Standards and Recommended Practices for Instrumentation and Control

10 Edition, 1989–Volume Two

**Reference Guides
for Instrumentation and Control**

Instrument Society of America

Copyright © Instrument Society of America 1989

All rights reserved

Printed in the United States of America

No part of this publication may be reproduced, stored in a retrieval system, or transmitted, in any form or by any means, electronic, mechanical, photocopying, recording or otherwise, without the prior written permission of the publisher.

INSTRUMENT SOCIETY OF AMERICA
67 Alexander Drive
P.O. Box 12277
Research Triangle Park
North Carolina 27709

Library of Congress Catalog Card No. 71-129243
ISBN: 1-55617-170-6
ISSN: 1042-6019

Contents*

Instrument Society of America Standards and Recommended Practices
- Title Listing v
- Abstracts 46
- Complete Texts for RP60.3 through ANSI C100.6 ix

Additional National and International Developers of Standards for Instrumentation and Control
- Developer Listing vii
- Abstracts and Contact Information 1

Agents for Standards Throughout the World 93

Subject Index 97

*Pages i through ix are located at the front of Volume Two
 Pages 1 through 176 are located at the back of Volume Two

Instrument Society of America Standards and Recommended Practices Title Listing*

VOLUME ONE

- **RP2.1** Manometer Tables
- **S5.1** Instrumentation Symbols and Identification
- **S5.2** Binary Logic Diagrams for Process Operations
- **S5.3** Graphic Symbols for Distributed Control/Shared Display Instrumentation, Logic and Computer Systems
- **S5.4** Instrument Loop Diagrams
- **S5.5** Graphic Symbols for Process Displays
- **RP7.1** Pneumatic Control Circuit Pressure Test
- **S7.3** Quality Standard for Instrument Air
- **S7.4** Air Pressure for Pneumatic Controllers, Transmitters, and Transmission Systems
- **RP7.7** Recommended Practice for Producing Quality Instrument Air
- **RP12.1** Electrical Instruments in Hazardous Atmospheres
- **S12.4** Instrument Purging for Reduction of Hazardous Area Classification
- **RP12.6**** Installation of Intrinsically Safe Systems for Hazardous (Classified) Locations
- **S12.10**** Area Classification in Hazardous (Classified) Dust Locations
- **S12.11** Electrical Instruments in Hazardous Dust Locations
- **S12.12** Electrical Equipment for Use in Class I, Division 2 Hazardous Classified Locations
- **S12.13, Part I** Performance Requirements, Combustible Gas Detectors
- **S12.13, Part II**** Installation, Operation and Maintenance of Combustible Gas Detectors
- **RP16.1,2,3** Terminology, Dimensions, and Safety Practices for Indicating Variable Meters (Rotameters, Glass Tube, Metal Tube, Extrusion Type Glass Tube)
- **RP16.4** Nomenclature and Terminology for Extension Type Variable Area Meters (Rotameters)
- **RP16.5** Installation, Operation, Maintenance Instructions for Glass Tube Variable Area Meters (Rotameters)
- **RP16.6** Methods and Equipment for Calibration of Variable Area Meters (Rotameters)
- **S18.1** Annunciator Sequences and Specifications
- **S20** Specification Forms for Process Measurement and Control Instruments, Primary Elements and Control Valves
- **S26** Dynamic Response Testing of Process Control Instrumentation
- **RP31.1** Specification, Installation, and Calibration of Turbine Flowmeters
- **S37.1** Electrical Transducer Nomenclature and Terminology
- **RP37.2** Guide for Specifications and Tests for Piezoelectric Acceleration Transducers for Aerospace Testing
- **S37.3** Specifications and Tests for Strain Gage Pressure Transducers
- **S37.5** Specifications and Tests for Strain Gage Linear Acceleration Transducers
- **S37.6** Specifications and Tests for Potentiometric Pressure Transducers
- **S37.8** Specifications and Tests for Strain Gage Force Transducers
- **S37.10** Specifications and Tests for Piezoelectric Pressure and Sound-Pressure Transducers
- **S37.12** Specifications for Tests for Potentiometric Displacement Transducers
- **RP42.1** Nomenclature for Instrument Tube Fittings

*Arranged in numerical order and individually page numbered within the ISA Complete Texts section, the document titles appear in the upper right hand corner and document numbers appear in the upper left corner of each odd-numbered page.

**New or revised since the 9th edition of *Standards and Practices for Instrumentation*.

ISA Standards Title Listing

S50.1 Compatibility of Analog Signals for Electronic Industrial Process Instruments
S51.1 Process Instrumentation Terminology
RP52.1 Recommended Environments for Standards Laboratories
RP55.1 Hardware Testing of Digital Process Computers

VOLUME TWO

RP60.3 Human Engineering for Control Centers
RP60.6 Nameplates, Labels and Tags for Control Centers
RP60.8 Electrical Guide for Control Centers
RP60.9 Piping Guide for Control Centers
S61.1 Industrial Computer System FORTRAN Procedures for Executive Functions, Process Input-Output and Bit Manipulation
S61.2 Industrial Computer System FORTRAN Procedures for File Access and the Control of File Contention
S67.01 Transducer and Transmitter Installation for Nuclear Safety Applications
S67.02 Nuclear-Safety-Related Instrument Sensing Line Piping and Tubing Standards for Use in Nuclear Power Plants
S67.03 Standard for Light Water Reactor Coolant Pressure Boundary Leak Detection
S67.04** Setpoints for Nuclear Safety-Related Instrumentation
S67.06 Response Time Testing of Nuclear Safety-Related Instrument Channels in Nuclear Power Plants
S67.10 Sample-Line Piping and Tubing Standard for Use in Nuclear Power Plants
S67.14 Qualifications for Certification of Instrumentation and Control Technicians in Nuclear Power Plants
S71.01 Environmental Conditions for Process Measurement and Control Systems: Temperature and Humidity
S71.04 Environmental Conditions for Process Measurement and Control Systems: Airborne Contaminants
S72.01 PROWAY—LAN Industrial Data Highway
RP74.01 Application and Installation of Continuous-Belt Weighbridge Scales
S75.01 Flow Equations for Sizing Control Valves
S75.02** Control Valve Capacity Test Procedure
S75.03 Face-to-Face Dimensions for Flanged Globe-Style Control Valve Bodies (ANSI Classes 125, 150, 250, 300 and 600)
S75.04 Face-to-Face Dimensions for Flangeless Control Valves (ANSI Classes 150, 300, and 600)
S75.05 Control Valve Terminology
RP75.06 Control Valve Manifold Designs
S75.07 Laboratory Measurement of Aerodynamic Noise Generated by Control Valves
S75.08 Installed Face-to-Face Dimensions for Flanged Clamp or Pinch Valves
S75.11 Inherent Flow Characteristic and Rangeability of Control Valves
S75.12 Face-to-Face Dimensions for Socket Weld-End and Screwed-End Globe-Style Control Valves (ANSI Classes 150, 300, 600, 900, 1500, and 2500)
S75.14 Face-to-Face Dimensions for Buttweld-End Globe-Style Control Valves (ANSI Class 4500)
S75.15 Face-to-Face Dimensions for Buttweld-End Globe-Style Control Valves (ANSI Classes 150, 300, 600, 900, 1500, and 2500)
S75.16 Face-to-Face Dimensions for Flanged Glove-Style Control Valve Bodies (ANSI Classes 900, 1500, and 2500)
S77.42 Fossil-Fuel Plant Feedwater Control System—Drum Type
S82.01** Safety Standard for Electrical and Electronic Test, Measurement, Controlling and Related Equipment—General Requirements—1988
S82.02** Safety Standard for Electrical and Electronic Test, Measuring, Controlling, and Related Equipment—Electrical and Electronic Test and Measuring Equipment—1988
S82.03** Safety Standard for Electrical and Electronic Test, Measuring, Controlling, and Related Equipment—Electrical and Electronic Process Measurement and Control Equipment—1988
MC96.1 Temperature Measurement Thermocouples
ANSI C100.6 3 Voltage or Current Reference Devices: Solid State Devices

Additional National and International Developers of Standards for Instrumentation and Control

DEVELOPER	ABSTRACTS PAGE
Acoustical Society of America (ASA)	1
Air-Conditioning and Refrigeration Institute (ARI)	3
Air Movement and Control Association (AMCA)	3
Aluminum Association (AA)	4
American Association of Textile Chemists and Colorists (AATCC)	4
American Boiler Manufacturers Association (ABMA)	4
American Chemical Society (ACS)	5
American Conference of Governmental Industrial Hygienists (ACGIH)	5
American Gas Association (AGA)	5
American Leather Chemists Association (ALCA)	7
American National Standards Institute (ANSI)	7
American Nuclear Society (ANS)	7
American Petroleum Institute (API)	8
American Society for Quality Control (ASQC)	12
American Society for Testing and Materials (ASTM)	13
American Society of Agricultural Engineers (ASAE)	19
American Society of Heating, Refrigerating, and Air-Conditioning Engineers, Inc. (ASHRAE)	19
American Society of Mechanical Engineers (ASME)	20
American Vacuum Society (AVS)	23
American Water Works Association (AWWA)	24
American Welding Society (AWS)	25
Anti-Friction Bearing Manufacturers Association, Inc. (AFBMA)	26
Association for the Advancement of Medical Instrumentation (AAMI)	26
Association of Official Analytical Chemists (AOAC)	27
The Chlorine Institute (CI)	27
Cooling Tower Institute (CTI)	28
Dairy and Food, 3A Sanitary Standards Committees (DFSSC)	28
Electronic Industries Association (EIA)	28
Factory Mutual System (FM)	32
Fluid Controls Institute, Inc. (FCI)	34
Gas Processors Association (GPA)	35
Industrial Fasteners Institute (IFI)	36
Industrial Risk Insurers (IRI)	36
Institute of Electrical and Electronics Engineers (IEEE)	36
Institute of Environmental Sciences (IES)	45
Institute of Interconnecting and Packaging Electronic Circuits (IPC)	45
Insulated Cable Engineers Association, Inc. (ICEA)	53
International Association of Plumbing and Mechanical Officials (IAPMO)	54
International Conference of Building Officials (ICBO)	54
International Electrotechnical Commission (IEC)	54
International Organization for Standardization (ISO)	59

Additional Standards Developers

Manufacturers Standardization Society of the Valve and Fitting Industry (MSS)	61
Metal Powder Industries Federation (MPIF)	62
National Association of Pipe Coating Applicators (NAPCA)	62
National Association of Relay Manufacturers (NARM)	63
National Board of Boiler and Pressure Vessel Inspectors (NBBI)	63
National Cable Television Association (NCTA)	64
National Council of Radiation Protection and Measurements (NCRP)	64
National Electrical Manufacturers Association (NEMA)	65
National Environmental Balancing Bureau (NEBB)	67
National Fire Protection Association (NFPA)	67
National Fluid Power Association (NFLDP)	69
National Institute of Standards and Technology (NIST)	63
Pipe Fabrication Institute (PFI)	76
Plumbing and Drainage Institute (PDI)	77
Radio Technical Commission for Aeronautics (RTCA)	77
Range Commanders Council (RCC)	81
Resistance Welder Manufacturers Association (RWMA)	83
Scientific Apparatus Makers Association (SAMA)	83
Society of Automotive Engineers, Inc. (SAE)	85
The Society of Naval Architects and Marine Engineers (SNAME)	89
Spring Manufacturers Institute, Inc. (SMI)	90
Technical Association of the Pulp and Paper Industry (TAPPI)	90
Ultrasonic Industry Association, Inc. (UIA)	90
Underwriters Laboratories, Inc. (UL)	90

Instrument Society of America Standards and Recommended Practices Complete Texts for RP60.3 ANSI C100.6

The Standards and Practices Department has been an integral part of the Instrument Society of America for forty-three years. To accomplish its primary goal of uniformity in the field of instrumentation, the Standards Department relies on the work of volunteer committee members whose efforts are coordinated and directed by volunteer Committee Chairmen and Standards and practices Department Directors. These volunteer members provide the time and technical expertise needed to develop and update standards in the field of instrumentation and control.

The Society also participates in many national and international standards activities. On March 26, 1976, the Instrument Society of America was approved as an Accredited Standards Writing organization by the American National Standards Institute (ANSI). Through strict adherence to our ANSI approved procedures, ISA has maintained its ANSI accreditation and continues to submit new and revised ISA standards to ANSI for approval as American National Standards. ISA is currently represented on three ANSI Standards Management Boards and is actively involved with the work of ISO and IEC. Through ANSI, ISA holds the Secretariat of ISO/TC 10/SC 3, "Graphical Symbols for Instrumentation," and IEC/TC SC65B, "Industrial-Process Measurement and Control: Elements of Systems." In addition twelve ISA standards committees serve as the advisory groups for U.S. representation to ISO and IEC committees and working groups.

The products of the ISA Standards program are well-surveyed, carefully written standards and recommended practices that outline accepted procedures in instrumentation throughout U.S. industry. To maintain their value, these standards and practices are reviewed and modified where necessary by the committee responsible for them, at intervals after publication. Toward the goal of improving its standards program, ISA welcomes criticism of its work. Correspondence should be directed to the attention of Standards Board Secretary, Instrument Society of America, 67 Alexander Drive, Research Triangle Park, North Carolina 27709.

The following ISA standards and recommended practices are arrange in numerical order and are individually page numbered. The document titles appear in the upper right corner and the document numbers appear in the upper left corner of each odd-numbered page.

Lois M. Ferson
Manager of Standards Services

ISA-RP60.3-1985

Recommended Practice

Human Engineering for Control Centers

Instrument Society of America

Instrument Society of America

ISBN 0-87664-897-9

ISA-RP60.3 Human Engineering for Control Centers

Copyright © 1985 by the Instrument Society of America. All rights reserved. Printed in the United States of America. No part of this publication may be reproduced, stored in a retrieval system, or transmitted, in any form or by any means (electronic, mechanical, photocopying, recording, or otherwise), without the prior written permission of the Publisher.

INSTRUMENT SOCIETY OF AMERICA
67 Alexander Drive
P.O. Box 12277
Research Triangle Park, NC 27709

PREFACE

This preface is included for informational purposes and is not part of ISA-RP60.3.

This recommended practice has been prepared as part of the service of the Instrument Society of America (ISA) toward a goal of uniformity in the field of instrumentation. To be of real value, this document should not be static, but should be subject to periodic review. Toward this end, the Society welcomes all comments and criticisms, and asks that they be addressed to the Secretary, Standards and Practices Board, Instrument Society of America, 67 Alexander Drive, P.O. Box 12277, Research Triangle Park, NC 27709, Telephone (919) 549-8411.

The ISA Standards and Practices Department is aware of the growing need for attention to the metric system of units in general, and the International System of Units (SI) in particular, in the preparation of instrumentation standards. The Department is further aware of the benefits to U.S.A. users of ISA standards of incorporating suitable references to the SI (and the metric system) in their business and professional dealings with other countries. Toward this end, the Department will endeavor to introduce SI-acceptable metric units in all new and revised standards to the greatest extent possible. The Metric Practice Guide, which has been published by the Institute of Electrical and Electronics Engineers as ANSI/IEEE Std. 268-1982, and future revisions will be the reference guide for definitions, symbols, abbreviations, and conversion factors.

It is the policy of the Instrument Society of America to encourage and welcome the participation of all concerned individuals and interests in the development of ISA standards. Participation in the ISA standards-making process by an individual in no way constitutes endorsement by the employer of that individual, of the Instrument Society of America, or of any of the standards that ISA develops.

The information contained in the preface, footnotes, and appendices is included for information only and is not a part of the recommended practice.

The SP60 Committee is preparing a series of recommended practices on control centers. ISA-RP60.3 is the fourth of this series to be published. The published recommended practices and drafts in preparation are listed below.

RECOMMENDED PRACTICE

SECTION	TITLE
dRP60.1*	Control Center (C.C.) Facilities
dRP60.2*	C.C. Design Guide and Terminology
RP60.3	Human Engineering for Control Centers
dRP60.4*	Documentation for Control Centers
dRP60.5*	Control Center Graphic Displays
RP60.6	Nameplates, Labels and Tags for Control Centers
dRP60.7*	Control Center Construction
RP60.8	Electrical Guide for Control Centers (published 1978)
RP60.9	Piping Guide for Control Centers (published 1981)
dRP60.10*	Control Center Inspection and Testing
dRP60.11*	Crating, Shipping and Handling for C.C.

*Draft Recommended Practice. For additional information on the status of this document, contact ISA Headquarters.

The persons listed below served as active members of the ISA Control Centers Committee, SP60, for the major share of its working period.

NAME	COMPANY
R. W. Borut, Chairman	The M. W. Kellogg Company
A. R. Alworth	Consultant
C. D. Armstrong	Tennessee Valley Authority
F. Aured	Engineering Enterprises
B. W. Ball	*Brown & Root Inc.
J. M. Fertitta, Secretary	The Foxboro Company
C. Goding	BIF Sanitrol, Unit of Gen. Signal
T. P. Holland	Johnson Controls, Inc. — Panel Unit
J. F. Jordan	Monsanto Company
J. G. McFadden	Public Service Electric & Gas
R. Munz	Mundix Control Systems, Inc.
R. F. Rossbauer	Fischer & Porter
†H. R. Solk	Consultant
A. Stockmal	Contraves
M. J. Walsh	*PROCON Inc.
R. L. Welch	Aramco Services Company

The persons listed below served as corresponding members of the ISA Control Centers Committee for the major share of its working period.

NAME	COMPANY
J. Cerretani	Detroit Edison Company
N. L. Conger	*Conoco
L. Corsetti	Crawford & Russell, Inc.
T. J. Crosby	Robertshaw Controls Company
C. R. Davis	Engineer
J. M. Devenney	Retired
F. L. Durfee	Swanson Engineering & Manufacturing
H. P. Fabisch	Fluor Engineering & Constructors, Inc.
J. Farina	Gismo Division of Guarantee Electric Company
H. L. Faul	H. L. Faul & Associates
M. E. Gunn	Swanson Engineering & Manufacturing
R. E. Hetzel	Stauffer Chemical Company
R. I. Hough	Hough Associates
J. R. Jordon	International Paper Company
H. Kamerer	Hough Associates
A. Kayser	Data Products New England
J. Kern	Aramco c/o Bechtel
R. W. Kief	Emanon Company, Inc.
A. L. Kress	3M Company
R. A. Landthorn	Ataboy Manufacturing
A. J. Langelier	Engineer
†C. S. Lisser	Consultant
S. F. Luna	General Atomics Company
R. G. Marvin	Dow Chemical Company

---

*Employer during development and ballot period
† Chairman Emeritus

NAME	COMPANY
A. P. McCauley, Jr.	Chagrin Valley Controls, Inc.
W. B. Miller	Moore Products Company
C. W. Moehring	Bechtel
D. P. Morrison	BIF/General Signal
R. L. Nickens	Reynolds Metal Company
F. W. Reichert	Engineer
I. Stubbs	Panelmatic, Inc.
J. F. Walker	Honeywell, Inc.
G. Walley	The N/P Company
S. J. Whitman	Acco/Bristol Division
W. T. Williams	Midwest Tech, Inc.
W. J. Wylupek	Moore Products Company

This recommended practice was approved for publication by the ISA Standards and Practices Board in June 1985.

NAME	COMPANY
N. Conger, Chairman	Fisher Controls Company
P. V. Bhat	Monsanto Company
W. Calder III	The Foxboro Company
R. S. Crowder	Ship Star Associates
B. Feikle	Bailey Controls Company
H. S. Hopkins	Westinghouse Electric Company
J. L. Howard	Boeing Aerospace Company
R. T. Jones	Philadelphia Electric Company
R. Keller	The Boeing Company
O. P. Lovett, Jr.	ISIS Corporation
E. C. Magison	Honeywell, Inc.
A. P. McCauley	Chagrin Valley Controls, Inc.
J. W. Mock	Bechtel Corporation
E. M. Nesvig	ERDCO Engineering Corporation
R. Prescott	Moore Products Company
D.-E. Rapley	Rapley Engineering Services
C. W. Reimann	National Bureau of Standards
J. Rennie	Factory Mutual Research Corporation
W. C. Weidman	Gilbert Commonwealth Inc.
K. Whitman	Consultant
*P. Bliss	Consultant
*B. A. Christensen	Continental Oil Company
*L. N. Combs	Retired
*R. L. Galley	Consultant
*T. J. Harrison	IBM Corporation
*R. G. Marvin	Roy G. Marvin Company
*W. B. Miller	Moore Products Company
*G. Platt	Retired
*J. R. Williams	Stearns Catalytic Corporation

*Director Emeritus

TABLE OF CONTENTS

Section   **Title**   **Page**

1. Scope 9
2. Physical Aspects 9
   2.1 Static Anthropometric Data 9
   2.2 Dynamic Anthropometric Data 9
3. Psychological Aspects 14
   3.1 General 14
   3.2 Information Sensing (Inputs) 14
      3.2.1 Visual Inputs 14
      3.2.2 Auditory Inputs 14
   3.3 Information Storage 14
   3.4 Information Processing 14
4. General Design 14
   4.1 Objective 14
   4.2 Implementation Guidelines 15
      4.2.1 Equipment Arrangement 15
      4.2.2 Accessibility 15
      4.2.3 Pattern Recognition 15
      4.2.4 Shape or Type 15
      4.2.5 Visibility and Readability 16
      4.2.6 Illumination 16
      4.2.7 Color Coding 16
      4.2.8 Auditory Techniques 16
      4.2.9 Safety Considerations 17
   4.3 Maintenance 17
5. Bibliography 17

LIST OF ILLUSTRATIONS

Figure   **Title**   **Page**

1. Standing Body Dimensions 10
2. Seated Body Dimensions 12

LIST OF TABLES

Table   **Title**   **Page**

1. Standing Body Dimensions 11
2. Seated Body Dimensions 13

1 SCOPE

The intent of this recommended practice is to present design concepts which are compatible with the physical and mental capabilities of the control center operator while recognizing any of the operator's limitations.

The control center can be designed for efficient functioning of the man-machine system after one first defines the information the operator needs to control the process, and the controls to be provided.

This recommended practice is limited to those aspects of human engineering that will affect the layout of and the equipment selection for the control center. It is recognized that some of the human factors discussed in this document are also used in the design and manufacture of instruments.

2 PHYSICAL ASPECTS

2.1 Static Anthropometric Data

In the dimensional design of a control center, one should consider the physical characteristics of the plant operators. The possible range in these characteristics is rather wide, but can be narrowed for a specific design if knowledge of such factors as age, sex, and physical qualities of the expected group is applied. Useful data may be obtained from the references listed in the bibliography for this recommended practice. It is important to note that in application of these data, one should not design the control center for average human characteristics but rather for the normally expected extremes of the subject group. Figures 1 and 2 and Tables 1 and 2 provide static anthropometric data taken from Military Standard (MIL. STD) 1472.

2.2 Dynamic Anthropometric Data

2.2.1 While the basic layout for a control center may be designed using static anthropometric data, it should be checked and refined using dynamic data. If this is done, the range to be covered by one operator can be described by physical reach and reasonable lateral movement. Included in the relevant dynamic information are the ranges of eye and head movement. If there is no head movement, the visual range is a cone whose angle is approximately 60°. However, discomfort results if the eye must be positioned off the standard line of sight. Therefore, head movement is used to accommodate or assist any scanning requirement. The result, then, is a control center whose dimensions and shape match the process and the operator. The standard profiles detailed in ISA-dRP60.7, "Control Center Construction," represent industry practice in the application of anthropometric data.

2.2.2 Displays should be arranged and positioned so that they are perpendicular to the line of sight when the eyes and head are in a comfortable position, and in any case should not be less than 45° from the normal line of sight. This is particularly important for displays which require constant attention.

The use of manually operated equipment located in positions which require frequent extreme physical movements, such as reaching, stooping, or squatting, will produce fatigue; thus, these locations should be avoided.

Consideration should be given to techniques or features in the control center which will help minimize fatigue when it is a requirement that the operator must stand most of the time. Arm rests or hand rails reduce upper back fatigue, and a foot rest or foot rail aids in easing lower back strain or fatigue.

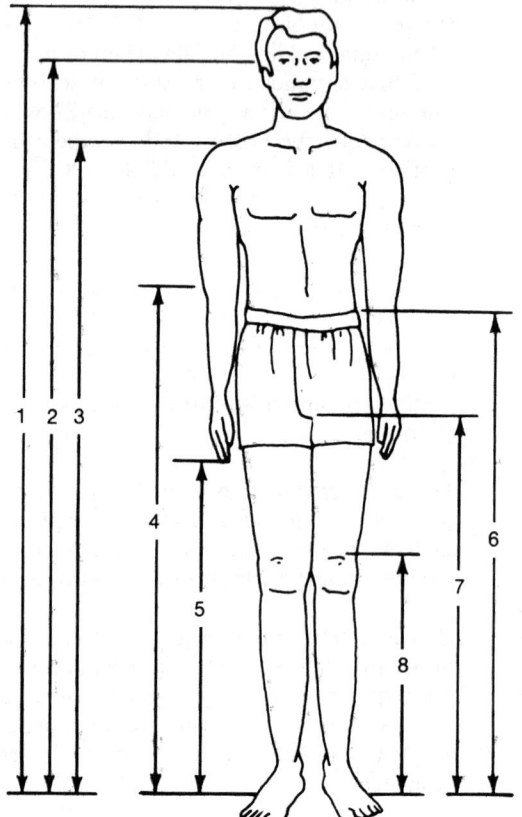
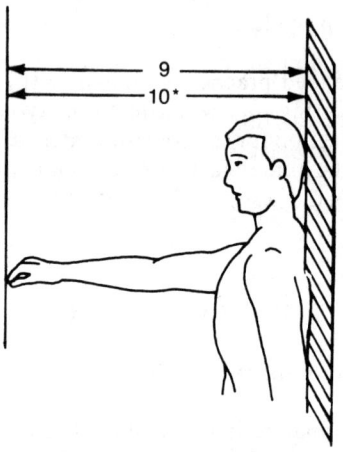

*SAME AS 9. HOWEVER, RIGHT SHOULDER IS EXTENDED AS FAR FORWARD AS POSSIBLE WHILE KEEPING THE BACK OF THE LEFT SHOULDER FIRMLY AGAINST THE BACK WALL.

Figure 1. Standing Body Dimensions

TABLE 1

Standing Body Dimensions

	\multicolumn{6}{c}{Percentile values in centimeters}					
	\multicolumn{3}{c}{5th percentile}			\multicolumn{3}{c}{95th percentile}		
	Ground troops	Aviators	Women	Ground troops	Aviators	Women
Weight (kg)	55.5	60.4	46.4	91.6	96.0	74.5
Standing body dimensions						
1 Stature	162.8	164.2	152.4	185.6	187.7	174.1
2 Eye height (standing)	151.1	152.1	140.9	173.3	175.2	162.2
3 Shoulder (acromiale) height	133.6	133.3	123.0	154.2	154.8	143.7
4 Elbow (radiale) height	101.0	104.8	94.9	117.8	120.0	110.7
5 Fingertip (dactylion) height		61.5			73.2	
6 Waist height	96.6	97.6	93.1	115.2	115.1	110.3
7 Crotch height	76.3	74.7	68.1	91.8	92.0	83.9
8 Kneecap height	47.5	46.3	43.8	58.6	57.8	52.5
9 Functional reach	72.6	73.1	64.0	90.9	87.0	80.4
10 Functional reach, extended	84.2	82.3	73.5	101.2	97.3	92.7
	\multicolumn{6}{c}{Percentile values in inches}					
Weight (lb)	122.4	133.1	102.3	201.9	211.6	164.3
Standing body dimensions						
1 Stature	64.1	64.6	60.0	73.11	73.9	68.5
2 Eye height (standing)	59.5	59.9	55.5	68.2	69.0	63.9
3 Shoulder (acromiale) height	52.6	52.5	48.4	60.7	60.9	56.6
4 Elbow (radiale) height	39.8	41.3	37.4	46.4	47.2	43.6
5 Fingertip (dactylion) height		24.2			28.8	
6 Waist height	38.0	38.4	36.6	45.3	45.3	43.4
7 Crotch height	30.0	29.4	26.8	36.1	36.2	33.0
8 Kneecap height	18.7	18.4	17.2	23.1	22.8	20.7
9 Functional reach	28.6	28.8	25.2	35.8	34.3	31.7
10 Functional reach, extended	33.2	32.4	28.0	39.8	38.3	36.5

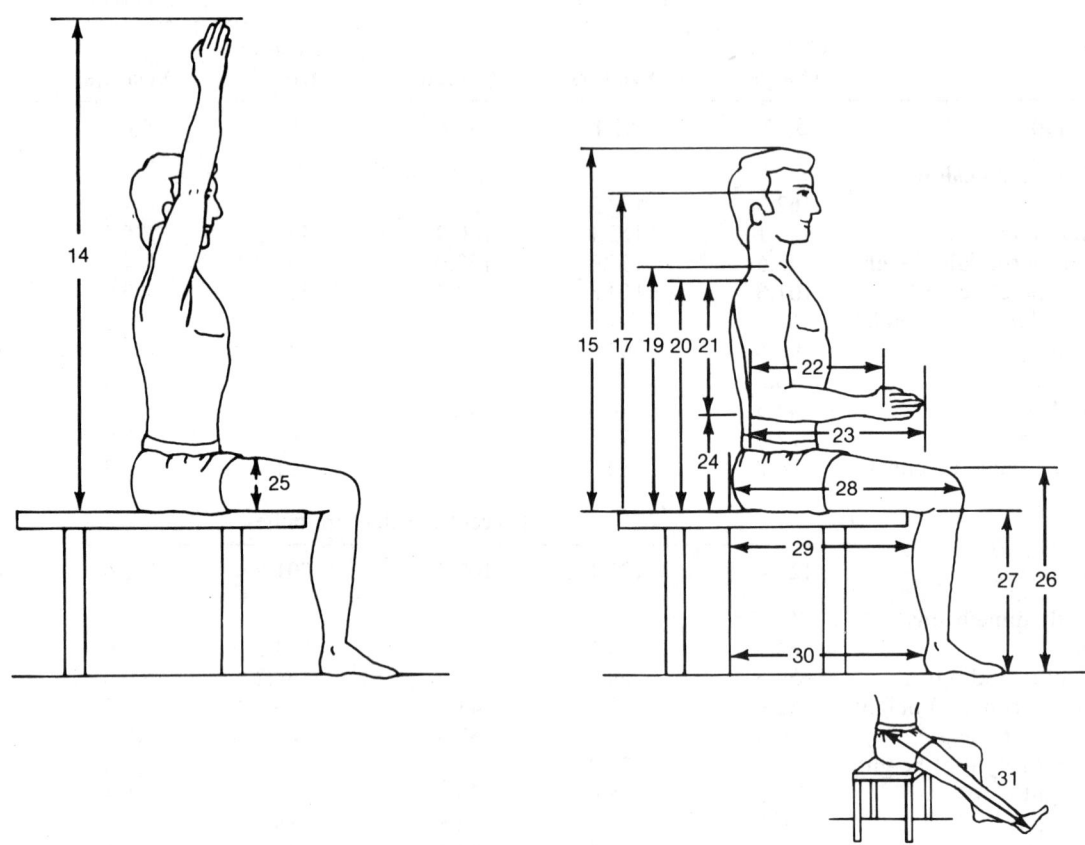

Figure 2. Seated Body Dimensions

TABLE 2
Seated Body Dimensions

| | Percentile values in centimeters ||||||
| | 5th percentile ||| 95th percentile |||
	Ground troops	Aviators	Women	Ground troops	Aviators	Women
Seated body dimensions						
14 Vertical arm reach, sitting	128.6	134.0	117.4	147.8	153.2	139.4
15 Sitting height, erect	83.5	85.7	79.0	96.9	98.6	90.9
16 Sitting height, relaxed	81.5	83.6	77.5	94.8	96.5	89.7
17 Eye height, sitting erect	72.0	73.6	67.7	84.6	86.1	79.1
18 Eye height, sitting relaxed	70.0	71.6	66.2	82.5	84.0	77.9
19 Mid-shoulder height	56.6	58.3	53.7	67.7	69.2	62.5
20 Shoulder height, sitting	54.2	54.6	49.9	65.4	65.9	60.3
21 Shoulder-elbow length	33.3	33.2	30.8	40.2	39.7	36.6
22 Elbow-grip length	31.7	32.6	29.6	38.3	37.9	35.4
23 Elbow-fingertip length	43.8	44.7	40.0	52.0	51.7	47.5
24 Elbow rest height	17.5	18.7	16.1	28.0	29.5	26.9
25 Thigh clearance height		12.4	10.4		18.8	17.5
26 Knee height, sitting	49.7	48.9	46.9	60.2	59.9	55.5
27 Popliteal height	39.7	38.4	38.0	50.0	47.7	45.7
28 Buttock-knee length	54.9	55.9	53.1	65.8	65.5	63.2
29 Buttock-popliteal length	45.8	44.9	43.4	54.5	54.6	52.6
30 Buttock-heel length		46.7			56.4	
31 Functional leg length	110.6	103.9	99.6	127.7	120.4	118.6

	Percentile values in inches					
Seated body dimensions						
14 Vertical arm reach, sitting	50.6	52.8	46.2	58.2	60.3	54.9
15 Sitting height, erect	32.9	33.7	31.1	38.2	38.8	35.8
16 Sitting height, relaxed	32.1	32.9	30.5	37.3	38.0	35.3
17 Eye height, sitting erect	28.3	30.0	26.6	33.3	33.9	31.2
18 Eye height, sitting relaxed	27.6	28.2	26.1	32.5	33.1	30.7
19 Mid-shoulder height	22.3	23.0	21.2	26.7	27.3	24.6
20 Shoulder height, sitting	21.3	21.5	19.6	25.7	25.9	23.7
21 Shoulder-elbow length	13.1	13.1	12.1	15.8	15.6	14.4
22 Elbow-grip length	12.5	12.8	11.6	15.1	14.9	14.0
23 Elbow-fingertip length	17.3	17.6	15.7	20.5	20.4	18.7
24 Elbow rest height	6.9	7.4	6.4	11.0	11.6	10.6
25 Thigh clearance height		4.9	4.1		7.4	6.9
26 Knee height, sitting	19.6	19.3	18.5	23.7	23.6	21.8
27 Popliteal height	15.6	15.1	15.0	19.7	18.8	18.0
28 Buttock-knee length	21.6	22.0	20.9	25.9	25.8	24.9
29 Buttock-popliteal length	17.9	17.7	17.1	21.5	21.5	20.7
30 Buttock-heel length		18.4			22.2	
31 Functional leg length	43.5	40.9	39.2	50.3	47.4	46.7

3 PSYCHOLOGICAL ASPECTS

3.1 General

3.1.1 This section deals with how an operator distinguishes, comprehends, and reacts to the information displayed in a control center. Measurement and control systems may contain thousands of devices representing functions or bits of information displayed, each potentially requiring operator actions. Considering the tremendous quantity of information confronting the operator, a high level of mental concentration is demanded.

3.1.2 Psychological aspects are thought processes which are difficult to measure but which must be considered in any control center design. The importance of these aspects increases rapidly with the complexity of the process and the quantity of the control center instrumentation. Psychological aspects also affect the selection and arrangement of the display devices.

3.1.3 Too often, control centers are designed without considering the principles of human engineering design. To the casual observer, the resulting control center may be an impressive display of lights and instruments. To the operator who must live with such a presentation, it can become bewildering. Operator confusion can lead to costly operational mistakes.

3.1.4 The operator can be considered a system — complete with sensory inputs, data processing capability (including information storage and processing), and responsive outputs. The following sections show how human beings fit this information processing system analogy. This information should help the designer incorporate good human engineering concepts into the control center design.

3.2 Information Sensing (Inputs)

3.2.1. Visual Inputs

Visual inputs provide 80 to 90 percent of the information required to operate a control center. The operator interprets:

1. Color, and changes of color

2. Position, and change of position (dials and switch handles)

3. Patterns, and changes of pattern, e.g., on cathode ray tube (CRT) displays

4. Digital values

5. Nameplate data

3.2.2 Auditory Inputs

Auditory input makes up the remaining 10 to 20 percent of the information required to operate a control center. This input reinforces the visual information by adding sounds from attention-getting devices such as horns, sirens, and other audible devices.

3.3 Information Storage

Memory is the information storage unit which makes it possible to integrate previous experience with present activity. The ability to perform the job's requirements depends on how well the operator can process the information presented and relate this to previously stored information.

3.4 Information Processing

The way in which the information is presented determines the operator's ability to recognize, comprehend, and react to a situation. It is evident that the operator can be overloaded with information, which leads to inefficiency and inaccuracy. The ability to recognize, comprehend, and react correctly to a situation is usually determined by the manner and speed with which information is presented. Information theory may be used as a tool to measure the optimum combination of display type and speed to avoid the operator's becoming overloaded and confused. For more information, refer to McCormick, *Human Factors Engineering*, Chapter 5 (see Section 5).

4 GENERAL DESIGN

4.1 Objective

4.1.1 This section will describe specific features and techniques that should be used in the design and layout of a control center, implementing the human engineering factors treated in Sections 2 and 3. Not all of the following recommendations will apply to every application. The designer should include only those features that are compatible with the overall design requirements of the control center and facility, taking into consideration hardware availability, cost, and delivery time.

4.1.2 Information display devices include analog indicators, recorders, indicating lights, backlighted nameplates, annunciator windows, digital displays, CRT displays, projection screens, and printers or typers. Control devices include selector switches, pushbutton switches, keyboards, and analog control stations (manual/automatic, manual loading, setpoint, and other types).

4.1.3 The use of computers to process data and generate displays should be considered where large amounts of data and/or complex operating data must be monitored and controlled. The computer can assist the operator in performing the necessary control functions. A computer can operate display devices such as CRTs, indicating lights, projection screens, digital displays, analog indicators, and printers or typers.

4.2 Implementation Guidelines

4.2.1 Equipment Arrangement

Control devices should be located on the control center in the same general sequence that the operator will follow in any of the operations of systems or subsystems, or located in the same relative positions that the actual equipment is located. The same concept may be applied to the location of individual sections of a control center.

Those devices which are interrelated or are used to determine the operating condition and status of a system or subsystem may be grouped together.

To assist the operator in identifying and locating a particular group of devices, space may be left between adjacent groups, different colors may be used, or groups may be outlined with tape or painted stripe.

Consistent criteria should be used for the relative location of specific status conditions where multiple status display devices are used.

Mounting the devices within graphic sections on the control center can result in easier identification by the operator.

Use consistent criteria for the location of nameplates in relation to their devices.

Layouts for duplicate units should be identical to improve operator proficiency as long as there is proper unit identification. Arrangements utilizing reverse or mirror images in the design of control centers for duplicate units should be avoided.

4.2.2 Accessibility

Control devices should be mounted on sections of the control center that are within the normal reach of the operator when the operator is in normal position, either standing or sitting.

Control devices should be mounted on the control center with sufficient clearance, so that the operator can conveniently operate them without interference from other devices or adjacent control center surfaces. Consideration should also be given to:

1. Plant conditions that might require gloves
2. The size or strength of an operator's hand
3. Hindrances due to safety clothing
4. Other operational hindrances

Frequently adjusted controls and often-referenced visual indicators should be located at the most convenient elevation. The height is related to the operator's normal position of standing or sitting. ISA-dRP60.7, "Control Center Construction," shows a typical profile, with an operator's average standing or sitting line of elevation. These typical profiles are based on the physical aspects outlined in Section 2.

Space should be provided for convenient storage of and access to operating checklists and instructions.

4.2.3 Pattern Recognition

Pattern recognition and symmetry may be used to aid the operator in the detection of abnormal conditions. Mounting several edgewise indicators adjacent to each other is one way to use this technique.

The direction of motion of the pointers for all indicators should be established and used consistently. Common practice is for the pointer to move from the bottom of the scale to the top, from left to right or clockwise as the displayed variable increases in value.

The direction of operation of control switches and selector switches should be chosen consistently for the functions that they perform. For example, all equipment starting functions on a particular control center may be performed when the switch handle is turned in a clockwise direction and stopped when the switch handle is turned in the counterclockwise direction. For another example, valves may open when the switch handle is turned in the clockwise direction, close when turned in the counterclockwise direction, and stop when in a neutral, or center position.

4.2.4 Shape or Type

The use of particular types of display or control devices in the control center facility should be standardized to increase the operator's efficiency and possibly improve the aesthetic appearance of the control center.

A specific size or shape of switch handle should be used as a means to identify a specific function; this will increase operator accuracy during emergencies.

The type of switch handle that will be most comfortable and convenient for the operator to use for the particular application should be selected. For the smaller-sized selector switches, use of "bat lever knobs" and "gloved hand operators" reduces the strain on the operator's hand. This is particularly true with spring return switches, for which the operator may be required to hold the switch in the momentary position for relatively long periods of time.

Where illuminated pushbuttons are used, select an arrangement or type of pushbutton that does not produce large amounts of heat and cause the pushbutton to become hot to the touch.

4.2.5 Visibility and Readability

Locate the display devices on the control center above the associated controls so that no part of the operator's body will obstruct the view of the related display devices during operation of the controls.

Mount those devices that have hinged doors or sections, such as recorders, so that operation or observation is not blocked from critical displays or control devices when these doors or sections are swung out from their normal operating positions.

When selecting a minimum size for the letters, numbers, and symbols on legends, scales, engraved nameplates, and other displays, take into consideration the distance from which the device will be read by the operator in the normal operating position; also consider the significance of the device. See ISA-RP60.6, "Nameplates, Labels and Tags for Control Centers," for recommended letter sizes.

Abbreviations should be avoided, but may be used when necessary. Consistent terminology and abbreviations should be used.

Requirements for the resolution of readings may determine the length of scales and the type and size of indicators and recorders.

Those devices which have graduated scales and require reading by the operator should be mounted on a section of the control center that is within the line of sight of the operator when in the normal operating position (standing or sitting) at the control center.

Glare on the surface of the display device may be reduced by:

1. Use of antiglare glass or materials for the windows of display devices

2. Use of hoods that extend over the top of the display device — especially for CRTs, digital displays, and projection screens

3. Proper selection of the mounting angle for the display device in relation to both the operator and the source of reflection

4. Use of adjustable lighting intensity in the area of the control center

5. Use of indirect or diffused lighting. Glare on the surfaces of the control centers and nameplates should be minimized through the proper selection of surface finish.

4.2.6 Illumination

Consider the use of internal illumination for indicators and recorders either (1) to increase readability, or (2) to call the operator's attention to an off-normal variable (by having the illumination flash or otherwise change from the continuously lighted condition).

Variation of light intensity and flash frequency can be used to draw the operator's attention to a display device.

Higher flash rates can be used in identifying more important devices.

When selecting indicating lights, backlighted nameplates, CRT displays, digital displays, and projection screens, consider the light intensity of the device, the lighting levels in the control center facility, and the effects of deterioration of the brightness of the device with age.

Some devices, such as CRT displays, have an adjustment for brightness. Status light power supplies may be designed to provide different intensities. See ISA-dRP60.1, "Control Center Facilities," for the recommended lighting levels inside the control room.

4.2.7 Color Coding

The establishment of a color coding system can assist in the identification of those devices that are associated with a particular system, subsystem, function, or piece or type of equipment. Nameplates, bezels, control center surfaces, graphic materials, switch rings, and switch actuators can be provided in different colors.

Use a consistent color designation scheme for all indicating lights and backlighted displays in the control room. Each color should designate a specific status condition or priority.

Color coding the indicating scale backgrounds by group or function can result in easier identification by the operator.

Contrasting colors for the background, lettering, and pointers of indicators and recorders should aid in their readability. Care should be taken in the selection of color for background and lettering of backlighted devices so that they will be readable in both the illuminated and extinguished state, if this is the desired effect.

4.2.8 Auditory Techniques

Audible devices are frequently used with display devices (such as annunciator windows, CRT displays, and printers or typers) to alert the operator to a condition that requires attention. Audible devices may also be used to direct attention to a particular control center or piece of equipment. Types of audible devices include: horns, bells, buzzers, sirens, gongs, chimes, klaxons, and pre-programmed verbal messages.

Audible devices should be chosen so that each device has a distinct and recognizable sound. This can be accomplished by use of different types of devices, tone sequences, pulse frequencies, and by varying the pitch of each signal. All devices in the control room, including telephones, should be considered.

Each audible device should have the capability of being adjusted for loudness so that the device can be heard, at the operator's location, over the background noise while at the same time not being offensively loud.

4.2.9 Safety Considerations

In critical applications such as emergency shutdown and trip functions, either guarded or protected pushbuttons or dual operation devices should be used to minimize inadvertent actuation.

Removable handles and key-operated switches may be used to prevent unauthorized operation.

"Panic Buttons" for emergency trip actuations should use mushroom head type pushbuttons or other readily operated devices. They should not be located where they could be accidentally actuated.

4.3 Maintenance

4.3.1 Each device should be located and mounted on the control center so that it is readily identified and accessible for maintenance, testing, or calibration without loss of operability. Provisions for easy removal of each device should be included. Interference from other devices, framing, reinforcing members, and surfaces of the control center, or from adjacent wiring, tubing, and piping should be taken into consideration during the design and layout.

4.3.2 Internal surfaces of the control center should be colored a flat white to improve visibility during maintenance.

4.3.3 Reduce to the maximum extent possible the interference with the functions of the operator during the performance of maintenance, testing, or calibration of the control center devices. This may be done by mounting the devices so that they can be quickly removed, and by locating auxiliary devices away from the immediate operating area. Locate the test and calibration connections so that they do not interfere with operations.

4.3.4 A means should be provided for readily identifying faulty display devices on the control center. This can be done by providing:

1. A lamp test feature and/or dual lamps

2. A dim-bright operating feature

3. An upscale or downscale burnout

4. A live zero

5. An out-of-service indication for critical items

6. Three displays or channels — to avoid the ambiguity of dual indications in critical services

4.3.5 A means should be provided for disconnecting the power source to the devices mounted on the control center. This may be done by the use of circuit breakers, disconnect switches, or individual plugs and receptacles.

4.3.6 If custom changes have been made to standard equipment in order to provide desired human engineering features, a complete documentation of these changes should be maintained.

5 BIBLIOGRAPHY

1. N. Diffrient, A. R. Tilley, and J. Bardaguy, "Human Scale 1-2-3," Massachusetts Institute of Technology Press, Cambridge, MA, 1974.

2. E. J. McCormick, *Human Factors Engineering*, 2d ed., McGraw-Hill Book Company, New York, NY, 1964.

3. W. E. Woodson and D. W. Conover, *Human Engineering Guide for Equipment Designers*, 2d ed., University of California Press, Berkeley, CA, 1964.

4. Lockheed Missiles and Space Company, Inc., for the Electric Power Research Institute, "Human Factors Review of Nuclear Power Plant Control Room Design," Project 501, EPRI NP-309, March 1977, Electric Power Research Institute, Palo Alto, CA.

5. Naval Publications Information Center, MIL Std. 1472, "Human Engineering Design Criteria for Military Systems, Equipment and Facilities," 10 May 1984, Naval Publications Information Center, Philadelphia, PA.

6. Naval Publications Information Center, MIL Std. H-46855, "Human Engineering Requirements for Military Systems, Equipment and Facilities," 5 April 1984, Naval Publications Information Center, Philadelphia, PA.

7. R. A. Behan and H. W. Wendhausen, NASA SP-5117, "Some NASA Contributions to Human Factors Engineering — A Survey," National Aeronautics and Space Administration, Washington, D.C., 1973.

ISA-RP60.6-1984

Recommended Practice

Nameplates, Labels and Tags for Control Centers

Instrument Society of America

ISBN 0-87664-813-8

ISA-RP60.6 Nameplates, Labels and Tags for Control Centers

Copyright ©1984 by the Instrument Society of America. All rights reserved. Printed in the United States of America. No part of this publication may be reproduced, stored in a retrieval system, or transmitted, in any form or any means electronic, mechanical, photocopying, recording or otherwise without the prior written permission of the publisher.

INSTRUMENT SOCIETY OF AMERICA
67 Alexander Drive
P.O. Box 12277
Research Triangle Park, North Carolina 27709

Copyright ©1984 by the Instrument Society of America

Nameplates, Labels and Tags for Control Centers

PREFACE

This preface is included for information purposes and is not a part of ISA-RP60.6.

This Recommended Practice has been prepared as a part of the service of the Instrument Society of America (ISA) toward a goal of uniformity in the field of instrumentation. To be of real value, this document should not be static, but should be subject to periodic review. Toward this end, the Society welcomes all comments and criticisms, and asks that they be addressed to the Secretary, Standards and Practices Board, Instrument Society of America, 67 Alexander Drive, P.O. Box 12277, Research Triangle Park, NC 27709, Telephone (919) 549-8411.

The ISA Standards and Practices Department is aware of the growing need for attention to the metric system of units in general, and the International System of Units (SI) in particular, in the preparation of instrumentation standards. The Department is further aware of the benefits to U.S.A. users of ISA standards of incorporating suitable references to the SI (and the metric system) in their business and professional dealings with other countries. Toward this end, this Department will endeavor to introduce SI-acceptable metric units in all new and revised standards to the greatest extent possible. The Metric Practice Guide, which has been published by the Institute of Electrical and Electronic Engineers as ANSI/IEEE Std. 268-1982, and future revisions will be the reference guide for definitions, symbols, abbreviations, and conversion factors.

It is the policy of the Instrument Society of America to encourage and welcome the participation of all concerned individuals and interests in the development of ISA standards. Participation in the ISA standards-making process by an individual in no way constitutes endorsement by the employer of that individual, of the Instrument Society of America, or of any of the standards that ISA develops.

The SP60 Committee is preparing a series of recommended practices on control centers. ISA-RP60.6 is the third of this series to be published. The published recommended practices and drafts in preparation are listed below:

RECOMMENDED PRACTICE

SECTION	TITLE
dRP60.1*	Control Center (C.C.) Facilities
dRP60.2*	C.C. Design Guide and Terminology
dRP60.3*	Human Engineering for Control Centers
dRP60.4*	Documentation for Control Centers
dRP60.5*	Control Center Graphic Displays
RP60.6	Nameplates, Labels and Tags for Control Centers
dRP60.7*	Control Center Construction
RP60.8	Electrical Guide for Control Centers (published 1978)
RP60.9	Piping Guide for Control Centers (published 1981)
dRP60.10*	Control Center Inspection and Testing
dRP60.11*	Crating, Shipping and Handling for C.C.

The persons listed below served as active members of the ISA Control Centers Committee for the major share of its working period.

NAME	COMPANY
H. R. Solk, Chairman	Consultant
J. M. Fertitta, Secretary	The Foxboro Company
A. R. Alworth	Shell Oil Company
C. D. Armstrong	Tennessee Valley Authority
F. Aured	Panels, Inc.
B. W. Ball	Brown & Root, Inc.
**R. W. Borut	The M. W. Kellogg Company
C. Goding	BIF Sanitrol, Unit of Gen. Signal
T. P. Holland	Johnson Controls, Inc.-Panel Unit
J. F. Jordon	Monsanto Company
J. G. McFadden	Public Service Electric & Gas
R. Munz	Mundix Control Systems, Inc.
R. F. Rossbauer	Fischer & Porter
A. Stockmal	Contraves
M. J. Walsh	Procon, Inc.
R. L. Welch	El Paso Natural Gas Company

*Draft Recommended Practice, for additional information on the status of this document contact ISA Headquarters.
**Chairman Emeritus

Instrument Society of America

The persons listed below served as corresponding members of the ISA Control Centers Committee for the major share of its working period:

NAME	COMPANY
J. Cerretani	Detroit Edison Company
N. L. Conger	Conoco
L. Corsetti	Crawford & Russell, Inc.
T. J. Crosby	Robertshaw Controls Company
C. R. Davis	Engineer
J. M. Devenney	Retired
F. L. Durfee	Swanson Engineering & Manufacturing
H. P. Fabisch	Fluor Engineering & Constructors, Inc.
J. Farina	Gismo Div. of Guarantee Electric Co.
H. L. Faul	A-E Development Corporation
M. E. Gunn	Swanson Engineering & Manufacturing
R. E. Hetzel	Stauffer Chemical Company
R. I. Hough	Hough Associated
J. R. Jordon	International Paper Company
H. Kamerer	Wilmington Steel Metal Ind.
A. Kayser	Consolidated Controls Corporation
J. L. Kern	Weyerhauser Company
R. W. Kief	Emanon Company
A. L. Kress	3M Company
R. A. Landthorn	Panels, Inc.
A. J. Langelier	Engineer
*C. S. Lisser	Oak Ridge National Laboratory
S. F. Luna	General Atomics Company
R. G. Marvin	Dow Chemical Company
A. P. McCauley, Jr.	Chagrin Valley Controls, Inc.
W. B. Miller	Moore Products Company
C. W. Moehring	Bechtel
D. P. Morrison	BIF/General Signal
R. L. Nickens	Reynolds Metal Company
F. W. Reichert	Engineer
I. Stubbs	Panelmatic, Inc.
J. F. Walker	Honeywell, Inc.
G. Walley	The N/P Company
S. J. Whitman	Acco/Bristol Division
W. T. Williams	Instrumentation Engineer
W. J. Wylupek	Moore Products Company

*Chairman Emeritus

This Recommended Practice was approved for publication by the ISA Standards and Practices Board in February 1984.

NAME	COMPANY
W. Calder III, Chairman	The Foxboro Company
N. L. Conger	Conoco
B. Feikle	Bailey Controls Company
T. J. Harrison	IBM Corporation
H. S. Hopkins	Westinghouse Electric Company
J. L. Howard	Boeing Aerospace Company
R. T. Jones	Philadelphia Electric Company
R. Keller	The Boeing Company
O. P. Lovett, Jr.	Isis Corporation
E. C. Magison	Honeywell, Inc.
A. P. McCauley	Chagrin Valley Controls, Inc.
J. W. Mock	Bechtel Corporation

E. M. Nesvig	ERDCO Engineering Corporation
R. Prescott	Moore Products Company
D. Rapley	Stearns Catalytic Corporation
W. C. Weidman	Gilbert Commonwealth, Inc.
K. A. Whitman	Allied Chemical Corporation
* P. Bliss	
* B. A. Christensen	
* L. N. Combs	
* R. L. Galley	
* R. G. Marvin	
* W. B. Miller	Moore Products Company
* G. Platt	Bechtel Power Corporation
* J. R. Williams	Stearns Catalytic Corporation

* Director Emeritus

TABLE OF CONTENTS

Section **Title** **Page**

1. Purpose 9
2. Scope 9
3. Functional Definitions 9
4. Identification 9
5. Types of Identification 9
6. Materials 9
7. Methods of Fabrication 10
8. Methods of Attachment 11
9. Typical Locations 11

APPENDICES

Section **Page**

Appendix A, Construction and Examples 13
Appendix B, Abbreviations 17
Appendix C, Character Heights and Suggested Viewing Distance 33
Appendix D, Gothic Lettering Examples 34

LIST OF TABLES

1. Identification Types 10

1 PURPOSE

The purpose of this Recommended Practice is to assist the designer or engineer in choosing and specifying the method of identifying items mounted on a control center or associated with a control center facility.

2 SCOPE

This Recommended Practice is intended to summarize identification methods and to suggest the uses of nameplates, labels, and tags. Examples are included for guidance in preparing drawings and specifications. (See Appendix A.) Since materials will determine the tools and techniques used, savings may be realized by using the materials and symbols standardized by the control center manufacturer selected — provided that the materials fully satisfy the application and purpose intended.

3 FUNCTIONAL DEFINITIONS

3.1 Nameplate - Used to display basic information, including function.

EXAMPLE:

| UNIT AND/OR FUNCTION | FRC-104 ○ OIL TO BURNER 4 ○ FLOW |

3.2 Label - Used to inform of detailed instructions about item identified.

EXAMPLE:

| NOTICE AND/OR TECHNICAL INFORMATION | CAUTION CONTAINS STATIC SENSITIVE COMPONENTS |

3.3 Tags - Used to display specific information about item identified.

EXAMPLE:

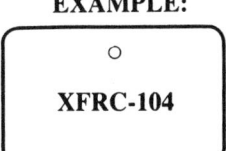

| INSTRUMENT WIRE OR LOOP NUMBER | XFRC-104 |

4 IDENTIFICATION

4.1 The effectiveness of controls and displays depends upon identification. Therefore, this identification should be positioned as close as is practical to its respective item. The location of the identification in relation to the item should be consistent throughout the control center to minimize confusion for operation, maintenance, and testing.

4.2 Identification lettering should be oriented horizontally and centered or left-justified. The minimum spacing between words should be the width of the letter "E." The spacing between the bottom of a line of lettering and the top of the following line should be a minimum of 40 percent and a maximum of 66 percent of the letter height. (See Appendix A for recommended spacing.)

4.3 Similar types of identification on controls and displays should have the same height letters with uniform stroke width. (See Appendix B for abbreviations and Appendix C for letter height recommendations.)

4.4 Letters should preferably be black or white on a sharply contrasting background, upper case (capital letters are preferred). Dark letters on light background will minimize illegibility from dirt buildup in the engraved letters. The condensed gothic style of letters is usually the easiest to read. (See Appendix D.)

4.5 Background color may be used to emphasize separate but related functions. (See ISA-RP60.3 Human Engineering for Control Centers.) The most legible combination of letter color and background color should be used in lieu of using one letter color on all background colors.

4.6 When manufacturers' standards are not suitable, the material, type, and method of attachment of the identification should be specified.

5 TYPES OF IDENTIFICATION

5.1 The type and method of identification should be standardized early in the design to ensure consistency.

5.2. Nameplates are one of the most common types of identification and are used mostly on panels and displays. Nameplates are normally adjacent to the instruments to be identified. Nameplates may also be used in the back of panels to aid assembly, maintenance, and troubleshooting.

5.3 Labels made of flexible material may be used in all of the above situations and are especially useful on irregular surfaces (instrument bodies, etc.).

5.4 Tags are suitable for identifying interconnecting piping and wiring. Tags may be fastened with wire or plastic ties when other methods of attachment are not acceptable. (See Appendix A.)

5.5 Terminal identification for wiring or tubing is necessary for installation, maintenance, and troubleshooting. See Table 1 for a summary of processes used in terminal identification.

6 MATERIALS

6.1 Most materials that can be cut or shaped can be used for identification purposes. Some of the most common materials used are metals, plastics, and laminated stocks. Nameplates are frequently made from laminated engraving stock.

6.2 Metals are used in making nameplates and tags. Aluminum, brass, steel, tinplate, and stainless steel are commonly used in embossing, engraving, stamping, and photoetching processes. Metals can be painted many different colors with chemical-resistant, baked enamel or epoxy paints.

6.3 Plastics are used in making nameplates, tags, and terminal identifications. A transparent plastic engraving stock, such as acrylic, rigid vinyl, and polycarbonate, is generally coated with paint on the back surface and engraved from the back side. The engraved area is then filled with a contrasting color paint.

6.4 Laminated engraving stocks have a constrasting color center core. A wide range of colors are available. Using an engraving machine, the characters are transferred to the stock by routing out the top colored layer to expose the contrasting color core. Laminated engraving stocks are available in various thicknesses and may be cut to the final nameplate size from sheet stock.

6.5 Heavy paper or fabric tags with handwritten or typed descriptions have uses as temporary identification if permanent methods are not available or desirable at the time of their installation.

6.6 The choice of materials used for identification may depend on the environment of the service. For instance, ultraviolet radiation from direct sunlight and heat can cause certain types of plastics to deteriorate.

7 METHODS OF FABRICATION

7.1 There are various acceptable methods of making nameplates, labels, and tags for identification of controls and displays, including the following processes:

7.1.1 Etching is a process of engraving lines in plates by means of acid. Etchings can be made on copper, steel, and glass, with copper being the most frequently used material.

7.1.2. Engraving is a method of carving words or designs on the surface of metal or plastics.

7.1.3 Painting is a process utilizing any kind of paint in various colors to cover almost any surface with designs and words.

7.1.4 Stenciling is a process of reproducing designs or letters on most surfaces by passing a brush wet with ink or paint over a stencil. A stencil is a thin sheet of metal or other material with lines and dots cut into it to form a pattern.

7.1.5 Silk screening is a means of printing with stencils on almost any surface. Almost any kind of ink or paint can be used in the process. Bolting cloth (stencil silk) is stretched across a frame, and a stencil pattern is applied to this supporting surface. A squeegee is used to force the ink or paint through the open areas of the stencil.

7.1.6 Typing is a common method used to produce printed characters on paper or other material.

7.1.7 Photo-etching is a photographic process of superimposing lines or designs, in color, permanently to anodized aluminum.

7.1.8 Embossing is a process in which a raised design is stamped or pressed by mechanical means on the surface of metal, paper, or plastic.

7.1.9 Stamping is the process of imprinting or making impressions by the use of a die or other stamping tools.

TABLE 1

IDENTIFICATION TYPES

PROCESS	NAMEPLATES	LABELS	TAGS	TERMINAL IDENTIFICATION*
Etching	X		X	
Engraving	X	X	X	
Painting	X	X	X	X
Silk Screening		X		X
Stenciling		X		X
Typing	X	X	X	X
Photo-etching	X	X		
Lithography		X	X	
Embossing	X		X	
Stamping		X	X	X

*See Section 5 TYPES OF IDENTIFICATION for terminal identification

7.2 Each of the above processes has a certain cost vs. longevity ratio and should be selected on the basis of best serving the purpose desired.

8 METHODS OF ATTACHMENT

8.1 The most common methods of attachment are mechanical and adhesive.

8.2 Mechanical methods of attachment of nameplates, labels, and tags directly to the surface of an item by non-corrosive rivets, screws, brads, or clamps can be considered permanent. Wrap-around loops of chain, wire, and plastic ties can also be used for securing identification directly to the item to be identified.

8.3 Adhesives are available to attach nameplates to clean, smooth surfaces. The same purpose may be served by using double-faced pressure-sensitive tape. The permanence of adhesives is dependent upon the conditions to which they are subjected. A hostile and changing environment is the natural enemy of the adhesive method of attaching nameplates, and the permanence of the adhesive method is directly related to environmental exposure.

8.4 One of the considerations in choosing the method of attachment may be the desired degree of permanence. There may be instances where anticipated future revisions would sway the selection to a method of attachment that allows easy removal.

9 TYPICAL LOCATIONS

9.1 Identification of instruments and controls on the face of a control center may be used to make clear the relationship between systems and their subsystems. An example is shown in Appendix A (A-1), page 13.

9.2 Typical nameplates are shown in Appendix A (A-2), page 13. After the identification wording is selected and the size of lettering and number of lines are chosen, the overall dimensions of the nameplate may be selected using Appendix A (A-3), page 15. To aid in operator recognition, it may be desirable to identify similar types of devices or functions with the same size of nameplate. It is not recommended, however, that one size of nameplate be routinely used for all devices and functions without regard to the space actually required by the lettering in each case.

9.3 Typical wire terminal identifications are shown in Appendix A (A-4), page 16. Wiring schedules should establish the identification method used.

9.4 Control Center components that may be removed for maintenance should, as a minimum, have their permanent mounting locations identified. The removable component may also be identified.

RP60.6

Nameplates, Labels and Tags for Control Centers

APPENDIX A
CONTROL CENTER EXAMPLES
A-1

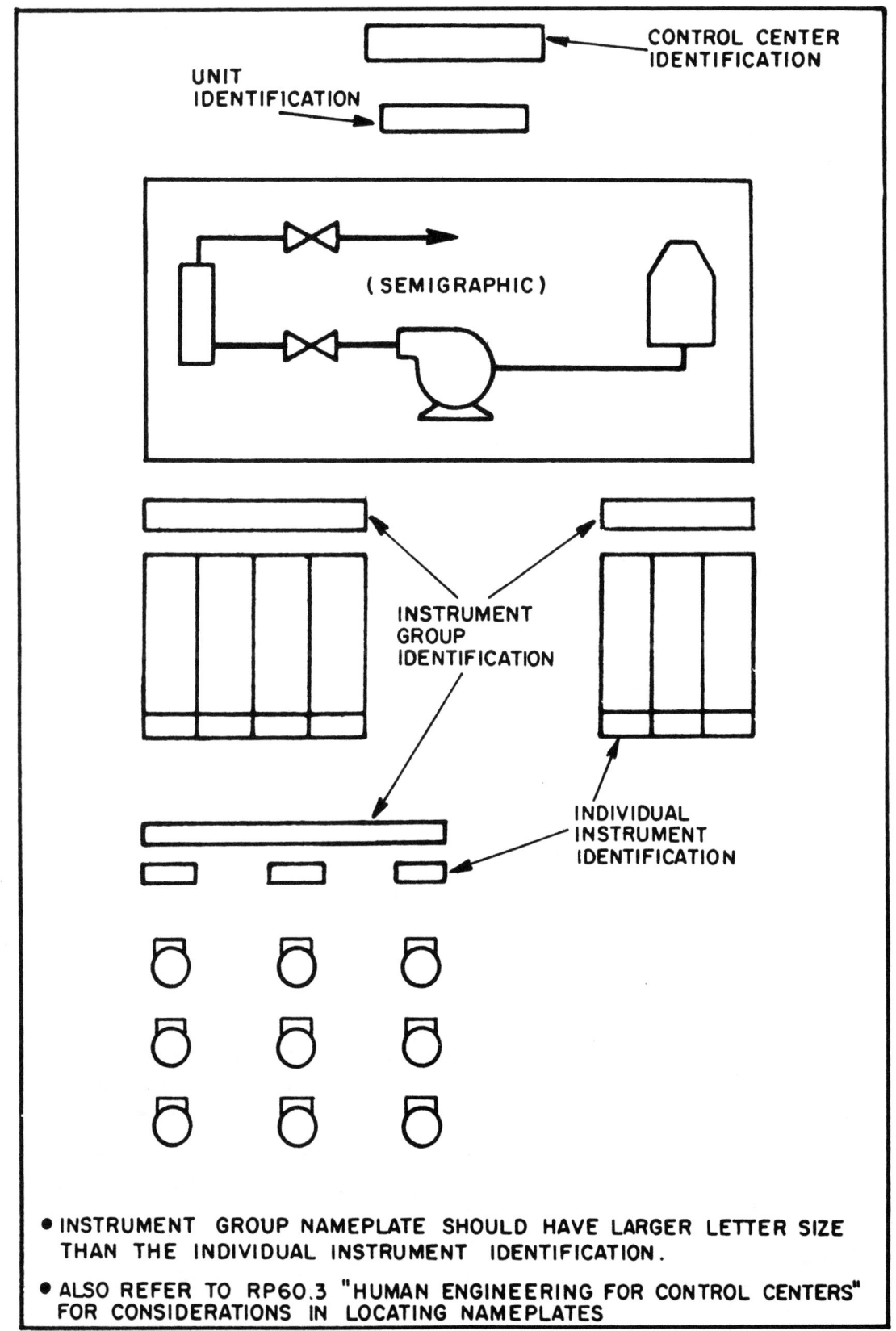

**APPENDIX A
NAMEPLATES
(USABLE AREA)
A-2**

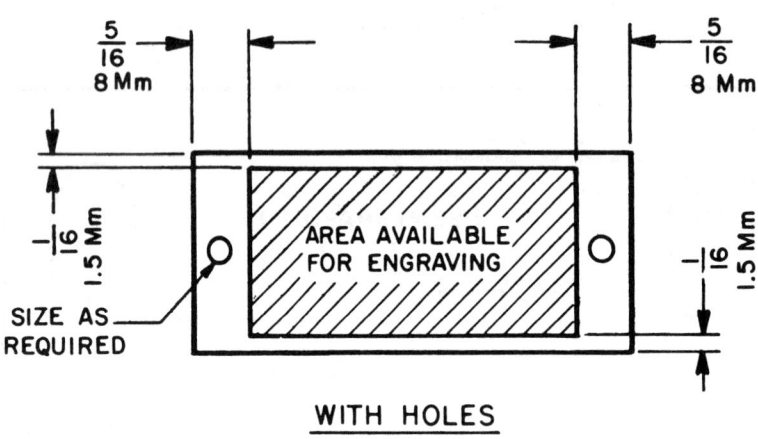

WITH HOLES

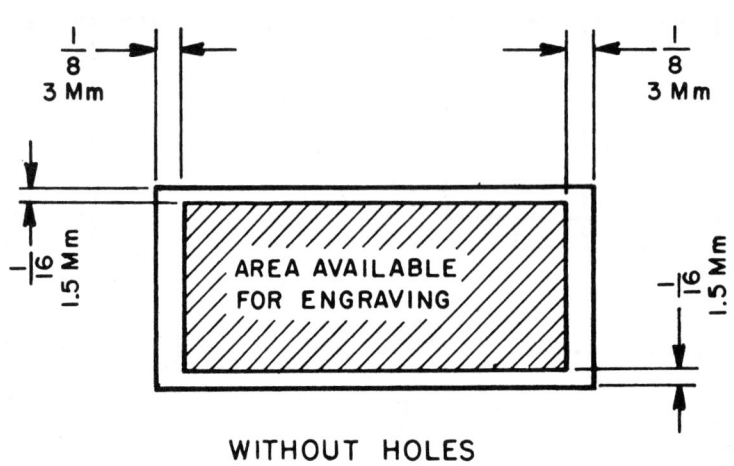

WITHOUT HOLES

OVERALL HEIGHT OF LETTERS	LETTER SPACING AVERAGE NUMBER OF LETTERS PER INCH *	SPACE BETWEEN LINES OF LETTERS
1/4 "	5	7/64 "
3/16 "	7	5/64 "
5/32 "	8	1/16 "
1/8 "	10	3/64 "
3/32 "	12	1/32 "

* 1 BLANK SPACE = 1 LETTER

APPENDIX A
NAMEPLATE
LETTER SIZE AND SPACING GUIDE
A-3

Based on gothic style letters with stroke width of 1/6th to 1/8th of character height.

AVERAGE LETTERS PER LINE

LETTER HEIGHT		NAMEPLATE LENGTH 1	1 1/4	1 1/2	1 5/8	2	2 1/4	2 1/2	2 3/4	3	3 1/4	3 1/2	3 3/4	4	4 1/2	5	5 1/2	6	6 1/2	7	7 1/2	8	8 1/2
1/4 in.	WITH HOLES	1	3	4	5	6	8	9	10	11	13	14	15	16	19	21	24	26	29	31	34	36	39
	W/O HOLES	3	5	6	6	8	10	11	12	13	15	16	17	18	21	23	26	28	31	33	36	38	41
3/16 in.	WITH HOLES	2	4	6	7	9	11	12	15	16	18	20	21	23	27	30	34	37	41	44	48	51	55
	W/O HOLES	5	7	8	9	12	14	15	17	19	21	22	24	26	29	33	36	40	43	47	50	54	57
5/32 in.	WITH HOLES	3	5	7	8	11	13	15	17	19	21	23	25	27	31	35	39	54	57	61	66	69	63
	W/O HOLES	6	8	10	11	14	16	18	20	22	24	26	28	30	34	38	42	46	50	54	58	62	66
1/8 in.	WITH HOLES	3	6	8	10	13	16	18	21	23	26	28	31	33	38	43	48	53	58	63	68	73	78
	W/O HOLES	7	10	12	13	17	20	22	25	27	30	32	35	37	42	47	52	57	62	67	72	77	82
3/32 in.	WITH HOLES	4	7	10	12	16	19	22	25	28	31	34	37	40	46	52	58	64	70	76	82	88	94
	W/O HOLES	9	12	15	16	21	24	27	30	33	36	39	42	46	52	58	64	70	76	82	88	94	100

MAXIMUM LINES PER NAMEPLATE

LETTER HEIGHT		NAMEPLATE HEIGHT 1/2	5/8	3/4	7/8	1	1 1/4	1 1/2	1 3/4	2	2 1/4	2 1/2	2 3/4	3									
1/4 in.		1	1	2	2	3	3	4	4	5	6	7	7	8									
3/16 in.		1	2	2	3	3	4	5	6	7	8	9	10	11									
5/32 in.		2	2	3	3	4	5	6	7	9	10	11	12	13									
1/8 in.		2	3	4	4	5	6	8	9	11	12	14	15	17									
3/32 in.		3	4	5	6	7	9	11	13	15	17	19	21	23									

APPENDIX A
WIRE TERMINAL IDENTIFICATION
A-4

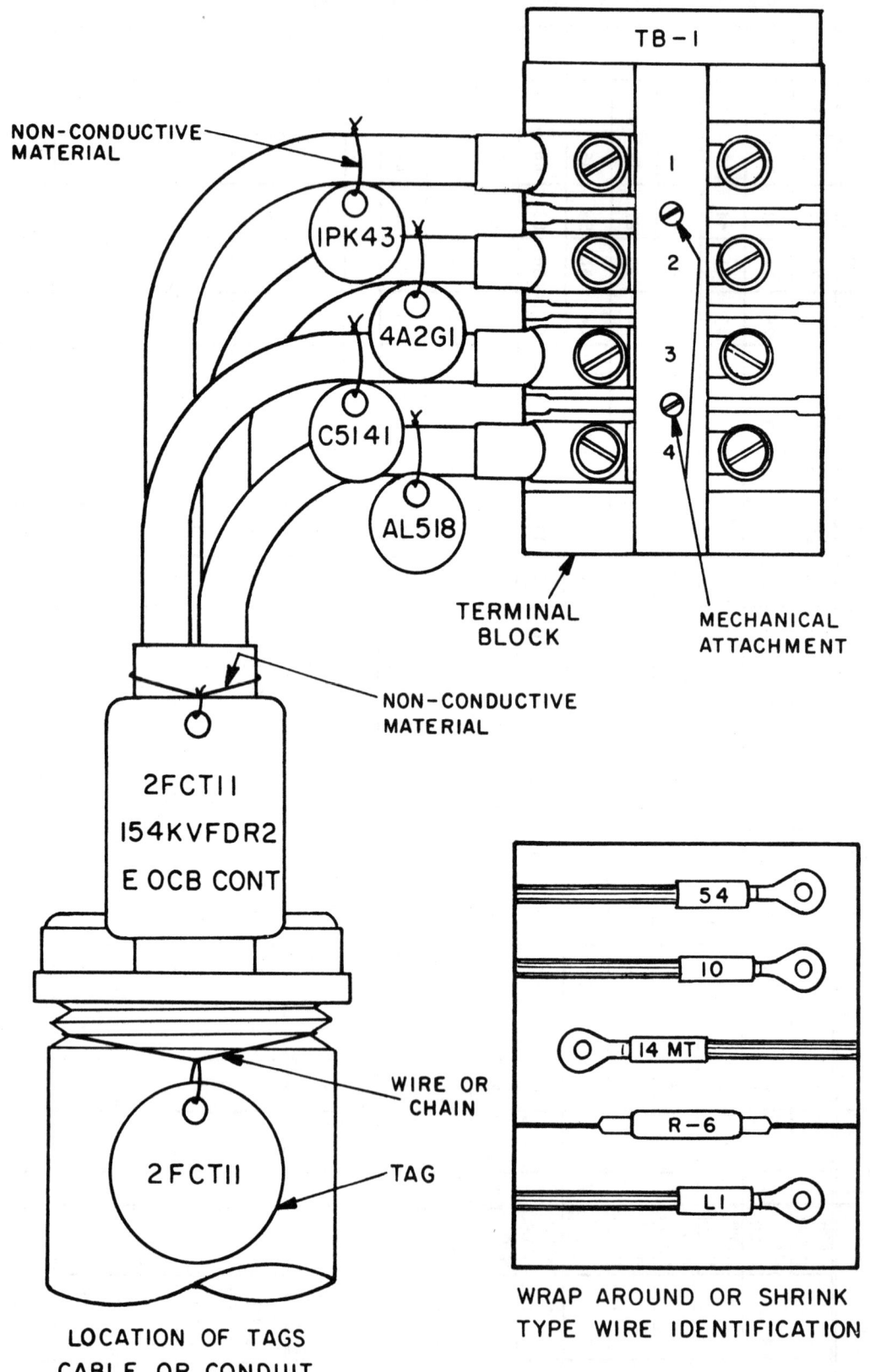

APPENDIX B
ABBREVIATIONS

Where possible, abbreviations should be avoided. Where abbreviations are used they should be consistent throughout the control center. One abbreviation should not have more than one meaning.

The following Table of Abbreviations is representative only and is not intended to be either mandatory or complete.

- A -

Abandoned	ABND
Abbreviate	ABBR
Abnormal	ABNL
About	ABT
Above	ABV
Absolute	ABS
Absorber	ABR
Absorptive	ABSV
Abutment	ABUT
Accelerate	ACCEL
Acceleration (due to gravity)	G
Acceleration (in general)	A
Access Opening	AO
Access Panel	AP
Accessory	ACCESS
Account	ACCT
Accumulate	ACCUM
Accumulator	ACC
Acetylene	ACET
Acidproof	AP
Acoustic	ACST
Acre	AC
Acre-foot	AC-FT
Activation	ACTN
Actual	ACTL
Actual Cubic Foot (per min.)	ACFM
Actuate	ACTE
Actuating	ACTG
Actuator	ACTR
Adapter	ADPT
Addition	ADD
Additive	ADDT
Adhesive	ADH
Adjust	ADJ
Advance	ADV
After	AFT
Afternoon	PM
Aggregate	AGGR
Agitator	AGTR
Air Blast	AB
Air Blast Circuit Breaker	ABCB
Air Blast Transformer	ABT
Air Break Switch	ABS
Air Circuit Breaker	ACB
Air Condition	AIR COND, A/C
Air Conditioning (Cooling-Heating System	A/CS
Air Conditioning Unit	A/CU
Air Cooled	AIR CLD
Air Handling Unit	AHU
Airborne	ABN
Airplane	APL
Air Supply	A/S
Airtight	AT
Alarm	ALM
Alarm Check Valve	ACV
Alcohol	ALC
Alkaline	ALK
Alloy	ALY
Alphanumeric Keyboard	ANK
Alteration	ALTRN
Alternate	ALT.
Alternating Current	AC
Alternator	ALTNTR
Altitude	ALT
Aluminum	AL
Amber	AMB
Ambient	AMB
American Wire Gage	AWG
Ammeter	AM
Amount	AMT
Ampere	A, AMP
Ampere-hour	AH
Amperes per Centimeter per °F	A/CM/°/F
Ampere per Meter	A/M
Ampere Turn	AT
Amplifier	AMPL
Amplitude	AM
Amplitude Modulation	AM
Analog Control Output	ACO
Analog Status Input	ASI
Analog Status Output	ASO
Analog-to-Digital	A/D
Analog-to-Digital Converter	ADC
Analysis	ANAL
Analyzer	ANALZ
And So Forth	ETC
Angstrom	A
Anhydrous	ANHYD
Annulus	ANNS
Annunciator	ANN
Anode	A
Antenna	ANT
Apartment	APT
Apparatus	APP
Area	A
Armature	ARM
Armored	ARMD
Arrange	ARR
Arrangement	ARRGT
Arrester	ARR
Artificial	ART
Assembly	ASSY
Assistant	ASST

Associate	ASSOC
Atmosphere	ATM
Atomic	AT
Attach	ATT
Attention	ATTN
Attenuator	ATTEN
Audible	AUD
Audio Frequency	AF
Authorized	AUTH
Automatic	AUTO
Automatic Brightness Control	ABC
Automatic Door Seal	ADS
Automatic Check Valve	ACV
Automatic Dispatch System	ADS
Automatic Frequency Control	AFC
Automatic Gain Control	AGC
Automatic Noise Limiter	ANL
Automatic Sensitivity Control	ASC
Automatic Volume Control	AVC
Automatic Volume Expansion	AVE
Autotransformer	AUTOXFMR
Auxiliary	AUX
Auxiliary Area	AA
Auxiliary Building	AUXBLDG
Auxiliary Control Panel	ACP
Auxiliary Control Room	ACR
Auxiliary Feedwater	AFW
Auxiliary Feedwater Pump	AFWP
Auxiliary Oil Pump	AOP
Auxiliary Relay	AR
Average	AVG
Average Power Range Monitoring	APRM
Aviation	AVI
Avoirdupois	AVDP
Azimuth	AZ

- B -

Back Flow Prevention Valve	BPVLV
Back Gear	BG
Back Pressure	BP
Back Pressure Control	BPC
Back Pressure Valve	BPV
Back Wash	BW
Back Water Valve	BWV
Bacteriological	BACT
Baffle	BAF
Balance	BAL
Balance of Plant	BOP
Ballast	BALL
Bandwidth	B
Bank	BK
Barometer	BAR
Barrel	BBL
Barrels Per Day	BPD
Barrels Per Hour	BPH
Barrier	BARR
Base Group	BG
Basement	BSMT
Basic Network	BAS NET
Battery	BAT
Baume	BE

Beacon	BCN
Beam	BM
Bearing	BRG
Beat Frequency Oscillator	BFO
Bedplate	BDPLT
Bell Alarm Switch	BASW
Below	BEL
Belt Conveyor	BC
Bending Moment	M
Bessemer	BESS
Between	BET
Binary Coded Decimal	BCD
Binary Input Multiplexer	BIM
Binary Output Multiplexer	BOM
Bituminous	BIT
Black	BLK, BK
Blank	BLK
Blasting Powder	BLSTGPWD
Block	BLK
Blower	BLO
Blow Down	BLWDN
Blowoff	BO
Blowout Coil	BOC
Blue	BLU, BL
Blueprint	BP
Board	BD
Boiler	BLR
Boiler Feed	BF
Boiler Feed Booster Pump	BFBP
Boiler Feed Pump	BFP
Boiler Feed Pump Turbine	BFPT
Boiler Feed Water	BFW
Boiler Horsepower	BHP
Boiler House	BH
Boiler Pressure	BP
Boiling Point	BP
Boiler Water Reactor	BWR
Booster	BSTR
Borrow	BOR
Borrowed Light	BLT
Bose Chaudhuri-Hocquenghem	BCH
Bottled	BTL
Bottom	BOT
Bottom Layer	BL
Boulder	BLDR
Boundary	BDY
Bracket	BRKT
Brake	BK
Brake Horsepower	BHP
Brake Horsepower-Hour	BHP-HR
Brake Mean Effective Pressure	BMEP
Branch	BR
Break	BRK
Breaker	BKR
Breeder	BDR
Bridge	BRDG
Brightness	BRT
British Thermal Units	BTU
British Thermal Units per Hour	BTUH
British Thermal Units per Hour (thousand)	MBH

Brown	BRN, BR
Brush	BR
Bubbler	BUBLR
Buffer	BUF
Building	BLDG
Building Heating System	BHS
Bulkhead	BHD
Bundle	BDL
Bus Tie	BT
Bushing Current Transformer	BCT
Butterfly Valve	BTFLY VLV
Button	BUT
Buzzer	BUZ
By-pass	BYP
By-pass Handwheel Motor	BPHM
By-pass Handwheel Switch	BPHS

- C -

Cabinet	CAB
Cable	CA
Cadmium Plate	CD PL
Calculate	CALC
Caliber	CAL
Calibrate	CAL
Calorie	CAL
Camber	CAM
Camera	CAMR
Candlepower	CP
Capacitor	CAP
Capacity	CAP
Capital	CAP
Carbon Dioxide	CO_2
Carbon Dioxide Storage, Fire Protection and Purging Sys.	CO_2S
Carbon Monoxide	CO
Carload	CL
Carriage	CRG
Carrier	CARR
Carrier Current	CARR CUR
Carton	CTN
Casing	CSG
Cast (used with other materials)	C
Cast Iron	CI
Cast Iron Pipe	CIP
Cast Iron Soil Pipe	CISP
Cast Steel	CS
Casting	CSTG
Catalogue	CAT
Catch Basin	CB
Cathode-ray Tube	CRT
Caustic	CAUS
Cavitation	CAVIT
Cavity	CAV
Ceiling	CLG
Celsius	C
Cement	CEM
Center	CTR
Center Line	CL
Center Matched	CM
Center of Buoyancy	CB
Center of Floatation	CF
Center of Gravity	CG
Center of Pressure	CP
Center Tap	CT
Centigrade (obsolete, use Celsius)	C
Centigram	CG
Centimeter	CM
Centimeter-gram-second System	CGS
Central Lubricating Oil System	CLOS
Central Processing Unit	CPU
Centrifugal	CNTFGL
Centrifugal Force	CF
Ceramic	CER
Chain	CH
Chain Grate	CG
Chamber	CHBR
Chamfer	CHFR
Change	CHG
Change Notice	CN
Change Order	CO
Channel	CHAN
Charge	CHG
Charger	CHGR
Check	CHK
Check Valve	CV
Chemical	CHEM
Chemical Addition Tank	CAT
Chemical Cleaning System	CHCS
Chemically Pure	CP
Chilled Water	CHW
Chilled Water Return	CWR
Chilled Water Supply	CWS
Chlorination	CLN
Circle	CIR
Circuit	CKT
Circuit Breaker	CB
Circular	CIR
Circular Mil	CMIL
Circular Mils, Thousands	KCMIL
Circular Pitch	CP
Circulating Water Pump	CWP
Clamp	CLP
Class	CL
Cleanout	CO
Clear	CLR
Clearance	CL
Clevis	CLV
Clockwise	CW
Closed	CLSD
Closed-Circuit Television	CCTV
Closing	CL
Closing Coil	CC
Clutch	CL
Coated	CTD
Coaxial	COAX
Code, Alarm, Paging	CAP
Code Call	CC
Code Call Alarm and Paging	CCAP
Coefficient	COEF
Cold Water	CW
Collector	COLL

Column	COL
Combination	COMB
Combine	COMB
Combustion	COMB
Commercial	COML
Common	COM
Common Battery	CB
Communication	COMM
Commutator	COMM
Commutator End	CE
Company	CO
Compare	CMPR
Compartment	COMPT
Compensate	COMP
Complete	COMPL
Component	CMPNT
Component Cooling	CC
Component Cooling System	CCS
Composite	CX
Composition	COMP
Compound	CMPD
Compress	CPRS
Compressed Air	CA
Compressor	CPRSR
Computation	COMP
Computer	COMP
Concentrate	CONC
Concrete	CONC
Condensate	CNDS
Condensate System	CS
Condenser	COND
Condenser Circulating Water	CCW
Condenser Circulating Water Pump	CCWP
Condenser Circulating Water Pump Station	CCWPS
Condenser Circulating Water System	CCWS
Condition	COND
Conductivity	CNDCT
Conductor	CNDCT
Configuration	CFGN
Connector	CONN
Constant	CONST
Constant Current Transformer	CCT
Construction	CONSTR
Contact	CONT
Contactor	CNTOR
Container	CNTNR
Containment	CNTMT
Contaminated	CONTAM
Continue, Continuous	CONT
Contract	CONTR
Contractor	CONTR
Contrast	CTRS
Control	CONT
Control Building	CONT BLDG
Control Element Assembly	CEA
Control Relay	CR
Control Rod Drive	CRD
Control Rod Drive System	CRDS
Control Station	CONT STA
Control Switch	CS
Control Valve	CV
Controller	CONT
Converter	CONV
Conveyor	CNVR
Coolant	COOL
Cooled	CLD
Cooler	CLR
Cooling	CLG
Cooling Coil	CC
Cooling Towers	CT
Cooling Water	CW
Corner	COR
Correct, Correction	CORR
Correspond	CRSP
Cotton	COT
Coulomb	C
Counter	CNTR
Counterclockwise	CCW
Counterbalance	CBAL
Counterweight	CTWT
Coupling	CPLG
Cover	COV
Crank	CRK
Crankcase	CRKC
Critical Path Method	CPM
Cross Arm	XARM
Cross Beam	XBEAM
Cross Connection	XCONN
Cross Road	XRD
Crossbar	XBAR
Crosstie	XTIE
Cruising	CRUIS
Crushed	CR
Crystal	XTAL
Cubic	CU
Cubic Centimeter	CM^3, CC
Cubic Feet	FT^3, CU FT
Cubic Feet Per Minute	CFM
Cubic Feet Per Second	CFS
Cubic Inch	IN^3, CU IN.
Cubic Meter	M^3, CU M
Cubic Micrometer	CU MU
Cubic Millimeter	MM^3, CU MM
Cubic Yard	YD^3, CU YD
Current	CUR
Current Directional Relay	CDR
Current Transformer	CT
Current to Current	I/I
Current to Pneumatic	I/P
Current to Voltage	I/E
Customer	CUST
Cut Down	CD
Cut Out	CO
Cycle	C
Cycles Per Minute	CPM
Cylinder	CYL

- D -

Damper	DMPR
Dash Pot	DP

Deadweight	DWT
Deaerating Feed Tank	DFT
Decay	DCA
Decibel	DB
Decimal	DEC
Decimeter	DM
Deck	DK
Decrease	DECR
Deflect	DEFL
Degree	°, DEG
Dehumidifier	DHMR
Delayed Automatic Volume Control	DAVC
Delineation	DEL
Deliver	DELIV
Demand Indicator	DI
Demand Meter	DM
Demineralizer	DMNRLZR
Demodulator	DEM
Density	D
Describe	DESCR
Designation	DESIG
Detector	DET
Develop	DEV
Development	DEV
Deviation	DEVN
Device	DV
Dew Point	DP
Dew Point Temperature	DPT
Diagonal	DIAG
Diagram	DIAG
Diameter	DIA
Diaphragm	DIAPH
Diesel	DSL
Diesel Generator	DG
Diesel Generator Building	DGB
Diesel Oil	DO
Differential	DIFF
Differential Expansion Detector	DXD
Differential Pressure	DP
Differential Temperature	DIFF T
Differential Time Relay	DIFF TR
Digital-to-Analog Converter	DAC
Dimension	DIM
Direct Connected	DIR CONN
Direct Current	DC
Direction	DIR
Director	DIR
Discharge	DISCH
Disconnect	DISC
Discriminator	DISCR
Dispatch	DISP
Display Generator	DG
Disposal	DPSL
Dissolved Oxygen	DO
Distance	DIST
Distill	DSTL
Distillate	DSTLT
Distilled Water	DW
Distribute	DISTR
District	DIST
District Engineer	DE

Ditto	DO
Diverter	DIV
Divide	DIV
Division	DIV
Dolomite	DOLM
Door	DR
Door Post	DP
Double	DBL
Double Acting	DA
Double Feeder	DF
Double Glass	DG
Double Groove (insulators)	DG
Double Pole Double Throw	DPDT
Double Pole, Single Throw	DPST
Doubler	DBLR
Down	DN
Downscale	DNS
Downspout	DS
downstream	DNSTR
Drain	DR
Drainage	DR
Drawing	DWG
Drawing List	DL
Drawn	DRWN
Drive	DR
Drive FIT	DF
Drop	D
Drop Manhole	DMH
Dry Bulb	DB
Dry Bulb Temperature	DBT
Dry Pipe Valve	DPV
Ductile Iron	DI
Duplex	DX
Duplicate	DUP
Dynamic	DYN
Dynamic Snubber	DYN S
Dynamotor	DYNM
Dynamo	DYN
Dynamometer	DYNMT

- E -

Each	EA
East	E
Eccentric	ECC
Eccentricity Detector	ED
Economizer	ECON
Eductor	EDUC
Effective	EFF
Effective Horsepower	EHP
Efficiency	EFF
Effluent	EFL
Ejector	EJCTR
Elastic Limit	EL
Elbow	ELL
Electric	ELEC
Electric Motor Operated	EMO
Electric Control Room	ECR
Electrical Hydraulic Controls	EHC
Electrohydraulic	ELYHD
Electrolyte	ELECT
Electrolytic	ELECTC

Electromechanical	ECMCH
Electromotive Force	EMF
Electrostatic	ES
Element	ELEM
Elementary	ELEM
Elevate	ELEV
Elevation	EL
Elevator	ELEV
Elongation	ELONG
Emergency	EMER
Emergency Bearing Oil Pump	EBOP
Emergency Feedwater System	EFS
Emergency Gas Treatment System	EGTS
Emergency Oil Pump	EOP
Emergency Power	EP
Enclose	ENCL
End Cell Switch	EC SW
Engine	ENG
Engineer	ENGR
Engineering	ENGRG
Entainment	ENTM
Entrance	ENT
Envelope	ENV
Environmental System	ES
Equal	EQ
Equalizer	EQL
Equation	EQ
Equipment	EQPT
Equipment Drain	ED
Equipment Drain Tank	EDT
Equivalent	EQUIV
Erection	EREC
Escape	ESC
Essential	ESSN
Essential Control Instrumentation	ECI
Estimate	EST
Evacuation	EVAC
Evaporator	EVAP
Excavate	EXC
Excavation	EXCAV
Excessive	EXC
Exchange	EXCH
Excitation	EXC
Exciter	EXC
Exclusive	EXCL
Executive	EXEC
Exhaust	EXH
Exhaust Hood Thermostat	EHT
Existing	EXIST
Expand	EXP
Expansion	EXP
Expansion Valve	EXP V
Experiment	EXP
Explosion Proof	EP
Exploration	EXPL
Expose	EXP
Expulsion	EXP
Extension	EXT
Exterior	EXT
External	EXT
Extinguish	EXT
Extra High Voltage	EHV
Extraction	EXTR
Extraction Steam System	ESS
Extreme High Water	EHW
Extreme Low Water	ELW
Extrude	EXTR

- F -

Facility	FACIL
Fahrenheit	F
Fail As Is	FAI
Fail Close	FC
Fail Open	FO
Failure	FAIL
Failsafe	FLSF
Farad	F
Feature	FEA
Federal	FED
Feed	FD
Feeder	FDR
Feedwater	FW
Feedwater Control System	FWCS
Feet	FT
Feet Per Day	FT/D, FPD
Feet Per Minute	FT/MIN, FPM
Feet Per Second	FT/SEC, FPS
Feet Per Year	FT/YR, FPY
Female	FEM
Fence	FEN
Fiber	FBR
Field	FLD
Field Multiplex	FM
Figure	FIG
Filament	FIL
Filament Center Tap	FCT
Fillet	FIL
Filling	FILL
Filter	FLTR
Finger	FGR
Fire	F
Fire Control	FC
Fire Department Connection	FDC
Fire Door	F DR
Fire Extinguisher Cabinet	FEC
Fire Hose	FH
Fire Hose Cabinet	FHC
Fire Hose Rack	FHR
Fire Hydrant	FH
Fire Main	FM
Fire Protection System	FPS
Fireproof	FPRF
Fireproofing	FPRFG
Fixture	FIX
Flameproof	FP
Flametight	FT
Flange	FLG
Flashing	FL
Flashless	FLHLS
Flat	F
Flat Bar	FB
Flexible	FLEX

Float	FLT
Float Switch	FS
Flooding	FLDNG
Floor	FL
Floor Drain	FD
Flow Control Valve	FCV
Flow Diagram	FD
Flow Switch	FWS
Flowmeter	FM
Fluid	FL
Fluorescent	FLUOR
Flush	FL
Flywheel Effect	WR²
Focus	FOC
Foot	FT
Footcandle	FC
Footlambert	FL
Foot-Pound	FT-LB
Foot-Pound Second	FPS
For Example	EG
Force	F
Forced Draft	FD
Forced Draft Blower	FDB
Forced Oil Air	FOA
Forenoon	AM
Forged Steel	FST
Forging	FORG
Fork	FK
Forward	FWD
Foundation	FDN
Foundry	FDRY
Four-wire	4W
Fractional	FRAC
Fractional Horsepower	FHP
Frame	FR
Framework	FRWK
Framing	FMG
Free on Board	FOB
Freeboard	FREEBD
Freezing Point	FP
Freight	FRT
Frequency	FREQ
Frequency Changer	FREQ CH
Frequency Shift	FS
Frequency Shift Keying	FSK
Frequency, High	HF
Frequency, Low	LF
Frequency, Medium	MF
Frequency Modulation	FM
Frequency, Super High	SHF
Frequency, Ultra High	UHF
Frequency, Very High	VHF
Frequency, Very Low	VLF
Fresh Water	FW
From Below	FR BEL
Front	FR
Front Connected	FC
Front View	FV
Fuel	F
Fuel Building	FB
Fuel-Air Ratio	F/A RATIO
Fuel Oil	FO
Fuel Oil System	FOS
Furnish	FURN
Fuse	FU
Fuse Block	FB
Fusible	FSBL
Fusion Point	FNP
Future	FUT

- G -

Gage or Gauge	GA
Gallery	GALL
Gallon	GAL
Gallons Per Acre Per Day	GPAD
Gallons Per Hour	GPH
Gallons Per Minute	GPM
Gallons Per Second	GPS
Galvanize	GALV
Galvanize Iron	GI
Galvanized Steel	GS
Galvanized Steel Wire Rope	GSWR
Garage	GAR
Gas	G
Gas Analyzer	GA
Gas Circuit Breaker	GCB
Gas Pressure Gage	GPG
Gas Stripper	GS
Gas Turbine	GTRB
Gasoline	GASO
Gate Input Card	GIC
Gate Output Register	GOR
Gate Valve	GTV
General	GEN
General Purpose Register	GPR
Generator	GEN
Generator Cooling System	GCS
Giga	G
Giga Hertz	GHZ
Gigavar	GVAR
Gigavolt-Ampere	GVA
Gigawatt	GW
Girder	G
Glass	GL
Glaze	GL
Globe Valve	GLV
Governor Load Change Motor	GLCM
Grade	GR
Grading	GRG
Graduation	GRAD
Grains Per Gallon	GPG
Gram	G
Gram Calorie	G-CAL
Gram-meter	G-M
Graphic	GRAPH
Graphite	GPH
Gravel	GVL
Gravity	G
Grey	GY
Grease Trap	GT
Green	G, GRN
Grid	G

Grind	GRD
Grommet	GROM
Groove	GRV
Gross Weight	GRWT
Ground	GRD
Group	GR
Guard	GD
Gypsum	GYP
Gyroscope	GYRO

- H -

Hand Control	HC
Hand Generator	HG
Hand Reset	HR
Hand Switch	HS
Hand Wheel	HD WHL
Hand Hole	HH
Handle	HDL
Hard	H
Head	HD
Head Water Gage Well	HWGW
Header	HDR
Headless	HDLS
Headquarters	HQ
Head Water	HW
Heat	HT
Heat Exchanger	HX
Heat Rejection System	HRS
Heater	HTR
Heater Drain & Vent System	HDVS
Heating	HTG
Heating Boiler System	HBS
Heating Cabinet	HC
Heating Coil	HC
Heating, Ventilating, and Air Conditioning	HVAC
Heavy	HVY
Height	HGT
Henry	H
Hertz	HZ
Hexagon	HEX
High	H
High Frequency	HF
High-High	HH
High-Low	HL
High Pressure	HP
High Pressure Fire Protection System	HPFPS
High Pressure Injection	HPI
High Pressure Injection System	HPIS
High Speed	HS
High Temperature	HT
High Temperature Water	HTW
High Tension	HT
High Voltage	HV
Highway	HWY
Holdup Pump	HUP
Holdup Tank	HUT
Holding	HLDG
Holding Coil	HC
Hollow	HOL
Horizontal	HORIZ
Horsepower	HP
Hot Water	HW
Hot Water, Circulating	HWC
Hot Well	HW
Hour	HR
House	HSE
Housing	HSG
Humidistat	HSTAT
Hundred	C
Hundredweight	CWT
Hybrid	HYB
Hydraulic	HYDR
Hydroelectric	HYDRELC
Hydrogen	H_2
Hydrogen (Ion Concentration)	pH
Hydrostatic	HYDRO
Hypochlorite	HYPCL

- I -

Identical	IDENT
Identify	IDENT
Ignition	IGN
Illuminate	ILLUM
Impact	IMP
Impedance	IMP
Imperial	IMP
Impulse	IMP
Inboard	INBD
Incandescent	INCAND
Inch	IN
Inch-pound	IN-LB
Inch Per Inch	IN/IN
Inches Per Revolution	IN/REV
Inches Per Second	IPS
Incinerator	INCIN
Include	INCL
Incoming	INC
Increase	INCR
Independent	INDEP
Indicate	IND
Indicated Horsepower	IHP
Indicating Lamp	IL
Indicator	IND
Induced Draft	ID
Inductance or Induction	IND
Information	INFO
Injection	INJ
Injection Water System	IWS
Inlet	INL
Input/Output	I/O
Inside	INS
Inspect	INSP
Install	INSTL
Instantaneous	INST
Instantaneous Relay	IR
Instruction	INSTR
Instruction Book	IB
Instrument	INSTR
Instruction Book	IB
Instrument	INSTR
Instrumentation	INSTM

Insulating Oil System	IOS
Insulting Transformer	IT
Intake	INTK
Integral	INT
Integrating	INT
Intercept	INTCP
Interceptor	INTCPR
Interchangeable	INTCHG
Intercommunication	INTERCOM
Intercooler	INCOLR
Integrated Control System	ICS
Interior	INT
Interior Communication	IC
Interlock	INTLK
Intermediate	INTMD
Intermediate Distributing Frame	IDF
Intermediate Frequency	IF
Intermediate Power Amplifier	IPA
Intermediate Pressure	IP
Intermediate Range Monitoring	IRM
Intermittent	INTMT
Internal	INT
Interphone Control Station	ICS
Interrupt	INTER
Interruptions Per Minute	IPM
Interruptions Per Second	IPS
Intersect	INT
Inverse	INVS
Inverse Time Relay	ITR
Invert	INV
Inverter	INVR
Ion Exchanger	IX
Irregular	IRREG
Island	I
Isolate	ISOL
Isolation	ISLN
Issue	ISS

- J -

Jack	J
Jacket Water	JW
Jacking Pump	JP
Joint	JT
Joule	J
Journal	JNL
Junction	JCT
Junction Box	JB

- K -

Kelvin	K
Key	K
Key Box	KBX
Keyseat	KST
Keyway	KWY
Kick Plate	KP
Kick Plate & Drip	KP&D
Kilo	K
Kilocalorie	KCAL
Kilocycle	KC
Kilocycles Per Second	KS
Kilogram	KG
Kilogram Meter	KG-M
Kilograms Per Cubic Meter	KG PER CU M
Kilograms Per Second	KGPS
Kilohertz	KHZ
Kiloliter	KL
Kilometer	KM
Kilometer Per Hour	KMPH
Kilometer Per Second	KMPS
Kilovar	KVAR
Kilovarhour	KVARH
Kilovolt	KV
Kilovolt-ampere	KVA
Kilovolt-ampere Hour	KVAH
Kilowatt	KW
Kilowatt-hour	KWH
Kinescope	KIN
Kip (1000 lb)	K
Kip-feet	K-FT
Knife Switch	KN SW
Knocked Down	KD
Knock-out	KO

- L -

Laboratory	LAB
Ladder	LAD
Lambert	L
Laminate	LAM
Lamp	L
Lamp, Green Indicating	GIL
Lamp, Red Indicating	RIL
Lamp, Yellow Indicating	YIL
Lamp, White Indicating	WIL
Landing	LDG
Latch Indicator Switch	LIS
Lateral	LAT
Latitude	LAT
Lavatory	LAV
Leakage	LKG
Leakoff	LOFF
Left	L
Left Hand	LH
Length	LG
Letdown	LTDN
Level	LVL
Level Controller	LC
Level Switch	LS
Light	LT
Lighting	LTG
Lighting Cabinet	LC
Lightning Arrester	LA
Limestone	LMST
Limit	LIM
Limit Switch	LIM SW
Limited	LTD
Limiter	LIM
Line	L
Linear	LIN
Linear Foot	LIN FT
Link	LK
Liquid	LIQ
Liquid Petroleum	LP

Liter	L
Live Load	LL
Load Bearing Switch	LBS
Load Dispatch System	LDS
Load Limit Motor	LLM
Load Limiting Resistor	LLR
Load Ratio	LR
Load Ratio Control	LRC
Load Shifting Resistor	LSR
Locked Closed	LC
Locked Open	LO
Logarithmic	LOG
Long	L
Longitude	LONG
Low	L
Low Frequency	LF
Low-Low	LL
Low Pass	LP
Low Point	LP
Low Pressure	LP
Low Pressure Injection	LPI
Low Pressure Injection System	LPIS
Low Speed	LPS
Low Temperature	LT
Low Tension	LT
Low Torque	LT
Low Vacuum Alarm	LVA
Low Voltage	LV
Lower	LWR
Lubricating Oil	LO
Lumen	LM
Lumens Per Watt	LPW

- M -

Machine	MACH
Magnet	MAG
Magnetic	MAG
Magneto	MAG
Main	MN
Main Control Room	MCR
Main Distributing Frame	MDF
Main Feedwater Pump	MFP
Main Feedwater Pump Turbine	MFPT
Main Steam Isolation Signal	MSIS
Main Steam Systems	MSS
Main Steam Vault	MSV
Maintenance	MAINT
Makeup	MKUP
Manhole	MH
Manifold	MANF
Manual	MNL
Manual Volume Control	MVC
Manufacture	MFR
Manufactured	MFD
Manufacturer	MFRR
Manufacturing	MFG
Marine	MAR
Master	MA
Master Oscillator	MO
Master Trip Solenoid	MTS
Masterswitch	MSW

Material	MATL
Maximum	MAX
Maximum Differential Pressure	MAX DIFF P
Maximum Differential Temperature	MAX DIFF T
Maximum Working Pressure	MWP
Mean Effective Difference	MED
Mean Effective Pressure	MEP
Mean Sea Level	MSL
Mechanical	MECH
Mechanism	MECH
Medical	MED
Medium	MED
Mega	M
Megacycles	MC
Megahertz	MHZ
Megavolt Ampere	MVA
Megawatt	MW
Megawatt Demand Setter	MDS
Megawatt Hour	MWH
Megohm	MΩ
Metal	MET
Meteorological	MET
Meter (instrument)	MTR
Meter (Instr, Comb Form & Unit of Length)	M
Metering	MTR
Meters Per Second	MPS
Micro	μ
Microampere	μA
Microfarad	μF
Microhenry	μH
Micrometer	μM
Micromho	μMHO
Micro-Micro (use Pico)	
Micromicrofarad (use Picofarad)	
Micron (use Micrometer)	
Microseconds	μSEC
Microvolt	μV
Microvolts Per Meter	μV/M
Microwatt	μW
Microwave	MW
Miles	MI
Miles Per Gallon	MPG
Miles Per Hour	MPH
Milli	M
Milliampere	MA
Milligram	MG
Millihenry	MH
Millilambert	ML
Milliliter	ML
Millimeter	MM
Million Gallons Per Day	MGD
Milliseconds	MS
Millivolt	MV
Milliwatt	MW
Minimum	MIN
Minimum Differential Pressure	MIN DIFF PRESS
Minimum Differential Temperature	MIN DIFF TEMP
Minute	MIN
Miscellaneous	MISC
Mixture	MIX
Modulator	MOD

Modulator and Demodulator	MODEM
Moisture	MSTRE
Molding	MLDG
Molecular Weight	MOL WT
Monitor	MON
Month	MO
Motor	MOT
Motor Control Center	MCC
Motor Field	MF
Motor Generator	MG
Motor Operated	MO
Motor Operated Disconnect	MOD
Motor Operated Valve	MOV
Mounted	MTD
Mounting	MTG
Multiple	MULT
Multiplexer	MPX

- N -

Nameplate	NP
Nano Second	NSEC
Negative	NEG
Net Positive Section Head	NPSH
Network	NET
Neutral	NEUT
Nitrogen	N_2
Nominal	NOM
Normal	NOR
Normal Pool Elevation	NOR POOL EL
Normal Power	NORP
Normally Closed	NC
Normally Open	NO
North	N
Northeast	NE
Northwest	NW
Nozzle	NOZ
Nuclear	NUC
Number	NO

- O -

Off-Gas	OG
Off-Gas System	OGS
Ohm	Ω
Oil Circuit Breaker	OCB
Oil Insulated	OI
Oil Level Gage	OLG
Oil Switch	OS
Open-Close-Open	O-C-O
Opening	OPNG
Operate	OPR
Operator	OPER
Optical	OPT
Orientation	ORNT
Orifice	ORF
Oscillograph	OSC
Oscillograph Test Block	OSTB
Oscilloscope	OSC
Oscilloscope Test Block	OSTB
Ounce	OZ
Ounce-foot	OZ-FT
Ounce-inch	OZ-IN

Outboard	OUTBD
Outgoing	OUT
Outgoing Repeater	OGR
Outgoing Trunk	OGT
Outlet	OUT
Output	OUT
Outside	OUT
Outside Diameter	OD
Overall	OA
Overcurrent	OC
Overflow	OVFL
Overhead	OVHD
Overload	OVLD
Overspeed	OVSP
Overvoltage	OVV
Oxidized	OXD
Oxygen	O, OXY

- P -

Packing	PKG
Page	P
Pages	PP
Pair	PR
Panel	PNL
Parallel	PAR
Part	PT
Part 1 of 2	P1/O2
Parts Per Billion	PPB
Parts Per Million	PPM
Partial	PART
Particulate	PART
Partition	PTN
Party	PTY
Passage	PASS
Passageway	PASSWY
Patent	PAT
Pattern	PATT
Peck	PK
Penetration	PEN
Percent	PCT
Percission	PERC
Perforate	PERF
Permanent	PERM
Permissive	PERM
Perpendicular	PERP
Personnel	PERS
Phase	PH
Photograph	PHOTO
Physical	PHYS
Pick Up	PU
Pico	P
Pico Coulomb	PC
Pico Farad	PF
Picture	PIX
Piece	PC
Piezometer	PIEZ
Pile	PL
Pilot	PLT
Pilot Relay	PR
Piston-Operated Check Valve	POCV
Plant	PLT

Plant Alarm and Display System	PADS
Plant Computer	PC
Plant Data Acquisition System	PDAS
Plant Monitoring System	PMS
Plant Protection System	PPS
Plastic	PLSTC
Platform	PLATF
Plotting	PLOT
Pneumatic	PNEU
Pneumatic Operated	PO
Point	PT
Polarized	POL
Pole	P
Portable	PORT
Position	POSN
Positive	POS
Potable Water	POT W
Potential	POT
Potential Difference	PD
Potential Transformer	PT
Potentiometer	POT
Pound	LB
Pound-foot	LB-FT
Pound-inch	LB-IN
Pounds Per Cubic Foot	LB/FT^3, PCF
Pounds Per Hour	LB/HR, PPH
Pounds Per Square Foot	LB/FT^2, PSF
Pounds Per Square Foot Absolute	PSFA
Pounds Per Square Foot Gage	PSFG
Pounds Per Square Inch	PSI
Pounds Per Square Inch Absolute	PSIA
Pounds Per Square Inch Gage	PSIG
Power	PWR
Power Amplifier	PA
Power Cabinet	PC
Power Circuit Breaker	PCB
Power Directional Relay	PDR
Power Factor	PF
Power Factor Meter	PFM
Power Line Carrier	PLC
Power Supply	PWR SPLY
Preamplifier	PREAMP
Precipitator	PPTR
Preferred	PFD
Premolded	PRMLD
Prepare	PREP
Press	PRS
Pressure	PRESS
Pressure Reducing Valve	PRV
Pressure Switch	PS
Pressurizer	PZR
Pressurizer Relief Tank	PRT
Primary	PRI
Primary Makeup Water System	PMWS
Priming	PRMG
Printer	PTR
Private Automatic Exchange	PAX
Private Branch Exchange	PBX
Process	PRCS
Product Detector	PD
Project	PROJ
Proof	PRF
Proportional	PROPNL
Protection	PROT
Public-Address System	PA
Pull Box	PB
Pull Button Switch	PULL B S
Pulsating Current	PC
Pulse Code Modulation	PCM
Pulse Duration Modulation	PDM
Pulse Time Modulation	PTM
Pulses Per Second	PPS
Pulverizer	PULV
Pump	PMP
Punch	PCH
Purification	PRFCN
Purging	PRNG
Purple	PRP, PR
Push Button	PB
Push Button Station	PB STA
Push-Pull	P-P
Pyrometer	PYR

- Q -

Quadrant	QUAD
Quantity	QTY
Quart	QT
Quarter	QTR
Quartz	QTZ

- R -

Radial	RDL
Radian	RAD
Radiation	RADN
Radio Frequency	RF
Radius	R
Range	RNG
Rankline	R
Ratio	R
Ratio Detector	RD
Raw Cooling Water	RCW
Raw Service Water	RSW
Raw Water	RW
Raw Water Chlorination System	RWCLS
Reactive	REAC
Reactive Factor Meter	RFM
Reactive Kilovolt Ampere	KVAR
Reactive Volt Ampere	VAR
Reactive Voltmeter	RVM
Reactor	REAC
Reactor Building	RB
Real Time Clock	RTC
Receive	RCV
Received	RCVD
Receiver	RCVR
Receiving	RCVG
Receptacle	RCPT
Reception	RCPN
Reciprocal	RECIP
Reciprocate	RECIP
Recirculate	RECIRC
Recirculation	RECIRCN

Reclosing	RECL
Recognition	RECOG
Record	RCD
Recorder	RCDR
Recording	RCDG
Rectifier	RECT
Red	R
Reduce	RED
Reduction	REDUC
Refinery	REF
Refractory	REFR
Refrigerant	REFRIG
Refrigerate	REFR
Regenerative	REGEN
Register	REG
Regular	REG
Regulate	REG
Reheat	RH
Reheat Stop Valve	RSV
Reheater	RHR
Relative Humidity	RH
Relay	RLY
Release	RLSE
Relief	RLF
Relief Valve	RV
Remote Control	RC
Required	REQD
Reserve	RSV
Reservoir	RSVR
Residual	RESID
Residual Heat Removal	RHR
Residual Heat Removal Pump	RHRP
Residual Heat Removal System	RHRS
Resistance	RES
Resistance Temperature Device	RTD
Retainer	RET
Retard	RET
Retractable	RETR
Return	RTN
Reverse	REV
Revolution	REV
Revolutions Per Minute	RPM
Revolutions Per Second	RPS
Rheostat	RHEO
Right	R
Rod Block Monitoring	RBM
Roller Bearing	RB
Roof	RF
Room	RM
Root Diameter	RD
Root Mean Square	RMS
Rotary	ROT
Rotate	ROT
Roughness Height Rating	RHR
Round	RD
Rubber	RUB
Rupture	RUPT

- S -

Safety	SAF
Safety Valve	SV
Salinometer	SAL
Salt Water	SW
Salvage	SALV
Sample	SMPL
Sampling	SMPLG
Sampling System	SS
Sanitary	SAN
Saturate	SAT
Saybolt Seconds Furol (Oil Viscosity)	SSF
Saybolt Seconds Universal (Oil Viscosity)	SSU
Scale	SC
Schematic	SCHEM
Scrubber	SCRB
Seal Oil Backup	SOBU
Sealed	SLD
Sea Level	SL
Second	SEC
Second-Feet (Cubic Feet Per Second)	CFS
Secondary	SEC
Section	SECT
Segment	SEG
Select	SEL
Selector	SEL
Semi-Automatic	SEMI AUTO
Sender	SDR
Separator	SEP
Sequence	SEQ
Sequential	SEQL
Serial	SER
Serial Channel Input/Output	SCIO
Serial Model Interface	SMI
Series	SER
Service	SERV
Service Air System	SAS
Service Building	SB
Set Point	SP
Settling	SET
Sewer	SEW
Shaft	SFT
Shaft Horsepower	SHP
Shield	SHLD
Short Circuit	SHR CKT
Short Circuit Ratio	SCR
Short Wave	SW
Shunt	SH
Shut-Off Valve	SOV
Shutdown	SD, SHTDN
Side	S
Signal	SIG
Silence	SIL
Simplex	SX
Single	S
Single Feeder	SF
Single Frequency	SF
Single Pole, Double Throw	SPDT
Single Pole, Single Throw	SPST
Single Sideband	SSB
Sink	SK
Skimmer	SKIM
Sleeve	SLV
Sleeve Bearing	SB

Slide	SL
Small	SM
Smoke	SMK
Socket	SOC
Solenoid	SOL
Solenoid Operated Valve	SOV
Solid State	SS
Soot Blower	SB
Sound	SND
South	S
Southeast	SE
Southwest	SW
Space	SP
Spacer	SPCR
Spare	SP
Speaker	SPKR
Special	SPL
Specific	SP
Specific Gravity	SP GR
Specific Heat	SP HT
Speech-Plus-Tone	S+T
Speed	SP
Speed No Load	SNL
Spherical	SPHER
Spherical Candle Power	SCP
Spindle	SPDL
Split Phase	SP PH
Square	SQ
Square Root	SQRT
Squirrel Cage	SQ CG
Stairway	STWY
Standard	STD
Standby	STBY
Standby Diesel Generator System	SDGS
Standpipe	SP
Starting	STG
Startup	SU, STUP
Static Pressure	SP
Station	STA
Station Drainage System	SDS
Station Service	SS
Stationary	STA
Stator	STAT
Steam	ST
Steam Generator	SG, STGEN
Steam Jet Air Ejector	SJAE
Steering	STEER
Stock	STK
Storage	STOR
Storm Water	ST W
Strainer	STR
Stream	STR
Strip	STR
Structural	STR
Structure	STR
Stuffing Box	SB
Submerged	SUB
Substation	SUBSTA
Suction	SUCT
Summary	SUM
Summer	SUMR

Supervise or Supervisor	SUPV
Supervisory Control and Data Acquisition	SCADA
Supplement	SUPP
Supply	SUP
Support	SUP
Suppression	SUPPR
Suspend	SUSP
Suspended	SUSP
Swing	SWG
Switch	SW
Switch and Relay Types	
Single Pole Switch	SP SW
Single Pole, Single Throw Switch	SPST SW
Single Pole, Double Throw Switch	SPDT SW
Double Pole Switch	DP SW
Double Pole, Single Throw Switch	DPST SW
Double Pole, Double Throw Switch	DPDT SW
Triple Pole Switch	3P SW
Triple Pole, Single Throw Switch	3PST SW
Triple Pole, Double Throw Switch	3PDT SW
4 Pole Switch	4P SW
4 Pole Single Throw Switch	4PST SW
4 Pole Double Throw Switch	4PDT SW
Switchboard	SWBD
Switcher	SWR
Switchgear	SWGR
Switchyard	SWYD
Synchroscope	SYNSCP
System	SYS

- T -

Tabulate	TAB
Tachometer	TACH
Tandem	TDM
Tangent	TAN
Tangent Spiral	TS
Tank	TK
Taper	TPR
Taper Per Foot	TPF
Taper Per Inch	TPI
Taper Shank	TS
Tarpaulin	TARP
Technical	TECH
Tee	T
Teeth	T
Teeth Per Inch	TPI
Telegraph	TLG
Telemeter	TLM
Telephone	TEL
Telephone Jack	TJ
Telephone Line Carrier	TLC
Teletypewriter Exchange	TWX
Television	TV
Temperature	TEMP
Temperature Absolute	T
Temperature Indicator	TEMP IND
Temperature Indicator Control	TIC
Temperature Meter	TM
Temperature Modifier	TM
Temperature Monitoring System	TMS
Temporary	TEMP

Tennessee	TN
Tensile Strength	TS
Tension	TENS
Tentative	TENT
Terminal	TERM
Terminal Board or Block	TB
Territory	TERR
Tertiary	TER
Test Block	TB
Test Link	TL
Texas	TX
Thermal	THRM
Thermal Element	TE
Thermal Switch	TS
Thermocouple	TC
Thermometer	THERM
Thermostat	THERMO
Thick	THK
Thousand	M
Thousand Circular Mills	KCMIL
Thousand Cubic Feet	MCF
Thousand Foot Pound	KIP-FT
Thousand Pound	KIP
Throttle	THROT
Through	THRU
Thrust	THR
Thursday	THU
Ticket	TKT
Tie Line	TL
Timber	TMBR
Time	T
Time Delay	TD
Time Delay Closing	TDC
Time Delay Opening	TDO
Toggle	TGL
Tolerance	TOL
Toll	T
Ton Mile	TON-MI
Tons of Refrigeration	TR
Tons Per Hour	TPH
Tons Per Square Foot	TSF
Torque	TOR
Total	TOT
Total Dynamic Head	TDH
Total Head, Feet	TH
Total Indicator Reading	TIR
Total Pressure	TP
Tower	TWR
Township	TWP
Train	TN
Training	TNG
Transceiver	XCVR
Transducer	XDCR
Transfer	TRANS
Transfer Switch	XS
Transfer Trip	TT
Transfer Trip Relay	TTR
Transformer	XFMR
Translator	XLTR
Transmission	XMSN
Transmit-Receive	TR

Transmitter	XMTR
Transmitting	XMTG
Transportation	TRANS
Transverse	TRANSV
Tread	TRD
Treatment	TRTMT
Trimmer	TRIM
Triode	TRI
Trip Coil	TC
Trip Cut Out	TCO
Trunk	TRK
Truss	T
Tubing	TUB
Tuesday	TUE
Tuned Radio Frequency	TRF
Tunnel	TNL
Turbine	TURB
Turbine Building	TB
Turbine Drive	TD
Turbine Extraction Taps and Drain System	TETDS
Turbine Generator	TURBOGEN, T-G
Turbogenerator Control System	TGCS
Turnout	TO
Turns Per Inch	TPI
Turret	TUR
Twisted	TW
Two Frequency	2F
2-way Trunk	2WT
Two Wires	2W

- U -

Ultimate	ULT
Ultra High Frequency	UHF
Under Voltage	UV
Under Voltage Device	UVD
Underground	UNDG
Underwater	UNDW
Unit	U
Unit Cooler	UC
Unit Heater	UH
Unit Operator	UO
Universal	UNIV
Upper	UPR
Upscale	UPS
Upstream	UPSTR
Utah	UT

- V -

Vacuum	VAC
Vacuum Priming System	VPS
Vacuum Tube	VT
Valve	V
Valve Box	VB
Valve Position Detector	VPD
Variable	VAR
Variable Frequency Oscillator	VFO
Variable Spring Support	VS
Varies	VRS
Vegetable	VEG
Velocity	VEL
Velocity Pressure	VP

Vent Pipe	VP
Vent Stack	VS
Ventilate	VENT
Ventilating System	VS
Ventilation	VENT
Venturi	VTI
Vermont	VT
Versus	VS
Vertical	VERT
Very-High Frequency	VHF
Very-Low Frequency	VLF
Vibrate	VIB
Vibration	VIB
Vibration Transmitter	VT
Video	VID
Video-Frequency	VDF
Violet	VIO, V
Virginia	VA
Viscosity	VISC
Vitreous	VIT
Voice Frequency	VG
Volt	V
Voltage to Current	E/I
Voltage to Pneumatic	E/P
Voltage Relay	VR
Voltampere	VA
Voltmeter	VM
Volts Per Mil	V/MIL
Volume	VOL
Volume Control Tank	VCT

- W -

Wall	W
Warehouse	WHSE
Washer	WASH
Washington	WA
Waste Disposal System	WDS
Waste Water & Oil	WWO
Water	WTR
Water Gage	WG
Water Heater	WH
Water Line	WL
Water, Oil or Gas	WOG
Water Surface Elevation	W S EL
Water Tank	WT
Water Treatment System	WTS
Waterproof	WPF
Watertight	WT
Waterworks	WW
Watt	W
Watthour	WH
Watthour Meter	WHM
Wattmeter	WM
Watts Per Candle	WPC
Weatherproof	WP
Wednesday	WED
Week	WK
Weight	WT
Weld Neck	WN
West	W
West Virginia	WV
Wet Bulb	WB
Wet Bulb Temperature	WBT
Wetted Surface	WS
White	WHT, W
Wide	W
Width	W
Wind	WD
Winding	WDG
Window	WDW
Wiring	WIR
Wisconsin	WI
With	W/
Without	W/O
Working Pressure	WPR
Working Steam Pressure	WSP
Wyoming	WY

- Y -

Yard	YD
Year	YR
Yellow	YEL, Y

- Z -

Zero Speed Switch	ZSS
Zone Switch	ZS

APPENDIX C

CHARACTER HEIGHTS

Intended Viewing Distance	Minimum Height	Nominal Height
1-2/3 ft. or less	0.09 in.	0.16 in.
1-2/3 ft. to 3 ft.	0.17 in.	0.28 in.
3 ft. to 6 ft.	0.34 in.	0.56 in.
6 ft. to 12 ft.	0.68 in.	1.12 in.
12 ft. to 20 ft.	1.13 in.	1.87 in.

APPENDIX D
GOTHIC LETTERING EXAMPLES

1/8" LETTER HEIGHT .015 LETTER STROKE

ABCDEFGHIJKLMONPQRSTUVW

XYZ-0123456789%?"'#$( )&

3/16" LETTER HEIGHT .020 LETTER STROKE

ABCDEFGHIJKLMNOPQRSTUVW

XYZ-0123456789%?"'#$( )&

1/4" LETTER HEIGHT .025 LETTER STROKE

ABCDEFGHIJKLMNOPQRSTUVW

XYZ-0123456789%? #$( )&

INSTRUMENT SOCIETY of AMERICA
Research Triangle Park, North Carolina

ISA-RP60.8-1978

Recommended Practice

Electrical Guide for Control Centers

Instrument Society of America

ISBN 0-87664-444-2

ISA-RP60.8 Electrical Guide for Control Centers

Copyright © 1978 by the Instrument Society of America. All rights reserved. Printed in the United States of America. No part of this publication may be reproduced, stored in a retrieval system, or transmitted, in any form or by any means (electronic, mechanical, photocopying, recording, or otherwise), without the prior written permission of the Publisher.

INSTRUMENT SOCIETY OF AMERICA
67 Alexander Drive
P.O. Box 12277
Research Triangle Park, North Carolina 27709

SUMMARY OF SP60

ISA S60 is an overall set of recommended practices being compiled by the ISA Control Centers Committee. RP60.8 is the first of this series to be issued.

SECTION	TITLE
RP60.1	Control Center (C.C.) Facilities
RP60.2	C.C. Design Guide and Terminology
RP60.3	Human Engineering for Control Centers
RP60.4	Documentation for Control Centers
RP60.5	Control Center Graphic Displays
RP60.6	Nameplates, Labels, Tags and Terminal Identification
RP60.7	Control Center Constructions
RP60.8	Electrical Guide for Control Centers
RP60.9	Piping Guide for Control Centers
RP60.10	Control Center Inspection and Testing
RP60.11	Crating, Shipping and Handling for C.C.

PREFACE

This Preface is included for information purposes and is not part of RP60.08.

This Standard has been prepared as a part of the service of the Instrument Society of America toward a goal of uniformity in the field of instrumentation. To be of real value, this document should not be static, but should be subject to periodic review. Toward this end, the Society welcomes all comments and criticisms, and asks that they be addressed to the Secretary, Standards and Practices Board, Instrument Society of America, 67 Alexander Drive, P.O. Box 12277, Research Triangle Park, North Carolina 27709, Telephone (919) 549-8411.

The ISA Standards and Practices Department is aware of the growing need for attention to the metric system of units in general, and the International System of Units (SI) in particular, in the preparation of instrumentation standards. The Department is further aware of the benefits to USA users of ISA Standards of incorporating suitable references to the SI (and the metric system) in their business and professional dealings with other countries. Towards this end this Department will endeavor to introduce SI and SI-acceptable metric units in all new and revised standards to the greatest extent possible. The Metric Practice Guide, which has been published by the American Society for Testing and Materials as ANSI designation Z210.1 (ASTM E380-76, IEEE Std. 268-1975), and future revisions, will be the reference guide for definitions, symbols, abbreviations, and conversion factors.

The persons listed below served as active members of the ISA Control Centers Committee for the major share of its working period.

H. R. Solk, Chairman	Comsip Customline Corp.
C. S. Lisser	Oak Ridge National Lab.
R. W. Borut	M. W. Kellogg Co.
J. M. Fertitta, Secretary	The Foxboro Co.
A. R. Alworth	Shell Oil Co.
M. Arevalo	Fischer & Porter Co.
C. D. Armstrong	Tennessee Valley Authority
B. Carlisle	E. I. duPont deNemours & Co., Inc.
J. E. Cerretani	Detroit Edison
J. M. Devenney	E. I. duPont deNemours & Co., Inc.
J. Farnia	Gismo Div. of GECO, Guarantee Electric Co.
C. H. Goding	BIF, Unit of General Signal
J. Gump	The Riley Co.
R. E. Hetzel	Stauffer Chemical Co.
T. P. Holland	Johnson Controls, Inc. — Panel Unit
A. A. Kayser	Bristol Division of ACCO
W. H. Kelley (Deceased)	Bechtel Associates Professional Corp.
J. L. Kern	Monsanto Co.
R. W. Kief	Emanon Co., Inc.
S. F. Luna	Gulf General Atomics
J. G. McFadden	Public Service Electric & Gas
R. Munz	The Mundix Co.
F. Nones	The Foxboro Co.
W. A. Richards	General Electric Co.
A. Stockmal	Owens-Illinois, Fecker Systems Div.
M. J. Walsh	Procon, Inc.
W. T. Williams	Robertshaw Controls Co.
B. W. Ball	Brown & Root, Inc.
C. F. Aured	Panels, Inc.
R. F. Rossbauer	Fischer & Porter Co.
R. A. Landthorn	Panels, Inc.

This Recommended Practice was approved by the ISA Standards and Practices Board on June 28, 1978.

A. P. McCauley, Chairman	SCM Corporation
P. Bliss	Pratt & Whitney Aircraft Group
E. J. Byrne	Brown & Root, Inc.
W. Calder	The Foxboro Company
B. A. Christensen	Continental Oil Company
R. Coel	Fluor Corporation
M. R. Gordon-Clark	Scott Paper Company
T. J. Harrison	IBM Corporation
H. T. Hubbard	Southern Company Services, Inc.
H. H. Koppel	Bailey Meter Company
P. S. Lederer	National Bureau of Standards
E. C. Magison	Honeywell, Inc.
R. L. Nickens	Reynolds Metals Company
G. Platt	Bechtel Power Corporation
G. Quigley	Moore Instrument Company, Limited
W. C. Weidman	Gilbert Associates, Inc.
K. A. Whitman	Allied Chemical Corporation
J. R. Williams	Stearns-Roger Inc.
W. E. Vannah	The Foxboro Company
M. T. Yothers, Secretary	Instrument Society of America
*L. N. Combs	E. I. duPont deNemours Company, Inc.
*R. L. Galley	Bechtel, Inc.
*R. G. Marvin	Dow Chemical Company
*W. B. Miller	Moore Products Company

*Director Emeritus

TABLE OF CONTENTS

Section		Page
1.	Scope	2
2.	General	2
	2.1 Application	2
	2.2 Compliance	2
	2.3 Wiring	2
	2.4 Terminations	3
	2.5 Sizing Conduit and Electrical Metallic Tubing	3
3.	General Purpose Open Wiring	3
	3.1 General Application	3
	3.2 Raceway	3
	3.3 Open Wiring	4
	3.4 Safety	4
	3.5 Working Space	4
4.	General Purpose Enclosed Wiring	4
	4.1 General Application	4
	4.2 Enclosed Wiring	4
5.	Hazardous Location Class I, Division 2 Open Wiring	4
	5.1 General Application	4
	5.2 Raceway	4
	5.3 Open Wiring	4
	5.4 Seals	4
6.	Hazardous Location Class I, Division 2 Enclosed Wiring	4
	6.1 General Application	4
	6.2 Enclosed Wiring	4
	6.3 Seals	5
7.	Hazardous Location Class I, Division 1 Wiring	5
	7.1 General Application	5
	7.2 Equipment	5
	7.3 Wiring	5
	7.4 Seals	5
	7.5 Terminal Box	5
	7.6 Purging	5
	Figure 1 Terminal Block Layout	6

1. SCOPE

1.1 Scope

This Section of ISA-RP60 is intended to assist the design engineer in establishing the electrical requirements of a control center; it is also intended to comply with the provisions of the National Electrical Code. What follows are the practices normally used by architects, engineers, user organizations and control center manufacturers. Special considerations which may apply to particular devices or circuits are not taken into account in this section.

2. GENERAL

2.1 Application

This Section of ISA-RP60 applies to the inter-connection wiring between devices within a control center, or between control centers. It does not apply to the internal wiring of an instrument or device, nor to the wiring between the control center and field-mounted instruments.

2.2 Compliance

All wiring recommendations in this Standard are intended to comply with the National Electrical Code. Compliance with other applicable local or industry codes requires specific reference in the control center specification. General references to these codes should be avoided.

2.3 Wiring

2.3.4 Power Wiring

Each device requiring control center power shall be wired so that when wires are removed from any one device, power will not be disrupted to any other device. It is recommended that no voltages greater than 125 Vdc or 120 Vac be used. If higher voltages are required, refer to color coding below. Provisions should be made for power wire segregation and identification to insure personnel safety.

The recommended minimum size for power wiring up to 15 A is #14 stranded copper conductor, 600 V, 60 C with thermo-plastic insulation. Different insulation material should be considered if the insulation has to be self-extinguishing or nontoxic. Refer to Table 310-13 "Conductor Application and Insulations" of the National Electrical Code, latest revision for particular provisions. Where power is 150 W or less and run to each instrument through a separate fuse or circuit breaker, the wire size can be reduced to #16 or #18. Refer to Article 310 — "Conductors for General Wiring" in the National Electrical Code, latest revision for currents greater than 15A.

2.3.1.1 Color Coding of ac Power Wiring

(1) 120 Vac, 2 wire, single phase:
Phase — Φ1 or L1, black
Neutral — N or L2, white

(2) 120/240 Vac or 120/208 Vac, 3 wire, single phase with neutral:
Phase 1 — Φ1 or L1, black
Phase 2 — Φ2 or L2, red
Neutral — N, white

(3) 240 Vac, 2 wire, single phase:
Φ1 or L1, black
Φ2 or L2, red

(4) Any 3-phase circuit regardless of voltage:
Phase 1 — Φ1 or L1, black
Phase 2 — Φ2 or L2, red
Phase 3 — Φ3 or L3, blue
Neutral in 4-wire system — N, white

(5) Interlock control circuits wired from an external power source:
yellow

(6) Case ground wire, if insulated — green or green with one or more yellow stripes.

2.3.1.2 Color Coding of dc Power Wiring

Due to numerous voltage levels and existing standards, it is recommended that distinctive colors be used which do not conflict with other colors being used.

2.3.1.3 Circuit Protection

In addition to circuit protection requirements of the National Electrical Code, special circuit considerations must be given. In particular, the solution to the problem of the failure of any device in a system which will cause loss of power for that system or other systems.

2.3.2 Grounding

Each device requiring power should be grounded to the control center by mounting to effect a conducting path or by a ground wire. Ground wire size should be no less than the supply conductor wire size. To determine wire size for grounding of control center to earth ground, refer to Table 250-95 — "Size of Equipment Grounding Conductors for Grounding Raceway and Equipment" of the National Electrical Code, latest revision.

2.3.3 Control Wiring

Control wiring is considered distinct from power or signal wiring. This type involves wiring of push buttons, pilot lights, selector switches, relay logic, timers and programming devices. Recommendations for wire sizes, fusing and color codes may be found in the Joint Industrial Council Electrical Standards for General Purpose Machine Tools EGP-1-67.

2.3.4 Signal Wiring

Normal usage for signal wiring is #18 AWG stranded copper conductor. The recommended minimum wire size is #24 AWG, which should be used in a multi-conductor cable or be twisted pair. For Class 1, 2 or 3 wiring, refer to Article 725 — "Class 1, 2 and 3 Remote-Control, Signalling, and Power Limited Circuits" of the

National Electrical Code, latest revision. Pulse type or low level signals may require twisted pair, shielded, or twisted pair and shielding wiring.

Wire type, signal grounding requirements, and shield requirements should be based on instrument manufacturer's recommendations. Wire carrying measurement signals associated with thermocouples, resistance thermometers, pH instruments and other low level signals are best run directly to the instrument without intermediate terminations. Isolated routing should be provided in the control center for running these wires.

To avoid inductive pickup, power wiring or control wiring should have maximum possible separation from signal wiring. A practical distance is not less than 150 mm (6 in.). If the power wiring has to cross the signal wiring, the crossing should be as close to a right angle as possible.

For intrinsically safe signal wiring, refer to ANSI/ISA RP12.6 "Installation of Intrinsically Safe Instrument Systems in Class I Hazardous Locations."

2.3.5 Wire Identification

All wiring is normally identified at both ends by suitable wire markers or color codes in accordance with the user's design documentation. See ISA-RP60.6 — "Nameplates, Labels, Tags and Terminal Identification" for marker materials and other details.

2.4 Terminations

2.4.1 Terminal Blocks

Each terminal block and its terminals should be identified, as suggested in ISA-RP60.6 — "Nameplates, Labels, Tags and Terminal Identification." Provisions should be made for spare terminations; the recommended minimum is ten percent. Normal practice is twenty percent.

Recommendations for minimum terminal block spacing are as follows:

(See Figure 1, page 6)

Depth of terminal box if used, 100 mm (4 in.). For control center wiring, 65 mm (2.5 in.) between terminal block and sidewall, 90 mm (3.5 in.) between parallel terminal blocks. For field wiring, 100 mm (4 in.) between terminal block and sidewall, 130 mm (5 in.) between terminal blocks.

If terminals are mounted on standoffs, wires can be routed behind the terminal blocks which make connections and wire markers more accessible.

Typical wire connections used at screw type terminal blocks are spade and ring lugs. Another type of terminal block utilizes compression fittings which accept striped wire. The ring lug and compression fitting terminal block are recommended for applications involving vibration.

Preferably, not more than two conductors should be terminated at each terminal connection. If two wire sizes must be terminated in a single compression type fitting, refer to terminal block manufacturer for the maximum difference between wire sizes.

2.4.2 Wiring Between Control Centers

Interconnection between control center sections, which will be separated for shipment, should be adequately prepared for convenient reassembly. Wire of adequate length for final connection should be disconnected, and pulled back in sections for shipment.

Possible ways of handling the interconnecting wiring are:

(1) Wiring of each section brought to terminal blocks in that section. Jumper wires used between terminal blocks of both sections.

(2) Wiring in one section, say section B, brought to terminal blocks in the section. Wires from other sections, say A and C, are brought directly to section B terminal blocks. For shipping, wires from A and C are removed from the terminal blocks in B and coiled back to their respective sections.

(3) As in (1) and (2) but use multi-wire connectors instead of terminal blocks.

2.4.3 Cable Connectors

Cables with multi-pin connectors are an acceptable form of interwiring control centers. The connectors can be mounted on special channels made by the connector manufacturer or control center supplier.

Suitable wire and cable identification should be provided as in 2.3.5. Also, the connector should be labeled. Use live female connector whenever circuit voltages could endanger personnel.

2.5 Sizing Conduit and Electrical Metallic Tubing

Consult Chapter 9 — "Tables and Examples" of the National Electrical Code, latest revision, to determine the maximum number of conductors in conduit or electrical metallic tubing if it is installed for the internal control center wiring.

3. GENERAL PURPOSE OPEN WIRING

3.1 General Application

General purpose open wiring — wiring carried in raceways, or bundled and open.

3.2 Raceway

Wires should be run in vented non-combustible or self-extinguishing non-metallic raceway. Metallic wireways are also acceptable.

3.3 Open wiring

Where the use of raceways is not practical, wires should be run open, bundled and bound with suitable cable ties at regular intervals not exceeding 300 mm (12 in.). All wires within a bundle should be run parallel to one another. Bundles should have a uniform appearance, a circular cross section, and be securely fastened to the control center framework.

3.4 Safety

Consideration should be given to live parts of terminal blocks, pilot lights and switches which may affect personnel safety or circuit integrity. If the control center itself does not constitute an acceptable enclosure, these live parts should be covered or enclosed with a housing. Where wires are to be run to a housing from either an open bundle or from raceway, the housing should be fitted with a suitable insulating bushing or grommet.

3.5 Working Space

Where working space is required within a control center, Article 110-16 — "Working Space About Electric Equipment (600 Volts or less, Nominal)" of the National Electrical Code, latest revision should be observed.

4. GENERAL PURPOSE ENCLOSED WIRING

4.1 General Application

Enclosed general purpose wiring — all wiring and terminals are enclosed.

4.2 Enclosed Wiring

All wiring should be run enclosed in electrical metallic tubing, flexible conduit with plastic covering or sheet metal wireway. Sharp edges, burrs, rough surfaces, or threads with which wire insulation may come in contact should be removed from conduit fitting, wireway, or any other parts.

Electrical metallic tubing, flexible conduit or raceway should be securely fastened to the control center. Pull boxes should be provided with pilot holes, plugs or knockouts for conduits.

5. HAZARDOUS LOCATION CLASS I, DIVISION 2 OPEN WIRING

5.1 General Application

Used when control center is made of components for which general purpose enclosures are acceptable under Article 501-3 (b-1, 2, 3) of the National Electrical Code, latest revision. In order to comply with the National Electrical Code, the control center construction must be considered a single general purpose enclosure. An example is a totally enclosed cabinet, including bottom and hinged, non-removable doors with locks.

Wiring and terminals need not be in explosion-proof enclosures. Arcing contacts are either hermetically sealed or enclosed in explosion-proof housing with sealed fittings or in circuits which under normal conditions do not release sufficient energy to ignite a specific hazardous atmospheric mixture. Any device operating at auto ignition temperature of flammable material is enclosed in an explosion-proof housing and provided with seals.

5.2 Raceway

Wires should be run in vented non-combustible or self-extinguishing non-metallic raceway. Metallic wireways are also accepted.

5.3 Open Wiring

Where the use of raceways is not practical, wires should be run open, bundled and bound with suitable cable ties or equal at regular intervals not to exceed 300 mm (12 in.). All wires within a bundle should be run parallel to one another. Bundles should have a uniform appearance, a circular cross section, and be securely fastened to the control center framework.

5.4 Seals

Seals are used at explosion-proof housings containing arcing devices that have been mounted in the control center. They may also be required where the field conduit enters the control center.

Seals should be poured in the field and each will bear a prominent tag or be painted to indicate this fact. Dams for these seals can be put in place prior to shipment of control center. For proper location and application of seals, refer to Article 501-5 — "Sealing and Drainage" of the National Electrical Code, latest revision.

6. HAZARDOUS LOCATION CLASS I, DIVISION 2 ENCLOSED WIRING

6.1 General Application

Control center construction must be considered a single general purpose enclosure. An example is an enclosed cabinet including bottom and hinged, non-removable doors with locks. All wiring and terminals are enclosed. Arcing contacts are either hermetically sealed or enclosed in explosion-proof housing with sealed fittings or in circuits which under normal conditions do not release sufficient energy to ignite a specific hazardous atmospheric mixture. Any device operating at auto ignition temperature of flammable material is enclosed in an explosion-proof housing and provided with seals.

For non-incendive, intrinsically safe system, refer to ANSI/ISA RP12.6 "Installation of Intrinsically Safe Instrument Systems in Class I Hazardous Locations."

6.2 Enclosed Wiring

All wiring should be run in accordance with Article 501-4b — "Wiring Methods — Class 1, Division 2" of the National Electrical Code, latest revision.

6.3 Seals

Seals are used at explosion-proof housings containing arcing devices that have been mounted in the control center. They may also be required where the field conduit enters the control center. Seals should be poured in the field and each will bear a prominent tag or be painted to indicate this fact. Dams for these seals can be put in place prior to shipment of control center. For proper location and application of seals, refer to Article 501-5 — "Sealing and Drainage" of the National Electrical Code, latest revision.

7. HAZARDOUS LOCATION CLASS 1, DIVISION 1 WIRING

7.1 General Application
All wiring and terminals are enclosed. Equipment, boxes and fittings are explosion-proof.

7.2 Equipment
Terminals, switches, relays, instruments, meters, etc. shall be provided with an enclosure approved for appropriate Class and Group.

7.3 Wiring
All Wiring must be enclosed in threaded rigid metal conduit or other types approved by the National Electrical Code, explosion-proof boxes, fittings and joints. Threaded joints shall be made up with at least the number of fully engaged threads required by the Area Classification. Sharp edges, burrs, rough surfaces, or threads with which insulation of conductors may come in contact, should be removed from conduit fittings and other parts.

7.4 Seals
Seals should be poured in the field and each will bear a prominent tag or be painted to indicate this fact. Dams for these seals can be put in place prior to shipment of control center. For proper location and application of seals, refer to Article 501-5 — "Sealing and Drainage" of the National Electrical Code, latest revision.

7.5 Terminal Box
Threaded holes with plugs for conduit fittings should be provided in each box for field wiring. Holes should be sized for conduit applicable to the number of existing terminal connections.

7.6 Purging
Where purging is required, ISA-S12.4 — "Instrument Purging for Reduction of Hazardous Area Classification" should be consulted, along with NFPA 496 — "Purge and Pressurized Enclosures for Electrical Equipment."

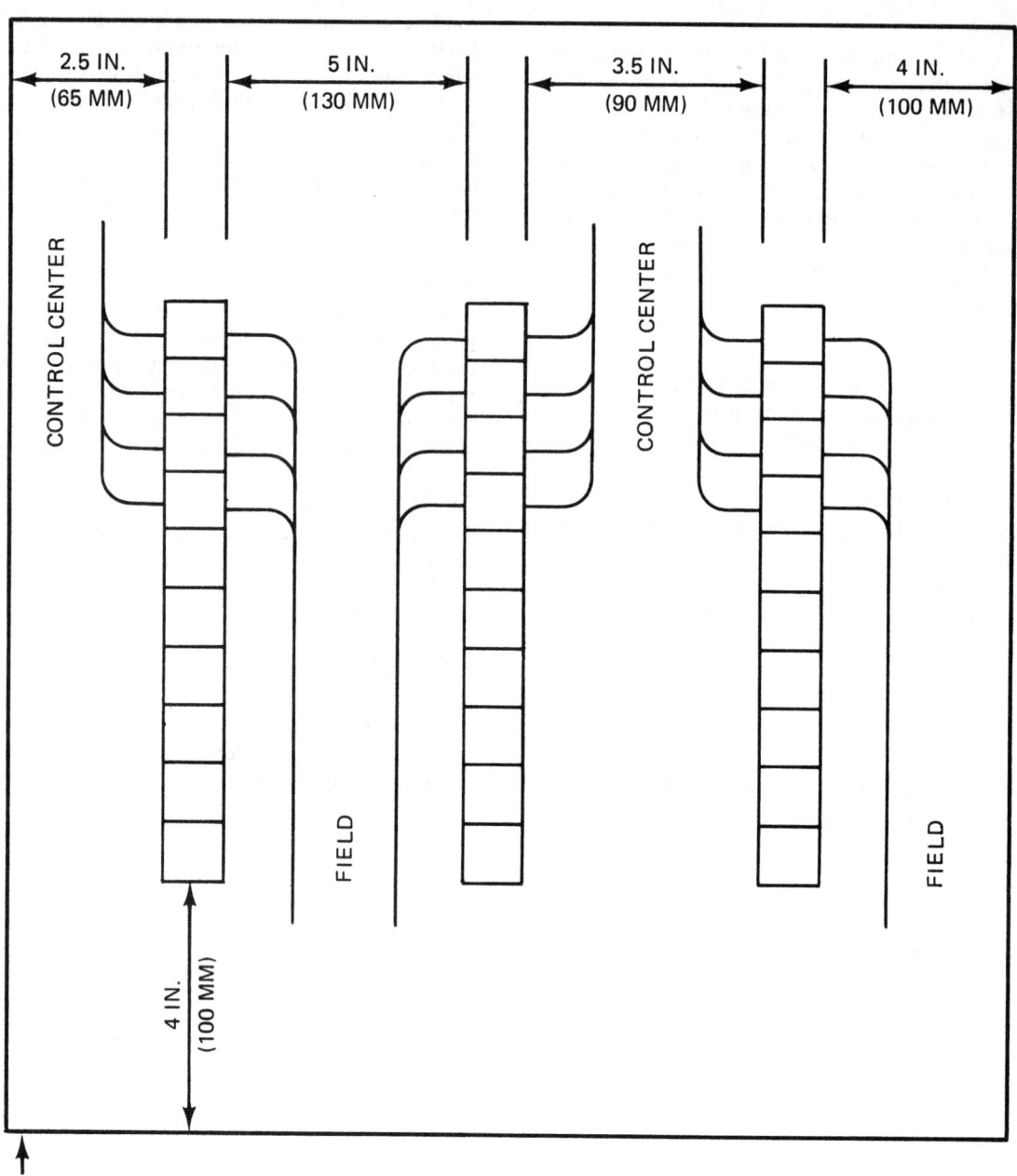

**FIGURE 1
TERMINAL BLOCK LAYOUT
(Minimum Dimensions Recommended)**

INSTRUMENT SOCIETY of AMERICA
Research Triangle Park, North Carolina

ISA-RP60.9-1981

Recommended Practice

Piping Guide for Control Centers

Instrument Society of America

ISBN 0-87664-556-2

ISA-RP60.9 Piping Guide For Control Centers

Copyright © 1981 by the Instrument Society of America. All rights reserved. Printed in the United States of America. No part of this publication may be reproduced, stored in a retrieval system, or transmitted, in any form or any means electronic, mechanical, photocopying, recording or otherwise without the prior written permission of the publisher.

INSTRUMENT SOCIETY OF AMERICA
67 Alexander Drive
P.O. Box 12277
Research Triangle Park, North Carolina 27709

Copyright © 1981 by the Instrument Society of America

PREFACE

This Preface is included for information purposes and is not part of RP60.9.

This Standard has been prepared as a part of the service of the Instrument Society of America toward a goal of uniformity in the field of instrumentation. To be of real value, this document should not be static, but should be subject to periodic review. Toward this end, the Society welcomes all comments and criticisms, and asks that they be addressed to the Secretary, Standards and Practices Board, Instrument Society of America, 67 Alexander Drive, P.O. Box 12277, Research Triangle Park, NC 27709, Telephone (919) 549-8411.

The ISA Standards and Practices Department is aware of the growing need for attention to the metric system of units in general, and the International System of Units (SI) in particular, in the preparation of instrumentation standards. The Department is further aware of the benefits to USA users of ISA Standards of incorporating suitable references to the SI (and the metric system) in their business and professional dealings with other countries. Towards this end this Department will endeavor to introduce SI — acceptable metric units in all new and revised standards to the greatest extent possible. The Metric Practice Guide, which has been published by the American Society for Testing and Materials as ANSI designation Z210.1 (ASTM E380-76, IEEE Std. 268-1975), and future revisions, will be the reference guide for definitions, symbols, abbreviation, and conversion factors.

It is the Policy of the Instrument Society of America to encourage and welcome the participation of all concerned individuals and interests in the development of ISA Standards. Participation in the ISA Standards making process by an individual in no way constitutes endorsement by the employer of that individual of the Instrument Society of America or any of the Standards which ISA develops.

The SP60 Committee is preparing a series of recommended practices on control centers. ISA-RP60.9 is the second of this series to be published. The published recommended practices and drafts in preparation are listed below:

RECOMMENDED PRACTICE

SECTION	TITLE
dRP60.1*	Control Center (C.C.) Facilities
dRP60.2*	C.C. Design Guide and Terminology
dRP60.3*	Human Engineering for Control Centers
dRP60.4*	Documentation for Control Centers
dRP60.5*	Control Center Graphic Displays
dRP60.6*	Nameplates, Tags and Labels for Control Centers
dRP60.7*	Control Center Construction
RP60.8	Electrical Guide for Control Centers (published 1978)
RP60.9	Piping Guide for Control Centers (published 1981)
dRP60.10*	Control Center Inspection and Testing
dRP60.11*	Crating, Shipping and Handling for C.C.

The persons listed below served as active members of the ISA Control Centers Committee for the major share of its working period:

NAME	COMPANY
H. R. Solk, Chairman	Consultant
C. S. Lisser - Past Chairman	Oak Ridge National Laboratory
R. W. Borut - Past Chairman	The M. W. Kellogg Company
J. M. Fertitta, Secretary	The Foxboro Company
A. R. Alworth	Shell Oil Company
M. Arevalo	Fischer & Porter Company
C. D. Armstrong	Tennessee Valley Authority
B. Carlisle	E. I. duPont deNemours & Company, Inc.
J. M. Devenney	E. I. duPont deNemours & Company, Inc.
J. Farina	Gismo Division of GECO, Guarantee Electric Company
C. H. Goding	BIF, Unit of General Signal
J. Gump	The Riley Company
R. E. Hetzel	Stauffer Chemical Company
T. P. Holland	Johnson Controls, Inc. — Panel Unit
A. A. Kayser	Bristol Division of ACCO
W. H. Kelley (Deceased)	Bechtel Associates Professional Corporation
J. L. Kern	Monsanto Company

*Draft Recommended Practice, for additional information on the status of this document contact ISA Headquarters.

Instrument Society of America

Name	Company
R. W. Kief	Emanon Company, Inc.
S. F. Luna	Gulf General Atomics
J. G. McFadden	Public Service Electric & Gas
R. Munz	The Mundix Company
F. Nones	The Foxboro Company
W. A. Richards (Deceased)	General Electric Company
A. Stockmal	Owen-Illinois, Fecker Systems Division
M. J. Walsh	Procon, Inc.
R. L. Welch	El Paso Natural Gas Company
W. T. Williams	Robertshaw Controls Company
B. W. Ball	Brown & Root, Inc.
C. F. Aured	Panels, Inc.
R. F. Rossbauer	Fischer & Porter Company
R. A. Landthorn	Panels, Inc.
J. F. Jordan	Monsanto Company

The persons listed below served as corresponding members of the ISA Control Centers Committee for the major share of its working period:

NAME	COMPANY
J. Cerretani	Detroit Edison Company
N. L. Conger	Continental Oil Company
L. Corsetti	Crawford & Russell, Inc.
T. J. Crosby	Robertshaw Controls Company
C. R. Davis	Engineer
H. L. Faul	A-E Development Corporation
L. Ferson	ISA Headquarters
M. E. Gunn	Swanson Engineering & Manufacturing
R. I. Hough	Hough Associated
W. B. Miller	Moore Products Company
H. Kamerer	Wilmington Sheet Metal Ind.
A. L. Kress	3M Company
A. J. Langelier	Engineer
R. G. Marvin	Retired
A. P. McCauley, Jr.	Diamond Shamrock Corp.
C. W. Moehring	Bechtel
D. P. Morrison	BIF/General Signal
R. L. Nickens	Reynolds Metal Company
H. P. Fabisch	Fluor Engr. & Constructors, Inc.
F. W. Reichert	Engineer
J. F. Walker	Honeywell, Inc.
G. Walley	The N/P Company
W. J. Wylupek	Industry Appl. Engineering
J. R. Jordan	International Paper Company

This Recommended Practice was approved for publication by the ISA Standards and Practices Board in May 1981.

NAME	COMPANY
T. J. Harrison, Chairman	IBM Corporation
P. Bliss	Consultant
W. Calder	The Foxboro Company
B. A. Christensen	Continental Oil Co.
M. R. Gorden-Clark	Scott Paper Company
R. T. Jones	Philadelphia Electric Company
R. Keller	Boeing Company
O. P. Lovett, Jr.	Jordan Valve
E. C. Magison	Honeywell, Inc.
A. P. McCauley	Diamond Shamrock Corporation
E. M. Nesvig	ERDCO Engineering Corporation
R. L. Nickens	Reynolds Metals Company
G. Platt	Bechtel Power Corporation

R. Prescott   Moore Products Company
R. W. Signor   General Electric Company
W. C. Weidman   Gilbert Associates, Inc.
K. A. Whitman   Allied Chemical Corporation
*L. N. Combs
*R. L. Galley
*R. G. Marvin
*W. B. Miller   Moore Products Company
*J. R. Williams   Stearns-Rogers, Inc.

*Director Emeritus

TABLE OF CONTENTS

Section	Title	Page
1.	Scope	9
2.	Definitions	9
	2.1 Piping	9
	2.2 Field Piping	9
	2.3 Direct Process Piping	9
	2.4 Signal Piping	9
	2.5 Air	9
	2.6 Pneumatic Supply	9
	2.7 Signal Air	9
3.	General	9
	3.1 Application	9
	3.2 Reference Standards or Practices	9
4.	Pneumatic Piping	9
	4.1 Pneumatic Supply	9
	4.2 Signal Piping	10
	4.3 Fittings	11
5.	Direct Process Piping	11

LIST OF ILLUSTRATIONS

Figure	Title	Page
1.	Installation of Dual Pressure Regulators and Dual Air Filters	10
2.	Installation of Combination Filter - Regulators	10
3.	Air Supply Header	10

1 SCOPE

This Recommended Practice is intended to assist the designer or engineer in the definition of piping requirements for pneumatic signals and supplies in control centers. This Recommended Practice is based on current practices. Because of the special nature of each control center, specific rules are not practical and accepted guidelines should take precedence. This Recommended Practice is a presentation of these guidelines.

Piping external to the control center (Field Piping) is beyond the scope of this document.

2 DEFINITIONS

2.1 Piping

For the purpose of this document, the term "piping" includes metal or plastic tube, pipe fittings, valves, and similar components, and the practice of assembling these items into a system.

2.2 Field Piping

That piping connecting the control center to items external to the control center.

2.3 Direct Process Piping

That piping between the process and the control center which contains process fluid.

2.4 Signal Piping

That piping interconnecting instruments, instrument devices or bulkhead fittings.

2.5 Air

For the purposes of this document, air implies use of any suitable, and normally clean, dry, safe gas.

2.6 Pneumatic Supply

Air at a nominally constant pressure used to operate pneumatic devices.

2.7 Signal Air

Air at varying pressure used to represent process or control information.

3 GENERAL

3.1 Application

Piping is employed in control centers to convey air for two basic reasons: (a), that of supplying energy for the operation of instruments and other devices, and (b), that of transmission of information between instruments. Integrity of the piping system is essential to avoid loss of the pneumatic supply and degradation of the transmitted signals.

3.2 Reference Standards or Recommended Practices

3.2.1 ISA-S7.3 Quality Standard for Instrument Air

3.2.2 ISA-RP42.1 Nomenclature for Instrument Tubing Fittings (Threaded)

3.2.3 ISA-RP7.1 Pneumatic Control Circuit Pressure Test

3.2.4 ISA-S7.4 (SAMA* RC2-5) Air Pressures for Pneumatic Controllers and Transmission Systems

3.2.5 SAMA* RC19-10 Tubing Connection Markings for Pneumatic Instruments.

3.2.6 ISA dRP60.6** Nameplates, Tags and Labels for Control Centers

3.2.7 API† RP 550, Part I, Section 12 Control Centers

4 PNEUMATIC PIPING

4.1 Pneumatic Supply

Supply air is usually delivered to the control center from external sources at pressures typically between 60 and 150 psi gage (400-1000 kPa). It should be clean, dry and suitable for the application and environment. Refer to ISA-S7.3, "Quality Standard for Instrument Air". A sufficient flow of supply air should be available to meet the control center requirements for transient as well as steady state conditions. Displacement of oxygen in the control center or control room, resulting from the use of bleed type instruments when gases other than air are used, should be considered.

4.1.1 Pressure Reducing Station

A pressure reducing station (sometimes termed as "air set") reduces and regulates the supply air pressure to a level suitable for the application. The reduced pressure is usually 20 psi gage (140 kPa) for systems employing 3-15 psi gage (20-100 kPa) signal ranges and 40 psi gage (270 kPa) for 6-30 psi gage (40-200 kPa) signal ranges.

Pressure reducing stations are normally located inside of the control center, but can be external if special access requirements or space limitations exist.

4.1.1.1 Sizing

The number of air users in the control center establishes the capacity requirements. It is a common practice to determine the total instantaneous air usage by adding together the maximum consumption of each pneumatic device and multiplying by a safety or sizing factor of 1.3 to 2.0.

Air filters and pressure regulators are selected from manufacturers' capacity data based on the total adjusted instantaneous usage and required regulated pressure.

4.1.1.2 Design

An installation of dual pressure regulators and dual air filters is generally employed. They are arranged in parallel with appropriate valving so that one system can be serviced while the other is in use. (See Figure 1.) When 3-way valves are used, they should not interrupt supply to the instruments during a transfer from one leg to the other. Pressure gages should be provided in each leg to indicate the pressure of the regulated air. A pressure gage

---

*Scientific Apparatus Makers Association
**Draft Recommended Practice, for additional information on the status of this document contact ISA Headquarters.
†American Petroleum Institute

may also be installed to indicate the pressure of the supply air. A pressure gage with a blow-out back or disc and a plastic lens is recommended. Low air pressure alarms are often initiated from suitable pressure switches at the pressure reducing station. Adequate clearance is necessary to provide for service and maintenance of the system.

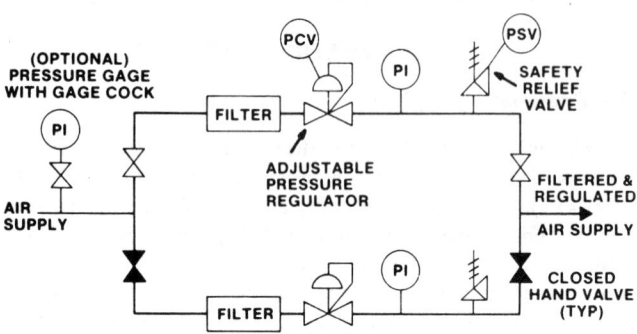

Figure 1. Installation of Dual Pressure Regulators and Dual Air Filters

Combination filter-regulators can be employed where the air usage or service requirements permit. (See Figure 2.) On control centers with a few air users, a separate air filter and pressure regulator may be employed for each air user.

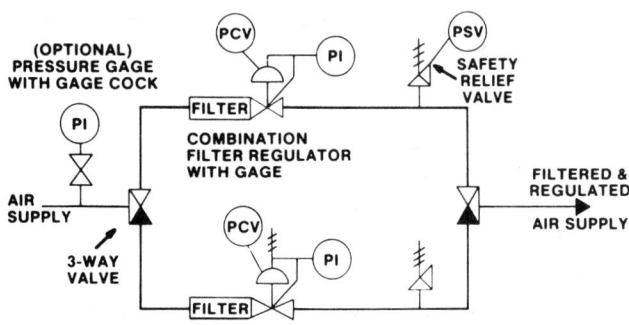

Figure 2. Installation of Combination Filter-Regulators

4.1.2 Regulated Air Supply Header

The regulated air is normally piped to a larger diameter plastic or non-ferrous metal pipe section acting as a manifold or header and reservoir from which each air user receives its supply. The headers are normally installed in a horizontal or vertical plane in the lower part of the control center and have individual shut-off valves at each air supply take-off point. Each shut-off valve should be identified with a non-corrosive metal or plastic tag. Refer to ISA-dRP60.6 "Nameplates, Tags and Labels for Control Centers".* (See Figure 3.) It is a good practice to provide up to 15 percent spare valved connections. The assembly is normally mounted in a rigid manner, sloping at least 1/8 inch per foot away from the supply end, down toward a drain cock. If there is the possibility that failure of the pressure reducing station could result in over pressure damage to the air users, a suitably sized safety relief valve should be installed. (See Figures 1, 2, and 3.) The safety relief valve should be sized to pass the rated flow capacity of one regulator failing full open.

The individual air users are connected to the header shut-off valves by tubing. This supply tubing is typically 1/4 inch outside diameter soft copper, plastic or stainless steel. Some devices with larger air usage may require supply tubing of 3/8 inch or 1/2 inch outside diameter, as suggested by the specifications for that device.

It is a general practice to increase the volume of the air supply header as the number of air users increase. To minimize the pressure drop across the air supply header during large transients in air demand, it is a common practice to use larger diameters and longer sections of pipe.

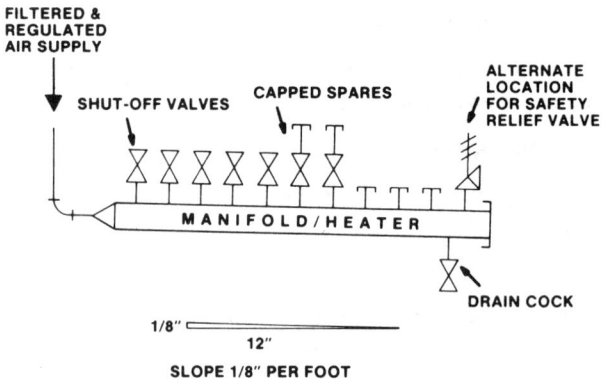

Figure 3. Air Supply Header

4.2 Signal Piping

Signal piping is typically 1/4 inch outside diameter (OD) metal or plastic tubing. The signal piping system often includes a test tee and isolating valve at each receiving device to facilitate testing, calibration, or maintenance.

4.2.1 Common Materials

(a) Soft Copper, 1/4 in. OD x 0.030 in. wall thickness
(b) Polyethylene, 1/4 in. OD x 0.040 in. wall thickness
(c) Aluminum, 1/4 in. OD x 0.032 in. wall thickness
(d) Stainless Steel, 1/4 in. OD x 0.030 in. wall thickness

Certain applications may require plated or plastic coated metal tubing.

---

*Draft Recommended Practice, for additional information on the status of this document contact ISA Headquarters.

4.2.2 Labeling and Color Coding

Plastic or noncorrosive metal nametags should be attached to bulkhead fittings to identify signal source, application, and/or other information. Refer to ISA-dRP60.6, "Nameplates, Tags and Labels for Control Centers".** Color coded plastic tube can identify use of application, i.e.:

- air supply — RED
- transmitted measurement — ORANGE
- controller output to valve, pneumatic set slave, etc., — YELLOW
- seal (to remote mounted controller) — PURPLE
- set (to remote mounted controller) — BLACK
- branch transmitted measurement to alarm element — GREEN
- branch transmitted measurement to readout element — BLUE
- all others — NATURAL*

4.2.3 Routing and Arrangement

All tubing installation should be in accordance with recognized good practices which generally provide that:

(a) Metal tube runs should be routed horizontally and vertically with diagonal routing minimized.
(b) All bends in metal tubing should be made with a tool designed for that purpose to prevent kinks or flattening.
(c) Tubing runs should be routed to provide maximum access to control center interior for ease of maintenance and equipment removal.
(d) Tubing runs should be grouped for mutual support and neat appearance and secured with noncorrosive metal gang straps. Plastic ties or straps can be used for bundling of plastic tubes or the tubes can be run in wireways.
(e) Sharp bends in plastic tubes should be supported or protected to avoid kinks.
(f) Test tees, isolating valves, and quick disconnects should be rigidly supported.
(g) Tube fittings should be made up or installed in accordance with the manufacturer's recommendations.
(h) An isolating valve should be provided at each receiver when one signal line serves two or more receiver instruments.

Piping and tubing terminations are normally accomplished through the use of fittings designed to provide leak free connections. Flared or compression type fittings are normally employed for tube terminations. Refer to ISA-RP42.1, "Nomenclature for Instrument Tubing-Fittings (Threaded)". Fittings are normally threaded for insertion into tapped pipe thread holes in instruments or other devices.

Plastic tubing terminations are normally the compression type. Special metallic tube fittings, including weld-type and solder-type, are commercially available for special applications.

4.3 Fittings

4.3.1 Bulkhead Fittings

Bulkhead fittings provide a means for connection of field piping to the appropriate tube run inside the control center enclosure. Bulkhead connections can be installed inside the control center enclosure or through the outer surface. The bulkhead fitting normally includes a means for rigid attachment to the control center enclosure and connectors for attachment of tubes at each end.

4.3.2 Tube Fittings

Tees, elbows, couplings, nipples, gage cocks, shut-off valves, and other tubing components, are commonly available in plastic, brass, certain alloys of stainless steel and other metals. Material selection should normally be based on the application, operating pressures, flowing media and ambient environmental conditions.

Pipe thread connections should be made leak proof by employing suitable sealing compounds or joint tape.

5 DIRECT PROCESS PIPING

Extra precautions are required when direct process piping is brought into a control center. Safety considerations should include reference to Federal, Provincial, State and Local Codes, restrictions of appropriate regulatory agencies, and other authorities. Material selection, pipe sizes, types of fittings, shut-off valves, blow-down facilities, and other features, should adhere to the requirements of the operating conditions such as temperature, pressure, and corrosiveness of the measured fluid.

---

*Uncolored or white pigmented plastic

**Draft Recommendation Practice, for additional information on the status of this document contact ISA Headquarters.

INSTRUMENT SOCIETY of AMERICA
Research Triangle Park, North Carolina

/ISA-S61.1-1976

Standard

Industrial Computer System FORTRAN Procedures for Executive Functions, Process Input/Output, and Bit Manipulation

Instrument Society of America

ISBN 0-87664-393-4

ISA-S61.1 Industrial Computer System FORTRAN
Procedures for Executive Functions,
Process Input/Output, and Bit Manipulation

Copyright © 1976 by the Instrument Society of America. All rights reserved. Printed in the United States of America. No part of this publication may be reproduced, stored in a retrieval system, or transmitted, in any form or by any means (electronic, mechanical, photocopying, recording, or otherwise), without the prior written permission of the Publisher.

INSTRUMENT SOCIETY OF AMERICA
67 Alexander Drive
P.O. Box 12277
Research Triangle Park, North Carolina 27709

PREFACE

This Preface, all footnotes and all Appendices are included for informational purposes and are not part of Standard ISA-S61.1.

This Standard has been prepared as a part of the service of the Instrument Society of America toward a goal of uniformity in the field of instrumentation. To be of real value this document should not be static but should be subjected to periodic review. Toward this end the Society welcomes all comments and criticisms and asks that they be addressed to the Standards and Practices Board Secretary, Instrument Society of America, 67 Alexander Drive, P.O. Box 12277, Research Triangle Park, North Carolina 27709.

The ISA Standards Committee on Industrial Computer System FORTRAN Procedures SP61, operates within the ISA Standards and Practices Department, W. B. Miller, Vice President. The SP61 Committee also is a working group of the FORTRAN Committee of the International Purdue Workshop on Industrial Computer Systems which provides the impetus for the development of this and related standards. The FORTRAN Committee is chaired by Ms. Maxine N. Hands; General Chairman of the International Purdue Workshop is Dr. T. J. Williams.

Committee SP61, Industrial Computer System FORTRAN Procedures was established in June 1971 as a working group of the FORTRAN Committee of the Purdue Workshop on Standardization of Industrial Computer Languages (later reorganized and renamed the International Purdue Workshop on Industrial Computer Systems). As a result of their activity, ISA Standard S61.1-1972, Industrial Computer System FORTRAN Procedures for Executive Functions and Process Input Output, was published by the ISA in 1972. The committee continued to work on additional draft standards as part of the International Purdue Workshop. In 1974, extensive use of S61.1-1972 and discussions with ANSI X3J3 FORTRAN Committee indicated a need for revision. This revision was accomplished by a working group of the Purdue Workshop FORTRAN Committee which constituted the SP61 Committee listed below:

M. R. Gordon-Clark, Chairman	Scott Paper Company
A. Arthur	IBM Corporation
F. E. Bearden	Modular Computer Systems
R. H. Caro	The Foxboro Company
L. M. Cartright	Inland Steel Company
W. Diehl	Hewlett-Packard Company
D. Frost	Honeywell, Inc.
M. N. Hands	Datum, Inc.
T. L. Luekens	Johnson Service Company
J. McGovney	Interdata, Inc.
O. Petersen	Norwegian Institute of Technology
W. A. Resch, III	Tennessee Eastman Company
D. W. Zobrist	ELDEC

The members of the SP61 Committee who developed the original S61.1-1972 Standard and their affiliations at the time of its approval are as follows:

E. A. Kelly, Chairman	Kaiser Aluminum & Chemical Corp.
A. J. Arthur	IBM Corporation
W. Diehl	Digital Equipment Corporation
M. R. Gordon-Clark	Scott Paper Company
M. N. Hands	Honeywell, Inc.
R. E. Hohmeyer	Control Data Corporation
P. H. Jarvis	General Electric Corporation
L. J. Lockwood	General Motors Corporation
P. K. Mattheiss	Sun Oil Company
R. S. Moser	Applied Automation, Inc.
G. W. Oerter	Leeds & Northrup Company
R. Osmundsen	Technical University of Norway
S. Taylor	Honeywell, Inc.

In preparing the revised S61.1-1976, the committee received extensive assistance and guidance from the members of the International Purdue Workshop and the ANSI X3J3 FORTRAN Committee. Well over 100 people contributed to this revised standard through technical discussions, critiques of drafts, by offering suggestions towards its improvement, and in other ways. The contributions of all of these people are greatly acknowledged. In addition to the SP61 Committee and International Purdue Workshop members, the following people have formally reviewed this revised standard as a Board of Review. They have indicated their general concurrence with this standard; however, it should be noted that they have acted as individuals and their approval does not necessarily constitute approval by their company or facility.

J. J. Anderson	Watermation
J. C. Archer	Gilbert Associates, Inc.
J. Barnard	NY State Agricultural Experiment Station
D. Braedley	Polysar Ltd.
J. A. Conover	Monsanto
R. L. Curtis	ALCOA
D. A. DeMattia	Kaiser Aluminum and Chemical Corporation
D. G. Dimmler	Brookhaven National Laboratory
C. L. Edwards	El Paso Products Company
R. Egger	Boeing Aerospace Company
V. Elischer	Lawrence Berkeley Laboratory
G. F. Erk	Sun Oil Company

R. W. Gellie	National Research Council of Canada
J. C. Houser	Simpson Timber Company
T. J. Harrison	IBM Corporation
W. R. Hodson	Leeds and Northrup Company
R. E. Hohmeyer	Control Data Corporation
D. L. Hutchins	Procter and Gamble Company
T. S. Imsland	Fischer Controls
P. Kushkowski	Northeast Utilities
R. F. Laird, Jr.	E. I. duPont de Nemours & Company
F. R. Lenkszus	Argonne National Laboratory
A. C. Lumb	Procter and Gamble Company
R. G. Marvin	Dow Chemical Company
B. B. Misra	Babcock and Wilcox Company
J. H. Morrison	Fuller Engineering
N. Moseley, Jr.	Merck & Company, Inc.
A. W. Petty	Procter and Gamble Company
D. Rosich	City College of New York
E. H. Spencer	Exxon Corporation
R. J. Spitznas	Baltimore Gas & Electric Company
N. G. Sutter	Honeywell, Inc.
R. F. Thomas	Los Alamos Scientific Laboratory
J. Tsing	Bechtel Power Corporation
J. White	University of Arizona
R. G. Wilhelm	Industrial Nucleonics Corporation
F. G. Willard	Westinghouse Electric Corporation

This Revised Standard was approved by the ISA Standards and Practices Board on June 1, 1976.

W. B. Miller, Chairman	Moore Products Company
P. Bliss	Pratt & Whitney Aircraft Company
E. J. Byrne	Brown & Root, Inc.
W. Calder	The Foxboro Company
B. A. Christensen	Continental Oil Company
L. N. Combs	Retired E. I. duPont deNemours & Company
R. L. Galley	Bechtel Corporation
R. G. Hand, Secretary	Instrument Society of America
T. J. Harrison	IBM Corporation
H. T. Hubbard	Southern Services, Inc.
T. S. Imsland	Fisher Controls Company
P. S. Lederer	National Bureau of Standards
O. P. Lovett, Jr.	E. I. duPont deNemours & Company
E. C. Magison	Honeywell Inc.
R. L. Martin	Tex-A-Mation Engineering Inc.
A. P. McCauley	Glidden Durkee Div. SCM Corporation
T. A. Murphy	The Fluor Corporation, Ltd.
R. L. Nickens	Reynolds Metals Company
G. Platt	Bechtel Power Corporation
A. T. Upfold	Polysar Ltd.
K. A. Whitman	Allied Chemical Corporation
J. R. Williams	Stearns-Roger Corporation

TABLE OF CONTENTS

Section	Page
1. Scope	5
2. Executive Interface	5
2.1 Control of Program Execution	5
2.2 Starting a Program Immediately or After a Specified Time Delay	5
2.3 Starting a Program At a Specified Time	5
2.4 Delaying Continuation of a Program	6
2.5 Program Termination Accomplished Using Present Standard	6
3. Process Input/Output Function Interfaces	6
3.1 Control of Analog and Digital Sensors and Outputs	6
3.2 Analog Inputs in a Sequential Order	6
3.3 Analog Inputs in Any Sequence	6
3.4 Analog Output	7
3.5 Digital Input	7
3.6 Momentary Digital Output	7
3.7 Latching Digital Output	8
4. Bit String Manipulation	8
4.1 Types of Manipulation	8
4.2 Logical Operations	8
4.3 Shift Operations	9
4.4 Bit Testing and Setting	9
5. Date and Time Information	9
5.1 Obtain Time of Day	9
5.2 Obtain Date	10
Appendix A	10
Appendix B	10

1 SCOPE

This standard presents external procedure references for use in industrial computer systems. These external procedure references permit interface of programs with executive routines, process input and output functions, allow manipulation of bit strings and provide access to time and date information. These procedures are intended for use with programs written in FORTRAN conforming to the International Standards Organization (ISO) Programming Language-FORTRAN R1539-1972.[1] expected to be executing both in a solitary and in a multiprogramming environment under the control of a real time executive routine.

2 EXECUTIVE INTERFACE

2.1 Control of Program Execution

Executive interfaces provide the facility to control operation of the programs within the system. Through these external procedures, one may start, stop or delay the execution of application programs.

The argument m, shown below, shall be set equal to or greater than two (2) in value for all instances in which the request is not accepted by the executive routine. Individual implementations may specify unique values of m within the allowable range to designate the specific reason for which the request was rejected.

2.2 Starting a Program Immediately or After a Specified Time Delay

Execution of a reference to the subroutine START shall, after the expiration of the specified time delay, cause the execution of the designated program. The actual time delay obtainable in a specific industrial computer system is subject to the resolution of that system's real time clock. Execution of the designated program will commence at the program's first executable statement. The form of this call is:

CALL START (i, j, k, m)

where:

i specifies the program to be executed.

The argument is either:
a) an integer expression
or b) an integer array name
or c) a procedure name

The processor shall define which of the above three forms is acceptable.

j specifies the minimum length of time, in units as specified by k, to delay before executing the program. If the value of j is zero or negative, the requested program will be run as soon as permissible. This argument shall be an integer expression.

k specifies the units of time as follows:

0 — Basic counts of the system's real time clock
1 — Milliseconds
2 — Seconds
3 — Minutes

This argument shall be an integer expression.

m is set on return to the calling program to indicate the disposition of the request as follows:

The value must be 1 or greater
1 — Request accepted
2 or greater — Request not accepted

This argument shall be an integer variable or integer array element.

2.3 Starting a Program at a Specified Time

Execution of a reference to the subroutine TRNON shall cause the designated program to be executed at a specified time of day. Execution of the designated program will commence at the program's first executable statement. The form of this call is

CALL TRNON (i, j, m)

where:

i specifies the program to be executed.

The argument is either:
a) an integer expression
or b) an integer array name
or c) a procedure name

The processor shall define which of the above three forms is acceptable.

j designates an array whose first three (3) elements contain the absolute time of day at which the specified program is to be executed. These elements are as follows:

First element — Hours (0 to 23)
Second element — Minutes (0 to 59)
Third element — Seconds (0 to 59)

This argument shall be an integer array name.

m is set on return to the calling program to indicate the disposition of the request as follows:

The value must be 1 or greater
1 — Request accepted
2 or greater — Requested rejected

This argument shall be an integer variable or integer array element.

2.4 Delaying Continuation of a Program

Execution of a reference to the subroutine WAIT shall return after a delay of a specified length of time. The form of this call is:

CALL WAIT (j, k, m)

where:

j specifies the length of time in units as specified by

---

[1] ANSI X3.9 — 1966 FORTRAN

k to delay before returning to the calling procedure. If the value is zero or negative, no delay will occur. Limitations of the implementation shall not cause the precise time to be less than requested. This argument shall be an integer expression.

k specifies units of time as follows:

0 — Basic counts of the system's clock
1 — Milliseconds
2 — Seconds
3 — Minutes

This argument shall be an integer expression.

m is set on return to the calling program to indicate the disposition of the request as follows:

The value must be 1 or greater
 1 — Request accepted
 2 or greater — Delay as specified has not occurred.

This argument shall be an integer variable or integer array element.

2.5 Program Termination Accomplished Using Present Standard

Termination of programs shall be accomplished using the STOP statement from (ISO) R1539 — 1972.

3 PROCESS INPUT/OUTPUT FUNCTION INTERFACES

3.1 Control of Analog and Digital Sensors and Outputs

Process input-output function interfaces allow access to data related to specific analog and digital sensors and outputs. The execution of references to these subroutines will result in return only being made to the requesting program when the requested data transfer is complete, or when the requested operation is terminated. The argument m, shown below, is to be interrogated by the requesting program in order to determine the status of the request. An individual processor may specify unique values for m within the allowable range to designate the specific error conditions.

The results of the input operation and the data presented for output operations are processor dependent. The operations described are intended to be unformatted transfers to and from the specified storage in the processor. Therefore, the representation in the storage of the processor is an image of the input or output device specified.

3.2 Analog Inputs in a Sequential Order

Execution of a reference to this subroutine allows the input of data from any number of analog points in a sequence which is specified by the processors input hardware interface. The argument j specifies the first point to be read and also the particular form of all the data placed in the array k. The form of this call is:

CALL AISQW (i, j, k, m)

where:

i specifies the number of analog points to be read. This argument shall be an integer expression.

j specifies hardware or software acquisition and conversion information. Specific information relevant to construction of this argument is processor dependent because various types of analog-to-digital systems are available. This argument shall be either an integer expression or an integer array name.

k designates an array to which the values are assigned. This argument shall be an integer array name, or an integer array element.

m indicates the disposition of the request as follows:

The value must be 1 or greater.
 1 — All data collected
 2 — Operation incomplete
 3 or greater — Error conditions

This argument shall be an integer variable or an integer array element.

3.3 Analog Inputs in any Sequence

Execution of references to this subroutine allows the input of data from analog points in a sequence which is independent of the hardware. This may also be referred to as the input of analog data in a random order. The array j controls the input sequence and allows the specification of hardware or software acquisition and conversion information on an individual point basis. The form of this call is:

CALL AIRDW (i, j, k, m)

where:

i specifies the number of analog points to be read. This argument shall be an integer expression.

j specifies hardware or software acquisition and conversion information for each analog input signal. Specific information relevant to construction of this argument is processor dependent. This argument shall be an integer array name, or an integer array element.

k designates an array to which the values are assigned. The order of the elements in k will correspond to the order in j. This argument shall be an integer array name, or an integer array element.

m indicates the disposition of the request as follows:

The value must be 1 or greater
 1 — All data collected
 2 — Operation incomplete
 3 or greater — Error conditions

This argument shall be an integer variable or integer array element.

3.4 Analog Output

Execution of references to this subroutine allows the output of analog signals in any sequence. The form of this call is:

CALL AOW (i, j, k, m)

where:

- i specifies the number of analog points to be written. This argument shall be an integer expression.

- j specifies hardware or software conversion and transmission information for the analog output point. Specific information relevant to construction of this argument is processor dependent because various types of digital-to-analog systems are available. This argument shall be an integer array name, or an integer array element.

- k designates an array from which the analog output values are taken. The order of the elements in k will correspond to the order j. This argument shall be an integer array name or integer array element.

- m indicates the disposition of the request as follows:

  The values must be 1 or greater
  1 — All data outputted
  2 — Operation incomplete
  3 or greater — Error conditions

  This argument shall be an integer variable or integer array element.

3.5 Digital Input

Execution of references to this subroutine causes the input of process information coded as a set of bits. A set of bits is typically organized and read as an external digital word whose length is processor dependent. The form of this call is:

CALL DIW (i, j, k, m)

where:

- i specifies the number of external digital words to be read. This argument shall be an integer expression.

- j specifies hardware or software acquisition and conversion information for each digital input word. Specific information relevant to construction of this argument is processor dependent because various types of digital inputs are available. This argument shall be an integer array name, or an integer array element.

- k designates an array to which the requested values are assigned. The order of the elements in k will correspond to the order in j. This argument shall be an integer array name, or an integer array element name.

- m indicates the disposition of the request as follows:

  The value must be 1 or greater.
  1 — All data collected
  2 — Operation incomplete
  3 or greater — Error conditions

  This argument shall be an integer variable or integer array element.

3.6 Momentary Digital Output

The execution of references to this subroutine causes the output of momentary digital signals. These signals consist of sets of bits typically organized as an external digital word whose length is processor dependent. This type of output is characterized by an action which momentarily sets individual outputs when a corresponding bit in the output data is set. These output bits will then be reset after a specific time period. The form of this call is:

CALL DOMW (i, j, k, n, m)

where:

- i specifies the number of external digital words to be written. This argument shall be an integer expression.

- j specifies hardware transmission information for each word of output. Specific information relevant to construction of this argument is processor dependent because various types of digital outputs are available. This argument shall be an integer array name, or an integer array element.

- k designates an array whose contents are the image words to be written. The order of the elements in k will correspond to the order in j. This argument shall be an integer array name, or an integer array element.

- n specifies the duration, measured in basic system clock counts, that the outputs are to remain set. If the processor does not allow selection of duration, this argument is ignored but must be present. This argument shall be an integer expression.

- m indicates the disposition of the request as follows:

  The value must be 1 or greater
  1 — All outputs accomplished
  2 — Operation incomplete
  3 or greater — Error conditions

  This argument shall be an integer variable or integer array element.

3.7 Latching Digital Output

Execution of references to this subroutine causes the output of digital signals which can be latched in either the set or reset state. These signals consist of sets or reset state. These signals consist of sets of bits typically organized as an external digital word whose length is processor dependent. This type of output is characterized by an action which sets individual outputs when a corresponding bit in the output is set and clears individual outputs when a corresponding bit is reset. The form of this call is:

CALL DOLW (i, j, k$_1$, k$_2$, m)

where:

i specifies the number of digital words to be written. This argument shall be an integer expression.

j specifies hardware transmission information for each word of digital output. Specific information relevant to construction of this argument is processor dependent. This argument shall be an integer array name, or an integer array element name.

k$_1$ designates an array whose values are the image words to be output. This argument shall be an integer array name, or an integer array element.

k$_2$ designates an array whose values define digital outputs which can be changed by the subroutine. A bit set in the k$_2$ array indicates that the digital output will be changed to the state defined by the corresponding bit position in the corresponding integer array element in k$_1$. The order of the elements in k$_1$ and k$_2$ will correspond to the order in j. This argument shall be an integer array name, or an integer array element.

m indicates the disposition of the request as follows:

The value must be 1 or greater
1 — All outputs accomplished
2 — Operation incomplete
3 or greater — Error conditions

This argument shall be an integer variable or integer array element.

4 BIT STRING MANIPULATION

4.1 Types of Manipulation

The subprograms which follow allow the programmer to view integer data as ordered sets of bits (a_n, a_{n-1}, ... a_0), where the set is a place positional binary representation of an integer value, thus permitting interrogation and manipulation of integers on a bit-by-bit basis. The value of n is processor defined.

4.2 Logical Operations

These operations are external functions. In the following functions, j and m are integer expressions. Operations are performed on all bits which represent the value of an integer internal to the processor. Operations are done bit-by-bit on corresponding bits, that is, the corresponding bits of the actual arguments j and m are used to generate the integer result.

4.2.1 Inclusive Or

The form of this function reference is:

IOR (j, m)

where the result of IOR (j, m) is:

$$\sum_{k=0}^{n} 2^k * (j_k + m_k - (j_k * m_k))$$

4.2.2 Logical Product

The form of this function reference is:

IAND (j, m)

where the result of IAND (j, m) is:

$$\sum_{k=0}^{n} 2^k * (j_k * m_k)$$

4.2.3 Logical Complement

The form of this function reference is:

NOT (j)

where the result of NOT (j) is:

$$\sum_{k=0}^{n} 2_k * (1-j_k)$$

4.2.4 Exclusive Or

The form of this function reference is:

IEOR (j, m)

where the result of IEOR (j, m) is:

$$\sum_{k=0}^{n} 2^k * (2 - (j_k + m_k)) * (j_k + m_k)$$

4.3 Shift Operations

This operation is an external function. In the following function j and m are integer expressions. Operations are performed on all bits which represent the value of an integer internal to the processor, and are used to generate an integer result.

The form of this function reference is:

ISHFT (j, m)

where the result of ISHFT (j, m) is:

If the value of m is positive or zero

$$\sum_{k=0}^{n-m} 2^{k+m} * j_k$$

If the value of m is negative

$$\sum_{k=m}^{n} 2^{k+m} * j_k$$

4.4 Bit Testing and Setting

These operations are external functions. In the following functions j and m are integer expressions.

4.4.1 Bit Test

This logical function tests a specified bit of an integer.

The form of this function reference is:

$$BTEST (j, m)$$

where the result of BTEST (j, m) is:

if IAND $(j, 2^m)$ = 0 then FALSE, else TRUE

4.4.2 Bit Set

This function sets a specified bit of an integer.

The form is this function reference is:

$$IBSET (j, m)$$

where the result of the function reference IBSET (j, m) is:

$$IOR (j, 2^m)$$

4.4.3 Bit Clear

This function clears a specified bit of an integer.

The form of this function reference is:

$$IBCLR (j, m)$$

where the result of the function reference IBCLR (j, m) is:

$$IAND (j, NOT(2^m))$$

5 DATE AND TIME INFORMATION

5.1 Obtain Time of Day

Execution of references to this subroutine allows a program to determine the current time of day. The form of this call is:

$$CALL\ TIME\ (j)$$

where:

j   designates an integer array into whose first three (3) elements the absolute time of day will be placed. The contents of these elements shall be as follows:

First Element   — Hours 0 to 23
Second Element — Minutes 0 to 59
Third Element   — Seconds 0 to 59

5.2 Obtain Date

Execution of references to this subroutine allows a program to determine the current calendar date. The form of this call is:

$$CALL\ DATE\ (j)$$

where:

j   designates an integer array into whose first three (3) elements the date will be placed. The contents of these elements shall be as follows:

First Element   — AD year since zero
Second Element — Month 1 to 12
Third Element   — Day 1 to 31

APPENDIX A

(This appendix is not part of the ISA Standard S61.1, but is included to facilitate its use.)

Considerations leading to "ISA-S61.1 Industrial Computer System FORTRAN Procedures for Executive Functions Process Input/Output, and Bit-Manipulation."

A.1 Historical Development

This standard is a direct outgrowth of the International Purdue Workshop on Industrial Computer Systems whose goals are:

> To make the definition, justification, hardware and software design, procurement, programming, installation, commissioning, operation and maintenance of industrial computer systems more efficient and economical through the development of standards and/or guidelines on an international basis.

The Workshop formed several committees to achieve their objectives. The FORTRAN Committee was charged with the task of preparing a set of Industrial Process Standards compatible with FORTRAN as defined by ISO-R1539-1972. This standard is a result of that committee's work. Additional standards are being developed.

A.2 Criteria Used in Developing This Standard

The committee assessed that FORTRAN was the most widely used language in the industrial environment. It thus used the following guidelines in the development of the standard.

1. The standard should cover features commonly used by existing industrial computer systems.
2. The standards should be easy to implement for most vendors.
3. The standards should follow the syntax and intent of FORTRAN as defined by ISO.
4. The standards should not restrict the future evolution of FORTRAN.

The development of the FORTRAN language is presently the responsibility of ISO who has delegated this responsibility to ANSI X3J3.

In order that these standards comply with the ISO-R1539-1972 standard as far as possible, external procedure references were used rather than direct changes or additions to the syntax of FORTRAN. This does not imply that this is the only way to provide these features nor does it exclude the possibility or desirability that the language will be developed through syntax changes so that these features and other related features can be performed.

APPENDIX B

(This appendix is not part of the ISA Standard S61.1, but is included to facilitate its use.)

NOTES BY SECTION

B.1 Section 2 Notes

This standard is a permissive standard in that it does not prescribe how the executive will respond to external procedure references nor does it describe how the information is passed to the executive routine. In particular, the argument "i" in CALL START (2.2) and CALL TRNON (2.3) has three forms, an integer expression, an integer array name or a program name, only one of which is permissible in any particular program. This restriction is a necessary consequence of the requirements of the FORTRAN language that any argument which is an external procedure reference be of a defined type; integer expressions, integer array names and program names are different types (Section 8.4.2 of ISO-R1539-1972). It is also a consequence of the FORTRAN language that if the argument is a program name, this name must appear in an EXTERNAL statement (Section 7.2.15 of ISO-R1539-1972).

Examples of the use of the argument i as an integer expression are:

```
        CALL START (7,0,0,M)

        DATA J/7/
        CALL START (J,0,0,M)

        DATA J/2HAB/
        CALL START (J,0,0,M)
```

An example of the use of the argument i as an integer array name is:

```
INTEGER XYZ
DIMENSION XYZ (3)
DATA XYZ(1), XYZ(2),XYZ(3)/2HAB,2HCD,2HEF/
CALL   START   (XYZ,0,0,M)
```

An example of the use of the argument i as a procedure name is:

```
        EXTERNAL  ABCD
        CALL   START   (ABCD,0,0,M)
```

Interchangeability of programs between different processors is reduced by permitting three possible types for this argument but the committee feels it is premature to standardize to achieve full interchangeability in this area.

B.2 Section 3 Notes

As an extension to the process I/O calls, six additional calls may be defined. They are AISQ, AIRD, AO, DI, DOM, DOL. Processors conforming to ISO-R1539-1972 require extensions to support these calls; namely, the FORTRAN processor must maintain association with the data block during execution of these subroutines by the executive routine. These calls accommodate executive routines which permit continuation of execution in the requesting program while the process input or output is being accomplished. These are related to the six standard calls as shown below:

STANDARD	EXTENSION
AISQW	AISQ
AIRDW	AIRD
AIW	AI
DIW	DI
DOMW	DOM
DOLW	DOL

wherein the arguments are defined identically as those of the corresponding standard calls.

For these extensions, the requesting program is required to have provision for periodically testing the status of the request. This is accomplished by use of the argument m.

All values returned by these extensions should not be considered as defined until such time as the parameter m indicates operation completed or error condition.

The process system designer must ensure the availability of these arguments to the process input/output interface subroutine in his total system. No method is given of how this will be achieved, although most existing industrial systems have features which can be used for this purpose, such as global common.

For the standard process I/O calls, the return to the calling program should not be made with m=2. On return m should equal 1 to indicate successful completion or m should be 3 or greater to indicate an error was found.

The argument j is defined to be processor dependent and can take several different forms. For example, in AISQW and AIRDW

1. An alias for the analog input similar to a logical unit number
2. A one word per analog input hardware address
3. A multiple word per analog input hardware address in which case the array j must be a multiple of the size of k.

B.3 Section 4 Notes

The basic unit of computer storage is the integer according to ISO FORTRAN. For the purposes of industrial computing it is necessary to view integer data as an ordered set of bits (a_n, o_{n-1}, ...a_0) where the set is a place positional binary representation of an integer value.

For reasons of mathematical exactitude it is necessary to express the bit functions as a summation of a series, but on a "normal" two's complement binary computer, they will operate in the expected manner for such functions. However, on a binary coded decimal machine the results of these functions may not be in the expected manner for such functions and it will not be possible to reference every 'bit' of a word.

Items such as difference in word lengths, one's complement versus two's complement, negative numbers and BCD versus pure binary will affect transportability. However, due to the differences in the computers themselves, these differences are to be expected.

These functions are not intended to discourage the introduction of syntactic changes to FORTRAN.

B.3.1 Examples of Section 4

The following examples illustrate the use of Section 4 in less mathematical terminology. The processor is assumed to be a seven (7) bit +, two's complement machine, with the sign bit on the extreme left.

```
Let   j = 10                (0001010)
      m =  3                (0000011)
Then  IOR (j,m) = 11        (0001011)
      IAND (j,m) = 2        (0000010)
      NOT (j) = -11         (1110101)
      IEOR (j,m) = 9        (0001001)
      ISHFT (j,m) = -48     (1010000)
      ISHFT (j,-m) = 1      (0000001)
      ISHFT (m,j) = 0       (0000000)
      BTEST (j,m) = .TRUE.
      IBSET (j,m) = 10      (0001010)
      IBCLR (j,m) = 2       (0000010)
```

B.4 Section 5 Notes

The format used for referencing time and date follows the ISO standard R2014-1971, writing of calendar dates in all-numeric form.

The standard ISA-S61.1 does not define the basis used or how the processor maintains date and time but does define how this information is made available to the program.

INSTRUMENT SOCIETY of AMERICA
Research Triangle Park, North Carolina

ANSI/ISA-S61.2-1978
Approved August 30, 1978

American National Standard

Industrial Computer System FORTRAN Procedures for File Access and the Control of File Contention

Instrument Society of America

ISBN87664-412-4

ANSI/ISA S61.2 Industrial Computer System FORTRAN Procedures for File Access and the Control of File Contention.

Copyright© 1978 by Instrument Society of America. All rights reserved. Printed in the United States of America. No part of this publication may be reproduced, stored in a retrieval system, or transmitted, in any form or any means, electronic, mechanical, photocopying, recording or otherwise without the prior written permission of the publisher:

INSTRUMENT SOCIETY OF AMERICA
67 Alexander Drive
P.O. Box 12277
Research Triangle Park, North Carolina 27709

Copyright© 1978 by Instrument Society of America.

PREFACE

This Preface, all footnotes, and all Appendices are included for informational purposes and are not part of Standard ISA-S61.2.

This Standard has been prepared as a part of the service of the Instrument Society of America toward a goal of uniformity in the field of instrumentation. To be of real value, this document should not be static but should be subjected to periodic review. Toward this end, the Society welcomes all comments and criticisms and asks that they be addressed to the Standards and Practices Board Secretary, Instrument Society of America, 67 Alexander Drive, P.O. Box 12277, Research Triangle Park, North Carolina 27709.

The ISA Standards Committee on Industrial Computer System FORTRAN Procedures SP61, operates within the ISA Standards and Practices Department, A. P. McCauley, Vice President. The SP61 Committee also is a working group of the FORTRAN Committee of the International Purdue Workshop on Industrial Computer Systems which provides the impetus for the development of this and related standards. The FORTRAN Committee is chaired by M. R. Gordon-Clark; General Chairman of the International Purdue Workshop is Dr. T. J. Williams.

Committee SP61, Industrial Computer System FORTRAN Procedures was established in June 1971 as a working group of the FORTRAN Committee of the Purdue Workshop on Standardization of Industrial Computer Languages (later reorganized and renamed the International Purdue Workshop on Industrial Computer Systems). As a result of their activity, ISA Standard S61.1-1972, Industrial Computer System FORTRAN Procedures for Executive Functions and Process Input Output, was published by the ISA in 1972. In 1974, extensive use of S61.1-1972 and discussions with ANSI X3J3 FORTRAN Committee indicated a need for revision and the standard was revised as S61.1-1976. The development of S61.2-1977 began in 1972 and the standard was prepared by a working group of the International Purdue Workshop FORTRAN Committee which constituted the SP61 committee listed below.

M. R. Gordon-Clark, Chairman	Scott Paper Company
A. Arthur	IBM Corporation
F. E. Bearden	Modular Computer Systems
R. H. Caro	The Foxboro Company
L. M. Cartright	Inland Steel Company
W. Diehl	Hewlett-Packard Company
M. N. Hands	Digital Equipment Corporation
C. C. Haskell	Union Carbide Company
T. L. Luekens	Johnson Service Company
O. Petersen	Norwegian Institute of Technology
W. A. Resch, III	Tennessee Eastman Company
R. W. Swearingen	Virtual Systems Inc.
D. W. Zobrist	ELDEC

This standard was approved for publication by the Standards and Practices Board on February 6, 1978.

A. P. McCauley, Chairman	SCM Corporation
P. Bliss	Pratt & Whitney Aircraft Group
E. J. Byrne	Brown & Root, Inc.
W. Calder	The Foxboro Company
B. A. Christensen	Continental Oil Company
R. Coel	Fluor Corporation
T. J. Harrison	IBM Corporation
H. T. Hubbard	Southern Company Services, Inc.
P. S. Lederer	National Bureau of Standards
E. C. Magison	Honeywell, Inc.
R. L. Nickens	Reynolds Metals Company
G. Platt	Bechtel Power Corporation
G. Quigley	Moore Instrument Company, Ltd.
H. G. Tobin	ITT Research Institute
W. C. Weidman	Gilbert Associates, Inc.
K. A. Whitman	Allied Chemical Corporation
J. R. Williams	Stearns-Roger, Inc.
W. E. Vannah	The Foxboro Company
M. T. Yothers, Secretary	Instrument Society of America
*L. N. Combs	Retired, E. I. du Pont de Nemours & Co.
*R. L. Galley	Bechtel, Inc.
*R. G. Marvin	Retired, Dow Chemical Company
*W. B. Miller	Moore Products Company

Director Emeritus

TABLE OF CONTENTS

Section   **Page**

1. Scope 1
   1.1 Definitions 1
   1.2 Background Information 1
   1.3 File System Environment for this Standard 1
2. Interfaces to Access Files 1
   2.1 General 1
   2.2 Creation of Files 1
   2.3 Deletion of Files 2
   2.4 Opening Files 2
   2.5 Closing Files 3
   2.6 Modify Access Privileges 3
3. Input/Output to Unformatted Direct Access Files 4
   3.1 Direct File Read 4
   3.2 Direct File Write 4

Appendix A 4
Appendix B 5
Appendix C 6

LIST OF TABLES

Table 1 — Features and Attributes of Files 2

1. SCOPE

This standard presents external procedure references for use in industrial computer control systems. These external procedure references provide means for accessing files, and also provide means for resolving problems of file access contention in a multiprogramming/multiprocessing environment. In such an environment, it is expected that concurrent programs will attempt to access the same file at the same time; therefore, the external procedure references defined in this standard provide the information necessary for the processor to resolve such simultaneous access in an orderly manner. The method for resolution of access control is left to the processor.

The procedure references in this standard are intended to provide the methods by which the program can inform the processor of the manner in which it intends to use the file but the references are not intended to require specific properties or attributes to be associated with the referenced files. This standard provides the means to avoid contention problems when used in conjunction with sound program design but the implementation of this standard is no assurance that such problems will not arise.

This standard is based on FORTRAN ISO R1539-1972 and is not based on FORTRAN ANSI X3.9-1978 (see Appendix C).

1.1 Definitions

Access Privilege: The right or permission to access (read or write) a file granted by the processor following a request for such permission.

File: A collection of related records treated as a unit. For the purpose of this standard, the records are viewed as being of fixed length. Record storage and access are independent of the internal format of records.

FORTRAN: The computing language defined as full FORTRAN ISO R1539-1972 which is the same as American National Standard FORTRAN ANSI X3.9-1966.

Other terminology used in this standard is defined in other references primarily in the document describing Standard FORTRAN.

1.2 Background Information

In computing systems, individual programs may have various relationships:
— Programs are executed sequentially.
— Several programs can be operating concurrently but with no shared resources.
— Several programs can be operating currently and these concurrent programs may share resources.

A file can be such a shared resource.

Files exist in all computing systems and can have various attributes and features, such as:
— A file can contain data, programs, or catalogue information.
— There can be a variety of ways for file access such as sequential, direct, and stream.
— A file can be created or deleted by a program, by a system utility, or at system generation time.
— A file can have security attributes associated with the file for the purpose of ensuring file privacy.
— When a file is associated with a program, this association can be restricted by the processor for reasons of privacy.
— A file can be associated with a set of related concurrent programs and this association can be restricted to assure orderly resolution of contention problems among the concurrent programs.
— A file can be internal or external to a program.
— A file can reside on fixed or removable media.
— A file can reside on main storage or backing storage.
— Restrictions for reasons of privacy or contention may apply to a file or a component of a file such as records and data items.

1.3 File System Environment for this Standard

In industrial computer systems, concurrent program operation with shared resources such as files is a common occurrence. This standard does not address all the areas of file management but is concerned with the problems that most commonly arise in industrial computer systems.

Table 1 shows those features covered by the standard and those excluded; however, the excluded features may affect the result of a request for association of a concurrent program to a file. Such restrictions on associations are processor-dependent and are outside the scope of this standard.

2 INTERFACES TO ACCESS FILES

2.1 General

The procedure references defined in this section of the standard are non-interruptible; that is, the processor will only execute one such procedure reference at a time. This requirement ensures that the features described are executed in an orderly manner.

The argument m, shown below, shall be set equal to or greater than two (2) in value when the request is not accepted by the executive routine. Individual implementation may specify unique values of m within the allowable range to designate the specific reason for which the request was rejected.

2.2 Creation of Files

Execution of a reference to the subroutine CFILW shall establish, but not open, a named file. Files established by CFILW do not have any privacy attribute to restrict a concurrent program from accessing the files. The contents of a newly created file are undefined by the standard. The form of call is:

$$\text{CALL CFILW } (j, n_1, n_2, m)$$

where:

j specifies the name of the file.

TABLE 1.
FEATURES AND ATTRIBUTES OF FILES

Included In The Standard	Excluded From The Standard
— Files whose contents are considered to be data.	— Files whose contents are not considered to be data by the accessing concurrent program.
— Files which exist on fixed media only or on removable media that are not removed.	— Files that exist on removable media which are removed.
— Files that reside in main storage or in backing storage.	
— Files that are external to a concurrent program.	— Files that are internal to a concurrent program.
— Creation and delection of files by a concurrent program.	— Creation and deletion of files by a system utility or at system generation.
— The association of a file to a concurrent program for both system created and for concurrent program created files.	
— Restrictions on file access as applied to the file.	— Restrictions on file access as applied to a component of a file.
— The association of a file to a concurrent program irrespective of the method of access (direct, sequential, or stream).	
— Read and write methods of access for direct access files only (sequential files are covered by standard FORTRAN ISO R1539-1972).	— Methods of file access except for direct access.
	— Attributes of a file for the purpose of ensuring file privacy.

The argument is either:
   a) an integer expression
or b) an integer array name
or c) a procedure name

The processor shall define which of the above three forms are acceptable.

n_1 specifies the number of storage units per record in this file. This argument shall be an integer expression.

n_2 specifies the number of records in this file. This argument shall be an integer expression.

m is set on return to the calling program to indicate the disposition of the request.

The value must be 1 or greater
   1 — File successfully created
   2 or greater — File not created

This argument shall be an integer variable name or integer array element name.

2.3 Deletion of Files

Execution of a reference to the subroutine DFILW shall remove a file from the file system. Any file created by the mechanism of Section 2.2 can be deleted by the execution of a reference to DFILW. A file currently open to another program must not be deleted. The form of this call is:

   CALL DFILW (j,m)

where:

i   specifies the name of the file.

   The argument is either:
      a) an integer expression
   or b) an integer array name
   or c) a procedure name

   The processor shall define which of the above three forms are acceptable.

m   is set on return to the calling program to indicate the disposition of the request.

   The value must be 1 or greater
      1 — File successfully deleted
      2 or greater — File not deleted

   This argument shall be an integer variable name or integer array element name.

2.4 Opening Files

Execution of a reference to the subroutine OPENW shall associate the unit specified by the program with the named file, and shall define the desired access privilege of that program to the file. The form of this call is:

   CALL OPENW (i,j,k,m)

where:

i specifies the unit by which the file, named by the argument j, is referenced in the program. This argument shall be an integer expression.

j specifies the name of the file.

   The argument is either:
   a) an integer expression
   or b) an integer array name
   or c) a procedure name

   The processor shall define which of the above three forms are acceptable.

k specifies the access privilege desired by the program. It is a declaration of the program's intended use of the file. This argument shall be an integer expression. The following values are defined:

   1. Read Only — The calling program can read but not write; other concurrent programs can read and write.

   2. Shared — The calling program can read or write; other concurrent programs can read or write.

   3. Protected Read — The calling program can only read; other concurrent programs can only read.

   4. Exclusive All — Only the calling program can access the file.

m set on return to the calling program to indicate the disposition of the request.

   The value must be 1 or greater
   1 — File successfully opened to the calling program.
   2 or greater — File not open to the calling program.

   This argument shall be an integer variable name or integer array element name.

Limitations

If the file is currently open to another program, the disposition of a possible request for a particular access privilege will be as follows:

— Read Only — Fails if another program currently has Exclusive All privilege, otherwise succeeds.

— Shared — Fails if another program currently has Exclusive All or Protected Read privileges, otherwise succeeds.

— Protected Read — Fails if another program currently has Exclusive All, or Shared privileges, otherwise succeeds.

— Exclusive All — Fails.

Any attempt to open a file will be successful only if the file exists. If the file was created by a mechanism outside of the standard, the attributes given to the file at its creation may restrict the granting of an access privilege to the program.

2.5 Closing Files

Execution of a reference to the subroutine CLOSEW shall end the program association of the specified logical unit with a named file. The form of the call is:

CALL CLOSEW (i,m)

where:

i specifies the unit. This argument shall be an integer expression.

m is set on return to the calling program to indicate the disposition of the request.

   The value must be 1 or greater
   1 — File successfully closed to the program.
   2 or greater — Non-performance.

   This argument shall be an integer variable name or integer array element name.

2.6 Modify Access Privileges

Execution of a reference to the subroutine MODAPW shall change the calling program's access privilege to a file previously opened by the calling program without closing and reopening the file.

If the calling program does not have access privileges to the file the request fails.

If the request for change cannot be granted, the previous access privilege remains in force. The form of this call is:

CALL MODAPW (i,k,m)

where:

i specifies the unit. This argument shall be an integer expression.

k specifies the access privilege desired. This argument shall be an integer expression. The following values are defined:

   1. Read Only — The calling program can read but not write; other concurrent programs can read or write.

   2. Shared — The calling program can read or write; other concurrent programs can also read or write.

   3. Protected Read — The calling program can only read; other concurrent programs can only read.

   4. Exclusive All — Only the calling program can access the file.

m is set on return to the calling program to indicate the disposition of the request.

   The value must be 1 or greater
   1 — Access privileges requested is granted to the program.
   2 or greater — Access privileges before the request remains in force.

   This argument shall be an integer variable name or integer array element name.

Limitations

If the file is currently open to another program, the disposition of a request for a particular access privilege will be as follows:

— Read Only — Will always succeed.
— Shared — Fails if another program currently has Protected Read privileges, otherwise succeeds.
— Protected Read — Fails if another program has Shared privileges, otherwise succeeds.
— Exclusive All — Fails

If the file was created by a mechanism outside of the standard, the attributes given to the file at its creation may restrict the granting of an access privilege to the program.

3. INPUT/OUTPUT TO UNFORMATTED DIRECT ACCESS FILES

Various methods of accessing files exist such as "sequential" and "direct." Direct access is a method in which items of information are stored and become available independently.

This section of the standard provides for unformatted direct access to files. Access to sequential files is defined by ISO R1539-1972. In this standard, direct access files are considered to consist of fixed length records. These files are considered to be resident in some mass memory that is permanently attached to the computer. The length of a record in these files is defined in terms of a storage unit.

3.1 Direct File Read

The execution of a reference to the subroutine RDRW results in the sequential transfer of one data record from a file. The file is treated as a direct access file for selection of the record. The calling program must have opened the file and must currently have access privileges. The form of the call is:

CALL RDRW (i,j,k,n,m)

where:

i   specifies the unit. This argument shall be an integer expression.

j   specifies the record number of the record to be read. This argument shall be an integer expression which evaluates to a positive integer.

k   designates the first variable into which information is to be placed.

     This argument shall be:
       a) an integer variable name
    or b) integer array element name
    or c) integer array name

n   specifies the maximum number of storage units that can be transferred.

m   is set on return to the calling program to indicate the disposition of the request.

The value is 1 or greater
   1 — Data transfer completed successfully
   2 or greater — Data transfer fails.

The argument shall be an integer variable name or integer array element name.

3.2 Direct File Write

The execution of a reference to the subroutine WRTRW writes unformatted direct accessed information into files that have previously been opened and access privileges assigned to the calling program. The request will only be successful if the program's access privilege is shared, or exclusive all. The form of the call is:

CALL WRTRW (i,j,k,n,m)

where:

i   specifies the unit. This argument shall be an integer expression.

j   specifies the record number of the record to be written. This argument shall be an integer expression which evaluates to a positive integer.

k   designates the first variable from which information is to be obtained.

     This argument shall be:
       a) an integer variable name
    or b) integer array element name
    or c) integer array name

n   specifies the maximum number of storage units that can be transferred.

m   is set on return to the calling program to indicate the disposition of the request.

The value is 1 or greater
   1 — Data transfer completed successfully
   2 or greater — Data transfer fails

The argument shall be an integer variable name or integer array element name.

APPENDIX A

This Appendix is not part of the ISA Standard S61.2 but is included to facilitate its use.

Considerations leading to Industrial Computer System FORTRAN Procedures for File Access.

A.1 Historical Development

This standard is a direct outgrowth of the International Purdue Workshop on Industrial Computer Systems whose goals are:

To make the definition, justification, hardware and software design, procurement, programming, installation, commissioning, operation, and maintenance of industrial computer system more efficient and economical through the development of standards and/or guidelines on an international basis.

The Workshop formed several committees to achieve its objectives. The FORTRAN Language Committee was charged with the task of preparing a set of Industrial Process standards compatible with standard FORTRAN.

This standard is the result of that committee's work.

A.2 Criteria Used in Developing FORTRAN Standards

The committee assessed the status of FORTRAN as used in the industrial environment and followed the guidelines below in the development of the standards:

(1) The standards should cover features commonly used by existing industrial computer systems.
(2) The standards should be easily implemented by most vendors.
(3) The standards should follow the syntax and intent of FORTRAN as defined by ISO R1539-1972.
(4) The standards should not restrict the future evolution of FORTRAN language.

The development of FORTRAN language standards is presently the responsibility of ANSI/X3J3. In order that ISA standards comply with the ANSI standards as far as possible, external-procedure references were used rather than direct changes or additions to the syntax of FORTRAN. This does not imply that this is the only way to provide these features, nor does it exclude the possibility or desirability that ANSI will develop the language syntax to perform these and other related forms.

APPENDIX B

This Appendix is not part of the ISA Standard S61.2 but is included to facilitate its use.

NOTES BY SECTION

B.1 Section 2 Notes

This standard is a permissive standard in that it does not prescribe how the executive will respond to external procedure references nor does it describe how the information is passed to the executive routine. In particular, the argument "j" in CALL CFILW (2.2), CALL DFILW (2.3) and CALL OPENW (2.4) has three forms: an integer expression, an integer array name or a program name, only one of which is permissible in any particular program. This restriction is a necessary consequence of the requirements of the FORTRAN language that any argument which is an external procedure reference be of a defined type; integer expressions, integer array names and program names are different types (Section 8.4.2 of ISO-R1539-1972). It is also a consequence of the FORTRAN language that if the argument is a program name, this name must appear in an EXTERNAL statement (Section 7.2.15 of ISO-R1539-1972).

Examples of the use of the argument j as an integer expression are:

```
        CALL CFILW (7,10,10,M)

    DATA J/7/
    CALL CFILW (J,10,30,M)

    DATA J/2HAB/
    CALL CFILW (J,10,30,M)
```

An example of the use of the argument j as an integer array name is:

```
INTEGER XYZ
DIMENSION XYZ (3)
DATA XYZ (1), XYZ (2), XYZ (3)/2HAB, 2HCD, 2HEF/
CALL CFILW (XYZ,10,10,M)
```

An example of the use of the argument j as a procedure name is:

```
        EXTERNAL ABCD
        CALL CFILW (ABCD,10,10M)
```

Interchangeability of programs between different processors is reduced by permitting three possible types for this argument but the committee feels it is premature to standardize to achieve full interchangeability in this area.

This standard does not cover all areas of file management. In particular, no attempt is made to consider file privacy. In an industrial computer, system privacy is not a significant problem because the intent is to provide a common data base on plant operation for control and information purposes for all users of the system, but the problem of contention is acute because of the asynchronous nature of the requests to the system.

File management can be very complex, especially with removable media because the system must check that the correct media is present for all accesses to the file. Such checking involves considerable system overhead which is much reduced if a restriction is made to non-removable media. The committee considers the restriction to non-removable media satisfactory for industrial systems, especially if adequate measures are taken by the operators of the equipment at those times when removable media are changed.

The access privilege states described in Sections 2.4 and 2.6 are not the only states that can exist but the committee considers states different from those described in Sections 2.4 and 2.6 to be of low utility.

B1.1 Example

Consider a file called FILNAM that contains data on a process. The file is eight (8) records long, and each record contains ten (10) integers. The first seven (7) records contain process information and the eighth record contains status information.

Program ABC reads process data from the first seven records of FILNAM and calculates status information which is written into the eighth record.

Program XYZ reads the status record of FILNAM, performs some actions, and resets the status record.

Program ABC is executed asynchronously when the process information changes. Program XYZ runs at regular intervals.

In this example, the file contention procedures are used to ensure an orderly use of the status information of the file FILNAM:

B.2 Section 3 Notes

The procedures for reading and writing direct access files described in Section 3 are limited in scope. The

EXAMPLE:

```
C
C     example program ABC
C
      INTEGER FILNAM,IVAL(10),JVAL(10)
C
C     open file logical unit number to 11
C        file name FILNAM
C        access privileges read only
C
      FILNAM=333
      CALL OPENW(11,FILNAM,1,M)
      IF(M.NE.1) GO TO 90
C
C     read record number 1
C
      CALL RDRW(11,1,IVAL,10,M)
      IF(M.NE.1) GO TO 90
C
C
C     continue processing data
C
C
C
C     change access privileges to
C        exclusive all
C        and update record number 8
C
   30 CALL MODAPW(11,4,M)
      IF(M.NE.1) GO TO 30
      CALL WRTRW(11,8,JVAL,10,M)
      IF(M.NE.1) GO TO 90
      CALL CLOSEW(11,M)
      IF(M.NE.1) GO TO 90
      STOP
C
C     error handling routines
C
   90 CONTINUE
```

```
C
C     example program XYZ
C
      INTEGER FILNAM,KVAL(10)
C
C     open file logical unit number to 7
C        file name FILNAM
C        access privileges exclusive all
C
      FILNAM=333
   20 CALL OPENW(7,FILNAM,4,M)
      IF(M.NE.1) GO TO 20
C
C     read record 8 and update contents
C
      CALL RDRW(7,8,KVAL,10,M)
      IF(M.NE.1) GO TO 90
C
C
C     continue processing data
C
C
C
C     write the updated record
C
      CALL WRTRW(7,8,KVAL,10,M)
      IF(M.NE.1) GO TO 90
C
C
      CALL CLOSEW(7,M)
      IF(M.NE.1) GO TO 90
      STOP
C
C     error handling routine
C
   90 CONTINUE
```

committee is aware of these limitations and considers that more comprehensive procedures for reading and writing files require changes to FORTRAN syntax. Such changes are the responsibility of the ANSI X3J3 committee.

APPENDIX C

This appendix is not part of the ISA Standard S61.2 but is included to facilitate its use.

Relationship of ISA S61.2-1978 to the American National Standard FORTRAN ANSI X3.9-1978.

C.1 ANSI X3.9-1978 FORTRAN

The standard ISA S61.2-1978 is based on the computing language full FORTRAN ISO R1539-1972 which is the same as American National Standard ANSI X3.9-1966, because this document defined standard FORTRAN during the development of ISA S61.2-1978. However, the ANSI X3J3 committee has released a revised version of standard FORTRAN which has been approved as an American National Standard ANSI X3.9-1978. In general, the changes, clarifications, deletions and additions to standard FORTRAN do not invalidate ISA S61.2-1978. ANSI X3.9-1978 consists of a full language and a subset. It is recommended that any supplier providing ANSI X3.9-1978 with ISA S61.2-1978 follow the procedures in sections C.2 and C.3.

C.2 Full FORTRAN ANSI X3.9-1978

ISA S61.2-1978 is compatible with ANSI X3.9-1978 with the following exceptions.

C.2.1 Use of Hollerith

In Appendix A, examples are given of the use of the argument "i" involving Hollerith data type. ANSI X3.9-1978 does not include a Hollerith data type and the equivalent examples would use the character data type.

C.2.2 Read/Write to Direct Access Media

ANSI X3.9-1978 provides for input/output to direct access media in more detail and with more features than ISA S61.2-1978 Section 3. It is recommended that the procedures of ANSI X3.9-1978 be used rather than Section 3 of ISA S61.2-1978. If Section 3 of ISA S61.2-1978 is implemented there is a possible problem in transportation of the subroutines with a standard conforming FORTRAN program; namely if the argument "k" is an integer variable name or integer array element name rather than an integer array name, there is no guarantee that consecutive memory locations will contain the desired data, although in many implementations of standard conforming FORTRAN this problem would not arise.

C.2.3 Extensions to ANSI X3.9-1978

Section 2 of ISA S61.2-1978 can be implemented by an extension to the input/output statements of ANSI X3.9-1978. The recommended method for such an extension is to use an additional keyword specifier.

PRIVILEGE — n where where n is an integer expression identical to the argument "k" in Section 2.

The actions performed by execution of a reference to the subroutines OPENW and MODAPW can be provided by this keyword specifier.

C.3 Subset of FORTRAN ANSI X3.9-1978

The comments of Section C.2 apply to the subset of FORTRAN ANSI X3.9-1978.

INSTRUMENT SOCIETY of AMERICA
Research Triangle Park, North Carolina

ANSI/ISA-S67.01-1979
Reaffirmed: July 29, 1987

Standard

Transducer and Transmitter Installation for Nuclear Safety Applications

Instrument Society of America

ISBN 0-87664-477-9

ISA-S67.01 Transducer and Transmitter Installation for Nuclear Safety Applications

Copyright © 1979 by the Instrument Society of America. All rights reserved. Printed in the United States of America. No part of this publication may be reproduced, stored in a retrieval system, or transmitted, in any form or by any means (electronic, mechanical, photocopying, recording, or otherwise), without the prior written permission of the Publisher.

INSTRUMENT SOCIETY OF AMERICA
67 Alexander Drive
P.O. Box 12277
Research Triangle Park, North Carolina 27709

Instrument Society of America

PREFACE

This Preface is included for informational purposes and is not a part of Standard S67.01.

This Standard has been prepared as a part of the service of the Instrument Society of America toward a goal of uniformity in the field of instrumentation. To be of real value, this document should not be static, but should be subjected to periodic review. Toward this end, the Society welcomes all comments and criticisms, and asks that they be addressed to the Standards and Practices Board Secretary, Instrument Society of America, 67 Alexander Drive, P.O. Box 12277, Research Triangle Park, NC 27709.

Begun in April 1974, under the directorship of Robert L. Galley and the assistance of H. C. Schmidt, W. M. Deutsch (deceased), J. A. Nay, and M. J. Kimbell, this Standard is one of the first ISA ventures directed specifically at the nuclear power industry. Shortly thereafter, the ISA Nuclear Power Plant Standards Committee (NPPSC) was formed within the Power Industries Division of ISA to oversee the development of standards for the nuclear power industry and to serve as the SP-67 committee for those standards.

The question of definitions between "transducer" and "transmitter" was raised repeatedly in the development of this standard. It was generally agreed that industry practice is to use "transmitter" for devices in which the values of the measurand are converted, operated upon, and scaled to a standardized output signal. In contrast, a "transducer" is commonly considered to be a fixed device for a single conversion of measurand value to some signal which is physically inherent to the "transducer" design, and which cannot be scaled or operated upon within the "transducer" itself. Thus, in common usage as seen by this subcommittee, a "transmitter" will contain at least one "transducer" (and often several) along with amplifiers and other devices. However, the subcommittee recognizes (with some reservation) that "transducer" can, through generic expansion, be used to designate devices commonly referred to as "transmitters." Because this standard is meant to apply to instruments included in both definitions, the word "transducer" has been selected for consistent use throughout. The user of this Standard is respectfully requested to include the instrument person's common usage of "transmitter" as part of the thought process when the single word "transducer" appears.

It is important to note that the installation of transducers, if not done properly, can negate the suitability of a device for its use in nuclear safety related systems. Since there are many different instrument service conditions and a wide variety of viable system and instrument designs, the user of this Standard will find that the design responsibilities, rather than the design itself, are sometimes delineated herein.

The ISA Standards and Practices Department is aware of the growing need for attention to the metric system of units in general, and the International System of Units (SI) in particular, in the preparation of instrumentation standards. The Department is further aware of the benefits to USA users of ISA Standards of incorporating suitable references to the SI (and the metric system) in their business and professional dealings with other countries. Towards this end, this Department will endeavor to introduce SI and SI-acceptable metric units as optional alternatives to English units in all new and revised standards to the greatest extent possible. The ANSI "Standards for Metric Practice"[1] will be the reference guide for definitions, symbols, abbreviations, and conversion factors. SI (metric) unit conversions in this Standard are given only to the precision intended in selecting the original numerical value. When working in the SI units system, the given SI value should be used. When working in customary U.S. units, the given U.S. value should be used.

The SP-67.01 members directly responsible for the development of this Standard are:

Name	Company
J. A. Nay (Chairman)	Westinghouse Nuclear Technology Division
L. E. Friedline	Bailey Controls Co.
J. P. Hanrihan	The Foxboro Co. (until 2/22/77)
H. S. Hopkins	Westinghouse Computer & Instrument Division
M. K. Kessie	Bechtel Power Corp.
J. R. Klingenberg	W. R. Holway Associates
R. C. LaSell	Rosemount, Inc.
J. V. Lipka	Gilbert Associates
J. Rowe	The Foxboro Co. (after 2/22/77)
A. J. Schager	Cleveland Electric Illuminating Co.
D. W. Stubley	Atomic Energy of Canada, Ltd.
B. W. Washburn	Los Alamos Scientific Laboratory

Approval by Topical Subcommittee 2 of the ISA NPPSC was obtained on October 10, 1977. Membership consisted of:

Name	Company
E. W. O'Neal (Chairman)	Sargent & Lundy
R. L. Boatright	Georgia Power Co.
L. C. Fron	Detroit Edison Co.
E. M. Good	Florida Power Co.

I. A. Pinkis	Commonwealth Associates Inc.
R. M. Rello	Air Products & Chemicals Inc.
S. J. Sims	U.S. Dept. of Energy
W. C. Weston	Stone & Webster Engineering

Approval by the ISA SP-67 Main Committee (NPPSC) was obtained on October 18, 1978, in accordance with NPPSC Procedures which do not require unanimity for approval. Membership consisted of:

Name	Company
H. T. Hubbard (Chairman)	Southern Company Services, Inc.
B. W. Ball	Brown & Root Inc.
R. L. Boatright	Georgia Power Co.
J. M. Dahlquist, Jr.	Baltimore Gas & Electric Co.
H. D. Foreman	Brown & Root Inc.
L. E. Friedline	Bailey Controls Co.
L. C. Fron	Detroit Edison Co.
E. M. Good	Florida Power Corp.
W. G. Gordon	Bechtel Power Corp.
L. F. Griffith	Ralph M. Parsons Co.
T. Grochowski	Babcock & Wilcox
H. S. Hopkins	Westinghouse Computer & Instrument Div.
R. J. Howarth	General Physics Corp.
R. N. Hubby	Leeds & Northrup Co.
R. T. Jones	Philadelphia Electric Co.
M. J. Kimbell	Kaiser Engineers Inc.
J. R. Klingenberg	W. R. Holway & Associates
R. C. LaSell	Rosemount Inc.
J. V. Lipka	Gilbert Associates
S. F. Luna	General Atomic Co.
D. W. Miller	Ohio State University
G. C. Minor	MHB Technical Associates
J. A. Nay	Westinghouse Nuclear Technology Division
E. W. O'Neal	Sargent & Lundy
A. F. Pagano, Jr.	Tennessee Valley Authority
E. F. Pain	Pyco Inc.
G. E. Peterson	Commonwealth Edison Co.
R. L. Phelps	Southern California Edison Co.
I. A. Pinkis	Commonwealth Associates Inc.
M. F. Reisinger	Combustion Engineering Inc.
R. M. Rello	Air Products & Chemicals Inc.
U. Shah	Washington Public Power Supply System
S. J. Sims	U.S. Dept. of Energy
J. Tana	Ebasco Services
R. J. Ungaretti	Philadelphia Electric Co.
K. Utsumi	General Electric Co.
R. C. Webb	Pacific Gas & Electric Co.
E. C. Wenzinger	U.S. Nuclear Regulatory Commission (NRC)
W. C. Weston	Stone & Webster Engineering

Instrument Society of America

TABLE OF CONTENTS

Section	Page
1 Scope	5
2 Purpose	5
3 Definition and Terminology	5
4 Applicability of the Code	5
4.1 In-Line Transducers	5
4.2 Off-Line Transducers	5
4.3 Mounting Structures	5
5 Equipment Mounting	5
5.1 Mounting of In-Line Transducers	5
5.2 Mounting of Off-Line Transducers	6
5.3 Mechanical Protection	7
5.4 Close-Coupled Transducers	7
5.5 Auxiliary Equipment	7
6 Location of Equipment	7
6.1 Selecting a Location	7
6.2 Separation of Redundant Transducers	7
6.3 Accessibility for Periodic Test and Service	8
6.4 Auxiliary Equipment	8
7 Environmental Considerations	8
7.1 Application of Safety Factors	8
7.2 Seismic Considerations	8
7.3 Operating Vibration	8
7.4 Normal Ambient Operating Conditions	8
7.5 Special Operating Conditions	8
8 Interface Connections	8
8.1 Process Fluid Connections	8
8.2 Types of Tubing and Piping Connections	9
8.3 Electrical Connections	10
9 Service, Calibration, and Test Facilities	11
9.1 Calibration Test Connections	11
9.2 Vents and Drains	11
9.3 Signal Test Connections	11
9.4 Communications	11
9.5 Labeling	11
10 Quality Assurance	11
11 References and Bibliography	11
11.1 References	11
11.2 Bibliography	12

APPENDICES

Section	Page
APPENDIX A, Seismic Guidance	13
APPENDIX B, Safety Classification	14
APPENDIX C, Normal Ambient Conditions	14

LIST OF ILLUSTRATIONS

Figure	Page
1 Interface Examples	9

1 SCOPE

This standard covers the installation of transducers for nuclear-safety-related applications, excepting those for measurands of liquid metals.

2 PURPOSE

This standard establishes requirements and recommendations for the installation of transducers and auxiliary equipment for nuclear power plant applications outside of the main reactor vessel. To be considered are:

Mounting Structures and Materials
Seismic Design
Separation
Protection from Mechanical Damage
Ambient Variations (environmental)
Signal Connections
Process Connections
Service and Test Provisions

3 DEFINITIONS AND TERMINOLOGY

ISA S51.1[2] is the basic reference for terms not defined herein.

ISA S37.1[3] is the reference for terms not included in S51.1.

auxiliary equipment — for the purposes of this standard these are separate devices, such as field mounted power supplies, which are appended to the basic transducer and are located in the same general area as the transducer. Equipment located away from the transducer (such as control-board-mounted controllers and rack-mounted power supplies) is not included in the definition as used in this standard.

Code — refers to the ASME Boiler and Pressure Vessel Code, Section III[4]; and other sections required to implement the requirements of Section III.

Code Class — a definition of the applicability of the Code, determined through consideration of pressure boundary integrity.

in-line — exposed directly to the process fluid in, or substantially in, the main flow paths or tankage of fluid systems (for example, thermowells, turbine flow meters, primary flow elements, float cages).

Nuclear-Safety-Class — any numeric or alphanumeric designation which indicates that the transducer or auxiliary equipment is nuclear-safety-related. (Guidance in selecting and determining appropriate Nuclear Safety Classes is given in Appendix B.)

Nuclear-Safety-Related (NSR) — That which is essential to:

(1) Emergency Reactor Shutdown
(2) Containment Isolation
(3) Reactor Core Cooling
(4) Containment or Reactor Heat Removal
(5) Prevention or mitigation of a significant release of radioactive material to the environment

or, is otherwise essential to provide reasonable assurance that a nuclear power plant can be operated without undue risk to the health and safety of the public.

off-line — not exposed to the main body of process fluid.

This includes devices such as temperature transducers in thermowells, pipe surface temperature transducers, and transducers connected to the main process fluid via sensing lines with one or more isolation valves (for example, pressure transducers, continuous sample analyzers).

pressure boundary integrity — a general consideration of the ability to retain the process fluid within the process piping and tankage.

qualified — demonstrated to be acceptable for the design requirements.

sensing line — piping or tubing connecting the transducer to the process.

4 APPLICABILITY OF THE CODE

4.1 In-Line Transducers

Installations of transducers in Code piping, vessels, or equipment shall comply with the Code.

Transducers for nuclear-safety-related applications shall be permitted to be installed in or connected to non-Code piping, vessels, or equipment, as necessary to obtain measurements from systems not having nuclear-safety-related requirements for pressure boundary integrity. The Code shall not apply to these installations.

4.2 Off-Line Transducers

The Code shall apply to the installation of the sensing lines for off-line transducers up to the point shown on system drawings or to the connection at the transducer, whichever is closer to the main process fluid. Sensing lines within the scope of this standard to which the Code does not apply shall be designed and installed per ANSI B31.1.[5]

4.3 Mounting Structures

Mounting structures which support Code equipment shall comply with the Code (Subsection NF).

5 EQUIPMENT MOUNTING

5.1 Mounting of In-Line Transducers

5.1.1 Flow-Through Devices

Flow-through transducers — such as rotameters, magnetic flow meters, turbine meters, certain radiation monitors, flow nozzles, and orifice plates should be flange-mounted to the piping, except in ASME Code Class 1 installations.

The piping shall not be supported by the instrument. Auxiliary supports shall be provided to restrain the instrument where the mass of the instrument puts unacceptable stress on the piping.

5.1.2 Flow-Past Devices

Flow-past devices (such as protective wells, target meters, and pitot-tube-type flowmeters), designed to be inserted into the flow path through the wall of a pipe or vessel, shall be installed and mounted through a piping penetration specifically designed to accommodate the transducer.

Where removal for service is required, one of the following alternatives to permanent welding of the transducer into a boss shall be used:

(1) The weld boss may be provided on a removable flange.

(2) For insertion shank diameters of 1-inch OD (25.4 mm) or less, mounting with a flareless fitting may be used. (See 8.2.4, Flareless Connections).

Mating fittings shall be welded into the pipe boss or fitting. The mating parts shall be obtained from the same manufacturer as the nuts and ferrules, or shall be proven by test to mate and seal with the nuts and ferrules to perform their intended function.

(3) Designs permitting removal of transducers from pressurized systems may be used. They shall provide positive non-friction restraint in both the inserted and in the withdrawn position.

(4) Straight threads with resilient seals, and conical or spherical seat unions may be used.

5.2 Mounting of Off-Line Transducers

5.2.1 Mounting Structure Design

The mounting structure shall not induce stresses greater than the allowable design stresses on the transducer or on the sensing lines, signal lines and electrical cables serving the transducer.

The structure shall be designed to allow the transducer to be mounted in the physical orientation(s) for which it is qualified. Only qualified mounting hardware shall be used. Preference should be given to use of hardware furnished and qualified as part of the transducer.

5.2.2 Fasteners

Bolts, screws or other fasteners used for mounting shall be chemically and metallurgically compatible with the equipment, the structure, and the environment. Fasteners shall be able to withstand seismic stresses, structural vibrations, and normal loads. Vibration-resistant fastening mechanisms shall be used.

5.2.3 Materials

The materials selected for the mounting structure shall be resistant to the design environment; for example, aluminum would be prohibited in caustic spray environments. Materials subject to corrosion shall be protected against corrosion.

Protective coatings shall be resistant to both normal and (where applicable) accident environments and the coatings shall be chemically and metallurgically compatible with the materials to be protected. Paints or paint-type coatings used on mounting structures shall be fire-retardant.

5.2.4 Mounting Structure Configurations

One of the following configurations shall be used for mounting:

(1) Wall bracket.

(2) Open rack — Partial combination with cabinet-type structure by the addition of flat plate side panels or cross-bracing is permitted to obtain satisfactory stiffness. Bracing to the wall or other structure is permitted.

(3) Cabinet (totally enclosed construction) — Unless cabinets are designed specifically to control the transducer environment, they shall be designed to allow natural air circulation.

Where a controlled environment is essential to the accuracy or continued function of the enclosed equipment, the auxiliary equipment for controlling the environment shall be designed and installed to the same requirements as the enclosed equipment. External indication of the controlled environmental parameters shall be provided.

(4) Panels — Flat plates attached to open racks, wall brackets, or cabinets, as defined above.

(5) Pipe stands or stanchions — A single length of 2-inch (nominal) pipe, mounted to the floor or wall.

Friction mounting of the bracket (such as that provided by U-bolt clamps around a pipe) shall not be used. Through-bolting or welding is permitted to avoid friction mounting.

5.2.5 Construction and Assembly Methods

Bolting methods shall meet the requirements of 5.2.2.

Where welding is used as an assembly method, all intersecting surfaces shall have a continuous butt or fillet weld. Tack welding shall not be used as a structural weld. For long joints between two members, fillet welds may be interrupted and appropriately spaced along the joint.

Attachments to ASME Code equipment shall conform to the requirements of the Code. For non-Code attachment or assembly, the weld shall be shown to have a stressed cross section sufficient to support the maximum design loads.

All welds shall be made in accordance with procedures and with filler materials approved by the designer. Welders shall be qualified in accordance with the installer's or constructor's welding certification program, including certification to Section IX of the Code where applicable.

Other assembly methods (such as bonding or adhesives) shall be qualified by test for all structural and environmental design conditions, including simulation of required service life.

5.2.6 Mounting Structure Attachment

Each mounting structure shall be attached to the floor or wall structure by bolting or welding in place.

Plates connected to embedded anchors are preferred for providing a structural member for attachment to concrete floors and walls. Expansion bolts, cement, or other means of attachment to blind holes drilled in concrete should be avoided in original designs, but are acceptable where modifications or additions are required after the concrete is in place.

Mounting structures and their attachments shall have a seismic capability not less than that required by any of the supported transducers or auxiliary equipment.

5.2.7 Inspection Requirements for Mountings

Inspection of the final assembly shall be performed to verify that:
  (1) Fasteners used are the size, type, and material specified
  (2) Welding is as specified
  (3) Paint and other protective coatings have been applied as specified to corrodible materials, and the coating remains intact or has been repaired
  (4) Transducers and auxiliary equipment items are installed in the orientation appropriate for seismic qualification and that the designated mounting brackets or supports shown on the approved design drawings have been properly installed
  (5) Piping, tubing, fittings, valves, and electrical connections have been installed and tested in accordance with the drawings or specifications
  (6) Non-destructive examination, if specified on the drawings or other documents, has been completed.

Inspections shall be documented.

5.3 Mechanical Protection

Impingement barriers shall be used where required to protect against damage from postulated missiles, fluid jets, and other identified moving objects. The barriers may also serve to separate redundant transducers and auxiliary equipment. The barriers shall be designed consistent with the seismic qualification requirements for the protected devices.

5.4 Close-Coupled Transducers

When transducers are coupled or attached directly to the process piping, the requirements of 5.1 and 5.2 (except 5.2.4, Mounting Structure Configuration) shall apply. Friction mounting is not permitted. When isolation valves from the process fluid are not provided, it shall be shown that transducer stability and reliability are sufficient to preclude the need for checks and calibrations at times other than during normal system shutdowns.

5.5 Auxiliary Equipment

The requirements of 5.2 and 5.3 shall apply to auxiliary equipment.

6 LOCATION OF EQUIPMENT

6.1 Selecting a Location

6.1.1 Transducers shall not be installed in locations with environments which may exceed the qualified capabilities of the transducer, except when it can be shown that a potentially adverse environment will exist only after the transducer's function is no longer required. For instance, transducers with no post-accident requirements may be located in the reactor containment without being qualified for post-accident environments, so long as their failure will not limit the availability of other equipment that is required to be operable. Selection of a location for transducers should consider the availability of transducers qualified to operate in the various design environments.

6.1.2 The transducers should be located or enclosed as necessary to protect them against physical damage, radiation, freezing, or rates of change in ambient conditions in excess of that for which they have been qualified.

6.2 Separation of Redundant Transducers

6.2.1 Transducers shall be located so that no credible single event can prevent the proper function of a nuclear safety related system. Piping, tubing and wiring associated with the transducers shall be placed with similar attention to separation of locations. A minimum of 18 inches (450 mm) separation shall be provided between surfaces of redundant transducers. Where analysis shows that specific postulated events could circumvent the protection afforded by separation, additional spacing or protective barriers shall be provided. Protective barriers may also be used in lieu of separation where it is impractical to maintain the required spacing. If barriers are used, a minimum of one inch (25 mm) of free space shall be provided on each side between transducers and the barrier.

6.2.2 Redundant transducers should be mounted on structures separate from the mounting structures that support their redundant counterparts. Grouping of non-redundant transducers on a common structure is permitted.

6.2.3 Where redundant transducers must be included on the same structure, service access to the transducers, or to the subassemblies of one transducer, shall be separated from access to components or subassemblies of their redundant counterpart(s). (For instance: with cabinet construction, separate access doors should be provided.)

6.3 Accessibility for Periodic Test and Service

6.3.1 The transducers should be located in an accessible place for ease of periodic testing, servicing, removal and replacement. Adequate lighting, electrical outlets, water, compressed air, and similar utility services shall be provided as required for testing and surveillance requirements.

6.3.2 Placement of equipment (with respect to the mounting structures, impingement barriers, and adjacent equipment) shall allow clearance for the removal of covers and normal use of hand tools for mounting, calibration, and servicing. Placement shall allow removal and replacement of a complete device without disturbing equipment not functionally associated with the instrument loop being serviced. (That is, it shall not be necessary to shut down one instrument loop to facilitate replacement or service of a transducer or auxiliary equipment in another instrument loop.)

6.4 Auxiliary Equipment

The requirements of section 6 shall also apply to auxiliary equipment.

7 ENVIRONMENTAL CONSIDERATIONS

7.1 Application of Safety Factors

Safety factors (beyond the identified uncertainties) may be added to the requirements for installations by adjusting the severity of calculated values for both normal and abnormal ambient conditions. Care should be exercised to assure that the adjustments are actually conservative and realistic. (For example, an increase in ambient temperature is normally considered to be in the conservative direction for design purposes, but a decrease in temperature may be considered to be more conservative where viscosity of a fill fluid affects an important time response.)

7.2 Seismic Considerations

The design of nuclear safety related transducer installations shall accommodate seismic disturbances such that the transducers are not subjected to seismic shock or vibration in excess of that for which they are qualified. (See Appendix A.) Addition, deletion, or relocation of equipment on mounting structures (beyond the original design considerations), shall require re-verification of the seismic suitability.

7.3 Operating Vibration

Exposure to vibrational excitation due to pumps, turbines, or other sources should be avoided. Where high vibration is unavoidable, the equipment should be mounted on an adjoining non-vibrating surface; or, if no other reasonable alternative exists, the equipment should be isolated by shock mounting for the expected vibration frequency. (Caution: Shock mounts are typically low-pass mechanical filters and will often amplify vibrations. Special care in design is essential to avoid this problem.)

7.4 Normal Ambient Operating Conditions

Normal and abnormal ambinet operating conditions shall be determined and documented. As a minimum, limits of the following ambient conditions shall be stated (See Appendix C.):

(1) temperature
(2) pressure
(3) humidity
(4) radiation

It is recognized that special containment designs (such as sub-atmospheric containments and ice containments) require special consideration of transducer ambients. When the conditions exceed plant normal ranges, the conditions shall be compared to the qualification of the transducer or auxiliary equipment on an individual basis, and shall be verified to be satisfactory for each transducer during any period of abnormal ambient conditions during which the transducer is required to operate.

7.5 Special Operating Conditions

7.5.1 Chemical Environments

When chemicals are transported in the lines being measured, the wetted materials of the transducer installation shall be compatible with the chemical. When the environment contains a chemical content, the installation shall be qualified for the environment.

7.5.2 Thermal Protection

If the installation is exposed to ambient temperatures below the freezing, condensation, or precipitation point of fluids, the transducer, lines, and other parts shall be heated by steam studs, electrical heat tracing, a radiant electric housing, or another suitable means. (See also 5.2.4.) Such provisions shall be treated as auxiliary equipment.

7.6 Auxiliary Equipment

The requirements of Section 7 also apply to auxiliary equipment.

8 INTERFACE CONNECTIONS

8.1 Process Fluid Connections

The term "process fluid connections" shall apply to all valves, fittings, attaching bosses, adapters, tubing, and piping used to connect transducers to process piping.

(See Figure 1.) Sensing lines are installed per other standards* from the process pipe, tank, or equipment to:

(1) The off-line-transducer connection, if neither bulkhead fittings nor service valves are provided (A), or
(2) The inlet connection of the service and calibration valve normally provided at the off-line transducer (B), or
(3) A bulkhead connection provided as part of an off-line-transducer mounting rack, cabinet, or other structure (C).

For in-line transducers, process connections are inherent in the mounting and shall be made as specified in 5.1, Mounting of In-Line Transducers.

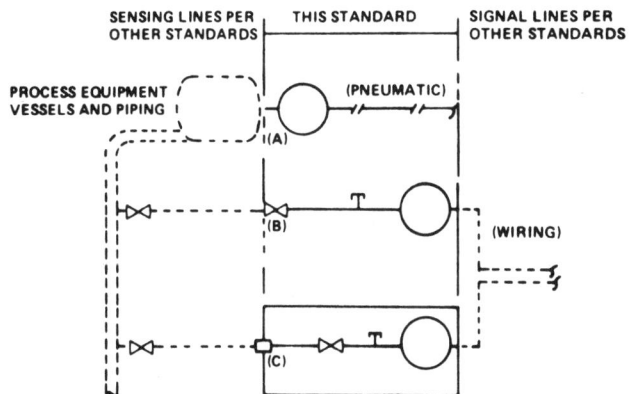

Fig. 1. Interface Examples

8.1.1 Performance Considerations

8.1.1.1 Temperature Measurements

Interface connection methods that do not involve the use of thermowells between the measurand and the transducer (that is: in-line installation) shall be used wherever necessary to achieve the response required by the system design. Installation designs should consider sources of error in accordance with ASME Performance Test Code, PTC 19.3, "Temperature Measurement" [6] as applicable to the measurand and selected transducer.

8.1.1.2 Pressure and Differential Pressure Measurements

Designers of the installation should consider the performance guides and sources of error given in ASME Performance Test Code, PTC 19.2, "Pressure Measurement," [7] as applicable to the measurand and method of measurement.

---

*See proposed Standard ISA-S67.02 "Instrument Sensing line Standards for Use in Nuclear Power Plants" (Draft 2) ANSI N677, 1978.

8.1.1.3 Flow Measurements

Designers of the installation should consider the general requirements, recommendations, conditions for proper installation and operation, errors, and other characteristics of the primary element or flowmeter, given in the ASME publication on "Fluid Meters." [8]

8.1.1.4 Dynamic Response

The effect of valves, fittings, tubing and the volumetric displacement of the transducer shall be included in determining the response of the installation.

8.2 Types of Tubing and Piping Connections

Each type of connection selected shall be qualified to the design conditions at the connection. Materials employed for valves, fittings, adapters, tubing, piping, thread lubricants, and seals shall be selected to meet the particular conditions of service required by the design: including material compatibility for welding, compatibility with the fluid chemistry, and material hardness requirements of mechanical joint fittings.

Tubing shall be routed or installed to permit access for adjustments on instruments.

8.2.1 Flanges

Proprietary flange designs (usually two-bolt manifold adaptors) are permitted as disassembly joints for the sensing lines at the transducer, when qualified for use with the transducer.

Other flanges shall comply with ANSI B16.5. [9] Due to their bulk, installation of other flanges in the immediate area of off-line transducers is not recommended.

8.2.2 Screwed Connections

Tapered pipe threads shall conform to ANSI B2.1. [10] However, tapered pipe threads shall not be used as take-down joints when repeated disassembly and re-assembly are planned. Straight thread fittings with metal-to-metal or resilient seals are permitted.

8.2.3 Welded Connections

Except as necessary to provide disassembly joints for instrument servicing, welded tubing and fittings are permitted for sensing tubing installations.

8.2.4 Flareless Connections

Disassembly joints, in tubing with diameters of 1-inch OD (25.4 mm) or less, may be made using flareless couplings. The joints shall be visually inspectable and shall provide permanent attachment of the ferrules to the tubing at installation by swaging or otherwise permanent mechanical upset (bite) on the tubing. All parts of the original fitting shall be designed and manufactured by the same company.

Replacement parts from different manufacturers shall be qualified to mate and seal properly at the design service conditions, or the complete fitting shall be replaced when service is needed.

8.2.5 Other Parts

8.2.5.1 Valves

Valves should be furnished in the form of factory-built manifolds (two, three, or more valves) as necessary and appropriate to provide for calibration, venting, draining, etc., with a minimum of tubing assembly at installation.

8.2.5.2 Calibration Test Connections

Connections provided at or near the transducer for the use of portable test and calibration equipment shall be provided with a plugged or capped connection designed for repeated assembly and disassembly.

8.2.6 Filled Systems

These systems include all filled capillary systems such as: chemical seal diaphragms with capillary tubing, pressure sensitive bellows with capillary tubing, temperature bulbs with capillary tubing, and so forth. Whether factory-filled or field-filled, the capillary tubing is an extension of the transducer, but shall be routed, separated, supported, and protected per sensing line installation standards. Armor provided as part of the capillary tubing is for protection during normal installation and use; it shall not alone be considered to meet separation or protective barrier requirements.

Shut-off and calibration valves in a sealed capillary line may seriously compromise the functional capability of the transducer and shall not be provided. Valves provided for capillary field-filling operations shall not be capable of interrupting the pressure-sensing path, and shall have the capability of being totally sealed against external leakage by welding after the capillary lines are filled. Where an all-welded capillary tubing (that is: no mechanical joints or seals) is provided, a double leakage barrier for the process fluid or containment atmosphere is inherent in the design; reactor containment penetrations, if any, shall not require additional valving in the capillary.

For piping design purposes, the connection interface shall be defined as the remote bellows, diaphragm, or bulb. The remote bellows, diaphragm, or bulb shall be part of the transducer; the housing for the remote bellows, diaphragm, or bulb and its connections to the process shall be part of the piping and shall meet any applicable codes or other piping design criteria.

8.3 Electrical Connections

8.3.1 Connection Boxes

Materials at connection boxes shall be compatible with the expected environment. Where electrical connections must be protected from the environment, such as would be experienced in postaccident service, the connection box design shall be qualified for the design conditions. Tapered pipe thread connections for use with conduit or cable adaptors are permitted. Connection boxes furnished as part of qualified transducers shall be installed in the same configurations for which they were qualified.

8.3.2 Electrical Terminations and Hook-up

8.3.2.1 Screw Terminals

When used, screw terminals on terminal boards or strips shall be qualified and shall have nominal ratings of at least twice the instrument signal voltage and current (or power supply voltage and current as applicable). Ring-tongue wire terminations are preferred to open tongue (spade lug) terminations.

8.3.2.2 Splices

Splices should be avoided. Use of screw terminals in electrical junction boxes is preferred. However, where necessary for proper connection to transducer pigtails, a single insulated splice, made only by soldering or crimping and suitably insulated, shall be used for each wire. Space shall be provided for neat and uncluttered coiling or lay-back of excess wire. Coiled or layed-back wire shall not overlap or be laced to the actual splice. The complete splicing system shall be qualified for the design conditions.

8.3.3 Quick Connectors

Quick connectors may be used where necessary for ease of removal or checking of transducers and their auxiliary equipment, or for guaranteeing proper multi-conductor hook-up between two devices where the devices and cable are pre-assembled and checked. The connectors and cable shall be matched in terms of size (wire connection size and overall cable size), cable retention, strain relief, and requirements for cable flexibility, pull space and other applicable factors. Solid wire shall not be used with the removable half of quick connectors. Connectors shall be mechanically retained in the connected configuration by screw threads with a minimum of 1-1/4 turns or equivalent mechanical detent. Both halves of each connector shall be furnished by the same manufacturer, or shall otherwise be qualified to mate properly and to provide an adequate electrical connection. The connector and cable combination shall be qualified for the design conditions.

8.3.4 Electrical Cable and Wire

Cables and wire furnished with racks, or special interconnecting cables furnished for hook-up between transducers and auxiliary equipment, shall be flame-retardant and qualified for the design conditions.

8.3.5 Shielding

Where shielding is specified by the system designer, it

shall be insulated against grounding. Shields shall be grounded only at the points specified by the system designer. Shielding continuity shall be carried ungrounded through all other connection boxes, penetrations and connectors. Braided or spiral exterior armor provided for mechanical protection should not be considered shielding.

9 SERVICE, CALIBRATION, AND TEST FACILITIES

9.1 Calibration Test Connections

Test connections shall be provided in each sensing line for the calibration and test of equipment in place. Test connections shall be capable of being isolated from the process pressure. Test connections shall be located to permit access, testing, and accurate calibration.

9.2 Vents and Drains

Capability for venting, draining, and flushing the transducer installation shall be provided where required. In addition, methods for capture and disposal of the drained, vented or flushed process fluid from radioactive systems shall be included as part of this capability.

9.3 Signal Test Connections

Test connections shall be provided for test and calibration of transducers and auxiliary equipment when the test connections will not adversely affect the transducer during those periods when it is required to function.

Test connections shall not interfere with the normal operation of any transducer not being tested.

9.4 Communications

Communication methods, for purposes of transducer calibration and service, should be coordinated with (and may be a part of) the overall plant communication network and systems. Due to the possibility of electromagnetic interference, portable radio transceivers shall not be the only method provided for transducer calibration and service.

9.5 Labeling

Permanent labels shall be provided to facilitate inspection of the installation, for warning purposes, and as guides to service and calibration. As a minimum, each transducer location shall be labeled with the transducer tag number. Where redundant groups or sets of transducers are identified by color coding, transducer and auxiliary equipment mounting labels shall be similarly color coded. Non-nuclear-safety transducer labels, and warning labels or service and calibration guide labels that are not unique to a redundant group or protection set, shall be of a neutral color.

Removable service panels or other features that may compromise the internal environment shall bear clear and permanent warning labels explaining the effects of their improper use. An example follows:

CAUTION

CABINET CONTAINS TEMPERATURE SENSITIVE INSTRUMENTS. THIS COVER MUST BE IN PLACE TO MAINTAIN SYSTEM ACCURACY.

10 QUALITY ASSURANCE

The installation of nuclear-safety-related transducers on nuclear facilities being licensed by the NRC under Title 10, Code of Federal Regulations, Part 50 (10CFR50) is subject to the requirements of the eighteen criteria of 10CFR50 Appendix B. Additional guidance can be found in the ANSI N 45.2 standards series.

It is not the intent of this standard to duplicate the work of ANSI N 45.2 or to state specific details or dicate the methods or procedures needed to ensure that nuclear-safety-related transducers are purchased, stored, installed, tested, inspected, and operated correctly. The requirements of this standard should be integrated into the user's total quality assurance program in the manner that best suits the individual's organization.

11 REFERENCES AND BIBLIOGRAPHY

11.1 References

(1) ANSI-Z210.1-1976, "Standard for Metric Practice," American National Standards Institute, New York, N.Y. 10017

(2) ISA-S51.1-1976, "Process Instrumentation Terminology," Instrument Society of America, Pittsburgh, Pa. 15222

(3) ISA-S37.1-1969, "Electrical Transducer Nomenclature and Terminology," Instrument Society of America, Pittsburgh, Pa. 15222

(4) ASME Boiler and Pressure Vessel Code, Section III, 1977, "Nuclear Power Plant Components," American Society of Mechanical Engineers, New York, N.Y. 10017

(5) ANSI-B-31.1-1977, "Power Piping," American Society of Mechanical Engineers, New York, N.Y. 10017.

(6) ASME-PTC-19.3-1974, "Temperature Measurement," American Society of Mechanical Engineers, New York, N.Y. 10017

(7) ASME-PTC-19.2-1964, "Instruments and Apparatus," Supplement to ASME Power Test Code, Part 2 — "Pressure Measurement" American Society of Mechanical Engineers, New York, N.Y. 10017

(8) Fluid Meters, Their Theory and Application. Sixth Edition, American Society of Mechanical Engineers, New York, N.Y. 10017, 1971.

(9) ANSI-B16.5-1973, "Steel Pipe Flanges, Flanged Valves, and Fittings," American Society of Mechanical Engineers, New York, N.Y. 10017

(10) ANSI-B2.1-1968, "Pipe Threads," American Society of Mechanical Engineers, New York, N.Y. 10017

11.2 Bibliography

The following documents may provide additional information, guidance or requirements for various aspects of the installation of transducers.

(A) ANSI-N18.2a (ANS 51.8) - 1975, "Nuclear Safety Criteria for the Design of Stationary Pressurized Water Reactor Plants," American Nuclear Society, Hinsdale, Ill. 60521

(B) ANSI-N212 (ANS 52.1) - 1978, "Nuclear Safety Criteria for the Design of Stationary Boiling Water Reactor Plants," American Nuclear Society, Hinsdale, Ill. 60521

(C) ANSI N213 (ANS 53.1) "Gas Cooled Reactor Design Criteria" (In preparation), American Nuclear Society, Hinsdale, Ill. 60521

(D) IEEE-323-1974, "IEEE Standard for Qualifying Class 1E Equipment for Nuclear Power Generating Stations," Institute of Electrical and Electronics Engineers, Inc., New York, N.Y. 10017

(E) ASTM D-2633-74, "Standard Methods of Testing Thermoplastic Insulated and Jacketed Wire and Cable," American Society for Testing and Materials, Philadelphia, Pa. 19103

(F) IEEE-383-1974, "IEEE Standard for Type Test of Class 1E Electric Cables, Field Splices, and Connections for Nuclear Power Generating Stations," Institute of Electrical and Electronic Engineers, Inc., New York, N.Y. 10017

(G) ANSI N45.2-1977, "Quality Assurance Program Requirements for Nuclear Facilities," American Society of Mechanical Engineers, New York, N.Y. 10017

(H) SNT-TC-1A (June 1975 edition), "Personnel Qualification and Certification in Nondestructive Testing," (Current Edition and Supplements as available) American Society for Nondestructive Testing, Evanston, Ill. 60202

APPENDIX A
SEISMIC GUIDANCE

A.1 This appendix provides additional information, for guidance only, regarding the installation of transducers which are to resist seismic forces.

A.2 While it is not within the scope of this document to define the qualification methods or requirements for the transducers themselves, it may be assumed that the following information about specific transducers will be available to the designers of their installation and mounting structures:

(1) Transducer test input acceleration spectra which define mounting-point accelerations in each major axis, and to which the transducer has been or will be qualified. These would normally be expected to fall in the region above the curve of Figure A-1 or Figure A-2 as applicable.

(2) Functional data, including output signal sensitivity to applied frequencies (signal off-sets, vibration-induced output "noise" and so forth), and identification of the most sensitive and least sensitive axes, if applicable.

(3) Structural transducer data, including natural or resonant frequencies of the transducer housing/mounting bracket combination, and identification of the most sensitive and least sensitive axes, if applicable.

(4) Detail drawings of each type of transducer, showing the qualified mounting configurations (orientations, brackets, fasteners) and the mass and center of gravity with respect to the qualified mounting configuration.

A.3 It should also be assumed that the following information may not be available for a particular transducer:

(1) Structural damping values of (or within) the transducer itself. (Due to the complexity and relatively small size of most transducers, damping values can seldom be determined.)

(2) Cross-coupling of multi-axis response of (or within) the transducer itself. (Again, due to the complexity and size, these effects can seldom be defined accurately.)

(3) Information on the vibration response of the test table mounting structure. (The test accelerations should be measured and reported, based on the acceleration at the transducer mounting point.)

A.4 Based upon the assumption in A.2 and A.3, and the "normal maximums" given below, it may thus not be necessary that the installation designer and the transducer designer have full knowledge of the "opposite" sides of their interfaces in order to achieve a qualified installation. Each may work independently, compare results with the peak acceleration interface curves of Figures A-1 or A-2 as appropriate, and require or perform special comparisons between actual data only for those cases where the interface is crossed by either party.

A.5 In-Line Transducers

The peak acceleration values of Figure A-1 should be considered the normal maximums for in-line or close-coupled installations.

Generic certification of the structural suitability of installations may be permitted for piping which falls below the boundary of Figure A-1.

When the piping design cannot be made to fall below the boundary of Figure A-1, the calculated piping response should be compared to the qualification of the transducer on an individual basis; it should be verified to be satisfactory for each transducer.

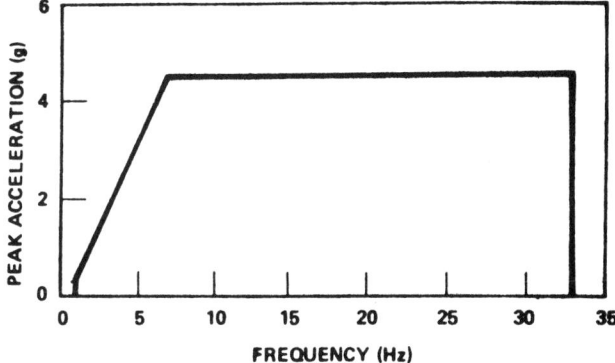

Fig. A-1. Transducer/Mounting Interface Normal Maximum Seismic Envelope for In-Line Transducers.

A.6 Off-Line Transducers and Auxiliary Equipment

The peak acceleration values of Figure A-2 should be considered the normal maximums for off-line installations.

Generic certification of the structural stability of installations may be permitted for installations which fall below the boundary of Figure A-2. Where the mounting design and location cannot be made to fall within the boundary of Figure A-2, the calculated response should be compared to the qualification of the transducer on an individual basis; it should be verified to be satisfactory for each transducer.

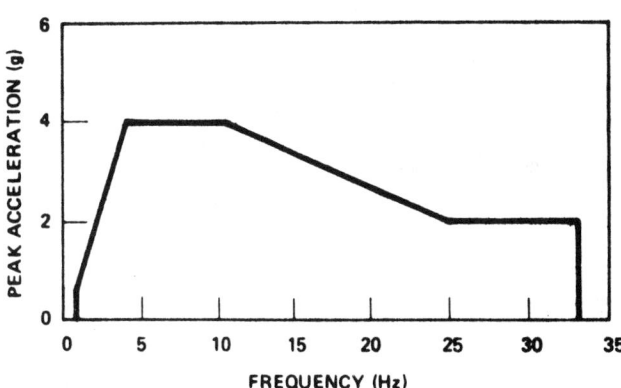

Fig. A-2. Transducer/Mounting Interface Seismic Envelope for Off-Line Transducers.

APPENDIX B
SAFETY CLASSIFICATION

This appendix provides additional information for guidance only, regarding some of the considerations appropriate to the selection of a safety class for transducers. This appendix does not define specific safety classes.

The application of a Safety Class to transducers requires a consideration of their functions and the results of a malfunction, as well as pressure boundary integrity. Code Class definition, while important, addresses only the fluid-retaining capabilities of the installation. Thus, Code Class and Safety Class are not mutually definitive for items within the scope of this standard.

Each classification action should be evaluated individually, and assignment should be made based upon the component functions and the classification system applicability. The category breakdown given below is intended to be consistent with the definitions of ANSI-N-18.2 (for PWRs), or ANSI-N-212 (for BWRs), or ANSI-N-213 (for GCRs). Reasoning similar to that promulgated by the above references should be used for nuclear reactor plant types not otherwise covered. In addition, various IEEE standards and international standards contain definitions which may apply under various circumstances. At the time of preparation of this standard, no single reference was available for universal application to instruments.

The following categories of functions are selected such that each group contains functions of approximately equal importance to nuclear safety:

B.1 A Safety Classification would apply to those transducers whose individual failure could cause system action which would directly result in a loss-of-coolant accident.

In general, applications requiring these transducers are not used.

B.2 A Safety Classification would apply to those transducers which are part of instrument systems that actively perform safety system functions, such as those that:

(1) Are required to shut down the reactor, isolate the reactor containment, cool the reactor core, cool the reactor containment, control the presence of chemically explosive gases (such as hydrogen) inside the reactor containment to within the acceptable limits, or control radioactivity inside and outside the reactor containment to within acceptable limits.

(2) Provide interlocks to prevent an operator error that could lead to incidents or events representing limiting plant design cases.

(3) Are required to maintain the plant in a safe and secure shutdown condition.

(4) Are required to enable the operator to take manual action essential to safety during the course of an accident or during postaccident operations.

B.3 A Safety Classification would apply to those transducers, not covered above, which are part of the instrument systems that passively monitor variables which are required to:

(1) Verify that plant operating conditions are within the normal operating bands assumed for the safety analysis of the plant

(2) Indicate the status of Safety System bypasses

(3) Monitor radioactive effluents to ensure that release rates and total releases are within the limits established for plant operation

(4) Monitor the removal of decay heat from spent fuel

B.4 Instruments not included in the above definitions do not have a function required for nuclear safety. They are designated NNS (Non-Nuclear-Safety). Installation of functionally-NNS transducers is not within the scope of this standard, except incidentally when they must be installed in ASME Code piping.

APPENDIX C
NORMAL AMBIENT CONDITIONS

This appendix is for guidance only. The following ambient conditions should be considered to be normal maximums and minimums:

Temperature: 35 to 160°F (2 to 70°C) maximum limits
25°F (14°C) span of normal variation

Pressure: 0.9 to 1.1 atm (90 to 110 kPa absolute)

Humidity: 95% maximum

Radiation: 0.5 rad/hr. (5×10^{-3} J/Kg-h) maximum rate for operation, and 0.5 millirems/hr (5×10^{-6} J/Kg-h) maximum rate when personnel access is required.

INSTRUMENT SOCIETY of AMERICA
Research Triangle Park, North Carolina

ANSI/ISA-S67.02-1980
Approved February 17, 1983

American National Standard

Nuclear-Safety-Related Instrument Sensing Line Piping and Tubing Standards for Use in Nuclear Power Plants

Instrument Society of America

ISBN 87664-489-2

ISA-S67.02 Nuclear-Safety-Related
Instrument Sensing Line Piping and Tubing Standards
for Use in Nuclear Power Plants, 1980

Copyright © 1980 by the Instrument Society of America. All rights reserved. Printed in the United States of America. No part of this publication may be reproduced, stored in a retrieval system, or transmitted, in any form or any means electronic, mechanical, photocopying, recording or otherwise without the prior written permission of the publisher.

INSTRUMENT SOCIETY OF AMERICA
67 Alexander Drive
P.O. Box 12277
Research Triangle Park, North Carolina 27709

Copyright © 1980 by the Instrument Society of America

PREFACE

This preface and appendices referring to other standards are included for informational purposes only and are not part of ISA-S67.02. Applicability of other standards or codes are as stated in the text. Where references are made to other standards, a particular paragraph reference is indicated for clarity where applicable.

Committee ISA SP67.02 formed in 1974, adopted its draft scope on September 19, 1974, and forwarded it to the Instrument Society of America's Standards and Practices Board for acceptance as part of the minutes of that meeting. On December 9, 1974, this committee received an approved Scope and Project Charter N677 from American National Standards Institute (ANSI).

It is the consensus of the committee that this standard addresses those portions of the safety-related instrument sensing line tubing (piping) runs that are unique to the nuclear power plant, concentrating therefore on meeting nuclear safety considerations as legislated by 10 CFR 50 (Code of Federal Regulations), Appendix A Criteria. The separation of redundant sensing lines as contained in this standard is predicated on the assumption that the equipment and instruments to which those sensing lines are connected are adequately separated.

The ISA Standards and Practices Department is aware of the growing need for attention to the metric system of units in general, and the International System of Units Scientific International (SI) in particular, in the preparation of instrumentation standards. The Department is further aware of the benefits to U.S.A. users of ISA standards of incorporating suitable references to the SI (and the metric system) in their business and professional dealings with other countries. Toward this end, this Department will endeavor to introduce SI and SI-acceptable metric units as optional alternatives to English units in all new and revised standards to the greatest extent possible. The ASTM Metric Practice Guide will be the reference guide for definitions, symbols, abbreviations, and conversion factors. SI (metric) unit conversions in this standard are given only to the precision intended in selecting the original numerical value. When working in the SI units system, the given SI value should be used. When working in customary U.S. units, the given U.S. value should be used.

Where the failure of instrument sensing lines from nuclear-safety-related processes to instruments that are not nuclear-safety-related is demonstrated not to produce either unacceptable leakage of process fluid or unacceptable flooding, jet impingement forces or other failure related hazard to nuclear-safety-related equipment, this standard does not apply.

Instrument sensing lines from nonnuclear-safety-related processes to nuclear-safety-related instruments are not in the scope of this standard.

Containment isolation requirements are not part of this standard.

COMMITTEE SP67.02

Name	Company
E. W. O'Neal, Chairman	Prosche & O'Neal Engineering, Inc.
W. G. Deutsch (Deceased)	Fluor Pioneer, Inc.
L. C. Fron, Asst. Chairman	Detroit Edison Company
G. E. Peterson, Secretary	Commonwealth Edison Company
R. L. Boatright	Georgia Power Company
F. Cunningham	NUPRO Company
A. DiPerna	Bechtel Power Company
E. M. Good	Florida Power Company
J. E. Ouzts	Nuclear Regulatory Commission
R. Shamarao	Anderson Greenwood & Company
J. Stevens	Gibbs & Hill, Inc.
M. Verdugo	General Atomic Company
W. C. Weston	Stone & Webster Engineering Corporation

This standard was approved by Topical Subcommittee 2 of the ISA NPPSC on December 15, 1977.

Name	Company
E. W. O'Neal, Chairman	Sargent & Lundy
R. L. Boatright	Georgia Power Company
L. C. Fron	Detroit Edison Company
E. M. Good	Florida Power Company
I. A. Pinkis	Commonwealth Associates, Inc.
R. M. Rello	Air Products & Chemicals, Inc.
S. J. Sims	U. S. Department of Energy
W. C. Weston	Stone & Webster Engineering Corporation

Instrument Society of America

The Standard was approved by ISA SP67 (NPPSC) Committee on October 19, 1978.

Name	Company
H. T. Hubbard, Chairman	Southern Company Services, Inc.
B. W. Ball	Brown & Root, Inc.
R. L. Boatright	Georgia Power Company
J. M. Dahlquist, Jr.	Baltimore Gas & Electric Company
H. D. Foreman	Brown & Root, Inc.
L. E. Friedline	Bailey Controls Company
L. C. Fron	Detroit Edison Company
E. M. Good	Florida Power Corporation
W. G. Gordon	Bechtel Power Corporation
L. F. Griffith	Ralph M. Parsons Company
T. Grochowski	Babcock & Wilcox
H. S. Hopkins	Westinghouse Computer & Instrument Division
R. J. Howarth	General Physics Corporation
R. N. Hubby	Leeds & Northrup Company
R. T. Jones	Philadelphia Electric Company
M. J. Kimbell	Kaiser Engineers, Inc.
J. R. Klingenberg	W. R. Holway & Associates
R. C. LaSell	Rosemount, Inc.
J. V. Lipka	Gilbert Associates, Inc.
S. F. Luna	General Atomic Company
D. W. Miller	Ohio State University
G. C. Minor	MHB Technical Associates
J. A. Nay	Westinghouse PWR Systems Division
E. W. O'Neal	Sargent & Lundy
A. F. Pagano, Jr.	Tennessee Valley Authority
E. F. Pain	Pyco Inc.
G. E. Peterson	Commonwealth Edison Company
R. L. Phelps	Southern California Edison Company
I. A. Pinkis	Commonwealth Associates, Inc.
M. F. Reisinger	Combustion Engineering, Inc.
R. M. Rello	Air Products & Chemicals, Inc.
U. Shah	Washington Public Power Supply System
S. J. Sims	U. S. Department of Energy
J. Tana	Ebasco Services, Inc.
R. J. Ungaretti	Philadelphia Electric Company
K. Utsumi	General Electric Company
R. C. Webb	Pacific Gas & Electric Company
E. C. Wenzinger	U.S. Nuclear Regulatory Commission (NRC)
W. C. Weston	Stone & Webster Engineering Corporation

This standard was approved for publication by the ISA Standards and Practices Board in June 1980.

Name	Company
E. C. Magison, Chairman	Honeywell, Inc.
P. Bliss	Pratt & Whitney Aircraft Group
W. Calder	The Foxboro Company
B. A. Christensen	Continental Oil Company
R. Coel	Fluor Corporation
J. E. French	U. S. Department of Commerce
M. R. Gordon-Clark	Scott Paper Company
T. J. Harrison	IBM Corporation
R. T. Jones	Philadelphia Electric Company
E. M. Nesvig	ERDCO Engineering Corporation
R. L. Nickens	Reynolds Metals Company
G. Platt	Bechtel Power Corporation
R. Prescott	Moore Products Company
W. C. Weidman	Gilbert Associates, Inc.
K. A. Whitman	Allied Chemical Corporation

*L. N. Combs
*R. L. Galley
*R. G. Marvin   Moore Products Company
*W. B. Miller
*J. R. Williams
*Director Emeritus

Nuclear-Safety-Related Instruments Sensing Line Piping and Tubing Standards for Use in Nuclear Power Plants

TABLE OF CONTENTS

Section	Page
1 Purpose	9
2 Scope	9
3 Definitions and Terminology	9
4 Pressure Boundary and Mechanical Design Requirements	9
4.1 Summary of Requirements	9
4.2 Mechanical Design Requirements	9
4.2.1 Instrument Sensing Lines in Accordance with ASME Boiler and Pressure Vessel Code Section III.	9
4.2.2 Instrument Sensing Lines in Accordance with ANSI B31.1 Power Piping	12
4.2.3 Media Isolation Devices Bellows or Diaphragms and Permanently Filled Capillaries	12
5 Protection of Nuclear-Safety-Related Instrument Sensing Lines	24
5.1 Redundant Instrument Taps	24
5.2 Routing Instrument Sensing Lines	24
5.2.1 General Considerations	24
5.2.2 Special Considerations	24
5.3 Identification and Color Coding	25
6 Auxiliary Devices, Fittings and Supports	25
6.1 Types of Sensing Line Connections within the Scope of this Standard	25
6.2 Root and Accessible Isolation Valves	25
6.3 Restriction Devices and Instrument Response	25
6.4 Automatic Reset Excess Flow Check Valves	26
6.5 Instrument Sensing Line Supports	26
7 Materials	26
8 Documentation and Quality Assurance	26
8.1 Documentation for Instrument Sensing Lines where ASME Section III is Required by this Standard	26
8.2 Documentation for Instrument Sensing Lines where ANSI B31.1 is Required by this Standard	26
8.3 Documentation for Permanently Filled Capillary Tubes	26
8.4 Quality Assurance Requirements for Instrument Sensing Lines for which ASME Section III is Required by this Standard	27
8.5 Quality Assurance Requirements for Instrument Sensing Lines for which ANSI B31.1 is Required by this Standard and for Seismic Category I Permanently Filled Capillary Tubes	27
APPENDIX A, Interface Standards and Documents	29

LIST OF ILLUSTRATIONS

Figure	Page
1 Water Filled Instrument Sensing Lines for Water and Steam Service ASME Class I Process, Instrument in Containment	13
2 Permanently Filled Instrument Sensing Lines for Water and Steam Service ASME Class I Process, Capillary Filled Instruments in Containment	14
3 Water Filled Instrument Sensing Lines for Water and Steam Service ASME Class 2 Process, Instruments in Containment	15
4 Permanently Filled Instrument Sensing Lines for Water and Steam Service ASME Class 2 Process, Capillary Filled Instruments in Containment	16
5 Water Filled Instrument Sensing Line for Water and Steam Service ASME Class 3 Process	17
6 Permanently Filled Instrument Sensing Line for Water and Steam Service ASME Class 3 Process, Capillary Filled Instruments	18
7 Water Filled Instrument Sensing Lines for Water and Steam Service ASME Class I Process, Instruments Outside Containment	19
8 Water Filled Instrument Sensing Lines for Water and Steam Service ASME Class I Process, Instruments Outside Containment	20
9 Containment Atmosphere Instrument Sensing Lines Instruments Outside Containment	21

10 Containment Atmosphere Instrument Sensing Lines Diaphragm with Filled Instruments
   Outside Containment.. 22
11 Containment Atmosphere Instrument Sensing Lines Capillary Filled Instruments
   Outside Containment.. 23

LIST OF TABLES

Table **Page**

1 Minimum Mechanical Design Requirements for Instrument Piping and Tubing that Do Not
  Penetrate the Primary Reactor Containment.. 10
2 Minimum Mechanical Design Requirements for Instrument Piping and Tubing that Penetrate
  the Primary Reactor Containment 11

1 PURPOSE

This standard establishes the applicable code requirements and code boundaries for the design and installation of instrument sensing lines interconnecting nuclear-safety-related power plant processes with both nuclear-safety-related and nonnuclear-safety-related instrumentation.

This standard addresses the difference between the pressure boundary integrity of an instrument sensing line in accordance with the appropriate parts of Section III, Boiler and Pressure Vessel Code, American Society of Mechanical Engineers (ASME) or American National Standards Instutite (ANSI) B31.1 as applicable and the ensurance that the protection function of nuclear-safety-related instruments is available.

2 SCOPE

This standard covers design, protection, and installation of nuclear-safety-related instrument sensing lines for light water cooled nuclear power plants. The standard covers the pressure boundary requirements for piping, capillary tubing, and tubing lines up to and including one inch (25.4 mm) outside diameter or three-quarter inch nominal pipe. The boundaries of this standard span from the process tap to the upstream side of the instrument panel, bulk head fitting or instrument shutoff valve.

This standard is also applicable to systems (except liquid metal) in other types of nuclear power plants.

3 DEFINITIONS AND TERMINOLOGY

accessible isolation valve—the isolation valve nearest the measured process on an instrument sensing line which is available to personnel during normal plant operation. The root valve may or may not perform the function of the accessible isolation valve, dependent on its location.

channel—a collection of instrument loops, including their sensing lines, that may be treated or routed as a group while being separated from instrument loops assigned to other redundant channels.

instrument shutoff valve—the valve or valve manifold nearest the instrument.

loop—a combination of one or more interconnected instruments arranged to measure or control a process variable, or both.

nonnuclear-safety (NNS)—any instrument not included in *nuclear-safety-related* below.

nuclear-safety-related (NSR)—that which is essential to:
(1) emergency reactor shutdown
(2) containment isolation
(3) reactor core cooling
(4) containment or reactor heat removal
(5) prevention or mitigation of a significant release of radioactive material to the environment or is otherwise essential to provide reasonable ensurance that a nuclear power plant can be operated without undue risk to the health and safety of the public.

Where the term "Nuclear-Safety-Related" is used in this standard, it refers to meeting the requirements of Title 10, Part 50, Code of Federal Regulations (10 CFR 50).

redundant sensing line(s)—sensing lines for redundant instruments as used in this standard are defined as a sensing line or group of lines that are provided to duplicate the function of another sensing line (e.g., sensing lines that transfer pressure energy for measurement of the same pressure energy for the same process).

root valve—the first valve located on the instrument sensing line after it taps off the main process.

sensing line—for the purpose of this standard, a pipe or tube of relatively static fluid that connects the process being sensed to the sensor (transducer) and is filled with the process fluid, or is a fluid filled capillary.

4 PRESSURE BOUNDARY AND MECHANICAL DESIGN REQUIREMENTS

4.1 Summary of Requirements

Tables 1 and 2 summarize the minimum pressure boundary and mechanical design requirements for nuclear-safety-related instrument sensing lines utilized in nuclear power plants. Table 1 applies to instrument sensing lines that do *not* penetrate the primary reactor containment and Table 2 applies to instrument sensing lines that penetrate the primary reactor containment.

Tables 1 and 2 are divided into four columns. Column 1 refers the user to the applicable figure or figures which graphically show the mechanical design requirements specified in columns 2, 3, and 4. The figure referenced in column 1 also indicates the pressure boundary scope of this standard. Column 2 indicates the process system code classification. Column 3 indicates the instrument sensing line seismic category. Where more than one seismic category is listed, the seismic category change is shown on the figure or figures referenced in column 1. Column 4 indicates the applicable design code for the instrument sensing line. Where more than one design code is listed, the design code change is shown on the figure or figures referenced in column 1.

4.2 Mechanical Design Requirements

The design of components, parts, and appurtenances utilized in the instrument sensing lines under the scope of this standard shall, as a minimum, be in accordance with the design code(s) specified in column 4 of Tables 1 and 2. Figures 1 through 11 illustrate typical applications of these requirements.

4.2.1 Instrument Sensing Lines in Accordance With ASME Boiler and Pressure Vessel Code Section III

Where identified in this standard, ASME Section III requirements shall apply for materials, design, fabrication, examination, testing, marking, stamping, documentation by the certificate holder and quality assurance.

Design and service limits for instrument sensing lines identified as ASME Class 1, 2, or 3 and Seismic Category I by

TABLE 1

MINIMUM MECHANICAL DESIGN REQUIREMENTS FOR INSTRUMENT PIPING AND TUBING THAT DO NOT PENETRATE THE PRIMARY REACTOR CONTAINMENT[a]

	Illustration	Process Piping ASME Code Class	Instrument Sensing Line Seismic Category	Applicable Design Code as Invoked by This Standard in Section 4
Figure 1	Water Filled Instrument Sensing Lines for Water and Steam Service ASME Class 1 Process, Instruments in Containment	1	Category 1[b,c]	ASME Section III, Class 1 & 3; ANSI B31.1[d,e]
Figure 2	Permanently Filled Instrument Sensing Lines for Water and Steam Service ASME Class 1 Process, Capillary Filled Instruments in Containment	1	See Figure 2[b,c]	ASME Section III, Class 1 & 3; ANSI B31.1[d,e]
Figure 3	Water Filled Instrument Sensing Lines for Water and Steam Service ASME Class 2 Process, Instruments in Containment[f]	2	Category 1[b,c]	ASME Section III, Class 2 & 3; ANSI B31.1[e,g]
Figure 4	Permanently Filled Instrument Sensing Lines for Water and Steam Service ASME Class 2 Process, Capillary Filled Instruments in Containment[f]	2	See Figure 4[b,c]	ASME Section III, Class 2 & 3; ANSI B31.1[e,g]
Figure 5	Water Filled Instrument Sensing Lines for Water and Steam Service ASME Class 3 Process[h]	3[i]	Category 1[b,c]	ASME Section III, Class 3; ANSI B31.1[e]
Figure 6	Permanently Filled Instrument Sensing Lines for Water and Steam Service ASME Class 3 Process, Capillary Filled Instruments[h]	3[i]	See Figure 6[b,c]	ASME Section III, Class 3; ANSI B31.1[e]

a. This table and associated figures do not contain physics of application or containment isolation requirements.

b. Design and service limits for instrument sensing lines identified as ASME Class 1, 2, or 3 and Seismic Category I by this standard, shall be in compliance with ASME Section III, Sub-subarticles NB3650, NC3650, and ND3650. Moments due to earthquakes shall be included. These requirements shall apply to media isolation devices and filled capillaries where Seismic Category I requirements are invoked by this standard. (Reference paragraph 4.2.1.)

c. Design and service limits for instrument sensing lines identified as ANSI B31.1 and Seismic Category I by this standard shall be in accordance with ANSI B31.1, paragraph 104.8. Moments due to earthquakes shall be included. These requirements shall apply to media isolation devices and filled capillary tubes where Seismic Category I requirements are invoked by this standard. (Reference paragraphs 4.2.2 and 4.2.3.)

d. A flow restrictor upstream of the Code classification change shall be provided such that loss of reactor coolant through one ruptured instrument sensing line shall not prevent orderly reactor shutdown assuming makeup is only by the normal make-up systems. (Reference paragraph 6.3, this standard.)

e. See referenced figure or figures for seismic category, and design code boundaries.

f. Figures 3 and 4 shall also be used for applications where the process pipe or equipment and instrument sensing lines are all outside the primary containment wall. When used in this manner, the shield wall and penetration may or may not apply.

g. A flow restrictor upstream of the Code classification change shall be provided such that the blowdown from one ruptured instrument sensing line shall not prevent the process system from performing its intended safety function. (See paragraph 6.3.)

h. Figures 5 and 6 shall be used for applications where the process instrument, process instrument tap, and instrument-sensing-line are located completely inside or outside the primary containment as long as they do not penetrate the primary containment.

i. The process system is classified as Seismic Category I.

TABLE 2

MINIMUM MECHANICAL DESIGN REQUIREMENTS FOR INSTRUMENT PIPING AND TUBING THAT PENETRATE THE PRIMARY REACTOR CONTAINMENT[a]

Illustration		Process Piping ASME Code Class	Instrument Sensing Line Seismic Category	Applicable Design Code as Invoked by This Standard in Section 4
Figure 7	Water Filled Instrument Sensing Lines for Water and Steam Service ASME CLASS 1 Process Instruments Outside Containment	1	Category I[b,c]	ASME Section III, Class 1,2,3; ANSI B31.1[d,e,f]
Figure 8	Water Filled Instrument Sensing Lines for Water and Steam Service ASME Class 1 Process Instruments Outside Containment	1	Category I[b,c]	ASME Section III, Class 1,2,3; ANSI B31.1[d,e,f]
Figure 9	Containment Atmosphere Instrument Sensing Lines Instruments Outside Containment	3	Category I[b,c]	ASME Section III, Class 2 & 3; ANSI B31.1[e,f]
Figure 10	Containment Atmosphere Instrument Sensing Lines Diaphragm with Filled Instruments Outside Containment	3	Category I[b]	ASME Section III, Class 2; See Paragraph 4.2.3
Figure 11	Atmosphere Instrument Sensing Lines Capillary Filled Instruments Outside Containment	3	Category I[b]	See Paragraph 4.2.3

a. This table and associated figures do not contain physics of application or containment isolation requirements.

b. Design and service limits for instrument sensing lines identified as ASME Class 1, 2, or 3 and Seismic Category I by this standard, shall be in compliance with ASME Section III. Sub-subarticles NB3650, NC3650, and ND3650. Moments due to earthquakes shall be included. These requirements shall apply to media isolation devices and filled capillaries where Seismic Category I requirements are invoked by this standard. (Reference paragraph 4.2.1.)

c. Design and service limits for instrument sensing lines identified as ANSI B31.1 and Seismic Category I by this standard shall be in accordance with ANSI B31.1, paragraph 104.8. Moments due to earthquakes shall be included. These requirements shall apply to media isolation devices and filled capillary tubes where Seismic Category I requirements are invoked by this standard. (Reference paragraphs 4.2.2 and 4.2.3.)

d. A flow restrictor upstream of the Code Classification change shall be provided such that loss of reactor coolant through one ruptured instrument sensing line shall not prevent orderly reactor shutdown assuming makeup is only by the normal make-up systems. (Reference paragraph 6.3, this standard.)

e. See referenced figure or figures for seismic category, and design code boundaries.

f. Instrument sensing lines shall be sized or restricted so that the potential offsite exposure would be substantially below the guidelines of 10 CFR 100 if a failure of the instrument sensing line outside the primary containment occurs during an accident. (Reference paragraph 6.3.)

this standard, shall be in compliance with ASME Section III, Sub-subarticles NB3650, NC3650, and ND3650. Moments due to earthquakes and other transient dynamic loadings shall be included.

(1) ASME Class 1.

Where instrument sensing lines are identified as ASME Class 1 by this standard, the applicable requirements of ASME Section III, Subsections NCA and NB, shall apply. The user is specifically referenced to Paragraph NB3676.

(2) ASME Class 2.

Where instrument sensing lines are identified as ASME Class 2 by this standard, the applicable requirements of ASME Section III, Subsections NCA and NC, shall apply. The user is specifically referenced to Paragraph NC3676. Primary containment wall penetrations are ASME Section III, Class 2. The user is specifically referenced to Paragraph NE-1132, Subsection NE, Division 1, ASME Section III, for the applicable requirements.

(3) ASME Class 3.

Where instrument sensing lines are identified as ASME Class 3 by this standard, the applicable requirements of ASME Section III, Subsections NCA and ND, shall apply. The user is specifically referenced to Paragraph ND3676.

4.2.2 Instrument Sensing Lines in Accordance with ANSI B31.1 Power Piping

Where ANSI B31.1 is required by this standard in Section 4, the user shall comply with ANSI B31.1, Paragraph 122.3, requirements for materials, design, fabrication, examination, and testing.

Where instrument sensing lines identified as ANSI B31.1 are interconnected with process piping systems classified as ASME Class 1, 2, or 3 and are identified as Seismic Category I in Tables 1 and 2, the following additional requirements shall apply:

(1) A material manufacturer's certificate of compliance with the material specification shall be furnished for all pressure boundary items.

(2) All pressure boundary items shall be pressure tested to 1.5 times the process system design pressure in accordance with Paragraphs 137.3 and 137.4, ANSI B31.1.

(3) Design and service limits for instrument sensing lines identified as ANSI B31.1 and Seismic Category I by this standard shall be in accordance with ANSI B31.1, Paragraph 104.8. Moments due to earthquakes shall be included.

(4) The connection between ASME Section III and ANSI B31.1 components shall be in accordance with Paragraph 127, Chapter 5, ANSI B31.1.

4.2.3 Media Isolation Devices Bellows or Diaphragms and Permanently Filled Capillaries

(1) Bellows or Diaphragms

Where bellows and diaphragms are identified as ASME Class 1, 2, or 3 by this standard in Section 4, these devices shall be designed in compliance with ASME Section III, Paragraphs NB3649, NC3649, or ND3649, as applicable. The requirements of paragraph 4.2.1 of this standard shall also apply.

Where bellows or diaphragms are identified as ANSI B31.1 by this standard, these devices shall be designed in accordance with ANSI B31.1, paragraph 104.7. The requirements of paragraph 4.2.2 of this standard shall also apply.

(2) Permanently Filled Capillary Tubes

Where permanently filled capillary tubes are identified as ASME Class 1, 2, or 3 by this standard, the pressure design and minimum wall thickness shall be established in accordance with ASME Section III or the pressure design shall be established by proof tests in accordance with ANSI B16.9.

Where permanently filled capillary tubes are identified as ANSI B31.1 by this standard, the pressure design and minimum wall thickness shall be established in accordance with ANSI B31.1 or the pressure design shall be established by proof tests in accordance with ASME Section 1, paragraph A-22.

(3) Each permanently filled capillary tube and media isolation device shall be pressure tested to 1.5 times the process system design pressure in accordance with paragraphs 137.3 and 137.4, ANSI B31.1.

(4) The fill fluid used shall not shorten the life of, or prevent the piping or tubing wetted parts from performing their required functions. Mercury shall not be used as a fill fluid.

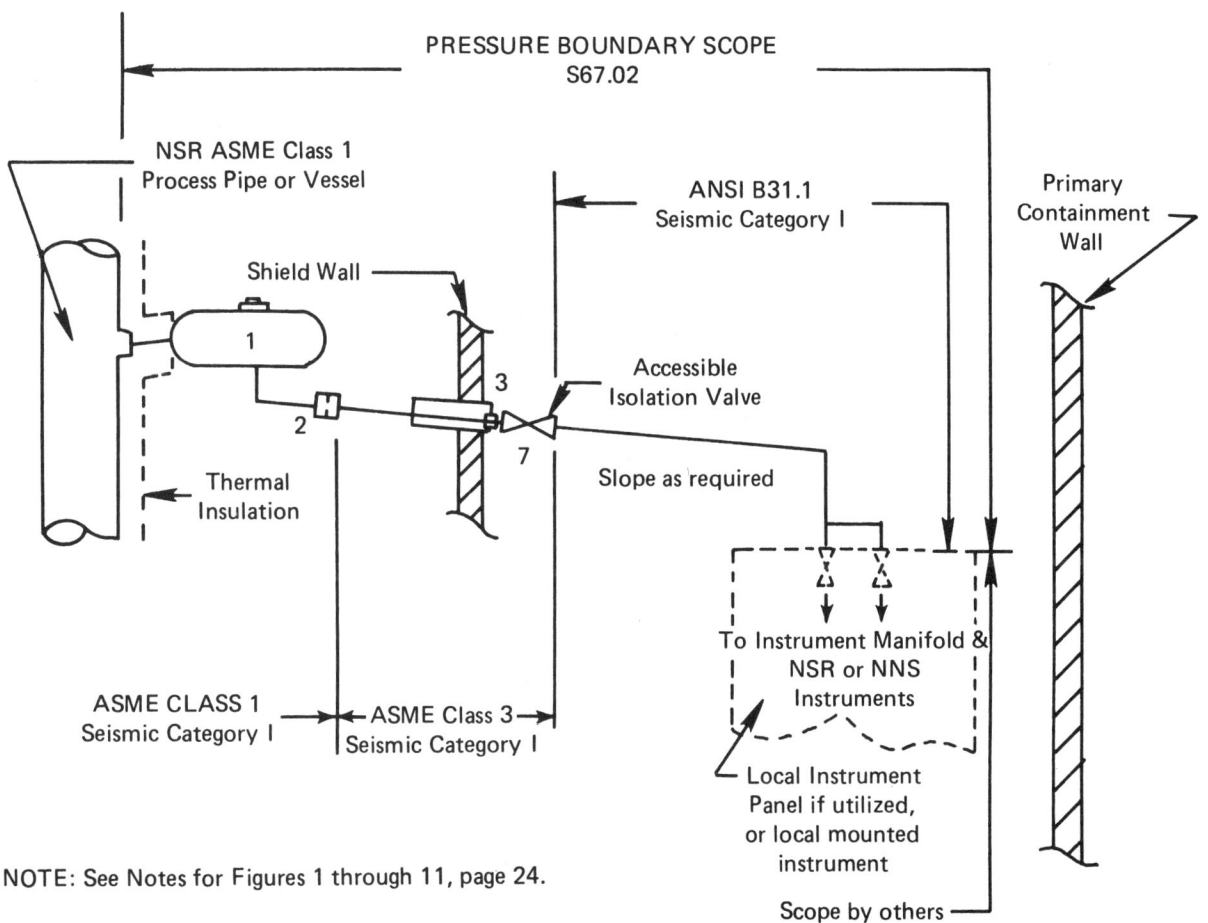

NOTE: See Notes for Figures 1 through 11, page 24.

Figure 1 : Water Filled Instrument Sensing Lines for Water and Steam Service ASME Class 1 Process, Instruments in Containment

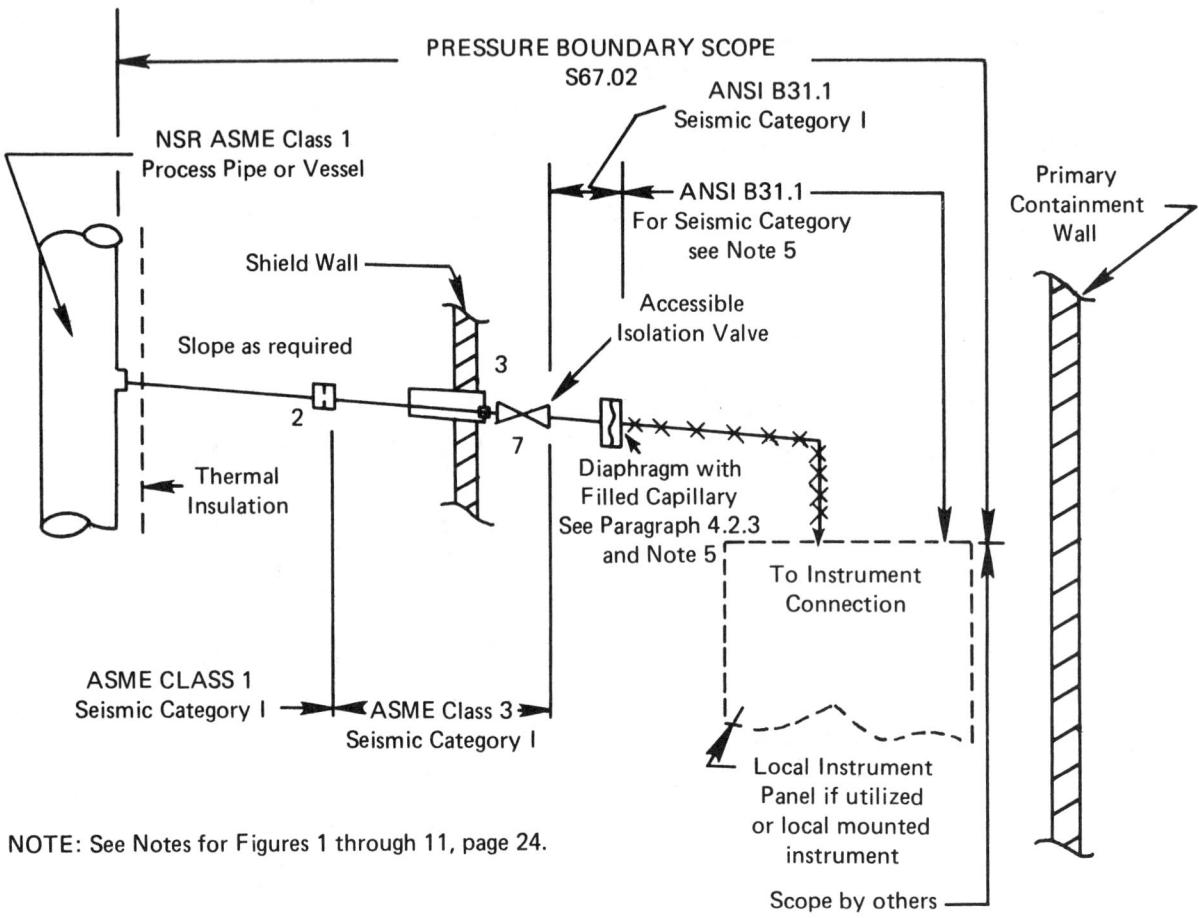

Figure 2: Permanently Filled Instrument Sensing Lines for Water and Steam Service ASME Class 1 Process, Capillary Filled Instruments in Containment

S67.02

Nuclear-Safety-Related Instruments
Sensing Line Piping and Tubing
Standards for Use in Nuclear Power Plants

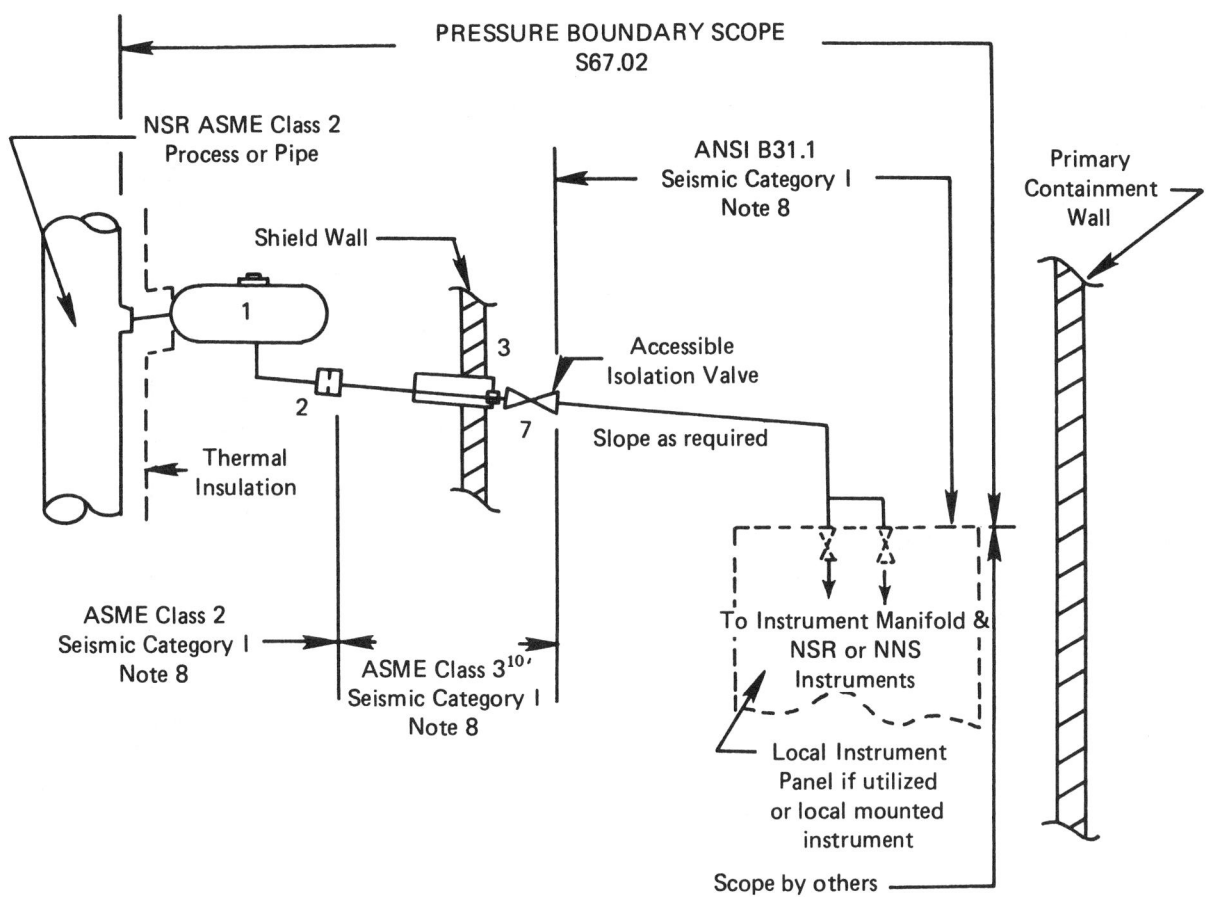

NOTE: See Notes for Figures 1 through 11, page 24.

Figure 3[6]: Water Filled Instrument Sensing Lines for Water and Steam Service ASME Class 2 Process, Instruments in Containment

Instrument Society of America

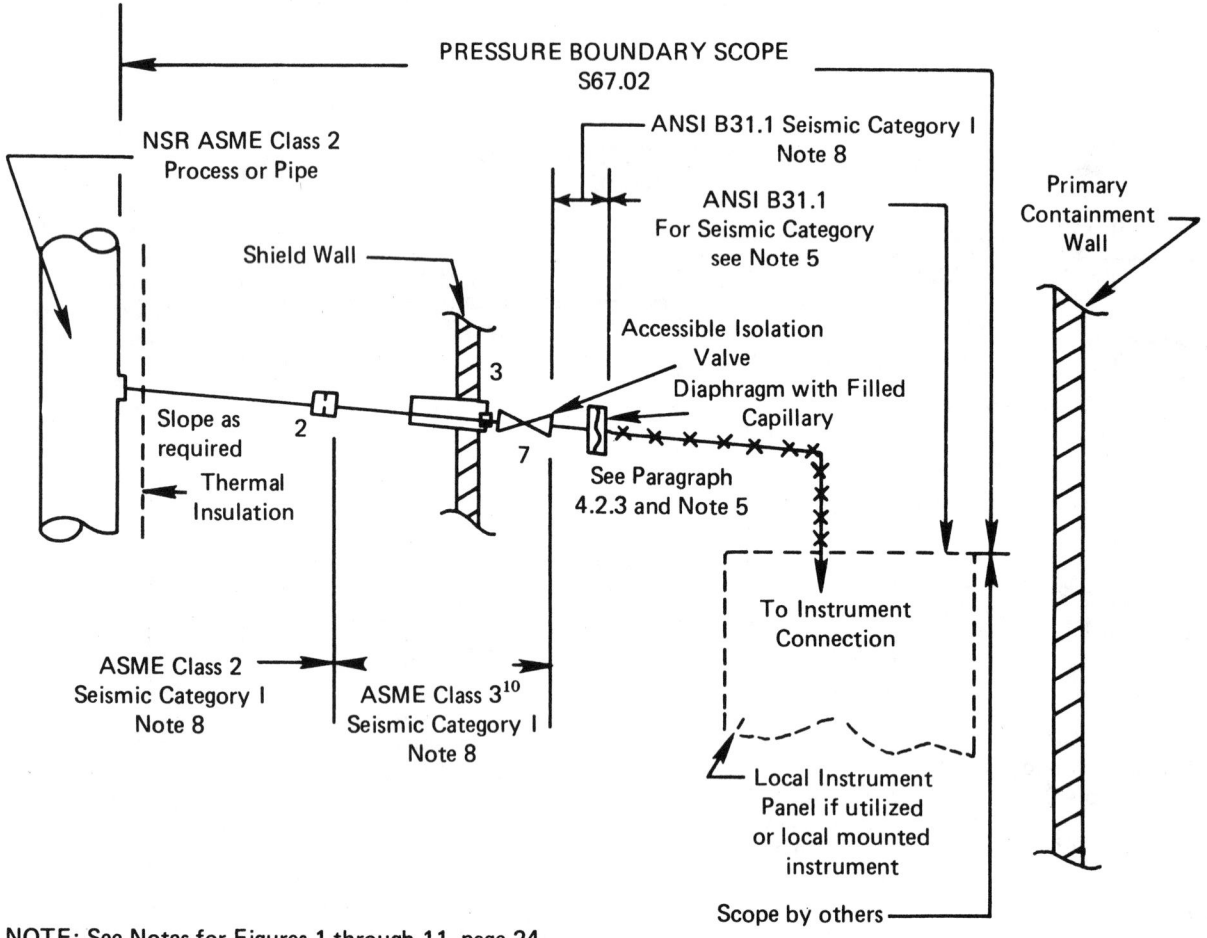

NOTE: See Notes for Figures 1 through 11, page 24.

Figure 4[6]: Permanently Filled Instrument Sensing Lines for Water and Steam Service ASME Class 2 Process, Capillary Filled Instruments in Containment

S67.02

Nuclear-Safety-Related Instruments
Sensing Line Piping and Tubing
Standards for Use in Nuclear Power Plants

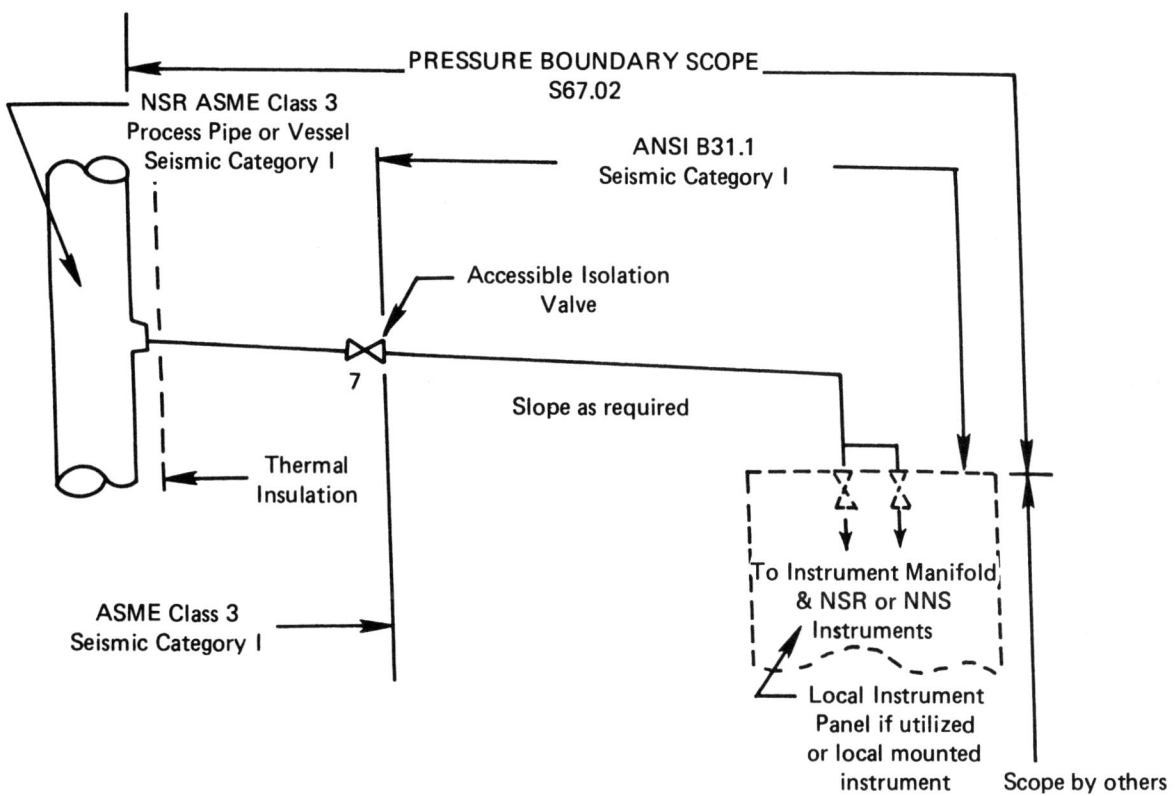

NOTE: See Notes for Figures 1 through 11, page 24.

Figure 5[9]: Water Filled Instrument Sensing Line for Water and Steam Service ASME Class 3 Process

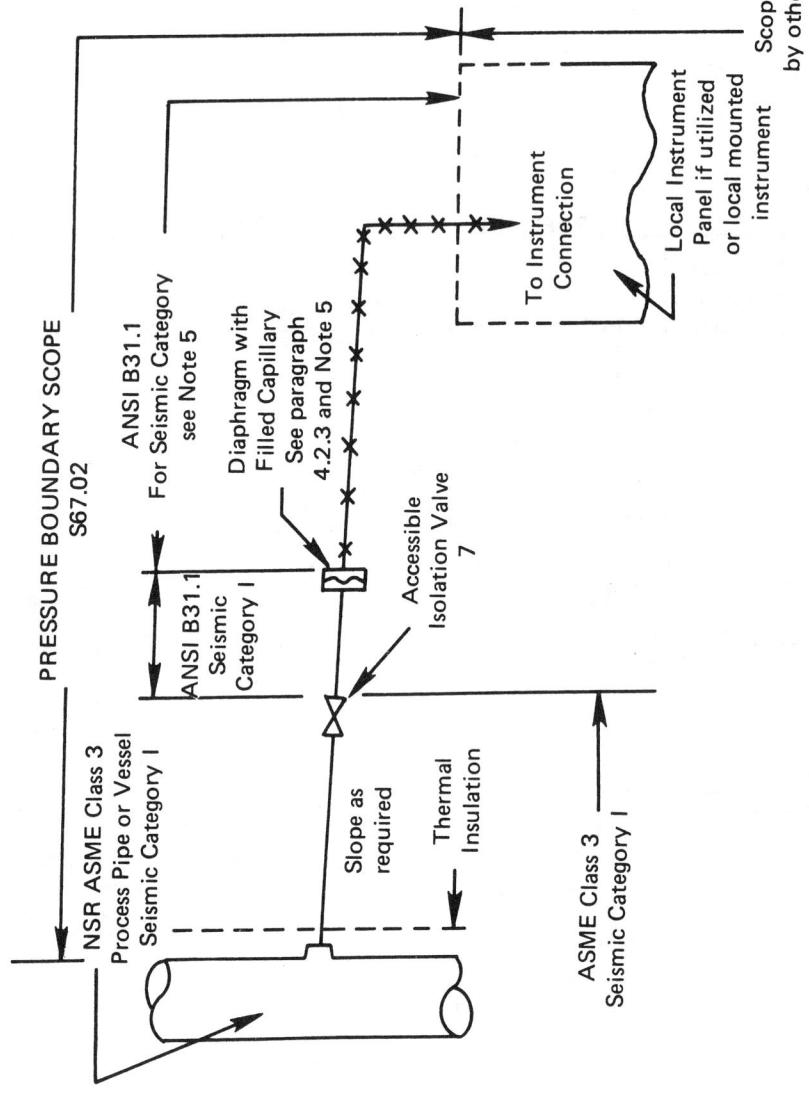

NOTE: See Notes for Figures 1 through 11, page 24.

Figure 6[9]: Permanently Filled Instrument Sensing Lines for Water and Steam Service ASME Class 3 Process, Capillary Filled Instruments

Nuclear-Safety-Related Instruments
Sensing Line Piping and Tubing
Standards for Use in Nuclear Power Plants

S67.02

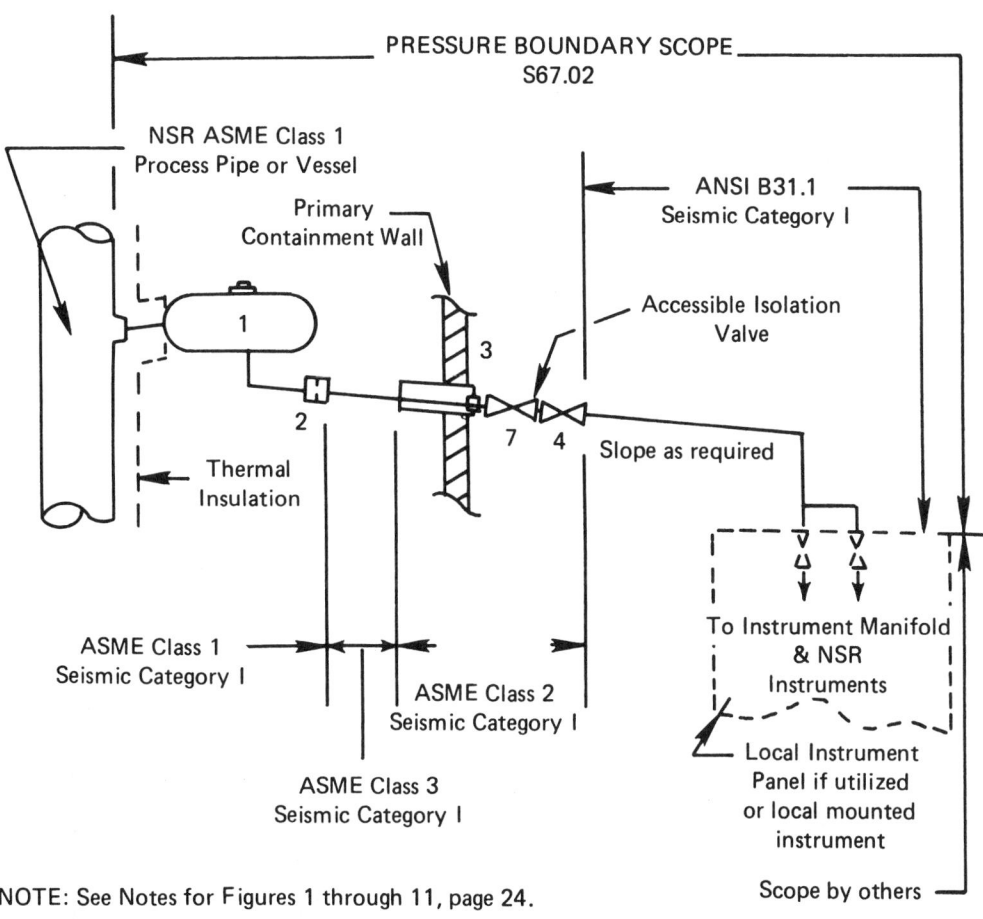

NOTE: See Notes for Figures 1 through 11, page 24.

Figure 7: Water Filled Instrument Sensing Lines for Water and Steam Service ASME Class 1 Process, Instruments Outside Containment

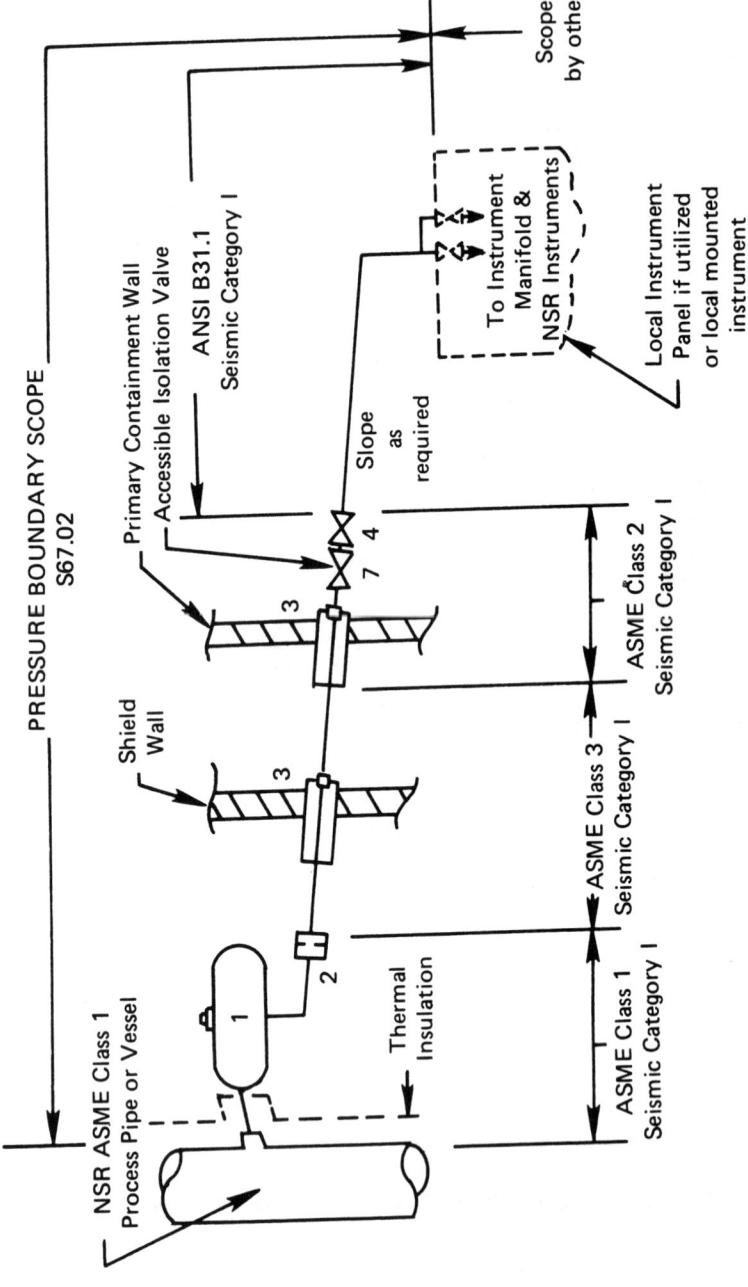

NOTE: See Notes for Figures 1 through 11, page 24.

Figure 8: Water Filled Instrument Sensing Lines for Water and Steam Service ASME Class 1 Process, Instruments Outside Containment

S67.02

Nuclear-Safety-Related Instruments Sensing Line Piping and Tubing Standards for Use in Nuclear Power Plants

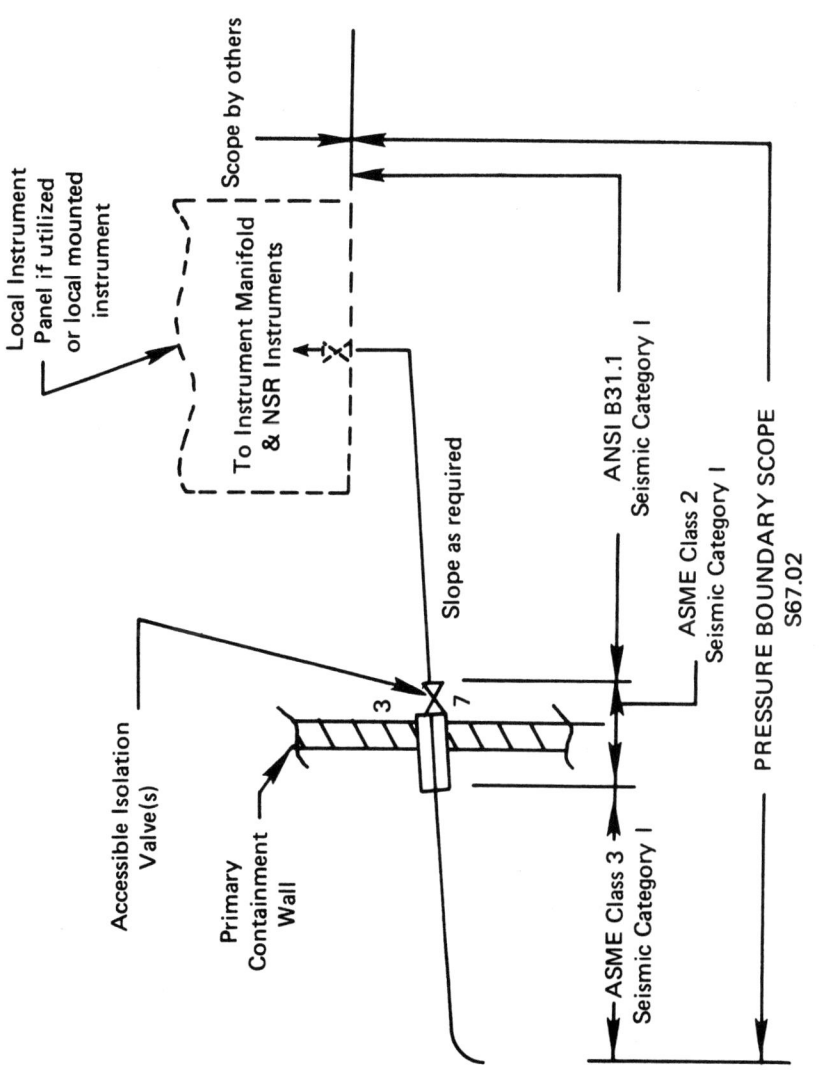

NOTE: See Notes for Figures 1 through 11, page 24.

Figure 9: Containment Atmosphere Instrument Sensing Lines, Instruments Outside Containment

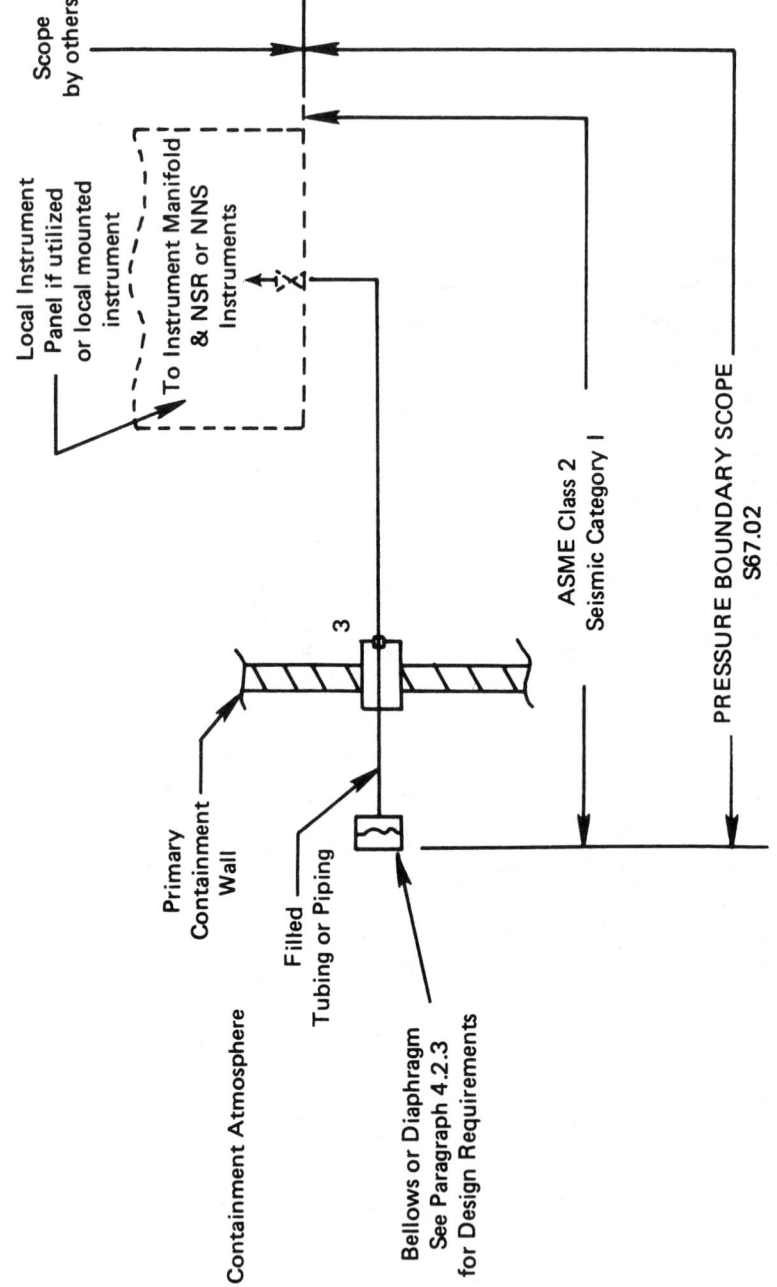

Figure 10: Containment Atmosphere Instrument Sensing Lines Diaphragm with Filled Instruments Outside Containment

NOTE: See Notes for Figures 1 through 11, page 14.

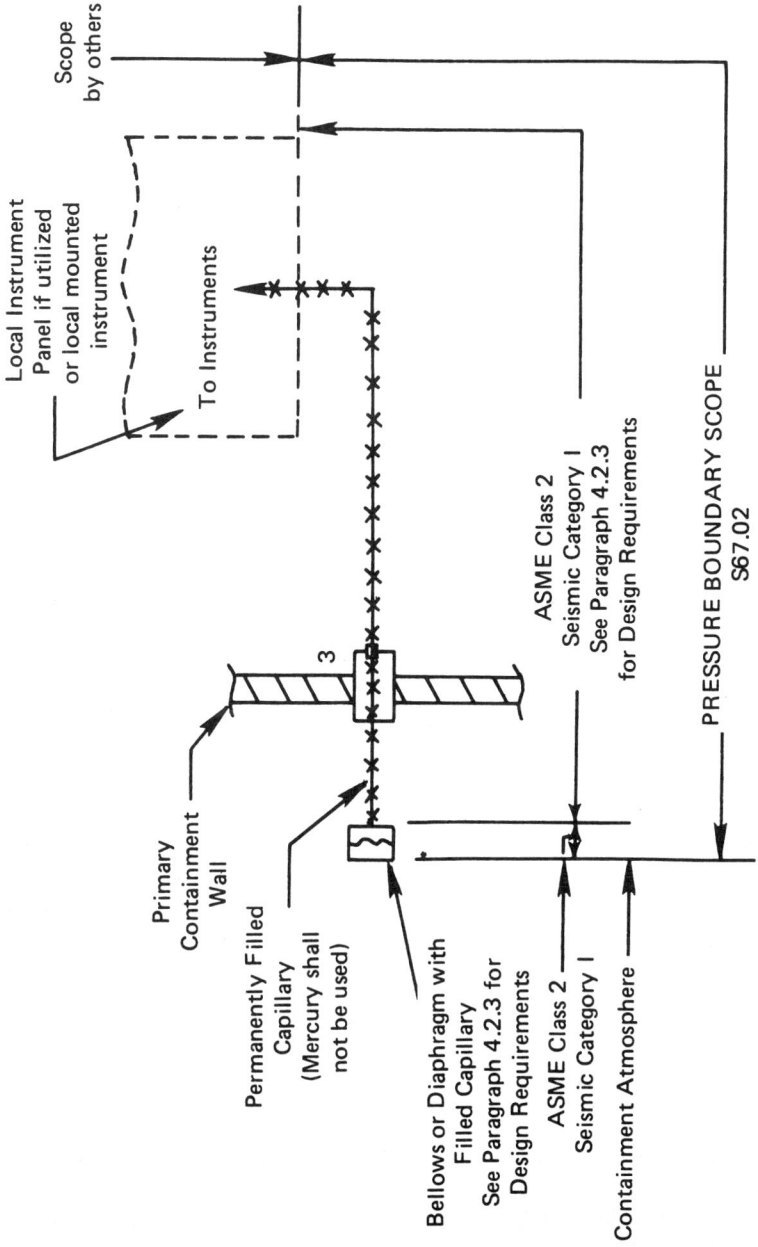

NOTE: See Notes for Figures 1 through 11, page 24.

Figure 11: Containment Atmosphere Instrument Sensing Lines Capillary Filled Instruments Outside Containment

Notes for Figures 1 through 11

These notes are part of this standard.

1. Condensate pot, as required.

2. Restricting device. (See paragraph 6.3)

3. Penetration.

4. Self-Actuating Excess Flow Check Valve: The self-actuating excess flow check valve is one method of meeting containment isolation requirements. Self-actuating excess flow check valves, if used, shall be designed in accordance with the rules of ASME Section III, Subarticle NC-3500. The valves shall be provided with automatic reset and indication of open and closed position, and shall have the capability of being periodically tested. See paragraph 6.4 and ANSI N271 for containment isolation requirements.

5. Seismic Category I when instrument is NSR. Non-Seismic when the instrument is NNS. (See Section 3)

6. Figures 3 and 4 shall also be used for applications where the process pipe or equipment and instrument sensing line are all outside the primary containment wall. When used in this manner, the shield wall and penetration (Note 3) above shall not apply.

7. Accessible isolation valve (may be the root valve, as well). (See paragraph 6.2)

8. Where the instrument sensing lines are carbon steel, restricting devices shall not be used and paragraph 6.3.1 shall apply.

9. Figures 5 and 6 shall be used for applications where the instrument, process instrument tap, and instrument-sensing line are located completely inside or outside the primary containment as long as they do not penetrate the primary containment.

10. Where the sensing line connects to nonnuclear-safety-related instruments only this portion of the sensing line may be ANSI B31.1, Seismic Category I.

5 PROTECTION OF NUCLEAR-SAFETY-RELATED INSTRUMENT SENSING LINES

Redundant instrumentation sensing lines shall be so routed and protected so that any credible effects (consequences) of any design basis event that is to be mitigated by signals sensed through those sensing lines shall not render any of these redundant sensing lines inoperable, unless it can be demonstrated that the protective function is still accomplished. This level of protection shall ensure that, after the event, a single failure shall not prevent mitigation of that event. Credible effects of design basis events that do not depend on a given group of redundant instrument sensing lines for mitigation or accident prevention may render inoperable any or all of that group of sensing lines without violating this criterion provided that the overall protective function is accomplished. All nuclear-safety-related instrument sensing lines should be protected from damage during normal operation occurrences.

Protection may be achieved by physical separation (barriers or spatial, with documented analysis or calculations, where required (see paragraph 5.2 and Section 3) or demonstrated by documented analysis or calculations that protection is achieved by other means. The analysis or calculations shall be maintained as part of the plant design records.

5.1 Redundant Instrument Taps

A single process pipe tap to connect process signals to redundant instruments shall not be used.

5.2 Routing Instrument Sensing Lines

5.2.1 General Considerations

(1) Instrument sensing lines shall be routed to prevent violating required separation between redundant instrument channels. See paragraph 5.2.1(2). Separation between redundant instrument sensing lines shall be provided by free air space or barriers, or both, such that no single failure can cause the failure of more than one redundant sensing line. See paragraph 5 above.

(2) The minimum separation between redundant instrument sensing lines shall be at least eighteen inches (450 mm) in air, in nonmissile, nonhigh energy jet stream, nonpipe whip or nonhostile areas. As an alternative, a suitable steel or concrete barrier shall be used. When a suitable barrier is used, it shall extend at least one inch (25 mm) beyond the line of sight between redundant sensing lines and shall be designed and mounted to Seismic Category I requirements. In hostile areas subject to high energy jet stream, missiles, and pipe whip, the separation shall be provided by space in air, steel or concrete barriers, or both, and documented with analysis or calculations as necessary to prove that the separation protects the redundant sensing lines from failure due to a common cause. All barriers shall be designed and mounted to Seismic Category I requirements.

(3) Instrument sensing lines shall be run along walls, columns, or ceilings whenever practical, avoiding open or exposed areas, to decrease the likelihood of persons supporting themselves on the lines, or of damage to the sensing lines by pipe whip, missiles, jet forces, or falling objects.

(4) Supports, brackets, clips, or hangers shall not be fastened to the instrument sensing lines for the purpose of supporting cable trays or any other equipment.

(5) Routing of the nuclear-safety-related instrument sensing lines shall ensure that the function of the lines is not affected by vibration, abnormal heat, or stress. (Refer to Section 6 of this standard, which covers installation hardware.)

5.2.2 Special Considerations

(1) Tubing penetrating walls and floors: Where redundant instrument sensing lines penetrate a wall or floor, the required separation shall be maintained. See paragraph 5.2.1(2) above. Care shall be taken to ensure the tubing or pipe does not rest on or against any abrasive surface.

(2) Tubing, piping, and capillary tubes penetrating shielding walls: Care shall be taken to avoid personnel exposure to radiation "streaming" from radioactive sources to the surrounding areas through instrument sensing line penetrations in the shield walls. One of the following methods shall be used:

   (a) All instrument sensing line penetrations shall be located at a minimum height of seven feet (2.2 meters) above floor level or,

   (b) The tubing penetrations shall be pitched toward an inner corner of the operating compartment, avoiding a direct radiation streaming path, or,

   (c) The tubing penetration shall be arranged as a labyrinth pattern, or,

   (d) If neither method a, b, or c above is practical, then the sensing lines penetrating the shield wall shall be surrounded by a pipe sleeve and the open space between the sensing line and sleeve, filled with a suitable radiation absorbing material. The sensing line shall make a right angle bend after it leaves the sleeve and a radiation shield placed in front of the line where it exits the pipe sleeve if the radiation absorbing filler material is not adequate or cannot be used.

(3) Any taps, piping, and tubing provided for testing shall comply with this standard.

5.3 Identification and Color Coding

The instrument sensing tubing or piping runs pertaining to a nuclear-safety-related-instrument channel shall be identified and color coded so as to identify its channel. Each instrument sensing line in this channel shall have an identification tag showing the color, channel, and unique line identification number.

A list of these line numbers and associated channel number shall be kept for record. Each instrument sensing line as a minimum, shall be tagged at its process line root valve connection and also at the instrument, and at any point in between where the sensing line passes through a wall or a floor (on both sides of such penetrations).

6 AUXILIARY DEVICES, FITTINGS AND SUPPORTS

6.1 Types of Sensing Line Connections within the Scope of this Standard

Flareless or welded tube or pipe fittings may be used for tube sizes not exceeding one inch (25.4 mm) outside diameter or three-quarter inch (nominal pipe) and with the requirements listed below:

6.1.1 Fittings shall be of a compatible material with the tubing or pipe material on which they are used to avoid electrolysis and to provide acceptable weld joints.

6.1.2 Tube fittings shall be used at pressure-temperature ratings not exceeding the recommendation of the tube fitting manufacturer and to meet the environmental and process system requirements.

6.1.3 Piping, tubing, tube fittings, pipe fittings, valves, restrictor devices, and other appurtenances shall meet the same requirements as defined in Section 4.

6.1.4 Flareless tube fittings, if used, shall be made of a design in which the sealing member or members shall grip the outer surface of the tube to prevent blowout, but without seriously deforming the inside diameter. The sealing member or members shall form a pressure seal against the fitting body with a full surface seal.

6.1.5 Tube fittings shall be installed in accordance with manufacturer's recommendations.

6.1.6 The mating parts of the original flareless tube fittings shall be designed and manufactured by the same company. Replacement parts from different manufacturers shall be qualified to mate and seal properly at the design service conditions, or the complete fitting shall be replaced when service is needed.

6.1.7 In the absence of any existing standards, the designer shall determine that the type of fitting selected is qualified for design conditions including vibration, pressure, and thermal shock and applicable environmental conditions, or demonstrated by test to be able to perform its intended function. The fittings selected shall not degrade the inherent strength of the tubing specified.

6.1.8 Screwed joints in which pipe threads provide the only seal shall not be used.

6.2 Root and Accessible Isolation Valves.

It is recommended that the root valves be three-quarter inch unless special requirements necessitate a different size. Isolation valves shown in Figures 1 through 11 of this standard refer to the first accessible isolation valve available to operating personnel during normal plant operation.

The root valve may or may not perform the function of the accessible isolation valve, dependent on its location. The valves shall be in compliance with ASME Boiler and Pressure Vessel Code Section III, paragraphs NB3676, NC3676, ND3676, or paragraph 122.3, ANSI B31.1, as identified in Section 4, this standard, is required. Isolation valves shall be located as close as practical to the process line, consistent with accessibility.

6.3 Restriction Devices and Instrument Response

6.3.1 Restriction devices shall be provided as shown in Figures 1 through 11 and in the footnotes to Tables 1 and 2. These restriction devices shall be sized in accordance with Table 1 (notes d and g) and Table 2 (notes d and f), as applicable. If restriction device sizing requirements prevent the achievement of the required response characteristic of a nuclear-safety-related instrument, or where drainback of condensate in a sensing line to the process is required for instrument operation, the restriction devices shall not be installed and the safety classification of the instrument sensing line shall remain the same as that of the process system.

6.3.2 When restriction devices are utilized, they shall be installed as close to the process as practical, and shall be upstream of the root valve. The preferred method of construction is a permanently welded fitting. Where the required diameter of the restriction is not less than that required for free drain-back to the process, the restrictor may be installed upstream of a condensate pot.

6.3.3 Where instrument sensing lines penetrate the primary containment wall, restriction devices shall be installed in addition to any self-actuated excess flow check valves unless covered otherwise in this standard. Refer to Section 4, this standard.

6.4 Automatic Reset Excess Flow Check Valves

The self-actuated automatic reset excess flow check valve is one acceptable method of meeting containment isolation requirements when dead-ended nuclear-safety-related instrument sensing lines penetrate the containment. Refer to Figures 7 and 8 for typical arrangements. The self-actuated automatic reset excess flow check valve shall be of the same material and Section III ASME Code Class as the line upstream in which it is installed and shall have the capability of being periodically tested and visually inspected.

6.4.1 Each valve shall have indication of open and closed positions, refer to ANSI N271 for containment isolation requirements.

6.4.2 Each valve may be designed with an integral orificed bleed flow when closed to refill the instrument sensing line downstream of the valve to equalize the pressure across the valve for automatic reset after testing valve operation. The bleed flow shall be small enough so that the requirements of 10 CFR 100 (Code of Federal Regulations) are not violated in case a nuclear-safety-related instrument sensing line break occurs downstream of the excess flow check valve during an accident.

6.5 Instrument Sensing Line Supports

6.5.1 Within Section 4 of this standard, where Seismic Category I hangers and supports are specified, they shall be anchored to Seismic Category I structures.

6.5.2 Hanger and support design shall include provisions for seismic, jet impingement, pipe whip, missiles, and thermal expansion of the process tap and instrument sensing line to which the hangers or supports may be subjected during normal operation, seismic, or other credible event.

6.5.3 Subsection NF, Section III, ASME Boiler and Pressure Vessel Code, shall apply to all hangers and support design for instrument sensing lines that come under ASME Section III by this standard. All others under ANSI B31.1 shall be designed to meet the load requirements.

6.5.4 Protection barriers shall be supported in accordance with paragraphs 6.5.1, 6.5.2, and 6.5.3 above.

6.5.5 Material selection for hangers, clamps, pads, and spacers in contact with the sensing lines shall be compatible to avoid corrosion.

7 MATERIALS

Materials of instrument sensing lines, valves, and fittings that are part of the pressure boundary shall be in accordance with paragraphs NB3676, NC3676, and ND3676, Section III, ASME Boiler and Pressure Vessel Code, paragraph 122.3 ANSI B31.1, and paragraph 4.2.2 of this standard. Material for sensing lines, valves, fittings, and supports shall be in accordance with Section 4 and paragraph 6.5 of this standard.

8 DOCUMENTATION AND QUALITY ASSURANCE

Sufficient documentation and quality control procedures shall exist or be implemented to assure a satisfactory nuclear-safety-related instrument sensing line installation in accordance with the provisions of this standard. As a minimum, the following is required:

8.1 Documentation for instrument sensing lines where ASME Section III is required by this standard:

All records and procedures that are required to be maintained by ASME Section III shall become part of the instrument sensing line documentation package. In addition, the following documents shall be maintained:

8.1.1 Analysis or qualification data showing that Seismic Category I requirements are satisfied.

8.1.2 Analysis or calculations proving physical protection (separation) where applicable.

8.2 Documentation for instrument sensing lines where ANSI B31.1 is required by this standard:

All records and procedures required to be maintained by ANSI B31.1 shall become part of the instrument sensing line documentation package. In addition, the following documents shall be maintained:

8.2.1 Material certification of compliance as required by paragraph 4.2.2 of this standard.

8.2.2 Pressure test reports (See paragraph 4.2.2)

8.2.3 Analysis or qualification data showing that Seismic Category I requirements are satisfied.

8.2.4 Analysis or calculations proving physical protection and/or separation where applicable.

8.2.5 Proof test reports, if proof tests are used to establish pressure ratings of media isolation devices.

8.3 Documentation for permanently filled capillary tubes:

The following documents shall be maintained for permanently filled capillary tubes:

8.3.1 Proof test reports, if proof test reports are used to establish pressure ratings.

8.3.2 Analysis or qualification data showing that Seismic Category I requirements are satisfied.

8.3.3 Pressure test reports (See paragraph 4.2.3)

8.3.4 Analysis or calculations proving physical protection (separation) where applicable.

8.4 Quality assurance requirements for instrument sensing lines for which ASME Section III is required by this standard.

Quality assurance requirements shall be in accordance with ASME Section III.

8.5 Quality assurance requirements for instrument sensing lines for which ANSI B31.1 is required by this standard and for Seismic Category I permanently filled capillary tubes.

Quality assurance procedures shall be established by the owner or his agent ensuring that instrument sensing lines, appurtenances, and supports are installed in accordance with the design drawings and documents.

APPENDIX A

INTERFACE STANDARDS AND DOCUMENTS

(Appendix A is not part of Standard ISA-S67.02.)

The documents listed in paragraphs A.1 through A.8 were considered in the development of this standard.

A.1 AMERICAN NATIONAL STANDARDS INSTITUTE (ANSI)

The following documents, approved by the American National Standards Institute, 1430 Broadway, New York, NY 10018, were considered:

B16.9 — Factory Made Wrought Steel Buttwelding Fittings, 1978.

B31.1 — Power Piping, Paragraphs 104.8 and 122.3 Through 122.3.9—1977 with Winter 1978 Addenda.

N18.2a — (ANS 51.8)—"Nuclear Safety Criteria for the Design of Stationary Pressurized Water Reactor Plants." 1975. American Nuclear Society, Hinsdale, IL 60521.

N45.2 — Quality Assurance Program Requirements for Nuclear Facilities, 1977.

N45.2.11 — Quality Assurance Requirements for Design of Nuclear Power Plants—1974.

N212 (ANS 52.1) — "Nuclear Safety Criteria for the Design of Stationary Boiling Water Reactor Plants." 1978. American Nuclear Society, Hinsdale, IL 60521.

N271 (ANS 56.2) — "Containment Isolation Provisions for Fluid Systems." 1976. American Nuclear Society, Hinsdale, IL 60521.

A.2 AMERICAN PETROLEUM INSTITUTE (API)

The following document is available from the American Petroleum Institute, 2101 L. Street, NW, Washington, DC 20037.

API RP 550, Part I, Paragraph 1.5b, March 1965.

A.3 AMERICAN SOCIETY OF MECHANICAL ENGINEERS (ASME)

The publications which follow can be obtained from the American Society of Mechanical Engineers, 345 East 47th Street, New York, NY 10017.

Boiler and Pressure Vessel Code, Section III—Nuclear Power Plant Components.

Division 1 and Division 2, Subsection NCA - 1977 with Summer 1978 Addenda.

Division 1, Subsection NB, Paragraph NB3676 - 1977 with Summer 1978 Addenda.

Division 1, Subsection NC, Paragraph NC3676 - 1977 with Summer 1978 Addenda.

Division 1, Subsection ND, Paragraph ND3676—1977 with Summer 1978 Addenda.

Division 1, Subsection NE, Paragraph NE1132 - 1977 with Winter 1977 Addenda.

Division 1, Subsection NF - 1977 with Summer 1978 Addenda.

Fluid Meters, Sixth Edition - 1971, Paragraph II-II-14.

Power Test Code PTC 19.2 - 1964, Paragraph 2.08 and Figure 2.4.

A.4 AMERICAN SOCIETY FOR TESTING AND MATERIALS (ASTM)

The following documents are available from the American Society for Testing and Materials, 1916 Race Street, Philadelphia, PA 19103.

Z210.1 ASTM/IEEE Standard for Metric Practice.

E-380-76-"Metric Practice Guide."

A.5 INSTITUTE OF ELECTRICAL AND ELECTRONIC ENGINEERS (IEEE)

The following documents are available from the Institute of Electrical and Electronic Engineers, 345 East 47th Street, New York, NY 10017.

279 — Criteria for Protection Systems for Nuclear Power Generating Stations - 1971.

336 — Installation, Inspection, and Testing Requirements for Instrumentation and Electrical Equipment During the Construction of Nuclear Power Generating Stations - 1977.

379 — Guide to Application of Single Failure Criterion to Nuclear Power Generating Stations Class 1E Systems - 1977.

384 — Criteria for Independence of Class 1E Equipment and Circuits - 1977.

603 — Trial Use—Standard Criteria for Safety Systems for Nuclear Power Generating Stations - April 15, 1977. (This standard is intended to replace IEEE-279 at the conclusion of the trial use period.)

A.6 INSTRUMENT SOCIETY OF AMERICA (ISA)

The following document can be obtained from the Instrument Society of America, 67 Alexander Drive, P.O. Box 12277, Research Triangle Park, NC 27709.

S67.01 (ANSI-N678) — "Transducer and Transmitter Installation for Nuclear Safety Applications."

A.7 U.S. CODE OF FEDERAL REGULATIONS LICENSINGS FOR PRODUCTION AND UTILIZATION FACITITIES

10 CFR 50, Appendix A - 1974.

Criterion 24 – Separation of Protection and Control Systems.

Criterion 33 – Reactor Coolant Makeup.

Criterion 53 – Provisions for Containment Testing and Inspection.

Criterion 54 – Systems Penetrating Containment.

Criterion 55 – Reactor Coolant Pressure Boundary Penetrating Containment.

Criterion 56 – Primary Containment Isolation.

Criterion 57 – Closed System Isolation Valves.

10 CFR 50, Appendix B - February 1971.

Quality Assurance Criteria for Nuclear Power Plants.

10 CFR 100 - Reactor Site Criteria - 1975.

A.8 U.S. NUCLEAR REGULATORY COMMISSION REGULATORY GUIDE

1.11 – Instrument Lines Penetrating Primary Reactor Containment - March 1971 and Supplement 1972.

1.26 – Quality Group Classification and Standards - February 1976.

1.29 – Seismic Design Classification - August 1973.

1.46 – Protection Against Pipe Whip Inside Containment - May 1973.

1.48 – Design Limits and Loading Combinations for Seismic Category I Fluid System Components - May 1973.

1.64 – Quality Assurance Requirements for the Design of Nuclear Power Plants - June 1976.

1.75 – Physical Independence of Electric Systems - January 1975.

INSTRUMENT SOCIETY of AMERICA
Research Triangle Park, North Carolina

ANSI/ISA-S67.03-1982
Approved February 8, 1984

American National Standard

Standard for Light Water Reactor Coolant Pressure Boundary Leak Detection

Instrument Society of America

ISBN 0-87664-734-4

ISA-S67.03 Standard for Light Water Reactor
Coolant Pressure Boundary Leak Detection

Copyright © 1982 by the Instrument Society of America. All rights reserved. Printed in the United States of America. No part of this publication may be reproduced, stored in a retrieval system, or transmitted, in any form or any means electronic, mechanical, photocopying, recording or otherwise without the prior written permission of the publisher.

INSTRUMENT SOCIETY OF AMERICA
67 Alexander Drive
P.O. Box 12277
Research Triangle Park, North Carolina 27709

Copyright © 1982 by the Instrument Society of America

PREFACE

This preface is included for information purposes and is not part of S67.03.

This Standard has been prepared as a part of the service of the Instrument Society of America toward a goal of uniformity in the field of instrumentation. To be of real value, this document should not be static, but should be subject to periodic review. Toward this end, the Society welcomes all comments and criticisms, and asks that they be addressed to the Secretary, Standards and Practices Board, Instrument Society of America, 67 Alexander Drive, P.O. Box 12277, Research Triangle Park, NC 27709, Telephone (919) 549-8411.

The ISA Standards and Practices Department is aware of the growing need for attention to the metric system of units in general, and the International System of Units (SI) in particular, in the preparation of instrumentation standards. The Department is further aware of the benefits to USA users of ISA standards of incorporating suitable references to the SI (and the metric system) in their business and professional dealings with other countries. Towards this end this Department will endeavor to introduce SI — acceptable metric units in all new and revised standards to the greatest extent possible. The Metric Practice Guide, which has been published by the American Society for Testing and Materials as ANSI designation Z210.1 (ASTM E380-76, IEEE Std. 268-1975), and future revisions, will be the reference guide for definitions, symbols, abbreviation, and conversion factors.

It is the policy of the Instrument Society of America to encourage and welcome the participation of all concerned individuals and interests in the development of ISA standards. Participation in the ISA standards making process by an individual in no way constitutes endorsement by the employer of that individual of the Instrument Society of America or any of the standards which ISA develops.

The American National Standards Institute (ANSI) assigned work on this standard to ISA Committee SP-67 "Nuclear Power Plant Standards" in December, 1973. The assignments, considered a priority project needing urgent and prompt action, was given to Subcommittee SP-67.03 chaired by M. J. Kimbell during the May 20, 1974 Boston ISA Power Conference. The subcommittee performed a literature search of leak test standards and current nuclear power plant practice in relation to reactor coolant leak detection for representative pressurized water and boiling water power reactors. This information was utilized during the preparation of this Standard together with comments received from concerned reviewers.

The information contained in this preface, the footnotes and attached Appendices A and B is included for information only and is not a part of the Standard.

The following individuals served as members of the ISA Subcommittee SP-67.03 which prepared this standard:

NAME	COMPANY
U. Shah, Chairman	Washington Public Power Supply System
M. J. Kimbell	Bechtel, Inc.
B. G. Atraz	General Electric Co.
J. Dodds	Bechtel Power Corporation
J. Hersey	Bechtel Power Corporation
M. Hildenbrand	Nuclear Measurements Corp.
R. Ulman	Victoreen Inst. Co.
L. S. Loomer	Bechtel Power Corporation
R. M. Norris	Washington Public Power Supply System
M. F. Reisinger	Combustion Engineering, Inc.
B. Segal	U.S. Nuclear Regulatory Commission
G. B. Stramback	General Electric Company
I. Sturman	Bechtel Power Corporation
T. N. Crawford	Pacific Gas and Electric Co.
J. H. Gebert	Iowa Electric Light & Power Co.

This standard was approved by ISA SP67 in January 1980.

NAME	COMPANY
B. W. Ball	Brown & Root, Inc.
G. G. Boyle	Honeywell
T. Crawford	Pacific Gas & Electric Co.
J. M. Dahlquist, Jr.	Baltimore Gas & Electric Co.
H. T. Evans	Pyco, Inc.
L. C. Fron	Detroit Edison Co.
R. L. Gavin	Sargent & Lundy
E. M. Good	Florida Power Corp.
W. G. Gordon	Bechtel Power Corporation

Instrument Society of America

S. C. Gottilla	Burns & Roe, Inc.
L. F. Griffith	Yarway
T. Grochowski	Babcock & Wilcox Co.
G. Harrington	Rosemount
H. S. Hopkins	Westinghouse Electric Corp.
R. J. Howorth	General Physics Corp.
R. N. Hubby	Leeds & Northrup Co.
R. T. Jones	Philadelphia Electric Co.
J. R. Karvinen	MERDI-CDIF
M. J. Kimbell	Bechtel, Inc.
J. R. Klingenberg	W. R. Holway & Assoc.
J. V. Lipka	Gilbert Assoc., Inc.
S. F. Luna	General Atomic Co.
D. W. Miller	Ohio State University
G. C. Minor	MHB Technical Associates
J. W. Mock	EG&G
J. A. Nay	Westinghouse Elec. Corp.
G. E. Peterson	Commonwealth Edison Co.
R. L. Phelps, Jr.	Southern Calif. Edison Co.
M. F. Reisinger	Combustion Engineering, Inc.
R. M. Rello	Air Products & Chemicals, Inc.
U. Shah	Washington Public Power Supply System
J. Tana	Ebasco Services, Inc.
R. J. Ungaretti	Philadelphia Electric Co.
K. Utsumi	General Electric Co.
R. C. Webb	Pacific Gas & Electric Co.
E. C. Wenzinger, Sr.	U.S. Nuclear Regulatory Comm.
W. C. Weston	Stone & Webster Engr. Corp.

This standard was approved for publication by the ISA Standards and Practices Board in October 1982.

NAME	COMPANY
T. J. Harrison, Chairman	IBM Corporation
P. Bliss	Consultant
W. Calder	The Foxboro Company
N. Conger	Continental Oil Co.
B. Feikle	Bailey Controls Co.
R. T. Jones	Philadelphia Electric Co.
R. Keller	Boeing Company
O. P. Lovett, Jr.	Isis Corp.
E. C. Magison	Honeywell, Inc.
A. P. McCauley	Diamond Shamrock Corp.
J. W. Mock	EG&G Idaho, Inc.
E. M. Nesvig	ERDCO Engineering Corp.
G. Platt	Bechtel Power Corp.
R. Prescott	Moore Products Company
W. C. Weidman	Gilbert Associates
K. A. Whitman	Allied Chemical Corp.
J. R. Williams	Stearns-Roger, Inc.
B. A. Christensen*	
L. N. Combs*	
R. L. Galley*	
R. G. Marvin*	
W. B. Miller*	Moore Products Company
R. L. Nickens*	

*Director Emeritus

TABLE OF CONTENTS

Section	Page
1 Introduction	7
2 Scope	7
3 Purpose	7
4 Definitions and Descriptions	7
5 Leakage Classifications and Sources	8
5.1 Leakage Classifications	8
5.2 Potential Identified Leakage Sources	8
6 General Design Requirements	9
6.1 Principal Monitoring Systems for Unidentified Leakage	9
6.2 Coolant Leakage Detection System Performance	9
6.3 Safety Classification	9
6.4 Collecting and Measuring Identified Leakages	9
6.5 Monitoring Intersystem Leakage	9
6.6 System Availability	9
6.7 Human Engineering and Operability Features	9
6.8 Power Sources	9
6.9 Design Basis Documentation	10
7 Specific Leakage Detection Methods and Requirements	10
7.1 Sump Level and Sump Pump Discharge Flow Monitoring Leakage Detection	10
7.2 Radiation Monitoring Leakage Detection	10
7.3 Containment Air Cooler Condensate Flow Collection for Leakage Detection	11
7.4 Reactor Coolant Inventory	11
7.5 Humidity Monitoring Leakage Detection	12
7.6 Temperature Monitoring Leakage Detection	12
7.7 Primary Containment Pressure Monitoring	13
7.8 Tape Moisture Sensors for Leakage Detection	13
7.9 Visual Observation	13
8 References	14
Appendix A Leakage Not Into the Containment	14
Appendix B Suggested Methods and Procedures for Leakage Detection Systems	16

LIST OF TABLES

Table	Page
1 Capabilities of Leakage Monitoring Methods	13

1 INTRODUCTION

Nuclear power plants vary widely with respect to size, capacity, and design details. Available leak detection methods must be individually examined by the designers to determine their suitability for a particular plant or system. The applicable federal regulation on the requirements for Reactor Coolant Pressure Boundary (RCPB) leak detection is specified in the Code of Federal Regulations, Title 10, Part 50 (10CFR50), Appendix A, Criterion 30.*

Detection of leakage from pressurized pipes and vessels is needed because small leaks may develop into larger leaks or ruptures. During reactor operation, detection of reactor coolant leakage from nonisolatable portions of the RCPB is important to allow early identification of minor flaws before they can develop into a pipe break or component rupture that could result in the accidental loss of coolant. Since the RCPB is housed in a containment structure, physical access is limited during power operation and remote indicating leakage detection systems are necessary.**

This standard defines design criteria that are intended to insure that adequate RCPB leak detection capabilities are provided to the nuclear plant operator and to meet the intent of the Code of Federal Regulations. However, the burden of proof of compliance with federal regulations shall remain the responsibility of the plant owner.

2 SCOPE

This standard covers identification and quantitative measurement of reactor coolant system leakage in light water cooled power reactors. Leak detection for gas and liquid metal cooled reactors and for containment building structures surrounding the reactor coolant pressure boundary is not covered in this standard.

3 PURPOSE

The purpose of this standard is to standardize criteria, methods, and procedures for assuring the design and operational adequacy of reactor coolant pressure boundary leak detection systems used in light water cooled nuclear power plants. A further objective is to encourage design improvements yielding increased utility and reliability of reactor coolant leakage detection systems.

4 DEFINITIONS AND DESCRIPTIONS

As used in this standard, the following definitions apply.

Accessible Area — An area routinely or periodically entered by plant personnel in the performance of routine functions during normal plant operation and in accordance with applicable health physics procedures.

Accuracy — ''Degree of conformity of an indicated value to a recognized accepted standard value, or ideal value''. Reference: Instrument Society of America (ISA) Standard S51.1-1979; ''Process Instrumentation Terminology.''

Boiling Water Reactor (BWR) — A nuclear steam supply system in which process steam is generated in the reactor vessel.

Calibration — ''The adjustment of device or series of devices, in order to bring the output to a desired value, within a specified tolerance, for a particular value of input.'' Reference: ISA S51.1-1979.

Coolant — The fluid contained within the reactor coolant pressure boundary.

Leak — An opening, however minute, that allows undesirable passage of a fluid from its containing boundaries.

Leakage — The fluid that passes through a leak. The fluid referred to in this Standard is the primary coolant water unless otherwise stated.

Abnormal Leakage — That leakage from the Reactor Coolant Pressure Boundary (RCPB) which is considered to be unusual, unexpected or in excess of technical specification allowances.

Allowable Leakage — That leakage value defined in plant operational technical specifications above which plant operation must be altered or interrupted as necessary to perform corrective actions to reduce the leakage to allowable values.

Identified Leakage — See Section 5.1.1.

Leakage Rate — Leakage expressed in volumetric units per unit of time at 20°C and one atmosphere pressure.

Unidentified Leakage — See Section 5.1.1 and 5.1.2.

Monitoring Instrument System — A system that provides information about RCPB leakage conditions so that the operator can take action.

Nuclear Safety Related (NSR) — Instrumentation ''which is essential to: 1) Emergency Reactor Shutdown; 2) Containment Isolation; 3) Reactor Core Cooling; 4) Containment or Reactor Heat Removal; 5) prevention or mitigation of a significant release of radioactive material to the environment or, is otherwise essential to provide reasonable assurance that a nuclear power plant can be operated without undue risk to the health and safety of the public.'' Reference: ISA S67.01-1979.

Non-Nuclear Safety (NNS) — Instrumentation not included in NSR.

Operating Basis Earthquake (OBE) — That earthquake which ''. . . could reasonably be expected to affect the plant site during the operating life of the plant; it is that earthquake which produces the vibratory ground motion for which those features of the nuclear power plant necessary for continued operation without undue risk to the health and safety of the public are designed to remain functional.'' Reference 10CFR100, Appendix A, III(d).

---

*Other related federal regulations not addressed in this standard cover the requirements for RCPB fracture prevention, inservice inspection, coolant makeup capability and quality assurance as specified in the Code of Federal Regulations, Title 10, Part 50 (10CFR50), References (1) and (2).

**Appendix A provides information on primary coolant leakage detection outside of the containment structure for boiling water reactors (BWR).

Pressurized Water Reactor (PWR) — A nuclear steam supply system in which the pressurized primary coolant fluid is heated by the reactor core, and the process steam is generated in a steam generator by heat transfer from the primary coolant.

Primary Containment — The structure that encloses the reactor coolant pressure boundary.

Reactor Coolant Pressure Boundary (RCPB) — "... all those pressure-containing components of boiling and pressurized water-cooled nuclear power reactors, such as pressure vessels, piping, pumps, and valves, which are:

(1) Part of the reactor coolant system, or

(2) Connected to the reactor coolant system, up to and including any and all of the following:

   (i) The outermost containment isolation valve in system piping which penetrates primary reactor containment.

   (ii) The second of two valves normally closed during normal reactor operation in system piping which does not penetrate primary reactor containment.

   (iii) The reactor coolant system safety and relief valves.

For nuclear power reactors of the direct cycle boiling water type, the reactor coolant system extends to and includes the outermost containment isolation valve in the main steam and feedwater piping." Reference: 10CFR50, Section 50.2(v).

Sensitivity — "... ratio of the change in output magnitude to the change of the input which causes it after the steady-state has been reached." Reference: ISA S51.1-1979.

Time Constant — "The time required for the output of a first-order system forced by a step change to complete 63.2 percent of the total rise or decay." Reference: ISA S51.1-1979.

Time Response of Instrumentation — "... an output expressed as a function of time, resulting from the application of a specified input under specified operating conditions." Reference: ISA S51.1-1979.

5 LEAKAGE CLASSIFICATIONS AND SOURCES

The significance of leakage from the RCPB will depend upon the leak location, the leakage rate, duration, and the nature of the flow path permitting the leakage. Through-wall cracks or flaws are the most difficult to detect and monitor because they can occur at any RCPB location. This type of leak is also of most concern because the leak may develop from some unpredicted combination of internal defects and external stresses in a nonisolatable portion of the RCPB.

5.1 Leakage Classifications

A principal concern in leakage monitoring is the capability to discriminate between unidentified leakage from the RCPB and leakage from identifiable sources into the containment. Being able to discriminate allows more rapid and reliable assessment of plant operating conditions. The following leakage classifications, as used in this standard, facilitate identification of leakage sources and interpretation of leakage data:

5.1.1 Identified Leakage

(A) Leakage into collection systems, e.g., pump seal or valve packing leakage that is collected and measured, or

(B) Leakage into the containment which meets all of the following conditions:

   (1) The leaks have been specifically located and the rate quantified.

   (2) The leaks are not cracks or flaws in the RCPB.

An example of (B) above is a quantified leakage of component cooling water into the containment.

5.1.2 Unidentified Leakage

Leakage into the containment which is not classified as identified leakage.

5.1.3 Intersystem Leakage

Coolant leakage across RCPB passive barriers such as heat exchanger tubes or tube sheets into other closed systems. Such leakage is not normally released to the containment atmosphere and is a separate classification.

5.1.4 Other Leakage

Any leakage not from the RCPB and outside of the reactor containment structure, if not covered by the above classifications. An example of such leakage could be leakage from steam or feedwater lines outside the containment structure of a BWR plant. A summary of such leakage sources and typical detection methods frequently used is given in Appendix A of this standard.

5.2 Potential Identified Leakage Sources

Variations in plant designs do not allow a single definitive check list of all potential leakage sources. However, probable leakage sources can be identified during plant design and appropriate leakage detection, measurement and collection (leakoff) systems provided. Collection and isolation, to the extent practical, of leakage from identified sources enhance the monitoring capability for unidentified leakage. The following are some of the more common types of leakage sources that can be easily identified:

(A) *Dynamic seals* such as valve stem packing, pump drive shaft seals and control rod drive gland seals.

(B) *Static seals* such as the reactor head pressure seals, equipment gaskets and valve seat seals in lines connected to the RCPB.

(C) *Pressure relief systems* such as pressure relief valves, rupture disks and safety relief valves.

(D) *Passive interface boundaries with the RCPB* such as instrument bellows, diaphragms and Bourdon tubes, thermometer wells and heat exchanger tubes.

6 GENERAL DESIGN REQUIREMENTS

This standard is not intended to replace applicable handbooks and texts, such as References (11) and (12), which provide detailed design and analytical techniques. Suggested methods and procedures for developing required design information are given in the references and appendices to this standard.

6.1 Principal Monitoring Systems for Unidentified Leakage

At least three dissimilar, diverse, and independent principal methods of monitoring coolant leakage from the RCPB to the containment shall be provided. One of these methods shall be sump level and/or sump flow monitoring. Other acceptable methods are identified in Section 7 and Table 1.

6.2 Coolant Leakage Detection System Performance

The sensitivity and response characteristics for each of three principal leak detection monitoring systems shall be shown by design calculations or performance tests to be capable of indicating and alarming a 1 gpm (3.8 liters/min) leakage increase within one hour. It is recognized that some systems other than sump monitoring may not be capable of meeting this requirement during certain normal plant operating conditions. In these cases, these systems shall be designed for leakage sensitivity that is as high as reasonably achievable. When identified leakages are superimposed on unidentified leakages the above sensitivity requirements shall apply also.

6.3 Safety Classification

The RCPB leak detection systems covered in this standard are non-nuclear safety systems or monitoring instrument systems.

6.4 Collecting and Measuring Identified Leakages

Seals, relief systems, and other probable sources of leakage shall be identified. Leakage collection and measurement systems shall be provided for sufficient identified sources to limit the expected leakage to the containment atmosphere to the extent practical. The residual uncollected liquid leakage shall not prevent unidentified liquid leakage monitoring systems from meeting Section 6.2 requirements.

Leakage to the primary reactor containment from identified sources shall be collected or otherwise isolated so that:

(A) The flow rates from identified leaks are monitored separately from unidentified leaks.

(B) The total flow rate from identified leaks can be established and monitored with a sensitivity capable of detecting a 1 gpm (3.8 liters/min) leakage increase within 1 hr for PWR plants and 2 gpm (7.6 liters/min) leakage increase within 1 hr for BWR plants.

6.5 Monitoring Intersystem Leakage

Provisions shall be made to monitor systems connected to the RCPB through passive barriers for indications of intersystem leakage. Acceptable methods include radioactivity monitoring and water inventory monitoring. See Appendix A for additional information.

6.6 System Availability

The RCPB leakage detection systems shall be designed to operate whenever the plant is not in cold shutdown condition.

6.6.1 Ambient Conditions

Monitoring system shall be designed to maintain specified accuracy and performance features for the range of ambient temperature, humidity, and radiation levels that are expected at the component locations during normal plant operations.

6.6.2 Seismic Events

The sump monitoring system and at least one of the other diverse monitoring channels provided shall be demonstrated to be acceptable for the design requirements after any seismic event for which plant shutdown is not required, i.e., less than an operating basis earthquake. The guidelines of IEEE Standard 344, Reference (6), may be used for seismic qualification.

Recorders need not function during or after seismic events provided alarm and indication capability remain available.

6.7 Human Engineering and Operability Features

6.7.1 Displays and alarms for all leakage detection systems shall be located in the main control room. In cases where additional process displays related to leakage are identified, these indications should be provided at one general display location. Leakage monitoring displays may also include computer functions and CRT displays. Quantitative measurements with procedures for locating leakage sources should be monitored and controlled from one general display location.

6.7.2 Capability for online calibration of leakage detection channels by simulated detector outputs or other means shall be provided. Readouts, alarm set points, and calibration factors shall be capable of periodic adjustment to compensate for changes in actual environmental or background conditions.

6.7.3 Leakage measuring system displays shall be expressed in volumetric units or if expressed in other units, conversion capability shall be provided. Capability for trend monitoring of measured leakage shall be provided.

6.7.4 Capability for entire channel calibration and maintenance during refueling outages shall be provided. The leakage detection systems shall be designed and located for ease of periodic testing, servicing, removal, and replacement.

6.8 Power Sources

Two of the three principal leakage detection systems shall be energized from separate power sources. Seismically qualified systems shall be powered from seismically qualified power sources.

6.9 Design Basis Documentation

Compliance with this standard shall be supported by design basis documentation to include the following:

(A) The data and design basis used for the design of each RCPB leak detection system, e.g., primary coolant temperature, pressure, radioactivity.

(B) A description of the analytical derivations and methods used to determine each system's sensitivity, response time, and alarm set point.

(C) The limitation and approximate accuracy of each leak detection method and its leakage measurement range in coolant volume per unit of time.

(D) Seismic qualification as appropriate.

(E) The procedures which describe the calibration and operation of the RCPB leakage detection systems.

(F) Identification and seismic classification of the power source for each monitoring system.

7 SPECIFIC LEAKAGE DETECTION METHODS AND THEIR REQUIREMENTS

The objective of leakage detection is to identify and quantify the leakage to such an extent that the seriousness of the leak can be determined. The following subsections present requirements and brief descriptions of methods that have been successfully applied to the detection, measurement, or location of leakage from the RCPB[3]. Methods considered to be developmental, such as acoustic[4,17] and ultrasonic, or not in general use are not discussed in this standard. However, this should not discourage the utilization of such methods if they meet the requirements of this standard. The calculational methods and procedures discussed hereafter and in the Appendix B represent idealized cases. These methods and equations can and should be modified and refined as needed to fit specific cases, e.g., purge versus nonpurge containment, containment atmosphere mixing and transport time, recirculation, filtration for calculation of estimated leakage rates.

Table 1 summarizes the capabilities of detection methods presented in this standard for detecting, measuring, and locating leakage from the RCPB. Current general practice is to provide at least three principal detection methods, one of which is sump monitoring.

7.1 Sump Level and Sump Pump Discharge Flow Monitoring Leakage Detection

Reactor coolant pressure boundary leakage can be detected and measured by monitoring open containment sump levels and/or sump pump discharge flow rates[10]. The following requirements apply:

(A) Identified equipment leakage from large valve stem packing glands and other readily identifiable sources shall be monitored by piping the flow to closed equipment drain tanks or sumps so that an average background identified leakage rate can be established.

(B) Open containment sumps shall collect unidentified containment leakage including containment cooler condensate (See Section 7.3.). Sensitivity and response time shall be such that a one gpm (3.8 liters/min) of liquid collected into the sump can be detected in less than one hour. For an instrumentation method to be acceptable for a given configuration, verification by calculation that the above requirements will be satisfied is required.

The leak location is not identifiable by this method unless relatively small areas of piping are draining into different sumps for each area. Both sump level change and sump discharge flow can be monitored to detect a leak.

Gaseous releases from identified leakage collection points shall be controlled so that they do not decrease the effectiveness of the radioactivity monitors.

7.2 Radiation Monitoring Leakage Detection

Monitoring the containment for radioactivity is a requirement specified in 10CFR50 Appendix A[7]. Monitors used to meet this requirement may also be used for RCPB leak detection provided the monitors meet the requirements of this standard. The response and sensitivity characteristics of monitor outputs shall be correlated to leakage rates and coolant activity.

7.2.1 Air Radioparticulate and Radiogas Activity Monitors

(A) Description

Air radiation monitors have the potential for detection of coolant leakages from the RCPB. The sensitivity and response time depends, among other things, on the sampling system design, containment atmosphere mixing characteristics, the radiation detector characteristics, the ambient radiation background, and the concentrations of detectable isotopes in the coolant and in the containment atmosphere.

The quantitative measurement of coolant leakage may sometimes be feasible by this method, but extensive current information about plant physical parameters and coolant radioisotope inventory are required in order to perform the computations. Graphical representation of the relationship of leakage rate to the principal parameters can be used to estimate leakage. However, plant process computers are a potentially more useful tool for rapid reduction and interpretation of the monitored data. Sufficient data and understanding of the principles are needed to properly interpret an increase in radiation monitor readout in terms of coolant leakage. For example, a decrease in reactor power level may cause an increase in the primary coolant radioactivity burden, and thus an apparent increase in leakage rate that is actually a false indication.

(B) Sampling System Design

The piping between the sample line inlet and the detector location shall be designed in accordance with ANSI N13.1, "Guide to Sampling Airborne Radioactive Materials in Nuclear Facilities", 1969.

(C) Assumptions for Design Computations

These assumptions provide a uniform design basis and shall be used to estimate the capabilities of radiation monitors to detect coolant leakage from RCPB and to show compliance with the design requirements of Section 6. A model to determine the radioactivity concentrations is given in Appendix B. Due to the range of operating conditions which may be expected from plants, several calculations will be necessary to cover most situations. These calculations will cover the range of expected equilibrium containment airborne concentrations. The following assumptions shall be used unless technical justification is provided.

(1) Expected coolant and RCPB leakage activity levels shall be taken from the American National Standard Source Term Specification (ANSI N237). If particulate monitors are used, the situation in which there is no failed fuel, i.e., only corrosion products in the coolant, shall be included. Coolant concentrations will vary by orders of magnitude within short time periods due to changes in operating conditions.

(2) For particulate monitors, owing to differences in source terms in PWRs, only Rb-88 which is in secular equilibrium with its parent isotope Kr-88 need be considered. For BWRs, a spectrum of isotopes shall be used. For the case with no fission products, several of the corrosion products will have to be considered since no single isotope dominates.

For gaseous monitors, Xe-133 gives over 95 percent of the dose and is sufficient for PWR analyses. For BWRs, a spectrum of isotopes shall again be used.

For iodine monitors, the five radioiodines I-131 through I-135 should be considered because of their abundance as fission products.

(3) Coolant leakage shall be assumed to be uniformly mixed in the appropriate containment or dry-well free volume, unless HVAC and building design indicate that uniform mixing will not occur within one hour. If the containment is continuously purged then this effect shall be included.

(4) Plateout factors of 0.999 for particulates, 0.99 for iodines, and zero for noble gases as a conservative design basis shall be applied to the activity in the coolant leakage. See Reference (16) for additional information.

(5) The equilibrium containment airborne activity levels shall be based on available operating data or other documented basis such as Reference (5). The concentrations at equilibrium in a containment from a continuous leak of 1 to 1.5 gpm approximate the observed data. These activity levels are in the range of the Maximum Permissible Concentrations (MPC) of 10CFR20, Table I for Xe-133 and I-131.

For maximum monitor sensitivity cases, e.g., when failed fuel levels are low and the containment airborne concentrations are low shortly after a startup, the airborne activity levels can be neglected. However, it should be realized that increases in containment activity may be due to temporary spiking increases in the coolant concentrations rather than due to increases in leakage following startups and power transients.

(6) Monitor background count rates shall be based on the activity and radiation levels expected at the detector location.

7.2.2 Intersystem Leakage Monitoring

Intersystem leakage of coolant through the RCPB into other systems, e.g., primary to secondary system leaks in heat exchangers, is detectable by secondary system liquid radiation monitoring or secondary system off-gas monitoring.

Radiation monitoring does provide the capability for detection of small intersystem leakages of coolant through the RCPB, but sensitivity requirements are not defined by this standard. Some of the factors affecting sensitivity and response time are the concentration of the detectable isotopes in the secondary fluid, proximity of the sampling point to the leak, and required cooling of high temperature samples for proper detection functioning.

The quantification of leakage can be accomplished if correlated with the known secondary system parameters such as flow, volume, activity, blowdown, and background activity along with the primary system activity.

Actual leak location can be identified only as being in the common barrier area between systems, or possibly, to a particular part of multibarrier systems by use of isolating valves or the type of detectable isotopes.

7.3 Containment Air Cooler Condensate Flow Collection for Leakage Detection

The condensate flow-monitoring method consists of measuring the flow rate of the liquid runoff from the drain pans under each containment air cooler unit. The increase of such condensate runoff can be indicative of increased vapor phase leakage into the containment[9].

The response and sensitivity characteristics of the instrumentation system used for condensate flow monitoring shall be estimated for determining the ability to detect one gpm leakage within one hour. The baseline of normal condensate flow shall be derived from the range of normal operating parameters anticipated including the expected normal leakage from both the RCPB and auxiliary systems which could affect normal condensate flow.

7.4 Reactor Coolant Inventory

The reactor coolant closed loop design of PWR plants

permits the maintenance of a coolant inventory which is constant except for controlled additions, controlled discharges, and uncontrolled leakage. Controlled coolant additions and discharges can be measured, recorded, and corrected to maintain the inventory balance. The resulting information is useful in evaluating the integrity of the RCPB. This surveillance method cannot generally be used on BWR plants with sufficient accuracy to be of value in detecting small RCPB leakage.

In establishing this method for PWR plants the following parameters must be at least considered:

(A) Density (temperature and pressure) of each fluid being measured.

(B) Water levels in pressurizer and all collection points.

(C) Duration of monitoring period.

7.5 Humidity Monitoring Leakage Detection

Humidity monitoring can detect the increased vapor content of air produced by the vapor phase portion of coolant leakage. Humidity detectors placed within the primary containment have the potential to detect leakage but suffer the quantitative uncertainty of the unknown proportion of liquid to vapor from any leak source. When used in large volume containment areas, the sensitivity may be on the order of several gpm[5,9]. These detectors cannot locate leakage except for area localization of the source, thus responding best in small contained volume areas. The response and sensitivity characteristics of the humidity detectors shall be considered in estimating system capability to meet the criterion of detecting a one gpm leakage within one hour. The baseline or normal specific humidity shall be based upon the range of normal operation parameters anticipated including the expected normal leakage rate.

7.6 Temperature Monitoring Leakage Detection

The sensitivity and system response time of this leakage detection method is highly dependent upon the following application conditions:

(A) The volume of space to be monitored by each temperature sensor.

(B) The thermal transport distance and conditions between the sensor and the potential leak locations.

(C) Potential heat losses from the measured volume.

(D) Normal temperature fluctuations expected in the absence of coolant leakage.

(E) The presence, or potential presence, of abnormal heat sources other than coolant leakage.

(F) The temperature sensor time constant (including the time interval between monitoring each point in multipoint sequential monitoring systems).

Multiple temperature sensor locations are usually necessary to monitor large volumes such as equipment rooms or containment building areas to provide useful leakage measurement sensitivity[8]. Temperature response and sensitivity can be optimized by mounting sensors in confined spaces such as relief valve or seal leakoff lines and contained spaces around piping and equipment, e.g., thermocouples attached to the metal sheathing of thermally insulated RCPB piping. False alarms are also minimized in this manner.

Differential temperature measurement may be used where ambient temperature can be measured entering and exiting from rooms or areas containing RCPB equipment and piping, e.g., with sensors located in heating and ventilating ducts. This application method minimizes the effects of inlet temperature variations and thus may increase measurement sensitivity, provided that measurement response time does not become too long. The same application principles described previously for absolute temperature measurement also apply to the differential temperature measurement.

7.6.1 Sensor Time Response Characteristics

Temperature sensor response characteristics expressed in time units that are quoted by instrument manufacturers are usually based on tests in moving fluids at stated flow conditions, most commonly, in water flowing at 3 ft/sec 0.914 m/s) past the sensor. Sensor response used for leakage detection shall be stated in terms of the anticipated fluid conditions that correspond to the intended measurement application. Sensor time response characteristics in a moving air stream will usually be several orders of magnitude longer than for the same sensor using 3 ft/sec 0.914 m/s) water flow velocity past the sensor.

7.6.2 Sensor Temperature Response to Coolant Leakage

Temperature sensor response to a one gpm coolant leakage rate shall be estimated for each unique measurement application by one of the following methods:

(A) Calculation of sensor response time from fabrication details of the sensor and the surrounding enclosed fluid conditions that result from coolant leakage.

(B) Correction and correlation of manufacturer sensor time response characteristics under stated fluid conditions to agree with the anticipated application fluid conditions.

(C) Test measurement of the overall temperature sensor system response under conditions which simulate the anticipated leakage detection application conditions.

Suggested procedures for methods (A) and (B) above are given in Reference (13) and Appendix B of this standard. Temperature sensors with remote electronic indication and trip devices shall have provisions for calibrating the trip device by insertion of a calibrating signal of known value. This method will usually permit ready identification of trip devices which are not functioning or are out of calibration.*

---

* A number of operating nuclear power plants have issued Abnormal Occurence Reports in which drift of setpoints for temperature switches (ranging from 0.5 to greater than 6 percent error) occurred in leakage detection applications. See USAEC office of Operations Evaluation Study: "Setpoint Drift in Nuclear Power Plant Safety-Related Instrumentation", OOE-ES-003, dated August 1974.

7.7 Primary Containment Pressure Monitoring

RCPB leakage will cause a pressure increase in the primary containment structure. The detection capability of a pressure monitoring system is on the order of large leakage because of the large size of the containment volume. Small leakage may cause pressure changes that fall within the range of normal containment pressure fluctuations for considerable periods of time.

Quantitative measurement of leakage by pressure monitoring methods is of questionable value for small leaks due to the large number of variables which can influence the measurement. Also this method provides no information on the leakage source location within the monitored containment volume.

7.8 Tape Moisture Sensors for Leakage Detection

The moisture sensitive tape method is a continuous monitoring system consisting of a sensing element which is normally placed next to the insulation of process piping. The element provides an electrical signal when activated by moisture (as produced by a leak) which may be used with an indicating device that generates an alarm signal. These sensors can quickly detect leakage from the piping on which they are mounted and fairly precisely localize the leak area. However, the amount of leakage cannot be measured. Typical design criteria are given in Appendix B.5.

7.9 Visual Observation

This is the most flexible of all monitoring methods, but detection sensitivity is heavily dependent upon the frequency of inspection and accessibility of equipment areas. The American Society of Mechanical Engineers Boiler and Pressure Vessel Code, Section XI, covers periodic mandatory inspection requirements of the RCPB. Inclusion of devices such as closed circuit television, temperature sensitive tapes and paint can be valuable aids in locating and identifying leakage sources. This method is not recommended as one of the principal monitoring systems; however, it can be used to augment other detection methods with respect to leak location.

TABLE 1
CAPABILITIES OF LEAKAGE MONITORING METHODS

Method	Leakage Detection Sensitivity	Leakage Measurement Accuracy	Leak Location
Sump Monitoring	G[a]	G	P[b]
Condensate Flow Monitors	G	F[c]	P
Radiogas Activity Monitor	F	F	F
Radioparticulate Activity Monitor	F	F	F
Primary Coolant Inventory[d] (Based on Makeup Flow Integrator)	G	G	P
Humidity-Dew Point	F	P	P
Tape Moisture Sensors	G	P	G
Temperature	F	P	F
Pressure	F	P	P
Liquid Radiation Monitor[e]	G	F	F
Visual[f]	F	P	G

[a] G (Good) - can generally be applied to meet intent of this standard if properly designed and utilized.
[b] F (Fair) - may be acceptable, marginal, or unable to meet intent of this standard depending upon application conditions and the number of measurement points or locations.
[c] P (Poor) - not normally recommended but might be used to monitor specific confined locations.
[d] For PWR during steady state conditions.
[e] For detection of intersystem leakage; may also be used for location function in sump or drain monitoring.
[f] Provided that the leakage area is visible.

8 REFERENCES

1. Code of Federal Regulations, Title 10, Part 50 - Licensing of Production and Utilization Facilities, Appendix A, "General Design Criteria for Nuclear Power Plants", Criteria 14, 31, 32, 33.
2. Code of Federal Regulations, Title 10, Part 50 - Licensing of Production and Utilization Facilities, Appendix B, "Quality Assurance Criteria for Nuclear Power Plants and Fuel Reprocessing Plants".
3. W. A. Maxwell, "A Survey of Leak Detection Methods for the On-Line Monitoring of BWR and PWR Loops", Southern Nuclear Engineering, Inc., SNE-36 (UC-37-4), October 1967.
4. R. L. Bell, "A Progress Report on the Use of Acoustic Emission to Detect Incipient Failure in Nuclear Pressure Vessel", Nuclear Safety, Vol. 15, No. 5, September-October 1974.
5. U.S. NRC Report NUREG-0017, "Calculation of Releases of Radioactive Materials in Gaseous and Liquid Effluents from Pressurized Water Reactors".
6. IEEE Std. 344, IEEE Recommended Practices for Seismic Qualification of Class IE Equipment for Nuclear Power Generating Stations, 1975.
7. Code of Federal Regulations, Title 10, Part 50 -Licensing of Production and Utilization Facilities, Appendix A, "General Design Criteria for Nuclear Power Plants", Criterion 64.
8. U.S. Atomic Energy Commission, Docket No. 50-254, November 29, 1973, Abnormal Occurence Report, B. B. Stephenson, Commonwealth Edison Company, Quad Cities Nuclear Station to J. F. O'Leary.
9. A. H. Klose and D. R. Miler, "Detection of Leaks in Inaccessible Areas of the Boiling Water Reactor Nuclear Power Plant", Conference Proceedings, U.S. Atomic Energy Commission CONF-671-11, January 1968.
10. U.S. Atomic Energy Commission, Docket No. 50-220, May 4, 1972, Description of Leak Detection System of Nine Mile Point Nuclear Station, F. J. Schneider, Niagara Mohawk Power Corporation to D. J. Skovholt.
11. J. M. Harrer, J. G. Becherly, "Nuclear Power Reactor Instrumentation Systems Handbook", Vol. 1 and Vol. 2, Office of Information Services, U.S. Atomic Energy Commission, 1973.
12. S. Glasstone and A. Sesonske, "Nuclear Reactor Engineering", D. Van Nostrand Company, Inc.
13. Rosemount Engineering Company Bulletin 9612, Rev. B, Appendices C and E.
14. A. J. Hornfeck, "Response Characteristics of Thermometer Elements", Transactions of the American Society of Mechanical Engineers, February, 1949.
15. W. H. McAdams, "Heat Transmission", McGraw-Hill Book Company, 1954.
16. USAEC Report WASH-1258, "Final Environmental Statement concerning proposed Rule-Making Action: Numerical guide for design objective and limiting conditions for operation to meet the criterion "as low as practicable" for radio-active material in light water cooled power reactor effluents". Wash. D.C., July, 1973.
17. US DOE COO/2974-2 UC-78, "Acoustic Monitoring Systems Tests at Indian Point Unit 1, Final Report" prepared by Westinghouse Electric Corporation (WCAP 9650).

THESE APPENDICES ARE INCLUDED FOR INFORMATION PURPOSES AND ARE NOT A PART OF THE STANDARD

APPENDIX A
LEAKAGE NOT INTO THE CONTAINMENT

A.1 Intersystem Leakage Detection

Process systems connected to the portion of the reactor coolant system pressure boundary that is inside containment up to the second system isolation valve should be monitored for the detection of intersystem leakage. If these systems or components are not isolated from the reactor coolant system during normal operation* the interface between components of these systems and secondary systems should be monitored for intersystem leakage (i.e., heat exchangers in reactor coolant cleanup or chemical and volume control systems). If these systems are isolated from the reactor coolant system during normal operation then the leakage into these systems (past the isolation valves) should be monitored. Table A-1 idenifies the systems or components for PWRs and BWRs connected to the reactor coolant system that should be monitored for the detection of intersystem leakage. Table A-2 identifies the typical methods used for intersystem leakage detection.

---

*As limited in Section 6.6 of the Standard.

TABLE A-1
SYSTEMS AND COMPONENTS CONNECTED TO REACTOR COOLANT PRESSURE BOUNDARY

A. Pressurized Water Reactors
1. Accumulators
2. Safety Injection Systems (High and Low Pressure)
3. Pressurizer Relief Tank
4. Secondary Side of Steam Generators
5. Residual Heat Removal System (Inlet and Outlet)
6. Secondary Side of Reactor Coolant Pump Thermal Barriers
7. Secondary Side of Residual or Decay Heat Removal Heat Exchangers
8. Secondary Side of Letdown Line Heat Exchangers
9. Secondary Side of Reactor Coolant Pump Seal Water Heat Exchangers

B. Boiling Water Reactors
1. Safety Injection Systems (High and Low Pressure Core Spray and Coolant Injection Systems)
2. Residual Heat Removal System (Inlet and Outlet)
3. Reactor Core Isolation Cooling System
4. Steam Side of High Pressure Coolant Injection (BWR-4 only)
5. Secondary Side of Reactor Water Cleanup System Heat Exchangers
6. Secondary Side of Reactor Coolant Pump Integral Heat Exchangers
7. Secondary Side of Residual Heat Removal Heat Exchangers

TABLE A-2
TYPICAL INTERSYSTEM LEAKAGE DETECTION METHODS

Methods
1. Lifting of Relief Valves
2. Leak-Off Temperature or ΔT
3. Tank or Sump Level Indication
4. Flow Rate or Δ Flow or Level Rate
5. Pressure or ΔP
6. Coolant Sampling
7. Radiation Monitoring
8. Cooling Water Temperature or ΔT

A.2 Primary Coolant Leak Detection Outside of the Containment

Boiling water reactor plants have some systems outside the containment, such as feedwater and main stream lines, which contain reactor coolant at or near reactor operating pressure and temperature. These systems are not a part of the RCPB, although they contain reactor coolant. Leakage in such connecting systems is presently detected by methods similar to those described in Section 7.0 of this standard. Such detection methods may initiate a safety related isolation from the reactor coolant system. Table A-3 indicates potential reactor coolant leak sources outside of the containment structure and leak detection methods used for their detection. This type of leakage is not covered by the scope of this standard (see definition of RCPB).

TABLE A-3
POTENTIAL LEAK SOURCES AND TYPICAL LEAK DETECTION METHODS

A. Potential Leak Sources

1. Residual Heat Removal System
2. Reactor Core Isolation Cooling System
3. Feedwater System Outside Containment
4. Main Steam Lines and Equipment Outside Containment

B. Typical Leak Detection Methods

1. Leakoff Temperature or ΔT
2. Airborne Radioactivity
3. Sump Level, Flow Rate, or Δ Flow or Level Rate
4. Humidity Measurement
5. Pressure or ΔP
6. Liquid Radiation Monitor
7. Cooling Water T or ΔT
8. Cooling Air T or ΔT

APPENDIX B
SUGGESTED METHODS AND PROCEDURES FOR LEAKAGE DETECTION SYSTEMS

B.1 Sump Level and Flow Measuring Methods

B.1.1 Purpose

This Appendix outlines suggested instrumentation methods for measuring sump level and leakage flow monitoring. It contains general equations for measurement resolution and response time. These equations are based on simplifying assumtions and may require some modification for a given configuration.

B.1.2 Methods

B.1.2.1 Analog Level Transmitters and Level Memory Method

(A) General Description

This method uses a sensing well as a hydraulic level memory device and compares present with previous values on timed cycles. Data recording and alarms for several measurements are provided. See Figures B-1 and B-2 and general equations in B.1.3.1 for this method.

(B) Details of Operation

Leakage inside the containment, other than identified leakage which is piped to the equipment drain sump, drains into the containment sump as unidentified leakage. The sump level is measured continuously by level transmitter LT-1A. The output of this transmitter is recorded on one pen of the dual pen recorder LR-1/FR-2.

A hydraulic level memory is provided by using a sensing well which is piped to the sump at an elevation below the lowest operating level in the sump. The well is isolated from the sump by solenoid-actuated valve, KV-3 which is operated by time-cycle controller, KC-3. Valve KV-3 is closed most of the time. However, it opens periodically to equalize the level in the sensing well with the level in the sump. The level in the sensing well is measured by level transmitter LT-1B. The time period between valve KV-3 closing and KV-3 opening is referred to hereafter as the measurement period, T. The output of level transmitter LT-1A is adjusted by electronic converter LY-1 to compensate for any nonlinear level-to-volume characteristics of the sump, if applicable.

The difference in the outputs from electronic converter LY-1 and level transmitter LT-1B is provided continuously by subtracting relay, LDY-2. The output of LDY-2 is transmitted to gain relay FY-2A, which is used to calibrate the loop by compensating for the measurement period and sump dimensions. The output of FY-2A, at the end of each measurement period, provides a pseudo-instantaneous leak rate measurement. This output is recorded on pen 2 of recorder LY-1/FR-2.

The sump level is controlled by electronic level

S67.03

Standard for Light Water
Reactor Coolant Pressure
Boundary Leak Detection

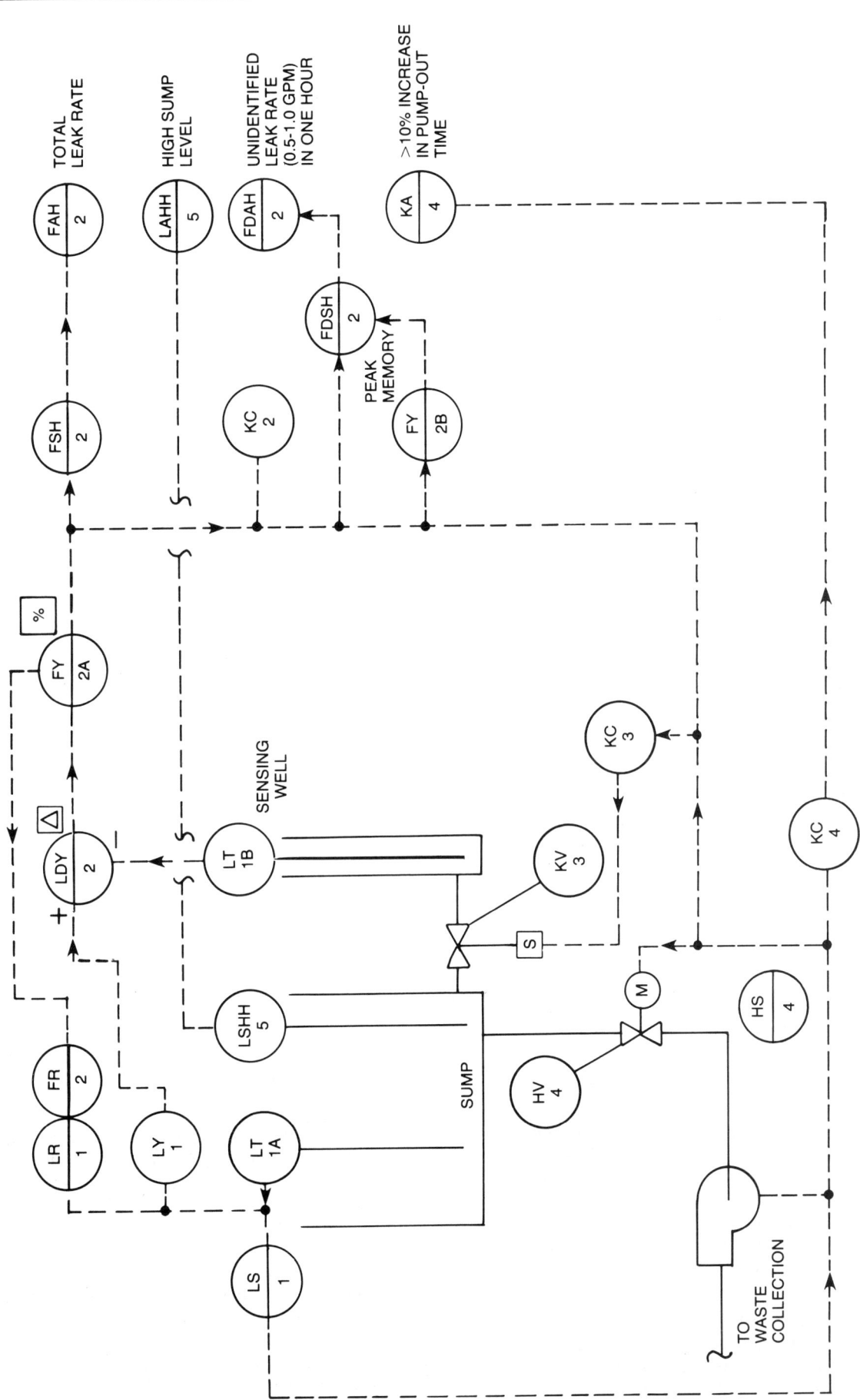

Figure B-1. Analog Level Transmitters and Level Memory

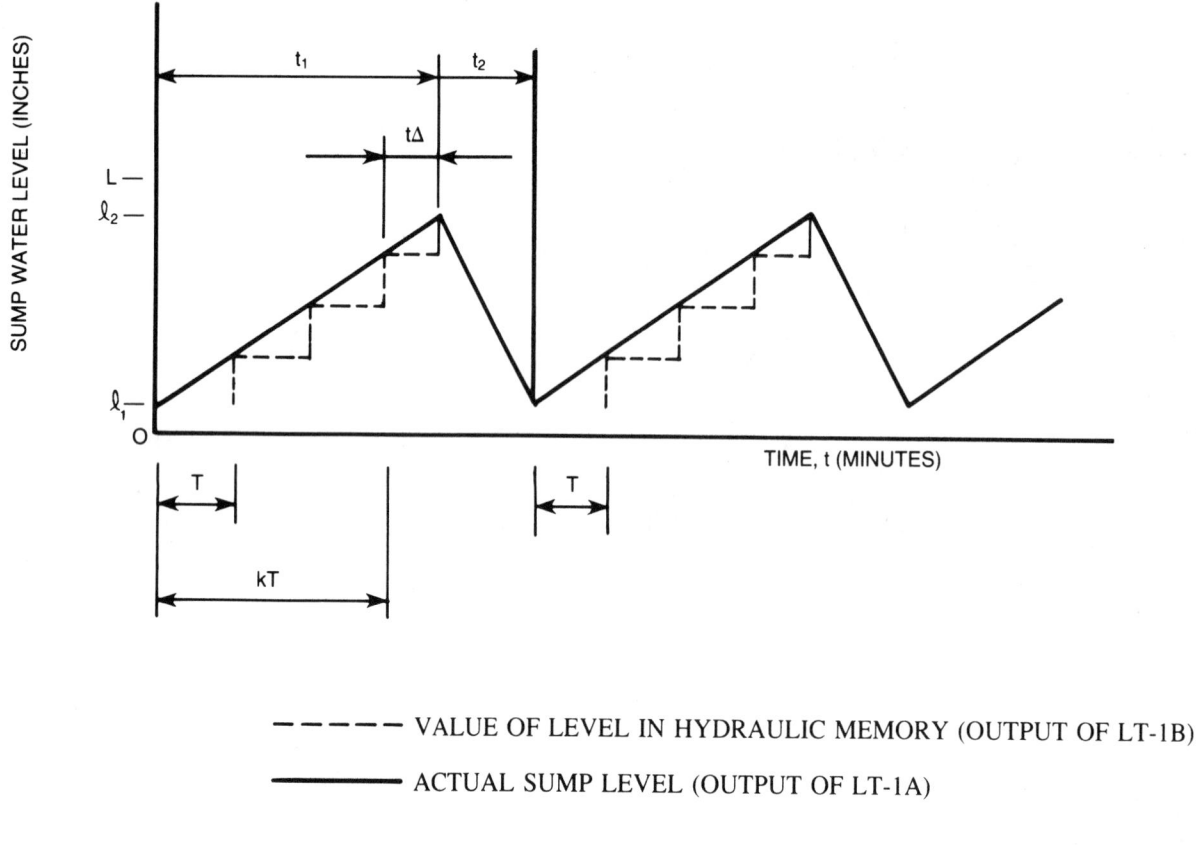

- - - - - VALUE OF LEVEL IN HYDRAULIC MEMORY (OUTPUT OF LT-1B)

────── ACTUAL SUMP LEVEL (OUTPUT OF LT-1A)

L = TOTAL SUMP CAPACITY
ℓ_2 = SUMP LEVEL AT WHICH LS-1 TURNS ON THE SUMP PUMP
ℓ_1 = SUMP LEVEL AT WHICH LS-1 TURNS OFF THE SUMP PUMP
T = MEASUREMENT PERIOD
t_1 = SUMP FILL TIME
t_2 = SUMP DRAIN TIME
$t\Delta$ = TIME FROM LAST LEVEL MEMORY RESET PRIOR TO BEGINNING OF SUMP PUMP DRAIN

Figure B-2. Typical Sump Fill/Drain Operation Curve

switch LS-1. When the level in the sump increases to the upper setpoint l_2 LS-1 opens valve HV-4 and turns the sump pump on. When the sump level decreases to low level l_1, LS-1 resets which turns the sump pump off, and momentarily opens valve KV-3, resetting the sensing well. Also, the peak-memory device, FY-2B, is reset at this time. Thus, the pseudo-instantaneous and increase leak rate measurement cycles are restarted. The sump and level memory levels are shown as a function of time in Figure B-2.

One alarm branch, FDAH-2, monitors the increase in the rate of unidentified leakage into the sump. This branch provides an alarm when the increase in leak rate exceeds 0.5 gpm to 1.0 gpm within one hour. The output of FY-2A is passed to peak-memory device FY-2B, which remembers the highest value attained by the input until its memory is reset. However, the input to FY-2B is interrupted by a time-delay relay that is activated once every 40 to 60 minutes by time-cycle controller KC-2. The same timer opens valve KV-3 momentarily, and then closes valve KV-3 to start a new measurement cycle. The time delay relay is adjusted so that as soon as the first measurement period is completed the input to FY-2B is interrupted. In this way the pseudo-instantaneous leak rate for the first measurement period is retained by FY-2B. The leak rate measuring cycle is repeated, during which time the output of

FY-2B remains constant at the highest value during the first measurement period. Receiver switch FDSH-2 substracts the output of FY-2B from the output of FY-2A. The resulting difference represents the increase in leak rate. If this difference exceeds the set point of 0.5 to 1.0 gpm, FDSH-2 actuates a high unidentified leak rate alarm, FDAH-2.

Another alarm branch, FAH-2, monitors the total rate of identified and unidentified leakage into the sump. This branch has receiver switch FSH-2, which receives the output from FY-2A. The set point for FSH-2 is set to correspond to the total leak rate limit defined in the Plant Technical Specification (typically 5 gpm). If this set point is exceeded, FSH-2 actuates a high leak rate alarm, FAH-2.

In addition to those described above, this system has the following features:

(1) The sump pump may be started and stopped by hand switch HS-4.

(2) An additional level switch, LSHH-5, which is independent of level transmitter LT-1A, actuates an alarm in case of pump failure.

(3) An additional timer, KC-4, is provided to generate an alarm in the event that the pump-out time exceeds the normal pump-out time plus 10 percent.

B.1.2.2 Integrated Level Switches Method

This method employs the use of several level switches with slightly different set points. An average leak rate is determined by measuring the time it takes for the sump level to increase from one set point to the next, taking into account the identified leak rate and the sump geometry. One of the following methods may be used:

(1) Level transducer with electronic switches

(2) Individual independent level switches, or

(3) Float-type transducers with magnetically coupled switches

The general equations related to this method are described in B.1.3.2.

B.1.2.3 Drain Pan with Monitoring of Flow into Sump Method

If the sump design can be coordinated before construction, it would be possible to install a shallow drain pan just below the floor level in which the sump is installed. This would collect all of the leakage coming into the sump and direct it to a central point in the drain pan that would allow the total sump flow to pass through a 0-10 gpm flowmeter. Depending on the type of flowmeter, it may be necessary to provide a trap so that the meter is filled. This method will provide an actual instantaneous measurement of the leakage flowing into the sump. With additional instruments such as those described above, the increase in leak rate can be determined and the sump emptied.

B.1.2.4 Pump Controlled by Timer and Level Switch Method

For this method it is necessary that the sump pump discharge flow be accurately measured. This method employs a timer to turn the sump pump on and a level switch to turn it off. By measuring the time it takes to pump the sump down and the pump discharge flow rate, an average leak rate can be determined. The pump frequency established by the timer must be selected such that the requirements of Section 7.1 of this standard can be satisfied. The general equations related to this method are described in B.1.3.3.

B.1.2.5 Sump Level or Flow Monitoring with Microprocessor Method

This method employs a single level or flow measuring device and a microprocessor. The microprocessor can be programmed to do all of the memory comparison, pump control, recording display, alarm functions described above. In addition, the microprocessor can simplify the calibration procedures for the leak detection systems.

NOTE
Presently supplied radiation monitors usually include field located microprocessors that can be programmed for these additional functions.

B.1.3 General Equations

B.1.3.1 Analog Level Transducer and Level Memory

(A) Definitions

Q	= Sump capacity: gallons
q	= Actual volume of liquid in the sump: gallons
L	= Sump level (depth of sump): inches
L_{f_s}	= Span of level device: inches
T	= Measurement period (period of time that the sump level is measured to determine an average leak rate): minutes
m	= Percent of level capacity to which sump is allowed to fill (between high & low set points): %
p	= Pump discharge flow rate: gpm
Δl_i	= Change in level during ith measurement period: inches
R	= Resolution (degree to which measuring system can detect leak rate): gpm
e	= Level measurement loop error: ±% of full scale
E	= Level measurement loop error: ±inches
x	= Leakage rate (when assumed constant): gpm
$\bar{x}_{im}$	= Measured average leak rate for the ith measurement period: gpm

x_i = Actual average leak rate for ith measurement period: gpm

t_1 = Sump fill time: minutes

t_2 = Sump drain time: minutes

k = Number of measurement periods that are completed before the pump starts running: integer

$t\Delta$ = Time from last level memory reset prior to beginning to drain the sump: minutes

NOTE
For metric units, substitute litres per second for gpm, meters for inches, and seconds for minutes in these definitions.

(B) Assumptions

(1) The measurement period starts immediately after the sump pump stops.

(2) The error in determining the measurement period, T, is not significant.

(C) Expression for Measurement Error

In general, the level measurement loop error, e, may be expressed as:

$$e = \left[\sum_{i=1}^{n} (e_i)^2\right]^{1/2} \quad (B-1)$$

where:

e_i = accuracy of the ith instrument used in the level measurement loop (± % of span)

n = number of instruments used in the level measurement loop (integer)

Since, for this method, the hydraulic level memory is allowed to equalize with the actual sump level when the sump is being pumped down, the accuracy of the level switch controlling the sump pump will have no effect on the total system error for average leak rate detection. Therefore, it is not necessary to include the accuracy of the level switch in the above equation for level measurement loop error.

(D) Resolution

The measured average leak rate for the ith measurement period may be expressed as follows:

$$\bar{x}_{im} = \bar{x}_i \pm R \quad (B-2)$$

where the resolution, R, may be expressed as

$$R = \frac{qe}{100T} \quad (B-3)$$

As indicated in this standard, a resolution of 1.0 gpm or better is required, therefore,

$$R \leq 1.0 \quad (B-4)$$

The maximum number of gallons that can be collected in the sump in one measurement period may be expressed by:

$$q_{max} = \frac{M_{max}Q}{100} \quad (B-5)$$

When the sump is filled to the high level set point in one measurement period, Equations (B-3) and (B-5) give a maximum value for resolution as follows:

$$R_{max} = \frac{M_{max}Qe}{10,000T} \quad (B-6)$$

Substituting Equation (B-6) into Equation (B-4) gives the following limiting equation for the measurement period, T:

$$T \geq \frac{M_{max}Qe}{10,000} \quad (B-7)$$

(E) Level Change to Measurement Error Relationship

The sump level for this method is allowed to increase until it reaches the high level, then the pump is activated and will run until the sump level drops down to the low level. Prior to primary pump activation, the change in sump level for a given measurement period may be expressed as:

$$\Delta l_i = \frac{\bar{x}_i \, TL}{Q} \quad (B-8)$$

The level measurement loop error may be expressed as:

$$E = \frac{eL_{fs}}{100} \quad (B-9)$$

Note that the above equation is equivalent to the resolution expressed in terms of inches. Therefore, to be detectable, a level change must be greater than the level measurement error.

$$\Delta l_i > E \quad (B-10)$$

Substituting Equations (B-8) and (B-9) into equation (B-10) gives:

$$\frac{\bar{x}_i \, TL}{Q} > \frac{eL_{fs}}{100} \quad (B-11)$$

or

$$T > \frac{eQL_{fs}}{\bar{x}_i L \, 100} \quad (B-12)$$

and

$$\bar{x}_i > \frac{eQL_{fs}}{LT \, 100} \quad (B-13)$$

(F) Determining the Measurement Period, T

In order to meet the requirements of this standard, the measurement period, T, must be determined carefully. The following equations are provided for this purpose. Referring to the Typical Sump Fill/Drain Operation (See Figure B. 2), it can be seen that $t\Delta$ is considered dead time so far as leak detection is concerned since it is not a part of a complete measurement cycle. Likewise, t_2 is leak detection dead time since the pump is draining the sump at this time. Any leakage that may occur during these two time periods ($t\Delta$, t_2) must be taken into account when determining the measurement period, T. To accomplish this, the dead time plus the measurement period must be less than 60 minutes to be able to detect a one gpm increase of leakage in one hour. This equation may be written as:

$$t\Delta + t_2 + T < 60 \quad (B\text{-}14)$$

or

$$T < 60 - t_2 - t\Delta \quad (B\text{-}15)$$

Assuming that the leak rate is constant (x = constant), we have:

$$t_1 = \frac{mQ}{100x} \quad (B\text{-}16)$$

$$t_2 = \frac{mQ}{(p-x)100} \quad (B\text{-}17)$$

Since $t\Delta = t_1 - kT$, Equation (B-15) may be expressed as:

$$T < 60 - \frac{mQ}{(p-x)100} - \frac{mQ}{100x} + kT \quad (B\text{-}18)$$

$$(1-k)T < \frac{6000(p-x)(x) - mQp}{(p-x)(x)100} \quad (B\text{-}19)$$

or

$$T < \frac{mQp - 6000x(p-x)}{x(p-x)(k-1)} \quad (B\text{-}20)$$

B.1.3.2 Integrated Level Switches

(A) Definitions

The definitions of Paragraph B.1.3.1A apply. The following definition also applies to this method:

N = Number of level switches employed in an integrated level swtich method.

(B) Assumptions

Assume that the leak rate is directly proportional to the associated change in sump level.

(C) Expression for Measurement Error

Equation (B-1) of Paragraph B.1.3.1C applies.

(D) Resolution

To detect an unidentified leak rate of one gpm within one hour, the difference in set points of adjacent level switches must not exceed the level change that would occur for a one gpm leak rate at the end of a one hour period.

$$\Delta l < \frac{(1.0 \text{ gpm})(60 \text{ min}) L}{Q} \quad (B\text{-}21)$$

or

$$\Delta l < \frac{60L}{Q} \quad (B\text{-}22)$$

(E) Level Change to Measurement Error Relationship

Equations of Paragraph B.1.3.1E apply.

(F) Determining the Number of Level Switches Required to Satisfy Standard

In order for this method to detect a one gpm leak rate within one hour, a sufficient number of switches must be provided such that the difference in set points of adjacent level switches is equal to the level change that would be caused by a one gpm leak rate in one hour.

$$N \geq \frac{Q}{60} \quad (B\text{-}23)$$

B.1.3.3 Equations for Pump Controlled by Timed and Level Switch

(A) Definitions

The definition of Paragraph B.1.3.1A apply.

(B) Assumptions

(1) The sump pump discharge flow rate is constant and is independent of sump level.

(2) The error in measuring pump down time is not significant.

(C) Expression for Measurement Error

The equations of Paragraph B.1.3.1C apply.

(D) Resolution

Resolution may be expressed as Equation (B-6) of Paragraph B.1.3.1.D.

$$R_{max} = \frac{mQe}{10,000 \, t_2} \quad (B\text{-}24)$$

for the sump pump controlled by level switches method. For the sump pump controlled by a timer and a level switch method, the sump could possibly fill completely. For this case, m = 1.

B.2 Condensate Flow Monitors

Containment air coolers have drain pans that duct containment atmosphere condensate to a sump. Other liquids coming to the same sump can come from liquid sources in the containment, e.g., service water. It is useful to measure the chiller condensate in order to distinguish it from these other liquids because the principle source of humidity in the containment is primary coolant flashing. One method of measuring condensate flow is to provide the drain pan with level switches. Then the measurement is made in the same manner as for the sumps described previously. The pans usually have gravity flow to the sump; thus, a valve can be used to control when the pan is emptied.

Depending on the cooler size and expected flow, it may be possible to find a flow meter for this application. However, it is cautioned that these are relatively small flows and the flow meters may be susceptible to becoming plugged.

B.3 Radioactivity Monitoring Methods

B.3.1 Airborne Radiation Monitoring Coolant Leakage Measurement

The instantaneous containment atmosphere radioactivity concentration due to reactor coolant leakage, for a single radioisotope, may be expressed as:

$$V\frac{dA}{dt} = LC - \lambda AV - P_f LC - QA \quad (B\text{-}25)$$

where
- V = containment free volume, cc
- A = concentration of radioisotope in the containment atmosphere, μCi/cc
- C = concentration of radioisotope in the reactor coolant, μCi/cc
- L = leakage rate of reactor coolant to containment atmosphere, liters/min.
- λ = decay constant of isotope, min
- P_f = plateout factor, fraction of leaking radioisotope that is removed by plateout, dimensionless
- Q = atmosphere removal rate (purge rate) of containment, cc/min (Q is zero for a containment without purge)

Simplifying equation (B-25) yields:

$$\frac{dA}{dt} = \frac{LC}{V}(1 - P_f) - \lambda A - \frac{QA}{V} \quad \text{(B-26)}$$

Solution of this differential equation gives the isotope activity concentration in the containment atmosphere as a function of time for a given leakage rate.

For $A = A_o$ at $t = 0$, the solution is

$$A = \frac{LC(1-P_f)}{\lambda V+Q} - \left[\frac{LC(1-P_f)}{\lambda V+Q} - A_o\right] e^{-(\lambda + \frac{Q}{V})t} \quad \text{(B-27)}$$

This equation may be used to estimate the radioactivity transient in the containment atmosphere due to a reactor coolant leak, and thus form the basis for estimating monitor response to a leak.

B.3.2 Airborne Radioactivity Monitor Sensitivity

Coolant leakage from the RCPB that does reach the containment atmosphere vaporizes and is diluted by the air in the containment free volume. The containment air is continuously sampled, ducted through detectors designed to measure radioactivity in the form of gases, particulates, iodides, or a combination of these and returned to containment. Typical minimum detectable concentrations at a 95 percent confidence level in a 1.0 mr/hr Co-60 gamma field for detectors currently available are as follows:

Detector Type	Concentration	Isotope***
a. Radiogas	1×10^{-6} μCi/cc	Kr-85
b. Radioparticulate	1×10^{-10} μCi/cc*	Cs-137
c. Radioiodine	1×10^{-9} μCi/cc**	I-131

B.3.3 Liquid Radiation Monitor Sensitivity

A typical liquid radiation monitor is capable of responding to a minimum detectable concentration of 10^{-5} μCi/cc of Zn-65 in a 1.0 mr/hr Co-60 gamma field with a 95 percent confidence level.

B.3.4 Typical System Factory Checkout Calibration and Field Installation Specification for Radiological Monitoring Systems

Before delivery, the complete radiological monitoring system should be calibrated and tested by the supplier in the order specified below to ensure that the complete system and all components conform to this specification.

The test procedure should provide a step-by-step method of verifying that all adjustments have been made, and that all functions operate properly. High and low limits shall be specified for all adjustments, and should be listed on a system checkout sheet. The actual value of the adjusted variable should be recorded on the system checkout sheet.

Any of the tests or calibrations may be witnessed by the reactor owner's representative at his discretion.

(A) ELECTRICAL CHECKOUT - The electrical checkout should include point-to-point continuity tests and electrical insulation tests in accordance with the requirements of Section 20-5.3.4.1 and 20-5.3.4.2 of ANSI C37.20, Standard for Switchgear Assemblies. However, coaxial and shielded cables should not have high voltages applied to them, nor should they be tested with megohmmeters. The manufacturer should be responsible for protecting instruments and devices that may be damaged by high voltage tests.

(B) OPERATIONAL TESTS - The completely assembled, piped, and wired racks should be tested at the factory in the presence of witnesses. (It is recommended that prior agreement be reached that any costs arising from these tests, including the repair of leaks or the replacement of defective materials, will be borne by the manufacturer).

(C) CALIBRATION - After the instruments have been accepted by the witness of (A) & (B) above, but before shipment, each sampler-detector-readout system should be factory calibrated with appropriate liquid or gas sources. Aerosol detectors are calibrated by cross-referenced standards. The sources should be traceable to the National Bureau of Standards and calibrated according to ASTM D 1690-67, D 2459-72, and D 2577-72, as applicable. This calibration should consist of:

(1) Operating the system to reach thermal stability.

(2) Recording the background counting rate in a specified Co-60 gamma field, usually 1.0 mR/hr.

(3) Introducing a laboratory-calibrated concentration of the reference or control isotope into the sampler, and recording the counting rate above background.

(4) Recording pertinent environmental conditions and operating parameters, such as detector high voltage, spectrometer and amplifier gain settings, ratemeter time constant, ambient temperature, sampling system temperature, sampling system pressure, and actual gamma field intensity.

The calibration procedure should be performed with three concentrations. The lowest should produce counting rate in the lowest decade of response, the median shall be approximately midscale between the lowest and highest concentrations, and the highest

---

* For a one hour collection time or less with a moving filter tape at 1 to 1.5 inches per hour tape speed.

** For a one hour collection time or less with a fixed filter.

***These referenced isotopes are used for instrument calibration, therefore, monitor sensitivity should be cross-calibrated to the isotopes listed in Paragraph 7.2.1C.

should produce a counting rate in the last decade of response.

A plot of the concentration of the reference isotope in terms of micro-curies/cc versus the net counting rate produced should then be drawn on log-log scales. This plot should be used as a calibration curve in the field. The three concentrations should lie on a straight line within one standard deviation of the net count rate. Following the dispersion of each concentration into the sampler, a sample of the concentration should be extracted into a glass vial for separate verification of each concentration, at a place designated by the reactor owner or his representative.

The following procedure should be used to demonstrate that the system meets radiological sensitivity specifications:

(1) Determine the standard deviation of the background counting rate under the specified operating conditions (including 1.0 mR/hr Co-60 gamma field) by taking the square root of the background counting rate.

(2) Multiply this standard deviation of the background counting rate by the specified factor, usually 2.56 (2.56 corresponds to 99 percent statistical confidence; for 95 percent confidence, multiply by 1.96).

(3) From the calibration curve plotted as described above, determine the net counting rate resulting from the specified minimum detectable concentration. The counting rate so determined must be greater than 2.56 (or 1.96 if so specified) times the standard deviation of the background counting rate for the system sensitivity to be acceptable.

An auxiliary Co-60 or Cs-137 calibration source should be supplied as a field calibration aid. For the scintillation detector instrument, the calibration sources should be taped to the detector crystal and the detector inserted into its shielded sampler that has been purged, and the resultant counting rate data recorded on the calibration curve.

The position of the calibration source should be clearly recorded by the use of identifying reference marks on the sampler and on the source. The system should be calibrated in the field by using the auxiliary calibration source data. Calibration data and decay curves for the auxiliary calibration sources should be supplied with the system manuals.

After the auxiliary calibration source is removed, the detector is replaced in the purged shielded chamber, and the check source is actuated. The counting rate data resulting from the actuation of check sources should also be recorded on the calibration curve.

(D) SYSTEM OPERATION - The radiological monitoring system shall be calibrated and operated as a system at the manufacturer's plant. The system should be operated for 200 hours continuously, and the last 100 hours must be without any failure or drift greater than specified. During the first 100 hours the system shall be thermally cycled twice each 24 hours to 120 ± 10°F and 70 ± 10°F with at least 2 hours at each nominal temperature. The detectors shall be placed in a radiation field that causes the readout to indicate in the second or third decade of the detector range. Compliance with the drift specification may be checked by recalibration.

(E) FIELD CHECKOUT AND CALIBRATION - The onsite checkout should consist of using the auxiliary calibration source and/or the check sources to check the calibration of each detector and, unless exception has been granted, to exercise all equipment functions such as alarms and remote readouts. When the system and accessory instruments meet the functional requirements and the performance guarantees, the system may be accepted. (It is recommended that the manufacturer be requested to submit a quotation to provide a field engineer, if needed, to repair any shipping damage, make final connections, and check out and calibrate the system.)

B.4 Humidity Monitoring

B.4.1 Calculation of Leakage Rate

The instantaneous change in containment atmosphere specific humidity due to coolant leakage can be expressed by the following relation:

$$M\left(\frac{dw}{dt}\right) = xL - \sum_{i=1}^{n} c_i \qquad (B-28)$$

where

M = total mass of atmosphere in containment
w = containment atmosphere specific humidity, mass of water vapor/mass of atmosphere
L = total leakage flow rate into the containment, mass/time
x = fraction of leak that flashes to vapor, lbm-vapor/lbm-liquid
c_i = condensation rate of "ith" containment air cooling unit, mass/time
n = number of operating containment air cooling units

The fraction of leak that flashes to vapor can be determined from the isenthalpic relation:

$$h = xh_g + (1 - x) h_f \qquad (B-29)$$

where

h = enthalphy of reactor coolant in RCPB, Btu/lbm, (J/Kg)
h_g = enthalpy of saturated vapor at containment temature, Btu/lbm (J/Kg)
h_f = enthalpy of saturated liquid at containment temperature, Btu/lbm (J/Kg)

B.4.2 Estimation of Humidity Transient

The following derivation is useful in estimating the humidity transient in a containment due to an RCPB leak and thus forms a basis for estimating monitor's response to a leak.

The condensation rate in a containment air cooling unit can be estimated by:

$$q = m(h_1 - h_2) - mh_{w2}(w_1 - w_2) \quad (B\text{-}30)$$

where

- q = heat removal rate of air cooling unit, Btu/min, (Watts)
- m = air mass flow rate in cooling unit, lbm-air/min, (Kg/sec)
- h_1 = enthalpy of air entering cooling unit, Btu/lbm, (J/Kg)
- h_2 = enthalpy of air at condensing temperature, Btu/lbm, (J/Kg)
- h_{w2} = enthalpy of saturated liquid (condensate) at leaving temperature, Btu/lbm, (J/Kg)
- w_1 = specific humidity of air entering air cooling unit, lbm-water vapor/lbm-air, (Kg vapor/Kg air)
- w_2 = specific humidity of air leaving air cooling unit, lbm-water vapor/lbm-air, (Kg vapor/Kg air)

This equation can be solved for w_1 by trial and error by assuming a final temperature, which determines the outlet conditions h_{w2}, h_2 and w_2, and seeking a balanced equation. Upon solution of this equation the condensation rate, c, assuming no reevaporation is:

$$c = m(w_1 - w_2) \quad (B\text{-}31)$$

It can be shown that over any practical range of concern the above relationship can be replaced by the mathematical approximation,

$$c = \alpha w_1^2 + \beta w_1 - \gamma \quad (B\text{-}32)$$

Where: α, β and γ are constant coefficients determined from a fit to parametric data of w_1 versus c, for the conditions of interest, calculated using Equations (B-30) and (B-31). These coefficients require derivation for each containment configuration.

Substituting Equation (B-32) into equation (B-28) with $w_1 = w$ and for equal capacity cooling unit gives:

$$M \frac{dw}{dt} = xL + \alpha w^2 - \beta w + \gamma \quad (B\text{-}33)$$

The solution to this equation for $w = w_0$ at $t = 0$ is,

$$w = \frac{\beta - K \left[\frac{1\beta + \sqrt{K}}{\Theta - \sqrt{K}} - e^{\left(\frac{K}{m}\right)t} \right]}{\left[\frac{1}{\Theta} - e^{\left(\frac{K}{m}\right)t} \right] 2\alpha} \quad (B\text{-}34)$$

for $w \leq \frac{\beta - \sqrt{K}}{2\alpha}$

where

$$k = \beta^2 - 4\alpha(\gamma + xL)$$

$$\Theta = \frac{2\alpha w_0 - \beta - \sqrt{k}}{2\alpha w_0 - \beta + \sqrt{k}}$$

Equation (B-34) may be used to estimate the specific humidity transient in the containment due to an RCPB leak. The designer is cautioned that where other effects, such as mixing, condensation, etc. are significant, these factors should also be included in the equations used.

B.5 Tape Moisture Sensor Design Criteria

The following typical design criteria are recommended for moisture sensitive tape leak detection systems:

(A) The tapes should be manufactured with halogen free chemicals and conductors should be stainless steel.

(B) The resistance between two strips of conductor should be at least 10 megohms when the tape is dry. The resistance of the tape should be not more than 100,000 ohms when moistened in one spot with one drop of water. The length of separately monitored detector strips should not exceed approximately 20 feet (6 meters).

(C) The tape shoule not be applied directly to an uninsulated surface if the surface temperature exceeds the boiling point of water. Weep holes in insulation and cover materials should be provided so that water leakage from the pipe or vessel will drain on the tape. The tape should be qualified for the ambient environmental conditions of the application.

(D) A control unit which performs a continuity check each 30 seconds or less and is capable of identifying any single fault and of annunciating a zone alarm condition should be provided.

Figure B-3 illustrates a typical installation method. Other sensor and system designs are available.

B.6 Estimating Temperature Sensor Time Constant from Sensor Configuration and Fluid Conditions

Realistic time constant estimates for thermocouples, resistance thermometers and other types of temperature sensors can be computed since the sensor time constant is the active sensor mass heat capacity divided by the heat transfer rate to the sensor mass. This relationship is given by the following expressions:

$$\tau = \frac{H}{Q} \quad (B\text{-}35)$$

$$\tau = \frac{MC_{pm}}{hA} \quad (B\text{-}36)$$

$$\tau = \frac{VC_{pv}}{hA} \quad (B\text{-}37)$$

where

- τ = time constant: h (hrs), s (sec)
- H = heat capacity: $\frac{BTU}{°F}$ (joules/°C, j/°C)
- Q = heat transfer: $\frac{BTU}{hr\,°F}$ (watts/°C, w/°C)
- h = heat transfer coefficient: $\frac{BTU}{ft^2\,°F\,hr}$ $\left(\frac{w}{m^2\,°C}\right)$
- A = heat transfer area: ft^2 (m^2)
- M = mass of junction or sensor material: lb, (kg)
- C_{pm} = mass basis specific heat of junction material: $\frac{BTU}{lb\,°F}$ $\left(\frac{J}{Kg\,°C}\right)$
- C_{pv} = volume basis specific heat of junction material: $\frac{BTU}{ft^3\,°F}$ $\left(\frac{J}{M^3\,°C}\right)$

S67.03

Standard for Light Water Reactor Coolant Pressure Boundary Leak Detection

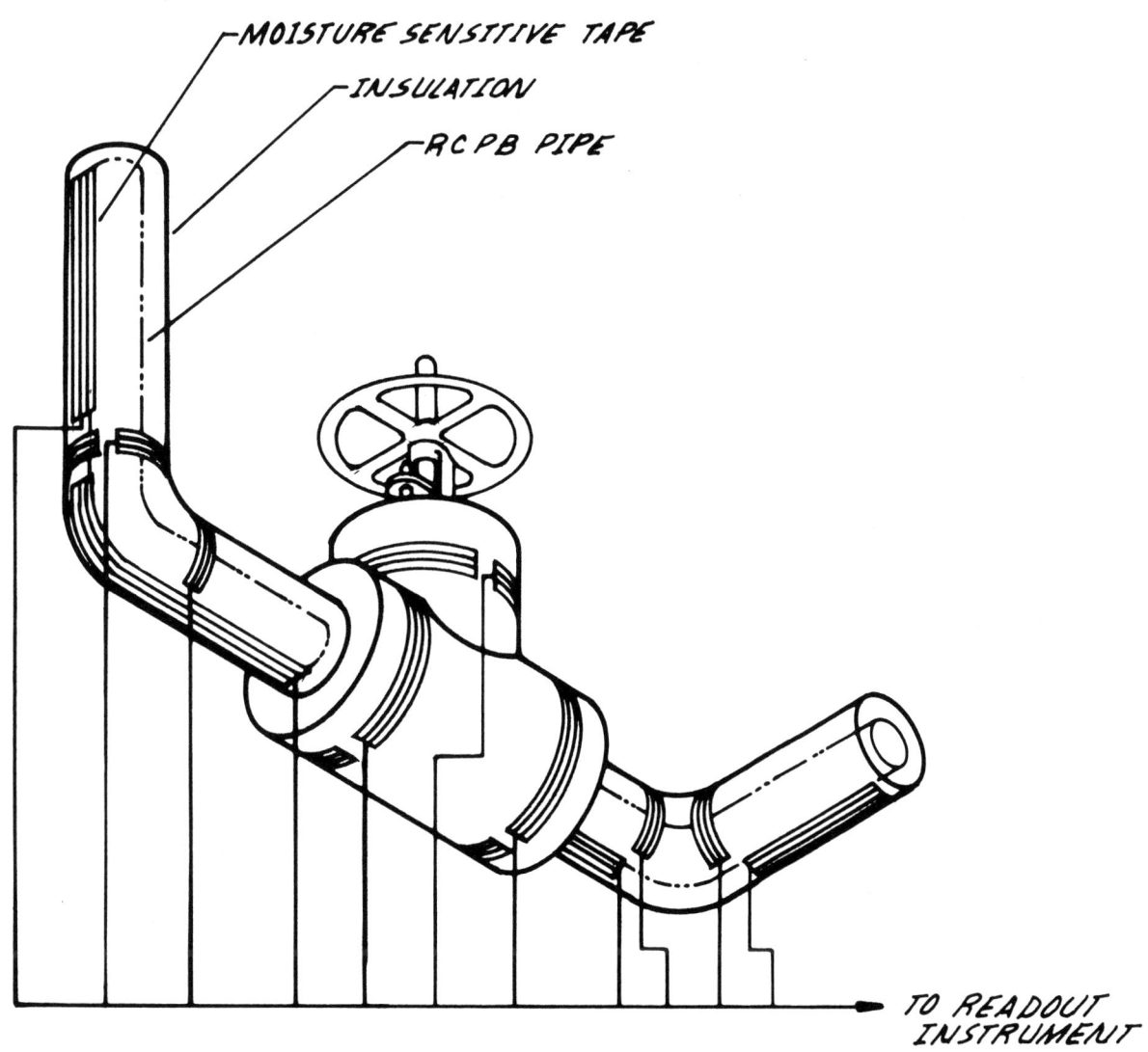

Figure B-3. Typical Moisture Sensitive Tape Installation

V = volume of junction or sensor material: ft³, (m³)

Satisfactory estimates of sensor time constants can usually be computed, where ΔT between sensor and surrounding fluid is sufficiently low to make radiation effects negligible, by assuming that the major resistance to heat flow lies in the film coefficient between the sensor and fluid medium. This assumption is valid for a single time constant system such as a bare thermocouple. When the sensing element is inside a well, there may be two or more films to consider. The heat transfer area is assumed to be the average film surface area.

The heat transfer film coefficient (h) is determined by the existing fluid conditions as expressed in the following relationship.

$$h = \frac{k N_u}{d} \qquad (B\text{-}38)$$

where

k = fluid conductivity: $\frac{BTU}{hr\ ft^{2}\,°F}\left(\frac{W}{m°C}\right)$

N_u = Nusselt number: dimensionless

d = sensor diameter: ft (meter)

One expression for the Nusselt number is:

$$N_u = 0.43 + X\ Re^Y\ Pr^{0.31} \qquad (B\text{-}39)$$

Table B-1 below gives values of X and Y as a function of Reynold's number that apply in Equation (B-39) above.

TABLE B-1
VALUES OF X AND Y AS FUNCTION OF REYNOLDS NUMBER

Reynold's Number	X	Y
0.4 - 4	0.989	0.330
4 - 40	0.911	0.385
40 - 4,000	0.683	0.466
4,000 - 40,000	0.193	0.618
40,000 - 400,000	0.0265	0.805

The Reynold's number (Re) and Prandtl number (Pr) are derived from the following expressions at the measurement fluid conditions.

$$Re = \frac{dv\rho}{\mu} \text{ (dimensionless)} \quad (B\text{-}40)$$

$$Pr = \frac{\mu C_p}{k} \text{ (dimensionless)} \quad (B\text{-}41)$$

where:

ρ = fluid density: $\frac{lb}{ft^3}\left(\frac{Kg}{m^3}\right)$

v = fluid velocity: $ft/hr\left(\frac{m}{s}\right)$

μ = fluid viscosity: $lb/ft\text{-}hr\left(\frac{Kg}{ms}\right)$

d = sensor diameter: ft (m)

k = fluid conductivity: $\frac{BTU}{hr\ ft^2\ °F}\left(\frac{W}{m^2\ °C}\right)$

C_p = specific heat: $\frac{BTU}{lb\ °F}\left(\frac{J}{K_g\ °C}\right)$

B.6.1 Converting Sensor Time Constant for Different Fluid Conditions

When the manufacturer of a temperature sensor provides the instrument time constant for a given set of fluid conditions, the data can be converted to a new set of fluid conditions. This method usually is capable of more accurate estimates of time constants because manufacturers' data is usually based on actual tests.

If the time constant τ_1 is measured under one set of conditions and the τ_2 is desired for another set of fluid conditions, then the relationship can be expressed as indicated in equation (B-42) since the heat capacity and transfer area of the thermometer remain unchanged.

$$\frac{\tau_1}{\tau_2} = \frac{h_2}{h_1} = \frac{k_2(0.43 + X_2 Re_2^Y Pr_2^{0.31})}{k_1(0.43 + X_1 Re_1^Y Pr_1^{0.31})} \quad (B\text{-}42)$$

Calculated values for sensor time constants should correlate well with experimental results. Otherwise significant deviations from reality will have been made in the assumptions used in calculations. The values for X and Y in Equation (B-42) above are derived from Table B-1, as a function of Reynold's number for the respective fluid conditions.

B.6.2 Temperature Sensor Response Characteristics

The response of a single time constant (τ) system to a step change input is given by the following expression:

$$T_t = T_f(1 - e^{\frac{t}{\tau}}) + T_i \quad (B\text{-}43)$$

where

T_t = temperature at a time t

T_f = final step-change temperature value

τ = sensor 63 percent response time constant

T_i = temperature before step change

When the response is exponential as for the above single time constant expression, it can be shown that the dynamic error of indicated temperature in degrees is simply the rate of change in temperature multiplied by the time constant (τ).

After a time interval corresponding to three time constants, a sensor will have attained 95 percent of the response to a step change. Therefore, for a ramp or constant rate of temperature change, the instrument lag time in seconds is approximately equal to the time constant after elapse of a time equal to three time constants.

Some temperature sensor designs, e.g., a sensor with thermowell, constitute a multiple time constant system that includes one or more thermal lag time constants. General practice, and the intent of this standard, is to permit lumping of these sensor delays into a single time constant, see Reference (14). This procedure is not strictly correct in the mathematical sense, but will give satisfactory estimates of sensor response over relatively small temperature increments.

B.6.3 Coolant Leakage Measurement Sensitivity Estimates

The rate of change of temperature for the process fluid in a system can be estimated for an assumed coolant leakage rate or heat input rate from the following generalized equation:

$$MC_p \left(\frac{dT}{dt}\right) = \Sigma\, Q(\text{input}) - \Sigma\, Q(\text{output}) \qquad (B\text{-}44)$$

Where M and C_p are the average mass and heat capacitance of the heat transfer fluid, dT/dt is the rate of change of the fluid temperature, $\Sigma\, Q$ (input) is the summation of heat inputs within measurement system boundaries and $\Sigma\, Q$ (output) is the summation of heat losses from the system boundary.

If initial conditions at t = o are assumed to be zero steady-state coolant leakage, constant environmental temperatures and the only new heat input to the system is a new coolant leakage flashing to steam vapor at a constant rate, then

$$MC_p \left(\frac{dT}{dt}\right) = Wh_g - \Sigma\, Q\,(\text{losses}) \qquad (B\text{-}45)$$

Where W = coolant leakage rate and h_g = flashing coolant or steam enthalpy.

If we substitute instrument indicated temperature rate of change, $\tau\, dT/dt$ for the equivalent fluid temperature rate of change after a time lapse of three sensor time constants (see Section B.6.2 above), and rearrange the equation, then an expression of coolant leakage rate in terms of rate of change of indicated temperature for finite time intervals is derived as follows:

$$W = \left(\frac{1}{h_g}\right)\left[MC_p\,\tau\left(\frac{dT}{dt}\right) + \Sigma\, Q\,(\text{losses})\right] \qquad (B\text{-}46)$$

The summation of heat losses from the measurement system limits will be a nonlinear function of time as the system fluid (usually air and water vapor) temperature increases and as the heat sink masses increase in temperature. However, for leakage sensitivity prediction estimates of respective exposed surface areas (A), overall heat transfer coefficients (U) and differences in fluid temperature and heat sink area temperature (ΔT) can be made for finite time increments so that the following equation for heat losses applies:

$$\Sigma\, Q\,(\text{losses}) = U_1 A_1 \Delta t_1 + \cdots U_n A_n \Delta t_n \qquad (B\text{-}47)$$

Where n corresponds to the number of different heat sinks considered in the calculation. When additional heat removal and mixing mechanisms are significant these should also be included in the estimates. For the purpose of these calculations, justifiable simplifying assumptions are permissible in arriving at heat transfer coefficients, area, etc. For longer periods of leakage, iterative methods may be used for adjusting heat loss coefficients and temperature differences. Using these methods, the leakage rate that will produce a finite and measurable temperature change (sensitivity) in a given unit of time can be estimated. With such backup design calculations, a coolant leakage detection system which alarms on rate of change of temperature rather than absolute or differential temperature may be justifiable. The above method of estimating sensitivity assumes complete and instantaneous mixing of flashing coolant and air volume in the measurement system. In cases where this assumption introduces large errors in the response calculations, i.e., large mixing volumes with considerable distances between sensor and postulated leakage, additional expressions defining the thermal transport mechanisms must be included in the calculations or the temperature sensor must be located closer to the postulated leakage in smaller mixing volumes.

INSTRUMENT SOCIETY of AMERICA
Research Triangle Park, North Carolina

ANSI/ISA-S67.04-1988
Approved February 4, 1988
Second Printing

American National Standard

Setpoints for Nuclear Safety-Related Instrumentation

Instrument Society of America

Instrument Society of America

ISBN 1-55617-083-1

ISA-S67.04 Setpoints for Nuclear Safety-Related Instrumentation

Copyright © 1987 by the Instrument Society of America. All rights reserved. Printed in the United States of America. No part of this publication may be reproduced, stored in a retrieval system, or transmitted, in any form or by any means (electronic, mechanical, photocopying, recording, or otherwise), without the prior written permission of the publisher.

INSTRUMENT SOCIETY OF AMERICA
67 Alexander Drive
P.O. Box 12277
Research Triangle Park, North Carolina 27709

PREFACE

This preface is included for informational purposes and is not part of ISA-S67.04.

This standard has been prepared as part of the service of the Instrument Society of America (ISA) toward a goal of uniformity in the field of instrumentation. To be of real value, this document should not be static, but should be subject to periodic review. Toward this end, the Society welcomes all comments and criticisms, and asks that they be addressed to the Secretary, Standards and Practices Board, Instrument Society of America, 67 Alexander Drive, P. O. Box 12277, Research Triangle Park, NC 27709, Telephone (919) 549-8411.

The ISA Standards and Practices Department is aware of the growing need for attention to the metric system of units in general, and the International System of Units (SI) in particular, in the preparation of instrumentation standards. The Department is further aware of the benefits to U.S.A. users of ISA standards of incorporating suitable references to the SI (and the metric system) in their business and professional dealings with other countries. Toward this end, this Department will endeavor to introduce SI-acceptable metric units in all new and revised standards to the greatest extent possible. The *Metric Practice Guide*, which has been published by the Institute of Electrical and Electronics Engineers as ANSI/IEEE Std. 268-1982, and future revisions will be the reference guide for definitions, symbols, abbreviations, and conversion factors.

It is the policy of the Instrument Society of America to encourage and welcome the participation of all concerned individuals and interests in the development of ISA standards. Participation in the ISA standards-making process by an individual in no way constitutes endorsement by the employers of the individual, of the Instrument Society of America, or of any of the standards that ISA develops.

The information contained in the preface, footnotes, and appendices is included for information only and is not a part of the standard.

Instrument setpoint drift is a problem which has led to numerous abnormal occurrence reports (now referred to as "Licensee Event Reports"). Section 50.36 "Technical Specifications" of 10 *CFR* 50* requires that, where a Limiting Safety System Setting (LSSS) is specified for a variable on which a safety limit has been placed, the setting be so chosen that automatic protective action will correct the most severe abnormal situation anticipated before a safety limit is exceeded. Protective instruments are provided with setpoints where specific actions are either initiated, terminated, or prohibited. Setpoints correspond to certain provisions of Technical Specifications that are incorporated into the facility operating license.

The single most prevalent reason for the drift of a setpoint out of compliance with a technical specification has been the selection of a setpoint that does not allow a sufficient margin between the setpoint and the technical specification limit to account for instrument accuracy, the expected environment, and minor calibration variations. In some cases the setpoint selected was numerically equal to the technical specification limit and stated as an absolute value, thus leaving no apparent margin for uncertainties. In other cases the setpoint was so close to the upper or lower limit of the instrument's range that instrument drift placed the setpoint beyond the instrument's range, thus nullifying the trip function. Other causes for drift of a setpoint out of conformity with the technical specification have been instrumentation design inadequacies and questionable calibration procedures.

The Instrument Society of America sponsored a review of the setpoint drift problem in April 1975 by establishing the SP67.4 committee (now renumbered as SP67.04).

The committee's review indicated that a more thorough consideration of setpoint drift was necessary in the design, test, purchase, installation, and maintenance of nuclear safety-related instrumentation.

The purpose of this revision is for clarification and to reflect current industry practice. The term "Trip Setpoint" is now consistent with the terminology used in the NRC Standard Technical Specifications and reflects what previously was known as "Upper Setpoint Limit."

This document has been developed to specifically address the establishment and maintenance of setpoints for individual safety-related instrument channels.

---

* Code of Federal Regulations, Title 10, Chapter 1, Part 50, Washington, D.C., 1987.

This standard is intended for use primarily by the owners of nuclear power plant facilities or their agents (Nuclear Safety Steam Suppliers, Architect/Engineers, etc.) in establishing procedures for determining setpoints, setpoint margins, and test routines in safety-related instrument channels.

Adherence to this standard will not in itself suffice to protect the public health and safety because it is the integrated performance of the structures, the mechanical systems, the fluid systems, the instrumentation, and the electric systems of the plant that limit the consequences of design basis events. On the other hand, failure to meet these requirements may be an indication of system inadequacy. Each application for a construction permit or an operating license for a nuclear power plant is required to develop these items to comply with Title 10, *Code of Federal Regulations,* Chapter 1, Part 50. Each applicant has the responsibility to assure himself and others that this integrated performance is adequate.

ISA Standards Committee SP67.04 operates as a subcommittee under SP67, the ISA Nuclear Power Plant Standards Committee, R. L. Phelps, Chairman.

The persons listed below served as active members of SP67.04 for the major share of its working period:

NAME	COMPANY
M. A. Widmeyer, Chairman	Washington Public Power Supply System
R. O. Allen, Secretary	Combustion Engineering, Inc.
J. Bauer	GA Technologies, Inc.
M. Belew, Vice Chairman	Tennessee Valley Authority
B. Beuchel	Yankee Atomic
W. Brown, Vice Chairman	ISD Corporation
J. Carolan	Philadelphia Electric Company
J. Fougere	Stone and Webster Engineering
J. Hill, Secretary	Northern States Power
S. Kincaid	Nuclear Energy Consultants, Inc.
J. Mauck	U. S. Nuclear Regulatory Commission
E. Schindhelm	Westinghouse Corporation
K. Utsumi	General Electric Company
G. D. Whitmore	Duquesne Light Company

The following people served as members of ISA Committee SP67:

NAME	COMPANY
R. L. Phelps, Chairman	Southern California Edison Company
R. O. Allen/P. C. Newcomb*	Combustion Engineering, Inc.
C. E. Andersen	Philadelphia Electric Company
R. W. Bailey	The Ohio State University
M. Belew	Tennessee Valley Authority
T. R. Bietsch	Advanced Technology Engineering Systems, Inc.
G. G. Boyle	Microswitch Division of Honeywell
S. Charnetski	Brown and Root, Inc.
T. N. Crawford, Sr./K. L. Hermann*	Pacific Gas and Electric Company
J. M. Dahlquist, Jr.	Baltimore Gas and Electric Company
A. Ellis	Western Electric Corporation
R. Estes	Nuclear Technical Consultant
H. T. Evans	PYCO, Inc.
L. C. Fron	Detroit Edison Company
R. L. Givan	Sargent and Lundy
W. Gordon	Bechtel Power Corporation
S. C. Gotilla	Gibbs and Hill
T. Grochowski	UNC Engineering Services, Inc.
H. Hopkins (Managing Director)	Consultant
R. N. Hubby	Leeds and Northrup Company
J. R. Karvinen	Mountain States Energy, Inc.
J. V. Lipka	Gilbert/Commonwealth, Inc.
L. R. McNeil	INPO
G. C. Minor	MHB Tech. Associates
J. A. Nay	Westinghouse Electric Corporation
R. Naylor	NPS
G. E. Peterson, Vice Chairman	Commonwealth Edison Company
I. Sturman	Bechtel Power Corporation
J. Tana	EBASCO Services, Inc.
K. Utsumi	General Electric Company
E. C. Wenzinger, Sr./F. Rosa*	U. S. Nuclear Regulatory Commission
M. A. Widmeyer	Washington Public Power Supply System

*One vote

Instrument Society of America

This standard was approved for publication by the ISA Standards and Practices Board in September 1987.

NAME	COMPANY
D. E. Rapley, Chairman	Rapley Engineering Services
D. N. Bishop	Chevron U.S.A. Inc.
W. Calder III	The Foxboro Company
N. Conger	Fisher Controls International, Inc.
R. S. Crowder	Ship Star Associates
C. R. Gross	Eagle Technology
H. S. Hopkins	Utility Products of Arizona
R. T. Jones	Philadelphia Electric Company
R. Keller	The Boeing Company
A. P. McCauley	Chagrin Valley Controls, Inc.
E. M. Nesvig	ERDCO Engineering Corporation
R. Prescott	Moore Products Company
C. W. Reimann	National Bureau of Standards
R. H. Reimer	Allen-Bradley Company
J. Rennie	Factory Mutual Research Corporation
W. C. Weidman	Gilbert/Commonwealth Inc.
K. A. Whitman	Farleigh Dickinson University
*P. Bliss	Consultant
*B. A. Christensen	Continental Oil Company
*L. N. Combs	Consultant
*R. L. Galley	Consultant
*T. J. Harrison	Florida State University
*O. P. Lovett, Jr.	ISIS Corporation
*E. C. Magison	Honeywell, Inc.
*R. G. Marvin	Roy G. Marvin Company
*W. B. Miller	Moore Products Company
*J. W. Mock	Bechtel Western Power Corporation
*G. Platt	Consultant
*J. R. Williams	Stearns Catalytic Corporation

*Director Emeritus

TABLE OF CONTENTS

Section	Title	Page
1. Purpose		9
2. Scope		9
3. Definitions		9
4. Establishment of Setpoints		10
4.1 Safety Limits		10
4.2 Safety Analysis		10
4.3 Limiting Safety System Setting		10
4.4 Combination of Uncertainties		12
5. Instrument Performance and Trip Setpoint Setting		12
6. Qualification Record Keeping		12
7. Maintenance of Safety-Related Setpoints		12
7.1 Initial Calibration		12
7.2 Operation		12
7.3 Replacement		13
References		13
Informative References		13
Figure 1. Nuclear Safety-Related Setpoint Relationships		10

1 PURPOSE

The purpose of the standard is to develop a basis for establishing nuclear safety-related instrumentation setpoints for actions determined by the design basis. This standard accounts for instrument errors and drift in the channel from the sensor, including the primary element, through and including the bistable trip device.

2 SCOPE

This standard defines minimum requirements for assuring that setpoints are established and held within specified limits in nuclear safety-related instruments in nuclear power plants.

3 DEFINITIONS

3.1 Reference Accuracy—In process instrumentation, a number or quantity that defines a limit that error will not exceed when a device is used under specified operating conditions. Error represents the difference between the measured value and the standard or ideal value. [1]

3.2 Allowable Value—The limiting value that the trip setpoint can have when tested periodically, beyond which the instrument channel is declared inoperable and corrective action must be taken.

3.3 Analytical Limit—Limit of a measured or calculated variable established by the safety analysis to ensure that a safety limit is not exceeded.

3.4 Design Basis—The design basis for protection systems for nuclear power generating stations is delineated in IEEE Standard 603-1980, "IEEE Standard Criteria for Safety Systems for Nuclear Power Generating Stations," Part 4, Safety System Design Basis.

3.5 Drift—An undesired change in output over a period of time, which change is unrelated to the input, environment, or load. [1]

3.6 Dynamic Response—The behavior of the output of a device as a function of the input, both with respect to time. [1]

3.7 Foldover—A device characteristic exhibited when a further change in the input produces an output signal which reverses its direction from the specified input-output relationship.

3.8 Hysteresis—That property of an element evidenced by the dependence of the value of the output, for a given excursion of the input, upon the history of prior excursions and the direction of the current traverse. [1]

3.9 Instrument Channel—An arrangement of components and modules as required to generate a single protective action signal when required by a generating station condition. A channel loses its identity where single protective action signals are combined.

3.10 Instrument Range—The region between the limits within which a quantity is measured, received, or transmitted, expressed by stating the lower and upper range values. [1]

3.11 Limiting Safety System Setting (LSSS)—Limiting Safety System Settings for nuclear reactors are settings for automatic protective devices related to those variables having significant safety functions. [2]

NOTE: For the purposes of this standard, the phrase "nuclear reactors" used in this definition should be understood to mean "nuclear power generating stations."

3.12 Repeatability—The closeness of agreement among a number of consecutive measurements of the output for the same value of the input under the same operating conditions, approaching from the same direction, for full-range traverses. [1]

3.13 Nuclear Safety-Related Instrumentation—That which is essential to:

1. Provide emergency reactor shutdown

2. Provide containment isolation

3. Provide reactor core cooling

4. Provide for containment or reactor heat removal

5. Prevent or mitigate a significant release of radioactive material to the environment; or is otherwise essential to provide reasonable assurance that a nuclear power plant can be operated without undue risk to the health and safety of the public.

3.14 Safety Limit—A limit on an important process variable that is necessary to reasonably protect the integrity of physical barriers that guard against uncontrolled release of radioactivity. [2]

3.15 Saturation—A device characteristic exhibited when a further change in the input signal produces no significant additional change in the output.

3.16 Sensor—The portion of a channel which responds to changes in a plant variable or condition, and converts the measured process variable into an electric or pneumatic signal. [3]

3.17 Trip Setpoint—A predetermined value at which a bistable device changes state to indicate that the quantity under surveillance has reached the selected value.

3.18 Test Interval—The elapsed time between the initiation (or successful completion) of tests on the same sensor, channel, load group, safety group, safety system, or other specified system or device. [4]

4 ESTABLISHMENT OF SETPOINTS

Trip setpoints in nuclear safety-related instruments shall be selected to provide sufficient allowance between the trip setpoint and the safety limit to account for uncertainties and dynamic responses. Detailed requirements for safety-related instrument setpoint relationships are given in the sections which follow, as illustrated in Figure 1.

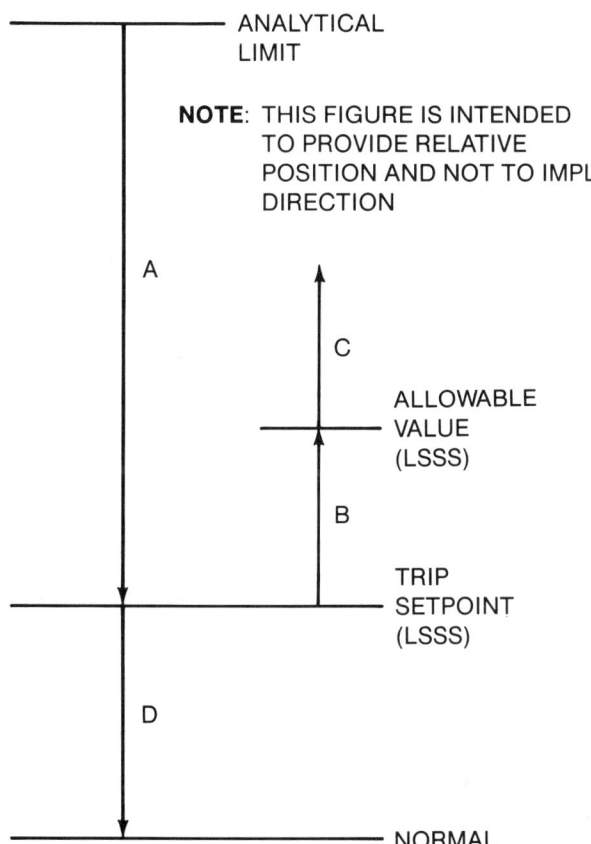

A. ALLOWANCES DESCRIBED IN PARAGRAPH 4.3.1
B. ALLOWANCES DESCRIBED IN PARAGRAPH 4.3.2
C. REGION OF TECHNICAL SPECIFICATION REPORTABLE OCCURRENCE
D. PLANT OPERATING MARGIN

Figure 1. Nuclear Safety-Related Setpoint Relationships

4.1 Safety Limits

Physical barriers are designed into nuclear power plants to prevent the uncontrolled release of radioactivity. Safety limits are chosen to maintain the integrity of these physical barriers. For the purpose of this standard, design limits for Engineered Safety Features are treated the same as safety limits. Safety limits can be defined in terms of directly measured process variables such as pressure or temperature. Safety limits can also be defined in terms of a calculated variable involving two or more measured process variables. An example of a calculated variable is Departure from Nucleate Boiling Ratio.

4.2 Safety Analysis

The safety analysis establishes an analytical limit in terms of a measured or calculated variable and a specific time after that value is reached to begin protective action. Satisfying these two constraints will ensure that the safety limit of 4.1 will not be exceeded during anticipated operational occurrences and design basis events. Choosing a limiting safety system setting (LSSS) to begin protective action before the analytical limit is reached will ensure that the consequences of a design basis event are not more severe than the safety analysis predicted. An LSSS, derived from an analytical limit, is published in the technical specifications and maintained by plant operating procedures.

4.3 Limiting Safety System Setting

Each LSSS normally has two components, called a trip setpoint and its allowable value. Figure 1 illustrates the relationships between an analytical limit and an LSSS. Detailed requirements for developing trip setpoints and allowable values are given in the sections that follow.

4.3.1 Trip Setpoint

The bases for selection of a trip setpoint shall be documented and shall include the data, assumptions, and methods used. The data used shall be taken from operating experience, equipment qualification tests, vendor design specifications, engineering analyses, laboratory tests, and certified engineering drawings. Any assumptions used, such as ambient temperature during equipment calibration and operation, shall be clearly identified.

An allowance shall be provided between the trip setpoint and the analytical limit to ensure a trip before the analytical limit is reached. The allowance used shall account for all applicable design basis events and the process instrument uncertainties listed below unless they were included in the determination of the analytical limit. A repeatable offset of fixed magnitude shall be accounted for in the setpoint calculation or during instrument calibration.

The uncertainties listed below may not be all-inclusive. Additional uncertainties that may apply to a particular instrument channel shall be accounted for in determining the trip setpoint allowance. Not all the uncertainties listed below apply to every measurement channel.

1. Instrument Calibration Uncertainties Caused by the:

   a. Calibration Standard

   b. Calibration Equipment

   c. Calibration Method

2. Instrument Uncertainties During Normal Operation Caused by:

   a. Reference Accuracy [1], including:

      (i) Conformity (Linearity)

      (ii) Hysteresis

      (iii) Dead Band

      (iv) Repeatability

   b. Power Supply Voltage Changes

   c. Power Supply Frequency Changes

   d. Temperature Changes

   e. Humidity Changes

   f. Pressure Changes

   g. Vibration (In-service and Seismic)

   h. Radiation Exposure

   i. Analog-to-Digital Conversion

3. Instrument Drift

All instruments may not have the same calibration interval. The drift used shall be based on instrument-specific calibration intervals.

4. Instrument Uncertainties Caused by Design Basis Events

Only uncertainties specific to the event and required period of service need be used. The use of different uncertainty components for the same process equipment for different events is permitted. Any residual effects of a design basis event shall also be included.

   a. Temperature Effects

The uncertainties associated with event-specific temperature profiles shall be used where possible. If these are not available, use the uncertainty associated with a imitiing temperature.

   b. Radiation Effects

The uncertainties associated with event-specific temperature profiles shall be used where possible. If these are not available, use the uncertainty associated with a lmiting temperature.

   c. Seismic/Vibratory Effects

The uncertainties associated with a safe shutdown or operating basis earthquake shall be used.

5. Process-Dependent Effects

The determination of the trip setpoint allowance shall account for uncertainties associated with the process variable. Examples include the effect of fluid stratification on temperature measurement, the effect of changing fluid density on level measurements, and process oscillations or noise.

6. Calculation Effects

The determination of the trip setpoint allowance shall account for uncertainties resulting from the use of a mathematical model to calculate a variable from measured process variables. For example, the use of differential pressure to determine flow.

7. Dynamic Effects

The determination of the trip setpoint allowance shall allow for response delays in the instrument channel. The instrument channel response time shall be no more than the limiting response time required by the safety analysis.

4.3.2 Allowable Value

The uncertainties of that portion of the instrument channel being tested to be used to determine the allowance between the trip setpoint and the Allowable Value are:

1. Instrument Calibration Uncertainties

2. Instrument Uncertainties During Normal Operation

3. Instrument Drift

The assumptions, data, and methods used to determine the allowable value shall be documented and shall be consistent with those used to determine the trip setpoint.

4.4 Combination of Uncertainties

The following methods are acceptable for combining uncertainties. Alternative methods can be used whenever documented justification is provided.

4.4.1 Square-Root Sum-of-Squares Method

Uncertainties that are random and independent can be combined by the square-root sum-of-squares method. This method is a direct application of the central limit theorem and is valid whenever a common uncertainty source does not exist. When two independent uncertainties, $(\pm a)$ and $(\pm b)$, are combined by this method, the resulting uncertainty is $(\pm c)$ where $c = (a^2 + b^2)^{1/2}$.

4.4.2 Algebraic Method

Uncertainties that are not random or independent shall be combined by the algebraic method. In this method, the combination of two dependent uncertainties, $(+a, -0)$ and $(+0, -b)$, results in a third uncertainty distribution with limits $(+a, -b)$.

4.4.3 Probabilistic and Statistical Methods

The interaction of dependent and independent uncertainties can be simulated by developing a stochastic model and inferring the combined uncertainties.

5 INSTRUMENT PERFORMANCE AND TRIP SETPOINT SETTING

Trip setpoints and safety analysis limits shall be located in the portion of the instrument's range where the accuracy is within the accuracy required by the setpoint analysis.

Instrument performance requirements shall be specified so that saturation, foldover, or any other cause will not cause misoperation for expected values of the process variable.

All components of error that are calibrated out shall be documented. Correction factors used to determine the trip setpoint (for example, to compensate for pressure differences between the measurement point and the sensor location) shall be separately identified. The relationship between instrument and process units shall also be identified.

6 QUALIFICATION RECORD KEEPING

Nuclear safety-related instrument seismic and environmental qualification shall be documented and available to verify parameters used in determining the setpoints.

7 MAINTENANCE OF SAFETY-RELATED SETPOINTS

Maintenance of setpoints shall include all actions taken to assure that instrumentation is installed and continues to operate within the design requirements used to establish the setpoints. The following sections address those aspects of safety-related instrument setpoint maintenance necessary to support the allowable values and trip setpoints as described in Section 4. Information in this section is supplemental to other industry standards that give guidance in maintenance of safety-related setpoints (see references).

7.1 Initial Calibration

All instruments and instrument channels should have some form of preoperational testing, using written procedures, performed upon them as soon as practical after installation and prior to being placed in service. Inability to perform such testing should be documented and justified. If this preoperational test is to satisfy the requirements of a surveillance test program, the test shall be consistent with the requirements of that program. The results of preoperational tests should be made available to preparers of procedures for further testing of the equipment, such as surveillance procedures, and to the personnel who evaluate the test data.

7.2 Operation

Testing and maintenance is performed to ensure that trip setpoints remain within their established limits during operation. Formal documentation is necessary to support the investigation and documentation of any occurrence where a limit is exceeded, in the high or low direction as applicable.

7.2.1 Periodic channel tests shall be performed to ensure that the instrument channel is functioning in compliance with the safety analysis. Channel tests on the trip setpoint shall be made by:

1. Varying the monitored variable (the same or a substitute process variable) and recording the point at which the channel trips, or

2. Substituting a known signal into the instrument channel as close to the monitored process variable as permitted by plant operating procedures and recording the point at which the channel trips.

7.2.2 Tests performed to satisfy Technical Specification requirements shall use written procedures to verify the proper operation of instruments and instrument channels. This verification shall be achieved by recording of sufficient "as found" data to determine the setpoint in terms of the measured or derived process variables, prior to any adjustment. "As Found" data shall be the data taken during the first traverse in the direction of concern during test.

If "as found" data indicate that no instrument adjustment is necessary, documentation of the testing and "as found" data is all that is required. If there is a need for adjustment, record the "as found" and "as left" data.

If "as found" data show that an allowable value was exceeded, the appropriate action for reporting the event shall take place. This action shall include investigation to determine the cause of the event, and appropriate corrective action to prevent a reoccurrence. Possible items of consideration are:

1. Testing frequency

2. Setpoint revision in the conservative direction

3. Reevaluation of the allowable value

4. Equipment installation and environment

5. Calibration (equipment and technique)

7.3 Replacement

Replacement materials, parts, and components shall be identical to the original equipment, or evaluated and documented to be equivalent or better in performance to that called for in the basis for the trip setpoint. Where it is impractical to comply with the original design basis requirements for the trip setpoint, the trip setpoint shall be reevaluated to take the replacement equipment's performance into account.

REFERENCES

1. Definition per ANSI/ISA-S51.1-1979, "Process Instrumentation Terminology."

2. Definition per "Code of Federal Regulations" Title 10, Chapter 1, Part 50, dated January 1, 1987, Paragraph 50.36.

3. Definition per IEEE Standard 603-1980 "IEEE Standard Criteria for Safety Systems for Nuclear Power Generating Stations."

4. IEEE Standard 338-1977, "IEEE Standard Criteria for the Periodic Testing of Nuclear Power Generating Station Safety Systems."

5. IEEE Standard 323-1983, "IEEE Standard for Qualifying Class 1E Equipment for Nuclear Power Generating Stations."

INFORMATIVE REFERENCES

The Instrument Society of America (ISA) has developed standards for the nuclear industry through the SP67 Nuclear Power Plant Standards Committee (NPPSC).

ANSI/ASME NQA-1-1983, "Quality Assurance Program Requirements for Nuclear Facilities."

ANSI/ISA-67.01-1981 (R 1987), "Transducer and Transmitter Installation for Nuclear Safety Applications."

IEEE Standard 352-1975 (R 1980), "IEEE Guide for General Principles of Reliability Analysis of Nuclear Power Generating Station Protection Systems."

IEEE Standard 498-1985, "IEEE Standard Requirements for the Calibration and Control of Measuring and Test Equipment Used in the Construction and Maintenance of Nuclear Power Generating Stations."

ANSI/ISA-S67.02-1980, "Nuclear-Safety-Related Instrument Sensing Line Piping and Tubing Standards for Use in Nuclear Power Plants."

ANSI/ISA-S67.06-1984, "Response Time Testing of Nuclear Safety-Related Instrument Channels in Nuclear Power Plants."

These standards may be obtained by writing the publishers at the following addresses:

American National Standards Institute (ANSI), 1430 Broadway, New York, New York 10018.

Institute of Electrical and Electronics Engineers (IEEE), 345 E. 47th St., New York, New York 10017.

Instrument Society of America (ISA), 67 Alexander Drive, P.O. Box 12277, Research Triangle Park, N.C. 27709.

ANSI/ISA-S67.06-1984
Approved August 29, 1986

American National Standard

Response Time Testing of Nuclear Safety-Related Instrument Channels in Nuclear Power Plants

Instrument Society of America

ISBN 0-87664-847-2

ISA-S67.06 Response Time Testing of
Nuclear Safety-Related Instrument Channels
in Nuclear Power Plants

Copyright ©1984 by the Instrument Society of America. All rights reserved. Printed in the United States of America. No part of this publication may be reproduced, stored in a retrieval system, or transmitted, in any form or any means (electronic, mechanical, photocopying, recording or otherwise) without the prior written permission of the publisher.

INSTRUMENT SOCIETY OF AMERICA
67 Alexander Drive
P.O. Box 12277
Research Triangle Park, North Carolina 27709

Copyright ©1984 by the Instrument Society of America

PREFACE

This preface is included for information purposes and is not a part of ISA-S67.06

This standard has been prepared as part of the service of the Instrument Society of America (ISA) toward a goal of uniformity in the field of instrumentation. To be of real value, this document should not be static, but should be subject to periodic review. Toward this end, the Society welcomes all comments and criticisms, and asks that they be addressed to the Secretary, Standards and Practices Board, Instrument Society of America, 67 Alexander Drive, P.O. Box 12277, Research Triangle Park, NC 27709, Telephone (919) 549-8411.

The ISA Standards and Practices Department is aware of the growing need for attention to the metric system of units in general, and the International System of Units (SI) in particular, in the preparation of instrumentation standards. The Department is further aware of the benefits to U.S.A. users of ISA standards of incorporating suitable references to the SI (and the metric system) in their business and professional dealings with other countries. Toward this end, this Department will endeavor to introduce SI-acceptable metric units in all new and revised standards to the greatest extent possible. The Metric Practice Guide, which has been published by the Institute of Electrical and Electronics Engineers as ANSI/IEEE Std. 268-1982, and future revisions will be the reference guide for definitions, symbols, abbreviations, and conversion factors.

It is the policy of the Instrument Society of America to encourage and welcome the participation of all concerned individuals and interests in the development of ISA standards. Participation in the ISA standards-making process by an individual in no way constitutes endorsement by the employer of that individual, of the Instrument Society of America, or of any of the standards that ISA develops.

The information contained in the preface, footnotes, and appendices is included for information only and is not a part of the standard.

The committee has determined that the terms "sensor" and "transducer" will be interchangeable throughout this standard. The term "sensor" is preferred due to its wider application. A sensor is considered to be the device which transforms the monitored variable into an intelligible signal.

The SP67.06 members directly responsible for the development of this standard are:

NAME	COMPANY
A. E. Ellis, Chairman	Westinghouse Combustion Control Division
R. J. Ungaretti, Past Chairman	EMC Control, Inc.
K. V. Allen	Entor Corporation
D. S. Berto	Combustion Engineering
D. G. Cain	Electric Power Research Institute
W. Ciaramitaro	Westinghouse Water Reactors Division
A. S. Hintze (for Lou Lewis)	United States Nuclear Regulatory Commission
T. Kerlin	University of Tennessee
D. W. Miller	Ohio State University
C. H. Neuschaefer	Combustion Engineering, Inc.
W. H. Sides	Oak Ridge National Laboratory
E. C. Wenzinger, Sr.	U.S. Nuclear Regulatory Commission

Previous members who have contributed to the development of this standard are:

NAME	COMPANY
D. Burton	Duke Power Company
C. G. Foster	Industrial Design and Engineering
D. G. Harding	Yankee Atomic Electric
R. Hedtke	Rosemount, Inc.
C. W. Mayo	Babcock & Wilcox
J. W. Mock	EG&G, Idaho
J. Ouzts (alternate)	U. S. Nuclear Regulatory Committee
A. Pagano	Tennessee Valley Authority
T. Swain	Technology of Energy Corporation
A. Wade	Yankee Atomic Electric
C. Weber	Power Technical Associates

Approval by SP67 was obtained on October 21, 1980. Membership consisted of:

NAME	COMPANY
H. S. Hopkins, Chairman	Westinghouse Combustion Control Division
B. W. Ball	Brown & Root, Inc.
G. G. Boyle	Micro Switch Division
T. Crawford	Pacific Gas & Electric Company
J. M. Dahlquist, Jr.	Baltimore Gas & Electric Company
H. T. Evans	PYCO, Inc.
L. C. Fron	Detroit Edison Company
R. L. Givan	Sargent & Lundy

NAME	COMPANY
W. G. Gordon	Bechtel Power Corporation
S. C. Gottilla	Burns & Roe, Incorporated
L. F. Griffith	Yarway Corporation
T. Grochowski	Babcock and Wilcox Company
G. Harrington	Rosemount, Inc.
R. J. Howarth	General Physics Corporation
R. N. Hubby	Leeds & Northrup Company
R. T. Jones	Philadelphia Electric Company
J. R. Karvinen	Mountain States Energy, Inc.
J. R. Klingenberg	W. R. Holway & Associates
J. V. Lipka	Gilbert Commonwealth, Inc.
S. F. Luna	General Atomic Company
D. W. Miller	The Ohio State University
M. V. Miller	Tennessee Valley Authority
G. C. Minor	MHB Technical Associates
J. W. Mock	EG&G, Idaho

J. A. Nay	Westinghouse Advance Rx Division
G. E. Peterson	Commonwealth Edison Company
R. L. Phelps, Jr.	Edison Company
M. F. Reisinger	Electronic Associates, Inc.
R. M. Rello	Air Products & Chemicals, Inc.
U. Shah, P.E.	Washington Public Power
H. G. Shugars	Electric Power Research Institute
J. Tana	Ebasco Services, Inc.
R. J. Ungaretti	EMC Controls, Inc.
K. Utsumi	General Electric Company
E. C. Wenzinger, Sr.	U. S. Nuclear Regulatory Commission
W. C. Weston	Stone & Webster Engineering Corporation

This standard was approved for publication by the ISA Standards and Practices Board in October 1984.

NAME	COMPANY
W. Calder III, Chairman	The Foxboro Company
P. V. Bhat	Monsanto Company
N. L. Conger	Conoco
B. Feikle	Bailey Controls Company
H. S. Hopkins	Westinghouse Electric Company
J. L. Howard	Boeing Aerospace Company
R. T. Jones	Philadelphia Electric Company
R. Keller	The Boeing Company
O. P. Lovett, Jr.	ISIS Corporation
E. C. Magison	Honeywell, Inc.
A. P. McCauley	Chagrin Valley Controls, Inc.
J. W. Mock	Bechtel Corporation
E. M. Nesvig	ERDCO Engineering Corporation
R. Prescott	Moore Products Company
D. Rapley	Stearns Catalytic Corporation
W. C. Weidman	Gilbert Commonwealth, Inc.
K. A. Whitman	Consultant
*P. Bliss	Consultant
*B. A. Christensen	Continental Oil Company
*L. N. Combs	Retired
*R. L. Galley	Consultant
*T. J. Harrison	IBM Corporation
*R. G. Marvin	Roy G. Marvin Company
*W. B. Miller	Moore Products Company
*G. Platt	Bechtel Power Corporation
*J. R. Williams	Stearns Catalytic Corportion

*Director Emeritus

TABLE OF CONTENTS

Section	Title	Page
1.	Scope	9
2.	Purpose	9
3.	Definitions and Terminology	9
4.	General Criteria	10
5.	Test Boundaries	10
6.	General Requirements for Testing	10
7.	Test Methods	10
8.	Acceptance Criteria for Test Methods	10
9.	Test Equipment - General Requirements	10
10.	Test Results	10
11.	Maintenance	11
12.	Test Methods - Specific Requirements	11
	12.1 Instrument Channels Utilizing Pressure Sensors	11
	12.1.1 Substitute Process Perturbation	11
	(1) Ramp Input Signal	11
	(2) Other Input Signals	11
	12.1.2 Noise Analysis	11
	(1) Program Description	11
	(2) Test Methods	12
	12.1.3 Impulse Lines	12
	(1) Liquid	12
	(2) Gas	12
	(3) Remainder of Channel	12
	12.2 Instrument Channels Utilizing Temperature Sensors	12
	12.2.1 Resistance Temperature Detectors (RTDs)	12
	12.2.2 Thermocouples	13
	12.2.3 Bypass Lines	13
	12.3 Instrument Channels Utilizing Neutron Sensors	13
	12.4 Multiple Input Testing	14
	12.5 Remainder of Channel Testing	14
	12.5.1 Step Inputs	14
	12.5.2 Ramp Inputs	14
13.	Design Requirements	14
14.	References and Bibliography	14

APPENDICES

Section	Title	Page
APPENDIX A, Noise Analysis Techniques		17

1 SCOPE

This standard delineates requirements and methods for determining the response time characteristics of nuclear safety-related instrument channels. The standard applies only to those instrument channels whose primary sensors measure pressure, temperature, or neutron flux.

2 PURPOSE

The purpose of this standard is to provide the nuclear power industry with requirements and acceptable methods for response time testing of nuclear safety-related instrument channels.

3 DEFINITIONS AND TERMINOLOGY

*Allowable Response Time-*The limiting *response time* established in the safety analysis and documented in the plant's Technical Specifications.

*Channel-*An arrangement of components and modules as required to generate a single protective action signal when required by a generating station condition. A *channel* loses its identity where single-action signals are combined. (See Reference 1.)

*Impulse Line-*Piping or tubing connecting the process to the *sensor*. (See Reference 2.)

*Indirect Test-*A test that measures a quantity other than response time. The actual response time is determined using this quantity and previous measurements of this quantity which have a known relationship to the actual response times.

*Instrument Channel, Response Time-*The time interval from the time when the monitored variable exceeds its trip setpoint until the time when a protective action is initiated.

*Nuclear Safety-Related (NSR)-*That which is essential to provide for:

(1) Emergency reactor shutdown
(2) Containment isolation
(3) Reactor core cooling
(4) Containment or reactor heat removal
(5) Prevention or mitigation of significant release of radioactive material into the environment, or that which is otherwise essential to provide reasonable assurance that a nuclear power plant can be operated without undue risk to the health and safety of the public.

*Response Time Characteristics-*Those properties (e.g., transfer function, *time constant*, delay time, power spectral density) of the equipment from which its response time can be determined.

*Response Time, Fluid Transport-*The response time associated with *fluid transport* from the location at which a property is to be measured to the sensor location. This delay may include contributions from both the transport time associated with fluid velocity and mixing times determined by mass flow rate and system configuration.

*Sensor-*That portion of a channel which responds to changes in a plant variable or condition, and which converts the measured process variable into an instrument signal. (See Reference 3.)

*Setpoint-*A predetermined level at which a bistable device changes state to indicate that the quantity under surveillance has reached the selected value. (See Reference 4.)

*Step Response Time-*Of a system or a component, the time required for an output to go through a specified percentage of the total excursion either before, or (in the absence of overshoot) as a result of a step change to the input.

NOTE: This is usually stated for 90, 95, or 99 percent change.

(See "time constant" for the use of a 63.2 percent value; see also Reference 4.)

*Test Interval-*The elapsed time between the performance of tests.

*Time Constant-*The value T in an exponential response term $A^{(-t/T)}$ or in one of the transform factors $1 + sT$, $1 + jwT$, $1/(1 + sT)$, $1/(1 + jwT)$ where:

s = complex variable
t = time, seconds
T = time constant
j = $\sqrt{-1}$
w = angular velocity, radians per second.

NOTE: For the output of a first-order system forced by a step or an impulse, T is the time required to complete 63.2 percent of the total rise or decay; at any instant during the process, T is the quotient of the instantaneous rate of change divided into the change still to be completed. In higher order systems, there is a time constant for each of the first-order components of the process. In a Bode diagram, break points occur at $w = 1/T$. (See Reference 4.)

*Unterminated Ramp-*A ramp that starts at the variable's initial value, becomes linear, and continues to a higher or lower value beyond the setpoint of interest, such that the instrument's or channel's desired output signal is obtained while the input ramp is still linear.

*White Noise-*Random noise that has a constant energy per unit bandwidth at every frequency in the range of interest.

4 GENERAL CRITERIA

Periodic testing shall be conducted to verify that response time characteristics of nuclear safety-related systems are within the limit assumed in the plant safety analysis and as defined in the plant's Technical Specifications. Tests to verify response time characteristics shall be performed in accordance with written procedures and the test results documented as specified in IEEE 338-1977, Section 6.6. (See Reference 5.)

5 TEST BOUNDARIES

As presented herein, a response time verification test encompasses the instrument channel portion of the overall safety system. These tests' boundaries include impulse lines, thermowells, and all other components that affect the instrument channel response time.

6 GENERAL REQUIREMENTS FOR TESTING

6.1 The total instrument channel should be tested in a single test. When the total channel is not tested in a single test, separate tests on groups of components and/or on single components encompassing the total instrument channel shall be combined to verify total channel response. All active and passive components in the instrument channel shall be included to determine the overall channel response time. The response time obtained by adding the individual response times of each component or groups of components will be greater than or equal to the actual instrument channel response time.

6.2 All testing shall be performed in situ. Equipment removal may be considered as an alternative only if it can be shown that such removal will not result in the elimination of testing of any portion of the channel that has an effect on the response time.

6.3 Calibration verification of instrument channels need not be performed in conjunction with response time testing if the instrument channel is within the required calibration interval. If at the time of response time testing the instrument channel exceeds its required calibration interval, or is found to be out of tolerance, the need to perform channel calibration prior to the response test shall be evaluated and documented. The evaluation shall include whether or not calibration adjustments could affect response times.

6.4 The test interval shall be established to detect an unacceptable response time. The test interval is determined by three factors: (1) the margin between the present test value of the response time and the allowable response time; (2) the time rate of change of response time; and (3) the reliability and qualification of as-built equipment.

6.5 Environmental or ambient effects on response time shall be covered in design qualification tests and need not be simulated during response time testing.

7 TEST METHODS

Response time tests shall be conducted using direct or indirect response time measurements. Where indirect methods are used, a known quantitative relationship between the measured quantity and response time shall be established and periodically verified by direct measurement of response time.

8 ACCEPTANCE CRITERIA FOR TEST METHODS

All test methods for response time measurement shall be validated by: (1) comparison with other direct methods in suitable laboratory or in situ tests; (2) through theoretical justification for the procedure; (3) through specification of assumptions and conditions that must be satisfied to ensure validity of the test; and (4) through verification that essential conditions for validity of the test exist during the in situ tests.

To be acceptable, indirect tests shall provide test results equal to or more conservative than direct response time tests.

9 TEST EQUIPMENT - GENERAL REQUIREMENTS

The calibration of test equipment used in verifying response time characteristics shall be traceable to the National Bureau of Standards.

The response time characteristics and accuracies of the dynamic test and recording equipment used in determining equipment response time characteristics shall be known and accounted for in determining test results. Test equipment shall have a known frequency response bandwidth that encompasses the allowable response time being verified. Test equipment accuracy shall be equal to or better than the required accuracy width and accuracy which are required to minimize the effect of test equipment characteristics on the test results.

10 TEST RESULTS

Test results shall be compared to the allowable response time. If the results are found to exceed this limit, an investigation shall be performed to determine the cause. Repair or replacement shall be performed, as required. (See also Section 11.)

Where testing indicates a rate of change in response time characteristics such that the allowable response time may be exceeded prior to the next test, degradation is indicated and shall be investigated to determine the cause, and the appropriate action shall be taken.

Test results shall be documented and filed to ensure recoverability.

11 MAINTENANCE

After any repair or replacement of material, parts, or components, the response time characteristics of that equipment shall be verified by test and test results documented, unless it is shown and documented that the repair or replacement cannot affect response time.

12 TEST METHODS - SPECIFIC REQUIREMENTS

Acceptable response time tests may be classified according to whether they are perturbation type or passive in nature. Perturbation tests require some direct means of stimulating the sensor or channel. Passive methods monitor inherent process fluctuations through the sensing system, and are referred to as "noise analysis." Both methods are discussed in the following subsections.

12.1 Instrument Channels Utilizing Pressure Sensors

This section includes all pressure-sensing applications, such as those for absolute pressure, differential pressure, level, flow, etc.

Caution: Entrapped air in liquid pressure sensors can cause scatter in or unrepeatable test results. If acceptance criteria are exceeded, the cause of the scatter shall be investigated.

12.1.1 Substitute Process Perturbation

Where perturbation of the actual process variable is not a practical method of testing pressure channels, (e.g., reactor high pressure) channel perturbation shall be accomplished by utilizing a substitute pressure input test signal. The input test signal(s) shall simulate design basis event pressure transient(s) unless this is not practicable. If not practicable, one of the following alternatives shall be used:

(1) Ramp Input Signal

This is a direct method of determining response time. To a first-order approximation, an unterminated ramp satisfies a majority of applications. (See References 6, 7.)

In order to bound the response time of the channel in anticipation of potential degradation modes, apply to the transducer two ramps, as defined below. Do not deliberately pressure cycle the instrument prior to performing the tests.
Caution: Try to avoid any inadvertent pressure cycling prior to the test.

Apply the slow ramp first.

(a) The first (slow) ramp shall be selected based on the slowest transients for which automatic protective action is required by design.

(b) The second (fast) ramp shall be selected based on the fastest transient for which automatic protective action is required by design.

(2) Other Input Signals

In some cases, alternate test input pressure perturbations which do not simulate the design basis event pressure transient are acceptable; for example, sinusoidal variations for frequency response or step inputs for time constant determinations. These specific examples are direct methods of determining response time. In these specific cases, the linearity property must be verified for the components tested. Test results shall then be converted to the equivalent response time for the components tested.

12.1.2 Noise Analysis

This is an indirect method of determining response time.

(1) Program Description

The program described in this section is designed to detect a possible change in sensor response time of nuclear safety-related (NSR) sensors using the noise analysis technique. The normal fluctuations in the process variables are used as input to the sensor system. The response of the sensor can be analyzed in the time domain or the frequency domain. If the bandwidth over which the input noise spectrum appears is white and stationary and encompasses the sensor bandwidth, information on the sensor dynamics can be obtained. The sampling and statistical estimation schemes used shall be those necessary to provide valid results for the sensor bandwidth and to ensure statistically significant conclusions. If the input noise is not white, measurement of sensor output may be used to derive information about the change in the sensor's response time by detection of change in frequency content.

The program has two phases: the baseline measurement and the periodic surveillance phase. The baseline phase establishes the reference for the surveillance phase. Acquisition of data for the baseline phase of the program is normally limited to early operation of the plant or when new sensors are installed. The surveillance phase of the program is performed during normal operation of the plant, and evaluations are made by determining changes in response time in comparison with previous measurements.

A limitation of noise analysis for some pressure sensors is evidence showing that the sensors may respond differently for large and small perturbations. That is, some sensors exhibit a nonlinear dynamic behavior. If this is the case, monitoring of the process noise level will not be valid. It shall be the responsibility of the user to assure that dynamic linearity methods can be applied. Details of this program are outlined in Appendix A.

Baseline Phase

The objective of this phase of the program is to establish the relationship between the actual measurement response time and the response at a baseline reference. This will be performed when the required process noise for the baseline measurement is present. In addition, acceptance criteria must be established for estimated changes in response time. These criteria must be consistent with the requirements in Section 8.

Surveillance Phase

The objective of the surveillance phase of the program is to periodically determine whether changes in the response have occurred beyond acceptable limits. This may be done by either periodic noise measurements and analysis or by a suitable continuous surveillance monitoring system. This program is carried out over the lifetime of the equipment.

To determine if changes have occurred, it is necessary to reestablish the equipment at the baseline reference point, repeat the test, and perform the analysis which was carried out in the baseline phase.

The changes observed in the surveillance phase shall be compared to the previously established acceptance criteria from the baseline phase. Failure to meet acceptance criteria shall be investigated, including a direct measurement of response time.

Apparent changes in response time, indicated by these methods, do not mean that significant degradation necessarily occurred. Long-term, nonstationary process effects may be responsible. A change in sensor characteristics which did not produce significant response time degradation also might have occurred. At any time during the sensor's installed life, the baseline can be updated by repeating the previous steps in conjunction with actual measurement of response time.

(2) Test Methods

The test method for power spectral density, autoregressive analyses, and zero-crossing analyses consists of baseline and surveillance phases. These test phases are the same as those described in this section and in Appendix A.

12.1.3 Impulse Lines

This is an indirect method for determining response time which has the validity of a direct method, and therefore the completion of Section 7 is not required.

(1) Liquid

(a) Verify that impulse line sizing is properly matched with the transmitter used and with the length of line required.

(b) Verify that an appropriately conservative assumption on the allowance for unobstructed impulse line delays has been made in the value of response time (RT) used in the safety analysis for the overall channel response.

(c) Verify during start-up testing and at the intervals required by the plant's Technical Specifications that the impulse line is not blocked to a specified extent. This may be accomplished by an examination of the line flow in the forward and reverse direction or in the direction required to initiate a safety action. If a line is suspected of having a partial blockage, a further investigation should be conducted to attempt to restore the line to its unobstructed condition.

(2) Gas

Gas impulse lines with accessible process connections (for example, containment pressure) shall be response time tested using the methods outlined in Section 12.1.1. This test shall include the entire impulse line and primary sensor in one test. Gas impulse lines with inaccessible process connections shall be flow tested as described in Section 12.1.3. (See References 1, 6.)

(3) Remainder of Channel

The methods of Sections 12.1.1 and 12.5 can be used to test the remainder of the channel after it has been established that no significant degradation in impulse line flows has occurred.

12.2 Instrument Channels Utilizing Temperature Sensors

12.2.1 Resistance Temperature Detectors (RTDs)

(1) Loop Current Step Response

This is an indirect method for determining response time which has the validity of a direct method. In the loop current step response test, the sensor is heated internally by passing a current through the normal sensor leads. The resulting transient is analyzed to give the response time characteristics of the sensor. Analysis methods shall meet the requirements of Section 8. (See References 8, 10, 11.)

(2) Noise Analysis Techniques

This is an indirect method for determining response time.

(a) Program Description

Noise analysis methods can be used to determine sensor degradation from a baseline reference. The general approach for using noise analysis in this manner is the same as that for pressure sensors. (See Section 12.1.2.) Unlike for pressure sensors, however, it is not necessary to verify that RTDs respond linearly for perturbations of different magnitudes.

(b) Test Methods

The test methods for power spectral density, autoregressive analysis, and zero-crossing analysis consists of baseline and surveillance phases. Those phases are the same as those described in Section 12.1.2 and in Appendix A.

(3) Self Heating

This is an indirect method for determining response time.

The self-heating method for RTDs can be used to determine sensor degradation from a baseline reference. The physical basis for the self-heating tests is that the under steady-state process conditions the difference in RTD temperatures for different internal heat generation rates is inversely proportional to the overall heat transfer coefficient. (See References 8, 9.)

12.2.2 Thermocouples

(1) Loop Current Step Response

This is an indirect method for determining response time which has the validity of direct method. In the loop current step response test, the sensor is heated internally by passing a current through the normal sensor leads. The resulting transient is analyzed to give the response time characteristics of the sensor. Analysis methods shall meet the requirements of Section 8. (See References 8, 10, 11.)

(2) Noise Analysis Techniques

This is an indirect method for determining response time.

(a) Program Description

Noise analysis methods can be used to determine sensor degradation from a baseline reference. The general approach for using noise analysis in this manner is the same as that for using pressure sensors. (See Section 12.1.2.) Unlike for pressure sensors, however, it is not necessary to verify that thermocouples respond linearly for perturbations of different magnitudes.

(b) Test Methods

The test method for power spectral density, autoregressive analysis, and zero-crossing analysis consists of baseline and surveillance phases. These test phases are the same as those described under Section 12.1.2 and in Appendix A.

12.2.3 Bypass Lines

For temperature measurement systems that use bypass lines for sampling the process fluid, it is necessary to consider the effect of fluid transport in the bypass line on the channel response time. The contribution of fluid transport effects to the channel response time is a monotonically decreasing function of fluid mass flow rate.

The mass flow rate in the bypass line shall be verified to be greater than or equal to the flow rate value that corresponds to the allowed contribution to the channel response time from fluid transports effects. That is, any response time measurements for systems employing bypass lines must include, at least, a comparative flow measurement to determine whether flow in the bypass line is greater than or less than the value that corresponds to the allowed fluid transport response time. When a comparative flow measurement is used, the allowed fluid transport response time is added to the other components of the channel response time to assure a conservative total channel response time.

12.3 Instrument Channels Utilizing Neutron Sensors

There are several methods presently available in research and development for measuring these response times or determining degradation. These methods have not yet proven acceptable and therefore are not included herein.

12.4 Multiple Input Testing

For instrument channels with multiple inputs, the response time test shall be performed in accordance with IEEE 388-1977, Section 6.3.4, Paragraph 5. (See Reference 5.) The response time shall be determined for each process variable and shall be verified to be less than its allowable response time.

12.5 Remainder of Channel Testing

The remainder of the channel test shall be performed where the total channel (as described in Section 5) was not tested in a single test. This test will typically start at the sensor output and go to the channel output as required in Section 6.1.

Typically, the output of the sensor is transmitted to a signal conditioner. Signal conditioners are those components that receive the sensor output (RESISTANCE, MV, MA) and modify it. When the signal conditioning is integral to the bistable unit, it shall be tested as one assembly.

12.5.1 Step Inputs

For bistables, the step input shall start at the initial value of the process variable (0 percent). The final value (100 percent point) of the step input shall be selected such that the trip point of the bistable occurs at greater than or equal to the 63 percent point. This way, the delay time at the trip point will be no less than the actual response time.

12.5.2 Ramp Inputs

For ramp inputs, the response time shall be determined as follows:

(1) A ramp test signal that simulates the sensor output for limiting design basis event transient shall be applied to the input of the signal converter.

(2) The remainder of channel components shall be exercised through those set points identified in the technical specification which required initiation of a protective function.

(3) The input and output signals shall be recorded. The response time shall be determined by measuring the time differential between the input and output signals at the channel trip point.

13 DESIGN REQUIREMENTS

Special provisions shall be made in the design to facilitate channel response time testing in accordance with this standard.

14 REFERENCES AND BIBLIOGRAPHY

14.1 References

1. IEEE 279-1971, *IEEE Standard: Criteria for Protection Systems for Nuclear Power Generating Stations*, (R 1978) (ANSI/IEEE). New York, NY: The Institute of Electrical and Electronics Engineers, Inc.

2. ISA-S67.02-1980, *Nuclear-Safety-Related Instrument Sensing Line Piping and Tubing Standards for Use in Nuclear Power Plants*. Research Triangle Park, NC: Instrument Society of America.

3. ISA-S67.04-1982, *Setpoints for Nuclear Safety-Related Instrumentation Used in Nuclear Power Plants*. Research Triangle Park, NC: Instrument Society of America.

4. ISA-S51.1-1979, *Process Instrumentation Terminology*, (ANSI/ISA). Research Triangle Park, NC: Instrument Society of America.

5. IEEE 338-1977, *Standard Criteria for the Periodic Testing of Nuclear Power Generating Station Safety Systems*, (ANSI/IEEE). New York, NY: The Institute of Electrical and Electronics Engineers, Inc.

6. EPRI Report NP-267, Project 503-1, October 1976, *Sensor Response Time Verification*. Palo Alto, CA: Electric Power Research Institute, p. 120.

7. Cain, D. G., and Foster, C. G., "A Practical Means for Pressure Transducer Response Verification." *Nuclear Technology* 36 (Mid-December 1977), LaGrange Park, IL: American Nuclear Society.

8. EPRI Report NP-834, Project 503-3, July 1978, *In Situ Response Time Testing of Platinum Resistance Thermometers*. Palo Alto, CA: Electric Power Research Institute, p. 122.

9. Burton, Dale A. "Resistance Temperature Detector Time Response Testing by the 'Self-Heating' Method." *Instrumentation in the Power Industry* 23 (1980): 155-59. Research Triangle Park, NC: Instrument Society of America.

10. ORNL-TM-4912, March 1976, *Analytical Methods for Interpreting In-Situ Measurements of Response Times in Thermocouples and Resistance Thermometers*. Oak Ridge, TN: Oak Ridge National Laboratories.

11. Kerlin, T. W.; Miller, L. F.; and Hashemian, H. M. "In-Situ Response Time Testing of Platinum Resistance Thermometers." *ISA Transactions* 17 (No. 4, 1978): 71-88. Research Triangle Park, NC: Instrument Society of America.

14.2 Bibliography

The following documents may provide additional information, guidance, or requirements for various aspects of response time testing.

American National Standards Institute. "Validation of Techniques for Response Time Testing of Temperature Sensor in PWR's." *ANSI Transactions* (June 1980).

"Comparative Study of On-Line Response Time Measurement Methods for Platinum Resistance Thermometers." International Measurement Confederation IMEKO, Moscow, U.S.S.R., June 1979.

EPRI Report NP-459, Project 503-3, January 1977, *In Situ Response Time Testing of Platinum Resistance Thermometers."* Palo Alto, CA: Electric Power Research Institute, p. 121.

EPRI Report NP-834, Vol. 2, July 1978, *In-Situ Response Time Testing of Platinum Resistance Thermometers —Noise Analysis Method."* Palo Alto, CA: Electric Power Research Institute, p. 122.

EPRI Report NP-1166, Project 503-2, May 1980, *ARMA Sensor Response Time Analysis.* Palo Alto, CA: Electric Power Research Institute, p. 122.

EPRI Report NP-1486, Project 1161-1, August 1980, *Temperature Sensor Response Characterization.* Palo Alto, CA: Electric Power Research Institute, p. 122.

Hashemian, H. M.; Kerlin, T. W.; and Upadhyaya, B. R. *"Apparatus for Measuring the Degradation of a Sensor Time Constant."* Patent application filed with U.S. Patent Office.

NUREG 0809, "Safety Evaluation Report Review of Resistance Temperature Detector Time Response Characteristics." Office of Nuclear Reactor Regulation, 1981.

Poore, W. P. "Resistance Thermometer Characteristics and Time Response Testing." Thesis, December 1979, Nuclear Engineering Department, University of Tennessee, Knoxville, Tennessee.

Upadhyaya, B. R., and Kerlin, T. W. "Estimation of Response Time Characteristics of Platinum Resistance Thermometers by the Noise Analysis Technique." *ISA Transactions* 17 (No. 4, 1978): 21-28. Research Triangle Park, NC: Instrument Society of America.

APPENDIX A

Noise Analysis Techniques

A.1 Power Spectral Density

Power spectral density analysis involves the determination of signal power per unit frequency as a function of frequency. For sensor response evaluation, the noise signal analyzed is the sensor output that results from normal process fluctuations. If the process fluctuations have a constant power spectrum (white noise over the nominal sensor bandwidth), then the power spectral density of the sensor output signal is proportional to the square of the frequency response gain of the sensor. Consequently, the sensor response characteristics can be evaluated by fitting a transfer function to the measured power spectral density if the white noise assumption is valid. This empirically determined transfer function then may be used to predict the response of the sensor to any input of interest.

If the white noise assumption is not valid, then the above procedure cannot be used. However, changes in the sensor response characteristic may alter a measured power spectral density.

A.2 Autoregressive Analysis

Autoregressive analysis involves fitting a simple formula to the measured data. The formula has the form

$$\gamma_k = \sum_{i=1}^{i=N} a_i \gamma_{k-i}$$

where:

γ_k = sample k of the output
N = order of the fit
a_i = an autoregressive coefficient

The fit provides estimates of the a_i (usually obtained by least squares fitting techniques). Once the a_i are known, the autoregressive model may be used to evaluate sensor response characteristics. As with the power spectral density approach, the results are quantitative only if the process fluctuations have white noise characteristics.

A.3 Zero-Crossing

The rate at which a sensor output crosses its average value in response to a specific fluctuating input decreases as the sensor time constant increases. Consequently, a device that monitors the crossing rate can be used to detect changes in sensor time constant and/or changes in input fluctuations. Masking of any effects due to changes in sensor time constant by exactly compensating changes in input fluctuations is implausible. Therefore, measuring the crossing rate will detect changes in the sensor time constant if the sensitivity of the crossing rate to changes in the time constant is large enough. For temperature sensors (where the response is governed by heat diffusion), the sensitivity is unity (an x percent increase in time constant causes an x percent decrease in crossing rate). The usual practice is to remove the average value of the signal and measure the rate of crossing of the zero value in the remaining signal. Consequently, the method is often called the zero-crossing technique.

INSTRUMENT SOCIETY of AMERICA
Research Triangle Park, North Carolina

ANSI/ISA–S67.10–1986
Approved August 7, 1986

American National Standard

Sample-Line Piping and Tubing Standard for Use in Nuclear Power Plants

Instrument Society of America

Instrument Society of America

ISBN 0-87664-974-6

ISA–S67.10 Sample-Line Piping and Tubing Standard for Use in Nuclear Power Plants

Copyright © 1986 by the Instrument Society of America. All rights reserved. Printed in the United States of America. No part of this publication may be reproduced, stored in a retrieval system, or transmitted, in any form or by any means (electronic, mechanical, photocopying, recording, or otherwise), without the prior written permission of the publisher.

INSTRUMENT SOCIETY OF AMERICA
67 Alexander Drive
P.O. Box 12277
Research Triangle Park, North Carolina 27709

PREFACE

This preface and the appendix referring to other standards and codes are included for informational purposes only and are not part of ISA-S67.10. Applicability of referenced standards and codes is as stated in the text of this standard. Where reference is made to a standard or code, a particular paragraph is indicated for clarity, as applicable.

This standard has been prepared as part of the service of the Instrument Society of America (ISA) toward a goal of uniformity in the field of instrumentation. To be of real value, this document should not be static, but should be subject to periodic review. Toward this end, the Society welcomes all comments and criticisms, and asks that they be addressed to the Secretary, Standards and Practices Board, Instrument Society of America, 67 Alexander Drive, P. O. Box 12277, Research Triangle Park, NC 27709, Telephone (919) 549-8411.

The ISA Standards and Practices Department is aware of the growing need for attention to the metric system of units in general, and the International System of Units (SI) in particular, in the preparation of instrumentation standards. The Department is further aware of the benefits to U.S.A. users of ISA standards of incorporating suitable references to the SI (and the metric system) in their business and professional dealings with other countries. Toward this end, this Department will endeavor to introduce SI-acceptable metric units as optional alternatives to English units in all new and revised standards to the greatest extent possible. The *Metric Practice Guide*, which has been published by the Institute of Electrical and Electronics Engineers as ANSI/IEEE Std. 268-1982, and future revisions will be the reference guide for definitions, symbols, abbreviations, and conversion factors. SI (metric) conversions in this standard are given only to the precision intended in selecting the original numerical value. When working in the SI units, the given SI value should be used; when working in customary U.S. units, the given U.S. value should be used.

It is the policy of the Instrument Society of America to encourage and welcome the participation of all concerned individuals and interests in the development of ISA standards. Participation in the ISA standards-making process by an individual in no way constitutes endorsement by the employer of that individual, of the Instrument Society of America, or of any of the standards that ISA develops.

The information contained in the preface, footnotes, and appendices is included for information only and is not a part of the standard.

ISA Committee SP67.10 was formed in October 1978, after approval of its purpose and scope by the ISA Standards and Practices Board. Approval of the project charter was granted by the American National Standards Institute (ANSI) on March 18, 1981.

The intent of this standard is to provide criteria for the design and installation of lines through which liquid or gaseous samples flow. This standard has been developed to complement ISA Standards S67.01 and S67.02, which address instrument installation and sensing lines, respectively. The format of these standards was adhered to and their contents were reviewed closely while this standard was being written.

Instrument Society of America

The following people served as members of ISA Committee SP67.10:

NAME	COMPANY
T. R. Bietsch, Chairman	Nuclear Power Services, Inc.
R. P. Cuilwik, Cochairman	Coastal Consulting Services, Inc.
L. McBride*	Stock Equipment Company
P. McCarthy	Cygna Energy Services
J. Minnicks*	Arizona Public Service
K. Grimm	Brown and Root, Inc.

The following people are former ISA Committee SP67.10 members who have contributed to this standard:

NAME	COMPANY
R. M. Rello	Air Products & Chemicals, Inc.
A. M. Romano	Stone and Webster Engineering Corporation
D. Simko	Crawford Fitting Company
R. C. Taggert	Bendix Process Instruments Division
F. Zikas	Parker-Hannifin Company

The following people served as members of ISA Committee SP67 during the appproval of SP67.10:

NAME	COMPANY
R. L. Phelps, Jr., Chairman	Southern California Edison Company
G. E. Peterson, Vice-Chairman	Commonwealth Edison Company
C. E. Andersen	Philadelphia Electric Company
R. W. Bailey	The Ohio State University
M. Belew	Tennessee Valley Authority
T. R. Bietsch	Nuclear Power Services, Inc.
G. G. Boyle	Micro Switch Division of Honeywell
S. Charnetski	Consultant
T. N. Crawford	Pacific Gas and Electric Company
J. M. Dahlquist, Jr.	Baltimore Gas and Electric Company
A. Ellis	Western Electric Corporation
R. Estes	Fluid Components, Inc.
H. T. Evans	PYCO, Inc.
L. C. Fron	Detroit Edison Company
R. L. Givan	Sargent and Lundy Company
W. Gordon	Bechtel Power Corporation
S. C. Gottilla	Gibbs & Hill
T. Grochowski	Babcock & Wilcox Company
H. Hopkins	Utility Products of Arizona
R. Hubby	Leeds and Northrup Company
J. R. Karvinen	Mountain States Energy, Inc.
J. V. Lipka	Gilbert/Commonwealth, Inc.
L. R. McNeil	INPO
G. C. Minor	MHB Technical Associates
J. A. Nay	Westinghouse Electric Corporation
P. C. Newcomb	Combustion Engineering, Inc.
I. Sturman	Bechtel Power Corporation
J. Tana, Jr.	EBASCO Services, Inc.
K. Utsumi	General Electric
F. Rosa	Nuclear Regulatory Commission
M. Widmeyer	Washington Public Power Supply

*Corresponding member

This standard was approved for publication by the ISA Standards and Practices Board in April 1986.

NAME	COMPANY
N. Conger, Chairman	Fisher Controls Company
W. Calder, III	The Foxboro Company
R. S. Crowder	Ship Star Associates
H. S. Hopkins	Utility Products of Arizona
J. L. Howard	Boeing Aerospace Company
R. T. Jones	Philadelphia Electric Company
R. Keller	The Boeing Company
O. P. Lovett, Jr.	ISIS Corporation
E. C. Magison	Honeywell, Inc.
A. P. McCauley	Chagrin Valley Controls, Inc.
J. W. Mock	The Bechtel Group, Inc.
E. M. Nesvig	ERDCO Engineering Corporation
R. Prescott	Moore Products Company
D. E. Rapley	Rapley Engineering Services
C. W. Reimann	National Bureau of Standards
J. Rennie	Factory Mutual Research Corporation
W. C. Weidman	Gilbert/Commonwealth, Inc.
K. Whitman	Fairleigh Dickinson University
*P. Bliss	Consultant
*B. A. Christensen	Continental Oil Company
*L. N. Combs	Consultant
*R. L. Galley	Consultant
*T. J. Harrison	IBM Corporation
*R. G. Marvin	Roy G. Marvin Company
*W. B. Miller	Moore Products Company
*G. Platt	Consultant
*J. R. Williams	Stearns Catalytic Corporation

*Director Emeritus

TABLE OF CONTENTS
Title

Section		Page
1	Scope	9
2	Purpose	9
3	Definitions and Terminology	9
4	Pressure Boundary and Mechanical Design Requirements	10
	4.1 General	10
	4.2 Mechanical Design	10
5	Fabrication, Routing, Installation, and Protection	11
	5.1 General	11
	5.2 Selection of Piping Versus Tubing	12
	5.3 Assembly	12
	5.4 Sampling Taps	12
	5.5 Routing	13
	5.6 Fittings and Connections	15
	5.7 In-Line Components	16
	5.8 Valves	16
	5.9 Sample-Line Termination	17
	5.10 Supports and Mounting Structures	17
	5.11 Sample-Line Taps	18
	5.12 Flush and Backflush	18
	5.13 Common Sample Lines	19
	5.14 Bypass Lines	19
	5.15 Personnel Protection	19
Appendix A—Interface Standards and Documents		20

1 SCOPE

This standard covers design, protection, and installation of sample lines for light-water-cooled nuclear power plants and the pressure boundary requirements for piping and tubing. The boundaries of this standard span from the process tap to the upstream side of the sample panel, bulkhead fitting, or analyzer shutoff valve, and include in-line sample probes.

2 PURPOSE

This standard establishes the applicable requirements and limits for the design and installation of sample lines interconnecting nuclear safety-related power plant processes with sampling instrumentation.

This standard addresses the maintenance of the pressure boundary integrity of a sample line (in accordance with the appropriate parts of the American Society of Mechanical Engineers [ASME], *Boiler and Pressure Vessel Code,* Section III or American National Standards Institute [ANSI] B31.1* as applicable) and provides assurance that the protection function of nuclear safety-related sample instruments is available.

3 DEFINITIONS AND TERMINOLOGY

3.1 *Backflush*—The injecting of a fluid in a reverse flow manner to remove sample-line fluid or obstructions.

3.2 *Channel*—A collection of instrument loops, including their sample lines, which may be treated or routed as a group while being separated from instrument loops assigned to other redundant channels. (See ANSI/ISA-S67.02-80.)

3.3 *Flush*—The injecting of a fluid into the sample line at an upstream point to remove sample fluid from the downstream line.

3.4 *Grab-Sample Point*—The point in the sample line where the flow of sample fluid can be directed to a portable container. It may be referred to as "sample point."

3.5 *Instrument*—For the purpose of this standard, the device that performs some analysis of the sample fluid and for which a sample line is required and connected. Also referred to as "analyzer" or "monitor."

3.6 *Instrument Shutoff Valve*—The valve or valve manifold of the sample line located nearest the instrument. Also referred to as "component isolation valve."

3.7 *Isolation Valve*—The isolation valve nearest the instrument, grab-sample point, or in-line component which is available to personnel during normal plant operation. The root valve may or may not perform the function of the isolation valve, depending on its location.

3.8 *Lag Time*—An interval of time between the initiation of a discrete sample (particle, molecule, atom) at the sample tap to termination at a specific volumetric flow rate through the sample line.

3.9 *Main-Line Class*—Refers to the pressure and temperature ratings, the material from which the pipe is constructed, and the appropriate code, such as ANSI B31.1.

3.10 *Nuclear Safety-Related (NSR)*—Activation or control of systems or components that are essential to emergency reactor shutdown, containment isolation, reactor core-cooling, and containment and reactor heat removal, or are otherwise essential in preventing or mitigating a significant release of radioactive material to the environment, or are otherwise used to provide reasonable assurance that a nuclear power plant can be operated without any risk to the health and safety of the public.

3.11 *Purge*—Increasing the sample flow above normal for the purpose of replacing current sample-line fluid or removing deposited or trapped materials.

3.12 *Reach Rod*—A valve extension mechanism used to provide for manual operation of valves which are inaccessible.

3.13 *Root Valve*—The first valve located in a sample line after it taps off the process. It is typically located in close proximity to the sample tap. (See ANSI/ISA-S67.02-80.)

3.14 *Sample Line*—A piping and/or tubing system which removes fluid from a process either continuously or periodically for the purpose of determining the constituents or the physical properties of the process fluid. The sample line begins at the process tap or nozzle used for sampling, and terminates where the flow of sample fluid ends as a discrete and controlled entity.

3.15 *Sample Sink*—An installed device with controlled drainage and/or ventilation at which a grab sample may be obtained.

3.16 *Sample Tap*—The point where the sample line taps into the process line (pipe, duct, container) and the point where sample flow begins. It may also be referred to as "sample connections," "sample nozzle," or "process tap."

3.17 *Sample Vessel*—An integrally valved, portable sample container designed to obtain pressurized samples at process pressure.

---

*The titles of standards referenced within are listed in Appendix A.

3.18 *Inaccessible Area*—An area for which the radiation level, as defined by the Architect Engineer, precludes personnel entry during power operations and other operational situations. These areas are typically indicated by "zones," which depict accessibility based on various plant evolutions.

4 PRESSURE BOUNDARY AND MECHANICAL DESIGN REQUIREMENTS

4.1 General

4.1.1 All sample pressure boundary connections, tubing, piping, fittings, valves, and in-line sampling devices and equipment shall be in accordance with the requirements of ASME Section III, the requirements of ANSI B31.1, and this standard, as appropriate. Additional installation requirements are made by ASME *Performance Test Codes* PTC 19.2 and 19.11 and ASTM Standard D1192 for water and steam sampling. Additional installation requirements are made by ANSI N13.1-1969 for airborne radioactive materials sampling.

4.1.2 Sample-line classifications are derived from the process (source) pipe class. ASME Section III classes or other standards defined for the process pipe are matched in the sample line up through and including the root valve or other device(s) which permit a change of the code classification. ISA Standard S67.02 provides guidance in the determination of code classification.

4.1.3 Sample lines shall meet, as a minimum, ANSI B31.1 requirements from the root valve until sample-line termination up through the last in-line grab-sample isolation and/or throttling valve.

4.1.4 All lines classified as "ASME Section III" shall be designed and classified as "Seismic Category I." Seismic Category I may apply to other classifications (see ANSI B31.1). It is the responsibility of the Architect Engineer to establish the Seismic Category I applicability on a case-by-case basis.

4.1.5 Attachments to ASME Code equipment shall conform to the requirements of the ASME Code. For non-Code attachment or assembly, the weld shall be shown to have a stressed cross section sufficient to support the maximum design loads without exceeding allowable stresses.

4.2 Mechanical Design

The design of components, parts, and appurtenances utilized in the instrument sample lines under the scope of this standard shall be, as a minimum, in accordance with the stipulated design codes specified in this section.

4.2.1 Sample Lines in Accordance with ASME Section III

Where ASME *Boiler and Pressure Vessel Code* Section III requirements are identified as applicable by this standard, they shall apply to design, materials, fabrication, examination, testing, marking, stamping, and documentation of the sample line and all components included in the sample line.

Design and service limits for instrument sampling lines identified as ASME Class 1, 2, or 3 and Seismic Category I by this standard, shall be in compliance with ASME Section III. Moments due to earthquakes and other transient dynamic and static loadings shall be included.

1. ASME Class 1
Where instrument sampling lines are identified as ASME Class 1 by this standard, the applicable requirements of ASME Section III, Subsections NCA and NB, shall apply.

2. ASME Class 2
Where instrument sampling lines are identified as ASME Class 2 by this standard, the applicable requirements of ASME Section III, Subsections NCA and NC, shall apply.

3. ASME Class 3
Where instrument sampling lines are identified as ASME Class 3 by this standard, the applicable requirements of ASME Section III, Subsections NCA and ND, shall apply.

4.2.2 Sample Lines in Accordance with ANSI B31.1

Where ANSI B31.1 *Power Piping Code* requirements are identified as applicable by this standard, they shall comply with ANSI B31.1, with respect to materials, design, fabrication, examination, and testing.

The Quality Assurance Program requirements that are implemented should provide control over activities affecting quality to the extent consistent with the importance to safety of the sample lines.

Where instrument sample lines identified as ANSI B31.1 are interconnected with process piping systems classified as ASME Class 1, 2, or 3 and are identified as Seismic Category I, the following additional requirements shall apply:

1. A material manufacturer's certificate of compliance with the material specification shall be furnished for all pressure boundary items.

2. All pressure boundary items shall be pressure-tested in accordance with ANSI B31.1.

3. Design and service limits for instrument sampling lines identified as ANSI B31.1 and Seismic Category I by this standard shall be in accordance with ANSI B31.1.

4. The connection between ASME Section III and ANSI B31.1 components shall be in accordance with ANSI B31.1.

4.2.3 Temperature and Pressure Design

A sampling line and all its associated wetted parts shall be suitable for use at the design temperature and the design pressure of the process system from which the sample is drawn, except as follows:

1. If a reliable means of cooling is provided, then the sampling system downstream of the cooler may be designed to the lowest reduced temperature.

2. If a reliable means of pressure reduction is provided, then the sampling system downstream of the pressure-reduction means may be designed to the lowest reduced pressure.

A "reliable means" denotes a method that satisfies the single-failure criterion. For temperature, a passive cooling device, such as a length of pipe of suitable design transferring heat to the atmosphere, may serve. For pressure, a fixed-restriction orifice, though passive, if used alone is not acceptable. For temperature reduction and pressure reduction, active devices may be used provided that safeguard instrumentation is provided to protect the sample system in the event of failure of the active device.

4.2.4 Sample Cooler Design

Sample coolers, heat exchangers, and associated pressure boundary equipment for sampling shall conform to the following applicable requirements:

1. American Society of Mechanical Engineers (ASME), Section VIII of the *Boiler and Pressure Vessel Code*

2. American Society for Testing and Materials (ASTM), D1192

3. Tubular Exchanger Manufacturers Association (TEMA), Standards

4.2.5 Sample-Line Diameter Sizing

Line size should be based upon process requirements such as pressure and temperature, flow and fluid state, and installation requirements such as mechanical support strength and routing. To assist in the design of diameter sizing the following should be considered and documented.

4.2.5.1 Small (less than or equal to 3/4-inch [19-mm] nominal pipe size) sample lines provide advantages where space is limited and routing is complex. Due to the low mass, small sample lines provide an advantage where rapid cooling or heating of the line is required, and their low internal volume minimizes radioactive shielding. Small sample lines may also reduce class requirements in ASME Section III applications.

4.2.5.2 Large (greater than 3/4-inch [19-mm] nominal pipe size) sample lines provide high mechanical strength and large flow capabilities for viscous or particulate fluid. The increased mass provides an advantage where slow cooling or heating of the line is required.

4.2.6 Restriction Orifices

Restriction orifices may be installed in sample lines as a passive flow limiting device. Such devices should not be installed in applications where constituents of the sample flow may collect and release from the restriction. Restriction orifices may be used to change the Code classification of the sample line by limiting maximum sample flow should the sample line break. (For the use of a restriction orifice as a class break device, see ANSI/ISA-S67.02.)

The application of orifices for the purpose of downgrading the classification of a system shall be in concert with the other design considerations necessary to allow this action.

It is recommended that wherever practical the restriction orifice be installed as close as possible to the process piping. Restriction orifices shall not be used where fluid may plug orifice bore and should not be utilized where accumulation of radioactive particles is possible.

5 FABRICATION, ROUTING, INSTALLATION, AND PROTECTION

5.1 General

5.1.1 Sample lines and components shall be fabricated, routed, installed, and protected in the most efficient manner possible in accordance with this standard and related documents (see Appendix A).

5.1.2 Sample lines shall be as short as possible, consistent with design requirements and personnel safety. Sample lines shall have no unnecessary in-line components and shall have as few taps and dead legs as possible, consistent with required function.

5.1.3 Lag times should be as short as possible, as determined by the requirements of the sample. The desired sample time can be obtained by adjusting the flow rate and by proper choice of the sample line I.D.

5.1.4 Sampling lines shall be protected against mechanical loads; otherwise, additional wall thickness or support may be required for mechanical strength.

5.1.5 Any permanent taps, piping, and tubing provided for testing or calibration shall comply with this standard.

5.1.6 Where samples are being taken to measure particulates or other impurities which are expected to stratify, a multi-port-type sampling tap extending across the pipe diameter should be provided. Further, sample flow rates should be adjusted so that fluid velocity through the sample nozzles is the same as that which exists in the process lines (isokinetic sample rate).

The isokinetic sample rate should normally be adjusted for the flow rate expected at 100 percent load.

Where isokinetic sampling is required, it is important that there be sufficient upstream straight run ahead of the sample tap to assure a stable, predictable flow profile across the line. Straight-run requirements vary with the upstream configuration.

5.1.7 When sampling water, samples should be taken at a point where the fluid is turbulent. If one of the constituents to be measured is considerably heavier than water, or if there is two-phase flow, the points where centrifugal action may cause concentration of any constituents shall be avoided. (Water samples do not present much of a problem where the flow rate assures turbulent flow.) A Reynolds number of 4000 or greater is usually considered sufficient to assure turbulent flow, although there is some variation depending upon pipe size and wall roughness. Even with turbulent flow, the sample should be taken at a distance from the pipe wall (to avoid sampling a stagnant wall film).

5.1.8 In nonradioactive sample lines, the sample-line length should be kept to a minimum for applications where dissolved solids or suspended particulates are the constituent of interest in the sample. Local grab-sample points shall be as close to the process-sampling connection as practical.

5.2 Selection of Piping Versus Tubing

Tubing is generally preferred over piping because of lower initial cost and greater ease of handling. However, piping should be used in the following cases:

1. Where rigidity is required, as for line-mounted instruments which generally do not have other supports.

2. Where required as follows:

    a. To permit welding of austenitic steel to ferritic steel, utilizing an austenitic steel union with a dissimilar weld process

    b. To avoid overstressing tubing under the design conditions of pressure and temperature, where tubing of greater wall thickness does not meet design requirements and exceeds tubing fitting manufacturer recommendations

Wherever practical, stainless steel tubing should be used for sample lines. Except where rigidity is required, tubing of main-line material and of suitable strength may be substituted for main-line class piping.

5.3 Assembly

Sample lines shall be assembled using approved methods as required by codes, standards, and manufacturers' recommendations. Moisture, dirt, and other foreign material shall be cleaned from the sample-line interior.

5.3.1 Welding

Welding procedures and welding performance qualifications shall be in accordance with the latest edition and addenda of ASME Section IX at the time of qualification. Filler materials used for qualification and/or installations shall meet the requirements of ASME Sections IX and XI, Part C, as applicable.

5.3.2 Assembly of Fittings

Flareless fittings, flexible metal-hose fittings, threaded fittings, and other joining devices shall be assembled and installed as recommended by the manufacturer (see Section 5.6).

5.3.3 Other assembly methods shall be qualified by test for all structural and environmental design conditions, including vibration and mechanical cycling.

5.4 Sampling Taps

Sample taps shall be installed to provide a representative sample flow. Sample nozzles shall be used when a surface tap does not provide representative flow.

5.4.1 Sample taps shall be installed far enough downstream of injection points to the process so that the additional material will be well mixed in the process fluid.

5.4.2 Liquid sample connections should be made on the side of horizontal pipe runs. Steam and gas sample connections should be made on vertical pipe or duct runs wherever practical. Steam and gas sample connections to horizontal pipe or duct runs should be made on the top of the pipe or duct run.

5.4.3 An individual sampling tap shall not be required for a sampling line if there is a continuous source of system blowdown and if the location and design of the blowdown tap meets requirements for a sampling tap. In this case, the sample line may tap into the blowdown line or the blowdown line may be used as a continuous-flow sampling line if the sampling function does not interfere with the blowdown function and if the blowdown cannot damage the sensing element or interfere with the elements' function.

5.4.4 A sample line shall have its own dedicated sample tap from the main process connection.

5.4.5 Small-diameter sample taps into larger lines or equipment should be located away from areas where inspection or maintenance are necessary. Minimizing the number and locating such penetrations away from maintenance or inspection work areas can reduce radiation exposure.

5.5 Routing

5.5.1 General Routing Considerations

All sample lines including trays, supports, instrumentation, valving, and other in-line devices shall be installed with good engineering practices in order to avoid contact interferences caused by relative motion between the sample line and other adjacent equipment or devices. Sources of relative motion which shall be considered are thermal expansion, seismic motions, vibrations, and design basis accidents or events. The Code classification of the sample line will determine the requirements for relative motion that shall be considered.

5.5.1.1 Bends rather than fittings should be used to change the direction of a run of piping or tubing. The minimum bending radius for cold-bending of tubing using a bending tool shall be established by the designer. Larger-radius bends should be considered for slurry samples to lessen the possibility of blockage, and for radioactive air samples to lessen the possibility of particle plateout.

5.5.1.2 In the absence of further engineering analysis, sample lines including trays, supports, instrumentation, valving, and other in-line devices shall be installed with the following minimum contact surface separations:

1. Three-inch (75-mm) separation for adjacent large piping (2½ inches [65 mm] and larger), including insulation, appurtenances, and hardware attachments

2. Two-inch (50-mm) separation for adjacent small piping (2 inches [50 mm] and smaller), including insulation, appurtenances, and hardware attachments

3. Two-inch (50-mm) separation for other adjacent equipment and devices, such as cable trays, cable tray supports, mechanical equipment, electrical equipment, ventilation ductwork, and duct supports

5.5.1.3 Sample lines shall be routed to avoid environmental extremes wherever practical. Sample lines routed through outside or low-temperature areas shall be heat-traced as necessary to prevent freezing, condensing, or solidification with loss of sample flow. Sample lines shall be routed away from significant sources of heat if the source of heat will have a detrimental effect on the sample or if the rating of the sample line may be exceeded by additional heating.

5.5.1.4 Gas-sampling lines should have a continuous downward slope toward the source to promote their being kept free of liquid. The slope is preferred to be 1 inch, or more, per foot (80 millimeters, or more, per meter) of run. Where the preferred minimum slope cannot be obtained, the sample lines shall be installed to the maximum slope available and in no case shall be less than ¼ inch per foot (20 millimeters per meter). Liquid-sampling lines need not be sloped; however, standard process piping installation standards shall be used. High-point vent and drains shall be used as necessary to provide system drainage.

5.5.1.5 Process or sampling fluids shall not be piped to an emergency shutdown panel, technical support center, or a control room.

5.5.1.6 Sampling lines should be run along walls, columns, or ceilings, avoiding open or exposed areas to decrease the likelihood of damage to the sample lines by pipe whip, missiles, jet forces, or falling objects.

5.5.1.7 Sample lines should be routed in such a manner that they do not obstruct normal personnel passage within the plant. Additionally, if the sample line is for atmospheric samples, the line should be directed downward to prevent the entry of foreign matter.

5.5.1.8 Valves, orifices, and instrument taps shall not be located on any piping inside an inaccessible area. Where this cannot be achieved, follow the guidelines of Section 5.8.6.

5.5.1.9 Portions of the pipe run which require relatively frequent inspection or maintenance should be grouped and arranged to be readily accessible. The intent is to provide optimum access, thereby minimizing the time required for performance of activities in radiation exposure areas.

5.5.1.10 Orientation of process taps for liquid service is preferred on the horizontal side of the process pipe; for gas service, vertical top side of the process pipe is preferred. Orientation of process taps for liquid or gas shall never be located on the vertical bottom side of the process pipe.

5.5.2 Routing for Lines Containing Radioactive Fluids

5.5.2.1 Shielding for sample lines containing radioactive fluids should be used when practicable.

5.5.2.2 Lines containing radioactive fluids shall not be routed through areas requiring low background radiation levels, such as laboratory counting rooms, without being properly shielded.

5.5.2.3 Before routing a line containing radioactive fluids through a low-radiation zone, determine that the line can be properly shielded along its entire length.

5.5.2.4 Lines should be routed taking into consideration the necessity for portable or temporary shielding.

5.5.2.5 Lines containing radioactive fluids should be routed away from components which require frequent maintenance. Additionally, lines containing radioactive fluids shall be away from doorways, accessways, labyrinths, stairways, or ladders.

5.5.2.6 Lines containing radioactive fluids shall not be routed with dead-legs or low points which cannot be readily drained or flushed. Draining and flush connections should be placed at appropriate places accessible from outside the pipeway for all radioactive lines in the pipeway.

5.5.2.7 For tubing and piping penetrating shield walls, care shall be taken to minimize personnel exposure to radiation "streaming" from radioactive sources to the surrounding areas through instrument sampling-line penetrations in the shield walls and the sample line itself. One of the following methods should be used:

1. All instrument sampling-line penetrations shall be located at a minimum height of 7 feet (2.5 m) above floor level, or

2. The tubing penetrations shall be pitched toward an inner corner of the operating compartment, avoiding a direct radiation-streaming path, and

3. The tubing penetration shall be arranged as a labyrinth pattern, or

4. If methods A, B, or C above are not practical, the sampling lines penetrating the shield wall shall be surrounded by a pipe sleeve, and the open space between the sampling line and sleeve shall be filled with a radiation-absorbing material determined by the type of radiation expected. The sampling line shall make a right angle bend after it leaves the sleeve, and a radiation shield shall be placed in front of the line where it exits.

5.5.3 Safety-Related Sample-Line Routing

5.5.3.1 Redundant instrumentation sample lines shall be routed and protected so that any credible effect (consequence) of a design basis event which is to be mitigated through those sample lines shall not render redundant sample lines inoperable, unless it can be demonstrated and documented that the protective function is still accomplished. This level of protection shall assure that, after the event, a single failure shall not prevent mitigation of that event. Credible effects of design basis events which do not depend on a given group of redundant instrument sample lines for mitigation or accident prevention may render inoperable any or all of that group of sample lines without violating this criterion, provided that the overall protective function is accomplished. All nuclear safety-related instrument sample lines should be protected from damage during normal operating occurrences.

5.5.3.2 The minimum separation between redundant instrument sample lines and taps shall be at least 18 inches (500 mm) in air, in nonmissile, nonhigh-energy jet stream, nonpipe, whip, or nonhostile areas. As an alternative, a suitable steel or concrete barrier shall be used. When a suitable barrier is used, it shall extend at least 1 inch (25 mm) beyond the line of sight between redundant sample lines and shall be designed and mounted to Seismic Category I requirements. In hostile areas subject to high-energy jet stream, missiles, and pipe whip, the separation shall be provided by space in air, steel or concrete barriers, or both, and documented with analysis or calculations as necessary to prove that the separation protects the redundant sample lines from failure due to a common cause. All barriers shall be designed and mounted to Seismic Category I requirements.

5.5.3.3 Where redundant instrument sample lines penetrate a wall or floor, the required separation or barriers shall be maintained.

5.5.3.4 Safety-related redundant sample lines shall not share the same containment penetration. Wherever practical, redundant sample lines shall be routed to separate penetrations, pipe enclosures, or other passageways. Where practical, sample lines should be routed on opposite sides of installed piping or equipment and on opposite walls of piping enclosures or other passageways. Existing unrelated equipment may be used to provide adequate separation.

5.5.4 Post-Accident Sample-Line Routing

Post-accident sampling lines shall be routed, installed, and shielded such that under accident conditions the personnel radiation exposure for a grab-sample collection will be minimized. This exposure limitation includes entry, collection, flush, purge, and exit times as well as any time required to manipulate shielded sample containers and other equipment.

5.6 Fittings and Connections

5.6.1 In the absence of any existing standards, the designer shall determine that the type of fitting selected is qualified for the design conditions, including vibration, temperature, pressure, thermal shock, material compatibility and applicable environmental conditions, or demonstrated by test to be able to perform its intended function.

5.6.2 Welded Fittings

5.6.2.1 Piping and Tubing

Welded fittings should be used in the cases listed below. Butt-welded fittings should be considered for large (over 1-inch [25-mm]) radioactive sample lines where socket-weld fittings would create radioactive particle traps.

1. If severe erosion, crevice corrosion, shock, or vibration is expected to occur

2. Radioactive systems

3. Hazardous fluid systems

4. Post-accident sampling systems

Instruments and in-line components which require frequent removal for maintenance should not have welded fittings.

Welding procedures and materials shall meet the requirements of this standard. (See Section 5.3.1.)

5.6.2.2 Tubing

For tubing, flareless fittings may be used except in the cases listed below, which require the use of welded fittings:

1. Where a sample line must be protected against inadvertent disassembling

2. For those portions of sample lines that are inaccessible or may be hidden from inspection

The mating parts of the original flareless tube fittings shall be designed and manufactured by the same company. Replacement parts from different manufacturers may be used, provided both of the manufacturers' guidelines for the use of their fittings is met concerning design service conditions, or provided the complete fitting shall be replaced when service is needed. The connection between the root valve and the sample tap shall be welded.

In the interest of leak integrity, care should be exercised to protect mating seal surfaces before and after original makeup. In particular, care should be exercised during maintenance; body and nut, ferrule, and tube assembly should be protected from dirt and physical damage.

5.6.3 Threaded Pipe Fittings

5.6.3.1 Threaded pipe fittings are not recommended for general use in nuclear power plant sampling systems. Threaded pipe fittings may be used as necessary with instruments or components which have threaded connections as supplied or when a welded attachment cannot be made.

5.6.3.2 Threaded fittings should be seal-welded. (Exception is made for instruments with integral electronics or other heat-sensitive components.)

5.6.3.3 National Pipe Thread (NPT) connectors may be used; however, the assembly shall comply with the requirements of ANSI B2.1 and B31.1.

5.6.3.4 Tapered pipe threads shall conform to ANSI B2.1. However, tapered pipe threads shall not be used as takedown joints when repeated disassembly and reassembly are planned. Straight thread fittings with metal-to-metal or resilient seals are permitted.

5.6.3.5 Appropriate thread lubricants or compounds should be used in the assembly. For threaded connections of stainless steel to stainless steel, lubrication is required to prevent seizing and galling and to promote sealing. The lubricants shall meet ASME NQA-1 for chloride and halogen content.

5.6.4 Flanges

5.6.4.1 Flanges may be used for in-line components which must be periodically removed. Flanges must be design-rated for the component as well as for the sample line at the point of insertion.

5.6.4.2 Flanges shall comply with ANSI B16.5 unless they are of proprietary design and are provided as an integral part of an instrument.

5.6.4.3 Flange gaskets shall be rated for sample-line pressure and temperature and should be of the spiral round design.

5.6.4.4 Flanges shall be installed so that they can be inspected visually and are readily accessible for maintenance.

5.6.5 Flexible Metal Hose

5.6.5.1 Shielded flexible metal hose may be used in sampling lines to accommodate thermal, seismic, and vibrational motions if its ratings equal or exceed the design requirements, including service life for the sampling line. It shall be installed in accordance with the manufacturer's instructions. Flexible metal hose shall be used within the limits of the manufacturer's ratings. The internal wetted surface shall be compatible with the sample fluid. Flexible metal hose should not be used with highly corrosive or highly radioactive fluids. Any barriers or screens used for protective purposes shall allow for the design motions of the flexible metal hose without contact. The manufacturer's fittings and adapters shall be used at each end to connect to the sample line or component.

5.6.5.2 For sample connections to high-pressure (>900 psi [6200 kPa]) high-temperature (>260° C [500° F]), radioactive, or otherwise hazardous fluids, flexible tubing or hose shall not be used between the process tap and the root valve. Nonshielded flexible hose or tubing may be used at a sample sink for low-pressure, low-temperature, and nonhazardous applications.

5.7 In-Line Components

In-line components include all devices through which the sample flows from process connection to sample-line termination. It includes all pressure boundary components except as otherwise covered by this standard. Examples are coolers, degassers, filters, in-line instrument probes, strainers, expansion coils, delay coils, etc.

5.7.1 Expansion Coils

Expansion coils which allow for thermal, seismic, or vibrational motion may be used if the expansion coil equals or exceeds the design requirements for the sampling line. (**Note:** Expansion coils increase sample-line volume and lag time.)

5.7.2 Sample Coolers

Sample coolers should be installed such that the sample flow enters the tube from the top and exits the bottom, and the cooling fluid enters the shell from the bottom, and exits the top.

5.7.3 Sample Pumps

Sample pumps should be properly sized to provide the required flow. Sample pumps shall have a pressure-relief capability from discharge to suction, or discharge to a suitable container.

5.7.4 Strainers and Filters

5.7.4.1 Strainers or filters should not be used in sample lines unless

1. The strainer or filter is necessary for the protection of an in-line instrument, or

2. The strainer or filter is necessary for sample preparation or conditioning.

5.7.4.2 If a filter or strainer is installed in a sample line for a startup service, then there shall be a direct or straight bypass line, and the filter or strainer shall be tapped off the main sampling line. Two isolation valves shall be provided to isolate the filter. These isolation valves shall be installed as close as practical to the sample line to reduce the volume of dead leg.

5.7.5 Moisture Traps

Use of moisture traps should be limited to those applications where they do not have an adverse effect on the sample.

5.8 Valves

5.8.1 Sampling systems shall have as few valves installed as practical, consistent with code, design, and functional requirements.

5.8.2 Valves shall be installed with the stem oriented in the most accessible direction, consistent with other requirements.

5.8.3 Where practical, valves shall be located in low-radiation zones or in locations where radiation exposure to personnel will be minimal.

5.8.4 If a reach rod must be used, it shall be less than 10 feet (3 m) long and should have no more than one gear box.

5.8.5 The pressure-temperature ratings (class) shall be as required by the main-line class until such point that the sample line is reduced to a lower class.

5.8.6 Root Valves

Root valves shall be installed as close as practical to the process connection. The root-valve stem and handle shall be clear of any main-line insulation. The root valve shall be made accessible. If a root valve close to the process connection is not accessible during normal operations due to location or radiation exposure rate, then the root-valve function shall be made accessible by one of the following procedures:

1. Relocate the process connection to an accessible location.

2. Install a reach rod.

3. Install a second root valve at the point that the sample line becomes accessible.

5.8.7 Component Isolation Valves

5.8.7.1 Valving shall be installed to allow for the isolation of in-line components or instrumentation as necessary for operation, maintenance, or calibration.

5.8.7.2 When the sample line terminates at an instrument or equipment (dead-end service), an upstream component isolation valve shall be installed.

5.8.7.3 In multicomponent sample lines which require continuous flow, individual component isolation and bypass valves may be installed.

5.8.7.4 Component isolation valves should be installed so that they are accessible from the component location.

5.8.8 Local Grab-Sample Valves

Local grab-sample valves shall be accessible from floor or platform and should be of convenient height. Sample-station collection nozzles shall be directed so that their discharge cannot inadvertently be directed at the operator.

5.8.9 Sample-Line Grab-Sample Valves

In welded systems, the last grab-sample valve may be installed with flareless connections if the sample line at that point does not require welded connections. This would allow for greater ease of replacement. Sample-station collection nozzles shall be oriented so that their discharge cannot inadvertently be directed at the operator.

5.9 Sample-Line Termination

5.9.1 Grab-Sample Termination

Grab-sample lines shall terminate as close as practical to a drain, drain funnel, sink, or collection sump suitable for their disposal. Adequate spacing shall be provided for the collection of samples.

5.9.2 Sample Sinks, Panels, and Stations

Arrangements should be provided for obtaining samples at central locations where sampling hoods are used to control the release and dispersal of airborne or gaseous radioactivity. Arrangements should be provided for radioactive drain capability and recirculation of the samples. The use of central locations reduces the time required to obtain radioactive samples and to transport the samples to the laboratory. Sample stations should be located such that representative samples can be obtained. In the event that a sampling station is used infrequently, its location in terms of accessibility and radiation dose takes on less significance. Generally, sampling stations should not be located in areas greater than 2.5 mrem/h (0.025 mSv). Criteria governing the *location* of sample stations are as follows:

1. Sample stations should be located to allow for collection of a representative sample for analysis.

2. The location assigned for sample stations should be within a readily accessible area.

3. Where practical, sample stations should be located in low-radiation zones where radiation exposure to personnel will be minimal.

4. Where possible, the sample stations should be close to the chemical laboratories.

5.9.3 Sampling Return to Process

If sampling lines are routed back to process or to where the discharge cannot be observed (including any flushing or calibration fluids used), they shall have flow monitoring in order to determine proper sampling.

5.9.3.1 Sample lines and components shall meet the main-line class requirements of the termination process pipe, duct, tank, or containment.

5.9.3.2 Sample-flow conditioning shall not degrade the process.

5.10 Supports and Mounting Structures

5.10.1 General Requirements

5.10.1.1 Mounting structures and their attachments shall have a seismic capability not less than that required by any of the supported transducers or auxiliary equipment.

5.10.1.2 Supports, brackets, clips, or hangers shall not be fastened to the sample line or components for the purpose of supporting cable trays or any other equipment.

5.10.1.3 The sample line shall not be supported by the instrument. Auxiliary supports shall be provided to restrain the instrument where the mass of the instrument puts unacceptable stress on the piping.

5.10.1.4 Hanger and support design shall include provisions for seismic, jet impingement, pipe whip, missiles, and thermal expansion of the process tap and instrument sampling line to which the hangers or supports may be subjected during normal operation, seismic or other credible event.

5.10.1.5 Sample lines shall be supported with anchors and guides. An anchor serves to clamp the line in place, while a guide will allow axial motion of the line. No more than one anchor point will be used in any straight run of line, and sufficient offset shall be provided on both sides of the anchor point to allow for the design thermal and seismic motions. If a vertical straight run of sample line has dead weight approaching or exceeding the compressive strength of the line, then

1. The line shall be bent or curved to allow for another anchor point with sufficient offset for design motions, or

2. An expansion loop shall be installed to allow for another anchor point with sufficient offset for design motions.

Such vertical runs should be avoided where practical.

5.10.2 Component Supports

5.10.2.1 Sample-line valves should not be supported by the sample-line tubing in an open run of tubing. Sample-isolation valves may be supported by tubing if the tubing is supported within 3 inches (75 mm) from the valve (as by a bulkhead union in a sampling panel) and if the appropriate code permits.

5.10.2.2 Sample-line root valves may be supported by the process pipe or duct as allowed by the appropriate code.

5.10.2.3 In-line isolation and the throttling valves may be supported by sample-line pipe as allowed by the appropriate code.

5.10.2.4 Valves on flexible metal hose shall be independently supported.

5.10.2.5 The weight of insulation, heat tracing, etc., shall be considered.

5.10.3 Operating Vibration

CAUTION

Shock mounts are typically low-pass mechanical filters and will often amplify vibrations. Special care in design is essential to avoid this problem.

Exposure to vibrational excitation due to pumps, turbines, or other sources should be avoided. Where high vibration is unavoidable, the equipment should be mounted on an adjoining nonvibrating surface; or, if no other reasonable alternative exists, the equipment should be isolated by shock mounting for the expected vibration frequency.

5.11 Sample-Line Taps

A tap made on a sample line and the lines and components connected to the tap shall meet the code and design requirements of the sample line until reduced by in-line components (class break). The main-line class of the sample line shall be maintained through an ancillary tap made on the sample line.

5.12 Flush and Backflush

5.12.1 Flushing lines shall have at least two means of isolation from the sample line to prevent sample flow through the purge or flushing line. The requirements of the sample line shall apply through the first form of isolation. The first flush valve shall be capable of throttling. It shall be readily accessible from any grab-sample point in high-temperature, high-pressure, or hazardous-fluid applications.

5.12.2 Where a comparison of design pressures between the sampling line and the flushing line indicates that contamination of the flushing fluid source is probable and if such contamination is not tolerable, then the second form of isolation should be a check valve.

5.12.3 Where practical, flush and backflush lines should be installed such that the flushing flow is upward in liquid systems and is downward in gas or vapor systems, for the purpose of flushing out unwanted liquid or gas.

5.12.4 Sample lines which have a flushing requirement to control radiation exposure shall have an isolation valve as close as possible to the last place of adequate shielding, and the flushing connection shall be made as near to this isolation valve as possible. Where practical, a manufactured manifold should be used.

5.12.5 Flush connections which could act as a trap or settling point for radioactive solids should be located on the top of the pipe or tube.

5.12.6 Backflush capability shall be installed for post-accident sampling lines to remove sample-line blockages.

5.13 Common Sample Lines

Sampling lines used to sample two or more nonidentical processes shall have valving to ensure selection of the desired sample source and positive isolation of the remaining sources. Where practical, this valving shall be accessible from the sampling-instrument or the grab-sample point. Positive indication of the sample source shall be provided at the grab-sample point and in the area of the sampling instrument. Any alarms or indications derived from the sampling instrument shall clearly indicate the sample source.

5.14 Bypass Lines

5.14.1 Sampling instruments shall always be installed in the main sampling line (main flow path). They shall never be installed in the bypass line.

Radiation-monitoring sample lines which are designed for continuous sampling, but which require periodic stopping of sample flow for calibration, checking, or maintenance shall have bypass capability, with the bypass flow routed to a safe sample termination point.

5.15 Personnel Protection

5.15.1 Hot-sample lines in liquid or gas service shall not be insulated except where sample cooling is undesired. Screens or covered trays shall be provided for personnel protection in normally accessible areas.

5.15.2 High-pressure components shall be installed behind a sample sink or panel for personnel safety in the event of component rupture.

5.15.3 Remotely operated or automatic sampling stations should be used in applications where accumulated personnel dose can be significantly reduced by avoiding exposing personnel to the fluid being sampled.

APPENDIX A
INTERFACE STANDARDS AND DOCUMENTS

This appendix is included for information only and is not a part of the standard.

A.1 American National Standards Institute (ANSI)

The following documents, approved by the American National Standards Institute, 1430 Broadway, New York, NY 10018, were considered.

- B2.1 (1968), "Pipe Threads," (Except Dryseal) Specifications, Dimensions, and Gaging for Taper and Straight Pipe Threads, Including Certain Special Applications," (ASME).
- B16.5 (1981), "Pipe Flanges and Flanged Fittings" (ASME).
- B31.1 (1983), "Power Piping," with 1984 Addenda for Parts A, B, and C" (ASME).
- N13.1 (1969), "Guide to Sampling Air Borne Radioactive Materials in Nuclear Facilities."

A.2 American Society of Mechanical Engineers (ASME)

The publications which follow can be obtained from the American Society of Mechanical Engineers, 345 East 47th Street, New York, NY 10017.

- Section III, *Boiler and Pressure Vessel Code*, "Nuclear Power Plant Components."
- Section VIII, *Boiler and Pressure Vessel Code*, "Nuclear Power-Plant Components."
- Section IX, *Boiler and Pressure Vessel Code*, "Nuclear Power Plant Components."
- Performance Test Codes PTC 19.2 (1964), "Pressure Measurement Instruments and Apparatus."
- Performance Test Codes PTC 19.11 (1970), "Part 11. Water and Steam in the Power Cycle (Purity and Quality, Leak Detection and Measurement) Instruments and Apparatus (R 1984)."
- Performance Test Codes PTC 23 (1958), "Atmospheric Water-Cooling Equipment."
- NQA-1 (1983), "Quality Assurance Program Requirements for Nuclear Facilities."

A.3 Instrument Society of America (ISA)

The following documents can be obtained from the Instrument Society of America, 67 Alexander Drive, P.O. Box 12277, Research Triangle Park, NC 27709.

- ANSI/ISA-S67.01 (1979), "Transducer and Transmitter Installation for Nuclear Safety Applications."
- ANSI/ISA-S67.02 (1980), "Nuclear-Safety-Related Instrument Sensing Line Piping and Tubing Standards for Use in Nuclear Power Plants."

A.4 *U.S. Code of Federal Regulations*, Title 10, Part 50 (1974), "Domestic Licensing of Production and Utilization Facilities," (revised January 1986).

A.5 American Society for Testing and Materials (ASTM)

The following document is available from the American Society for Testing and Materials, 1916 Race Street, Philadelphia, PA 19103.

- D1192 (1970), "Specification for Equipment for Sampling Water and Steam" (R 1977).

A.6 Tubular Exchanger Manufacturers Association (TEMA)

The following documents are available from the Tubular Exchanger Manufacturers Association, 25 North Broadway, Tarrytown, NY 10591.

- Standards of Tubular Exchanger Manufacturers Association (Includes 1982 Addenda)
- Sec. 4E-78, Installation, Operation, and Maintenance
- Sec. 12RGP-78, Recommended Good Practice

ANSI/ISA-S67.14-1983
Approved November 17, 1983

American National Standard

Qualifications and Certification of Instrumentation and Control Technicians in Nuclear Power Plants

Instrument Society of America

ISBN 0-87664-788-3

ISA-S67.14 Qualifications and Certification of
Instrumentation and Control Technicians
in Nuclear Power Plants

Copyright © 1983 by the Instrument Society of America. All rights reserved. Printed in the United States of America. No part of this publication may be reproduced, stored in a retrieval system, or transmitted, in any form or any means electronic, mechanical, photocopying, recording or otherwise without the prior written permission of the publisher.

INSTRUMENT SOCIETY OF AMERICA
67 Alexander Drive
P.O. Box 12277
Research Triangle Park, North Carolina 27709

Copyright © 1983 by the Instrument Society of America

PREFACE

This Preface is included for informational purposes and is not part of ISA-S67.14.

This standard has been prepared as a part of the service of the Instrument Society of America toward a goal of uniformity in the field of instrumentation. To be of real value, this document should not be static, but should be subjected to periodic review. Toward this end, the Society welcomes all comments and criticisms, and asks that they be addressed to the Secretary, Standards and Practices Board, Instrument Society of America, 67 Alexander Drive, P.O. Box 12277, Research Triangle Park, NC 27709, Telephone (919) 549-8411.

The ISA Standards and Practices Department is aware of the growing need for attention to the metric system of units in general, and the International System of Units (SI) in particular, in the preparation of instrumentation standards. The Department is further aware of the benefits to USA users of ISA standards of incorporating suitable references to the SI (and the metric system) in their business and professional dealings with other countries. Toward this end, this Department will endeavor to introduce SI - acceptable metric units in all new and revised standards to the greatest extent possible. The Metric Practice Guide, which has been published by the Institute of Electrical and Electronics Engineers, Inc. as ANSI/IEEE 268-1982, and future revisions, will be the reference guide for definitions, symbols, abbreviations, and conversion factors.

It is the policy of the Instrument Society of America to encourage and welcome the participation of all concerned individuals and interests in the development of ISA standards. Participation in the ISA standards making process by an individual in no way constitutes endorsement by the employer of that individual of the Instrument Society of America or any of the standards which ISA develops.

The purpose of this standard is to provide the nuclear power industry with bases for certifying the qualifications of instrumentation and control technicians who work on equipment that is important to safety. This certification is intended to reduce the possibility that unqualified personnel could perform improper maintenance on plant equipment important to safety.

The standard does not intend to say what those qualifications should be since they will differ by plant I&C Department organization and company job descriptions. Appendix A is meant to be a guideline only.

This standard is meant to be a document subordinate to and elaborating on both:

- Qualifications of Inspection, Examination and Testing Personnel for Nuclear Power Plants, ANSI/ASME N45.2.6 - 1978; and
- Selection, Qualification and Training of Personnel for Nuclear Power Plants, ANSI/ANS 3.1 - 1981.

These documents (and their newer draft versions) tell what must be done; this standard, how to certify it. The standard committee was aware of the The Institute of Nuclear Power Operations (INPO) "Guidelines for Instrumentation and Control Technician Qualification July 1981" as a document describing possible technical content descriptions for qualifications and/or training, but did not endorse its application to every plant or situation. This standard tells users how to certify competency and proficiency to whatever qualification specifications a plant or company wishes to cite as applicable in their case.

This standard describes four (4) Technician categories or levels which are meant to be functional descriptions of typical skill proficiency and competency levels identified at numerous plants as well as empirically in job analyses. They are essentially points of reference for application of this standard to a specific plant situation. No specific number of categories, levels or classifications are required by this standard. Each plant or company has their own organizational structure which works best for them. All they need to do is establish a certification program, procedure and/or plan which cross-references how they will relate their job descriptions/categories to the four (4) functional levels. This standard has no intention to change organizational arrangements already in place. It is recommending a functional organization only. Utilities may also take exception to our description by categorizing technicians horizontally as Computer Technicians or Repairmen, etc. The key to this interface is the existence of a good representative job description of each type of technician at a station or plant.

The following individuals served as members of the ISA Subcommittee SP67.14 which prepared this standard:

NAME	COMPANY
R. W. Bailey, Chairman	Transportation Research Center of Ohio/The Ohio State University
R. E. Bebermeyer	Consumers Power Company
K. R. Carlton	Johnson Controls
S. Ferguson	NUS Corporation
S. B. Fowler	Westinghouse Electric Company
T. Gogniat, retired	Westinghouse Electric Company
L. Goodrum	Plant Instrument Testing, Inc.
M. Griffis	Georgia Power Company
R. Hinson	South Carolina Electric and Gas

R. Howarth	General Physics Corporation
K. C. Jones	Pennsylvania Power and Light Company
J. R. Karvinen	Mountain States Energy
L. McNeil	General Physics Corporation
W. R. Meyer	Public Service Electric and Gas
D. W. Miller	The Ohio State University
O. G. Parker	State Technical Institute at Memphis
R. T. Profant, Jr.	Department of Energy Savannah River Operations Office
R. G. Profeta	General Electric Instrument and Control Training
J. M. Reed	Consumers Power Company
L. H. Rice, Jr.	Public Service Electric and Gas
S. Richardson	United States Nuclear Regulatory Commission
A. I. Rylander	Houston Lighting and Power
J. C. Stultz	Transportation Research Center of Ohio/The Ohio State University
K. P. Voigt III	Philadelphia Electric Company

This standard was approved by ISA SP67 in January 1982.

NAME	COMPANY
H. Hopkins, Chairman	Westinghouse Electric Co.
B. W. Ball	Brown & Root Inc.
G. G. Boyle	Micro Switch Division
T. N. Crawford	Pacific Gas & Electric Co.
J. M. Dahlquist	Baltimore Gas & Electric
H. T. Evans	Baltimore Gas & Electric
L. C. Fron	Detroit Edison
R. L. Givan	Sargent & Lundy Co.
L. F. Griffith	Yarway
T. Grochowski	Babcock & Wilcox Co.
R. J. Howarth	General Physics Corp.
R. N. Hubby	Leeds & Northrup Co.
J. R. Karvinen	Mountain States Energy
J. V. Lipka	Gilbert Associates
S. F. Luna	General Atomic Co.
F. K. Luteran	Weston Components & Controls
D. W. Miller	The Ohio State University
G. C. Minor	MHB Technical Assoc.
J. W. Mock	Bechtel
J. A. Nay	Westinghouse Electric Corp.
P. C. Newcomb	Combustion Engineering Inc.
G. E. Peterson	Commonwealth Edison
R. L. Phelps	Edison Company
M. F. Reisinger	Electronics Assoc. INc.
U. Shah	Alyeska Pipeline Service Co.
H. G. Shugars	Electric Power Research Inst.
J. Tana	Ebasco Services
R. J. Ungaretti	EMC Controls
K. Utsumi	Ebasco Services Inc.
E. C. Wenzinger	Nuclear Regulatory Comm.
W. C. Weston	Stone & Webster Eng. Corp.

This standard was approved for publication by the ISA Standards and Practices Board in June 1983.

NAME	COMPANY
W. Calder III, Chairman	The Foxboro Co.
N. Conger	Conoco
B. Feikle	Bailey Controls
T. J. Harrison	IBM Corp.
H. S. Hopkins	Westinghouse Electric Co.
J. L. Howard	Boeing Aerospace Co.
R. T. Jones	Philadelphia Electric Co.
R. Keller	The Boeing Company
O. P. Lovett	Retired
A. P. McCauley	Diamond Shamrock
E. C. Magison	Honeywell Inc.
J. W. Mock	Bechtel
E. M. Nesvig	ERDCO
R. Prescott	Moore Products Co.
D. E. Rapley	Stearns-Roger Inc.
W. C. Weidman	Gilbert Associates Inc.
K. A. Whitman	Allied Chemical Corp.
J. R. Williams	Stearns-Roger Inc.
P. Bliss*	
B. Christensen*	
L. N. Combs*	
R. L. Galley*	Consultant
R. G. Marvin*	
W. B. Miller*	Moore Products Co.
R. L. Nickens*	
G. Platt*	Bechtel Power Corp.

*Directors Emeritus

TABLE OF CONTENTS

SECTION   **PAGE**

1 Scope 9
2 Purpose 9
3 Definitions 9
4 Technician Categories 9
5 General Criteria 10
6 Certification of Qualifications 10
7 Education and Experience Requirements 10
8 Recertification 11
9 Bibliography 11
Appendix A (Typical Knowledge and Skills List) 13

1 SCOPE

This standard identifies the criteria for certification of instrumentation and control technicians in nuclear power plants. These criteria address qualifications based on education, experience, training, and job performance.

2 PURPOSE

The purpose of this standard is to provide the nuclear power industry with bases for certifying the qualifications of instrumentation and control technicians who work on equipment that is important to safety.

3 DEFINITIONS

3.1 Control Loop

Two or more devices processing a single variable which may provide an input signal to a control system.

3.2 Control System

A system in which deliberate guidance or manipulation is used to achieve a prescribed value of a variable. (Ref:ANSI/ISA-S51.1-1979)

3.3 Device

An apparatus for performing a prescribed function. (Ref: ANSI/ISA-S51.1-1979)

3.4 Direction

Having the person who is qualified to perform the task physically present when the task is performed or in continuous communication with the person performing the task.

3.5 Experience

Applicable work in design, construction, pre-operational and startup testing activities, operation, maintenance, on-site activities, or technical services. Observation of others performing work in the above areas is not experience. This experience can be obtained during startup or operations in a nuclear power plant, in fossil power plants, in other industries, or in the military.

3.6 Group Leader

Shall be the person in the highest level of supervision whose responsibilities are oriented solely toward instrumentation and control.

3.7 Important to Safety

Those structures, systems, and components that provide reasonable assurance that the facility can be operated without undue risk to the health and safety of the public. (Ref: 10 CFR Part 50, Appendix A)

3.8 Instrumentation

A collection of instruments or their application for the purpose of observation, measurement, or control. (Ref: ANSI/ISA-S51.1-1979)

3.9 Knowledge

Familiarity with theory and concepts, and detailed understanding of job-related topics.

3.10 Shall, Should and May

The word "shall" is to be understood as a requirement, the word "should" as a recommendation, the word "may" as permissive, neither mandatory nor recommended. (Ref: ISA-RPA-la-1970)

3.11 Skill

The ability to demonstrate the practical application of knowledge.

3.12 Technical Supervision

Providing guidance as needed to a subordinate in the performance of an assigned task.

4 TECHNICIAN CATEGORIES

4.1 Introduction

This section describes four (4) Technician categories or levels which are meant to be functional descriptions. No specific number of categories, levels, or classifications are required.

4.2 Technician I

4.2.1 Shall be a person who performs simple tasks without direction on devices which perform basic functions such as temperature, pressure, flow, or level measurement.

4.2.2 Shall be a person who can perform Technician II level tasks under the direction of a Technician II.

4.2.3 Shall be a person who can provide technical assistance or instruction or both during on-the-job training to a Technician I.

4.3 Technician II

4.3.1 Shall be a person who can perform the tasks of a Technican I and tasks such as troubleshooting, calibration, and repair without direction on instrumentation and control loops.

4.3.2 Shall be a person who can perform Technician III level tasks under the direction of a Technician III.

4.3.3 Shall be a person who can provide technical assistance or instruction or both during on-the-job training to a Technician I or Technician II.

4.4 Technician III

4.4.1 Shall be a person who can perform the tasks of a Technician II and can perform complex troubleshooting, calibration, and repair without direction on instrumentation and control systems and equipment.

4.4.2 Shall be a person who can perform Technician IV level tasks under the direction of a Technician IV.

4.4.3 Shall be a person who can provide technical assistance or instruction or both during on-the-job training to a Technician I, Technician II, or Technician III.

4.5 Technician IV

4.5.1 Shall be a person who can perform the tasks of a Technician III and can provide technical supervision to other technicians.

4.5.2 Shall be a person who can provide technical assistance or instruction or both during on-the-job training to a Technician I, Technician II, Technician III, or Technician IV.

4.6 The terms "Technician I", "Technician II", "Technician III", "Technician IV" and "Group Leader" need not be used as job titles at a specific plant. However, the job titles and job descriptions shall be correlated with the above terms in the plant certification program.

5 General Criteria

5.1 The Instrumentation and Control (I&C) Group in a nuclear power plant shall consist of a Group Leader and one or more persons who are certified in the categories of Technician I, Technician II, Technician III, and/or Technician IV. The I&C Group Leader's qualifications are defined in ANSI/ANS-3.1-1981, "Selection and Training of Personnel for Nuclear Power Plants." The Group Leader may also be a Technician IV. Personnel who are not qualified in one of the above categories may be a part of the I&C Group provided they do not work on equipment important to safety without direction.

5.2 A program to certify the knowledge and skills qualifications that are not plant specific shall be instituted by the employer of personnel in the technician categories. A program to certify the knowledge and skills qualifications that are plant specific shall be instituted by the Licensee of the plant in which personnel in the technician categories work. The Licensee has ultimate responsibility for assuring that every technician performing work on plant equipment is properly certified at the appropriate level. The program requirements are specified in Section 6. For example, a contractor providing the technician would certify the technician in non-plant specific items while the Licensee who is operating the plant would certify the contractor's technician in plant specific items only. The Licensee would certify its own technician in both non-plant specific and plant specific items. The qualifications for each technician category shall be based upon the job description of each category in a specific plant. The job description shall be a list which may not be complete but shall clearly identify the types of tasks each level of technician can perform without direction.

5.3 In addition to the knowledge and skill items, there shall be minimum education and experience requirements for each technician category. These requirements are specified in Section 7.

5.4 In order for each person to meet the qualifications in the technician categories, a training program should be an integral part of the certification program. An example of the skill and knowledge subjects for a training program is provided in Appendix A.

6 Certification of Qualifications

6.1 Each technician shall be tested on at least a representative sample of the knowledge and skill items on the job description of the category for which the technician is to be certified.

6.2 Knowledge proficiency shall be demonstrated and documented by written and/or oral examination.

6.3 Skill proficiency shall be demonstrated and documented by observation of the technician's performance of actual or simulated tasks.

6.4 A technician may be certified to individually perform tasks of the next higher technician category by successful completion of the knowledge and/or skill proficiency examination(s) directly related to those tasks.

6.5 The Licensee shall maintain documentation, in accordance with existing plant and personnel procedures, on each technician to show the following items:

(1) Name of employer and certifying agency
(2) Name of certified individual
(3) Level of technician certification
(4) Dates of effective period of certification
(5) Signature of employer's or certifying agency's designated representative
(6) Training records to support certification
(7) Education and experience from outside sources
(8) Evaluation by the immediate supervisor

6.6 The Licensee shall maintain documentation, in accordance with existing plant and personnel procedures, on skill proficiency, written, and oral examinations to show the following:

(1) Name of technician
(2) Name of evaluator
(3) Date of examination
(4) Pass/fail criteria
(5) Pass/fail status of the technician
(6) Questions and answer keys for written examinations
(7) Evaluation criteria and results for skill proficiency examinations
(8) Questions and pass/fail criteria for each question on oral examinations

6.7 The evaluation by the immediate supervisor shall include, but not be limited to comment on the technician's:

(1) Technical competence
(2) Quality of work

7 Education and Experience Requirements

7.1 The minimum education and experience requirements are specified below.

7.2 Technician I

(1) High school diploma or equivalent, as specified in Section 4.1 of ANSI/ANS-3.1-1981
(2) Six (6) months experience
(3) Certification of Technician I knowledge and skill qualifications

7.3 Technician II

(1) High school diploma or equivalent, as specified in Section 4.1 of ANSI/ANS-3.1-1981
(2) Three (3) years of working experience in his/her specialty
(3) Certification of Technician II knowledge and skills qualifications

or

(1) Associate degree or higher in electronics, instrumentation, or a related field
(2) One (1) year of working experience in his/her specialty
(3) Certification of the Technician II knowledge and skills qualifications

7.4 Technician III

(1) Education and experience requirements for Technician II
(2) Certification of the Technician III knowledge and skills qualifications

7.5 Technician IV

(1) Education and experience requirements of Technican II
(2) Certification of the Technician IV knowledge and skills qualifications

8 RECERTIFICATION

8.1 A technician must be recertified under the following conditions:

(1) Every two years
(2) When the Technician transfers to another plant or unit on which the Technician is not certified
(3) If the Technician does not work in his/her certification level for more than 180 consecutive days

8.2 Recertification requirements shall be equivalent in scope and difficulty to the current certification requirements. Satisfactory evaluation by the immediate supervisor and/or examination results may be the basis for recertification.

8.3 Failure to meet the recertification criteria shall cause the Technician's certification to be immediately removed and the cause shall be evaluated and documented by the immediate supervisor. The following are possible courses of action.

8.3.1 A retraining program may be initiated, with emphasis on the weak areas.

8.3.2 Recertification to a lower level may be performed.

8.3.3 Reassignment to work not requiring certification.

9 BIBLIOGRAPHY

Process Instrumentation Terminology; ANSI/ISA S51.1-1979

Qualifications of Inspection, Examination and Testing Personnel for Nuclear Power Plants; ANSI/ASME N45.2.6-1978

Selection, Qualification and Training of Personnel for Nuclear Power Plants; ANSI/ANS-3.1-1981

10CFR Part 50, Appendix A

Style Manual for Standards and Recommended Practices; ISA RPA-la-1970

APPENDIX A
TYPICAL KNOWLEDGE AND SKILL LIST

This appendix is not part of the standard but is attached to provide information to facilitate determination of the knowledge and skills an Instrumentation and Control Technician should have to work safely and effectively in a nuclear power plant. The knowledge and skills listed here are typically found in I&C training programs. Each plant should tailor its training program content and technician skill requirements to the specific needs of the nuclear power plant. It is expected this may be accomplished by adding to or subtracting from these lists.

A.1 Knowledge

(1) Mathematics

- Arithmetic Operations
- Ratio and Proportions
- Algebra
- Geometry
- Trigonometry
- Fundamental Calculus

(2) Physical Sciences

- Physics
- Mechanics
- Heat Transfer Fundamentals
- Nuclear Physics
- Health Physics
- Chemistry Fundamentals
- Pneumatics/Hydraulics
- Basic Water Chemistry

The following topics will include Theory of Operation and Failure Mechanisms and Modes

(3) Electricity and Electronics

- Basic Electricity
- Basic Electronics
- Analog Electronics
- Digital Electronics
- Fundamental Computer Theory
- Electrical/Electronic Measurement and Test Equipment Theory of Operation
- Electrical/Electronic Maintenance and Repair Techniques

(4) Process Measurement and Instrumentation

- Basic Concepts and Terminology
- Temperature Measurement Fundamentals and Applications
    Fluid Thermometry
    Thermocouples
    Bimetallic Devices
    Resistance Temperature Detectors
    Thermistors
- Pressure Measurement Fundamentals and Applications
    Bourdon Tube Devices
    Diaphragms
    Bellows
    Capsules
    Strain Gauges
- Level Measurement Fundamentals and Applications
    Float Systems
    Displacer Systems
    Differential Pressure Devices
    Radiation Level Systems
    Sonic Level Systems
- Flow Measurement Fundamentals and Applications
    Primary Elements
    Differential Pressure Devices
    Magnetic Flow Measurement
    Turbine Flowmeters
    Rotometers
- Miscellaneous Transducers
    Fire Detection Transducers
    Security System Transducers
    Meteorologic Transducers
    Environmental Transducers
    Seismic Transducers
- Analytical Transducers
    In-line and Laboratory pH Meters
    O_2 Analyzers
    H_2 Analyzers
    Conductivity/Salinity Cells

(5) Signal Processing/Conditioning Equipment

- Transducers
    Pneumatic/Electronic
    Force Balance
    Motion Balance
    Moment Balance
- Signal Conditioners
    Integrators/Ratemeters, Function Generators, and Square Rooters
- Signal Converters
    P/I & I/P & E/P & P/E & E/I & A-D/D-A, etc.
- Summers
- Amplifiers
- Bistables
- Control Relays/Solenoids
- Indicators
    Visual
    Audible, etc.
    Annunciators
- Recorders

(6) Process Control Instrumentation

- Automatic Control Theory
- Controllers
    Pneumatic
    Electronic
- Analog Control Systems
- Digital Control Systems
- Final Control Elements
    Positioners
    Actuators

(7) Radiation Detection Instrumentation

- Portable Radiac
- Area Radiation Monitor
- Portal Radiation Monitors
- Process Radiation Monitors
- Reactor/Nuclear Monitoring Systems

(8) Nuclear Power Plant Systems

- Reactor Theory/Construction
- Primary/Secondary Systems and Flowpaths
- Auxiliary Systems
- Turbine-Generator and Auxiliary Systems
- Emergency/Safeguard Systems
- Reactor Control and Protection Systems
- Major Electrical Distribution

(9) Basic Plant Operation

- Plant Layout
- Plant Procedures
- Company Policies

(10) Advanced Plant Instrumentation

- Computer Systems
- Multivariable Control Systems
- Multiplexed Control and Information Systems
- Microprocessor/Minicomputer Based Systems

A.2. Skills

(1) Proper use of the following technical information:

- Maintenance/Instrumentation/Administrative Procedures
- Technical Manuals
- Wiring and Termination Drawings
- Design Specification Sheets
- Solid-State Schematics
- Power Distribution Drawings
- Assembly Drawings
- Unit Conversion Tables
- Functional Control Drawings
- Integrated Circuit Schematics
- Electro-Mechanical Drawings
- P & ID's
- Electronic Control Drawings
- Piping and Elementary Drawings
- Relay Logic Drawings
- Complex Logic Diagrams
- Computer Flow Charts

(2) Proper use of the following measurement and test equipment:

- Miscellaneous Hand Tools
- Digital VOM
- VOM
- Electrometer
- Function Generators
- Differential (Null) VM
- Power Supplies
- Calibrators-Pneumatic/Electronic
- Calibrated Resistances/Decade Boxes
- Oscilloscope/Oscillograph
- Calculator
- Calibration Standards
- Various Gauges
- XY Plotters/Recorders
- Wheatstone Bridge
- N_2 Pressure Regulator Calibration Unit
- Strip Chart Recorder/Visicorder
- Millivolt Potentiometer
- Kelvin Bridge
- Vacuum Testers
- Deadweight Tester/Comparator
- Capacitance Bridge
- Transistor Tester
- Noise/Vibration Analyzers
- Megohm Bridge
- Oscilloscope Photography/Camera
- Tubing Bender
- Frequency Spectrum Analyzer
- Pulse Counter
- Frequency Counter
- Pico Amp Source
- Pico Ammeter
- High Speed Event Recorders
- Meter Shunts
- Inductance Bridge
- Stroboscope

(3) Align, Calibrate, Test, Troubleshoot, Adjust, and Repair

- Typical Analog and Digital Circuitry
- Process Measurement Sensing Elements and Transducers
- Signal Processing/Conditioning Equipment
- Process Control Loops
- Complex Measurement and Control Systems
- Nuclear Radiation Instrumentation Systems
- Reactor Plant Control/Instrumentation Systems
- Advanced Plant Instrumentation
- Proper Valving Techniques

(4) Proper use of the following plant administrative documents or systems:

- I & C Shop Practices/Procedures/Forms
- Information/Library/Records Management Systems
- Equipment Utilization Forms
- Maintenance Orders/Work Orders
- Bypass Control/Jumper/Wire Removal Forms
- Plant Tagging System
- FSAR and Technical Specifications
- Radiation Work Permits/Forms
- Channel Check Procedures/Forms
- Functional Test Procedures/Forms
- Calibration Procedures/Forms
- Alignment Procedures/Forms
- Receipt/Inspection Procedures/Forms
- Document Control Systems/Forms
- Surveillance Test Procedures/Forms
- Confined Space Entry Techniques/Forms
- Nonconformance Items

- Review, Approval and Control of Procedures/Changes
- Training and Certification Procedures/Forms
- Temporary Procedure Changes/Forms
- Facility/Equipment Change or Modification Procedures/Forms

(5) Proper Plant Safety Practices
- First Aid/CPR
- Electrical Safety Techniques
- Handling of Toxic Materials
- Handling of Radioactive Sources/Contaminated Materials
- Use of Safety Equipment
- Use of High Pressure/Temperature/Pneumatic/Hydraulic and Fluid Systems

INSTRUMENT SOCIETY OF AMERICA
Research Triangle Park, North Carolina 27709

ANSI/ISA-S71.01-1985
Approved August 15, 1986

American National Standard

Environmental Conditions for Process Measurement and Control Systems: Temperature and Humidity

Instrument Society of America

Instrument Society of America

ISBN 0-87664-894-4

ISA-S71.01 Environmental Conditions for Process Measurement and Control Systems: Temperature and Humidity

Copyright © 1985 by the Instrument Society of America. All rights reserved. Printed in the United States of America. No part of this publication may be reproduced, stored in a retrieval system, or transmitted, in any form or by any means (electronic, mechanical, photocopying, recording, or otherwise), without the prior written permission of the publisher.

INSTRUMENT SOCIETY OF AMERICA
67 Alexander Drive
P.O. Box 12277
Research Triangle Park, North Carolina 27709

PREFACE

This preface is included for informational purposes and is not part of ISA-S71.01.

This standard has been prepared as part of the service of the Instrument Society of America (ISA) toward a goal of uniformity in the field of instrumentation. To be of real value, this document should not be static, but should be subject to periodic review. Toward this end, the Society welcomes all comments and criticisms, and asks that they be addressed to the Secretary, Standards and Practices Board, Instrument Society of America, 67 Alexander Drive, P.O. Box 12277, Research Triangle Park, NC 27709, Telephone (919) 549-8411.

The ISA Standards and Practices Department is aware of the growing need for attention to the metric system of units in general, and the International System of Units (SI) in particular, in the preparation of instrumentation standards. The Department is further aware of the benefits to U.S.A. users of ISA standards of incorporating suitable references to the SI (and the metric system) in their business and professional dealings with other countries. Toward this end, this Department will endeavor to introduce SI-acceptable metric units in all new and revised standards to the greatest extent possible. The Metric Practice Guide, which has been published by the Institute of Electrical and Electronics Engineers as ANSI/IEEE Std. 268-1982, and future revisions will be the reference guide for definitions, symbols, abbreviations, and conversion factors.

It is the policy of the Instrument Society of America to encourage and welcome the participation of all concerned individuals and interests in the development of ISA standards. Participation in the ISA standards-making process by an individual in no way constitutes endorsement by the employer of that individual, of the Instrument Society of America, or of any of the standards that ISA develops.

The information contained in the preface, footnotes, and appendices is included for information only and is not a part of the standard.

This document is one of several standards covering various environmental conditions affecting process measurement and control systems. In developing this standard, the committee goals included the following:

1. To provide a practical standard that can be applied with a minimum of research and technical effort by the user.

2. To provide a concise method of stating environmental classifications for convenient communication between all users of the standard.

3. To cover real-world ranges of each classified parameter.

In order to be compatible with international standards, the SP71 committee used the same limit values, wherever appropriate, as presented in Publication 654-1, First edition (1979), of the International Electrotechnical Commission: "Operating Conditions for Industrial-Process Measurement and Control Equipment, Part 1: Temperature, Humidity and Barometric Pressure."

For Classes B3 and B4 described in this standard, the committee specified limits of 5 to 90 percent relative humidity instead of 5 to 95 percent relative humidity as specified by the International Electrotechnical Commission. The committee concluded that for this class (Class B), relative humidity values above 90 percent should be covered in Severity Level X.

Instrument Society of America

The persons listed below served as members of ISA Committee SP71, which prepared this standard.

NAME	COMPANY
W. Holway, Chairman	The Foxboro Company
D. Boyle	National Bureau of Standards (retired)
D. Cummins	Purafil, Inc.
J. Duffy	Fisher Controls Company
K. Gulick	Digital Equipment Corporation
M. Huza	Circul-Aire Inc.
F. Kent	Fischer and Porter Company
E. J. Laderoute	The Foxboro Company
M. Lombardi	Honeywell, Inc.
R. Magnuson	Hewlett Packard
W. T. Mitchell	Dow Chemical, USA
R. Prescott (Director)	Moore Products
E. Rasmussen	Fluor Engineers & Constructors, Inc.
W. T. Rhodes	CONOCO, Inc.
R. H. Walton	Exxon Company, USA

This standard was approved for publication by the ISA Standards and Practices Board in February 1985.

NAME	COMPANY
N. Conger, Chairman	Fisher Controls Company
P. V. Bhat	Monsanto Company
W. Calder III	The Foxboro Company
R. S. Crowder	Ship Star Associates
B. Feikle	Bailey Controls Company
H. S. Hopkins	Westinghouse Electric Company
J. L. Howard	Boeing Aerospace Company
R. T. Jones	Philadelphia Electric Company
R. Keller	The Boeing Company
O. P. Lovett, Jr.	ISIS Corporation
E. C. Magison	Honeywell, Inc.
A. P. McCauley	Chagrin Valley Controls Inc.
J. W. Mock	Bechtel Corporation
E. M. Nesvig	ERDCO Engineering Corporation
R. Prescott	Moore Products Company
D. E. Rapley	ISTS
C. W. Reimann	National Bureau of Standards
J. Rennie	Factory Mutual Research Corporation
W. C. Weidman	Gilbert Commonwealth Inc.
K. Whitman	Consultant
*P. Bliss	Consultant
*B. A. Christensen	Continental Oil Company
*L. N. Combs	Retired
*R. L. Galley	Consultant
*T. J. Harrison	IBM Corporation
*R. G. Marvin	Roy G. Marvin Company
*W. B. Miller	Moore Products Company
*G. Platt	Retired
*J. R. Williams	Stearns Catalytic Corporation

*Director Emeritus

S71.01 — Environmental Conditions for Process Measurement and Control Systems: Temperature and Humidity

TABLE OF CONTENTS

Section	Title	Page
1.	Purpose	7
2.	Scope	7
3.	Introduction	7
4.	Definitions	8
4.1	Normal Operating Conditions, Operative Limits, and Transportation and Storage Conditions	8
4.2	Maintenance Conditions	8
4.3	Maintenance	8
5.	Location Classifications	8
5.1	Air-Conditioned Locations (Class A)	8
5.2	Enclosed Temperature-Controlled Locations (Class B)	8
5.3	Sheltered Locations (Class C)	8
5.4	Outdoor Locations (Class D)	8
5.5	Special Locations (Class X)	8

LIST OF TABLES

Table	Title	Page
1.	Location, Class, and Severity Levels	9

LIST OF APPENDICES

Appendix	Title	Page

Appendix A: Psychrometric Charts

Chart 1:	Air-Conditioned Locations, Class A1	11
Chart 2:	Air-Conditioned Locations, Class A2	12
Chart 3:	Enclosed Temperature-Controlled Locations, Class B1	13
Chart 4:	Enclosed Temperature-Controlled Locations, Class B2	14
Chart 5:	Enclosed Temperature-Controlled Locations, Class B3	15
Chart 6:	Enclosed Temperature-Controlled Locations, Class B4	16
Chart 7:	Sheltered Locations, Class C1	17
Chart 8:	Sheltered Locations, Class C2	18

1 PURPOSE

The purpose of this standard is to establish uniform classifications of temperature and humidity conditions for industrial process measurement and control systems. This document is one of a series of standards on environmental conditions for process measurement and control systems.

2 SCOPE

2.1 This standard covers temperature and humidity environmental conditions for industrial process measurement and control equipment. Specifications for other environmental conditions are beyond the scope of this standard.

2.2 This standard establishes temperature and humidity classes for fixed (non-mobile) installations during normal operation (nonemergency conditions) or during transportation and storage.

2.3 The classes of temperature and humidity conditions stated in this standard are suitable for use in activities related to process instrumentation, including design, manufacturing, sales, installation, test, use, and maintenance. These classes may also be used as a guide when establishing requirements for environmental control of buildings or other protective housings for industrial process measurement and control systems.

2.4 These classifications pertain only to the environment external to the equipment which may affect the equipment externally or internally.

2.5 The effects of environmental conditions on safety, comfort, and performance of operating and maintenance personnel are not considered in this standard.

3 INTRODUCTION

3.1 Environmental classifications have been established according to the type of location. Within each classification, severity levels have also been established. Parameter limit values are tabulated for each classification and severity level of the location. These values are shown in Table 1 of this standard. The classification consists of a class location letter followed by a severity identification numeral.

EXAMPLE: Temperature and Humidity Classification A2 would represent Class A Location and Level 2 Severity.

3.2 The manufacturer and/or user should specify the equipment performance in a stated environmental Class and Severity Level. The following example shows how a manufacturer or user might specify several sets of environmental classes for operating or maintaining the same equipment.

EXAMPLE NO. 1

Conditions	Temperature and Humidity Class
Normal Operating Conditions*	A2
Operative Limit*	B2
Transportation & Storage Conditions*	C2
Maintenance Conditions†	B3
Shutdown Conditions‡	C1

* These terms are defined in ISA-S51.1, "Process Instrumentation Terminology."
† See Section 4, this standard, for definition.
‡ Specified separately only when Shutdown Conditions differ from Transportation and Storage Conditions. See also Section 4.1, this standard.

The above example may also be specified as follows:

EXAMPLE NO. 2.

Condition	Temperature Range (°C)	Control Tolerance (°C)	Max. Rate of Change (°C/Hour)	Humidity Range (% R.H.)*	Control Tolerance (% R.H.)	Max. Moisture Content (Kg/Kg Dry Air)
Normal Operating	18 to 27	±2	±5	20 to 80	±10	N.A.†
Operative Limit	5 to 40	±3	±10	10 to 75	N.A.	0.020
Transportation and Storage	–40 to 85	N.A.	±10	5 to 100	N.A.	0.028
Maintenance	5 to 40	±10	±20	5 to 90	N.A.	0.028
Shutdown	–25 to 55	N.A.	±5	5 to 100	N.A.	0.028

NOTES: * R.H. = relative humidity
† N.A. = Not Applicable

4 DEFINITIONS

4.1 Normal Operating Conditions, Operative Limits, and Transportation and Storage Conditions

These terms are defined in the ISA-S51.1 Standard "Process Instrumentation Terminology." The ISA-S51.1 definition of "Transportation and Storage Conditions" includes "Shutdown." If the shutdown conditions are different from transportation and storage conditions, the shutdown environment shall be specified separately.

4.2 Maintenance Conditions

Conditions under which maintenance is performed.

4.3 Maintenance

Any activity intended to keep equipment in satisfactory working condition, including tests, measurements, replacements, adjustments, and repairs. (Refer to the Scientific Apparatus Manufacturers Association Standard PMC 32.1, "Process Instrumentation Reliability Terminology."

5 LOCATION CLASSIFICATIONS

5.1 Air-Conditioned Locations (Class A)

Class A locations are locations where both air temperature and relative humidity are controlled. These locations are usually provided for computers and other electronic equipment requiring a controlled air environment.

Special consideration should be given where hygroscopic materials, such as punched cards and chart paper, will be used. These materials, depending on the manufacturer's recommendation, may require a relative humidity less than the maximum given in Table 1. The special requirements shall be described by use of Severity Level "X" of Table 1.

5.2 Enclosed Temperature Controlled Locations (Class B)

Class B locations are locations where air temperature is controlled but relative humidity is not controlled. These locations are usually provided where continuous operator surveillance is required. This class may also represent storage and occasionally transportation conditions.

5.3 Sheltered Locations (Class C)

Class C locations are locations protected from direct exposure to the climatic elements, such as sunlight, rain and other precipitation, and full wind pressure.

Neither heating nor cooling is normally provided. Ventilation, if any, may be either natural or forced. Minimum air temperature inside the enclosure may be as low as the outdoor air temperature. Maximum air temperature inside the enclosure may be considerably greater than the outdoor air temperature due to solar radiation heating of the shelter surfaces. The air inside the shelter is the environment for the equipment. Condensation may occur on surfaces within the shelter or within the equipment enclosure due to temporary excursions below the local dew point.

Sheltered locations are provided where minimum protection is required for operators, maintenance personnel, or equipment. Examples of equipment shelters range from box enclosures to equipment "shacks."

5.4 Outdoor Locations (Class D)

Class D locations are locations where there is no specific protection from the environment.

Equipment in these locations may be subjected to sudden and severe changes of environment due to weather or other factors. Minimum temperature of the equipment may be as low as the outdoor air temperature. Maximum temperature of the equipment may be considerably greater than the outdoor air temperature due to solar radiation heating. Differential temperature conditions may also exist in the equipment when part of the equipment is exposed to direct heat radiation with the remaining surface shaded, or by other circumstances of this type. Condensation may occur due to temporary excursions below the local dew point. In addition to the effect of ambient air temperature, the effect of radiated heat from the sun or other sources should be considered for selecting severity levels in Table 1.

5.5 Special Locations (Class X)

It is recognized that extreme or special service conditions exist in which the excursions of temperature or humidity differ from the previously mentioned classes. To accommodate this situation, a special Class "X" is included in Table 1. Specifications for equipment in Class X are a matter of negotiation between user and supplier.

Table 1 [a,b,c]

Location, Class, and Severity Levels

Location	Class	Severity Level	Temperature Limits (°C)	Control Point Tolerance (°C)	Maximum Rate of Change (°C/Hour)	Humidity Limits (% Relative Humidity)	Control Point Tolerance (% Relative Humidity)	Maximum Moisture Content (Kg/Kg Dry Air)
Air Conditioned	A	1	18 to 27[d]	±2[e]	±5[f]	35 to 75[d]	±5[e]	N.A.
		2	18 to 27[d]	±2[e]	±5[f]	20 to 80[d]	±10[e]	N.A.
		X	T.B.S.[d]	T.B.S.[e]	T.B.S.[f]	T.B.S.[d]	T.B.S.[e]	T.B.S.
Enclosed Temperature Controlled	B	1	15 to 30[d]	±2[e]	±5[f]	10 to 75[d]	N.A.	N.A.
		2	5 to 40[d]	±3[e]	±10[f]	10 to 75[d]	N.A.	0.020
		3	5 to 40[d]	±10[e]	±20[f]	5 to 90[d]	N.A.	0.028
		4	5 to 50[d]	±10[e]	±20[f]	5 to 90[d]	N.A.	0.028
		X	T.B.S.[d]	T.B.S.[e]	T.B.S.[f]	T.B.S.[d]	N.A.	T.B.S.
Sheltered	C	1	−25 to 55	N.A.	±5	5 to 100	N.A.	0.028
		2	−40 to 85	N.A.	±10	5 to 100	N.A.	0.028
		X	T.B.S.	N.A.	T.B.S.	5 to 100	N.A.	T.B.S.
Outdoor	D	1	−25 to 70	N.A.	±10	5 to 100	N.A.	N.A.
		2	−40 to 85	N.A.	±20	5 to 100	N.A.	N.A.
		3	−55 to 65	N.A.	±20	5 to 100	N.A.	N.A.
		X	T.B.S.	N.A.	T.B.S.	T.B.S.	N.A.	N.A.

NOTES:
[a] This table applies for atmospheric pressures between 86 kPa and 108 kPa
[b] N.A. = Not Applicable
[c] T.B.S. = To Be Specified
[d] Operating temperature/humidity to be selected from within temperature/humidity limits
[e] Allowable variation from the selected operating temperature/humidity control point
[f] Maximum rate of change within the control tolerance

APPENDIX A
PSYCHROMETRIC CHARTS

Chart 1: Air-Conditioned Locations, Class A1

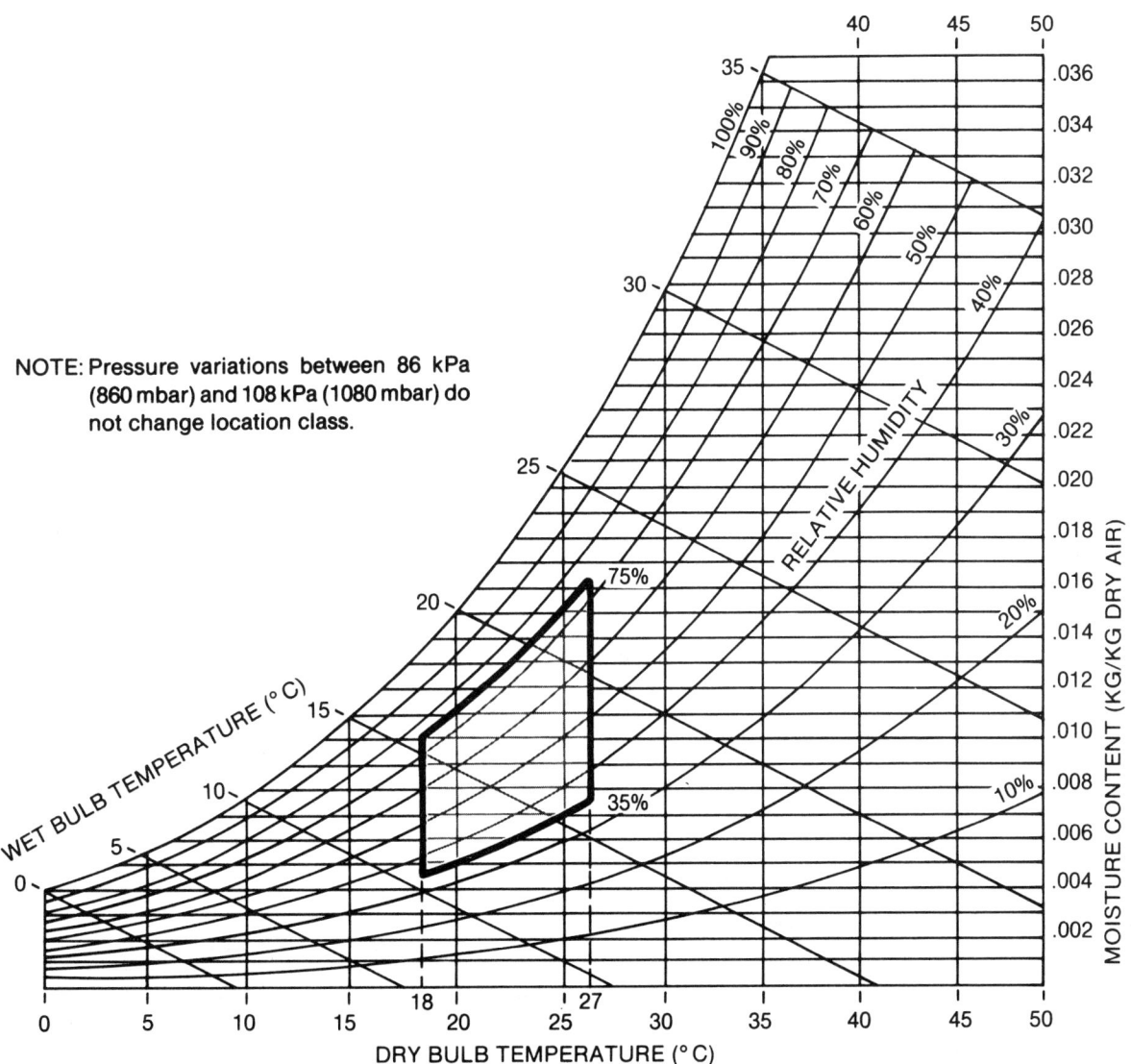

APPENDIX A
PSYCHROMETRIC CHARTS

Chart 2: Air-Conditioned Locations, Class A2

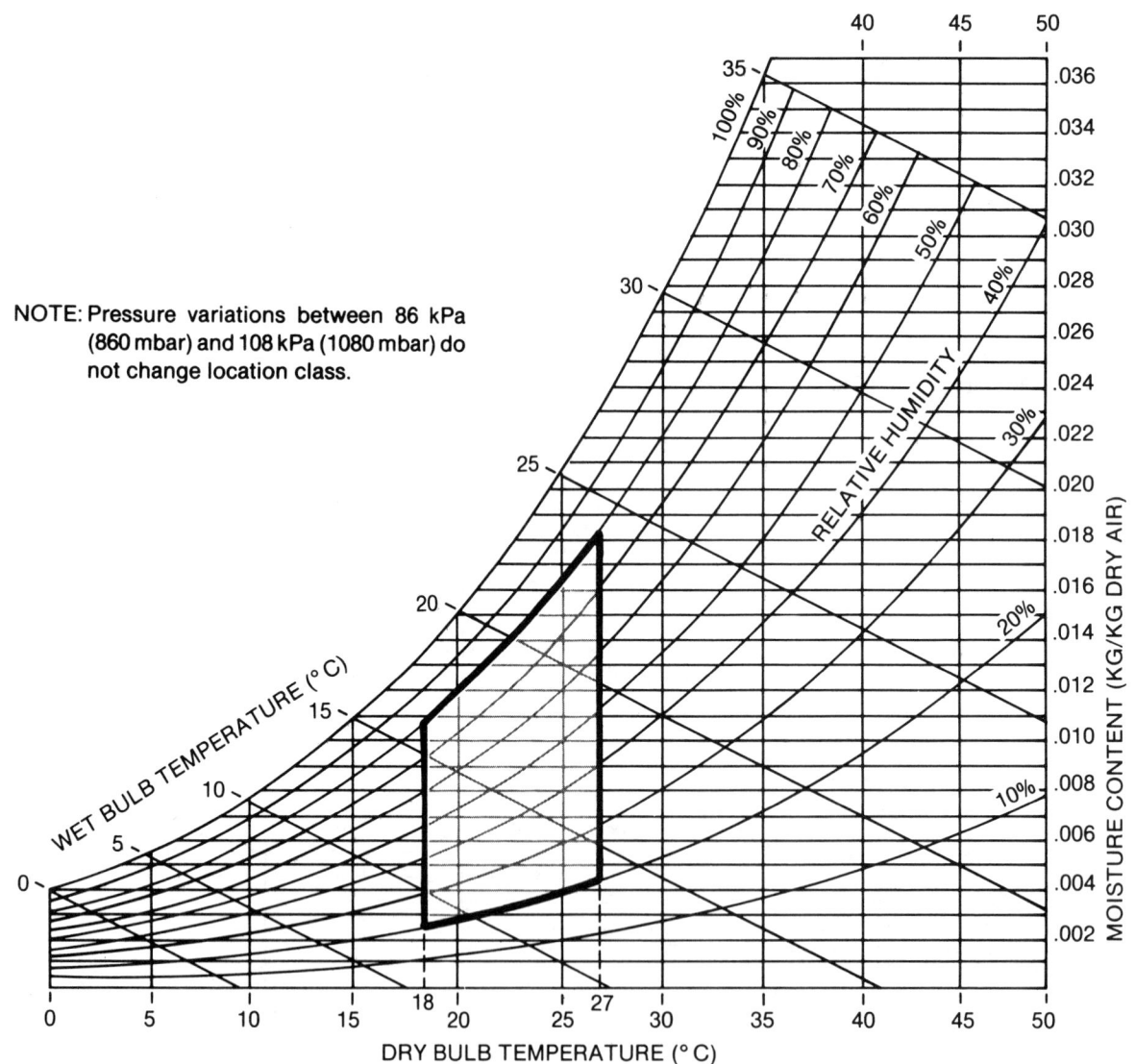

APPENDIX A
PSYCHROMETRIC CHARTS

Chart 3: Enclosed Temperature-Controlled Locations, Class B1

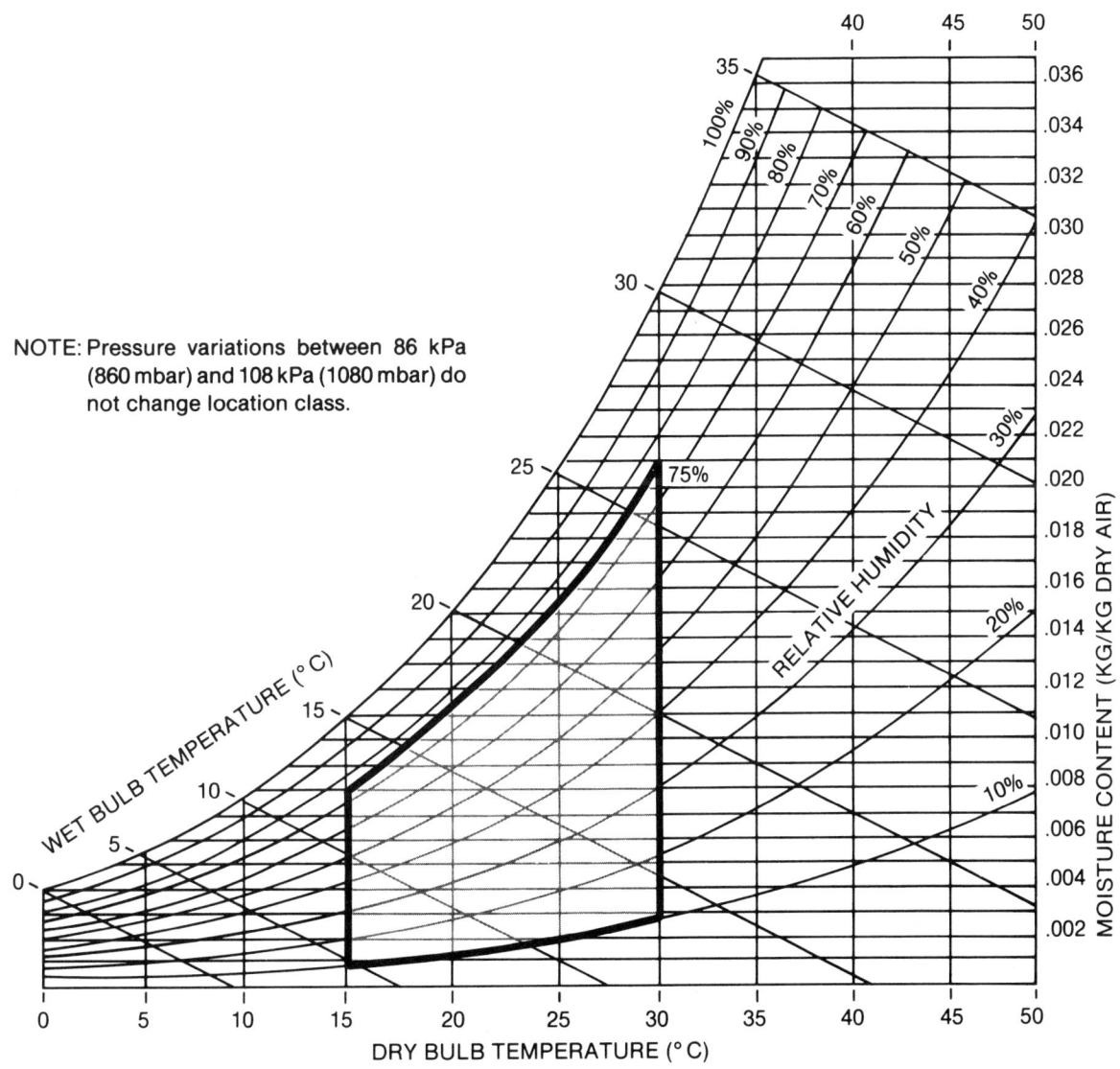

Instrument Society of America

APPENDIX A
PSYCHROMETRIC CHARTS

Chart 4: Enclosed Temperature-Controlled Locations, Class B2

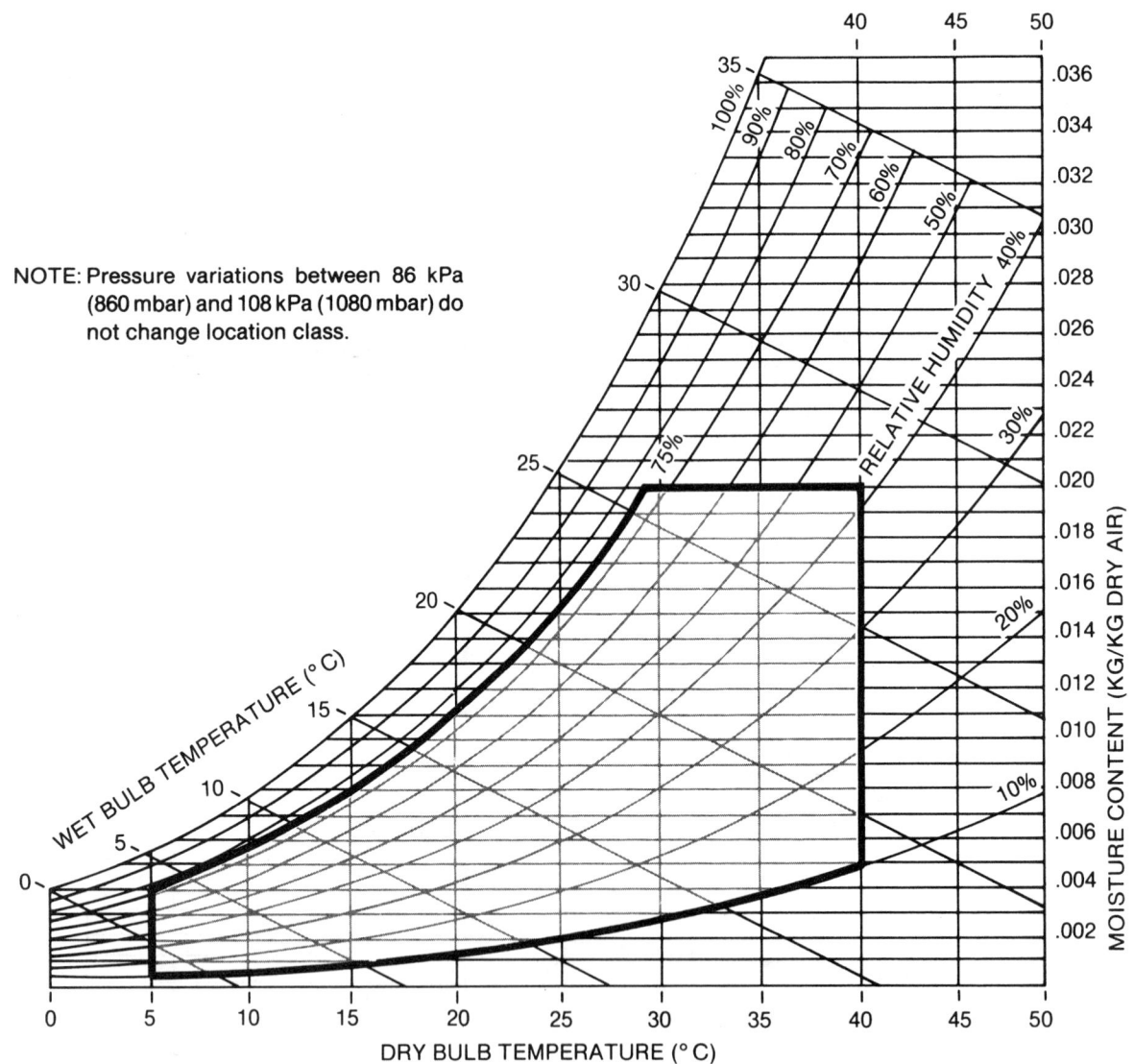

APPENDIX A
PSYCHROMETRIC CHARTS

Chart 5: Enclosed Temperature-Controlled Locations, Class B3

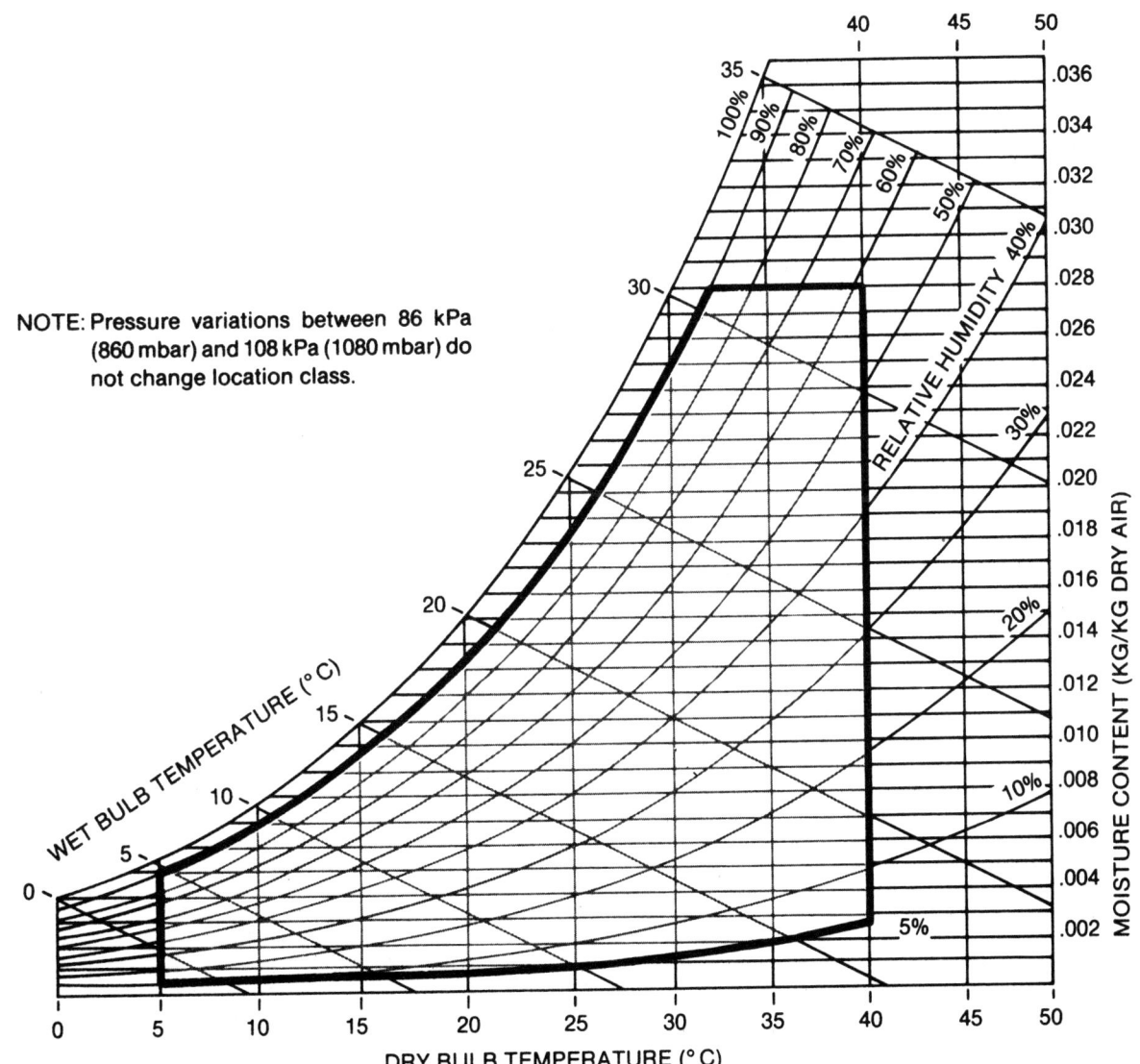

APPENDIX A
PSYCHROMETRIC CHARTS

Chart 6: Enclosed Temperature-Controlled Locations, Class B4

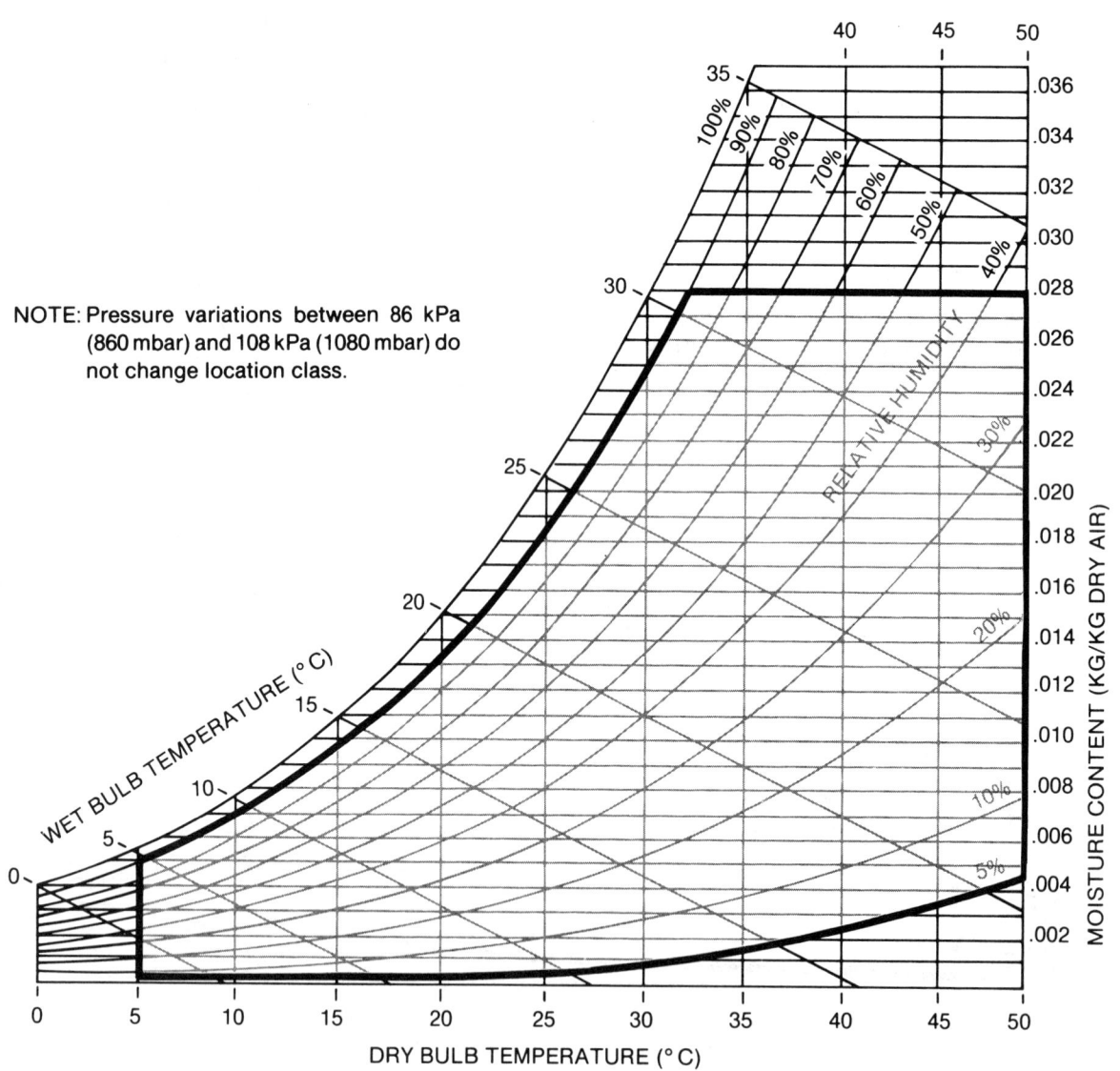

S71.01 — Environmental Conditions for Process Measurement and Control Systems: Temperature and Humidity

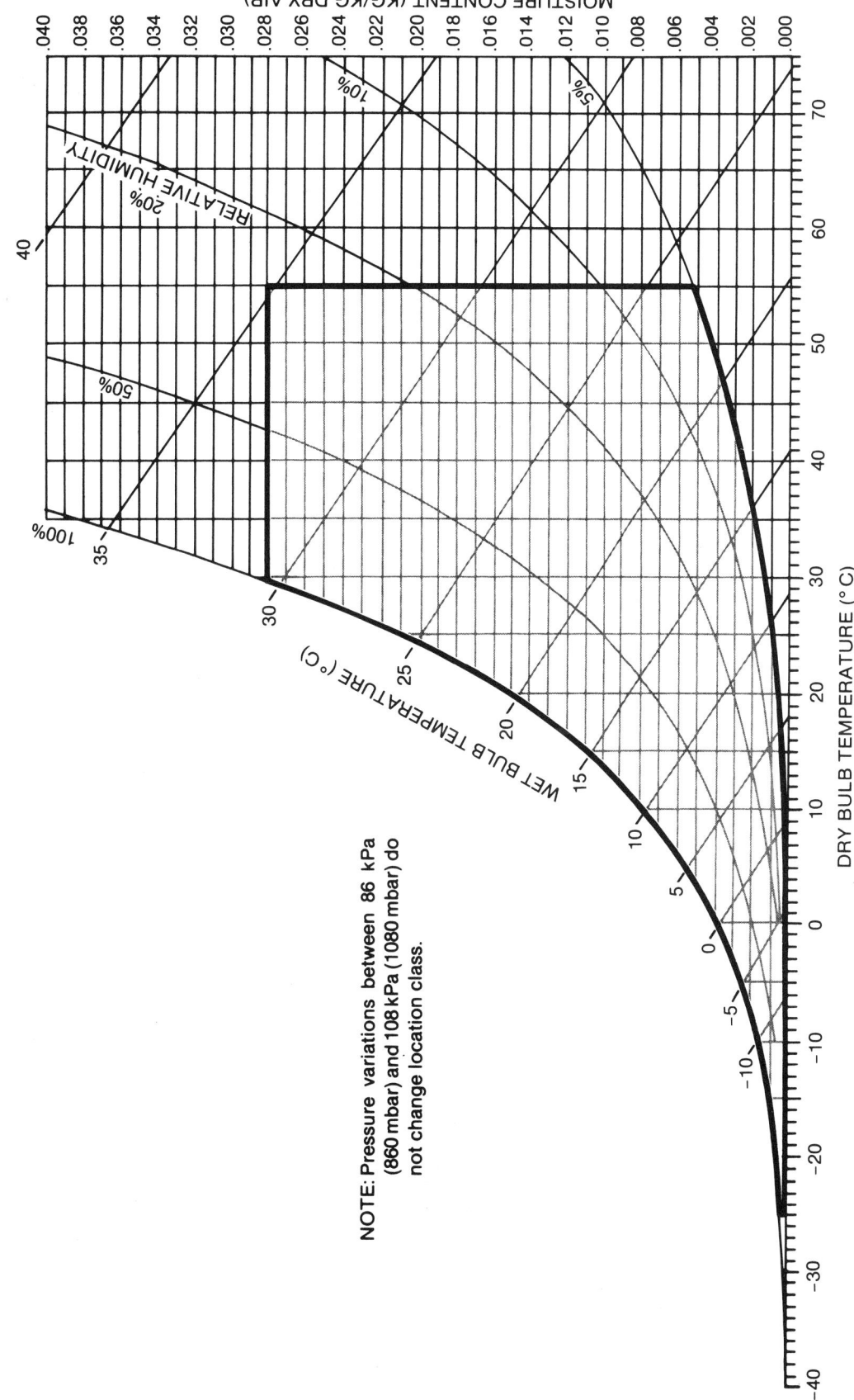

APPENDIX A
PSYCHROMETRIC CHARTS

Chart 7: Sheltered Locations, Class C1

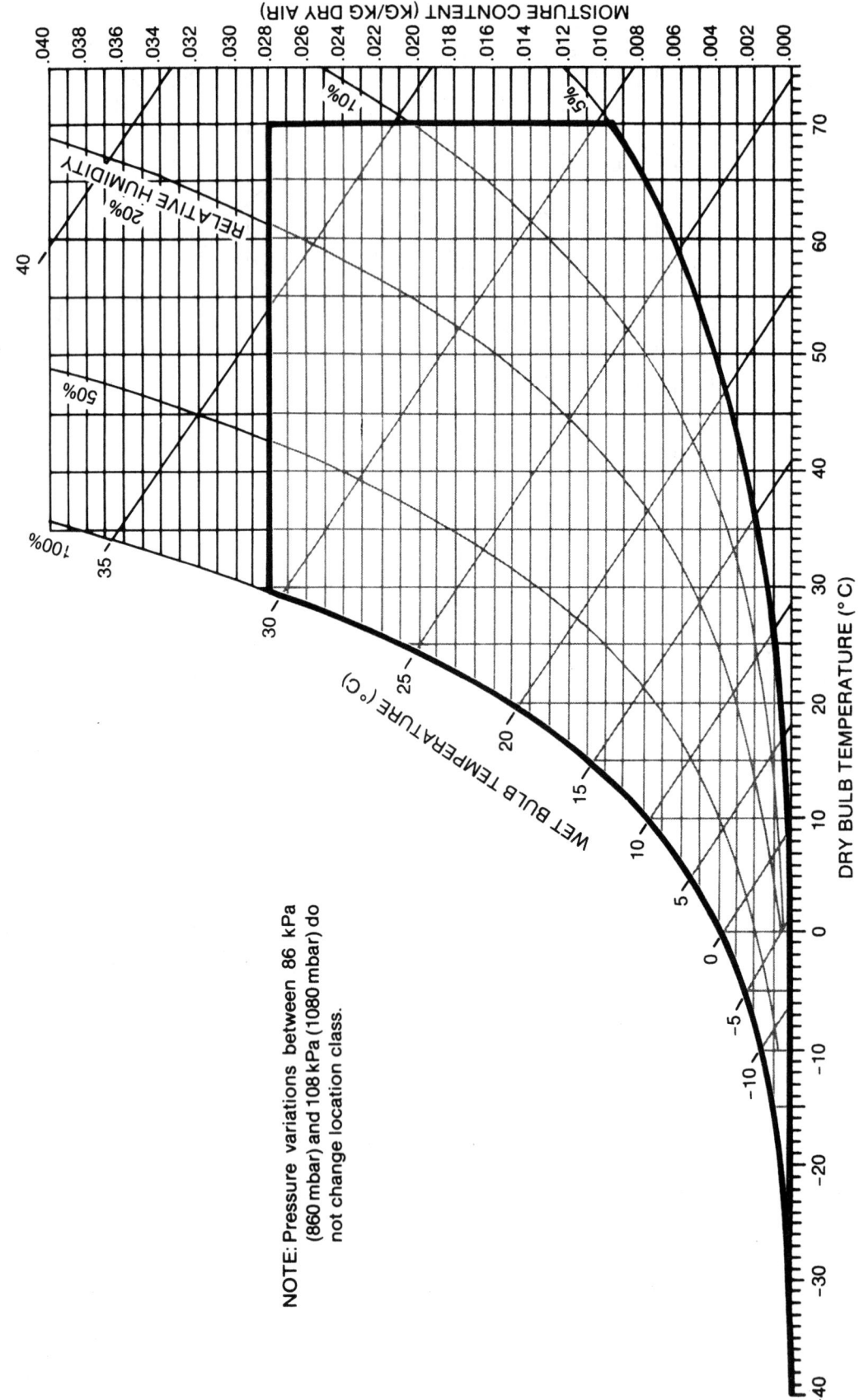

APPENDIX A
PSYCHROMETRIC CHARTS

Chart 8: Sheltered Locations, Class C2

ANSI/ISA-S71.04-1985
Approved February 3, 1986

American National Standard

Environmental Conditions for Process Measurement and Control Systems: Airborne Contaminants

Instrument Society of America

Instrument Society of America

ISBN 0-87664-895-2

ISA-S71.04 Environmental Conditions for Process Measurement and Control Systems: Airborne Contaminants

Copyright © 1985 by the Instrument Society of America. All rights reserved. Printed in the United States of America. No part of this publication may be reproduced, stored in a retrieval system, or transmitted, in any form or by any means (electronic, mechanical, photocopying, recording, or otherwise), without the prior written permission of the publisher.

INSTRUMENT SOCIETY OF AMERICA
67 Alexander Drive
P.O. Box 12277
Research Triangle Park, North Carolina 27709

PREFACE

This preface is included for informational purposes and is not part of ISA-S71.04.

This standard has been prepared as part of the service in the Instrument Society of America (ISA) toward a goal of uniformity in the field of instrumentation. To be of real value, this document should not be static, but should be subject to periodic review. Toward this end, the Society welcomes all comments and criticisms, and asks that they be addressed to the Secretary, Standards and Practices Board, Instrument Society of America, 67 Alexander Drive, P.O. Box 12277, Research Triangle Park, NC 27709, Telephone (919) 549-8411.

The ISA Standards and Practices Department is aware of the growing need for attention to the metric system of units in general, and the International System of Units (SI) in particular, in the preparation of instrumentation standards. The Department is further aware of the benefits to U.S.A. users of ISA standards of incorporating suitable references to the SI (and the metric system) in their business and professional dealings with other countries. Toward this end, this Department will endeavor to introduce SI-acceptable metric units in all new and revised standards to the greatest extent possible. The Metric Practice Guide, which has been published by the Institute of Electrical and Electronics Engineers as ANSI/IEEE Std. 268-1982, and future revisions will be the reference guide for definitions, symbols, abbreviations, and conversion factors.

It is the policy of the Instrument Society of America to encourage and welcome the participation of all concerned individuals and interests in the development of ISA standards. Participation in the ISA standards-making process by an individual in no way constitutes endorsement by the employer of that individual, of the Instrument Society of America, or of any of the standards that ISA develops.

The information contained in the preface, footnotes, and appendices is included for information only and is not a part of the standard.

This document is one of several standards covering various environmental conditions affecting process measurement and control systems. In developing this standard, the committee goals included the following:

1. To provide a practical standard that can be applied with a minimum of research and technical effort by the user.

2. To provide a concise method of stating environmental classifications for convenient communication between users of the standard.

3. To cover real-world ranges of each classified parameter.

This standard is limited to airborne contaminants and biological influences only, covering contamination influences that affect industrial process measurement and control systems.

The persons listed below served as members of ISA Committee SP71, which prepared this standard.

NAME	COMPANY
W. Holway, Chairman	The Foxboro Company
D. Boyle	National Bureau of Standards (retired)
D. Cummins	Purafil, Inc.
J. Duffy	Fisher Controls Company
K. Gulick	Digital Equipment Corporation
M. Huza	Circul-Aire Inc.
F. Kent	Fischer and Porter Company
E. J. Laderoute	The Foxboro Company
M. Lombardi	Honeywell, Inc.
R. Magnuson	Hewlett Packard
W. T. Mitchell	Dow Chemical, USA
R. Prescott (Director)	Moore Products
E. Rasmussen	Fluor Engineers & Constructors, Inc.
W. T. Rhodes	CONOCO, Inc.
R. H. Walton	Exxon Company, USA

Instrument Society of America

This standard was approved for publication by the ISA Standards and Practices Board in February 1985.

NAME	COMPANY
N. Conger, Chairman	Fisher Controls Company
P. V. Bhat	Monsanto Company
W. Calder III	The Foxboro Company
R. S. Crowder	Ship Star Associates
B. Feikle	Bailey Controls Company
H. S. Hopkins	Westinghouse Electric Company
J. L. Howard	Boeing Aerospace Company
R. T. Jones	Philadelphia Electric Company
R. Keller	The Boeing Company
O. P. Lovett, Jr.	ISIS Corporation
E. C. Magison	Honeywell, Inc.
A. P. McCauley	Chagrin Valley Controls Inc.
J. W. Mock	Bechtel Corporation
E. M. Nesvig	ERDCO Engineering Corporation
R. Prescott	Moore Products Company
D. E. Rapley	ISTS
C. W. Reimann	National Bureau of Standards
J. Rennie	Factory Mutual Research Corporation
W. C. Weidman	Gilbert Commonwealth Inc.
K. Whitman	Consultant
*P. Bliss	Consultant
*B. A. Christensen	Continental Oil Company
*L. N. Combs	Retired
*R. L. Galley	Consultant
*T. J. Harrison	IBM Corporation
*R. G. Marvin	Roy G. Marvin Company
*W. B. Miller	Moore Products Company
*G. Platt	Retired
*J. R. Williams	Stearns Catalytic Corporation

*Director Emeritus

TABLE OF CONTENTS

Section	Title	Page
1.	Purpose	7
2.	Scope	7
3.	Introduction	7
4.	Airborne Contaminants — Liquids	7
5.	Airborne Contaminants — Solids	8
6.	Airborne Contaminants — Gases	9
7.	Biological Influences	11

LIST OF TABLES

Table	Title	Page
1.	Classification of Chemically Active Contaminants — Liquid Aerosols	8
2.	Classification of Airborne Particulates	9
3.	Classification of Reactive Environments	11

LIST OF APPENDICES

Appendix	Title	Page
Appendix A	Terminology	13
Appendix B	Section 1 Some Common Sources of Reactive Environmental Constituents	13
	Section 2 Some Common Emissions of Natural and Industrial Processes	14
Appendix C	Copper Reactivity Samples	15

1 PURPOSE

The purpose of this standard is to classify airborne contaminants that may affect process measurement and control instruments.

The classification system provides user and manufacturers of instruments with a means of specifying the type and concentration of airborne contaminants to which a specified instrument may be exposed.

This document is one of a series of standards on environmental conditions for process measurement and control systems.

2 SCOPE

2.1 This standard covers airborne contaminants and biological influences that affect industrial process measurement and control equipment. Specifications for other environmental conditions, including nuclear radiation and hazardous atmospheres, are beyond the scope of this standard.

2.2 This standard establishes airborne contaminant classes for fixed (non-mobile) installations during normal operation (nonemergency conditions) or during transportation and storage.

2.3 The classes of environmental conditions stated in this standard are suitable for use in activities related to process instrumentation, including design, manufacturing, sales, installation, test, use, and maintenance. These classes may also be used as a guide when establishing requirements for environmental control of buildings or other protective housings for industrial process measurement and control systems.

2.4 These classifications pertain only to the environment external to the equipment which may affect the equipment externally or internally.

2.5 The effects of environmental conditions on safety, comfort, and performance of operating and maintenance personnel are not considered in this standard.

2.6 CAUTION — Airborne or biological contaminants not listed in this document could cause equipment damage. Caution should be used when a combination of factors approach or surpass class "X". Obtaining the guidance of a chemical specialist is suggested when this condition occurs.

3 INTRODUCTION

3.1 Environmental classifications have been established according to the type of contaminant. Within each classification, severity levels have also been established. Parameter limit values are tabulated for each classification and severity level of the contaminant. The classification consists of a class contaminant letter followed by a severity identification numeral.

3.2 Some of these contaminants may appear in more than one form, i.e., gas, liquid, or solid. For the purpose of this document, a particular contaminant will be classified in the form in which it most often exists under ambient conditions.

3.3 The user or equipment manufacturer should specify the equipment performance in the stated environmental class and severity level. It is possible to specify several sets of contaminant classes and severity levels for the same equipment.

3.4 Chemical/temperature effects may cause the rates of destructive chemical reactions to more than double for every 10° C increase in temperature.

3.5 High or variable relative humidity and contaminant mixtures may accelerate corrosive effects. These concerns are addressed in Section 6 and Table 3.

3.6 Contaminants are listed as Class A, Class B, etc., or as Special Class X with increasing Severity Levels 1, 2, 3, and X.

The contaminants are listed with a prefix of "L" for liquid, "G" for gas and "S" for solids.

An example of table usage would be:

EXAMPLE

Suitable to operate for normal life in the following Airborne Contaminant Severity Levels:

Contaminant	Concentration	Reference Class/ Severity	Table No.
Liquids: Trichlorethylene	<5 µg/kg	LA2	1
Oils	<100 µg/kg	LB3	1
Sea salt mist	Within 0.5 km inland	LC2	1
Solids: Particle size	Concentration Level		
>1 mm	<1000 µg/m^3	SA1	2
100 to 1000 µm	<3000 µg/m^3	SB2	2
1 to 100 µm	< 350 µg/m^3	SC3	2
<1 µm	< 350 µg/m^3	SD3	2
Gases: Harsh: >2000 angstroms film formation on exposed copper coupon after one month exposure		G3	3

4 AIRBORNE CONTAMINANTS — LIQUIDS
(Refer to Table 1)

4.1 Liquids — This refers to liquids that will corrode unprotected equipment. They are transported to the equipment by condensation, rain, splashing liquids, or cleaning fluids sprayed from hoses. The majority of these are not classified, but should be specified to the manufacturers of equipment by special classification *LX*.

4.2 Vapors — Solvents sometimes occur as vapors, which may condense and form puddles that become corrosive to instruments and controls.

4.3 Aerosols — Aerosols are liquids carried in gas or air in the form of small droplets generating mists.

Aerosols can vary in composition and are a major source of chemical contamination to equipment.

4.4 Sea Salt Mist (Refer to Table 1)

Example: Class LC1: Inland more than 0.5 km from shore
Class LC2: Inland less than 0.5 km from shore
Class LC3: Offshore installations (oil rigs, etc.)

4.5 General Examples (Refer to Table 1)

Contaminant	Contaminant Classes
Trichlorethylene	LA2
Oils (engine rooms, compressor station)	LB1
Specials (contaminant must be specified)	LX3

Table 1

Classification of Chemically Active Contaminants: Liquid Aerosols (Measured in $\mu g/kg$ except as specified)

Contaminant	Class	Severity Level 1 Value	Severity Level 2 Value	Severity Level 3 Value	Severity Level X (special) Value
Vapors*	LA	< 1.0	< 5.0	< 20.0	≥ 20.0
Oils	LB	< 5.0	< 50.0	<100.0	≥100.0
Sea salt mist	LC	More than 0.5 km inland	Within 0.5 km inland	Offshore installation	T.B.S.
Special T.B.S.	LX	T.B.S.	T.B.S.	T.B.S.	T.B.S.

*For example, trichlorethylene ($CHClCCl_2$)

NOTES: $1.0 \mu g/kg = 1.0$ part per billion ($p/10^9$)
T.B.S. = To Be Specified
< is defined as "less than"
> is defined as "more than"
≥ is defined as "greater than or equal to"

5 AIRBORNE CONTAMINANTS — SOLIDS
(Refer to Table 2)

5.1 General

Dust is a universal contaminant and is a cause of environmentally induced equipment failures. Failure modes may be mechanical, chemical, electrical, thermal, or magnetic. To maximize equipment reliability and life, every effort should be made to minimize exposure to airborne particulates. The sensitivity of control equipment to different types of particulates varies widely. Some of the major effects are discussed in Section 5.2. Specifications should include a description of these characteristic types of particulates if they are relevant. Particle size and concentration classifications are given in Table 2.

5.2 Particulate Properties that Affect Equipment

5.2.1 Magnetic Permeability — Magnetically permeable substances can accumulate in magnetic fields; for example, the movement of forcecoils or galvanometer movements can be severely restricted or entirely demobilized by magnetic substances accumulating in air gaps of the permanent magnets. Likewise, electrical motors can be seriously damaged by magnetic materials accumulating between rotor and stator.

5.2.2 Thermal Conductivity — The thermal insulating properties of some solid particles can cause overheating of cooling systems (which become insulated by surface deposits of these substances). For example, the cooling fins of power electronics can be seriously insulated by textile fibers.

5.2.3 Electrical Conductivity — Solid substances are divided into two groups, the good electrical conductors and the highly insulating substances.

Electrical conductors–such as metals, carbon blacks, and coal dusts–can cause short circuits when settling between terminals.

Insulating substances can accumulate static charges that upset the functioning of computers and integrated circuits. Some insulators adsorb moisture under conditions of high relative humidity. This causes an increase in conductivity and can result in equipment failures due to electrical leakage.

5.2.4 Adhesiveness — This characteristic causes a contaminant to adhere to and accumulate on surfaces. This intensifies undesirable effects such as thermal insulation, high voltage discharge, and bearing failures. Adhesive qualities may be inherent to the contaminant, such as tobacco smoke, which contains sticky tars.

5.2.5 Chemical — Airborne particulate matter varies from hard crystalline structures such as metallic ores to soft porous structures such as atmospheric dust, fly ash, and smoke. Porous particles with sizes less than one micrometer may adsorb gaseous contaminants and moisture. This can cause equipment failure due to accelerated corrosion.

5.2.6 Abrasiveness — Abrasiveness is a significant factor in mechanical erosion by high velocity solid contaminants. It also contributes to the accelerated wear of moving parts.

5.3 Explanation of Table 2

Solid particulates are classified by size. The environment should be described in terms of concentration severity level for each class, Classes SA through SD.

Table 2

Classification of Airborne Particulates

Particle Size	Class	Severity Level (Concentration measured in $\mu g/m^3$)			
		1	2	3	X
>1 mm	SA	< 1000	< 5000	<10 000	≥10 000
100 μm to 1000 μm	SB	< 500	< 3000	< 5000	≥ 5000
1 μm to 100 μm	SC	< 70	< 200	< 350	≥ 350
<1 μm	SD	< 70	< 200	< 350	≥ 350

NOTES: μm = micrometer = 0.001 millimeter
μm/m^3 = micrograms per cubic meter

6 AIRBORNE CONTAMINANTS — GASES
(Refer to Table 3)

6.1 Reactivity

Two methods have been used for environmental characterization. One is a direct measure of selected gaseous air pollutants. The other, which can be termed "reactivity monitoring," provides a quantitative measure of the overall corrosion potential of an environment.

Pollution analysis may provide short-term estimates for specific sites. High values will confirm that a severe environment exists. The reverse, however, is not necessarily true. Industrial environments may contain a complex mixture of contaminants that interact to greatly accelerate (or retard) the corrosive action of individual gas species.

To avoid these practical difficulties, the nature of industrial environments is defined in terms of the rate at which they react with copper. As a direct measure of overall corrosion potential, reactivity monitoring involves the placement of specially prepared copper coupons in the operating environments. Copper has been selected as the coupon material because data exists which correlates copper film formation with reactive (corrosive) environments. It has proven to be particularly useful for environmental characterization. Analyses may consist of measurements of film thickness, film chemistry, or weight loss. Sensitivity of reported techniques is well within the range required for meaningful application data.

Four levels of corrosion severity are established in Table 3. Concentration levels of some gases that contribute to these reactivity rates are also cited.

6.2 Contamination Effects

Each site may have different combinations and concentration levels of corrosive gaseous contaminants. Performance degradation can occur rapidly or over many years, depending on the particular concentration levels and combinations present at a site. The following paragraphs describe how various pollutants contribute to equipment performance degradation.

6.2.1 Relative Humidity

High relative humidity accelerates the corrosion caused by gaseous contaminants in an exponential manner. Equally important is the recognition of the fact that temperature fluctuations dramatically affect relative humidity and often induce local condensation. Although water is universally present in industrial atmospheres, the concentration varies widely. It promotes the corrosive degradation of equipment in the three major ways described below.

1. Directly, as a reactive chemical attacking metals and plastics.

2. Interactively with other atmospheric constituents, in most cases forming a more reactive combination. An example of this is sulfur dioxide, SO_2, which combines with water to form sulfurous acid.

3. Electrochemically: Many species when dissolved in water form a conductive solution. When electric potential differences exist between two dissimilar metals, the conditions for electrolytic or galvanic corrosion processes are set up. These are different phenomena, but both are caused by and/or promoted by an electrolyte.

6.2.2 Inorganic Chlorine Compounds
(Expressed as Cl_2 in Table 3)

This group includes chlorine, chlorine dioxide, hydrogen chloride, etc., and reactivity will depend upon the specific gas composition. In the presence of moisture, these gases generate chloride ions which react readily with the copper, tin, silver, and iron alloys. These reactions are significant even when the gases are present at low parts per billion levels. For example, the corrosivity of air containing 1 part per billion of chlorine would probably place that environment in the "Moderate" Class G2 category described in 6.3.2. A concentration of 10 parts per billion would probably increase the severity level to Class G3 or GX. These reactions are attenuated in dry atmospheres. At higher concentrations, many elastomers and some plastics are oxidized by exposure to chlorinated gases. Particular care must be given to equipment which is exposed to atmospheres which contain chlorinated contaminants. Sources of chloride ions, such as cleaning compounds and cooling tower vapors, etc., should be considered when classifying industrial environments. They are seldom absent in major installations.

6.2.3 Active Sulfur Compounds (Expressed as H_2S in Table 3)

This group includes hydrogen sulfide, elemental sulfur, and organic sulfur compounds such as the mercaptans. When present at low parts per billion levels, they rapidly attack copper, silver, aluminum, and iron alloys. The presence of moisture and small amounts of inorganic chlorine compounds greatly accelerate sulfide corrosion. Note, however, that attack still occurs in low relative humidity environments. Active sulfurs rank with inorganic chlorides as the predominant cause of atmospheric corrosion in the process industries.

6.2.4 Sulfur Oxides (Expressed as SO_2 and SO_3 in Table 3)

Oxidized forms of sulfur (SO_2, SO_3) are generated as combustion products of sulfur-bearing fossil fuels. Low parts per billon levels of sulfur oxides can passivate reactive metals and thus retard corrosion. At higher levels they attack certain types of masonry, metals, elastomers, and plastics. The reaction with masonry and metals normally occurs when these gases dissolve in water to form sulfurous and sulfuric acid.

6.2.5 Nitrogen Oxides (Expressed as NO_X in Table 3)

NO_X compounds (NO, NO_2, N_2O_4) are formed as combustion products of fossil fuels and have a critical role in the formation of ozone in the atmosphere. They are also believed to have a catalytic effect on corrosion of base metals by chlorides and sulfides. In the presence of moisture, some of these gases form nitric acid which, in turn, attacks most common materials.

6.2.6 Hydrogen Fluoride (Expressed as HF in Table 3)

This compound is a member of the halogen family and reacts like inorganic chloride compounds.

6.2.7 Ammonia and Derivatives (Expressed as NH_3 in Table 3)

Reduced forms of nitrogen (ammonia, amines, ammonium ions) occur mainly in fertilizer plants, agricultural applications, and chemical plants. Copper and copper alloys are particularly susceptible to corrosion in ammonia environments.

6.2.8 Photochemical Species (Expressed as O_3 in Table 3)

The atmosphere contains a wide variety of unstable, reactive species which are formed by the reaction of sunlight with moisture and other atmospheric constituents. Some have lifetimes measured in fractions of a second as they participate in rapid chain reactions. In addition to ozone, a list of examples would include the hydroxyl radical as well as radicals of hydrocarbons, oxygenated hydrocarbons, nitrogen oxides, sulfur oxides, and water. Because of the transient nature of most of these species, their primary effect is on outdoor installations and enclosures. In general, plastics and elastomers are more susceptible than metals to photochemical effects.

6.2.9 Strong Oxidants

This includes ozone plus certain chlorinated gases (chlorine, chlorine dioxide). Ozone (O_3) is an unstable form of oxygen which is formed from diatomic oxygen by electrical discharge or by solar radiation in the atmosphere. These gases are powerful bleaching and oxidizing agents. They attack the surface of many elastomers and plastics. Photochemical oxidation–the combined effect of oxidants and ultraviolet light (sunlight)–is particularly potent. Ozone may also function as a catalyst in sulfide and chloride corrosion of metals, but its precise role is unclear.

6.3 Explanation of Contaminant Severity Levels

There is a broad distribution of contaminant concentrations and reactivity levels existing within industries using process measurement and control equipment. Some environments are severely corrosive, while others are mild.

The purpose of the contaminant classes is to define environments on the basis of corrosion rate of oxygen-free high conductivity copper, which is prepared and tested as described in Appendix C.

6.3.1 Severity Level G1

Mild — An environment sufficiently well-controlled such that corrosion is not a factor in determining equipment reliability.

6.3.2 Severity Level G2

Moderate — An environment in which the effects of corrosion are measurable and may be a factor in determining equipment reliability.

6.3.3 Severity Level G3

Harsh — An environment in which there is a high probability that corrosive attack will occur. These harsh levels should prompt further evaluation resulting in environmental controls or specially designed and packaged equipment.

6.3.4 Severity Level GX

Severe — An environment in which only specially designed and packaged equipment would be expected to survive. Specifications for equipment in this class are a matter of negotiation between user and supplier.

7 BIOLOGICAL INFLUENCES

7.1 Biological Influences

Flora and fauna are important constituents of the environment in which industrial process measurement and control equipment is expected to function properly. Usually a tropical climate has more living contaminants, but other climates can have similar problems.

For example, insects can cause unexpected shutdowns of pneumatic equipment by blocking off all breather openings with a clay-like cement which they use to form their nests. Also, insulating material is often subject to damage by cockroaches and rodents, etc., which simply remove the insulation by nibbling it off the wires. The accumulation of fungi, molds, dead animals or insects can cause mechanical, electrical, or thermal equipment failures.

The subject of "flora and fauna" is a general classification for plant growth and animal (insect) life and is not specific enough to be useful, Therefore, any plant growth or animal life which may affect equipment performance should be considered.

Table 3

Classification of Reactive Environments

Severity Level	G1 Mild	G2 Moderate	G3 Harsh	GX Severe
Copper Reactivity Level (in angstroms)*	< 300	< 1000	< 2000	≥ 2000

The gas concentration levels shown below are provided for reference purposes. They are believed to approximate the Copper Reactivity Levels stated above, providing the relative humidity is less than 50%. For a given gas concentration, the Severity Level (and Copper Reactivity Level) can be expected to be increased by one level for each 10% increase in relative humidity above 50% or for a relative humidity rate of change greater than 6% per hour.

	Contaminant	Gas			Concentration†				
Reactive Species†,‡	Group A	H_2S	<	3	<	10	<	50	≥ 50
		SO_2, SO_3	<	10	<	100	<	300	≥ 300
		Cl_2	<	1	<	2	<	10	≥ 10
		NO_x	<	50	<	125	<	1250	≥ 1250
	Group B§	HF	<	1	<	2	<	10	≥ 10
		NH_3	<	500	<10 000		<25 000		≥25 000
		O_3	<	2	<	25	<	100	≥ 100

NOTES: *Measured in angstroms after one month's exposure. See Appendix C, Item Numbers 2, 3.
†mm^3/m^3 (cubic millimeters per cubic meter) parts per billion average for test period for the gases in Groups A and B
‡The Group A contaminants often occur together and the reactivity levels include the synergistic effects of these contaminants.
§The synergistic effects of Group B contaminants are not known at this time.

APPENDIX A

TERMINOLOGY

Corrosion — Deterioration of a substance (usually a metal) because of a reaction with its environment.

Contaminant — That which contaminates to make impure or corrupt by contact or mixing.

Electrochemical Corrosion — Corrosion of metal caused by current flowing through an electrolyte between anode and cathode areas.

Erosion — Deterioration by the abrasive action of fluids, usually accelerated by the presence of solid particles in suspension.

Halide — Compound containing fluorine, bromine, chlorine, or iodine.

Halogen — Bromine, chlorine, fluorine, or iodine.

Hygroscopic — Having a tendency to absorb water.

Oxidation — Loss of electrons by a constituent of a chemical reaction.

Oxide — Chemical compound of an element, usually metal, with oxygen.

Reduction — Gain of electrons by a constituent of a chemical reaction.

APPENDIX B — Section 1

Some Common Sources of Reactive Environmental Constituents

Category	Symbol	Constituent	Some Common Sources
Gas	H_2S	Hydrogen sulfide	Geothermal emissions, microbiological activities, fossil fuel processing, wood pulping, sewage treatment, combustion of fossil fuel, auto emissions, ore smelting, sulfuric acid manufacture
Gas	SO_2, SO_3	Sulfur dioxide	Combustion of fossil fuel, auto emissions, ore smelting, sulfuric acid manufacture, tobacco smoke
Gas	S_8, R-SH	Mercaptans	Foundries, sulfur manufacture
Gas	HF	Hydrogen fluoride	Fertilizer manufacture, aluminum manufacture, ceramics manufacture, steel manufacture, electronic device manufacture, fossil fuel
Gas	NO_x	Oxides of nitrogen	Automobile emissions, fossil fuel combustion, microbes, chemical industry
Gas	N_2	Active organic nitrogen	Automobile emissions, animal waste, vegetable combustion, sewage, wood pulping
Gas	NH_3	Ammonia	Microbes, sewage, fertilizer manufacture, geothermal steam, refrigeration equipment, cleaning products, reproduction (blueprint) machines
Solid	C	Carbon	Incomplete combustion (aerosol constituent), foundry
Gas	CO	Carbon monoxide	Combustion, automobile emissions, microbes, trees, wood pulping
Gas	Cl_2, ClO_2	Chlorine, Chlorine dioxide	Chlorine manufacture, aluminum manufacture, paper mills, refuse decomposition, cleaning products
Gas	HCl	Hydrogen chloride	Automobile emissions, combustion, oceanic processes, polymer combustion
Gas	HBr, HI	Halogen compounds	Automotive emissions
Liquid	Cl	Chloride ions	Aerosol content, oceanic processes, ore processing
Gas	O_3	Ozone	Atmospheric photochemical processes mainly involving nitrogen oxides and oxygenated hydrocarbons, automotive emissions, electrostatic filters
Gas	C_nH_n	Hydrocarbons	Automotive emissions, fossil fuel processing, tobacco smoke, water treatment, microbes. Many other sources, both natural and industrial, paper mill
Solid	—	Inorganic dust	Crystal rock, rock and ore processing, combustion, blowing sand and many industrial sources

APPENDIX B — Section 2

Some Common Emissions of Natural and Industrial Processes

Natural Processes	Emissions
Microbes	H_2, NH_3, NO_X, H_2S, CO, large variety of organics of many types
Sewage	NH_3, aldehydes, many organics, H_2S, mercaptans, H_2, S, CO
Geothermal	H_2, H_2S, SO_2
Marshy area	H_2S, NH_3, SO_2
Animal matter	Many organics, mainly oxygenated
Forest fire	HCl, CO, CO_2
Oceans	NaCl, chloride ions

Industrial Processes	Emissions
Power generation	SO_2, C, CO, NO_X, hydrocarbons, organics
Automotive combustion	SO_2, SO_3, HCl, HBr, NO_X, hydrocarbons, organics, CO, HBr
Diesel combustion	CO, NO_X, many organics
Fossil fuel processing	H_2S, S, SO_2, NH_3, hydrocarbons, other organics, mercaptans
Plastic manufacture	All organics, aldehydes, alcohols, NH_3, SO_2
Cement plants	SO_3, dust, SO_2, NO_X, CO
Steel blast furnaces	H_2S, SO_2, CO, HF, coal dust
Steel electric furnace	H_2S, SO_2, C, CO
Coke plants	H_2S, CO, HCN, carbon, dust
Pulp manufacture	Cl_2, SO_2, H_2S, CO, wood fibers, dust
Chlorine plants	Chlorine, chlorine compounds, NaCl
Fertilizer manufacture	HF, NH_3, CH_4, gas, liquids, dust, acids
Food processing	Hydrocarbons, many organics
Rubber manufacture	H_2S, S_8, R-SH
Paint manufacture	C, hydrocarbons, oxygenated hydrocarbons, dust
Aluminum manufacture	HF, SO_2, C, dust
Ore smelting	SO_2, CO, H_2, dust
Tobacco smoke	H_2S, SO_2, HCN, CO, tars and particulates
Gasoline and fuel vapors	Hydrocarbons, oxygenated hydrocarbons
Battery manufacture	SO_2, acids, dust

APPENDIX C

Copper Reactivity Samples

1. Sample Preparation — Copper samples (nominal size 15 cm^2) should be prepared from 99.99 purity, oxygen-free high conductivity (OFHC), 0.635 mm thick sheet; ½ - ¾ hard.

Prepare as follows:

(1) Abrade with 240X metallograph paper using a wax lubricant.
(2) Abrade with 400X metallograph paper as in Step (1).
(3) Abrade with 600X metallograph paper as in Step (1).*
(4) Scrub with cotton soaked in hot reagent grade acetone.
(5) Dip in hot reagent grade isopropyl alcohol.
(6) Store in glass containers purged with dry nitrogen.

*Steps (3) through (5) should be done as near to placement time as possible.

2. Sample Exposure — Three copper coupons should be placed vertically at the site being monitored. Particular care should be taken to avoid surface contamination such as fingerprints. Installation should be in an area which has air flow rates that are characteristic of the site.

Corrosion is defined in terms of the corrosion film thickness which builds up with one month of exposure. It is recognized that film buildup will be quite slow in mild areas but rapid in severe sites. To facilitate film thickness measurements in these extreme conditions, test times can be extended to three months at mild sites or reduced to two weeks in harsh environments. Copper corrosion is non-linear, so changes of this type must be made with great care. Experience has shown that measurements over longer or shorter test times can be reduced to a normalized one month value by using the relationship

$$x_1 = x\,(t_1/t)^A$$

where x_1 is the equivalent film thickness after one month

x is the measured film thickness after time t

t_1 is thirty days

t is the actual test time (days)

A is equal to 0.3 for G1, 0.5 for G2, 1 for G3 and GX

3. Sample Analysis — Film thickness should be determined by cathodic reduction using the method of W. E. Campbell and U. B. Thomas, "Tarnish Studies," Bell Telephone System Technical Publications, Monograph 13, 1170 (1939).

4. References

Abbott, W. H. "The Effects of Operating Environments on Electrical and Electronic Equipment Reliability in the Pulp and Paper Industry." Paper presented at the IEEE Industry Applications Society 1983 Pulp and Paper Technical Conference, May 1983. *IEEE Conference Record.* New York: Institute of Electrical and Electronics Engineers, Inc., 1983.

Rice, D. W., et al. "Atmospheric Corrosion of Copper and Silver." *Electrochem. Soc.* 128, no. 2 (February 1981): 275-284.

ANSI/ISA-S72.01-1985
Approved February 3, 1986
Second Printing 1987

American National Standard

PROWAY-LAN
Industrial Data Highway

Instrument Society of America

ISBN 0-87664-896-0

ISA-72.01 PROWAY-LAN Industrial Data Highway

Copyright © 1985 by the Instrument Society of America. All rights reserved. Printed in the United States of America. No part of this publication may be reproduced, stored in a retrieval system, or transmitted, in any form or by any means (electronic, mechanical, photocopying, recording, or otherwise), without the prior written permission of the publisher.

INSTRUMENT SOCIETY OF AMERICA
67 Alexander Drive
P.O. Box 12277
Research Triangle Park, North Carolina 27709

Second Printing, 1987

PREFACE

This standard has been prepared as part of the service of the Instrument Society of America (ISA) toward a goal of uniformity in the field of instrumentation. To be of real value, this document should not be static, but should be subject to periodic review. Toward this end, the Society welcomes all comments and criticisms, and asks that they be addressed to the Secretary, Standards and Practices Board, Instrument Society of America, 67 Alexander Drive, P.O. Box 12277, Research Triangle Park, NC 27709, Telephone (919) 549-8411.

The ISA Standards and Practices Department is aware of the growing need for attention to the metric system of units in general, and the International System of Units (SI) in particular, in the preparation of instrumentation standards. The Department is further aware of the benefits to U.S.A. users of ISA standards of incorporating suitable references to the SI (and the metric system) in their business and professional dealings with other countries. Toward this end, this Department will endeavor to introduce SI-acceptable metric units in all new and revised standards to the greatest extent possible. The Metric Practice Guide, which has been published by the Institute of Electrical and Electronics Engineers as ANSI/IEEE Std. 268-1982, and future revisions will be the reference guide for definitions, symbols, abbreviations, and conversion factors.

It is the policy of the Instrument Society of America to encourage and welcome the participation of all concerned individuals and interests in the development of ISA standards. Participation in the ISA standards-making process by an individual in no way constitutes endorsement by the employer of that individual, of the Instrument Society of America, or of any of the standards that ISA develops.

The information contained in the preface, footnotes, and appendices is included for information only and is not a part of the standard.

Foreword

Efforts to develop a standard for data communications to support process and industrial control began in 1976. This work was centered in Working Group 6 of Subcommittee 65C of the International Electrotechnical Committee, supported by SP72 of the Instrument Society of America and committees in Japan and Germany. This work led to development of the PROWAY A&B (token-passing bus) draft standards.

Through ANSI, members of ISA's SP72 committee continue to serve as members of the International Electrotechnical Commission (IEC) Subcommittee 65C Working Group 6 on intersubsystem computer communications. These experts have made significant contributions to the development of a series of IEC draft standards currently under consideration for worldwide acceptance. ISA-S72.01-1985 is in complete harmony with the comparable IEC document, Draft IEC Publication 955.

In 1980, Project 802 of the Institute of Electrical and Electronics Engineers began work on a general standard for Local Area Network (LAN) communications. Several members of SP72 joined the 802 committee and thus many 802 concepts were drawn from the PROWAY effort.

In January 1983, SP72 evaluated the Token-Passing Bus proposal of the IEEE 802.4 committee and concluded that it could be made suitable for industrial control. SP72 also concluded that both the general welfare and standards harmonization would be enhanced by minimizing the number of Token Bus LAN standards.

In April 1983, SP72 began a cooperative effort with the IEEE 802 committee. This effort led to development of the ISA S72.01-1985 PROWAY-LAN Industrial Data Highway standard for industrial process control. This standard is known as PROWAY-C by the IEC.

PROWAY-LAN is compatible with (but more restrictive than) the IEEE 802.2 and 802.4 standards for general LANs. Significant enhancements were made to both the IEEE 802.2 and 802.4 standards to allow a compatible implementation of features that provide the reliability and timeliness required for industrial control.

The following is an alphabetical list of those who participated in developing the PROWAY-LAN standard. Those marked * were active members of SP72 as of the date shown on this document.

Mark Bauer	Simon Korowitz*
William Brown*	Bob Lawler
Anthony Capel*	Jim Lindgren
Richard Caro*	Laurie Lindsay
Fred Casadei	Gunther Martin
Barry Cole	Ronald Martin
Steven Cooper	Samuel Miles
Robert Crowder* — Chairman	Gene Nines*
Dorel Damsker*	Toshio Ogawa
Stephen Dana*	Thomas Phinney
Emanuel Delahostria*	Dennis Quy*
Bruce Dilger	John Ricketson
Cesar Doreza*	Fred Schwierske
Jack Dorsey* — Treasurer	Paul Senechal
Robert H. Douglas*	Harvey Shepherd*
Robert Eskeridge	Don Smith*
Gerhard Funk	Mark Steiglitz
Maris Graube	Garry Stephens
Klaus Grund	Ken Still
Robert Guzik	Kathleen Sturgis*
Mel Hagar*	David Sweeton*
Jeffrey Hale	Allen Tanzman
Robert Harold*	Seppo Turunen
Thomas Harrison	Michael Ullman
Wade Harsy* — Secretary	Jay Warrior*
Philip Jacobs	Earl Whitaker*
Dittmar Janetzky*	Chris Wilmering
Kenneth Jones	Graeme Wood
	Prentiss Yates*

Special thanks are due to Anthony C. Capel, Robert S. Crowder, Dittmar Janetzky, Gunther Martin, Thomas L. Phinney, and David C. Sweeton for technical contributions and to David E. Carlson, Robert H. Douglas, Robert S. Crowder and Eugene Nines for coordinating the efforts of the IEEE 802.2 and 802.4, ISA SP72 and IEC/SC65C/W66 committees during the joint development of this standard for industrial applications of Token Bus Local Area Networks.

This standard was approved for publication by the ISA Standards and Practices Board in June 1985.

N. Conger, Chairman	C. W. Reimann
P. V. Bhat	J. Rennie
W. Calder III	W. C. Weidman
R. S. Crowder	K. Whitman
B. Feikle	*P. Bliss
H. S. Hopkins	*B. A. Christensen
J. L. Howard	*L. N. Combs
R. T. Jones	*R. L. Galley
R. Keller	*T. J. Harrison
O. P. Lovett, Jr.	*R. G. Marvin
E. C. Magison	*W. B. Miller
A. P. McCauley	*G. Platt
J. W. Mock	*J. R. Williams
E. M. Nesvig	
R. Prescott	*Director Emeritus
D. E. Rapley	

Contents

SECTION			PAGE
I.	Introduction		1
	I.1	Scope	1
	I.2	Introduction	1
		I.2.1 Layering	1
		I.2.2 Precedence	2
		I.2.3 Definitions	2
		I.2.4 References	2
	I.3	Device Types	2
		I.3.1 Classes of Stations	2
		I.3.2 Compliance Requirements	2
	I.4	PROWAY Organization	3
	I.5	Overview of the PROWAY Link Control (PLC) Layer	3
		I.5.1 PROWAY Link Control Services	3
		I.5.2 PROWAY Link Control Sublayer Organization	3
		I.5.2.1 Local State Machine Functions	3
		I.5.2.2 Remote State Machine Functions	3
	I.6	Overview of the Token Bus Access Method	4
		I.6.1 The Essence of the Token Bus Access Method	4
		I.6.2 General MAC Sublayer Functions	4
	I.7	MAC Layer Internal Structure	5
		I.7.1 Interface Machine (IFM)	5
		I.7.2 Access Control Machine (ACM)	6
		I.7.3 Receive Machine (RxM)	6
		I.7.4 Transmit Machine (TxM)	6
	I.8	Token Bus Access Method Characteristics	6
	I.9	Overview of the Physical Layer and Media	7
		I.9.1 Summary of Phase Continuous FSK (Frequency Shift Keying) Physical Layer	7
		I.9.2 Summary of Single-Channel Coaxial Cable Bus Medium	7
		I.9.3 Alternate Physical and Media Implementation	7
	I.10	Standard Organization	8
	I.11	Correspondence between the PROWAY and IEEE 802 Standards	9
	I.12	Compatibility with IEEE 802	10
I-A	Appendix — References		10
1.	Functional Requirements		12
	1.1	Overview	12
	1.2	Application Environment and Main Features	13
		1.2.1 Application Characteristics	13
		1.2.2 Economic versus Technical Factors	13
		1.2.3 Main Features of PROWAY	13
	1.3	Device Types	13
		1.3.1 Communications between Control Devices	13
		1.3.2 Communications with Other Devices	13
		1.3.3 Classes of Stations	13
	1.4	System Structure	14
		1.4.1 Control System Structure	14
		1.4.2 Data Highway Mode of Operation	14
		1.4.3 Configuration Classes	14

SECTION			PAGE
	1.5	Maintenance and Service Features	14
		1.5.1 Testing and Fault Diagnosis	14
		1.5.2 Effect of State Transitions	14
	1.6	Safety	14
		1.6.1 Electrical Faults	14
		1.6.2 Intrinsic Safety	15
	1.7	Performance in Industrial Environment	15
		1.7.1 Industrial Environment	15
		1.7.1.1 Induced Noise	15
		1.7.1.2 Electromagnetic Environment	15
		1.7.1.3 Differences in Earth Potential	15
		1.7.2 Data Circuit Bit Error Rate	15
		1.7.3 Residual Error Rate	15
		1.7.4 Information Transfer Rate	15
		1.7.5 Media Access and Highway Transaction Times	16
		1.7.5.1 Media Access Time	16
		1.7.5.2 Highway Transaction Time	16
		1.7.5.3 Definitions	16
	1.8	System Availability	16
		1.8.1 Effect of Failures	16
		1.8.2 Internal Status and Error Reporting	16
		1.8.3 Automatic Recovery	16
		1.8.4 Control of Stations	16
2.	The Interface to PROWAY		17
	2.1	Organization	17
2A.	User-PLC Interface and Service Specification		17
	2A.1	Scope and Field of Application	17
	2A.2	Overview of PROWAY Services	17
		2A.2.1 General Description of Services Provided	17
		2A.2.2 Model Used for the Service Specification	17
		2A.2.3 Overview of Services	18
		2A.2.3.1 Send Data with Acknowledge (SDA)	18
		2A.2.3.2 Send Data with No Acknowledge (SDN)	18
		2A.2.3.3 Request Data with Reply (RDR)	18
		2A.2.4 Overview of Interactions	18
		2A.2.5 Overview of Data Highway Operation	19
	2A.3	Mandatory Features	19
	2A.4	Detailed Specification of PLC Interactions with the User of PROWAY	19
		2A.4.1 Send Data with Acknowledge (SDA)	19
		2A.4.1.1 Description of Operation	19
		2A.4.1.2 Definition of Primitives of Local User-Highway Interface	20
		2A.4.2 Send Data with No Acknowledge	23
		2A.4.2.1 Description of Operation	23
		2A.4.2.2 Inter-operability	23
		2A.4.2.3 Definition of Primitives at Local User-Highway Interface	23
		2A.4.3 Request Data with Reply (RDR)	26
		2A.4.3.1 Description of Operation	26
		2A.4.3.2 Definition of Primitives at Local User-PLC Interface	26

SECTION	PAGE

2B.	User-Management Interface and Service Specification		31
	2B.1	Scope and Field of Application	31
		2B.1.1 Mandatory Features	31
	2B.2	Overview of Management Services	31
		2B.2.1 General Description of Services Provided at the User-Management Interface	31
		2B.2.2 Model Used for the Service Specification	31
		2B.2.3 Overview of Interactions	33
	2B.3	Detailed Specification of Management Interactions with the User of PROWAY	33
		2B.3.1 L_MGMT.request	33
		2B.3.1.1 Function	33
		2B.3.1.2 Semantics	33
		2B.3.1.3 When Generated	33
		2B.3.1.4 Effect of Receipt	33
		2B.3.1.5 Additional Comments	
		2B.3.2 L_MGMT.confirm	33
		2B.3.2.1 Function	33
		2B.3.2.2 Semantics	33
		2B.3.2.3 When Generated	34
		2B.3.2.4 Effect of Receipt	34
		2B.3.2.5 Additional Comments	34
		2B.3.3 L_MGMT.indication	35
		2B.3.3.1 Function	35
		2B.3.3.2 Semantics	35
		2B.3.3.3 When Generated	35
		2B.3.3.4 Effect of Receipt	35
2-A.	Appendix — Overview of Data Highway Operation		35
	2-A.1	Organization of Data Highway Overview	35
	2-A.2	Send Data with Acknowledge (SDA) Service Diagram	37
		2-A.2.1 Topological and Sequential Relationships	37
		2-A.2.2 Service State Diagrams	38
	2-A.3	Send Data with No Acknowledge	40
		2-A.3.1 Topological and Sequential Relationships	40
		2-A.3.2 Service State Diagrams	41
	2-A.4	Request Data With Reply (RDR) Service Diagram	43
		2-A.4.1 Topological and Sequential Relationships	43
		2-A.4.2 Service State Diagrams	43
3.	The PROWAY Link Control (PLC) Sublayer		46
3A.	PROWAY Link Control (PLC) Sublayer Definitions and Mandatory Features		47
	3A.1	Notations Used in PLC Machine State Tables	47
		3A.1.1 Notations Used for User-PLC Interface Parameters	47
		3A.1.2 Notations Used for Link_protocol Data Unit Parameters	47
		3A.1.3 Notations Used for PLC-MAC Interface Parameters	47
	3A.2	Variable Definitions for the PLC Machine State Tables	48
		3A.2.1 Global PLC Variables	48
		3A.2.2 Local PLC Variables	48
		3A.2.3 Remote PLC Variable	48
	3A.3	Parameter Definitions for the PLC Machine State Tables	48
		3A.3.1 Local PLC Parameters	48
		3A.3.2 Remote PLC Parameters	48
	3A.4	Constants Used in the PLC Machine State	49

SECTION				PAGE
	3A.5	Functions and Procedures — Definitions for the PLC State Machines		49
		3A.5.1	LOCAL_STATUS?	49
		3A.5.2	UPDATE_CONTEXT	49
		3A.5.3	VALIDATE?	49
		3A.5.4	SEQUENCE	49
		3A.5.5	BUILD_PDU	49
		3A.5.6	EXTRACT_CONTEXT	50
		3A.5.7	NOTIFY_MGT	50
		3A.5.8	REMOTE_STATUS?	50
		3A.5.9	RESOURCES?	50
		3A.5.10	DUPLICATE?	50
		3A.5.11	ACTIVATED?	50
		3A.5.12	UPDATE_HISTORY	50
		3A.5.13	REQUEST_DATA_AREA	50
		3A.5.14	RELEASE_DATA_AREA	51
		3A.5.15	UPDATE_DATA_AREA	51
		3A.5.16	RESPONSE_TYPE	51
		3A.5.17	BUILD_RPDU	51
		3A.5.18	EVEN	51
		3A.5.19	UPDATE_SEQ	51
	3A.6	Mandatory Features		51
		3A.6.1	Validity of Response Frames	51
		3A.6.2	PLC_station_delay	51
		3A.6.3	Mandatory State Machines and Features	52
			3A.6.3.1 Local Machine	52
			3A.6.3.2 Remote Machine	52
		3A.6.4	Labeling	52
3B.	PLC Machine Formal Description			53
	3B.1	Overview		53
		3B.1.1	State Machine Invocations	53
		3B.1.2	State Diagrams	53
	3B.2	Techniques Used in the PLC Machine State Descriptions		53
	3B.3	State Transition Table for Local PLC State Machine		55
	3B.4	State Transition Table for the Remote PLC State Machine		58
3C.	PLC Protocol Data Unit Format			61
	3C.1	Function of the PLC_Protocol_Data_Unit		61
	3C.2	Structure of the PLC_Protocol_Data_Unit		61
		3C.2.1	PLC Header Composition	62
		3C.2.2	Service_Access Points	62
			3C.2.2.1 LSAP Representation	63
			3C.2.2.2 LSAP Field Composition	63
			3C.2.2.3 LSAP Address Usage	63
			3C.2.2.4 Global LSAP Assignments	63
			3C.2.2.5 Locally Administered LSAP Assignments	64
		3C.2.3	Specification of Shared Data Areas	64
		3C.2.4	Implied SAP	64
		3C.2.5	L_PDU_TYPE Definitions	64
			3C.2.5.1 Sequence_numbers	64
			3C.2.5.2 R_Status Definition	64
	3C.3	L_Data_Unit		64

SECTION				PAGE
		3C.3.1	Data Unit Size	64
		3C.3.2	Bit Order	65
	3C.4	Invalid L_pdu		65
3-A.	Appendix — Prevention of Loss and Duplication for the PROWAY Confirmed Data Transfer Services			66
	3-A.1	Introduction		66
	3-A.2	Overview of the PROWAY Confirmed Services		66
	3-A.3	Sources of Frame Loss		66
	3-A.4	Protection against Loss and Creation of Duplicate Frames		67
	3-A.5	Protection against Duplication		67
	3-A.6	Operation with Non-PROWAY MAC Layers		67
	3-A.7	An Example of Erroneous Detection of Duplication when Sequence Information Is Not Used		68
4.	PLC-MAC Interface and Service Specification			68
	4.1	Scope and Field of Application		68
	4.2	Overview of the PLC-MAC Data Transfer Service		68
		4.2.1	General Description of Services Provided	68
		4.2.2	Model Used for the Service Specification	69
		4.2.3	Notation	69
		4.2.4	Overview of Interactions	69
		4.2.5	Mandatory Features	69
	4.3	Detailed Interactions with the PLC Entity		69
		4.3.1	MA_DATA.request	70
			4.3.1.1 Function	70
			4.3.1.2 Semantics	70
			4.3.1.3 When Generated	70
			4.3.1.4 Effect of Receipt	70
			4.3.1.5 Additional Comments	70
		4.3.2	MA_DATA.indication	71
			4.3.2.1 Function	71
			4.3.2.2 Semantics	71
			4.3.2.3 When Generated	71
			4.3.2.4 Effect of Receipt	71
			4.3.2.5 Additional Comments	71
		4.3.3	MA_DATA.confirm	71
			4.3.3.1 Function	71
			4.3.3.2 Semantics	71
			4.3.3.3 When Generated	72
			4.3.3.4 Effect of Receipt	72
			4.3.3.5 Additional Comments	72
5.	The Medium Access Control (MAC) Layer			73
5A.	Informal Description of MAC Sublayer Operation			74
	5A.1	Token Ring Steady State Operation		74
		5A.1.1	Slot_Time	75
		5A.1.2	Token Passing	75
		5A.1.3	Adding New Stations to the Token Ring	76
			5A.1.3.1 Bound on Token Rotation Time	77
			5A.1.3.2 User Notification of Ring Membership	77
		5A.1.4	Token Ring Initialization	78
			5A.1.4.1 Leaving the Token Ring	78
		5A.1.5	Priorities	78

SECTION			PAGE
	5A.1.6	MAC Confirmed Data Transfer Service	80
	5A.1.7	Randomized Variables	80
5A.2	Access Control Machine (ACM) States		80
	5A.2.0	Offline	80
	5A.2.1	Idle	81
	5A.2.2	Demand In	81
	5A.2.3	Demand Delay	82
	5A.2.4	Claim Token	83
	5A.2.5	Use Token	83
	5A.2.6	Await IFM Response	84
	5A.2.7	Check Access Class	84
	5A.2.8	Pass Token	85
	5A.2.9	Check Token Pass	85
	5A.2.10	Await Response	85
5B.	MAC Sublayer Definitions and Requirements		86
5B.1	MAC Definitions		86
	5B.1.1	Immediate_response	86
	5B.1.2	MAC-symbols	86
	5B.1.3	MAC_symbol_time	86
	5B.1.4	Octet_time	87
	5B.1.5	PHY-symbols	87
	5B.1.6	Transmission_path_delay	87
	5B.1.7	MAC_station_delay	87
	5B.1.8	PLC_station_delay	87
	5B.1.9	Safety_margin	87
	5B.1.10	Slot_time	87
	5B.1.11	Response_time	87
	5B.1.12	Response_window	87
	5B.1.13	Maximum_retry_limit	88
5B.2	Transmission Order		88
5B.3	Delay Labeling		88
5B.4	Miscellaneous Requirements		88
	5B.4.1	Station Initialization	88
	5B.4.2	Token-Passing Order	88
	5B.4.3	Station Receipt of Its Own Transmission	90
	5B.4.4	Token Holding Time	90
	5B.4.5	Address Lengths	90
	5B.4.6	Randomized Variables	91
	5B.4.7	Contention Delay	91
	5B.4.8	Token Claiming	91
5B.5	Use of Address Bits in Contention Algorithm		91
	5B.5.1	Claim_token Frame Length	91
	5B.5.2	Demand Delay Time Interval	92
5B.6	Priority of Transmitted Frames		92
	5B.6.1	Access Classes	92
	5B.6.2	Priority to Access_Class Mapping	92
	5B.6.3	Token Rotation Timers	92
	5B.6.4	Recommended Priority Assignments in PROWAY System	93
5B.7	Delegation of Right to Transmit		93
5B.8	Notification of Logical Ring Modification		93
5B.9	Station Addresses		93

SECTION			PAGE
	5B.10	Required MAC Machines	93
		5B.10.1 Mandatory Access Control Machine (ACM) Functions	93
	5B.11	Interface Machine	93
		5B.11.1 Interface Machine (IFM) Description	93
		5B.11.2 Mandatory Interface Machine (IFM) Functions	94
	5B.12	Receive Machine	94
		5B.12.1 Receive Machine Description	94
		5B.12.2 Mandatory Receive Machine (RxM) Functions	95
	5B.13	Transmit Machine	95
		5B.13.1 Transmit Machine Description	95
		5B.13.2 Mandatory Transmit Machine (TxM) Functions	95
5C.	Access Control Machine		95
	Formal Description		95
	5C.1	Variables and Functions	95
		5C.1.1 Station Management Variables	95
		5C.1.2 Interface Machine Variables and Functions	97
		5C.1.3 Note on Logical Ring Membership Control	97
		5C.1.4 Timers	97
		5C.1.4.1 Slot_time Interval Timers	97
		5C.1.4.2 Octet_time Interval Timers	98
		5C.1.5 Receive Machine Variables and Functions	99
		5C.1.6 Other Variables and Functions	99
	5C.2	Access Control Machine Formal Description	103
		5C.2.1 List of Unique ACM Words	103
5D.	Frame Formats		104
	5D.1	Frame Components	105
		5D.1.1 Preamble	105
		5D.1.2 Start Delimiter	105
		5D.1.3 Frame Control Field	106
		5D.1.3.1 MAC_control_frame	106
		5D.1.3.2 Data_frames	106
		5D.1.4 Address Fields	107
		5D.1.4.1 Destination Address Field Composition	107
		5D.1.4.2 Individual Addresses	108
		5D.1.4.3 Source Address Field	109
		5D.1.4.4 Numerical Interpretation of Addresses	109
		5D.1.5 MAC Data Unit Field	109
		5D.1.6 Frame Check Sequence (FCS) Field	109
		5D.1.7 End Delimiter	110
		5D.1.8 Abort Sequence	111
	5D.2	Enumeration of Frame Types	111
		5D.2.1 MAC Control Frame Formats	111
		5D.2.1.1 Claim_token	111
		5D.2.1.2 Solicit_successor_1	111
		5D.2.1.3 Solicit_successor_2	111
		5D.2.1.4 Who_follows	111
		5D.2.1.5 Resolve_contention	112
		5D.2.1.6 Pass Token	112
		5D.2.1.7 Set_successor	112
		5D.2.2 Data Frame Formats	112
		5D.2.2.1 Request_with_no_response Data Frame	112

SECTION				PAGE
		5D.2.2.2	Request_with_response Data Frame	112
		5D.2.2.3	Response Data Frame	112
	5D.2.3	Invalid Frames		113
5-A.	Appendix — An Example PROWAY Implementation at $1*10^6$ Bits/Sec			113
5-B.	Appendix — Examples of Individual PROWAY Station Addresses			114
6.	MAC-Physical Layer Interface Specification			117
6A.	MAC-Physical Layer Interface Service Specification			118
	6A.1	Scope and Field		118
	6A.2	Overview of Physical Layer Service		118
		6A.2.1	General Description of Services Provided by the PHY Layer	118
		6A.2.2	Model Used for the Service Specification	118
		6A.2.3	Overview of Interactions	118
		6A.2.4	Basic Services and Options	119
	6A.3	Detailed Specifications of Interactions with the Physical Layer Entity		119
		6A.3.1	PHY_DATA.request	119
			6A.3.1.1 Function	119
			6A.3.1.2 Semantics	119
			6A.3.1.3 When Generated	119
			6A.3.1.4 Effect of Receipt	119
			6A.3.1.5 Constraints	119
			6A.3.1.6 Additional Comments	120
		6A.3.2	PHY_DATA.indication	120
			6A.3.2.1 Function	120
			6A.3.2.2 Semantics	120
			6A.3.2.3 When Generated	120
			6A.3.2.4 Effect of Receipt	120
			6A.3.2.5 Additional Comments	120
6B.	MAC-Physical Layer Interface Implementation Specification			121
	6B.1	Scope and Field of Application		121
	6B.2	Terminology		121
		6B.2.1	Model Used for Specification	121
	6B.3	Overview of Interactions		121
		6B.3.1	Signals to the PHY Entity	121
		6B.3.2	Signals to the MAC Entity	121
	6B.4	Detailed Specification of MAC Entity-Physical Entity Interface Signals		122
		6B.4.1	Physical_Send Signal (PSC0, PSC1, PSC2)	122
			6B.4.1.1 Physical_Send Encoding	122
			6B.4.1.2 Requirements	122
		6B.4.2	Physical_Receive Signals (PRC0, PRC1, PRC2)	122
			6B.4.2.1 Physical_Receive Encoding	122
			6B.4.2.2 Requirements	123
		6B.4.3	Timing Signals	123
			6B.4.3.1 Physical_Send_Timing (PST)	123
			6B.4.3.2 Physical_Receive_Timing (PRT)	123
			6B.4.3.3 Physical_Transmit_Timing (PTT)	124
		6B.4.4	Management Signals	124
			6B.4.4.1 Physical_Line_Disconnect (PLD)	124
			6B.4.4.2 Physical_Watchdog_Status (PWS)	124
			6B.4.4.3 Physical_Primary_Signal_Source (PPS)	124
	6B.5	MAC-PHY Interface Realization		125
		6B.5.1	Grounding	125

SECTION				PAGE
		6B.5.1.1	Signal Ground	125
		6B.5.1.2	Protective Ground	125
		6B.5.1.3	External Ground Connections	125
		6B.5.1.4	Circuit Ground Connections	126
		6B.5.1.5	Cable Shield Connection	126
	6B.5.2	MAC-PHY Interface Connector		126
		6B.5.2.1	MAC-PHY Interface Connector Mechanical Requirements	126
		6B.5.2.2	Assignment of MAC-PHY Connector Pin Numbers	127
		6B.5.2.3	MAC-PHY Interface Connector Electrical Requirements	127
	6B.5.3	MAC-PHY Interconnection Cable		128
		6B.5.3.1	Cable Length	128
		6B.5.3.2	Cable Requirements	128
		6B.5.3.3	Precaution for Electromagnetic Radiation	128
7.	Alternate Physical Layers and Media			128
8.	Specification of the Single-Channel Phase-Continuous-FSK Physical (PHY) Layer and its Interface to the Medium			128
8.1	Scope and Field of Application			128
	8.1.1	Nomenclature		128
	8.1.2	Object		129
	8.1.3	Compatibility Considerations		130
	8.1.4	Operational Overview of the Single-Channel Coaxial-Cable-Bus Medium		130
8.2	Overview of the Phase-Continuous-FSK Physical (PHY) Layer			130
	8.2.1	General Description of Functions		130
		8.2.1.1	Symbol Transmission and Reception Functions	130
		8.2.1.2	Jabber Inhibit Function	130
		8.2.1.3	Local Administrative Functions	130
	8.2.2	Model Used for the Functional Specification		131
	8.2.3	Required Functions		131
8.3	Application of Station Management-PHY Layer Interface Specification			131
8.4	Single-Channel Phase-Continuous-FSK Physical Layer Functional, Electrical and Mechanical Specifications			131
	8.4.1	Data Signaling Rates		131
	8.4.2	Symbol Encoding		131
	8.4.3	Modulated Line Signal		132
		8.4.3.1		132
		8.4.3.2		132
		8.4.3.3		132
		8.4.3.4		133
		8.4.3.5		133
	8.4.4	Jabber Inhibit		133
	8.4.5	Coupling to the Medium		133
	8.4.6	Receiver Sensitivity and Selectivity		133
	8.4.7	Symbol Timing		133
	8.4.8	Symbol Decoding		133
	8.4.9	Received Signal Source Selection		134
	8.4.10	Transmitter Enable/Disable		134
	8.4.11	Redundant Media Considerations		134
	8.4.12	Reliability		134

SECTION			PAGE
	8.5	Environmental Specifications	134
		8.5.1 Safety Requirements	134
		8.5.2 Electromagnetic and Electric Environment	135
		8.5.2.1 Electromagnetic Environment	135
		8.5.2.2 Differences in Earth Potential	135
		8.5.3 Temperature and Humidity	135
		8.5.4 Regulatory Requirements	135
	8.6	Labeling	135
9.	Single-Channel Coaxial-Cable-Bus Medium "Layer" Specification		136
	9.1	Scope and Field of Application	136
		9.1.1 Nomenclature	136
		9.1.2 Object	137
		9.1.3 Compatibility	137
	9.2	Overview of the Coaxial-Cable-Bus Medium "Layer"	137
		9.2.1 General Description of Functions	137
		9.2.1.1 Overview of the Single-Channel Coaxial-Cable-Bus Medium	137
		9.2.2 Model Used for the Functional Specification	138
		9.2.3 Mandatory Characteristics and Options	138
	9.4	Single-Channel Coaxial-Cable-Bus Medium "Layer" Functional, Electrical and Mechanical Specifications	138
		9.4.1 Coupling to the Station	138
		9.4.2 Characteristic Impedance and Impedance Matching	138
		9.4.3 Signal Level	138
		9.4.4 Distortion	139
		9.4.5 Noise Floor and Signal-to-Noise (S/N) Ratio	139
		9.4.6 Cable Shielding	139
		9.4.7 Compatibility at the Station Interface	139
		9.4.8 Redundancy Considerations	139
		9.4.9 Reliability	139
	9.5	Installation Requirements	139
		9.5.1 Sound Installation Practice	139
		9.5.2 Grounding	139
		9.5.3 Termination of Drop Cables and Taps	140
		9.5.4 Regulatory Requirements	140
		9.5.5 Installation and Maintenance Guidelines	140
		9.5.6 Notice	140
	9.6	Environmental Specifications	140
		9.6.1 Electromagnetic and Electric Environment	140
		9.6.1.1 Electromagnetic Fields	141
		9.6.1.2 Differences in Earth Potentials	141
		9.6.2 Temperature and Humidity	141
	9.7	Transmission Path Delay Considerations	141
	9.8	Documentation	141
	9.9	Network Configuration	142
9-A.	Appendix — Guidelines for Configuring the Medium for a Single-Channel Coaxial-Cable-Bus Local Area Network		142
	9-A.1	Network Configuration	142
	9-A.2	Medium Components	142
		9-A.2.1 Coaxial Cable	142
		9-A.2.1.1 Trunk Cables	142

SECTION			PAGE
		9-A.2.1.2 Drop Cables	143
	9-A.2.2	Splitters	143
	9-A.2.3	Trunk Connection (Transformer Tap)	143
9-A.3	Example Network Configuration		144
	9-A.3.1	Single Control Room Application	144
	9-A.3.2	Complex Plant Area Application	145
9-A.4	Installation Guidelines		146
	9-A.4.1	Grounding	146
	9-A.4.2	Surge Protection	146
	9-A.4.3	Termination	147
	9-A.4.4	Joining Cable Sections	147
	9-A.4.5	Pretested Cable	147
10.	PROWAY Management		147
10A.	Station Management Activities		148
10A.1	Scope and Field of Application		148
10A.2	Overview of Station Management Activities		149
	10A.2.1	User Requests	149
	10A.2.2	User Notification of Changes	149
10A.3	Compliance		149
10A.4	Detailed Description of Station Management Activities in Response to Local User Requests		149
	10A.4.1	Return Retry Counter	149
	10A.4.2	Enter Line_Connect State	150
	10A.4.3	Enter Line_Disconnect State	150
	10A.4.4	Enter Configure State	150
	10A.4.5	Activate L_SAP	150
	10A.4.6	Activate RSAP	151
	10A.4.7	Deactivate SAP	151
10A.5	Detailed Description of Station Management Activities in Notifying the User of Changes in Station Status		151
10B.	Station Management-PLC Sublayer Interface Service Specification		152
10B.1	Scope and Field of Application		152
10B.2	Overview of Station Management-PLC Services		152
	10B.2.1	General Description of Services Provided	152
	10B.2.2	Model Used for the Service Specification	152
	10B.2.3	Overview of Interactions	152
		10B.2.3.1 L_RESET	153
		10B.2.3.2 L_STATUS	153
		10B.2.3.3 L_SAP_ACTIVATE	153
		10B.2.3.4 L_RSAP_ACTIVATE	153
		10B.2.3.5 L_SAP_ACTIVATE	153
		10B.2.3.6 Services to Test Local and Remote PLC Entities	153
		10B.2.3.7 Compliance	153
10B.3	Detailed Specifications of Interactions with the Station Management Entity		153
	10B.3.1	L_RESET.request	153
		10B.3.1.1 Function	153
		10B.3.1.2 Semantics	153
		10B.3.1.3 When Generated	154
		10B.3.1.4 Effect of Receipt	154
	10B.3.2	L_STATUS.request	154
		10B.3.2.1 Function	154

SECTION			PAGE
	10B.3.2.2	Semantics of the Service Primitive	154
	10B.3.2.3	When Generated	154
	10B.3.2.4	Effect of Receipt	154
10B.3.3	L_STATUS.indication		154
	10B.3.3.1	Function	154
	10B.3.3.2	Semantics of the Service Primitive	154
	10B.3.3.3	When Generated	155
	10B.3.3.4	Effect of Receipt	155
10B.3.4	L_STATUS.confirm		155
	10B.3.4.1	Function	155
	10B.3.4.2	Semantics	155
	10B.3.4.3	When Generated	155
	10B.3.4.4	Effect of Receipt	155
10B.3.5	L_SAP_ACTIVATE.request		155
	10B.3.5.1	Function	155
	10B.3.5.2	Semantics	155
	10B.3.5.3	When Generated	156
	10B.3.5.4	Effect of Receipt	156
	10B.3.5.5	Additional Comments	156
10B.3.6	L_SAP_ACTIVATE.confirm		156
	10B.3.6.1	Function	156
	10B.3.6.2	Semantics	157
	10B.3.6.3	When Generated	157
	10B.3.6.4	Effect of Receipt	157
10B.3.7	L_SAP_DEACTIVATE.request		157
	10B.3.7.1	Function	157
	10B.3.7.2	Semantics of the Service Primitive	157
	10B.3.7.3	When Generated	157
	10B.3.7.4	Effect of Receipt	157
	10B.3.7.5	Additional Comments	157
10B.3.8	L_SAP_DEACTIVATE.confirm		157
	10B.3.8.1	Function	157
	10B.3.8.2	Semantics	157
	10B.3.8.3	When Generated	158
	10B.3.8.4	Effect of Receipt	158
10B.3.9	L_RSAP_ACTIVATE.request		158
	10B.3.9.1	Function	158
	10B.3.9.2	Semantics	158
	10B.3.9.3	When Generated	158
	10B.3.9.4	Effect of Receipt	158
	10B.3.9.5	Additional Comments	158
10B.3.10	L_RSAP_ACTIVATE.confirm		158
	10B.3.10.1	Function	158
	10B.3.10.2	Semantics	158
	10B.3.10.3	When Generated	159
	10B.3.10.4	Effect of Receipt	159
10C. Station Management-MAC Interface Service Specification			159
10C.1 Scope and Field of Application			159
10C.2 Overview of the Station Management-MAC Service			160
10C.2.1	General Description of Services Provided		160
10C.2.2	Model Used for the Service Specification		160

SECTION			PAGE
	10C.2.3	Overview of Interactions	160
		10C.2.3.1 MA_INITIALIZE_PROTOCOL	160
		10C.2.3.2 MA_SET_VALUE	160
		10C.2.3.3 MA_READ_VALUE	160
		10C.2.3.4 MA_EVENT	161
		10C.2.3.5 MA_FAULT_REPORT	161
		10C.2.3.6 MA_GROUP_ADDRESS	161
	10C.2.4	Compliance	161
10C.3	Detailed Interactions with the Station Management Entity		161
	10C.3.1	MA_INITIALIZE_PROTOCOL.request	161
		10C.3.1.1 Function	161
		10C.3.1.2 Semantics	161
		10C.3.1.3 When Generated	161
		10C.3.1.4 Effect of Receipt	161
		10C.3.1.5 Additional Comments	161
	10C.3.2	MA_INITIALIZE_PROTOCOL.confirm	162
		10C.3.2.1 Function	162
		10C.3.2.2 Semantics	162
		10C.3.2.3 When Generated	162
		10C.3.2.4 Effect of Receipt	162
	10C.3.3	MA_SET_VALUE.request	162
		10C.3.3.1 Function	162
		10C.3.3.2 Semantics	162
		10C.3.3.3 When Generated	162
		10C.3.3.4 Effect of Receipt	163
	10C.3.4	MA_SET_VALUE.confirm	163
		10C.3.4.1 Function	163
		10C.3.4.2 Semantics	163
		10C.3.4.3 When Generated	163
		10C.3.4.4 Effect of Receipt	163
	10C.3.5	MA_READ_VALUE.request	163
		10C.3.5.1 Function	163
		10C.3.5.2 Semantics	163
		10C.3.5.3 When Generated	163
		10C.3.5.4 Effect of Receipt	164
	10C.3.6	MA_READ_VALUE.confirm	164
		10C.3.6.1 Function	164
		10C.3.6.2 Semantics	164
		10C.3.6.3 When Generated	164
		10C.3.6.4 Effect of Receipt	164
	10C.3.7	MA_EVENT.indication	164
		10C.3.7.1 Function	164
		10C.3.7.2 Semantics	164
		10C.3.7.3 When Generated	164
		10C.3.7.4 Effect of Receipt	164
	10C.3.8	MA_FAULT_REPORT.indication	164
		10C.3.8.1 Function	164
		10C.3.8.2 Semantics	165
		10C.3.8.3 When Generated	165
		10C.3.8.4 Effect of Receipt	165
		10C.3.8.5 Additional Comments	165

SECTION			PAGE
	10C.3.9	MA_GROUP_ADDRESS.request	165
		10C.3.9.1 Function	165
		10C.3.9.2 Semantics	165
		10C.3.9.3 When Generated	165
		10C.3.9.4 Effect of Receipt	165
		10C.3.9.5 Additional Comments	166
	10C.3.10	MA_GROUP_ADDRESS.confirm	166
		10C.3.10.1 Function	166
		10C.3.10.2 Semantics	166
		10C.3.10.3 When Generated	166
		10C.3.10.4 Effect of Receipt	166
10D.	Station Management-Physical Layer Interface Service Specification		166
10D.1	Scope and Field of Application		166
10D.2	Overview of the Physical Layer-Management Service		166
	10D.2.1	General Description of Services Provided at Interface	166
	10D.2.2	Model Used for the Service Specification	167
	10D.2.3	Overview of Interactions	167
		10D.2.3.1 PHY_RESET Request and Confirm	167
		10D.2.3.2 PHY_MODE_SELECT Request and Confirm	168
		10D.2.3.3 PHY_MODE_CHANGE.indication	168
	10D.2.4	Compliance	168
10D.3	Detailed Specifications of Interactions with the Station Management Entity		168
	10D.3.1	PHY_RESET.request	168
		10D.3.1.1 Function	168
		10D.3.1.2 Semantics	168
		10D.3.1.3 When Generated	168
		10D.3.1.4 Effect of Receipt	168
	10D.3.2	PHY_RESET.confirm	168
		10D.3.2.1 Function	168
		10D.3.2.2 Semantics	168
		10D.3.2.3 When Generated	168
		10D.3.2.4 Effect of Receipt	169
	10D.3.3	PHY_MODE_SELECT.request	169
		10D.3.3.1 Function	169
		10D.3.3.2 Semantics	169
		10D.3.3.3 When Generated	169
		10D.3.3.4 Effect of Receipt	169
	10D.3.4	PHY_MODE_SELECT.confirm	169
		10D.3.4.1 Function	169
		10D.3.4.2 Semantics	170
		10D.3.4.3 When Generated	170
		10D.3.4.4 Effect of Receipt	170
	10D.3.5	PHY_MODE_CHANGE.indication	170
		10D.3.5.1 Function	170
		10D.3.5.2 Semantics	170
		10D.3.5.3 When Generated	170
		10D.3.5.4 Effect of Receipt	170
		10D.3.5.5 Additional Comments	170
10E.	Token Ring Management		170
10E.1	Scope and Field of Application		170
10E.2	Overview of Token Ring Management		171
	10E.2.1	Management of Ring Participation	171

SECTION	PAGE

	10E.2.2	Maintenance of the Active_Station_List	171
	10E.2.3	Report of Active_Station_List Membership	172
	10E.2.4	Report of Spontaneous Changes in Active_Station_List Membership	172
10E.3	Compliance		172
10E.4	Maintenance of the Active_Station_List		172
	10E.4.1	Informal Overview of Active_Station_List Maintenance Procedures	172
	10E.4.2	Detailed Active_Station_List Maintenance Procedures	172
	10E.4.3	Communications with the Management Entities of Other Stations	174
		10E.4.3.1 Transmission of Active_Station_List and Change_Notifications	174
		10E.4.3.2 Transmission of Active_Station_List Requests	174
10F.	Reset Management		174
10-A.	Appendix — Examples of Active Station List Maintenance		175
10-B.	Appendix — Recommendations for Management of Ring Participation and Redundant Media		177
	10-B.1	Recommendations for Ring Participation Management	177
	10-B.1.1	Management of Ring Participation	177
	10-B.1.2	Ring Participation Management Station Descriptions	177
	10-B.1.3	Line_Disconnect State	178
	10-B.1.4	Line_Connect State	178
	10-B.1.5	Active_Station_List_Receive State	179
	10-B.1.6	Line_Connect State	179
	10-B.1.7	Active_Station_List_Update Substate	180
	10-B.1.8	Configure State	181
	10-B.1.9	Line_Disconnect State	181
	10-B.2	Recommendation for Management of Redundant Media	181

FIGURES		PAGE
Fig I-1	LAN Model	1
Fig I-2	Logical Ring on Physical Bus	4
Fig I-3	MAC Layer Functional Partitioning	5
Fig 2A-1	Relationship to LAN Model	17
Fig 2B-1	Relationship to LAN Model	31
Fig 2-2	Local Station Interaction Diagram	36
Fig 2-3	Remote Station Interaction Diagram	37
Fig 2-4	Topological Behavior of the send_data_with_acknowledge Service	37
Fig 2-5	Sequential Relationship of the send_data_with_acknowledge Service	37
Fig 2-6	Local User State Diagram for the send_data_with_acknowledge Service	38
Fig 2-7	Remote User State Diagram for the send_data_with_acknowledge Service	38
Fig 2-8	Local Highway Entity State Diagram of the send_data_with_acknowledge Service	39
Fig 2-9	Remote Highway Entity State Diagram for the send_data_with_acknowledge Service	39
Fig 2-10	Topological Behavior of the send_data_with_no_acknowledge Service	40
Fig 2-11	Sequential Relationship of send_data_with_no_acknowledge Service	40
Fig 2-12	Local User State Diagram for the send_data_with_no_acknowledge Service	41
Fig 2-13	Remote User State Diagram for the send_data_with_no_acknowledge Service	41
Fig 2-14	Local Highway Entity State Diagram for the send_data_with_no_acknowledge Service	42

FIGURES		PAGE
Fig 2-15	Remote Highway Entity State Diagram for the send_data_with_no_acknowledge Service	42
Fig 2-16	Topological Behavior of the request_data_with_reply Service	43
Fig 2-17	Sequential Relationship of the request_data_with_reply Service	43
Fig 2-18	Local User State Diagram for the request_data_with_reply Service	44
Fig 2-19	Remote User State Diagram for the request_data_with_reply Service	44
Fig 2-20	Local Highway Unit State Diagram for the request_data_with_reply Service	45
Fig 2-21	Remote Highway Entity State Diagram for the request_data_with_reply Service — Data Transmission	45
Fig 3-1	Relationship to LAN Model	46
Fig 3-2	PLC Local Machine State Diagram	54
Fig 3-3	PLC Remote Machine State Diagram	54
Fig 4-1	Relationship to LAN Model	69
Fig 5-1	Relationship to LAN Model	73
Fig 5-2	Token Rotation Time Example	116
Fig 5-3	MAC Finite State Machine Diagram	81
Fig 5-7	PROWAY PLC and MAC_protocol_data and Transmission Order Unit Composition Order	89
Fig 5-8	Logical Token Passing Ring	90
Fig 6-1	Relationship to LAN Model	118
Fig 6-2	MAC-PHY Entity Grounding	125
Fig 6-3	MAC-Physical Interface Connector	126
Fig 8-1	Relationship to LAN Model and Medium	129
Fig 8-2	Manchester Data Encoded Physical Signal Encodings	132
Fig 9-1	Relationship to LAN Model	136
Fig 9-2	Nondirectional Transformer Tap	144
Fig 9-3	Example Cable Configuration — Simple Control Room Application	145
Fig 9-4	Example Cable Configuration — Complex Plant Area Application	146
Fig 10A-1	Relationship to LAN Model	148
Fig 10B-1	Relationship to LAN Model	152
Fig 10C-1	Relationship to LAN Model	159
Fig 10D-1	Relationship to LAN Model	167
Fig 10E-1	Relationship to LAN Model	171
Fig 10-2	Ring Participation Management	178
Fig 10-3	Line_Connect State Diagram	179
Fig 10-4	Line_Disconnect State Diagram	180

TABLES		PAGE
Table 2-1	Management Request Codes and Arguments	33
Table 2-2	User-Management Confirmation Codes and Results	34
Table 2-3	User-Management Indication Codes and Reason	35
Table 3-1	State Transition Table for Local PLC State Machine	55
Table 3-2	State Transition Table for the Remote PLC State Machine	58
Table 3-3	PLC Protocol Data Unit Structure	62
Table 6-1	Physical_Send Encoding	122
Table 6-2	Physical_Receive Encoding	123
Table 6-3	Physical_Interface Pin Assignments	127
Table 9-1	Example Coaxial Cable Types	143
Table 9-2	Example Specifications of Single Drop Nondirectional Transformer Taps	144

Standard

PROWAY-LAN
Industrial Data Highway

I. Introduction and Overview

I.1 Scope. This standard specifies those elements which are required for compatible interconnection of stations by way of a Local Area Network (LAN) using the Token-Bus access method in an industrial environment. These elements include:

(1) The electrical and physical characteristics of the transmission medium.
(2) The electrical transmission signaling method used by the Physical protocol.
(3) The services and signals provided at interface between the Physical and Media Access Control entities.
(4) The frame formats transmitted.
(5) The Token Bus Media Access protocol.
(6) The services provided at the conceptual interface between the Medium Access Control (MAC) sublayer and the PROWAY Link Control (PLC) sublayer above it.
(7) The PROWAY Link Control (PLC) protocol.
(8) The services provided at the conceptual interface between the User of PROWAY and the PROWAY Link Control (PLC) sublayer.
(9) Management of PROWAY stations.

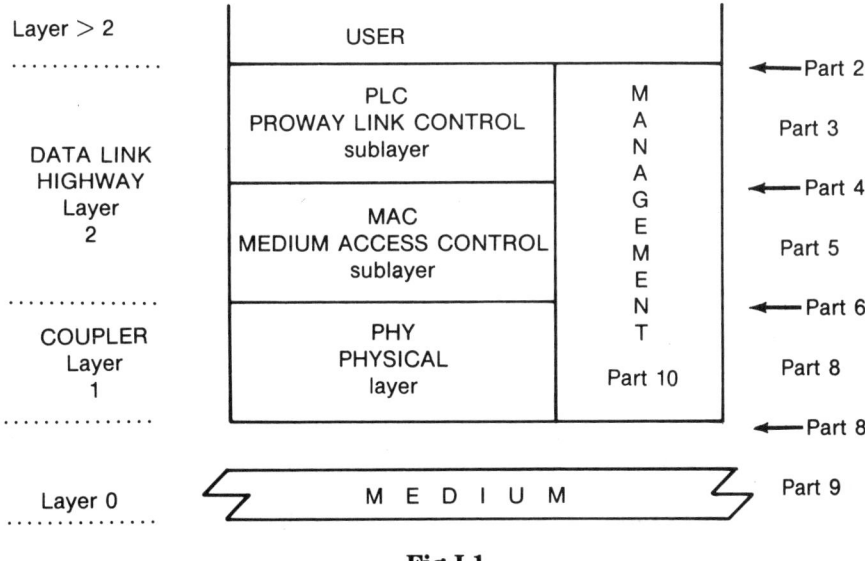

Fig I-1
LAN Model

I.2 Introduction

I.2.1 Layering. Within this standard the operation of a station is specified in terms of the layered model shown in Figure I-1. The figure also shows which parts of the standard specify interfaces between layers and which parts specify the operation of the layers themselves. This figure is repeated in each part of the standard to indicate the relationship of that part to the remaining parts of the standard and to the LAN model.

I.2.2 Precedence. Parts I and 1 are introductory and explanatory in nature. Parts I and 1 are not intended to convey precise specifications. In case of a conflict between Parts I and 1 and any other part of the standard, the specifications of the other part shall take precedence.

I.2.3 Definitions. Terms that are used in a specific or specialized manner in any part of this standard are defined at the beginning of that part. General definitions are given in References 1 and 2, Appendix I-A.

I.2.4 References. Documents referenced by this standard are listed in Appendix I-A.

I-3 Compliance
I.2.3 Classes of Stations. This standard defines three classes of stations:

INITíATOR stations perform all functions of the Local state machine and the reception functions of the Remote state machine.

RESPONDER stations perform all the functions of the Remote state machine and none of the functions of the Local state machine. A simple device (such as an instrument) can be connected to the Local Area Network as a **RESPONDER**.

INITIATOR/RESPONDER stations perform all functions of both the Local and Remote state machines.

I.3.2 Compliance Requirements. Implementations may claim compliance with this standard as one of the classes of stations given in I.3.1. Implementations that claim compliance with this station shall

(1) Offer the mandatory PLC-user interface and services specified in this standard for that class of station.
(2) Support the PROWAY Link Control protocol specified in this standard for that class of station.
(3) Support the mandatory PLC-MAC interface and services specified in this standard for that class of station.
(4) Offer the mandatory station management services specified in this standard for that class of station.
(5) Support the mandatory features of the Token Bus Medium Access protocol specified in this standard for that class of station.
(6) Support the MAC-Physical interface services specified in this standard.

It is strongly recommended that **INITIATOR** stations support reporting of the list of stations participating in the logical token ring.

A number of options are specified in this standard. An implementation must indicate which, if any, of these options are supported and the supported values of each station attribute.

INDUSTRIAL DATA HIGHWAY

I.4 PROWAY Organization. The PROWAY Data Highway has three primary functional layers or entities:

- **PROWAY Link Control — PLC**
- **Media Access Control — MAC**
- **Physical Signaling — PHY**

Each of these layers is briefly introduced in the following sections.

The PLC and MAC sublayers together comprise the Data Link (Highway) level of the ISO Model. The PHY level comprises the Physical level of the ISO Model.

I.5 Overview of the PROWAY Link Control (PLC) Layer

I.5.1 PROWAY Link Control Services. PROWAY provides four basic services to its users:

- **Sending data using a confirmed** (immediate response) **protocol** from one local (originating) station to one remote (responding) station. This service is known as **Send Data with Acknowledge or SDA**.
- **Sending data without acknowledge or retry** from one local station to one, some, or all remote (receiving) stations. This less secure service is known as **Send Data with No Acknowledge or SDN**.
- One local station **requesting previously submitted information** from one remote station using the confirmed (immediate response) **protocol**. This service is known as **Request Data with Reply or RDR**.

I.5.2 PROWAY Link Control Sublayer Organization. The PLC sublayer functions are logically divided into two independent state machines:

(1) Local State Machine
(2) Remote State Machine

I.5.2.1 Local State Machine Functions. The Local state machine handles all requests from and confirmations to the local PLC user. These local requests result in transmission of request frames. Functions of the local PLC state machine include:

- Accepting local user requests (excluding update_requests)
- Generating request frames
- Receiving response frames
- Passing confirmations to the local PLC user (excluding update_confirmations)

I.5.2.2 Remote State Machine Functions. The remote state machine passes indications to the remote PLC user, manages the shared data areas, and returns requested data to the local machine. Functions of the remote PLC state machine include:

- Receiving request frames
- Passing indications to the remote PLC user
- Generating response frames
- Accepting remote user requests to update a shared_data_area
- Passing update confirmations to the remote PLC user

I.6 Overview of the Token Bus Access Method
I.6.1 The Essence of the Token Bus Access Method

(1) A token (or baton) controls the right of access to the physical medium; the station which holds (possesses) the token has control over the medium.
(2) The token is passed by stations residing on the medium. As the token is passed from station to station a logical ring is formed.
(3) Steady state operation consists of a data transfer phase and a token transfer phase.
(4) Ring maintenance functions within the stations provide for ring initialization, lost token recovery, new station addition to logical ring, and general housekeeping of the logical ring. The ring maintenance functions are replicated among all the token-using stations on the network.

Shared media generally can be categorized into two major types. These types are broadcast and sequential. This standard deals exclusively with the broadcast type. On a broadcast medium, every station may receive all signals transmitted. Media of the broadcast type are usually configured as a physical bus.

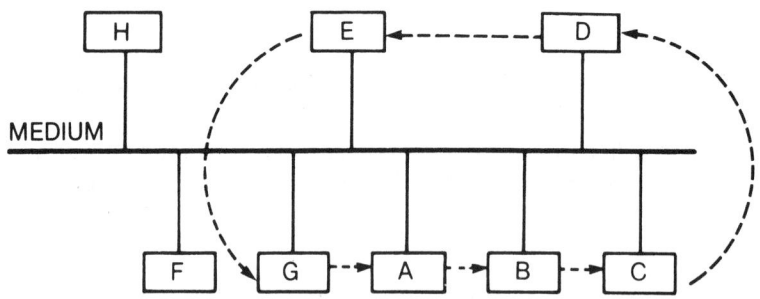

Fig I-2
Logical Ring on Physical Bus

In Figure I-2, note that the token bus access method is always sequential in a logical sense. That is, during normal steady state operation, the right to access the medium passes from station to station. Further, note that the physical connectivity has little impact on the order of the logical ring and that stations can respond to a query from the token holder even without being part of the logical ring. (For example, stations H and F can receive frames but cannot initiate a transmission since they will never be sent the token.)

The Medium Access Control (MAC) sublayer provides sequential access to the shared bus medium by passing control of the medium from station to station in a logically circular fashion. The MAC sublayer determines when the station has the right to access the shared medium by recognizing and accepting the token from the predecessor station, and it determines when the token must be passed to the successor station.

I.6.2 General MAC Sublayer Functions

(1) Lost token timer
(2) Distributed initialization
(3) Token holding timer (for multiple classes of service and ring maintenance)
(4) Token rotation timer (for multiple classes of service and ring maintenance)

INDUSTRIAL DATA HIGHWAY

ISA
S72.01-1985

(5) Limited data buffering
(6) Node address recognition
(7) Frame encapsulation (including token preparation)
(8) Frame Check Sequence (FCS) generation and checking
(9) Valid token recognition
(10) New ring member addition
(11) Node failure error recovery
(12) Allowing an immediate PLC response to or acknowledgement of a frame sent by the token holder
(13) Retry transmission for certain classes of service

I.7 MAC Layer Internal Structure. The MAC layer performs several functions which are loosely coupled. The descriptions and specifications of the MAC sublayer in this standard are organized in terms of one of several possible partitionings of these functions. The partitioning used here is illustrated in Figure I-3, which shows four asynchronous logical "machines," each of which handles some of the MAC functions, as discussed below.

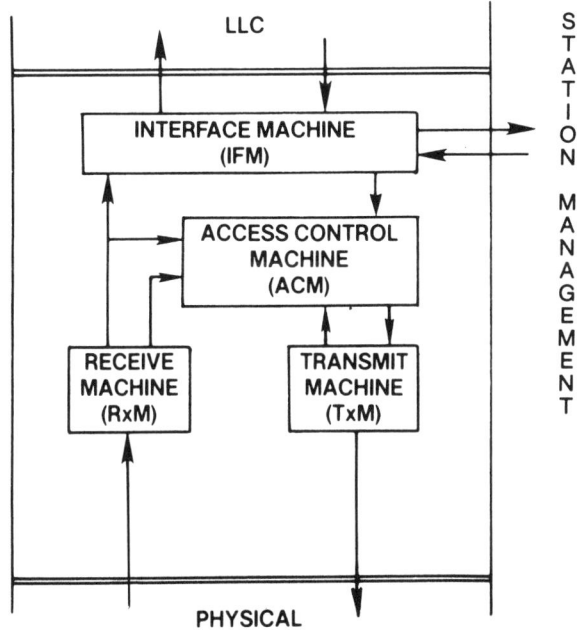

Fig I-3
MAC Layer Functional Partitioning

I.7.1 Interface Machine (IFM). This machine acts as an interface and buffer between the PLC and MAC sublayers and between Station Management and the MAC sublayer. It interprets all incoming service primitives and generates appropriate outgoing service primitives. This machine handles the mapping of "quality of service" parameters from the PLC view to the MAC view, and generation of a response frame when required. The IFM handles queuing of service requests sent to a PLC protocol data unit. It performs the "address recognition" function on received PLC frames, accepting only those addressed to this station. Finally, it generates an immediate response when requested by the initiating station.

5

I.7.2 Access Control Machine (ACM). This machine cooperates with the ACMs of all other stations to control transmission on the shared bus medium. The ACM manages multiple MAC access classes in order to provide different levels of "quality of service" to the PLC layer. When required, it will wait for a response or acknowledgment of a transmitted frame from the remote station. If the required response is not received, the local ACM will retry the transmission. The ACM is also responsible for initialization and maintenance of the logical ring, including the admission of new stations. Finally, it has responsibility for the detection of and, where possible, recovery from faults and failures in the token bus network and for informing station management of changes in the membership of the token ring. The ACM is not required in a **RESPONDER** station.

I.7.3 Receive Machine (RxM). This machine accepts atomic symbol inputs from the physical layer, assembles them into frames, which it validates and passes the frames to the IFM and ACM. The RxM accomplishes this by recognizing the frame Start Delimiter (SD) and the frame End Delimiter (ED), checking the Frame Check Sequence (FCS) and validating the frame's structure. The RxM also identifies and indicates the reception of **noise_bursts** and the **bus_quiet** condition.

I.7.4 Transmit Machine (TxM). This machine generally accepts a data frame from the ACM and transmits it, as a sequence of atomic symbols in the proper format, to the physical layer. The TxM builds a MAC protocol data unit by prefacing each frame with the required preamble and SD, and appending the FCS and ED.

I.8 Token Bus Access Method Characteristics. An understanding of the basic characteristics of the token bus access method assists in understanding where and when a token-passing bus is an appropriate LAN technology.

Some of the important features of the token bus access method are as follows:

(1) The method is efficient in the sense that under high offered load the coordination of the stations requires only a small percentage of the media's capacity.
(2) The method is fair in the sense that it offers each station an equal share of the media's capacity. It does not, however, require any station to use its full share.
(3) The method permits multiple priorities.
(4) The method supports open loop (no response) and closed loop (with response or acknowledgment) data transfers.
(5) The method coordinates station transmissions to minimize and control interference with other stations.
(6) The method imposes no additional requirements on the media and the modem capabilities over those necessary for transmission and reception of multi-bit, multi-frame sequences at the specified bit error rate.
(7) In the absence of system noise, the method provides computable, deterministic, worst case bounds on access delay, for the highest priority class of service, for any given network and loading configuration.
(8) Periods of controlled interference are distinguishable; system noise measurements are possible during the remaining periods.
(9) The method places minimal constraints on how a station which momentarily controls the media may use its share of the media's capacity. In particular, the access method allows a station to use request/reply access methods during that station's access periods.
(10) The method permits the presence of a large number of low-cost reduced-function **(RESPONDER)** stations in the network with one or more full-function **(INITIATOR)** stations. (It is assumed that at least one **INITIATOR** station is needed to make the system operational, for example, to initialize.) A **RESPONDER** station does not require Media Access Control logic.

(11) Use of the immediate response feature in conjunction with the confirmed (SDA or RDR) data transfer services allows protection against loss or duplication of frames at any desired level of assurance.

I.9 Overview of the Physical Layer and Media
I.9.1 Summary of Phase Continuous FSK (Frequency Shift Keying) Physical Layer

Transmit Level: +64 to +66 dB (1 mV, 75 ohm) (dBmV); i.e., approximately 2 V rms.

Receiver Sensitivity: +4 to +50 dB (1 mV, 75 ohm) (dBmV), and the ability to receive one's own transmissions.

Noise Floor: ≤ -15 dB (1 mV, 75 ohm) over a 3-7 MHz band.

Data Rate: 1 Mbits/sec.

Signaling: Manchester encoding of **data**, **non_data**, and **pad_idle** symbols. The transmission symbol representations are

{HL}   = **zero** }
{LH}   = **one** } **data**
{LL}{HH} = pairs of **non_data** symbols containing no frequency transitions for one full MAC symbol period
Octets of consecutive **ones** and **zeros** = **pad_idle** (preamble)

Modulation: Phase continuous FSK (Frequency Shift Keying) (a form of frequency modulation).

(1) Higher frequency = 6.25 ± 0.08 MHz = [H]
(2) Lower frequency = 3.75 ± 0.08 MHz = [L]

Clock Recovery: From transitions generated by the Manchester encoding.

I.9.2 Summary of Single-Channel Coaxial Cable Bus Medium

Topology: Omnidirectional bus

Cable: 75 ohm coaxial trunk cable, such as types RG-6 and semi-rigid, with flexible drop cables such as RG-59.

Recommended Cable Configuration: Semi-rigid trunk and flexible drop cable.

Trunk Connection (TAP): 75 ohm nondirectional passive impedance-matching tap.

Repeaters: Active repeaters and amplifiers are not used in a PROWAY System.

I.9.3 Alternate Physical and Media Implementation.
Additional physical layers, data rates, and compatible media (e.g., 5 Mbits/sec, phase coherent or fiber optics) as defined in Chapters 12/13 and 16/17 of Reference 4 are compatible implementations of this standard.

I.10 Standard Organization. This standard is organized in 10 parts, which are summarized below.

Part I serves as an introduction. It begins with a general discussion of the PROWAY link control services and token-passing bus access methods. The PLC and MAC sublayer functional partitioning used in subsequent parts is introduced here. Features of the token-passing bus access method are next reviewed. The Physical layer and Media are then surveyed.

Appendix I-A lists standards referenced by this standard.

Part 1 gives the functional requirements for industrial local area networks.

Part 2 outlines the user interface to the Process Data Highway (PROWAY).

Part 2A details the user interface for transmission of data through PROWAY; i.e., it details the user interface to the PROWAY Link Control (PLC) sublayer.

Part 2B defines the user interface to station management functions.

Appendix 2-A provides an overview of the operation of the Data Link layer (PLC and MAC sublayers combined) and the user interactions with PROWAY.

Part 3 specifies the PROWAY Link Control (PLC) sublayer.

Part 3A gives definitions that are applicable to the PLC state machines and specifies mandatory requirements of the PLC sublayer.

Part 3B defines the functionality of the PROWAY Link Control (PLC) sublayer by means of state tables.

Part 3C details the format of Protocol Data Units (L_pdu) exchanged between cooperating PLC entities.

Part 4 details the interfaces between PLC and MAC sublayers. It defines the services and interfaces provided to the PLC sublayer.

Part 5 specifies the Medium Access Control (MAC) sublayer and details the format of frames exchanged between cooperating MAC entities.

Part 5A discusses the basic concepts of the medium access protocol and provides an informal description of the actions in each state of the Access Control Machine.

Part 5B contains definitions of essential MAC terms and components and specifies those requirements for the MAC sublayer protocols which are not covered elsewhere. It specifies the Interface machine and references the Receive and Transmit machines in Reference 4, Appendix I-A.

Part 5C specifies the MAC Access Control Machine in Reference 4. This is the definitive specification of the token-passing bus-MAC cooperation.

It also describes the MAC layer variables, functions and procedures used in the state machine.

Part 5D details the MAC frame structure including delimiters, addressing, and the FCS. All of the frame formats which MAC handles, including MAC control frames, are enumerated.

Appendix 5-A gives an example configuration for a PROWAY network that meets the requirements of Part 1.

Appendix 5-B gives examples of 16 bit station addresses.

Part 6 details the interface between the MAC sublayer and the Physical layer.

Part 6A specifies the MAC-Physical interface services in an abstract manner and provides descriptions of the interface symbols, requests, and indications.

Part 6B defines the MAC-Physical interface implementation required when the MAC and Physical entities are separate pieces of equipment.

Part 7 is reserved for future use.

Part 8 specifies the Phase-Continuous-FSK physical layer entity and its interface to the medium.

Part 9 gives requirements for a single-channel coaxial cable bus medium.

Appendix 9-A gives guidance for configuring and installing single-channel coaxial cable bus networks.

Part 10 defines PROWAY station management.

Part 10A informally defines the activities of the station management entity with respect to local requests by means of tables of required actions.

Part 10B defines the interface between the station management entity and the PLC sublayer.

Part 10C details the interface between the station management entity and the MAC sublayer.

Part 10D details the interface between the station management entity and the Physical layer.

Part 10E specifies maintenance of the list of stations participating in the token ring (the active station list) and the format of Protocol Data Units (Mgt_pdu's) exchanged between cooperating management entities in maintaining this list.

Part 10F specifies a communications reset capability.

Appendix 10-A gives examples of active station list activity.

Appendix 10-B provides recommendations for preconditions to entering and leaving the logical token ring.

I.11 Correspondence between the PROWAY and IEEE 802 Standards. The parts of the PROWAY-LAN standard are related to the following sections of IEEE 802 standard.

PROWAY-LAN PART	IEEE STANDARD AND SECTION
PROWAY-LAN Part 1	802.4 — Section 1
PROWAY-LAN Part 2A	802.2 — Section 2
PROWAY-LAN Part 2B	—
PROWAY-LAN Part 3A	802.2 — Definitions
PROWAY-LAN Part 3B	802.2 — Section 6 and Single-Frame
PROWAY-LAN Part 3C	802.2 — Sections 3, 5
PROWAY-LAN Part 4	802.4 — Section 2
PROWAY-LAN Part 5A	802.4 — Section 5
PROWAY-LAN Part 5B	802.4 — Section 6
PROWAY-LAN Part 5C	802.4 — Section 7
PROWAY-LAN Part 5D	802.4 — Section 4
PROWAY-LAN Part 6A	802.4 — Section 8
PROWAY-LAN Part 6B	—
PROWAY-LAN Part 8	802.4 — Section 10
PROWAY-LAN Part 9	802.4 — Section 13
PROWAY-LAN Appendix 9-A	802.4 — Appendix 13A
PROWAY-LAN Part 10A	—
PROWAY-LAN Part 10B	802.2 — Management Proposal
PROWAY-LAN Part 10C	802.4 — Section 3
PROWAY-LAN Part 10D	802.4 — Section 9
PROWAY-LAN Part 10E	—

I.12 Compatibility with IEEE 802. This standard defines the features which are mandatory for conformance to PROWAY specifications. Conformance to IEEE 802 specifications requires additional features as defined in the IEEE 802.1, 802.2, and 802.4 standards (see References 2-4 of Appendix I-A).

Implementation specifics for Chapters 2, 3, 6, and 10 are subject to harmonization with the future work of the IEEE 802 committee.

In any case, where there is a direct conflict between this standard and a requirement of an equivalent part of one of the above IEEE 802 standards, the requirement of the IEEE 802 standard shall take precedence. Where this standard imposes additional requirements beyond those of the above IEEE 802 standards, the additional PROWAY requirements shall take precedence.

Appendix I-A
References

Ref. No.		Part No.
1	ISO 7498-1984. Information Processing Systems Interconnection — Basic Reference Model.	All
2	IEEE Std 802.1, Local Area Network Standard — Overview, Interworking, and Management. (Draft standard, publication pending)	I, 10
3	IEEE Std 802.2 (1985), Logical Link Control. IEEE 802.2 Proposed Draft Addendum — Acknowledged Connectionless Service, Draft 13, Oct. 1986.	I, 2, 3

4	IEEE Std 802.4 (1987), Token-Passing Bus Access Method and Physical Layer Specifications	I, 4, 5
5	Department of Defense, MIL Standard 1851A (1983), ADA Programming Language.	I
6	ADA™ Programming Language	I, 5
7	IEC Publication 79-10 (1972) Part 10: Classification of Hazardous Areas	1
8	IEC Publication 79-1 (1971) Part 1: Construction and Test of Flameproof Enclosures of Electrical Apparatus	1
9	IEC Publication 79-3 (1972) Part 3: Spark Test Apparatus for Intrinsically Safe Circuits	1
10	IEC Publication 79-11 (1984) Part 11: Construction and Test of Intrinsically Safe and Associated Apparatus	1
11	ISO 4902:1980, Data Communication — 37-Pin and 9-Pin DTE/DCE Interface Connectors and Pin Assignments	6
12	CCITT Yellow Book, Vol. II (1980), Characteristics for Balanced Double Current for General Use with Integrated Circuit Equipment in the Field of Data Communications	6
13	IEC Publication 348 (1978), Safety Requirements for Electronic Measuring Apparatus	6
14	FCC Docket 20780-1980 (Part 15) Technical Standards for Computing Equipment. Reconsidered First Report and Order, April 1980.	8
15	NFPA National Electric Code, Article 250: Grounding; Article 800: Communication Circuits; Article 820: Community Antenna Television and Radio Distribution Systems	8, 9
16	IEC 716-1983, Expression of the Properties of Signal Generators	
17	UL 94, Test for the Flammability of Plastic Materials for Parts in Devices and Appliances (Rated under V-0)	8, 9
18	UL 114, Office Appliances and Business Equipment	8, 9
19	UL 478, Electronic Data Processing Units and Systems	8, 9
20	CSA Standard C22.2 No. 154-M 1983, Data Processing Equipment	8, 9
21	IEC 435 (1983), Safety of Data Processing Equipment	8, 9
22	CCITT Yellow Book, Vol. 8, Fascicle VIII.1, Data Communications over the Telephone Network	8

23	CCITT Yellow Book, Vol. 8, Fascicle VIII.3, Data Communication Networks	8
24	CCITT Yellow Book, Vol. 8, Fascicle VIII.2, Data Communication Networks	8
25	CCITT Yellow Book, Vol. 3, Fascicle III.4, Line Transmission of Non-Telephone Signals	9
26	CCITT Yellow Book, Vol. 3, Fascicle III.4, Impedance Matching between Repeaters and Coaxial Pairs in Television Transmission	9
27	CCITT Yellow Book, Vol. 3, Fascicle III.4, Annex A to Recommendation J.73	9
28	General Motors Unified Communications Systems Task Force of MCC/CMC Computers in Manufacturing Subcommittee, 1978, General Broadband Coaxial Cable Networks for Digital, Video, and Audio Transmission	9
29	Rheinfelder, W. *CATV Circuit Engineering*, TAB Books, Blue Ridge Summit, PA	9
30	Rheinfelder, W. *CATV System Engineering*, TAB Books, Blue Ridge Summit, PA	9
31	RCA/Cablevision Systems: *Design and Construction of CATV Systems*, Van Nuys, CA	9
32	*Basic CATV Concepts*, Theta Com CATV/TEXCAN, Phoenix, AZ	9
33	EIA CB8-1981, Components Bulletin, List of Approved Agencies, US and Other Countries, Impacting Electronic Components and Equipment	8
34	IEC Publication 255-4 (1976), Single Input Energizing Quantity Measuring Relays with Dependent Specified Time	I
35	US Military Standard — MIL C 39012	8, 9

1. Functional Requirements

1.1 Overview. The PROWAY standard defines the protocols, interfaces, and media for layers 1 and 2 of the ISO reference model. Compliance with this standard, and with complementary standards at the higher layers will allow unambiguous communication between the devices that comprise a Distributed Industrial or Process Control System over a shared Process Data Highway (PROWAY). Compliance with these standards will enable devices from different manufacturers to cooperate in the same control system. The PROWAY-LAN standard is applicable to control systems for both continuous and discrete processes and to a wide range of factory automation systems. Industrial control systems are distinguished from other on-line, real-time computer networks in that control systems' outputs cause material or energy to move.

This standard applies to serial transmission over a single shared electrical transmission line; i.e., a coaxial cable. However, future revisions may define alternative transmission media, such as fiber optics.

INDUSTRIAL DATA HIGHWAY

ISA
S72.01-1985

The Functional Requirements given in Part 1 were developed prior to the other parts of the standard and were the primary basis for evaluating the technical merits of proposals for use in Parts 2 through 10 of this standard. If there are conflicts between Part 1 and other Parts of this standard, the requirements of the other Part take precedence.

1.2 Application Environment and Main Features

1.2.1 Application Characteristics. The characteristics of the data highway should be such that they provide optimum features for use in industrial control systems and shall be applicable to both continuous and discrete processes. An industrial data highway is characterized by the following:

(1) Event-driven communication which allows real-time response to events.
(2) Very high availability.
(3) Very high data integrity.
(4) Proper operation in the presence of electromagnetic interference and differences in earth potentials.
(5) Dedicated intra-plant transmission lines.

1.2.2 Economic versus Technical Factors. To achieve broad applicability it is essential that industrial data highways should be economically viable in control systems under the following conditions:

(1) With low or high information transfer requirements.
(2) Within a control room and/or while exposed to the plant environment.
(3) In geographically small or large plants.

The economic and technical factors may need to be reconciled to achieve a balance of transmission line length versus data signaling rate.

1.2.3 Main Features of PROWAY

Number of stations	≤ 100
Length of highway	≤ 2000 m
Data signaling rate	$\geq 1 \times 10^6$ bits/sec
Data circuit bit error rate	$< 1 \times 10^{-8}$
Residual error rate	$< 3 \times 10^{-15}$ at a bit error rate of $1*10^{-6}$
Maximum user data in frame	≤ 1000 octets
Information transfer rate	$\geq 3 \times 10^5$ bits/sec
High priority media access time	≤ 10 millisec

1.3 Device Types

1.3.1 Communications between Control Devices. Communications shall be provided among commonly used devices in process or industrial control applications.

It is intended that this standard will provide optimum performance when intelligent control devices are communicating.

1.3.2 Communications with Other Devices. An industrial data highway is not intended to provide an optimized interface for high-speed computer memories or peripherals. However, no devices or types of devices are excluded from exchanging data over an industrial data highway, provided they conform to the requirements of this standard.

1.3.3 Classes of Stations. The data highway shall support full function (**INITIATOR**) stations and reduced function (**RESPONDER**) stations. Simple devices may be directly connected to the data highway as **RESPONDERS**.

1.4 System Structure

1.4.1 Control System Structure. An industrial data highway shall be capable of supporting control systems with centralized intelligence, distributed intelligence, hierarchical intelligence and combinations thereof.

1.4.2 Data Highway Mode of Operation. The data highway shall be capable of supporting transmission of event oriented data in real time. The normal mode of operation uses transaction message pairs, such that each request message is followed by its related response or acknowledgment message; i.e., by an immediate response.

Any two stations on a single data highway shall perform direct data interchange without involving store and forward at a third station.

1.4.3 Configuration Changes. The data highway shall support reconfiguration of the control system while the process is operating. During reconfiguration, a transient disturbance to the exchange of frames is permitted, provided that the data highway is able to detect such disturbance and that it can recover full operation within a time appropriate to the application.

Examples of these configuration changes are as follows:

(1) Extending, shortening or rerouting transmission lines
(2) Connecting or disconnecting stations from the transmission line

1.5 Maintenance and Service Features

1.5.1 Testing and Fault Diagnosis. An industrial data highway must include means for carrying out test and fault diagnoses on line.

1.5.2 Effect of State Transitions. Any station shall be able to perform transitions from one state to another without generating bit errors between other stations. Examples of such transitions are as follows:

(1) On-line/off-line
(2) Power-on/power-off
(3) Ready/not ready
(4) Busy/not busy
(5) Local/remote

1.6 Safety

1.6.1 Electrical Faults. All devices used in the data highway shall be capable of withstanding the application of an allowable fault potential appropriate to the application. Application of this potential to the device's connection to the transmission line shall not damage the device nor cause it either to damage other devices, or become hazardous to personnel. Three classes of installation can be identified. They are as follows:

(1) The fault potential is the power mains voltage in the area traversed by the transmission line(s).
(2) The fault potential is represented by a pulse typically 2.5 kV peak with a 1 microsec rise time and 50 microsec decay time, the test procedure is given in Reference 34, Appendix I-A.
(3) The fault potential is that which is generated by a lightning strike to an arbitrary point near the transmission line(s). Such a fault is typically characterized by a 10 microsec rise time to 5,000 amps and a 20 microsec fall time to half that value. This is referred to as a 10/20 microsec pulse.

INDUSTRIAL DATA HIGHWAY

1.6.2 Intrinsic Safety. The design of the data highway shall include consideration of possible extensions to use the equipment, or sections of it, in hazardous atmospheres.

The supplier shall state which of the four categories below his equipment will meet:

(1) Not suitable for hazardous atmospheres (See Reference 7, Appendix I-A)
(2) Flameproof construction (See Reference 8, Appendix I-A)
(3) Intrinsically safe (See References 9 and 10, Appendix I-A)
(4) Claimed to meet the requirements for intrinsic safety certification but not actually certified.

For categories (2) and (3), the supplier shall give the name of the approving authority, the class of certification and the approval certificate number.

1.7 Performance in Industrial Environment
1.7.1 Industrial Environment
1.7.1.1 Induced Noise. The noise floor on a main trunk cable installed in a typical industrial environment in accordance with Part 9 may be as high as 0 dBmV measured over a 3 to 7 MHz bandwidth.

1.7.1.2 Electromagnetic Environment. The industrial environment may include an ambient plane wave field of

(1) 2 volts/meter from 10 kHz through 30 MHz
(2) 5 volts/meter from 30 MHz through 1 GHz.

1.7.1.3 Differences in Earth Potential. Typical differences in earth potential for an industrial environment are as follows:

(1) When the transmission medium is entirely contained in a protected area, this difference in earth potential is typically less than 10 V peak-to-peak at frequencies less than 400 Hz.
(2) When the transmission medium is exposed to the plant environment, this difference in earth potential is typically less than 50 V peak-to-peak at frequencies less than 400 Hz.
(3) When the transmission medium is exposed to a severe plant environment (for example a power station), this difference in earth potential may typically rise to 1000 V peak-to-peak at frequencies less than 10 MHz.

1.7.2 Data Circuit Bit Error Rate. The data highway, when installed in a typical industrial environment and in accordance with Part 9, shall exhibit a data circuit bit error rate of no more than 1×10^{-8}. The manufacturer shall provide a graph relating data circuit bit error rate to the noise floor, differences in earth potential, data rate, and other relevant factors.

1.7.3 Residual Error Rate. The data highway shall achieve a residual error rate of no more than 3×10^{-15} when the data circuit bit error rate is 1×10^{-6}. The manufacturer shall provide a graph which relates the residual error rate to the data circuit bit error rate.

NOTE: The corresponding rate of undetected frame errors is one error per 1000 years of data highway operation, assuming 100% utilization of a data signaling rate of 1×10^6 bits/sec.

1.7.4 Information Transfer Rate. The data highway shall achieve an information transfer rate of at least 3×10^5 bit/sec, when data circuit bit error rate is 1×10^{-6} and the data rate is 1×10^6 bits/sec. The manufacturer shall provide a graph which relates information transfer rate to the data circuit bit error rate and data signaling rate.

1.7.5 Media Access and Highway Transaction Times. The manufacturer shall provide information relating average and maximum media access time and Highway transaction time to highway configuration, loading, priority, and other relevant factors.

1.7.5.1 Media Access Time. An industrial control data highway shall have a maximum media access time of no more than **10 millisec** for a frame submitted at the highest priority by an arbitrary station under any set of conditions equivalent to those listed below.

- 2000 m highway length
- 1×10^6 bits/sec data signaling rate
- 20 stations participate in the token ring
- 10 stations initiate SDA messages with an average user data length of 16 octets simultaneously
- address length = 16 bits
- no more than one error occurs during each token rotation

One set of conditions under which this media access time may be achieved is given in Appendix 5-A.

At a given data rate, error rate, address length and propagation delay, the maximum media access time is determined by the number of token holding stations, the offered load at the highest priority, and the maximum frame length allowed.

1.7.5.2 Highway Transaction Time. The highway transaction time, when using the SDA or RDR services, is determined by the queuing delay within the initiating station, the media access time, the user data length and the number of retries required.

1.7.5.3 Definitions

- **Media access time** is defined as the period of time between a request becoming next to be transmitted and the time when the first bit of the SD for the corresponding request frame appears on the common bus medium.
- **Highway transmission time** is the period of time between submission of a request at the PLC-user interface and the appearance of the corresponding confirmation at that interface.

1.8 System Availability

1.8.1 Effect of Failures. No single failure of any part of any device used within or connected to the data highway shall cause failure of the entire control system, or of any function except those in which the failed device is directly involved.

It shall be possible to configure an industrial control system which can tolerate without the loss of communication function, changes of configuration, failure of any one transmission line, or failure of any one station.

1.8.2 Internal Status and Error Reporting. The data highway shall have an internal status and error reporting capability.

1.8.3 Automatic Recovery. The data highway shall be capable of automatic recovery after commonly occurring failures are corrected.

1.8.4 Control of Stations. The data highway shall support loading, starting, stopping, reloading, and resetting of any station.

2. The Interface to PROWAY

2.1 Organization. Part 2A specifies the data transfer services provided to the PROWAY user by the PROWAY Link Control sublayer of the Data Link (Highway) layer of the ISO reference model.

Part 2B specifies the administrative services provided to the PROWAY user by the station management entity.

2A. User-PLC Interface and Service Specification

2A.1 Scope and Field of Application. This part specifies the data transfer services provided to the user of PROWAY (PLC user) by the PROWAY Link Control (PLC) sublayer at the boundary between the user of PROWAY and the PROWAY Link Control sublayer of the Data Link (Highway) layer of the ISO reference model. This standard specifies these services in an abstract way. It does not specify or constrain the implementation entities and interfaces within a computer system. The relationship of this part to other parts of this standard and to LAN specifications is illustrated in Figure 2A-1.

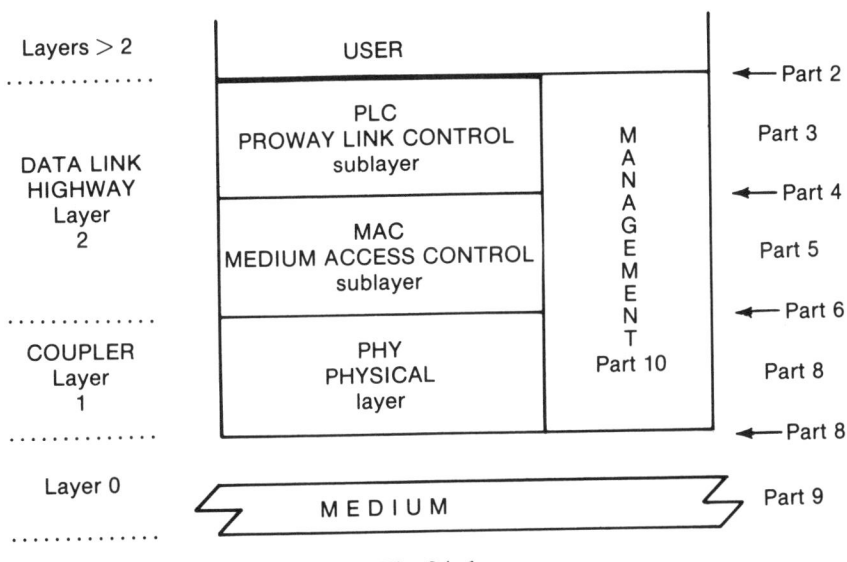

**Fig 2A-1
Relationship to LAN Model**

2A.2 Overview of PROWAY Services

2A.2.1 General Description of Services Provided. This section describes informally the services provided to the user of PROWAY by the PROWAY Link Control (PLC) sublayer of the Highway layer. These services provide data transfer and control services between peer PLC entities. They provide the means by which PROWAY Link Control entities can exchange Link service data units (L_sdu) over the shared Process Data Highway, i.e., PROWAY Local Area Network (LAN). The data transfer can be open-loop or closed-loop, point-to-point or multipoint.

2A.2.2 Model Used for the Service Specification. The general model and descriptive method are given in Reference I, Appendix I-A. The specific application of this model to each service is given in 2A.4.

2A.2.3 Overview of Services. The data transfer services provided to a user of PROWAY are:

- **Send Data with Acknowledge (SDA)**, compatible with the Single Frame service of Reference 3, Appendix I-A (when data is not requested).
- **Send Data with No Acknowledge (SDN)**, equivalent to and inter-operable with the Unacknowledged Connectionless service of Reference 3, Appendix I-A.
- **Request Data with Reply (RDR)**, compatible with the Single Frame service of Reference 3, Appendix I-A (when no data transmitted).

Note that a compatible Connection-Oriented service may be implemented in some stations on a PROWAY-LAN. This standard does not specify the Connection-Oriented service, which is specified in Reference 3, Appendix I-A.

The SDA and RDR services provide **confirmed** data transfers.

2A.2.3.1 Send Data with Acknowledge (SDA). This service allows a local user of PROWAY to send user-supplied data (Link service data unit, L_sdu) to a single remote station. The local user receives a confirmation of the receipt or nonreceipt of the data. At the remote station, this L_sdu (if received correctly) is passed to the remote user of PROWAY. If such an L_sdu is presented to a remote user, it is known to be correct and identical to the L_sdu presented at the local station. If an error in transmission occurs, the local PROWAY station will attempt to retransmit the data.

2A.2.3.2 Send Data with No Acknowledge (SDN). This service allows a local user of PROWAY to send user-supplied data (Link service data unit, L_sdu) to a single, a group of, or all remote stations. The local user receives confirmation of transmission completion but not of receipt of the data. At the remote station(s) this L_sdu (if received correctly) is passed to the remote user(s) of PROWAY. If such an L_sdu is presented to a remote user, it is known to be correct and identical to the L_sdu presented at the local station. However, there is no acknowledgment that such a delivery was made.

2A.2.3.3 Request Data with Reply (RDR). This service allows a local user of PROWAY to request data (Link service data unit, L_sdu) previously submitted by the user at a single remote station. The local user receives either the data requested or an indication that the data was not available. If the data is presented to the local user, it is known to be correct and identical to the L_sdu presented earlier at the remote station. In the event of transmission errors, the local PROWAY station will make additional requests for the data.

2A.2.4 Overview of Interactions. These services are accomplished using a set of primitives. A request primitive is used by the local user to request a service. A confirmation primitive is returned to the local user when the service is completed. An indication primitive is passed to the remote user whenever an unsolicited event is noted. The primitives used for each of the above services are:

- **Send Data with Acknowledge (SDA)**
    L_DATA_ACK.request
             .indication
             .confirm

- **Send Data with No Acknowledge (SDN)**
    L_DATA.request
          .indication
          .confirm

INDUSTRIAL DATA HIGHWAY

ISA
S72.01-1985

- **Request Data with Reply (RDR)**

    L_REPLY.request
    .indication
    .confirm

    L_REPLY_UPDATE.request
    .confirm

The local user of PROWAY can have only one outstanding request (i.e., awaiting a confirmation) for each invocation of the local PLC entity supported by this station. A station must support one invocation of the local PLC entity for each priority and SSAP that it supports (see Section 3B.1.1).

2A.2.5 Overview of Data Highway Operation. An overview of the combined operation of the Data Link layer (PLC and MAC sublayers) is given in Appendix 2-A.

2A.3 Mandatory Features. The primitives which are mandatory for **INITIATOR** and **RESPONDER** stations are listed as M in the table below. Optional primitives are listed as O, and non-applicable primitives are listed as –.

	INITIATOR	RESPONDER
Send Data with Acknowledge (SDA)		
L_DATA_ACK.request	M	–
.indication	M	M
.confirm	M	–
Send Data with No Acknowledge (SDN)		
L_DATA.request	M	–
.indication	M	M
.confirm	O	–
Request Data with Reply (RDR)		
L_REPLY.request	M	–
.indication	M	M
.confirm	M	–
L_REPLY_UPDATE.request	O	M
.confirm	O	M

2A.4 Detailed Specification of PLC Interactions with the User of PROWAY

2A.4.1 Send Data with Acknowledge (SDA)

2A.4.1.1 Description of Operation. The local PLC user prepares a Link service data unit (L_sdu) containing process information or commands, both transparent to PROWAY, for one remote PLC user. This data is passed to the local PLC entity by an L_DATA_ACK.request primitive over the local PLC-user interface. The local PLC entity accepts this service request and attempts to send the L_sdu to the remote PLC entity. The local PLC entity provides a confirmation to the local PLC user indicating the success or a failure of the data transfer.

A positive acknowledgment is required from the remote PLC entity before the local PLC entity returns a positive confirmation to the local PLC user. If this acknowledgment is not received in a timely manner, the local MAC entity will make up to a predefined number of attempts to pass the

L_sdu to the remote MAC. No other traffic occurs on the local area network between the original transmission of the data and its associated acknowledgment.

When the frame is correctly received, the remote PLC entity receives and passes the L_sdu to the remote PLC user over its PLC-user interface using a L_DATA_ACK.indication primitive.

2A.4.1.2 Definition of Primitives of Local User-Highway Interface
2A.4.1.2.1 L_DATA_ACK._request

2A.4.1.2.1.1 Function. This primitive is the service request primitive for the send_data_with_ acknowledge data transfer service.

2A.4.1.2.1.2 Semantics. The primitive shall provide parameters as follows:

L_DATA_ACK.request
  (SSAP,
  DSAP,
  L_sdu,
  remote_address,
  service_class)

The SSAP and DSAP parameters specify the local and remote user link service access points involved in the data transfer service, as defined in Part 3C.

The remote_address parameter specifies the MAC address of the remote station, as defined in Part 5D. The remote_address must be an individual address.

The L_sdu parameter specifies the link service data unit to be transferred by the PLC entity.

The service_class parameter specifies the MAC access_class and thus the MAC priority desired for the data transfer, as defined in Part 5D. Recommended access_class assignments for PROWAY systems are as follows:

MAC Priority	Access_class	Usage
Highest	6	Urgent messages, i.e., those performing critical alarm, interlock, and control coordination functions
	4	Normal control actions and ring maintenance functions
	2	Routine data gathering and display, and data base update functions
Lowest	0	File and program transfers

2A.4.1.2.1.3 When Generated. This primitive is passed from the local PLC user to the local PLC entity to request a L_sdu be sent to one remote PLC user using send_data_with_acknowledge procedures.

2A.4.1.2.1.4 Effect of Receipt. Receipt of this primitive causes the local PLC entity to send an L_sdu using send_data_with_acknowledge procedures.

INDUSTRIAL DATA HIGHWAY

ISA
S72.01-1985

2A.4.1.2.1.5 Additional Comments. Only one SDA, SDN, or RDR request may be concurrently outstanding (i.e., awaiting a confirmation) for each invocation, as defined in Part 3B, of the local PLC entity supported by this station.

2A.4.1.2.2 L_DATA_ACK.indication

2A.4.1.2.2.1 Function. This primitive is the service indication for the send_data_with_acknowledge data transfer service.

2A.4.1.2.2.2 Semantics. The primitive shall provide parameters as follows:
L_DATA_ACK.indication
  (SSAP,
  DSAP,
  local_address,
  remote_address,
  L_sdu,
  service_class)

The SSAP and DSAP parameters specify the local and remote user link service_access points involved in the data transfer service, as defined in Part 3C.

The local_address and remote_address parameters specify the source address and the destination address of the corresponding SDA frame, as defined in Part 5D.

The L_sdu parameter specifies the link service data unit of the corresponding SDA frame.

The service_class parameter specifies the actual MAC priority of the corresponding SDA frame, as defined in Part 5D.

2A.4.1.2.2.3 When Generated. The primitive is passed from the remote PLC entity to the remote PLC user when a MA_DATA.indication with L_pdu_type = **SDA** is received and the received L_pdu is not a duplicate.

2A.4.1.2.2.4 Effect of Receipt. The effect of receipt of this primitive on the remote PLC user is unspecified.

2A.4.1.2.2.5 Additional Comments. the contents of the L_sdu parameter are logically complete and unchanged relative to the L_sdu parameter in the associated L_DATA_ACK.request primitive.

2A.4.1.2.3 L_DATA_ACK.confirm

2A.4.1.2.3.1 Function. This primitive is the service confirmation primitive for the send_data_with_ acknowledge data transfer service.

2A.4.1.2.3.2 Semantics. The primitive shall provide parameters as follows:
L_DATA_ACK.confirm
  (SSAP,
  DSAP,
  remote_address,
  L_sdu,
  service_class,
  L_status)

The SSAP and DSAP parameters specify the local and remote user link service_access points involved in the data transfer service. They are identical to the SSAP and DSAP parameters of the corresponding L_DATA_ACK.request primitive.

The link service data unit is **null**.

The remote_address parameter specifies the destination_address parameter of the MA_DATA.confirmation primitive.

The service class parameter specifies the priority parameter of the MA_DATA.confirmation primitive.

The L_status parameter indicates the success or failure of the previous associated send_data_with_acknowledge data transfer request and whether any error condition is **temporary** or **permanent**.

The values that the Command_status components of the L_status parameter can assume are as follows:

Code	Meaning	Temporary/Permanent
OK =	Command L_sdu accepted	-
TE =	No acknowledgement from remote station after specified retries (possibly due to a non-PROWAY station)	T
DS =	Local station disconnected from line	P
WD =	Local station watchdog timed out	P
IV =	Invalid parameters in request	P
RS =	Request for an unimplemented or unactivated service at remote DSAP: no action taken	P
LS =	Unimplemented service at local SSAP	P
UN =	Resources not available to remote PLC or MAC entity: no action taken	T
UE =	User-PLC interface error	P
PE =	Protocol error	T/P
IP =	Permanent implementation dependent error	P
IT =	Temporary implementation dependent error	T

In Reference 3, Appendix I-A, an additional Response_status component of L_status may be present.

2A.4.1.2.3.3 When Generated. This primitive is passed from the local PLC entity to the local PLC user to indicate the success or failure of the previous associated send_data_with_acknowledge data transfer request.

2A.4.1.2.3.4 Effect of Receipt. The effect of receipt of this primitive on the local PLC user is unspecified.

2A.4.1.2.3.5 Additional Comments. If the transfer was unsuccessful, this primitive indicates that the remote PLC sublayer entity received and positively acknowledged the L_sdu. If an error in transmission occurs, the local MAC entity will make up to a predefined number of attempts to retransmit the data.

If L_status indicates a **Temporary** error, the local PLC-user entity may assume that a future retry of the associated request may be successful.

If L_status indicates a **Permanent** error, the local PLC-user entity should assume that management intervention may be required before a retry of the associated request may be successful.

It is assumed that sufficient information is available to the local PLC user to associate this confirmation with the corresponding request.

2A.4.2 Send Data with No Acknowledge (SDN)
2A.4.2.1 Description of Operation. The local PLC user prepares data (process information or commands, both transparent to PROWAY) for any or all remote users. This data is passed by the L_DATA.request primitive over the PLC-user interface. The local PLC entity accepts this service request, attempts to send this data to the specified remote PLC entities, and returns a local transmission confirmation to the local PLC user. This confirmation reports only local failures.

There is no guarantee of delivery to the remote PLC entities addressed, since no remotely generated acknowledgments or local retries are employed. The data is transmitted once on the line and is received (subject to the propagation delay of the line) simultaneously by all addressed stations. Each remote PLC entity which receives the data passes it to the remote PLC user by an L_DATA.indication primitive.

2A.4.2.2 Inter-operability. Inter-operability, with any station conforming to References 3 and 4, Appendix I-A, is achieved using SDN procedures.

2A.4.2.3 Definition of Primitives at Local User-Highway Interface
2A.4.2.3.1 L_DATA.request
2A.4.2.3.1.1 Function. This primitive is the service request primitive for the send_data_with_no_acknowledge data transfer service.

2A.4.2.3.1.2 Semantics. The primitive shall provide parameters as follows:
L_DATA.request
 (SSAP,
  DSAP,
  remote_address,
  L_sdu
  service_class)

The SSAP and DSAP parameters specify the local and remote users' link service_access points involved in the data transfer service.

The remote_address parameter specifies the MAC address of the remote station as specified in Part 5D. The remote_address may be either an individual address, a group address, or a broadcast address.

The L_sdu parameter specifies the link service data unit to be transferred by the local PLC entity.

The service_class parameter specifies the MAC priority desired for the data transfer as specified in Part 5D.

MAC Priority	Access_class	Usage
Highest	6	Urgent messages, i.e., those performing critical alarm, interlock and control coordination functions
	4	Normal control actions and ring maintenance functions
	2	Routine data gathering and display, and data base update functions
Lowest	0	File and program transfers

2A.4.2.3.1.3 When Generated. This primitive is passed from the local PLC user to the local PLC entity to request that a L_sdu be sent to one, a group, or all remote users using send_data_with_no_acknowledge procedures.

2A.4.2.3.1.4 Effect of Receipt. Receipt of this primitive causes the local PLC entity to send a L_sdu using send_data_with_no_acknowledge procedures.

2A.4.2.3.1.5 Additional Comments. Only one SDA, SDN, or RDR request may be concurrently outstanding (i.e., awaiting a confirmation) for each invocation, as defined in Part 3B, of the local PLC entity supported by this station).

2A.4.2.3.2 L_DATA.indication
2A.4.2.3.2.1 Function: This primitive is the service indication primitive for the send_data_with_no_acknowledge data transfer service.

2A.4.2.3.2.2 Semantics. The primitive shall provide parameters as follows:
L_DATA.indication
  (SSAP,
  DSAP,
  local_address,
  remote_address,
  L_sdu,
  service_class)

The SSAP and DSAP parameters specify the local and remote user link service_access points involved in the data transfer.

the local address and remote_address parameters specify the source_address and destination_address of the corresponding SDN frame, as defined in Part 5D.

The L_sdu parameter specifies the link service data unit of the corresponding SDN frame.

The service_class parameter specifies the actual MAC priority parameter of the corresponding SDN frame, as defined in Part 5D.

INDUSTRIAL DATA HIGHWAY

2A.4.2.3.2.3 When Generated. The primitive is passed from the remote PLC entity to the remote PLC user when a MA_DATA.indication L_pdu_type = **SDN** is received.

2A.4.2.3.2.4 Effect of Receipt. The effect of receipt of this primitive on the remote PLC user is unspecified.

2A.4.2.3.2.5 Additional Comments. The contents of the L_sdu parameter are logically complete and unchanged relative to the L_sdu parameter in the associated L_DATA.request primitive.

2A.4.2.3.3 L_DATA.confirm
2A.4.2.3.3.1 Function. This primitive is the confirmation of the send_data_with_no_acknowledge data transfer service.

2A.4.2.3.3.2 Semantics. The confirmation shall provide parameters as follows:
L_DATA.confirm
  (SSAP,
  DSAP,
  remote_address,
  service_class
  L_status)

The SSAP and DSAP parameters specify the local and remote user link service_access points involved in the data transfer service. They are identical to the SSAP and DSAP parameters of the corresponding L_DATA.request primitive.

The remote_address specifies the destination_address parameter of the MA_DATA.confirm primitive.

The service_class parameter specifies the priority parameter of the MA_DATA.confirm primitive.

The L_status parameter indicates the local success or failure of the previous associated send_data_with_no_acknowledge data transfer request.

The values that the Command_status component of the L_status parameter can assume are:

```
OK = Transmission complete at local station
DS = Local station disconnected from line
WD = Local station watchdog timed out
LE = Invalid parameters (locally detected)
LS = Unimplemented service at local station
```

2A.4.2.3.3.3 When Generated. This confirmation is passed from the local PLC entity to the local PLC user to indicate the local success or failure of the previous associated send_data_with_no_acknowledge data transfer request.

2A.4.2.3.3.4 Effect of Receipt. The effect of receipt of this confirmation by the local PLC user is unspecified.

ISA
S72.01-1985

PROWAY-LAN

2A.4.2.3.3.5 Additional Comments. It is assumed that sufficient information is available to the local PLC user to associate this confirmation with the corresponding request.

2A.4.3 Request Data with Reply (RDR)
2A.4.3.1 Description of Operation. The local PLC user requests data from a remote PLC user by a L_REPLY.request primitive passed to the local PLC entity over the PLC-user interface. The local PLC entity accepts this service request and sends a request for the data to the remote PLC entity. The local PLC entity provides a L_REPLY.confirm to the local PLC user with the requested data or a failure indication.

The remote PLC entity receives the request for data and immediately responds by transmitting a copy of data that had been previously submitted by the remote PLC user with an L_REPLY_UPDATE.request for the corresponding DSAP. The remote PLC user is informed of this reply transmission by an L_REPLY indication.

If a response is not received in a timely manner, the local MAC entity may make up to a predefined number of attempts to obtain the requested data. Between the original transmission of the request and the associated response, no other traffic occurs on the local area network.

The remote PLC user is responsible for maintaining valid data available to the remote PLC entity. The remote PLC user may make a L_REPLY_UPDATE.request to update the shared_data_area, and a L_REPLY_UPDATE.confirm is returned by the remote PLC after the update is complete. Note that the data updating procedures occur asynchronously to any data transfer requests.

2A.4.3.2 Definition of Primitives at Local User-PLC Interface
2A.4.3.2.1 L_REPLY.request
2A.4.3.2.1.1 Function. This primitive is the service request primitive for the request_data_with_reply (RDR) service.

2A.4.3.2.1.2 Semantics. The primitive shall provide parameters as follows:
L_REPLY.request
 (SSAP,
  DSAP,
  remote_address,
  L_sdu,
  service_class)

The SSAP parameter specifies the link service_access point of the local user requesting the data.

The DSAP parameter specifies which shared_data_area is requested from the remote PLC and thus which remote user will receive the indication primitive.

The remote address parameter specifies the MAC address of the remote station, as defined in Part 5D.

The link service data unit is **null**.

The service_class specifies the MAC priority of the request as specified in Part 5D. Recommended access class assignments for PROWAY systems are as follows:

26

MAC Priority	Access_class	Usage
Highest	6	Urgent messages, i.e., those performing critical alarm, interlock and control coordination functions
	4	Normal control actions and ring maintenance functions
	2	Routine data gathering and display, and data base update functions
Lowest	0	File and program transfer

2A.4.3.2.1.3 When Generated. This primitive is passed from the local PLC user to the local PLC entity to request data from one remote PLC entity.

2A.4.3.2.1.4 Effect of Receipt. Receipt of this primitive causes the local PLC entity to request the data specified by the DSAP parameter using the request_data_with_response procedures.

2A.4.3.2.1.5 Additional Comments. The remote PLC user is responsible for maintaining valid data available to the remote PLC entity in the shared_data_area. This data has been previously transferred to the remote PLC entity by an L_REPLY_UPDATE.request primitive with an identical DSAP parameter. The underlying PLC entity must prevent the remote PLC user and the remote PLC entity itself from accessing (one writes while the other reads) the shared_data_area at the same time.

Only one SDA, SDN, or RDR request may be concurrently outstanding (i.e., awaiting a confirmation) for each invocation, as defined in Part 3B, of the local PLC entity supported by this station.

2A.4.3.2.2 L_REPLY.indication
2A.4.3.2.2.1 Function. This primitive is the service indication primitive for the request_data_with_reply service.

2A.4.3.2.2.2 Semantics. The primitive shall provide parameters as follows:
L_REPLY.indication
  (SSAP,
  DSAP)
  local_address,
  L_sdu,
  remote_address,
  service_class)

The SSAP parameter specifies the link service_access point of the local PLC user requesting the data.

The DSAP parameter specifies the remote PLC-user link service_access point that receives this indication. It also specifies which shared_data_area was transmitted as a response to an RDR frame.

The local_address and remote_address parameters specify the source_address and destination_address of the corresponding RDR frame.

The link service data unit is **null**.

The service_class parameter specifies the actual MAC priority of the corresponding RDR frame, as defined in Part 5D.

2A.4.3.2.2.3 When Generated. This primitive is passed by the remote PLC entity to the remote PLC user when a shared_data_area is transmitted as a response to an RDR frame.

2A.4.3.2.2.4 Effect of Receipt. The effect of receipt of this primitive on the remote PLC user is unspecified.

2A.4.3.2.3 L_REPLY.confirm
2A.4.3.2.3.1 Function. This primitive is the service confirmation primitive for the request_data_with_reply service.

2A.4.3.2.3.2 Semantics. The primitive shall provide parameters as follows:
L_REPLY.confirm
  (SSAP,
  DSAP,
  remote_address,
  L_sdu,
  service_class,
  L_status)

The SSAP and DSAP parameters specify the local and remote user link service_access points involved in the data transfer service. They are identical to the SSAP and DSAP parameters of the corresponding L_REPLY.request primitive. In addition the DSAP identifies the remote shared_data_area whose contents are the L_sdu that accompanies this confirmation.

The remote_address specifies the destination_address parameter of the MA_DATA.confirmation primitive.

The L_sdu parameter identifies the data returned by the remote PLC entity.

The service_class parameter specifies the priority parameter of the MA_DATA.confirmation primitive.

The L_status parameter indicates the success or failure of the previous associated request_data_with_response data transfer request and whether any error condition is temporary or permanent.

The values that the Response_status component of the L_status parameter can assume are as follows:

Code	Meaning	Temporary/ Permanent
OK =	Response L_sdu returned	-
TE =	No response from remote station after specified retries (possibly due to a non-PROWAY station)	T
DS =	Local station disconnected from line	P
WD =	Local station watchdog timed out	P
IV =	Invalid parameters in request	P
RS =	Request for an unimplemented or unactivated service at remote DSAP: no action taken	P
LS =	Unimplemented service at local SSAP	P
UN =	Resources not available to remote PLC or MAC entity: no action taken	T
UE =	User-PLC interface error	P
PE =	Protocol error	T/P
IP =	Permanent implementation dependent error	P
IT =	Temporary implementation dependent error	T
NE =	Response L_sdu never submitted at destination	P

In Reference 3, Appendix I-A, an additional Command_status component of L_status may be present.

2A.4.3.2.3.3 When Generated. This primitive is passed from the local PLC entity to the local PLC user to indicate the success or failure of the previous associated request_data_with_reply data transfer request and to pass the requested data if the transfer was successful.

2A.4.3.2.3.4 Effect of Receipt. The effect of receipt of this primitive on the local PLC user is unspecified.

2A.4.3.2.3.5 Additional Comments. The local PLC entity delivers either the requested data or the reason for failure to the local PLC user. If an error in transmission occurs, the local MAC entity will make up to a predefined number of requests for the data.

If L_status indicates a **Temporary** error, the local PLC-user entity may assume that a future retry of the associated request may be successful.

If L_status indicates a **Permanent** error, the local PLC-user entity should assume that management intervention may be required before a retry of the associated request may be successful.

It is assumed that sufficient information is available to the local PLC user to associate this confirmation with the corresponding request.

2A.4.3.2.4 L_REPLY_UPDATE.request

2A.4.3.2.4.1 Function. This primitive is the update_request primitive of the request_data_with_ reply service.

2A.4.3.2.4.2 Semantics. The primitive shall provide parameters as follows:
L_REPLY_UPDATE.request
 (DSAP,
  L_sdu)

The DSAP parameter specifies the remote PLC-user link service_access point making this update_request and which shared_data_area is to be updated.

The L_sdu parameter specifies the new contents of the shared_data_area identified by the DSAP parameter.

2A.4.3.2.4.3 When Generated. The primitive is passed by the remote PLC user to the remote PLC entity to request update of a shared_data_area.

2A.4.3.2.4.4 Effect of Receipt. Receipt of this primitive causes the remote PLC entity to attempt to update the specified shared_data_area.

2A.4.3.2.4.5 Additional Comments. The shared_data_area can be updated only if the remote PLC entity is not attempting to generate a response by accessing the same shared_data_area.

This primitive has significance only to the station containing the shared_data_area, i.e., the remote station.

2A.4.3.2.5 L_REPLY_UPDATE.confirm
2A.4.3.2.5.1 Function. This primitive is the update_confirmation primitive of the request_data_with_reply service.

2A.4.3.2.5.2 Semantics. This primitive shall provide parameters as follows:
L_REPLY_UPDATE.confirm
 (DSAP,
  L_status)

The DSAP parameter specifies the remote PLC-user link service_access point that receives this update_confirmation and which shared_data_area was the subject of the attempted update.

The L_status parameter specifies success or failure of the corresponding update_request. The values that L_status can assume are:

```
Code              Meaning
OK = shared data area updated
UN = shared data area busy and not updated
```

2A.4.3.2.5.3 When Generated. This primitive is passed from the remote PLC entity to the remote PLC user to indicate success or failure of the corresponding L_REPLY_UPDATE.request.

2A.4.3.2.5.4 Effect of Receipt. The effect of receipt of this primitive on the remote PLC user is unspecified.

2B. User-Management Interface and Service Specification

2B.1 Scope and Field of Application. This section specifies the administrative services related to layers 1 and 2 of the ISO reference model that are provided by the Management entity of each station at the PROWAY management interface between the user of PROWAY and that management entity. This standard specifies these services abstractly. It does not specify or constrain the implementation of entities or interfaces within a computer system. The relationship of this part to other parts of this standard and to LAN specifications is illustrated in Figure 2B-1.

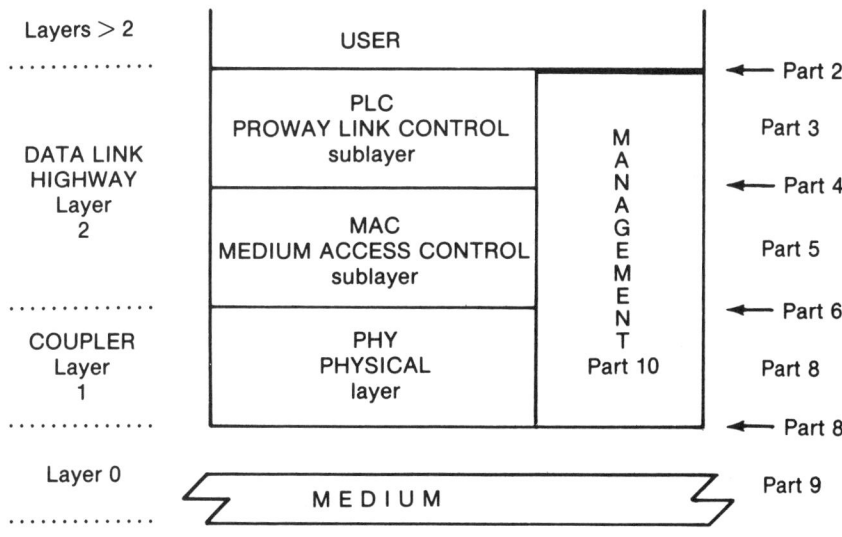

**Fig 2B-1
Relationship to LAN Model**

2B.1.1 Mandatory Features. The management functions described in Part 2B are mandatory if the corresponding PLC, MAC or PHY function is implemented. The specific primitives and codings given in Part 2B are subject to further study and harmonization with the work of the IEEE 802 committee.

2B.2 Overview of Management Services

2B.2.1 General Description of Services Provided at the User-Management Interface. This section informally describes the L_MGMT services related to layers 1 and 2 of the ISO reference model provided to the user of PROWAY by each station's management entity. These services include the following:

(1) Changes in the configuration or status of the station
(2) Requests for information on token ring performance or membership
(3) Notifications of unsolicited changes in station or token ring status
(4) Request for the current value of PLC, MAC, or PHY entity parameters

The L_MGMT service has only local significance.

2B.2.2 Model Used for the Service Specification. The model and descriptive method are given in Reference 1, Appendix I-A.

2B.2.3 Overview of Interactions. One set of primitives is provided at the user-management interface:

L_MGMT.request
       .confirm
         .indication

L_MGMT.request primitive is passed from the local user of PROWAY to the local management entity to request a change in the station's configuration or status or current information on token ring performance or membership.

L_MGMT.indication primitive is passed from the local management entity to a designated local user to indicate a change in station status or token ring membership.

L_MGMT.confirm primitive is passed from the local management entity to the local user of PROWAY to convey the results of the corresponding L_MGMT.request primitive and any requested information.

2B.3 Detailed Specification of Management Interactions with the User of PROWAY
2B.3.1 L_MGMT.request
2B.3.1.1 Function. This primitive is the service request primitive of the L_MGMT service.

2B.3.1.2 Semantics. The primitive shall provide parameters as follows:

L_MGMT.request
  (SSAP,
    code_type,
    arguments)

SSAP specifies the link service_access point of the entity requesting the corresponding L_MGMT service.

The code_type parameter can assume any one of the values shown in Table 2-1.

The arguments are dependent upon the code_type chosen and are listed in Table 2-1 along with the corresponding code_type. The allowed format and range of values for each parameter are specified in Part 10B, 10C, or 10D, as appropriate.

2B.3.1.3 When Generated. This primitive is passed from the local user of PROWAY to the local management entity to request a management action.

2B.3.1.4 Effect of Receipt. Receipt of this primitive by the management entity causes the management entity to perform the requested action. This action may require the cooperation of the PLC, MAC, or PHY entity.

2B.3.1.5 Additional Comments

2B.3.2 L_MGMT.confirm
2B.3.2.1 Function. This primitive is the service confirmation primitive of the L_MGMT service.

2B.3.2.2 Semantics. This primitive shall provide parameters as follows:

Table 2-1
Management Request Codes and Arguments

Code_Type	Action Requested	Arguments
1	Return list of all stations participating in the token ring	None
2	Return value of PLC, MAC, or PHY parameter	L_user identifier, Parameter identifier, access_control information
3	Activate LSAP (except for RDR response component	SSAP, service activated at this SSAP, role in each service activated, maximum L_sdu length for each service activated
4	Activate RSAP (RDR response component of this LSAP	SSAP to receive L_REPLY. indications when an RDR request is received at this DSAP, DSAP associated with this shared_buffer_area, shared_buffer_area identification (format is implementation dependent)
5	Deactivate SAP	SAP
6	Enter line_disconnect state	None
7	Enter line_connect state	None
8	Enter configure state	None

L_MGMT.confirmation
 (SSAP,
  local_address,
  code_type,
  results,
  Mgt_status)

SSAP specifies the link service_access point of the local user of PROWAY involved in the corresponding L_MGMT service.

Local_address specifies this station's MAC address as defined in Part 5D.

Code_type can assume any of the values indicated in Table 2.2.

Results provide information dependent on the code_type as listed in Table 2-2. The allowed range of values for each parameter is specified in Part 10B, 10C, or 10D, as appropriate.

Table 2-2
User-Management Confirmation Codes and Results

Code_Type	Action Attempted	Results
1	Return list of all stations participating in the token ring	Active_stations list. (format is implementation dependent
2	Value of specified PLC, MAC, or PHY layer parameter	Status, parameter_value
3	Activate LSAP	Services activated at this SSAP, role in each active service, maximum L_sdu length for each service activated, shared buffer area identification, user to receive L_REPLY. indication
5	Deactivate SAP	SAP
6	Entry to line_disconnect state	None
7	Entry to line_connect state	None
8	Entry to configure state	None

L_MGT_status indicates success or failure of the action requested in the corresponding L_MGMT.request. Mgt_status may assume the following values:

Code	Meaning
OK =	Requested action performed
CE =	Unimplemented code type
IP =	Invalid parameters
RJ =	Unable to perform requested action

2B.3.2.3 When Generated. This primitive is passed from the local management entity to the requesting entity after the requested management operation is complete.

2B.3.2.4 Effect of Receipt. The effect of receipt of this primitive on the user of PROWAY is unspecified.

2B.3.2.5 Additional Comments. This primitive has local significance only.

2B.3.3 L_MGMT.indication

2B.3.3.1 Function. This primitive is the indication primitive of the L_MGMT service.

2B.3.3.2 Semantics. The primitive shall provide parameters as follows:
L_MGMT.indication
 (SAP,
 code_type)

SAP is the user notification service_access point of the entity that receives this service indication as defined in 3C.2.3.

Code_type specifies the change that has occurred. Table 2-3 itemizes each reported change.

Reason indicates the underlying cause of L_MGMT.indication.

Table 2-3
User-Management Indication Codes and Reason

Code_Type	Associated Change	Reason
1	A change has occurred to the active stations list	Defined in 10E
2	The station has entered the line_disconnect state spontaneously	Defined in 10A

Additional changes to be reported are under study.

2B.3.3.3 When Generated. This primitive is passed from the local management entity to the entity identified by the user_notification_SAP after the management entity has been notified of a change by the PLC, MAC or PHY layer.

2B.3.3.4 Effect of Receipt. The effect of receipt of this primitive on the receiving entity is unspecified.

Appendix 2-A
Overview of Data Highway Operation

2-A.1 Organization of Data Highway Overview. Figures 2-2 and 2-3 show the overall diagrams of the local and remote highway entities of a station that supports a single invocation of the local PLC entity. These figures show the combined actions of the PLC and MAC sublayers and the PHY layer as seen from the perspective of the local and remote users of PROWAY.

Figures 2-4 through 2-21 describe the operation of each service provided by the PROWAY data highway.

The description of each of the services begins with a topological and a sequential diagram which depicts the operation of the entire PROWAY data highway from the perspective of the PROWAY users. They are followed by example state diagrams for the local and remote users of this service. (These diagrams are included to clarify PROWAY-user interactions. They are not referenced in this standard.) The local and remote highway entity state diagrams show the operation of the Data Link (Highway) layer (PLC and MAC sublayers combined) as seen by the users of PROWAY. These diagrams show specific PLC-user and PLC-MAC interactions for this service.

The specifics of the PLC sublayer are given in Part 3, while the specifics of the MAC sublayer are given in Part 5.

In all highway state diagrams, the PLC states are identical to those of the same name defined in Part 3. In the topological diagrams, the solid arcs within PROWAY represent logical paths and the dotted lines represent frame transmissions. The value of the L_status parameter in a configuration depends on the logical path taken. In sequential diagrams, the time axis is shown as a pair of vertical lines with time increasing toward the bottom of the page and a malfunction indicated by an "X".

All diagrams given in Appendix 2-A are for explanation only and are not a requirement of this standard. In case of conflict between Appendix 2-A and other parts of this standard, the other parts take precedence.

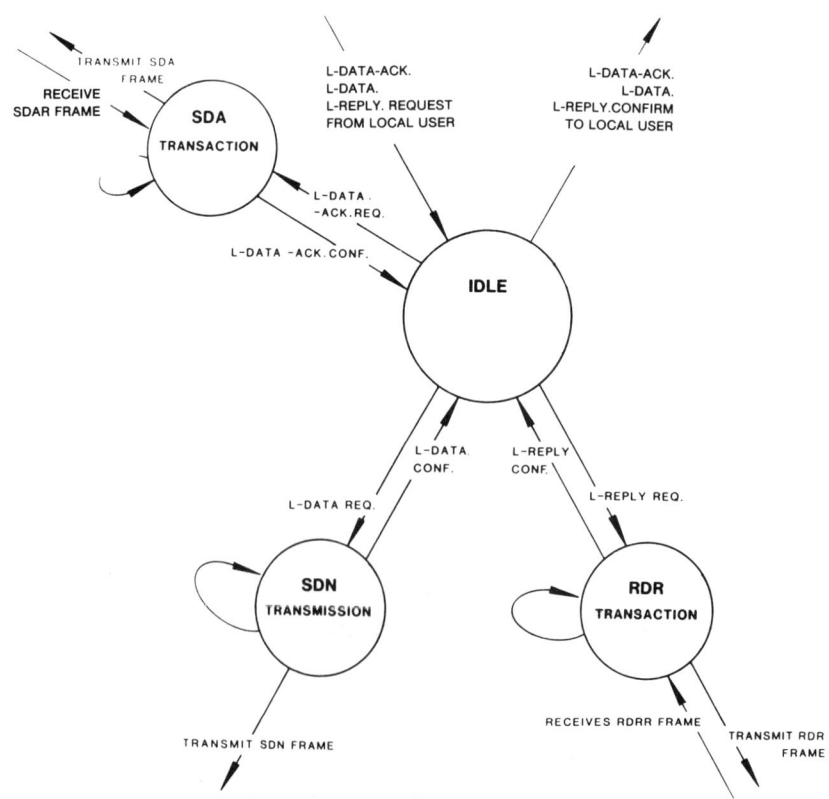

Fig 2-2
Local Station Interaction Diagram

INDUSTRIAL DATA HIGHWAY

ISA
S72.01-1985

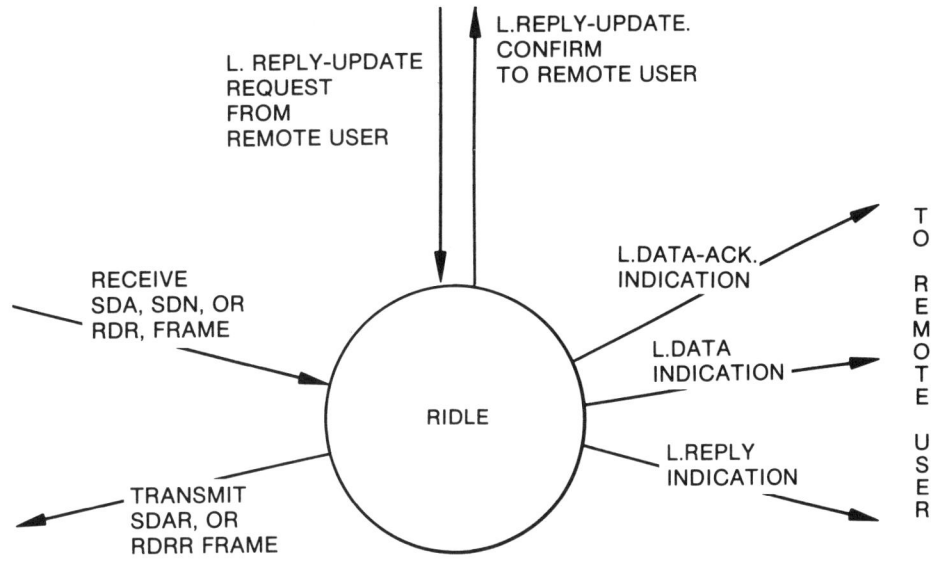

Fig 2-3
Remote Station Interaction Diagram

2-A.2 Send Data with Acknowledge (SDA) Service Diagram

2-A.2.1 Topological and Sequential Relationships. The topological behavior of the send_data_with_acknowledge service is shown in Figure 2-4.

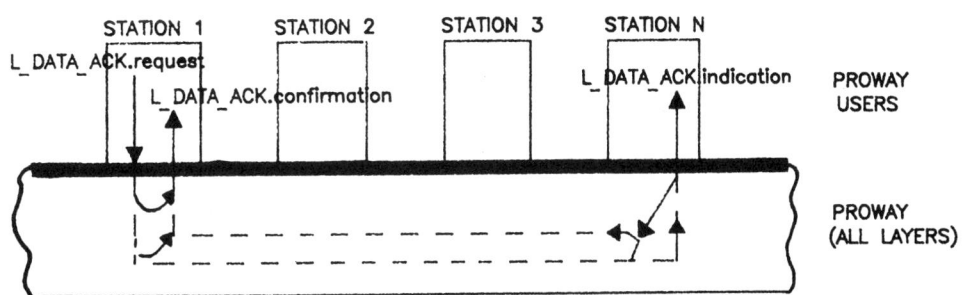

Fig 2-4
Topological Behavior of the Send_Data_With_Acknowledge Service

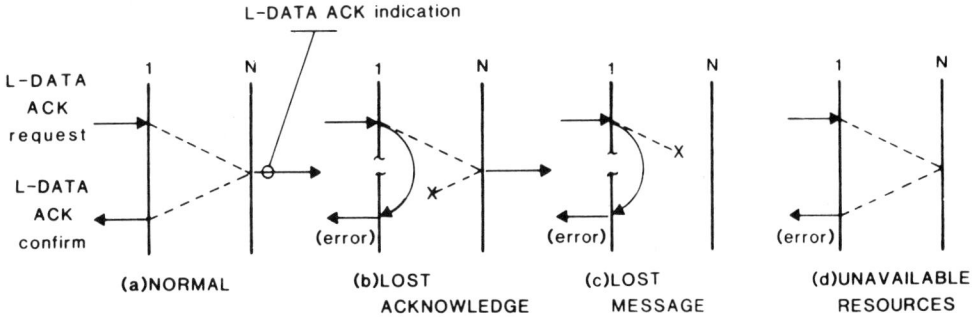

Fig 2-5
Sequential Relationship of the Send_Data_With_Acknowledge Service

37

ISA
S72.01-1985

PROWAY-LAN

2-A.2.2 Service State Diagrams. the state diagrams for the send_data_with_acknowledge service are shown in Figures 2-6 through 2-9. Figures 2-6 and 2-7 show local and remote PLC-user state diagrams which indicate how PLC users interact at the PLC-user interface. Figures 2-8 and 2-9 show local and remote highway state diagrams to indicate how PROWAY processes the user service requests. Figures 2-8 and 2-9 correspond to sections of the overall highway entity state machines shown in Figures 2-2 and 2-3.

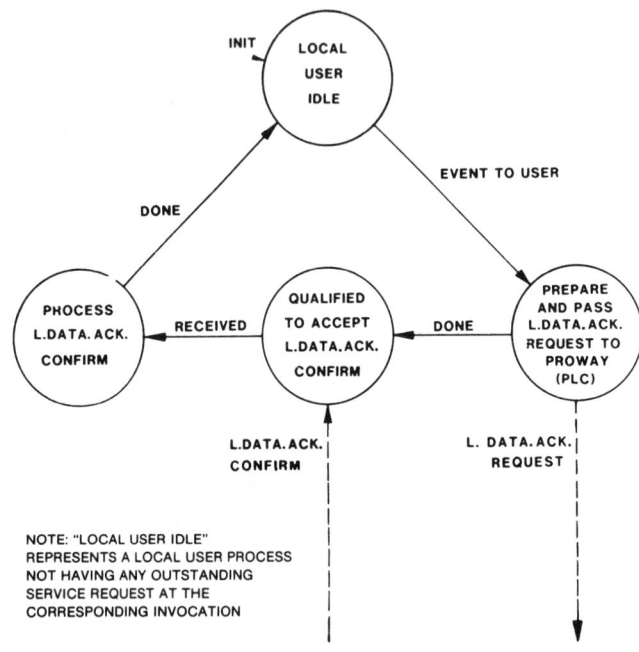

Fig 2-6
Local User State Diagram for the Send_Data_With_Acknowledge Service

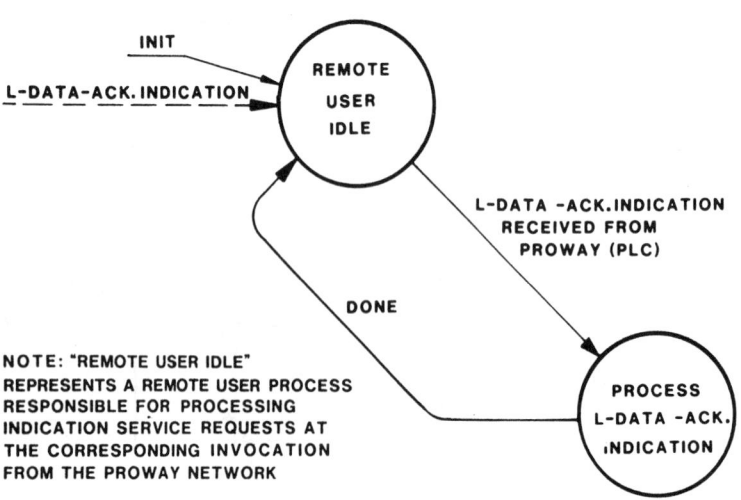

Fig 2-7
Remote User State Diagram for the Send_Data_With_Acknowledge Service

38

INDUSTRIAL DATA HIGHWAY

ISA
S72.01-1985

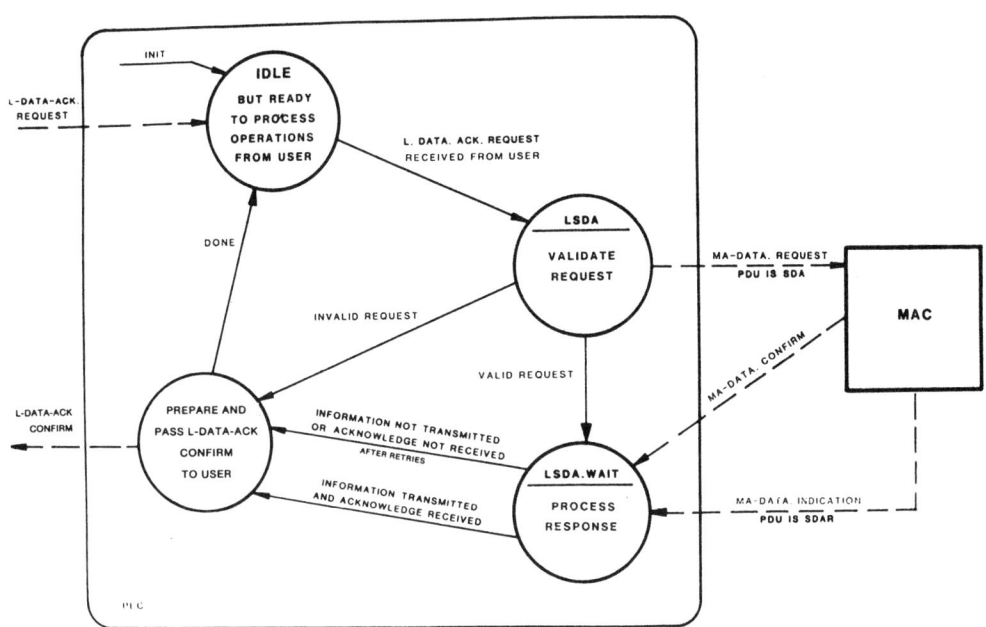

Fig 2-8
Local Highway Entity State Diagram of the Send_Data_With_Acknowledge Service

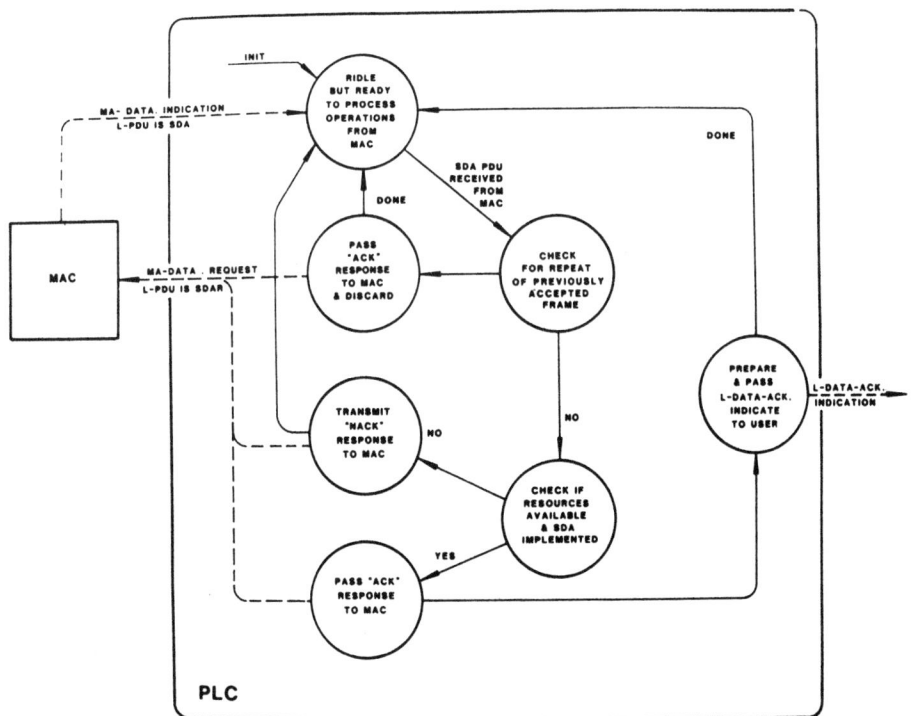

Fig 2-9
Remote Highway Entity State Diagram for the Send_Data_With_Acknowledge Service

2-A.3 Send Data with No Acknowledge (SDN) Service Diagram

2-A.3.1 Topological and Sequential Relationships.
The topological behavior of the send_data_with_no_acknowledge service is shown in Figure 2-10.

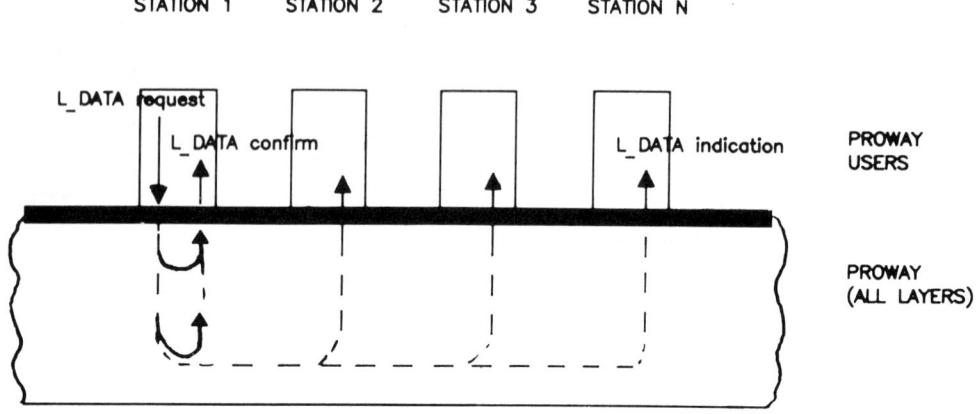

Fig 2-10
Topological Behavior of the Send_Data_With_No_Acknowledge Service

The sequential relationship of the send_data_with_no_acknowledge service is shown in Figure 2-11.

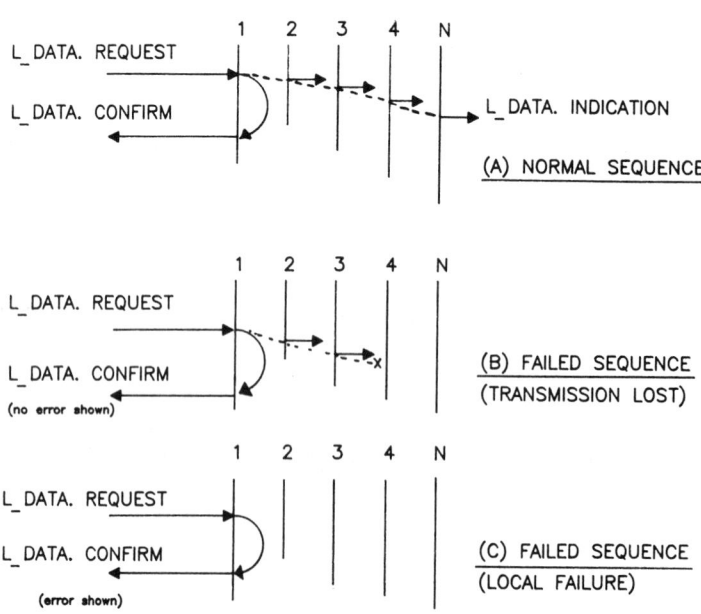

Fig 2-11
Sequential Relationship of the Send_Data_With_No_Acknowledge Service

INDUSTRIAL DATA HIGHWAY

ISA
S72.01-1985

2-A.3.2 Service State Diagrams. The state diagrams for the send_data_with_no_acknowledge service are shown in Figures 2-12 through 2-15. Figures 2-12 and 2-13 show local and remote PLC-user state diagrams to indicate how the PLC user interacts at the PLC-user interface. Figures 2-14 and 2-15 show local and remote highway entity state diagrams to indicate how PROWAY processes the user service requests. These last diagrams correspond to sections of the overall highway entity state machines shown in Figures 2-2 and 2-3.

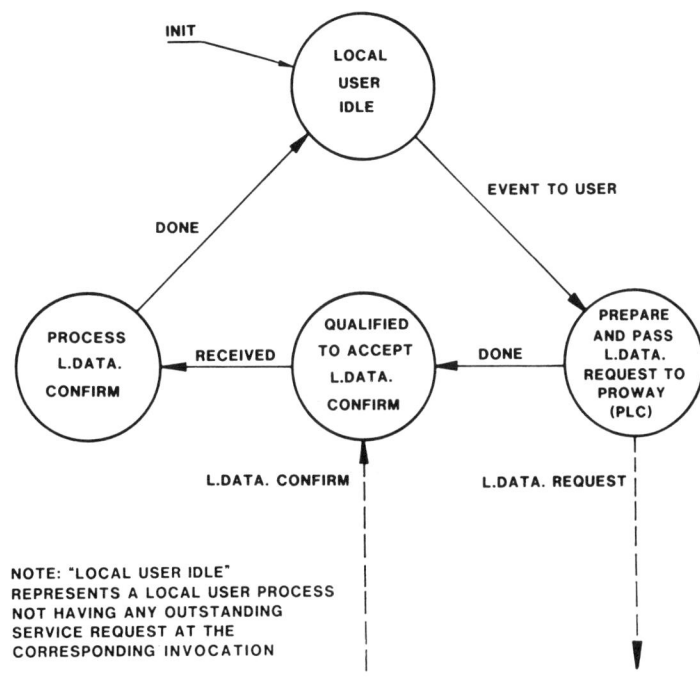

Fig 2-12
Local User State Diagram for the Send_Data_With_No_Acknowledge Service

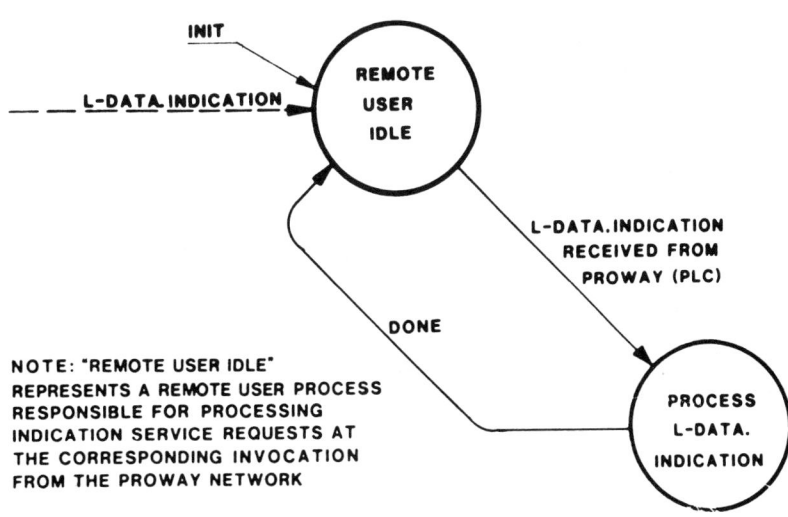

Fig 2-13
Remote User State Diagram for the Send_Data_With_No_Acknowledge Service

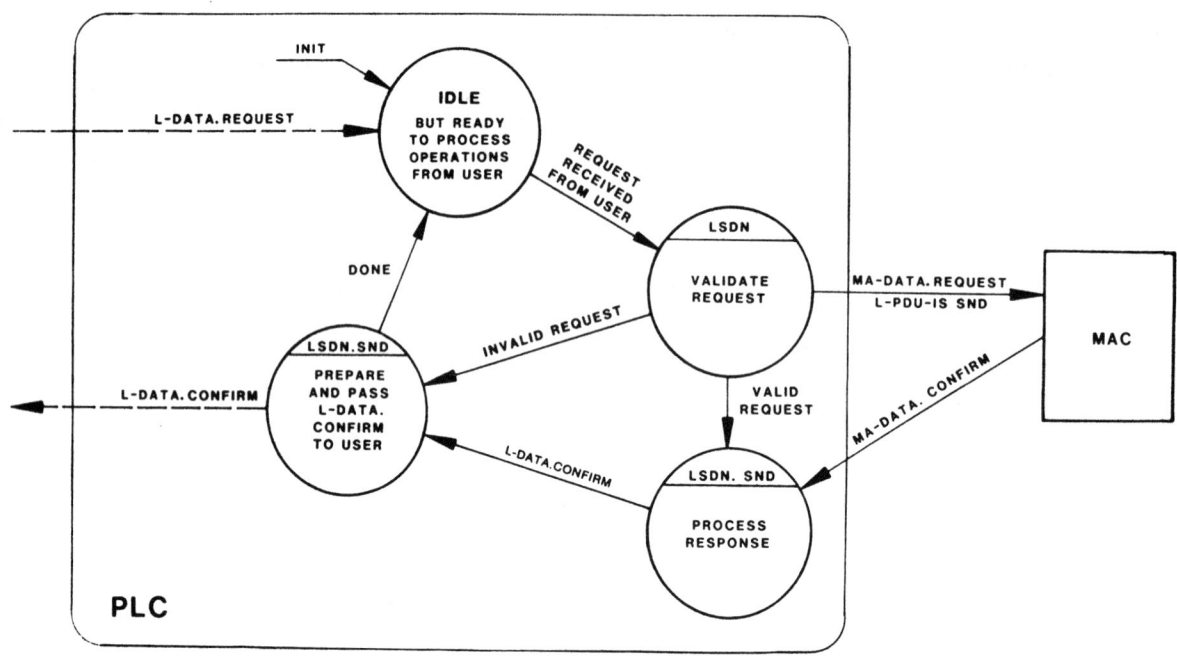

Fig 2-14
Local Highway Entity State Diagram for the Send_Data_With_No_Acknowledge Service

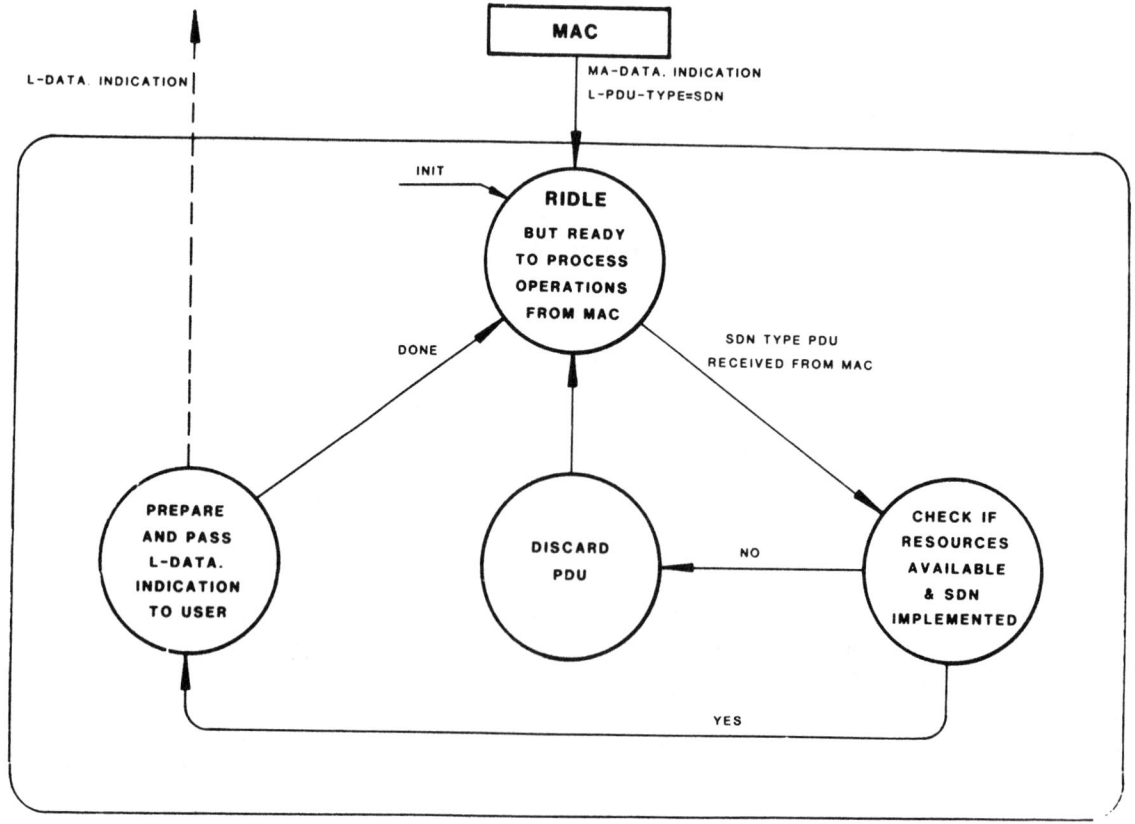

Fig 2-15
Remote Highway Entity State Diagram for the Send_Data_With_No_Acknowledge Service

INDUSTRIAL DATA HIGHWAY

ISA
S72.01-1985

2-A.4 Request Data With Reply (RDR) Service Diagram

2-A.4.1 Topological and Sequential Relationships. The topological behavior of the request_data_with_reply service is shown in Figure 2-16.

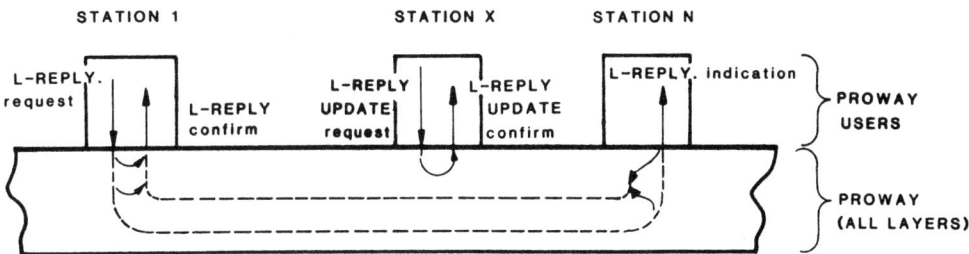

Fig 2-16
Topological Behavior of the Request_Data_With_Reply Service

The sequential relationship of the request_data_with_reply service is shown in Figure 2-17.

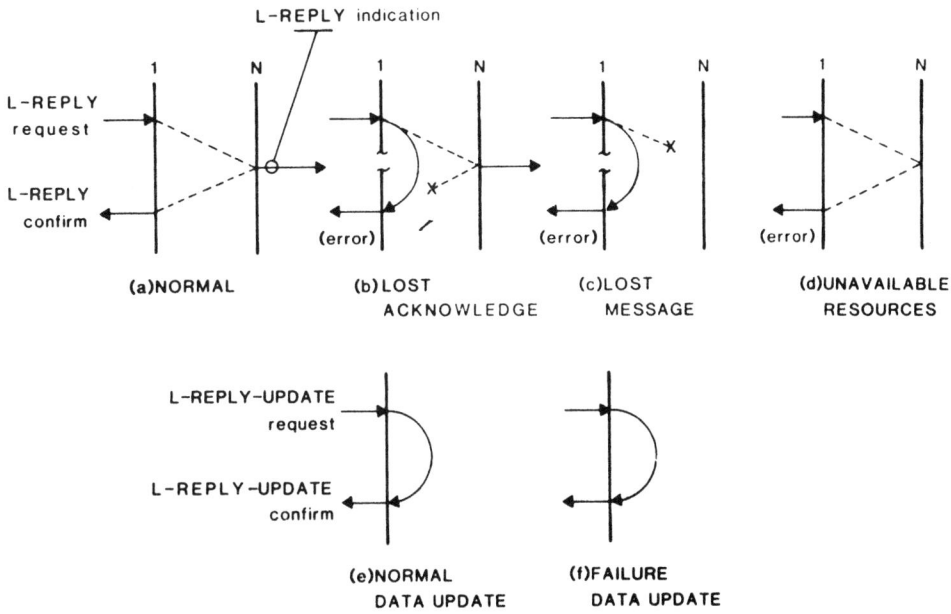

Fig 2-17
Sequential Relationship of the Request_Data_With_Reply Service

2-A.4.2 Service State Diagrams. The state diagrams for the request_data_with_reply service are shown in Figures 2-18 through 2-21. Figures 2-18 and 2-19 show local and remote PLC-user state diagrams to indicate how the user interacts at the PLC-user interface. Figures 2-20 and 2-21 show local and remote highway entity state diagrams to indicate how PROWAY processes the user service requests. These last diagrams correspond to sections of the overall highway entity state machines shown in Figures 2-2 and 2-3.

43

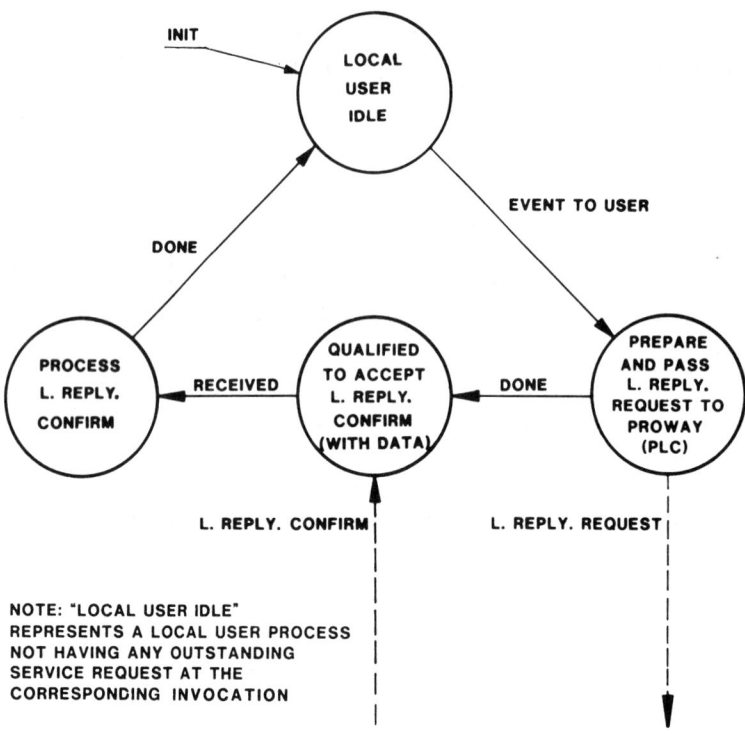

Fig 2-18
Local User State Diagram for the Request_Data_With_Reply Service

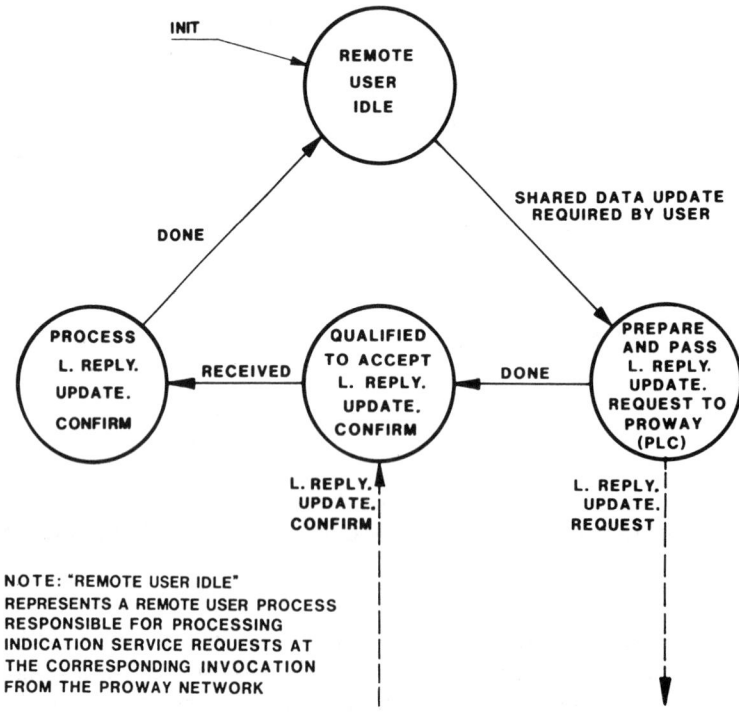

Fig 2-19
Remote User State Diagram for the Request_Data_With_Reply Service

INDUSTRIAL DATA HIGHWAY

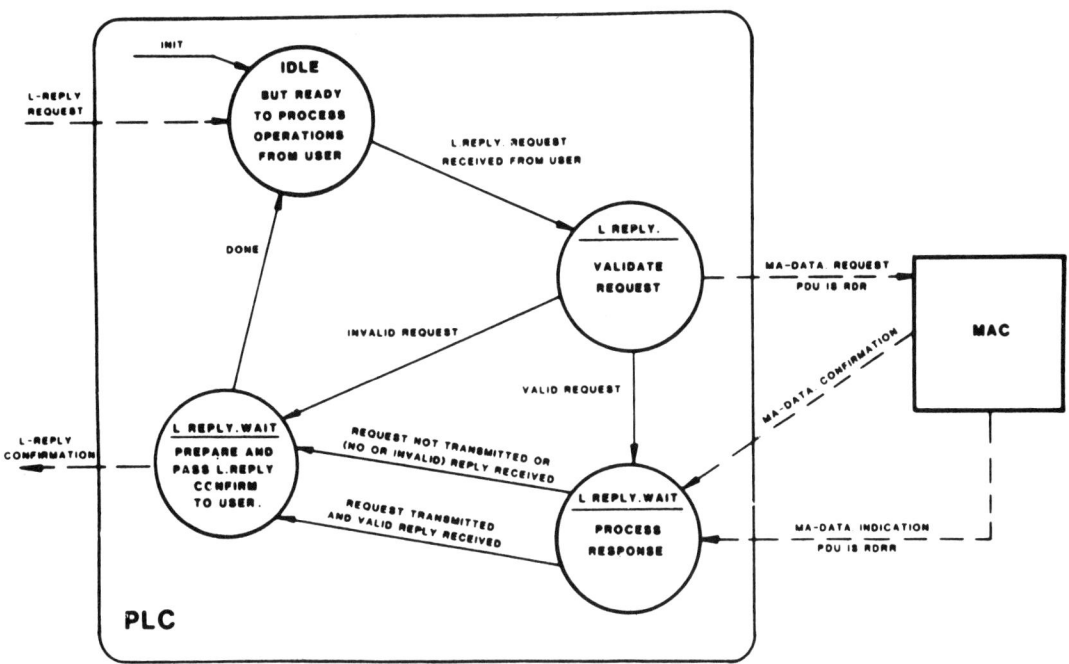

Fig 2-20
Local Highway Unit State Diagram for the Request_Data_With_Reply Service

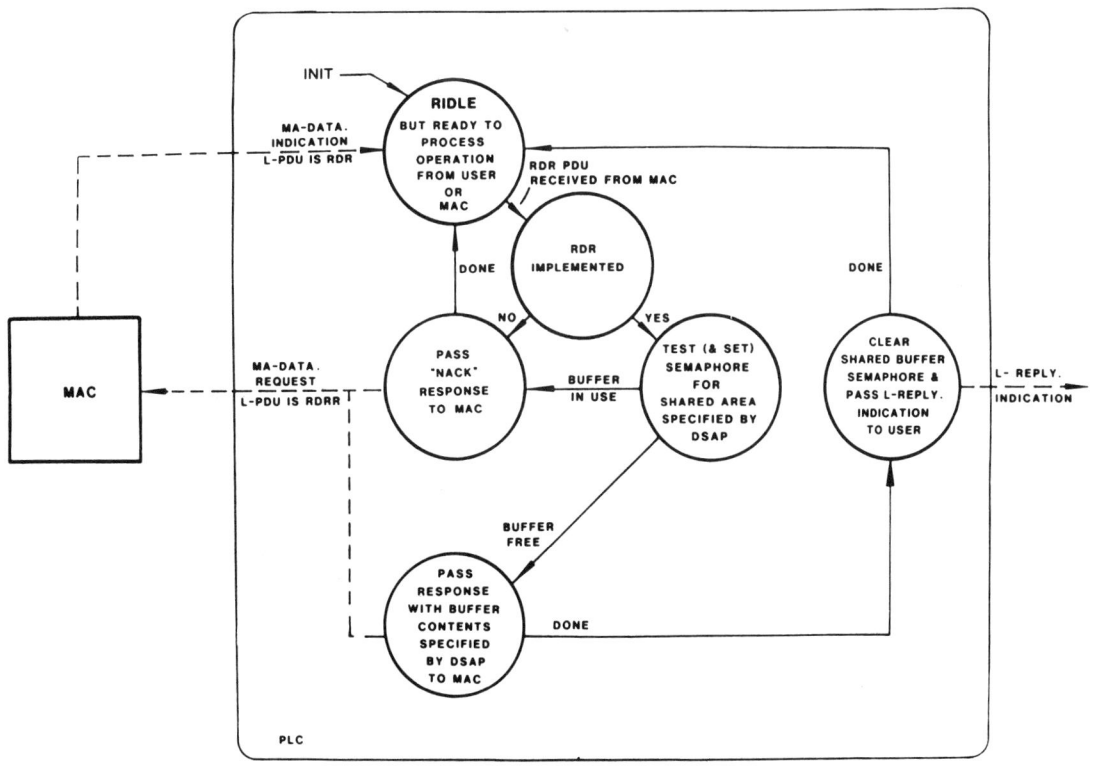

Fig 2-21
Remote Highway Entity State Diagram for the Request_Data_With_Reply Service —
Data Transmission

3. The PROWAY Link Control (PLC) Sublayer

Part 3 specifies the operation of the PROWAY Link Control (PLC) sublayer of the Data Link (Highway) layer of the ISO reference model in an abstract way. It does not specify or constrain the implementation of PLC sublayer entities or interfaces within a computer system.

An introduction to the interactions of the PLC sublayer with the user of PROWAY and with the MAC sublayer of the Data Link layer is given in Appendix 2-A.

Part 3A gives precise definitions of terms and procedures used in the PLC machine state tables and specifies mandatory features of the PLC mechanism.

Part 3B, PLC state machines and descriptions, gives the formal definition of the PLC local and remote state machines using state transition tables and state diagrams.

Part 3C defines the PLC Link_protocol data unit (L_pdu) by which a PLC entity communicates with cooperating PLC entities.

The relationship of this part to other parts of this standard and to LAN specifications is illustrated in Figure 3-1.

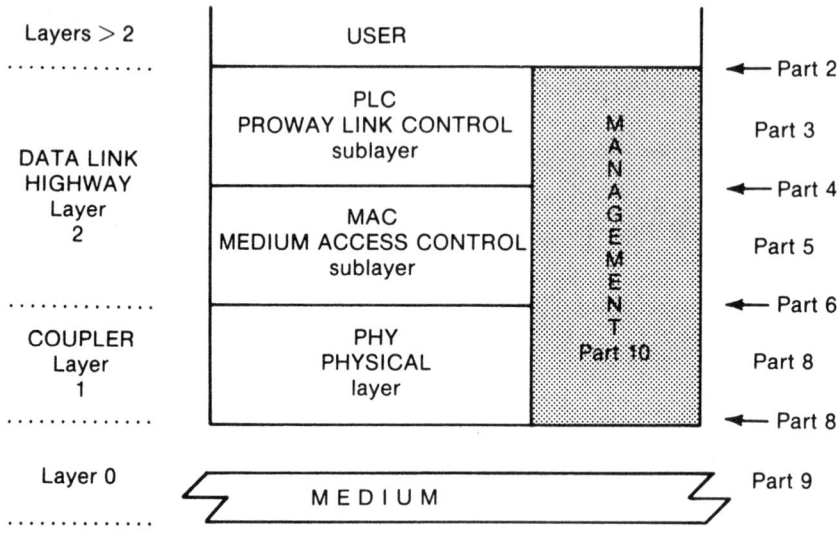

**Fig 3-1
Relationship to LAN Model**

3A. PROWAY Link Control (PLC) Sublayer Definitions and Mandatory Features

3A.1 Notations Used in PLC Machine State Tables

3A.1.1 Notations Used for User-PLC Interface Parameters. This section defines abbreviations used in the PLC machine state tables for primitives and parameters passed across the user-PLC interface. For precise definitions, see Part 2A.

```
SSAP     = local user link service_access point parameter in a primitive
           at the user-PLC interface
LA       = local address parameter in a primitive at the user-PLC
           interface
DSAP     = remote user link service_access point parameter in a primitive
           at the user-PLC interface
RA       = remote address parameter in a primitive at the user-PLC
           interface
SC       = service_class parameter in a primitive at the user-PLC
           interface
L_Status = L_status parameter in a primitive at the user-PLC interface
L_sdu    = the link service data unit parameter in a primitive at the
           user-PLC interface
Confirm  = confirmation
```

3A.1.2 Notations Used for Link_protocol Data Unit Parameters. This section defines abbreviations used in the PLC machine state table for parameters of L_pdu's. For precise definitions and coding, see Part 3C. All of the parameters are transmitted as coded in Part 3C:

```
DSAP     = DSAP field of an L_pdu
SSAP     = SSAP number and L/G bit of the SSAP field of an L_pdu
L_pt     = L_pdu_type parameter of the L_pdu_type field of an L_pdu
R_Status = Response_status parameter of the R_status field of an L_pdu
L_du     = L_data_unit parameter of an L_pdu
SEQ      = the sequence_number bit of the L_pdu_type field of an L_pdu.
           Sequence_number is type BOOLEAN and takes on the values 0 or 1
```

3A.1.3 Notations Used for PLC-MAC Interface Parameters. This section defines abbreviations used in the PLC machine state tables for parameters passed across the PLC-MAC interface. For precise definitions, see Part 4. Where noted these parameters are transmitted as coded at the PLC-MAC interface.

```
M_sdu    = MAC service data unit parameter of a primitive at the PLC-MAC
           interface and in a MAC frame
SA       = the source_address parameter of a primitive at the PLC-MAC
           interface and in a MAC frame
DA       = the destination_address parameter of a primitive at the PLC-MAC
           interface and in a MAC frame
CC       = the confirmation_class parameter of a primitive at the PLC-MAC
           interface
PR       = the priority parameter in a primitive at the PLC-MAC interface
M_status = the M_status parameter of a primitive at the PLC-MAC interface
```

3A.2 Variable Definitions for the PLC Machine State Tables. This section defines variables used by the PLC state machines. The formats used are implementation-dependent.

3A.2.1 Global PLC Variables. These variables are shared by all PLC state machines in one station:

RHIS contains the L_pdu_type, source_address, priority and sequence_number corresponding to the latest received L_pdu. Usage: Remote PLC machine.

STAT contains the R_status value of last transmitted response frame. Usage: Remote PLC machine.

3A.2.2 Local PLC Variables. These variables are specific to the local PLC state machine which corresponds to each PLC invocation (see 3B.1.1).

CONTEXT	= The SSAP, DSAP, remote_address, service class and priority of the request currently being processed by this PLC invocation
current_priority	= the priority component of CONTEXT
current_RA	= the remote_address component of CONTEXT
current_DSAP	= the DSAP component of CONTEXT
TSEQ (destination station)	= the latest sequence_number (0 or 1) transmitted to each supported destination station at each supported priority. The total number of TSEQ = stations_supported.

3A.2.3 Remote PLC Variable. This variable is specific to the remote PLC machine which corresponds to each supported DSAP:

SEM = the semaphore which controls access to the shared_data_area associated with this machine. SEM assumes the values **busy** and **not_busy**. The shared_data_area may be accessed only when SEM = **not busy**.

3A.3 Parameter Definitions for the PLC Machine State Tables. This section defines parameters used in the PLC state machines. The formats used are implementation-dependent.

3A.3.1 Local PLC Parameters. These parameters are specific to the local PLC state machine which corresponds to each PLC invocation. These parameters are established by the station management entity when this SAP is activated as described in Part 10A and 10B:

services for which the **INITIATOR** role has been activated.

maximum_L_sdu_length for each service activated as an **INITIATOR**.

3A.3.2 Remote PLC Parameters. These parameters are specific to the remote PLC state machine which corresponds to each DSAP. These parameters are established by the station management entity when this SAP is activated as described in Part 10A and 10B:

services for which the **RESPONDER** role of this DSAP has been activated.

maximum_L_sdu_length for each service for which the **RESPONDER** role has been activated.

INDUSTRIAL DATA HIGHWAY

shared_data_area identified when this DSAP's RDR response role was activated.

NSAP = the SSAP specified to receive L_REPLY.indications when this DSAP's RDR response component was activated

3A.4 Constants Used in the PLC Machine State. This section defines constants used in the PLC state machines. The formats used are implementation-dependent.

priorities_supported = the number of priorities supported in this station = **4**

stations_supported = the number of stations with which this station exchanges SDA, or RDR L_pdu's $\leq$ **256**

3A.5 Functions and Procedures — Definitions for the PLC State Machines. This section defines all functions and procedures (other than the issuance of service primitives) which are performed by the PLC state machines. The service primitives are defined in Part 2A and Part 4.

3A.5.1 LOCAL_STATUS?

    Returns: **Active** IF the INITIATOR role of one or more services of the SSAP underlying this machine are now activated as described in Parts 10A and 10B
    Returns: **Inactive** OTHERWISE

3A.5.2 UPDATE_CONTEXT

    Returns: None
    Function: Save SSAP, DSAP, remote_address and service_class of the request referenced on this arc in CONTEXT
    current_priority = **EVEN** (service_class)

3A.5.3 VALIDATE?

    Returns: **IP** IF the parameters of the current request primitive do not meet the prescription of Part 2A
    Returns: **LS** IF NOT **IP** and the INITIATOR role of the service specified by the current request primitive is not currently activated as described in Part 10A and 10B
    Returns: **Valid** OTHERWISE

Additional Comments: The parameters of the current request primitive are referenced on any arc that invokes the VALIDATE? function.

3A.5.4 SEQUENCE (destination_address)

    Returns: Sequence_number = the TSEQ value corresponding to this MAC destination_address and priority

3A.5.5 BUILD_PDU (L_pdu_type, sequence_number, SSAP, DSAP, L_du)

    Returns: An L_pdu
    Function: Builds a command L_pdu of the format specified in Part 3C with values as supplied in the calling arguments.

The sequence_number parameter is merged with the L_pdu_type parameter to form the L_pdu type field of the L_pdu.

The C/R bit of the SSAP field is set = **0**. This L_pdu becomes a parameter of the MA.DATA.request generated on the arc which references **BUILD_PDU**.

3A.5.6 EXTRACT_CONTEXT

Returns: The SSAP, DSAP, remote_address, and service_class now contained in CONTEXT. The order of these parameters is as shown above.
Function: Sets CONTEXT to null

3A.5.7 NOTIFY_MGT

Returns: None
Function: Notifies the station management entity of a Protocol error and the conditions under which it occurred

3A.5.8 REMOTE_STATUS?

Returns: **Active** IF the RESPONDER role in one or more services of the DSAP underlying this machine is currently activated as described in Parts 10A and 10B
Returns: **Inactive** OTHERWISE

3A.5.9 RESOURCES?

Returns: **Available** IF PLC resource is available and no MAC parameter indicated that MAC was without resources
Returns: **Unavailable** OTHERWISE

3A.5.10 DUPLICATE?

Returns: **Yes** IF the L_pdu_type, source_address, priority, and sequence_number associated with the current MA_DATA.indication all agree with the values previously saved in RHIS
Returns: **No** OTHERWISE

3A.5.11 ACTIVATED?

Returns: **Yes** IF the RESPONDER role for the service specified in the current MA_DATA.indication is currently activated for the DSAP underlying this machine
Returns: **No** OTHERWISE

3A.5.12 UPDATE_HISTORY

Returns: None
Function: Save L_pdu_type, source_address, priority and sequence_number associated with the current MA_DATA.indication in RHIS

3A.5.13 REQUEST_DATA_AREA

Returns: **Not_busy** IF SEM = **not_busy**

INDUSTRIAL DATA HIGHWAY

ISA
S72.01-1985

Returns: **Busy** OTHERWISE
Function: Set SEM = **busy** IF **not_busy** is returned

3A.5.14 RELEASE_DATA_AREA

Returns: None
Function: Set SEM = **not_busy**

3A.5.15 UPDATE_DATA_AREA

Returns: None
Function: The L_sdu of the current L_REPLY_UPDATE.request replace the contents of the shared_data_area associated with this machine

3A.5.16 RESPONSE_TYPE (L_pdu_type)

Returns: **SDAR** IF L_pdu_type = SDA
RDRR IF L_pdu_type = RDR
Returns: Error indication OTHERWISE

3A.5.17 BUILD_RPDU (L_pdu_type, sequence_number, SSAP, DSAP, R_status, L_du)

Returns: An L_pdu
Function: Builds a response L_pdu of the format specified in Part 3C with values as supplied in the calling arguments.
The C/R bit of the SSAP field is set = **1**.
This L_pdu becomes a parameter of the MA.DATA.request generated on the arc which references **BUILD_RPDU**.

3A.5.18 EVEN (service_class)

Returns: **0** IF service_class = 0 or 1
2 IF service_class = 2 or 3
4 IF service_class = 4 or 5
6 IF service_class = 6 or 7

3A.5.19 UPDATE_SEQ (destination_address)

Returns: None
Function: Complements TSEQ (destination_address)

3A.6 Mandatory Features

3A.6.1 Validity of Response Frames. The remote PLC entity shall provide a valid response for each SDA and RDR L_pdu received. Specifically the UN status must be returned in all cases where resources are not available. Updates of the shared_data_areas shall only influence responses to RDR type L_pdus.

3A.6.2 PLC_station_delay. The remote PLC entity shall provide a valid response L_pdu when it receives a MA_DATA.indication containing an SDA and RDR L_pdu. **This response shall be provided in a timely manner so that the remote station complies with the PLC_station_delay requirements of 5B.1.8.**

3A.6.3 Mandatory State Machines and Features

3A.6.3.1 Local Machine. The local machine including all of its areas and features is mandatory for **INITIATOR** stations. The local machine is not required in **RESPONDER** stations. The minimum required attributes of the local PLC entity are given in the table below.

Attribute	Minimum INITIATOR Requirement
Individual Locally Administered SSAPS	4
Globally Administered SSAPS	01110001 01000000 11000000 01110000 01110010
Priorities supported	4
Minimum, maximum L_sdu_length supported	512 octets

3A.6.3.2 Remote Machine. The remote machine is mandatory for all stations. The arcs of this machine and related machine features are categorized as Mandatory = M or – for Not required for **INITIATOR** and **RESPONDER** machines in the table below. Any variable or function referenced by a mandatory arc is itself mandatory.

Arc	INITIATOR	RESPONDER
1, 3, 4, 6, 7, 8, 9, 10 and 15	M	M
2, 11, 12, 13	–	M

The minimum required attributes of the remote PLC entity for **INITIATOR** and **RESPONDER** stations are given in the table below.

Attribute	Minimum INITIATOR Requirement	Minimum RESPONDER Requirement
Individually Locally Administered DSAPS	4	1
Group Locally Assigned DSAPS	0	0
Shared data areas	0	1
Global DSAPS	01110001 01000000 11000000 01110000 01110010	01000000 01110000
Priorities supported	4	4
Minimum, maximum L_sdu_length supported	512 octets	16 octets

3A.6.4 Labeling. The vendor shall label each station to show which PLC options are supported and the supported value of each PLC attribute.

3B. PLC Machine Formal Description

3B.1 Overview. The PLC sublayer implementation in each station with Initiator functions contains one local machine for each invocation supported by this station's PLC implementation. An invocation is defined as a specific combination of one SSAP and one priority. The local state machine handles all requests from and confirmations to the local user at the priority and SSAP corresponding to this invocation.

The PLC sublayer implementation in each station with Responder functions contains one remote machine for each DSAP supported by this station's PLC implementation. This remote state machine handles all indications to the remote user at this DSAP and manages the shared_data_area associated with this DSAP. These indications usually arise as a result of frames received over the local area network.

Each state machine describes the set of operations performed to support one of these PLC invocations or DSAPs. It is defined using state machine descriptive techniques. These state machines do not specify particular implementation techniques; rather they are intended to describe the external characteristics of the PLC entity as perceived by the corresponding PLC entity in another station or by a higher layer, i.e., the user, in the same station.

3B.1.1 State Machine Invocations. Each machine is shown for the support of a single invocation. A separate local state machine is invoked for each priority/SSAP combination that is supported by this station's implementation of the PLC sublayer. A separate remote machine is invoked for each DSAP that is supported by this station implementation of the PLC sublayer.

Stations that are members of a single network need not support the same number of PLC invocations.

3B.1.2 State Diagrams. Figures 3-2 and 3-3 diagram the local and remote state machines, respectively. These state transitions are detailed in Tables 3-1 and 3-2, which are constructed according to the techniques given in 3B.2.

3B.2 Techniques Used in the PLC Machine State Descriptions. This section provides guidance in interpreting the PLC machine state tables.

Tables 3-1 and 3-2 display the state transitions for the local and remote PLC machines. Each includes columns for the current state, the event which causes state transition, any action(s) taken before the transition, and the next state. This combination of a current state, an event causing a transition, some actions, and the next state is known as an arc. These tables define and number all valid arcs.

The following points apply to the interpretation of the state tables:

(1) There is no implied ordering to the arcs. Also arcs are mutually exclusive. Thus the first event that is satisfied causes the corresponding arc, and no other arc from the current state, to be executed.
(2) An arc may terminate in the same or a different state.
(3) Events which are not listed as valid inputs to the current state shall not cause state transitions.
(4) Actions specified for an arc are executed in the order that they appear in the state table. No other actions are taken on the transition.
(5) Functions, procedures, variables, and constants defined in Part 3A are in these tables.
(6) Primitives defined in Parts 2A, 4 and 10B are referenced.

ISA
S72.01-1985

PROWAY-LAN

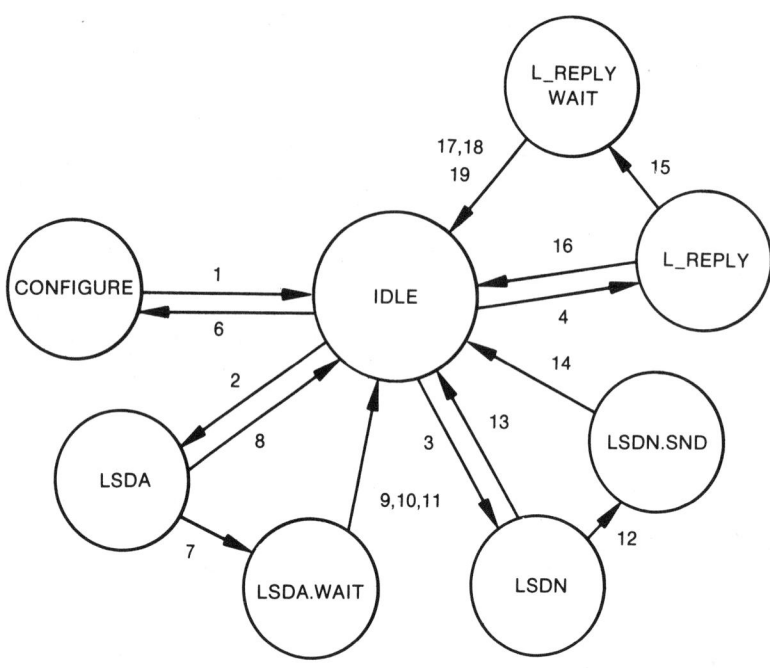

**Fig 3-2
PLC Local Machine State Diagram**

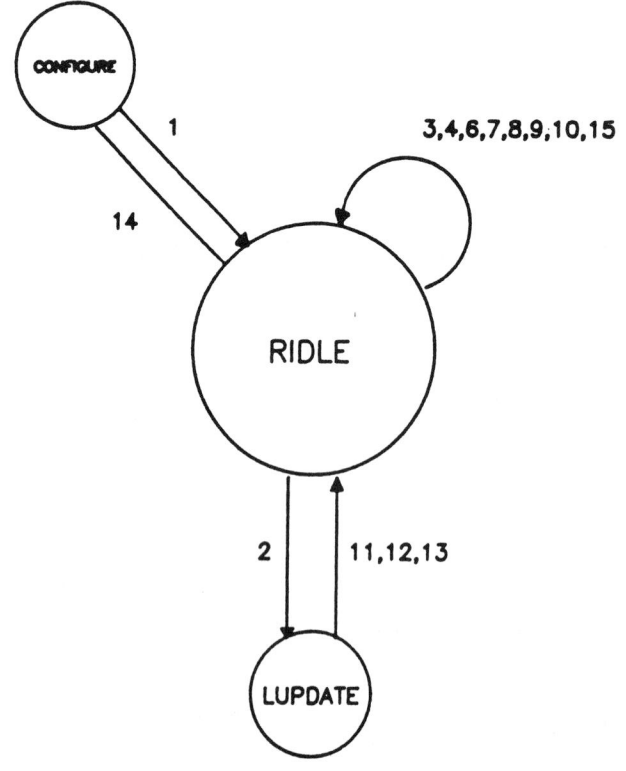

**Fig 3-3
PLC Remote Machine State Diagram**

54

(7) By convention if one and only one primitive at a given interface is mentioned in the definition of an arc, then all parameters related to that interface are parameters of that primitive. If none or more than one primitive at an interface are mentioned in the definition of an arc, then each parameter is explicitly related to the appropriate primitive.

(8) When request and confirmation primitives at the same interface occur in the same arc, the confirmation parameters are those of the request unless otherwise stated.

(9) The parameters listed for a primitive are the only parameters whose value are germane in deciding which (if any) arc to take.

(10) Activation and configuration of LSAP's are described in Parts 10A and 10B and Appendix 10-B.

(11) Error recovery is for further study.

3B.3 State Transition Table for Local PLC State Machine. Table 3-1 describes the local (originating) PLC state machine. The local machine handles service requests (except Update requests) originating from the local PROWAY user and confirmation to that local user.

Table 3-1
State Transition Table for the Local PLC State Machine

CURRENT STATE	EVENT	ACTION	NEXT STATE	ARC #
CONFIGURE	LOCAL_STATUS?=Active	None	IDLE	1
IDLE	L_DATA_ACK.request	UPDATE_CONTEXT	LSDA	2
IDLE	L_DATA.request	UPDATE_CONTEXT	LSDN	3
IDLE	L_REPLY.request	UPDATE_CONTEXT	L_REPLY	4
IDLE	LOCAL_STATUS?=Inactive	None	CONFIGURE	6
LSDA	VALIDATE?=Valid	BUILD_PDU (L_pt:=SDA, SEQ:=SEQUENCE (RA), SSAP:=SSAP, DSAP:=DSAP, L_du:=L_sdu) MA_DATA.request (DA:=RA, M_sdu:=L_pdu, PR:=current_priority, CC:=RR)	LSDA.WAIT	7
LSDA	VALIDATE?≦Valid	L_DATA_ack.confirm (EXTRACT_CONTEXT !SSAP, !DSAP, !RA, !SC),	IDLE	8
LSDA.WAIT	MA_DATA.confirm (M_status≦OK)	L_DATA_ack.confirm (EXTRACT_CONTEXT !SSAP, !DSAP,	IDLE	9

**Table 3-1
State Transition Table for the Local PLC State Machine**
(Continued)

CURRENT STATE	EVENT	ACTION	NEXT STATE	ARC #
		!RA, !SC, L_sdu:=null, L_status:=M_status		
LSDA.WAIT	(MA_DATA.indication (SA=current_RA AND SSAP=CURRENT_DSAP AND L_pt=SDAR) AND MA_DATA.confirm (M_status=OK))	L_DATA.ACK.confirm (EXTRACT_CONTENT !SSAP, !DSAP, !RA, !SC, L_sdu:=null, L_status) UPDATE_SEQ (current_RA)	IDLE	10
LSDA.WAIT	(MA_DATA.indication (SA≤current_RA OR SSAP≤current_DSAP OR L_pt≤SDAR or L_du≠null))	NOTIFY_MGT.	IDLE	11
LSDN	VALIDATE?=Valid	BUILD_PDU (L_pt:=SDN, SEQ:=0 SSAP:=SSAP, DSAP:=DSAP, L_du:=L_sdu) MA_DATA.request (DA:=RA, M_sdu:=L_pdu, CC:=RQ, PR:=current_priority)	LSDN.SND	12
LSDN	VALIDATE?≤Valid	L_DATA.confirm (EXTRACT_CONTEXT !SSAP, !DSAP, !RA, !SC, L_STATUS:=VALIDATE)	IDLE	13
LSDN.SND	MA_DATA.confirm	L_DATA.confirm (EXTRACT_CONTEXT !SSAP, !DSAP, !RA, !SC, L_status:=M_status)	IDLE	14

Table 3-1
State Transition Table for the Local PLC State Machine
(Continued)

CURRENT STATE	EVENT	ACTION	NEXT STATE	ARC #
LREPLY	VALIDATE?=Valid	BUILD_PDU (L_pt:=RDR, SEQ:SEQUENCE(RA) SSAP:=SSAP DSAP:=DSAP L_du:=null) MA_DATA.request (DA:=RA, M_sdu:=L_pdu, PR:=current_priority, CC:=RR)	LREPLY.WAIT	15
LREPLY	VALIDATE?≤Valid	L_REPLY.confirm (EXTRACT_CONTEXT !SSAP, !DSAP, !RA, !SC, L_sdu=null, L_status:=VALIDATE)	IDLE	16
LREPLY.WAIT	MA_DATA.confirm (M_status≤OK)	L_REPLY.confirm (EXTRACT_CONTEXT !SSAP, !DSAP, !RA, !SC, L_sdu:=null, L_status:=M_status)	IDLE	17
LREPLY.WAIT	(MA_DATA.indication (SA=current_RA AND SSAP=current_DSAP AND L_pt=RDRR) AND MA_DATA.confirm (M_status=OK))	L_REPLY.confirm (EXTRACT_CONTEXT !SSAP, !DSAP, !RA, !SC, L_sdu, L_status), UPDATE_SEQ (current_RA)	IDLE	18
LREPLY.WAIT	(MA_DATA.indication (SA≤current_RA OR SSAP≤current_DSAP OR L_PT≤RDRR))	NOTIFY_MGT	IDLE	19

The following arcs describe conditions that are implied in the state description of Reference 3: 2, 3, 4, 5, 8, 11, 13, 14, 16, 19.

ISA
S72.01-1985

PROWAY-LAN

3B.4 State Transition Table for the Remote PLC State Machine. Table 3-2 describes the remote PLC state machine which handles service indications originating from the MAC layer and manages shared_data_areas.

Table 3-2
State Transition Table for the Remote PLC State Machine

CURRENT STATE	EVENT	ACTION	NEXT STATE	ARC #
CONFIGURE	REMOTE_STATUS?=Active	None	RIDLE	1
RIDLE	REMOTE_STATUS?=Inactive	None	CONFIGURE	14
RIDLE	L_REPLY_UPDATE.request	None	LUPDATE	2
RIDLE	(MA_DATA.indication (L_pt=SDN) AND RESOURCES?=Available AND ACTIVATED?=Yes	L_DATA.indication (SSAP:=SSAP, DSAP:=DSAP, LA:=SA, RA:=DA, SC:=PR, L_sdu:=L_du)	RIDLE	3
RIDLE	(MA_DATA.indication (L_pt=SDA), AND L_du≤null AND RESOURCES?=Available AND ACTIVATED?=Yes AND DUPLICATE?=No	BUILD_RPDU (L_pt:=SDAR, SEQ:=NOT(SEQ), SSAP:=DSAP, DSAP:=SSAP R_status:=OK, L_du:=null) MA_DATA.request (M_sdu:=L_pdu, DA:=SA, CC:=RS, PR:=PR) L_DATA_ACK.indication (SSAP:=SSAP, DSAP:=DSAP, LA:=SA, RA:=DA, L_sdu:=L_du) UPDATE_HISTORY STAT:=OK	RIDLE	4
RIDLE	(MA_DATA.indication (L_pt=SDA) AND DUPLICATE?=Yes AND ACTIVATED?=Yes	BUILD_RPDU (L_pt:=SDAR, SEQ:=NOT(SEQ), SSAP:=DSAP, DSAP:=SSAP, R_status:=STAT, L_du:=null)	RIDLE	6

Table 3-2
State Transition Table for the Remote PLC State Machine
(Continued)

CURRENT STATE	EVENT	ACTION	NEXT STATE	ARC #

		MA_DATA.request
		(M_sdu:=L_pdu,
		  DA:=SA,
		  CC:=RS,
		  PR:=PR)		

RIDLE	(MA_DATA.indication
	(L_pt=RDR) AND
	REQUEST_DATA_AREA=
	  Not Busy AND
	ACTIVATED?=Yes)
		BUILD_RPDU
		(L_pt:=RDRR,
		 SEQ:=NOT(SEQ),
		 SSAP:=DSAP,
		 DSAP:=SSAP,
		 R_status:=OK,
		 L_du:=shared_data_
		 area contents)
		MA_DATA.request
		(M_sdu:=L_pdu,
		  DA:=SA,
		  CC:=RS,
		  PR:=PR)
		
		L_REPLY.indication
		(SSAP:=NSAP,
		 DSAP:=DSAP,
		 LA:=SA,
		 RA:=DA,
		 SC:=PR,
		 L_sdu:=null)
		
		RELEASE_DATA_AREA
		
		UPDATE_HISTORY
		
		STAT:=OK
			RIDLE	7

RIDLE	(MA_DATA.indication
	(L_pt=RDR) AND
	(REQUEST_DATA_AREA
	=Busy OR
	RESOURCES?=
	  Unavailable AND
	ACTIVATED?=Yes)
		BUILD_RPDU
		(L_pt:=RDRR,
		 SEQ:=NOT(SEQ),
		 SSAP:=DSAP,
		 DSAP:=SSAP,
		 L_status:=UN,
		 L_du:=null)
		MA_DATA.request
		(M_sdu:=L_pdu,
		  DA:=SA,
		  CC:=RS,
		  PR:=PR)
		
		UPDATE_HISTORY
		
		STAT:=UN
			RIDLE	8

Table 3-2
State Transition Table for the Remote PLC State Machine
(Continued)

CURRENT STATE	EVENT	ACTION	NEXT STATE	ARC #
RIDLE	(MA_DATA.indication (L_pt=SDA AND L_du= null AND RESOURCES?= Available AND ACTIVATED?=Yes AND DUPLICATE?=No	BUILD_RPDU (L_pt:=SDAR, SEQ:=NOT(SEQ), SSAP:=DSAP, DSAP:=SSAP, R_status:=OK, L_du:=null) MA_DATA.request (M_sdu:=L_pdu, DA:=SA, CC:=RS, PR:=PR) UPDATE_HISTORY STAT:=OK	RIDLE	15
RIDLE	(MA_DATA.indication (CC=RR) AND ACTIVATED?=No	BUILD_RPDU (L_pt:=RESPONSE_TYPE (L_pdu_type), SEQ:=NOT(SEQ), SSAP:=DSAP, DSAP:=SSAP, R_status:=RS, L_du:=null) MA_DATA.request (M_sdu:=L_pdu, DA:=SA, CC:=RS, PR:=PR) UPDATE_HISTORY STAT:=RS	RIDLE	9
RIDLE	(MA_DATA.indication (L_pt=SDA) AND DUPLICATE?=No AND ACTIVATED?=Yes RESOURCES?= Unavailable)	BUILD_RPDU (L_pt:=SDAR, SEQ:=NOT(SEQ), SSAP:=DSAP, DSAP:=SSAP, R_status:=UN, L_du:=null MA_DATA.request (M_sdu:=L_pdu, DA:=SA, CC:=RS, PR:=PR) STAT:=UN	RIDLE	10

Table 3-2
State Transition Table for the Remote PLC State Machine
(Continued)

CURRENT STATE	EVENT	ACTION	NEXT STATE	ARC #
LUPDATE	(RVALIDATE?=Valid AND REQUEST_DATA_AREA= Not Busy)	UPDATE_DATE_AREA L_REPLY_UPDATE.confirm (DSAP:=DSAP, L_status:=OK) RELEASE_DATE_AREA	RIDLE	11
LUPDATE	(RVALIDATE?≤Valid)	L_REPLY_UPDATE.confirm (DSAP:=DSAP L_status:=RVALIDATE)	RIDLE	12
LUPDATE	(VALIDATE?=Valid AND REQUEST_DATA_AREA= Busy)	L_REPLY_UPDATE.confirm (DSAP:=DSAP, L_status:=UN	RIDLE	13

The following arcs describe conditions that are implied in the state descriptions of Reference 3: 1, 11, 12, 13, 14.

3C. PLC Protocol Data Unit Format

3C.1 Function of the PLC_Protocol_Data_Unit. The PLC_protocol_data_unit (L_pdu) is used to transfer data, commands or status information between cooperating PLC entities.

The **send_data_with_no_acknowledge (SDN) L_pdu** is used by the local PLC to convey an L_du to one or more remote stations without requiring an acknowledge. The SDN L_pdu is invoked by the L_DATA.request primitive.

The **send_data_with_acknowledge (SDA) L_pdu** is used by the local PLC to convey an L_pdu to one remote station and to request an acknowledge from that remote station. The SDA L_pdu is invoked by the L_DATA_ACK.request primitive.

The **request_data_with_response (RDR) L_pdu** is used by the local PLC to pass a request for data to a remote station. The RDR L_pdu is invoked by the L_REPLY.request primitive.

3C.2 Structure of the PLC_Protocol_Data_Unit. Each PLC_protocol_data_unit (L_pdu) must contain a 3 or 4 octet PLC header. In addition the L_pdu may contain a Link_data_unit (L_du), i.e., information field.

The format of the PLC L_pdu is:

PLC_header	L_du
3 or 4 octets	0 to 1000 octets

↑ First field delivered to or received from the MAC sublayer.

The structure of the L_pdu depends on the service that is requested. The structure of the L_pdu for each PROWAY service is given in Table 3-3.

Table 3-3
PLC Protocol Data Unit Structure

L_pdu_type	L_pdu_type Coding	R_status Present	Command/ Response	L_du Present
SND=Send data with no ack.	11000000	No	Command	Yes
SDA=Send data with ack.	1110011S	No	Command	Yes
RDR=Request data with resp.	1110111S	No	Command	No
SDAR=SDA Response	1110011S	Yes	Response	No
RDRR=RDR Response	1110111S	Yes	Response	Yes
	↑ 1st bit delivered_to or received_from MAC			

S = Sequence_number

NOTE: The above bit patterns are equivalent to those for the type 1 and type 3 services of Reference 3 of Appendix I-A.

3C.2.1 PLC Header Composition. The PLC header specifies the PLC L_pdu_type of this L_pdu and provides additional information that depends on the L_pdu_type. The PLC header consists of:

a DSAP field
a SSAP field
a L_pdu_type field
a R_status field for response L_pdu_types as shown in Table 3-3.

The format of the PLC_header is as shown below.

─── 1st octet delivered to or received from MAC

DSAP Field	SSAP Field	L_pdu_type	R_status
1 octet	1 octet	1 octet	0 or 1 octet

3C.2.2 Service_Access Points. Each PLC L_pdu shall contain two link service_access point (LSAP) fields: the destination service_access point (DSAP) field and the source service_access point (SSAP) field. The DSAP shall identify the one or more service_access points for which the PLC L_du is intended. The SSAP shall identify the specific access point from which the L_du was initiated.

3C.2.2.1 LSAP Representation. The LSAP representation shall be as shown below.

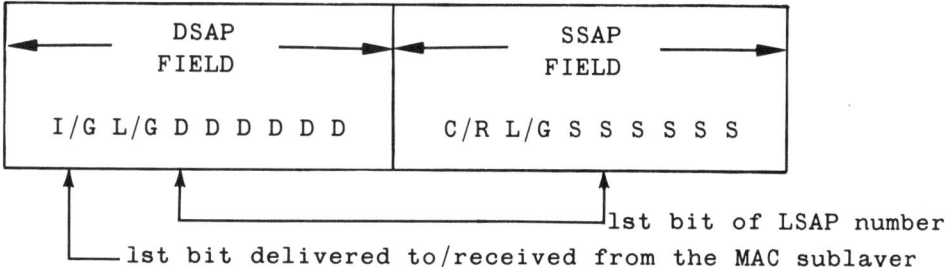

```
I/G = 0   INDIVIDUAL DSAP
I/G = 1   GROUP DSAP
C/R = 0   COMMAND PDU
C/R = 1   RESPONSE PDU
L/G = 0   LOCALLY ASSIGNED LSAP
L/G = 1   LSAP IS RESERVED FOR DEFINITIONS BY STANDARDS BODIES

LSAP_NUMBER RANGE = 0-63
```

3C.2.2.2 LSAP Field Composition
3C.2.2.2.1 Each LSAP field shall contain one octet.

3C.2.2.2.2 Each LSAP field shall contain 6 bits of the actual LSAP number.

3C.2.2.2.3 The I/G bit shall be the first bit delivered to the MAC of the DSAP field. If this bit is "0", it shall indicate that the address is an individual DSAP address. If this bit is "1", it shall indicate that the address is a group DSAP address that identifies none, one or more, or all of the service_access points that are serviced by the PLC entity. Group DSAP addresses are allowed only if the L_pdu_type is SDN.

3C.2.2.2.4 The C/R bit shall be the first bit delivered to the MAC of the SSAP field. If this is "0", it shall indicate that the L_pdu is a command. If this bit is "1", it shall indicate that the L_pdu is a response as shown in Table 3-3.

3C.2.2.3 LSAP Address Usage. An individual LSAP shall be usable as both an SSAP and a DSAP address. A group or the global LSAP shall be used only as a DSAP address and only in conjunction with SDN L_pdu's.

3C.2.2.4 Global Administered LSAP Assignments
3C.2.2.4.1 Global DSAP. DSAP fields = 11111111 (all "1s") is the Global DSAP. This DSAP designates an SDN L_pdu as destined for all DSAPs actively being serviced by the underlying MAC entity.

3C.2.2.4.2 PROWAY Application LSAP. The LSAP = 01110010 is reserved for higher layer PROWAY applications.

3C.2.2.4.3 Station Management LSAPs. LSAP field = 01110000 is designated as the individual LSAP for this station's station management entity.

3C.2.2.4.4 Active_Station_List LSAP. LSAP field = 01110001 is designated as the SSAP and DSAP used to exchange Mgt_pdu's related to maintenance of the active_station_list as defined in Part 10E.

3C.2.2.5 Locally Administered LSAP Assignments. Recommended values of locally administered individual and group LSAP numbers in PROWAY systems are for further study.

3C.2.3 Specification of Shared Data Areas. For the RDR service an individual DSAP also designates a particular shared_data_area. This shared_data_area is updated by L_UPDATE.request primitives designating this LSAP.

3C.2.4 Implied SAP. The user identified by the user notification LSAP will receive notifications of changes in ring membership as defined in Part 10E and changes in station status as defined in Part 10A.

3C.2.5 L_PDU_TYPE Definitions. The L_pdu_type specifies the underlying service to be provided by the PLC layer. It also specifies the composition of the L_pdu as shown in Table 3-3 and carries sequence_number information to assist in detection of duplicate frames.

3C.2.5.1 Sequence_numbers. SDA, RDR, SDAR and RDRR L_pdu's have the sequence_number bit set as specified in Part 3B. The sequence_number bit is reserved and set equal zero for SDN L_pdu's.

3C.2.5.2 R_Status Definition. R_status conveys the disposition of the immediately preceding command L_pdu by the remote MAC and/or PLC entities.

The format of the R_status field is shown below.

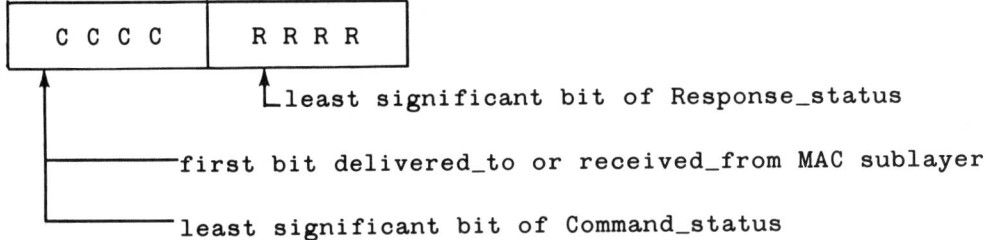

```
T = Temporary error.  A future retry of the associated transmission is
                      likely to succeed.
P = Permanent error.  Management intervention may be required before a
                      retry of the associated transmission is likely to
                      succeed.
```

3C.3 L_Data_Unit

3C.3.1 Data Unit Size. The L_du shall consist of any integral number (including zero) of octets. The maximum length of an L_du in a PROWAY system is 1000 octets.

Command_Status				
Value	Code	Meaning		T/P
0	OK	Command LSDU accepted		-
1	RS	Request for an unimplemented or unactivated service or receipt of an RDR pdu containing a non-null data unit at remote DSAP: no action taken		P
5	UE	User-PLC interface error		P
6	PE	Protocol error		P
7	IP	Permanent implementation dependent error		P
9	UN	Resources not available to remote PLC or MAC entity: no action taken		T
15	IT	Temporary implementation dependent error		T
Others		Reserved		-

Response_Status				
Value	Code	Meaning		T/P
0	OK	Response LSDU present		-
1	RS	Request for an unimplemented service: no action taken		P
3	NE	Response LSDU never submitted to remote PLC		P
4	NR	Response LSDU not requested		-
5	UE	User-PLC interface error		P
6	PE	Protocol error		P
7	IP	Permanent implementation dependent error		P
9	UN	Resources not available to remote PLC or MAC entity: no action taken		T
15	IT	Temporary implementation dependent error		T
Others		Reserved		-

3C.3.2 Bit Order. The information field shall be delivered to the source MAC sublayer in the same bit order as received from the source user of PLC. The information field shall be delivered to the destination user of PLC in the same bit order as received from the destination MAC sublayer.

3C.4 Invalid L_pdu. An invalid L_pdu shall be defined as one which meets at least one of the following conditions:

(1) It is identified as such by the Physical layer or the Medium Access Control (MAC) sublayer.
(2) It is not an integral number of octets in length.
(3) It does not contain a properly formatted PLC header and an L_du corresponding to Table 3-3. Note that the L_du length may be 0 octets in an L_pdu that contains an L_du.
(4) Its length is less than 3 octets.

Invalid L_pdu's shall be ignored by the destination PLC entity.

Appendix 3-A
Prevention of Loss and Duplication for the PROWAY Confirmed Data Transfer Services

3-A.1 Introduction. This appendix is explanatory in nature. It is not a requirement of this standard.

The PROWAY **confirmed services** (SDA and RDR) are designed to prevent loss and duplication of messages. This appendix provides a general overview of how this protection is achieved. It first describes the underlying characteristics of the confirmed services. This is followed by a detailed examination of loss and duplication scenarios. The analysis shows that loss prevention adds a source of duplication, which must be accommodated by a sequence_number generated by the local (originating) PLC entity and checked by the remote (responding) PLC entity.

3-A.2 Overview of the PROWAY Confirmed Services. The PROWAY confirmed services are provided by the PLC sublayer in conjunction with the token bus MAC sublayer. When the local user issues a request for an SDA or RDR service, the local PLC queues a request to the MAC layer at the specified priority. The sequence_number bit of the PLC L_pdu_type field is toggled, as described below, in order to prevent frame duplication. The local MAC composes the required request_with_response frame.

When the local MAC obtains the token, it sends the request_with_response frame and waits for an immediate response. It does not transmit any other frames nor pass the token while waiting. If a timer expires, the local MAC will resend the original request_with_response frame. This is repeated until a response is received or the retry counter exceeds the maximum_retry_limit which had been previously set by station management.

When the remote MAC receives any frame, it passes the received frame up to the remote PLC. If the frame is of L_pdu_type = SDA or RDR, it is the responsibility of the remote PLC to make the determination as to whether the data unit is a duplicate. If the frame is not a duplicate or is for the SDN service, the data unit contained in the frame is passed to the remote user.

For all request_with_response frames (L_pdu_type = SDA or RDR) the remote PLC immediately generates the appropriate response (acknowledgment or reply) and directs the remote MAC to send this response to the source of the request_with_response frame. The sequence_number bit of the request_with_response frame is returned in the response frame to maintain compatibility with other standards.

When the originating MAC receives the response frame or the retry counter is exhausted, it notifies the local PLC of completion of the originating request.

3-A.3 Sources of Frame Loss. Two potential sources of frame loss in the confirmed services are:

First. Loss of frames due to line errors which cause the Request frame or its associated response frame to be lost.

Second. Loss of a request_with_response frame because the remote station incorrectly decided that frame is a repeat, discards it and repeats its previous response. This may cause the originating station to incorrectly assume that the request_with_response frame was properly received.

3-A.4 Protection against Loss and Creation of Duplicate Frames. The prevention mechanism for the first type of loss is the MAC level retry mechanism at the originating station. If a response is not detected before the timeout expires, the frame is retransmitted and the timer is restarted. This is repeated until the retry counter exceeds the limit established by station management or a response is heard. If the retry counter is exhausted, local user is notified that the link has failed. In this case the only unknown is whether the remote station correctly received any of the frames transmitted.

If the remote station has received a frame that the local station thinks was lost, a potential for duplication on the next token round exists. The level of protection against duplication is R, the maximum_retry_limit, since R request_with_response or response frames must be lost to create the above condition.

The retry mechanism prevents loss except in the case where the retry counter is exhausted. In this case, the local user must send an SDA frame with a null data unit to regain synchronization with the remote user.

There is a failure scenario with such a scheme if the response delays are not bounded. If a frame is sent and the timeout expires before the response is returned, a second copy of the frame is sent, which is also confirmed. When the first acknowledge is received, the originator sends a new frame, which may get lost. However, the response from the retransmission of the first frame may be received and incorrectly associated with the second data frame. In the PROWAY specification, the reply must be sent by the remote station within 2 octet times. So long as the retry timeout is set longer than 3 slot times, the scenario described above will not occur.

3-A.5 Protection against Duplication. Duplication occurs at the remote station if it cannot differentiate between a retry of a previously received frame and a new frame. The PROWAY specification uses a single sequence_number bit to solve the duplication problem.

The remote PLC makes the duplicate determination based on the last received source_address, SSAP, priority, and the state of the sequence_number bit. If all fields match and the previous frame was accepted, then the frame will be rejected as a duplicate and the previous status is returned as the response. Otherwise the frame is treated as a new frame. Note that the response to this frame may be generated immediately by PLC prior to checking the address fields if buffer space is available to store the incoming frame. This alleviates the processing load required before generating the response.

Each time the local PLC entity queues a request_with_response frame to the remote station at a given priority, it will have to toggle the sequence_number bit of the PLC L_pdu_type field to allow the remote PLC to differentiate a new request_with_response frame from a retry.

The local PLC must maintain a history and state of the sequence_number bit for the previous request_with_response frame queued to each destination station for each priority.

3-A.6 Operation with Non-PROWAY MAC Layers. The confirmed service depends on the token bus request_with_response procedure which prevents independent traffic from intervening between the request frame and its response. A confirmed connectionless service can prevent loss of frames using non-PROWAY MACs by implementing Link level retry procedures. Prevention of frame duplication using a non-PROWAY MAC requires maintenance of state information for each pair of communicating LSAPs.

3-A.7 An Example of Erroneous Detection of Duplication when Sequence Information Is Not Used. Another loss occurs if the originating station queues two different request_with_response frames with identical source address, destination address and priority while the destination station sees nothing between them. This can be caused by the destination station missing the intervening frames or the source failing to insert frames.

A station may send two request_with_response frames with the same source address, destination address, and frame control field sequentially.

If toggling of the sequence bit was not used, duplicating could occur (due to lost frames at destination station). As shown in the following example, if station C misses the request_with_response and response between B and A, it will assume that the next message it receives is a duplicate. Note that the transmissions between B and A could be a pair of token passes or a request_with_response.

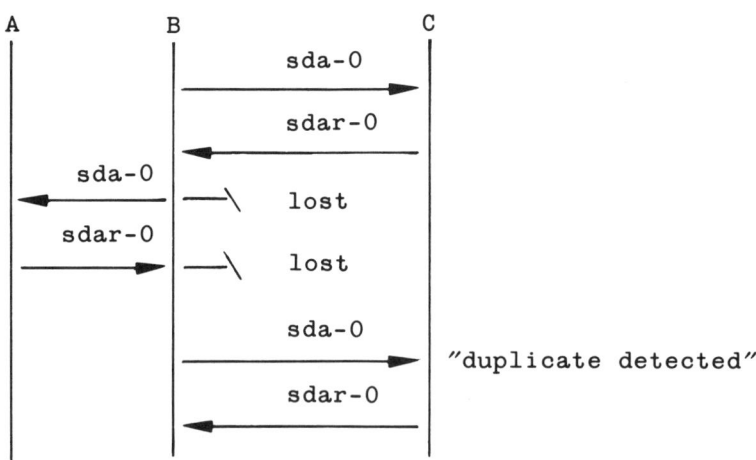

4. PLC-MAC Interface and Service Specification

4.1 Scope and Field of Application. This section specifies the services provided to the PROWAY Link Control (PLC) sublayer at the boundary between the Link Control sublayer and the Medium Access sublayer of the Data Link (HIGHWAY) layer of the ISO reference model. This standard specifies these services in an abstract way. It does not specify or constrain the implementation entities and interfaces within a computer system. The relationship of this section to other sections of this standard and to LAN specifications is illustrated in Figure 4-1.

4.2 Overview of the PLC-MAC Data Transfer Service
4.2.1 General Description of Services Provided. This section describes informally the MAC_DATA transfer service provided to the PLC sublayer by the MAC sublayer. This service supports data transfer between peer PLC entities. It provides the means by which PLC entities can exchange Link_protocol_data_units (L_pdu). The data transfer can be open-loop, closed-loop, point-to-point or multi-point.

INDUSTRIAL DATA HIGHWAY

ISA
S72.01-1985

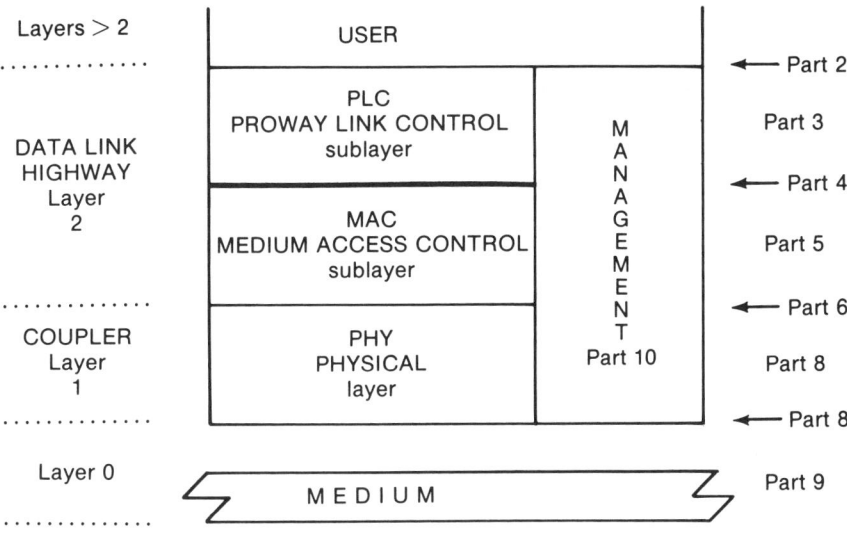

**Fig 4-1
Relationship to LAN Model**

4.2.2 Model Used for the Service Specification. The model and descriptive method are provided in Reference 1, Appendix I-A.

4.2.3 Notation. In this part only, stations, MAC entities and PLC entities are frequently identified by their relationship to the associated frame; i.e., as source or destination. This notation avoids the ambiguities that arise when both the local (originating) and remote (receiving) PLC entities submit a MA_DATA.request for the same service, e.g., SDA, RDR, RSR.

4.2.4 Overview of Interactions. The primitives associated with the MA_DATA data transfer service are:

MA_DATA.request
MA_DATA.indication
MA_DATA.confirm

MA_DATA.request primitive is passed from the source PLC entity to the source MAC entity to request that an M_sdu be sent to one, a group of, or all destination stations.

MA_DATA.indication primitive is passed from the destination MAC entity to the destination PLC entity to indicate the arrival of an M_sdu.

MA_DATA.confirm primitive is passed from the source MAC entity to the source PLC entity to convey the results of the corresponding MA_DATA.request primitive.

4.2.5 Mandatory Features. The MA_DATA service and all its primitives are mandatory and are required in all implementations.

4.3 Detailed Interactions with the PLC Entity. This section describes in detail the primitives and parameters associated with the MA_DATA data transfer service. Note that the service is specified in an abstract sense. The parameters specify the information that must be available to the sending entity. A specific implementation is not constrained in the method of making this information available. For example, the M_sdu parameter associated with some of the data transfer service primitives may be provided by actually passing the MAC_service_data_unit, by passing a descriptor, or by other means. The values of some selection parameters may also be implied by an implementation.

4.3.1 MA_DATA.request

4.3.1.1 Function. This primitive is the service request primitive for the MA_DATA data transfer service.

4.3.1.2 Semantics. The primitive shall provide parameters as follows:

MA_DATA.request
   (destination_address,
     M_sdu,
     confirmation_class,
     priority)

The **destination_address parameter** specifies the MAC entity address of the destination station or stations, as defined in Part 5D. The destination address may be either an individual address, a group address, or the broadcast address.

The **M_sdu parameter** specifies the MAC_service_data_unit to be transmitted by the source MAC entity for the source PLC entity.

The **confirmation_class parameter** specifies whether a response is required from the remote MAC entity and identifies such a response. The possible values of the confirmation class parameter are:

RQ=request_with_no_response
RR=request_with_response
RS=response

The **priority parameter** specifies the desired MAC priority, as defined in Part 5D.

4.3.1.3 When Generated. This primitive is passed from a source PLC entity to the source MAC entity to request that the MAC entity compose and transmit a frame with the specified priority and confirmation class on the local area network.

4.3.1.4 Effect of Receipt. Receipt of this primitive causes the source MAC entity to attempt to compose and transmit the specified frame.

4.3.1.5 Additional Comments. Group and broadcast destination_addresses may be used only with confirmation_class = RQ.

A value of request_with_response for the confirmation_class parameter indicates that the next MA_DATA.indication should have a confirmation_class of response, in which case that next MA_DATA.indication shall be associated with this MA_DATA.request.

A value of response for the confirmation_class parameter indicates that the immediately prior MA_DATA.indication shall have had a confirmation_class of request_with_response.

4.3.2 MA_DATA.indication

4.3.2.1 Function. This primitive is the service indication primitive for the MA_DATA data transfer service.

4.3.2.2 Semantics. The primitive shall provide parameters as follows:

MA_DATA.indication
 (destination_address,
  source_address,
  M_sdu,
  confirmation_class,
  priority)

The **destination_address and source_address parameters** specify the DA and SA fields, as defined in Part 5D, of a frame received by the destination MAC entity.

The **M_sdu parameter** specifies the MAC_service_data_unit received by the destination MAC sublayer entity.

The **confirmation_class and priority parameters** specify the quality of service of the frame received by the destination MAC entity. The semantics are identical to 4.3.1.2.

4.3.2.3 When Generated. This primitive is passed from the destination MAC entity to the destination PLC entity to indicate the arrival of a frame from the PHY entity of the destination station. Such frames are reported only when they are free of detected errors and their (individual, group, or broadcast) destination address designates the destination MAC entity.

4.3.2.4 Effect of Receipt. The effect of receipt of this primitive by the destination PLC entity is specified in Part 3B.

4.3.2.5 Additional Comments. If delivered, the contents of the M_sdu parameter are logically complete and unchanged relative to the M_sdu parameter in the associated MA_DATA.request at the source station.

NOTE: This is a guarantee of transparency.

A value of request_with_response for the confirmation_class parameter indicates that the receiving PLC entity should immediately respond with an MA_DATA.request which itself has a confirmation_class of response.

A value of response for the confirmation_class parameter indicates that this MA_DATA.indication may be associated with a prior MA_DATA.request which itself had a confirmation_class of request_with_response, and which was issued by the same PLC entity.

4.3.3 MA_DATA.confirm

4.3.3.1 Function. This primitive is the service confirmation primitive for the MA_DATA data transfer service.

4.3.3.2 Semantics. This primitive shall provide parameters as follows:

MA_DATA.confirm
 (confirmation_class,
  priority,
  M_status)

The **confirmation_class and priority parameters** specify the quality of service actually provided. The semantics are identical to 4.3.1.2.

The **M_status parameter** specifies the success or failure of the service associated with the corresponding MA_DATA.request. M_status may assume values as listed below.

```
Code    Meaning
 OK  =  Requested service performed
        {Frame transmitted for confirmation_class = RQ or RS}
        {Response frame received for confirmation_class = RR}
 DS  =  Source station disconnected from line
 WD  =  Source station watchdog timed out
 IP  =  Invalid parameter (detected at source)
 TE  =  No acknowledgment from remote station after specified retries
        (possibly due to a non-PROWAY station)
```

4.3.3.3 When Generated. This primitive is passed from the source MAC entity to the source PLC entity to confirm the success or failure of the service associated corresponding MA_DATA.request.

Thus when confirmation_class = RQ or RS, the MA_DATA.confirm is passed immediately after the attempt to transmit the RQ or RS frame.

When confirmation_class = RR, the MA_DATA.confirm is passed:

(1) immediately if the RR frame cannot be transmitted, or
(2) when the requested RS frame is received, or
(3) when all predefined retries have been exhausted.

4.3.3.4 Effect of Receipt. The effect of receipt of this primitive by the local PLC entity is specified in Part 3B. Receipt of this primitive by the remote PLC has no effect.

4.3.3.5 Additional Comments. Success indicates that the requested M_sdu has been transmitted correctly to the best of the source MAC entity's knowledge.

In the case that the corresponding MA_DATA.request had a confirmation_class parameter specifying request_with_response, the MA_DATA.confirm is associated with the MA_DATA.indication conveying the response, if any such occurred.

5. The Medium Access Control (MAC) Sublayer

Part 5 specifies the Medium Access Control sublayer of the Data Link (Highway) layer of the ISO reference model.

Part 5A provides a description of the token bus medium access control mechanism including its operational and exception recovery functions. This part is intended to assist the reader in understanding the MAC sublayer and its operation. Where statements in this part conflict with Part 5B, 5C, 5D, Reference 4, Appendix I-A, or are incomplete, the other part shall take precedence.

Part 5B, MAC Sublayer Definitions and Requirements, contains precise definitions of MAC-specific terms and mandatory aspects of the medium access mechanism which are not included in Parts 5D or 5C.

Part 5C, ACM Formal Description, describes the required behavior of the access control machine of the MAC sublayer. The state tables that specify the ACM are given in Reference 4.

Part 5D defines the required MAC frame formats. This includes all allowed frame formats and the arrangement of all frame subfields.

The relationship of this part to other parts of this standard and to LAN specifications is illustrated in Figure 5-1.

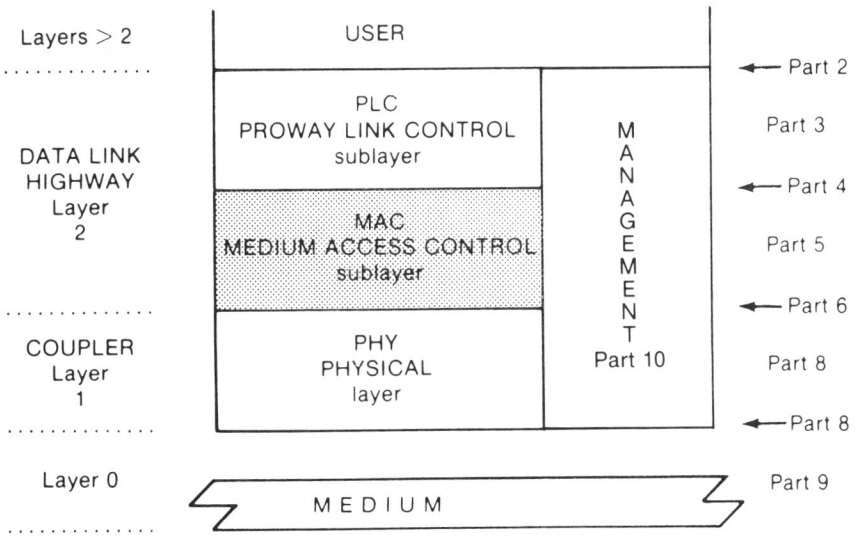

Fig 5-1
Relationship to LAN Model

5A. Informal Description of MAC Sublayer Operation

Specific responsibilities of the Medium Access Control sublayer for a broadcast medium involve managing ordered access to the medium, providing a means for admission and deletion of stations (adjustment of logical ring membership), and handling fault recovery.

The faults considered here are those caused by communications errors or station failures. These faults include the following:

(1) Multiple tokens
(2) Lost tokens
(3) Token-pass failure
(4) "Deaf" station (that is, a station with an inoperative receiver)
(5) Duplicate station addresses

This medium access protocol is intended to be robust, in the sense that it should tolerate and survive multiple concurrent errors.

Some basic observations are useful in understanding the operation of tokens on a broadcast medium.

(1) Stations are connected in parallel to the medium. Thus, when a station transmits, its signal is received (or "heard") by all stations on the medium. Other stations can interfere with the first station's transmission but cannot predictably alter its contents.
(2) When a station transmits, it may assume that all other stations hear something (though not necessarily what was transmitted).
(3) When a station receives a valid frame (properly formed and delimited and containing a correct frame check sequence), it may infer that some station transmitted the frame and, therefore, that all stations heard something.
(4) When a station receives something other than a valid frame (that is, noise), it may make no inference about what the other stations on the medium might have heard.
(5) Not all stations need to be involved in token passing (only those which desire to initiate transmissions).
(6) Multiple tokens and lost tokens may be detected by any station. There are no special "monitor" stations that perform token recovery functions.
(7) Due to spatial separation, stations cannot be guaranteed to have a common perception of the system state at any instant. (The medium access protocol described herein accounts for this.)

5A.1 Token Ring Steady State Operation. Steady state operation (the network condition where a logical ring has been established and no error conditions are present) simply requires the sending of the token to a specific successor station as each station is finished transmitting (see Figure I-2).

Other essential and more difficult tasks are establishing the logical ring (at initialization or re-establishing it in the case of a catastrophic error) and maintaining the logical ring (allowing stations to enter and leave the logical ring without disrupting the other stations in the network.

The right to transmit, the token, passes among all stations in the logical ring. Each participating station knows the address of its predecessor (the station that transmitted the token to it), referred to as Previous Station or PS. It knows its successor (which station the token should be sent to next), referred to as Next Station or NS. It knows its own address, referred to as This Station or TS. These predecessor and successor stations are dynamically determined and maintained by the algorithms described.

INDUSTRIAL DATA HIGHWAY

The following introduce major elements and features of the token bus access protocol.

NOTE: For the purpose of description, all state machines are presumed to be instantaneous with respect to external events.

5A.1.1 Slot_Time. In describing the access operations, the term slot_time is used to refer to the maximum time any station need wait for an immediate medium access level response from another station. Slot_time is precisely defined in 5B.1.10.

The slot_time (along with the station's address and several other station management parameters) must be known to the station before it may attempt to transmit on the network. If all stations in a network are not using the same value for slot_time, the medium access protocol may not operate properly. The method of setting these parameters in each station is outside the scope of this standard.

5A.1.2 Token Passing. The token (right to transmit) is passed from station to station in numerically descending, station-address order. When a station hears a token frame addressed to itself, it "has the token" and may transmit data frames.

When a station has the token it may temporarily delegate its right to transmit to another station by sending a request_with_response data frame. When a station hears a request_with_response data frame addressed to itself it must respond with a response data frame, if the request with response option is implemented. The response data frame causes the right to transmit to revert back to the station which sent the request_with_response data frame.

After each station has completed transmitting any data frames it may have and has completed other maintenance functions (described in 5A.1.3), the station passes the token to its successor by sending a "token" MAC_control frame.

After sending the token frame the station listens to make sure that its successor hears the token frame and is active. If the sender hears a valid frame following the token, it assumes that its successor has the token and is transmitting. If the token sender does not hear a valid frame following its token pass, it must attempt to assess the state of the network.

If the token-sending station hears a noise_burst or frame with an incorrect FCS, it cannot be sure from the source address which station sent the frame. The medium access protocol treats this condition in a way which minimizes the chance of the station causing a serious error. If a noise_burst is heard, the token-sending station sets an internal indicator and continues to listen in the check_token_pass state for up to four more slot times. If nothing more is heard, the station assumes that it heard its own token that had been garbled and so repeats the token transmission. If anything is heard during the following four slot time delay, the station assumes its successor has the token.

If the token holder does not hear a valid frame after sending the token the first time, it repeats the token pass operation once, performing the same monitoring as during the first attempt.

If the successor does not transmit after a second token frame, the sender assumes that its successor has failed. The sender then sends a who_follows frame with its successor's address in the data field of the frame. All stations compare the value of the data field of a who_follows frame with the address of their predecessor (the station that normally sends them the token). The station whose predecessor is the successor of the sending station responds to the who_follows frame by sending its address in a set successor frame. The station holding the token thus establishes a new successor, bridging the failed station out of the logical ring.

If the sending station hears no response to a who_follows frame, it repeats the frame a second time. If there is still no response, the station tries another strategy to re-establish the logical ring. The station now sends a solicit_successor_2 frame with its own address as both DA and SA, asking any station in the system to respond to it. Any operational station that hears the request and needs to be part of the logical ring responds, and the logical ring is re-established using the response window process discussed next.

If all attempts at soliciting a successor fail, the station assumes that a fault may have occurred; either all other stations have failed, all stations have left the logical ring, the medium has broken, or the station's own receiver has failed so that it cannot hear other stations who have been responding to its requests. Under such conditions the station quits attempting to maintain the logical ring. If the station has no frames to send, it listens for some indication of activity from other stations. If the station has data frames to send, it sends its remaining data frames and then repeats the token pass process. Once the station has sent its frames and still cannot locate a successor it will become silent, listening for another station's transmissions.

In summary, the token is normally passed from station to station using a short token pass frame. If a station fails to pick up the token, the sending station uses a series of recovery procedures that grow increasingly more drastic as the station repeatedly fails to find a successor station.

5A.1.3 Adding New Stations to the Token Ring. New stations are added to the logical ring through a controlled contention process using "response windows". A response window is a controlled interval of time (equal to one slot_time) after transmission of a MAC control frame in which the station sending the frame pauses and listens for a response. If the station hears a transmission start during a response window, the station continues to listen to the transmission, even after the response window time expires, until the transmission is complete. Thus the response windows define the time interval during which a station must hear the beginning of a response from another station.

The two frame types, solicit_successor_1 and solicit_successor_2, indicate the opening of response windows for stations wishing to enter the logical ring. The solicit_successor frame specifies a range of station addresses between the frame source and destination addresses. Stations whose addresses fall within this range and who wish to enter the logical ring respond to the frame.

The sender of a solicit_successor frame transmits the frame and then waits, listening for a response in the response window following the frame. Responding stations send the frame sender their set_successor frame which requests inclusion in the logical ring. If the frame sender hears a valid request, it allows the new station to enter the logical ring by changing the address of its successor to the new station and passing its new successor the token.

In any response window there exists the possibility that more than one station will simultaneously desire logical ring entry. To minimize contention when this happens, the token pass sequence is limited by requiring that a station only request admission when a window is opened for an address range that spans its address.

There are two solicit_successor frames. Solicit_successor_1 has one response window following. Solicit_successor_2 has two response windows. Solicit_successor_1 is sent when the station's successor's address is less than the station's address. This is the normal case when the token is being passed from higher to lower addressed stations. Solicit_successor_1 allows only stations whose address is in the range between the token sender and the token destination to respond, thus limiting the possible contenders and preserving the descending order of the logical ring.

INDUSTRIAL DATA HIGHWAY

Exactly one station in the logical ring has its station's address below that of its successor, that is, the unique station having the lowest address which must send the token to the "top" of the address-ordered logical ring. When soliciting successors, this station must open two response windows, one for those stations with addresses below its own, and one for stations with addresses above its successor. The station with the lowest address sends a solicit_successor_2 frame when opening response windows. Stations having an address below the sender respond in the first response window, while stations having an address higher than the sender's successor respond in the second response window.

In any response window, when the soliciting station hears a valid set_successor frame, it has found a new successor. When multiple stations simultaneously respond, only unrecognizable noise may be heard during the response period. The soliciting station then sequences through an arbitration algorithm to identify a single responder, by sending a resolve_contention frame. The stations which had responded to the earlier solicit_successor frame and which have not yet been eliminated by the iterative resolve responders algorithm, choose a two-bit value from the station's address and listen for 0, 1, 2, or 3 slot_times as determined by that listen delay value. (This listen delay value is further described later.) If these contending stations hear anything (that is, non-silence) while listening, they eliminate themselves from the arbitration. If they hear only silence, they continue to respond to further resolve_contention requests from the soliciting stations.

NOTE: By knowing and controlling the frequency with which response windows are opened and by virtue of the finite length of the response resolution algorithm, hard bounds on the access delay can always be calculated (determinism). For 16-bit addresses, the response resolution cycle needs to be run a maximum of nine times (16/2+1, for 16 address bits taken two bits at a time, plus one pair of "random" bits).

5A.1.3.1 Bound on Token Rotation Time. A maximum token rotation time for ring maintenance is established with the ring maintenance timer, similar to the rotation times established for the differing access classes of data transmissions, as described in 5A.1.5. If the token appears to rotate slower than the time established for this ring maintenance timer, a station defers the solicit successor procedure until a later token pass. When the network is less heavily loaded on the next or succeeding token pass, the station performs the ring maintenance function of soliciting new successors.

The ring maintenance timer gives station management control over whether the station solicits successors immediately on entering the token ring or defers for one (or more) token pass(es). The ring_maintenance_initial value is set into the ring maintenance timer when the station enters the ring. If this value is large, the station will not find the timer expired and will solicit successors immediately. If this value is zero, the station will find the timer expired and will pass the token.

When a station gets the token, it services the four access_class data queues and then performs ring maintenance. If the inter_solicit_count is zero, the station should solicit its successors. The do_solicit_successor arc is taken if the ring maintenance token rotation timer has not expired. If either the timer has expired or the inter_solicit_count is not yet zero, the do_pass_token arc is taken and the token is simply passed.

5A.1.3.2 User Notification of Ring Membership. It is necessary for the user of PROWAY to have access to a list of other stations which are active in the logical token ring. The creation of this live list is performed by station management as described in Part 10A.

Whenever a station changes its successor in the logical ring, the station's own station management entity will be notified. That station management entity will then read the new successor's address (NS) and inform other stations when appropriate.

5A.1.4 Token Ring Initialization. Initialization is primarily a special case of adding new stations; it is triggered by the exhaustion of an inactivity timer (bus_idle) in one station. If the inactivity timer expires, the station sends a claim_token frame. As in the response window algorithm, the initialization algorithm assumes that more than one station may try to initialize the network at a given instant. This is resolved by address sorting the initializers.

Each potential initializer sends a claim_token frame having an information field length that is a multiple of the system slot time (the multiple being 0, 2, 4, or 6 based on selected bits of the station address). Each initializing station then waits one slot time for its own transmission, and for other stations that chose the same frame length, to pass. The station then samples the state of the medium.

If a station senses non-silence, it knows that some other station(s) sent a longer length transmission. The station defers to those stations with the longer transmission and re-enters the idle state.

If silence was detected and unused bits remain in the address string, the station attempting initialization repeats the process using the next two bits of its address to derive the length of the next transmitted frame. If all bits have been used and silence is still sensed, the station has "won" the initialization contest and now holds the token.

Once there is a unique token in the network, the logical ring builds by way of the response window process previously described.

NOTE: A random pair of bits is used at the end of the address sort algorithm to ensure that two stations with the same address (which is a fault condition) will not permanently bring down an entire system. If the two stations don't separate (random choices identical), they both attempt to form a logical ring, and at most one of them will succeed. If they do separate (random choices are different), one will get in. In the latter case, the station which doesn't get in will hear a transmission from a station with an identical address and so will discover the error condition.

5A.1.4.1 Leaving the Token Ring. A station may remove itself from the logical ring at any time by waiting for the token, then sending a set_successor frame to its predecessor on the logical ring with the address of its successor. The exiting station then sends the token as usual to its successor. Re-admission to the logical ring requires one of the sequences described in 5A.1.3 and 5A.1.4.

5A.1.5 Priorities. The token-passing access method provides a priority mechanism by which higher layer data frames awaiting transmission are assigned to different "service classes", ranked or ordered by their desired transmission priority. The priority mechanism allows the MAC sublayer to provide four service classes to the PLC sublayer, and higher level protocols. The priority of each frame is determined by the "priority" specified in the request command to MAC.

The token bus access method distinguishes only four levels of priority, called "access classes". Thus there are four request queues to store frames pending transmission. The access classes are named 0, 2, 4, and 6, with 6 being the highest priority and 0 the lowest.

MAC maps the two most significant bits of the priority class requested by the PLC sublayer into a two bit priority value, which is included in the frame format field. The priority value is then mapped into the MAC access class by ignoring the least significant bit in the priority field. Thus service classes 0 and 1 correspond to access class 0, service classes 2 and 3 to access class 2, service classes 4 and 5 to access class 4, and service classes 6 and 7 to access class 6.

The service class value in the request to MAC shall be carried in the FC octet. For all stations, the rule governing the transmission of highest priority frames is that a station may not transmit consecutive frames for more than some maximum time set by station management. This time, called the hi_pri_token_hold_time, prevents any single station from monopolizing the network. If a station has more access class 6 data frames to send than it can transmit in one hi_pri_token_hold_time period, it is prohibited from sending additional frames, except for retries, after that time has expired. It then completes any required retries and must pass the token.

A station must complete all retries before passing the token to prevent the introduction of duplicate frames. For example, the remote station could correctly receive a frame but the "ACK" frame might be lost. If the local station terminated without retrying and receiving an ACK, then the remote station would have received and processed a frame that the local station thought was lost. On the next token round, the local station would most likely retry the original transmission and a duplicate frame would be created. For a more complete discussion of lost and duplicate frames see Appendix 3-A.

When a station has lower access class frames to send and has time available, it may only send these frames subject to the priority system rules described in these paragraphs.

The object of the priority system is to allocate network bandwidth to the higher pricrity frames and only send lower priority frames when there is sufficient unused bandwidth. The network bandwidth is allocated by timing the rotation of the token around the logical ring. Each access class is assigned a "target" token rotation time. For each access class the station measures the time it takes the token to circulate around the logical ring.

If the token returns to a station in less than the target rotation time, the station can send frames of that access class until the target rotation time has expired. If the token returns after the target rotation time has been reached, the station cannot send frames of that priority on this pass of the token.

Each station shall have three rotation timers for the three lower access classes. Each access class has a queue of frames to be transmitted. When a station receives the token it first services the highest access class queue, which uses the hi_pri_token_hold_time to control its operation. After having sent any frames of the highest priority, the station begins to service the rotation timers and queues, working from higher to lower access classes.

Each access class acts as a virtual substation in that the token is passed, internally, from the highest access class downward, through all access classes, before being passed to the station's successor.

The access class service algorithm consists of loading the residual value from a token rotation timer into a "token hold timer" and reloading the same rotation timer with the target rotation time for that access class. (Thus frames sent by a station, for this access class, are accounted for in the access class's next token rotation time computation.) If the "token hold timer" has a remaining positive value, the station can transmit frames at this access class until either the "token hold timer" times out or this access class's queue is empty. When either event occurs the station begins to service the next lower access class.

In all cases the station completes any required retries before moving to the next lower access class. When the lowest level is serviced the station performs any required logical ring maintenance and passes the token to its successor.

5A.1.6 MAC Confirmed Data Transfer Service. Immediate response procedures, coupled with appropriate PLC procedures, provide confirmed data exchanges. When a higher layer entity requests a confirmed data transmission, the PLC entity issues a MA_DATA.request with a request_with_response frame to the local MAC entity.

When the local MAC entity obtains the token, it sends the request_with_response frame and waits for a response frame. If a timer expires without a valid response, the local MAC entity will transmit the original frame. This is repeated until a response is received or the allowed number of retries is exhausted.

When the remote (responding) MAC entity receives the request_with_response frame, it passes the received frame to the remote PLC entity. The remote PLC entity generates an appropriate response and directs the remote MAC entity to send this response frame at once to the source of the request_with_response frame (i.e., the local station).

When the local (originating) MAC entity receives the response frame or when the retry count is exhausted, it associates the frame (if available) with the request_with_response frame now being processed and notifies the originating PLC entity of the completion of its original request.

The retry mechanism of the request_with_response procedure prevents loss of frames at any arbitrary level of confidence determined by the maximum_retry_limit. However, since the local station repeats the request_with_response frame when a response frame is not received, it is possible for the remote station to receive duplicates of the original request_with_response frame. The remote PLC entity eliminates duplicate frames using the procedures given in Part 3B. A more complete explanation of the causes for and prevention of loss and duplication is given in Appendix 3-A.

The local (originating) MAC engages in one request_with_response activity at a time. All retries and timeouts for that request are completed before the local MAC processes another request or passes the token.

5A.1.7 Randomized Variables. Several of the variables used by the medium access protocol have two-bit "random" values. Some of these randomized variables are used to improve error recovery probabilities under certain conditions. The randomization of max_inter_solicit_count forces stations to operate "out of step" when opening response windows.

5A.2 Access Control Machine (ACM) States. The medium access logic in a station is described here as a computation machine which sequences through a number of distinct phases, called states. These states are introduced in the following clauses. The states and transitions between them are illustrated in Figure 5-3. (The dashed lines group states into functional areas.) Part 5C contains the complete state transition table which provides a formal description of the token-passing bus access machine.

5A.2.0 Offline. "Offline" is the state the access machine is in immediately following power-up or following the detection, by the MAC sublayer, of certain fault conditions. After powering up, a station tests itself and its connection to the medium without transmitting on the medium. This "internal" self-testing is station implementation dependent and does not affect other stations on the network. Thus, the self-test procedure is beyond the scope of this standard.

INDUSTRIAL DATA HIGHWAY

ISA
S72.01-1985

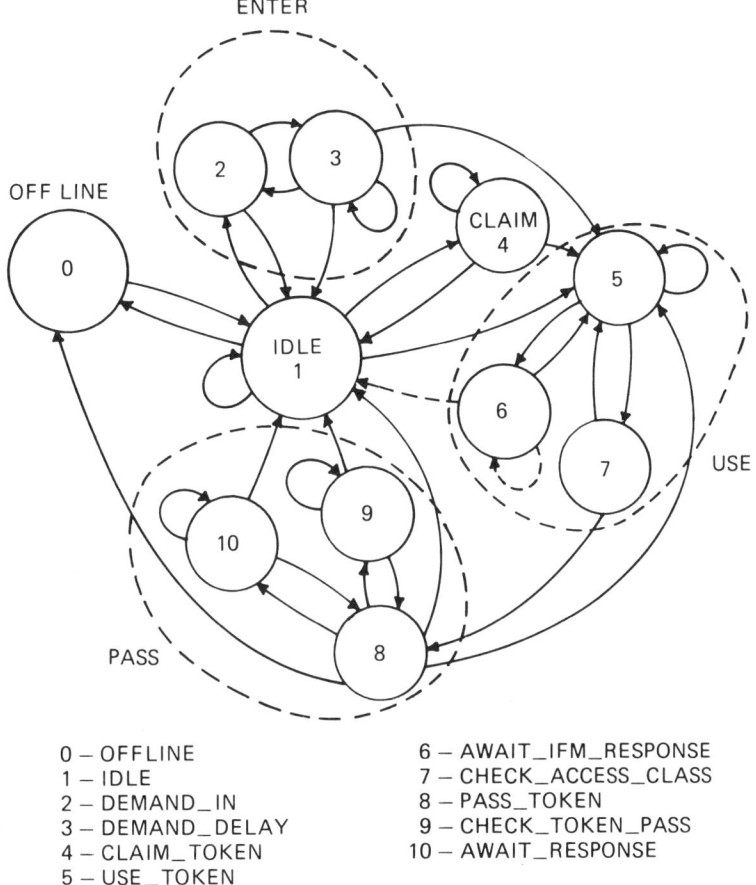

**Fig 5-3
MAC Finite State Machine Diagram**

After completing any power-up procedures, the station remains in the offline state until it has had all necessary internal parameters initialized and been instructed to go online.

5A.2.1 Idle. "Idle" is the state where the station is listening to the medium and not transmitting.

If a MAC_control frame is received for which the station must take action, the appropriate state is entered. For example, if a token frame, addressed to the station, is received, the station enters the "use token" state.

If the station goes for a long period of time (a defined multiple of the slot_time) without hearing any activity on the medium, it may infer that recovery of the logical ring is necessary. The station attempts to claim the token (enters the "claiming token" state) and (re)initialize the logical ring.

5A.2.2 Demand In. The "demand-in" state is entered from the idle state if a solicit_successor frame that spans the station's address is received by a station desiring logical ring entry. (The demand-in state is also entered from the demand-delay state during the contention resolution process discussed in 5A.1.3.) In the demand-in state the contending station sends the token holder a set_successor frame and goes to the demand-delay state to await a response.

If a station intends to respond to a solicit_successor frame or a who_follows in the first response window, the station enters the demand-in state from the idle state with a zero delay and so immediately transmits a set_successor response and goes to the demand-delay state. If the station intends to respond in the second response window after a frame or is participating in the contention resolution process, the station delays in the demand-in state before transmitting a set_successor frame.

While delaying in the demand-in state, if the station hears any transmissions, it must assume that another station with a higher numbered address is requesting the token and so it must return to the idle state.

5A.2.3 Demand Delay. "Demand-delay" is the state the station enters after having sent a set_successor frame in the demand-in state. In the demand-delay state a station can expect to hear:

(1) A token from the token holder indicating its set_successor frame was heard,
(2) A resolve_contention frame from the token holder indicating that all stations which are still demanding into the logical ring should perform another step of the contention resolution process, or
(3) Set_successor frames from other stations, which the station ignores.

If the station either hears nothing or hears a frame other than one of the above, the station must leave the demand-delay state. The station then abandons soliciting the token and returns to the idle state.

In case 1 above, the token holder has heard the soliciting station and sent the token. The contention resolution process is over. The soliciting station upon receiving the token goes to the use token state and begins transmitting.

In case 2, the token holder has heard responses from multiple stations soliciting the token and sent a resolve_contention frame. All stations currently in the demand-delay state respond to this frame. The responding stations first set a delay and return to the demand-in state where they listen for other contenders. If no other contenders have been heard by the time the delay period has expired, the station then sends another set_successor frame to the token holder.

The contention for the token by multiple stations is resolved by having each station delay a period of time in the demand-in state before transmitting another set_successor frame. The delay interval is chosen by taking the station's (unique) address and using two bits from that address to determine the delay interval. The first pass of the resolution process uses the most significant two address bits; the next pass the next two address bits; etc. Thus stations will delay 0, 1, 2, or 3 slot times when entering the demand-in state before transmitting.

When multiple stations request to enter the logical ring, the desired result is for the token holding station to pass the token to the highest addressed station. In order to select the highest addressed contending station from multiple contenders, the one's complement of the station's address is used to determine the delay in the demand-in state. Thus, stations with numerically higher addresses delay shorter intervals and send their set_successor messages sooner than stations with lower addresses. Stations with numerically lower addresses hear the transmissions of the higher addressed stations and drop out of the contention process.

If two contending stations have the same value for the selected two address bits, they delay the same amount of time and transmit more or less simultaneously. If the token holder hears multiple responses and does not hear a valid set_successor from any station, it sends another resolve_contention frame, starting another step of the contention resolution process.

The contention resolution process can take at most nine passes (16/2+1) for 16-bit addresses of the following cycle:

(1) All remaining demanding stations send set_successor frames to the token holder.
(2) They all listen for a response from the token holder and ignore other set_successor frames.
(3) They all hear a resolve_contention frame from the token holder.
(4) They all delay a number of slot times based on the next two bits of their own addresses.
(5) If they hear another frame during the delay, they drop from contention.

The contention resolution process should resolve so that the contending station with the highest address is heard by the token holder and receives the token. However, if two stations are erroneously assigned the same station address, they will both sequence through the contention process using the same delays and may not resolve.

To permit eventual resolution in this error condition, a final resolution pass is taken, using a two-bit random number, after the station's address bits have all been used and the contention is still unresolved. If both stations choose the same random value or another error prevents resolution, then the token holder and the contending stations abandon the resolution process until the next response window is opened.

(Thus, two stations with the same address which consistently choose the same random number value may never be able to enter the logical ring.)

5A.2.4 Claim Token. The "claim_token" state is entered from the idle state after the inactivity timer expires (and the station desires to be included in the logical ring). In this state, the station attempts to initialize or reinitialize the logical ring by sending claim_token frames.

To resolve multiple simultaneous stations sending claim_token frames, each station delays for one slot_time after sending the claim_token frame, and then monitors the medium as previously described. If after this delay the bus is quiet, the station sends another claim_token frame.

If the station sends max_pass_count (where max_pass_count is half the number of bits in the station's address plus one, since the address bits are used in pairs) claim_token frames without hearing other transmissions, the station has successfully "claimed" the token, and goes to the use-token state.

5A.2.5 Use_Token. Use_token is the state the station enters after just receiving or claiming a token. This is the state in which a station can send data frames.

Upon entering this state just after receiving or claiming a token, the station starts the "token holding" timer, which limits the amount of time the station may begin transmission of frames before passing to the next access class. The expiration of the token holding timer does not prevent retries, and all retries must be completed before passing to the next access class. The value initially loaded in the holding timer, hi_pri_token_hold_time, is a system-imposed parameter.

After each data frame is sent, the ACM enters the await_IFM_response state. It will return to the use-token state as soon as the confirmation requirements of the frame last sent are satisfied.

When a station's token holding timer has expired and all required retries are initiated and any transmission in progress is complete, or when the station has no more to transmit, it enters the check_access_class state.

When a station sends a data frame it sets the response_window_timer to 3 slot times and enters the await_IFM_response state.

5A.2.6 Await_IFM_Response. This state is entered when a data frame has been sent. The ACM may wait for the Interface Machine (IFM) to signal the reception of a response.

If the frame sent in the use-token state was a request_with_no_response frame, no response is expected. The use-token state is again entered to check for another frame or holding timer timeout. If the frame sent was a request_with_response frame, the station waits in the await_IFM_response state for one of the following:

(1) A response frame addressed to the requestor
(2) Any other valid frame
(3) A timeout

If a response frame addressed to the requestor is heard, the use_token state is again entered to check for another frame or holding-timer timeout. (The IFM passes the response frame to the PLC entity as it does all other data frames addressed to the station. The IFM also associates the response frame with the request_with_response frame just previously transmitted.)

If any other valid frame is heard, an error has occurred. The station returns to the idle state and processes the received frame.

If a timeout occurs before a valid frame is heard, the station repeats sending the request_with_ response data frame. If the station repeats sending that frame the number of times specified by the maximum_retry_limit, a parameter established by station management, then the request is abandoned, and the IFM notifies the PLC entity that no response to the frame was received. The use-token state is entered to check for another frame or token-holding-timer timeout.

5A.2.7 Check_Access_Class. "Check_access_class" controls the transmission of frames for different access classes. If the priority option is not implemented, all frames are considered to be high priority and the check_access_class state only serves to control entry to token passing.

A station may send frames for lower access classes before passing the token. Each access class other than the highest has a target token rotation time. At the time that the station has the token and begins to consider transmitting frames for that access class, the residual time left in the target rotation timer is loaded into the token holding timer and the station transitions back to the use_ token state. At this time the target rotation timer is also reloaded to its initial value.

Thus the station will alternate between use_token and check_access_class states for each access class. If time is available, data frames will be sent in the use_token state. When the lowest priority access class has been checked, the station will proceed to the pass_token process, described next.

After a station has completed sending data frames, it must enter the token passing state. Three conditions can occur:

(1) The station knows its successor, so simply passes the token and enters the check_token_pass state.
(2) The station knows its successor but must first check if new stations desire to enter the logical ring. The station sends a solicit_successor frame and enters the await_response state.
(3) The station does not know its successor. (This condition occurs after initialization and under error conditions.) The station sends a solicit_successor frame, opening response windows for all stations in the system. and enters the await-response state.

5A.2.8 Pass_Token. "Pass_token" is the state in which a station attempts to pass the token to its successor.

Before passing the token, the station will, when its inter_solicit_count value is zero and the time remains on its ring_maintenance timer, allow new stations to enter the logical ring. The token holding station does this by sending a solicit_successor_1 or solicit_successor_2 frame, as appropriate, and enters the await-response state. (See the description of response windows for the details of this operation.

Following any new successor solicitation if the address of the successor, NS, is known, the station performs a simple token pass. (See the description of token passing for details of this operation.) If the successor responds and the station hears a valid frame, the station has completed its token passing obligations.

If NS is not known, the station sends a solicit_successor_2 frame to itself. Since this frame has two response windows and identical source and destination addresses, it forces all stations on the network which desire to be in the logical ring (whether or not they were previously) to respond. Those stations whose addresses are lower than the sender of the token frame transmit in the first window; those with addresses higher transmit in the second.

The station monitors the response windows for set_successor frames from potential successors, exactly as for the token pass. If no responses are heard, the station stops trying to maintain the logical ring and listens for transmissions from any other station. (See the description of token passing for details of this operation.)

5A.2.9 Check_Token_Pass. "Check_token_pass" is the state in which the station waits for a reaction from the station to which it just passed the token.

The station sending the token waits one slot time for the station receiving the token to transmit. The one slot time delay accounts for the time delay between receiving a frame and the recipient taking the response action.

If a valid frame is heard which started during the response window, the station assumes the token pass is successful. The frame is processed as if it were received in the idle state.

If nothing is heard in one slot time, the station sending the token assumes the token pass was unsuccessful and returns to the pass_token state to either repeat the pass or try another strategy.

If noise or an invalid frame is heard, the station will continue to listen for additional transmissions, as described in detail in 5A.1.2.

5A.2.10 Await_Response. In the await_response state the station attempts to sequence candidate successors through a distributed contention resolution algorithm until one of those successors' set_successor frames is correctly received or until no successors appear. The state is entered from the pass_token state whenever the station determines it is time to open a response window or if the station does not know its successor (as in initialization or when a token pass fails).

The station waits in the await_response state for a number of response window times. If nothing is heard for the entire duration of the window(s) opened, the station goes to the pass_token state, either to pass the token to its known successor or to try a different token-pass strategy.

If a set_successor frame is received, the station waits for the rest of the response window time to pass. The station then enters the pass_token state and sends the token to the new successor.

If the received frame is other than a set_successor frame, the station drops the token (since someone else must have it to be able to send any other frame type, thus a duplicate token situation) and re-enters the idle state.

If noise is heard during the response windows, the station cycles through a procedure of sending resolve_contention frames which open four response windows each, and waiting for a distinguishable response that began in a response window. The loop repeats a maximum of max_pass_count times, each time instructing contending stations to select a different two bits of their address to determine which of the four opened windows to transmit in.

5B. MAC Sublayer Definitions and Requirements

Part 5B specifies mandatory aspects of the MAC sublayer operation and mechanism which are not specified in Parts 5C or 5D. All specifications in Part 5B are required for conformance to this standard.

5B.1 MAC Definitions. The following paragraphs define critical MAC parameters, which are constrained by this specification.

5B.1.1 Immediate_response. Immediate_response is defined as the immediate transmission of a response to a received frame. This assumes that no other transmissions or actions were intervening.

5B.1.2 MAC-symbols. MAC-symbols are defined as the smallest unit of information exchanged between MAC sublayer entities. The six MAC-symbols are as follows:

Name	Abbreviation
zero	0
one	1
non_data	N
pad_idle	P
silence	S
bad_signal	B

Where binary 0 and 1 data bits are discussed, they are sent and received as **zero** and **one** MAC-symbols, respectively.

5B.1.3 MAC_symbol_time. MAC_symbol_time is the time required to send a single MAC_symbol. This is the inverse of the LAN data rate.

Nominal Data Rate	Nominal MAC_symbol_time
1 M bits/sec	1 microsec
5 M bits/sec	200 nanosec
10 M bits/sec	100 nanosec

5B.1.4 Octet_time. The term octet_time shall be understood to correspond to the time interval required to transmit eight (8) MAC-symbols.

5B.1.5 PHY-symbols. Physical layer symbols (PHY-symbols) correspond to the waveforms impressed on the physical medium. See Section 8.2.1.1 for the PHY-symbol definitions).

5B.1.6 Transmission_path_delay. Transmission_path_delay is the worst case delay which transmissions experience going through the physical medium from a transmitter to a receiver. The following formula is a general definition of transmission_path_delay:

Transmission_path_delay = worst_case physical_medium_delay

NOTE: Refer to Section 9.7 for a detailed discussion of transmission_path_delay.

5B.1.7 MAC_station_delay. MAC_station_delay is the time from the receipt of the PHY-symbols corresponding to the last MAC-symbol of the received ED at the receiving station's physical medium interface to the impression of the first immediate_response PHY-symbols onto the physical medium by that station's transmitter when the frame is a MAC frame.

5B.1.8 PLC_station_delay. PLC_station_delay is the time from the receipt of the PHY-symbols corresponding to the last MAC-symbol of the received ED at the receiving station's physical medium interface to the impression of the first immediate_response PHY-symbols onto the physical medium by that station's transmitter when the frame is a request_with_response data frame.

PLC_station_delay shall be $\leq$ 16 microseconds.

5B.1.9 Safety_margin. Safety_margin is defined as a time interval no less than one MAC_symbol_time.

Safety_margin $\geq$ MAC_symbol_time

5B.1.10 Slot_time. Slot_time is the maximum time any station need wait for an immediate_response from another station. Slot_time is measured in octet_times, and is defined as

Slot_time = INTEGER ({ [2*(Transmission_path_delay + MAC_station_delay)
        + Safety_margin]/MAC_symbol_time
        + 7}/8)

5B.1.11 Response_time. Response_time is the anticipated time any station need wait for a response_frame from another station. Response_time is defined as

Response_time = slot_time + PLC_station_delay + nominal preamble

The response_time determines the maximum amount of time the ACM may spend in state six (await_IFM_response) as shown in Figure 5-3.

5B.1.12 Response_window. A response_window is the basic time interval which the MAC protocol allows, following certain MAC_control frames, for an immediate_response from another station. This interval is one slot_time long.

Response_window duration = slot_time

If a station, waiting for a response, hears a transmission start during a response_window, that station shall not transmit again at least until the received transmission terminates.

5B.1.13 Maximum_retry_limit. This limit is the maximum number of retries a MAC sublayer will make in attempting to secure a response to a request_with_response frame.

5B.2 Transmission Order. The frame formats used by the MAC sublayer and the detailed contents of the octets of those frames are specified in Part 5D. The octets of a frame and the MAC-symbols of an octet shall be transmitted from the PLC to the MAC sublayer, and from the MAC sublayer to the Physical layer, and vice versa, in the order specified in Figure 5-7; the first octet of the frame shall be transmitted first and the first MAC-symbol of each octet shall be transmitted first. The first octet and first MAC-symbol correspond to the top octet and the left-most MAC-symbol shown in Figure 5-7. Those notations within octets of Figure 5-7 which are not MAC-symbols are casual descriptions.

Figure 5-7 describes a complete general frame or MAC protocol_data_unit (M_pdu) containing a PLC protocol_data_unit (L_pdu).

5B.3 Delay Labeling. Vendors shall provide a worst case value for MAC and PLC station delay times. Vendors must also specify a minimum network slot time when their equipment anticipates some minimum delay in order to function correctly. When uncertain of the exact value of the delay, vendors shall state an upper bound. Vendors of equipment conforming to this standard shall label the equipment with that equipment's contribution to the station delay. A vendor of a complete station would label the station with the total station delay. A vendor of a component, intended to be assembled by an end user into a station, would label the component or otherwise document the delays that contribute to station delay.

5B.4 Miscellaneous Requirements
5B.4.1 Station Initialization. On power-up, the station shall enter the offline state. While in the offline state the station shall not impress any signaling on the LAN medium.

The station shall progress from the offline state to the idle state only when it has been loaded with the basic station operating parameters needed for correct operation of the MAC protocols. These operating parameters include at least the following:

(1) TS (station address)
(2) address_length (implicit in TS)
(3) slot_time
(4) hi_pri_token_hold_time
(5) max_ac_4_rotation_time
(6) max_ac_2_rotation_time
(7) max_ac_0_rotation_time
(8) max_inter_solicit_count
(9) max_ring_maintenance_rotation_time
(10) ring_maintenance_timer_initial_value
(11) maximum_retry_limit
(12) in_ring_desired
(13) min_post_silence_preamble_length

5B.4.2 Token-Passing Order. The token shall be passed from station to station in numerically descending station address order, except that the station with the lowest address shall pass the

INDUSTRIAL DATA HIGHWAY

ISA
S72.01-1985

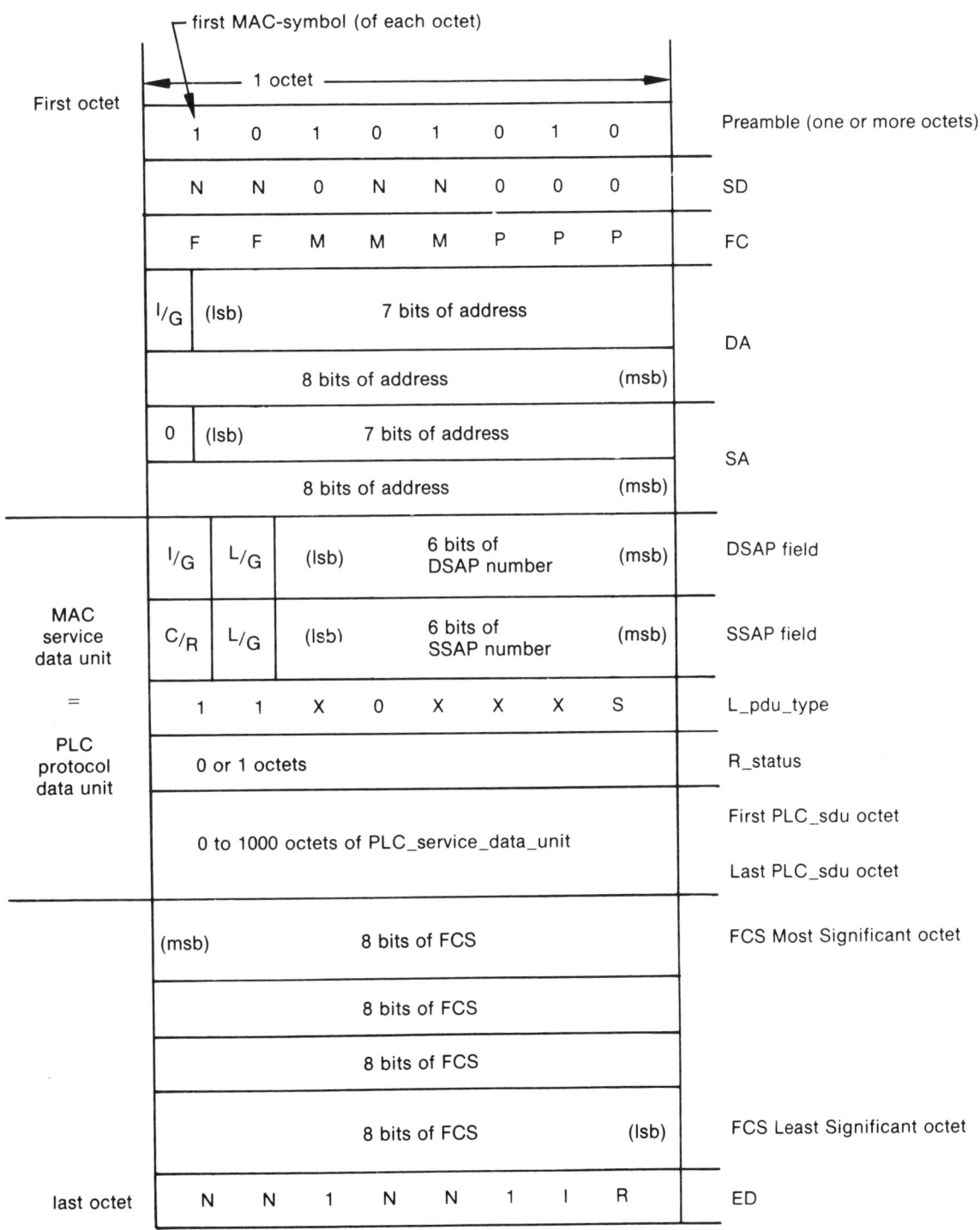

**Fig 5-7
PROWAY PLC and MAC Protocol_Data_Unit
Composition and Transmission Order with 16-Bit Address**

token to the station with the highest address, in order to close the logical ring. Figure 5-8 illustrates the address-ordered logical ring and shows the logical relationships which hold between addresses of adjacent stations in a logical ring with three or more members.

5B.4.3 Station Receipt of Its Own Transmission. In systems with significant transmission_path_ delay, a transmitting station may receive its own transmissions after some small but significant delay. The MAC access mechanism of such a transmitting station shall not be misled by the receipt of its own transmissions. The state diagrams in Part 5C specify where a station should ignore its own transmissions.

5B.4.4 Token Holding Time. The token holding station shall only begin transmitting a non-retry frame when there is time remaining on the token_hold_timer. A transmission may run past the expiration of the token_hold_timer by up to the time necessary to transmit a maximum length data frame. Waiting for any response frame and any required retries may also cause a token holding station to exceed the token_hold_timer.

5B.4.5 Address Lengths. Addresses shall be two octets (16 bits) or optionally 6 octets (48 bits) long. Address composition is specified in Part 5D, and examples are given in Appendix 5-B. Note that use of 48-bit addresses may reduce system throughput and responsiveness.

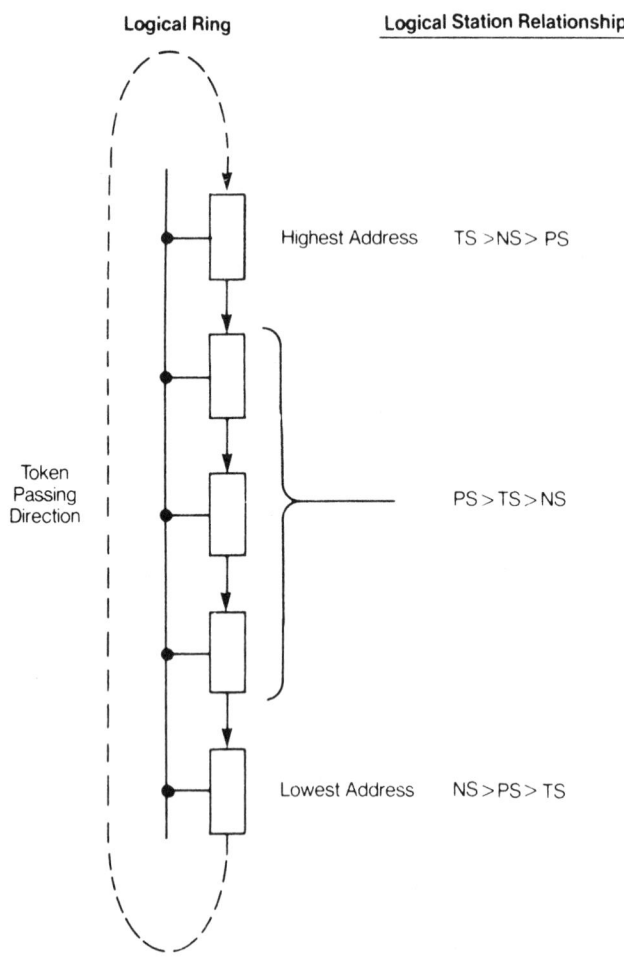

**Fig 5-8
Logical Token-Passing Ring**

5B.4.6 Randomized Variables. The station must provide a two-bit (that is, four valued) random variable for use in the MAC protocols. For the medium access protocol to benefit from the randomization, the technique used to create the random values must be statistically independent between stations. Thus random number generators tied in any way to the network data clock, for example, would not produce statistically independent variable values.

The variables shall be re-randomized "periodically". Periodically shall be interpreted to mean either an interval not to exceed 50 milliseconds, or every use of the random variable.

5B.4.7 Contention Delay. If the station hears a solicit_successor or who_follows frame, it determines which response window in which to contend based on the station's address and the SA/DA addresses in the frame. If the station wants to contend in the first window, it loads the contention timer with zero, so the station proceeds immediately to the demand_in state. If the station wants to contend in the second window, the contention timer is loaded with one, so the station listens during the first window.

Following receiving a resolve_contention frame, if the station is contending, it loads the contention timer with the one's complement of two bits selected from its own address as indexed by the resolution pass count. The station thus listens zero, one, two, or three slot times before again contending.

5B.4.8 Token Claiming. If the bus_idle_timer expires, a station may transmit a claim_token frame and set the claim_timer. When the claim_timer expires, if no transmissions are present at that instant, the station sends an additional claim_token frame and repeats the delay and transmission check. This procedure is repeated until either transmissions from another station are heard or the value of the claim_pass_count equals max_claim_pass_count.

The length of the claim_token frames are 0+, 2+, 4+, or 6+ slot_times as a function of two bits of the station's address. Indexing through the address performs an address sort in the claim process, leaving the station with the highest address claiming the token.

5B.5 Use of Address Bits in Contention Algorithms. The contention processes used to claim a new token or demand logical ring entry both use the bits of the station address to accomplish a sorting-like resolution in which the station having the numerically largest or highest address value wins. The following paragraphs treat the address as an array of binary (0/1) values or address bits; for notational purposes, "address(i)" indicates the ith binary bit of the station's address, with address(1) the most significant bit. These address_bits are used two at a time, starting with the most significant address_bits.

5B.5.1 Claim_token Frame Length. A station which is attempting to claim a new token first determines that no other station is transmitting, then transmits a claim_token frame containing a data_unit with a length equal to 0, 2, 4, or 6 slot_times. It then waits or delays one slot_time before again listening for other transmissions. The token claiming contention process shall consist of N cycles of listening, transmitting and delaying, where N is a function of the station's address length in bits:

N = (address_length / 2) + 1

The length, L, of the nth claim_token frame's data_unit, in octet_times (for the nth cycle of the token claiming process), shall be determined as follows:

for $1 \leq n < N$
        L := 2*slot_time*bit_value
   bit_value := (2*address_bit((2*n)-1) + address_bit(2*n))

where bit_value equals 0, 1, 2, or 3 as a function of the two bits used in cycle n.

for $n = N$
$$:= 2*\text{slot\_time}*\text{random\_4}$$

where random_4 = a random number equal to 0, 1, 2 or 3.

5B.5.2 Demand Delay Time Interval. A station which is demanding entry into the logical ring first listens for other transmissions, delaying its next transmission for 0, 1, 2 or 3 slot_times; then, if it has heard no other transmissions, it transmits a fixed length set_successor frame. This delay preceding set_successor frame transmissions is called the demand_delay. The contention process for demanding logical ring entry shall consist of at most N cycles of transmission and listening delays, where N is a function of the address length in bits:

N = (address_length / 2) + 1

The number of slot_times, D, to delay after the nth transmission (for the nth cycle of the contention resolution process) shall be determined as follows:

for $1 \leq n < N$
$$D := 2*\text{slot\_time}*\text{bit\_value}$$
$$\text{bit\_value} := (2*\text{address\_bit}((2*n)-1) + \text{address\_bit}(2*n))$$

where bit_value equals 0, 1, 2, or 3 as a function of the two bits used in cycle n.

for $n = N$
$$D := \text{random\_4}$$

where random_4 = a random number equal to 0, 1, 2, or 3.

5B.6 Priority of Transmitted Frames

5B.6.1 Access Classes. The priority mechanism shall provide four levels of service with respect to a frame's priority of access to the medium; these levels are called access_classes. The access_classes shall be identified as 0, 2, 4 and 6, and access_class 6 shall be the highest priority or most favored level of service.

5B.6.2 Priority to Access_Class Mapping. The PLC request priority contained in the request to MAC is eventually satisfied through use of the access_classes. The priority request shall first be satisfied by assignment of the request to one of eight MAC service_classes. These MAC service_classes shall then be mapped into MAC access_classes as described in the following table:

service_class	access_class	priority
0, 1	0	lowest
2, 3	2	
4, 5	4	
6, 7	6	highest

5B.6.3 Token Rotation Timers. A station shall provide three (actual or virtual) token_rotation_timers, one for each access_class. These timers shall all run concurrently, counting downward from an initial value to zero, at which point they shall stop counting and their status shall be "expired".

These timers shall count in units of octet_times, and shall otherwise be managed as specified in Part 10C.

5B.6.4 Recommended Priority Assignments in PROWAY System. The following priority assignments are recommended for all PROWAY systems.

access_class	usage
6	Urgent messages, i.e., those performing critical alarm, interlock and coordination functions.
4	Normal control actions and ring maintenance functions.
2	Routine data gathering and display and data base updates.
0	File and program transfer.

5B.7 Delegation of Right to Transmit. A station holding a token may request a second station to transmit a response without the second station holding the token. The first station, in effect, delegates the authority to transmit to a secondary station.

The secondary station shall conform to all of the requirements imposed by this standard on the token holder except for participating in the logical token passing ring and associated protocol mechanisms (unless the secondary is or needs to be in_ring). The secondary station need not be in the logical ring from an address sense. The secondary shall not transmit on the network unless

(1) Delegated as a transmitter by a token holder, or
(2) Transmitting is authorized by the procedures specified in Part 5C.

5B.8 Notification of Logical Ring Modification. A change in the Next_Station (NS) variable requires that the MAC sublayer generate an indication to station management. Station management typically uses this to participate in the maintenance of a list of active stations.

5B.9 Station Addresses. Station addresses shall be defined as given in 5D.1.4.1.2 and 5D.1.4.2.

5B.10 Required MAC Machines. All MAC machines are required for **INITIATOR** stations. All MAC machines except the ACM are required for **RESPONDER** stations.

5B.10.1 Mandatory Access Control Machine (ACM) Functions. All ACM functions are mandatory in all implementations of **INITIATOR** stations.

5B.11 Interface Machine
5B.11.1 Interface Machine (IFM) Description. The Interface Machine (IFM) acts as an intermediary between the other functional machines of the MAC sublayer and the PLC and LLC sublayers and station management entity with which MAC must communicate.

The IFM has eight primary functions:

(1) Accepting or generating the service primitives supported at the MAC-MAC_user interface.
(2) Mapping higher layer requests between MAC_user terms of quality of service (priority and confirmation class) and MAC terms (access_class and MAC_action).

(3) Queuing pending service requests into one of four queues, separated by access_class, so that the requests may be handled according to access_class, as well as according to arrival order. Requests from PLC, LLC and station management at a given access_class share the same queue and are serviced on a first-in first-out basis.

(4) Recognizing the individual destination address of data frames destined for this station and the broadcast address.

(5) Recognizing group addresses.

(6) Notifying the ACM that a response frame has been received (for request_with_response frames) and passing this frame to the local (originating) PLC entity in association to the corresponding confirmation.

(7) Passing received request_with_no_response and request_with_response frames to the destination MAC_user entity specified by frame_type.

(8) Accepting a response frame from the remote (responding) PLC entity following receipt of a request_with_response frame, and sending this response frame.

5B.11.2 Mandatory Interface Machine (IFM) Functions. IFM functions that are mandatory for **INITIATOR** and **RESPONDER** stations are indicated by an M in the appropriate column in the table below. Optional functions are indicated by an O. A hyphen (-) in this table indicates that the associated function is not required for the corresponding type of station. These functions are fully described in Section 5B.11.1.

FUNCTION	INITIATOR	RESPONDER
1. Accepting or generating supported service primitives	M	M
2. Mapping between MAC_user and MAC terms	M	M
3. Queuing service requests into queues separated by access_class	M	-
4. Recognizing this station's individual destination address and the broadcast address	M	M
5. Recognizing group addresses	M	O
6. Receiving, associating, and passing response frames to the originating MAC_user	M	-
7. Passing request_with_response and request_with_no_response frames to the specified MAC_user	M	M
8. Accepting and transmitting response frames	M	M

The minimal attributes of MAC entities for **INITIATOR** and **RESPONDER** stations are:

Priorities supported	4	4
Group addresses recognized	16	0

5B.12 Receive Machine

5B.12.1 Receive Machine Description. The receive machine accepts MAC_symbols (see 5B.1.2) from the PHY layer and generates high-level data structures and signals for the MAC access control machine and the MAC interface machine (IFM).

INDUSTRIAL DATA HIGHWAY

The receive machine is specified in Reference 4, Appendix I-A.

The interface between the PHY layer and the MAC access machine is the PHY_DATA.indication primitive specified in Part 6. The description of the receive machine in Reference 4 embodies the PHY_DATA.indication primitive as an encoded MAC-symbol and an associated PHY_clk.

5B.12.2 Mandatory Receive Machine (RxM) Functions. All RxM functions are mandatory in all implementations.

5B.13 Transmit Machine

5B.13.1 Transmit Machine Description. The media_access_control machine forwards frames for transmission to the TxM as a unit (at least for the purposes of this description). The TxM then passes the data frame, along with appropriate delimitation, to the physical layer, one MAC-symbol at a time, for transmission on the physical medium. The transmit machine is responsible for sending the proper amount of preamble, computing the FCS and including it within the transmitted frame, and delimiting the frame with SD and ED.

The transmit machine is specified in Reference 4, Appendix I-A.

5B.13.2 Mandatory Transmit Machine (TxM) Functions. All TxM functions are mandatory in all implementations.

5C. Access Control Machine (ACM)

Formal Description. This part defines the token bus medium access control mechanism, and also defines the ACM, variables and functions used in the definition of the ACM. A formal state machine description of the access control mechanism using the variables and functions discussed here is given in Reference 4, Appendix I-A.

5C.1 Variables and Functions. The variables and functions of the state machine description given in Reference 4, Appendix I-A are grouped into categories as follows:

(1) Variables defined by station management
(2) Variables defined by the interface machine
(3) Timers
(4) Variables defined by the receive machine
(5) Other ACM variables and functions

5C.1.1 Station Management Variables. Station management provides the MAC machine with the station's address bit string (and thus implicitly with the length of all addresses). Also supplied by station management are other network parameters:

TS: This Station's address. A bit-string variable set to the value of the station's 16-bit or 48-bit address.

slot_time: An integer in the range of 1 to $(2^{13}-1)$ octet_times. See 5A.1.1 and 5B.1.10.

max_pass_count: An integer equal to half the station's address length in bits, plus one. (Thus equal to 9 for a 16-bit address length.)

The value of max_pass_count limits loops in the ACM. The value is used to limit the token contention process. After cycling through max_pass_count contention cycles the process must be stopped if, due to an error, a single contender cannot be resolved.

The value of max_pass_count is also used to stop the token claiming process. After sending max_pass_count claim token frames, if no other station is heard, a station can claim the token.

min_post_silence_preamble_length: An integer equal to the minimum number of octets of preamble to be transmitted at the beginning of a transmission after the station has been silent. The value of min_post_silence_preamble_length is determined by the type of physical layer used in the station.

max_inter_solicit_count: An integer number of token possessions within the range 16 to 255. The value, in addition to the ring_maintenance_timer, determines how often a station opens response windows. Normally, a station opens response windows prior to every Nth pass of the token, where N is the value of max_inter_solicit_count. The action is taken only when token rotation time has not exceeded the ring_maintenance_target_time.

If all stations in the ring used the same max_inter_solicit_count, they would all consistently open response windows on the same token rotation. This action could lead to rapid token rotations where no response windows were opened, and occasional rotations where every station opened a response window before passing the token.

To avoid all stations in the ring having the same value of max_inter_solicit_count, the least significant two bits of the value shall be chosen randomly. The actual value used for the max_inter_solicit_count shall be changed by each station by re-randomizing the least significant two bits of the variable at least every 50 milliseconds or on every use.

target_rotation_time (access_class): An array of integers in the range 0 to $2^{21}-1$ octet_times, used with the priority option and with the ring_maintenance timer. (See 5C.1.4 for a discussion of the variable's function.)

ring_maintenance_timer_initial_value: An integer in the range 0 to $2^{21}-1$ octet_times, used to determine the initial value of the ring_maintenance token_rotation_timer upon entry to the ring. A large value will cause the station to solicit successors immediately upon entry to the ring; a value of zero will cause the station to defer this solicitation for at least one rotation of the token.

hi_pri_token_hold_time: An integer in the range of 0 to $2^{16}-1$ octet_times. Used to control the maximum time a station can transmit frames at access class 6.

in_ring_desired: A Boolean variable which determines the access control machine's steady-state condition when it has no queued transmission requests. If the variable is true, the station should be in_ring (a participant in the token-passing logical ring). If false, the station should be out_of_ring (an observer of the token-passing logical ring).

maximum_retry_limit: An integer constant in the range of 0 to 7 specifying the maximum number of attempts that will be made to send a request_with_response frame.

5C.1.2 Interface Machine Variables and Functions

get_pending_frame(access_class): A function provided by the interface machine of MAC. This function pops the first frame off the pending frame queue for the indicated access class, and returns it to the access control machine for transmission.

any_send_pending: A Boolean variable reflecting the logical OR of all the Boolean variables send_pending(access_class). Any_send_pending is true if at least one of the pending frame queues is non-empty. If all queues are empty, the variable's value is false.

power_OK: A Boolean variable indicating that the ACM may begin operation. Provided by station management hardware.

5C.1.3 Note on Logical Ring Membership Control. The two Boolean variables in_ring_desired and any_send_pending determine the operation of the ACM with respect to contending for the token and being in the logical ring as follows:

Variables and States		ACM Actions
in_ring_desired	any_send_pending	
false	false	Do not contend for token. Drop out if currently in token-passing logical ring.
false	true	Contend for token. Send data, which may empty the ending frame queues and take any_send_pending false. Exit logical ring if any_send_pending becomes false.
true	false	Contend for token if not sole active station. Remain in token-passing logical ring even without data to send.
true	true	Contend for token. Remain in token-passing logical ring and send data.

5C.1.4 Timers. A number of timers are used in the description of the state machine. A timer is expressed as a set of procedures and a Boolean variable. The procedures are named xx_timer.start(value), where xx is the timer name and value is an integer that sets the timer delay. xx_timer.value returns the current value of the counter. The Boolean variables are named xx_timer.expired and have a value of false while the timer is running and true when the timer has expired.

For example, the bus_idle timer would be set to a value of one (slot_time) by executing bus_idle_timer.start(1). The variable bus_idle_timer.expired would then be false for one slot time.

5C.1.4.1 Slot_time Interval Timers. The first five timers (bus_idle_timer, contention_timer, claim_timer, response_timer and token_pass_timer) work in integral multiples of the network slot_time. (The first five timers are not used concurrently, thus they could be implemented in a single hardware timer.)

bus_idle_timer: Controls how long a station listens in the idle state for any data on the medium before entering the claiming token state and reinitializing the network. Most stations wait seven slot times. The one station in the network having lowest_address true waits six slot times. The function max_bus_idle returns the value 6 or 7 depending on the state of lowest_address.

claim_timer: Controls how long a station listens between sending claim_token frames. The claim_timer is always loaded with the value 1.

response_window_timer: Controls how long a station which has opened response windows listens before transmitting its next frame.

When sending a solicit_successor, who_follows or resolve_contention frame, this timer controls the length of time a station solicits responses. After sending a solicit_successor frame, the sending station loads the response_window timer with the number of windows opened. The timer thus determines how long the station remains in the await_response state listening for stations to respond. If the timer expires and nothing is heard, the station goes to the pass_token state and passes the token to its successor.

When sending a request_with_response data frame, the response_window timer controls how long a station waits for a response frame before repeating the request_with_response frame.

contention_timer: Controls how long a station listens in the demand_in state after hearing a resolve_contention, solicit_successor or who_follows frame when the station wants to contend for the token. If the station hears a transmission while listening, it has lost the contention and must return to the idle state.

token_pass_timer: Controls how long a station listens after passing the token to its successor.

If any frame is heard before the token_pass_timer expires, the station assumes that its successor has accepted the token. If the timer expires and a frame is not heard, the station assumes its successor did not accept the token and sequences to the next stage of the pass token procedure.

5C.1.4.2 Octet_time Interval Timers. The remaining timers have a granularity of one octet transmission time, rather than one network slot_time. They are used to implement the access class structure and limit the time during which a station may start to transmit frames for each access class.

token_rotation_timer(access_class): This is a set of four timers, one for each of the lower three access classes and one for ring maintenance.

When the station begins processing the token at a given access class the associated timer is reloaded with the value of the target_rotation_time for that level. When the station again receives the token it may send data of that access class until the residual time in the associated token_rotation_timer has expired.

Upon initial entry to the ring, the first three priority timers are set to a value of zero (expired), and the ring_maintenance timer is set to the value ring_maintenance_timer_initial_value.

token_hold_timer: The residual time from the current token_rotation_timer is loaded into the token_hold_timer just before the token_rotation_timer is reloaded. The station may send data frames of the corresponding access class as long as the token_hold_timer has not expired.

When the station is sending highest access class messages, the value of hi_pri_token_hold_time is loaded into the token_hold_timer. Thus highest access class messages are limited to only a fixed number of bytes regardless of current network loading.

5C.1.5 Receive Machine Variables and Functions. The outputs of the receive machine are several state variables and a data frame, as described in the following paragraphs.

bus_quiet: A Boolean variable which is true whenever the physical layer is reporting that silence is being received. False when something other than silence is being received. bus_quiet is set and reset by the receiver machine and is only read by the ACM.

Rx_frame: A record written by the receiver machine. The record is updated to reflect the contents of the most recently received valid frame. The major fields in the record are:

FC: The one-octet frame control field
DA: The two-octet destination address field
SA: The two-octet source address field
data_unit: The multi-octet data unit field
FCS: The four-octet frame check sequence field.

Rx_protocol_frame: This signal indicates that a valid frame has been received, and that the frame type is one of the MAC protocol frame types. This signal is set by the receive machine, and it is read and cleared by the access machine.

Rx_data_frame: This signal indicates that a valid frame has been received, and that the frame type is a LLC frame or station_management frame. This signal is set by the receive machine and read by both the access control machine and the interface machine; it is cleared only by the interface machine.

noise_burst: A Boolean variable set by the receiver machine when bus_quiet goes true (the bus goes from non-silence to silence) and neither Rx_protocol_frame nor Rx_data_frame were set during the transmission (that is, no valid frame was heard). It is reset by the access control machine when the noise burst condition has been processed.

5C.1.6 Other Variables and Functions. The following are internal to the MAC access control machine (ACM).

TH — Token Holder's address: The address of the current token holder. A temporary buffer loaded from the SA field of a solicit_successor, who_follows or resolve_contention frame. If a set_successor frame is sent by the station as part of the contention process, the DA address is taken from TH.

NS — Next Station's address: The address of a station's successor in the logical ring. NS is set when a station that does not know its successor hears a solicit_successor frame and contends for the token. The station sets NS to the value of the destination address field of the frame. (If the station successfully contends in a response window, it will receive the token and eventually pass it to the station whose address was loaded into NS.)

For example: Suppose a station with address 25 is not in the logical ring, and wants to enter. If this station hears a solicit_successor frame sent by station 30 with a DA address of 20, it will set NS to 20, the DA address in the frame. If the station contends in the window and is heard by station 30, it will be passed a token. When the station has completed sending data frames it passes the token to its successor station 20.

The NS variable is also loaded whenever the station receives a set_successor frame addressed to it.

NOTE that wherever the value of NS is changed, a MA_EVENT.indication is given to station management.

NOTE: Once a station thinks it knows the value of NS it no longer reloads NS when a contention window is opened spanning the station's address. The reason is that under error recovery conditions, stations will send solicit_successor_2 frames addressed to themselves that open response windows for all stations. If all stations reset their NS variables at this point, any logical ring that existed would collapse.

NS_known: A Boolean variable that indicates whether the station thinks it knows the address of its successor. NS_known is set true whenever the station receives a set_successor frame addressed to it. Normally the set_successor frame follows a successful contention, as in the previous example.

NS_known is set false whenever the station leaves the logical ring.

PS — Previous Station's address: The variable is set to the value of the source address of the last token addressed to the station.

If a who_follows frame is heard, the contents of the data field of the frame are compared with the contents of PS. If they are equal, the station responds to the who_follows request with a set_successor frame.

An example will clarify the use of PS. If a logical ring contains stations with addresses 30, 25, and 20, the station with address 20 will have 25 in its PS register, since this is the address of the station that sends it the token. If station 25 fails, when station 30 tries to send the token to station 25 it will get no response. After two tries at passing the token, station 30 sends a "who follows 25?" frame. Station 20 responds by sending a "set your successor to 20" frame. In this manner the failed station, 25, is quickly patched out of the ring.

max_access_class: An integer constant used to initialize the sequencing of the processing of the pending frame queues. The value of max_access_class is 6, the highest priority access class.

access_class: An integer which is used to sequence through the access classes while transmitting data frames.

The first, or highest priority, access_class equals the value of max_access_class (that is, 6). The variable access_class is then decremented (by 2) through all classes until less than zero, and then the station performs its ring maintenance functions and passes the token.

in_ring: A Boolean variable set true when the station receives a token frame addressed to it or when the station successfully completes the claiming token process. Set false if the station sets itself out of the ring.

sole_active_station: A Boolean variable used to mute stations having defective receivers. If a station's receiver becomes inoperative in an undetected manner, the station otherwise would disrupt the operation of the system by continually claiming the token and then soliciting a successor station.

If the sole_active_station variable is true, a station is prevented from entering the claiming token process until it has data to send. Thus a station with an inoperative receiver and no data to send will remain passively out of the ring.

If a station is a member of the ring and its receiver fails, it will be unable to hear its successor claiming the token. The station will cycle through the token passing recovery algorithm, quickly reaching the point where it has sent a solicit_successor_2 frame addressed to itself and received no response. At this point, the station sets sole_active_station true and becomes passive.

Sole_active_station is set false whenever the station hears a valid frame from another station.

lowest_address: A Boolean variable set true if the station's successor address is greater than the station's address.

At any one time there should be only one station in the logical ring with lowest_address set true. This is the station with the lowest_address of all those currently in the logical ring. When this station opens response windows during a token pass, it must open two windows. The first window is used by stations having an even lower address that wish to enter the ring. The second window is used by stations having a higher address than the recipient of the token, the station currently with the highest address in the ring.

Lowest_address is computed and set by a station whenever NS is changed.

Lowest_address is used for a second purpose unrelated to token passing. If the token-holding station fails, another station must recover the token. The bus_idle_timer is a "watchdog" timer. If no frames are heard by a station for an interval greater than this timer, the claim token process is started.

In an effort to minimize interference during the claiming process, one station is selected to use a shorter bus_idle_timer value than the other stations. This station recovers all lost token failures, except one it causes. The station with lowest_address true is always unique, so it is assigned this role.

just_had_token: A Boolean variable set true when the station passes the token and set false if the station hears a valid frame from another station.

Just_had_token is used to detect duplicate addressing failures on the network. If a station hears a valid frame with a source address equal to its own address and just_had_token is false, the station cannot have sent the frame. If such a frame is heard, the MAC sublayer notifies station management of the detection of another station on the network using the same MAC address; the MAC sublayer then enters the offline state.

heard: A three-state variable used in the await_response state. The states are:

nothing: The station has heard nothing (except its own transmissions) since beginning the resolve process.
collision: A noise burst has been heard.
successor: A valid set_successor frame has been received. At the end of the resolution period, the station will send the token to the station whose address was in the protocol data unit field of (one of) the valid set_successor frame(s).

claim_pass_count: An integer with a range from 0 to max_pass_count. Used as an index to TS to select two bits from the station's address. The value of the selected bits (times twice the slot time) determines the length of the information field of the claim_token frame to be sent. After each claim_token frame the value of the variable claim_pass_count is incremented by one.

contend_pass_count: An integer with a range from 0 to max_pass_count. Used as an index to TS to select two bits from the station's address. The one's complement of the selected bits (times the slot time) determines the length of time a station delays in the demand-in state after receiving a resolve_contention frame. If no other frames are heard before the contention_timer expires, the station sends a set_successor frame to the token holder, increments the value of contend_pass_count, goes to the demand-delay state, and waits for the token or for another resolve_contention frame.

contention_delay(cycle): An integer function which returns the value 0, 1, 2 or 3. This value is based upon the one's complement of a pair of address bits which are indicated by the cycle in the address sort. The value is used to control the number of slot_times the station delays before transmitting when demanding entry to the logical ring.

resolution_pass_count: An integer with a range from 0 to max_pass_count. Used to count the number of resolve contention passes the token-holding station makes. If the counter reaches the value of max_pass_count, the resolution process is abandoned and the token passed to the station's successor.

inter_solicit_count: An integer in the range of 0 to 255. Determines when a station must open a response window. Before passing the token, the value of inter_solicit_count is checked. If the value is zero, a new successor is solicited by opening a response window. If the value is not zero, the counter is decremented and the token is passed. Whenever anything is heard during the response windows following the solicit_successor frame, the counter value is set to zero so that it will again be zero when the station next receives and passes the token. Thus receipt of a set_successor frame during a response window causes the station to reopen the response window before that next token pass.

remaining_retries: An integer in the range of 0 to maximum_retry_limit. Used to control the number of retransmissions upon timeout of a request_with_response frame.

suppress_FCS: A Boolean variable used within the ACM to indicate that the current frame should be transmitted without having the transmitter state machine append a FCS.

transmitter_fault_count: An integer in the range of 0 to 7. Used to infer that the station's transmitter has probably failed and thus that the station's transmissions cannot be heard correctly by other stations on the network.

The value of transmitter_fault_count is incremented each time the station sequences to the end of the token contention process or fails to pass the token to any successor. Neither of these failures occurs during normal operation. The value of transmitter_fault_count is reset to zero if the station either wins the demand-in token contention process or successfully passes the token, since such an event indicates that another station correctly heard a transmission from the station.

If the value of transmitter_fault_count is incremented to max_transmitter_fault_count, the station reports a faulty_transmitter to station management and enters the offline state. The value of max_transmitter_fault_count is set to a maximum of 7, allowing for an occasional protocol sequencing impasse due to noise. If the station cannot enter the ring or pass the token (if already in the logical ring) seven times in a row, the inference made is that something has failed in the station, probably in the transmitter, and so the station removes itself from the logical ring.

INDUSTRIAL DATA HIGHWAY

ISA
S72.01-1985

first_time: A Boolean variable that controls processing of noise bursts in the await_response state. Set true upon entry to the state. Set false when the first noise burst is heard. If a noise burst is heard when first_time is false, the station returns to the idle state.

pass_state: A multistate variable used to control the operation of the pass_token substates. The action taken in the state depends on the value of the variable pass_state as follows (the actions are listed in the order taken by a station soliciting successors and failing:

pass_state value	action
solicit_successor	Send solicit_successor frame. Enter await_response state.
pass_token	Send token to successor. Enter check_token_pass_ state.
repeat_pass_token	Same action as pass_token substate.
who_follows	Send who_follows frame. Enter await_response state.
repeat_who_follows	Same action as who_follows substate.
solicit_any	Send solicit_successor_2 frame with DA = TS, opening 2 response windows that span all other stations. Enter await_response state.
total_failure	Set sole_active_station true and either silently pass the token back to itself (if the station has more data to send) or enter the idle state. This station will not transmit again unless it has data to send or it hears a valid frame from another station.

5C.2 Access Control Machine Formal Description. The access control machine (ACM) is described formally in Reference 4, Appendix I-A.

5C.2.1 List of Unique ACM Words. This is a list of unique words appearing in the MAC ACM state transition tables of Reference 4, Appendix I-A. This list includes only words from the EXIT CONDITION and ACTION TAKEN parts of these tables; CURRENT STATE, transition name, NEXT STATE and comments (strings beginning with —) were excluded.

```
access_class                          claim_timer.start
any_send_pending                      claim_token
bus_idle_timer                        collision
bus_idle_timer.expired                contend_pass_count
bus_idle_timer.start                  contention_delay
bus_quiet                             contention_timer
cdu                                   contention_timer.expired
claim_data_unit                       contention_timer.start
claim_pass_count                      DA
claim_timer                           data_unit
claim_timer.expired                   destination
```

duplicate_address	MA_FAULT_REPORT.indication
faulty_transmitter	MA_INITIALIZE_PROTOCOL.request
FC	noise_burst
FCS_suppression	nothing
first_time	NS
frame_control	NS_known
get_pending_frame	octet_time
heard	pass_state
hi_pri_token_hold_time	pass_token
in_ring	power_ok
in_ring_desired	ring_maintenance_timer
inter_solicit_count	remaining_retries
just_had_token	token_pass_timer.expired
lowest_address	token_pass_timer.start
max_access_class	token_rotation_timer
max_bus_idle	tok_pass_substate'succ
max_inter_solicit_count	total_failure
max_pass_count	transmitter_fault_count
maximum_retry_limit	TS
max_transmitter_fault_count	who_follows

5D. Frame Formats

This part defines the required MAC frame formats. This includes all allowed frame formats and the arrangement of all frame subfields. The term frame as used here refers to the protocol_data_units exchanged by MAC sublayer entities. The MAC_service_data_units received from the PLC sublayer are contained within these MAC frames.

NOTE that certain frame types exchanged with an alternate Link Control entity (LLC) are shown for completeness. This standard does not specify the formats or use of frames exchanged with LLC entities.

This part describes the frame components and formats used by medium access control. The MAC level transmit frames and abort sequences are described in the following clauses. First the components of the frames are discussed, followed by the definition of the valid frame formats. All frames sent or received by the MAC sublayer shall conform to the following general format:

PREAMBLE	SD	FC	DA	SA	DATA_UNIT ...	FCS	ED

where PREAMBLE = pattern sent to set receiver's modem clock and level (1 or more octets)
      SD = start delimiter (1 octet)
      FC = frame control (1 octet)
      DA = destination address (2 octets)
      SA = source address (2 or 6 octets)
DATA_UNIT = information (0 or more octets)
      FCS = frame check sequence (4 octets)
      ED = end delimiter (1 octet)

INDUSTRIAL DATA HIGHWAY

The number of octets between SD and ED, exclusive, shall be 1023 or fewer. The abort sequence shall conform to the following format:

| SD | ED |

```
where SD = start delimiter (1 octet)
      ED = end delimiter (1 octet)
```

Within this part the following acronyms are used for the addresses of the station under discussion, its successor and its predecessor in the logical ring:

TS — this station's address.
NS — next station's address.
PS — previous station's address.

5D.1 Frame Components. This clause describes the frame components which are shown in the previous illustrations in greater detail.

5D.1.1 Preamble. The preamble pattern precedes every transmitted frame. Preamble is sent by MAC as an appropriate number of pad_idle symbols. Preamble may be decoded by the receiver as arbitrary data symbols that occur outside frame delimiters. Preamble is primarily used by the receiving modem to acquire signal level and phase lock by using a known pattern. The preamble pattern is chosen for each modulation scheme and data rate for this purpose. Consult the parts on the physical layer for details.

A secondary purpose for the preamble is to guarantee a minimum ED to SD time period to allow stations to process the frame previously received. The minimum amount of preamble transmitted is a function of the data rate. This standard requires that the duration of the preamble shall be at least 2 microseconds, regardless of data rate, and that an integer number of octets shall be sent. At a data rate of 1 Mb/s one octet of preamble is required to meet the integer number of octets requirement. The maximum amount of preamble is constrained by the "jabber" control in the physical layer. Additionally, for claim token frames, all stations shall use the minimum number of preamble octets to ensure that all frames are of uniform specified length.

5D.1.2 Start Delimiter. The frame structure requires a start delimiter, which begins the frame. The start delimiter consists of signaling patterns that are always distinguishable from data.

The start delimiter is coded as follows (see Part 8 for representations of the symbol coding as present on the medium):

```
                           ┌── First MAC-symbol transmitted
                           ↓
Start Delimiter (SD):   │N N 0 N N 0 0 0│
                         1 2 3 4 5 6 7 8   ←── bit positions

where N = non_data MAC-symbol
      0 = zero MAC-symbol
```

5D.1.3 Frame Control Field. The frame control octet (FC) determines what class of frame is being sent among the following general categories:

(1) MAC control
(2) LLC data
(3) Station management data
(4) PLC data

5D.1.3.1 MAC_control_frame

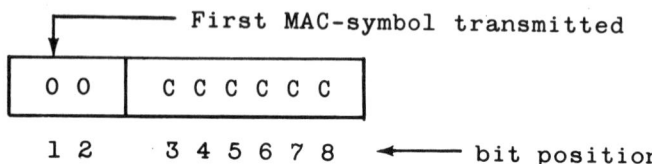

where CCCCCC = type of MAC_control frame as follows:

```
C C C C C C
3 4 5 6 7 8      bit positions

0 0 0 0 0 0      Claim_token
0 0 0 0 0 1      Solicit_successor_1  (has 1 response window)
0 0 0 0 1 0      Solicit_successor_2  (has 2 response windows)
0 0 0 0 1 1      Who_follows          (has 3 response windows)
0 0 0 1 0 0      Resolve_contention   (has 4 response windows)
0 0 1 0 0 0      Token
0 0 1 1 0 0      Set_successor
```

5D.1.3.2 Data_frames

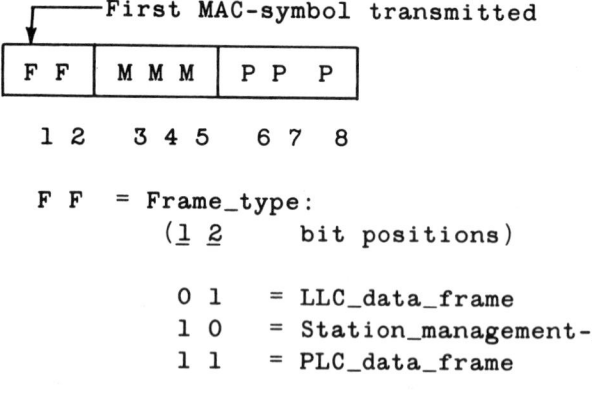

```
F F = Frame_type:
        (1 2      bit positions)

        0 1   = LLC_data_frame
        1 0   = Station_management-data_frame
        1 1   = PLC_data_frame

M M M = MAC-confirmation_class:

        (3 4 5     bit positions)

        0 0 0 = Request_with_no_response
        0 0 1 = Request_with_response
        0 1 0 = Response
```

INDUSTRIAL DATA HIGHWAY

ISA
S72.01-1985

```
P P P = priority:
        (6 7 8    bit positions)

        1 1 X = highest priority
        1 0 X = second highest priority
        0 1 X = third highest priority
        0 0 X = lowest priority
              └── least significant bit of priority class

    X = value has no significance to MAC but may contain information of
        significance to higher layers
```

Other bit patterns in the frame control octet are reserved for future study. The action of a station upon receiving an FC value not defined in this standard is not specified.

5D.1.4 Address Fields. Each frame shall contain two address fields: the destination address field and the source address field, in that order. Addresses shall be 16 bits or optionally 48 bits in length.

Note: **Appendix 5-B** shows examples of individual MOWAY address assignments.

5D.1.4.1 Destination Address Field Composition. The following illustration shows the possible representations of destination addresses.

5D.1.4.1.1 Generic Address Form

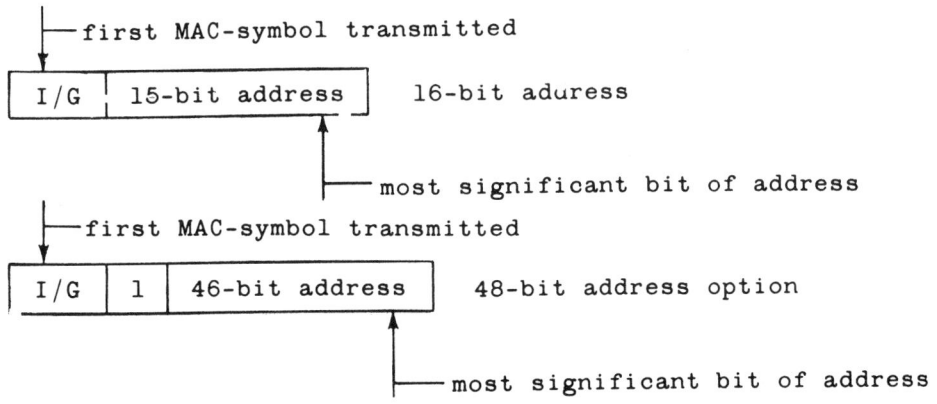

where I/G = Individual/Group address indication bit

The first MAC-symbol transmitted of the destination address (the I/G bit) distinguishes individual addresses from group addresses:

0 = individual address
1 = group address

5D.1.4.1.2 Individual PROWAY Station Addresses

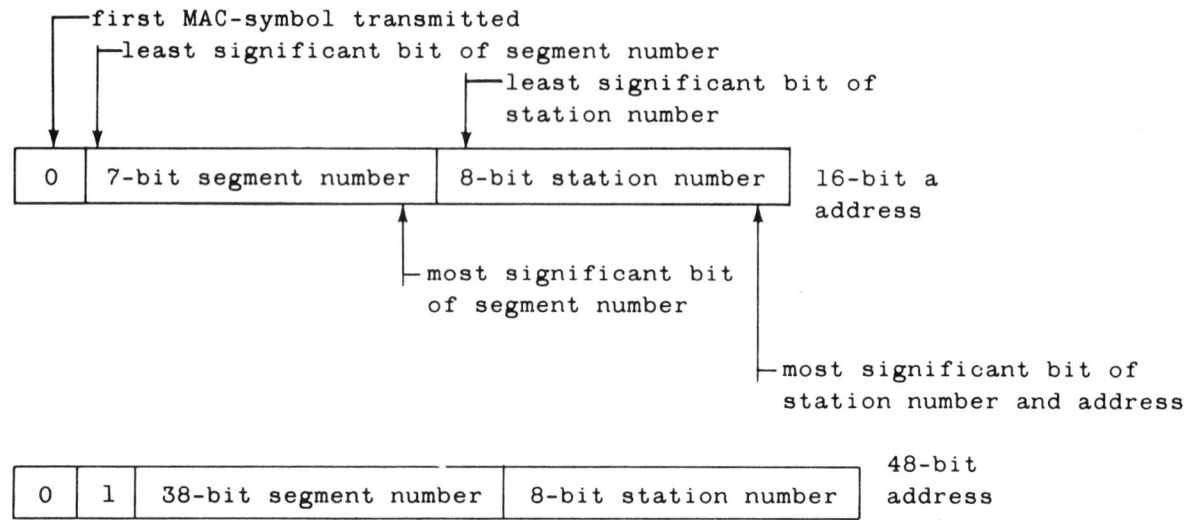

Station number identifies a unique station within a subnetwork, i.e., data highway. Station number may be used for maintenance of the active_station_list.

The lowest allowed individual station number is 0, and the highest is 255. The lowest allowed segment number is 0, and the highest is 127 for 16-bit address and $2^{38}-1$ for 48-bit addresses.

Segment number identifies the particular subnetwork within a LAN on which this station resides (see Appendix 5-A). When a LAN contains no subnetworks, it is recommended that segment number = 0.

5D.1.4.1.3 Group Addresses

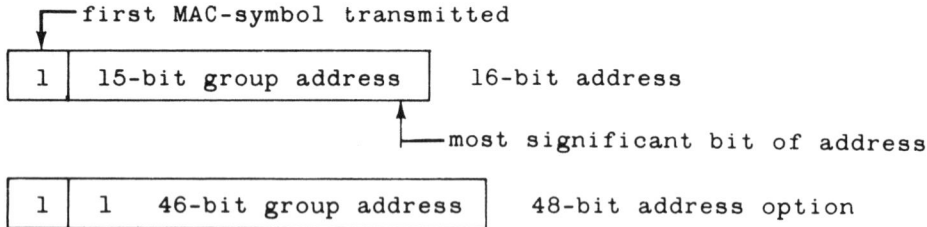

5D.1.4.2 Individual Addresses. An individual address identifies a particular station on the LAN and shall be distinct from all other individual station addresses on the same LAN. It consists of the identifying station number assigned to that station by the local administrative authority, preceded by an I/G bit = 0, a "0" (for 48-bit addresses), and a segment number assigned by the local administrative authority.

Group Addresses. A group address is used to address a frame to multiple destination stations. Group addresses may be associated with zero, one, or more stations on a given LAN. In particular, a group address is an address associated by convention with a group of logically related stations.

A group address shall not be used in a request_with_response data frame.

INDUSTRIAL DATA HIGHWAY

Recommended assignments of group addresses in PROWAY networks is for future study.

Broadcast Addresses: The group address consisting of all ones (that is, 16 ones for 16-bit addressing or 48 ones for 48-bit addressing, respectively) shall constitute the broadcast address, denoting the set of all stations on the given LAN.

A broadcast address shall not be used in a request_with_response data frame.

NOTE: For some of the frame types used by the token bus MAC procedures, the contents of the destination address field is irrelevant. In such cases, the originating station's own address or any other properly formed address can be sent in this field.

5D.1.4.3 Source Address Field. The source address identifies the station originating the frame and has the same format and length as the destination address in a given frame, except that the individual/group bit shall be set to 0; the significance of this bit being set to 1 is a subject for future study.

5D.1.4.4 Numerical Interpretation of Addresses. Strictly speaking, addresses are bit strings which serve as unique station identifiers or group identifiers. For the purpose of the MAC address comparison within the token bus MAC sublayer, as used in ordering the logical ring and as expressed in the formal access control machine of 5C, each MAC_address bit string is interpreted as if it were an unsigned integer value sent least significant bit first, and thus as if the last bit transmitted had the highest numeric significance.

Additionally the address bits are used in determining delays in the contention process and transmission lengths in the token claiming process. These processes start with the most significant address bits using two bits at a time. Thus the internal processing order is reversed from the serial transmission order on the medium.

5D.1.5 MAC Data Unit Field. Depending on the bit pattern specified in the frame's frame control octet, the MAC data unit field can contain either of the following:

(1) A PLC protocol data unit, as specified in Part 3, which is used to exchange PLC information between PLC entities.
(2) An LLC protocol data unit which is used to exchange LLC information between LLC entities. This standard does not specify the details of exchanges between LLC entities.
(3) A MAC management data frame which is used to exchange MAC management information between MAC management entities.
(4) A value specific to one of the MAC control frames.

Each octet of the MAC data unit field shall be transmitted low-order bit first.

The question of exchanging PLC data units using the LLC data unit type is for further study and harmonization with the IEEE 802 committee.

5D.1.6 Frame Check Sequence (FCS) Field. The FCS is a 32-bit frame checking sequence, based upon the following standard generator polynomial of degree 32:

$$X^{32} + X^{26} + X^{23} + X^{22} + X^{16} + X^{12} + X^{11} + X^{10} + X^8 + X^7 + X^5 + X^4 + X^2 + X + 1$$

The FCS is the one's complement of the sum (modulo 2) of:

(1) The remainder of

$$X^K * (X^{31} + X^{30} + X^{29} + ... + X^2 + X + 1)$$

divided (modulo 2) by the standard 32-bit generating polynomial, where K is the number of bits in the frame control, address (SA and DA), and MAC data_unit fields, and

(2) The remainder of the division (modulo 2) by the standard generating polynomial of the product of X^{32} by the content of the frame control, address (SA and DA), and MAC data_unit fields. The FCS is transmitted commencing with the coefficient of the highest degree term.

As a typical implementation, at the transmitter, the initial content of the register of the device computer the remainder of of the division is preset to all ones and is then modified during division of the frame format, address and information fields by the generator polynomial (as described above). The one's complement of the resulting remainder is transmitted as the 32-bit FCS sequence.

At the receiver, the initial content of the register of the device computing remainder is preset to all ones. The serial incoming protected bits and the FCS, when divided by the generator polynomial, results, in the absence of transmission errors, in a unique non-zero remainder value. The unique remainder value for the 32-bit FCS is the polynomial:

$$X^{31} + X^{30} + X^{26} + X^{25} + X^{24} + X^{18} + X^{15} + X^{14} + X^{12} + X^{11} + X^{10} + X^8 + X^6 + X^5 + X^4 + X^3 + X + 1$$

NOTE: In order to test the FCS generation and checking logic in a station, an implementation should provide a means of bypassing the FCS generation circuitry and providing an FCS from an external source. The ability to pass frames that have FCS errors along with the received FCS value and an error indication to higher levels of the protocol is another desirable testability feature.

5D.1.7 End Delimiter. The frame structure requires an end delimiter (ED), which ends the frame and determines the position of the frame check sequence. The data between the SD and ED must be an integral number of octets. All bits between the start and end delimiters are covered by the frame check sequence.

The end delimiter consists of signaling patterns that are always distinguishable from data. The end delimiter also contains bits of information that are not error checked. The end delimiter is coded as follows:

```
                              ┌─── First MAC-symbol transmitted
                              ↓
End Delimiter (ED):     │ N N 1 N N 1 I R │

                          1 2 3 4 5 6 7 8  ←─ bit positions
```

where N = non_data MAC-symbol
      1 = one MAC-symbol
      R = reserved and set = 0. Non-PROWAY stations may set this bit equal
          one if they detect an error in this frame.
      I = intermediate bit

The seventh ED MAC-symbol is called the **intermediate** bit. If **one**, it indicates that more transmissions from the station follow. If **zero**, it indicates that this is the last frame transmitted by the station and silence follows the ED. The I bit will be **zero** for request_with_response and response frames.

5D.1.8 Abort Sequence. This pattern prematurely terminates the transmission of a frame. The abort sequence is sent by a station that does not wish to continue to send a frame it has already begun.

```
            ┌── First MAC-symbol transmitted
            ▼
   ┌─────────────────┬─────────────────┐
   │ N N 0 N N 0 0 0 │ N N 1 N N 1 I R │
   └─────────────────┴─────────────────┘
         S D                 E D
```

where N = non_data MAC-symbol
      0 = zero MAC-symbol
      1 = one MAC-symbol
      R = reserved
      I = intermediate

5D.2 Enumeration of Frame Types. This clause shows how the components of the frames are arranged in the various frame types transmitted by the MAC sublayer. Part 3 discusses the frames and terminology used here.

5D.2.1 MAC Control Frame Formats. The following frames are sent and received by the MAC sublayer and are not passed to higher layers.

5D.2.1.1 Claim_token. The frame has a data_unit whose value is arbitrary and whose length in octets (between addresses and FCS exclusive) is 0, 2, 4, or 6 times the system's slot_time also measured in octets.

PREAMBLE	SD	00000000	DA	SA	arbitrary value, length = (0, 2, 4, 6) slot_time octets	FCS	ED

5D.2.1.2 Solicit_successor_1. The frame has a DA = the contents of the station's NS register and a null data unit. One response window always follows this frame.

PREAMBLE	SD	00000001	DA	SA	FCS	ED

                                                one response window

5D.2.1.3 Solicit_successor_2. The frame has DA = the contents of the station's NS or TS register and a null data unit. Two response windows always follow this frame.

PREAMBLE	SD	00000010	DA	SA	FCS	ED

                                                two response windows

5D.2.1.4 Who_follows. The frame has a data_unit = the value of the station's NS register. The format and length of the data_unit are the same as a source address. Three response windows always follow this frame. (This gives receivers two extra slot_times to make a comparison with an address other than TS.)

ISA
S72.01-1985

PROWAY-LAN

| PREAMBLE | SD | 00000011 | DA | SA | value NS | FCS | ED |

three response windows

5D.2.1.5 Resolve_contention. The frame has a null data_unit. Four response windows always follow this frame.

| PREAMBLE | SD | 00000100 | DA | SA | FCS | ED |

four response windows

5D.2.1.6 Token. The frame has DA = the contents of the station's NS register, and has a null data_unit.

| PREAMBLE | SD | 00001000 | DA | SA | FCS | ED |

5D.2.1.7 Set_successor. The frame has DA = the SA of the last frame received, and data_unit = the value of the station's NS or TS register. The format and length of the data_unit are the same as that of a source address.

| PREAMBLE | SD | 00001100 | DA | SA | new value of NS | FCS | ED |

5D.2.2 Data Frame Formats. Data frames have a DA and data_unit specified by the source station's link control or station management entity (as indicated by the FF field). Valid data frames with non-null data_units are delivered to the destination link control or station management entity specified by the FF field by the destination MAC entity.

There are three possible types of data frames.

5D.2.2.1 Request_with_no_response Data Frame. For a request_with_no_response data frame, the destination MAC entity does not respond immediately to the frame.

| PREAMBLE | SD | FF000PPP | DA | SA | Data_unit | FCS | ED |

5D.2.2.2 Request_with_response Data Frame. For a request_with_response data frame, the destination (responding) MAC entity passes the data unit to the destination PLC entity. The destination PLC entity immediately returns a response data_unit to the destination MAC entity, and the destination MAC entity immediately sends the corresponding response frame.

A station shall not send a request_with_response frame using a group or broadest destination address.

Conceptually a request_with_response frame delegates to the recipient the right to transmit a single frame addressed to the token holder.

| PREAMBLE | SD | FF001PPP | DA | SA | Data_unit | FCS | ED |

5D.2.2.3 Response Data Frame. A response data frame is sent by a responding station to the originating (requesting) station upon receipt of a request_with_response frame from the requesting station.

Upon receipt of the response data frame at the requesting station, the originating MAC entity passes the data_unit of the response frame to the originating PLC entity.

| PREAMBLE | SD | FF010PPP | DA | SA | Data_unit | FCS | ED |

5D.2.3 Invalid Frames. An invalid frame is defined as one which meets at least one of the following five conditions:

(1) It is identified as such by the Physical layer (for example, it contains **non_data** or **bad_signal** symbols).
(2) It is not an integral number of octets in length.
(3) It does not consist of a start delimiter, one frame control field, two properly formed address fields, one data unit field of appropriate length (dependent on the bit pattern specified in the frame control field), one FCS field, and an end delimiter, in that order.
(4) The FCS computation, when applied to all octets between the SD and the ED, fails to yield the unique remainder specified in 5D.1.6.
(5) It is recommended that implementations treat a frame with the intermediate bit of the end delimiter equal to **zero** as an invalid frame unless the ED is followed by at least two MAC_symbol_ times of silence.

NOTE that the intermediate bit equals **zero** for all request_with_response and response frames.

Implementations may also treat a frame meeting any of the following additional conditions as an invalid frame:

(1) The frame control field contains an undefined bit pattern.
(2) The reserved bit within the end delimiter of the frame is asserted. This may indicate that a non-PROWAY station has detected an error.

Invalid frames shall be treated as noise. Their existence, as noise bursts, is relevant at some points in the token bus elements of procedure.

Appendix 5-A
An Example PROWAY Implementation at $1*10^6$ Bits/Sec

Part 5C allows for ranges of parameter values in implementations of token bus media access control methods. This appendix gives one set of values which meet the access time specifications of Part 1.

MAC_station_delay = 2 octet times

PLC_station_delay = 2 octet times

hi_pri_token_hold_time = 64 octet times

Max_AC_4_rotation_time = 6000 octet times

Max_AC_2_rotation_time = 4000 octet times

Max_AC_0_rotation_time = 2000 octet times

maximum_ring_maintenance_timer = 25000 octet times

maximum_retry_limit = 4

transmission_path_delay = 10 microseconds

max_inter_solicit_count = 255

Typical preamble transmitted = 1 or 2 octet times

ring_maintenance_timer_initial_value = zero

maximum_L_pdu size for SSAP's which use access_class = $6 \leq 16$ octets

address length = 16 bits

data rate = $1*10^6$ bits/sec

Appendix 5-B
Examples of Individual PROWAY Station Addresses

This appendix illustrates the coding of 16-bit station addresses in PROWAY_LAN. It is not a mandatory requirement of this standard.

Segment #/Station #

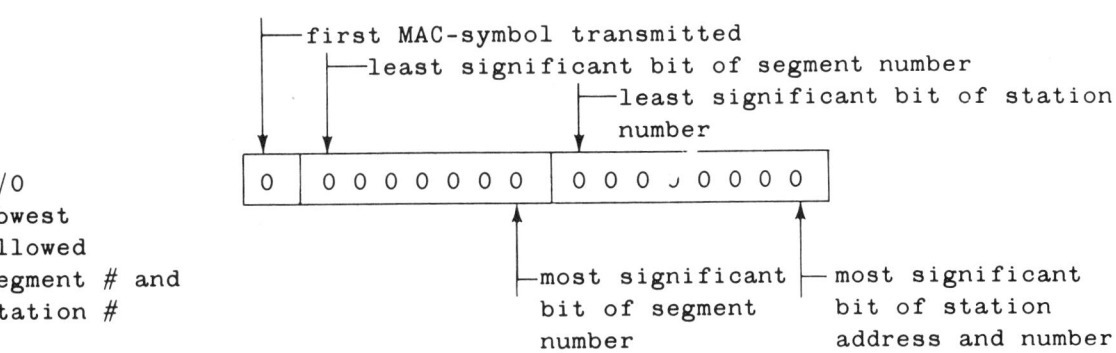

Address Coding

0/0 lowest allowed segment # and station #

1/1

65/130

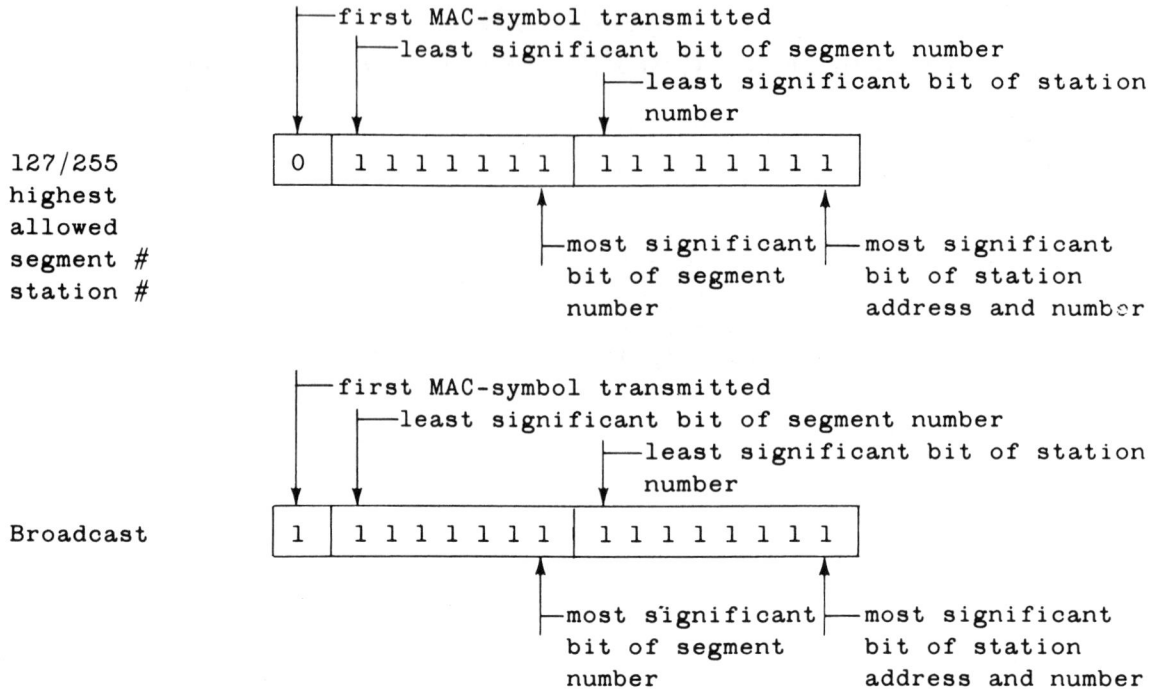

An integer representation of a station's address can be calculated using the following formula for 16-bit addresses:

station_address = 2*(segment_number) + 256*(station_number)

6. MAC-Physical Layer Interface Specification

This part specifies the interface between the MAC sublayer and the Physical layer.

Part 6A specifies the services at MAC-Physical interface in an abstract way. Part 6A is mandatory for all PROWAY Systems.

Part 6B specifies the implementation of MAC-Physical interface. Part 6B applies only to those stations in which MAC and Physical layers are embodied in two separate pieces of equipment.

The relationship of this part to other parts of this standard and to LAN specifications is illustrated in Figure 6-1.

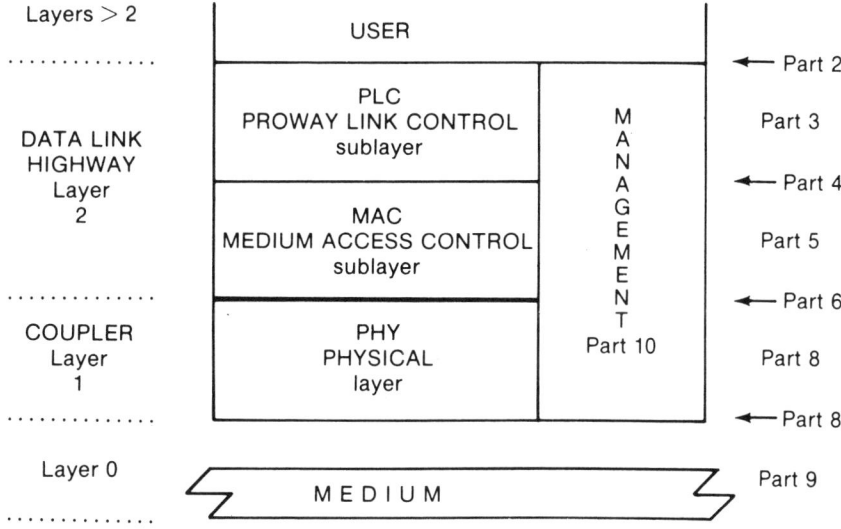

**Fig 6-1
Relationship to LAN Model**

ISA
S72.01-1985

PROWAY-LAN

6A. MAC-Physical Layer Interface Service Specification

6A.1 Scope and Field of Application. This part specifies the services provided to the MAC sublayer by the Physical layer of all stations conforming to this standard. It specifies these services in an abstract way. It does not specify or constrain the implementation entities and interfaces within a computer system.

6A.2 Overview of the Physical Layer Service

6A.2.1 General Description of Services Provided by the PHY Layer. These paragraphs describe informally the services provided by the Physical layer. These services provide for the transmission and reception of MAC-symbols, each with a duration of one MAC_symbol_period. Jointly, they provide the means by which cooperating MAC entities can coordinate their transmissions and exchange information by way of a shared communications medium.

6A.2.2 Model Used for the Service Specification. See Reference 1, Appendix I-A.

6A.2.3 Overview of Interactions. The primitives associated with symbol transmission and reception are PHY_DATA.request and PHY_DATA.indication.

The **PHY_DATA.request** primitive is passed to the physical layer to request that a symbol be impressed on the local area network's communications medium. Only one such request is accepted per MAC_symbol_period.

The **PHY_DATA.indication** primitive is passed from the physical layer to indicate the reception of a MAC symbol from the medium.

6A.2.4 Basic Services and Options. All PHY_DATA services are required in all implementations, and both of the PHY_DATA primitives are mandatory.

6A.3 Detailed Specifications of Interactions with the Physical Layer Entity. This part describes in detail the primitives and parameters associated with the Physical layer services. The parameters specify the information that must be available to the receiving (MAC or Physical) layer entity. A specific implementation is not constrained in the method of making this information available.

6A.3.1 PHY_DATA.request
6A.3.1.1 Function. This primitive is the service request primitive for the MAC symbol transfer service.

6A.3.1.2 Semantics. This primitive shall provide parameters as follows:

PHY_DATA.request (MAC-symbol)

MAC-symbol may specify one of:

(1) **zero** — corresponds to a binary 0
(2) **one** — corresponds to a binary 1 (**data** is the collective name for **zero** and **one**)
(3) **non_data** — used in delimiters, always sent in pairs, and always in octets with the form:
　　　　　　　　non_data　non_data　data　non_data　non_data　data　data　data
(4) **pad_idle** — send one symbol of preamble/inter_frame_idle (preamble is a physical-entity sequence of ones and zeros).
(5) **silence** — send silence for a duration of one MAC-symbol period. It is defined as the absence of carrier.

6A.3.1.3 When Generated. This primitive is passed from the MAC sublayer to the Physical layer to request that the specified symbol be transmitted on the local area network medium. This primitive shall be passed to the Physical layer once for each PHY_DATA.indication that the MAC sublayer receives from the Physical layer. There shall be an implementation-dependent constant phase relationship, determined by MAC, between a PHY_DATA.indication and the next subsequent PHY_DATA.request.

6A.3.1.4 Effect of Receipt. Receipt of this primitive causes the Physical layer to attempt to encode and transmit the symbol using the signaling appropriate to the local area network medium. The Physical layer signals its acceptance of the primitive by responding with a locally defined confirmation primitive.

6A.3.1.5 Constraints. **pad_idle** symbols, which are referred to collectively as **preamble**, are transmitted at the start of each MAC frame, both to provide a training signal for receivers and to provide a non-zero minimum separation between consecutive frames. The following constraints apply:

- An originating station must transmit a minimum number of octet multiples of **pad_idle** such that their duration is at least eight microseconds, and after completing transmission of the last required octet, it may (but need not) transmit more octets of **pad_idle** symbols before the first frame delimiter.

- **non_data** symbols shall be used only within frame delimiters, where they always shall be requested in pairs. The symbol sequences of those frame delimiters shall be: **non_data non_data data non_data non_data data data data** where each **data** symbol is either the symbol **zero** or the symbol **one**.

- When **data** symbols are transmitted between frame delimiters, the number of **data** symbols transmitted, not including those **data** symbols within the eight-symbol frame delimiter sequences, shall always be a multiple of eight. (That is, only complete octets of data symbols may be transmitted between frame delimiters.) When **pad_idle** symbols are transmitted between frame delimiters, the number of **pad_idle** symbols transmitted shall always be a multiple of eight. Octets of **pad_idle** symbols and octets of **data** symbols shall always be separated by frame delimiter octets, or a sequence of **silence** symbols, or both. (That is, **pad_idle** octets and **data** octets cannot be intermixed.)

- The jitter in the implementation-dependent constant phase relationship between consecutive PHY_DATA.indication and PHY_DATA.request primitives shall not be greater than 2 percent.

6A.3.1.6 Additional Comments. The confirmation of this request is a timed confirmation, which can only be made once per transmitted MAC-symbol period. Consequently, this request will only be repeated once per transmitted MAC-symbol period.

6A.3.2 PHY_DATA.indication
6A.3.2.1 Function. This primitive is the service indication primitive for the symbol transfer service.

6A.3.2.2 Semantics. This primitive shall provide parameters as follows:

PHY_DATA.indication (MAC-symbol)

MAC-symbol may specify one of:

(1) **zero** — corresponds to a binary 0
(2) **one** — corresponds to a binary 1
(3) **non_data** — used in delimiters, always sent in pairs
(4) **pad_idle** — corresponds to one MAC-symbol period during which preamble/inter_frame_idle was received and reported as a **one**
(5) **silence** — corresponds to one MAC-symbol period of received silence (or pseudo-silence)
(6) **bad_signal** — corresponds to one MAC-symbol period during which inappropriate signaling was received or when implementation-dependent receiver checks fail (refer to Part 8)

6A.3.2.3 When Generated. This primitive is passed from the Physical layer to the MAC sublayer to indicate that the specified symbol was received from the local area network medium.

6A.3.2.4 Effect of Receipt. The effect of receipt of this primitive by the MAC sublayer entity is defined in Part 5. If a **bad_signal** symbol is received during the reception of a frame, i.e., prior to the receipt by MAC of the end delimiter, MAC will treat this as an error and abort the frame.

6A.3.2.5 Additional Comments. This indication is a timed indication, which can only be made once per received MAC-symbol period. Consequently, this indication will only be repeated once per received MAC-symbol period.

Each transmission begins with **pad_idle** symbols, and it is expected that some, but not all, of these initial symbols may be "lost in transit" between the transmitting station and the receiving stations, and consequently reported as **silence**.

Where the Physical layer encoding for successive **pad_idle**s is a sequence of **ones** and **zeros**, receivers are permitted to decode such a transmitted sequence of **pad_idle**s as a sequence of **ones** and **zeros** and report them as such to the MAC entity. In other words, a receiver need not have the ability to detect and report **pad_idle** as such; rather it may report the corresponding signaling as **data**.

In the absence of errors or colliding transmissions, and with the above two exceptions for symbols transmitted as **pad_idle**, the sequence of symbols reported is identical to the sequence of symbols transmitted by the associated PHY_DATA.requests.

6B. MAC-Physical Layer Interface Implementation Specification

6B.1 Scope and Field of Application. This part recommends an implementation of MAC-Physical interface. Part 6B applies only to those stations in which the MAC and Physical layers are embodied in two separate pieces of equipment. This part provides a recommended implementation of the abstract primitives found in Part 6A. This part describes these primitives as a set of interface signals which define the MAC-Physical interface. It specifies the requirements for the interface signals and the interactions between the signals. It also defines the physical interface requirements including electrical signal levels, grounding technique and connectors. Finally, this part assigns connector pin numbers to the interface signals.

6B.2 Terminology
6B.2.1 Model Used for Specification. References 11, 12 and 13 of Appendix I-A.
6B.2.2 Compliance. The functions of this section that apply to features implemented in this station are mandatory: however, the open assignments and coding defined are for further study and harmonization with the IEEE 802 committee and ISO TC97/SC6.

6B.3 Overview of Interactions
6B.3.1 Signals to the PHY Entity. The following signals originate in the MAC entity:

PSC0 Physical_Send_Code_Weight_0
PSC1 Physical_Send_Code_Weight_1
PSC2 Physical_Send_Code_Weight_2
PST Physical_Send_Timing
PLD Physical_Line_Disconnect
PPS Physical_Primary_Signal_Source

6B.3.2 Signals to the MAC Entity. The following signals originate in the PHY entity:

PR0 Physical_Receive_Code_Weight_0
PR1 Physical_Receive_Code_Weight_1
PR2 Physical_Receive_Code_Weight_2
PRT Physical_Receive_Timing
PWS Physical_Watchdog_Status

ISA
S72.01-1985

PROWAY-LAN

6B.4 Detailed Specification of MAC Entity-Physical Entity Interface Signals
6B.4.1 Physical_Send Signal (PSC0, PSC1, PSC2)
Direction: to PHY entity

The Physical_Send coded signal implements the possible values which may be taken on by the PHY_DATA.request primitive.

6B.4.1.1 Physical_Send Encoding. The atomic signals comprising the Physical_Send signal are:

PSC0 — Physical_Send_Code_Weight 0
PSC1 — Physical_Send_Code_Weight 1
PSC2 — Physical_Send_Code_Weight 2

The coded value of the Physical_Send signal at the ON to OFF transition of the Physical_Send_Timing (PST) signal defines the value of the PHY_DATA.request primitive for the current MAC-symbol period according to Table 6-1.

Table 6-1
Physical_Send_Encoding

PHY.request Value	Physical_Send Encoding		
	PSC2	PSC1	PSC0
silence	0	0	X
pad_idle	0	1	X
non_data	1	0	X
zero	1	1	0
one	1	1	1
bad_signal		Not Used	
1 = ON 0 = OFF	X = NOT SIGNIFICANT		

6B.4.1.2 Requirements
(1) All of the Physical_Send atomic signals (PSC0, PSC1, PSC2) shall be valid at the ON to OFF transition of the Physical_Send_Timing (PST) signal.
(2) The MAC entity shall change the state of the Physical_Send atomic signals (PSC0, PSC1, PSC2) in synchronism with an OFF to ON transition of the Physical_Send_Timing (PST) signal.
(3) The open circuit receiver condition of the Physical_Send atomic signals (PSC0, PSC1, PSC2) shall be OFF.

6B.4.2 Physical_Receive Signals (PRC0, PRC1, PRC2)
Direction: From PHY entity

the Physical_Receive coded signal implements the possible values which may be taken on by the PHY_DATA.indication primitive.

6B.4.2.1 Physical_Receive Encoding. The atomic signal comprising the Physical_Receive signal are:

PRC0 — Physical_Receive_Code_Weight_0
PRC1 — Physical_Receive_Code_Weight_1
PRC2 — Physical_Receive_Code_Weight_2

The value of the Physical_Receive signal at the ON to OFF transition of the Physical_Receive_Timing (PRT) signal defines the value of the PHY_DATA.indication primitive for the current MAC-symbol period, according to Table 6-2.

Table 6-2
Physical_Receive Encoding

PHY.indication Value	Physical_Receive Encoding		
	PRC2	PRC1	PRC0
silence	0	0	X
pad_idle	Not Used		
non_data	1	0	X
bad_signal	0	1	X
zero	1	1	0
one	1	1	1
1 = ON 0 = OFF	X = NOT SIGNIFICANT		

6B.4.2.2 Requirements
(1) All of the Physical_Receive atomic signals (PRC0, PRC1, PRC2) shall be valid at the ON to OFF transition of the Physical_Receive_Timing (PRT) signal.
(2) The MAC entity shall change the state of the Physical_Receive atomic signals (PRC0, PRC1, PRC2) in synchronism with an OFF to ON transition of the Physical_Receive_Timing (PRT) signal.
(3) The open circuit receiver condition of the Physical_Receive atomic signals (PRC0, PRC1, PRC2) shall be OFF.

6B.4.3 Timing Signals
6B.4.3.1 Physical_Send_Timing (PST)
Direction: To PHY entity
6B.4.3.1.1 Function. The Physical_Send_Timing (PST) signal establishes the transmission bit rate of the PHY entity and provides the PHY entity with signal element timing information.

6B.4.3.1.2 Requirements
(1) The Physical_Send_Timing (PST) signal shall have ON and OFF states for nominally equal periods of time at a frequency corresponding to the bit rate of the MAC entity.
(2) The OFF to ON transition of the Physical_Send_Timing (PST) signal shall cause the PHY entity to interpret the current value of the PHY_DATA.request primitive in accordance with Section 6B.4.1.
(3) The open circuit receiver condition of Physical_Send_Timing (PST) shall be OFF.
(4) The MAC entity shall maintain a constant phase relationship between the PST and PTT signals.

6B.4.3.2 Physical_Receive_Timing (PRT)
Direction: From PHY entity
6B.4.3.2.1 Function. The Physical_Receive_Timing (PRT) signal provides the MAC entity with signal_element_timing information.

6B.4.3.2.2 Requirements
(1) The ON to OFF transitions of Physical_Receive_Timing (PRT) signal shall cause the MAC entity to interpret the current value of the PHY_DATA.indication primitive in accordance with Section 6B.4.3.
(2) The open circuit receive condition of the Physical_Receive_Timing (PRT) signal shall be off.

6B.4.3.2.3 Additional Comments. The Physical_Receive_Timing (PRT) signal is derived by the PHY entity from state transitions on the transmission line including **preambles**.

6B.4.3.3 Physical_Transmit_Timing (PTT)
Direction: From PHY entity

6B.4.3.3.1 Function. The Physical_Transmit_Timing (PTT) signal provides the MAC entity with accurate timing information and is used to establish signal_element_timing with the MAC entity.

6B.4.3.3.2 Requirements
(1) The Physical_Transmit_Timing (PTT) signal is derived by the PHY entity from a frequency source which meets the requirements of 8.5.1.
(2) The open circuit receiver condition of the Physical_Transmit_Timing (PTT) signal shall be OFF.
(3) The PHY entity shall begin to provide timing information on the PTT signal when the power supply to the PHY entity is switched on.

6B.4.4 Management Signals
6B.4.4.1 Physical_Line_Disconnect (PLD)
Direction: To PHY entity

6B.4.4.1.1 Function. The Physical_Line_Disconnect (PLD) signal implements the line_disconnect value of PHY_RESET.request primitive. This enables and disables the PHY entity transmitters.

6B.4.4.1.2 Requirements
(1) When the Physical_Line_Disconnect (PLD) signal is set to ON, the PHY entity's transmitters will be unconditionally disconnected from the line.
(2) When the Physical_Line_Disconnect (PLD) signal is set to OFF, all of the PHY entity's transmitters shall be enabled for normal operation.
(3) The ON to OFF transition of the Physical_Line_Disconnect (PLD) signal shall set the watchdog function of its "normal" state which is actively checking for correct PHY entity performance.
(4) This ON to OFF transition of the Physical_Line_Disconnect (PLD) signal shall reset the jabber circuit.
(5) The open circuit condition of the Physical_Line_Disconnect (PLD) signal shall be on.

6B.4.4.2 Physical_Watchdog_Status (PWS)
Direction: From PHY entity

6B.4.4.2.1 Function. The Physical_Watchdog_Status (PWS) signal implements the **jabber_inhibit** value of the PHY_MODE_CHANGE.indication. Thus it indicates the state of the PHY entity watchdog.

6B.4.4.2.2 Requirements
(1) The ON condition of the Physical_Watchdog_Status (PWS) signal shall be maintained whenever the PHY entity's watchdog is in the "normal" state.
(2) The OFF condition of the Physical_Watchdog_Status (PWS) signal shall indicate that the PHY entity watchdog has detected a fault state.

NOTE: The PWS = OFF condition causes the PHY entity to be disconnect itself from the line.

6B.4.4.3 Physical_Primary_Signal_Source (PPS)
Direction: To PHY entity

6B.4.4.3.1 Function. the Physical_Primary_Signal_Source (PPS) signal directs the PHY entity to receive signals from the primary medium or the alternate medium.

INDUSTRIAL DATA HIGHWAY

6B.4.4.3.2 Requirements
(1) The ON condition of this signal shall cause the primary receiver to be the source of receive signal.
(2) The OFF condition shall cause the alternate receiver to be the source of receiving signal.
(3) In conjunction with the PLD signal, loopback testing of the primary and alternate receivers and transmitters may be performed. Note that loopback testing requires that the station go offline through the activation of the PLD signal.
(4) The open circuit condition of Physical_Primary_Signal_Source (PPS) shall be ON.

6B.5 MAC-PHY Interface Realization
6B.5.1 Grounding
6B.5.1.1 Signal Ground. The signal ground conductor shall connect the MAC entity circuit common (ground) to the PHY entity circuit common (ground) so as to provide a conductive path directly between the two circuit common. See Figure 6-2.

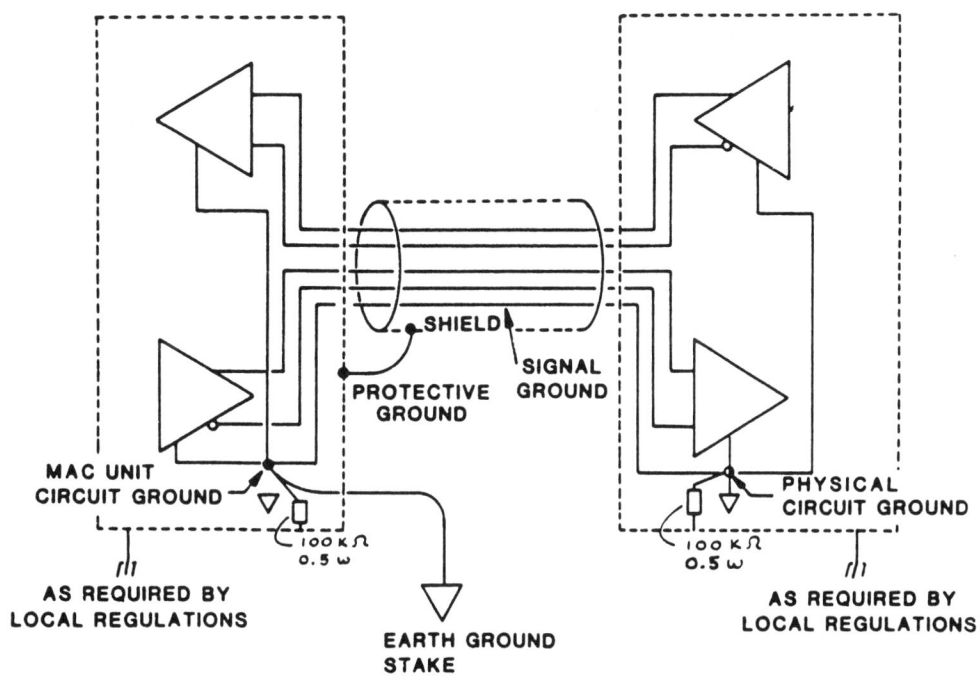

Fig 6-2
MAC Physical Grounding

6B.5.1.2 Protective Ground. The protective ground, alternatively termed frame ground, is defined for the purpose of this standard, as the electrical bonding of both the MAC and PHY entity unit to the respective equipment frame. A protective ground shall be provided and the appropriate requirements in Reference 13 of Appendix I-A, shall apply.

6B.5.1.3 External Ground Connections. Existing National safety or other National regulations shall be obesrved when any external ground is connected to the system protective ground.

NOTE: For example, some countries may have regulations concerning connections to the "Earth" line of the power supply.

6B.5.1.4 Circuit Ground Connections. The MAC entity and the PHY entity circuit commons (grounds) should each be connected in series with a resistor to their respective protective ground. When so connected, the nominal value of the resistance shall be 100K ohms with a power rating of not less than 0.5 W. Only the circuit common (ground) of the station shall be connected to real earth ground. An example of grounding arrangements is given in Figure 6-2.

6B.5.1.5 Cable Shield Connection. The shield of the MAC-PHY interconnecting cable shall be connected to the protective ground. This connection shall be made only at the MAC entity end of the cable by means of a connection between contact 1 of the unit's mating connector and protective ground leaving unconnected contact 1 of the mating connector of the opposite unit to minimize susceptibility to and generation of external noise.

WARNING: Significantly different frame potentials should be avoided. The interface connection system may not be capable of handling excessive ground currents.

6B.5.2 MAC-PHY Interface Connector

6B.5.2.1 MAC-PHY Interface Connector Mechanical Requirements. The mechanical requirements of the 37-pin interface connector to be used on the MAC and PHY entities and also at each end of the interface interconnecting cable shall be as specified in Reference 11, Appendix I-A. The MAC and PHY entities shall have fitted to each of them fixed interface connectors which have male contacts and female shells and each connector shall be equipped with two latching blocks (shown in Figure 6-3), as specified in Reference 11, Appendix I-A. Each end of the interface interconnecting cable shall be fitted with free connectors which have female contacts and male shells. These free connectors shall be equipped with means for latching to the blocks on the fixed connectors.

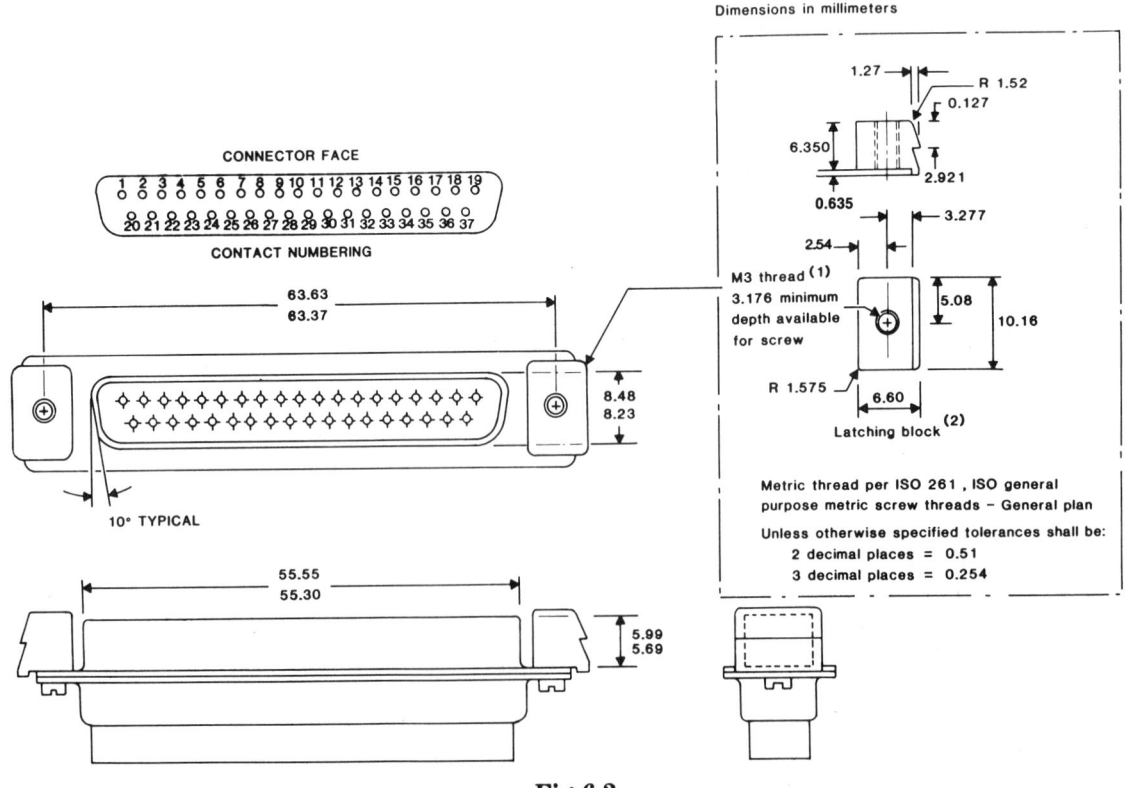

Fig 6-3
MAC-Physical Interface Connector

INDUSTRIAL DATA HIGHWAY

ISA
S72.01-1985

NOTE: The means for latching the free connectors to the blocks of the fixed connectors may be subject to National regulations.

6B.5.2.2 Assignment of MAC-PHY Connector Pin Numbers. The pin assignments for the interface signals and other necessary connections are shown in Table 6-3.

Table 6-3
MAC-Physical Interface Pin Assignments

Pin Number	Signal Connection	Pin Number	Signal Connection
1	Shield		
2	Reserved	20	Reserved
3	PSC1	21	PSC1
4	PSC2	22	PSC2
5	PST	23	PST
6	PRC2	24	PRC2
7	PSC0	25	PSC0
8	PRT	26	PRT
9	PWS	27	PWS
10	PLD	28	PLD
11	Reserved	29	Reserved
12	PRC1	30	PRC1
13	PRC0	31	PRC0
14	PPS	32	PPS
15	Reserved	33	Reserved
16	Reserved	34	Reserved
17	PTT	35	PTT
18	Reserved	36	Reserved
19	Signal Ground	37	Reserved

Pins 2-18 inclusive are A-A' interchange points and pins 20-36 inclusive are B-B' interchange points in accordance with Reference 12, Appendix I-A.
The pins designated **Reserved** shall not be used for signal or power connections that have not been defined in this standard.

6B.5.2.3 MAC-PHY Interface Connector Electrical Requirements. The MAC-PHY interface connector shall meet the electrical performance requirements as follows:

Voltage rating: 60 V

Testing voltage: 500 V; the voltage tests shall be carried out, as appropriate, in accordance with Clause 9.7.4 of IEC Publication 348

Contact rating: 5 A per contact

Contact resistance: Less than 20 milliohms

Endurance: After more than 1000 insertions the contact resistance shall not exceed 20 milliohms

Insulation resistance: Higher than 5×10^8 ohms

Contact material: Gold-plated alloy

6B.5.3 MAC-PHY Interconnection Cable

6B.5.3.1 Cable Length. The maximum length of the MAC-PHY interconnection cable used shall be such that the signal propagation time through this cable does not exceed 20% of nominal MAC_symbol_time as defined in 5B.1. The recommended maximum length is 25 meters at data rate of 1 Mbits/sec.

6B.5.3.2 Cable Requirements. The MAC-PHY interconnection cable shall include 17 twisted pairs for the balanced signal interchange circuits and a signal ground conductor. This cable shall embody an electrically conducting shield overall.

6B.5.3.3 Precaution for Electromagnetic Radiation. In installations prone to excessive electromagnetic radiation, special precautions shall be taken to ensure that the interference emf induced in the MAC-PHY interconnection cable is brought within acceptable limits.

7. Alternate Physical Layers and Media

Alternate physical layer and media specifications are given in Chapters 12/13 and 16/17 of Reference 4.

8. Specification of the Single-Channel Phase-Continuous-FSK Physical (PHY) Layer and its Interface to the Medium

8.1 Scope and Field of Application. This section specifies the functional, electrical and mechanical characteristics of the Physical (PHY) layer of this standard. This specification defines the Physical layer embodiments found in stations which could attach to a single-channel coaxial-cable-bus local area network. The relationship of this section to other sections of this standard and to LAN specifications is illustrated in Figure 8-1.

This standard specifies these Physical layer entities only insofar as necessary to insure:

(1) the interoperability of implementations conforming to this specification, and
(2) the protection of the local area network itself and those using it.

8.1.1 Nomenclature. This paragraph defines some terms used in this section whose meanings within the section are more specific than indicated in the glossary of this standard.

single-channel FSK system: a system whereby information is encoded, frequency modulated onto a carrier, and impressed on the coaxial transmission system. At any point on the medium, only one information signal at a time can be present within the channel without disruption.

Manchester encoding: a means by which separate data and clock signals can be combined into a single, self-synchronizable data stream, suitable for transmission on a serial channel. Within each data bit cell there are always two states, each with a width D/2. If the data bit is a one, the sequence of states low:high is inserted to represent the data value. If the data bit is a zero, the sequence of states high:low is inserted to represent the data value. This method creates a transition in the middle of a bit cell which is retrieved from the signal for use as a clock.

drop cable: a 75 ohm flexible coaxial cable which connects the station to the tap on the trunk cable.

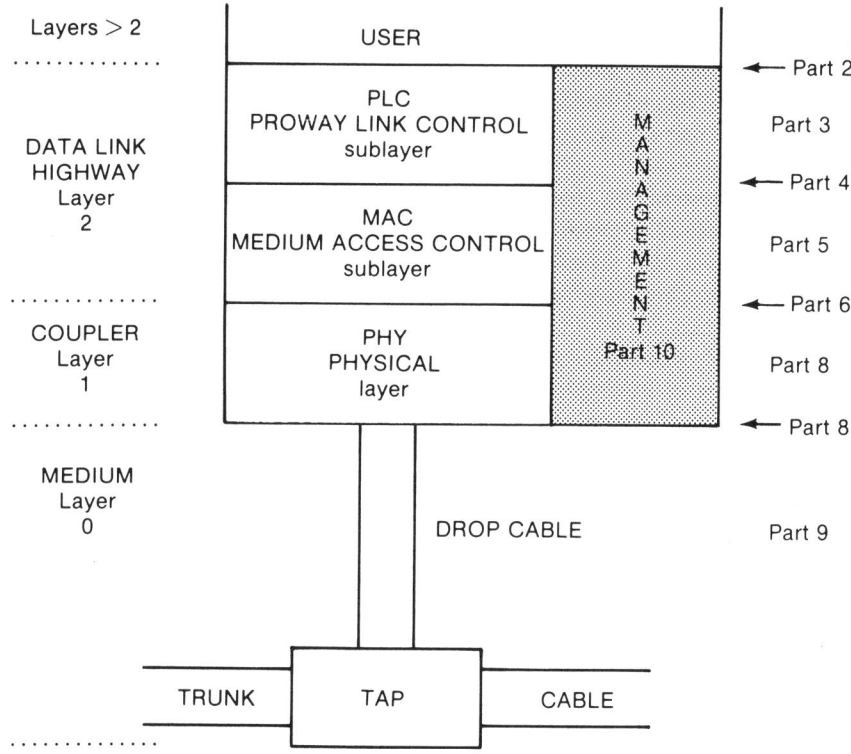

**Fig 8-1
Relationship to LAN Model and Medium**

FSK: frequency shift keying, a modulation technique whereby information is impressed upon a carrier by shifting the frequency of the transmitted signal to one of a small set of frequencies.

phase-continuous FSK: a particular form of FSK where the translations between signaling frequencies is accomplished by a continuous change of frequency (as opposed to the discontinuous replacement of one frequency by another, such as might be accomplished by a switch). Thus it is also a form of frequency modulation.

station: equipment connected to the local area network.

trunk cable: the main 75 ohm coaxial cable of a single-channel coaxial-cable-bus system.

8.1.2 Object. The object of this specification is to:

(1) Provide the physical means necessary for communication between local area network stations conforming to this standard that are connected by a single-channel coaxial-cable-bus medium.
(2) Define a physical interface that can be implemented independently among different manufacturers of hardware and achieve compatibility when interconnected by a common single-channel coaxial-cable-bus medium.
(3) Provide a communication channel capable of high bandwidth and low bit-error-rate performance. The resultant mean bit-error-rate at the MAC-PHY service interface (see Part 6) shall be less than 1×10^{-8}, with a mean undetected-bit-error rate of less than 1×10^{-9} at that interface.
(4) Provide for ease of installation and service in a wide range of environments.
(5) Provide for high network availability.

8.1.3 Compatibility Considerations. This standard applies to Physical layer entities which are designed to operate on a 75 ohm coaxial-cable-bus configured in a trunk and drop cable structure. All single-channel phase-continuous-FSK stations that signal at the same data rate shall be compatible at the medium interface specified in this part. Specific implementations based on this standard may be conceived in different ways provided compatibility at the medium interface is maintained.

8.1.4 Operational Overview of the Single-Channel Coaxial-Cable-Bus Medium. The communications medium specified in Part 9 consists of a trunk cable and drop cable structure, with branching (splitters) possible in both the trunk and drop cables by non-directional passive impedance-matching networks (**taps**). The drop cables are connected to the stations.

8.2 Overview of the Phase-Continuous-FSK Physical (PHY) Layer

8.2.1 General Description of Functions. These paragraphs describe informally the functions performed by the single-channel phase-continuous-FSK Physical layer entity. Jointly these physical entities provide a means whereby symbols presented at the MAC interface of one Physical layer entity can be conveyed to all of the Physical layer entities on the bus for presentation to their respective MAC interfaces.

8.2.1.1 Symbol Transmission and Reception Functions. Successive symbols presented to the Physical layer entity at its MAC-PHY interface are applied to an encoder which produces as output a three state PHY-symbol code: {H}, {L}, {0}.

That output is then applied to a two-tone FSK modulator which represents each transmitted {H} as the higher frequency tone, each {L} as the lower frequency tone. {0} represents the OFF condition of the transmitter output.

Each receiver is also coupled to the single-channel coaxial-cable-bus medium. It bandpass filters the received signal to reduce received noise, demodulates the filtered signal and then infers the transmitted PHY-symbol from the presence of carrier and the frequency of the received signal. It then decodes that inferred PHY-symbol by an approximate inverse of the encoding process and presents the resultant decoded MAC-symbols at its MAC-PHY interface.

For all MAC-symbols except **pad_idle**, this decoding process is an exact inverse of the encoding process in the absence of errors. The **pad_idle** symbols, which are referred to collectively as **preamble**, are transmitted at the start of each MAC frame, both to provide a training signal for receivers and to provide a non-zero minimum separation between consecutive frames. Since each transmission begins with **pad_idle** symbols, it is expected that some of these initial symbols may be "lost in transit" between the transmitter and receivers. Additionally, in phase-continuous-FSK systems, the encoding for the MAC-symbols **pad_idle** is a sequence of **ones** and **zeros** and receivers are permitted to decode the transmitted representation of **pad_idle** as a sequence of **ones** and **zeros** and report it as such to the MAC entity.

8.2.1.2 Jabber Inhibit Function. To protect the local area network from most faults in a station, each station contains a jabber-inhibit function. This function serves as a "watchdog" on the transmitter; if the station does not turn off its transmitter after a prolonged time (roughly 0.1 second), then the transmitter output must be automatically disabled for at least the remainder of the transmission.

8.2.1.3 Local Administrative Functions. These are functions which select various modes of operation. They are activated either manually, or by way of the Physical layer entity's station management interface, or both. They are:

INDUSTRIAL DATA HIGHWAY

(1) Enabling or disabling each transmitter output (a redundant medium configuration would have two or more transmitter outputs).
(2) Selecting the received signal source: any single medium (if redundant media are present), or any available loopback point.

NOTE: If a loopback point is selected, then all transmitter outputs must be inhibited.

8.2.2 Model Used for the Functional Specification. See Reference 22 of Appendix I-A.

8.2.3 Required Functions. All functions and requirements are mandatory in all implementations. However the limits appropriate to sections 8.5.2 and 8.5.3 shall be stated by the manufacturer.

8.3 Application of Station Management-PHY Layer Interface Specification. Refer to Part 10D.

8.4 Single-Channel Phase-Continuous-FSK Physical Layer Functional, Electrical and Mechanical Specifications. Unless otherwise stated, all voltage and power levels specified are in rms and dB (1 mV, 75 ohm) rms, respectively, based on transmissions of arbitrary data patterns.

8.4.1 Data Signaling Rates. The standard data signaling rate for phase-continuous-FSK systems is 1 Mbits/sec. The permitted tolerance for this signaling rate is ±0.01 percent for a transmitting station.

8.4.2 Symbol Encoding. The Physical layer entity transmits MAC-symbols presented to it at its MAC-PHY interface by the MAC entity. The possible MAC-symbols are (see 6A.3.1):

zero
one
non_data
pad_idle
silence

The transmission symbols are

{H}
{L}
{$\underline{0}$}

These transmission symbols are applied to a modulator for ½ of the bit period and transmitted.

The encoding for each of the input MAC-symbols is

silence   — Each **silence** symbol shall be encoded as the sequence {$\underline{0}$-$\underline{0}$}.
pad_idle — Each pair of **pad_idle** symbols shall be encoded as the sequence {L}{H}{H}{L}.
zero   — Each **zero** symbol shall be encoded as the sequence {H}{L}.
one   — Each **one** symbol shall be encoded as the sequence {L}{H}.
non_data — Non_data symbols are transmitted by the MAC sublayer entity in pairs. Each such pair of consecutive **non_data** symbols shall be encoded as the sequence {L}{L}{H}{H}.

Thus the **start_frame_delimiter** sequence will be encoded as

non_data	non_data	zero	non_data	non_data	zero	zero	zero
{LL},	{HH}	{HL}	{LL}	{HH}	{HL}	{HL}	{HL}

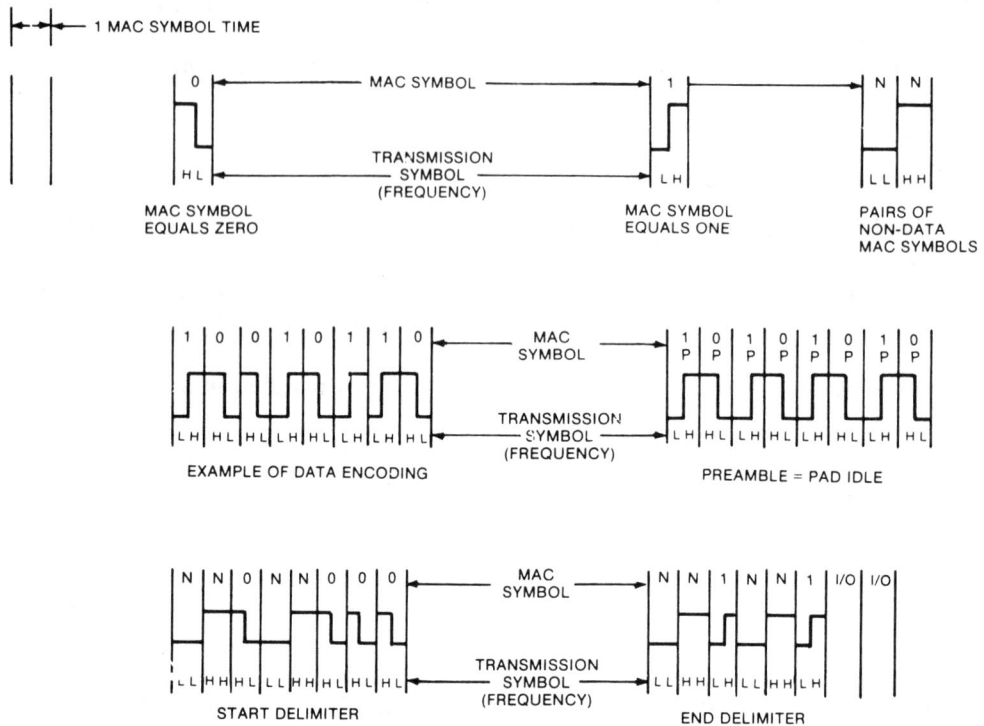

**Fig 8-2
Manchester Data Encoded Physical Signal Encodings**

The **end frame delimiter** sequence will be enclosed as

non_data	non_data	one	non_data	non_data	one	one/zero	one/zero
{LL},	{HH},	{LH},	{LL},	{HH},	{LH}		

8.4.3 Modulated Line Signal (at the line output of the station). The result of the transmission encoding step of 8.4.2 shall be applied to an FSK modulator with the result that each {H} shall be represented by the higher of the modulator's two signaling frequencies, each {L} shall be represented by the lower of the modulator's two signaling frequencies, and each {0} shall be represented by the absence of both carrier and modulation. The resultant modulated carrier shall be coupled to the single-channel coaxial-cable-bus medium as specified in 8.4.5.

8.4.3.1 The line signal shall correspond to an FSK signal with its carrier frequency at 5.00 MHz, varying smoothly between the two signaling frequencies of 3.75 MHz ±80 kHz and 6.25 MHz ±80 kHz.

8.4.3.2 Each of the transmission symbols resulting from the transmission encoding step of 8.4.2 shall be transmitted for a period equal to one-half of the inter-arrival time of the MAC-symbols which the MAC entity presents at the MAC interface. The maximum jitter in this periodicity shall be less than one percent of that MAC-symbol inter-arrival time.

8.4.3.3 When transitioning between the two signaling frequencies, the FSK modulator shall change its frequency in a continuous and monotonic manner within 100 ns, with amplitude distortion of at most ten percent.

8.4.3.4 The output level of the transmitted signal at the modulated carrier frequency into a 75 ohm resistive load shall be between +64 and +66 dB (1 mV, 75 ohm) (dBmV).

8.4.3.5 The residual or leakage transmitter-off output signal (that is, while "transmitting" the PHY-code {0}) shall be no more than -22 dB (1 mV, 75 ohm) (dBmV).

8.4.4 Jabber Inhibit. Each Physical layer entity shall have a self-interrupt capability to inhibit modulation from reaching the local area network medium. Hardware within the Physical layer shall monitor the output-on condition of the transmitter and shall provide a nominal window of 0.1 second ±25 percent during which time data link transmission may occur. If a transmission is in excess of this duration, the jabber inhibit function shall operate to inhibit any further output from reaching the medium. Reset of this jabber inhibit function shall occur upon receipt of a station management PHY_RESET.request (see Section 10D). Additional resetting means are permitted.

8.4.5 Coupling to the Medium. The Physical layer functions are intended to operate satisfactorily over a medium consisting of a 75 ohm bidirectional coaxial trunk cable, nondirectional impedance matching taps, and 75 ohm drop cables. The mechanical coupling of the station to the medium shall be to a drop cable by way of a connector on the station, as specified in Part 9.

The maximum Voltage-Standing-Wave-Ratio (VSWR) at the receiver connector shall be 1.5:1 or less when that connector is terminated with a 75 ohm resistive load. The VSWR is as measured over the spectral range of 3-7 MHz.

Both the transmitter and the receiver shall be transformer coupled to the center conductor of one of the medium's drop cables. The breakdown voltage between the windings shall be at least 500 volts ac rms. The shield of the coaxial cable medium may optionally be connected to chassis ground, and the impedance of that connection shall be less than 0.1 ohm.

8.4.6 Receiver Sensitivity and Selectivity. The Physical layer entity shall be capable of providing an undetected bit error rate of 1×10^{-9} or lower, and a detected bit error rate of 1×10^{-8} or lower, when receiving signals at 4 dBmV to +50 dBmV and with a signal to noise ratio (SNR) of 20 dB or greater. The noise shall be as measured over a spectral range of 3 to 7 MHz as described in Section 8.4.5.

In addition, each receiver must be able to properly interpret its own station's transmissions.

8.4.7 Symbol Timing. Each Physical layer entity shall recover the PHY-symbol timing information contained within the transitions between signaling frequencies of the received signal, and shall use this recovered timing information to determine the precise rate at which MAC-symbols should be delivered to the MAC interface. The jitter in this reported MAC-symbol timing relative to the PHY-symbol timing within the received signaling shall be less than eight percent. (When receiving **silence** from the medium, it shall be reported at the MAC interface at the nominal rate determined by 8.4.1 within ±25%).

8.4.8 Symbol Decoding. After demodulation and determination of each received PHY-symbol, that PHY-symbol shall be decoded by the process inverse to that described in 8.4.2, and the decoded MAC-symbols shall be reported at the MAC interface. (As noted in 8.2.1.1, receivers are permitted to decode the transmitted representation of **pad_idle** as a sequence of **ones** and **zeros**.)

Whenever a PHY-symbol sequence is received for which the encoding process has no inverse, those PHY-symbols shall be decoded as an appropriate number of **bad_signal** MAC-symbols and reported as such at the MAC-PHY interface. In such cases, the receiving entity should resynchronize the decoding process as rapidly as possible.

8.4.9 Received Signal Source Selection. The ability to select the source of received signaling, either a loopback point within the Physical layer entity or (one of) the (possibly redundant) media, as directed by the station management entity, is required. When the selected source is other than (one of) the media, the PHY entity shall disable transmission to all connected bus media automatically while such selection is in force.

8.4.10 Transmitter Enable/Disable. The ability to enable and disable the transmission of modulation onto the single-channel bus medium as directed by the station management entity is mandatory.

8.4.11 Redundant Media Considerations. Embodiments of this standard which can function with redundant media are encouraged, provided that the embodiment **as delivered** will function correctly in a non-redundant single-cable environment. Where redundant media are employed, separate N connectors and jabber-inhibit monitoring shall exist for each medium (although common inhibition is permissible), receiver signal source selection shall be provided capable of selecting any one of the redundant media, and it shall be possible to enable or disable each single transmitter independently of all other redundant transmitters when the source of received signaling is one of the redundant media.

8.4.12 Reliability. The Physical layer entity shall be designed such that its probability of causing a communication failure among other stations connected to the medium is less than 1×10^{-6} per hour of continuous (or discontinuous) operation.

Connectors and other passive components comprising the means of connecting the station to the coaxial cable medium shall be designed to minimize the probability of total network failure.

8.5 Environmental Specifications

8.5.1 Safety Requirements. This clause sets forth a number of recommendations and guidelines related to safety concerns. The list is incomplete; neither does it address all possible safety concerns. The designer and installer are urged to consult the relevant local, national, and international safety regulations to assure compliance with the appropriate standards. Reference 33 of Appendix I-A provides additional guidance on many relevant regulatory requirements.

Local area network cable systems as described in Part 9 are subject to at least four direct electrical safety hazards during their use, and designers of connecting equipment should be aware of these hazards. The hazards are:

(1) Direct contact between local network components and power or lighting circuits.
(2) Static charge buildup on local network cables and components.
(3) High-energy transients coupled onto the local network cabling system.
(4) Potential differences between safety grounds to which various network components are connected.

These electrical safety hazards, to which all similar cabling systems are subject, should be alleviated properly for a local network to perform properly. In addition to provisions for properly handling these faults in an operational system, special measures must be taken to ensure that the intended safety features are not negated when attaching or detaching equipment from the local area network medium of an existing network.

INDUSTRIAL DATA HIGHWAY

Sound installation practice is defined in Reference 15, Appendix I-A.

8.5.2 Electromagnetic and Electric Environment. Sources of interference from the environment include electromagnetic fields, electrostatic discharge, transient voltages between earth connections, and so forth. Several sources of interference will contribute to voltage buildup between the coaxial cable and the earth connection, if any, of the station.

The Physical layer entity shall meet its specifications when operating in an industrial environment.

8.5.2.1 Electromagnetic Environment. The industrial environment may include an ambient plane wave field of

(1) 2 volts/meter from 10 kHz through 30 MHz
(2) 5 volts/meter from 30 MHz through 1 GHz

8.5.2.2 Differences in Earth Potential. Typical differences in earth potential for an industrial environment are:

(1) When the medium is entirely contained in a protected area, this difference in earth potential is typically less than 10 V peak-to-peak at frequencies less than 400 Hz.
(2) When the medium is exposed to the plant environment, this difference in earth potential is typically less than 50 V peak-to-peak at frequencies less than 400 Hz.
(3) When the medium is exposed to a severe plant environment (for example a power station), this difference in earth potential may typically rise to 1000 V peak-to-peak at frequencies less than 10 MHz.

8.5.3 Temperature and Humidity. Any embodiment of this standard is expected to operate over a reasonable range of environmental conditions related to temperature, humidity, and physical handling such as shock and vibration. Specific requirements and values for these parameters are considered to be beyond the scope of this standard. Manufacturers are to indicate in the literature associated with system components and equipment (and on the components if possible) the operating environment specifications to facilitate selection, installation and maintenance of these components.

8.5.4 Regulatory Requirements. Regulatory requirements that may apply to local area network equipment and media include but may not be limited to those listed as References 15 through 20 and 33 of Appendix I-A.

8.6 Labeling. It is required that each embodiment (and supporting documentation) of a Physical layer entity conformant to this standard be labeled in a manner visible to the user with at least these parameters:

(1) Data rate capability in Mbits/sec (that is, 1 Mbit/sec).
(2) Worst-case round-trip delay which this equipment induces on a two-way transmission exchange between stations, as specified in 5B.1.6.
(3) Operating modes and selection capabilities as defined in 10D.
(4) When the station has multiple N-series connectors (for example, for redundant media) the role of each such connector shall be designated clearly by markings on the station in the vicinity of that connector.

9. Single-Channel Coaxial-Cable-Bus Medium "Layer" Specification

9.1 Scope and Field of Application. This section specifies the functional, electrical and mechanical characteristics of the medium "layer" (single-channel coaxial-cable-bus) of the PROWAY standard. This specification defines the medium "layer" embodiment of a single-channel coaxial-cable-bus local area network. The relationship of this section to other sections of this standard and to LAN specifications is illustrated in Figure 9-1.

This standard specifies the medium "layer" only insofar as necessary to insure:

(1) the interoperability of Physical layer entities conforming to this standard when connected to a medium layer conformant to this section, and
(2) the protection of the local area network itself and those using it.

9.1.1 Nomenclature

single-channel FSK system: a system whereby information is encoded, frequency modulated onto a carrier, and impressed on the coaxial transmission medium. At any point on the medium, only one information signal at a time can be present in the channel without disruption.

drop cable: a flexible coaxial cable of the single-channel coaxial-cable-bus medium which connects to a station.

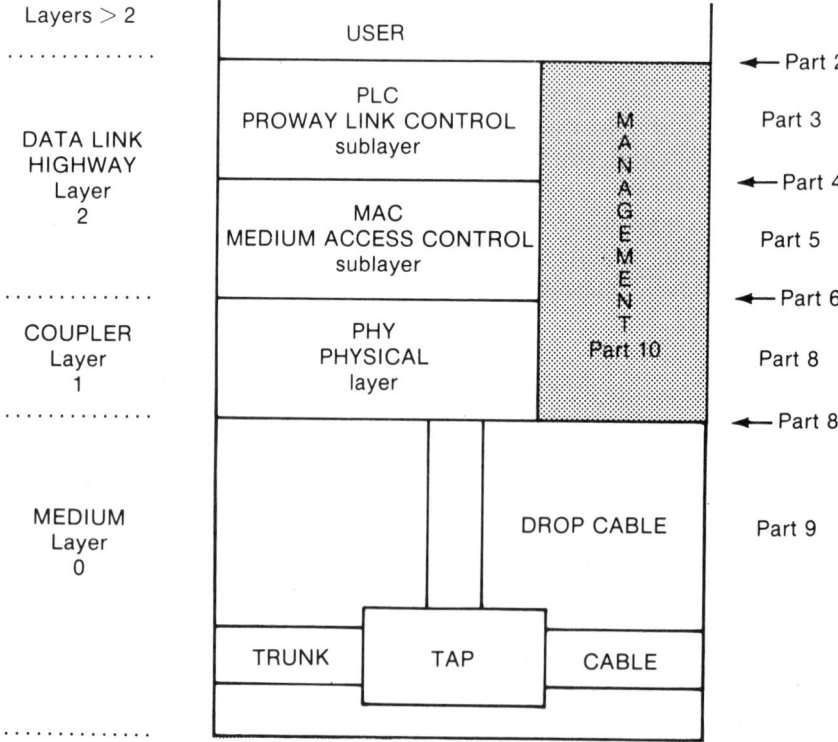

**Fig 9-1
Relationship to LAN Model**

INDUSTRIAL DATA HIGHWAY

ISA
S72.01-1985

connector: a coaxial connector.

FSK: frequency shift keying, a modulation technique whereby information is impressed upon a carrier by shifting the frequency of the transmitted signal to one of a small set of frequencies.

(impedance-matching) power splitter: a small module which electrically and mechanically couples one large diameter trunk cable to other large diameter trunk cables, providing a branching topology for each single-channel coaxial cable trunk. A power splitter combines signal energy received at its ports, splitting any signal energy received from a trunk symmetrically among the other trunks. It contains only passive electrical components (R, L, C).

(impedance-matching) splitter: a version of the power splitter used to couple drop cables together symmetrically.

station: equipment connected to the local area network.

(impedance-matching) tap: a module which electrically and mechanically couples the trunk cable to drop cables. It splits the signal energy received from each trunk cable very asymmetrically, with the bulk of that signal energy passed to the other trunk cable and only a small percentage going to the drop cables. It combines any signal energy received from the drop cables, splits a small part of that signal energy equally among the trunk cables, and passes most of the rest of that combined signal energy back to the drop cables. It contains only passive electrical components (R, L, C).

trunk cable: the main, usually larger diameter, semi-rigid, coaxial cable of a single-channel coaxial-cable-bus system.

9.1.2 Object. The object of this specification is to:

(1) specify the physical medium necessary for communication between stations conforming to this standard,
(2) provide for high network availability, and
(3) provide for ease of installation and service in a wide range of environments.

9.1.3 Compatibility. All implementations of the medium "layer" conformant to this standard shall be compatible at the interface of the drop cable to the station. Specific implementations based on this standard may be executed in different ways provided compatibility at the actual drop cable to station interface is maintained. The manufacturer shall state the values applicable to Section 9.6.

9.2 Overview of the Coaxial-Cable-Bus Medium "Layer"

9.2.1 General Description of Functions. These paragraphs describe informally the functions performed by the single-channel coaxial-cable-bus medium "layer". Jointly these functions provide a means whereby signals presented at the station interfaces to the drop cables can be combined and conveyed to all of the stations connected to any of the medium's drop cables. Thus all stations connected to this medium can communicate.

9.2.1.1 Overview of the Single-Channel Coaxial-Cable-Bus Medium. Stations are connected to the larger diameter **trunk** cable(s) of single-channel coaxial-cable-bus systems by smaller diameter **drop** cables and impedance-matching **taps**. These taps are passive devices which are nondirectional (that is, omni-directional) with regard to signal propagation. The nondirectional characteristics of the tap permits the station's signal to propagate in both directions along the trunk cable. The taps also minimize the effects of reflections due to any impedance mismatches along the trunk or on drop cables.

The topology of the single-channel coaxial-cable-bus system is that of a highly branched tree without a root. The stations are connected as leaves to the tree's branches. Branching is accomplished in the trunk itself by way of **power splitters**, which provide nondirectional coupling of the signals carried on the trunk cables similar to that of the just described taps. Like the taps, the power splitters employ only passive electrical components (R, L, C only).

Branching in the drop cables is provided by (drop cable) splitters which also employ only passive electrical components.

9.2.2 Model Used for the Functional Specification. See Reference 25, Appendix I-A.

9.2.3 Basic Characteristics and Options. All characteristics are mandatory and required in all medium implementations that claim conformance to this standard. However, the manufacturer shall state the values applicable to Section 9.6.

9.4 Single-Channel Coaxial-Cable-Bus Medium "Layer" Functional, Electrical and Mechanical Specifications. The single-channel coaxial-cable-bus medium "layer" is an entity whose sole function (relative to this standard) is signal transport between the stations of a local area network that are connected by the single-channel coaxial-cable-bus medium. Consequently only those characteristics of the medium "layer" which impinge on station-to-station signal transport, or on human and equipment safety, are specified in this standard.

An implementation of the medium "layer" shall be deemed conformant to this standard if it provides the specified signal transport services and characteristics for the stations of a single-channel coaxial-cable-bus local area network, and if it meets the relevant safety and environmental codes.

All measurements specified in the following paragraphs are to be made at the point of station connection to the medium (i.e., drop cable). Unless otherwise stated, all voltage and power levels specified are in rms and dB (1 mV, 75 ohm) [dBmV], respectively, and are based on transmissions of arbitrary data patterns.

9.4.1 Coupling to the Station. The connection of the single-channel coaxial-cable-bus medium to the station shall be by way of a flexible 75 ohm drop cable terminated in a male connector.

This combination shall mate with a female connector mounted on the station. The center conductor of the female connector shall be transformer coupled to the station's electronics. The shell of the female connector shall be electrically isolatable from the station by removal of a ground strap.

9.4.2 Characteristic Impedance and Impedance Matching. The characteristic impedance of the single-channel coaxial-cable-bus medium shall be 75 ± 3 ohm.

The maximum VSWR at each of the medium's N-connectors shall be 1.5:1 or less when the N-connector is terminated with a 75 ohm resistive load. The VSWR shall be measured over a spectral range of 3 to 7 MHz.

9.4.3 Signal Level. When receiving the signal of a single station whose transmit level is as specified in 8.4.3.4, the single-channel coaxial-cable-bus medium shall present those signals to each connected station at an amplitude of between +4 and +50 dBmV.

9.4.4 Distortion. The maximum group delay distortion shall be 25 nanosec over the spectral range of 3–7 MHz.

9.4.5 Noise Floor and Signal-to-Noise (S/N) Ratio. It is recommended that the in-band (3–7 MHz) noise floor be −15 dB (1 mV, 75 ohm) [dBmV] or less. In all cases the minimum received signal shall exceed the actual noise floor by at least 20 dBmV.

9.4.6 Cable Shielding. The shields of all trunk and drop cables shall provide an effective shielding factor of at least 90%.

9.4.7 Compatibility at the Station Interface. An embodiment of a single-channel coaxial-cable-bus "medium" is deemed to support a specific single-channel coaxial-cable-bus local area network if the requirements of 9.4.1 through 9.4.7 (inclusive) are met when measured from each point of station connection to the medium, independent of which one of the points of station connection is chosen for test signal origination.

9.4.8 Redundancy Considerations. As implied by 8.4.11, redundant single-channel coaxial-cable-bus media are encouraged by this standard. Where redundant media are employed, the provisions of 9.4.1 to 9.4.7 shall apply separately and independently to each single non-redundant medium interface.

9.4.9 Reliability. Connectors, taps and other passive components comprising the means of connecting the station to the coaxial cable medium or comprising a part of the medium shall be designed to minimize the probability of total network failure.

9.5 Installation Requirements. This clause sets forth a number of recommendations and guidelines related to safety concerns. The list is incomplete; neither does it address all possible safety concerns. The designer is urged to consult the relevant local, national, and international safety regulations to assure compliance with the appropriate standards; see Reference 33, Appendix I-A.

Local area network cable systems are subject to at least four direct electrical safety hazards during their use, and designers of connecting equipment should be aware of these hazards. The hazards are:

(1) Direct contact between local network components and power or lighting circuits
(2) Static charge buildup on local network cables and components
(3) High-energy transients coupled onto the local network cabling system
(4) Potential differences between safety grounds to which various network components are connected.

These electrical safety hazards, to which all similar cabling systems are subject, should be alleviated properly for a local area network to perform correctly. In addition to provisions for properly handling these faults in an operational system, special measures must be taken to ensure that the intended safety features are not negated when attaching or detaching equipment from the local area network medium of an existing network.

9.5.1 Sound Installation Practice. See Reference 15, Appendix I-A. Applicable local codes and regulations shall be followed in every instance in which such practice is applicable.

9.5.2 Grounding
9.5.2.1 The shields of the trunk cable segments and the tap housings of all connected taps shall be connected in series in each branch of a single-channel coaxial-cable-bus medium. This connected

conductive path shall be effectively grounded at several points along the length of the trunk cable, and at every point where the cable enters or leaves a building structure. The suggested practice is to effectively ground each tap housing, and to effectively ground the cable shield at least once per hundred meters on long cable runs between tap housings. However where large differences in ground potential exist between grounding points, other practices may be necessary. Effectively grounded means permanently connected to earth through a ground connection of sufficiently low impedance and having sufficient current-carrying capacity to prevent the building up of voltages that may result in undue hazard to connected equipment or to persons.

9.5.2.2 The shields of all drop coaxial cable segments shall be effectively grounded to the tap housings to which they connect.

9.5.2.3 Where there is reason to believe that the ground potential of an exposed shield of a coaxial cable or the housing of a connector or tap differs from the ground potential in the vicinity of that component by more than a few volts, an insulating sleeve or boot or cover should be affixed to that equipment in such a manner as to ensure that users (not installers) of the equipment will not inadvertently complete a circuit between the exposed shield or housing and the local ground through body contact.

9.5.3 Termination of Drop Cables and Taps. All taps and drop cables shall be terminated in a 75 ohm load, except that a small number of drop cables may be temporarily open circuited during maintenance procedures.

9.5.4 Regulatory Requirements. The regulatory requirements that may apply to local area network equipment and media include but may not be limited to those given in References 14-20 and 33, Appendix I-A.

In particular, the FCC requirements for radiation from the coaxial-cable-bus medium and connected equipment do apply, and must be considered in any embodiment of the medium "layer" specified in this standard.

9.5.5 Installation and Maintenance Guidelines. Installation and maintenance guidelines developed within the CATV industry for inter- and intra-facility installation of coaxial cable systems shall be followed where applicable. In addition the following caution shall be observed:

CAUTION: At no time should the shield of any portion of the coaxial trunk cable be permitted to float without an effective ground. If a section of floating cable is to be added to an existing cable system, the installer shall take care not to complete the circuit between the shield of the floating cable section and the grounded cable section through body contact.

9.5.6 Notice. The installation instructions for single-channel coaxial-cable-bus coaxial cable networks and components shall contain language which familiarizes the installer with the cautions and guidelines mentioned in Section 9.5.

9.6 Environmental Specifications

9.6.1 Electromagnetic and Electric Environment. Sources of interference from the environment include electromagnetic fields, electrostatic discharge, transient voltages between earth connections, and so forth. Several sources of interference will contribute to voltage buildup between the coaxial cable and the earth connection, if any, of the station.

The medium "layer" entity embodiment shall meet its specifications when operating in an industrial environment.

9.6.1.1 Electromagnetic Fields. The industrial environment may include an ambient plane wave field of:

(1) 2 volts/meter from 10 kHz through 30 MHz
(2) 5 volts/meter from 30 MHz through 1 GHz

9.6.1.2 Differences in Earth Potentials. Typical differences in earth potentials for an industrial environment are:

(1) Typically less than 10 V peak-to-peak at frequencies less than 400 Hz when the medium is entirely contained in a protected area.
(2) Typically less than 50 V peak-to-peak at frequencies less than 400 Hz when the medium is exposed to the plant environment.
(3) Typically rise to 1000 V peak-to-peak at frequencies less than 10 MHz when the medium is exposed to a severe plant environment (for example, a power station).

9.6.2 Temperature and Humidity. Any embodiment of this standard is expected to operate over a reasonable range of environmental conditions related to temperature, humidity, and physical handling such as shock and vibration. Specific requirements and values for these parameters are considered to be beyond the scope of this standard. Manufacturers are to indicate in the literature associated with system components and equipment (and on the components if possible) the operating environment specifications to facilitate selection, installation and maintenance of these components.

9.7 Transmission Path Delay Considerations. When specifying an embodiment of a medium "layer" which conforms to this standard, a vendor shall state the transmission_path_delay (see 5B.1) as the maximum **one-way** delay which the single-channel coaxial-cable-bus medium could be expected to induce on a transmission from any connected station to any other station. The delays induced by the transmitting and receiving stations themselves should not be included in the transmission_path_delay.

For each potentially worst-case path through the medium, a path delay is computed as the sum of the medium-induced delay in propagating a signal from one station to another. The transmission_path_delay used for determining the network's slot_time (see 5B.1) shall be the largest of these path delays for the cable system.

These path delay computations shall take into account all circuitry delays in all medium "layer" splitters or other components as well as all signal propagation delays within the cable segments themselves.

The transmission_path_delay shall be expressed in terms of the network's symbol signaling rate on the medium. When not an integral number of signaled symbols, it shall be rounded up to such an integral number. When uncertain of the exact value of the delay, vendors shall state an upper bound for the value.

9.8 Documentation. It is mandatory that each vendor of an embodiment of a medium "layer" entity conformant to this standard provide to the user supporting documentation with at least these parameters:

(1) Specific sections of this standard to which the embodiment conforms
(2) The transmission_path_delay, as specified in 9.6 and 5B.1

ISA
S72.01-1985

PROWAY-LAN

9.9 Network Configuration. The medium must provide a transmission path from each station to all other stations with an attenuation of at least 16 decibels (dB) and no more than 60 dB. This is accomplished by attaching each station to the main trunk using a drop cable and tap and calculating two worst-case attenuation values: the first a minimum, generally between the two closest stations; and a second a maximum, often between the two furthest stations. An acceptable design will place both attenuation values within the range given above. Optimum medium performance (i.e., immunity to noise and interference) will occur for designs having minimum attenuation values within the range given.

Since attenuation is dependent on cable length, cable type, and the taps used, it may be necessary to select specific medium components to arrive at an acceptable solution for any given application.

This topic is further discussed in Appendix 9-A.

Appendix 9-A
Guidelines for Configuring the Medium for a
Single-Channel Coaxial-Cable-Bus Local Area Network

This appendix is not a mandatory requirement of this standard. It is included for information only.

The following recommendations for designing and installing local area networks using coaxial cable transmission media are the result of practical experience, and correspond to typical field conditions that are encountered.

9-A.1 Network Configuration. The medium must provide a transmission path from each station to all other stations with an attenuation of at least 16 decibels (dB) and no more than 60 dB. This is accomplished by attaching each station to the main trunk using a drop cable and tap and calculating two worst-case attenuation values: the first a minimum, often between the two closest stations; and a second a maximum, generally between the two furthest stations. An acceptable design will place both attenuation values given within the range given above. Optimum medium performance (i.e., immunity to noise and interference) will occur for designs having minimum attenuation values within the range given.

Since attenuation is dependent on cable length, cable type, and the taps used, it may be necessary to select specific medium components to arrive at an acceptable solution for any given application.

9-A.2 Medium Components

9-A.2.1 Coaxial Cable. Main trunk and drop cables must be selected to meet both the physical and electrical environments and the cable system attenuation objectives. All cables must provide in excess of 90% coverage of the center conductor by the shield. Certain noisy environments may require special cable and shielding considerations.

9-A.2.1.1 Trunk Cables. Several example trunk cable types in common use by the CATV industry are given in Table 9-1. Larger diameter cables are normally lower in attenuation but less convenient to install. Cables using lower loss dielectrics are also available but some of these may be more susceptible to physical and environmental abuse. Cables with supporting steel (messenger) wire, armoring, steel tape, flooding compounds, or extra insulation sleeves are available for special applications.

Table 9-1
Example Coaxial Cable Types

Type Cable	Typical Attenuation dB/100 m at 10 MHz	V max kV(rms)	Typical Application
RG-59*	4.8	3.5	Drop cable
RG-6*	3.2	3.5	Trunk or drop cable
RG-11*	1.7	2.0	Trunk cable
0.412	1.1	TBD	Trunk cable
0.500		TBD	Trunk cable
0.750	0.6	TBD	Main trunk

* RG series type but with full shield coverage.

9-A.2.1.2 Drop Cables. The drop cables specified in this section are variable lengths of flexible 75 ohm cable, typically not to exceed 30 meters (100′) so that loss in the drop cable is less than 1 dB. This length of drop cable permits relative freedom in routing the trunk cable and locating the station. Since the drop cable length is not negligible, the drop cable must be terminated in its characteristic impedance of 75 ohm to preserve the impedance-matched conditions at the tap. Typical drop cable types are shown in Table 9-1.

9-A.2.2 Splitters. The simplest network topology is a long unbranched trunk, requiring the trunk cable to be routed near each station site, in turn. Branched topologies may be implemented by impedance-matched nondirectional splitters, which are three-port passive networks that divide the signal incident at one port into two equal parts that are transmitted to the other two ports. The insertion loss between any two ports of a typical, commercially available nondirectional splitter is 6.1 dB. When branches are implemented by way of such splitters, a separate loss budget must be calculated for each possible end-to-end path, so that the highest loss path can be used to select the trunk cable.

9-A.2.3 Trunk Connection (Transformer Tap). The drop cable is coupled to the trunk cable through a passive, nondirectional, coupling network, the transformer tap, that is impedance-matched at all ports. A fixed small fraction of the signal traveling in either direction on the trunk cable is transferred to the drop cable. A signal originating on the drop cable is attenuated and then propagates out equally in both directions on the trunk. A tap with example attenuation values is shown in Figure 9-2. Taps supporting several drop cables from a single physical housing and supporting differing trunk to drop cable attenuation values are available. The insertion losses shown represent the way the incident power is divided between the ports; they are not due to dissipation in the network. Typical attenuation values are given in Table 9-2.

In every case, all ports must be matched for proper operation. This means that a 75 ohm termination must be connected to every unused port, and that the cable attached to any port must be properly terminated.

The coupling networks consist only of passive R, L and C elements. Power connections are not required.

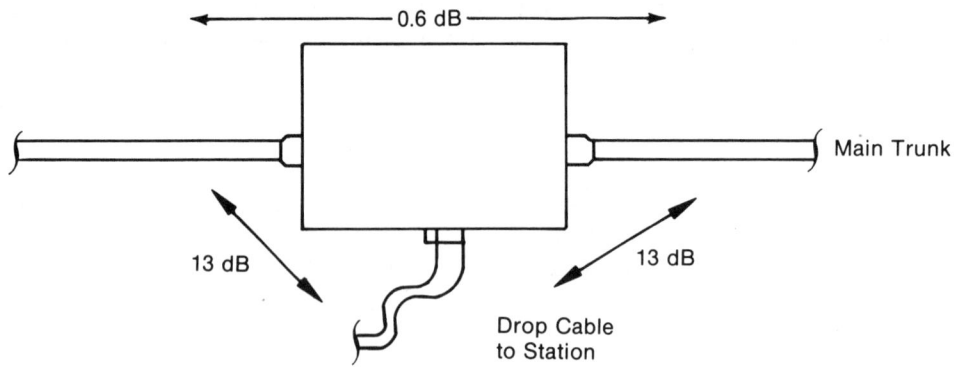

**Fig 9-2
Nondirectional Transformer Tap**

**Table 9-2
Example Specifications of
Single Drop Nondirectional Transformer Taps**

Drop Attenuation	Through Attenuation
7 dB	1.8 dB
10	1.0
13	0.6
16	0.4
19	0.3
22	0.2

The coupling networks are enclosed by sealed housings to provide both environmental protection and electrical shielding. Tap housings are available for indoor, outdoor, and below-ground applications. The housings, typically metal castings, include integral connectors for the trunk cable and drop cable(s). The housings for semi-rigid cable are designed such that the coupling network may be removed from the housing without disturbing the trunk connections.

9-A.3 Example Network Configurations

9-A.3.1 Single Control Room Application. A simple application where only ten PROWAY stations need to be supported over a 100 meter distance is shown in Figure 9-3. An acceptable design using 13 dB single drop taps and RG-11 type coaxial cable is verified by calculating minimum and maximum attenuation values.

Minimum attenuation is between adjacent stations and is 26 dB. Maximum attenuation is between stations 1 and 10 and is 33.3 dB:

Attenuation at tap 1	13
Attenuation of RG-11 main trunk (100 m)	1.5 (@ 7 MHz)
Attenuation of 8 intervening taps	4.8
Attenuation at tap 10	13
Attenuation of cable drop (2 × 40 m)	1.3 (@ 7 MHz)
Total Attenuation	33.6

INDUSTRIAL DATA HIGHWAY

ISA
S72.01-1985

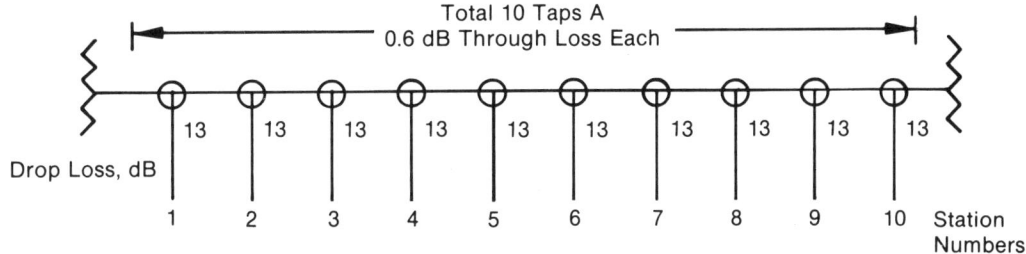

Fig 9-3
Example Cable Configuration — Simple Control Room Application

9-A.3.2 Complex Plant Area Application. A complex application where 100 PROWAY stations need to be supported over a 2 km distance is considered. The exact design will depend upon the physical topology of the application. However, for the purposes of this example it is assumed that half the stations are within 1 km of each other while the remainder are anywhere within the 2 km limit. Each station requires its own tap and up to a 40 meter drop cable between the tap and the station itself.

An initial design using 13 dB taps and 0.412 coaxial cable failed when the maximum attenuation was calculated to be 104.1 dB:

Attenuation at first tap	13
Attenuation of 0.412 main trunk (2 km)	18
Attenuation of 98 taps	58.8
Attenuation of last tap	13
Attenuation of cable drop (2 × 40 m)	1.3
Total Attenuation	104.1

An acceptable design was obtained by using lower attenuation coaxial cable by grading the tap to drop attenuation values as shown in Figure 9-4. The worst-case attenuation for the entire network is within the central 1 km area, which contains 50 taps. Maximum attenuation is between the furthest stations in the core area (stations 26 and 75 in Figure 9-4) and is 58.9 dB:

Attenuation at tap 25	22
Attenuation of 1 km 0.750 main trunk	4
Attenuation of 48 intervening 0.2 taps	9.6
Attenuation of tap 75	22
Attenuation of cable drop (2 × 40 m)	1.3
Total Attenuation	58.9

The worst case for the outlying 50 stations, considering the total 2 km area, occurs when the remaining 1 km of main trunk is arbitrarily placed between stations 25 and 26 or between stations 75 and 76. Assuming this worst-case situation, worst-case attenuation is between stations 1 and 76 and is 57.3 dB:

Attenuation at tap 1	10
Attenuation of 74 intervening taps	19
Attenuation of 2 km main trunk	8
Attenuation at tap 76	19
Attenuation of cable drops (2 × 40 m)	1.3
Total Attenuation	57.3

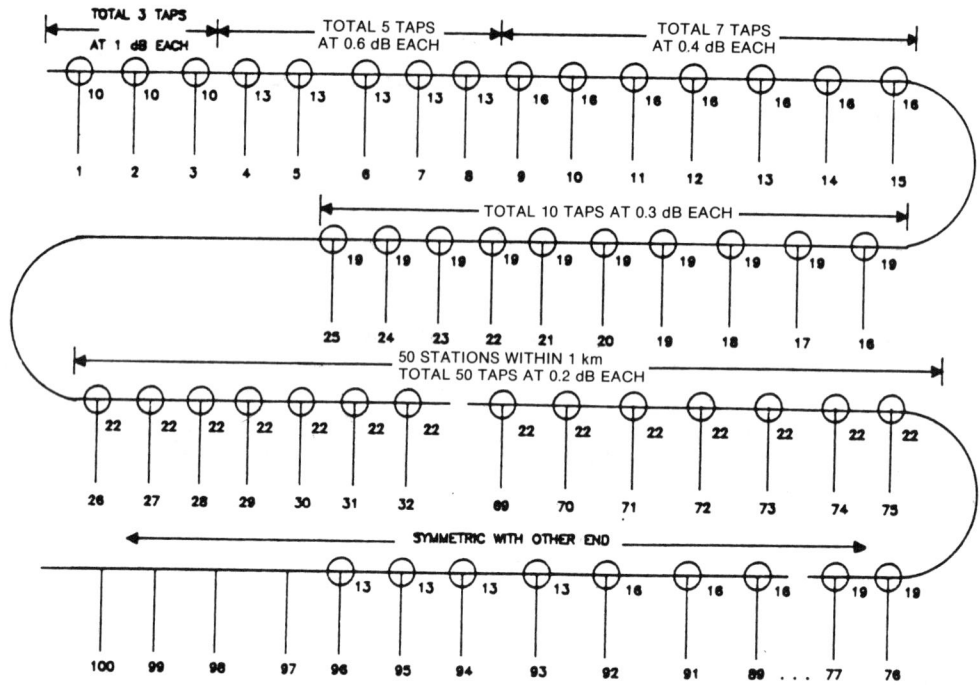

Fig 9-4
Example Cable Configuration — Complex Plant Area Application

The end-to-end attenuation between stations 1 and 100 is 56.1 dB:

Attenuation of taps 1 and 100	20
Attenuation of 98 intervening taps	26.8
Attenuation of 2 km main trunk cable	8
Attenuation of cable drops (2 × 40 m)	1.3
Total Attenuation	56.1

9-A.4 Installation Guidelines

9-A.4.1 Grounding. The trunk cable shield may be floating, single-point, or multiple-point grounded as far as signal is concerned. Grounds thus may be installed in compliance with EMI and safety codes and other regulations applicable to the particular installation. This usually means grounding where the cable enters or leaves a building, and at intervals not exceeding approximately 100 meters within the building. Grounds should be applied carefully by a clamp that does not crush or damage the cable, because such cable damage causes serious reflections. Suitable clamps are available from suppliers of CATV system hardware.

9-A.4.2 Surge Protection. It is good practice to protect cable against ground surges due to lightning. Suitable protectors that meet the IEEE-472 requirements should be at each end of the cable. The capacitive loading of protectors must be small to avoid affecting physical entity (modem) performance, and must not exceed values for a standard tap. For maximum surge protection, a low impedance, heavy-duty ground connection is required.

9-A.4.3 Termination. The trunk cable must be properly terminated at both ends. All drop cables must be properly terminated at the station end. All unused tap ports must be terminated. Shielded 75 ohm coaxial terminations with broadband characteristics are commercially available for most coaxial cables. Since transmit levels on the trunk cable are approximately 66 dB (1 mV, 75 ohm) [dBmV], power ratings of ¼ watt are sufficient.

9-A.4.4 Joining Cable Sections. In general, the trunk cable will consist of a number of separate sections of coaxial cable. Some sections will be joined by connections to tap housings, while others may be joined by splicing connectors (for semi-rigid cable) or straight-through connectors (for flexible cable). Flexible cables will be fitted with matched connectors at each end, while semi-rigid cables will simply have their ends properly prepared to mate with corresponding connectors.

A good engineering practice is to maintain constant impedance between cable sections by using one cable type from one manufacturer for the entire trunk. This practice will avoid significant reflections where cables join. When dissimilar types must be joined, it is suggested that a lossy (attenuating) impedance-matched connector, such as a tap, be employed to reduce reflections.

9-A.4.5 Pretested Cable. It is a good practice to pretest all trunk cable before installation. The objectives are to insure that the attenuation does not exceed the expected values at frequencies of interest, and to insure that concealed (that is, internal) discontinuities that can cause reflections do not exist. Most cable suppliers will pretest and certify all cable before shipment for a nominal charge.

On-site testing after installation is also recommended, since any damage may degrade operating margins or cause outright failure. A recommended method for testing the installed cable for damage, improper termination, shorts, or discontinuities is to use a time domain reflectometer, which is available from various instrument manufacturers.

10. PROWAY Management

Within the PROWAY architecture, the needs to initiate, terminate and monitor activities within a station, to monitor local area network status, and to recover from abnormal conditions are handled by the management entity of the ISO reference model. Management activities related to layers 1 and 2 of the ISO reference model fall into the following categories:

(1) Activation, maintenance and termination of LAN resources at this local station including parameter initialization and modification
(2) Monitoring of this local station's status including the reporting of statistics
(3) Error control for this local station including the performance of diagnostic, reconfiguration and restart functions
(4) Monitoring of participants in the logical token passing ring, i.e., maintenance of the active_station_list

The first three categories are station management activities. They involve exchanges of information between the local user and the local station management entity at the PROWAY management interface (Part 2B) and between the local station management entity and this local station's PLC, MAC and Physical entities at the interfaces defined in Parts 10B, 10C and 10D. These local activities are detailed in Part 10A.

The last activity requires communication with management entities of the remote station as well as with the local user and the local MAC entities. The remote communication is accomplished using the MAC_DATA service. Token ring maintenance activities are detailed in Part 10E.

10A. Station Management Activities

10A.1 Scope and Field of Application. This part specifies the local station management activities related to layers 1 and 2 of the ISO reference model provided by the management entity of this standard. This standard specifies these activities in an abstract way. It does not specify or constrain the implementation of station management entities and interfaces within a computer system. The relationship of this part to other parts of this standard and to the LAN specifications is illustrated in Figure 10A-1.

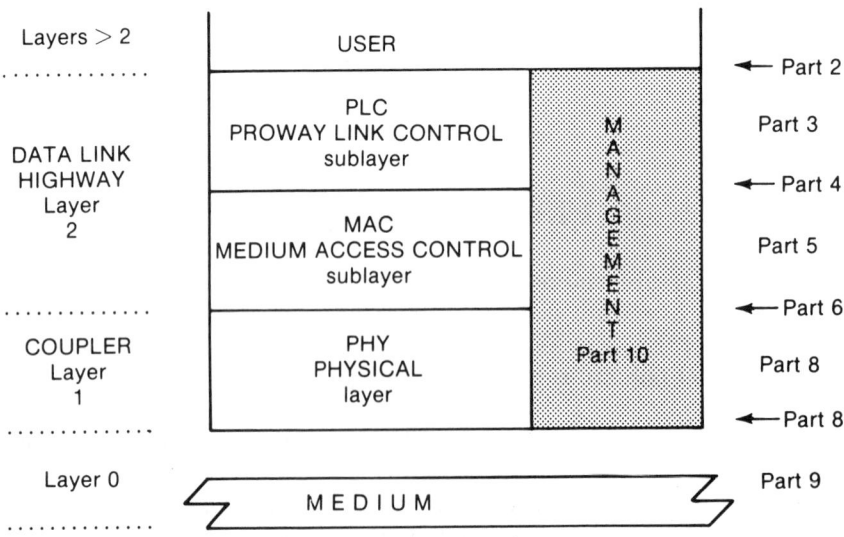

**Fig 10A-1
Relationship to LAN Model**

INDUSTRIAL DATA HIGHWAY

ISA
S72.01-1985

10A.2 Overview of Station Management Activities

10A.2.1 User Requests. The local user prepares a request for station management and passes this request to the local station management entity over the management interface defined in Part 2B. The local station management entity accepts this service request, performs any specified actions and immediately responds with a confirmation and any requested results. The local station management entity does not exchange information with remote station management entities in satisfying the user requests.

User requests to station management are:

Return Retry Counter
Enter Line Connect State
Enter Line Disconnect State
Enter Configure State
   Activate LSAP
   Activate RSAP
   Deactivate SAP

Activation of the PLC service_access_points (SAPs) is allowed only while the associated PLC state machines are in the Configure state.

Additional user requests to station management are for further study.

10A.2.2 User Notification of Changes. Important changes in the status of the local station may generate spontaneous indications to the station management entity from the PLC, MAC, or Physical entities. The station management entity conveys this information to a defined local user as an indication.

10A.3 Compliance. The functions of this section that apply to features implemented in this station are mandatory; however, the primitives and coding defined are for further study and harmonization with the IEEE 802 committee.

10A.4 Detailed Description of Station Management Activities in Response to Local User Requests. The local station management entity provides services to the local user in response to user requests.

User requests to station management are made using the L_MGMT.request primitive defined in Part 2B. Station management performs the appropriate actions and responds to the user with an L_MGMT.confirmation, also defined in Part 2B.

The following descriptions detail for each L_MGMT.request the station management actions performed and the conditions which are prerequisite to the action being performed (the states of the management entity referenced are described informally in Appendix 10-B). If the specified conditions are not met, the station management entity shall return a L_MGMT.confirmation with Mgt_status indicating the cause of failure. The value of results is unspecified in the event of failure. The arguments for each L_MGMT.request and the results returned with the corresponding L_MGMT.confirmation are given in Part 2B.

10A.4.1 Return Retry Counter

Conditions: Local station management entity in Line_Connect state.

Action: Issue MA_READ_VALUE.request (# retries of request_with_response_frames)

Get MA_READ_VALUE.confirm (# retries of request_with_response_frames)

Results: # retries of request_with_response_frames

10A.4.2 Enter Line_Connect State

Conditions: Local station management entity must be in Line_Disconnect state.

Action: Local station management entity initializes the PLC, MAC and PHY entities as specified in Parts 10B.3.1, 10C.3.3, 10C.3.9, 10D.3.1 and 10D.3.7 and then enters the Line_Connect state. Refer to Appendix 10-B for recommendations of actions the local station management entity should take in transitioning to the Line_Connect state.

No response is returned to the local user until the local station management entity reaches the Line_Connect state.

10A.4.3 Enter Line_Disconnect State

Conditions: Local station management entity must be in Line_Connect state

Action: Local station management entity exits from the token ring gracefully by setting in_ring_desired=**false** and enters the Line_Disconnect state.

Refer to Appendix 10-B for recommendations of actions local station management entity should take in transitioning to the Line_Disconnect state.

No response is made to the local user until the local station management entity reaches the Line_Disconnect state.

10A.4.4 Enter Configure State

Conditions: Local station management entity must be in the Line_Connect state with no PLC SAP components activated.

Action: Local station management entity enters the Configure state for the purpose of accepting SAP activation requests. This corresponds to entry to the Configure state of the local and remote PLC state machines described in Part 3B.

10A.4.5 Activate L_SAP

Conditions: Local and remote PLC state machines for this SAP must be in the Configure state.

Action: Issue L_SAP_ACTIVATE.request
  (SSAP,
   Services activated,
   Role for each service activated,
   Maximum L_sdu_length for each service activated)

Receive L_SAP_ACTIVATE.confirm
  (SSAP,
   Services_activated,
   Role for each service activated,
   Maximum L_sdu_length for each service activated)

Local station management entity enters Line_Connect state relative to this SAP when any associated RSAP activation for this SAP is complete. This corresponds to entry to the IDLE and RIDLE states of the local and remote state machines described in Part 3B.

10A.4.6 Activate RSAP

Conditions: Remote PLC state machine for this SAP must be in the Configure state.

Action: Issue L_RSAP_ACTIVATE.request
(SSAP to receive L_REPLY.indication,
 DSAP associated with shared_buffer_area,
 shared_buffer_identification)

Receive L_RSAP_ACTIVATE.confirm
(SSAP to receive L_REPLY.indication,
 DSAP associated with shared_buffer_area,
 shared_buffer_identification)

Local station management entity enters Line_Connect state relative to this SAP when any activation of this SAP is completed.

10A.4.7 Deactivate SAP

Conditions: Local station management entity must wait until there are no outstanding request_with_response frames for this SAP.

Action: Issue L_SAP_DEACTIVATE.request (SAP),
Receive L_SAP_DEACTIVATED.confirm (SAP).

Local station management entity enters Configure state relative to this SAP. This corresponds to entry to the Configure state of the local and remote PLC state machines described in Part 3B.

10A.5 Detailed Description of Station Management Activities in Notifying the User of Changes in Station Status.
The station management entity shall report all

PHY_MODE_CHANGE.indications,
MA_FAULT_REPORT.indications,
MA_EVENT.indications with no successor

to the user identified by SSAP=01100000 using an L_MGMT.indication as described in Part 2B. This indication shall identify the cause of the failure as reported by the PHY or MAC indication.

10B. Station Management-PLC Sublayer Interface Service Specification

10B.1 Scope and Field of Application. This part specifies the services provided by the station management entity to the PROWAY link control (PLC) sublayer of this standard. This standard specifies these services in an abstract way. It does not specify or constrain the implementation entities and interfaces within a computer system. The relationship of this to other parts of this standard and to LAN specifications is illustrated in Figure 10B-1.

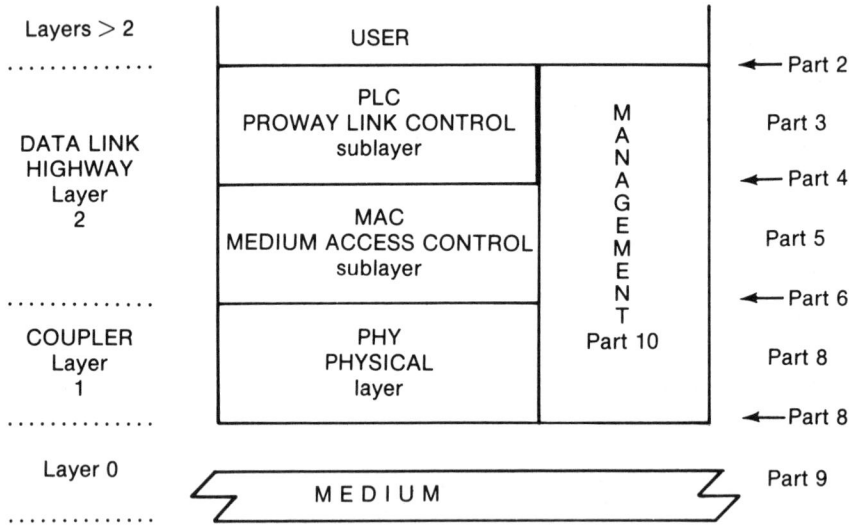

**Fig 10B-1
Relationship to LAN Model**

10B.2 Overview of Station Management-PLC Services

10B.2.1 General Description of Services Provided. This section informally describes the services provided to the station management entity by the PROWAY Link Control (PLC) sublayer. These services are local administrative services between the PLC sublayer and its manager. These services provide the means of

(1) Resetting the PLC entity
(2) Activating and configuring the PLCs service_access_points, i.e., SAPs
(3) Specifying the values of constants appropriate for the network
(4) Notifying the station management entity of relevant changes in the PLC entity's status.

10B.2.2 Model Used for the Service Specification. Reference 1, Appendix I-A.

10B.2.3 Overview of Interactions. The primitives associated with these local administrative services are:

L_RESET.request
L_STATUS.request

L_STATUS.indication
L_STATUS.confirm
L_SAP_ACTIVATE.request
L_SAP_ACTIVATE.confirm
L_RSAP_ACTIVATE.request
L_RSAP_ACTIVATE.confirm
L_SAP_DEACTIVATE.request
L_SAP_DEACTIVATE.confirm

10B.2.3.1 L_RESET. The L_RESET.request primitive is passed to the PLC entity to reset PLC entity.

10B.2.3.2 L_STATUS. The L_STATUS.request primitive is passed to the PLC entity to request a report of the status of an SAP component of a local or remote PLC entity. The L_STATUS.indication primitive is passed from the PLC entity as a result of a remote request for status information or a local change of status. The L_STATUS.confirm primitive is passed from the PLC entity to convey the results of the previous associated L_STATUS.request.

10B.2.3.3 L_SAP_ACTIVATE. The L_SAP_ACTIVATE.request primitive is passed to the PLC entity to request the activation and configuration of an LSAP component in the PLC ENTITY. This primitive does not activate or configure the RDR response component of this LSAP. The L_SAP_ACTIVATE.confirm primitive is passed from the PLC entity to convey the results of the previous associated L_SAP_ACTIVATE.request.

10B.2.3.4 L_RSAP_ACTIVATE. This primitive is passed to the PLC entity to request the activation and configuration of the component of that SAP that responds to RDR requests. The L_RSAP.confirm is passed from the PLC entity to convey the result of the previously associated L_RSAP_ACTIVATE.request.

10B.2.3.5 L_SAP_DEACTIVATE. The L_SAP_DEACTIVATE.request primitive is passed to the PLC entity to request the deactivation of an SAP component in the PLC entity. The L_SAP_DEACTIVATE.confirm primitive is passed from the PLC entity to convey the results of the previous associated L_SAP_DEACTIVATE.request.

10B.2.3.6 Services to Test Local and Remote PLC Entities. For future study are services to configure active SAP's and to test local and remote PLC entities.

10B.2.3.7 Compliance. The functions of this section that apply to features implemented in this station are mandatory; however, the primitives and coding defined are for further study and harmonization with the IEEE 802 committee.

10B.3 Detailed Specifications of Interactions with the Station Management Entity. This section describes in detail the primitives and parameters associated with the identified services. Note that the parameters are specified in an abstract sense. The parameters specify the information that must be available to the receiving entity. A specific implementation is not constrained in the method of making this information available.

10B.3.1 L_RESET.request

10B.3.1.1 Function. This primitive is the service request primitive for the PLC sublayer reset service.

10B.3.1.2 Semantics. This primitive shall be parameterless as follows:

ISA
S72.01-1985

PROWAY-LAN

L_RESET.request

10B.3.1.3 When Generated. This primitive is passed from the station management entity to the PLC sublayer entity to request that the PLC sublayer entity reset itself.

10B.3.1.4 Effect of Receipt. Receipt of this primitive causes the PLC sublayer entity to reset itself exactly as at power-on. All LSAPs, except management LSAPs, are deactivated and all management LSAPs are activated by this request.

10B.3.2 L_STATUS.request
10B.3.2.1 Function. This primitive is the service request primitive for the L_status service.

10B.3.2.2 Semantics of the Service Primitive. The primitive shall provide parameters as follows:

L_STATUS.request
  (SSAP,
  DSAP,
  remote_address
  status_type)

SSAP specifies the local LSAP associated with the status request.

DSAP and remote_address specify the remote LSAP for which status is requested. If the status request is local, these parameters are null.

The status_type specifies the type of status being requested.

10B.3.2.3 When Generated. This primitive is passed from the station management entity to the PLC entity to request local or remote status of an SAP component.

10B.3.2.4 Effect of Receipt. Receipt of this primitive specifying local status will cause the local PLC entity to indicate the requested local status. Actions of the local PLC entity on receiving a request for remote status are for future study.

10B.3.3 L_STATUS.indication
10B.3.3.1 Function. This primitive is the service indication primitive for L_STATUS service.

10B.3.3.2 Semantics of the Service Primitive. The primitive shall provide parameters as follows:

L_STATUS.indication
  (SSAP,
  DSAP,
  remote_address,
  status_type)

SSAP specifies the local LSAP associated with the status indication.

DSAP and remote_address specify the remote LSAP for which status is requested. If the status request is local or the primitive is initiated by local spontaneous action, these parameters are null.

Status_type conveys the local or remote status depending upon whether the primitive is initiated by local spontaneous action, or by receipt of a status update from a remote LSAP.

10B.3.3.3 When Generated. This primitive is passed from the PLC entity to the station management entity to inform the station management entity of the receipt of a remote status update, or to inform the station management entity of spontaneous changes within the PLC sublayer which require interaction with the upper layers and/or the station management entity.

10B.3.3.4 Effect of Receipt. The effect of receipt of this primitive by the station management entity is for future study.

10B.3.4 L_STATUS.confirm

10B.3.4.1 Function. This primitive is the service confirmation primitive for the L_STATUS service.

10B.3.4.2 Semantics. The primitive shall provide parameters as follows:

L_STATUS.confirmation
  (SSAP,
   DSAP,
   remote_address,
   status_type,
   status_value)

SSAP specifies the local LSAP associated with the status confirmation.

DSAP and remote_address specify the remote LSAP for which status was requested. If the status request is local, these parameters are null.

Status_type specifies the type of status being returned.

Status_value conveys the requested status information.

10B.3.4.3 When Generated. This primitive is passed from the PLC entity to the station management entity to convey local or remote SAP status in response to an earlier request for status.

10B.3.4.4 Effect of Receipt. The effect of receipt of this primitive by the station management entity is unspecified.

10B.3.5 L_SAP_ACTIVATE.request

10B.3.5.1 Function. This primitive is the service request primitive of LSAP activation and configuration service. This primitive deals with all components of the LSAP except those that respond to an RDR request.

10B.3.5.2 Semantics. The primitive shall provide parameters as follows:

L_SAP_ACTIVATE.request
  (SSAP,
   Services activated,
   Role in each service activated,
   Maximum L_sdu_length in each service activated)

SSAP specifies the local LSAP which is to be activated and configured.

Services activated specifies the services be supported by the SAP being activated. Allowed services are:

A given SAP may be configured to support one or more of these services in any combination.

SDA
SDN
RDR
non-PROWAY

Role specifies separately the role of this LSAP for each activated service. Allowed roles are:

INITIATOR
RESPONDER
BOTH = INITIATOR + RESPONDER

NOTE that the RESPONDER role for the RDR service is configured with the L_RSAP_ACTIVATE.request.

Maximum L_sdu_length specifies separately for each activated service the maximum size of the L_data_unit exchanged across the PLC-user interface (Part 2A). The range of maximum L_sdu_length is 1 to 1000 octets.

10B.3.5.3 When Generated. This primitive is passed from the station management entity to the PLC entity to activate and configure an LSAP, excluding its RDR response component.

10B.3.5.4 Effect of Receipt. The receipt of this primitive by the PLC entity will cause activation of an LSAP with the specified LSAP address and configuration.

10B.3.5.5 Additional Comments. The details of the "activation" are implementation-dependent.

Care should be taken that all SAPs specified in group DSAPs have been activated for the SDN service.

For efficiency, and for a well-structured system, a user should not be required to process the indication primitives from communication services it does not support. Hence the L_SAP_Activate.request specifies the type(s) of service that a user entity expects to receive from PLC. This implies that if the L_SAP_Activate.request specifies only SDA/RESPONDER and a later SDN L_pdu addressed to that LSAP is received by this (remote) PLC entity, the remote PLC entity will return an L_pdu with L_status = RS to the originating station. The remote PLC will not inform the remote user of the arrival of the SDN L_pdu.

NOTE: Management LSAPs must always be able to receive SDN Management pdus. These management LSAPs are made active when the PLC entity is activated.

In Reference 3, Appendix I-A, an additional parameter may be supplied with the L_SAP_Activate.request to support a Link level retry procedure which is not used in PROWAY. For compatibility a PROWAY station may specify:

maximum_number_of_transmissions = 1

The L_SAP_Activate.request and the L_RSAP_Activate.request for an SAP must be associated to allow coordinated exits from the Configure state as described in Part 10A and Appendix 10-B.

10B.3.6 L_SAP_ACTIVATE.confirm
10B.3.6.1 Function. This primitive is the service confirmation primitive for the SAP activation service.

10B.3.6.2 Semantics. The primitive shall provide parameters as follows:

L_SAP_Activate.confirm
 (SSAP,
  results)

SSAP indicates the local SAP for which the results are being conveyed.

The results parameter conveys the results of the previous associated L_SAP_Activate.request.

10B.3.6.3 When Generated. This primitive is passed from the PLC entity to the station management entity to convey the results of the previous associated L_SAP_Activate.request primitive. The results indicate either that the SAP activation attempt was successful or that an activation of one or more of the services requested could not be achieved. In case of failure, results indicates those services whose activation failed.

10B.3.6.4 Effect of Receipt. If the activation request was issued on behalf of a user, the receipt of this primitive by the station management causes it to issue an L_MGMT.confirm notifying the user of the success or failure of the activation.

10B.3.7 L_SAP_DEACTIVATE.request
10B.3.7.1 Function. This primitive is the service request primitive for the LSAP deactivation service.

10B.3.7.2 Semantics of the Service Primitive. The primitive shall provide parameters as follows:

L_SAP_Deactivate.request
       (SAP)

The SAP parameter specifies the LSAP address which is to be deactivated.

10B.3.7.3 When Generated. This primitive is passed from the station management entity to the PLC entity to deactivate an LSAP.

10B.3.7.4 Effect of Receipt. The receipt of this primitive by the PLC sublayer will cause the deactivation of an SAP with the given LSAP address. It may optionally cause the purging of all outstanding service requests for this LSAP.

10B.3.7.5 Additional Comments. The details of "deactivation" are implementation-dependent.

Deactivation will not occur while this LSAP has a request_with_response frame outstanding.

10B.3.8 L_SAP_DEACTIVATE.confirm
10B.3.8.1 Function. This primitive is the service confirmation primitive for the SAP deactivation service.

10B.3.8.2 Semantics. The primitive shall provide parameters as follows:

L_SAP_Deactivate.confirm
 (SAP
  results)

ISA
S72.01-1985

PROWAY-LAN

SAP indicates the local SAP for which the results are being conveyed.

The results parameter conveys the results of the previous associated L_SAP_Deactivate.request.

10B.3.8.3 When Generated. This primitive is passed from the PLC entity to the station management entity to convey the results of the previous associated L_SAP_Deactivate.request primitive. The results indicate either that the LSAP deactivation attempt was successful or that a deactivation of one or more of the services requested could not be achieved. In case of failure, results indicates those services whose deactivation failed.

10B.3.8.4 Effect of Receipt. If the deactivation request was issued on behalf of a user, the receipt of this primitive by the station management causes it to issue an L_MGMT.confirm notifying the user of the success or failure of the activation.

10B.3.9 L_RSAP_ACTIVATE.request
10B.3.9.1 Function. This primitive is the service request primitive for the RDR response component activation and configuration service.

10B.3.9.2 Semantics. The primitive shall provide parameters as follows:

L_SAP_Activate.request
  (SSAP,
   DSAP,
   shared_buffer_identification)

SSAP specifies the user which is to receive L_REPY.indications when an RDR request is received at this DSAP.

DSAP specifies the DSAP number associated with the shared_buffer_area.

Shared_buffer_identification indicates the shared buffer which is made available for the RDR service. Its format is implementation-dependent.

10B.3.9.3 When Generated. This primitive is passed from the station management entity to the PLC entity to activate and configure the RDR response component of an LSAP.

10B.3.9.4 Effect of Receipt. The receipt of this primitive by the PLC entity will initiate the activation and configuration of the indicated RDR response component.

10B.3.9.5 Additional Comments. The details of the "activation" are implement-dependent. The L_SAP_Activate.request and the L_RSAP_Activate.request for an SAP must be associated to allow coordinated exit from into the Configure state as described in Part 10A and Appendix 10-B.

10B.3.10 L_RSAP_ACTIVATE.confirm
10B.3.10.1 Function. This primitive is the service confirmation primitive for the RDR response component activation and configuration service.

10B.3.10.2 Semantics. The primitive shall provide parameters as follows:

L_RSAP_Activate.confirm
(SSAP,
DSAP,
results)

SSAP indicates the user which receives the L_REPY.indication for when an RDR request is received at this DSAP.

DSAP indicates the DSAP for which the results are being conveyed.

The results parameter conveys the results of the previous associated L_RSAP_Activate.request.

10B.3.10.3 When Generated. This primitive is passed from the station entity to the station management entity to convey the results of the previous associated L_RSAP_Activate.request primitive. The results indicate either that the DSAP configuration attempt was successful or that the configuration of the SAP could not be achieved.

10B.3.10.4 Effect of Receipt. If the activation request was issued on behalf of a user, the receipt of this primitive by the station management entity causes it to issue an L_MGMT.confirm notifying the user of the success or failure of the activation.

10C. Station Management-MAC Interface Service Specification

10C.1 Scope and Field of Application. This part specifies the services provided to the station management entity by the MAC sublayer of this standard. This standard specifies these services in an abstract way. It does not specify or constrain the implementation entities and interfaces within a computer system. The relationship of this part to other parts of this standard and to LAN specifications is illustrated in Figure 10C-1.

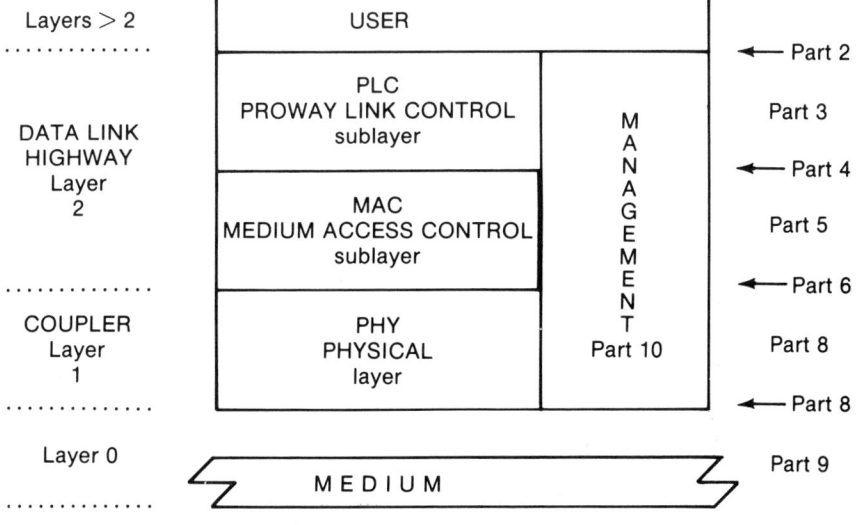

**Fig 10C-1
Relationship to LAN Model**

10C.2 Overview of the Station Management-MAC Service

10C.2.1 General Description of Services Provided. This clause describes informally the services provided to and by the station management functions by and to the token-passing Medium Access Control sublayer of the Highway layer. These services are local administrative services between the MAC sublayer and its manager. They provide the means and method of

(1) Resetting the MAC entity, and selecting the MAC entity's MAC address (and implicitly the length of all MAC addresses on the network).
(2) Specifying the values of timer and counter preset constants appropriate for the network (for example, slot_time).
(3) Determining whether the MAC entity should be a member of the token-passing ring, and whether its address appears to be unique to the local network.
(4) Reading the current values of some of the MAC entity's parameters.
(5) Notifying the station management entity of relevant changes in the MAC entity's parameters.
(6) Specifying the set of group addresses which the MAC entity recognizes.

10C.2.2 Model Used for the Service Specification. The model and descriptive method are detailed in Reference 1, Appendix I-A.

10C.2.3 Overview of Interactions. The primitives associated with local administrative services (items 1-6 above) are:

MA_INITIALIZE_PROTOCOL.request
MA_INITIALIZE_PROTOCOL.confirmation

MA_SET_VALUE.request
MA_SET_VALUE.confirmation

MA_READ_VALUE.request
MA_READ_VALUE.confirmation

MA_EVENT.indication
MA_FAULT_REPORT.indication

MA_GROUP_ADDRESS.request
MA_GROUP_ADDRESS.confirmation

10C.2.3.1 MA_INITIALIZE_PROTOCOL. The MA_INITIALIZE_PROTOCOL.request primitive is passed to the MAC sublayer to reset the entire MAC sublayer and to select the MAC entity's MAC address (and implicitly the length of all MAC addresses on the network), the token-passing protocol appropriate for the network (that is, token_bus) and the station's role in that network (i.e., originate_only). It elicits an immediate MA_INITIALIZE_PROTOCOL.confirm specifying whether the desired protocol is available.

10C.2.3.2 MA_SET_VALUE. The MA_SET_VALUE.request primitive is passed to the MAC sublayer from the station management entity to set the value of a MAC variable. The MA_SET_VALUE.confirm returns a success or failure indication for the associated primitive.

10C.2.3.3 MA_READ_VALUE. The MAC_READ_VALUE.request primitive is passed to the MAC sublayer from the station management entity to request the value of a MAC variable. The MA_SET_VALUE.confirm returns the value of the requested MAC variable.

10C.2.3.4 MA_EVENT. The MA_EVENT.indication primitive is passed to the station management entity to indicate that the value of a significant MAC variable has changed.

10C.2.3.5 MA_FAULT_REPORT. The MA_FAULT_REPORT.indication primitive is passed to the station management entity to indicate that an error has been inferred by the MAC entity.

10C.2.3.6 MA_GROUP_ADDRESS. The MA_GROUP_ADDRESS.request primitive is passed to the MAC sublayer to specify the set of MAC-level group addresses which the MAC entity should recognize, so that valid MAC-level frames with these destination addresses will be passed to the MAC user entity by way of the appropriate indicate primitive. The MA_GROUP_ADDRESS.confirm primitive returns a success or failure indication for the associated request primitive.

10C.2.4 Compliance. The functions of this section that apply to features implemented in this station are mandatory; however, the primitives and coding defined are for further study and harmonization with the IEEE 802 committee.

10C.3 Detailed Interactions with the Station Management Entity. This clause describes in detail the primitives and parameters associated with the MAC administrative services. Note that the parameters are specified in an abstract sense. The parameters specify the information that must be available to the receiving entity. A specific implementation is not constrained in the method of making this information available. For example, the station address (TS) or set-of-group-addresses parameters associated with some of the administrative service primitives may be provided by actually passing MAC addresses, by passing descriptors, or by other means. The values of some selection parameters may be implied by an implementation. The MAC sublayer may also provide local confirmation mechanisms for all request type primitives.

10C.3.1 MA_INITIALIZE_PROTOCOL.request

10C.3.1.1 Function. This primitive is the service request primitive for the protocol initialization service. It also functions as a RESET service request primitive for the entire MAC sublayer.

10C.3.1.2 Semantics. This primitive shall provide parameters as follows:

MA_INITIALIZE_PROTOCOL.request
 (desired_protocol)

Desired_protocol specifies the token-passing protocol which the MAC entity should implement; i.e., simple token_bus.

NOTE that PROWAY stations always use the simple token_bus protocol.

10C.3.1.3 When Generated. This primitive is passed from the station management entity to the MAC sublayer entity to request that the MAC sublayer entity reset itself and reconfigure itself as specified by the parameters.

10C.3.1.4 Effect of Receipt. Receipt of this primitive causes the MAC entity to reset itself exactly as at power-on, select the desired protocol, and generate a MA_INITIALIZE_PROTOCOL.confirmation to indicate the availability of the desired protocol.

10C.3.1.5 Additional Comments. This primitive causes the MAC entity to be a non-member of the token-passing ring.

10C.3.2 MA_INITIALIZE_PROTOCOL.confirmation

10C.3.2.1 Function. This primitive is the service confirmation primitive for the protocol initialization service.

10C.3.2.2 Semantics. This primitive shall provide parameters as follows:

MA_INITIALIZE_PROTOCOL.confirm
  (status)

Status indicates the success or failure of the initialization request.

10C.3.2.3 When Generated. This primitive is passed from the MAC sublayer entity to the station management entity to indicate the success or failure of the previous associated protocol initialization request.

10C.3.2.4 Effect of Receipt. The effect of receipt of this primitive is unspecified.

10C.3.3 MA_SET_VALUE.request

10C.3.3.1 Function. This primitive is the service request primitive for the setting of values of MAC sublayer variables.

10C.3.3.2 Semantics. This primitive shall provide parameters as follows:

MA_SET_VALUE.request
  (variable name,
   desired_value)

Variable_name indicates which MAC sublayer variables are to be assigned the specified desired_value. The following variables shall be settable by way of this primitive:

(1) TS (station address, see 5C.1.1)
(2) slot_time
(3) hi_pri_token_hold_time
(4) max_ac_4_rotation_time
(5) max_ac_2_rotation_time
(6) max_ac_0_rotation_time
(7) max_ring_maintenance_rotation_time
(8) ring_maintenance_initial_value
(9) max_inter_solicit_count
(10) in_ring_desired
(11) event_enable_maxk (see 10C.3.7)
(12) maximum_retry_limit

The ability to set other MAC variables is for future study.

The range of desired_value is the range defined in Part 5 for the specified MAC variable.

10C.3.3.3 When Generated. This primitive is passed from the station management entity to the MAC sublayer entity to request that the MAC sublayer entity change the value of the specified variable.

10C.3.3.4 Effect of Receipt. The receipt of this primitive causes the variable value to be changed and to generate the associated MA_SET_VALUE.confirm primitive.

10C.3.4 MA_SET_VALUE.confirm
10C.3.4.1 Function. This primitive is the service confirmation primitive for the MAC sublayer set value service.

10C.3.4.2 Semantics. This primitive shall provide parameters as follows:

MA_SET_VALUE.confirm
 (status)

Status indicates the success or failure of the set value request. If the MAC variable of the MAC_SET_VALUE.request is not implemented, status shall be returned as failure.

10C.3.4.3 When Generated. This primitive is passed from the MAC sublayer entity to the station management entity to indicate the success or failure of the previous associated MA_SET_VALUE.request.

10C.3.4.4 Effect of Receipt. The effect of receipt of this primitive is unspecified.

10C.3.5 MA_READ_VALUE.request
10C.3.5.1 Function. This primitive is the service request primitive for the reading of values of MAC sublayer variables.

10C.3.5.2 Semantics. This primitive shall provide parameters as follows:

MA_READ_VALUE.request
 (variable_name)

Variable_name indicates which of the MAC sublayer variables is to be read. Readable MAC sublayer variables are:

(1) Address of successor, NS (see 5D)
(2) Address of predecessor, PS (see 5D)
(3) Number of retries on request_with_response frames
(4) In-ring

Additional candidate variables under study are:

(1) Number of stations in ring
(2) Measured token rotation time
(3) Number of valid received frames
(4) Number of received frames with FCS errors

The ability to read other variable names is for future study.

The range of values associated with each variable is given in Part 5.

10C.3.5.3 When Generated. This primitive is passed from the station management entity to the MAC sublayer entity to request that the MAC sublayer entity return the value of the specified variable.

10C.3.5.4 Effect of Receipt. The receipt of this primitive causes the value of the specified variable to be returned by way of the associated MA_READ_VALUE.confirm primitive.

10C.3.6 MA_READ_VALUE.confirm
10C.3.6.1 Function. This primitive is the service confirmation primitive for the read value service.

10C.3.6.2 Semantics. This primitive shall provide parameters as follows:

MA_READ_VALUE.confirm
  (variable_name,
   current_value,
   status)

If status equals success, the MAC variable, variable_name, has taken on the current_value specified in the associated read value primitive. If status equals failure, the MAC variable is not readable or not implemented and the value of current_value is not specified.

10C.3.6.3 When Generated. This primitive is passed from the MAC sublayer entity to the station management entity to indicate the success or failure of the previous associated MA_READ_VALUE.request.

10C.3.6.4 Effect of Receipt. The effect of receipt of this primitive is unspecified.

10C.3.7 MA_EVENT.indication
10C.3.7.1 Function. This primitive is the service indication primitive by which station management is informed of significant events within the MAC sublayer.

10C.3.7.2 Semantics. This primitive shall provide parameters as follows:

MA_EVENT.indication
  (event)

Event identifies the event that has occurred with the MAC sublayer. The events that cause an MA_EVENT.indication are:

(1) Change of successor address
(2) Change of successor address to null

Additional events which lead to MA_EVENT.indications are for future study.

10C.3.7.3 When Generated. This primitive is passed from the MAC sublayer entity to the station management entity to indicate the occurrence of an enabled significant event within the MAC sublayer. It is enabled when an MA_SET_VALUE.request has set the event_enable mask to a\non-zero value.

10C.3.7.4 Effect of Receipt. The effect of receipt of this primitive is on the station management entity and is specified in Part 10A.

10C.3.8 MA_FAULT_REPORT.indication
10C.3.8.1 Function. This primitive is the service indication primitive for MAC failure indications.

10C.3.8.2 Semantics. This primitive shall provide parameters as follows:

MA_FAULT_REPORT.indication
  (fault_type)

Fault_type identifies the particular fault condition the MAC layer has detected. The events which cause an MA_FAULT_REPORT.indication are:

(1) Duplicate_address
(2) Faulty_transmitter

The duplicate_address fault is indicated when the MAC sublayer entity has inferred that another MAC entity on the network which has the same MAC address as the current value of the variable TS.

The faulty_transmitter fault is indicated when the MAC sublayer entity has inferred evidence that the station's transmitter is not being received correctly by other stations in the network.

10C.3.8.3 When Generated. This primitive is passed from the MAC sublayer entity to the station management entity to indicate that the MAC entity has detected a failure condition within the protocol. The indication is passed when the access control machine transitions to the offline state.

The situations in which an MA_FAULT_REPORT.indication is passed are defined in the ACM state tables in Part 5C.

10C.3.8.4 Effect of Receipt. The effect of receipt of this primitive is on the station management entity and is specified in Part 10A.

10C.3.8.5 Additional Comments. This primitive is generated in response to detection of a network administration or hardware fault which may be due to failure of circuitry in either the detecting station or another station.

10C.3.9 MA_GROUP_ADDRESS.request

10C.3.9.1 Function. This primitive is the service request primitive for the protocol's group address activation service.

10C.3.9.2 Semantics. This primitive shall provide parameters as follows:

MA_GROUP_ADDRESS.request
  (set of group-addresses)

Set of group-addresses specifies a group of zero or more MAC entity addresses.

10C.3.9.3 When Generated. This primitive is passed from the station management entity to the MAC sublayer entity to request that the MAC sublayer entity recognize the specified set of group addresses, so that valid MAC-level frames with these destination addresses will be passed to LLC, PLC or station management by way of the appropriate indication primitive.

10C.3.9.4 Effect of Receipt. Receipt of this primitive causes the MAC sublayer entity to load the desired set of group addresses for comparison with the destination address of LLC, PLC and station management frames and to return a MA_GROUP_ADDRESS.confirm primitive. The last set of group addresses loaded are the only ones recognized. Loading zero (no) group addresses will deactivate the group address recognition service.

10C.3.9.5 Additional Comments. The predefined broadcast group address (all address bits = one) is always recognized and cannot be disabled.

10C.3.10 MA_GROUP_ADDRESS.confirm

10C.3.10.1 Function. This primitive is the confirmation primitive for the protocol's group address activation service.

10C.3.10.2 Semantics. This primitive shall provide parameters as follows:

MA_GROUP_ADDRESS.confirm
(status)

Status indicates the success or failure of the request.

10C.3.10.3 When Generated. This primitive is passed from the MAC sublayer entity to the station management entity to indicate the success or failure of the previous associated MA_GROUP_ADDRESS.request.

10C.3.10.4 Effect of Receipt. The effect of receipt of this primitive is unspecified.

NOTE: To provide a basis for network monitoring and analysis, an implementation also may support the following:

(1) Reception of all data_frames, independent of destination address
(2) Reception of all frames, including MAC_control frames, as specified in 5D.1.3.

10D. Station Management-Physical Layer Interface Service Specification

10D.1 Scope and Field of Application. This section specifies the services provided to the station management entity by the Physical layer entity of this standard. The relationship of this section to other sections of this standard and to LAN specifications is illustrated in Figure 10D-1.

10D.2 Overview of the Physical Layer-Management Service
10D.2.1 General Description of Services Provided at Interface. These paragraphs describe informally the services provided to the station management entity by the Physical layer entity. These services are all local administrative services between the Physical layer and its manager. This set of services provides the means and method of:

(1) Resetting the Physical layer entity
(2) Determining the available and current operating modes of the Physical layer entity and selecting the appropriate operating modes. Modal choices can include:

INDUSTRIAL DATA HIGHWAY

(a) Transmitter output disable/enable (per drop cable)
(b) Received signal source (either a drop cable or a specified loopback point)
(c) Received signal level reporting

(3) Notifying station management of changes in current operating modes not caused by a PHY_MODE.select request.

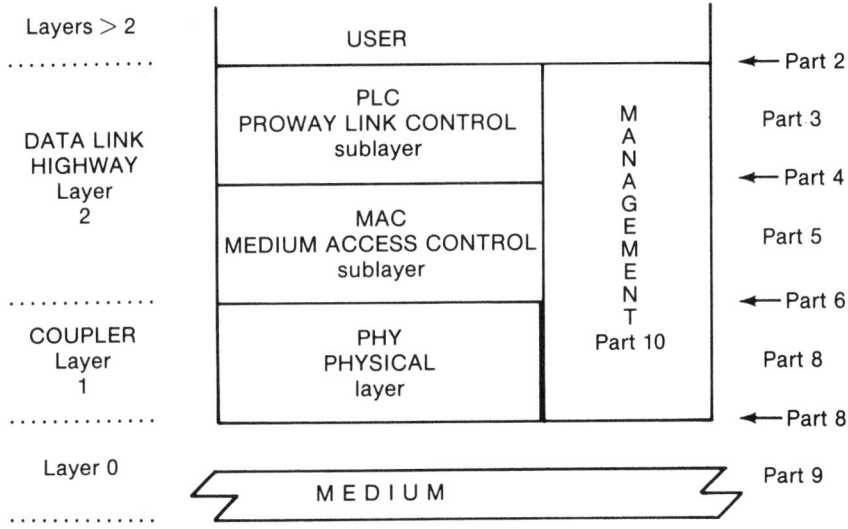

**Fig 10D-1
Relationship to LAN Model**

10D.2.2 Model Used for the Service Specification. The model and descriptive method are detailed in Reference 1, Appendix I-A.

10D.2.3 Overview of Interactions. The primitives associated with local administrative services are

PHY_RESET.request
PHY_RESET.confirm
PHY_MODE_SELECT.request
PHY_MODE_SELECT.confirm
PHY_MODE_CHANGE.indication

10D.2.3.1 PHY_RESET Request and Confirm. The PHY_RESET.request primitive is passed to the Physical layer to reset the physical entity. It elicits an immediate PHY_RESET.confirm indicating success or failure of the reset.

167

10D.2.3.2 PHY_MODE_SELECT Request and Confirm. The PHY_MODE_SELECT.request primitive is passed to the Physical layer to select the Physical layer entity's mode of operation with respect to a specified class of modal operation. It elicits an immediate PHY_MODE_SELECT.confirm primitive specifying the success or failure of the selection request.

10D.2.3.3 PHY_MODE_CHANGE.indication. The PHY_MODE_CHANGE.indication primitive is passed to the station management entity to notify it of the occurrence of a non-commanded change in one of the Physical layer entity's modes of operation (for example, transmitter disabling by way of activation of the Physical layer entity's jabber inhibit function).

10D.2.4 Compliance. The functions of this section that apply to features implemented in this station are mandatory; however, the primitives and coding defined are for further study and harmonization with the IEEE 802 committee.

10D.3 Detailed Specifications of Interactions with the Station Management Entity. This subsection describes in detail the primitives and parameters associated with the Physical layer administrative services. Note that the parameters are specified in an abstract sense. The parameters specify the information that must be available to the station management entity. A specific implementation is not constrained in the method of making this information available. For example, the representation of received signal source is not specified.

10D.3.1 PHY_RESET.request
10D.3.1.1 Function. This primitive is the service request primitive for the Physical layer reset service.

10D.3.1.2 Semantics. This primitive shall be parameterless, as follows:

PHY_RESET.request

10D.3.1.3 When Generated. This primitive is passed from the station management entity to the Physical layer entity to request that the Physical layer entity reset itself.

10D.3.1.4 Effect of Receipt. Receipt of this primitive causes the Physical layer entity to reset itself exactly as at power-on.

10D.3.2 PHY_RESET.confirm
10D.3.2.1 Function. This primitive is the service confirmation primitive for the Physical layer reset service.

10D.3.2.2 Semantics. This primitive shall provide parameters as follows:

PHY_RESET.confirm
  (LAN_topology_type,
   PHY_role)

LAN_topology_type specifies that the associated LAN is a token_bus (on a broadcast medium).

PHY_role specifies the Physical layer entity should function as an originate_only station.

10D.3.2.3 When Generated. This primitive is passed from the Physical layer entity to the station management entity in confirmation to the previous associated physical reset request.

10D.3.2.4 Effect of Receipt. The station management entity uses the LAN topology type and PHY role information to determine which protocol the associated MAC entity should employ.

10D.3.3 PHY_MODE_SELECT.request
10D.3.3.1 Function. This primitive is the service request primitive for the Physical layer current_mode select service.

10D.3.3.2 Semantics. This primitive shall provide parameters as follows:

PHY_MODE_SELECT.request
 (mode_class,
  new_mode

mode_class may specify one of the classes listed below. Other values of the mode_class parameter will elicit a failure confirmation status.

(1) transmitter_outputs_inhibit
(2) received_signal_source

new_mode specifies the desired mode of the designated mode_class. Allowed values for each mode_class:

(1) transmitter_output_inhibits:

True — transmitter is inhibited from outputting to its associated medium
False — transmitter is enabled to output to its associated medium

NOTE: In a PHY entity with multiple transmitter this primitive indicates the desired state of each transmitter.

(2) received_signal_source:

Primary — station is to listen to the primary medium (primary receiver = ON, alternate = OFF)
Alternate — station is to listen to the alternate medium (primary receiver = OFF, alternate receiver = ON)
Looped — station is to looping back from within PHY layer (both receivers = OFF)

NOTE: In a PHY entity with multiple loopback points this primitive indicates the loopback point desired.

10D.3.3.3 When Generated. This primitive is passed from the station management entity to the Physical layer entity to request that the Physical layer entity change its current operating mode of the designated mode_class to the mode specified by new_mode.

10D.3.3.4 Effect of Receipt. Receipt of this primitive causes the Physical layer entity to attempt to change its operating mode of the designated mode_class to the specified mode, and to generate a PHY_MODE_SELECT.confirm primitive to indicate the status of the requested change.

NOTE: When a looped back condition exists in the PHY entity, all transmitters must be inhibited.

10D.3.4 PHY_MODE_SELECT.confirm
10D.3.4.1 Function. This primitive is the service confirmation primitive for the Physical layer current_mode select service.

10D.3.4.2 Semantics. This primitive shall provide parameters as follows:

PHY_MODE_SELECT.confirm
  (mode_class,
   status)

mode_class may take any of the values specified in 10D.3.3. all other values of the mode_class parameter shall elicit a **failure** confirmation status.

status indicates the **success** or **failure** of the operating mode selection request.

10D.3.4.3 When Generated. This primitive is passed from the Physical layer entity to the station management entity in confirmation to the previous associated mode select request to indicate the success or failure of that request.

10D.3.4.4 Effect of Receipt. The effect of receipt of this primitive by the station management entity is unspecified.

10D.3.5 PHY_MODE_CHANGE.indication

10D.3.5.1 Function. This primitive is the service indication primitive for the Physical layer non-commanded mode change notification service.

10D.3.5.2 Semantics. This primitive shall be parameterless, as follows:

PHY_MODE_CHANGE.indication

10D.3.5.3 When Generated. This primitive is passed from the Physical layer entity to the station management entity to indicate that one or more of the Physical layer entity's current modes has changed since the previous PHY_MODE_SELECT.confirm or PHY_MODE_CHANGE.indication primitive.

10D.3.5.4 Effect of Receipt. The effect of receipt of this primitive by the station management entity is unspecified.

10D.3.5.5 Additional Comments. This primitive is used to provide timely notification of non-commanded mode changes, such as automatic transmitter_output_inhibit due to the activation of the Physical layer's jabber inhibit function.

10E. Token Ring Management

10E.1 Scope and Field of Application. This part specifies those activities of the management entity of a station that participates in the token ring that relate to maintenance and reporting of the active_station_list. These activities are referred to as Token Ring Management.

This standard specifies these activities informally and in an abstract way. It does not specify or constrain the implementation of entities and interfaces within a computer system. The relationship of this section to other sections of this standard and to LAN specifications is illustrated in Figure 10E-1.

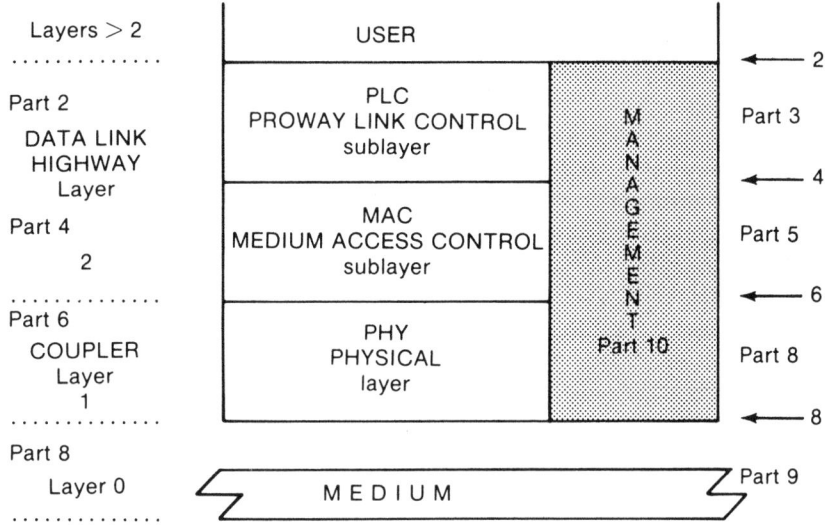

Fig 10E-1
Relationship to LAN Model

10E.2 Overview of Token Ring Management. The management entity of a PROWAY station that participates in the logical token ring has responsibilities related to ring participation. These are:

(1) Managing the station's participation in the logical token ring
(2) Maintenance of the active_station_list
(3) Reporting membership and changes in the active_station_list to PROWAY users

10E.2.1 Management of Ring Participation. The station's management of participation in the logical token ring is described in Part 10A. Recommendations to supplement Part 10A are informally given in Appendix 10B.

The PROWAY user may request management transitions to the configure, line_connect and the line_disconnect states by primitives passed across the user-management interface described in Part 2B. The configure state is a stable, non-communicating state of an LSAP component of the PLC entity which allows setting of parameters which affect the station's interaction with other participating stations. The line_connect state provides a means for an initialization of the PLC, MAC and PHY entities and for entry into the logical token ring. The line_disconnect state provides a means for orderly exit from the token ring.

10E.2.2 Maintenance of the Active_Station_List. The management entity of each station that participates in the token ring maintains the list of stations participating in that logical token ring. This list is known as the active_station_list. A station which is currently cooperating in the token-passing process, i.e., one for which the MAC variable **in-ring** is true, is known as an active station and should be included in the active_station_list. Each station's management entity maintains an active_station_list. It does this in cooperation with the management entities of all other stations participating in the token ring by receiving, adjusting, recording, and passing on the current active_station_list and active_station_change notifications. The methods for creating and maintaining this list, and the responsibilities of each station's management entity, are specified in 10E.4.

10E.2.3 Report of Active_Station_List Membership. A user may submit an L_MGMT.request for a report of all stations currently participating in the token-passing process as described in Part 2B. The list will be returned by an L_MGMT.confirm as described in Part 2B.

10E.2.4 Report of Spontaneous Changes in Active_Station_List Membership. A particular designated user identified by SSAP=01100000 will be notified by an L_MGMT.indication of all changes in active_station_list membership as described in Part 2B. The management activities in generating the L_MGMT.indication are described in 10E.4.

10E.3 Compliance. Implementation of the protocol and frame formats of Part 10E are strongly recommended for all PROWAY stations that participate in the logical token ring, i.e., all **INITIATORS**.

NOTE: Proper operation of this protocol requires participation by all token-holding stations.

10E.4 Maintenance of the Active_Station_List
10E.4.1 Informal Overview of Active_Station_List Maintenance Procedures
Change Notifications. When a station detects a change in its NS, it notifies its new NS of the change. This initiates a series of change notifications that travel in turn to each station in the token ring. At each station the notification is used to update that station's active_station_list. The notification is also passed to that station's NS if required. Change notifications for more than one change may be circulating around the token ring simultaneously.

Creation of Active_Station List. The active_station_list is initially created by a bootstrap procedure involving the change notifications.

Maintenance of Active_Station_List. The active_station_list is maintained by each station as it receives change notifications.

Receiving an Initial Active_Station_List. Stations which have never been in the ring or which have left the ring have no active_station_list except TS. A station entering (or re-entering) the ring receives an initial active_station_list from its PS.

Removal of Change Notifications. When a station receives a change notification which is redundant (i.e., its list is already correct with respect to the change) it does not continue to circulate the change. This insures that the change circulates to each station only once. The redundant notification will usually be removed by its initiator but may be removed by a successor if the initiator leaves the ring before the notification makes one round of the ring.

Transmission of Active_Station_Lists and Change Notifications. Request_with_response procedures are used to prevent aliasing of the time order of change notifications. Priority = 4 is used to allow timely notification of changes.

Interactions with Participation in the Logical Token Ring. Recommended interactions with entry into and exiting fromi the token ring are given in Appendix 10-B.

10E.4.2 Detailed Active_Station_List Maintenance Procedures. The management entity of each station that participates in the logical token ring must cooperate in maintaining the active_station_list. The following protocol must be followed by each individual station's management entity:

(1) When the management entity enters the line_connect state it must initialize its active_station_list to TS only.
(2) When the management entity receives an MA_EVENT.indication indicating that NS has changed:

IF event = NS_changed_to_null [i.e., to out-of-ring] THEN initialize active_station_list to TS only

ELSE [i.e., NS changed to non-null address]

IF current active_station_list $\leq$ TS [i.e., now in-ring]

IF (new_NS > old_NS) OR
(new_NS < any station_address in current active_station_list) [i.e., new station added to ring]

(a) Add address of new_NS to current active_station_list.
(b) Send complete active_station_list to new_NS.
(c) Generate an L_MGMT.indication of the change in the active_station_list.

ELSE [i.e., station or stations removed from ring]

(a) Send active_station_delete notification to new NS. This notification includes the addresses of all stations in current active_station_list for which TS > station_address > new_NS. No notification is sent if no stations meet this new criteria.
(b) Remove all stations from active_station_list for which TS > station_address > new_NS.
(c) Generate an L_MGMT.indication of the change in the active_station_list.

END IF

ELSE [i.e., not now in-ring]

(a) Start an active_station_list_timer which will expire in approximately 1 second.

END IF

END IF

(3) When an active_station_add_notification is received from PS FOR each address contained in the active_station_add_notification,

IF this station is NOT in the current active_station_list,

(a) Add station to active_station_list.
(b) Send an active_station_add_notification for this station to NS.
(c) Generate an L_MGMT.indication of the change in the active_station_list.

ELSE

(The notification is redundant and no action is taken.)

END IF

NEXT station_address

(4) When an active_station_delete_notification is received from PS FOR each address contained in the active_station_delete_notification,

IF this station is in the active_station_list,

(a) Delete station from active_station_list.
(b) Send an active_station_delete_notification for this station to NS.
(c) Generate an L_MGMT.indication of the change in the active_station_list.

ELSE

(The notification is redundant and no action is taken.)

END IF

NEXT station_address

(5) When an active_station_list is received from PS,

(a) Store the received list as the current active_station_list
(b) Send an active_station_add_notification for TS to NS.
(c) Stop the active_station_list_timer.

(6) When the active_station_list_timer expires [i.e., PS left the ring before transmitting an initial active_station_list],

(a) Determine address of PS using an MA_READ_VALUE.request and confirmation.
(b) Send an active_station_list_request to PS.
(c) Restart the active_station_list_timer.

10E.4.3 Communications with the Management Entities of Other Stations

10E.4.3.1 Transmission of Active_Station_List and Change_Notifications. The active_station_list and active_station_change_notifications are sent to this station's NS using the SDA service at Priority 4. The SSAP and DSAP for these transmissions shall equal 01110000.

10E.4.3.1.1 Format of Transmitted Active_Station_List. For the active_station_list, the first octet of the Mgt_pdu shall have a value of **0**. The remainder of the Mgt_pdu is a list of the station addresses of all stations in the current active_station_list of the initiating management entity. Station addresses shall be arranged in ascending numerical order to facilitate storage and examination of members of the list. Station addresses shall be in the format of Part 5D.

10E.4.3.1.2 Format of Transmitted Active_Station_Add_Notification. For active_station_add_notifications, the first octet of the Mgt_pdu shall have a value of **1**. The remainder of the Mgt_pdu shall contain station addresses of stations to be added to the active_station_list. Station addresses shall be arranged in ascending numerical order. Station addresses shall be in the format of Part 5D and Appendix 5-B.

10E.4.3.1.3 Format of Transmitted Active_Station_Delete_Notification. For active_station_delete_notifications, the first octet of the Mgt_pdu shall have a value of **2**. The remainder of the Mgt_pdu shall contain station addresses of stations to be deleted from the active_station_list. Station addresses shall be arranged in ascending numerical order. Station addresses shall be in the format of Part 5D.

10E.4.3.2 Transmission of Active_Station_List_Requests. Active_station_list_requests are sent to this station's PS using the SDN service at Priority 4. The SSAP and DSAP for these transmissions shall equal 01110000. The Mgt_pdu shall consist of one octet which shall have a value of **3**.

10F. Reset Management

The subject of resetting a station's communication function is for future study.

Appendix 10-A
Examples of Active Station List Maintenance

This appendix gives examples of the operation of active_station_list maintenance procedures as specified in Part 10E. This appendix is **not** mandatory.

```
Starting Condition - Station 3 only = on line
```

	6	5	4	3	2	1
NS =				3			MAC
LIST =				3			STA.MGT

Event - Station 2 comes on line

Resulting Condition

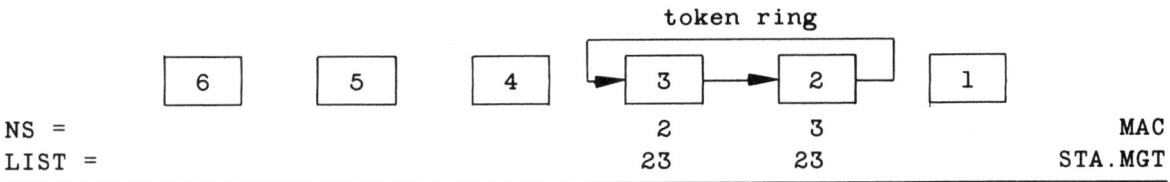

NS =				2	3		MAC
LIST =				23	23		STA.MGT

Description:

(a) Station 2 comes on line and is added to token ring
(b) Station 3 receives MA_EVENT.indication with new_NS = 2
(c) Station 3 adds 2 to active_station_list
(d) Station 3 sends active_station_list to its new_ NS (Station 2)
(e) Station 2 receives and saves the active_station_list
(f) Station 2 sends notification for station 2 to its NS (Station 3)
(g) Station 3 receives the notification and detects that it is redundant (i.e., Station 2 is already in the list) and so ends the cycle and the active_station_list is formed.

Starting Condition - All Stations Active

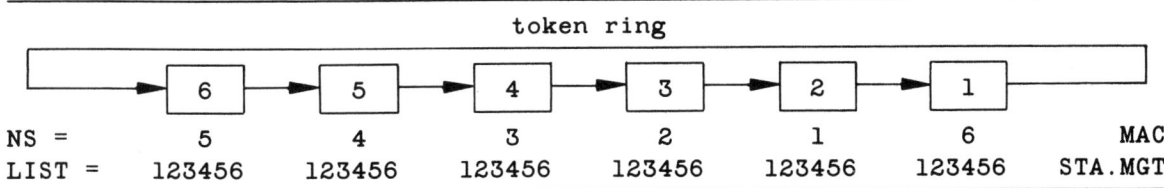

| NS = | 5 | 4 | 3 | 2 | 1 | 6 | MAC |
| LIST = | 123456 | 123456 | 123456 | 123456 | 123456 | 123456 | STA.MGT |

Event - Stations 2 and 3 leave the ring

Resulting Condition

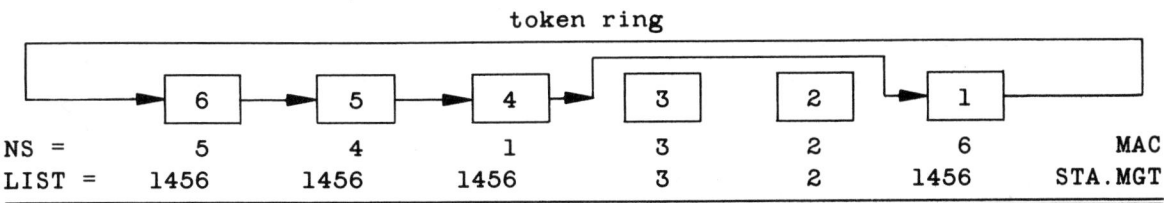

| NS = | 5 | 4 | 1 | 3 | 2 | 6 | MAC |
| LIST = | 1456 | 1456 | 1456 | 3 | 2 | 1456 | STA.MGT |

Description:

(a) Stations 2 and 3 leave the ring
(b) Station 4 management entity receives MA_EVENT.indication with new_NS = 1. This condition is new_NS < old_NS indicating a station or stations were deleted from the ring
(c) Station 4 deletes from its active_station_list all station addresses such that TS > station address > new_NS (i.e., 2 and 3)
(d) Station 4 sends an active_station_delete_notification to its new_NS listing all stations from its active_station_list such that TS > address > new_NS (i.e., 2 and 3)
(e) Station 1 receives the active_station_delete_notification from 4 and deletes the listed stations from its active_station_list
(f) Station 6 receives the active_station_delete_notification from 1 and deletes the listed stations from its active_station_list
(g) Station 5 receives the active_station_delete_notification from 6 and deletes the listed stations from its active_station_list
(h) Station 4 receives the active_station_delete_notification from 5, detects that the listed deletions are redundant, and ends the cycle by ignoring the notification

Starting Condition — Stations 1, 2, 4, 6 Active

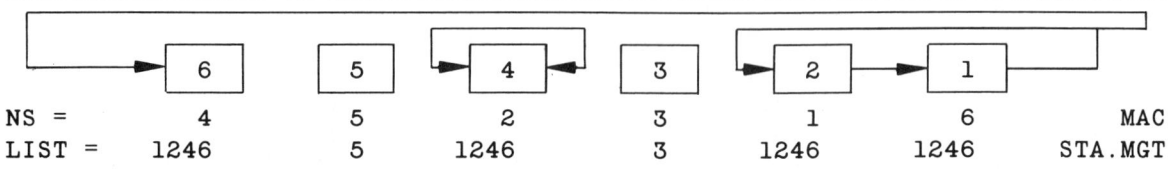

Event — Stations 3 and 5 enter the ring simultaneously

Resulting Condition — All Stations Active

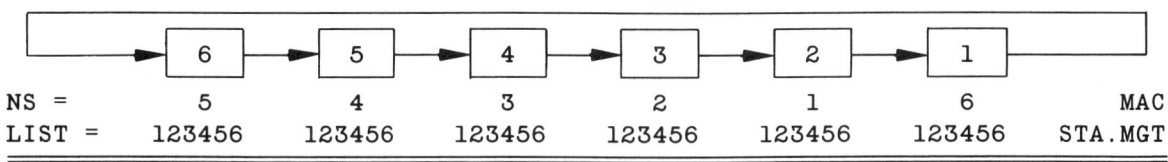

Description:

Notes: (a) is the common activity of both changes occurring, simultaneously (b) and all items at left margin are results of 5 entering the ring (c) and all indented items are the result of 3 entering the ring

(a) Stations 3 and 5 enter the ring at essentially the same instant
(b) Station 6 receives a MA_EVENT.indication with new_NS = 5 and begins a recycle by sending its active_station_list to station 5
(c) Station 4 receives MA_EVENT.indication with = 3 and begins a cycle by sending its active_station_list to station 3
(d) Station 5 receives the list from 6 and saves it

```
(e) Station 5 sends active_station_add_notification to station 4 to add 5
(f) Station 3 receives the active_station_list from 4 and saves it
(g) Station 3 sends an active_station_add_notification to station 2 to
add 3
(h) Station 4 adds 5 to its active_station_list and passes on the add 5
notification to 3
(i) Station 2 adds 3 to its active_station_list and passes on the add 3
notification to 1
(j) Station 3 adds 5 to active_station_list and passes on the add 5
notification to 2
(k) Station 1 adds 3 to active_station_list and passes on the add 3
notification to 6
(l) Station 2 adds 5 to active_station_list and passes on the add 5
notification to 1
(m) Station 6 adds 3 to active_station_list and passes on the add 3
notification to 5
(n) Station 1 adds 5 to active_station_list and passes on the add 5
notification to 6
(o) Station 5 adds 3 to active_station_list and passes on the add 3
notification to 4
(p) Station 6 receives the add 5 notification and stops the cycle because
the notification is redundant
(q) Station 4 receives the add 3 notification and stops the cycle because
the notification is redundant
```

Appendix 10-B
Recommendations for Management of
Ring Participation and Redundant Media

This appendix gives recommendations for management of a station's participation in the token ring and for management of redundant media. These recommendations are **not** mandatory.

10-B.1 Recommendations for Ring Participation Management
10-B.1.1 Management of Ring Participation.
The station's management of participation in the logical token ring is described in Part 10A. Recommendations to supplement Part 10A are informally given in Appendix 10-B.

The PROWAY user may request management transitions to the configure, line_connect and the line_disconnect states by primitives passed across the user-management interface described in Part 2B. The configure state is a stable, non-communicating state of an LSAP component of the PLC entity which allows setting of parameters which affect the station's interaction with other participating stations. The line_connect state provides a means for an initialization of the PLC, MAC and PHY entities and for entry into the logical token ring. The line_disconnect state provides a means for orderly exit from the token ring.

10-B.1.2 Ring Participation Management Station Descriptions.
This section provides an informal description of the state diagram of Figure 10-2.

ISA
S72.01-1985

PROWAY-LAN

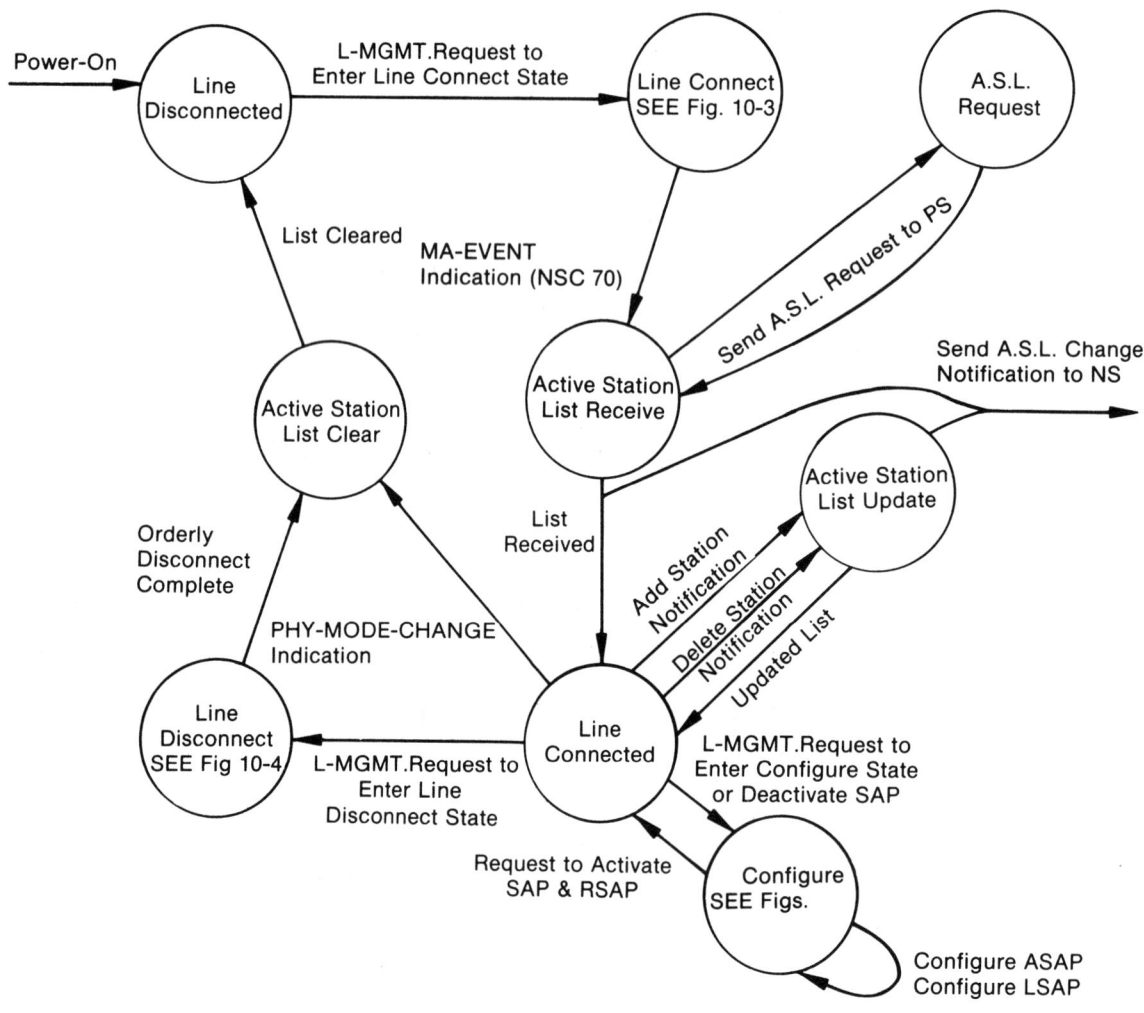

**Fig 10-2
Ring Participation Management**

10-B.1.3 Line_Disconnect State. The line_disconnect state is entered after the completion of an orderly disconnect in response to a user L_MGMT.request to enter line_disconnect state. The transitions required to achieve an orderly disconnection are described in Figure 10-3.

The line_disconnect state exits to the line_connect state in response to a user L_MGMT.request to enter line_connect state as described in Part 10A.

10-B.1.4 Line_Connect State. The line_connect state is entered from the line_disconnect state in response to a user L_MGMT.request to enter the line_connect state. This is the user's method of requesting connection to the network. Entry to the line_connect state starts the process of establishing an orderly connection to the network. Prior to attempting a connection, the PLC, MAC and PHY entities are initialized as described in Part 10A.4.2. The transitions required to achieve an orderly connection are described in Figure 10-4.

INDUSTRIAL DATA HIGHWAY

ISA
S72.01-1985

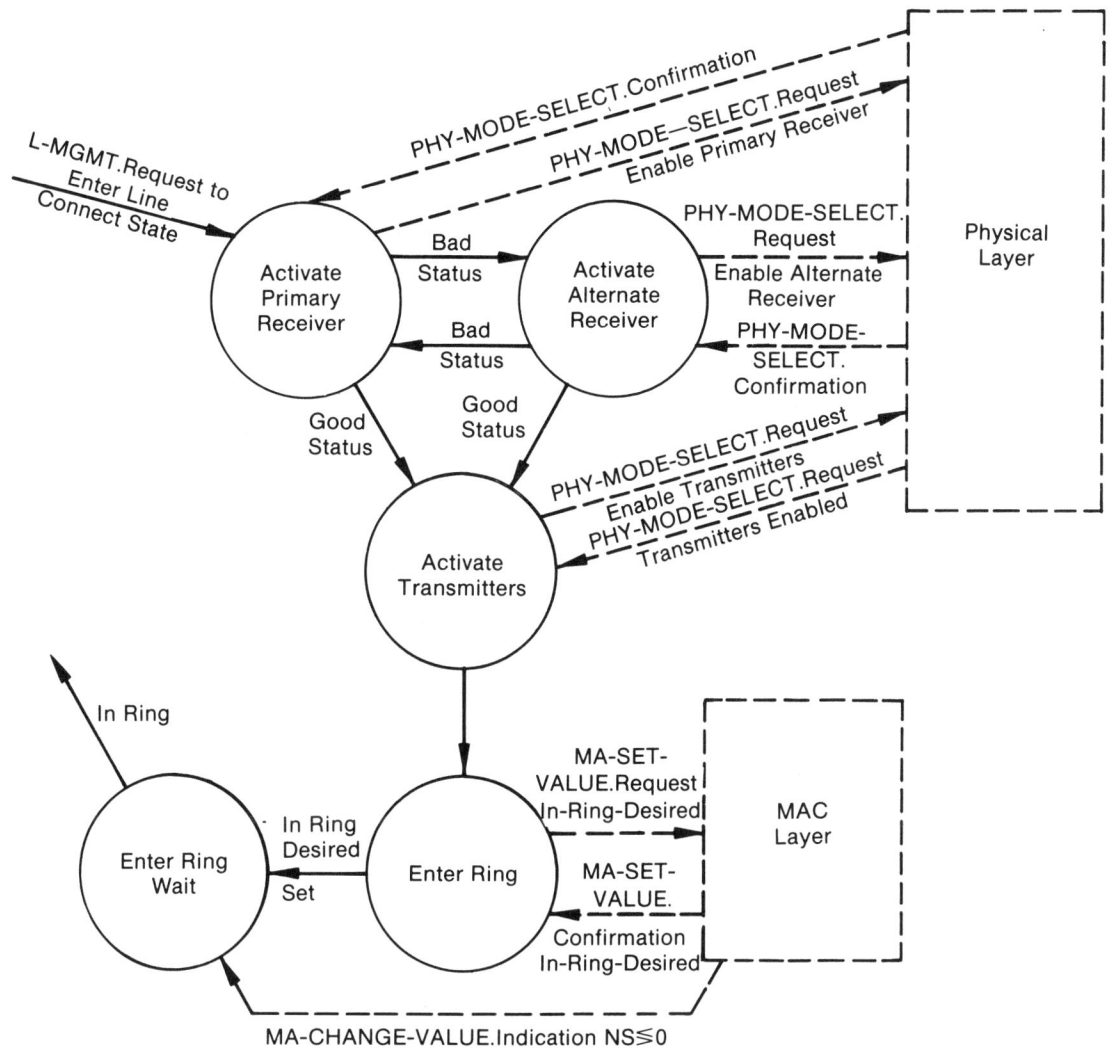

Fig 10-3
Line_Connect State Diagram

The line_connect state exits to the active_station_list_receive state after the orderly connection to the network is complete. The details of this condition are described in Figure 10-4.

10-B.1.5 Active_Station_List_Receive State. The active_station_list_receive state is entered from the line_connect state after the orderly connection of the station as detailed in Figure 10-4. The station waits in this state for an active_station_list to be received from the station's PS as described in Part 10E.

The active_station_list_receive substate is exited to the line_connect state upon receipt of an active_station_list.

10-B.1.6 Line_Connect State. The line_connect state is entered from the active_station_receive state upon receipt of an active_station_list from the station's PS. This is the nominal state allowing communication by this station on the local area network.

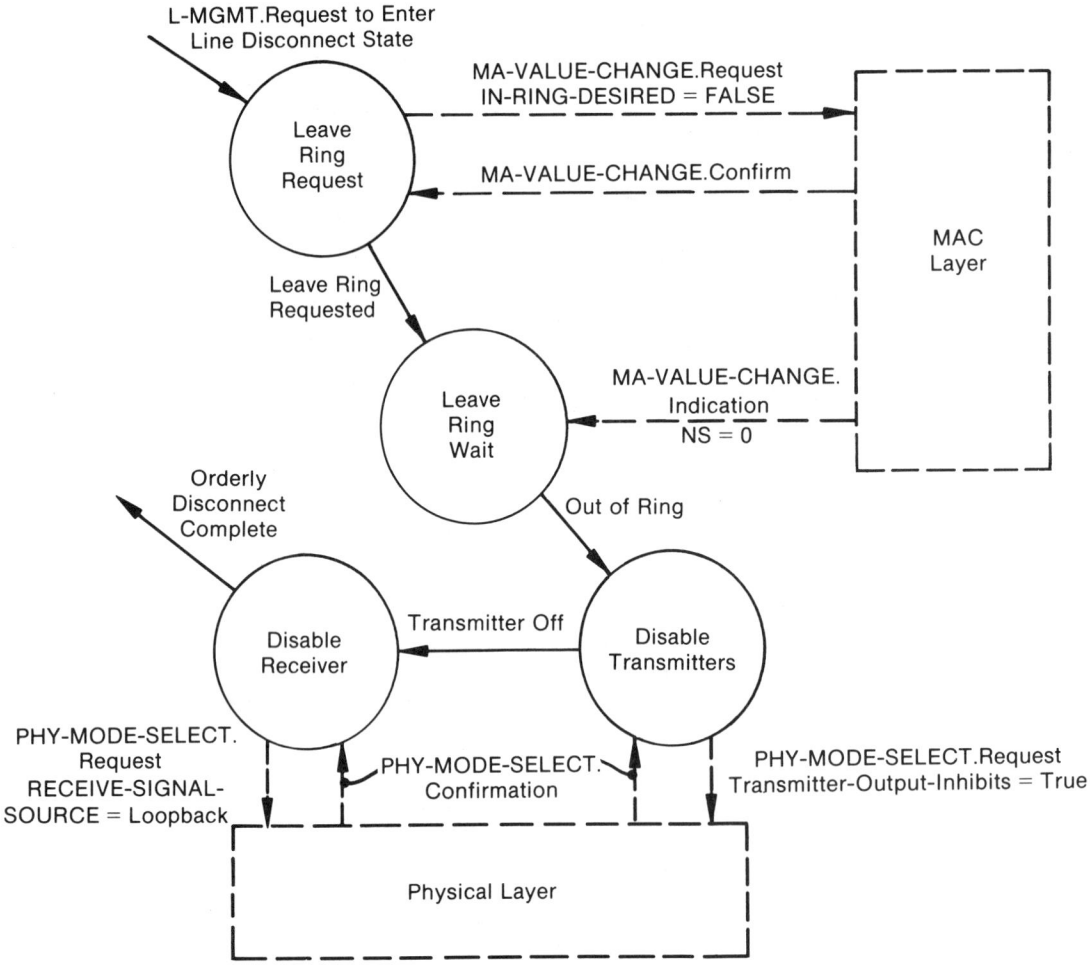

**Fig 10-4
Line_Disconnect State Diagram**

The line_connect state may exit to one of four states as described below.

(1) Receipt of an active_station_add_notification or active_station_delete_notification will cause an exit from the line_connect state to the active_station_list update substate.
(2) Receipt of a PHY_MODE_CHANGE.indication will cause an exit from the line_connect state to the active_station_list_clear state. This indication is received due to a "jabber halt" condition as described in Part 10D.
(3) The user may cause an exit from the line_connect state by issuing an L_MGMT.request to enter the line_disconnect state or configure state.
(4) The user may request deactivation of a PLC SAP component by issuing an L_SAP_Deactivate. request.

10-B.1.7 Active_Station_List_Update Substate. The station enters the active_station_list_update substate from the line_connect state on receipt of an active_station_add_notification or active_ station_delete_notification from the station's PS as described in Part 10E. Note that entry to or exit from the active_station_list_update substate does not affect operation of the PLC entity.

The station's copy of the active_station_list is updated while in this state according to the protocol described in Part 10E. The station then returns to the line_connect state.

10-B.1.8 Configure State. The configure state is entered for all supported specified PLC SAP components on receipt from the user of an L_MGMT.request to enter configure state. It is entered for the specified PLC SAP component on receipt of an L_SAP_Deactivate.request. These user requests are only allowed in the line_connect state. The enter_configure state command is used after initial entry to the line_connect state as described in 10-B.1.6 above. L_SAP_Deactivate.requests are used to disable or reconfigure a currently active SAP component.

PLC SAPs component may be activated while in the configure state relative to that SAP component. The configure state exits to the line_connect state for each SAP when that SAP has been activated by the L_MGMT.requests described in Parts 10A and 10B. The configure state of the station management entity corresponds to the configure state of the local and remote PLC state machines described in Part 3B.

10-B.1.9 Line_Disconnect State. The line_disconnect state is entered in response to a user L_MGMT.request to enter line_disconnect state. This initiates an orderly disconnect as described in Figure 10-4. This includes a notification to PS that this station desires to leave the ring. Notification of PS is achieved by issuing an MA_VALUE_CHANGE.request for in_ring_desired equals **false**.

10-B.2 Recommendation for Management of Redundant Media. Redundant media are best managed by transmitting on all available media while selecting one functional medium for reception. State transitions to accomplish this goal are outlined in Figures 10-3 and 10-4.

In addition, media not currently in use for reception should be tested and the user notified of any failures. These activities are for further study.

ISA-RP74.01-1984

Recommended Practice

Application and Installation of Continuous-Belt Weighbridge Scales

Instrument Society of America

ISBN 0-87664-814-6

ISA-RP74.01 Application and Installation of
Continuous-Belt Weighbridge Scales

Copyright © 1984 by the Instrument Society of America. All rights reserved. Printed in the United States of America. No part of this publication may be reproduced, stored in a retrieval system, or transmitted, in any form or any means electronic, mechanical, photocopying, recording or otherwise without the prior written permission of the publisher.

INSTRUMENT SOCIETY OF AMERICA
67 Alexander Drive
P.O. Box 12277
Research Triangle Park, North Carolina 27709

Copyright © 1984 by the Instrument Society of America

PREFACE

This preface is included for information purposes and is not a part of ISA-RP74.01.

This recommended practice has been prepared as a part of the service of the Instrument Society of America (ISA) toward a goal of uniformity in the field of instrumentation. To be of real value, this document should not be static, but should be subject to periodic review. Toward this end, the Society welcomes all comments and criticisms, and asks that they be addressed to the Secretary, Standards and Practices Board, Instrument Society of America, 67 Alexander Drive, P.O. Box 12277, Research Triangle Park, NC 27709, Telephone (919) 549-8411.

The ISA Standards and Practices Department is aware of the growing need for attention to the metric system of units in general, and the International System of Units (SI) in particular, in the preparation of instrumentation standards. The Department is further aware of the benefits to U.S.A. users of ISA standards of incorporating suitable references to the SI (and the metric system) in their business and professional dealings with other countries. Toward this end, this Department will endeavor to introduce SI-acceptable metric units in all new and revised standards to the greatest extent possible. The Metric Practice Guide, which has been published by the Institute of Electrical and Electronic Engineers as ANSI/IEEE Std. 268-1982, and future revisions will be the reference guide for definitions, symbols, abbreviations, and conversion factors.

It is the policy of the Instrument Society of America to encourage and welcome the participation of all concerned individuals and interests in the development of ISA standards. Participation in the ISA standards-making process by an individual in no way constitutes endorsement by the employer of that individual, of the Instrument Society of America, or of any of the standards that ISA develops.

The information contained in the preface, footnotes, and appendices is included for information only and is not a part of the recommended practice.

Prior to issuance of this recommended practice, NBS Handbook 44 (Specification Tolerances and Other Technical Requirements for Commercial Weighing and Measuring Devices) and OIML IR50 (Laws of the Member States' Journal of European Communities Relating to Continuous Totalizing Weighing Machines) were available to guide specifiers and buyers of weighbridge-type, continuous-belt weigh scales. These two documents are intended for use on the regulation of weighing and apply to sales approved by weights and measures personnel.

The purpose of this recommended practice is to furnish design criteria to simplify the specification of weighbridge-type, continuous-belt weigh scales and to provide recommendations for their installation, calibration, and maintenance. It is recognized that nuclear-radiation-type belt weigh scales and "bolt-on" sensors perform a valuable service in the industry and can provide Class III accuracy (1-percent error). This recommended practice, however, addresses only the weighbridge-type scale, leaving the application and installation of nuclear scales and other weighing devices to subsequent consideration.

The persons listed below served as members of ISA Committee SP74:

NAME	COMPANY
N. L. Kautsky, Chairman	Stearns Catalytic Corporation
R. W. Brockmeyer	AMAX Corporation
P. Chase	Ramsey Engineering Company
S. Ciesiewski	Peabody Coal Company
W. J. Coulam	Kennecott Research Center
E. Duncan	K & P, Inc.
D. R. Gilmore	Thayer Scale Company
M. R. Gruber	Southern Weighing & Inspection Bureau
W. Heerman	Tech Sales, Inc.
R. Horne	Ralph M. Parsons Company
G. Kachel	Riede Systems, Inc.
O. C. Murdoch	Duval Corporation
R. L. Nickens (deceased)	Reynolds Metals Company
R. Osgood	K & P, Inc.
L. J. Walker	Merrick Scale Company

Instrument Society of America

This recommended practice was approved for publication by the ISA Standards and Practices Board in March 1984.

NAME	COMPANY
W. Calder III, Chairman	The Foxboro Company
N. L. Conger	Conoco
B. Feikle	Bailey Controls Company
T. J. Harrison	IBM Corporation
H. S. Hopkins	Westinghouse Electric Company
J. L. Howard	Boeing Aerospace Company
R. T. Jones	Philadelphia Electric Company
R. Keller	The Boeing Company
O. P. Lovett, Jr.	Isis Corporation
E. C. Magison	Honeywell, Inc.
A. P. McCauley	Chagrin Valley Controls, Inc.
J. W. Mock	Bechtel Power Corporation
E. M. Nesvig	ERDCO Engineering Corporation
R. Prescott	Moore Products Company
D. Rapley	Stearns Catalytic Corporation
W. C. Weidman	Gilbert Commonwealth
K. A. Whitman	Allied Chemical Corporation
*P. Bliss	
*B. A. Christensen	
*L. N. Combs	
*R. L. Galley	
*R. G. Marvin	
*W. B. Miller	Moore Products Company
*G. Platt	Bechtel Power Corporation
*J. R. Williams	Stearns Catalytic Corporation

*Director Emeritus

TABLE OF CONTENTS

Section	Title	Page
1.	Purpose and Scope	7
2.	Definitions	7
3.	Conveyor Design Recommendations	9
3.1	Conveyor Length	9
3.2	Maximum Angle of Inclination	9
3.3	Belt Speed	9
3.4	Troughing Angle Carrying Idlers	9
3.5	Idler Spacing	9
3.6	Takeup	9
3.7	Concave/Convex Curves	9
3.8	Conveyor Stringers	9
3.9	Conveyor Vertical Supports	9
3.10	Support Footing	9
3.11	Training Idlers	9
3.12	Weigh Idlers	10
3.13	Conveyor Load Point	10
3.14	Wind Loading	10
3.15	Material Transition	10
3.16	Idler Alignment	10
3.17	Belt Tracking	10
3.18	Calibration Chain Storage	10
4.	Calibration	10
4.1	Methods	10
4.1.1	Material Run—Preferred Method	10
4.1.2	Simulated Test Calibration with a Test Chain	10
4.1.3	Simulated Test Calibration with a Static Weight	10
4.1.4	Simulated Test with Electronic Calibration	10
4.2	Frequency of Alignment, Calibration, and Zero Balancing	11
4.2.1	Zero Balancing Interval	11
4.2.2	Span Calibration Interval	11
5.	Accuracy	11
6.	Belt Weigh Scale Components	11
6.1	Load Sensor and Reactor	11
6.2	Speed Measurement System	11
6.3	Multiplier	11
6.4	Information Display	12
6.5	Transmission	12
6.6	Power Supply	12
6.7	Enclosure	12
7.	Scale Maintenance	12
7.1	Cleanliness	12
7.2	Alignment	12
7.3	Tension Adjustment	13
7.4	Monitoring Accuracy	13
8.	Bibliography	13

APPENDICES

Section	Title	Page
APPENDIX A,	OIML Recommendation No. 50, Chapter II, Metrological Requirements (Metrological Measurements)	15
APPENDIX B,	Sample Form for a Test Report	19

LIST OF ILLUSTRATIONS

Figure	Title	Page
1.	Belt Conveyor and Weighbridge Assembly	8
2.	Sample Form for a Test Report	19

1 PURPOSE AND SCOPE

The purpose of this recommended practice is to furnish design criteria inducive to simplified specifications and to provide recommendations for installation, calibration, and maintenance of continuous-belt, weighbridge-type scales. The recommended practice will provide an effective base of comparison of scale suppliers, will establish minimum values, and will ensure that a scale specification and purchase incorporates the essentials to satisfy a particular weighing job. It will permit full knowledge of the weigh scale configuration (regardless of the manufacturer) early in the design stages of the belt conveyor.

2 DEFINITIONS

The following definitions apply to this recommended practice. (See also Figure 1.)

Approach Idler - The last *idler* passed before the material on a belt reaches the weighbridge.

Belt Conveyor - An endless fabric, rubber, plastic, leather, or metal belt operating over suitable drive, tail-end, and bend terminals; belt idlers; or slider beds for handling bulk materials, packages, or objects placed directly upon the belt.

Belt-Conveyor Scale - A device installed on a belt-conveyor structure that continuously weighs the material being conveyed.

Belt-Speed Sensor - A device that generates a signal as a function of belt speed.

Belt-Speed Transmitter - A device that transmits a belt speed signal to a receiver.

Bend Pulley - Any pulley used to change the direction of travel of a belt.

Calibration Test - A test using known weights and forces to load the scale in order to determine the performance of a *belt-conveyor scale*.

Concave Curve - A change in the angle of inclination of a *belt conveyor* where the center of the curve is above the conveyor.

Convex Curve - A change in the angle of inclination of a *belt conveyor* where the center of the curve is below the conveyor.

Conveyor Stringers - Support members for the conveyor on which the *idlers* are mounted.

Design Capacity - The maximum weight load that the scale is designed to weigh in one hour within the designated class accuracy. It is customarily 125 percent of normal capacity and is also known as "scale capacity."

Head Pulley - The pulley at the discharge end of the *belt conveyor*. The power drive for the belt is generally applied to the end pulley.

Idler Spacing - The center-to-center distance between consecutive *idler rollers*, measured parallel to the belt.

Idlers (Idler Rollers) - Freely-turning cylinders mounted on a frame to support the conveyor belt. For a flat belt, the *idlers* may consist of one or more horizontal cylinders transverse to the direction of belt travel. For a troughed belt, the *idlers* may consist of one or more horizontal cylinders with one or more additional cylinders at an angle that lifts the sides of the belt to form a trough.

Impact Idler - A belt *idler* incorporating resilient roll coverings to absorb large amounts of shock at the *loading point*.

Load Reactor - A device that generates a signal proportional to the force imposed upon it by the *load sensor*.

Load Sensor - See "weigh carriage." Also called "load receiving element."

Loading Point - The location at which material to be conveyed is applied to the conveyor.

Normal Capacity - Normal capacity is 80 percent of design capacity.

Retreat Idler - The first *idler* reached after the material on the belt leaves the weigh carriage. Also called "departure idler."

Service Idlers - Those *idlers* in the weighing area, including scale-borne *idlers* and several *idlers* on either side of the scale-borne *idlers*. They must be of the same type and grade and receive maintenance as *weigh idlers*.

Simulation Tests - Tests on the weighing unit in which either the movement of the belt, the effect of the material thereon, or both are simulated by using known weights and forces.

Skirting - Stationary sideboards or sections of the *belt conveyor* attached to the conveyor support frame or other stationary support to prevent the bulk material from falling off the side of the belt.

Snub Pulley - Any pulley used to increase the arc of contact between the belt and the drive pulley.

String Lines - Wires, piano wire, or monofilament line of suitable tensile strength and visibility, strung over each of the three rolls of the *weigh idlers* to confirm idler alignment and elevation (three-wire line alignment).

Tail Pulley - The pulley at the opposite end of the conveyor from the *head pulley*.

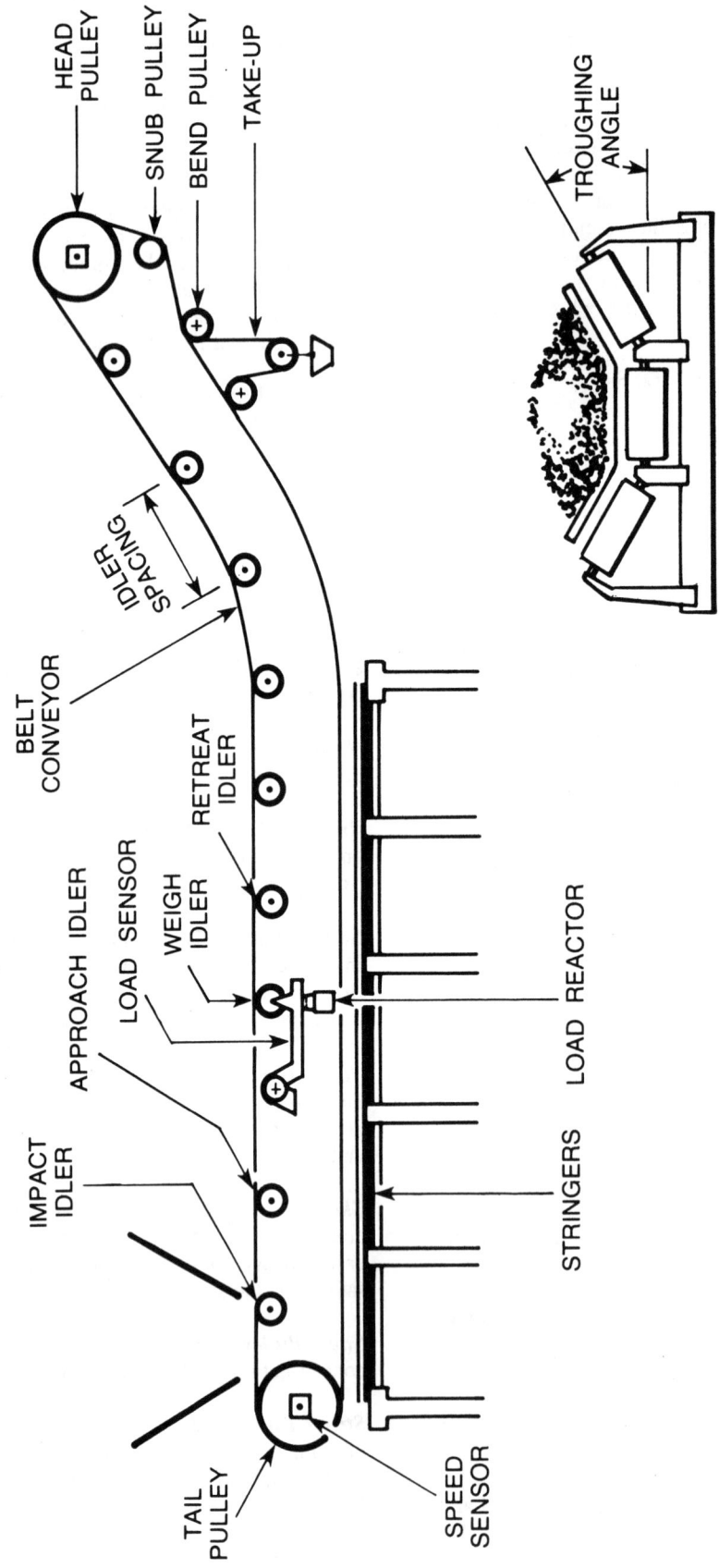

Figure 1. Belt Conveyor and Weighbridge Assembly

Takeup (Gravity) - A device plus a calculated quantity of dead weight to provide sufficient tension in a conveyor belt to ensure that the belt will be positively driven by the drive pulley. A counter-weighted *takeup* consists of a horizontal pulley free to move in either the vertical or horizontal direction, with dead weights applied to the pulley shaft to provide the tension required.

Test Chain - A calibrating device consisting of a series of rollers or wheels linked together to ensure that their weight is uniform and they move freely (so that chain weight loss due to wear is minimized).

Totalizer - A device used with a *belt-conveyor scale* to indicate the total weight of the material that has been conveyed over the scale. The master weight *totalizer* is the primary indicating element of the *belt-conveyor scale*. An auxiliary vernier counter used for scale calibration should not be part of the master weight *totalizer*. *Totalizers* can be remote auxiliary *totalizers* as well as local masters. The *totalizer* shows the accumulated weight; a *totalizer* may be nonresettable or resettable to zero to measure a definite amount of conveyed material.

Training Idlers - *Idlers* of special design or mounting that are intended to counteract any tendency of the belt to shift sideways.

Tripper - A device for unloading a *belt conveyor* at a point between the loading point and the head pulley.

Weigh Carriage - A structure supporting the *weigh idlers*, which in turn transmits weight to the load reactor.

Weigh Idlers - *Idlers* positioned in the *weigh carriage* assembly so that they sense the weight of the material on the conveyor belt and transmit the weight through the carriage to the *load reactor*.

Weighbridge-Type Belt Scale - A scale mounted above or below a *belt conveyor* that supports a section of the conveyor belt via a structural suspension system *(weigh carriage)* and *weigh idlers*.

3 CONVEYOR DESIGN RECOMMENDATIONS

The design and installation of the conveyor leading to and from the weigh scale are crucial to enable the scale to perform within accuracy specifications.

3.1 Conveyor Length

The maximum conveyor length may be dictated by the desire to make one belt revolution within a given time period or a given tonnage. The minimum conveyor length is 40 ft (12 m), unless the conveyor is specifically designed as a weighing conveyor and certified by the scale manufacturer.

3.2 Maximum Angle of Inclination

The maximum angle of inclination is a function of material properties and belt speed. The primary objective is that the material must not slide back on the belt.

3.3 Belt Speed

Belt speed is a function of conveyor length and application. In general, accuracy can be maintained in the range of 100 to 1,000 ft/min (30 to 305 m/min).

3.4 Troughing Angle Carrying Idlers

The maximum troughing angle for carrying idlers is 35°.

3.5 Idler Spacing

Idler spacing is a function of conveyor load, troughing, and belt construction, e.g., stiffness, and belt width. The scale can function as long as the conveyor does not spill the load between idlers.

3.6 Takeup

Where space permits, the takeup is vertical gravity type, preferably near the drive or head pulley.

3.7 Concave/Convex Curves

No vertical curve in the belt should be located between the loading point and the scale. The scale should be located not less than 40 ft (12 m) from the convex curve and not less than 70 ft (21.3 m) from the concave curve. The belt should be in contact with at least eight idlers on either side of the scale.

3.8 Conveyor Stringers

Conveyor stringers should be continuous to accommodate scale weigh idlers, with at least four idlers on either side of the scale, to minimize deflection under load. If the stringers are not continuous, welding them together can be helpful. (See Section 3.9.)

3.9 Conveyor Vertical Supports

Conveyor vertical supports are located at weigh-section, load-bearing points, with additional supports spaced at 10 ft (3.1 m) to span an area equal to at least four idlers on each side of the weigh idlers. Relative deflection between the idlers (eight idlers plus weigh idlers) should not exceed ±0.010 in. (0.0254 cm) under load.

3.10 Support Footing

Preferably, support footings are located on concrete foundations, but they may be located on a suitably reinforced concrete floor.

3.11 Training Idlers

Training idlers shall not be allowed within 40 ft (12 m) on

either side of the scale or within 10 idler spacings, whichever is greater.

3.12 Weigh Idlers

Weigh idlers shall be evenly spaced and normal (square) to the conveyor. At least four idlers on either side of the weighbridge shall be similarly aligned.

3.13 Conveyor Load Point

Conveyor loading shall be preferred at one point only. If multiple load points are required, they should be grouped in close proximity to one another. Where a load point is within eight approach idlers of the weighbridge, multiple impact idlers are recommended. The conveyor loading mechanism should be designed to provide uniform belt loading.

3.14 Wind Loading

The conveyor in the scale area should be provided with wind shields on the sides, ends, and top and bottom of each end of the scale for a distance of 30 ft (9 m) measured along the belt, where exposed to outdoor environment. The wind screen should protect against wind, at least 6 in. (15 cm) below the return belt.

3.15 Material Transition

The distance from the loading point to the scale shall allow a minimum settling time of 2 s for the material on the belt before it is weighed. Appropriate apron feeders, profilers, etc., should be used to produce as uniform a loading as possible.

3.16 Idler Alignment

The manufacturer's installation instructions should be followed for the stringline alignment of weighing, approach, and retreat idlers.

3.17 Belt Tracking

The belt should have sufficient flexibility to ensure contact with all weighing idler rolls and be centered when it is running empty. The belt should track within ±1/2 in. (1.25 cm) from empty to fully loaded. This can be aided by loading material onto the belt in the direction of travel rather than transverse to it. Lagging on the head pulley is advisable; it will grip the belt and carry it into a well-centered position. It will also permit belt tension to be reduced to a minimum, which will allow better weighing accuracy.

3.18 Calibration Chain Storage

A suitable protective support should be provided to allow storage of the test chain, where used, to facilitate handling of the chain onto the conveyor belt.

4 CALIBRATION

4.1 Methods

Operate the conveyor under load for a sufficient time, depending upon weather conditions, to ensure normal operating conditions and instrument warmup. In temperatures below freezing, it is advisable to operate with a load at least 1 hour prior to starting calibration.

Assure that the belt is clear, that all idlers are free, and that the belt is in alignment. To account for variations in belt thickness, zero-balance the weigh signal transmitter with an integrated average belt weight. To determine the average belt weight, run the belt for at least five integral revolutions or 3 min, whichever is greater (to reduce the error caused by inaccurate observations). Begin and end zero balance while the conveyor is running. On long belts (e.g., 1,000 ft (305 m), three revolutions should be sufficient for Class II accuracy). (See Section 5.)

4.1.1 Material Run—Preferred Method

Material is passed over the scale at or near the normal load rate (80 percent of the design capacity) for at least five revolutions of the belt (three revolutions for Class II accuracy). It is collected in a collecting bin with an integral (reference) scale. A second method is to collect material that has passed over the scale into a truck or railcar and weighed on a certified scale. (See Section 5.) The reference scale shall have an error no greater than 25 percent of the smallest tolerance to be applied when the standard is used.

4.1.2 Simulated Test Calibration with a Test Chain

The chain should permit calibration from 65 to 75 percent of the design capacity. The pounds-per-foot rating of the chain should be equal to the desired conveyor loading and be certified to 0.5 percent accuracy of the total weight. All links of the chain should weigh within 0.5 percent of the mean link weight. The chain shall have sufficient length to span the second fixed idler on both sides of the scale. It should have free rollers or wheels and a pitch that divides the idler spacing evenly — no longer than 6 in. Suitable support above the belt shall be provided to house and protect the chain from dust and moisture.

4.1.3 Simulated Test Calibration with a Static Weight

A static weight suspended from the center of the carriage may be used for an interim check on calibration. Failure of the check to fall within the predetermined limits would indicate a need for recalibration by the test chain or the material run.

4.1.4 Simulated Test with Electronic Calibration

The effect of the built-in electronic weight simulator,

if available, should permit calibration of the scale system between 60 and 80 percent of its design capacity.

The scale manufacturer should be consulted if the capacity or the type of load reactor is changed after the initial commissioning of the scale system.

After calibration, report accuracy as "as found" and "as left."

4.2 Frequency of Alignment, Calibration, and Zero Balancing

4.2.1 Zero Balancing Interval

The interval is determined by sensitivity to ambient conditions, stability of structure, rate of dust buildup, or spill on scale suspension. Zero should be checked every other day for several weeks after the installation and the results observed. The period can be lengthened thereafter.

4.2.2 Span Calibration Interval

Calibration with a built-in weigh signal, if available, should be done weekly for the Class I conveyor and monthly for the Class II conveyor. Tests with material and/or test chains should be done at regular intervals, not to exceed 90 days.

Accuracy classification and operating experience should guide the user when determining the allowable intervals between calibrations and the frequency of zero balancing.

Realignment with three-wire lines should be done once each year.

Test chains should be certified once every 2 yr at a minimum. As an alternative, a correction factor can be determined for the chain when material tests are run. (See Section 4.1.1.)

5 ACCURACY

Class I accuracy is defined as permitting no more than ±1/4 percent error of totalized weight when operated over a design capacity of 33-1/3 percent to 100 percent.

Class II accuracy is defined as permitting no more than ±1/2 percent maximum error of totalized weight when operated over a design capacity of 33-1/3 percent to 100 percent.

The accuracy classification should take into consideration the following:

(1) Custody transfer scales require high accuracy, as they are used for billing and accounting purposes.
(2) Control system scales, such as those used in ratio blending and general bulk transfers, require less accuracy, but repeatability must be assured.

Accuracy should be established based on an appropriate test condition (e.g., a material run, Section 4.1.1). Requirements for the determination of accuracy, as found in Chapter II of OIML* Recommendation No. 50 (June 1980) may be applicable to a specific installation. (See Appendix A.)

For the best accuracy, belt scales should operate at a minimum load of 7 lb/ft (10.4 kg/m) of the belt width. A counterbalance device that offsets the tare weight of the belt can be used to reduce the minimum load to 3 lb/ft (4.46 kg/m).

6 BELT WEIGH SCALE COMPONENTS

6.1 Load Sensor and Reactor

Each scale should include a load sensor (weighbridge) and a load reactor.

The weighbridge should utilize as many suspended idlers as required to meet the accuracy requirements of the installation.

The load reactor should measure the forces generated by the load sensor.

Strain gage load cells, LVDTs (linear variable differential transformers), and mechanical reactors are typical load reactors.

The load reactor and the sensor should incorporate not less than 150 percent of the design capacity overload protection. It should include temperature compensation over the temperature range specified and suitable environmental protection.

6.2 Speed Measurement System

A belt speed sensor should be employed in all scale systems. It should be designed such that there is no possibility that a slip will affect the results, regardless of whether the belt is loaded or unloaded.

The belt speed sensor transmitter should provide a continuous speed signal to the multiplier. It may be analog or digital. For troubleshooting purposes, it is desirable to have the speed signal available at the scale readout.

6.3 Multiplier

The totalized weight may be obtained either by multiplying the speed and weight information and integrating it with respect to time, or by integrating the weight information directly with respect to distance traveled.

---

*Organisation Internationale de Métrologie Légale

6.4 Information Display

At a minimum, all information to be displayed or transmitted should be available at one location.

Span and zero adjustments of the display should be non-interacting.

A weigh scale integrator/totalizer should be supplied to record the weight crossing the scale. The weight should be displayed in engineering units. The totalizer should contain a minimum of six digits and should display the total in accordance with the accuracy designation. Weight may be expressed in kilograms, pounds, tons, or multiples or subdivisions thereof. The value of the smallest unit should not be greater than 1/800 of the total number of counts produced on the master totalizer during the calibration test interval.

If the totalizer depends on electrical power, a backup power supply (e.g., a battery) should be considered in order to preclude rerunning the material upon loss of power.

The instantaneous weight flow rate should be displayed on all scales. The display should be indicated in engineering units or as design capacity (0 to 100 percent). It should be displayed in such a manner that the error of the indicator shall not be greater than twice that permitted in the designated accuracy class.

6.5 Transmission

When the instantaneous weight information is transmitted in pulse form, the signal may be produced by a dry contact or a solid-state circuit closure, or by an internally generated signal at a pulse rate proportional to the flow rate of the material, electrically isolated from the input signal and the ground. A local indication of contact closure should be supplied to facilitate troubleshooting.

When transmitted in analog form, a 4 to 20 mA dc signal may be used to a 0 to 600 Ω receiver to represent 0 to 100 percent of the scale's design capacity. The signal should be electrically isolated from the input signals and the ground. Alternatively, the signal may be transmitted via a solid-state device remotely powered from a 24 or 48 V dc source and be capable of driving loads of 100Ω to 1,000Ω. The accuracy of the flow signal should not exceed twice the class accuracy.

All transmission signals shall be available and clearly marked at a central termination junction.

6.6 Power Supply

The measuring system should maintain accuracy upon a ±10 percent deviation of the rated voltage.

6.7 Enclosure

The manufacturer may supply an enclosure suitable for the environment to house the above components. A key lock is advisable to secure calibration. The following information should be supplied on a permanent tag on the outside of the enclosure:

(1) Manufacturer

(2) Serial number

(3) Design capacity in units of weight per hour

(4) Design speed at which the belt will deliver design capacity, in feet per minute (meters per second)

(5) Units of registration, in tons, kilograms, etc.

(6) Number of weight units totalized for the load simulator for a specific number of feet of belt travel

(7) Power requirements

(8) Output signal, e.g., 4-20 mA, pulses/ton, etc.

(9) Hazardous area classification

(10) Identification of laboratory test approval, e.g., UL (Underwriters Laboratories) or FM (Factory Mutual).

The enclosure should allow the scale to operate within specification over the ambient temperature range expected, augmented if necessary by heaters.

In hazardous locations, the instrument housing should be located outside the hazardous area to allow meter reading and calibration adjustment under a "power on" condition.

All electronic equipment mounted in the enclosure and not interconnected with printed circuit boards should have crimped, soldered, or screw terminal interconnections. Suitable test points should be provided on all printed circuit boards to enable connection of the test equipment.

7 SCALE MAINTENANCE

7.1 Cleanliness

If it is impractical to keep the weight scale continuously clean of dust, dust can be allowed to build up on the weighbridge until it will no longer accumulate. The effect can then be zeroed out. Scales weighing large solid particles should be cleaned as necessary to prevent weighbridge binding.

A history of scale maintenance and process stability should be used to determine frequency of cleaning.

7.2 Alignment

Alignment should be made by using measurements of

scale carriage in relation to the conveyor frame. The need for "wire line" alignment checks should be considered any time conveyor work is performed in the scale area.

7.3 Tension Adjustment

Too much tension will cause excessive wear of both the belt and the scale components. It may also make the scale read light since the belt will not conform to weigh idlers properly.

Too little tension will cause the belt to sag and/or slip. If the scale is not speed-compensated, weighing errors will occur.

Proper tension should prevent slippage, but should not impede contact of the belt with idlers in the weigh area. As a guide, the tension should be tight enough to limit the sag at maximum load to less than 2 percent of the weigh idler pitch.

The calibration of the scale should be checked after mechanical maintenance, such as tension adjustment.

7.4 Monitoring Accuracy

Test chains should be clean, in good repair, and certified at a minimum of once every 2 yr. As an alternative, a correction factor can be determined for the chain when material tests (Section 4.1.1) are run.

Records of calibration and maintenance, including conveyor alignment and belt tension adjustment, should be maintained to develop a history of scale performance. The use of forms is recommended to provide a comprehensive, repeatable history log. (See Appendix B for sample form.)

8 BIBLIOGRAPHY

The following references are applicable to certifying a scale for billing applications:

1. *Association of American Railroads Scale Handbook,* 1983, Association of American Railroads, American Railroads Building, 1920 L Street, NW, Washington, DC 20036.

2. "Requirements for the Approval of Belt Conveyor Scales," Circular 9585-5, Southern Weighing and Inspection Bureau, Suite 306, Transportation Building, 151 Ellis Street, NE, Atlanta, GA 30303.

3. "An Explanation of Rules for the Inspection, Testing, and Calibration of Belt Conveyor Scales," Circular 9585-5, Southern Weighing and Inspection Bureau, Suite 306, Transportation Building, 151 Ellis Street, NE, Atlanta, GA 30303.

4. "Specifications, Tolerances, and Other Technical Requirements for Weighing and Measuring Devices, *Conveyor Belt Scale Handbook, Handbook 44,* 1980, National Bureau of Standards, U.S. Department of Commerce, Washington, DC 20234.

Additional references applicable to belt conveyors and belt-conveyor scales include the following:

1. *Belt Conveyors for Bulk Materials,* Conveyor Equipment Manufacturers Association, 1979, Van Nostrand Reinhold Publishing Company, 135 West 50th Street, New York, NY 10020.

2. *Continuous Totalizing Weighing Machines,* OIML IR50, June 1980, International Organization of Legal Metrology (Organisation Internationale de Métrologie Légale), 11 rue Turgot, 75009 Paris, France.

3. "Weighing Scales," PTC 19.5:1-1964, Supplement of ASME Power Test Codes, American Society of Mechanical Engineers, 345 East 47th Street, New York, NY 10017.

4. "Safety Standards for Conveyors and Related Equipment," ANSI B20.1-1976, American Society of Mechanical Engineers, 345 East 47th Street, New York, NY 10017.

APPENDIX A
OIML* RECOMMENDATION NO. 50, CHAPTER II
METROLOGICAL REQUIREMENTS

4 ACCURACY CLASSES

Belt weighers are divided into two accuracy classes: class 1 and class 2.

4.1 Characteristics of belt weighers in class 1

4.1.1 Totalisation scale interval (d_t)

The totalisation scale interval of the belt weigher must be:

— less than or equal to 0.05% of the load totalised in one hour at maximum flowrate (C_{max})
— more than or equal to 0.002% of this load (C_{max}).

4.1.2 Scale interval of the zero indicator (d_o)

The scale interval of the zero indicator must not exceed the following value of the load totalised in one hour at the maximum flowrate:

0.005 % for continuous (analogue) indication
0.0025% for discontinuous (digital) indication,

and must not be greater than the totalisation scale interval (d_t).

4.1.3 Scale interval of the test indicator

The scale interval of the test indicator must not exceed the following value of the minimum totalised load:

0.2% for continuous (analogue) indication
0.1% for discontinuous (digital) indication

and must not be greater than the totalisation scale interval (d_t).

4.2 Characteristics of belt weighers in class 2

4.2.1 Totalisation scale interval (d_t)

The totalisation scale interval of the belt weigher must be:

— less than or equal to 0.1% of the load totalised in one hour at the maximum flowrate (C_{max})
— more than or equal to 0.004% of this load (C_{max}).

4.2.2 Scale interval of the zero indicator (d_o)

The scale interval of the zero indicator must not exceed the following value of the load totalised in one hour at the maximum flowrate:

---

*Organisation International de Métrologie Légale

0.01 % for continuous (analogue) indication
0.005% for discontinuous (digital) indication

and must not be greater than the totalisation scale interval (d_t).

4.2.3 Scale interval of the test indicator

The scale interval of the test indicator must not exceed the following value of the minimum totalised load:

0.4% for continuous (analogue) indication
0.2% for discontinuous (digital) indication

and must not be greater than the totalisation scale interval (d_t).

*Organisation International de Métrologie Légale

4.3. Form of scale interval

The scale interval must be equal to a number of units of mass expressed by one of the following formulae:

1×10^n, 2×10^n or 5×10^n, where n represents a positive or negative whole number or zero.

However, the scale intervals of the zero indicator and those of the test indicator need not comply with this requirement.

4.4 Minimum flowrate

The minimum flowrate must be equal to 20 % of the maximum flowrate.

5 MAXIMUM PERMISSIBLE ERRORS

After the belt weigher has been set correctly to zero with no load, the maximum permissible errors, positive or negative, must be equal to the values specified below, for any totalised load equal to or greater than the minimum totalised load.

5.1 Maximum permissible errors on initial verification

5.1.1 Belt weighers in class 1

0.5% of the totalised load for any flowrate between 20 and 100% of the maximum flowrate.

5.1.2 Belt weighers in class 2

1% of the totalised load for any flowrate between 20 and 100% of the maximum flowrate.

5.2 Maximum permissible errors in service

5.2.1 Belt weighers in class 1

1% of the totalised load for any flowrate between 20 and 100% of the maximum flowrate.

5.2.2 Belt weighers in class 2

2% of the totalised load for any flowrate between 20 and 100% of the maximum flowrate.

6 VARIATION OF MAXIMUM PERMISSIBLE ERRORS

The maximum permissible errors specified above are applicable only to results of totalisation under the following conditions:

6.1 Discontinuous (digital) test indicating device

When the indicating device used for testing is discontinuous (digital), the maximum permissible errors must be increased by one scale interval of this device.

6.2 Belt weighers fitted with several totalisation indicating or printing devices

The errors on the indications or printed results provided for the same totalised load by different totalisation indicating or printing devices on the same belt weigher must not exceed the maximum permissible errors.

The difference between these indications or printed results, taken two by two, must not be greater than:

— one discontinuous (digital) scale interval when the results are supplied by two discontinuous (digital) indicating devices having the same scale interval,

— the absolute value of the maximum permissible error, when the results are provided by two continuous (analogue) indicating devices,

— the greater of the two values:
absolute value of the maximum permissible error and one discontinuous (digital) scale interval,
when the results are provided by a continuous (analogue) indicating device and a discontinuous (digital) indicating device.

6.3 Simulation tests

6.3.1 Maximum permissible errors during simulation tests

The maximum permissible errors, positive or negative, must be equal to the values specified below:

6.3.1.1 Belt weighers in class 1

For any flowrate between 5 and 20% of the maximum flowrate : 0.07% of the load totalised at the maximum flowrate during the duration of the test.

For any flowrate between 20 and 100% of the maximum flowrate : 0.35% of the totalised load.

6.3.1.2 Belt weighers in class 2

For any flowrate between 5 and 20% of the maximum flowrate : 0.14% of the load totalised at the maximum flowrate during the duration of the test.

For any flowrate between 20 and 100% of the maximum flowrate : 0.7% of the totalised load.

6.3.2 Displacement simulation error

The relative error due to the simulated displacement of the belt must be less than 20% of the maximum permissible error for the totalised load. This error is included in the maximum permissible error.

6.3.3 Difference between two results obtained due to a variation in the simulated speed

For any variation in the speed of the displacement simulation device corresponding to a variation of ± 10% of the speeds of the conveyor belt, the variation in the relative error of the results of the simulation tests shall not exceed 20% of the maximum permissible error in point 6.3.1.

6.3.4 Difference between two results obtained by varying the point of application of a load

When the point of application of the same load is varied, in a manner compatible with the design of the load receptor, the difference between the two results must not be greater than the absolute value of the maximum permissible error.

6.3.5 Zero-setting

For any load within the range of the zero-setting device, the results, after setting the machine to zero, shall comply with the maximum permissible errors for the totalised load.

6.3.6 Influence factors

6.3.6.1 Temperature

Belt weighers must comply with the requirements relating to the maximum permissible errors at all practically constant temperatures between −10°C and +40°C after adjustment of zero. However, for special applications the belt weighers may have different temperature ranges. In that case, the interval must be at least 30°C and must be indicated on the data plate.

For a temperature variation of 10°C at a rate not exceeding 5°C per hour, the zero indiction or, in the case of belt weighers fitted with a zero checking device with additional mass, the control value, must not vary by more than:

0.07% for class 1
0.14% for class 2

of the load totalised at the maximum flowrate for the duration of the test.

6.3.6.2 Power supply

Belt weighers must comply with the requirements relating to the maximum permissible errors, without adjustment of zero, for the following variations of the power supply:

15% to + 10% of the nominal voltage and
 2% to + 2% of the nominal frequency.

6.3.6.3 Other influence factors

Belt weighers must under normal conditions of use comply with the requirements relating to the maximum permissible errors when they are submitted to the effects of influence factors other than those referred to in points 6.3.6.1 and 6.3.6.2, and resulting from the conditions of their installation.

6.3.7 Metrological characteristics

6.3.7.1 Repeatability

The difference between two results obtained for the same load placed under the same conditions on the load receptor must not be greater than the absolute value of the maximum permissible error.

6.3.7.2 Discrimination of the totalisation indicating device

At any flowrate, between the minimum and maximum flowrate, the difference between the indications obtained for two totalised loads, differing from each other by a value equal to the maximum permissible error, must be at least equal to one half of the calculated value corresponding to the difference between these totalised loads.

6.3.7.3 Discrimination of the indicator used for zero-setting

For tests of a duration of three minutes there must be a clearly visible difference between the indications of the zero indicator obtained at no load and for a load, deposited or removed equal to the following percentages of the maximum capacity:

0.1% for class 1
0.2% for class 2.

6.3.7.4 Stability of zero

6.3.7.4.1 Short-term stability

The difference between the smallest and largest indications of the zero indicator obtained in five tests of three minutes duration must not exceed the following percentages of the load totalised in one hour at the maximum flowrate:

0.0025% for class 1
0.005 % for class 2.

6.3.7.4.2 Long-term stability

When the short-term stability tests are repeated after 3 hours of operation and without zero adjustment:

— the results must satisfy the requirements laid down in point 6.3.7.4.1 and
— the difference between the smallest and largest of all indications at the zero indicator must not exceed the following percentages of the load totalised in one hour at the maximum flowrate:

0.0035% for class 1
0.007 % for class 2.

6.3.7.5 Supplementary totalisation indicating devices

Supplementary totalisation indicating devices must not affect the operation of the belt weigher.

6.3.7.6 Zero checking device with additional mass

The requirements in points 6.3.7.3 and 6.3.7.4 apply also to testing of belt weighers fitted with zero checking devices with additional mass.

APPENDIX B

TEST REPORT

TEST DATE -	REPORTED BY-

1-Length Conveyor Belt-
2-Weight Test Chain P/Ft-
3-Lbs. For 1 Circuit of Conv-
4-Wgt. For 1 Circuit-
   (Same As Counter)

5-Registration In Unit Of-
6-One Rev. of Dial-
7-Wgt. For 1 Ft. Travel-
   (Same as Counter)
8-Wgt. For 1 In Travel-

	Zero	Adjustments	1	2	3
No. of Circuits Over-Under Travel Reading After Reading Before					
Registered Wgt. Correction + –					
Corrected Wgt. True Wgt.					
Diff Error in % Adjusted					

Chain Factor	Parts Used

Does Overloading Occur-

Conveyor Speed-
Peak Cap. Rating

Minimum Load Tonnage-

Normal Operating Load-
   (Tonnage)

COMMENTS:

Weight Registration To Be Net.
After Completion of Test

Scale
Reading As Left-
Reading as Start-   _____
Deduct

Control Room Counter

Reading After-
Reading Before-   _____

Figure 2. Sample Form of a Test Report

INSTRUMENT SOCIETY of AMERICA
Research Triangle Park, North Carolina

ANSI/ISA-S75.01-1985
Approved August 15, 1986

American National Standard

Flow Equations for Sizing Control Valves

Instrument Society of America

Instrument Society of America

ISBN 0-87664-899-5

ISA-S75.01 Flow Equations for Sizing Control Valves

Copyright © 1985 by the Instrument Society of America. All rights reserved. Printed in the United States of America. No part of this publication may be reproduced, stored in a retrieval system, or transmitted, in any form or by any means (electronic, mechanical, photocopying, recording, or otherwise), without the prior written permission of the publisher.

INSTRUMENT SOCIETY OF AMERICA
67 Alexander Drive
P.O. Box 12277
Research Triangle Park, North Carolina 27709

PREFACE

This preface is included for informational purposes and is not part of ISA-S75.01.

This standard has been prepared as part of the service of the Instrument Society of America (ISA) toward a goal of uniformity in the field of instrumentation. To be of real value, this document should not be static, but should be subject to periodic review. Toward this end, the Society welcomes all comments and criticisms, and asks that they be addressed to the Secretary, Standards and Practices Board, Instrument Society of America, 67 Alexander Drive, P.O. Box 12277, Research Triangle Park, NC 27709, Telephone (919) 549-8411.

The ISA Standards and Practices Department is aware of the growing need for attention to the metric system of units in general, and the International System of Units (SI) in particular, in the preparation of instrumentation standards. The Department is further aware of the benefits to U.S.A. users of ISA standards of incorporating suitable references to the SI (and the metric system) in their business and professional dealings with other countries. Toward this end, this Department will endeavor to introduce SI-acceptable metric units in all new and revised standards to the greatest extent possible. The Metric Practice Guide, which has been published by the Institute of Electrical and Electronics Engineers as ANSI/IEEE Std. 268-1982, and future revisions will be the reference guide for definitions, symbols, abbreviations, and conversion factors.

It is the policy of the Instrument Society of America to encourage and welcome the participation of all concerned individuals and interests in the development of ISA standards. Participation in the ISA standards-making process by an individual in no way constitutes endorsement by the employer of that individual, of the Instrument Society of America, or of any of the standards that ISA develops.

The information contained in the preface, footnotes, and appendices is included for information only and is not a part of the standard.

The following people served as members of ISA Committee SP75.05, which prepared this standard:

NAME	COMPANY
L. R. Driskell, Chairman	Consultant
J. B. Arant	E. I. du Pont de Nemours and Company, Inc.
H. D. Baumann	H. D. Baumann Associates, Ltd.
*C. S. Beard	
G. Borden	Bechtel Power Corporation
L. Griffith	Consultant
F. P. Harthun	Fisher Controls International, Inc.
R. B. Jones	Upjohn Company
A. P. McCauley	Chagrin Valley Controls, Inc.
J. Ozol	Omaha Public Power Company
R. A. Quance	Walsh Inc.
W. Rahmeyer	Colorado State University
K. Schoonover	Con-Tek
J. M. Simonsen	Valtek, Inc.
H. Sonderregger	ITT Grinnell Corporation
F. Volpe	Masoneilan Division, McGraw-Edison Company
W. C. Weidman	Gilbert Commonwealth, Inc.
L. Zinck	Union Carbide Corporation

The following people served as members of ISA Committee SP75 during the review of this standard:

NAME	COMPANY
L. R. Driskell, Chairman	Consultant
J. B. Arant	E. I. du Pont de Nemours and Company, Inc.
H. E. Backinger	John F. Kraus & Company

---

*Deceased

Instrument Society of America

NAME	COMPANY
G. Barb	Muesco, Inc.
H. D. Baumann	H. D. Baumann Associates, Ltd.
*C. S. Beard	
N. Belaef	Consultant
G. Borden	Bechtel Power Corporation
**R. Brodin/G. F. Stiles	Fisher Controls International, Inc.
E. H. C. Brown	Dravo Engineers, Inc.
E. J. Cooney	Air Products & Chemicals, Inc.
W. G. Dewart	Rockwell International
J. T. Emery	Honeywell, Inc.
H. J. Fuller	Worcester Controls Corporation
L. Griffith	Consultant
A. J. Hanssen	Fluid Controls Institute, Inc.
F. P. Harthun	Fisher Controls International, Inc.
H. P. Illing	Kieley & Mueller, Inc.
R. B. Jones	Upjohn Company
M. W. Kaye	M. W. Kellogg Company
R. Louviere	Creole Engineering
O. P. Lovett, Jr.	ISIS Corporation
A. P. McCauley	Chagrin Valley Controls, Inc.
T. V. Molloy	Pacific Gas & Electric
H. R. Nickerson	Resistoflex Company
J. Ozol	Omaho Public Power Company
R. A. Quance	Walsh Inc.
W. Rahmeyer	Colorado State University
J. N. Reed	Masoneilan Division, McGraw-Edison Company
G. Richards	Jordan Valve Div., Richards Industries, Inc.
J. Rosato	Rawson Company
K. Schoonover, Secretary	Con-Tek
H. Schwartz	Flexible Valve Corporation
**W. L. Scull/J. T. Muller	Leslie Company
F. O Seger	Willis Division, Smith International, Inc.
J. M. Simonsen	Valtek, Inc.
H. R. Sonderregger	ITT Grinnell Corporation
N. D. Sprecher	DeZurik
R. U. Stanley	Retired
R. E. Terhune, Vice-Chairman	Consultant
R. F. Tubbs	Copes-Vulcan
W. C. Weidman	Gilbert Commonwealth, Inc.
R. L. Widdows	Cashco, Inc.
L. Zinck	Union Carbide Corporation

This standard was approved for publication by the ISA Standards and Practices Board in May 1985.

NAME	COMPANY
N. Conger, Chairman	Fisher Controls Company
P. V. Bhat	Monsanto Company
W. Calder III	The Foxboro Company
R. S. Crowder	Ship Star Associates
B. Feikle	Bailey Controls Company
H. S. Hopkins	Westinghouse Electric Company

*Deceased
**One vote

NAME	COMPANY
J. L. Howard	Boeing Aerospace Company
R. T. Jones	Philadelphia Electric Company
R. Keller	The Boeing Company
O. P. Lovett, Jr.	ISIS Corporation
E. C. Magison	Honeywell, Inc.
A. P. McCauley	Chagrin Valley Controls, Inc.
J. W. Mock	Bechtel Corporation
E. M. Nesvig	ERDCO Engineering Corporation
R. Prescott	Moore Products Company
D. E. Rapley	ISTS
C. W. Reimann	National Bureau of Standards
J. Rennie	Factory Mutual Research Corporation
W. C. Weidman	Gilbert Commonwealth, Inc.
K. Whitman	Consultant
†P. Bliss	Consultant
†B. A. Christensen	Continental Oil Company
†L. N. Combs	Retired
†R. L. Galley	Consultant
†T. J. Harrison	IBM Corporation
†R. G. Marvin	Roy G. Marvin Company
†W. B. Miller	Moore Products Company
†G. Platt	Consultant
†J. R. Williams	Stearns Catalytic Corporation

†Director Emeritus

TABLE OF CONTENTS

Section	Title	Page
1.	Scope	9
2.	Introduction	9
	2.1 Flow Variables and Fluid Properties	9
3.	Nomenclature	9
4.	Incompressible Fluid — Flow of Nonvaporizing Liquid	10
	4.1 Equations for Turbulent Flow	10
	4.2 Numerical Constants N	10
	4.3 Piping Geometry Factor F_p	10
	4.4 Equations for Nonturbulent Flow	11
5.	Incompressible Fluid — Choked Flow of Vaporizing Liquid	12
	5.1 Liquid Choked Flow Equations	12
	5.2 Liquid Pressure Recovery Factor F_L	13
	5.3 Combined Liquid Pressure Recovery Factor F_{LP}	13
6.	Compressible Fluid — Flow of Gas and Vapor	14
	6.1 Equations for Turbulent Flow	14
	6.2 Numerical Constants N	14
	6.3 Expansion Factor Y	14
	6.4 Choked Flow	14
	6.5 Pressure Drop Ratio Factor x_T	15
	6.6 Pressure Drop Ratio Factor with Reducers or Other Fittings	15
	6.7 Ratio of Specific Heats Factor F_k	15
	6.8 Compressibility Factor Z	15
7.	References	27

LIST OF APPENDICES

Appendix	Title	Page
A.	Use of Flow Rate Equations for Sizing Valves	17
B.	Derivation of Factors F_p and F_{LP}	17
C.	Control Valve-Piping System Head Changes	18
D.	Representative Values of Valve Capacity Factors	19
E.	Reynolds Number Factor F_R	19
F.	Equations for Nonturbulent Liquid Flow	22
G.	Liquid Critical Pressure Ratio Factor F_F	24
H.	Derivation of Factor x_{TP}	24
I.	Control Valve Flow Equations — SI Notation (International System of Units)	25

LIST OF ILLUSTRATIONS

Figure	Title	Page
1.	Reynolds Number Factor	12
2.	Liquid Flow Rate Versus Pressure Drop for a Typical Valve (Constant Upstream Pressure and Vapor Pressure)	13
C-1.	Head Changes in a Control Valve-Piping System	19
E-1.	Reynolds Number Factor for Valve Sizing	20

LIST OF TABLES

Table	Title	Page
1.	Numerical Constants for Liquid Flow Equations	10
2.	Numerical Constants for Gas and Vapor Flow Equations	14
C-1.	Definitions of Head Terms	18
D-1.	Representative Values of Valve Capacity Factors	20
E-1.	Reynolds Number Factor F_R for Transitional Flow	21

1 SCOPE

This standard presents equations for predicting the flow of compressible and incompressible fluids through control valves. The equations are not intended for use when mixed-phase fluids, dense slurries, dry solids, or non-Newtonian liquids are encountered. In addition, the prediction of cavitation, noise, or other effects is not a part of this standard.

2 INTRODUCTION

The equations of this standard are based on the use of experimentally determined capacity factors obtained by testing control valve specimens according to the procedures of ANSI/ISA S75.02, "Control Valve Capacity Test Procedure" (Reference 1).

The equations are used to predict the flow rate of a fluid through a valve when all the factors, including those related to the fluid and its flowing condition, are known. When the equations are used to select a valve size, it is often necessary to use capacity factors associated with the fully open or rated condition to predict an approximate required valve flow coefficient (C_v). This procedure is further explained in Appendix A.

2.1 Flow Variables and Fluid Properties

The flow rate of a fluid through a control valve is a function of the following (where applicable):

- Inlet and outlet conditions

    Pressure
    Temperature
    Piping geometry

- Liquid properties

    Composition
    Density
    Vapor pressure
    Viscosity
    Surface tension
    Thermodynamic critical pressure

- Gas and vapor properties

    Composition
    Density
    Ratio of specific heats

- Control valve properties

    Size
    Valve travel
    Flow path geometry

3 NOMENCLATURE

Symbol	Description
C_v	Valve flow coefficient
d	Valve inlet diameter
D	Internal diameter of the pipe
F_d	Valve style modifier
F_F	Liquid critical pressure ratio factor, dimensionless
F_k	Ratio of specific heats factor, dimensionless
F_L	Liquid pressure recovery factor of a valve without attached fittings, dimensionless
F_{LP}	Product of the liquid pressure recovery factor of a valve with attached fittings (no symbol has been identified) and the piping geometry factor, dimensionless
F_P	Piping geometry factor, dimensionless
F_R	Reynolds number factor, dimensionless
F_s	Laminar, or streamline, flow factor, dimensionless
g	Local acceleration of gravity
G_f	Liquid specific gravity at upstream conditions [ratio of density of liquid at flowing temperature to density of water at 60° F (15.6° C)], dimensionless
G_g	Gas specific gravity (ratio of density of flowing gas to density of air with both at standard conditions, which is equal to the ratio of the molecular weight of gas to the molecular weight of air), dimensionless
k	Ratio of specific heats, dimensionless
K	Head loss coefficient of a device, dimensionless
K_B	Bernoulli coefficient, dimensionless
K_i	Velocity head factors for an inlet fitting, dimensionless
M	Molecular weight, atomic mass units
$N_1, N_2,$ etc.	Numerical constants for units of measurement used
p_1	Upstream absolute static pressure, measured two nominal pipe diameters upstream of valve-fitting assembly
p_2	Downstream absolute static pressure, measured six nominal pipe diameters downstream of valve-fitting assembly
Δp	Pressure differential, $p_1 - p_2$
p_c	Absolute thermodynamic critical pressure

Instrument Society of America

Symbol	Description
p_r	Reduced pressure, dimensionless
p_v	Absolute vapor pressure of liquid at inlet temperature
p_{vc}	Apparent absolute pressure at vena contracta
q	Volumetric flow rate
q_{max}	Maximum flow rate (choked flow conditions) at a given upstream condition
Re_v	Valve Reynolds number, dimensionless
T_r	Reduced temperature, dimensionless
T_c	Absolute thermodynamic critical temperature
T_1	Absolute upstream temperature (in degrees K or R)
U_1	Velocity at valve inlet
w	Weight or mass flow rate
x	Ratio of pressure drop to absolute inlet pressure ($\Delta p/p_1$), dimensionless
x_T	Pressure drop ratio factor, dimensionless
x_{TP}	Value of x_T for valve-fitting assembly, dimensionless
Y	Expansion factor, ratio of flow coefficient for a gas to that for a liquid at the same Reynolds number, dimensionless
Z	Compressibility factor, dimensionless
γ_1 (gamma)	Specific weight, upstream conditions
μ (mu)	Viscosity, absolute
ν (nu)	Kinematic viscosity, centistokes
ρ (rho)	Density
Subscripts	
1	Upstream conditions
2	Downstream conditions
s	Nonturbulent
t	Turbulent

4 INCOMPRESSIBLE FLUID — FLOW OF NONVAPORIZING LIQUID

The flow rate of a liquid through a given control valve at a given travel is a function of the differential pressure ($p_1 - p_2$) when the liquid does not partially vaporize between the inlet and outlet of the valve. If vapor bubbles form either temporarily (cavitation) or permanently (flashing), this relationship may no longer hold. (Refer to Section 5 for choked flow equations that apply when extensive vaporization occurs.) In the transitional region between nonvaporizing liquid flow and fully choked flow, the actual flow rate is less than that predicted by either the equations in this section or those in Section 5. Cavitation that occurs in this transitional region can produce physical damage to the valve and/or to the downstream piping and equipment.

4.1 Equations for Turbulent Flow

The equations for determining the flow rate of a liquid through a valve under turbulent, nonvaporizing flow conditions are:

$$\left. \begin{array}{l} q = N_1 F_p C_v \sqrt{\dfrac{p_1 - p_2}{G_f}} \\ \text{or} \\ C_v = \dfrac{q}{N_1 F_p} \sqrt{\dfrac{G_f}{p_1 - p_2}} \end{array} \right\} \quad (1)$$

$$\left. \begin{array}{l} w = N_6 F_p C_v \sqrt{(p_1 - p_2)\gamma_1} \\ \text{or} \\ C_v = \dfrac{w}{N_6 F_p \sqrt{(p_1 - p_2)\gamma_1}} \end{array} \right\} \quad (2)$$

4.2 Numerical Constants N

The numerical constants N are chosen to suit the measurement units used in the equations. Values for N are listed in Table 1.

**TABLE 1
NUMERICAL CONSTANTS
FOR LIQUID FLOW EQUATIONS**

Constant		Units Used in Equations					
N	w	q	$p, \Delta p$	d, D	γ_1	ν
N_1	0.0865	—	m³/h	kPa	—	—	—
	0.865	—	m³/h	bar	—	—	—
	1.00	—	gpm	psia	—	—	—
N_2	0.00214	—	—	—	mm	—	—
	890	—	—	—	in	—	—
N_4	76 000	—	m³/h	—	mm	—	centistokes*
	17 300	—	gpm	—	in	—	centistokes*
N_6	2.73	kg/h	—	kPa	—	kg/m³	—
	27.3	kg/h	—	bar	—	kg/m³	—
	63.3	lb/h	—	psia	—	lb/ft³	—

*To convert m²/s to centistokes, multiply m²/s by 10⁶. To convert centipoises to centistokes, divide centipoises by G_f.

4.3 Piping Geometry Factor F_p

The piping geometry factor F_p accounts for fittings attached to either the valve inlet or the outlet that disturb the flow to the extent that valve capacity is affected. F_p is actually the

10

ratio of the flow coefficient of a valve with attached fittings to the flow coefficient (C_v) of a valve installed in a straight pipe of the same size as the valve.

For maximum accuracy, F_p must be determined by the test procedures specified in Reference 1. Where estimated values are permissible (Reference 2), F_p may be determined by using the following equation:

$$F_p = \left(\frac{\Sigma K C_v^2}{N_2 d^4} + 1 \right)^{-1/2} \qquad (3)$$

(See Appendix B for the mathematical derivation of F_p.)

In many instances, the nominal sizes for valve and pipe (d and D) may be used in Equations 3, 5, 6, and 7 without significant error.

The factor ΣK is the algebraic sum of the effective velocity head coefficients of all fittings attached to but not including the valve. For instance,

$$\Sigma K = K_1 + K_2 + K_{B1} - K_{B2} \qquad (4)$$

where K_1 and K_2 are the resistance coefficients of the inlet and outlet fittings, respectively, and K_{B1} and K_{B2} are the Bernoulli coefficients for the inlet and outlet fittings, respectively. The Bernoulli coefficients compensate for the changes in pressure resulting from differences in stream area and velocity.

When the diameters of the inlet and outlet fittings are identical, $K_{B1} = K_{B2}$, both factors drop out of the equation. When the diameters of the inlet and outlet fittings are different, K_B is calculated as follows:

$$K_B = 1 - \left(\frac{d}{D} \right)^4 \qquad (5)$$

The fittings most commonly encountered are standard, short-pattern concentric pipe reducers. These fittings have little taper, and their pressure loss will not exceed that of an abrupt contraction with a slightly rounded entrance. On that basis, if experimental values for the resistance coefficients K_1 and K_2 are unavailable, estimated values may be computed as follows:

Inlet reducer only:

$$K_1 = 0.5 \left(1 - \frac{d^2}{D_1^2} \right)^2 \qquad (6)$$

Outlet increaser only:

$$K_2 = 1.0 \left(1 - \frac{d^2}{D_2^2} \right)^2 \qquad (7)$$

When the reducer and increaser are the same size:

$$K_1 + K_2 = 1.5 \left(1 - \frac{d^2}{D^2} \right)^2 \qquad (8)$$

(See Appendix C for a graphic representation of system head changes around a valve with attached reducers.)

4.4 Equations for Nonturbulent Flow

Nonturbulent flow occurs at high fluid viscosities and/or low velocities. In these circumstances, the flow rate through a valve is less than for turbulent flow, and the Reynolds number factor F_R must be introduced. F_R is the ratio of nonturbulent flow rate to the turbulent flow rate predicted by Equations 1 or 2. The corresponding nonturbulent equations then become, respectively:

$$\left. \begin{array}{l} q = N_1 F_R C_v \sqrt{\dfrac{p_1 - p_2}{G_f}} \\[1em] \text{or} \\[1em] C_v = \dfrac{q}{N_1 F_R} \sqrt{\dfrac{G_f}{p_1 - p_2}} \end{array} \right\} \qquad (9)$$

$$\left. \begin{array}{l} w = N_6 F_R C_v \sqrt{(p_1 - p_2) \gamma_1} \\[1em] \text{or} \\[1em] C_v = \dfrac{w}{N_6 F_R \sqrt{(p_1 - p_2) \gamma_1}} \end{array} \right\} \qquad (10)$$

Note the absence of the piping geometry factor in the above equations. For nonturbulent flow, the effect of close-coupled reducers or other flow-disturbing fittings is unknown. Thus, Equation 3 applies to turbulent flow only.

Tests (References 3 and 4) show that F_R can be found by using the valve Reynolds number and Figure 1. The shading around the central curve indicates the scatter of test data and the range of uncertainty of flow rate prediction in the nonturbulent regimes.

The valve Reynolds number is defined as:

$$\text{Re}_v = \frac{N_4 F_d q}{\nu F_L^{1/2} C_v^{1/2}} \left(\frac{F_L^2 C_v^2}{N_2 d^4} + 1 \right)^{1/4} \qquad (11)$$

The valve style modifier F_d in Equation 11 correlates data from tests of several valve styles with different hydraulic radii, so that a single curve represents all the styles tested. (See Appendix D for representative values of F_d.) Caution must be used in applying the curve in Figure 1 to valve styles for which F_d has not been established.

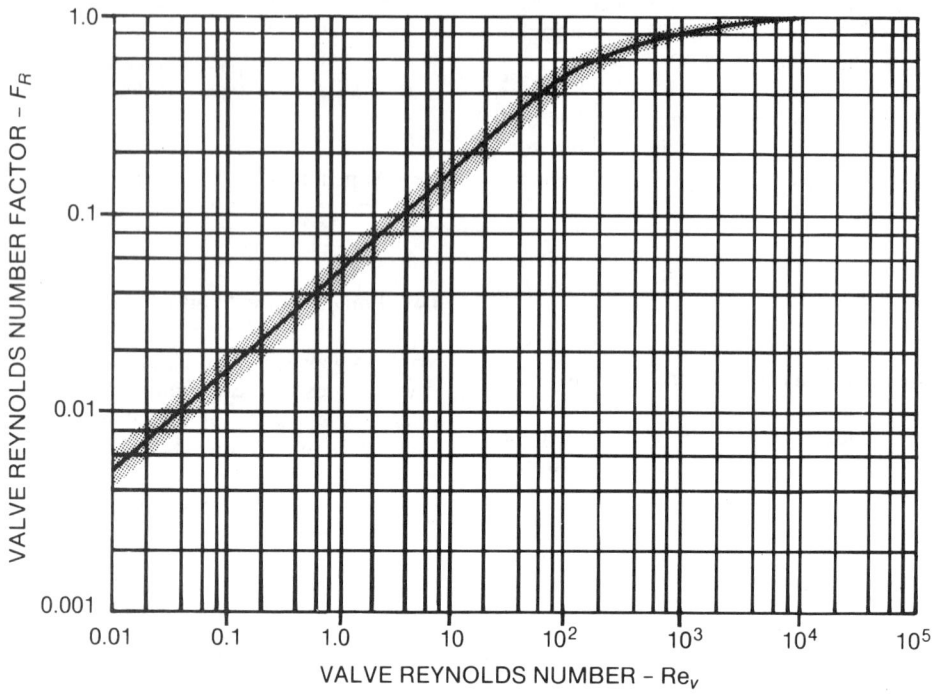

Figure 1. Reynolds Number Factor

The bracketed term in Equation 11 accounts for the *velocity of approach*.* Except for wide-open ball or butterfly valves, this term has only a slight effect on the Re_v calculation and can generally be omitted.

Most flow streams in process plant control valves are turbulent, with valve Reynolds numbers in excess of 10^4, where the Reynolds number factor is 1.0. When the flow regime is questionable, Equation 11 should be used to find Re_v. For additional information on nonturbulent flow, see Appendices E and F.

5 INCOMPRESSIBLE FLUID — CHOKED FLOW OF VAPORIZING LIQUID

Choked flow is a limiting, or maximum, flow rate. With fixed inlet (upstream) conditions, it is manifested by the failure of decreasing downstream pressure to increase the flow rate. With liquid streams, choking occurs as a result of vaporization of the liquid when the pressure within the valve falls below the vapor pressure of the liquid. Choked flow will be accompanied by either cavitation or flashing. If the downstream pressure is greater than the vapor pressure of the liquid, cavitation occurs. If the downstream pressure is equal to or less than the vapor pressure of the liquid, flashing occurs. This relationship between flow rate and pressure drop for a typical valve is shown in Figure 2.

*The flow rate through a valve is a function of the velocity of the jet stream at the vena contracta and the area of the jet at that location. This velocity is a function of the pressure drop across the valve orifice and also the valve inlet velocity, or *velocity of approach*. The velocity of approach factor is included in the valve flow coefficient C_v.

5.1 Liquid Choked Flow Equations

The equations for determining the maximum flow rate of a liquid under choked conditions for valves in straight pipes of the same size are as follows:

$$q_{max} = N_1 F_L C_v \sqrt{\frac{p_1 - p_{vc}}{G_f}}$$

or

$$C_v = \frac{q_{max}}{N_1 F_L} \sqrt{\frac{G_f}{p_1 - p_{vc}}} \quad (12a)$$

where

$$p_{vc} = F_F p_v \quad \text{(see Appendix G for } F_F) \quad (13a)$$

giving

$$q_{max} = N_1 F_L C_v \sqrt{\frac{p_1 - F_F p_v}{G_f}}$$

or

$$C_v = \frac{q_{max}}{N_1 F_L} \sqrt{\frac{G_f}{p_1 - F_F p_v}} \quad (14a)$$

The equations for determining the maximum flow rate of a liquid under choked conditions for valves with attached fittings are:

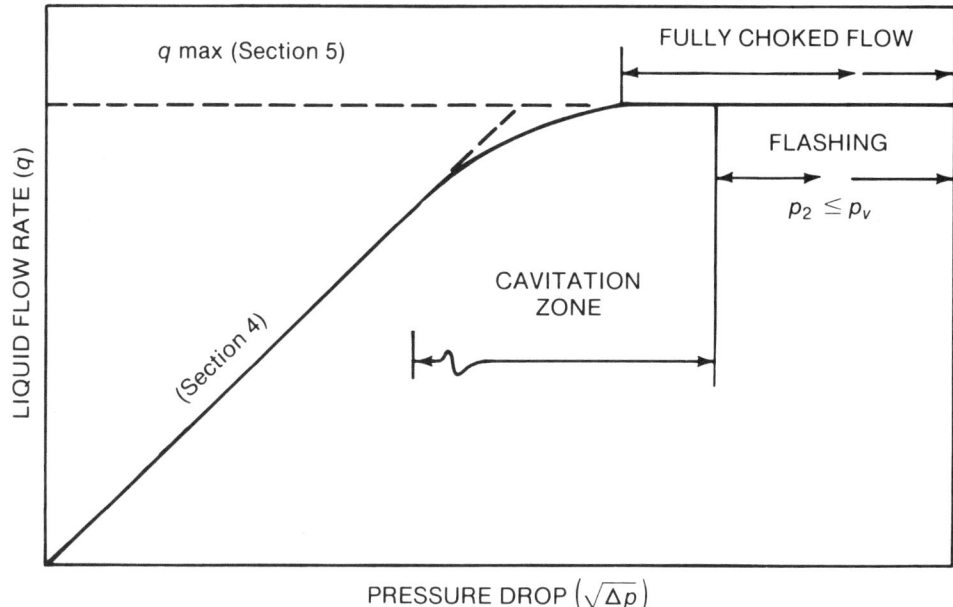

Figure 2. Liquid Flow Rate Versus Pressure Drop for a Typical Valve (Constant Upstream Pressure and Vapor Pressure)

$$q_{max} = N_1 F_{LP} C_v \sqrt{\frac{p_1 - p_{vc}}{G_f}}$$

or

$$C_v = \frac{q_{max}}{N_1 F_{LP}} \sqrt{\frac{G_f}{p_1 - p_{vc}}}$$

(12b)

where

$$p_{vc} = F_F p_v \quad \text{(see Appendix G for } F_F\text{)} \quad (13b)$$

giving

$$q_{max} = N_1 F_{LP} C_v \sqrt{\frac{p_1 - F_F p_v}{G_f}}$$

or

$$C_v = \frac{q_{max}}{N_1 F_{LP}} \sqrt{\frac{G_f}{p_1 - F_F p_v}}$$

(14b)

5.2 Liquid Pressure Recovery Factor F_L

The liquid pressure recovery factor F_L applies to valves without attached fittings (Reference 5). This factor accounts for the influence of the internal geometry of the valve on its capacity at choked flow. Under nonvaporizing flow conditions, it is defined by the equation:

$$F_L = \sqrt{\frac{p_1 - p_2}{p_1 - p_{vc}}} \quad (15a)$$

Representative F_L values for various valve styles are listed in Appendix D.

5.3 Combined Liquid Pressure Recovery Factor F_{LP}

When a valve is installed with reducers or other attached fittings, the liquid pressure recovery of the valve-fitting combination is not the same as that for the valve alone. For calculations involving choked flow, it is convenient to treat the piping geometry factor (F_p) and the F_L factor for the valve-fitting combination as a single factor, F_{LP}. The value of F_L for the combination is then F_{LP}/F_p, where

$$\frac{F_{LP}}{F_p} = \sqrt{\frac{p_1 - p_2}{p_1 - p_{vc}}} \quad (15b)$$

(Refer to Section 4.3 and Appendix B.)

For maximum accuracy, F_{LP} must be determined by using the test procedures specified in Reference 1. When estimated values are permissible, reasonable accuracy may be obtained by using the following equation to determine F_{LP}:

$$F_{LP} = F_L \left(\frac{K_i F_L^2 C_v^2}{N_2 d^4} + 1 \right)^{-1/2} \quad (16)$$

In this equation, K_i is the head loss coefficient ($K_1 + K_{B1}$) of any fitting between the upstream pressure tap and the inlet face of the valve only. (See Appendix B for the mathematical derivation of F_{LP}.)

6 COMPRESSIBLE FLUID — FLOW OF GAS AND VAPOR

The flow rate of a compressible fluid varies as a function of the ratio of the pressure differential to the absolute inlet pressure ($\Delta p/p_1$), designated by the symbol x. At values of x near zero, the equations in this section can be traced to the basic Bernoulli equation for Newtonian incompressible fluids. However, increasing values of x result in expansion and compressibility effects that require the use of appropriate correction factors (References 6 and 7).

6.1 Equations for Turbulent Flow

The flow rate of a gas or vapor through a valve may be calculated by using any of the following equivalent forms of the equation:

$$\left. \begin{aligned} w &= N_6 F_p C_v Y \sqrt{x p_1 \gamma_1} \\ \text{or} & \\ C_v &= \frac{w}{N_6 F_p Y \sqrt{x p_1 \gamma_1}} \end{aligned} \right\} \quad (17)$$

$$\left. \begin{aligned} q &= N_7 F_p C_v p_1 Y \sqrt{\frac{x}{G_g T_1 Z}} \\ \text{or} & \\ C_v &= \frac{q}{N_7 F_p p_1 Y} \sqrt{\frac{G_g T_1 Z}{x}} \end{aligned} \right\} \quad (18)$$

$$\left. \begin{aligned} w &= N_8 F_p C_v p_1 Y \sqrt{\frac{xM}{T_1 Z}} \\ \text{or} & \\ C_v &= \frac{w}{N_8 F_p p_1 Y} \sqrt{\frac{T_1 Z}{xM}} \end{aligned} \right\} \quad (19)$$

$$\left. \begin{aligned} q &= N_9 F_p C_v p_1 Y \sqrt{\frac{x}{M T_1 Z}} \\ \text{or} & \\ C_v &= \frac{q}{N_9 F_p p_1 Y} \sqrt{\frac{M T_1 Z}{x}} \end{aligned} \right\} \quad (20)$$

Note that the numerical value of x used in these equations must not exceed the choking limit ($F_K x_{Tp}$), regardless of the actual value of x. (See Section 6.4.)

6.2 Numerical Constants N

The numerical constants N are chosen to suit the measurement units used in the equations. Values for N are listed in Table 2.

TABLE 2
NUMERICAL CONSTANTS FOR GAS AND VAPOR FLOW EQUATIONS

Constant	Units Used in Equations					
N	w	q^*	$p, \Delta p$	γ_1	T_1	d, D
N_5 0.00241	—	—	—	—	—	mm
1000	—	—	—	—	—	in
N_6 2.73	kg/h	—	kPa	kg/m³	—	—
27.3	kg/h	—	bar	kg/m³	—	—
63.3	lb/h	—	psia	lb/ft³	—	—
N_7 4.17	—	m³/h	kPa	—	K	—
417	—	m³/h	bar	—	K	—
1360	—	scfh	psia	—	°R	—
N_8 0.948	kg/h	—	kPa	—	K	—
94.8	kg/h	—	bar	—	K	—
19.3	lb/h	—	psia	—	°R	—
N_9 22.5	—	m³/h	kPa	—	K	—
2250	—	m³/h	bar	—	K	—
7320	—	scfh	psia	—	°R	—

*q is in cubic feet per hour measured at 14.73 psia and 60 °F, or cubic meters per hours measured at 101.3 kPa and 15.6 °C.

6.3 Expansion Factor Y

The expansion factor Y accounts for the change in density of a fluid as it passes from the valve inlet to the vena contracta and for the change in area of the vena contracta as the pressure drop is varied (contraction coefficient). Theoretically, Y is affected by all of the following:

1. Ratio of port area to body inlet area
2. Internal geometry of the valve
3. Pressure drop ratio, x
4. Reynolds number
5. Ratio of specific heats, k

The influence of items 1, 2, and 3 are defined by the factor x_T. Test data (Reference 7) indicate that Y may be taken as a linear function of x, as shown in the following equation for a valve without attached fittings:

$$Y = 1 - \frac{x}{3 F_k x_T} \quad \text{(Limits } 1.0 \geq Y \geq 0.67\text{)} \quad (21)$$

For a valve with attached fittings, x_{TP} shall be substituted for x_T.

For all practical purposes, Reynolds number effects may be disregarded in the case of compressible fluids. The effect of the ratio of specific heats is considered in Section 6.7.

6.4 Choked Flow

If all inlet conditions are held constant and the differential pressure ratio (x) is increased by lowering the downstream pressure (p_2), the mass flow rate will increase to a maximum limit. Flow conditions where the value of x exceeds this limit are known as *choked flow*. Choking occurs when the jet stream at the vena contracta attains its maximum cross-sectional area at sonic velocity. This occurs at pressure ratios (p_1/p_{vc}) greater than about 2.0.

The value of x at the inception of choked flow conditions varies from valve to valve. It also varies with the piping geometry and with the thermodynamic properties of the flowing fluid. The factors involved are x_T (Section 6.5), x_{TP} (Section 6.6), and F_k (Section 6.7).

Choking affects the use of Equations 17 through 21 in the following manner: The value of x used in the equations must not exceed $F_k x_T$ or $F_k x_{TP}$, regardless of the actual value of x. The expansion factor Y at choked flow ($x \geq F_k x_{TP}$) is then at its minimum value of $\frac{2}{3}$.

6.5 Pressure Drop Ratio Factor x_T

For maximum accuracy, the pressure drop ratio factor x_T must be established by using the test procedures specified in Reference 1. Representative x_T values for valves are tabulated in Appendix D. These representative values are not to be taken as actual. Actual values must be obtained from the valve manufacturer.

6.6 Pressure Drop Ratio Factor with Reducers or Other Fittings x_{TP}

When a valve is installed with reducers or other fittings, the pressure drop ratio factor of the assembly (x_{TP}) is different from that of the valve alone (x_T). For maximum accuracy, x_{TP} must be determined by test (Reference 1). When estimated values are permissible, the following equation may be used to determine x_{TP}:

$$x_{TP} = \frac{x_T}{F_p^2} \left(\frac{x_T K_i C_v^2}{N_5 d^4} + 1 \right)^{-1} \qquad (22)$$

In this equation, x_T is the pressure drop ratio factor for a given valve installed without reducers or other fittings, and K_i is the sum of the inlet velocity head coefficients ($K_1 + K_{B1}$) of the reducer or other fitting attached to the valve inlet. This correction to x_T is usually negligible if d/D is greater than 0.5 and C_v/d^2 is less than 20, where d is in inches.

See Appendix H for the mathematical derivation of x_{TP}.

6.7 Ratio of Specific Heats Factor F_k

The ratio of specific heats of a compressible fluid affects the flow rate through a valve. The factor F_k accounts for this effect. F_k has a value of 1.0 for air at moderate temperatures and pressures, where its specific heat ratio is about 1.40. Both theoretical and experimental evidence indicate that for valve sizing purposes, F_k may be taken as having a linear relationship to k. Therefore:

$$F_k = \frac{k}{1.40} \qquad (23)$$

6.8 Compressibility Factor Z

Equations 18, 19, and 20 do not contain a term for the actual specific weight of the fluid at upstream conditions. Instead, this term is inferred from the inlet pressure and temperature based on the laws of ideal gases. Under some conditions, real gas behavior can deviate markedly from the ideal. In these cases, the compressibility factor Z shall be introduced to compensate for the discrepancy. Z is a function of both reduced pressure and reduced temperature. For use in this section, reduced pressure p_r is defined as the ratio of the actual inlet absolute pressure to the absolute thermodynamic critical pressure for the fluid in question. The reduced temperature is defined similarly. That is:

$$p_r = \frac{p_1}{p_c} \qquad (24)$$

$$T_r = \frac{T_1}{T_c} \qquad (25)$$

Absolute thermodynamic critical pressures and temperatures for most fluids, and curves from which Z may be determined, can be found in many reference handbooks of physical data.

APPENDIX A
USE OF FLOW RATE EQUATIONS FOR SIZING VALVES

Laboratory tests are conducted on actual valves in a prescribed test setup (Reference 1). The test fluid is usually water or air. The flow coefficient C_v and the factors F_L, x_T, etc., are determined at the rated valve travel. These data, along with factors to account for the actual fluid and the pipe configuration (F_k, F_F, F_p, etc.), are used in the equations of this standard to predict the flow rate with the valve fully open.

The principal use of the flow equations is to aid in the selection of an appropriate valve size for a specific application. In this procedure, the numbers in the equations consist of known values for the fluid and flow conditions and known values for the selected valve type at its rated opening. With these factors in the equation, the unknown (or product of unknowns, e.g., $F_p C_v$) can be computed. Although these computed numbers are often suitable for selecting a valve from a series of discrete sizes, they do not represent a true operating condition, because the factors are mutually incompatible. Some of the factors used in the equation are for the wide-open valve while others relating to the operating conditions are for the partially open valve.

Once a valve size has been selected, the remaining unknowns, such as F_p, can be computed and a judgment can be made as to whether the valve size is adequate. It is not usually necessary to carry the calculations further to predict the exact valve opening. To do this, all the pertinent sizing factors must be known at fractional valve openings.

Additional information on the use of the flow equations, along with example problems, is available in References 8 and 9.

APPENDIX B
DERIVATION OF FACTORS F_p AND F_{LP}

If a valve is installed between reducers, the C_v of the entire assembly is different from that of the valve alone. If the inlet and outlet reducers are the same size, the only effect is the added resistance of the fittings, which creates an additional pressure drop. If there is only one reducer or if there are reducers of different sizes, there will be an additional effect on the pressure due to the difference in velocity between the inlet and outlet streams.

The velocity head expressed in feet of fluid equals $U_2/2g$, where U is the velocity of the stream and g is the acceleration of gravity. Expressed in U.S. customary units, psi, gpm, and inches, the velocity pressure becomes:

$$p = \frac{q^2 G_f}{890 \, d^4} \quad (B1)$$

For a resistance coefficient K, the pressure difference then becomes:

$$\Delta p = K \left(\frac{q^2 G_f}{890 \, d^4} \right) \quad (B2)$$

From Equations 1 and B2, the resistance coefficient for a valve is:

$$K_{\text{valve}} = \frac{890 \, d^4}{C_v^2} \quad (B3)$$

The change in velocity pressure across a reducer with diameters d and D is:

$$\frac{q^2 G_f}{890 \, d^4} - \frac{q^2 G_f}{890 \, D^4} = \frac{q^2 G_f}{890 \, d^4} \left(1 - \frac{d^4}{D^4} \right) \quad (B4)$$

From Equations B2 and B4, we have the factor K_B, which has been called the Bernoulli coefficient.

Here:

$$K_B = \left(1 - \frac{d^4}{D^4} \right) \quad (B5)$$

By definition:

$$(F_P C_v)^2 = \frac{q^2 G_f}{\Delta p} \quad (B6)$$

From Equations B2 and B6, adding all K factors:

$$(F_P C_v)^2 = \frac{890 \, d^4}{K_{\text{valve}} + K_1 + K_2 + K_{B1} - K_{B2}} \quad (B7)$$

Substitute K_{valve} from Equation B3:

$$(F_P C_v)^2 = \frac{890 \, d^4}{\frac{890 \, d^4}{C_v^2} + \Sigma K} \quad (B8)$$

where

$$\Sigma K = K_1 + K_2 + K_{B1} - K_{B2} \quad (B9)$$

Then, rearranging Equation B8, we have:

$$F_p = \left(\frac{\Sigma K C_v^2}{890 \, d^4} + 1 \right)^{-1/2} \quad (B10)$$

It should be noted in Equation B9 that ΣK is the sum of all the effective velocity head coefficients. If the inlet and outlet reducers are the same size, $K_{B1} = K_{B2}$, and in Equation B9 both drop out because of the difference in their sign. For K_1 and K_2, see Equations 6 and 7.

By definition, from Equation 15:

$$F_L^2 = \frac{p_1 - p_2}{p_1 - p_{vc}} = \frac{\Delta p_a}{\Delta p_{vc}} \tag{B11}$$

where Δp_a is the pressure drop across the valve, and Δp_{vc} is the drop to the vena contracta.

Also, from Equation 1:

$$q^2 = (F_P C_v)^2 \frac{\Delta p_b}{G_f} = C_v^2 \frac{\Delta p_a}{G_f} \tag{B12}$$

where Δp_b is the drop across the valve with reducers.

From Equation B12:

$$\Delta p_a = F_P^2 \Delta p_b \tag{B13}$$

Substituting this expression into Equation B11, we have:

$$F_L^2 = F_P^2 \frac{\Delta p_b}{\Delta p_{vc}} \tag{B14}$$

By definition,

$$(F_L)_P^2 = \frac{\Delta p_b}{\Delta p_{vc} + \Delta p_i} \tag{B15}$$

where $(F_L)_P$ is the pressure recovery factor for the valve with reducers, and Δp_i is the drop across the inlet reducer.

From Equation B2:

$$\Delta p_i = \frac{K_i q^2 G_f}{890 \, d^4} \tag{B16}$$

where $K_i = K_1 + K_{B1}$.

Substituting the expression for q^2 from Equation B12 into Equation B16, we have:

$$\Delta p_i = \frac{K_i F_P^2 C_v^2 \Delta p_b}{890 \, d^4} \tag{B17}$$

Substituting Equations B14 and B17 into B15, we have the following development:

$$(F_L)_P^2 = \frac{\Delta p_b}{\dfrac{F_P^2 \Delta p_b}{F_L^2} + \dfrac{K_i F_P^2 C_v^2 \Delta p_b}{890 \, d^4}}$$

$$(F_L)_P = \frac{1}{F_P}\left(\frac{1}{F_L^2} + \frac{K_i C_v^2}{890 \, d^4}\right)^{-1/2}$$

$$F_{LP} = (F_L)_P F_P = \left[\frac{1}{F_L^2} + \frac{K_i}{890}\left(\frac{C_v}{d^2}\right)^2\right]^{-1/2}$$

$$F_{LP} = F_L \left[\frac{F_L^2 K_i}{N_2}\left(\frac{C_v}{d^2}\right)^2 + 1\right]^{-1/2} \tag{B18}$$

APPENDIX C
CONTROL VALVE-PIPING SYSTEM HEAD CHANGES

An understanding of the various loss mechanisms involved in a control valve-piping system can be obtained by looking at the energy grade lines and the hydraulic grade lines for a liquid flow system containing abrupt contractions and expansions in the form of concentric reducers. These are shown schematically in Figure C-1. For ease of comprehension, the curves are displayed as straight line segments. The energy grade line includes only the available energy and excludes internal energy. Each point of pressure change associated with this figure is defined below. Some of the pressure drops are nonrecoverable and some are recoverable, as shown in the hydraulic grade line. The terms below also define the various coefficients associated with the system. The Bernoulli coefficients, K_{B1} and K_{B2}, account for the change in the velocity pressure of the stream and relate the total kinetic energy to that calculated with the valve inlet velocity U_1.

TABLE C-1
DEFINITIONS OF HEAD TERMS
(Refer to Figure C-1)

Reference Letter (See Fig. C-1)	Head Terms*	U.S. Units	SI Units
A	Inlet pressure head	p_1/γ	$p_1/\rho g$
B	Inlet velocity head	$(d/D_1)^4(U_1^2/2g)$	$(d/D_1)^4(U_1^2/2g)$
C	Reducer drop	$(K_1+K_{B1})(U_1^2/2g)$	$(K_1+K_{B1})(U_1^2/2g)$
D	Differential to vena contracta	$(E)/(1-F_L^2)$	$(E)/(1-F_L^2)$
E	Pressure recovery at valve	$(D)-(H)$	$(D)-(H)$
F	Increaser recovery	$(K_{B2}-K_2)(U_1^2/2g)$	$(K_{B2}-K_2)(U_1^2/2g)$
G	Reducer loss	$K_1(U_1^2/2g)$	$K_1(U_1^2/2g)$
H	Valve loss	$N_2(d^4/C_v^2)(U_1^2/2g)$	$N_2(d^4/C_v^2)(U_1^2/2g)$
I	Increaser loss	$K_2(U_1^2/2g)$	$K_2(U_1^2/2g)$
J	Outlet pressure head	p_2/γ	$p_2/\rho g$
K	Outlet velocity head	$(d/D_2)^4(U_1^2/2g)$	$(d/D_2)^4(U_1^2/2g)$
L	Total head loss	$(p_1-p_2)/\gamma$	$(p_1-p_2)/\rho g$

*All units are absolute and consistent: pound, foot, and second in U.S. customary units; SI for metric units.

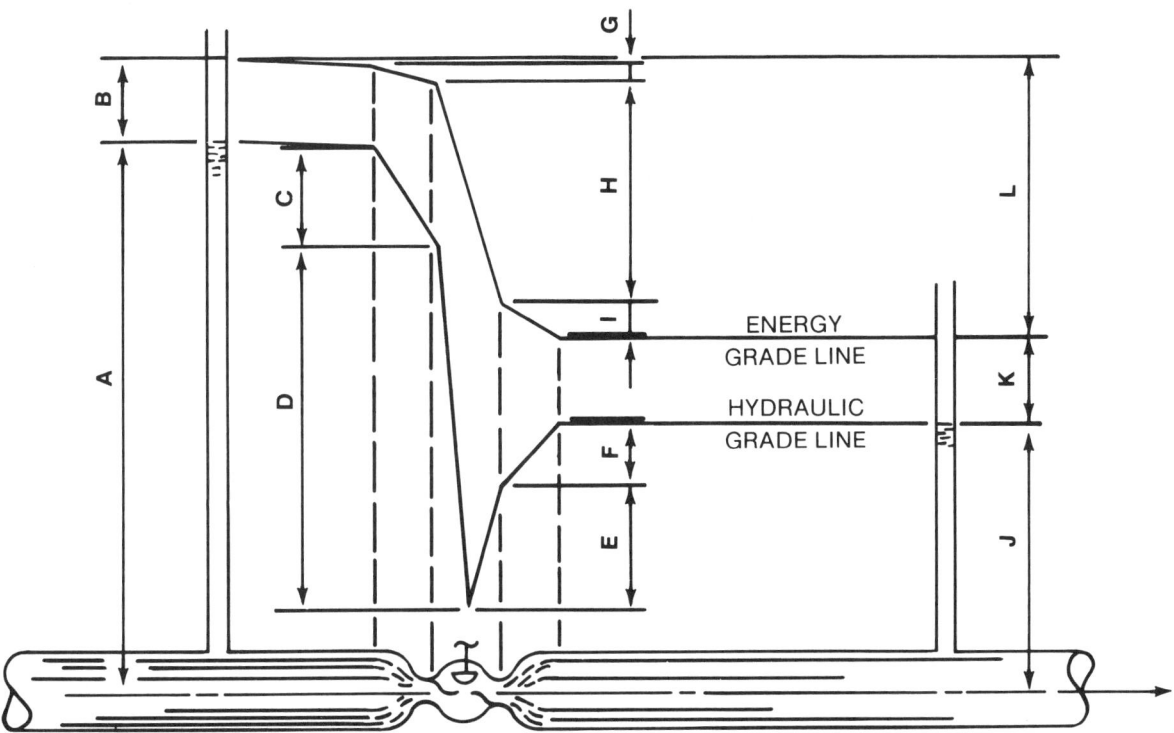

Figure C-1. Head Changes in a Control Valve-Piping System

APPENDIX D
REPRESENTATIVE VALUES
OF VALVE CAPACITY FACTORS

The values in Table D-1 are typical only for the types of valves shown at their rated travel for full-size trim. Significant variations in value may occur because of any of the following reasons: reduced travel, trim type, reduced port size, and valve manufacturer.

APPENDIX E
REYNOLDS NUMBER FACTOR F_R

The information contained in this appendix is an elaboration of the discussion presented in Section 4.4. It presents a method used for resolving laminar and transitional flow problems.

Figure E-1 shows the relationships between F_R and the valve Reynolds number Re_v for the three types of problems that may be encountered with viscous flow. These are:

1. Determining the required flow coefficient when selecting a control valve size

2. Predicting the flow rate that a selected valve will pass

3. Predicting the pressure differential that a selected valve will exhibit

In Figure E-1, the straight diagonal lines extending downward at an F_R value of approximately 0.3 indicate conditions under which laminar flow exists. At a valve Reynolds number of 40 000, all three curves in Figure E-1 reach an F_R value of 1.0. At this number and at all higher Re_v values, fully turbulent flow conditions exist. Between the laminar region, indicated by the straight diagonal lines of Figure E-1, and the turbulent region, where $F_R = 1.0$, the flow regime is transitional (i.e., neither laminar nor turbulent).

Equation 11 for determining the valve Reynolds number Re_v is:

$$Re_v = \frac{N_4 F_d q}{\nu F_L^{1/2} C_v^{1/2}} \left(\frac{F_L^2 C_v^2}{N_2 d^4} + 1 \right)^{1/4} \qquad (11)$$

F_R values and the solutions to the three classes of problems may be obtained by using the following procedures.

TABLE D-1
REPRESENTATIVE VALUES OF VALVE CAPACITY FACTORS

Valve Type	Trim Type	Flow Direction*	x_T	F_L	F_s	F_d**	C_v/d^2†
GLOBE							
Single port	Ported plug	Either	0.75	0.9	1.0	1.0	9.5
	Contoured plug	Open	0.72	0.9	1.1	1.0	11
		Close	0.55	0.8	1.1	1.0	11
	Characterized cage	Open	0.75	0.9	1.1	1.0	14
		Close	0.70	0.85	1.1	1.0	16
	Wing guided	Either	0.75	0.9	1.1	1.0	11
Double port	Ported plug	Either	0.75	0.9	0.84	0.7	12.5
	Contoured plug	Either	0.70	0.85	0.85	0.7	13
	Wing guided	Either	0.75	0.9	0.84	0.7	14
Rotary	Eccentric spherical plug	Open	0.61	0.85	1.1	1.0	12
		Close	0.40	0.68	1.2	1.0	13.5
ANGLE	Contoured plug	Open	0.72	0.9	1.1	1.0	17
		Close	0.65	0.8	1.1	1.0	20
	Characterized cage	Open	0.65	0.85	1.1	1.0	12
		Close	0.60	0.8	1.1	1.0	12
	Venturi	Close	0.20	0.5	1.3	1.0	22
BALL	Segmented	Open	0.25	0.6	1.2	1.0	25
	Standard port (diameter $\cong 0.8d$)	Either	0.15	0.55	1.3	1.0	30
BUTTERFLY	60-Degree aligned	Either	0.38	0.68	0.95	0.7	17.5
	Fluted vane	Either	0.41	0.7	0.93	0.7	25
	90-Degree offset seat	Either	0.35	0.60	0.98	0.7	29

* Flow direction tends to *open* or *close* the valve, i.e., push the closure member away from or towards the seat.
** In general, an F_d value of 1.0 can be used for valves with a single flow passage. An F_d value of 0.7 can be used for valves with two flow passages, such as double-ported globe valves and butterfly valves.
† In this table, d may be taken as the nominal valve size, in inches.

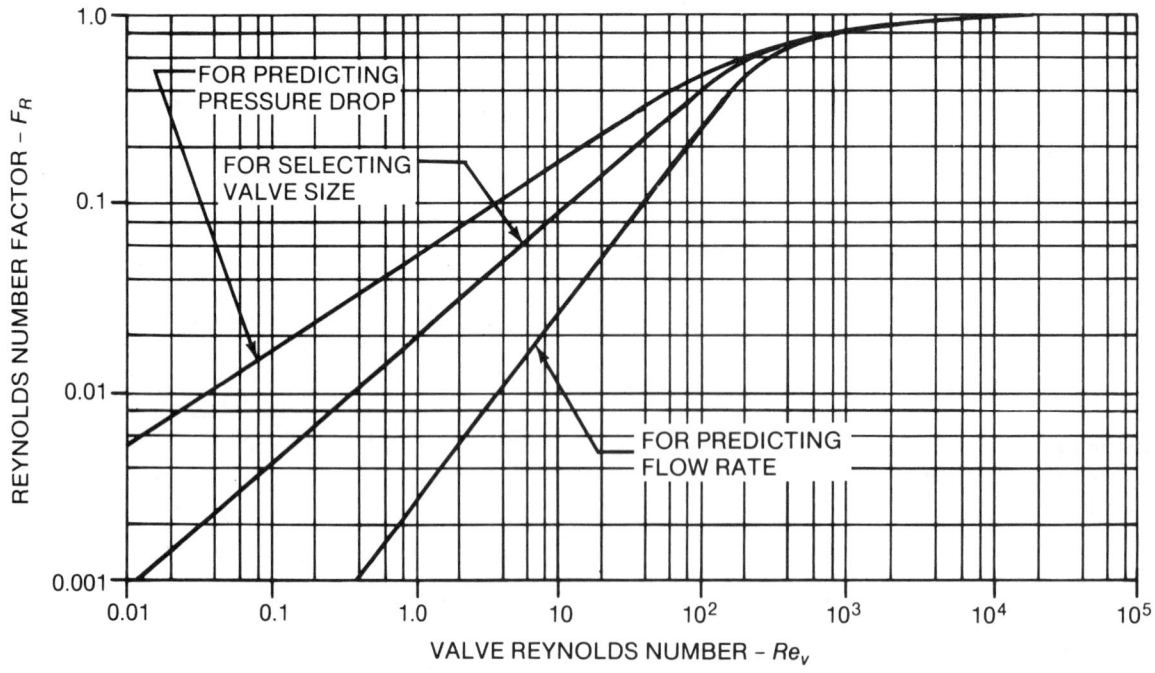

Figure E-1. Reynolds Number Factor for Valve Sizing
(See Figure 1 for the range of uncertainty.)

Determining Required Flow Coefficient (Selecting Valve Size)

The following treatment is based on valves without attached fittings; therefore, $F_p = 1.0$.

1. Calculate a pseudo valve flow coefficient C_{vt}, assuming turbulent flow, using:

$$C_{vt} = \frac{q}{N_1 \sqrt{\frac{p_1 - p_2}{G_f}}} \quad (E1)$$

2. Calculate Re_v by using Equation 11, substituting C_{vt} from Step 1 for C_v. For F_L, select a representative value for the valve style desired.

3. Find F_R as follows:

   a. If Re_v is less than 56, the flow is laminar, and F_R may be found by using either the curve in Figure E-1 labeled "For Selecting Valve Size" or by using the equation

$$F_R = 0.019 \, (Re_v)^{0.67} \quad (E2)$$

   b. If Re_v is greater than 40 000, the flow may be taken as turbulent, and $F_R = 1.0$.

   c. If Re_v lies between 56 and 40 000, the flow is transitional, and F_R may be found by using either the curve in Figure E-1 or Table E-1 in the column headed "Valve Size Selection."

4. Obtain the required C_v from:

$$C_v = \frac{C_{vt}}{F_R} \quad (E3)$$

5. After determining C_v, check the F_L value for the selected valve size and style. If this value is significantly different from the value selected in Step 2, use the new value and repeat Steps 1 through 4.

Predicting Flow Rate

1. Calculate q_t, assuming turbulent flow, using:

$$q_t = N_1 C_v \sqrt{\frac{p_1 - p_2}{G_f}} \quad (E4)$$

2. Calculate Re_v by using Equation 11, substituting q_t for q from Step 1.

3. Find F_R as follows:

   a. If Re_v is less than 106, the flow is laminar, and F_R may be found by using the curve in Figure E-1 labeled "For Predicting Flow Rate" or by using the equation

$$F_R = 0.0027 \, Re_v \quad (E5)$$

   b. If Re_v is greater than 40 000, the flow may be taken as turbulent, and $F_R = 1.0$.

   c. If Re_v lies between 106 and 40 000, the flow is transitional, and F_R may be found by using the curve in Figure E-1 or Table E-1 in the column headed "Flow Rate Prediction."

4. Obtain the predicted flow rate from:

$$q = F_R q_t \quad (E6)$$

Predicting Pressure Drop

1. Calculate Re_v according to Equation 11.

2. Find F_R as follows:

   a. If Re_v is less than 30, the flow is laminar, and F_R may be found by using the curve in Figure E-1 labeled "For Predicting Pressure Drop" or by using the equation

$$F_R = 0.052 \, (Re_v)^{0.5} \quad (E7)$$

   b. If Re_v is greater than 40 000, the flow may be taken as turbulent, and $F_R = 1.0$.

   c. If Re_v lies between 30 and 40 000, the flow is transitional, and F_R may be found by using the curve in Figure E-1 or Table E-1 in the column headed "Pressure Drop Prediction."

3. Obtain the predicted pressure drop from:

$$\Delta p = G_f \left(\frac{q}{N_1 F_R C_v} \right)^2 \quad (E8)$$

TABLE E-1
REYNOLDS NUMBER FACTOR
***FR* FOR TRANSITIONAL FLOW**

	Valve Reynolds Number, Re_v*		
F_R*	Valve Size Selection	Flow Rate Prediction	Pressure Drop Prediction
0.284	56	106	30
0.32	66	117	38
0.36	79	132	48
0.40	94	149	59
0.44	110	167	74
0.48	130	188	90
0.52	154	215	113
0.56	188	253	142
0.60	230	298	179
0.64	278	351	224
0.68	340	416	280
0.72	471	556	400
0.76	620	720	540
0.80	980	1100	870
0.84	1560	1690	1430
0.88	2470	2660	2300
0.92	4600	4800	4400
0.96	10 200	10 400	10 000
1.00	40 000	40 000	40 000

*Linear interpolation between listed values is satisfactory.

APPENDIX F
EQUATIONS FOR
NONTURBULENT LIQUID FLOW

The following method for handling liquid nonturbulent flow permits a direct solution for the unknown — flow rate, C_v, or Δp — without using tables or curves and without first computing a Reynolds number. It is especially useful with programmable calculators or computers. The results are in conformance with Section 4.4.

Figure 1 in Section 4.4 has the following features:

1. A straight horizontal line at $F_R = 1.0$, representing the turbulent flow region. Here, the flow rate varies as the square root of differential pressure (Equation 1).

2. A straight diagonal line, representing the laminar flow region. Here, the flow rate varies directly with the differential pressure.

3. A curved portion, representing the transitional flow region.

4. A shaded envelope to indicate the scatter of the test data (References 3 and 4) and the uncertainty to be expected in the nonturbulent flow region.

From Equation 9:

$$q = N_1 F_R C_v \sqrt{\frac{p_1 - p_2}{G_f}} \qquad (9)$$

and Equation 11:

$$Re_v = \frac{N_4 F_d q}{\nu F_L^{1/2} C_v^{1/2}} \left(\frac{F_L^2 C_v^2}{N_2 d^4} + 1 \right)^{1/4} \qquad (11)$$

For the laminar flow region, an equation can be written for the straight line found in Figure 1, such that:

$$F_R = \left(\frac{Re_v}{370} \right)^{1/2} \qquad (F1)$$

Combining these three equations, we obtain:

$$\left. \begin{array}{l} q = N_s (F_s C_v)^{3/2} \dfrac{\Delta p}{\mu} \\[2ex] \text{or} \\[1ex] C_v = \dfrac{1}{F_s} \left(\dfrac{q \mu}{N_s \Delta p} \right)^{2/3} \end{array} \right\} \qquad (F2)$$

where,

$$F_s = \frac{F_d^{2/3}}{F_L^{1/3}} \left(\frac{F_L^2 C_v^2}{N_2 d^4} + 1 \right)^{1/6} \qquad (F3)$$

and

μ = absolute viscosity, centipoises
N_s = a constant that depends on the units used, i.e.,

N_s	q	Δp
47	gpm	psi
1.5	m³/hr	kPa
15	m³/hr	bar

F_s is generally a function of a specific manufacturer's valve style and varies little from size to size. This variation is usually no greater than the uncertainty in the value of the factor F_d that accounts for the hydraulic radius. Representative values of F_s are listed in Appendix D. Once a particular valve has been selected, the actual values of F_d, F_L, and C_v/d^2 may be used to compute F_s.

Equation F2 may be solved directly for the unknown if the flow is fully laminar. In the transitional region, to avoid using a curve or table, the following equations have been established for determining F_R:

$$F_R = 1.044 - 0.358 \left(\frac{C_{vs}}{C_{vt}} \right)^{0.655} \qquad (F4)$$

$$F_R = 1.084 - 0.375 \left(\frac{\Delta p_s}{\Delta p_t} \right)^{0.336} \qquad (F5)$$

$$F_R = 1.004 - 0.358 \left(\frac{q_t}{q_s} \right)^{0.588} \qquad (F6)$$

In these equations, the subscript s denotes a value computed from Equation F2 assuming laminar flow conditions, and the subscript t denotes a value computed from Equation 9 assuming turbulent flow conditions ($F_R = 1.0$).

When the value F_R calculated by the above equations is less than 0.48, the flow may be taken as laminar, and Equation F2 governs. When F_R is greater than 0.98, the flow may be taken as turbulent, and Equation 9 governs ($F_R \cong 1.0$). The piping geometry factor F_p should not be used in either Equation 9 or Equation F2, because the effect that close-coupled fittings have on nonturbulent flow through control valves has not been established. Also, the equation used in this standard for F_p is based on turbulent flow only. For maximum accuracy, a valve must be installed with a straight inlet pipe the same size as the valve. The length of the straight pipe should be sufficient for the stream to attain its normal velocity profile, a condition upon which the research data are based.

The following examples demonstrate how problems may be solved.

PROBLEM 1. Find the valve size.

Given: $q = 500$ gpm, $G_f = 0.9$, $\Delta p = 20$ psi, $\mu = 20\,000$ cp

Selected valve: Butterfly, $C_v/d^2 = 19$, $F_s = 0.93$ (from a manufacturer's catalog or Appendix D)

Using Equation 9 for turbulent flow:

$$q = N_1 F_R C_{vt} \left(\frac{\Delta p}{G_f}\right)^{1/2}$$

$$500 = (1.0)(1.0) C_{vt} \left(\frac{20}{0.90}\right)^{1/2}$$

$$C_{vt} = 106$$

Using Equation F2 for laminar flow:

$$C_{vs} = \frac{1}{F_s}\left(\frac{q\mu}{N_s \Delta p}\right)^{2/3}$$

$$C_{vs} = \frac{1}{0.93}\left[\frac{500(20\,000)}{47(20)}\right]^{2/3} = 520$$

Using Equation F4 for transitional flow:

$$F_R = 1.044 - 0.358 \left(\frac{520}{106}\right)^{0.655} = 0.03$$

This value for F_R is less than the 0.48 limit for transitional flow, so the flow is laminar. The C_v required is 520. To meet this requirement, a representative 6-inch valve has a $C_v = 19d^2 = 684$, or as listed in the manufacturer's catalog.

PROBLEM 2. Find the differential pressure.

Given: $q = 1070$ gpm, $G_f = 0.84$, $\mu = 5900$ cp, $C_v = 400$, $F_s = 1.25$

Using Equation 9 assuming turbulent flow:

$$q = N_1 (1.0) C_v \left(\frac{\Delta p_t}{G_f}\right)^{1/2}$$

$$1070 = (1.0)(1.0)400 \left(\frac{\Delta p_t}{0.84}\right)^{1/2}$$

$$\Delta p_t = 601 \text{ psi}$$

Using Equation F2 assuming laminar flow:

$$q = N_s (F_s C_v)^{3/2} \frac{\Delta p_s}{\mu}$$

$$1070 = 47[1.25(400)]^{3/2} \frac{\Delta p_s}{5900}$$

$$\Delta p_s = 12.0 \text{ psi}$$

Using Equation F5 for transitional flow:

$$F_R = 1.084 - 0.375 \left(\frac{12.0}{6.01}\right)^{0.336} = 0.61$$

Because F_R is between 0.48 and 0.98, the flow is transitional.

Find the pressure drop using Equation 9:

$$q = N_1 F_R C_v \left(\frac{\Delta p}{G_f}\right)^{1/2}$$

$$1070 = 1.0(0.61)(400)\left(\frac{\Delta p}{0.84}\right)^{1/2}$$

$$\Delta p = 16 \text{ psi}$$

Note that the pseudo values of Δp, assuming turbulent (6 psi) or laminar flow (12 psi), are not applicable, because the flow is actually transitional.

PROBLEM 3. Find the valve size.

Given: $q = 17$ m³/h, $\rho = 1100$ kg/m³, $\Delta p = 69$ kPa, $\mu = 1000$ N·s/m² (or 10^6 cp)

Selected valve: Ball, $C_v/d^2 = 30$, $F_s = 1.3$

Using Equation 9 for turbulent flow:

$$q = N_1 F_R C_{vt} \left(\frac{\Delta p}{G_f}\right)^{1/2}$$

$$17 = 0.0865 (1.0) C_{vt} \left(\frac{69}{1.1}\right)^{1/2}$$

$$C_{vt} = 24.8$$

Using Equation F2 for laminar flow:

$$C_{vs} = \frac{1}{F_s}\left(\frac{q\mu}{N_s \Delta p}\right)^{2/3}$$

$$C_{vs} = \frac{1}{1.3}\left[\frac{17(10^6)}{1.5(69)}\right]^{2/3}$$

$$C_{vs} = 2310$$

For transitional flow:

$$F_R = 1.044 - 0.358 \left(\frac{2310}{24.8}\right)^{0.655} = -5.9$$

A value less than 0.48 indicates laminar flow. Therefore, the required C_v is 2310. To meet this requirement, a 250-mm (10-in) valve has a $C_v = 30(10)^2 = 3000$.

APPENDIX G
LIQUID CRITICAL PRESSURE RATIO FACTOR F_F

Flow rate is a function of the pressure drop from the valve inlet to the vena contracta. Under nonvaporizing liquid flow conditions, the apparent vena contracta pressure (p_{vc}) can be predicted from the downstream pressure (p_2), because the pressure recovery is a consistent fraction of the pressure drop to the vena contracta. The effect of this pressure recovery is recognized in the valve flow coefficient (C_v).

Under choked flow conditions, there is no relationship between p_2 and p_{vc} because vaporization affects pressure recovery. The liquid critical pressure ratio factor is used to predict p_{vc}. It is the ratio of the apparent vena contracta pressure under choked flow conditions to the vapor pressure of the liquid at its inlet temperature.

An equation for predicting F_F has been published in previous standards. A theoretical equation based on the assumption (Reference 10) that the fluid is always in thermodynamic equilibrium states that:

$$F_F = 0.96 - 0.28 \left(\frac{p_v}{p_c}\right)^{1/2} \qquad (G1)$$

Because a liquid does not remain in thermodynamic equilibrium as it flashes across a valve (Reference 11), the actual flow rate will be greater than that predicted by the use of Equation G1.

In experiments with nonvalve restrictions (Reference 12), the following equation for F_F was derived:

$$F_F = 1 - \frac{\sigma}{F_o} \qquad (G2)$$

where σ is the surface tension of the liquid in N/m and F_o is an experimentally determined orifice factor for the restriction or valve in the same units. This equation allows for the fact that liquids vaporizing across a restriction are not in thermodynamic equilibrium, but become metastable and choke at a critical vena contracta pressure. The equation has been tested only for deaerated water. Limited data indicate that values of F_o for values at rated travel range from around 0.2 N/m for a streamlined angle valve to nearly 1.0 for a more tortuous double-ported globe valve. The surface tension of water in N/m can be approximated based on the Othmer equation:

$$\sigma = \left[\frac{(374 - °C)}{4080}\right]^{1.05} \qquad (G3)$$

or

$$\sigma = \left[\frac{(705 - °F)}{7340}\right]^{1.05} \qquad (G4)$$

APPENDIX H
DERIVATION OF FACTOR x_{TP}

The slope of the Y versus x curve for any specific valve is determined using air or gas as the test fluid, and is designated by the value of x at $Y = 2/3$. This value, known as x_T, is the pressure drop ratio factor. For most valves, it is less than 1.0, but it may be greater for some valve styles.

If a valve is installed with a fitting at the inlet and/or outlet, the pressure drop ratio factor for the combination of the valve plus the fitting (x_{TP}) usually differs from that of the valve alone.

Let us consider a valve with reducers operating at choked flow

$$[x = x_{TP}, Y = Y_T \text{ for an ideal gas } (Z = 1)]$$

From Equation 18, the volumetric valve flow equation (in U.S. customary units is:

$$q_T = 1360 \, F_P C_v p_1 Y_T \sqrt{\frac{x_{TP}}{G_g T_1}} \qquad (H1)$$

where the subscript T indicates the terminal or choked condition. For the valve alone at choked flow, the equation is:

$$q_T = 1360 \, C_v p_i Y_T \sqrt{\frac{x_T}{G_g T_1}} \qquad (H2)$$

where p_i is the valve inlet pressure. From Equations H1 and H2, we have:

$$p_i = F_p p_1 \sqrt{\frac{x_{TP}}{x_T}} \qquad (H3)$$

From the gas laws, the mean specific weight across the inlet reducer is:

$$\gamma_1 = \left(\frac{p_1 - p_i}{2}\right) \frac{M}{RT_1} = \frac{144 \, (p_1 + p_i)}{2} \left(\frac{28.97 \, G_g}{1545 \, T_1}\right)$$

$$\gamma_1 = 1.350 \, (p_1 + p_i) \frac{G_g}{T_1} \qquad (H4)$$

Since the pressure drop, expressed in feet of head, is $K(U^2/2g)$,

$$\frac{144(p_1 - p_i)}{\gamma} = \frac{K}{2g} U^2 \text{ or } \frac{144 \, (p_1 - p_i)}{1.350 \, (p_1 + p_i) \frac{G_g}{T_1}}$$

$$= \frac{K}{2g} \left[\frac{q}{3600} \frac{14.73}{0.5 \, (p_1 + p_i)} \frac{T_1}{519.69} \frac{4(144)}{\pi d^2}\right]^2$$

Simplifying:
$$p_1^2 - p_i^2 = 1.214(10^{-9}) K G_g T_1 q^2 d^{-4} \quad \text{(H5)}$$

Substituting the expression for p_i from Equation H3, we have:
$$p_1^2 - F_p^2 p_1^2 \left(\frac{x_{TP}}{x_T}\right) = 1.214(10^{-9}) K G_g T_1 q^2 d^{-4} \quad \text{(H6)}$$

From Equation H1:
$$q_T^2 G_g \frac{T_1}{p_1^2} = (1360 F_p C_v Y_T)^2 x_{TP} \quad \text{(H7)}$$

Substituting this into Equation H6, with $q = q_T$ and $K = K_i$, we have:
$$1.214(10^{-9})(1360 F_p C_v Y_T)^2 K_i \frac{x_{TP}}{d^4}$$
$$= 1 - F_p^2 \frac{x_{TP}}{x_T} \quad \text{(H8)}$$

Solving for x_{Tp}, with $Y_T = 2/3$, we have:
$$x_{TP} = \frac{x_T}{F_p^2} \left(\frac{K_i x_T}{1000} \frac{C_v^2}{d^4} + 1\right)^{-1} \quad \text{(H9)}$$

APPENDIX I
CONTROL VALVE FLOW EQUATIONS—
SI NOTATION (International System of Units)

The valve flow coefficient that is compatible with SI units is A_v (Reference 13). At the present time, A_v does not have wide acceptance by the technical community. This appendix has been included for the benefit of those who wish to use pure, coherent SI units.

In the following equations, certain symbols commonly associated with SI practice differ from those listed in Section 3. These are:

A_v Valve flow coefficient, m^2 $[A_v = C_v \times 24(10^{-6})]$
ζ (Zeta)   Head loss coefficient ($\zeta = K$), dimensionless
ρ (Rho)   Density, kg/m³

Liquid Equations

Turbulent flow:
$$q = F_p A_v \left(\frac{\Delta p}{\rho}\right)^{1/2} \quad \text{(I1)}$$

$$w = F_p A_v (\Delta p \rho)^{1/2} \quad \text{(I2)}$$

$$F_p = \left[\left(\frac{\Sigma \zeta A_v^2}{1.23 d^4}\right) + 1\right]^{-1/2} \quad \text{(I3)}$$

Choked flow:
$$q = F_{LP} A_v \left[\frac{p_1 - p_{vc}}{\rho}\right]^{1/2} \quad \text{(I4)}$$

$$w = F_{LP} A_v \left[\rho(p_1 - p_{vc})\right]^{1/2} \quad \text{(I5)}$$

$$F_{LP} = F_L \left[\left(\frac{\zeta_1 F_L^2 A_v^2}{1.23 d^4}\right) + 1\right]^{-1/2} \quad \text{(I6)}$$

where $\zeta_i = \zeta_1 + \zeta_{B1}$

$$p_{vc} = F_F p_v \quad \text{(See Appendix G for } F_F\text{)} \quad \text{(I7)}$$

Laminar flow (see Appendix F):
$$q_s = (F_s A_v)^{3/2} \frac{\Delta p}{280 \mu} \quad \text{(I8)}$$

$$w_s = (F_s A_v)^{3/2} \frac{\Delta p \rho}{280 \mu} \quad \text{(I9)}$$

$$F_s = \frac{F_d^{2/3}}{F_L^{1/3}} \left(\frac{F_L^2 A_v^2}{1.23 d^4} + 1\right)^{1/6} \quad \text{(I10)}$$

Transitional flow:
$$q = F_R A_v \left(\frac{\Delta p}{\rho}\right)^{1/2} \quad \text{(I11)}$$

$$w = F_R A_v (\Delta p \rho)^{1/2} \quad \text{(I12)}$$

$$F_R = 1.044 - 0.358 \left(\frac{A_{vs}}{A_{vt}}\right)^{0.655} \quad \text{(I13)}$$

$$F_R = 1.084 - 0.375 \left(\frac{\Delta p_s}{\Delta p_t}\right)^{0.336} \quad \text{(I14)}$$

$$F_R = 1.004 - 0.358 \left(\frac{q_t}{q_s}\right)^{0.588} \quad \text{(I15)}$$

Limits for $F_R = 0.48$ to 1.0.

Gas and Vapor Equations

Turbulent flow:
$$w = F_p A_v Y (x p_1 \rho_1)^{1/2} \quad \text{(I16)}$$

$$q = 0.246 F_p A_v p_1 Y \left(\frac{x}{M T_1 Z}\right)^{1/2} \quad \begin{array}{l}\text{(Normal m}^3 \\ \text{at 0°C and} \\ \text{101.3 kPa)}\end{array} \quad \text{(I17)}$$

Limit: $x \leq F_k x_{TP}$ (in equation only)

$$x_{TP} = \frac{x_T}{F_p^2}\left[\left(0.72 x_T \zeta_i \frac{A_v^2}{d^4}\right) + 1\right]^{-1} \quad \text{(I18)}$$

where
$$\zeta_i = \zeta_1 + \zeta_{B1} \quad \text{(I19)}$$

7 REFERENCES

1. Instrument Society of America, ANSI/ISA S75.02, *Control Valve Capacity Test Procedure*, Instrument Society of America, Research Triangle Park, NC, 1981.

2. H. D. Baumann, "Effect of Pipe Reducers on Control Valve Capacity," *Instruments and Control Systems*, December 1968, pp. 99–102.

3. G. F. Stiles, "Liquid Viscosity Effects on Control Valve Sizing," Technical Manual TM 17A, October 1967, Fisher Governor Company, Marshalltown, IA.

4. E. B. McCutcheon, "A Reynolds Number for Control Valves," *Symposium on Flow, Its Measurement and Control in Science and Industry*, Vol. I, Part 3, Instrument Society of America, Pittsburgh, PA, 1974, pp. 1087–90.

5. H. D. Baumann, "The Introduction of a Critical Flow Factor for Valve Sizing," *ISA Transactions*, Vol. 2, 1963, pp. 107–11.

6. J. F. Buresh and C.B. Schuder, "The Development of a Universal Gas Sizing Equation for Control Valves," *ISA Transactions*, Vol. 3, 1964, pp. 322–28.

7. L. R. Driskell, "New Approach to Control Valve Sizing," *Hydrocarbon Processing*, July 1969, pp. 111–14.

8. Instrument Society of America, *ISA Handbook of Control Valves*, 2d ed., Instrument Society of America, Pittsburgh, PA, 1976.

9. L. R. Driskell, *Control Valve Selection and Sizing*, Instrument Society of America, Research Triangle Park, NC, 1983.

10. W. F. Allen, Jr., "Flow of a Flashing Mixture of Water and Steam through Pipes and Valves," *Journal of Basic Engineering*, April 1951, pp. 357–65.

11. J. F. Bailey, "Metastable Flow of Saturated Water," *Journal of Basic Engineering*, November 1951, pp. 1109–16.

12. J. G. Burnell, "Flow of Boiling Water through Nozzles, Orifices and Pipes," *Engineering*, December 12, 1947, pp. 572–76.

13. International Electrotechnical Commission (IEC) Standards: *Industrial Process Control Valves*, International Electrotechnical Commission, Geneva, Switzerland.
    534-1 (1976), Part 1: General Considerations.
    534-2 (1978), Part 2: Flow Capacity. Section One — Sizing Equations for Incompressible Fluid Flow under Installed Conditions.
    534-2-2 (1980), Part 2: Flow Capacity. Section Two — Sizing Equations for Compressible Fluid Flow under Installed Conditions.
    534-2-3 (1983), Part 2: Flow Capacity. Section Three Test Procedures.

ANSI/ISA-S75.02-1988
Approved November 29, 1988

American National Standard

Control Valve Capacity Test Procedure

Instrument Society of America

Instrument Society of America

ISBN 1-55617-120-X

ISA-S75.02 Control Valve Capacity Test Procedure

Copyright ©1988 by the Instrument Society of America. All rights reserved. Printed in the United States of America. No part of this publication may be reproduced, stored in a retrieval system, or transmitted, in any form or by any means (electronic, mechanical, photocopying, recording, or otherwise), without the prior written permission of the publisher.

INSTRUMENT SOCIETY OF AMERICA
67 Alexander Drive
P.O. Box 12277
Research Triangle Park, North Carolina 27709

Preface

This preface is included for informational purposes and is not part of ISA-S75.02.

This standard has been prepared as part of the service of the Instrument Society of America (ISA) toward a goal of uniformity in the field of instrumentation. To be of real value, this document should not be static, but should be subject to periodic review. Toward this end, the Society welcomes all comments and criticisms, and asks that they be addressed to the Secretary, Standards and Practices Board, Instrument Society of America, 67 Alexander Drive, P.O. Box 12277, Research Triangle Park, NC 27709, telephone (919) 549-8411.

The ISA Standards and Practices Department is aware of the growing need for attention to the metric system of units in general and the International System of Units (SI) in particular, in the preparation of instrumentation standards. The Department is further aware of the benefits to the U.S.A. users of ISA standards of incorporating suitable references to the SI (and the metric system) in their business and professional dealings with other countries. Toward this end, this Department will endeavor to introduce SI-acceptable metric units in all new and revised standards to the greatest extent possible. *The Metric Practice Guide*, which has been published by the Institute of Electrical and Electronics Engineers as ANSI/IEEE Std. 268-1982 and future revisions will be the reference guide for definitions, symbols, abbreviations, and conversion factors.

It is the policy of the Instrumentation Society of America to encourage and welcome the participation of all concerned individuals and interests in the development of ISA standards. Participation in the ISA standards-making process by an individual in no way constitutes endorsement by the employer of that individual, of the Instrument Society of America, or of any of the standards that ISA develops.

The information contained in the preface, footnotes, and appendices is included for information only and is not a part of the standard.

The following people served as members of ISA Committee SP75.06, which prepared this standard.

NAME	COMPANY
F. Harthun, Chairman	Fisher Controls International Incorporated
G. E. Barb	Anchor/Darling
R. Barnes	Valtek, Incorporated
G. Borden	Stone and Webster
L. Driskell	Consultant
H. Illing	DeZurik
M. Kirik	Rockwell International Corporation
G. Kovecses	Yarway Corporation
O. P. Lovett	Consultant
A. P. McCauley	Chagrin Valley Controls, Inc.
J. Ozol	Omaha Public Power Company
R. A. Quance	Consultant
G. Richards	Richards Industries, Incorporated
F. O. Seger	Masoneilan/Dresser
W. C. Weidman	Gilbert/Commonwealth, Inc.
L. R. Zinck	Union Carbide Corporation

The following people served as members of ISA Committee SP75 During the review of this standard.

NAME	COMPANY
L. R. Driskell, Chairman	Consultant
J. B. Arant	E. I. du Pont de Nemours & Company (retired)
H. E. Backinger	Kraus Company, Inc.
G. E. Barb	Anchor/Darling Valve Company
R. W. Barnes	Valtek, Inc.

Instrument Society of America

H. D. Baumann	H. D. Baumann Associates, Ltd.
G. Borden, Jr.	Stone and Webster
* R. Brodin/F. Harthun	Fisher Controls International, Inc.
E. H. C. Brown	Dravo Engineers, Inc.
E. J. Cooney	Air Products & Chemicals, Inc.
W. G. Dewart	Rockwell International
J. T. Emery	Honeywell, Inc.
H. J. Fuller	Consultant
L. F. Griffith	Consultant
A. J. Hanssen	Fluid Controls Institute, Inc.
B. Hart	M. W. Kellogg Company
H. P. Illing	DeZurik
J. D. Leist	Dow Chemical U.S.A.
R. Louviere	Creole Engineering
O. P. Lovett, Jr.	Consultant
P. C. Martin	Hammel Dahl & Jamesbury
A. P. McCauley	Chagrin Valley Controls, Inc.
H. Miller	Control Components Incorporated
T. V. Molloy	Pacific Gas & Electric
J. Ozol	Omaha Public Power Company
R. A. Quance	Consultant
W. Rahmeyer	Utah State University
J. N. Reed	Masoneilan/Dresser
G. Richards	Richards Industries, Inc.
J. Rosato	Rawson Company
K. Schoonover, Secretary	Con-Tek
H. Schwartz	Flexible Valve Corporation
W. L. Scull	Everlasting Valve Company
M. Shah	McEvoy-Willis
H. R. Sonderegger	Grinnell Corporation
R. U. Stanley	Retired
R. E. Terhune	Consultant
R. F. Tubbs	Copes-Vulcan-Charlotte
W. C. Weidman	Gilbert/Commonwealth, Inc.
R. L. Widdows	Cashco Inc.
L. R. Zinck, Vice Chairman	Union Carbide Corporation

*One vote per company

This standard was approved for publication by the ISA Standards and Practices Board in September 1988.

NAME	COMPANY*
D. E. Rapley, Chairman	Rapley Engineering Services
D. N. Bishop	Chevron U.S.A. Inc.
W. Calder III	The Foxboro Company
N. Conger	Fisher Controls International, Inc.
R. S. Crowder	Ship Star Associates
C. R. Gross	Eagle Technology
H. S. Hopkins	Utility Products of Arizona
R. T. Jones	Philadelphia Electric Company
A. P. McCauley	Chagrin Valley Controls, Inc.

E. M. Nesvig	ERDCO Engineering Corporation
R. Prescott	Moore Products Company
R. H. Reimer	Allen-Bradley Company
J. Rennie	Factory Mutual Research Corporation
W. C. Weidman	Gilbert/Commonwealth, Inc.
** P. Bliss	Consultant
** B. A. Christensen	Continental Oil Company
** L. N. Combs	Consultant
** R. L. Galley	Consultant
** T. J. Harrison	Florida State University
** R. Keller	The Boeing Company
** O. P. Lovett, Jr.	Consultant
** E. C. Magison	Honeywell Inc.
** R.G. Marvin	Roy G. Marvin Company
** W. B. Miller	Moore Products Company
** J. W. Mock	Bechtel Western Power Corporation
** G. Platt	Consultant
** C. W. Reimann	National Institute of Standards and Technology
** J. R. Williams	Stearns Catalytic Corporation

* Employer at time of approval
** Director Emeritus

Instrument Society of America

TABLE OF CONTENTS

Section	Page
1. Scope	9
2. Nomenclature	9
3. Test System	9
3.1 General Description	9
3.2 Test Specimen	10
3.3 Test Section	10
3.4 Throttling Valves	10
3.5 Flow Measurement	10
3.6 Pressure Taps	10
3.7 Pressure Measurement	10
3.8 Temperature Measurement	12
3.9 Installation of Test Specimen	12
3.10 Accuracy of Test	12
4. Test Fluids	12
4.1 Incompressible Fluid	12
4.2 Compressible Fluid	12
5. Test Procedure-Incompressible Fluids	12
5.1 C_v Test Procedure	12
5.2 F_L Test Procedure	14
5.3 F_P Test Procedure	14
5.4 F_{LP} Test Procedure	14
5.5 F_R Test Procedure	14
5.6 F_F Test Procedure	14
6. Data Evaluation Procedure-Incompressible Fluids	15
6.1 C_v Calculation	15
6.2 F_L Calculation	15
6.3 F_P Calculation	15
6.4 F_{LP} Calculation	15
6.5 F_R Calculation	15
6.6 F_F Calculation	15
7. Test Procedure -- Compressible Fluids	15
7.1 C_v Test Procedure	16
7.2 x_T Test Procedure	16
7.3 Alternative Test Procedure for C_v and x_T	16
7.4 F_P Test Procedure	17
7.5 x_{TP} Test Procedure	17
8. Data Evaluation Procedure -- Compressible Fluids	17
8.1 C_v Calculation	17
8.2 x_T Calculation	17
8.3 F_P Calculation	17
8.4 x_{TP} Calculation	18
9. Numerical Constants	18

LIST OF ILLUSTRATIONS

Figure   **Page**

1. Basic Flow Test System 10
2. Recommended Pressure Connection 11
3. Reynolds Number Factor 19

LIST OF TABLES

Table   **Page**

1. Piping Requirements-Standard Test Section 11
2. Minimum Upstream Test Pressure 13
3. Numerical Constants 18

1. Scope

This test standard utilizes the mathematical equations outlined in ANSI/ISA-S75.01, "Flow Equations for Sizing Control Valves," in providing a test procedure for obtaining the following:
(1) Valve flow coefficient, C_v
(2) Liquid pressure recovery factors, F_L and F_{LP}
(3) Reynolds Number factor, F_R
(4) Liquid critical pressure ratio factor, F_F
(5) Piping geometry factor, F_P
(6) Pressure drop ratio factor, x_T and x_{TP}

This standard is intended for control valves used in flow control of process fluids and is not intended to apply to fluid power components as defined in the National Fluid Power Association Standard NFPA T.3.5.28-1977.

2. Nomenclature

Symbol   **Description**

Symbol	Description
C_v	Valve flow coefficient
d	Valve inlet diameter
D	Internal diameter of the pipe
F_d	Valve style modifier
F_F	Liquid critical pressure ratio factor, dimensionless
F_k	Ratio of specific heats factor, dimensionless
F_L	Liquid pressure recovery factor of a valve without attached fittings, dimensionless
F_{LP}	Product of the liquid pressure recovery factor of a valve with attached fittings (no symbol has been identified) and the piping geometry factor, dimensionless
F_P	Piping geometry factor, dimensionless
F_R	Reynolds Number factor, dimensionless
G_f	Liquid specific gravity at upstream conditions [ratio of density of liquid at flowing temperature to density of water at 15.6°C (60°F)], dimensionless
G_g	Gas specific gravity (ratio of flowing gas to density of air with both at standard conditions, which is equal to the ratio of the molecular weight of gas to the molecular weight of air), dimensionless
k	Ratio of specific heats, dimensionless
m	The number of similar flow paths (i.e., $m = 1$ for single-ported valves, $m = 2$ for double-ported, etc.)
N_1, N_2, etc.	Numerical constants for units of measurement used
p_1	Upstream absolute static pressure, measured two nominal pipe diameters upstream of valve-fitting assembly
p_2	Downstream absolute static pressure, measured six nominal pipe diameters downstream of valve-fitting assembly
Δp	Pressure differential, $p_1 - p_2$
p_v	Absolute vapor pressure of liquid at inlet temperature
q	Volumetric flow rate
q_{max}	Maximum flow rate (choked flow conditions) at a given upstream condition
Re_v	Valve Reynolds Number, dimensionless
T_1	Absolute upstream temperature (in K or degrees R)
x	Ratio of pressure drop to absolute inlet pressure ($\Delta p/p_1$), dimensionless
x_T	Pressure drop ratio factor of the valve without attached fittings, dimensionless
x_{TP}	Value of x_T for valve-fitting assembly, dimensionless
Y	Expansion factor, ratio of flow coefficient for a gas to that for a liquid at the same Reynolds Number, dimensionless
$\nu(nu)$	Kinematic viscosity, centistokes

Subscripts:
1   Upstream conditions
2   Downstream conditions

3. Test System

3.1 General Description. A basic flow test system as shown in Figure 1 includes:
(1) Test specimen
(2) Test section
(3) Throttling valves
(4) Flow-measuring device
(5) Pressure taps
(6) Temperature sensor

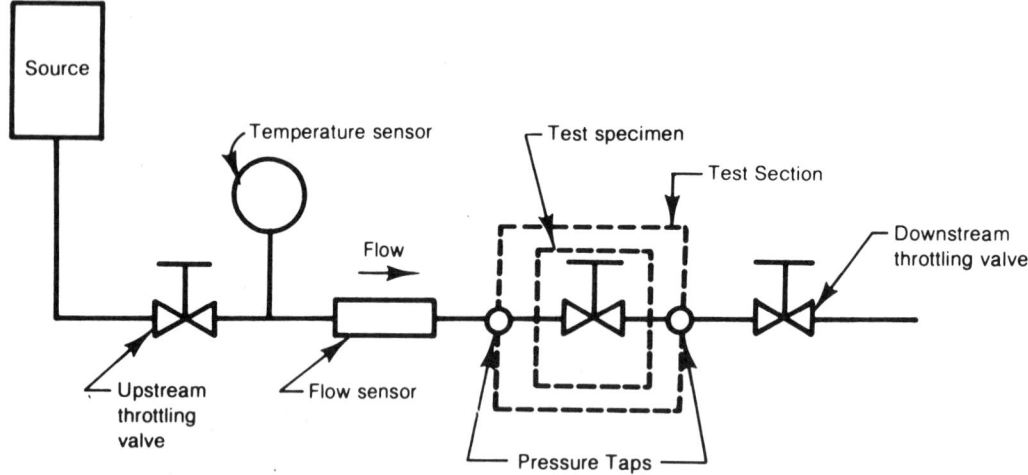

Figure 1. Basic Flow Test System

3.2 Test Specimen. The test specimen is any valve or combination of valve, pipe reducer, and expander or other devices attached to the valve body for which test data are required. Modeling of valves to a smaller scale is an acceptable practice in this standard, although testing of full-size valves or models is preferable. Good practice in modeling requires attention to significant relationships such as Reynolds Number, the Mach number where compressibility is important, and geometric similarity.

3.3 Test Section. The upstream and downstream piping adjacent to the test specimen shall conform to the nominal size of the test specimen connection and to the length requirements of Table 1.

The piping on both sides of the test specimen shall be Schedule 40 pipe for valves through 250-mm (10-in.) size having a pressure rating up to and including ANSI Class 600. Pipe having 10-mm (0.375-in.) wall may be used for 300-mm (12-in.) through 600-mm (24-in.) sizes. An effort should be made to match the inside diameter at the inlet and outlet of the test specimen with the inside diameter of the adjacent piping for valves outside the above limits.

The inside surfaces shall be reasonably free of flaking rust or mill scale and without irregularities that could cause excessive fluid frictional losses.

3.4 Throttling Valves. The upstream and downstream throttling valves are used to control the pressure differential across the test section pressure taps and to maintain a specific downstream pressure. There are no restrictions as to style of these valves. However, the downstream valve should be of sufficient capacity to ensure that choked flow can be achieved at the test specimen for both compressible and incompressible flow. Vaporization at the upstream valve must be avoided when testing with liquids.

3.5 Flow Measurement. The flow-measuring instrument may be any device that meets specified accuracy. This instrument will be used to determine the true time average flow rate within an error not exceeding ± 2 percent of the actual value. The resolution and repeatability of the instrument shall be within ± 0.5 percent. The measuring instrument shall be calibrated as frequently as necessary to maintain specified accuracy.

3.6 Pressure Taps. Pressure taps shall be provided on the test section piping in accordance with the requirements listed in Table 1. These pressure taps shall conform to the construction illustrated in Figure 2.

Orientation:
Incompressible fluids — Tap center lines shall be located horizontally to reduce the possibility of air entrapment or dirt collection in the pressure taps.

Compressible fluids — Tap center lines shall be oriented horizontally or vertically above pipe to reduce the possibility of dirt or condensate entrapment.

Multiple pressure taps can be used on each test section for averaging pressure measurements. Each tap must conform to the requirements in Figure 2.

3.7 Pressure Measurement. All pressure and pressure differential measurements shall be made to an error not exceeding ± 2 percent of actual value. Pressure-measuring devices shall be calibrated as frequently as necessary to maintain specified accuracy.

Pressure differential instruments are required in the measurement of the pressure differential across the test specimen to avoid additional inaccuracies resulting from taking the difference of two measurements. Exceptions to this are the procedures in Sections 5.2 and 7.2 for determin-

Table 1
Piping Requirements, Standard Test Section

A *,**	B	C	D	Standard Test Section Configuration
At least 18 nominal pipe diameters of straight pipe	2 nominal pipe diameters of straight pipe	6 nominal pipe diameters of straight pipe	At least 1 nominal pipe diameter of straight pipe	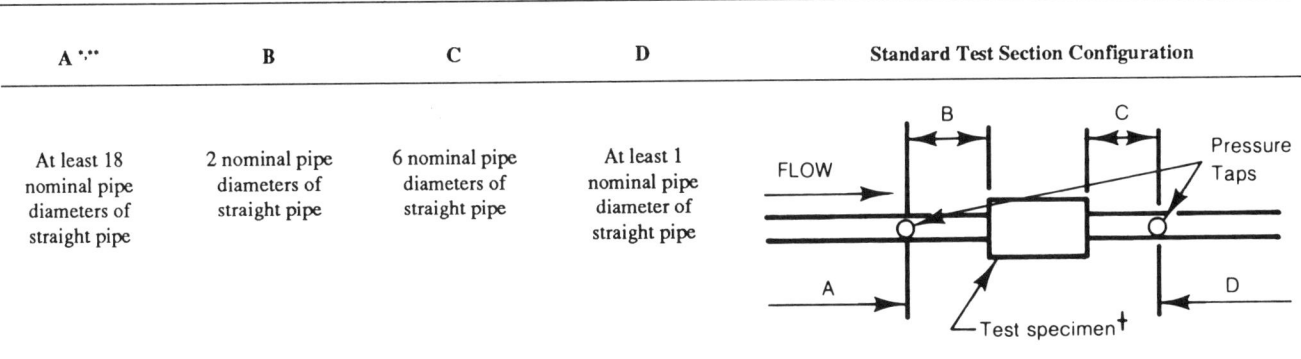

* Dimension "A" may be reduced to 8 nominal diameters if straightening vanes are used.

  Information concerning the design of straightening vanes can be found in ASME Performance Test Code PTC 19.5-1972, "Applications." Part II of "Fluid Meters, Interim Supplement on Instruments and Apparatus."

** If an upstream flow disturbance consists of two ells in series and they are in different planes, dimension "A" must exceed 18 nominal pipe diameters unless straightening vanes are used.

† See Section 3.2 for definition of the test specimen.

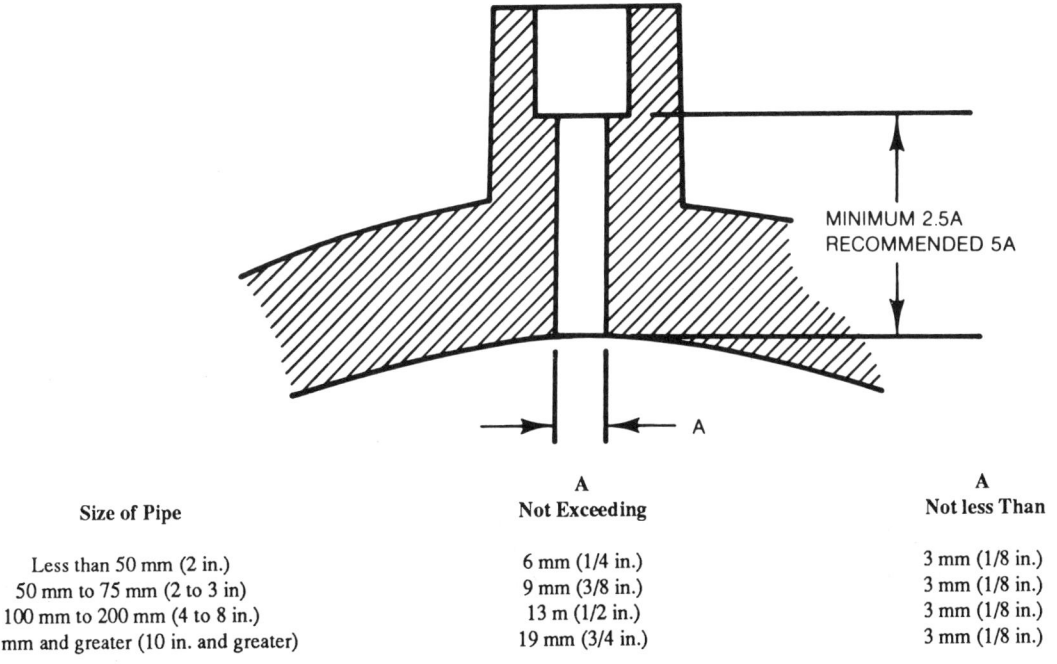

Size of Pipe	A Not Exceeding	A Not less Than
Less than 50 mm (2 in.)	6 mm (1/4 in.)	3 mm (1/8 in.)
50 mm to 75 mm (2 to 3 in)	9 mm (3/8 in.)	3 mm (1/8 in.)
100 mm to 200 mm (4 to 8 in.)	13 m (1/2 in.)	3 mm (1/8 in.)
250 mm and greater (10 in. and greater)	19 mm (3/4 in.)	3 mm (1/8 in.)

* Edge of hole must be clean and sharp or slightly rounded, free from burrs, wire edges or other irregularities. In no case shall any fitting protrude inside the pipe.
  Any suitable method of making the physical connection is acceptable if above recommendations are adhered to.

  Reference: ASME Performance Test Code PTC 19.5-1972, "Applications. Part II of Fluid Meters, Interim Supplement on Instruments and Apparatus."

Figure 2. Recommended Pressure Connection

ing maximum flow rates for incompressible and compressible flow, respectively.

3.8 Temperature Measurement. The fluid temperature shall be measured to an error not exceeding ± 1°C (± 2°F) of actual value.

The inlet fluid temperature shall remain constant within ± 3°C (± 5°F) during the test run to record data for each specific test point.

3.9 Installation of Test Specimen. The alignment between the center line of the test section piping and the center line of the inlet and outlet of the test specimen shall be as follows:

Pipe Size	Allowable Misalignment
15 mm thru 25 mm	0.8 mm
(1/2 in. thru 1 in.)	(1.32 in.)
32 mm thru 150 mm	1.6 mm
(1-1/4 in. thru 6 in.)	(1/16 in.)
200 mm and larger	1 percent of the diameter
(8 in. and larger)	

When rotary valves are being tested, the valve shafts shall be aligned with test section pressure taps.

Each gasket shall be positioned so that it does not protrude into the flow stream.

3.10 Accuracy of Test. Valves having an $N_3 C_v / d^2$ ratio of less than 30 will have a calculated flow coefficient, C_v, of the test specimen within a tolerance of ± 5 percent.

4. Test Fluids

4.1 Incompressible Fluids. Fresh water shall be the basic fluid used in this procedure. Inhibitors may be used to prevent or retard corrosion and to prevent the growth of organic matter. The effect of additives on density or viscosity shall be evaluated by computation using the equations in this standard. The sizing coefficient shall not be affected by more than 0.1 percent. Other test fluids may be required for obtaining F_R and F_F.

4.2 Compressible Fluids. Air or some other compressible fluid shall be used as the basic fluid in this test procedure. Vapors that may approach their condensation points at the vena contracta of the specimen are not acceptable as test fluids. Care shall be taken to avoid internal icing during the test.

5. Test Procedure — Incompressible Fluids

The following instructions are given for the performance of various tests using incompressible fluids.

The procedures for data evaluation of these tests follow in Section 6.

5.1 C_v Test Procedure. The following test procedure is required to obtain test data for the calculation of the flow coefficient C_v. The data evaluation procedure is provided in Section 6.1

5.1.1 Install the test specimen without reducers or other attached devices in accordance with piping requirements in Table 1.

5.1.2 Flow tests shall include flow measurements at three widely spaced pressure differentials within the turbulent, non-vaporizing region. The suggested differential pressures are:

(a) just below the onset of cavitation or the maximum available in the test facility, whichever is less;

(b) about 50% of the pressure differential of (a); and

(c) about 10% of the pressure differential of (a) and shall be measured across the test section pressure taps with the valve at the rated travel.

The line velocity should not exceed 13.7 m/s (45 ft/s).

A minimum valve Reynolds Number, Re_v, of 10^5 is recommended (see Equation 5). Deviations from standard requirements shall be recorded.

For large valves where flow source limitations are reached, lower pressure differentials may be used optionally as long as turbulent flow is maintained. Deviations from standard requirements shall be recorded.

5.1.3 In order to keep the downstream portion of the test section liquid filled and to prevent vaporization of the liquid, the upstream pressure must be maintained equal or greater than the minimum values in Table 2. This minimum upstream pressure is dependent on the liquid pressure recovery factor, F_L, of the test specimen. If F_L is unknown, a conservative estimate for the minimum inlet pressure should be made using the values in Table 2.

5.1.4 The valve flow test shall be performed at 100 percent of rated valve travel. Optional tests may be performed at each 10 percent of rated valve travel or any other points to more fully determine the inherent characteristic of the specimen.

5.1.5 The following data shall be recorded:

(1) Valve travel (measurement error not exceeding ±0.5 percent of rated travel)

(2) Upstream pressure (p_1) (measurement error not exceeding ± 2 percent of actual value)

(3) Pressure differential (Δp) across test section pressure taps (measurement error not exceeding ± 2 percent of actual value)

(4) Volumetric flow rate (q) (measurement error not exceeding ± 2 percent of actual value)

(5) Fluid inlet temperature (T_1) (measurement error not exceeding ± 1°C (± 2°F))

(6) Barometric pressure (measurement error not exceeding ± 2 percent of actual value)

Table 2
Minimum Upstream Test Pressure
for a Tempeature Range of 5°C to 40°C (41°F to 104°F)

Δp	kPa	35	70	100	140	350	700	1400
	bar	0.35	0.7	1.0	1.4	3.5	7	14
	psi	5	10	15	20	50	100	200

Absolute Upstream Pressure, p_1

F_L								
0.5	kPa	280	550	830	1100	2800	5500	11000
	bar	2.8	5.5	8.3	11.0	28.0	55.0	110.0
	psia	40	80	120	160	400	800	1600
0.6	kPa	190	380	570	760	1900	3900	7700
	bar	1.9	3.8	5.7	7.6	19.0	39.0	77.0
	psia	28	56	83	110	280	560	1110
0.7	kPa	150	280	420	560	1400	2800	5700
	bar	1.5	2.8	4.2	5.6	14.0	28.0	57.0
	psia	22	41	61	82	200	410	820
0.8	kPa	150	220	320	430	1100	2100	4300
	bar	1.5	2.2	3.2	4.3	11.0	21.0	43.0
	psia	22	31	47	62	160	310	620
0.9	kPa	150	180	260	340	830	1700	3400
	bar	1.5	1.8	2.6	3.4	8.3	17.0	34.0
	psia	22	27	37	49	120	250	490

NOTES:
- (a) Minimum upstream pressures have been calculated to provide a downstream gage pressure of at least 14 kPa (0.14 bar) (2 psig) above atmospheric pressure.

- (b) Upstream pressures were calculated using $p_1 \min = 2\Delta p/F_L^2$.

- (c) Upstream pressures were rounded to 2 significant digits while still maintaining a minimum pressure as specified in note (a).

- (d) Example: Estimated F_L for valve is 0.7.
  Pressure differential is 10 psi.
  From table: Minimum upstream pressure is 41 psia.

(7) Physical description of test specimen (i.e., type of valve, flow direction, etc.)

(8) Physical description of test system and test fluid

5.2 F_L Test Procedure. The maximum flow rate, q_{max}, is required in the calculation of the liquid pressure recovery factor, F_L. For a given upstream pressure, the quantity q_{max} is defined as that flow rate at which a decrease in downstream pressure will not result in an increase in the flow rate. The test procedure required to determine q_{max} is included in this section. The data evaluation procedure including the calculation of F_L is contained in Section 6.2. The test for F_L and corresponding C_v must be conducted at identical valve travels. Hence, the tests for both these factors at any valve travel shall be made while the valve is locked in a fixed position.

5.2.1 Install the test specimen without reducers or other attached devices in accordance with piping requirements in Table 1. The test specimen shall be at 100 percent of rated travel.

5.2.2 The downstream throttling valve shall be in the fully open position. Then, with a preselected upstream pressure, the flow rate will be measured and the downstream pressure recorded. Table 2 has been provided to assist the user in selecting an upstream pressure. This test establishes a "maximum" pressure differential for the test specimen in this test system. A second test run shall be made with the pressure differential maintained at 90 percent of the pressure differential determined in the first test with the same upstream pressure. If the flow rate in the second test is within 2 percent of the flow rate in the first test, the "maximum" or choked flow rate has been established. If not, the test procedure must be repeated at a higher upstream pressure. If choked flow cannot be obtained, the published value of F_L must be based on the maximum measurement attainable, with an accompanying notation that the actual value exceeds the published value, e.g., $F_L > 0.87$. Note that values of upstream pressure and pressure differential used in this procedure are those values measured at the pressure taps.

5.2.3 Record the following data:

(1) Valve travel (measurement error not exceeding ±0.5 percent of rated travel)

(2) Upstream pressure (p_1) and downstream pressure (p_2) (measurement error not exceeding ± 2 percent of actual value)

(3) Volumetric flow rate (q) (measurement error not exceeding ± 2 percent of actual value)

(4) Fluid temperature (measurement error not exceeding ± 1°C (± 2°F))

(5) Barometric pressure (measure error not exceeding ± 2 percent of actual value)

5.3 F_P Test Procedure. The piping geometry factor, F_P, modifies the valve sizing coefficient for reducers or other devices attached to the valve body that are not in accord with the test section. It is the ratio of the installed C_v with these reducers or other devices attached to the valve body to the rated C_v of the valve installed in a standard test section and tested under identical service conditions. This factor is obtained by replacing the valve with the desired combination of valve, reducers, and/or other devices and then conducting the flow test outlined in Section 5.1, treating the combination of the valve and reducers as the test specimen for the purpose of determining test section line size. For example, a 100-mm (4-in.) valve between reducers in a 150-mm (6-in.) line would use pressure tap locations based on 150-mm (6-in.) nominal diameter. The data evaluation procedure is provided in Section 6.3.

5.4 F_{LP} Test Procedure. Perform the tests outlined for F_L in Section 5.2, replacing the valve with the desired combination of valve and pipe reducers or other devices and treating the combination of valve and reducers as the test specimen. The data evaluation procedure is provided in Section 6.4.

5.5 F_R Test Procedure. To produce values of the Reynolds Number factor, F_R, nonturbulent flow conditions must be established through the test valve. Such conditions will require low pressure differentials, high viscosity fluids, small values of C_v, or some combination of these. With the exception of valves with very small values of C_v, turbulent flow will always exist when flowing tests are performed in accordance with the procedure outlined in Section 5.1, and F_R under these conditions will have the value of 1.0.

Determine values of F_R by performing flowing tests with the valve installed in the standard test section without reducers or other devices attached. These tests should follow the procedure for C_v determination except that:

(1) Test pressure differentials may be any appropriate values provided that no vaporization of the test fluid occurs within the test valve.

(2) Minimum upstream test pressure values shown in Table 2 may not apply if the test fluid is not fresh water at 20°C ± 14°C (68°F ± 25°F).

(3) The test fluid should be a Newtonian fluid having a viscosity considerably greater than water unless instrumentation is available for accurately measuring very low pressure differentials.

Perform a sufficient number of these tests at each selected valve travel by varying the pressure differential across the valve so that the entire range of conditions, from turbulent to laminar flow, is spanned. The data evaluation procedure is provided in Section 6.5.

5.6 F_F Test Procedures. The liquid critical pressure ratio factor, F_F, is ideally a property of the fluid and its temperature. It is the ratio of the apparent vena contracta pressure at choked flow conditions to the vapor pressure of liquid at inlet temperature.

The quantity of F_F may be determined experimentally

by using a test specimen for which F_L and C_v are known. The standard test section without reducers or other devices attached will be used with the test specimen installed. The test procedure outlined in Section 5.2 for obtaining q_{max} will be used with the fluid of interest as the test fluid. The data evaluation procedure is in Section 6.6.

6. Data Evaluation Procedure — Incompressible Fluids

The following procedures are to be used for the evaluation of the data obtained using the test procedures in Section 5. The pressure differentials used to calculate the flow coefficients and other flow factors were obtained using the test section defined in Table 1. These pressure measurements were made at the pressure taps and include the test section piping between the taps as well as the test specimen.

6.1 C_v Calculation

6.1.1 Using the data obtained in Section 5.1, calculate C_v for each test point at a given valve travel using the equation:

$$C_v = \frac{q}{N_1}\left(\frac{G_f}{\Delta p}\right)^{1/2} \quad (1)$$

Round off the calculated value to no more than three significant digits.

6.1.2 The rated C_v of the valve is the arithmetic average of the calculated values for 100 percent of rated travel obtained from the test data in Section 5.1.5. A critical examination of the individual values calculated should reveal equal values of C_v within the tolerance given in Section 3.10.

6.2 F_L Calculation

$$F_L = \frac{q_{max}}{N_1 C_v \left[(p_1 - 0.96 p_v)/G_f\right]^{1/2}} \quad (2)$$

where P_1 is the pressure at the upstream pressure tap for the q_{max} determination (Section 5.2).

6.3 F_P Calculation. Calculate F_P as follows at rated valve travel:

$$F_P = \frac{q}{N_1 C_v (\Delta p / G_f)^{1/2}} \quad (3)$$

6.4 F_{LP} Calculation

$$F_{LP} = \frac{q_{max}}{N_1 C_v \left[(p_1 - 0.96 p_v)/G_f\right]^{1/2}} \quad (4)$$

where p_1 is the pressure at the upstream pressure tap for the q_{max} determination (Section 5.2).

6.5 F_R Calculation.
Use test data, obtained as described under Section 5.5 and in Equation (1), Section 6.1, to obtain values of an apparent C_v. This apparent C_v is equivalent to $F_R C_v$. Therefore, F_R is obtained by dividing the apparent C_v by the experimental value of C_v determined for the test valve under standard conditions at the same valve travel. Although data may be correlated in any manner suitable to the experimenter, a method that has proven to provide satisfactory correlations involves the use of the valve Reynolds Number, which may be calculated from:

$$\text{Re}_v = \frac{N_4 F_d q}{\nu \, F_L^{1/2} C_v^{1/2}} \left(\frac{F_L^2 C_v^2}{N_2 D^4} + 1\right)^{1/4} \quad (5)$$

where:
 F_d = valve style modifier, accounts for the effect of geometry on Reynolds Number. F_d has been found to be proportional to $1/m^{1/2}$.
 m = the number of similar flow paths (i.e., $m = 1$ for single-ported valves, $m = 2$ for double-ported, etc.).
 ν = kinematic viscosity in centistokes.

Plotting values of F_R versus Re_v will result in the curve that appears as Figure 3 in ANSI/ISA-S75.01-1985, "Flow Equations for Sizing Control Valves".

6.6 F_F Calculation. Calculate F_F as follows:

$$F_F = \frac{1}{p_v}\left[p_1 - G_f \left(\frac{q_{max}}{N_1 F_L C_v}\right)^2\right] \quad (6)$$

where p_v is the fluid vapor pressure at the inlet temperature. $F_L C_v$ is determined for the test specimen by the standard method (Section 5.2 — F_L Test Procedure).

7. Test Procedure — Compressible Fluids

The following instructions are given for the performance of various tests using compressible fluids.
The procedures for data evaluation of these tests follow in Section 8.

7.1 C_v Test Procedure. The determination of the flow coefficient, C_v, requires flow tests using the following procedure to obtain the necessary test data. The data evaluation procedure is in Section 8.1. An alternative procedure for calculating C_v is provided in Section 7.3.

7.1.1 Install the test specimen without reducers or other devices in accordance with the piping requirements in Table 1.

7.1.2 Flow tests will include flow measurements at three pressure differentials. In order to approach flowing conditions that can be assumed to be incompressible, the pressure drop ratio ($x = \Delta p/p_1$) shall be ≤ 0.02.

7.1.3 The valve flow test shall be performed at 100 percent of rated valve travel. Optional tests may be performed at each 10 percent of rated valve travel or any other points to more fully determine the inherent characteristic of the specimen.

7.1.4 The following data shall be recorded:

(1) Valve travel (measurement error not exceeding ±0.5 percent of rated valve travel)

(2) Upstream pressure (p_1) (measurement error not exceeding ±2 percent of actual value)

(3) Pressure differential (Δp) across test section pressure taps (measurement error not exceeding ± 2 percent of actual value)

(4) Volumetric flow rate (q) (measurement error not exceeding ± 2 percent of actual value)

(5) Fluid temperature (T_1) upstream of valve (measurement error not exceeding ± 1°C (± 2°F))

(6) Barometric pressure (measurement error not exceeding ± 2 percent of actual value)

(7) Physical description of test specimen (e.g., type of valve, flow direction, etc.) and test fluid

7.2 x_T Test Procedure. The maximum flow rate, q_{max}, (referred to as choked flow) is required in the calculation of x_T, the pressure drop ratio factor. This factor is the terminal ratio of the differential pressure to absolute upstream pressure, ($\Delta p /p_1$), for a given test specimen installed without reducers or other devices. The maximum flow rate is defined as that flow rate at which, for a given upstream pressure, a decrease in downstream pressure will not produce an increase in flow rate. The test procedure required to obtain q_{max} is contained in this section with the data evaluation procedure in Section 8.2. An alternative procedure for determining x_T is provided in Section 7.3.

7.2.1 Install the test specimen without reducers or other attached devices in accordance with piping requirements in Table 1. The test specimen shall be at 100 percent of rated travel.

7.2.2 Any upstream supply pressure sufficient to produce choked flow is acceptable, as is any resulting pressure differential across the valve, provided that the criteria for determination of choked flow specified in Section 7.2.3 are met.

7.2.3 The downstream throttling valve will be in the wide-open position. Then, with a preselected upstream pressure the flow rate will be measured and the downstream pressure recorded. This test establishes the maximum pressure differential for the test specimen in this test system. A second test shall be conducted using the downstream throttling valve to reduce the pressure differential by 10 percent of the pressure differential determined in the first test (with the same upstream pressure). If the flow rate of this second test is within 0.5 percent of the flow rate for the first test, then the maximum flow rate has been established.

Although the absolute value of the flow rate must be measured to an error not exceeding ± 2 percent, the repeatability of the tests for x_T must be better than ± 0.5 percent in order to attain the prescribed accuracy. This series of tests must be made consecutively, using the same instruments, and without alteration to the test setup.

7.2.4 Record the following data:

(1) Valve travel (measurement error not exceeding ± 0.5 percent of rated travel)

(2) Upstream pressure (p_1) (measurement error not exceeding ± 2 percent of actual value)

(3) Downstream pressure (p_2) (measurement error not exceeding ± 2 percent of actual value)

(4) Volumetric flow rate (q) (measurement error not exceeding ± 2 percent of actual value)

(5) Fluid temperature upstream (T_1) of valve (measurement error not exceeding ± 1°C (± 2°F))

(6) Barometric pressure (measurement error not exceeding ± 2 percent of actual value)

(7) Physical description of test specimen (e.g., type of valve, flow direction, etc.) and test fluid

7.3 Alternative Test Procedure for C_v and x_T

7.3.1 Install the test specimen without reducers or other attached devices in accordance with piping requirements in Table 1. The test specimen shall be at 100 percent of rated travel.

7.3.2 With a preselected upstream pressure, p_1, measurements shall be made of flow rate, q, upstream fluid temperature, T_1, downstream pressure, p_2, for a minimum of five well-spaced values of x (the ratio of pressure differential to absolute upstream pressure).

7.3.3 From these data points calculate values of the product YC_V using the equation:

$$YC_v = \frac{q}{(N_7 P_1)} \left[\frac{(G_g T_1)}{x} \right]^{1/2} \qquad (7)$$

where Y is the expansion factor defined by:

$$Y = 1 - \frac{x}{3F_k x_T}$$

where:

$$F_k = \frac{k}{1.40}$$

7.3.4 The test points shall be plotted on linear coordinates as (YC_v) vs. x and a linear curve fitted to the data. If any point deviates by more than 5 percent from the curve, additional test data shall be taken to ascertain if the specimen truly exhibits anomalous behavior.

7.3.5 At least one test point, $(YC_v)_1$, must fulfill the requirement that:

$$(YC_v)_1 \geq 0.97(YC_v)_0$$

where $(YC_v)_0$ corresponds to $x \simeq 0$.

7.3.6 At least one test point, $(YC_v)_n$, must fulfill the requirement that:

$$(YC_v)_n \leq 0.83(YC_v)_0$$

7.3.7 The value of C_v for the specimen shall be taken from the curve at $x = 0$, $Y = 1$.

The value of x_T for the specimen shall be taken from the curve at $YC_v = 0.667 C_v$.

7.4 F_P Test Procedure. The piping geometry factor, F_P, modifies the valve sizing coefficient for reducers or other devices attached to the valve body that are not in accord with the test section. The factor F_P is the ratio of the installed C_v with the reducers or other devices attached to the valve body to the rated C_v of the valve installed in a standard test section and tested under identical service conditions. This factor is obtained by replacing the valve with the desired combination of valve, reducers, and/or other devices and then conducting the flow test outlined in Section 7.1, treating the combination of valve and reducers as the test specimen for the purpose of determining test section line size. For example, a 100-mm (4-inch) valve between reducers in a 150-mm (6-inch) line would use pressure tap locations based on a 150-mm (6-inch) nominal diameter. The data evaluation procedure is provided in Section 8.3.

7.5 x_{TP} Test Procedure. Perform the tests outlined for x_T in Section 7.2, replacing the valve with the desired combination of valve and pipe reducers or other devices and treating the combination of valve and reducers as the test specimen. The data evaluation procedure is provided in Section 8.4.

8. Data Evaluation Procedure — Compressible Fluids

The following procedures are to be used for the evaluation of the data obtained using the test procedures in Section 7. The pressure differentials used to calculate the flow coefficients and other flow factors were obtained using the test section defined in Table 1. These pressure measurements were made at the pressure taps and include the test section piping between the taps as well as the test specimen.

8.1 C_v Calculation. Using the data obtained in Section 7.1 and assuming the expansion factor $Y = 1.0$, calculate the flow coefficient, C_v for each test point using:

$$C_v = \frac{q}{N_7 p_1} \left(\frac{T_1 G_g}{x}\right)^{1/2} \qquad (8)$$

Calculate the arithmetic average of the three test valves obtained at rated travel to obtain the rated C_v.

8.2 x_T Calculation. Calculate x_T as follows:
From Equation (7):

$$q = N_7 Y C_v p_1 \left(\frac{x}{G_g T_1}\right)^{1/2}$$

When $x = F_k x_T$, then $q = q_{max}$:

$$q_{max} = N_7 Y C_v p_1 \left(\frac{F_k x_T}{G_g T_1}\right)^{1/2}$$

and:

$$x_T = \left(\frac{q_{max}}{N_7 Y C_v p_1}\right)^2 \frac{G_g T_1}{F_k} \qquad (9)$$

Assuming air as test fluid and substituting $Y = 0.667$, $G_g = 1.0$, and $F_k = 1.0$:

$$x_T = \left(\frac{q_{max}}{0.667 N_7 C_v p_1}\right)^2 T_1 \qquad (10)$$

8.3 F_P Calculation. Calculate F_P at rated valve travel:

$$F_P = \frac{q}{N_7 p_1 \left(\frac{x}{T_1 G_g}\right)^{1/2} C_{v\,rated}} \qquad (11)$$

8.4 x_{TP} Calculation. Calculate x_{TP} as follows:
From Equation (7):

$$q = N_7 F_p Y C_v p_1 \left(\frac{x_{TP}}{G_g T_1}\right)^{1/2} \quad (12)$$

with F_p added to account for reducers and other devices.
When $x = x_{TP}$, $q = q_{max}$

$$q_{max} = N_7 F_p Y C_v p_1 \left(\frac{x_{TP}}{G_g T_1}\right) \quad (13)$$

Assuming air as the test fluid:
$Y = 0.667$
$G_g = 1.0$
$F_k = 1.0$

$$x_{TP} = \left(\frac{q_{max}}{0.667 N_7 F_p C_v p_1}\right)^2 T_1 \quad (14)$$

9. Numerical Constants

The numerical constants, N, depend on the measurement units used in the general sizing equations. Values for N are listed in Table 3.

Table 3
Numerical Constants

	N	q*	p	v	T	d, D
N_1	0.0865	m³/h	kPa	--	--	--
	0.865	m³/h	bar	--	--	--
	1.00	gpm	psia	--	--	--
N_2	0.00214	--	--	--	--	mm
	890	--	--	--	--	in.
N_3	645	--	--	--	--	mm
	1.00	--	--	--	--	in.
N_4	76,000	m³/h	--	Centistoke**	--	mm
	17,300	gpm	--	Centistoke**	--	in.
N_7	4.17	m³/h	kPa	--	K	--
	417	m³/h	bar	--	K	--
	1,360	scfh	psia	--	°R	--

* The standard cubic foot is taken at 14.73 psia and 60°F and the standard cubic meter at 101.3 kPa and 15.61°C.

** Centistoke = 10^{-6} m²/sec.

All pressures are absolute.

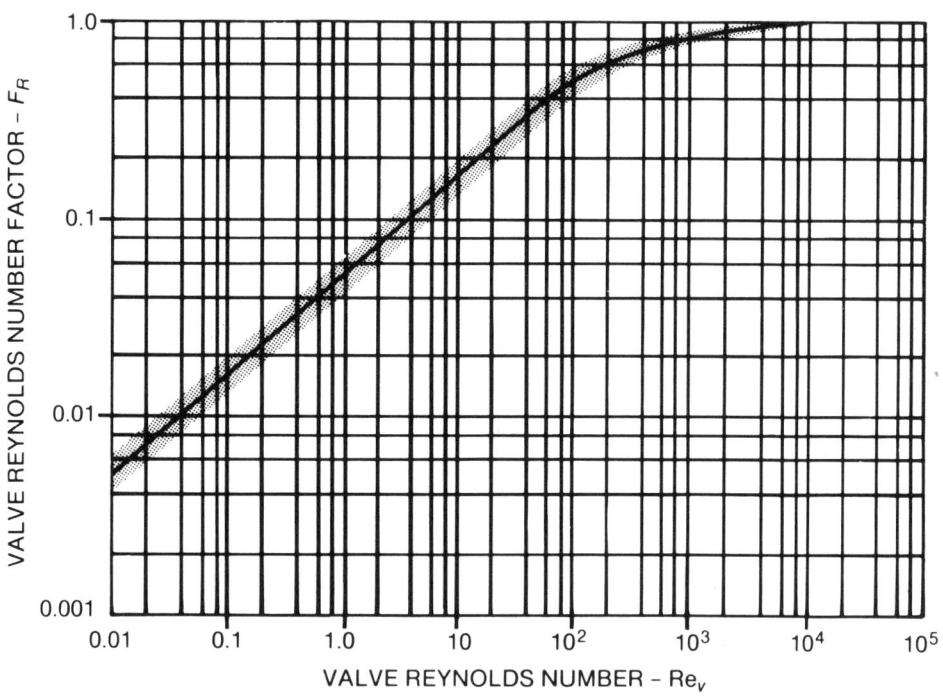

Figure 3. Reynolds Number Factor

ANSI/ISA-S75.03-1985
Approved April 10, 1985
(Formerly ISA-S4.01.1)

American National Standard

Face-to-Face Dimensions for Flanged Globe-Style Control Valve Bodies (ANSI Classes 125, 150, 250, 300 and 600)

Instrument Society of America

Instrument Society of America

ISBN 0-87664-841-3

ISA-S75.03 Face-to-Face Dimensions for
Flanged Globe-Style Control Valve Bodies

Copyright © 1984 by the Instrument Society of America. All rights reserved. Printed in the United States of America. No part of this publication may be reproduced, stored in a retrieval system, or transmitted, in any form or any means (electronic, mechanical, photocopying, recording or otherwise) without the prior written permission of the publisher.

INSTRUMENT SOCIETY OF AMERICA
67 Alexander Drive
P.O. Box 12277
Research Triangle Park, North Carolina 27709

Copyright © 1984 by the Instrument Society of America

PREFACE

This preface is included for information purposes and is not part of ISA-S75.03.

This standard has been prepared as part of the service of the Instrument Society of America (ISA) toward a goal of uniformity in the field of instrumentation. To be of real value, this document should not be static, but should be subject to periodic review. Toward this end, the Society welcomes all comments and criticisms, and asks that they be addressed to the Secretary, Standards and Practices Board, Instrument Society of America, 67 Alexander Drive, P.O. Box 12277, Research Triangle Park, NC 27709, Telephone (919) 549-8411.

The ISA Standards and Practices Department is aware of the growing need for attention to the metric system of units in general, and the International System of Units (SI) in particular, in the preparation of instrumentation standards. The Department is further aware of the benefits to U.S.A. users of ISA standards of incorporating suitable references to the SI (and the metric system) in their business and professional dealings with other countries. Toward this end, this Department will endeavor to introduce SI-acceptable metric units in all new and revised standards to the greatest extent possible. The Metric Practice Guide, which has been published by the Institute of Electrical and Electronic Engineers as ANSI/IEEE Std. 268-1982, and future revisions will be the reference guide for definitions, symbols, abbreviations, and conversion factors.

It is the policy of the Instrument Society of America to encourage and welcome the participation of all concerned individuals and interests in the development of ISA standards. Participation in the ISA standards-making process by an individual in no way constitutes endorsement by the employer of that individual, of the Instrument Society of America, or of any of the standards that ISA develops.

The information contained in the preface, footnotes, and appendices is included for information only and is not a part of the standard.

The following people served as members of ISA Committee SP75.08 (formerly SP4.1):

NAME	COMPANY
W. C. Weidman, Chairman	Gilbert Commonwealth, Inc.
H. D. Baumann	H. D. Baumann Associates, Ltd.
G. Borden	Bechtel Power Corporation
R. Chown	OTEC
E. J. Cooney	Air Products & Chemicals, Inc.
L. R. Driskell	Consultant
J. T. Emery	Honeywell, Inc.
A. J. Hanssen	Fluid Controls Institute
H. Illing	Kieley & Mueller, Inc.
M. W. Kaye	M. W. Kellogg Company
O. P. Lovett, Jr.	ISIS Corporation
H. R. Nickerson	Resistoflex Company
A. L. Pratt	ITT Hammel Dahl
R. A. Quance	Walsh Inc.
J. N. Reed	Masoneilan Division, McGraw Edison Co.
H. Schwartz	Flexible Valve Corporation
F. O. Seger	Willis Division, Smith International Inc.
J. M. Simonsen	Valtek, Inc.
R. U. Stanley	Retired
G. F. Stiles	Fisher Controls International, Inc.
R. F. Tubbs	Copes-Vulcan
S. Weiner	Monsanto Company
P. Wing, Jr.	Retired

Instrument Society of America

The following people served as members of ISA Committee SP75:

NAME	COMPANY
L. R. Driskell, Chairman	Consultant
J. B. Arant	E. I. duPont deNemours and Company, Inc.
H. E. Backinger	John F. Kraus & Company
G. Barb	Muesco, Inc.
H. D. Baumann	H. D. Baumann Associates, Ltd.
C. S. Beard	Mutual Stamp Company
N. Belaef	Consultant
G. Borden	Bechtel Power Corporation
D. E. Brown	R. Conrader Company
E. H. C. Brown	Dravo Engineers, Inc.
E. C. Cooney	Air Products & Chemicals, Inc.
W. G. Dewart	Rockwell International
J. T. Emery	Honeywell, Inc.
H. J. Fuller	Worcester Controls Corporation
L. Griffith	Consultant
A. J. Hanssen	Fluid Controls Institute, Inc.
F. P. Harthun	Fisher Controls International, Inc.
H. P. Illing	Kieley & Mueller, Inc.
R. B. Jones	Upjohn Company
M. W. Kaye	M. W. Kellogg Company
R. Louvere	Creole Corporation
O. P. Lovett, Jr.	ISIS Corporation
A. P. McCauley	Chagrin Valley Controls, Inc.
T. V. Molloy	Pacific Gas & Electric
J. T. Muller	Leslie Company
H. R. Nickerson	Resistoflex Company
J. Ozol	Omaha Public Power
R. A. Quance	Walsh Inc.
W. Rahmeyer	Colorado State University
J. N. Reed	Masoneilan Division, McGraw Edison Co.
G. Richards	Jordan Valve
J. Rosato	Rawson Company
K. Schoonover, Secretary	Con-Tek
H. Schwartz	Flexible Valve Corporation
F. O. Seger	Willis Division, Smith International, Inc.
J. M. Simonsen	Valtek, Inc.
H. Sonderregger	ITT Grinnell Corporation
N. Sprecher	DeZurik
R. U. Stanley	Retired
*G. F. Stiles/Roy Brodin	Fisher Controls International, Inc.
R. Terhune, Vice-Chairman	Exxon Company USA
R. F. Tubbs	Copes-Vulcan
W. C. Weidman	Gilbert Commonwealth, Inc.
R. L. Widdows	Cashco, Inc.
P. Wing	Retired
L. Zinck	Union Carbide

*One vote

This standard was approved for publication by the ISA Standards and Practices Board in July 1984.

NAME	COMPANY
W. Calder III, Chairman	The Foxboro Company
P. V. Bhat	Monsanto Company
N. L. Conger	Conoco
B. Feikle	Bailey Controls Company
T. J. Harrison	IBM Corporation
H. S. Hopkins	Westinghouse Electric Company
J. L. Howard	Boeing Aerospace Company
R. T. Jones	Philadelphia Electric Company
R. Keller	The Boeing Company
O. P. Lovett, Jr.	ISIS Corporation
E. C. Magison	Honeywell, Inc.
A. P. McCauley	Chagrin Valley Controls, Inc.
J. W. Mock	Bechtel Corporation
E. M. Nesvig	ERDCO Engineering Corporation
R. Prescott	Moore Products Company
D. Rapley	Stearns Catalytic Corporation
W. C. Weidman	Gilbert Commonwealth, Inc.
K. A. Whitman	Allied Chemical Corporation
*P. Bliss	
*B. A. Christensen	
*L. N. Combs	
*R. L. Galley	
*R. G. Marvin	
*W. B. Miller	Moore Products Company
*G. Platt	Bechtel Power Corporation
*J. R. Williams	Stearns Catalytic Corporation

*Director Emeritus

1 SCOPE

1.1 This standard applies to flanged globe-style control valves, sizes 1/2 inch through 16 inches, having top, top and bottom, port, or cage guiding.

2 PURPOSE

2.1 The purpose of this standard is to aid users in their piping design by providing ANSI Class 125 flat face, and ANSI Classes 150, 250, 300, and 600 raised face, flanged control valve dimensions, without giving special consideration of the equipment manufacturer to be used.

3 DEFINITION

3.1 For definitions of terms used in this standard, see ANSI/ISA-S75.01-1983, "Control Valve Terminology."

4 BIBLIOGRAPHY

The following bibliography is included for the definition of pressure classes and flange dimensions, material identification, and cross reference information. Items 4.1 through 4.5 were published by ASME (American Society of Mechanical Engineers).

4.1 American National Standards Institute, Inc. (ANSI) Standard B16.34-1981, "Valves—Flanged and Buttwelding End-Steel, Nickel Alloy, and other Special Alloys, 1981."

4.2 American National Standards Institute, Inc. (ANSI) Standard B16.1-1975, "Cast Iron Pipe Flanges and Flanged Fittings, Class 25, 125, 250 and 800, 1975."

4.3 American National Standards Institute, Inc. (ANSI) Standard B16.5-1981, "Pipe Flanges and Flanged Fittings— Steel, Nickel Alloy, and other Special Alloys, 1981."

4.4 American National Standards Institute, Inc. (ANSI) Standard B16.10-1973, "Face-to-Face and End-to-End Dimensions of Ferrous Valves, 1973."

4.5 American National Standards Institute, Inc. (ANSI) Standard B16.24-1979, "Bronze Pipe Flanges and Flanged Fittings, Class 150 and 300, 1979."

4.6 International Electrotechnical Commission (IEC) Publication 534, Industrial-Process Control Valves, 534-3(1976) Part 3: Dimensions. Section One: "Face-to-Face Dimensions for Flanged, Two-Way, Globe-Type Control Valves."

5 DIMENSIONAL DATA

5.1 Face-to-Face Dimensions for Flanged Globe-Style Control Valves. See Table 1.

Instrument Society of America

TABLE 1
FACE-TO-FACE DIMENSIONS FOR FLANGED GLOBE-STYLE CONTROL VALVES

Nominal Valve Size		PN 20* (ANSI Classes 125 and 150)		PN 50* (ANSI Classes 250 and 300)		PN 100 (ANSI Class 600)		Tolerance	
		Dimension "A"		Dimension "A"		Dimension "A"				
mm	inches	mm	inches	mm	inches	mm	inches	mm	inches
15	1/2	184	7.25	190	7.50	203	8.00	±1.6	±0.062
20	3/4	184	7.25	194	7.62	206	8.12	±1.6	±0.062
25	1	184	7.25	197	7.75	210	8.25	±1.6	±0.062
40	1-1/2	222	8.75	235	9.25	251	9.88	±1.6	±0.062
50	2	254	10.00	267	10.50	286	11.25	±1.6	±0.062
65	2-1/2	276	10.88	292	11.50	311	12.25	±1.6	±0.062
80	3	298	11.75	318	12.50	337	13.25	±1.6	±0.062
100	4	352	13.88	368	14.50	394	15.50	±1.6	±0.062
150	6	451	17.75	473	18.62	508	20.00	±1.6	±0.062
200	8	543	21.38	568	22.38	610	24.00	±1.6	±0.062
250	10	673	26.50	708	27.88	752	29.62	±1.6	±0.062
300	12	737	29.00	775	30.50	819	32.25	±3.2	±0.125
350	14	889	35.00	927	36.50	972	38.25	±3.2	±0.125
400	16	1016	40.00	1057	41.62	1108	43.62	±3.2	±0.125

*International standards groups have not finalized PN 20 and PN 50 as being designated for Class 125 and Class 250, respectively. However, dimensionally, the flanges are compatible.

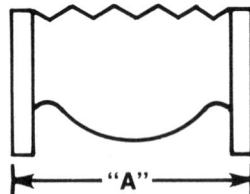
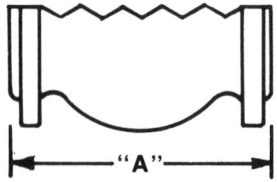

APPENDIX

This Appendix is not part of ISA Standard S75.03, but it is included to facilitate its use.

The ASME standards listed in the bibliography contain valve body design information in addition to face-to-face dimensions. Dimensions for metrically sized valves are nominal conversions that are conventionally used in documents by the Manufacturers Standardization Society (MSS) of the Valve and Fitting Industry (MSS-SP86-1981), by the International Organization for Standardization (ISO), and by the International Electrotechnical Commission (IEC).

Control valve initial dimensions, 10 inches-16 inches, were listed in the Fluid Control Institute (FCI) Standard, FCI-65-2-1975, which has been withdrawn from publication.

INSTRUMENT SOCIETY of AMERICA
Research Triangle Park, North Carolina

ANSI/ISA-S75.04-1985
Approved April 10, 1985
(Formerly ISA-S4.01.2)

American National Standard

Face-to-Face Dimensions for Flangeless Control Valves (ANSI Classes 150, 300, 600)

Instrument Society of America

Instrument Society of America

ISBN 0-87664-842-1

ISA-S75.04 Face-to-Face Dimensions for Flangeless
Control Valves

Copyright © 1984 by the Instrument Society of America. All rights reserved. Printed in the United States of America. No part of this publication may be reproduced, stored in a retrieval system, or transmitted, in any form or any means (electronic, mechanical, photocopying, recording or otherwise) without the prior written permission of the publisher.

INSTRUMENT SOCIETY OF AMERICA
67 Alexander Drive
P.O. Box 12277
Research Triangle Park, North Carolina 27709

Copyright © 1984 by the Instrument Society of America

PREFACE

This preface is included for information purposes and is not part of ISA-S75.04.

This standard has been prepared as part of the service of the Instrument Society of America (ISA) toward a goal of uniformity in the field of instrumentation. To be of real value, this document should not be static, but should be subject to periodic review. Toward this end, the Society welcomes all comments and criticisms, and asks that they be addressed to the Secretary, Standards and Practices Board, Instrument Society of America, 67 Alexander Drive, P.O. Box 12277, Research Triangle Park, NC 27709, Telephone (919) 549-8411.

The ISA Standards and Practices Department is aware of the growing need for attention to the metric system of units in general, and the International System of Units (SI) in particular, in the preparation of instrumentation standards. The Department is further aware of the benefits to U.S.A. users of ISA standards of incorporating suitable references to the SI (and the metric system) in their business and professional dealings with other countries. Toward this end, this Department will endeavor to introduce SI-acceptable metric units in all new and revised standards to the greatest extent possible. The Metric Practice Guide, which has been published by the Institute of Electrical and Electronic Engineers as ANSI/IEEE Std. 268-1982, and future revisions will be the reference guide for definitions, symbols, abbreviations, and conversion factors.

It is the policy of the Instrument Society of America to encourage and welcome the participation of all concerned individuals and interests in the development of ISA standards. Participation in the ISA standards-making process by an individual in no way constitutes endorsement by the employer of that individual, of the Instrument Society of America, or of any of the standards that ISA develops.

The information contained in the preface, footnotes, and appendices is included for information only and is not a part of the standard.

The following people served as members of ISA Committee SP75.08 (formerly SP4.1):

NAME	COMPANY
W. C. Weidman, Chairman	Gilbert Commonwealth, Inc.
H. D. Baumann	H. D. Baumann Associates, Ltd.
G. Borden	Bechtel Power Corporation
R. Chown	OTEC
E. J. Cooney	Air Products & Chemicals, Inc.
L. R. Driskell	Consultant
J. T. Emery	Honeywell, Inc.
A. J. Hanssen	Fluid Controls Institute
H. Illing	Kieley & Mueller, Inc.
M. W. Kaye	M. W. Kellogg Company
O. P. Lovett, Jr.	ISIS Corporation
H. R. Nickerson	Resistoflex Company
A. L. Pratt	ITT Hammel Dahl
R. A. Quance	Walsh Inc.
J. N. Reed	Masoneilan Division, McGraw Edison Co.
H. Schwartz	Flexible Valve Corporation
F. O. Seger	Willis Division, Smith International, Inc.
J. M. Simonsen	Valtek, Inc.
R. U. Stanley	Retired
G. F. Stiles	Fisher Controls International, Inc.
R. F. Tubbs	Copes-Vulcan
S. Weiner	Monsanto Company
P. Wing, Jr.	Retired

Instrument Society of America

The following people served as members of ISA Committee SP75:

NAME	COMPANY
L. R. Driskell, Chairman	Consultant
J. B. Arant	E. I. duPont deNemours and Company, Inc.
H. E. Backinger	John F. Kraus & Company
G. Barb	Muesco, Inc.
H. D. Baumann	H. D. Baumann Associates, Ltd.
C. S. Beard	Mutual Stamp Company
N. Belaef	Consultant
G. Borden	Bechtel Power Corporation
D. E. Brown	R. Conrader Company
E. H. C. Brown	Dravo Engineers, Inc.
E. C. Cooney	Air Products & Chemicals, Inc.
W. G. Dewart	Rockwell International
J. T. Emery	Honeywell, Inc.
H. J. Fuller	Worcester Controls Corporation
L. Griffith	Consultant
A. J. Hanssen	Fluid Controls Institute, Inc.
F. P. Harthun	Fisher Controls International, Inc.
H. P. Illing	Kieley & Mueller, Inc.
R. B. Jones	Upjohn Company
M. W. Kaye	M. W. Kellogg Company
R. Louvere	Creole
O. P. Lovett, Jr.	ISIS Corporation
A. P. McCauley	Chagrin Valley Controls, Inc.
T. V. Molloy	Pacific Gas & Electric
J. T. Muller	Leslie Company
H. R. Nickerson	Resistoflex Company
J. Ozol	Omaha Public Power
R. A. Quance	Walsh Inc.
W. Rahmeyer	Colorado State University
J. N. Reed	Masoneilan Division, McGraw Edison Co.
G. Richards	Jordan Valve
J. Rosato	Rawson Company
K. Schoonover, Secretary	Con-Tek
H. Schwartz	Flexible Valve Corporation
F. O. Seger	Willis Division, Smith International Inc.
J. M. Simonsen	Valtek, Inc.
H. Sonderregger	ITT Grinnell Corporation
N. Sprecher	DeZurik
R. U. Stanley	Retired
*G. F. Stiles/Roy Brodin	Fisher Controls International, Inc.
R. Terhune, Vice-Chairman	Exxon Company USA
R. F. Tubbs	Copes-Vulcan
W. C. Weidman	Gilbert Commonwealth, Inc.
R. L. Widdows	Cashco, Inc.
P. Wing	Retired
L. Zinck	Union Carbide

*One vote

This standard was approved for publication by the ISA Standards and Practices Board in July 1984.

NAME	COMPANY
W. Calder III, Chairman	The Foxboro Company
P. V. Bhat	Monsanto Company
N. L. Conger	Conoco
B. Feikle	Bailey Controls Company
T. J. Harrison	IBM Corporation
H. S. Hopkins	Westinghouse Electric Company
J. L. Howard	Boeing Aerospace Company
R. T. Jones	Philadelphia Electric Company
R. Keller	The Boeing Company
O. P. Lovett, Jr.	ISIS Corporation
E. C. Magison	Honeywell, Inc.
A. P. McCauley	Chagrin Valley Controls, Inc.
J. W. Mock	Bechtel Corporation
E. M. Nesvig	ERDCO Engineering Corporation
R. Prescott	Moore Products Company
D. Rapley	Stearns Catalytic Corporation
W. C. Weidman	Gilbert Commonwealth, Inc.
K. A. Whitman	Allied Chemical Corporation
*P. Bliss	
*B. A. Christensen	
*L. N. Combs	
*R. L. Galley	
*R. G. Marvin	
*W. B. Miller	Moore Products Company
*G. Platt	Bechtel Power Corporation
*J. R. Williams	Stearns Catalytic Corporation

*Director Emeritus

1 SCOPE

1.1 This standard applies to flangeless control valves utilizing a full ball or a segment of a ball and other rotary-stem or sliding-stem flangeless control valves, sizes 3/4 inch through 16 inches for ANSI Classes 150 through 600.

1.2 The face-to-face dimensions listed apply only to control valves which will be bolted between flanges.

1.3 This standard is not intended to include butterfly valves.

2 PURPOSE

2.1 The purpose of this standard is to aid users in their piping designs for flangeless control valves by providing valve face-to-face dimensions without giving special consideration to the equipment manufacturer to be used.

3 DEFINITIONS

3.1 For definitions of terms used in this standard, see ANSI/ISA-S75.05, "Control Valve Terminology."

4 BIBLIOGRAPHY

4.1 International Electrotechnical Commission (IEC), Publication 534, Industrial-Process Control Valves, Part 3, Section Two-1983, "Face-to-Face Dimensions of Flangeless Control Valves Except Wafer Butterfly Valves."

5 DIMENSIONAL DATA

5.1 Face-to-Face Dimensions for Flangeless Control Valves. See Table 1.

TABLE 1
FACE-TO-FACE DIMENSIONS FOR FLANGELESS CONTROL VALVES

Nominal Valve Size		PN 20, 50 & 100 ANSI Classes 150, 300 & 600		Tolerance	
		Dimension "A"			
mm	inches	mm	inches	mm	inches
20	3/4	76	3.00	±1.6	±0.062
25	1	102	4.00	±1.6	±0.062
40	1-1/2	114	4.50	±1.6	±0.062
50	2	124	4.88	±1.6	±0.062
80	3	165	6.50	±1.6	±0.062
100	4	194	7.62	±1.6	±0.062
150	6	229	9.00	±1.6	±0.062
200	8	243	9.56	±1.6	±0.062
250	10	297	11.69	±1.6	±0.062
300	12	338	13.31	±3.2	±0.125
350	14	400	15.75	±3.2	±0.125
400	16	400	15.75	±3.2	±0.125

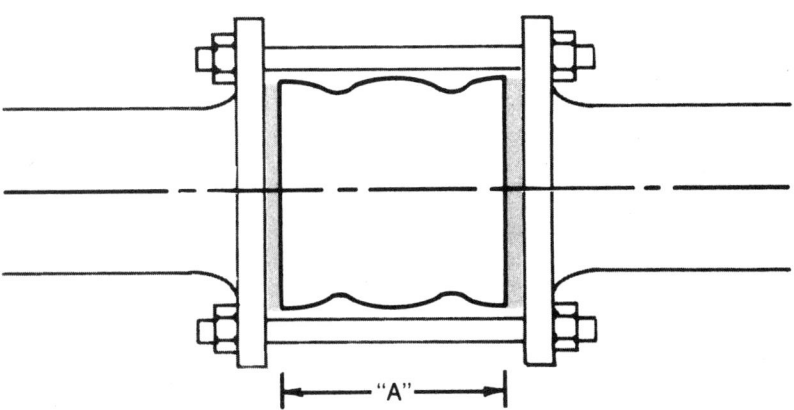

NOTE: Dimension "A" does not include line gaskets, and is a dimension for the valve itself.

APPENDIX

This appendix is not part of ISA-S75.04, but it is included to facilitate its use.

Face-to-face dimensions to start this standard were obtained from Scientific Apparatus Makers Association (SAMA), Proposed Standard PMC 23.3e-1974, "Face-to-Face Dimensions of Flangeless Control Valves." Dimensions for metrically-sized valves are nominal conversions that are conventionally used in the Manufacturers Standardization Society (MSS) of the Valve and Fitting Industry (MSS-SP86-1981), International Standards Organization (ISO), and International Electrotechnical Commission (IEC).

INSTRUMENT SOCIETY OF AMERICA
Research Triangle Park, North Carolina 27709

ANSI/ISA-S75.05-1983
Approved September 29, 1986

American National Standard

Control Valve Terminology

Instrument Society of America

ISBN 0-87664-753-0

ISA-S75.05 Control Valve Terminology

Copyright © 1983 by the Instrument Society of America. All rights reserved. Printed in the United States of America. No part of this publication may be reproduced, stored in a retrieval system, or transmitted, in any form or any means electronic, mechanical, photocopying, recording or otherwise without the prior written permission of the publisher.

INSTRUMENT SOCIETY OF AMERICA
67 Alexander Drive
P.O. Box 12277
Research Triangle Park, North Carolina 27709

Copyright © 1983 by the Instrument Society of America

PREFACE

This Preface is included for informational purposes and is not part of ISA-S75.05.

This standard has been prepared as a part of the service of the Instrument Society of America toward a goal of uniformity in the field of instrumentation. To be of real value, this document should not be static, but should be subject to periodic review. Toward this end, the Society welcomes all comments and criticisms, and asks that they be addressed to the Secretary, Standards and Practices Board, Instrument Society of America, 67 Alexander Drive, P.O. Box 12277, Research Triangle Park, NC 27709, Telephone (919) 549-8411.

The ISA Standards and Practices Department is aware of the growing need for attention to the metric system of units in general, and the International System of Units (SI) in particular, in the preparation of instrumentation standards. The Department is further aware of the benefits to USA users of ISA Standards of incorporating suitable references to the SI (and the metric system) in their business and professional dealings with other countries. Towards this end this Department will endeavor to introduce SI - acceptable metric units in all new and revised standards to the greatest extent possible. The Metric Practice Guide, which has been published by the Institute of Electrical and Electronics Engineers, Inc. as ANSI/IEEE 268-1982, and future revisions, will be the reference guide for definitions, symbols, abbreviations, and conversion factors.

It is the policy of the Instrument Society of America to encourage and welcome the participation of all concerned individuals and interests in the development of ISA standards. Participation in the ISA standards making process by an individual in no way constitutes endorsement by the employer of that individual of the Instrument Society of America or any of the standards which ISA develops.

Prior to the issue of this standard, there had been no standard which provided terminology for control valve functions, types of valves and parts of valves used for control of fluid flow. This standard provides terminology for control valves of seven different types and also for common types of actuators used with these valves. This standard names individual valve parts, defines assemblies of parts, and provides terminology for part and assembly functions. Terminology is provided for differently shaped parts performing the same function. Operating terminology is furnished for the complete valve-actuator unit as well as limited terminology for the function of auxiliary equipment used with control valves. Type of construction terminology is provided where there may be a reason to specify a particular construction.

This standard is provided with a subject index for cross-reference and a glossary to define commonly used control valve terms. Defined terms, where used as a part of other definitions, are set in italics to provide a ready cross reference.

Valve classification charts are provided to show the control valve function within the entire grouping of valves and the charts are extended to show the relationship of control valves by type.

This standard replaces ASME Standard 112, issued in 1961, "Diaphragm Actuated Control Valve Standard", which is now inactive and is to be withdrawn. For additional information see the ISA Handbook of Control Valves.

The following individuals served as members of ISA Subcommittee 75.01 which prepared this standard:

NAME	COMPANY
M. L. Freeman, Chairman (Deceased)	Ametek, Inc.
R. E. Terhune, Chairman	Exxon Company, U.S.A.
L. E. Brown, Secretary	Retired
N. Belaef	Retired
J. B. Hills	Lasko Corp.
R. E. Pfeiffer	Union Carbide Co.
C. A. Prior	Prior Associates, Inc.
G. Richards	Jordan Valve Co.
N. D. Sprecher	DeZurik - A Unit of General Signal
F. C. Sullivan	E.I. duPont de Nemours & Co.

The standard was approved by ISA SP75 in August 1981.

NAME	COMPANY
L. R. Driskell, Chairman	Consultant
J. Arant	E. I. duPont de Nemours & Co.
H. E. Backinger	John F. Kraus & Co.
G. E. Barb	Bechtel Corp.
H. D. Baumann	H. D. Baumann Associates, Inc.
C. S. Beard	Retired

Instrument Society of America

N. Belaef	Retired
G. Borden	Bechtel Power Corp.
D. Brown	Conrader Valve
E. H. C. Brown	Dravo Engineers and Construction
P. S. Buckley	E. I. duPont de Nemours & Co.
J. Curran	Emigrant Savings Bank
J. T. Emery	Honeywell, Inc.
H. J. Fuller	Worcester Controls Corp.
L. F. Griffith	Yarway Corp.
A. J. Hanssen	Fluid Controls Institute
F. Harthun	Fisher Controls International Inc.
H. Illing	Kieley & Mueller, Inc.
R. Jones	Upjohn Company
M. Kaye	Pullman Kellog Co.
O. P. Lovett	Isis Corporation
J. Manton	PEDCO
H. P. Meissner	Kieley & Mueller Inc.
T. V. Molloy	Pacific Gas & Electric Co.
J. T. Muller	Leslie Company
H. R. Nickerson	Resistoflex Corp.
R. E. Pfeiffer	Union Carbide Corp.
C. A. Prior	Prior Associates
R. A. Quance	Comeau, Boyle, Inc.
J. N. Reed	Masoneilan International
G. Richards	Jordan Valve Division
J. Rosato	Rawson Co.
K. Schoonover	Steam Systems & Services
F. O. Seger	Willis Oil Tool Co.
J. M. Simonsen	Valtek, Inc.
H. R. Sonderregger	ITT Grinnell Corp.
N. D. Sprecher	DeZurik - A Unit of General Signal
R. V. Stanley	Retired
G. F. Stiles	Fisher Controls International Inc.
R. E. Terhune	Exxon Co. USA
R. F. Tubbs	Copes-Vulcan
W. C. Weidman	Gilbert Associates, Inc.
R. L. Widdows	CASHCO
P. Wing	Retired
L. R. Zinck	Union Carbide

This standard was approved by the ISA Standards and Practices Board in April 1983.

NAME	COMPANY
W. Calder, III, Chairman	The Foxboro Company
N. Conger	CONOCO
B. Feikle	Bailey Controls Co.
T. J. Harrison	IBM Corporation
H. S. Hopkins	Westinghouse Electric Corp.
J. L. Howard	Boeing Aerospace Co.
R. T. Jones	Philadelphia Electric Co.
Richard Keller	The Boeing Co.
O. P. Lovett	Isis Corp.
E. C. Magison	Honeywell, Inc.
A. P. McCauley	Diamond Shamrock Corp.
J. W. Mock	Measurement Services
E. M. Nesvig	ERDCO Engineering Corp.
R. Prescott	Moore Products Co.
D. Rapley	Stearns-Roger Eng. Corp.
W. C. Weidman	Gilbert Associates, Inc.
K. A. Whitman	Allied Chemical Corp.
J. R. Williams	Stearns-Rogers Inc.

P. Bliss*
B. Christensen*
L. N. Combs*
R. L. Galley*
R. G. Marvin*
W. B. Miller*   Moore Products Company
R. L. Nickens*
G. Platt*

*Director Emeritus

TABLE OF CONTENTS

Section	Page
1 Scope and Purpose	9
2 Basic Definitions	9
3 Classification	9
4 Definitions of Parts Common to Many Types of Valves	12
5 Linear Motion Control Valve Type	13
5.1 Globe Valve	13
5.2 Gate Valve	15
5.3 Diaphragm Valve	16
5.4 Pinch or Clamp Valve	16
6 Rotary Motion Control Valve Types	17
6.1 Ball Valve	17
6.2 Butterfly Valve	17
6.3 Plug Valve	18
7 Control Valve Actuators	19
8 Auxiliary Equipment	22
9 Glossary	23
Appendix A Hard Facing of Control Valve Trim	27
Appendix B Butterfly Valve Liners and Seals	29
Subject Index	31

SCOPE AND PURPOSE

1.1 To provide terminology and classification for the following types of *control valves*.

1.1.1 *Linear motion type control valves* covered are: (Section 5)
   Globe (Subsection 5.1)
   Gate (Subsection 5.2)
   Diaphragm (Subsection 5.3)
   Pinch or Clamp (Subsection 5.4)

1.1.2 *Rotary motion type control valves* covered are: (Section 6)
   Ball (Subsection 6.1)
   Butterfly (Subsection 6.2)
   Plug (Subsection 6.3)

1.2 To provide terminology and classification for the following types of *control valve actuators*.

1.2.1 *Actuator types* covered are: (Section 7)
   Diaphragm (Subsection 7.1.1.1)
   Piston (Subsection 7.1.1.2)
   Vane (Subsection 7.1.1.3)
   Bellows (Subsection 7.1.1.4)
   Fluid Motor (Subsection 7.1.1.5)
   Electro Mechanical Type (Subsection 7.1.2.1)
   Electro Hydraulic Type (Subsection 7.1.2.2)
   Hydraulic (Subsection 7.1.3)

1.3 To provide the description and classification of *auxiliary equipment* which may be used with *control valves*. (Section 8)

1.4 To provide a glossary to define other terms commonly used in the control valve industry.

2 BASIC DEFINITIONS

2.1 Control Valve[*] — A power operated device which modifies the fluid flow rate in a process control system. It consists of a *valve* connected to an *actuator* mechanism that is capable of changing the position of a flow controlling element in the *valve* in response to a signal from the controlling system.

2.1.1 Valve — A *valve* is a device used for the control of fluid flow. It consists of a fluid retaining assembly, one or more ports between end openings and a movable *closure member* which opens, restricts or closes the *port(s)*.

2.1.2 Actuator — An *actuator* is a fluid powered or electrically powered device which supplies force and motion to a *valve closure member*.

2.1.3 Motion Conversion Mechanism — A mechanism between the *valve* and the power unit of the *actuator* to convert between linear and rotary motion. The conversion can be from *linear actuator* action to *rotary valve* operation or from *rotary actuator* action to *linear valve* operation.

3 CLASSIFICATION

The classification of *valves*, *actuators*, and *valve positioners* is illustrated in the following charts:

3.1 Valves
3.2 Control Valves
3.3 Actuators
3.4 Valve Positioners

---

[*]The following types of valves are excluded from this standard:

1) Regulator — A regulator, whether for flow, level, pressure or temperature is a valve with a positioning actuator using a self-generated power signal for moving the closure member relative to the valve port or ports in response and in proportion to the changes in energy of the controlled variable. The force to position the closure member is derived from the same fluid. Source: "Standard Classification and Terminology for Power Actuated Valves", (Fluid Controls Institute) FCI 55-1 (1962).

2) Relief Valve — A generic term applying to relief valves, safety valves or safety relief valves. Source: ANSI B95.1 or API RP520.

3) Hand Valve — A generic term applying to valves used in process piping to provide shut off or isolation.

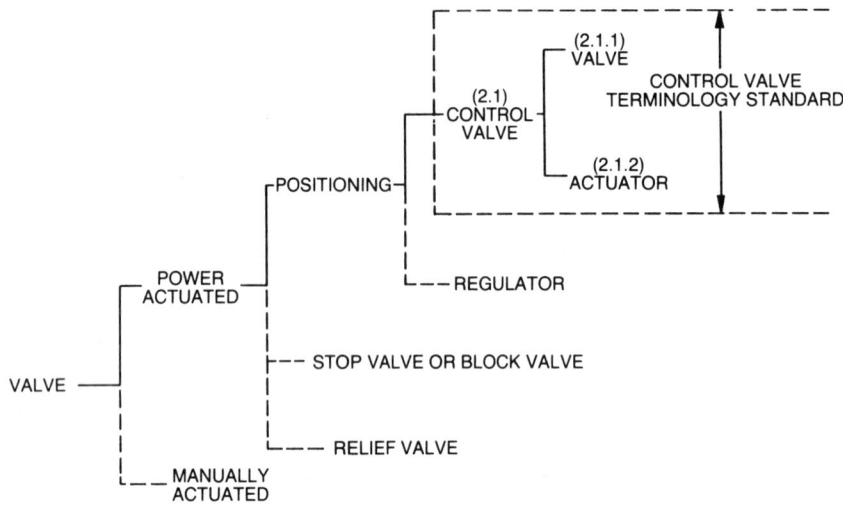

CHART 3.1 VALVES [a]

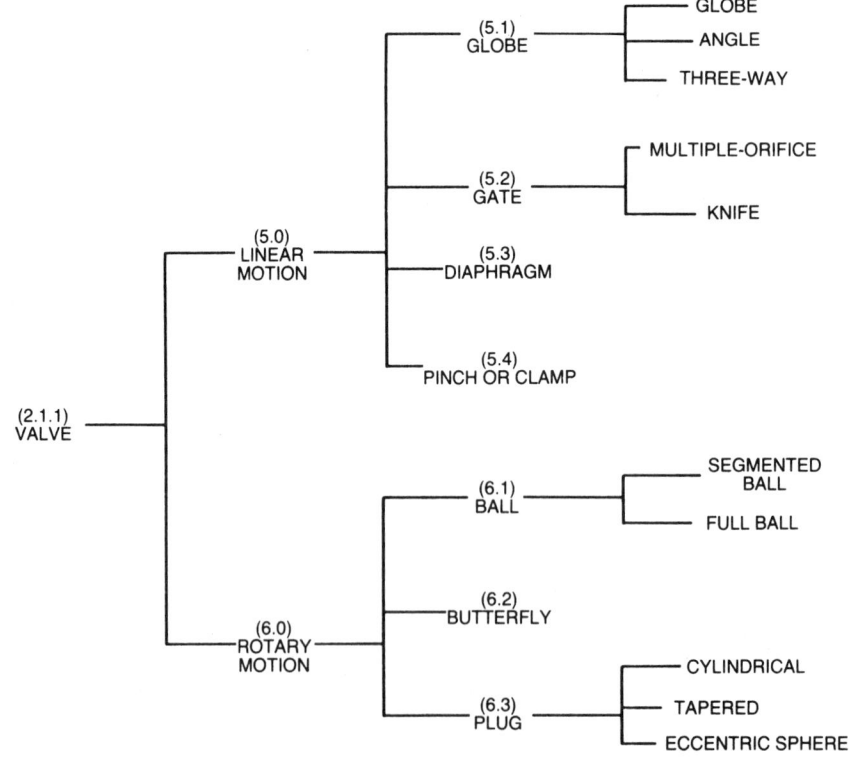

CHART 3.2 CONTROL VALVES

(a) Taken in part from FCI 55-1, 1962 Std.

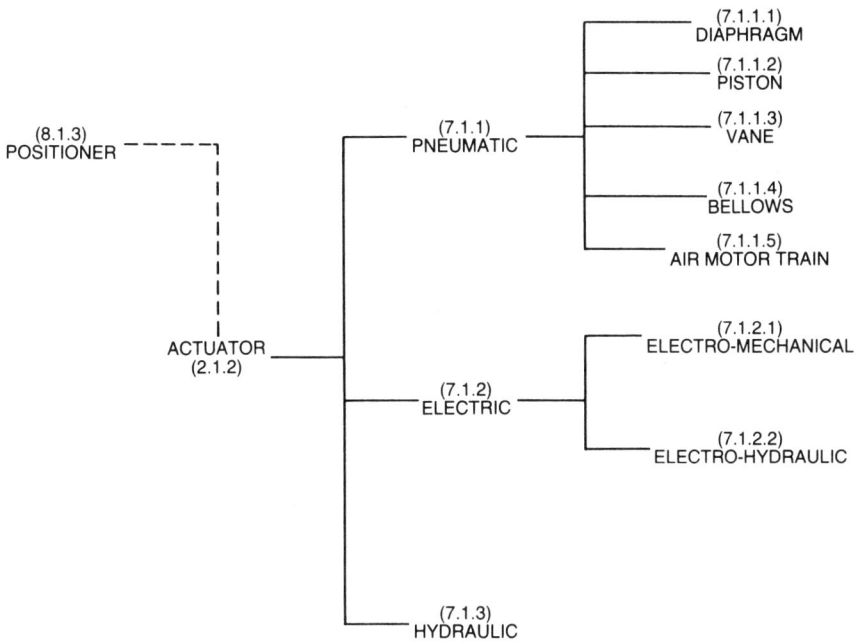

CHART 3.3 ACTUATORS

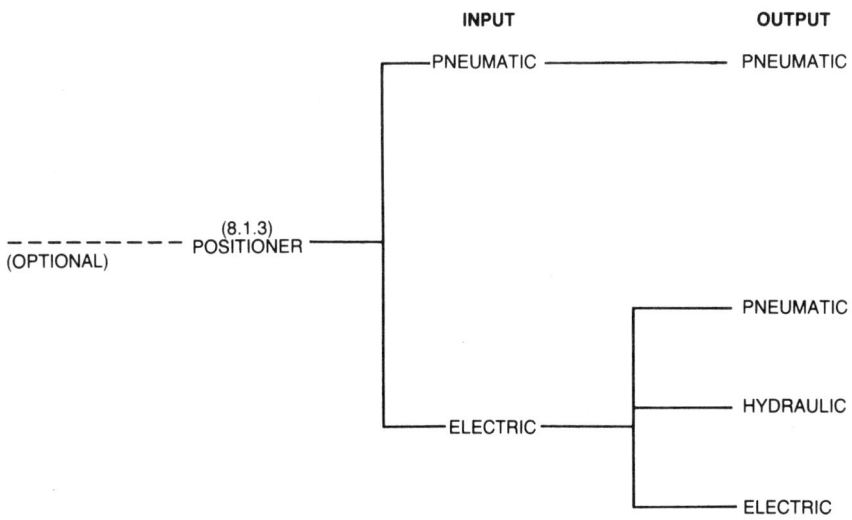

CHART 3.4 VALVE POSITIONERS

4 DEFINITIONS OF PARTS COMMON TO MANY TYPES OF VALVES

4.1 Body — The part of the *valve* which is the main pressure boundary. The *body* also provides the pipe connecting ends, the fluid flow passageway, and may support the seating surfaces and the *valve closure member*.

4.2 Bonnet — That portion of the valve pressure retaining boundary which may guide the *stem* and contains the *packing box* and *stem seal*. It may also provide the principal opening to the *body* cavity for assembly of internal parts or be an integral part of the *valve body*. It may also provide for the attachment of the *actuator* to the *valve body*.

4.2.1 Bonnet Types — Typical bonnets are bolted, threaded, or welded to or integral with the *body*. Other types sometimes used are defined below.

4.2.1.1 Extension Bonnet — A *bonnet* with a *packing box* that is extended above the *bonnet* joint of the *valve body* so as to maintain the temperature of the *packing* above or below the temperature of the process fluid. The length of the extension *bonnet* is dependent upon the difference between the fluid temperature and the *packing* design temperature limit as well as upon the *valve body* design.

4.2.1.2 Seal Welded Bonnet — A *bonnet* welded to a *body*, at assembly, to provide a zero leakage joint. This construction consists of a low-strength weld with the *bonnet* retained to the *body* by other means to withstand the *body* pressure load acting on the *bonnet* area.

4.2.2 Bonnet Gasket — A deformable sealing element between the mating surfaces of the *body* and *bonnet*. It may be deformed by compressive stress or energized by fluid pressure within the valve *body*.

4.2.3 Bonnet Bolting — A means of fastening the *bonnet* to the *body*. It may consist of studs with nuts for a flanged *bonnet* joint, studs threaded into the bonnet neck of the *body*, or bolts through the *bonnet* flange.

4.3 Closure Member — A movable part of the *valve* which is positioned in the flow path to modify the rate of flow through the *valve*.

4.3.1 Closure Member Types

4.3.1.1 Ball — A spherically shaped part which uses a portion of a spherical surface or an internal path to modify flow rate with rotary motion.

4.3.1.2 Disk — An essentially flat, circular shaped part which modifies the flow rate with either linear or rotary motion.

4.3.1.3 Gate — A flat or wedge-shaped sliding element that modifies flow rate with linear motion across the flow path.

4.3.1.4 Plug — A cylindrical part which moves in the flow stream with linear motion to modify the flow rate and which may or may not have a contoured portion to provide flow characterization. It may also be a cylindrical or conically tapered part, which may have an internal flow path, that modifies the flow rate with rotary motion.

4.4 Flow Control Orifice — The part of the flow passageway that, with the *closure member*, modifies the rate of flow through the *valve*. The orifice may be provided with a seating surface, to be contacted by or closely fitted to the *closure member*, to provide tight shutoff or limited leakage.

4.4.1 Seat Ring — A part that is assembled in the *valve body* and may provide part of the *flow control orifice*. The *seat ring* may have special material properties and may provide the contact surface for the *closure member*. Typical *seat ring* hard facing is shown in Appendix A.

4.4.2 Cage — A part in a *globe valve* surrounding the *closure member* to provide alignment and facilitate assembly of other parts of the *valve trim*. The *cage* may also provide flow characterization and/or a seating surface for *globe valves* and flow characterization for some *plug valves*.

4.4.3 Integral Seat — A *flow control orifice* and seat that is an integral part of the *body* or *cage* material or may be constructed from material added to the *body* or *cage*.

4.5 Stem — The rod, shaft or spindle which connects the valve *actuator* with the *closure member*.

4.6 Stem Seals — The part or parts needed to effect a pressure-tight seal around the *stem* while allowing movement of the *stem*.

4.6.1 Packing — A sealing system consisting of deformable material of one or more mating and deformable elements contained in a *packing box* which may have an adjustable compression means to obtain or maintain an effective pressure seal.

4.6.1.1 Packing Box — The chamber, in the *bonnet*, surrounding the *stem* and containing *packing* and other *stem* sealing parts.

4.6.1.2 Packing Follower — A part which transfers mechanical load to the *packing* from the packing flange or nut.

4.6.1.3 Lantern Ring — A rigid spacer assembled in the *packing box* with *packing* normally above and below it and designed to allow lubrication of the *packing* or access to a leak-off connection.

4.6.2 Pressure Energized Stem Seal — A part and/or *packing* material deformable by fluid pressure that bears against the *stem* to make a tight seal.

4.6.3 Bellows Stem Seal — A thin wall, convoluted, flexible member which makes a seal between the *stem* and *bonnet* or *body* and allows *stem* motion while maintaining a positive seal.

4.6.4 Back Seat — A seating surface in the *bonnet* area that mates with the *closure member* or *valve stem* in the extreme open position to provide pressure isolation of the *stem seal*.

4.7 Bushing — A fixed member which supports and/or guides the *closure member*, *valve stem* and/or *actuator stem*. The bushing supports the nonaxial loads on these parts and is subject to relative motion of the parts.

5 LINEAR MOTION CONTROL VALVE TYPES

Types of valves with a *closure member* that moves with a linear motion to modify the rate of flow through the *valve*.

5.1 Globe Valve — A *valve* with a linear motion *closure member*, one or more *ports* and a *body* distinguished by a globular shaped cavity around the *port* region. Typical *globe valve* types are illustrated below. Flow arrows shown indicate a commonly used flow direction.

TWO-WAY BODIES

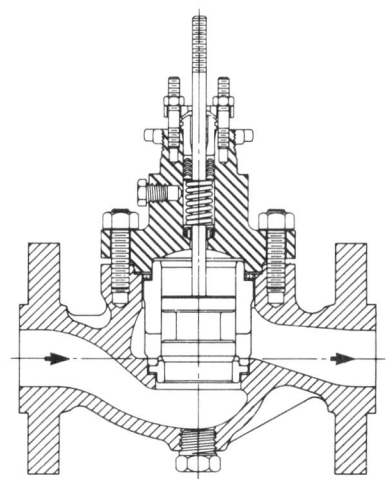

5.1 (a) Cage Guided

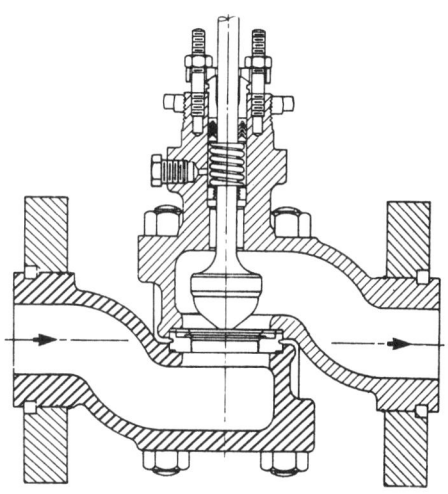

5.1 (b) Split Body Stem Guided

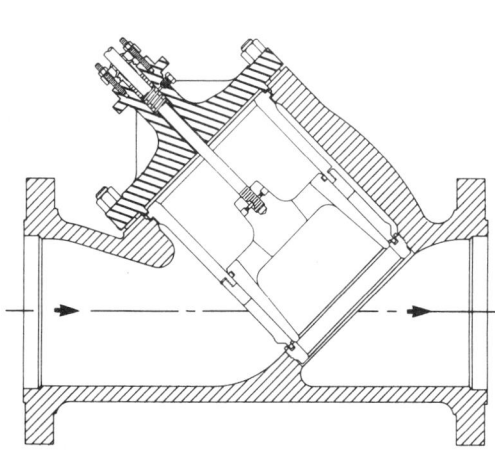

5.1 (c) Y Type Cage Guided

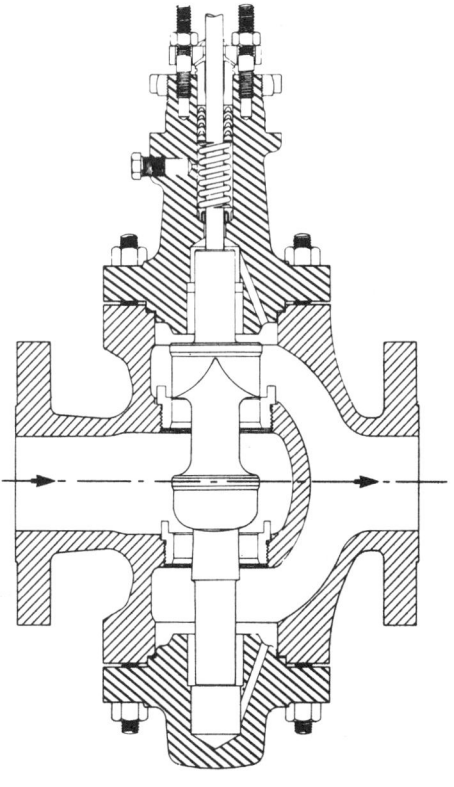

5.1 (d) Double Ported Post (or Top & Bottom) Guided

Globe Body Types (continued)

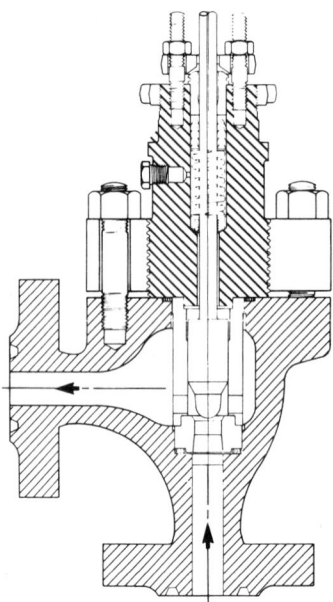

5.1 (e) Angle Body

THREE-WAY BODIES

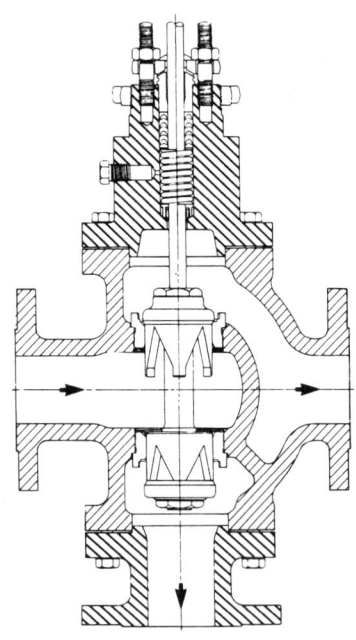 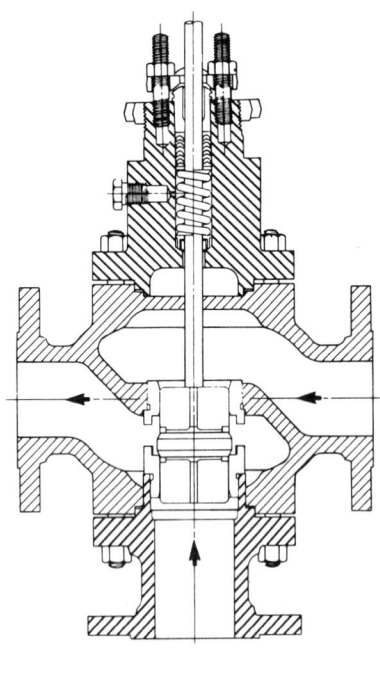 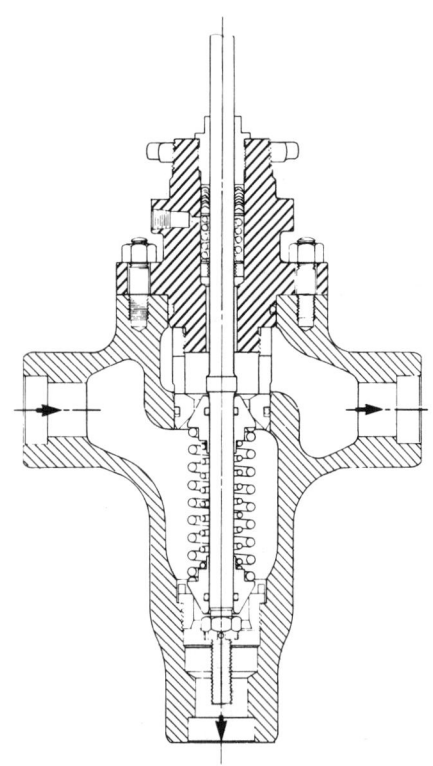

5.1 (f) Diverging   **5.1 (g) Converging**   **5.1 (h) Three Position**

5.1.1 Bonnet (See 4.2)

5.1.2 Bottom Flange — A part which closes a *valve body* opening opposite the *bonnet* opening. It may include a *guide bushing* and/or serve to allow reversal of the *valve* action. In *three-way valves* it may provide the lower flow connection and its *seat*.

5.1.3 Globe Valve Trim — The internal parts of a *valve* which are in flowing contact with the controlled fluid. Examples are the *plug, seat ring, cage, stem* and the parts used to attach the *stem* to the *plug*. The *body, bonnet, bottom flange*, guide means and *gaskets* are not considered as part of the *trim*.

5.1.3.1 Plug (See 4.3.1.4)

5.1.3.2 Cage (See 4.4.2)

5.1.3.3 Stem (See 4.5)

5.1.3.4 Anti-Noise Trim — A combination of *plug* and *seat ring* or *plug* and *cage* that by its geometry reduces the noise generated by fluid flowing through the *valve*.

5.1.3.5 Anti-Cavitation Trim — A combination of *plug* and *seat ring* or *plug* and *cage* that by its geometry permits non-cavitating operation or reduces the tendency to cavitate, thereby minimizing damage to the *valve* parts, and the downstream piping.

5.1.3.6 Balanced Trim — An arrangement of *ports* and *plug* or combination of *plug, cage, seals* and *ports* that tends to equalize the pressure above and below the *valve plug* to minimize the net static and dynamic fluid flow forces acting along the axis of the *stem* of a *globe valve*.

5.1.3.7 Erosion Resistant Trim — *Valve trim* which has been faced with very hard material or manufactured from very hard material to resist the erosive effects of the controlled fluid flow. See Appendix A.

5.1.3.8 Soft Seated Trim — *Globe valve trim* with an elastomeric, plastic or other readily deformable material used either in the *valve plug* or *seat ring* to provide tight shutoff with minimal *actuator* forces. See ANSI B16.104 for leakage classifications.

5.1.3.9 Seat Ring (See 4.4.1)

5.1.4 Globe Valve Plug Guides — The means by which the *plug* is aligned with the *seat* and held stable throughout its travel. The *guide* is held rigidly in the *body* or *bonnet*.

5.1.4.1 Stem Guide — A *guide bushing* closely fitted to the *valve stem* and aligned with the *seat*.

5.1.4.2 Post Guide — *Guide bushing* or *bushings* fitted to posts or extensions larger than the *valve stem* and aligned with the *seat*.

5.1.4.3 Cage Guide — A *valve plug* fitted to the inside diameter of the *cage* to align the *plug* with the *seat*.

5.1.4.4 Port Guide — A *valve plug* with wings or a skirt fitted to the *seat ring* bore.

5.2 Gate Valve — A *valve* with a linear motion *closure member* that is a flat or wedge shaped gate which may be moved in or out of the flow stream. It has a straight-through flow path.

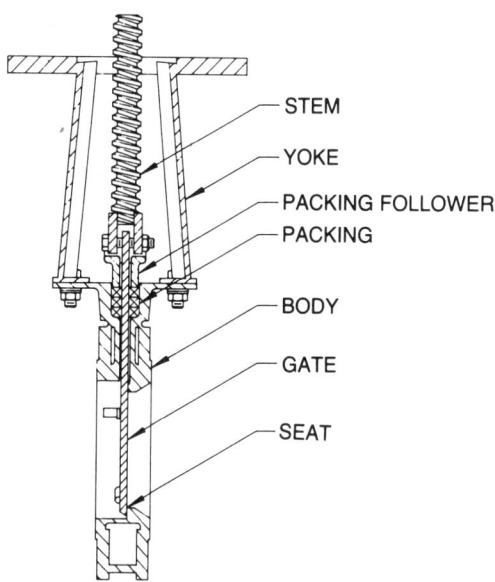

5.2.1 Types

5.2.1.1 Bonnetted — *Gate valve* having a *bonnet* which encloses the *gate* within the pressure boundary when in the open position. *Packing* is provided at the *stem*.

5.2.1.2 Bonnetless — *Gate valve* which has *packing* between the *gate* and *body*, such that the *gate* extends outside the pressure boundary in the open position.

5.2.2 Body (See 4.1)

5.2.2.1 Single Flange (Lugged) — A thin annular section *body* whose end surfaces mount between the pipeline flanges, or may be attached to the end of a pipeline without any additional flange or retaining parts, using either thru bolting and/or tapped holes.

5.2.2.2 Flanged Body — *Valve body* with full flanged *end connections*.

5.2.3 Bonnet (See 4.2)

5.2.4 Flow Control Orifice — (See 4.4)

5.2.4.1 Vee Orifice — "V" shaped *flow control orifice* which allows a characterized flow control as the gate moves in relation to the fixed Vee opening.

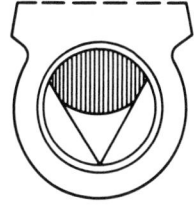

5.2.4.2 Multiple Orifice — A *flow control orifice* consisting of a moving member *(gate)* which slides reciprocally against a stationary member *(plate)*. Both elements contain several matching orifices and the flow area is changed as the *gate* slides.

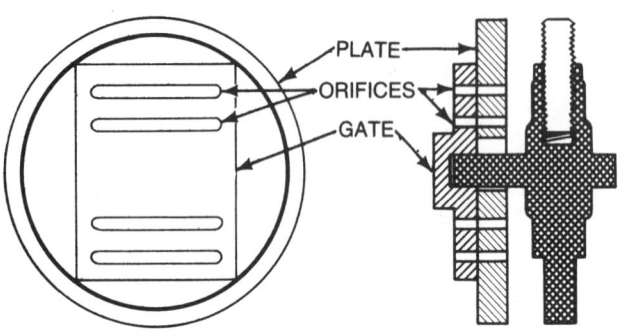

5.2.5 Stem — (See 4.5)

5.2.6 Packing Follower — (See 4.6.1.2)

5.2.7 Packing — (See 4.6.1)

5.2.8 Seat Joint — (See Glossary)

5.3 Diaphragm Valve — A *valve* with a flexible linear motion *closure member* that is forced into the internal flow passageway of the *body* by the *actuator*.

5.3.2 Adaptor Bushing — The part which attaches a close coupled diaphragm *actuator* to the *bonnet* of the *diaphragm valve body*.

5.3.3 Valve Diaphragm — A flexible member which is moved into the fluid flow passageway of the *body* to modify the rate of flow through the *valve*.

5.3.4 Compressor — A device which the *valve stem* forces against the backside of the diaphragm to cause the diaphragm to move toward and seal against the internal flow passageway of the *valve body*.

5.3.5 Finger Plate — A plate used to restrict the upward motion of the diaphragm and prevent diaphragm extrusion into the *bonnet* cavity in the full open position.

5.4 Pinch or Clamp Valve — A *valve* consisting of a flexible elastomeric tubular member connected to two rigid flow path ends whereby modulation and/or shut off of flow is accomplished by squeezing the flexible member into eventual tight sealing contact. The flexible member may or may not be reinforced. The flexible member may or may not be surrounded by a pressure retaining boundary consisting of a metal housing with *stem packing box*. Squeezing of the flexible member may be accomplished by: 1) single *stem* and leverage acting from both sides so that the total collapse and sealing occurs along the horizontal center line of the flexible member; 2) double *stem* action involving two separate *actuator* assemblies diametrically opposed, or 3) a separate source of fluid pressure applied to an annulus surrounding the flexible member. A *clamp valve* is a *pinch valve* but with clamps and shaped inserts used to provide stress relief in the creased area of the tubular member.

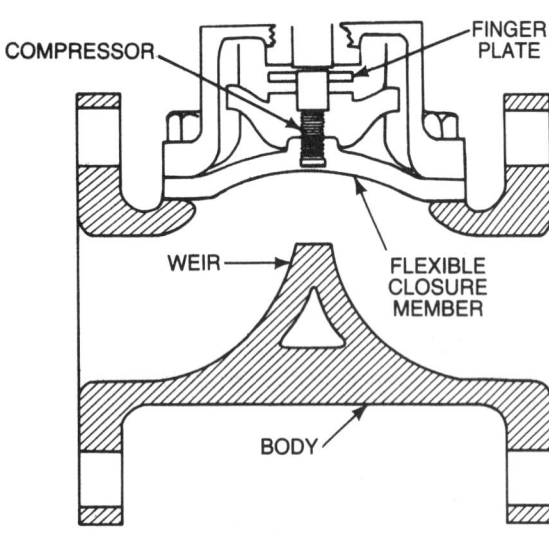

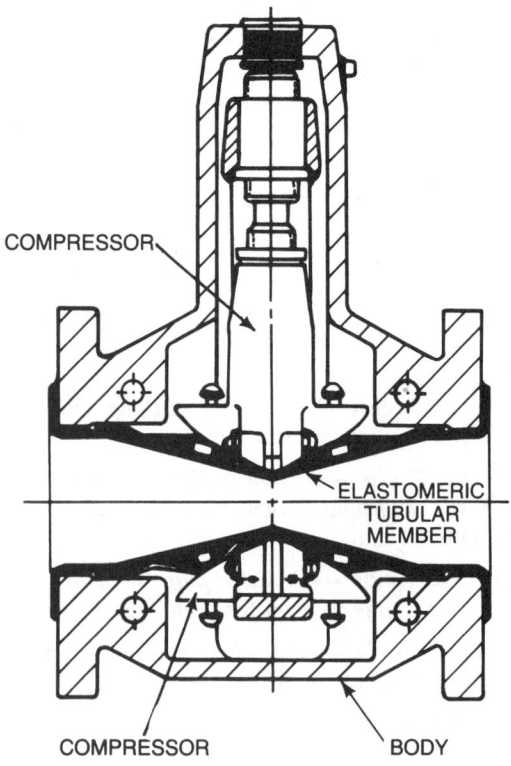

5.3.1 Body — (See 4.1).

5.3.1.1 Weir Type — A *body* having a raised contour contacted by a diaphragm to shut off fluid flow.

6 ROTARY MOTION CONTROL VALVE TYPES

Types of *valves* with a *closure member* that moves with a rotary motion to modify the rate of flow through the *valve*.

6.1 Ball Valve — A *valve* which modifies flow rates with rotary motion of the *closure member*, which is either a sphere with an internal passage or a segment of a spherical surface.

6.1.1 Body — (See 4.1) General body constructions are illustrated below:

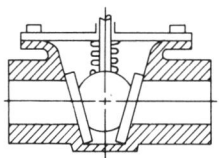 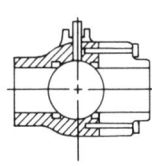 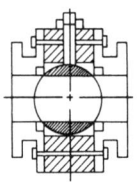

(a) One Piece   (b) Two Piece   (c) Three Piece

6.1.2 Typical Ball Types

6.1.2.1 Segmented Ball — A *closure member* that is a segment of a spherical surface which may have one edge contoured to yield a desired *flow characteristic*.

(a) Segmented Ball with Contoured Edge   (b) Segmented Ball with "V" Edge

6.1.2.2 Full Ball — A *closure member* that is a complete spherical surface with a flow passage through it. The flow passage may be round, contoured or otherwise modified to yield a desired *flow characteristic*.

6.1.2.3 Three-Way Ball — A *closure member* that is a spherical surface with one or more flow passages through it. The passages may be round, contoured or otherwise modified to yield a desired *flow characteristic*.

 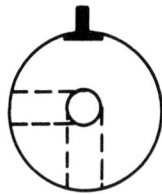

6.1.2.4 Floating Ball — A full ball positioned within the valve that contacts either of two *seat rings* and is free to move toward the *seat ring* opposite the pressure source when in the closed position to effect tight shutoff.

6.1.3 Seat Ring — (See 4.4.1)

6.1.3.1 Downstream Seating — Seating is accomplished by pressure differential thrust across the ball in the closed position, moving the ball slightly downstream into tighter contact with the *seat ring* seal which is supported by the *body*.

6.1.3.2 Upstream Seating — A *seat* on the upstream side of the ball, designed so that the pressure of the controlled fluid causes the *seat* to move toward the ball.

6.1.3.3 Spring Loaded Seat — A *seat* design that utilizes a mechanical means, such as a spring, to exert a greater force at the point of ball contact to improve the sealing characteristics, particularly at low pressure differential. The spring action may be accomplished by a metal spring arrangement or a compressed elastomer.

6.1.4 Stem — (See 4.5)

6.1.4.1 Loose Stem — A design in which the *stem* is not physically or mechanically attached to the ball, but drives the ball through intimate contact of surfaces. Typical Loose *Stem* Drives are:
 a) Tang
 b) Pin
 c) Splined

6.1.4.2 Integral Stem — A design in which the *stem* is either physically a part of the ball or mechanically made part of the ball. Some integral *stems* are designed to perform a turning and then lifting action.

6.1.4.3 Trunnion — Extensions of the ball used to locate, support and turn the ball within the *valve body*. May be integral or attached to the ball.

6.1.5 Stem Seal — (See 4.6)

6.2 Butterfly Valve — A *valve* with a circular body and a rotary motion disk *closure member*, pivotally supported by its *stem*.

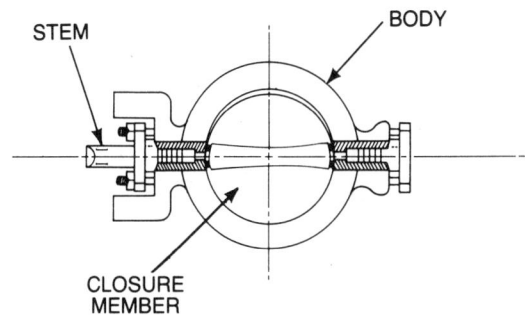

6.2.1 Body Types

6.2.1.1 Wafer Body — A *body* whose end surfaces mate with the pipeline flanges. It is located and clamped between the piping flanges by long bolts extending from flange to flange. A *wafer body* is also called a *flangeless body*.

6.2.1.2 Split Body — A *body* divided in half by a plane containing the longitudinal flow path axis.

6.2.1.3 Unlined Body — A *body* without a lining.

6.2.1.4 Lined Body — A *body* having a lining which makes an interference fit with the *disk* in the closed position thus establishing a seal. See Appendix B.

6.2.1.5 Single Flanged Body (Lugged) — (See 5.2.2.1)

6.2.2 Typical Disk Orientations

6.2.3 Typical Disk Shapes

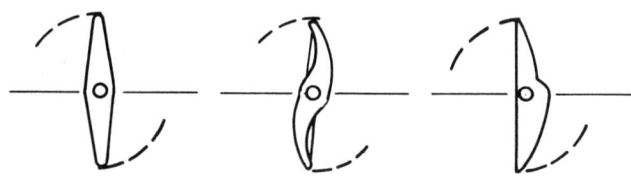

(a) Flat   (b) Cambered   (c) Nonsymmetrical Edge

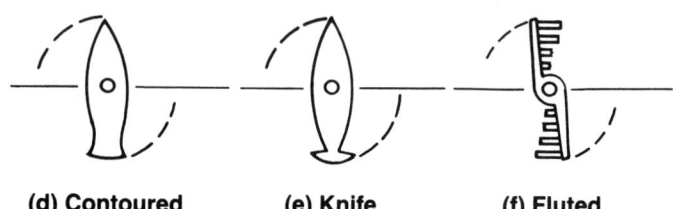

(d) Contoured   (e) Knife   (f) Fluted

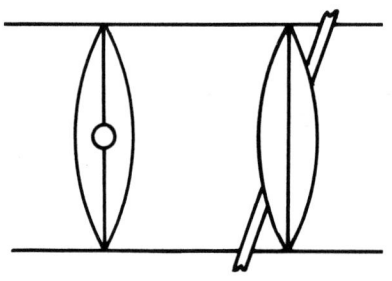

(a) Aligned   (b) Aligned with Canted Stem

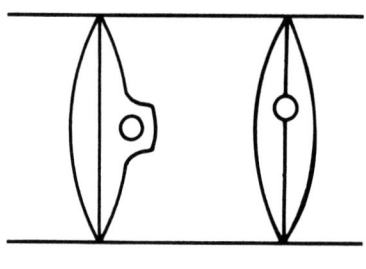

(c) Offset   (d) Cammed

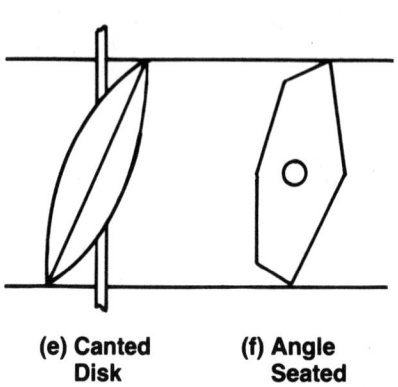

(e) Canted Disk   (f) Angle Seated

6.2.4 Seal on Disk — A seal ring located in a groove in the *disk* circumference. The *body* is unlined in this case.

6.2.5 Stem — (See 4.5)

6.2.6 Stem Bearings — Butterfly *stem* bearings are referred to as either the outboard or the inboard type, depending on their location, outside or inside of the *stem seals*.

6.2.7 Stem Seal — (See 4.6)

6.3 Plug Valve — A *valve* with a *closure member* that may be cylindrical, conical or a spherical segment in shape. It is positioned, open to closed, with rotary motion.

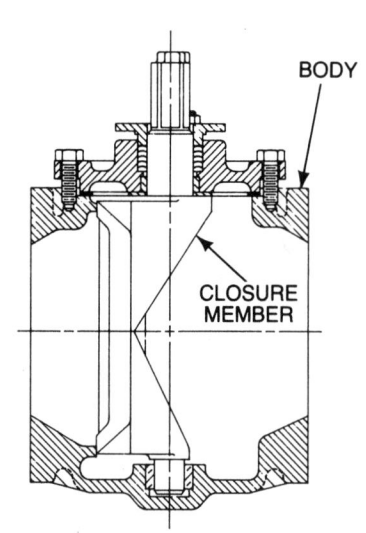

18

6.3.1 Body — (See 4.1)

6.3.2 Plug Configurations

(a) Cylindrical: Plug is cylindrical, with a flow passage through it, or is a partial cylinder.

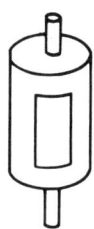

(b) Tapered: Plug is tapered and may be lifted from seating surface before rotating to close or open.

(c) Eccentric: Plug face is not concentric with plug *stem* centerline and moves into *seat* when closing.

(d) Eccentric Spherical Disk: *Disk* is spherical segment, not concentric with the *disk stem*.

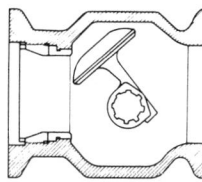

(e) Characterized Plug: Plug with contoured face, such as the "vee" plug shown, to provide various flow characteristics.

6.3.3 Characterized Sleeve — A part added to a *plug valve* to provide various flow characteristics.

6.4 Disk Valve — A *valve* with a *closure member* that consists of a *disk* which moves with a rotary motion against a stationary *disk*, each *disk* having flow passages through it.

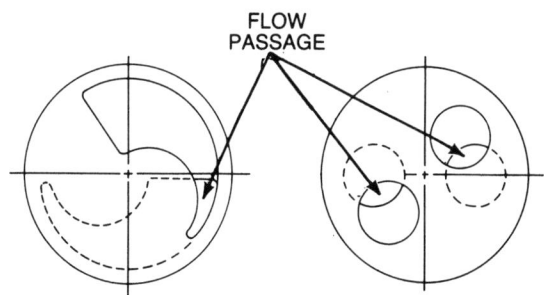

7 CONTROL VALVE ACTUATORS

An *actuator* consists of the complete assembly of parts required to operate a specific *valve* (See 2.1.2).

7.1 Power Unit — The portion of the *actuator* which converts fluid, electrical or mechanical energy into *stem* motion to develop thrust or torque.

7.1.1 Pneumatic — A device which converts the energy of a compressible fluid, usually air, into motion.

7.1.1.1 Diaphragm Type — A fluid powered device in which the fluid acts upon a flexible member, the diaphragm, to provide linear motion to the *actuator stem*.

7.1.1.2 Piston Type — A fluid powered device in which the fluid acts upon a movable cylindrical member, the piston, to provide linear motion to the *actuator stem*.

SINGLE ACTING DIAPHRAGM ACTUATORS

SINGLE ACTING PISTON ACTUATORS

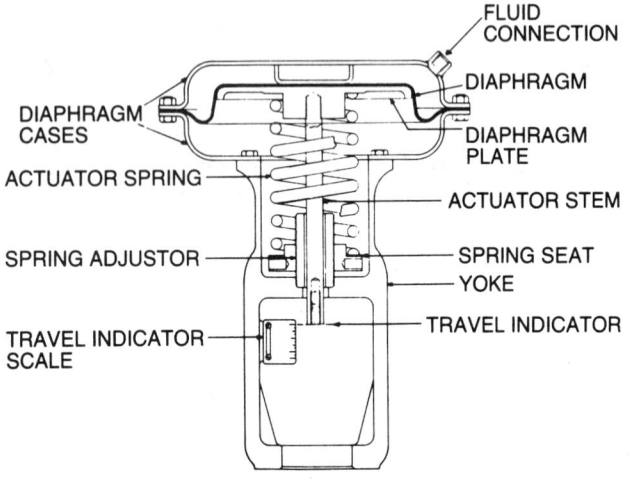

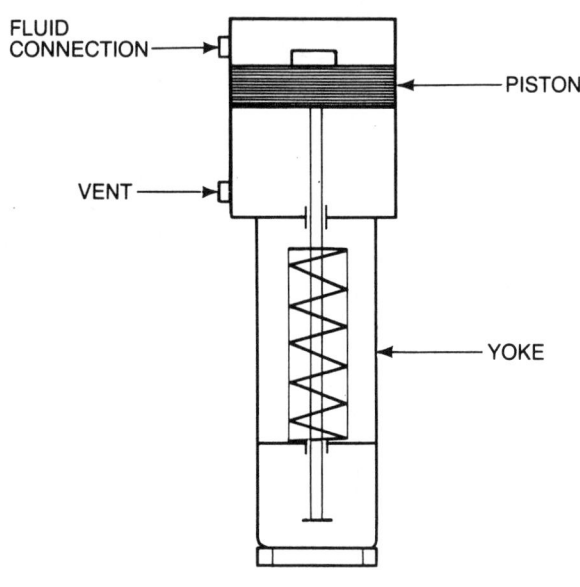

(a) Fluid-to-Extend Stem

(a) Fluid-to-Extend Stem

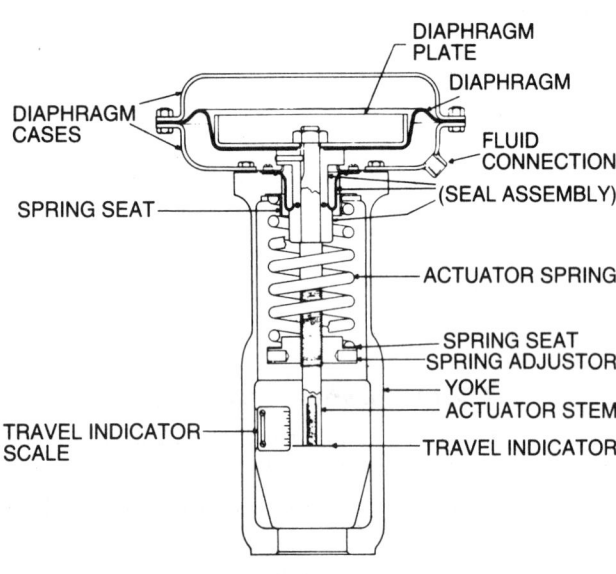

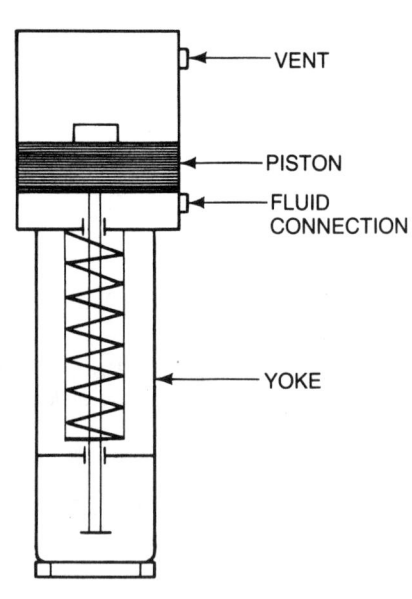

(b) Fluid-to-Retract Stem

(b) Fluid-to-Retract Stem

7.1.1.3 Vane Type — A fluid powered device in which fluid acts upon a movable pivoted member, the vane, to provide rotary motion to the *actuator stem*.

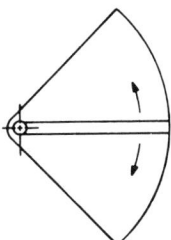

(a) Single Vane

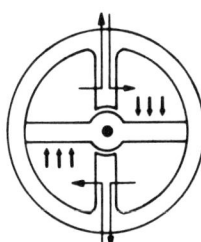

(b) Multiple Vane

(c) Rolling Diaphragm Vane

7.1.1.4 Bellows Type — A fluid powered device in which the fluid acts upon a flexible convoluted member, the bellows, to provide linear motion to the *actuator stem*.

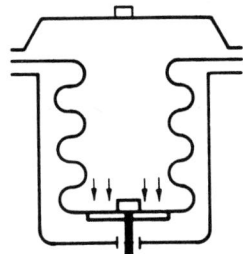

7.1.1.5 Fluid Motor Type — A fluid powered device which uses a rotary motor to position the *actuator stem*.

7.1.2 Electric — A device which converts electrical energy into motion.

7.1.2.1 Electro Mechanical Type — A device which uses an electrically operated motor driven gear train or screw to position the *actuator stem*. Such *actuators* may operate in response to either analog or digital electrical signals. The electro-mechanical actuator is also referred to as a motor gear train *actuator*.

7.1.2.2 Electro Hydraulic Type — A self contained device which responds to an electrical signal, positioning an electrically operated hydraulic pilot valve to allow pressurized hydraulic fluid to move an actuating *piston*, *bellows*, *diaphragm* or *fluid motor* to position a *valve stem*.

7.1.3 Hydraulic — A fluid powered device which converts the energy of an incompressible fluid into motion.

7.2 Yoke — The structure which rigidly connects the *actuator power unit* to the *valve*.

7.3 Stem — The part, usually a rod or shaft, which connects to the *valve stem* and transmits motion (force) from the *actuator* to the *valve*. The *actuator stem* delivering an output thrust may or may not be the same *stem* as that on the power unit *stem*.

7.4 Stem Connector — The device which connects the *actuator stem* to the *valve stem*.

7.5 Position Indicator — The device, such as a pointer and scale, which indicates the position of the *closure member*.

7.6 Motion Conversion Mechanism — (See 2.1.3). Device needed on some, but not all, assemblies to convert linear action to rotary *valve* operation.

7.6.1 Diaphragm Type Power Unit

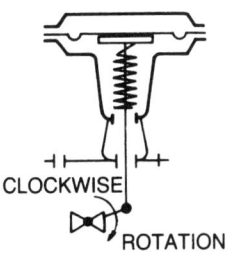   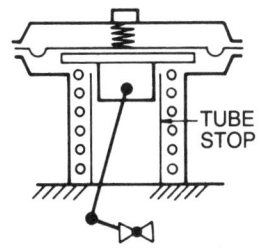

(a) Double Linkage   (b) Rocking Stem

7.6.2 Piston Type Power Unit

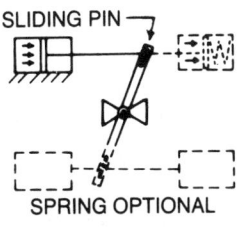   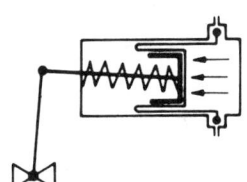

(a) Scotch Yoke   (b) Rocking Stem

7.6.2 Piston Type Power Unit (continued)

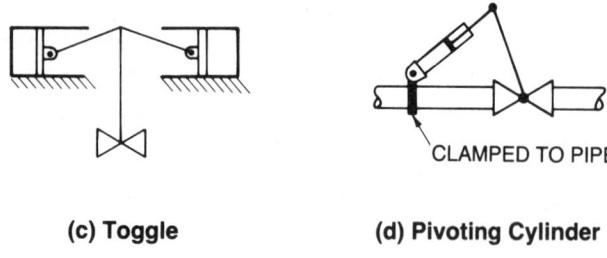

(c) Toggle   (d) Pivoting Cylinder

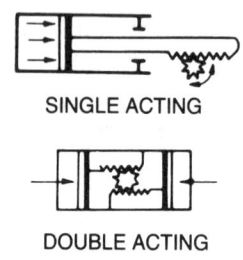

(e) Rack and Pinion

7.7 Action

7.7.1 Single Acting — An *actuator* in which the power supply acts in only one direction. In a spring and diaphragm *actuator*, for example, the spring acts in a direction opposite to the diaphragm thrust. Single acting spring and *diaphragm actuators* may be further classified as to direction of *stem* movement on increasing fluid pressure: a) air to extend *actuator stem*, b) air to retract *actuator stem*.

7.7.2 Double Acting — An *actuator* in which the power supply acts both to extend and retract the *actuator stem*.

8 AUXILIARY EQUIPMENT

8.1 Attached Equipment — The auxiliary equipment which must be located on the *valve* or *actuator*.

8.1.1 Manual Override — A device to manually impart motion in either one or two directions to the *valve stem*. It may be used as a limit stop. (Also see "Handwheel" in glossary.)

8.1.2 Snubber — A device which is used to damp the motion of the *valve stem*. This is usually accomplished by an oil filled cylinder/piston assembly. The *valve stem* is attached to the piston and the flow of hydraulic fluid from one side of the piston to the other is restricted.

8.1.3 Positioner — A position controller, which is mechanically connected to a moving part of a final control element or its *actuator,* and automatically adjusts its output pressure to the *actuator* in order to maintain a desired position that bears a predetermined relationship to the input signal. The *positioner* can be used to modify the action of the *valve* (reversing *positioner*), extend the stroke/controller signal (split range *positioner*), increase the pressure to the *valve actuator* (amplifying *positioner*) or modify the *control valve flow characteristic* (characterized *positioner*).

8.1.3.1 Single Acting Positioner — A *positioner* is single acting if it has a single output.

8.1.3.2 Double Acting Positioner — A *positioner* is double acting if it has two outputs, one with "direct" action and the other with "reversed" action.

8.1.3.3 Positioner Types — *Positioners* characterized by their input and output are available as:

  (a) pneumatic/pneumatic
  (b) electric/pneumatic
  (c) electric/hydraulic
  (d) electric/electric

8.1.4 Mechanical Limit Stop — A mechanical device to limit the *valve stem* travel.

8.1.5 Stem Anti-Rotation Device — A mechanical means of preventing rotation of the linear *actuator stem* and/or *valve stem*.

8.1.6 Position Transmitter — The position transmitter is a device that is mechanically connected to the *valve stem* and generates and transmits a pneumatic or electrical signal representing the *valve stem* position.

8.1.7 Position Switch — A position switch is a pneumatic, hydraulic or electrical device which is linked to the *valve stem* to detect a single, preset *valve stem* position.

8.2 Adjacent Equipment — The auxiliary equipment which may be located adjacent to the *valve* or *actuator*.

8.2.1 Air Set — A device which is used to control the supply air pressure to the *valve actuator* and its auxiliaries.

8.2.2 Signal Booster Relay — A pneumatic relay that is used to reduce the time lag in pneumatic circuits by reproducing pneumatic signals with high volume and/or high pressure output. These relays may be either volume boosters, amplifying or a combination of both.

8.2.3 Transducer — A device to convert one form of signal to another.

8.2.4 Control Signal Override Device — A device which overrides the effect of the control signal to the *valve actuator* to cause the *closure member* to remain stationary or assume a pre-selected position.

9 GLOSSARY

Accuracy — The degree of conformity of an indicated value to a recognized accepted standard value or ideal value. (Source: ISA S51.1)

Action, Air-To-Close — See *"Fail-Open"*.

Action, Air-To-Open — See *"Fail-Close"*.

Actuator Effective Area — The net area of *piston*, the *bellows*, *vane* or *diaphragm* acted on by fluid pressure to generate *actuator* output thrust. It may vary with relative *stroke* position depending upon the *actuator* design.

Actuator Environment — The temperature, pressure, humidity, radioactivity and corrosiveness of the atmosphere surrounding the *actuator*. Also, the mechanical and seismic vibration transmitted to the *actuator* through the piping or heat radiated toward the *actuator* from the *valve body*.

Actuator Travel Time — See *"Stroke Time"*.

Back Face — The machined surface on the side of a through-bolted flange, opposite the gasket face, that is provided for nut seating.

Back Seat — (See 4.6.4)

Bench Set — The shop calibration of the *actuator* spring range of a *control valve*, to account for the in service process forces.

Body Cavity — The internal chamber of the *valve body* including the *bonnet* zone and excluding the *body* ends.

Boss — A localized projection on a *valve* surface provided for various purposes, such as attachment of drain connections, or other accessories.

Bubble Tight — A nonstandard term. Refer to ANSI B16.104 for specification of leakage classifications.

Capacity — The rate of flow through a *valve* under stated test conditions.

Cage — (See 4.4.2.)

Cavitation — A two-stage phenomenon of liquid flow. The first stage is the formation of voids or cavities within the liquid system; the second stage is the collapse or implosion of these cavities back into an all-liquid state.

Center-To-End Dimension — The distance from the center line of a *valve body* to the extreme plane of a specific end connection. See *"Face-to-Face Dimension"* and *"End-to-End Dimension"*.

Characteristic, Equal Percentage — The *inherent flow characteristic* which, for equal increments of *rated travel*, will ideally give equal percentage changes of the existing *flow coefficient* (C_v).

Characteristic, Flow — Indefinite term, see *Characteristic, Inherent Flow* and *Characteristic, Installed Flow*.

Characteristic, Inherent Flow — The relationship between the flow rate through a *valve* and the travel of the *closure member* as the *closure member* is moved from the closed position to *rated travel* with constant pressure drop across the *valve*.

Characteristic, Installed Flow — The relationship between the flow rate through a *valve* and the travel of the *closure member* as the *closure member* is moved from the closed position to *rated travel* when the pressure drop across the *valve* varies as influenced by the system in which the *valve* is installed.

Characteristic, Linear Flow — An *inherent flow characteristic* which can be represented by a straight line on a rectangular plot of *flow coefficient* (C_v) versus per cent *rated travel*. Therefore, equal increments of travel provide equal increments of *flow coefficient* (C_v) at constant pressure drop.

Characteristic, Modified Parabolic Flow — An *inherent flow characteristic* which provides fine throttling action at low *valve plug* travel and approximately a *linear characteristic* for upper portions of *valve* travel. It is approximately midway between *linear* and *equal percentage*.

Characteristic, Quick Opening Flow — An *inherent flow characteristic* in which there is a maximum flow with minimum travel.

Clearance Flow — That flow below the minimum controllable flow with the *closure member* not seated.

Coefficient, Flow — A constant (C_v), related to the geometry of a *valve*, for a given *valve* opening, that can be used to predict flow rate. See ANSI/ISA S75.01 "Control Valve Sizing Equations" and ANSI/ISA S75.02 "Control Valve Capacity Test Procedure".

Coefficient, Rated Flow — The *flow coefficient* (C_v) of the *valve* at *rated travel*.

Coefficient, Relative Flow — The ratio of the *flow coefficient* (C_v) at a stated travel to the *flow coefficient* (C_v) at *rated travel*.

Coefficient, Valve Recovery — See *"Liquid Pressure Recovery Factor"*.

Cold Working Pressure — The maximum pressure rating of a *valve* or fitting coincident with ambient temperature, generally the range from −20F to +100F (−29C to +38C).

Common Port — The *port* of a *three-way valve* that connects to the other two flow paths.

Control Valve Gain — The change in the flow rate as a function of the change in *valve* travel. It is the slope of the *installed* or of the *inherent valve flow characteristic* curve and must be designated as installed or inherent.

Cycling Life — The specified minimum number of full scale excursions or specified partial range excursions over which a *control valve* will operate as specified without changing its performance beyond specified tolerance.

Dashpot — A mechanical damping device consisting of a cylinder and piston apparatus arranged so as to dampen the movement of a *valve stem*. A less preferred term. See *"Snubber"* (8.1.2).

Data Plate — A plate bearing the name of the manufacturer and other information related to the product as may be required by various regulations or codes. See also *"Nameplate"*.

Dead Band — The range through which an input can be varied without initiating an observable response.

Dead End Shut Off — A nonstandard term. Refer to ANSI B16.104 for specification of leakage classifications.

Drift — A change in output over a period of time with constant input.

Drip Tight — A nonstandard term. Refer to ANSI B16.104 for specification of leakage classifications.

Drop Tight — A nonstandard term. Refer to ANSI B16.104 for specification of leakage classifications.

Dual Sealing Valve — A *valve* which uses a resilient seating material for the primary seal and a metal-to-metal *seat* for a secondary seal.

End Connection — The configuration provided to make a pressure tight joint to the pipe carrying the fluid to be controlled.

End-To-End Dimension — See *"Face-to-Face Dimension"* and *"Center-to-End Dimension"*.

Face-To-Face Dimension — The dimension from the face of the inlet opening to the face of the outlet opening of a *valve* or fitting. See *"End-to-End Dimension"* and *"Center-to-End Dimension"*. See ANSI/ISA S75.03 "Uniform Face-to-Face Dimensions for Flanged Globe Style Control Valve Bodies" and ANSI/ISA S75.04 "Face-to-Face Dimensions of Flangeless Control Valves".

Facing, Flange — The finish on the end connection gasket surfaces of *flanged* or *flangeless valves*.

Fail-Close — A condition wherein the *valve closure member* moves to a closed position when the actuating energy source fails. See *"Normally Closed"*.

Fail-Open — A condition wherein the *valve closure member* moves to an open position when the actuating energy source fails. See *"Normally Open"*.

Fail-Safe — A characteristic of a particular *valve* and its *actuator*, which upon loss of actuating energy supply, will cause a *valve closure member* to fully close, fully open or remain in fixed position. Fail-safe action may involve the use of auxiliary controls connected to the *actuator*.

Flanged Ends — *Valve* end connections incorporating flanges which allow pressure seals by mating with corresponding flanges on the piping.

Flangeless Control Valve — A *valve* without integral line flanges, which is installed by bolting between companion flanges, with a set of bolts, or studs, generally extending through the companion flanges.

Full Face Gasket — A flat gasket which contacts the entire flat contact surface of two mating flanges, extending past the bolt holes. This term applies to flat face flanges only.

Gland — See *"Packing Follower"* (4.6.1.2) or *"Lantern Ring"* (4.6.1.3).

Grease Seal Ring — A nonstandard term. See *"Lantern Ring"* (4.6.1.3).

Hand Jack — A lesser used term. See *"Handwheel"*.

Handwheel — A manual override device to stroke a *valve* or limit its *travel*.

Types of Handwheels

Top Mounted — The *handwheel* is mounted on top of the *valve actuator* case. This type of *handwheel* does not have a clutch and is usually used to restrict the motion of the *valve stem* in one direction only.

Side-Mounted — Bellcrank lever types are externally mounted on the *control valve yoke*. They can provide a limit to the extent a *valve stem* will travel in either direction, but not in both directions.

In-Yoke Mounted — In-yoke gear types are designed with a worm gear drive which is contained in a lubricated housing. The gear box is integral with the *yoke* which is usually elongated to provide space for the worm gear assembly. With this type of *handwheel*, stops may be set in either or both directions to limit the travel of the *valve stem*. This type of *handwheel* is declutchable.

Shaft Mounted, Declutchable — A shaft mounted worm gear drive that can be declutched from the power *actuator*.

Hysteresis — The maximum difference in output value for any single input valve during a calibration cycle, excluding errors due to dead band. See ISA S51.1 "Process Instrumentation Terminology".

Identification Plate — See *"Data Plate"*.

Inlet — The *body* end opening through which fluid enters the *valve*.

In-Line Valve — A *valve* having a piston actuated *closure member* shaped like a *globe valve plug* which moves to seat axially in the direction of the flow path. In-Line valves are normally operated by a fluid energy source but may be operated mechanically.

Jacketed Valves — A *valve body* cast with a double wall or provided with a double wall by welding material around the body so as to form a passage for a heating or cooling medium. Also refers to *valves* which are enclosed in split metal jackets having internal heat passageways or electric heaters. Also referred to as "Steam Jacketed" or "Vacuum Jacketed". In a vacuum jacketed *valve*, a vacuum is created in the space between the *body* and secondary outer wall to reduce the transfer of heat by convection from the atmosphere to the internal process fluid, usually cryogenic.

Lapped-In — Mating contact surfaces that have been refined by grinding and/or polishing together or separately in appropriate fixtures.

Leakage — The quantity of fluid passing through a *valve* when the *valve* is in the fully closed position under stated closure forces, with the pressure differential and temperature as specified. Leakage is usually expressed as a percentage of the *valve* capacity at full *rated travel*. Refer to ANSI B16.104 for specification of leakage quantity.

Leak-Off Gland — A *packing box* with *packing* above and below the *lantern ring* so as to provide a sealed low pressure leak collection point for fluid leaking past the primary seal (lower packing).

Lens Joint Ends — *Valves* with the ends prepared for lens ring gaskets.

Lift — A nonstandard term. See *"Travel"*.

Limit Switch — See *"Position Switch"* (8.1.7)

Linearity — The closeness to which a curve approximates a straight line. See ISA S51.1 "Process Instrumentation Terminology".

Lined Valve Body — A *valve body* to which a protective coating or liner has been applied to internal surfaces of pressure containing parts or to the surfaces exposed to the fluid.

Liquid Pressure Recovery Factor — The ratio (F_l) of the *valve flow coefficient* (C_v) based on the pressure drop at the vena contracta, to the usual *valve flow coefficient* (C_v) which is based on the overall pressure drop across the *valve* in non-vaporizing liquid service. These coefficients compare with the orifice metering coefficients of discharge for vena contracta taps and pipe taps, respectively. See ANSI/ISA-S75.01 "Control Valve Sizing Equations."

Lubricant Ring — A nonstandard term. See *"Lantern Ring"* (4.6.1.3.)

Lubricated Packing Box — A *packing* arrangement consisting of a *lantern ring* with *packing* rings above and below with provision to lubricate the *packing*.

Lubricator Isolating Valve — In a *control valve*, an isolating *valve* is a small hand operated *valve* located between the *packing* lubricator assembly and the *packing box* assembly. It shuts off the fluid pressure from the lubricator assembly.

Modulating — The actions to keep a quantity or quality in proper measure or proportion. Also see *"Throttling"*.

Mounting Position — The location and orientation of an *actuator* or auxiliary component relative to the *control valve*. This can apply to the *control valve* itself relative to the piping.

Nameplate — A plate attached to a *control valve* bearing the name of the manufacturer. It may also contain specification and limitation information. See also *"Data Plate"*.

Needle Point Valve — A type of *valve* having a needle point plug.

Noise — *Control Valve* noise can be caused by:
  (1) Turbulent flow of liquid.
  (2) Aerodynamic flow.
  (3) Liquid cavitation flow.
  (4) Mechanical vibration.

Normally Closed Valve — A *valve* with means provided to move to and/or hold in its closed position without *actuator* energy supply. See *"Fail-Close"*.

Normally-Open Valve — A *valve* with means provided to move to and/or hold in its wide-open position without *actuator* energy supply. See *"Fail-Open"*.

Port — The *flow control orifice* of a *control valve*. It is also used to refer to the inlet or outlet openings of a *valve*.

Precision — A nonstandard term. See *"Repeatability"*.

Purged Packing Box — A packing arrangement consisting of a *lantern ring* inside the packing rings to permit introduction of a purge fluid to continually flush the space between the *stem* and *body*. It is usually used to purge, admit cooling fluid or detect *stem seal* leakage.

Rangeability, Inherent — The ratio of the largest *flow coefficient* (C_v) to the smallest *flow coefficient* (C_v) within which the deviation from the specified *inherent flow characteristic* does not exceed the stated limits.

Rated Travel — The amount of movement of the *valve closure member* from the closed position to the rated full open position.

Relative Travel — The ratio of the travel at a given opening to the *rated travel*.

Repeatability — The closeness of agreement among a number of consecutive measurements of the output for the same value of input under the same operating conditions, approaching from the same direction, for full range traverse. It does not include *hysteresis*. (Source: ISA S51.1).

Reproducibility — The closeness of agreement among repeated measurements of the output for the same value of input made under the same operating conditions over a period of time, approaching from both directions. It includes *hysteresis, dead band, drift* and *repeatability*. See ISA S51.1 "Process Instrumentation Terminology".

Resolution — The least interval between two adjacent discrete details which can be distinguished one from the other. See ISA S51.1 "Process Instrumentation Terminology".

Reversible Seat — Refers to a *seat ring* with *seating surfaces* on both sides such that when one surface has worn, the ring may be reversed to present a new surface to contact the *closure member*.

Seat Angle — The angle between the axis of the *seat orifice* and the *seating surface*. A flat seated *valve* has a seat angle of 90°. The *seat angle* of the *closure member* and *seat* may differ slightly to provide line contact.

Seat Joint — The area of contact between the *closure member* and the *valve seat* which establishes the sealing action.

Seat Load — The total net contact force between the *closure member* and *seat* with stated static conditions.

Sensitivity — The ratio of change in output magnitude to the change of the input which causes it after the steady-state has been reached. See ISA S51.1 "Process Instrumentation Terminology".

Shaft — (See 4.5 and 7.3)

Split Clamp Ends — *Valve* end connections of various proprietary designs using split clamps to apply gasket loading.

Spot Face — A machined annular surface around a bolt hole on the side of a through bolted flange, opposite the gasket face, that is provided for nut seating.

Spring Rate — The force change per unit change in length. This is usually expressed as pounds per inch or Newtons per millimeter.

Stem — (See 4.5 and 7.3)

Stem Boot — A protective device similar to a flexible bellows, used outside the *bonnet* to protect the *valve stem* from the surrounding atmosphere.

Stroke — See "*Travel*".

Stroke Cycle — *Travel* of the *closure member* from its closed position to the *rated travel* opening and return to the closed position.

Stroke Time — The time required for one-half a *stroke cycle* at specified conditions.

Swell Plug — Consists of a *piston actuator* coaxial with the flow-path axis and suspended from the *body* wall by radial fins. The piston compresses an annular elastomer member which forces it to close the annulus between the piston outside diameter and the *body* internal diameter.

Threaded Ends — *Valve* end connections incorporating threads, either male or female.

Three-Way Valve — A *control valve* with three end connections. (See 5.1 and 6.1.2.3)

Throttling — The actions to regulate fluid flow through a *valve* by restricting its orifice opening. Also see "*Modulating*".

Topworks — A nonstandard term. (See 7.)

Travel — The amount of movement of the *closure member* from the closed position to an intermediate or the rated full open position.

Travel Characteristic — The relationship between signal input and travel.

Travel Indicator — A means of externally showing position of the *closure member;* typically in terms of percent of or degrees of opening. Can be a visual indicator at or on the *valve* or a remote indicating device by means of transmitter or appropriate linkage. See 7.5.

Travel Indicator Scale — A scale or plate fastened to a *valve* and marked with graduations to indicate the *valve* opening position.

Trim — The internal parts of a *valve* which are in flowing contact with the controlled fluid.

Trim, Restricted — *Control valve trim* which has a flow area less than the full flow area for that *valve*.

Turndown — An obsolete term - See "*Rangeability, Inherent*".

Two-Way Valve — A *valve* with one inlet opening and one outlet opening.

Unbalance, Dynamic — The net force produced on the *valve stem* in any given open position by the fluid pressure acting on the *closure member* and *stem* within the pressure retaining boundary, with the *closure member* at a stated opening and with stated flowing conditions.

Unbalance, Static — The net force produced on the *valve stem* by the fluid pressure acting on the *closure member* and *stem* within the pressure retaining boundary with the fluid at rest and with stated pressure conditions.

Valve Plug — An obsolete term, see "*Closure Member*". (See 4.3.)

Vena Contracta — The location where cross-sectional area of the flowstream is at its minimum. The vena contracta normally occurs just downstream of the actual physical restriction in a *control valve*.

Weld Ends — *Valve* end connections which have been prepared for welding to the line pipe or other fittings. May be butt weld (BWE), or socket weld (SWE).

APPENDIX A
HARD FACING OF CONTROL VALVE TRIM
This appendix is nonmandatory. Use for reference only.

A.1 Hard Facing A material harder than the surface to which it is applied. Used to resist fluid erosion and/or to reduce the chance of galling between moving parts, particularly at high temperature.

A.1.1 Hard facing may be applied by:

a) Fusion welding, other welding, diffusion, or spray coating material.

b) Attaching a hard alloy insert to the surface to be protected by:

   1) Brazing.
   2) Electric (projection) welding.
   3) Bonding.

c) Interference Fit of an Insert to Part to be Protected.

   1) Press Fit.
   2) Shrink Fit.

A.1.2 Hard facing is not to be confused with "hard plating" which means an electro plated, or thin metal deposit, or induced surface hardening which is many orders of magnitude thinner than hard facing.

A.2 In addition to the use of the term "hard facing" the valve parts and specific areas of each part to be hard faced must be specified.

TABLE A-1
SPECIFIC LOCATION OF HARD FACING

	Linear Motion Valves			Rotary Motion Valves		
	Gate	Globe	Diaphragm	Ball	Butterfly	Plug
Seat Joint Area	✓	✓		✓	✓	✓
Complete Flow Control Area		✓		✓	✓	✓
Solid Plug Cap Brazed to Plug Body		✓				
Seat Joint Area		✓	✓	✓	✓	✓
Seat Insert Brazed in Bore		✓				
Complete Flow Control Orifice Area		✓		✓	✓	✓
Guide		✓				
Bearing				✓		✓
Cage		✓				
Body Bore Back Seat	✓					
Stem Back Seat	✓					

APPENDIX B
BUTTERFLY VALVE BODY LINERS AND SEALS
This appendix is nonmandatory. Use for reference only.

B.1 Contents
This appendix provides the recommended terminology for various types of *butterfly valve body* elastomeric and plastic liners and *body*-to-*disk* sealing means.

B.2 Liner Types

B.2.1 Slip-In — An annular shaped liner which makes a slight interference fit with the *body* bore and which may be readily forced into position through the *body* end. May be plain or reinforced.

B.2.2 Locked-In — A liner retained in the *body* bore by a key ring or other means.

B.2.3 Bonded — A liner vulcanized or cemented to the body *bore*.

B.2.4 Wrap-Around — A liner extending around the end faces of the *wafer body* to form a gasket seal with the pipe flanges. The liner may cover all or part of the flange contact area of the *wafer body*.

B.2.5 Flange Retained — A liner retained in the *body* by the pipe flanges or by a continuous or segmented ring. The segmented ring provides a means of adjusting the liner to *disk* interference to achieve improved sealing. The bore of the pipe flanges is smaller in diameter than the *body* bore, therefore the flanges retain the liner in the *body*.

B.2.6 Elastomeric Energized Liner — A resilient elastomeric ring under the main liner is compressed by the *disk* acting through the main liner, thus generating a resilient sealing action between the *disk* and the main liner.

B.2.7 Pressure Energized — A pressure source, either internal fluid pressure or an external fluid pressure source, energizes the liner forcing it into tighter contact with the disk.

B.2.8 Encapsulated Body — All surfaces of the *body* are covered by a continuous surface layer of a different material, usually an elastomeric or plastic material. A soft elastomer behind a harder encapsulating material may be used to provide interference for *disk* and *stem* sealing areas.

B.3 Seals

B.3.1 Flexible Lip Seal — A seal ring retained in the body bore with raised flexible lip which contacts an offset *disk* in the closed position yet is clear of the *disk* in other positions.

B.3.2 Pressure Energized Seal — A seal energized by interference fit between the disk groove and *valve* liner and also by differential pressure acting across the seal. The seal may be a solid section or have internal pressure *ports*.

B.3.3 Metal Piston Type Seal — A self-expandable metal seal ring installed in a groove on the *disk* circumference to block the clearance between the *disk* outer diameter and the liner bore with the *disk* in closed position.

SUBJECT INDEX
(Not including terms listed in the Glossary)

Term	Section	Term	Section
Acting, Single	7.7.1	Bonnetted	5.2.1.1
Acting, Double	7.7.2	Bottom Flange	5.1.2
Actuator	2.1.2	Box Packing	4.6.1.1
Bellows Type	7.1.1.4	Bushing	4.7
Classification Chart	3.3	Bushing, Adaptor	5.3.2
Diaphragm Type	7.1.1.1	Butterfly Liners	Appendix B
Double Linkage	7.6.1 (a)	Butterfly Valve	6.2
Electric	7.1.2	Cage	4.4.2, 5.1.3.2
Electro Hydraulic Type	7.1.2.2	Cage Guided	5.1 (a), 5.1.4.3
Electro Mechanical Type	7.1.2.1	Cage Trim	5.1.3.2
Fluid Motor Type	7.1.1.5	Cambered Disk	6.2.3 (b)
Hydraulic	7.1.3	Cammed Disk	6.2.2 (d)
Motor Gear Train	7.1.2.1	Canted Disk	6.2.2 (e)
Piston Type	7.1.1.2	Characterized Plug	6.3.2 (e)
Pneumatic	7.1.1	Characterized Sleeve	6.3.3
Position Indicator	7.5	Clamp Valve	5.4
Stem	7.3	Compressor	5.3.4
Stem Connector	7.4	Connector, Actuator Stem	7.4
Vane Type	7.1.1.3	Contoured Disk	6.2.3 (d)
Yoke	7.2	Control Signal	
Adaptor Bushing	5.3.2	Override Device	8.2.4
Air Set	8.2.1	Control Valve	
Aligned Disk	6.2.2 (a), 6.2.2 (b)	Classification Chart	3.2
Angle Body	5.1 (e)	Linear Motion	5
Angle Seated Disk	6.2.2 (f)	Rotary Motion	6
Anti-Cavitation Trim	5.1.3.5	Converging, Body	5.1 (g)
Anti-Noise Trim	5.1.3.4	Cylindrical Plug	6.3.2 (a)
Back Seat	4.6.4	Diaphragm Type Actuator	7.1.1.1
Balanced Trim	5.1.3.6	Diaphragm Valve	5.3
Ball Valve	6.1	Diverging Body	5.1 (f)
Floating	6.1.2.4	Disk, Aligned	6.2.2 (a), 6.2.2 (b)
Full	6.1.2.2	Angle Seated	6.2.2 (f)
Segmented	6.1.2.1	Cambered	6.2.3 (b)
Three-Way	6.1.2.3	Cammed	6.2.2 (d)
Bearings, Stem	6.2.6	Canted	6.2.2 (e)
Bellows Stem Seal	4.6.3	Contoured	6.2.3 (d)
Bellows Type Actuator	7.1.1.4	Eccentric Spherical	6.3.2 (d)
Body, Definition	4.1	Flat	6.2.3 (a)
Angle	5.1 (e)	Fluted	6.2.3 (f)
Converging	5.1 (g)	Knife	6.2.3 (e)
Diverging	5.1 (f)	Nonsymmetrical Edge	6.2.3 (c)
Flanged	5.2.2.2	Offset	6.2.2 (c)
Lined	6.2.1.4	Seal On	6.2.4
Single Flange	5.2.2.1	Valve	6.4
Split	5.1 (b), 6.2.1.2	Double Acting	7.7.2
Three Position	5.1 (h)	Double Linkage Power Unit	7.6.1 (a)
Three-Way	5.1 (f), (g), (h)	Double Ported	5.1 (d)
Two-Way	5.1 (a), (b), (c), (d), (e)	Downstream Seating	6.1.3.1
Unlined	6.2.1.3	Eccentric Plug	6.3.2 (c)
Wafer	6.2.1.1	Eccentric Spherical Disk	6.3.2 (d)
Bonnet	4.2	Electro Hydraulic Actuator	7.1.2.2
Bolting	4.2.3	Electro Mechanical Actuator	7.1.2.1
Extension	4.2.1.1	Erosion Resistant Trim	5.1.3.7
Gasket	4.2.2	Extension Bonnet	4.2.1.1
Seal Welded	4.2.1.2	Facing, Hard	Appendix A
Bonnetless	5.2.1.2	Finger Plate	5.3.5

SUBJECT INDEX
(Not including terms listed in the Glossary)

Term	Section	Term	Section
Flange, Bottom	5.1.2	Pinch Valve	5.4
Flange, Single	5.2.2.1	Piston Type Actuator	7.1.1.2
Flanged Body	5.2.2.2	Pivoting Cylinder Power Unit	7.6.2 (e)
Flat Disk	6.2.3 (a)	Plate, Finger	5.3.5
Floating Ball	6.1.2.4	Plug	4.3.1.4
Closure Member	4.3	Characterized	6.3.2 (e)
Closure Member		Cylindrical	6.3.2 (a)
Types		Eccentric	6.3.2 (c)
Ball	4.3.1.1	Eccentric Spherical Disk	6.3.2 (d)
Disk	4.3.1.2	Guide	5.1.4
Gate	4.3.1.3	Tapered	6.3.2 (b)
Plug	4.3.1.4	Valve	6.3
Flow Control Orifice	4.4	Pneumatic Actuator	7.1.1
Fluid Motor Actuator	7.1.1.5	Port Guide	5.1.4.4
Fluted Disk	6.2.3 (f)	Position Switch	8.1.7
Follower, Packing	4.6.1.2	Position Transmitter	8.1.6
Fulcrum-Lever Power Unit	7.6.1 (c)	Positioner	8.1.3
Full Ball	6.1.2.2	Positioner, Double Acting	8.1.3.2
Gasket, Bonnet	4.2.2	Positioner, Single Acting	8.1.3.1
Gate Valve	5.2	Positioner Types	8.1.3.3
Globe Valve	5.1	Positioners, Valve Classification Chart	3.4
Globe Valve Trim	5.1.3	Post Guide	5.1.4.2
Guide, Plug	5.1.4	Power Unit, Double Linkage	7.6.1 (a)
Guide, Port	5.1.4.4	Pivoting Cylinder	7.6.2 (e)
Guide, Stem	5.1.4.1	Rack and Pinion	7.6.2 (f)
Guided, Y Type Cage	5.1 (c) 5.1.4.3	Rocking Stem	7.6.1 (b)
Guided, Double Port Post	5.1 (d) 5.1.4.2		7.6.2 (b)
Guided, Split Body Stem	5.1 (b)	Scotch Yoke	7.6.2 (a)
Hard Facing	Appendix A	Toggle	7.6.2 (d)
Hydraulic Actuator	7.1.3	Variable Thrust	7.6.2 (c)
Integral Seat	4.4.3	Pressure Energized Stem Seal	4.6.2
Integral Stem	6.1.4.2	Rack and Pinion Power Unit	7.6.2 (f)
Indicator, Actuator Position	7.5	Regulator	2.1 footnote
Knife Disk	6.2.3 (e)	Relay, Signal Booster	8.2.2
Lantern Ring	4.6.1.3	Relief Valve	2.1 Footnote
Limit Stop, Mechanical	8.1.4	Ring, Lantern	4.6.1.3
Lined Body	6.2.1.4	Ring, Seat	4.4.1
Liners, Butterfly	Appendix B	Rocking Stem Power Unit	7.6.1 (b)
Loose Stem	6.1.4.1		7.6.2 (b)
Lugged Valve	5.2.2.1	Scotch Yoke Power Unit	7.6.2 (a)
Manual Override	8.1.1	Seal on Disk	6.2.4
Mechanical Limit Stop	8.1.4	Seal, Stem	4.6
Member Closure	4.3	Stem, Bellows	4.6.3
Motion Conversion Mechanism	2.1.3	Stem, Pressure Energized	4.6.2
Motor Gear Train Actuator	7.1.2.1	Seal, Welded Bonnet	4.2.1.2
Multiple Orifice	5.2.4.2	Seals, Butterfly	Appendix B
Nonsymmetrical Edge Disk	6.2.3 (c)	Seat, Back	4.6.4
Offset Disk	6.2.2 (c)	Integral	4.4.3
Orifice, Flow Control	4.4	Ring	4.4.1
Orifice, Multiple	5.2.4.2	Spring, Loaded	6.1.3.3
Orifice, Vee	5.2.4.1	Seating, Downstream	6.1.3.1
Override Device, Control Signal	8.2.4	Seating, Upstream	6.1.3.2
Override, Manual	8.1.1	Segmented Ball	6.1.2.1
Packing	4.6.1	Set, Air	8.2.1
Packing Box	4.6.1.1	Signal Booster Relay	8.2.2
Packing Follower	4.6.1.2, 5.2.6	Single Acting	7.7.1

SUBJECT INDEX
(Not including terms listed in the Glossary)

Term	Section	Term	Section
Single Flange Body	5.2.2.1, 6.2.1.5	Balanced	5.1.3.6
Single Flange	5.2.2.1	Cage	4.4.2, 5.1.3.2
Sleeve, Characterized	6.3.3	Erosion Resistant	5.1.3.7
Snubber	8.1.2	Plug	4.3.1.4
Soft Seated Trim	5.1.3.8	Soft Seated	5.1.3.8
Split Body	6.2.1.2	Stem	4.5
Stem	4.5	Trunnion	6.1.4.3
Actuator	7.3	Two-Way Bodies	5.1 (a), (b), (c), (d), (e)
Anti-Rotation Device	8.1.5		
Bearings	6.2.6	Unlined Body	6.2.1.3
Connector	7.4	Upstream Seating	6.1.3.2
Guide	5.1.4.1	Valve	2.1.1
Guided	5.1 (b)	Ball	6.1
Integral	6.1.4.2	Butterfly	6.2
Loose	6.1.4.1	Clamp	5.4
Seal	4.6, 6.2.7	Control	2.1
Seal, Bellows	4.6.3	Diaphragm	5.3
Seal, Pressure		Gate	5.2
Energized	4.6.2	Globe	5.1
Trim	4.5	Pinch	5.4
Stop, Mechanical Limit	8.1.4	Plug	6.3
Switch, Position	8.1.7	Relief	2.1 footnote
Tapered Plug	6.3.2 (b)	Valve, Classification Charts	3.1, 3.2
Three-Position	5.1 (h)	Valve Diaphragm	5.3.3
Three-Way Ball	6.1.2.3	Valve Disk	6.4
Three-Way Body	5.1 (f), (g), (h)	Vane Type Actuator	7.1.1.3
Toggle Power Unit	7.6.2 (d)	Vee Orifice	5.2.4.1
Transducer	8.2.3	Wafer Body	6.2.1.1
Transmitter, Position	8.1.6	Weir Type Diaphragm Valve	5.3.1.1
Trim, Globe Valve	5.1.3	Yoke, Actuator	7.2
Anti-cavitation	5.1.3.5	Y-Type Cage Guided Globe Valve	5.1 (c)
Anti-Noise	5.1.3.4		

INSTRUMENT SOCIETY of AMERICA
Research Triangle Park, North Carolina

ISA-RP75.06-1981

Recommended Practice

Control Valve Manifold Designs

Instrument Society of America

ISBN 0-87664-656-9

ISA-RP75.06 Control Valve Manifold Designs

Copyright © by the Instrument Society of America 1981. All rights reserved. Printed in the United States of America. No part of this publication may be reproduced, sorted in a retrieval system, or transmitted, in any form or any means electronic, mechanical, photocopying, recording or otherwise without the prior written permission of the publisher.

INSTRUMENT SOCIETY OF AMERICA
67 Alexander Drive
P.O. Box 12277
Research Triangle Park, North Carolina 27709

Copyright © by the Instrument Society of America 1981

PREFACE

This Preface is included for information purposes and is not part of RP75.06.

This Recommended Practice has been prepared as a part of the service of the Instrument Society of America toward a goal of uniformity in the field of instrumentation. To be of real value, this document should not be static, but should be subject to periodic review. Toward this end, the Society welcomes all comments and criticisms, and asks that they be addressed to the Secretary, Standards and Practices Board, Instrument Society of America, 67 Alexander Drive, P.O. Box 12277, Research Triangle Park, NC 27709, Telephone (919) 549-8411.

The ISA Standards and Practice Department is aware of the growing need for attention to the metric system of units in general, and the International System of Units (SI) in particular, in the preparation of instrumentation standards. The Department is further aware of the benefits to USA users of ISA Standards of incorporating suitable references to the SI (and the metric system) in their business and professional dealings with other countries. Towards this end this Department will endeavor to introduce SI – acceptable metric units in all new and revised standards to the greatest extent possible. The Metric Practice Guide, which has been published by the American Society for Testing and Materials as ANSI designation Z210.1 (ASTM E380-76, IEEE Std. 268-1975), and future revisions, will be the reference guide for definitions, symbols, abbreviation and conversion factors.

It is the policy of the Instrument Society of America to encourage and welcome the participation of all concerned individuals and interests in the development of ISA Standards. Participation in the ISA Standards making process by an individual in no way constitutes endorsement by the employer of that individual of the Instrument Society of America or any of the Standards or Recommended Practices which ISA develops.

From a survey initiated in 1947, Tentative Recommended Practice ISA-RP4.2, "Standard Control Valve Manifold Designs (Carbon Steel Control Valves Only)," (1956) was developed and published. It gave detailed dimensions for using ASA 300# flanged control valves 1-½ through 6 inch size, in 150# and 300# manifolds, 2 through 8 inch. Pressure-temperature ratings are now known as ANSI Class 150, Class 300, etc. rather than ASA 150#, 300#, etc. Dimensions have not changed since 1956 for these sizes and pressure ratings of control valves, block (gate) valves, and bypass valves (globe up through 4 inch, gate for 6 and 8 inch). Globe control valves still predominate in most industries, but other styles (flangeless, rotary, etc.) are being used and these have face-to-face or end-to-end dimensions that may be different than globe control valves (Bibliography, references 3, 4, and 5).

This document is not intended to promote the use of manifolds in piping systems nor to recommend the types shown as better than others; but if a designer has decided to install a manifold, *space estimates* may be obtained herein. Application information may be obtained in the *ISA Handbook of Control Valves*, reference 1 in the Bibliography.

Subcommittee SP75.08 (formerly Committee SP4)

NAME	COMPANY
W. C. Weidman (Chairman)	Gilbert Associates
G. Borden	Bechtel Power Corporation
R. Chown	Kamyr, Inc.
L. R. Driskell	Dravo Corporation
J. T. Emery	Honeywell, Inc.
D. G. Gaffney	Fisher Controls
M. Gugenheim	Bechtel Petroleum, Inc.
A. J. Hanssen	ITT Hammel-Dahl/Conoflow
M. Kaye	M. W. Kellogg Company
O. Lovett	E. I. duPont deNemours & Co.
M. May (Deceased)	W-K-M
H. P. Meissner	Kieley & Mueller, Inc.
H. R. Nickerson	Resistoflex Corporation
R. A. Quance	The SNC Group

Subcommittee SP75.08 (cont.)

NAME	COMPANY
J. Reed	Copes Vulcan, Inc.
F. O. Seger	Control Components, Inc.
J. M. Simonsen	Valtek, Inc.
R. U. Stanley	Bechtel Petroleum, Inc.
S. Weiner	Monsanto Company
P. Wing	Masoneilan International, Inc.

This Recommended Practice was approved for publication by the ISA Standards and Practices Board in April 1981.

NAME	COMPANY
T. J. Harrison, Chairman	IBM Corporation
P. Bliss	Consultant
W. Calder	The Foxboro Company
B. A. Christensen	Continental Oil Co.
M. R. Gorden-Clark	Scott Paper Company
R. T. Jones	Philadelphia Electric Company
R. Keller	Boeing Company
O. P. Lovett, Jr.	Jordan Valve
E. C. Magison	Honeywell, Inc.
A. P. McCauley	Diamond Shamrock Corporation
E. M. Nesvig	ERDCO Engineering Corporation
R. L. Nickens	Reynolds Metals Company
G. Platt	Bechtel Power Corporation
R. Prescott	Moore Products Company
R. W. Signor	General Electric Company
W. C. Weidman	Gilbert Associates, Inc.
K. A. Whitman	Allied Chemical Corporation
*L. N. Combs	
*R. L. Galley	
*R. G. Marvin	
*W. B. Miller	Moore Products Company
*J. R. Williams	Stearns-Rogers, Inc.

*Director Emeritus

TABLE OF CONTENTS

Section   Page

1. Scope 7
2. Purposes of these Manifold Designs 7
3. Application 7
4. Dimensions 7
5. Sizes 7
6. Manifold Type Details 7
   Bibliography 20

LIST OF ILLUSTRATIONS

Figure   Page

1. Type I Control Valve Manifold 8
2. Type II Control Valve Manifold 10
3. Type III Control Valve Manifold 12
4. Type IV Control Valve Manifold 14
5. Type V Control Valve Manifold 16
6. Type VI Control Valve Manifold 18

LIST OF TABLES

Table   Page

1. Type I Control Valve Manifold Dimensions for ANSI Class 300 9
2. Type II Control Valve Manifold Dimensions for ANSI Class 300 11
3. Type III Control Valve Manifold Dimensions for ANSI Class 300 13
4. Type IV Control Valve Manifold Dimensions for ANSI Class 300 15
5. Type V Control Valve Manifold Dimensions for ANSI Class 300 17
6. Type VI Control Valve Manifold Dimensions for ANSI Class 300 19

1 SCOPE

Control valves and bypass valves are sometimes manifolded in piping systems to allow manual manipulation of the flow through the systems in those situations in which (usually) the control valve is not in service. Six control valve manifold types are presented in this recommended practice with space estimates for various sizes. Each of these six types consists of a straight through globe control valve, isolating upstream and downstream block valves, and bypass piping with a manually operated valve.

2 PURPOSES OF THESE MANIFOLD DESIGNS

2.1 Provide space estimates for one popular pressure rating (American National Standards Institute (ANSI) Class 300) of several sizes of manifolds.

2.2 Suggest space estimates for other pressure ratings.

2.3 Suggest compact, model designs.

2.4 Reduce design expense.

2.5 Promote uniformity of practice.

2.6 Allow reuse of components.

3 APPLICATION

Application information and guidance may be found in the *ISA Handbook of Control Valves*, API RP550, (Bibliography, references 1 and 2), and other publications.

4 DIMENSIONS*

4.1 This revision of ISA-RP4.2 includes dimensions for flanged globe control valves of ANSI Class 300. (Class 600 control valves face-to-face dimensions are approximately 7 percent longer than Class 300; Class 150 are about 5 percent shorter.) Face-to-face dimensions of certain globe type valves are standardized per ISA (See Bibliography, references 3 & 4.). Dimensions for most gate valves, globe valves, and plug valves used as block and/or bypass may be obtained from Bibliography, reference 5.

4.2 Globe control valves have longer face-to-face dimensions than other styles. Use of flangeless valves may reduce dimension W significantly.

4.3 The user should verify dimensions of all components before releasing a manifold design for construction.

4.4 Actuator heights are not standardized; therefore the heights, for both plain and extended bonnet actuators, must be obtained from the chosen manufacturer.

4.5 Sizes are nominal for the control valves and both nominal and actual OD for the manifold pipe. Since space dimensions H and W are estimates, these are given to the next higher inch from a compact design. Millimeters are converted from inches and rounded to the next higher 10 mm increment for H, W, X, and Y.

4.6 Bibliography reference 1 contains manifolds with different details and dimensions. Reference 2 also contains several different designs. Reference 6 gives layout guidance.

5 SIZES

5.1 Manifold piping nominal sizes covered by the designs are 1, 1-½, 2, 3, 4, 6, and 8 inches (actual piping OD is 33.4, 48.3, 60.3, 88.9, 114.3, 168.3, 219.1 mm).

5.2 Control valve body nominal sizes covered by the designs are 1, 1-½, 2, 3, 4, and 6 inch. (25, 40, 50, 80, 100, and 150 mm).

6 MANIFOLD TYPE DETAILS

6.1 All the piping component pieces can be fabricated using conventional weld type fittings and pipe.

6.2 The six manifold designs are designated Type I through Type VI. The sequence in which these designs appear has no bearing on the preference or acceptability of the designs, nor does omission of other configurations indicate less acceptability.

6.3 Bypass valves are usually globe valves up through 4 inch (100 mm) size. Bypass valves are often gate valves in 6 and 8 inch sizes. The dimensions shown allow enough space for manifold size bypass valves.

6.4 Drain and vent connections may or may not be desirable. They are shown schematically on the type drawings. Reference 6 gives more details.

6.5 The clearances in the manifold designs are to allow for the removal of the actuator and the control valve plug from the bonnet side (top pull) or bottom (bottom pull) of a manifold-size valve with plain bonnet. The 12 inch (305 mm) clearance between the top of the actuator and the bypass (in Types I, III, and IV) as the yoke clears the valve stem is generally quite sufficient to facilitate disassembly. Dimension X allows clearance for actuator and plug removal outside the piping.

6.6 It is assumed that the piping system in which these manifolds will be used has been properly sized and designed so that the variation between main line size and the control valve size is not excessive. Where the main line size is more than two sizes larger than the control valve size, it is considered economical to reduce ahead of the manifold so that the manifold pipe size and the control valve fall within the above relationship. Reducing may also be considered at the manifold inlet to provide a manifold only one size larger than the control valve if dictated by economics or where the need for flexibility for installing a larger control valve is not likely. Because of this, references have been made to manifold sizes rather than to main line sizes.

---

*Refer to Figures for letter codes and to Tables for dimensions.

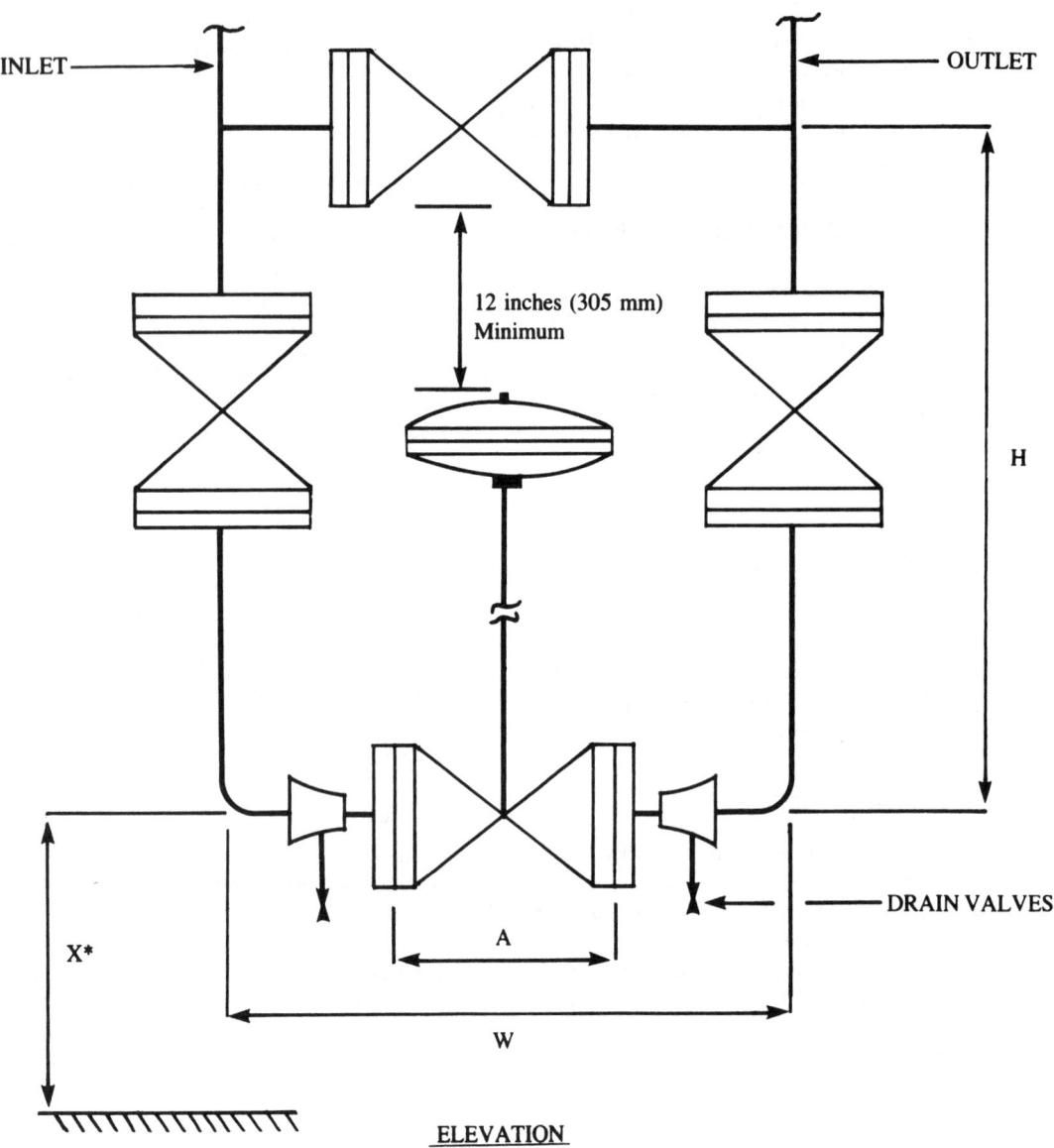

Figure 1. Type I Control Valve Manifold

TABLE 1
TYPE I CONTROL VALVE MANIFOLD DIMENSIONS FOR ANSI CLASS 300

INCHES							MILLIMETERS				
A*	W**	H**	X**	ACTUAL MANIFOLD PIPE OUTSIDE DIAMETER	NOMINAL SIZE MANIFOLD PIPE	NOMINAL SIZE CONTROL VALVE	NOMINAL SIZE CONTROL VALVE	ACTUAL MANIFOLD PIPE OUTSIDE DIAMETER	A*	W**	H**	X**
7¾	27	39	23	1.315	1	1	25	33.4	197	690	990	580
9¼	27	39	23	1.900	1½	1½	40	48.3	235	690	990	580
7¾	27	39	23	1.900	1½	1	25	48.3	197	690	990	580
10½	27	39	23	2.375	2	2	50	60.3	267	690	990	580
9¼	27	39	23	2.375	2	1½	40	60.3	235	690	990	580
7¾	27	39	23	2.375	2	1	25	60.3	197	690	990	580
12½	30	42	27	3.500	3	3	80	88.9	317	760	1070	690
10½	30	42	27	3.500	3	2	50	88.9	267	760	1070	690
9¼	30	42	27	3.500	3	1½	40	88.9	235	760	1070	690
14½	35	43	30	4.500	4	4	100	114.3	368	890	1090	760
12½	35	43	30	4.500	4	3	80	114.3	317	890	1090	760
10½	35	43	30	4.500	4	2	50	114.3	267	890	1090	760
14½	45	54	39	6.625	6	4	100	168.3	368	1140	1370	990
12½	45	54	39	6.625	6	3	80	168.3	317	1140	1370	990
18⅝	55	57	46	8.625	8	6	150	219.1	473	1400	1450	1170
14½	55	57	46	8.625	8	4	100	219.1	368	1400	1450	1170

*Actual dimensions from reference 3.
**Suggested dimensions

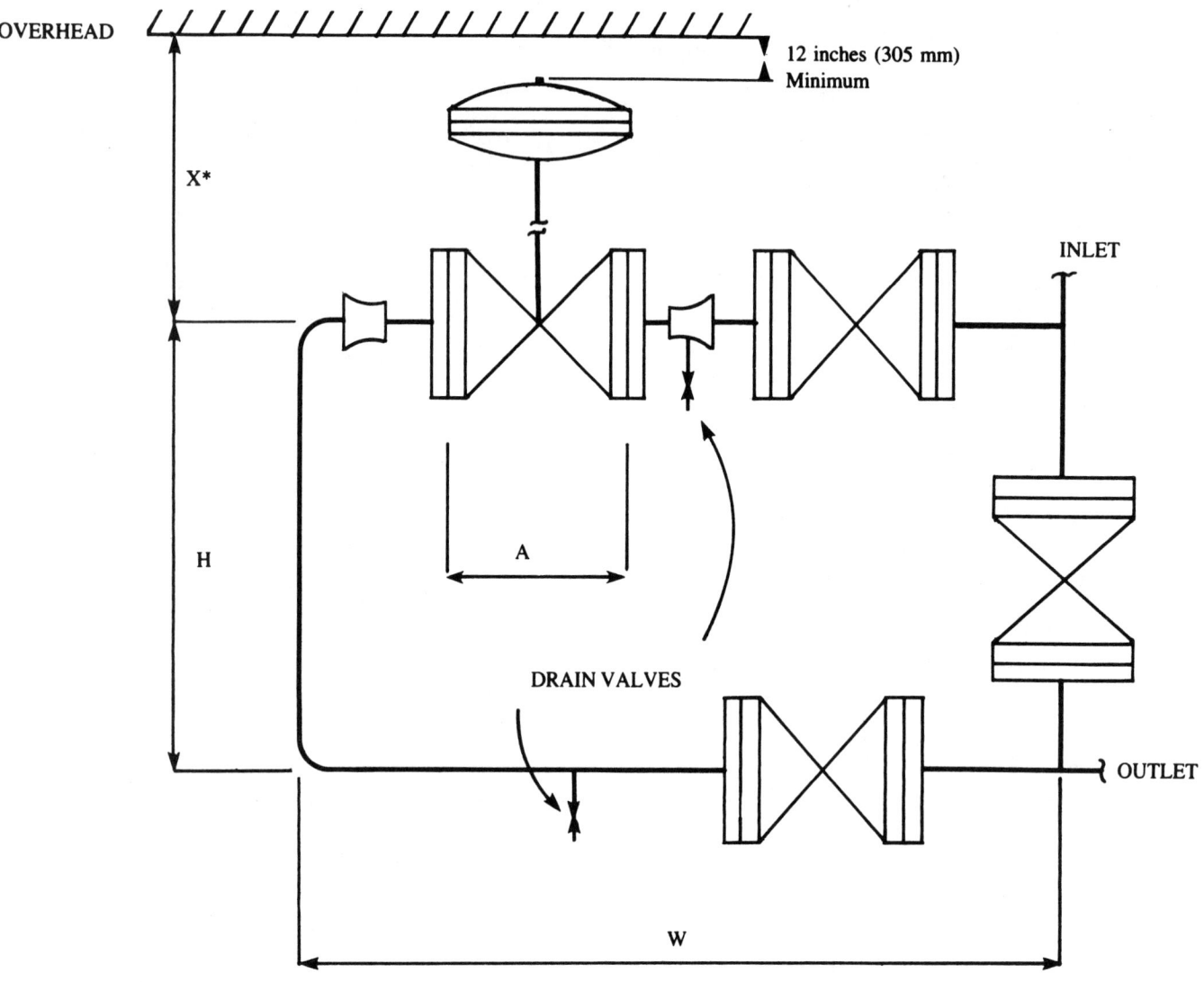

Figure 2. Type II Control Valve Manifold

TABLE 2
TYPE II CONTROL VALVE MANIFOLD DIMENSIONS FOR ANSI CLASS 300

		INCHES						MILLIMETERS				
A*	W**	H**	X**	ACTUAL MANIFOLD PIPE OUTSIDE DIAMETER	NOMINAL SIZE MANIFOLD PIPE	NOMINAL SIZE CONTROL VALVE	NOMINAL SIZE CONTROL VALVE	ACTUAL MANIFOLD PIPE OUTSIDE DIAMETER	A*	W**	H**	X**
7¾	44	25	37	1.315	1	1	25	33.4	197	1120	640	940
9¼	44	25	37	1.900	1½	1½	40	48.3	235	1120	640	940
7¾	44	25	37	1.900	1½	1	25	48.3	197	1120	640	940
10½	44	25	37	2.375	2	2	50	60.3	267	1120	640	940
9¼	44	25	37	2.375	2	1½	40	60.3	235	1120	640	940
7¾	44	25	37	2.375	2	1	25	60.3	197	1120	640	940
12½	48	29	39	3.500	3	3	80	88.9	317	1220	740	990
10½	48	29	39	3.500	3	2	50	88.9	267	1220	740	990
9¼	48	29	39	3.500	3	1½	40	88.9	235	1220	740	990
14½	56	33	40	4.500	4	4	100	114.3	368	1430	840	1020
12½	56	33	40	4.500	4	3	80	114.3	317	1430	840	1020
10½	56	33	40	4.500	4	2	50	114.3	267	1430	840	1020
14½	70	43	50	6.625	6	4	100	168.3	368	1780	1090	1270
12½	70	43	50	6.625	6	3	80	168.3	317	1780	1090	1270
18⅝	78	50	52	8.625	8	6	150	219.1	473	1990	1270	1320
14½	78	50	52	8.625	8	4	100	219.1	368	1990	1270	1320

*Actual dimensions from reference 3.
**Suggested dimensions

Instrument Society of America

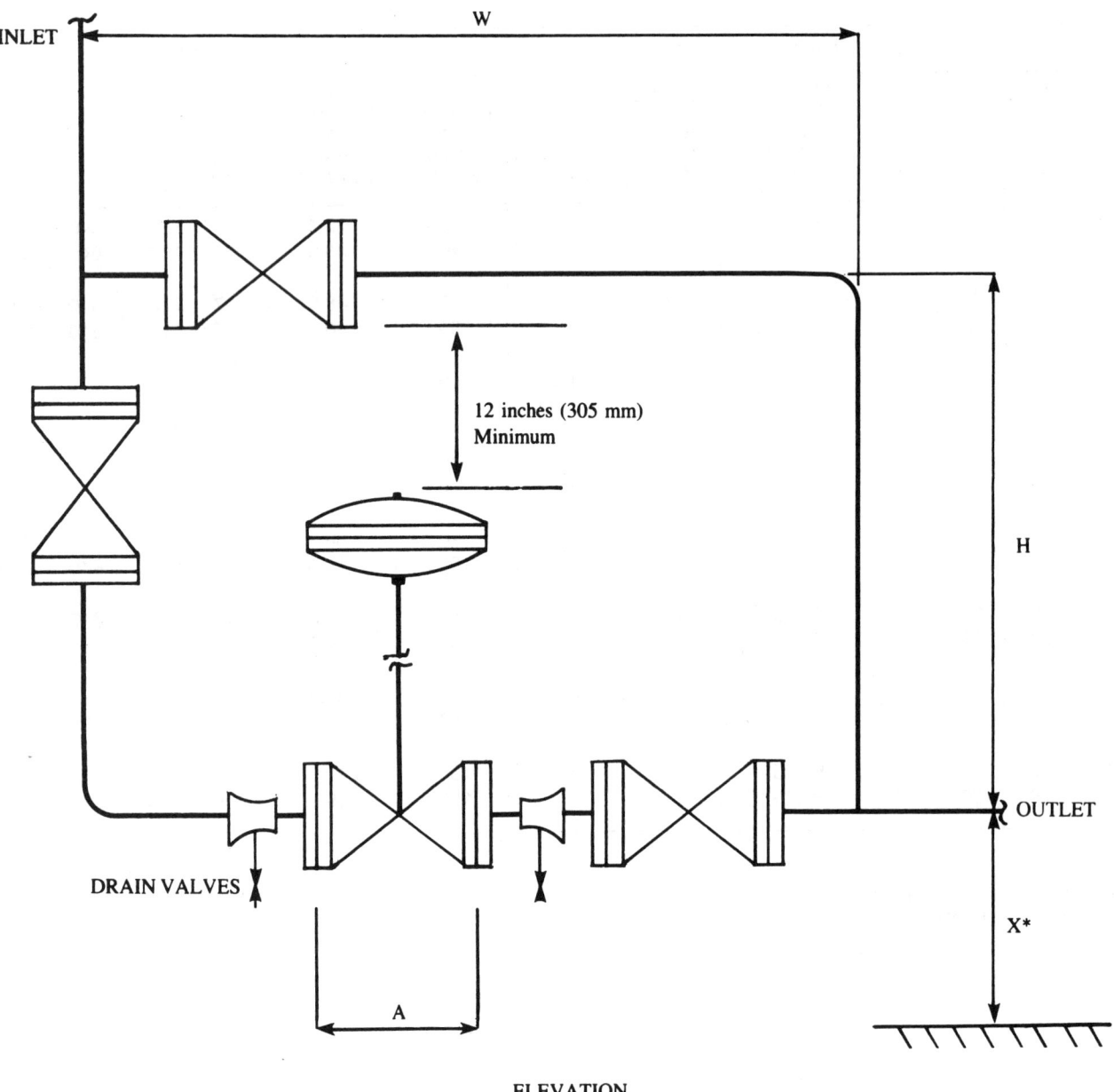

ELEVATION

NOTE: Dimensions shown are suggested piping dimensions and may vary depending upon actual dimensions of components being used.

*If Plug removes in this direction

Figure 3. Type III Control Valve Manifold

12

TABLE 3
TYPE III CONTROL VALVE MANIFOLD DIMENSIONS FOR ANSI CLASS 300

		INCHES							MILLIMETERS			
A*	W**	H**	X**	ACTUAL MANIFOLD PIPE OUTSIDE DIAMETER	NOMINAL SIZE MANI-FOLD PIPE	NOMINAL SIZE CONTROL VALVE	NOMINAL SIZE CONTROL VALVE	ACTUAL MANIFOLD PIPE OUTSIDE DIAMETER	A*	W**	H**	X**
7¾	44	39	23	1.315	1	1	25	33.4	197	1120	990	580
9¼	44	39	23	1.900	1½	1½	40	48.3	235	1120	990	580
7¾	44	39	23	1.900	1½	1	25	48.3	197	1120	990	580
10½	44	39	23	2.375	2	2	50	60.3	267	1120	990	580
9¼	44	39	23	2.375	2	1½	40	60.3	235	1120	990	580
7¾	44	39	23	2.375	2	1	25	60.3	197	1120	990	580
12½	48	42	27	3.500	3	3	80	88.9	317	1220	1070	690
10½	48	42	27	3.500	3	2	50	88.9	267	1220	1070	690
9¼	48	42	27	3.500	3	1½	40	88.9	235	1220	1070	690
14½	56	43	30	4.500	4	4	100	114.3	368	1430	1090	760
12½	56	43	30	4.500	4	3	80	114.3	317	1430	1090	760
10½	56	43	30	4.500	4	2	50	114.3	267	1430	1090	760
14½	70	54	39	6.625	6	4	100	168.3	368	1780	1370	990
12½	70	54	39	6.625	6	3	80	168.3	317	1780	1370	990
18⅝	78	57	46	8.625	8	6	150	219.1	473	1990	1450	1170
14½	78	57	46	8.625	8	4	100	219.1	368	1990	1450	1170

*Actual dimensions from reference 3.
**Suggested dimensions

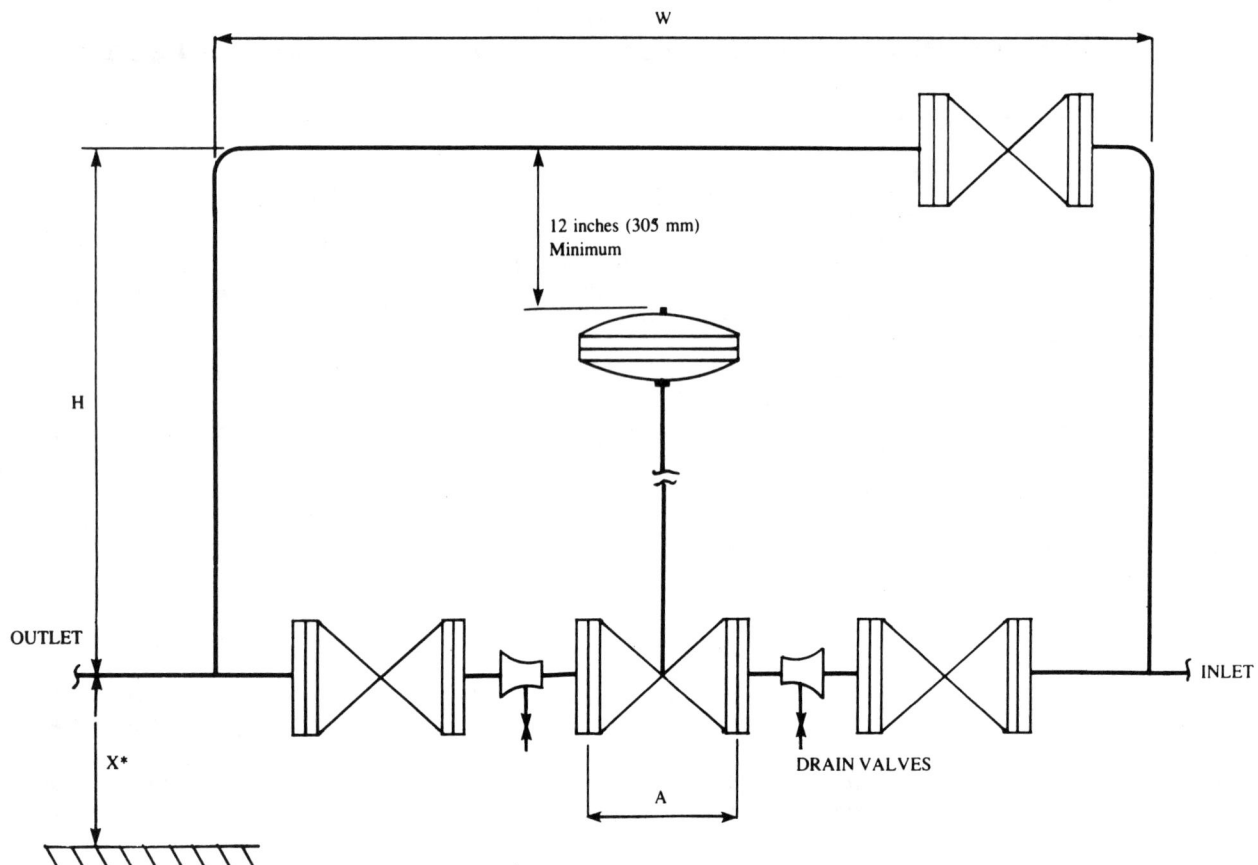

ELEVATION

NOTE: Dimensions shown are suggested piping dimensions and may vary depending upon actual dimensions of components being used.

Figure 4. Type IV Control Valve Manifold

TABLE 4
TYPE IV CONTROL VALVE MANIFOLD DIMENSIONS FOR ANSI CLASS 300

\multicolumn{7}{c}{INCHES}	\multicolumn{6}{c}{MILLIMETERS}

A*	W**	H**	X**	ACTUAL MANIFOLD PIPE OUTSIDE DIAMETER	NOMINAL SIZE MANIFOLD PIPE	NOMINAL SIZE CONTROL VALVE	NOMINAL SIZE CONTROL VALVE	ACTUAL MANIFOLD PIPE OUTSIDE DIAMETER	A*	W**	H**	X**
7¾	58	39	23	1.315	1	1	25	33.4	197	1470	990	580
9¼	58	39	23	1.900	1½	1½	40	48.3	235	1470	990	580
7¾	58	39	23	1.900	1½	1	25	48.3	197	1470	990	580
10½	58	39	23	2.375	2	2	50	60.3	267	1470	990	580
9¼	58	39	23	2.375	2	1½	40	60.3	235	1470	990	580
7¾	58	39	23	2.375	2	1	25	60.3	197	1470	990	580
12½	67	42	27	3.500	3	3	80	88.9	317	1700	1070	690
10½	67	42	27	3.500	3	2	50	88.9	267	1700	1070	690
9¼	67	42	27	3.500	3	1½	40	88.9	235	1700	1070	690
14½	74	43	30	4.500	4	4	100	114.3	368	1880	1090	760
12½	74	43	30	4.500	4	3	80	114.3	317	1880	1090	760
10½	74	43	30	4.500	4	2	50	114.3	267	1880	1090	760
14½	97	54	39	6.625	6	4	100	168.3	368	2460	1370	990
12½	97	54	39	6.625	6	3	80	168.3	317	2460	1370	990
18⅝	109	57	46	8.625	8	6	150	219.1	473	2770	1450	1170
14½	109	57	46	8.625	8	4	100	219.1	368	2770	1450	1170

*Actual dimensions from reference 3.
**Suggested dimensions

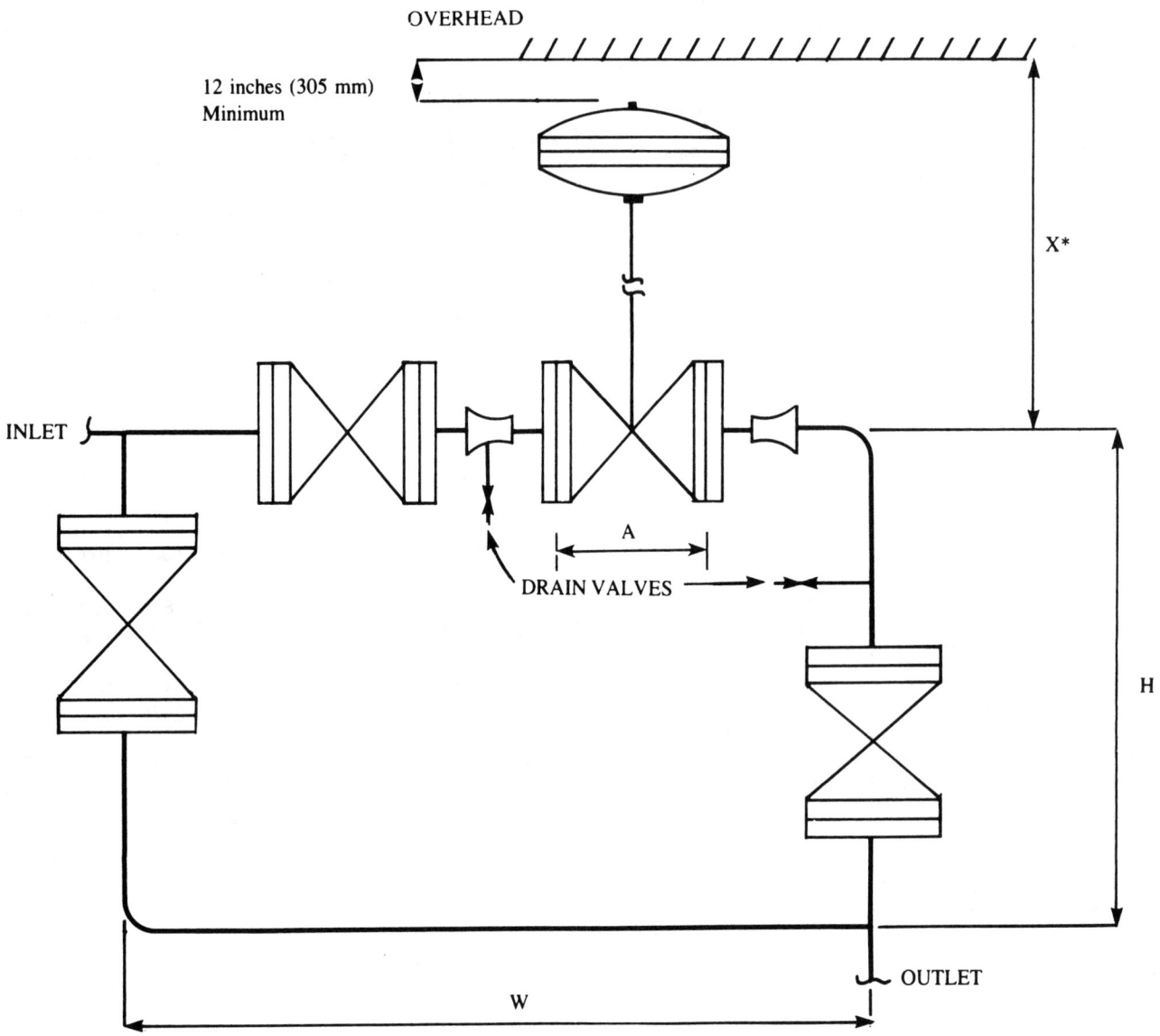

Figure 5. Type V Control Valve Manifold

RP75.06

Control Valve
Manifold Designs

TABLE 5
TYPE V CONTROL VALVE MANIFOLD DIMENSIONS FOR ANSI CLASS 300

\multicolumn{7}{c}{INCHES}	\multicolumn{6}{c}{MILLIMETERS}
				ACTUAL MANIFOLD PIPE OUTSIDE DIAMETER	\multicolumn{2}{c}{NOMINAL SIZE}	NOMINAL SIZE CONTROL VALVE	ACTUAL MANIFOLD PIPE OUTSIDE DIAMETER				
A*	W**	H**	X**		MANI-FOLD PIPE	CONTROL VALVE			A*	W**	H**	X**
7¾	44	36	37	1.315	1	1	25	33.4	197	1120	910	940
9¼	44	36	37	1.900	1½	1½	40	48.3	235	1120	910	940
7¾	44	36	37	1.900	1½	1	25	48.3	197	1120	910	940
10½	44	36	37	2.375	2	2	50	60.3	267	1120	910	940
9¼	44	36	37	2.375	2	1½	40	60.3	235	1120	910	940
7¾	44	36	37	2.375	2	1	25	60.3	197	1120	910	940
12½	48	39	39	3.500	3	3	80	88.9	317	1220	990	990
10½	48	39	39	3.500	3	2	50	88.9	267	1220	990	990
9¼	48	39	39	3.500	3	1½	40	88.9	235	1220	990	990
14½	56	39	40	4.500	4	4	100	114.3	368	1430	990	1020
12½	56	39	40	4.500	4	3	80	114.3	317	1430	990	1020
10½	56	39	40	4.500	4	2	50	114.3	267	1430	990	1020
14½	70	46	50	6.625	6	4	100	168.3	368	1780	1170	1270
12½	70	46	50	6.625	6	3	80	168.3	317	1780	1170	1270
18⅝	78	50	52	8.625	8	6	150	219.1	473	1990	1270	1320
14½	78	50	52	8.625	8	4	100	219.1	368	1990	1270	1320

*Actual dimensions from reference 3.
**Suggested dimensions

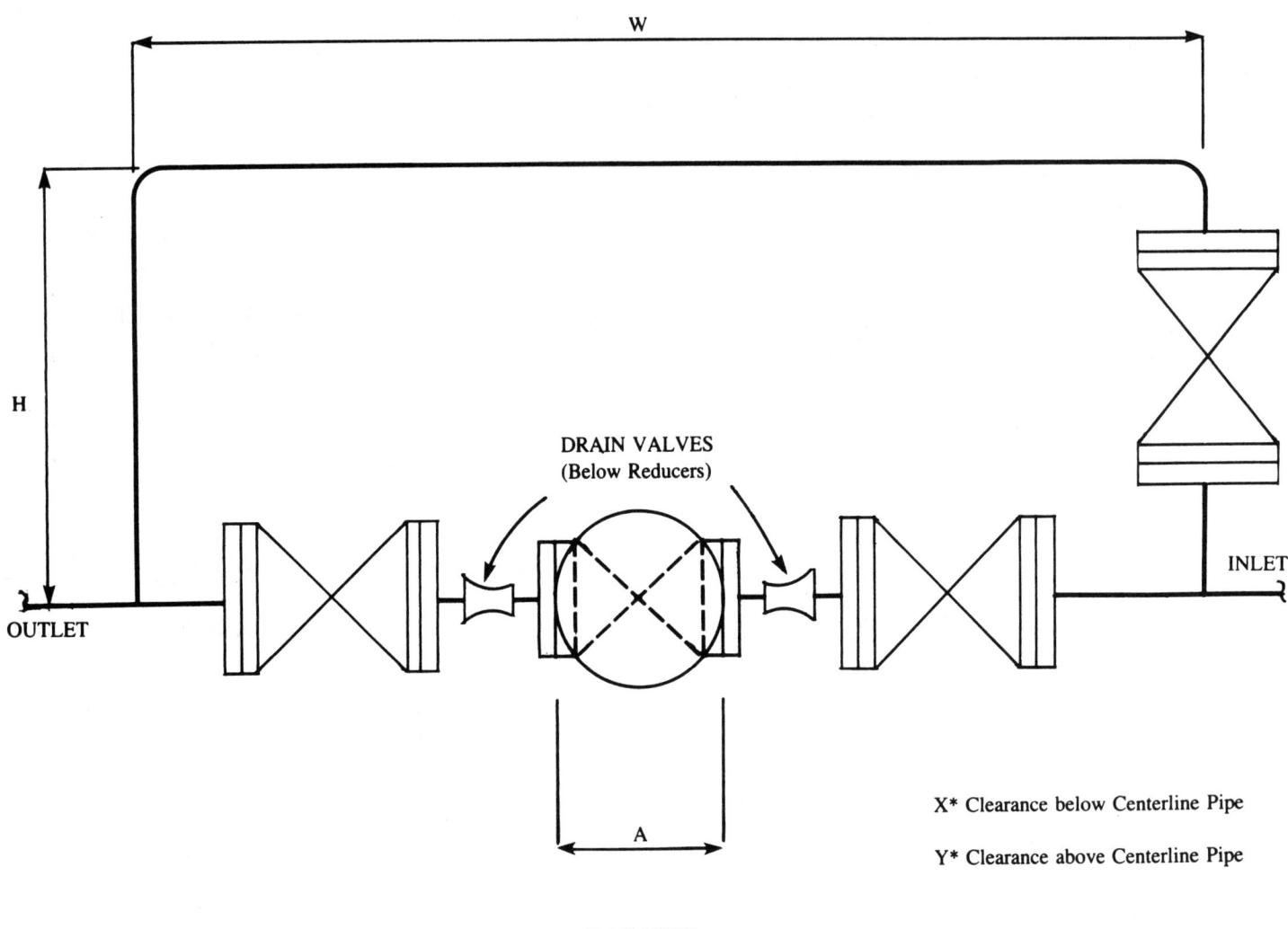

Figure 6. Type VI Control Valve Manifold

TABLE 6
TYPE VI CONTROL VALVE MANIFOLD DIMENSIONS FOR ANSI CLASS 300

		INCHES							MILLIMETERS					
					ACTUAL MANIFOLD PIPE OUTSIDE DIAMETER	NOMINAL SIZE		NOMINAL SIZE CONTROL VALVE	ACTUAL MANIFOLD PIPE OUTSIDE DIAMETER					
A*	W**	H**	X**	Y**		MANI-FOLD PIPE	CONTROL VALVE			A*	W**	H**	X**	Y**
7¾	58	21	23	37	1.315	1	1	25	33.4	197	1470	530	580	940
9¼	58	21	23	37	1.900	1½	1½	40	48.3	235	1470	530	580	940
7¾	58	21	23	37	1.900	1½	1	25	48.3	197	1470	530	580	940
10½	58	21	23	37	2.375	2	2	50	60.3	267	1470	530	580	940
9¼	58	21	23	37	2.375	2	1½	40	60.3	235	1470	530	580	940
7¾	58	21	23	37	2.375	2	1	25	60.3	197	1470	530	580	940
12½	67	26	27	39	3.500	3	3	80	88.9	317	1700	660	690	990
10½	67	26	27	39	3.500	3	2	50	88.9	267	1700	660	690	990
9¼	67	26	27	39	3.500	3	1½	40	88.9	235	1700	660	690	990
14½	74	29	30	40	4.500	4	4	100	114.3	368	1880	740	760	1020
12½	74	29	30	40	4.500	4	3	80	114.3	317	1880	740	760	1020
10½	74	29	30	40	4.500	4	2	50	114.3	267	1880	740	760	1020
14½	97	36	39	50	6.625	6	4	100	168.3	368	2460	910	990	1270
12½	97	36	39	50	6.625	6	3	80	168.3	317	2460	910	990	1270
18⅝	109	41	46	52	8.625	8	6	150	219.1	473	2770	1040	1170	1320
14½	109	41	46	52	8.625	8	4	100	219.1	368	2770	1040	1170	1320

*Actual dimensions from reference 3.
**Suggested dimensions

BIBLIOGRAPHY

1. Hutchison, J. W., *ISA Handbook of Control Valves*, Second Edition, Pittsburgh, Pa., Instrument Society of America, 1976
2. API RP550, *Manual on Installation of Refinery Instruments and Control Systems* Part 1 – Process Instrumentation and Control Section 6 – Control Valves and Accessories, Third Edition, Washington, D.C., American Petroleum Institute, 1976
3. ISA-S75.03, 1979, *Uniform Face-to-Face Dimensions for Flanged Globe Style Control Valve Bodies*, Pittsburgh, Pa., Instrument Society of America, 1979
4. ISA-S75.04, 1979, *Face-to-Face Dimensions of Flangeless Control Valves*, Pittsburgh, Pa., Instrument Society of America, 1979
5. ANSI Standard B16.10-1973, *Face-to-Face and End-to-End Dimensions of Ferrous Valves*, New York, N.Y., 1973
6. Weaver, Rip., *Process Piping Drafting*, Houston, Texas, Gulf Publishing Co., 1974

INSTRUMENT SOCIETY of AMERICA
Research Triangle Park, North Carolina

ISA-S75.07-1987

Standard

Laboratory Measurement of Aerodynamic Noise Generated by Control Valves

Instrument Society of America

Instrument Society of America

ISBN 1-55617-047-5

ISA-S75.07, Laboratory Measurement of Aerodynamic Noise Generated by Control Valves

Copyright © 1987 by the Instrument Society of America. All rights reserved. Printed in the United States of America. No part of this publication may be reproduced, stored in a retrieval system, or transmitted, in any form or by any means (electronic, mechanical, photocopying, recording, or otherwise), without the prior written permission of the publisher.

INSTRUMENT SOCIETY OF AMERICA
67 Alexander Drive
P. O. Box 12277
Research Triangle Park, North Carolina 27709

PREFACE

This preface is included for informational purposes and is not part of ISA-S75.07.

This standard has been prepared as part of the service of the Instrument Society of America (ISA) toward a goal of uniformity in the field of instrumentation. To be of real value, this document should not be static, but should be subject to periodic review. Toward this end, the Society welcomes all comments and criticisms, and asks that they be addressed to the Secretary, Standards and Practices Board, Instrument Society of America, 67 Alexander Drive, P.O. Box 12277, Research Triangle Park, NC 27709, Telephone (919) 549-8411.

The ISA Standards and Practices Department is aware of the growing need for attention to the metric system of units in general, and the International System of Units (SI) in particular, in the preparation of instrumentation standards. The Department is further aware of the benefits to U.S.A. users of ISA standards of incorporating suitable references to the SI (and the metric system) in their business and professional dealings with other countries. Toward this end, this Department will endeavor to introduce SI-acceptable metric units in all new and revised standards to the greatest extent possible. *The Metric Practice Guide*, which has been published by the Institute of Electrical and Electronics Engineers as ANSI/IEEE Std. 268-1982, and future revisions will be the reference guide for definitions, symbols, abbreviations, and conversion factors.

It is the policy of the Instrument Society of America to encourage and welcome the participation of all concerned individuals and interests in the development of ISA standards. Participation in the ISA standards-making process by an individual in no way constitutes endorsement by the employer of that individual, of the Instrument Society of America, or of any of the standards that ISA develops.

The information contained in the preface, footnotes, and appendices is included for information only and is not a part of the standard.

The following people served as members* of ISA Subcommittee SP75.07:

NAME	COMPANY
J. B. Arant, Chairman	E. I. du Pont de Nemours & Company
G. E. Barb	Bechtel, Inc. (now employed by Anchor/Darling Valve Company)
H. D. Baumann	H. D. Baumann Associates, Ltd.
C. S. Beard	Mutual Stamp Service (deceased)
S. J. Boyle	Masoneilan International, Inc. (now Masoneilan/Dresser)
T. Champlain	ITT Grinnell
A. Gharabegian	Fluor Engineers & Constructors
A. Glenn	Valtek, Inc.
J. Grosek	Honeywell Inc.
H. Illing	Kieley and Mueller, Inc. (now DeZurik/Kieley and Mueller Div.)
J. Kahrs	Leslie Company
R. Liebich	United Engineers & Constructors, Inc.
K. W. Ng	Naval Underwater Systems center
G. Reethof	Pennsylvania State University
G. Richards	Jordan Valve (now with Richards Industries, Inc.)
R. Rusali	Control Components International
F. O. Seger	Willis Oil Tool Company (now with Masoneilan/Dresser)
A. Shea	Copes-Vulcan, Inc.
A. Skovgaard	Yarway Corporation
J. Wang	Exxon Research & Engineering
W. C. Weidman	Gilbert Associates (now Gilbert/Commonwealth, Inc.)
R. J. Winkler	Fisher Controls Company (now Fisher Controls International, Inc.)

The following people served as alternates of ISA Subcommittee SP75.07:

NAME	COMPANY
R. Barnes	Valtek, Inc.
B. Broxterman	Jordan Valve Company
A. Fagerlund	Fisher Controls Company (now Fisher Controls International, Inc.)
C. Langford	E. I. du Pont de Nemours & Company
D. Minoofar	Control Components International
J. M. Simonsen	Valtek, Inc.
H. R. Sonderegger	ITT Grinnell Corporation
R. F. Tubbs	Copes-Vulcan, Inc.
A. Wolf	Exxon Research & Engineering

*Membership list and company at the time of the final vote (current company names are given in parentheses)

The following people served as members* of ISA Committee SP75:

NAME	COMPANY
L. R. Driskell, Chairman	Consultant
J. B. Arant	E. I. du Pont de Nemours & Company
H. E. Backinger	John F. Kraus & Company (Kraus Company, Inc.)
G. E. Barb	Muesco Inc. (now with Anchor/Darling Valve Company)
H. D. Baumann	H. D. Baumann Associates, Ltd.
C. S. Beard	Mutual Stamp Services (deceased)
N. Belaef	Retired
G. Borden, Jr.	Bechtel Power Corporation (retired)
D. E. Brown	R. Conrader Company
E. H. C. Brown	Dravo Engineers, Inc.
E. J. Cooney	Air Products & Chemicals, Inc.
W. G. Dewart	Rockwell International
J. T. Emery	Honeywell, Inc.
H. J. Fuller	Worcester Controls Corporation
L. Griffith	Consultant
A. J. Hanssen	Fluid Controls Institute, Inc.
F. Harthun	Fisher Controls Company (now Fisher Controls International, Inc.)
H. Illing	Kieley & Mueller Inc. (now DeZurik/Kieley and Mueller Div.)
R. B. Jones	Upjohn Company (now Dow Chemical USA)
M. W. Kaye	M. W. Kellogg Company
R. Louviere	Creole Engineering
O. P. Lovett, Jr.	ISIS Corporation (now retired)
J. Manton	Consultant
A. P. McCauley	Diamond Shamrock (now with Chagrin Valley Controls, Inc.)
T. V. Molloy	Pacific Gas & Electric
J. T. Muller	Leslie Company
H. R. Nickerson	Resistoflex Company
J. Ozol	Omaha Public Power Company
R. A. Quance	Comeau, Boyle Inc. (now consultant)
W. Rahmeyer	Colorado State University (now with Utah State University)
J. Reed	Masoneilan International, Inc. (now Masoneilan/Dresser
G. Richards	Jordan Valve (now with Richards Industries, Inc.)
J. Rosato	Rawson Company
K. Schoonover, Secretary	Con-Tek
H. Schwartz	Flexible Valve Corporation
F. O. Seger	Willis Oil Tool Company (now with Masoneilan/Dresser)
J. M. Simonsen	Valtek, Inc. (retired)
H. R. Sonderregger	ITT Grinnell Corporation
N. Sprecher	DeZurik
R. U. Stanley	Retired
G. Stiles/R. Brodin**	Fisher Controls Company (now Fisher Controls International, Inc.)
R. E. Terhune	Consultant
R. F. Tubbs	Copes-Vulcan, Inc.
W. C. Weidman	Gilbert Associates (Gilbert Commonwealth, Inc.)
R. L. Widdows	Cashco. Inc.
P. Wing	Masoneilan (retired)
L. Zinck	Union Carbide Corporation

*Membership list and company names at the time of the final vote (current company names are given in parentheses)
**One vote

Instrument Society of America

This standard was approved for publication by the ISA Standards and Practices Board in June 1987.

NAME	COMPANY
D. E. Rapley, Chairman	Rapley Engineering Services
D. N. Bishop	Chevron U.S.A. Inc.
W. Calder III	The Foxboro Company
N. Conger	Fisher Controls International, Inc.
R. S. Crowder	Ship Star Associates
C. R. Gross	Eagle Technology
H. S. Hopkins	Utility Products of Arizona
J. L. Howard	Boeing Aerospace Company
R. T. Jones	Philadelphia Electric Company
R. Keller	The Boeing Company
O. P. Lovett, Jr.	ISIS Corporation
E. C. Magison	Honeywell, Inc.
A. P. McCauley	Chagrin Valley Controls, Inc.
J. W. Mock	The Bechtel Group, Inc.
E. M. Nesvig	ERDCO Engineering Corporation
R. Prescott	Moore Products Company
C. W. Reimann	National Bureau of Standards
R. H. Reimer	Allen-Bradley Company
J. Rennie	Factory Mutual Research Corporation
W. C. Weidman	Gilbert/Commonwealth, Inc.
K. A. Whitman	Fairleigh Dickinson University
*P. Bliss	Consultant
*B. A. Christensen	Continental Oil Company
*L. N. Combs	Consultant
*R. L. Galley	Consultant
*T. J. Harrison	IBM Corporation
*R. G. Marvin	Roy G. Marvin Company
*W. B. Miller	Moore Products Company
*G. Platt	Consultant
*J. R. Williams	Stearns Catalytic Corporation

*Director Emeritus

TABLE OF CONTENTS

Section	Title	Page
1.	Purpose	9
2.	Scope	9
3.	Test System	9
4.	Testing Procedures	12
5.	Test Data	12

LIST OF FIGURES

Figure	Title	Page
1.	Control Valve Noise Test — System Components	9
2.	Test Arrangement for Test Specimen Outside Test Chamber	10
3.	Test Arrangement for Test Specimen Inside Test Chamber	11

1 PURPOSE

The purpose of this standard is to provide a procedure for testing, measuring, and reporting the aerodynamic noise-generating characteristics of a control valve and its associated piping.

2 SCOPE

This standard defines equipment, methods, and procedures for the laboratory testing and measurement of airborne sound radiated by a compressible fluid flowing through a control valve and its associated piping, including fixed-flow restrictions. The test may be conducted under any conditions mutually agreed upon by the user and the manufacturer. Although this standard is designed for measurement of the noise radiated from the piping downstream of the valve, other test variations are optional, including the use of insulation and nonstandard piping. (See Section 4.3.) Applications of this standard to control valves discharging directly to atmosphere are excluded from this standard.

3 TEST SYSTEM

The test system is shown in Figures 1, 2, and 3. The various parts are described below.

3.1 Throttling Valves

The upstream and/or downstream throttling valves (optional) are used to regulate the test pressures. Caution should be taken to avoid pressure drops which will create significant stream-borne noise. If such pressure drops are unavoidable, then silencers must be used.

3.2 Test Specimen

The test specimen is any valve, combination of valves, fixed restrictions, and associated piping components for which data are required. The test specimen and test section shall not be insulated, although optional tests may be conducted to determine the effect of insulation. (See Section 4.3.)

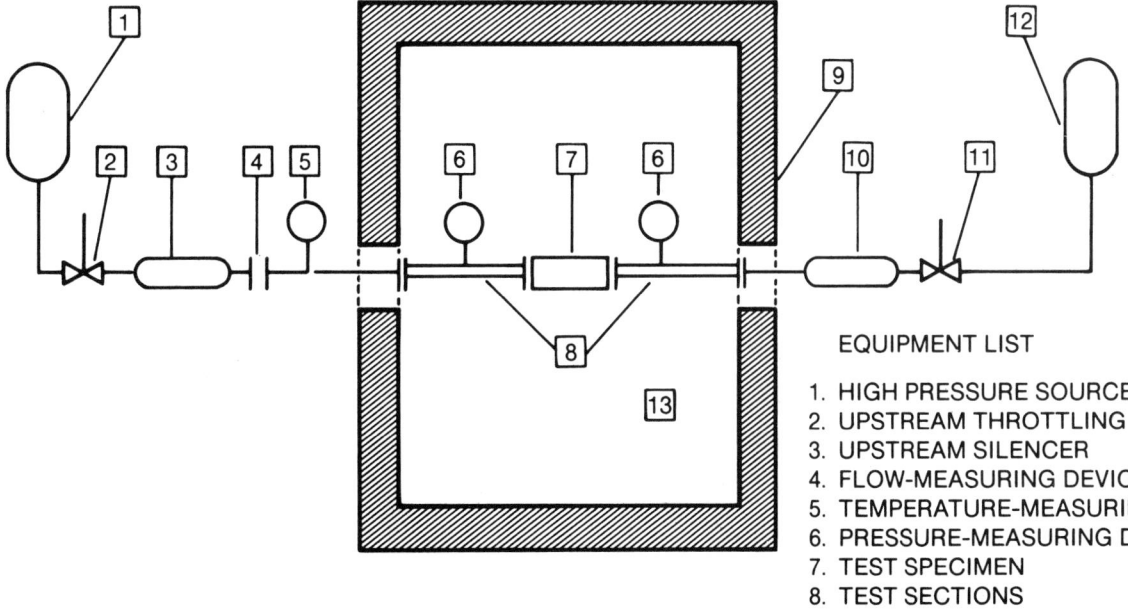

NOTES:
1. MICROPHONE PLACEMENT IS SHOWN IN FIGURES 2 AND 3
2. ITEMS 2, 3, 9, 10 AND 11 ARE OPTIONAL
3. ITEM 4 LOCATION IS OPTIONAL

EQUIPMENT LIST

1. HIGH PRESSURE SOURCE
2. UPSTREAM THROTTLING VALVE
3. UPSTREAM SILENCER
4. FLOW-MEASURING DEVICE
5. TEMPERATURE-MEASURING DEVICE
6. PRESSURE-MEASURING DEVICE
7. TEST SPECIMEN
8. TEST SECTIONS
9. TEST CHAMBER (ACOUSTIC ENVIRONMENT)
10. DOWNSTREAM SILENCER
11. DOWNSTREAM THROTTLING VALVE
12. LOW PRESSURE RECEIVER
13. MICROPHONE (NOTE 1)

Figure 1. Control Valve Noise Test — System Components

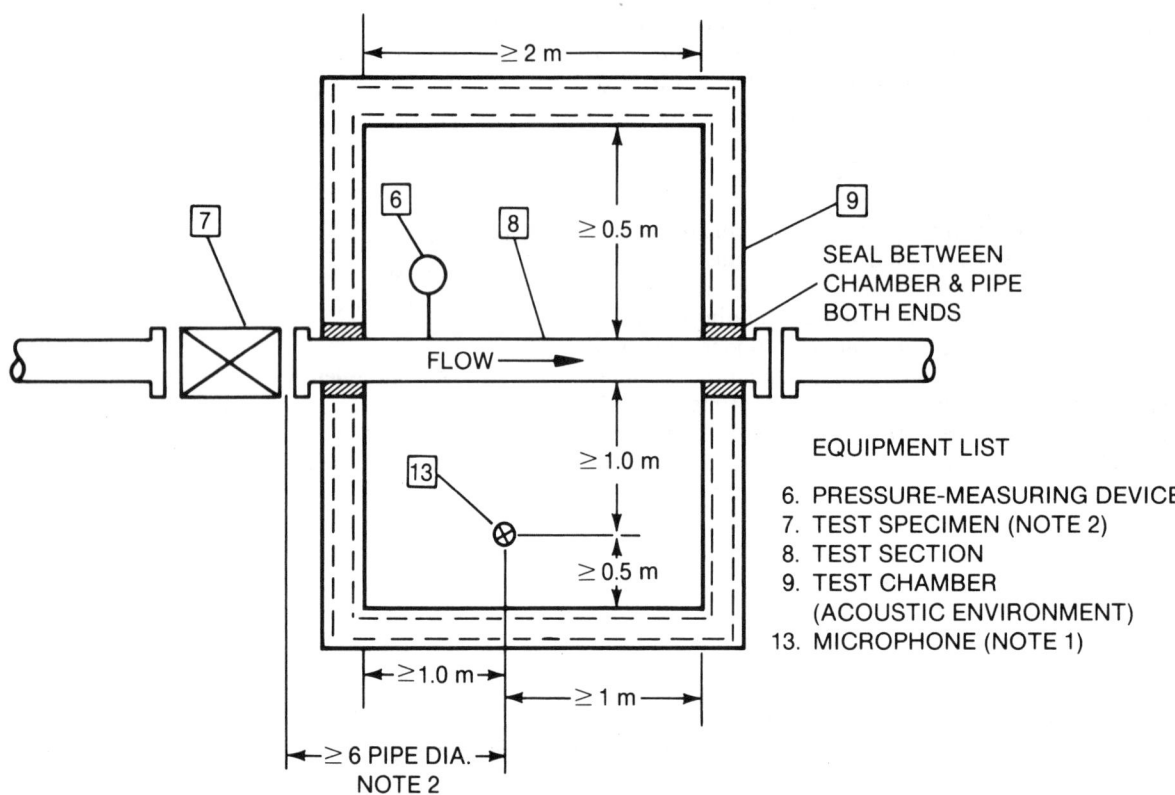

NOTES:
1. THE MICROPHONE MAY BE LOCATED ANYWHERE ON THE CIRCUMFERENCE OF THE CIRCLE HAVING A RADIUS OF PIPE OUTSIDE DIAMETER /2 + 1 m, PROVIDING IT IS NO CLOSER THAN 0.5 m TO THE NEAREST SURFACE.
2. SEE PARAGRAPH 4.2. THE TEST SECTION AND TEST SPECIMEN SHOULD BE AS CLOSELY COUPLED AS PHYSICALLY POSSIBLE.

Figure 2. Test Arrangement for Test Specimen Outside Test Chamber

3.3 Test Section Piping

There is no limitation concerning the maximum length of upstream and downstream piping connected to the test specimen. The exposed pipe within the acoustic environment shall be a minimum of 2 meters (m) in length, free of mechanical joints except for the connections between the test specimen and test section (upstream or downstream pipe, depending upon the test conducted). Piping for each side of the test section shall be Schedule 40 steel pipe for valves through 10-inch (in) body size (250 mm), having pressure ratings up to ANSI Class 600. Pipe having 0.375-in (10-mm) wall thickness shall be used for sizes ranging from 12 in (300 mm) through 24 in (600 mm).

An effort should be made to match the test specimen inlet and outlet inside diameters with those of the adjacent piping for valve sizes exceeding the above-mentioned limits. Uninsulated pipe shall be used. Other pipe schedules, pipe materials, or insulated piping may be used for optional tests. (See Section 4.3.)

3.4 Pressure Taps

Pressure taps shall be provided for the measurement of pressures, and the taps shall conform to ANSI/ISA S75.02-1982, "Control Valve Capacity Test Procedure," Paragraph 2.6.*

---

*Published by the Instrument Society of America, 67 Alexander Drive, P.O. Box 12277, Research Triangle Park, NC 27709.

S75.07

Laboratory Measurement of Aerodynamic
Noise Generated by Control Valves

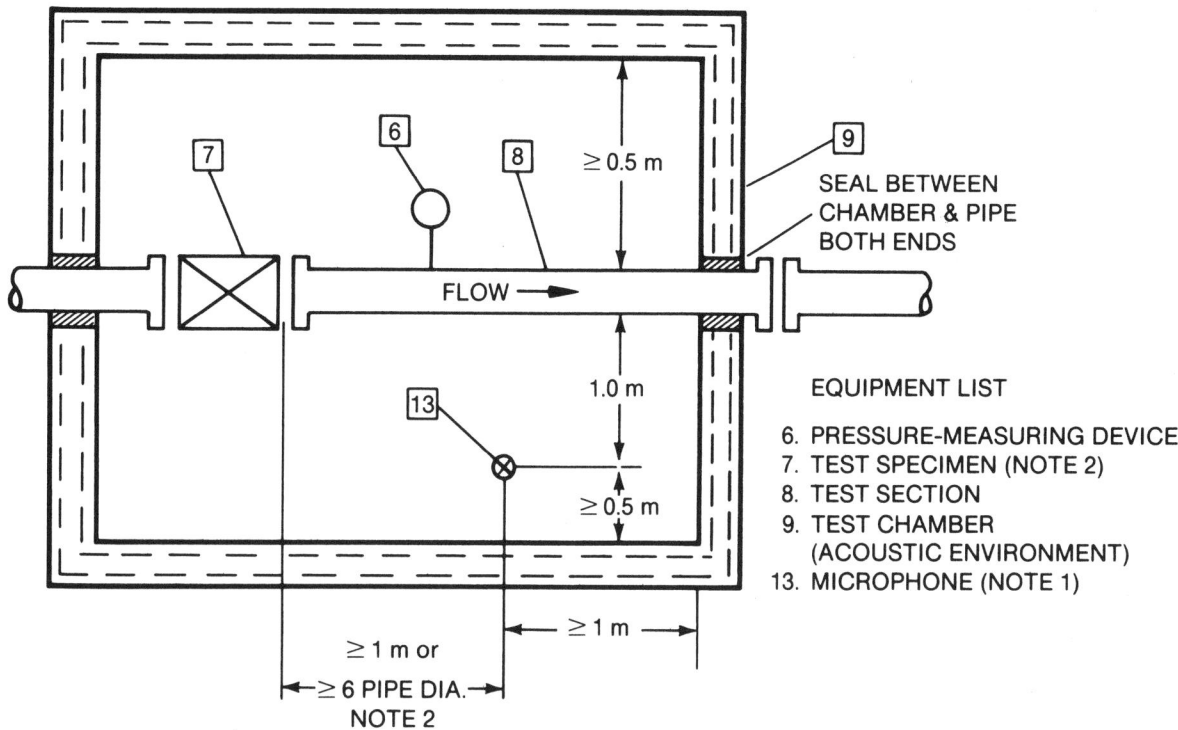

NOTES:
1. THE MICROPHONE MAY BE LOCATED ANYWHERE ON THE CIRCUMFERENCE OF THE CIRCLE HAVING A RADIUS OF PIPE OUTSIDE DIAMETER /2 + 1 m, PROVIDING IT IS NO CLOSER THAN 0.5 m TO THE NEAREST SURFACE.
2. SEE PARAGRAPH 4.2. THE TEST SECTION AND TEST SPECIMEN SHOULD BE AS CLOSELY COUPLED AS PHYSICALLY POSSIBLE.

Figure 3. Test Arrangement for Test Specimen Inside Test Chamber

3.5 Acoustic Environment

The test environment shall be controlled such that background noise, reflected noise, and other extraneous noise sources are at a minimum of 10 dB lower than the noise radiated by the test section. Depending upon the test system and the acoustic environment, upstream and/or downstream silencers may be necessary. General recommendations for the acoustic environment can be found in ANSI S1.13-1976, "Methods for the Measurement of Sound Pressure Levels."* No sound level correction shall be made for excessive sound levels caused by the test system, the acoustic environment, or other reasons.

3.6 Instrumentation

The instrumentation for sound level measurements shall conform to ANSI S1.13-1976, Section 5, entitled "Instrumentation for Noise Measurements." Specifications for sound level meters shall conform to ANSI S1.4-1983, "Specification for Sound Level Meters."** Calibration and sensitivity checks shall be corrected for atmospheric pressure to sea level conditions. Accuracy of flow, pressure, and temperature measurements shall conform to ANSI/ISA S75.02-1982, "Control Valve Capacity Test Procedure."

---

*Published by the American National Standards Institute, 1430 Broadway, New York, NY 10018.

**Ibid.

4 TESTING PROCEDURES

4.1 Fluid

Air is the preferred test fluid to be used, but other compressible fluids may be substituted where need and availability dictate. The fluid shall be sufficiently dry to ensure that any icing which may take place does not significantly affect the test results. Saturated vapors are not acceptable as test fluids, unless data are required for application to the particular saturated vapor.

4.2 Microphone Position

The microphone shall be located 1 m away from the nearest pipe surface. Downstream location shall be 1 m from the beginning of the exposed part of the test section, or six nominal pipe diameters downstream of the test specimen outlet, whichever is greater. For test specimens having multiple flow passages, the 6 pipe diameters may be changed to 10 hydraulic diameters of the largest single-flow passage of the test specimen. (For any circular cross section, Hydraulic Diameter D = 4 × Hydraulic Radius, or 4 × Area/Wetted Perimeter.)

4.3 Optional Tests

Additional tests may be conducted to evaluate nonstandard factors. These tests shall be conducted to otherwise concur with the parameters of this standard. Any exceptions shall be noted in the test data.

4.4 Blowdown Test Limitations

Blowdown test results are intended to simulate steady-state test results. As a guideline, the difference between the blowdown test results and the steady-state test results should not exceed 2 dB, for equivalent flow conditions. Such assurance of accuracy could be established mathematically, or through a small-scale demonstration test. The blowdown rate shall be limited to conform with ANSI S1.13-1976, "Methods for the Measurement of Sound Pressure Levels," Section 2, Definitions, and shall not exceed the transducer response and data acquisition capabilities of the instrumentation system.

In the blowdown method of testing, the inlet pressure to the test specimen decays during the test period. The blowdown rate is the rate at which the inlet pressure to the test specimen changes.

5 TEST DATA (State Units Used)

The minimum data to be recorded and reported are as follows:

1. Upstream pressure

2. Pressure drop (ΔP) and/or downstream pressure

3. Upstream fluid temperature

4. Flow rate

5. Valve travel (percent of full travel, ±2%)

6. Valve C_v at the test travel position(s)

7. Acoustic data consisting of the "A-weighted" sound level and either a 1/3-octave or full-octave band analysis, which shall be recorded over the frequency range of 180 Hz (250 Hz for full-octave band, or 200 Hz for 1/3-octave band center frequency) to 22 400 Hz (16 000 Hz for full-octave band, or 20 000 Hz for 1/3-octave band center frequency). Narrow band data may be obtained when furthur frequency resolution is desired.

8. Description of the complete test specimen

9. Description of the test facility, including:

    a. Piping and instrumentation schematic, including the pipe size, material, and wall thickness

    b. Description of the environmental chamber (if used)

    c. Dimension sketch of the test facility

10. Test fluid and its molecular weight or specific gravity

11. Instruments used (manufacturer, model number, span, accuracy specification)

12. Microphone position

13. Any deviations from this standard

RP2.1 — Manometer Tables
Reaffirmed-1978, 31 pp.

RP3.2 — Flange Mounted Sharp Edged Orifice Plates for Flow Measurement
1984, 52 pp.

S5.1 — Instrumentation Symbols and Identification (Formerly ANSI Y32.20)
1984, 55 pp.

S5.2 — Binary Logic Diagrams for Process Operations
ANSI/ISA-1976
(R 1981)
Reaffirmed-1981, 19 pp.

S5.2 — Diagramas Logicos Binarios Para Operaciones de Proceso
(Spanish Edition)

S5.3 — Graphic Symbols for Distributed Control/Shared Display Instrumentation, Logic and Computer Systems
1982, 16 pp.

S5.4 — Instrument Loop Diagrams
ANSI/ISA-1976 (R 1981)
Reaffirmed-1981, 11 pp.

S5.4 — Diagramos de Circuito de Instrumentos (Spanish Edition)

S5.5 — Graphic Symbols for Process Displays
ANSI/ISA-1986
1986, 44 pp.

RP7.1 — Pneumatic Control Circuit Pressure Test
1956, 6 pp.

S7.3 — Quality Standard for Instrument Air
ANSI/ISA-1975 (R 1981)
Reaffirmed-1981, 6 pp.

S7.4 — Air Pressures for Pneumatic Controllers, Transmitters, and Transmission Systems
1981, 4 pp.

RP7.7 — Recommended Practice for Producing Quality Instrument Air
1984, 16 pp.

RP12.1 — Electrical Instruments in Hazardous Atmospheres
1960, 7 pp.

S12.4 — Instrument Purging for Reduction of Hazardous Area Classification
1970, 12 pp.

RP12.6 — Installation of Intrinsically Safe Instrument Systems in Class I Hazardous Locations
ANSI/ISA-1977
1977, 12 pp.

S12.10 — Area Classification in Hazardous Dust Locations
1973, 23 pp.

S12.11 — Electrical Instruments in Hazardous Dust Locations
1973, 10 pp.

S12.12 — Electrical Equipment for Use in Class I, Division 2 Hazardous (Classified) Locations
1984, 24 pp.

RP16.1,2,3 — Terminology, Dimensions and Safety Practices for Indicating Variable Area Meters (Rotameters, Glass Tube, Metal Tube, Extension Type Glass Tube)
1959, 6 pp.

RP16.4 — Nomenclature and Terminology for Extension Type Variable Area Meters (Rotameters)
1960, 3 pp.

RP16.5 — Installation, Operations, Maintenance Instructions for Glass Tube Variable Area Meters (Rotameters)
1961, 6 pp.

RP16.6 — Methods and Equipment for Calibration of Variable Area Meters (Rotameters)
1961, 7 pp.

S18.1 — Annunciator Sequences and Specifications
ANSI/ISA-1979 (R 1985)
Reaffirmed-1985, 36 pp.

S20 — Specification Forms for Process Measurement and Control Instruments, Primary Elements and Control Valves
Reaffirmed-1981, 72 pp.

S26 — Dynamic Response Testing of Process Control Instrumentation
ANSI/MC4.1-1985
1975, 25 pp.

26F — Frequency Response from Pulse Test Data

RP31.1 — Specification, Installation, and Calibration of Turbine Flowmeters
ANSI/ISA-1977
1977, 21 pp.

S37.1 — Electrical Transducer Nomenclature and Terminology
ANSI/ISA-1975 (R 1982)
Reaffirmed-1982, 15 pp.

RP37.2 — Guide for Specifications and Tests for Piezoelectric Acceleration Transducers for Aerospace Testing
Reaffirmed-1982, 19 pp.

S37.3 — Specifications and Tests for Strain Gage Pressure Transducers
ANSI/ISA-1975 (R 1982)
Reaffirmed-1982, 22 pp.

S37.6 — Specifications and Tests of Potentiometric Pressure Transducers
ANSI/ISA-1976 (R 1982)
Reaffirmed-1982, 27 pp.

S37.8 — Specifications and Tests for Strain Gage Force Transducers
ANSI/ISA-1977 (R 1982)
Reaffirmed-1982, 15 pp.

S37.10 — Specifications and Tests for Piezoelectric Pressure and Sound-Pressure Transducers
ANSI/ISA-1975 (R 1982)
Reaffirmed-1982, 22 pp.

S37.12 — Specifications and Tests for Potentiometric Displacement Transducers
ANSI/ISA-1977 (R 1982)
Reaffirmed-1982, 21 pp.

RP42.1 — Nomenclature for Instrument Tube Fittings
1982, 12 pp.

S50.1 — Compatibility of Analog Signals for Electronic Industrial Process Instruments
ANSI/ISA-1982
Reaffirmed-1982, 11 pp.

S51.1 — Process Instrumentation Terminology
ANSI/ISA-1979
1979, 44 pp.

RP52.1 — Recommended Environments for Standards Laboratories
1975, 18 pp.

RP55.1 — Hardware Testing of Digital Process Computers
ANSI/ISA-1975 (R 1983)
Reaffirmed-1983, 54 pp.

RP60.3 — Human Engineering for Control Centers
1985, 16 pp.

RP60.6 — Nameplates, Labels and Tags for Control Centers
1984, 24 pp.

RP60.8 — Electrical Guide for Control Centers
1978, 6 pp.

RP60.9 — Piping Guide for Control Centers
1981, 12 pp.

S61.1 — Industrial Computer System FORTRAN Procedures for Executive Functions, Process Input-Output, and Bit Manipulation
ANSI/ISA-1977
1977, 11 pp.

S61.2 — Industrial Computer System FORTRAN Procedures for File Access and the Control of File Contention
ANSI/ISA-1978
1978, 7 pp.

S67.01 — Transducer and Transmitter Installation for Nuclear Safety Applications
ANSI/ISA-1979 (R 1986)
1979, 16 pp.

S67.02 — Nuclear Safety-Related Instrument Sensing Line Piping and Tubing Standards for Use in Nuclear Power Plants
ANSI/ISA-1980
1980, 32 pp.

S67.03 — Standard for Light Water Reactor Coolant Pressure Boundary Leak Detection
1982, 28 pp.

S67.04 — Setpoints for Nuclear Safety-Related Instrumentation Used in Nuclear Power Plants
1982, 16 pp.

S67.06 — Response Time Testing of Nuclear Safety-Related Instrument Channels in Nuclear Power Plants
ANSI/ISA-1986
1984, 20 pp.

S67.10 — Sample-Line Piping and Tubing Standard for Use in Nuclear Power Plants
1986, 20 pp.

S67.14 — Qualifications and Certification of Instrumentation and Control Technicians in Nuclear Power Plants
1983, 16 pp.

S71.01 — Environmental Conditions for Process Measurement and Control Systems: Temperature and Humidity
ANSI/ISA-1986
1985, 20 pp.

S71.04 — Environmental Conditions for Process Measurement and Control Systems: Airborne Contaminants
ANSI/ISA-1985
1985, 20 pp.

S72.01 — PROWAY-LAN Industrial Data Highway
ANSI/ISA-1985
1985, 186 pp.

RP74.01 — Application and Installation of Continuous-Belt Weighbridge Scales
1984, 28 pp.

S75.01 — Flow Equations for Sizing Control Valves
ANSI/ISA-1986
1977, 11 pp.

S75.02 — Control Valve Capacity Test Procedure
ANSI/ISA-1982
1981, 20 pp.

S75.03 — Face-to-Face Dimensions for Flanged Globe-Style Control Valve Bodies (ANSI Classes 125, 150, 250, 300, 600)
ANSI/ISA-1985
1984, 12 pp.

S75.04 — Face-to-Face Dimensions for Flangeless Control Valves (ANSI Classes 150, 300, 600)
ANSI/ISA-1985
1984, 8 pp.

S75.05 — Control Valve Terminology
ANSI/ISA-1986
1983, 33 pp.

S75.06 — Control Valve Manifold Designs
1981, 16 pp.

S75.08 — Installed Face-to-Face Dimensions for Flanged Clamp or Pinch Valves
ANSI/ISA-1985
1985, 14 pp.

S75.11 — Inherent Flow Characteristic and Rangeability of Control Valves
1984, 16 pp.

S75.12 — Face-to-Face Dimensions for Socket Weld-End and Screwed-End Globe-Style Control Valves (ANSI Classes 150, 300, 600, 900, 1500, and 2500)
1986, 12 pp.

S75.14 — Face-to-Face Dimensions for Buttweld-End Globe-Style Control Valves (ANSI Class 4500)
ANSI/ISA-1986
1984, 8 pp.

S75.15 — Face-to-Face Dimensions for Buttweld-End Globe-Style Control Valves (ANSI Classes 150, 300, 600, 900, 1500, and 2500)
1986, 12 pp.

S75.16 — Face-to-Face Dimensions for Flanged Globe-Style Control Valve Bodies (ANSI Classes 900, 1500, and 2500)
1986, 8 pp.

MC96.1 — American National Standard for Temperature Measurement Thermocouples
1982, 48 pp.

ANSI C100.6-3 — American National Standard for Voltage or Current Reference Devices: Solid State Devices
1984, 12 pp.

Call the Order Department at ISA Headquarters, (919) 549-8411, to request a complete catalog of ISA Publications and Training Aids. Prices are subject to change without notice. For more information, call toll-free, 1-800-334-6391.

ANSI/ISA-S75.08-1985
Approved: February 19, 1986

American National Standard

Installed Face-to-Face Dimensions for Flanged Clamp or Pinch Valves

Instrument Society of America

Instrument Society of America

ISBN 0-87664-907-X

ISA-S75.08 Installed Face-to-Face Dimensions for Flanged Clamp or Pinch Valves

Copyright © 1985 by the Instrument Society of America. All rights reserved. Printed in the United States of America. No part of this publication may be reproduced, stored in a retrieval system, or transmitted, in any form or by any means (electronic, mechanical, photocopying, recording, or otherwise), without the prior written permission of the Publisher.

INSTRUMENT SOCIETY OF AMERICA
67 Alexander Drive
P.O. Box 12277
Research Triangle Park, NC 27709

PREFACE

This preface is included for informational purposes and is not part of ISA-S75.08.

This standard has been prepared as part of the service of the Instrument Society of America (ISA) toward a goal of uniformity in the field of instrumentation. To be of real value, this document should not be static, but should be subject to periodic review. Toward this end, the Society welcomes all comments and criticisms, and asks that they be addressed to the Secretary, Standards and Practices Board, Instrument Society of America, 67 Alexander Drive, P.O. Box 12277, Research Triangle Park, NC 27709, Telephone (919) 549-8411.

The ISA Standards and Practices Department is aware of the growing need for attention to the metric system of units in general, and the International System of Units (SI) in particular, in the preparation of instrumentation standards. The Department is further aware of the benefits to U.S.A. users of ISA standards of incorporating suitable references to the SI (and the metric system) in their business and professional dealings with other countries. Toward this end, this Department will endeavor to introduce SI-acceptable metric units in all new and revised standards to the greatest extent possible. The Metric Practice Guide, which has been published by the Institute of Electrical and Electronics Engineers as ANSI/IEEE Std. 268-1982, and future revisions will be the reference guide for definitions, symbols, abbreviations, and conversion factors.

It is the policy of the Instrument Society of America to encourage and welcome the participation of all concerned individuals and interests in the development of ISA standards. Participation in the ISA standards-making process by an individual in no way constitutes endorsement by the employer of that individual, of the Instrument Society of America, or of any of the standards that ISA develops.

The information contained in this preface, footnotes, and appendices is included for information only and is not a part of the standard.

The following people served as members of ISA Subcommittee SP75.10, which prepared this standard:

NAME	COMPANY
H. R. Nickerson, Chairman	Resistoflex Company
G. Borden	Bechtel Power Corporation
C. Clarkson	Clarkson Company
F. Dicosimo	Robbin-Meyers
J. T. Emery	Honeywell, Inc.
B. Howes	Galigher-Ash Valve
R. B. Jones	Dow Chemical Company
M. W. Kaye	M. W. Kellogg Company
A. Levin	Red Valve Company, Inc.
E. Lincoln	Flow-Con Valve Corporation
O. P. Lovett, Jr.	ISIS Corporation
R. A. Quance	Walsh Inc.
H. Rich/R. Brodin	Fisher Controls International, Inc.
H. Schwartz	Flexible Valve Corporation
J. M. Simonsen	Valtek, Inc.
R. U. Stanley	Retired
W. C. Weidman	Gilbert/Commonwealth, Inc.

Instrument Society of America

The following people served as members of ISA Committee SP75 during the review of this standard:

NAME	COMPANY
L. R. Driskell, Chairman	Consultant
J. B. Arant	E. I. du Pont de Nemours and Company, Inc.
H. E. Backinger	John F. Kraus & Company
G. Barb	Muesco, Inc.
H. D. Baumann	H. D. Baumann Associates, Ltd.
*C. S. Beard	
N. Belaef	Consultant
G. Borden	Bechtel Power Corporation
**R. Brodin/G. F. Stiles	Fisher Controls International, Inc.
E. H. C. Brown	Dravo Engineers, Inc.
E. J. Cooney	Air Products & Chemicals, Inc.
W. G. Dewart	Rockwell International
J. T. Emery	Honeywell, Inc.
H. J. Fuller	Worcester Controls Corporation
L. Griffith	Consultant
A. J. Hanssen	Fluid Controls Institute, Inc.
F. P. Harthun	Fisher Controls International, Inc.
H. P. Illing	Kieley & Mueller, Inc.
R. B. Jones	Dow Chemical Company
M. W. Kaye	M. W. Kellogg Company
R. Louviere	Creole Engineering
O. P. Lovett, Jr.	ISIS Corporation
A. P. McCauley	Chagrin Valley Controls, Inc.
T. V. Molloy	Pacific Gas & Electric
H. R. Nickerson	Resistoflex Company
J. Ozol	Omaha Public Power Company
R. A. Quance	Walsh Inc.
W. Rahmeyer	Colorado State University
J. N. Reed	Masoneilan/Dresser
G. Richards	Jordan Valve Div., Richards Industries, Inc.
J. Rosato	Rawson Company
K. Schoonover, Secretary	Con-Tek
H. Schwartz	Flexible Valve Corporation
**W. L. Scull/J. T. Muller	Leslie Company
F. O. Seger	Willis Division, Smith International, Inc.
J. M. Simonsen	Valtek, Inc.
H. R. Sonderregger	ITT Grinnell Corporation
N. D. Sprecher	DeZurik
R. U. Stanley	Retired
R. E. Terhune, Vice-Chairman	Consultant
R. F. Tubbs	Copes-Vulcan
W. C. Weidman	Gilbert/Commonwealth, Inc.
R. L. Widdows	Cashco, Inc.
L. Zinck	Union Carbide Corporation

*Deceased
**One vote

This standard was approved for publication by the ISA Standards and Practices Board in August 1985.

NAME	COMPANY
N. Conger, Chairman	Fisher Controls Company
P. V. Bhat	Monsanto Company
W. Calder III	The Foxboro Company
R. S. Crowder	Ship Star Associates
B. Feikle	OTIS
H. S. Hopkins	Westinghouse Electric Company
J. L. Howard	Boeing Aerospace Company
R. T. Jones	Philadelphia Electric Company
R. Keller	The Boeing Company
O. P. Lovett, Jr.	ISIS Corporation
E. C. Magison	Honeywell, Inc.
A. P. McCauley	Chagrin Valley Controls, Inc.
J. W. Mock	Bechtel Corporation
E. M. Nesvig	ERDCO Engineering Corporation
R. Prescott	Moore Products Company
D. E. Rapley	Rapley Enginering Services
C. W. Reimann	National Bureau of Standards
J. Rennie	Factory Mutual Research Corporation
W. C. Weidman	Gilbert/Commonwealth, Inc.
K. Whitman	Consultant
*P. Bliss	Consultant
*B. A. Christensen	Continental Oil Company
*L. N. Combs	Retired
*R. L. Galley	Consultant
*T. J. Harrison	IBM Corporation
*R. G. Marvin	Roy G. Marvin Company
*W. B. Miller	Moore Products Company
*G. Platt	Retired
*J. R. Williams	Stearns Catalytic Corporation

*Director Emeritus

1 SCOPE

1.1 This standard applies to valves, sizes 1 inch through 8 inches, of the clamp or pinch valve design incorporating clamp or pinch elements.

2 PURPOSE

2.1 The purpose of this standard is to aid users in their piping design by providing installed face-to-face dimensions for control valves, incorporating clamp or pinch elements, which have flanges that mate with ANSI B16.1 Class 125 (PN20) and/or ANSI B16.5 Class 150 (PN20) flanges, without giving special consideration to the manufacturer of the equipment to be used.

2.2 This standard excludes solenoid-actuated valves and direct fluid-actuated valves.

2.3 In the absence of other standards, this standard can be applied to handwheel-operated valves.

3 DEFINITIONS

3.1 For definitions of terms used in this standard, see ISA-S75.05-1983, "Control Valve Terminology."

3.2 *One-Piece-Element Clamp* or *Pinch Valve* — A *one-piece-element clamp* or *pinch valve* is a valve consisting of a one-piece flexible element or liner installed in a body with the element or liner extending over the flange faces and acting as gaskets between the valve and connecting piping. See Table 1 for dimensions. "Long style" dimensions are the same as those for flanged-end control valves in ANSI B16.10 and ANSI/ISA-S75.03-1985 except for the 6-inch size. "Short style" dimensions are the same as those for flanged-end gate valves in ANSI B16.10 and MSS SP-72, except for the 6-inch and the 8-inch sizes.

3.3 *Two-Piece-Element Clamp* or *Pinch Valve* — A *two-piece-element clamp* or *pinch valve* is a valve consisting of two flexible elastomeric elements or liners installed between a two-piece flanged body. The flexible elements or liners also extend over the flange faces and act as gaskets between the valve and the connecting piping. See Table 2 for dimensions. No prior standards exist for these two-piece-element valves.

3.4 *Face-to-Face* — Each of the *face-to-face* dimensions in the tables is to be applied to the installed length of the valve after installation into a piping system. The free length face-to-face dimension of a valve before installation into a piping system may be slightly longer than dimensions shown in either Table 1 or Table 2.

4 BIBLIOGRAPHY

4.1 American National Standards Institute, Inc. (ANSI) Standard B16.10-1973, "Face-to-Face and End-to-End Dimensions of Ferrous Valves."

4.2 American National Standards Institute, Inc., ANSI/ISA-S75.03-1985, "Face-to-Face Dimensions for Flanged Globe-Style Control Valve Bodies."

4.3 Instrument Society of America (ISA) Standard ISA-S75.05-1983, "Control Valve Terminology."

4.4 Manufacturers Standardization Society of the Valve and Fittings Industry, Inc. (MSS), SP-72-1970, "Ball Valves with Flanged or Butt-Welding Ends for General Service."

5 DIMENSIONAL DATA OF INSTALLED VALVES

5.1 For face-to-face dimensions for flanged clamp or pinch valves, see Tables 1 and 2.

TABLE 1

One-Piece-Element Design

(See Paragraph 3.2)

Nominal Valve Size		Installed Face-to-Face Dimension "A"[a]			
		Long Style		Short Style	
mm	inches	mm[b,c]	inches[b,c]	mm[b,c]	inches[b,c]
25	1	184	7.25	127	5.00
40	1½	222	8.75	165	6.50
50	2	254	10.00	178	7.00
65	2½	276	10.88	190	7.50
80	3	298	11.75	203	8.00
100	4	352	13.88	229	9.00
125	5	381	15.00	—	—
150	6	457	18.00	406	16.00
200	8	543	21.38	483	19.00

[a] See Figure 1 for dimension "A"

[b] Tolerance for 25 through 125-mm (1 through 5-inch) size valves: ±1.5 mm (±0.06 inches)

[c] Tolerance for 150 and 200-mm (6 and 8-inch) size valves: ±3.0 mm (±0.12 inches)

TABLE 2

Two-Piece-Element Design

(See Paragraph 3.3)

Nominal Valve Size		Installed Face-to-Face Dimension "A"[a]	
mm	inches	mm[b]	inches[b]
25	1	133	5.25
40	1½	178	7.00
50	2	229	9.00
65	2½	254	10.00
80	3	305	12.00
100	4	381	15.00
150	6	495	19.50
200	8	625	24.63

[a] See Figure 1 for dimension "A"

[b] Tolerance for all size valves: ±1.5 mm (±0.06 inches)

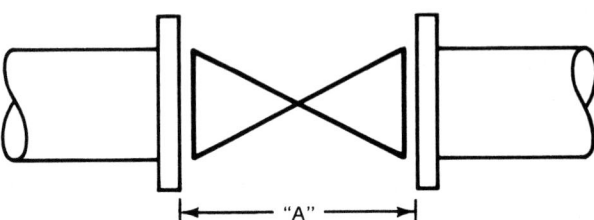

Figure 1. Installed Face-to-Face Dimension "A"

APPENDIX

This appendix is not part of the ISA Standard S75.08 but is included to facilitate its use.

Dimensions for metrically sized valves are nominal conversions that are conventionally used in documents by the Manufacturers Standardization Society (MSS) of the Valve and Fitting Industry (MSS-SP86-1981), by the International Organization for Standardization (ISO), and by the International Electrotechnical Commission (IEC).

ANSI/ISA-S75.11-1985
Approved April 3, 1985

American National Standard

Inherent Flow Characteristic and Rangeability of Control Valves

Instrument Society of America

ISBN 0-87664-835-9

ISA-S75.11 Inherent Flow Characteristic and
Rangeability of Control Valves

Copyright © 1984 by the Instrument Society of America. All rights reserved. Printed in the United States of America. No part of this publication may be reproduced, stored in a retrieval system, or transmitted, in any form or any means electronic, mechanical, photocopying, recording or otherwise without the prior written permission of the publisher.

INSTRUMENT SOCIETY OF AMERICA
67 Alexander Drive
P.O. Box 12277
Research Triangle Park, North Carolina 27709

Copyright © 1984 by the Instrument Society of America

PREFACE

This preface is included for information purposes and is not part of ISA-S75.11.

This standard has been prepared as part of the service of the Instrument Society of America (ISA) toward a goal of uniformity in the field of instrumentation. To be of real value, this document should not be static, but should be subject to periodic review. Toward this end, the Society welcomes all comments and criticisms, and asks that they be addressed to the Secretary, Standards and Practices Board, Instrument Society of America, 67 Alexander Drive, P.O. Box 12277, Research Triangle Park, NC 27709, Telephone (919) 549-8411.

The ISA Standards and Practices Department is aware of the growing need for attention to the metric system of units in general, and the International System of Units (SI) in particular, in the preparation of instrumentation standards. The Department is further aware of the benefits to U.S.A. users of ISA standards of incorporating suitable references to the SI (and the metric system) in their business and professional dealings with other countries. Toward this end, this Department will endeavor to introduce SI-acceptable metric units in all new and revised standards to the greatest extent possible. The Metric Practice Guide, which has been published by the Institute of Electrical and Electronic Engineers as ANSI/IEEE Std. 268-1982, and future revisions will be the reference guide for definitions, symbols, abbreviations, and conversion factors.

It is the policy of the Instrument Society of America to encourage and welcome the participation of all concerned individuals and interests in the development of ISA standards. Participation in the ISA standards-making process by an individual in no way constitutes endorsement by the employer of that individual, of the Instrument Society of America, or of any of the standards that ISA develops.

Prior to the issuance of this standard, there had been no standard which provided allowable deviations for control valve flow characteristics and which established criteria for rangeability of control valves.

In contrast to conventional globe valves, most rotary motion control valve types such as ball valves, butterfly valves or plug valves do not have a mathematically definable flow characteristic. The users of control valves, therefore, have to depend on the manufacturer to state the specific flow characteristic for a given style or size of valve either in graphic or tabular form. For sake of consistency, this method of presentation was also adapted for generic flow characteristics such as "equal-percentage" or "linear."

This standard states the limits within which a stated flow characteristic can be expected to be reproducible. Knowledge of specific flow coefficients (within allowable deviations) at stated travel positions will enable the user to calculate the installed flow characteristic for a specific control system.

The stated inherent rangeability of a specific control valve is related solely to the interaction between the closure member and the flow control orifice of a valve. This given value may not be applicable when the control valve is installed. Other factors such as the positioning accuracy of the actuator or the effects of hydraulic flow resistance of associated piping have to be considered when deriving the installed rangeability for a specific application.

The following individuals served as members of ISA Subcommittee SP75.11, which prepared this standard:

NAME	COMPANY
H. D. Baumann, Chairman	H. D. Baumann Assoc., Ltd.
M. Hellman, Secretary	Cashco, Inc.
J. B. Arant	E. I. duPont deNemours and Company, Inc.
H. Boger	Masoneilan Division, McGraw-Edison Co.
P. S. Buckley	E. I. duPont deNemours and Company, Inc.
J. F. Buresh	Retired
C. L. Crawford	Union Carbide Company
G. Keith (Alternate Member)	Masoneilan Division, McGraw-Edison Co.
R. E. Pfeiffer	Union Carbide Company
F. G. Shinskey	The Foxboro Company
J. M. Simonsen	Valtek, Inc.
G. Stiles	Retired
R. E. Terhune	Exxon Company, USA
S. Weiner	Monsanto Company
P. J. Schafbuch	Fisher Controls International, Inc.

Instrument Society of America

The following people served as members of ISA Committee SP75:

NAME	COMPANY
L. R. Driskell, Chairman	Consultant
J. B. Arant	E. I. duPont deNemours and Company, Inc.
H. E. Backinger	John F. Kraus & Company
G. Barb	Muesco, Inc.
H. D. Baumann	H. D. Baumann Assoc., Ltd.
C. S. Beard	
N. Belaef	
G. Borden	Bechtel Power Corporation
D. E. Brown	R. Conrader Company
E. H. C. Brown	Dravo Engineers, Inc.
E. C. Cooney	Air Products & Chemicals, Inc.
W. G. Dewart	Rockwell International
J. T. Emery	Honeywell, Inc.
H. J. Fuller	Worcester Controls Corporation
L. Griffith	
A. J. Hanssen	Fluid Controls Institute, Inc.
F. P. Harthun	Fisher Controls
H. P. Illing	Kieley & Mueller, Inc.
R. B. Jones	Upjohn Company
M. W. Kaye	M. W. Kellogg Company
R. Louvere	Creole
O. P. Lovett, Jr.	ISIS Corporation
J. Manton	
A. P. McCauley	Chagrin Valley Controls, Inc.
T. V. Molloy	Pacific Gas & Electric
J. T. Muller	Leslie Company
H. R. Nickerson	Resistoflex Company
J. Ozol	Omaha Public Power
R. A. Quance	Walsh Inc.
W. Rahmeyer	Colorado State University
J. N. Reed	Masoneilan
G. Richards	Jordan Valve
J. Rosato	Rawson Company
K. Schoonover	Con-Tek
H. Schwartz	Flexible Valve Corporation
F. O. Seger	Willis Oil Tool Company
J. M. Simonsen	Valtek, Inc.
H. Sonderregger	ITT Grinnell Corporation
N. Sprecher	DeZurik
R. U. Stanley	Retired
G. F. Stiles	Fisher Controls Company
R. Terhune	Exxon Company USA
R. F. Tubbs	Copes-Vulcan
W. C. Weidman	Gilbert Commonwealth
R. L. Widdows	Cashco, Inc.
P. Wing	Retired
L. Zinck	Union Carbide

Instrument Society of America

This standard was approved for publication by the ISA Standards and Practices Board in July 1984.

NAME	COMPANY
W. Calder III, Chairman	The Foxboro Company
P. V. Bhat	Monsanto
N. L. Conger	Conoco
B. Feikle	Bailey Controls Company
H. S. Hopkins	Westinghouse Electric Company
J. L. Howard	Boeing Aerospace Company
R. T. Jones	Philadelphia Electric Company
R. Keller	The Boeing Company
O. P. Lovett, Jr.	ISIS Corporation
E. C. Magison	Honeywell, Inc.
A. P. McCauley	Chagrin Valley Controls, Inc.
J. W. Mock	Bechtel Corporation
E. M. Nesvig	ERDCO Engineering Corporation
R. Prescott	Moore Products Company
D. Rapley	Stearns Catalytic Corporation
W. C. Weidman	Gilbert Commonwealth
K. A. Whitman	Allied Chemical Corporation
*P. Bliss	Pratt & Whitney
*B. A. Christensen	Continental Oil Company
*L. N. Combs	
*R. L. Galley	
*T. J. Harrison	IBM Corporation
*R. G. Marvin	
*W. B. Miller	Moore Products Company
*G. Platt	Bechtel Power Corporation
*J. R. Williams	Stearns Catalytic Corporation

*Director Emeritus

TABLE OF CONTENTS

Section	Title	Page
1.	Scope	9
2.	Basic Definitions	9
	2.1 Terminology	9
	2.2 Flow Coefficient	9
	2.3 Inherent Flow Characteristic	9
	2.4 Inherent Rangeability	9
	2.5 Relative Flow Coefficient (ϕ)	9
	2.6 Relative Travel (h)	9
3.	Typical Inherent Flow Characteristics	9
4.	Permissible Deviations Between Actual and Manufacturer-Stated Inherent Flow Characteristics	9

LIST OF ILLUSTRATIONS

Figure	Title	Page
1.	Example of Globe Valve Specimen Compared to Manufacturer-Specified Flow Characteristic	10
2.	Example of Butterfly Valve Specimen Compared to Manufacturer-Specified Flow Characteristic	11

LIST OF TABLES

Table	Title	Page
1.	Permissible Deviations Between Actual and Manufacturer-Stated Inherent Flow Characteristics	9

1 SCOPE

The scope of this standard is to define the statement of typical control valve inherent flow characteristics and inherent rangeabilities, and to establish criteria for adherence to manufacturer-specified flow characteristics.

2 BASIC DEFINITIONS

2.1 Terminology

Basic terminology used herein is based on definitions stated in "Control Valve Terminology" ISA Standard S75.05.

2.2 Flow Coefficient

A constant (Cv), related to the geometry of a valve, for a given valve opening, that can be used to predict flow rate. See ANSI/ISA S75.01 "Control Valve Sizing Equations" and ANSI/ISA S75.02 "Control Valve Capacity Test Procedure."

2.3 Inherent Flow Characteristic

The relationship between the flow rate through a valve and the travel of the closure member as the closure member is moved from the closed position to rated travel with constant pressure drop across the valve.

2.4 Inherent Rangeability

The ratio of the largest flow coefficient (Cv) to the smallest flow coefficient (Cv) within which the deviation from the specified inherent flow characteristic does not exceed the limits stated in Section 4.

2.5 Relative Flow Coefficient (ϕ)

The ratio of the flow coefficient (Cv) at a stated travel to the flow coefficient (Cv) at rated travel.

2.6 Relative Travel (h)

The ratio of the travel at a given opening to the rated travel.

3 TYPICAL INHERENT FLOW CHARACTERISTICS

3.1 The typical inherent flow characteristic for a specific size, type, and trim configuration of a control valve shall be specified by the manufacturers either graphically or in tabular form.

3.2 When tabulated, specific flow coefficients shall be stated for the following travel positions: at 5%, 10%, 20%, and every subsequent 10% of rated travel up to and including 100%.

3.3 The manufacturer may publish flow coefficients in addition to those at the above-stated travel positions.

3.4 In addition, the manufacturer is encouraged to specify the generic name of a specific flow characteristic such as "Linear," "Equal-Percentage," etc., if applicable, following the definitions in ISA Standard S75.05.

3.5 The manufacturer shall state the largest flow coefficient that meets the criteria of Section 4, if it is less than the rated flow coefficient. (See Figure 2.)

4 PERMISSIBLE DEVIATIONS BETWEEN ACTUAL AND MANUFACTURER-STATED INHERENT FLOW CHARACTERISTICS

4.1 When subjected to a flow test per ANSI/ISA S75.02, the individual test Cv values may not deviate by more than $\pm 10 \left\{ \frac{1}{\phi} \right\}^{0.2}$ percent from those values specified in the flow characteristic published by the manufacturer. Exceptions of this are Cvs at given travel positions falling below a Cv value of 5, or above a Cv value of $30d^2$. In the above relationships, d is the nominal valve size in inches, and ϕ is the relative flow coefficient based on published Cvs. Allowable deviations calculated by the above equation are listed in Table 1.

TABLE 1
PERMISSIBLE DEVIATIONS BETWEEN ACTUAL AND MANUFACTURER-STATED INHERENT FLOW CHARACTERISTICS

% Cv Rated	ϕ	Permitted +/− Deviation (%)	ϕ Range High	Low
5	0.05	18.2	0.0591	0.0409
10	0.1	15.8	0.116	0.0842
20	0.2	13.8	0.227	0.172
30	0.3	12.7	0.338	0.262
40	0.4	12.0	0.448	0.352
50	0.5	11.5	0.557	0.443
60	0.6	11.1	0.667	0.533
70	0.7	10.7	0.775	0.625
80	0.8	10.4	0.883	0.717
90	0.9	10.2	0.992	0.808
100	1.0	10.0	1.100	0.900

4.2 The inherent flow characteristic of a control valve, when plotted from test data at the stated increments of travel, shall show no major deviations in slope. By definition, a major deviation is when the slope of the line connecting two adjacent test points varies by more than 2 to 1 or 0.5 to 1 from the slope of a line drawn between flow coefficients specified by the manufacturer for the same travel positions. (See Figure 1.)

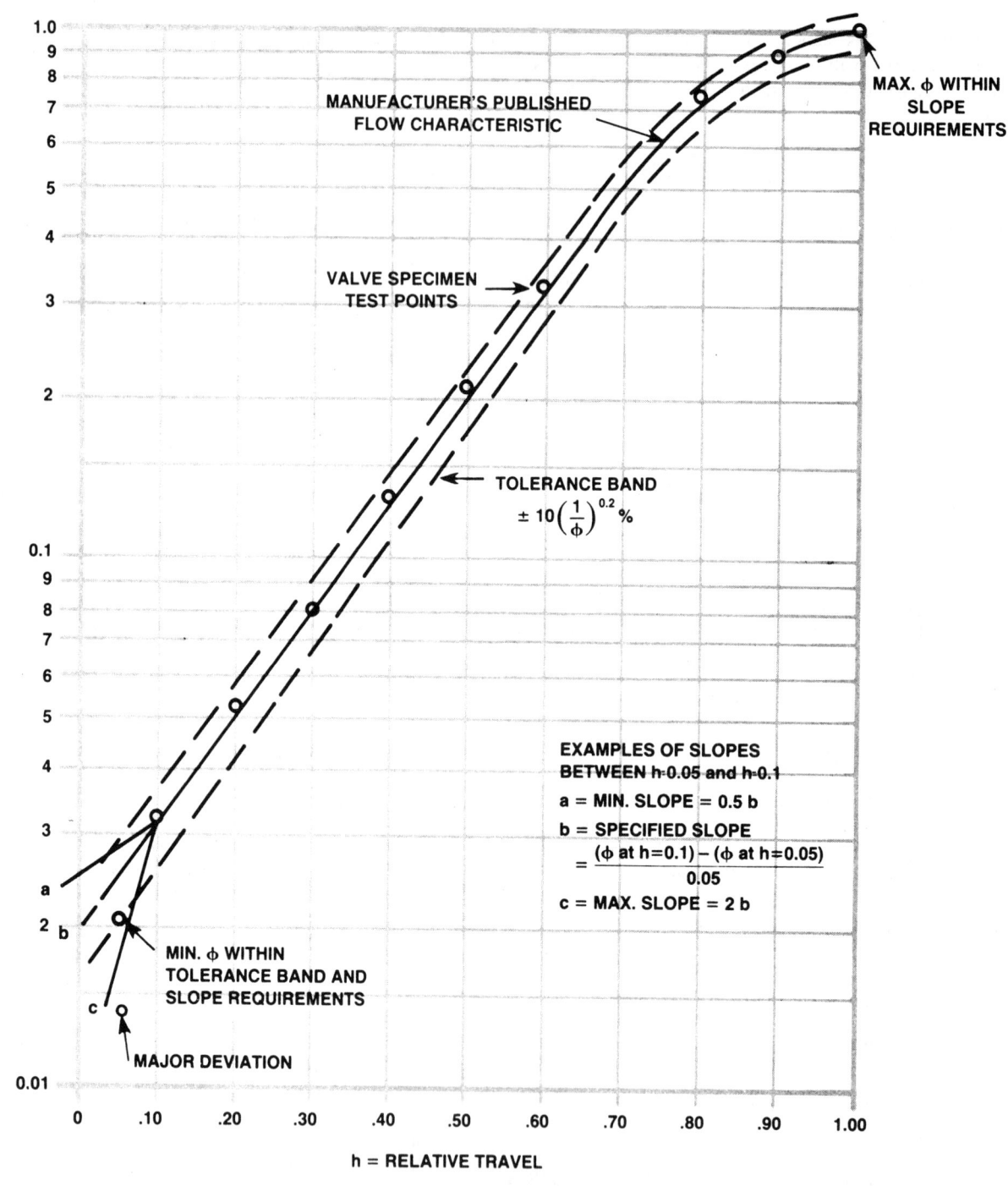

Figure 1. Example of Globe Valve Specimen Compared to Manufacturer-Specified Flow Characteristic

S75.11

Inherent Flow Characteristic
and Rangeability of Control Valves

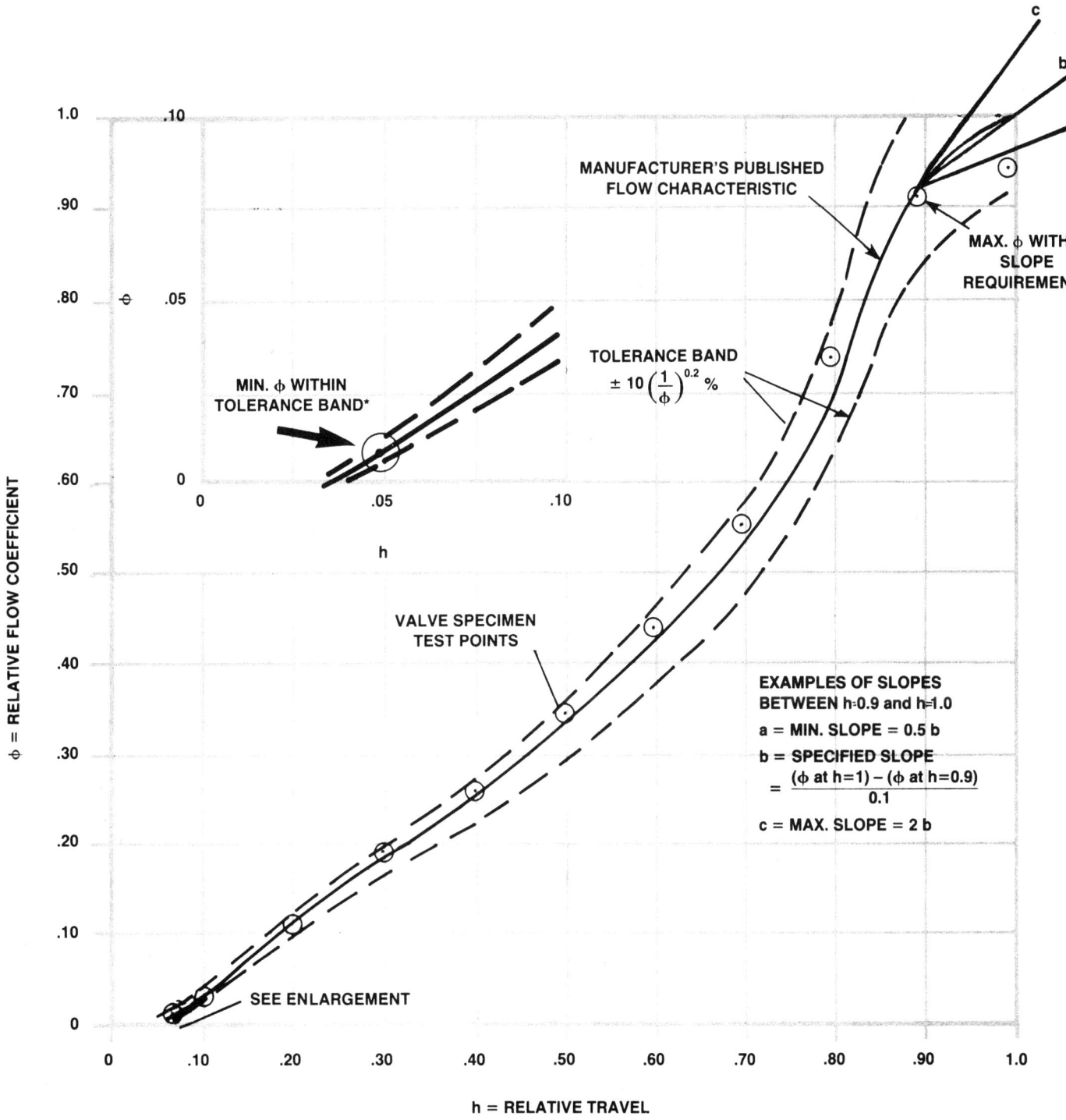

INHERENT RANGEABILITY OF TEST SPECIMEN (per Section 2.4): $\frac{\phi \text{ Max.}}{\phi \text{ Min.}} = \frac{\phi\ 0.91}{\phi\ 0.01} = 91$

The maximum flow coefficient ϕ meeting the requirements of Section 4 is 0.91; the minimum ϕ is 0.01.

*Tolerance band for $\phi = 0.01$ is $10 \left\{ \frac{1}{0.01} \right\}^{0.2} = \pm 25\%$ of $\phi = 0.01$.

Figure 2. Example of Butterfly Valve Specimen Compared to Manufacturer-Specified Flow Characteristic

INSTRUMENT SOCIETY OF AMERICA
Research Triangle Park, North Carolina 27709

ANSI/ISA-S75.12-1987
Approved March 18, 1987

American National Standard

Face-to-Face Dimensions for Socket Weld-End and Screwed-End Globe-Style Control Valves (ANSI Classes 150, 300, 600, 900, 1500, and 2500)

Instrument Society of America

Instrument Society of America

ISBN 0-87664-975-4

ISA-S75.12 Face-to-Face Dimensions for Socket Weld-End and Screwed-End Globe-Style Control Valves (ANSI Classes 150, 300, 600, 900, 1500, and 2500)

Copyright © 1986 by the Instrument Society of America. All rights reserved. Printed in the United States of America. No part of this publication may be reproduced, stored in a retrieval system, or transmitted, in any form or by any means (electronic, mechanical, photocopying, recording, or otherwise), without the prior written permission of the Publisher.

INSTRUMENT SOCIETY OF AMERICA
67 Alexander Drive
P.O. Box 12277
Research Triangle Park, North Carolina 27709

PREFACE

This preface is included for informational purposes and is not part of ISA-S75.12.

This standard has been prepared as part of the service of the Instrument Society of America (ISA) toward a goal of uniformity in the field of instrumentation. To be of real value, this document should not be static, but should be subject to periodic review. Toward this end, the Society welcomes all comments and criticisms, and asks that they be addressed to the Secretary, Standards and Practices Board, Instrument Society of America, 67 Alexander Drive, P.O. Box 12277, Research Triangle Park, NC 27709, Telephone (919) 549-8411.

The ISA Standards and Practices Department is aware of the growing need for attention to the metric system of units in general, and the International System of Units (SI) in particular, in the preparation of instrumentation standards. The Department is further aware of the benefits to U.S.A. users of ISA standards of incorporating suitable references to the SI (and the metric system) in their business and professional dealings with other countries. Toward this end, this Department will endeavor to introduce SI-acceptable metric units in all new and revised standards to the greatest extent possible. *The Metric Practice Guide*, which has been published by the Institute of Electrical and Electronics Engineers as ANSI/IEEE Std. 268-1982, and future revisions will be the reference guide for definitions, symbols, abbreviations, and conversion factors.

It is the policy of the Instrument Society of America to encourage and welcome the participation of all concerned individuals and interests in the development of ISA standards. Participation in the ISA standards-making process by an individual in no way constitutes endorsement by the employer of that individual, of the Instrument Society of America, or of any of the standards that ISA develops.

The information contained in this preface, footnotes, and appendices is included for information only and is not a part of the standard.

The following people served as members of ISA Committee SP75.08 (formerly SP4.1), which prepared this standard.

NAME	COMPANY
W. C. Weidman, Chairman	Gilbert/Commonwealth, Inc.
R. Barnes	Valtek, Inc.
H. D. Baumann	H. D. Baumann Associates, Ltd.
G. Borden, Jr.	Bechtel Power Corporation
R. Brodin	Fisher Controls International, Inc.
R. Chown	OTEC
E. J. Cooney	Air Products & Chemicals, Inc.
J. T. Emery	Honeywell Inc.
A. J. Hanssen	Fluid Controls Institute, Inc.
H. P. Illing	Kieley & Mueller, Inc.
R. B. Jones	Dow Chemical USA
M. W. Kaye	M. W. Kellogg Company
O. P. Lovett, Jr.	ISIS Corporation
P. C. Martin	Hammel Dahl & Jamesbury
H. R. Nickerson	Resistoflex Company
R. A. Quance	Walsh Inc.
J. N. Reed	Masoneilan/Dresser
J. Rosato	Rawson Company
H. Schwartz	Flexible Valve Corporation
F. O. Seger	Willis Division, Smith International, Inc.
J. M. Simonsen	Valtek, Inc.
H. R. Sonderregger	ITT Grinnell Corporation
R. U. Stanley	Retired
G. Stiles	Fisher Controls International, Inc.
R. F. Tubbs	Copes-Vulcan-Charlotte

Instrument Society of America

The following people served as members of ISA Committee SP75 during the review of this standard.

NAME	COMPANY
L. R. Driskell, Chairman	Consultant
J. B. Arant	E. I. du Pont de Nemours & Company
H. E. Backinger	Kraus Company, Inc.
G. E. Barb	Muesco Inc.
R. Barnes/J. M. Simonsen*	Valtek, Inc.
H. D. Baumann	H. D. Baumann Associates, Ltd.
N. Belaef	Consultant
G. Borden, Jr.	Bechtel Power Corporation
R. Brodin/F. Harthun/G. Stiles*	Fisher Controls International, Inc.
E. H. C. Brown	Dravo Engineers, Inc.
E. J. Cooney	Air Products & Chemicals, Inc.
W. G. Dewart	Rockwell International
J. T. Emery	Honeywell, Inc.
H. J. Fuller	Retired
L. F. Griffith	Consultant
A. J. Hanssen	Fluid Controls Institute, Inc.
H. P. Illing	Kieley & Mueller, Inc.
R. B. Jones	Dow Chemical USA
M. W. Kaye	M. W. Kellogg Company
R. Louviere	Creole Engineering
O. P. Lovett, Jr.	ISIS Corporation
P. C. Martin	Hammel Dahl & Jamesbury
A. P. McCauley	Chagrin Valley Controls, Inc.
T. V. Molloy	Pacific Gas & Electric
H. R. Nickerson	Resistoflex Company
J. Ozol	Omaha Public Power Company
R. A. Quance	Walsh Inc.
W. Rahmeyer	Utah State University
J. N. Reed	Masoneilan/Dresser
G. Richards	Richards Industries, Inc.
J. Rosato	Rawson Company
K. Schoonover, Secretary	Con-Tek
H. Schwartz	Flexible Valve Corporation
W. L. Scull/J. T. Muller*	Leslie Company
F. O. Seger	Willis Division, Smith International, Inc.
H. R. Sonderregger	ITT Grinnell Corporation
N. D. Sprecher	De Zurik
R. U. Stanley	Retired
R. E. Terhune, V. Chairman	Consultant
R. F. Tubbs	Copes-Vulcan-Charlotte
W. C. Weidman	Gilbert/Commonwealth, Inc.
R. L. Widdows	Cashco Inc.
L. R. Zinck	Union Carbide Corporation

*One vote

This standard was approved for publication by the ISA Standards and Practices Board in June 1986.

NAME	COMPANY
N. Conger, Chairman	Fisher Controls International, Inc.
W. Calder III	The Foxboro Company
R. S. Crowder	Ship Star Associates
H. S. Hopkins	Consultant
J. L. Howard	Boeing Aerospace Company
R. T. Jones	Philadelphia Electric Company
R. Keller	The Boeing Company
O. P. Lovett, Jr.	ISIS Corporation
E. G. Magison	Honeywell, Inc.
A. P. McCauley	Chagrin Valley Controls, Inc.
J. W. Mock	Bechtel Corporation
E. M. Nesvig	ERDCO Engineering Corporation
R. Prescott	Moore Products Company
D. E. Rapley	Rapley Engineering Company
C. W. Reimann	National Bureau of Standards
J. Rennie	Factory Mutual Research Corporation
W. C. Weidman	Gilbert/Commonwealth, Inc.
K. Whitman	Consultant
*P. Bliss	Consultant
*B. A. Christensen	Continental Oil Company
*L. N. Combs	Retired
*R. L. Galley	Consultant
*T. J. Harrison	IBM Corporation
*R. G. Marvin	Roy G. Marvin Company
*W. B. Miller	Moore Products Company
*G. Platt	Retired
*J. R. Williams	Stearns Catalytic Corporation

---

*Director Emeritus

1 SCOPE

1.1 This standard applies to socket weld-end globe-style control valves, sizes 1/2 inch through 4 inches, and screwed-end globe-style control valves, sizes 1/2 inch through 2-1/2 inches, having top, top and bottom, port, or cage guiding.

2 PURPOSE

2.1 The purpose of this standard is to aid users in their piping designs by providing ANSI Classes 150 through 2500 socket weld-end control valve dimensions and ANSI Classes 150 through 600 screwed-end control valve dimensions without giving special consideration to the equipment manufacturer to be used.

2.2 The short-long dimensions provided in Tables 1 and 2 clarify Section 2.1 by consolidating the diversity of existing manufacturers' lengths into two sets of dimensions for each size valve. Before using either the short or long dimensions, the piping designer should confirm with the selected valve manufacturer which dimension is correct for the valve(s) being supplied.

3 DEFINITIONS

3.1 For definitions of terms used in this standard, see ISA-S75.05, "Control Valve Terminology."

4 BIBLIOGRAPHY

4.1 Manufacturers Standardization Society (MSS) SP-84-1978, "Steel Valves — Socket Welding and Threaded Ends."

5 DIMENSIONAL DATA

5.1 For face-to-face dimensions for socket weld-end globe-style control valves, see Table 1.

5.2 For face-to-face dimensions for screwed-end globe-style control valves, see Table 2.

TABLE 1
FACE-TO-FACE DIMENSIONS FOR SOCKET WELD-END GLOBE-STYLE CONTROL VALVES

Nominal Valve Size		PN 20, 50, & 100 (ANSI Classes 150, 300, & 600)				PN 150 & 250 (ANSI Classes 900 & 1500)				PN 420 (ANSI Class 2500)				Tolerance	
		Dimension "A"				Dimension "A"				Dimension "A"					
		mm		inches		mm		inches		mm		inches			
mm	inches	Short	Long	Short	Long	Short	Long	Short	Long	Short	Long	Short	Long	mm	inches
15	1/2	165	206	6.69	8.12	194	279	7.00	11.00	216	318	8.50	12.50	± 6.4	±0.25
20	3/4	165	210	6.69	8.25	194	279	7.00	11.00	216	318	8.50	12.50	± 6.4	±0.25
25	1	197	210	7.75	8.25	197	279	7.00	11.00	216	318	8.50	12.50	± 6.4	±0.25
40	1-1/2	235	251	9.25	9.88	235	330	9.25	13.00	260	381	10.25	15.00	± 6.4	±0.25
50	2	267	286	10.50	11.25	292	375	11.50	14.75	318	400	12.75	15.75	± 6.4	±0.25
65	2-1/2	292	311	11.50	12.25	292	—	11.50	—	318	—	12.75	—	± 6.4	±0.25
80	3	318	337	12.50	13.25	318	533	12.50	21.00	381	660	15.00	26.00	± 6.4	±0.25
100	4	368	394	14.50	15.50	368	635	14.50	20.88	406	737	16.00	29.00	± 6.4	±0.25

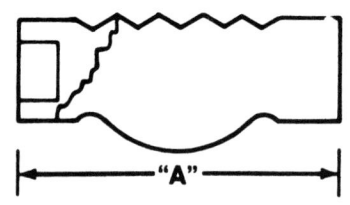

TABLE 2
FACE-TO-FACE DIMENSIONS FOR SCREWED-END GLOBE-STYLE CONTROL VALVES

| Nominal Valve Size || PN 20, 50, & 100 (ANSI Classes 150, 300, & 600) Dimension "A" |||| Tolerance ||
mm	inches	mm Short	mm Long	inches Short	inches Long	mm	inches
15	1/2	165	206	6.50	8.12	±1.6	±0.062
20	3/4	165	210	6.50	8.25	±1.6	±0.062
25	1	197	210	7.75	8.25	±1.6	±0.062
40	1-1/2	235	251	9.25	9.88	±1.6	±0.062
50	2	267	286	10.50	11.25	±1.6	±0.062
65	2-1/2	292	311	11.50	12.25	±1.6	±0.062

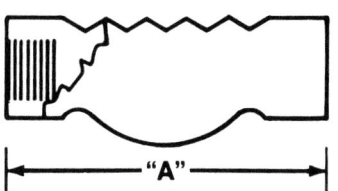

APPENDIX

This appendix is not part of ISA-S75.12, but it is included to facilitate its use.

Dimensions for metrically sized valves are nominal conversions that are conventionally used in the Manufacturers Standardization Society (MSS) of the Valve and Fitting Industry's Publication MSS-SP86-1981, and in International Standards Organization (ISO) and International Electrotechnical Commission (IEC) documents.

ANSI/ISA-S75.14-1984
Approved April 18, 1986

American National Standard

Face-to-Face Dimensions for Buttweld-End Globe-Style Control Valves (ANSI Class 4500)

Instrument Society of America

Instrument Society of America

ISBN 0-87664-848-0

ISA-S75.14 Face-to-Face Dimensions for Buttweld-End Globe-Style Control Valves, ANSI Class 4500

Copyright © 1984 by the Instrument Society of America. All rights reserved. Printed in the United States of America. No part of this publication may be reproduced, stored in a retrieval system, or transmitted, in any form or any means (electronic, mechanical, photocopying, recording or otherwise) without the prior written permission of the publisher.

INSTRUMENT SOCIETY OF AMERICA
67 Alexander Drive
P.O. Box 12277
Research Triangle Park, North Carolina 27709

Copyright © 1984 by the Instrument Society of America

PREFACE

This preface is included for information purposes and is not part of ISA-S75.14.

This standard has been prepared as part of the service of the Instrument Society of America (ISA) toward a goal of uniformity in the field of instrumentation. To be of real value, this document should not be static, but should be subject to periodic review. Toward this end, the Society welcomes all comments and criticisms, and asks that they be addressed to the Secretary, Standards and Practices Board, Instrument Society of America, 67 Alexander Drive, P.O. Box 12277, Research Triangle Park, NC 27709, Telephone (919) 549-8411.

The ISA Standards and Practices Department is aware of the growing need for attention to the metric system of units in general, and the International System of Units (SI) in particular, in the preparation of instrumentation standards. The Department is further aware of the benefits to U.S.A. users of ISA standards of incorporating suitable references to the SI (and the metric system) in their business and professional dealings with other countries. Toward this end, this Department will endeavor to introduce SI-acceptable metric units in all new and revised standards to the greatest extent possible. The Metric Practice Guide, which has been published by the Institute of Electrical and Electronics Engineers as ANSI/IEEE Std. 268-1982, and future revisions will be the reference guide for definitions, symbols, abbreviations, and conversion factors.

It is the policy of the Instrument Society of America to encourage and welcome the participation of all concerned individuals and interests in the development of ISA standards. Participation in the ISA standards-making process by an individual in no way constitutes endorsement by the employer of that individual, of the Instrument Society of America, or of any of the standards that ISA develops.

The information contained in the preface, footnotes, and appendices is included for information only and is not a part of the standard.

The following people served as members of ISA Committee SP75.08 (formerly ISA Committee SP4.1):

NAME	COMPANY
W. C. Weidman, Chairman	Gilbert Commonwealth, Inc.
H. D. Baumann	H. D. Baumann Associates, Ltd.
G. Borden	Bechtel Power Corporation
R. Brodin	Fisher Controls International, Inc.
R. Chown	OTEC
E. J. Cooney	Air Products & Chemicals, Inc.
L. R. Driskell	Consultant
J. T. Emery	Honeywell, Inc.
A. J. Hanssen	Fluid Controls Institute
H. Illing	Kieley & Mueller, Inc.
R. B. Jones	Upjohn Company
M. W. Kaye	M. W. Kellogg Company
O. P. Lovett, Jr.	ISIS Corporation
J. T. Muller	Leslie Company
H. R. Nickerson	Resistoflex Company
R. A. Quance	Walsh Inc.
J. N. Reed	Masoneilan Division, McGraw Edison Co.
J. Rosato	Rawson Company
H. Schwartz	Flexible Valve Corporation
F. O. Seger	Willis Division, Smith International, Inc.
J. M. Simonsen	Valtek, Inc.
H. R. Sonderregger	ITT Grinnell Corporation
R. U. Stanley	Retired
G. F. Stiles	Fisher Controls International, Inc.
R. F. Tubbs	Copes-Vulcan
S. Weiner	Monsanto Company
P. Wing, Jr.	Retired

Instrument Society of America

The following people served as members of ISA Committee SP75:

NAME	COMPANY
L. R. Driskell, Chairman	Consultant
J. B. Arant	E. I. duPont deNemours and Company, Inc.
H. E. Backinger	John F. Kraus & Company
G. Barb	Muesco, Inc.
H. D. Baumann	H. D. Baumann Associates, Ltd.
C. S. Beard	Mutual Stamp Company
N. Belaef	Consultant
G. Borden	Bechtel Power Corporation
D. E. Brown	R. Conrader Company
E. H. C. Brown	Dravo Engineers, Inc.
E. C. Cooney	Air Products & Chemicals, Inc.
W. G. Dewart	Rockwell International
J. T. Emery	Honeywell, Inc.
H. J. Fuller	Worcester Controls Corporation
L. Griffith	Consultant
A. J. Hanssen	Fluid Controls Institute, Inc.
F. P. Harthun	Fisher Controls International, Inc.
H. P. Illing	Kieley & Mueller, Inc.
R. B. Jones	Upjohn Company
M. W. Kaye	M. W. Kellogg Company
R. Louvere	Creole Corporation
O. P. Lovett, Jr.	ISIS Corporation
A. P. McCauley	Chagrin Valley Controls, Inc.
T. V. Molloy	Pacific Gas & Electric
J. T. Muller	Leslie Company
H. R. Nickerson	Resistoflex Company
J. Ozol	Omaha Public Power
R. A. Quance	Walsh Inc.
W. Rahmeyer	Colorado State University
J. N. Reed	Masoneilan Division, McGraw Edison Co.
G. Richards	Jordan Valve
J. Rosato	Rawson Company
K. Schoonover, Secretary	Con-Tek
H. Schwartz	Flexible Valve Corporation
F. O. Seger	Willis Division, Smith International, Inc.
J. M. Simonsen	Valtek, Inc.
H. Sonderregger	ITT Grinnell Corporation
N. Sprecher	DeZurik
R. U. Stanley	Retired
*G. F. Stiles/Roy Brodin	Fisher Controls International, Inc.
R. Terhune, Vice-Chairman	Exxon Company USA
R. F. Tubbs	Copes-Vulcan
W. C. Weidman	Gilbert Commonwealth, Inc.
R. L. Widdows	Cashco, Inc.
P. Wing	Retired
L. Zinck	Union Carbide

*One vote

This standard was approved for publication by the ISA Standards and Practices Board in September 1984.

NAME	COMPANY
W. Calder III, Chairman	The Foxboro Company
P. V. Bhat	Monsanto Company
N. L. Conger	Conoco
B. Feikle	Bailey Controls Company
H. S. Hopkins	Westinghouse Electric Company
J. L. Howard	Boeing Aerospace Company
R. T. Jones	Philadelphia Electric Company
R. Keller	The Boeing Company
O. P. Lovett, Jr.	ISIS Corporation
E. C. Magison	Honeywell, Inc.
A. P. McCauley	Chagrin Valley Controls, Inc.
J. W. Mock	Bechtel Corporation
E. M. Nesvig	ERDCO Engineering Corporation
R. Prescott	Moore Products Company
D. Rapley	Stearns Catalytic Corporation
W. C. Weidman	Gilbert Commonwealth, Inc.
K. A. Whitman	Consultant
*P. Bliss	Consultant
*B. A. Christensen	Continental Oil Company
*L. N. Combs	Retired
*R. L. Galley	Consultant
*T. J. Harrison	IBM Corporation
*R. G. Marvin	Roy G. Marvin Company
*W. B. Miller	Moore Products Company
*G. Platt	Bechtel Power Corporation
*J. R. Williams	Stearns Catalytic Corporation

*Director Emeritus

1 SCOPE

1.1 This standard applies to buttweld-end globe-style control valves, sizes 1/2 inch through 8 inches, having top and cage guiding.

2 PURPOSE

2.1 The purpose of this standard is to aid users in their piping designs by providing ANSI Class 4500 buttweld-end control valve dimensions, without giving special consideration to the equipment manufacturer to be used.

3 DEFINITION

3.1 For definitions of terms used in this standard, see ANSI/ISA-S75.05, "Control Valve Terminology."

4 BIBLIOGRAPHY

Both of the following references were published by ASME (The American Society of Mechanical Engineers).

4.1 American National Standards Institute, Inc. (ANSI) Standard B16.34-1981, "Valves—Flanged and Buttwelding End—Steel, Nickel Alloy, and other Special Alloys, 1981."

4.2 American National Standards Institute, Inc. (ANSI) Standard B16.25-1979, "Buttwelding Ends."

5 DIMENSIONAL DATA

5.1 For face-to-face dimensions for buttweld-end valves, see Table 1.

TABLE 1
FACE-TO-FACE DIMENSIONS FOR BUTTWELD-END GLOBE-STYLE CONTROL VALVES

Nominal Valve Size		ANSI Class 4500 Dimension "A"		Tolerance	
mm	inches	mm	inches	mm	inches
15	1/2	298	11.75	±1.6	±0.062
20	3/4	298	11.75	±1.6	±0.062
25	1	298	11.75	±1.6	±0.062
40	1-1/2	298	11.75	±1.6	±0.062
50	2	378	14.88	±1.6	±0.062
80	3	479	18.88	±1.6	±0.062
100	4	584	23.00	±1.6	±0.062
150	6	883	34.75	±2.4	±0.093
200	8	1,118	44.00	±2.4	±0.093

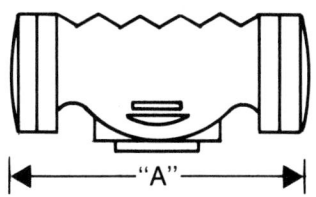

APPENDIX

This appendix is not part of ISA-S75.14, but it is included to facilitate its use.

The ANSI standards listed in the bibliography contain valve body design information in addition to face-to-face dimensions. Dimensions for metrically sized valves are nominal conversions that are conventionally used in the Manufacturers Standardization Society (MSS) of the Valve and Fitting Industry's Publication MSS-SP86-1981, and in International Organization (ISO) and International Electrotechnical Commission (IEC) documents.

INSTRUMENT SOCIETY of AMERICA
Research Triangle Park, North Carolina

ANSI/ISA-S75.15-1987
Approved March 18, 1987

American National Standard

Face-to-Face Dimensions for Buttweld-End Globe-Style Control Valves (ANSI Classes 150, 300, 600, 900, 1500, and 2500)

Instrument Society of America

Instrument Society of America

ISBN 0-87664-976-2

ISA-S75.15 Face-to-Face Dimensions for Buttweld-End Globe-Style Control Valves (ANSI Classes 150, 300, 600, 900, 1500, and 2500)

Copyright © 1986 by the Instrument Society of America. All rights reserved. Printed in the United States of America. No part of this publication may be reproduced, stored in a retrieval system, or transmitted, in any form or by any means (electronic, mechanical, photocopying, recording, or otherwise), without the prior written permission of the Publisher.

INSTRUMENT SOCIETY OF AMERICA
67 Alexander Drive
P.O. Box 12277
Research Triangle Park, North Carolina 27709

PREFACE

This preface is included for informational purposes and is not part of ISA-S75.15.

This standard has been prepared as part of the service of the Instrument Society of America (ISA) toward a goal of uniformity in the field of instrumentation. To be of real value, this document should not be static, but should be subject to periodic review. Toward this end, the Society welcomes all comments and criticisms, and asks that they be addressed to the Secretary, Standards and Practices Board, Instrument Society of America, 67 Alexander Drive, P.O. Box 12277, Research Triangle Park, NC 27709, Telephone (919) 549-8411.

The ISA Standards and Practices Department is aware of the growing need for attention to the metric system of units in general, and the International System of Units (SI) in particular, in the preparation of instrumentation standards. The Department is further aware of the benefits to U.S.A. users of ISA standards of incorporating suitable references to the SI (and the metric system) in their business and professional dealings with other countries. Toward this end, this Department will endeavor to introduce SI-acceptable metric units in all new and revised standards to the greatest extent possible. *The Metric Practice Guide*, which has been published by the Institute of Electrical and Electronics Engineers as ANSI/IEEE Std. 268-1982, and future revisions will be the reference guide for definitions, symbols, abbreviations, and conversion factors.

It is the policy of the Instrument Society of America to encourage and welcome the participation of all concerned individuals and interests in the development of ISA standards. Participation in the ISA standards-making process by an individual in no way constitutes endorsement by the employer of that individual, of the Instrument Society of America, or of any of the standards that ISA develops.

The information contained in the preface, footnotes, and appendices is included for information only and is not a part of the standard.

The following people served as members of ISA Committee SP75.08 (formerly SP4.1), which prepared this standard.

NAME	COMPANY
W. C. Weidman, Chairman	Gilbert/Commonwealth, Inc.
R. Barnes	Valtek, Inc.
H. D. Baumann	H. D. Baumann Associates, Ltd.
G. Borden, Jr.	Bechtel Power Corporation
R. Brodin	Fisher Controls International, Inc.
R. Chown	OTEC
E. J. Cooney	Air Products & Chemicals, Inc.
J. T. Emery	Honeywell Inc.
A. J. Hanssen	Fluid Controls Institute, Inc.
H. P. Illing	Kieley & Mueller, Inc.
R. B. Jones	Dow Chemical USA
M. W. Kaye	M. W. Kellogg Company
O. P. Lovett, Jr.	ISIS Corporation
P. C. Martin	Hammel Dahl & Jamesbury
H. R. Nickerson	Resistoflex Company
R. A. Quance	Walsh Inc.
J. N. Reed	Masoneilan/Dresser
J. Rosato	Rawson Company
H. Schwartz	Flexible Valve Corporation
F. O. Seger	Willis Division, Smith International, Inc.
J. M. Simonsen	Valtek, Inc.
H. R. Sonderregger	ITT Grinnell Corporation
R. U. Stanley	Retired
G. Stiles	Fisher Controls International, Inc.
R. F. Tubbs	Copes-Vulcan-Charlotte

Instrument Society of America

The following people served as members of ISA Committee SP75 during the review of this standard.

NAME	COMPANY
L. R. Driskell, Chairman	Consultant
J. B. Arant	E. I. du Pont de Nemours & Company
H. E. Backinger	Kraus Company, Inc.
G. E. Barb	Muesco Inc.
R. Barnes/J. M. Simonsen*	Valtek, Inc.
H. D. Baumann	H. D. Baumann Associates, Ltd.
N. Belaef	Consultant
G. Borden, Jr.	Bechtel Power Corporation
R. Brodin/F. Harthun/G. Stiles*	Fisher Controls International, Inc.
E. H. C. Brown	Dravo Engineers, Inc.
E. J. Cooney	Air Products & Chemicals, Inc.
W. G. Dewart	Rockwell International
J. T. Emery	Honeywell, Inc.
H. J. Fuller	Retired
L. F. Griffith	Consultant
A. J. Hanssen	Fluid Controls Institute, Inc.
H. P. Illing	Kieley & Mueller, Inc.
R. B. Jones	Dow Chemical USA
M. W. Kaye	M. W. Kellogg Company
R. Louviere	Creole Engineering
O. P. Lovett, Jr.	ISIS Corporation
P. C. Martin	Hammel Dahl & Jamesbury
A. P. McCauley	Chagrin Valley Controls, Inc.
T. V. Molloy	Pacific Gas & Electric
H. R. Nickerson	Resistoflex Company
J. Ozol	Omaha Public Power Company
R. A. Quance	Walsh Inc.
W. Rahmeyer	Utah State University
J. N. Reed	Masoneilan/Dresser
G. Richards	Richards Industries, Inc.
J. Rosato	Rawson Company
K. Schoonover, Secretary	Con-Tek
H. Schwartz	Flexible Valve Corporation
W. L. Scull/J. T. Muller*	Leslie Company
F. O. Seger	Willis Division, Smith International, Inc.
H. R. Sonderregger	ITT Grinnell Corporation
N. D. Sprecher	De Zurik
R. U. Stanley	Retired
R. E. Terhune, V. Chairman	Consultant
R. F. Tubbs	Copes-Vulcan-Charlotte
W. C. Weidman	Gilbert/Commonwealth, Inc.
R. L. Widdows	Cashco Inc.
L. R. Zinck	Union Carbide Corporation

---

*One vote

This standard was approved for publication by the ISA Standards and Practices Board in June 1986.

NAME	COMPANY
N. Conger, Chairman	Fisher Controls International, Inc.
W. Calder III	The Foxboro Company
R. S. Crowder	Ship Star Associates
H. S. Hopkins	Consultant
J. L. Howard	Boeing Aerospace Company
R. T. Jones	Philadelphia Electric Company
R. Keller	The Boeing Company
O. P. Lovett, Jr.	ISIS Corporation
E. G. Magison	Honeywell, Inc.
A. P. McCauley	Chagrin Valley Controls, Inc.
J. W. Mock	Bechtel Corporation
E. M. Nesvig	ERDCO Engineering Corporation
R. Prescott	Moore Products Company
D. E. Rapley	Rapley Engineering Company
C. W. Reimann	National Bureau of Standards
J. Rennie	Factory Mutual Research Corporation
W. C. Weidman	Gilbert/Commonwealth, Inc.
K. Whitman	Fairleigh Dickinson University
*P. Bliss	Consultant
*B. A. Christensen	Continental Oil Company
*L. N. Combs	Retired
*R. L. Galley	Consultant
*T. J. Harrison	IBM Corporation
*R. G. Marvin	Roy G. Marvin Company
*W. B. Miller	Moore Products Company
*G. Platt	Retired
*J. R. Williams	Stearns Catalytic Corporation

---

*Director Emeritus

1 SCOPE

1.1 This standard applies to buttweld-end globe-style control valves, sizes 1/2 inch through 18 inches, for ANSI Classes 150 through 2500, having top, top and bottom, port, or cage guiding.

2 PURPOSE

2.1 The purpose of this standard is to aid users in their piping designs by providing buttweld-end control valve dimensions, without giving special consideration to the equipment manufacturer to be used.

2.2 The short-long dimensions provided in Table 1 clarify Section 2.1 by consolidating the diversity of existing manufacturers' lengths into two sets of dimensions for each size valve. Before using either the short or long dimensions, the piping designer should confirm with the selected valve manufacturer which dimension is correct for the valve(s) being supplied.

3 DEFINITIONS

3.1 For definitions of terms used in this standard, see ANSI/ISA-S75.05, "Control Valve Terminology."

4 BIBLIOGRAPHY

The following standards were developed by the American Society of Mechanical Engineers (ASME).

4.1 American National Standards Institute, Inc., Standard ANSI B16.34-1981, "Valves — Flanged and Buttwelding End."

4.2 American National Standards Institute, Inc., Standard ANSI B16.25-1979, "Buttwelding Ends."

5 DIMENSIONAL DATA

5.1 For face-to-face dimensions for buttweld-end globe-style control valves, see Table 1.

TABLE 1
FACE-TO-FACE DIMENSIONS FOR BUTTWELD-END GLOBE-STYLE CONTROL VALVES

Nominal Valve Size		PN 20, 50, & 100 (ANSI Classes 150, 300, & 600)				PN 150 & 250 (ANSI Classes 900 & 1500)				PN 420 (ANSI Class 2500)				Tolerance	
			Dimension "A"				Dimension "A"				Dimension "A"				
		mm		inches		mm		inches		mm		inches			
mm	inches	Short	Long	Short	Long	Short	Long	Short	Long	Short	Long	Short	Long	mm	inches
15	1/2	187	203	7.38	8.00	194	279	7.62	11.00	216	318	8.50	12.50	±1.6	±0.062
20	3/4	187	206	7.38	8.25	194	279	7.62	11.00	216	318	8.50	12.50	±1.6	±0.062
25	1	187	210	7.38	8.25	197	279	7.75	11.00	216	318	8.50	12.50	±1.6	±0.062
40	1-1/2	222	251	8.75	9.88	235	330	9.25	13.00	260	359	10.25	14.12	±1.6	±0.062
50	2	254	286	10.00	11.25	292	375	11.50	14.75	318	400	12.50	15.75	±1.6	±0.062
65	2-1/2	292	311	11.50	12.25	292	375	11.50	14.75	318	400	12.50	15.75	±1.6	±0.062
80	3	318	337	12.50	13.25	318	460	12.50	18.12	381	498	15.00	19.62	±1.6	±0.062
100	4	368	394	14.50	15.50	368	530	14.50	20.88	406	575	16.00	22.62	±1.6	±0.062
150	6	451	508	17.75	20.00	508	768	24.00	30.25	610	819	24.00	32.25	±1.6	±0.062
200	8	543	610	21.38	24.00	610	832	24.00	32.75	762	1029	30.00	40.25	±1.6	±0.062
250	10	673	752	26.50	29.62	762	991	30.00	39.00	1016	1270	40.00	50.00	±1.6	±0.062
300	12	737	819	29.00	32.25	914	1130	36.00	44.50	1118	1422	44.00	56.00	±3.2	±0.125
350	14	851	1029	33.50	40.50	—	1257	—	49.50	—	1803	—	71.00	±3.2	±0.125
400	16	1016	1108	40.00	43.62	—	1422	—	56.00	—	—	—	—	±3.2	±0.125
450	18	1143	—	45.00	—	—	1727	—	68.00	—	—	—	—	±3.2	±0.125

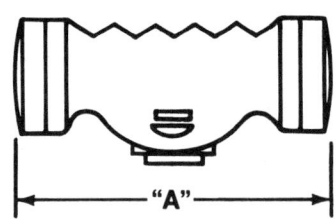

APPENDIX

This appendix is not part of ISA-S75.15, but it is included to facilitate its use.

The ANSI standards listed in the bibliography contain valve body design information in addition to face-to-face dimensions. Dimensions for metrically sized valves are nominal conversions that are conventionally used in the Manufacturers Standardization Society (MSS) of the Valve and Fitting Industry's Publication MSS-SP86-1981, and in International Standards Organization (ISO) and International Electrotechnical Commission (IEC) documents.

ANSI/ISA-S75.16-1987
Approved March 20, 1987

American National Standard

Face-to-Face Dimensions for Flanged Globe-Style Control Valve Bodies (ANSI Classes 900, 1500, and 2500)

Instrument Society of America

Instrument Society of America

ISBN 0-87664-977-0

ISA-S75.16 Face-to-Face Dimensions for Flanged Globe-Style Control Valve Bodies (ANSI Classes 900, 1500, and 2500)

Copyright © 1986 by the Instrument Society of America. All rights reserved. Printed in the United States of America. No part of this publication may be reproduced, stored in a retrieval system, or transmitted, in any form or by any means (electronic, mechanical, photocopying, recording, or otherwise), without the prior written permission of the Publisher.

INSTRUMENT SOCIETY OF AMERICA
67 Alexander Drive
P.O. Box 12277
Research Triangle Park, North Carolina 27709

PREFACE

This preface is included for informational purposes and is not part of ISA-S75.16.

This standard has been prepared as part of the service of the Instrument Society of America (ISA) toward a goal of uniformity in the field of instrumentation. To be of real value, this document should not be static, but should be subject to periodic review. Toward this end, the Society welcomes all comments and criticisms, and asks that they be addressed to the Secretary, Standards and Practices Board, Instrument Society of America, 67 Alexander Drive, P.O. Box 12277, Research Triangle Park, NC 27709, Telephone (919) 549-8411.

The ISA Standards and Practices Department is aware of the growing need for attention to the metric system of units in general, and the International System of Units (SI) in particular, in the preparation of instrumentation standards. The Department is further aware of the benefits to U.S.A. users of ISA standards of incorporating suitable references to the SI (and the metric system) in their business and professional dealings with other countries. Toward this end, this Department will endeavor to introduce SI-acceptable metric units in all new and revised standards to the greatest extent possible. *The Metric Practice Guide*, which has been published by the Institute of Electrical and Electronics Engineers as ANSI/IEEE Std. 268-1982, and future revisions will be the reference guide for definitions, symbols, abbreviations, and conversion factors.

It is the policy of the Instrument Society of America to encourage and welcome the participation of all concerned individuals and interests in the development of ISA standards. Participation in the ISA standards-making process by an individual in no way constitutes endorsement by the employer of that individual, of the Instrument Society of America, or of any of the standards that ISA develops.

The information contained in the preface, footnotes, and appendices is included for information only and is not a part of the standard.

The following people served as members of ISA Committee SP75.08 (formerly SP4.1), which prepared this standard.

NAME	COMPANY
W. C. Weidman, Chairman	Gilbert/Commonwealth, Inc.
R. Barnes	Valtek, Inc.
H. D. Baumann	H. D. Baumann Associates, Ltd.
G. Borden, Jr.	Bechtel Power Corporation
R. Brodin	Fisher Controls International, Inc.
R. Chown	OTEC
E. J. Cooney	Air Products & Chemicals, Inc.
J. T. Emery	Honeywell Inc.
A. J. Hanssen	Fluid Controls Institute, Inc.
H. P. Illing	Kieley & Mueller, Inc.
R. B. Jones	Dow Chemical USA
M. W. Kaye	M. W. Kellogg Company
O. P. Lovett, Jr.	ISIS Corporation
P. C. Martin	Hammel Dahl & Jamesbury
H. R. Nickerson	Resistoflex Company
R. A. Quance	Walsh Inc.
J. N. Reed	Masoneilan/Dresser
J. Rosato	Rawson Company
H. Schwartz	Flexible Valve Corporation
F. O. Seger	Willis Division, Smith International, Inc.
J. M. Simonsen	Valtek, Inc.
H. R. Sonderregger	ITT Grinnell Corporation
R. U. Stanley	Retired
G. Stiles	Fisher Controls International, Inc.
R. F. Tubbs	Copes-Vulcan-Charlotte

Instrument Society of America

The following people served as members of ISA Committee SP75 during the review of this standard.

NAME	COMPANY
L. R. Driskell, Chairman	Consultant
J. B. Arant	E. I. du Pont de Nemours & Company
H. E. Backinger	Kraus Company, Inc.
G. E. Barb	Muesco Inc.
R. Barnes/J. M. Simonsen*	Valtek, Inc.
H. D. Baumann	H. D. Baumann Associates, Ltd.
N. Belaef	Consultant
G. Borden, Jr.	Bechtel Power Corporation
R. Brodin/F. Harthun/G. Stiles*	Fisher Controls International, Inc.
E. H. C. Brown	Dravo Engineers, Inc.
E. J. Cooney	Air Products & Chemicals, Inc.
W. G. Dewart	Rockwell International
J. T. Emery	Honeywell, Inc.
H. J. Fuller	Retired
L. F. Griffith	Consultant
A. J. Hanssen	Fluid Controls Institute, Inc.
H. P. Illing	Kieley & Mueller, Inc.
R. B. Jones	Dow Chemical USA
M. W. Kaye	M. W. Kellogg Company
R. Louviere	Creole Engineering
O. P. Lovett, Jr.	ISIS Corporation
P. C. Martin	Hammel Dahl & Jamesbury
A. P. McCauley	Chagrin Valley Controls, Inc.
T. V. Molloy	Pacific Gas & Electric
H. R. Nickerson	Resistoflex Company
J. Ozol	Omaha Public Power Company
R. A. Quance	Walsh Inc.
W. Rahmeyer	Utah State University
J. N. Reed	Masoneilan/Dresser
G. Richards	Richards Industries, Inc.
J. Rosato	Rawson Company
K. Schoonover, Secretary	Con-Tek
H. Schwartz	Flexible Valve Corporation
W. L. Scull/J. T. Muller*	Leslie Company
F. O. Seger	Willis Division, Smith International, Inc.
H. R. Sonderregger	ITT Grinnell Corporation
N. D. Sprecher	De Zurik
R. U. Stanley	Retired
R. E. Terhune, V. Chairman	Consultant
R. F. Tubbs	Copes-Vulcan-Charlotte
W. C. Weidman	Gilbert/Commonwealth, Inc.
R. L. Widdows	Cashco Inc.
L. R. Zinck	Union Carbide Corporation

*One vote

This standard was approved for publication by the ISA Standards and Practices Board in June 1986.

NAME	COMPANY
N. Conger, Chairman	Fisher Controls International, Inc.
W. Calder III	The Foxboro Company
R. S. Crowder	Ship Star Associates
H. S. Hopkins	Consultant
J. L. Howard	Boeing Aerospace Company
R. T. Jones	Philadelphia Electric Company
R. Keller	The Boeing Company
O. P. Lovett, Jr.	ISIS Corporation
E. G. Magison	Honeywell, Inc.
A. P. McCauley	Chagrin Valley Controls, Inc.
J. W. Mock	Bechtel Corporation
E. M. Nesvig	ERDCO Engineering Corporation
R. Prescott	Moore Products Company
D. E. Rapley	Rapley Engineering Company
C. W. Reimann	National Bureau of Standards
J. Rennie	Factory Mutual Research Corporation
W. C. Weidman	Gilbert/Commonwealth, Inc.
K. Whitman	Consultant
*P. Bliss	Consultant
*B. A. Christensen	Continental Oil Company
*L. N. Combs	Retired
*R. L. Galley	Consultant
*T. J. Harrison	IBM Corporation
*R. G. Marvin	Roy G. Marvin Company
*W. B. Miller	Moore Products Company
*G. Platt	Retired
*J. R. Williams	Stearns Catalytic Corporation

---

*Director Emeritus

1 SCOPE

1.1 This standard applies to flanged control valves, sizes 1/2 inch through 18 inches, having top, top and bottom, port, or cage guiding.

2 PURPOSE

2.1 The purpose of this standard is to aid users in their piping design by providing ANSI Classes 900, 1500, and 2500, raised-face, flanged control valve dimensions, without giving special consideration to the equipment manufacturer to be used.

2.2 The short-long dimensions provided in Table 1 clarify Section 2.1 by consolidating the diversity of existing manufacturers' lengths into two sets of dimensions for each valve size. Before using either the short or long dimension, the piping designer should confirm with the selected valve manufacturer which dimension is correct for the valve(s) being supplied.

3 DEFINITIONS

3.1 For definitions of terms used in this standard, see ISA-S75.05, "Control Valve Terminology."

4 BIBLIOGRAPHY

The following references are included for the definition of pressure classes, flange dimensions, material identification, and cross-reference information. They were developed by the American Society of Mechanical Engineers (ASME).

4.1 American National Standards Institute, Inc. (ANSI), Standard ANSI B16.34-1981, "Valves — Flanged and Buttwelding End."

4.2 American National Standards Institute, Inc., Standard ANSI B16.5-1981, "Pipe Flanges and Flanged Fittings."

4.3 American National Standards Institute, Inc., Standard ANSI B16.10-1973, "Face-to-Face and End-to-End Dimensions of Ferrous Valves."

5 DIMENSIONAL DATA

5.1 For face-to-face dimensions for flanged globe-style control valves, see Table 1.

TABLE 1
FACE-TO-FACE DIMENSIONS FOR RAISED-FACE, FLANGED GLOBE-STYLE CONTROL VALVES

Nominal Valve Size		PN 150 (ANSI Class 900)				PN 250 (ANSI Class 1500)				PN 420 (ANSI Class 2500)				Tolerance	
		Dimension "A"				Dimension "A"				Dimension "A"					
		mm		inches		mm		inches		mm		inches			
mm	inches	Short	Long	Short	Long	Short	Long	Short	Long	Short	Long	Short	Long	mm	inches
15	1/2	273	292	10.75	11.50	273	292	10.75	11.50	308	318	12.12	12.50	±1.6	±0.062
20	3/4	273	292	10.75	11.50	273	292	10.75	11.50	308	318	12.12	12.50	±1.6	±0.062
25	1	273	292	10.75	11.50	273	292	10.75	11.50	308	318	12.12	12.50	±1.6	±0.062
40	1-1/2	311	333	12.25	13.12	311	333	12.25	13.12	359	381	14.12	15.00	±1.6	±0.062
50	2	340	375	13.38	14.75	340	375	13.38	14.75	—	400	—	16.25	±1.6	±0.062
65	2-1/2	—	410	—	16.12	—	410	—	16.12	—	441	—	17.38	±1.6	±0.062
80	3	387	441	15.25	17.38	406	460	16.00	18.12	498	660	19.62	26.00	±1.6	±0.062
100	4	464	511	18.25	20.12	483	530	19.00	20.87	575	737	22.62	29.00	±1.6	±0.062
150	6	600	714	21.87	28.12	692	768	24.00	30.25	819	864	32.25	34.00	±1.6	±0.062
200	8	781	914	30.75	36.00	838	972	33.00	38.25	—	1022	—	40.25	±1.6	±0.062
250	10	864	991	34.00	39.00	991	1067	39.00	42.00	1270	1372	50.00	54.00	±1.6	±0.062
300	12	1016	1130	40.00	44.50	1130	1219	44.50	48.00	1321	1575	52.00	62.00	±3.2	±0.125
350	14	—	1257	—	49.50	—	1257	—	49.50	—	—	—	—	±3.2	±0.125
400	16	—	1422	—	56.00	—	1422	—	56.00	—	—	—	—	±3.2	±0.125
450	18	—	1727	—	68.00	—	1727	—	68.00	—	—	—	—	±3.2	±0.125

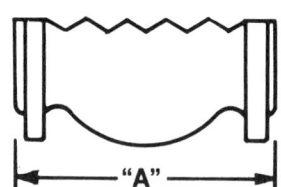

APPENDIX

This appendix is not part of ISA-S75.16, but it is included to facilitate its use.

The ANSI standards listed in the references contain valve body design information in addition to face-to-face dimensions. Dimensions for metrically sized valves are nominal conversions that are conventionally used in the Manufacturers Standardization Society (MSS) of the Valve and Fitting Industry's Publication MSS-SP86-1981, and in International Standards Organization (ISO) and International Electrotechnical Commission (IEC) documents.

ANSI/ISA-S77.42-1987
Approved: August 10, 1987

American National Standard

Fossil Fuel Power Plant Feedwater Control System — Drum Type

Instrument Society of America

Instrument Society of America

ISBN 1-55617-048-3

ISA-S77.42 Fossil Fuel Power Plant Feedwater Control System — Drum-Type

Copyright © 1987 by the Instrument Society of America. All rights reserved. Printed in the United States of America. No part of this publication may be reproduced, stored in a retrieval system, or transmitted, in any form or by any means (electronic, mechanical, photocopying, recording, or otherwise), without the prior written permission of the publisher.

INSTRUMENT SOCIETY OF AMERICA
67 Alexander Drive
P.O. Box 12277
Research Triangle Park, North Carolina 27709

PREFACE

This preface is included for informational purposes and is not part of ISA-S77.42.

This standard has been prepared as part of the service of the Instrument Society of America (ISA) toward a goal of uniformity in the field of instrumentation. To be of real value, this document should not be static, but should be subject to periodic review. Toward this end, the Society welcomes all comments and criticisms, and asks that they be addressed to the Secretary, Standards and Practices Board, Instrument Society of America, 67 Alexander Drive, P.O. Box 12277, Research Triangle Park, NC 27709, Telephone (919) 549-8411.

The ISA Standards and Practices Department is aware of the growing need for attention to the metric system of units in general, and the International System of Units (SI) in particular, in the preparation of instrumentation standards. The Department is further aware of the benefits to U.S.A. users of ISA standards of incorporating suitable references to the SI (and the metric system) in their business and professional dealings with other countries. Toward this end, this Department will endeavor to introduce SI-acceptable metric units in all new and revised standards to the greatest extent possible. The *Metric Practice Guide*, which has been published by the Institute of Electrical and Electronics Engineers as ANSI/IEEE Std. 268-1982, and future revisions will be the reference guide for definitions, symbols, abbreviations, and conversion factors.

It is the policy of the Instrument Society of America to encourage and welcome the participation of all concerned individuals and interests in the development of ISA standards. Participation in the ISA standards-making process by an individual in no way constitutes endorsement by the employer of that individual, of the Instrument Society of America, or of any of the standards that ISA develops.

The information contained in the preface, footnotes, and appendices is included for information only and is not part of the standard. Functional drawings in the appendix are provided using both ANSI/ISA-S5.1-1984 symbology and established power industry conventions.

This standard is part of a series of standards resulting from the efforts of the ISA SP77 Fossil Power Plant Standard Committee's Subcommittee SP77.40, "Boiler Controls." It should be used in conjunction with the other SP77 series of standards for safe, reliable, and efficient design, construction, operation, and maintenance of fossil-fired power plants. It is not intended that this standard establish any procedures or practices that are contrary to any other standard in this series.

A variety of feedwater control systems have been developed and used over the years to maintain drum level within limits at the required set point. This standard is intended to establish minimum requirements for feedwater control.

The following people served as members of ISA Subcommittee SP77.42, "Feedwater Committee."

NAME	COMPANY
G. R. McFarland, Chairman	Westinghouse Electric Corporation
W. T. Blazier	Illinois Power Company
N. Burris	Tennessee Valley Authority
R. Campbell	Salt River Project
J. Cooper	EBASCO Services, Incorporated
R. T. Criswell	Foster Wheeler Energy Corporation
T. Dimmery	Duke Power Company
W. Holland	Georgia Power Company
H. Hubbard	Southern Company Services, Incorporated
R. Hubby	Leeds & Northrup
J. R. Karvinen	Mountain States Energy, Incorporated
M. B. Laney	Duke Power Company
L. Martz	Westinghouse Electric Company
T. R. New	Southern Company Services, Incorporated
J. A. Rovnak	Stone & Webster Engineering Corporation
T. Russell	Fisher Controls International, Inc.
J. Tana	EBASCO Services, Incorporated
T. Toms	Carolina Power & Light Company
B. L. Traylor	Virginia Power Company
H. Wall	Duke Power Company

Instrument Society of America

The following people served as members of ISA Committee SP77.40, "Boiler Control Committee":

NAME	COMPANY
G. R. McFarland, Chairman/R. Spellman/ L. Martz*	Westinghouse Electric Corporation
E. J. Adamson/O. Taneja/J. A. Rovnak*	Stone & Webster Engineering Corporation
R. Anken	New York State Electric & Gas Company
C. Balasubramanian	United Engineers & Constructors, Inc.
M. A. Blaschke	Weyerhaeuser Company
W. T. Blazier	Illinois Power Company
R. Brodd	Public Service Company of New Mexico
N. Burris/J. Martin*	Tennessee Valley Authority
R. Campbell	Salt River Project
J. Cooper	EBASCO Services, Incorporated
R. L. Criswell	Foster Wheeler Energy Corporation
T. R. Dimmery/M. B. Laney/H. Wall*	Duke Power Company
J. Dole	Fox & Dole Company
G. Ellis/W. Holland*	Georgia Power Company
J. C. Flynn	Combustion Engineering, Incorporated
D. Hagerty	Bechtel Corporation
R. Hartswell	Long Island Lighting Company
R. Hicks	Consumers Power Company
C. F. Hildenbrand	Retired
H. Hubbard/T. R. New*	Southern Company Services, Incorporated
C. L. Hughart	Union Carbide Company
J. R. Karvinen	Mountain States Energy Incorporated
N. Kerman	Duquesne Light
W. Labos	Public Service Electric & Gas Company
I. Mazza	Consultant, Italy
A. Misra	Consultant, Mozambique, East Africa
W. S. Matz/R. Murphy*	Forney Engineering Company
J. R. McLemore	Alabama Power Company
E. McWilliams	Houston Lighting & Power Company
P. Papish	Pall Pneumatic Products Corporation
L. D. Rawlings II	Babcock & Wilcox
T. Russell	Fisher Controls International, Incorporated
L. Smith	Arkansas Power & Light Company
T. Toms	Carolina Power & Light Company
B. L. Traylor	Virginia Power Company
G. Vaccaro	Cleveland Electric Illuminating Company
R. Walker	Consultant

*One vote

This standard was approved for publication by the ISA Standards and Practices Board in June 1987.

NAME	COMPANY
D. E. Rapley, Chairman	Rapley Engineering Services
D. N. Bishop	Chevron U.S.A. Inc.
W. Calder III	The Foxboro Company
N. Conger	Fisher Controls International, Inc.
R. S. Crowder	Ship Star Associates
C. R. Gross	Eagle Technology
H. S. Hopkins	Utility Products of Arizona
J. L. Howard	Boeing Aerospace Company
R. T. Jones	Philadelphia Electric Company
R. Keller	The Boeing Company
O. P. Lovett, Jr.	ISIS Corporation
E. C. Magison	Honeywell, Inc.
A. P. McCauley	Chagrin Valley Controls, Inc.
J. W. Mock	The Bechtel Group, Inc.
E. M. Nesvig	ERDCO Engineering Corporation
R. Prescott	Moore Products Company
C. W. Reimann	National Bureau of Standards
R. H. Reimer	Allen-Bradley Company
J. Rennie	Factory Mutual Research Corporation
W. C. Weidman	Gilbert/Commonwealth Inc.
K. A. Whitman	Fairleigh Dickinson University
**P. Bliss	Consultant
**B. A. Christensen	Continental Oil Company
**L. N. Combs	Consultant
**R. L. Galley	Consultant
**T. J. Harrison	IBM Corporation
**R. G. Marvin	Roy G. Marvin Company
**W. B. Miller	Moore Products Company
**G. Platt	Consultant
**J. R. Williams	Stearns Catalytic Corporation

*One vote
**Director Emeritus

Instrument Society of America

TABLE OF CONTENTS

Section	Title	Page
1.	Purpose	7
2.	Scope	7
3.	Definitions	7
4.	Minimum Design Requirements for a Feedwater Control System	8
5.	References	15

Appendix A, Feedwater Control 17
Table A-1, Summary of Typical Control Elements 17

LIST OF ILLUSTRATIONS

Figure	Title	Page
1.	Typical Drum Level Differential Pressure Transmitter Connections	9
2.	Single-Element Feedwater Control (Functional Control Diagram)	10
2A.	Single-Element Feedwater Control (Functional Control Diagram Using the ANSI/ISA-S5.1-1984 Format)	10
3.	Two-Element Feedwater Control (Functional Control Diagram)	11
3A.	Two-Element Feedwater Control (Functional Control Diagram Using the ANSI/ISA-S5.1-1984 Format)	12
4.	Three-Element Feedwater Control (Functional Control Diagram)	13
4A.	Three-Element Feedwater Control Functional Control Diagram Using the ANSI/ISA-S5.1-1984 Format	14
5.	Water/Steam Density Relationship	19
6.	Level Measurements and Corrections	20
7.	Positive Static Head Curves	21
8.	Positive Static Head — Pumps in Parallel	22
9.	Positive Static Head — Pumps in Series	23
10.	Diagram Depicting Effects of Throttling and Speed Reduction	24

1 PURPOSE

The purpose of this standard is to establish minimum criteria for the control of levels, pressures, and flow for the safe and reliable operation of drum-type feedwater systems in fossil power plants.

2 SCOPE

The standard is intended to assist in the development of design specifications covering the measurement and control of feedwater systems in boilers with steaming capacities of 200 000 lb/hr (25 kg/s) or greater. The safe physical containment of the feedwater shall be in accordance with applicable piping codes and standards and is beyond the scope of this standard.

3 DEFINITIONS

The following definitions are included to clarify their use in this standard and may not correspond to the use of the word in other texts. For other definitions, reference ANSI/ISA S51.1-1979, "Process Instrumentation Terminology."

3.1 Boiler — The entire vessel in which steam or other vapor is generated for use external to itself, including the furnace, consisting of the following: waterwall tubes; the firebox area, including burners and dampers; the convection area, consisting of any superheater, reheater, and/or economizer sections, as well as drums and headers.

3.2 Cascade Control — Control action in which the output of one controller is the set point for another controller.

3.3 Controller — Any manual or automatic device or system of devices for the regulation of boiler systems to keep the boiler at normal operation. If automatic, the device or system is motivated by variations in temperature, pressure, water level, time, flow, or other influences.

3.4 Drum (Steam) — A closed vessel designed to withstand internal pressure. A device for collecting and separating the steam/water mixture circulated through the boiler.

3.5 Feedwater Control System — A control system using input signals derived from the process for the purpose of regulating feedwater flow to the boiler to maintain adequate drum level according to the manufacturer's recommendations.

3.6 Mass Feedwater Flow Rate — The mass flow rate of all water delivered to the boiler, derived either from direct process measurements and/or calculations from other parameters. When volumetric feedwater flow rate measurement techniques are employed and the feedwater temperature at the flow-measuring element varies 100° F (37.8° C), the measured (indicated) flow shall be compensated for flowing feedwater density to determine the true mass feedwater flow rate.

3.7 Mass Steam Flow Rate — The mass flow rate of steam from the boiler, derived either from direct process measurements and/or calculations from other parameters. If volumetric steam flow-rate measuring techniques are employed, the measured (indicated) flow shall be compensated for flowing steam density to determine the true mass steam flow rate.

3.8 Primary/Secondary Control Loop Controller — The controller which adjusts the set point for the secondary control loop controller in the cascade control action scheme.

3.9 Protective Logic Circuits — Logic circuits designed to prevent damage to equipment by related system equipment malfunctions, failure, or operator errors.

3.10 Pump Drive Control — A control component of the final device that translates a control system demand signal into an electronic, hydraulic, pneumatic, or mechanical signal which affects pump speed.

3.11 Redundant (Redundancy) — Duplication or repetition of elements in electronic or mechanical equipment to provide alternative functional channels in case of failure of the primary device.

3.12 Runback — An action by the boiler control system initiated by the loss of any auxiliary equipment that limits the capabilities of the unit to sustain the existing load. Upon runback initiation, the boiler demand signal is reduced at a preset rate to the capability of the remaining auxiliaries.

3.13 Rundown — An action by the boiler control system initiated by an unsafe operating condition — i.e., fuel/air limit (cross-limiting), temperature limits, etc. Upon rundown initiation, the boiler demand signal is reduced in a controlled manner to the load point where the unsafe operating condition is eliminated.

3.14 Shall, Should, and May — The word "SHALL" is to be understood as a REQUIREMENT, the word "SHOULD" as a RECOMMENDATION, and the word "MAY" as PERMISSIVE, neither mandatory nor recommended. (Reference: ISA-RPA-1a-1970, "Style Manual.")*

3.15 Shrinkage — A decrease (shrinkage) in drum level due to a decrease in steam bubble volume. This condition is due to a decrease in load (steam flow), with a resulting increase in drum pressure and a decrease in heat input.

3.16 Single-Element Feedwater Control — A control system whereby one process variable, drum level, is used as the input to the control loop that regulates feedwater flow to the drum to maintain the drum level at set point.

---

*See Section 5 for bibliographic information on references.

3.17 Steady-State — A characteristic of a condition, such as value, rate, periodicity, or amplitude, exhibiting only negligible change over a long (arbitrarily chosen) period of time.

NOTE: It may describe a condition in which some characteristics are static, others dynamic.

3.18 Swell — An increase (swell) in drum level due to an increase in steam bubble volume. This condition is due to an increase in load (steam flow), with a resulting decrease in drum pressure and an increase in heat input. Swelling also occurs during a cold start-up as the specific volume of the water increases.

3.19 Three-Element Feedwater Control — A control system whereby three process variables (steam flow, feedwater flow, and drum level) are used as inputs to the control loop that regulates feedwater flow to the drum to maintain the drum level at set point. This is a cascaded feedforward loop with drum level as the primary variable, steam flow as the feedforward input, and feedwater flow (feedback) as the secondary variable.

3.20 Transient — The behavior variable during the transition between two steady states.

3.21 Two-Element Feedwater Control — A control system whereby two process variables (steam flow and drum level) are used as inputs to the control loop that regulates feedwater flow to the drum to maintain the drum level at set point. The feedforward input is steam flow, with the output of the drum level controller as the primary control signal.

3.22 Two-Out-of-Three Logic Circuit (2/3 Logic Circuit) — A logic circuit that employs three independent inputs. The output of the logic circuit is the same state as any two matching input states.

4 MINIMUM DESIGN REQUIREMENTS FOR A FEEDWATER CONTROL SYSTEM

The control system shall meet operational requirements and correctly interface with the process. To accomplish this objective, the following requirements are defined for minimum system design:

1. Process measurement requirements
2. Control and logic requirements
3. Final control device requirements
4. System reliability and availability
5. Alarm requirements
6. Operator interface

4.1 Process Measurement Requirements

4.1.1 Instrument Installation for Feedwater Control

4.1.1.1 Instruments should be installed as close as is practical to the source of the measurement, with consideration being given to excessive vibration, temperature, and accessibility for periodic maintenance. Recommendations for the location of instrument and control equipment connections can be found in the joint publication by SAMA (Scientific Apparatus Makers Association) and ABMA (American Boiler Manufacturers Association), "Recommendations for Location of Instrument and Control Connections for the Operation and Control of Watertube Boilers." Specific requirements for the location of drum water level measuring devices are contained in Section 1 of the American Society of Mechanical Engineers' *Boiler and Pressure Vessel Code*.

4.1.1.2 Separate isolation valves, head chambers (when used), and impulselines shall be provided for each instrument. (See Figure 1.)

4.1.2 Drum Level Measurement

A drum level signal is required for single-element, two-element, and three-element feedwater control systems. (See Figures 1, 2, 3, and 4.) If the instruments used to measure drum level are sensitive to density variation, then density compensation techniques shall be employed.

4.1.3 Steam Flow Measurement

A mass steam flow signal is required for two-element and three-element feedwater control systems.

4.1.4 Feedwater Flow Measurement

A mass feedwater flow signal is required for three-element feedwater control systems.

4.2 Control and Logic Requirements

The function of the feedwater control system is to maintain drum water level within the boiler manufacturer's specified limits. The flow of feedwater to the drum is controlled by the variation of boiler feedpump speed and/or by the action of a control valve(s). Feedwater control can be accomplished by using the following control strategies:

1. Single-element control
2. Two-element control
3. Three-element control

4.2.1 Single-Element Feedwater Control

Single-element control is the minimum feedwater control system and shall be used for the following applications:

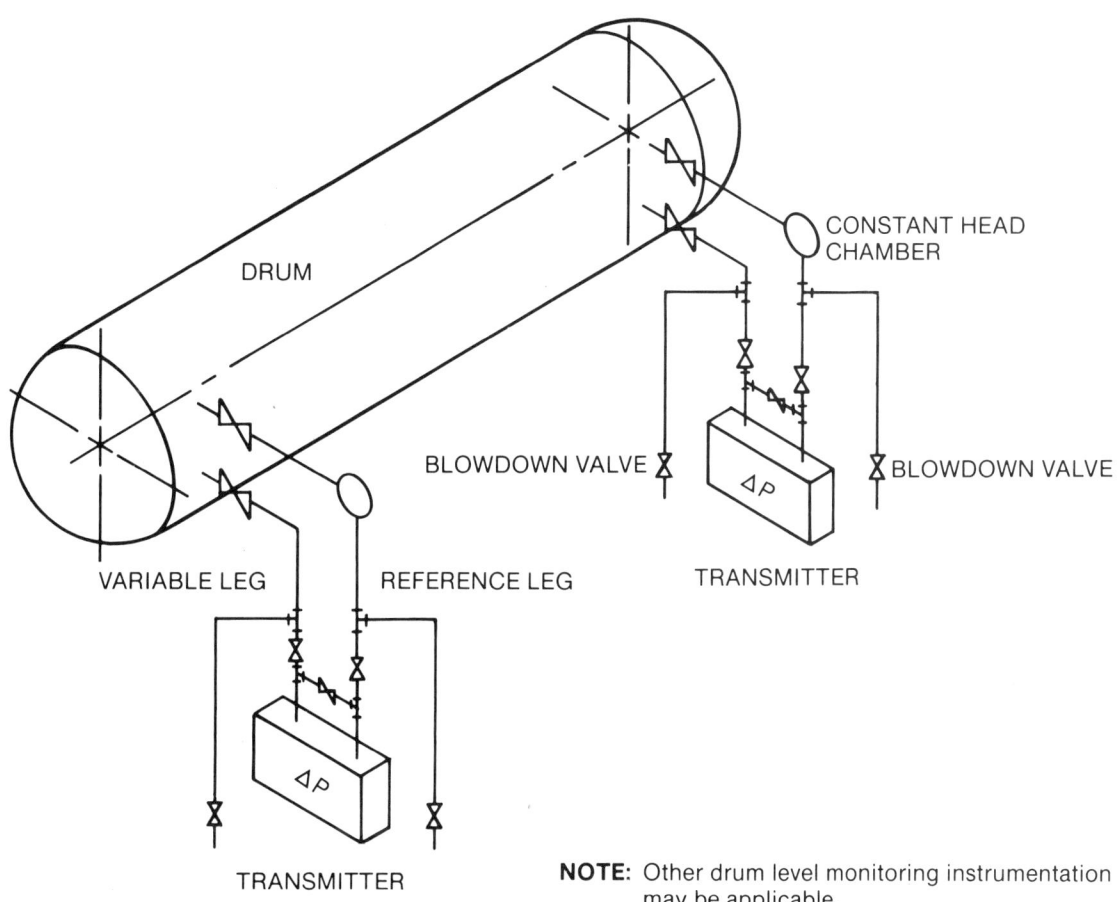

Figure 1. Typical Drum Level Differential Pressure Transmitter Connections

1. During start-up or at low-load operation, when flow measurements are generally not accurate.

2. When steam flow rate of change is nominal and feedwater supply pressure is essentially constant.

4.2.2 Two-Element Feedwater Control

Two-element control is the minimum feedwater control for a variable steam flow application and is not recommended for new applications.

4.2.3 Three-Element Feedwater Control

Three-element control shall be used for variable steam flow applications.

4.2.4 Feedwater Protective Logic

Requirements for protective logic signals shall be those determined by the specific equipment manufacturers. Protective logic signals to trip the fuel to the boiler may include high drum level and low drum level.

The following conditions shall produce a signal usable by other control systems, as covered under other standards within the SP77 series:

1. Loss of a boiler feedpump shall result in a unit load runback to the remaining on-line boiler feedpump capacity.

2. Exceeding the maximum capability limit of the feedwater system shall result in a unit load rundown to return the feedwater system to a controllable range.

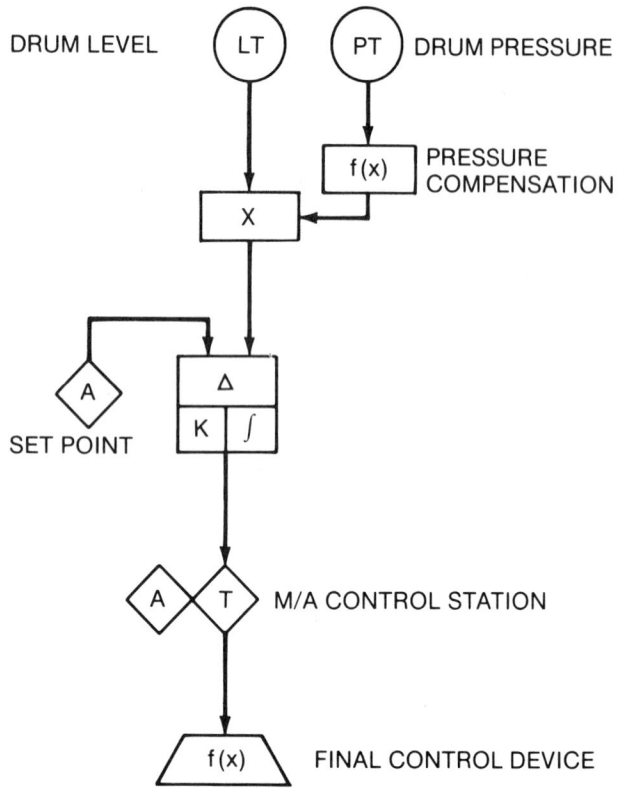

Typical single-drive control system. For simplicity, redundant transmitters have not been shown on this typical control drawing. See Figure 2A for ANSI/ISA-S5.1-1984 format.

Figure 2. Single-Element Feedwater Control
(Functional Control Diagram)

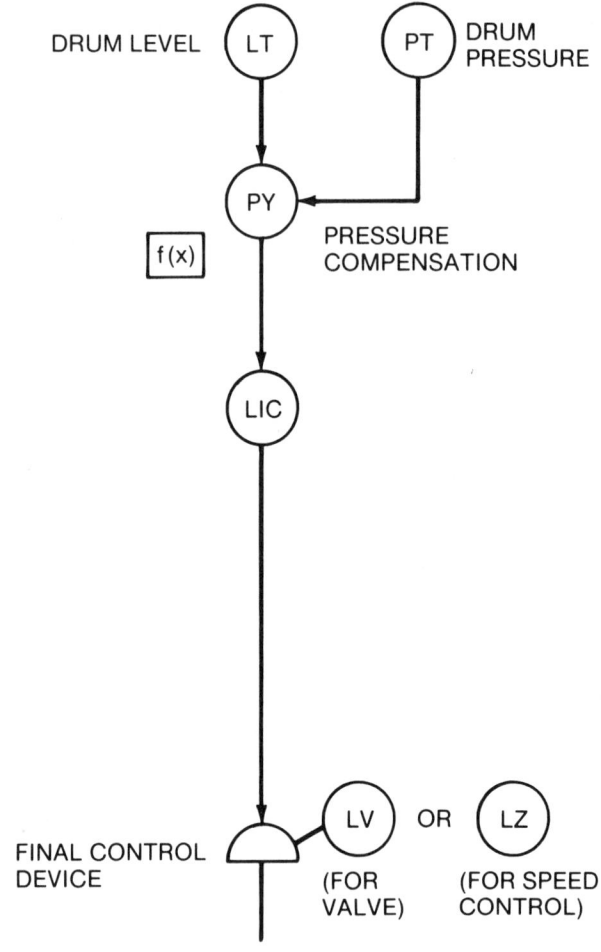

Figure 2A. Single-Element Feedwater Control
(Functional Control Diagram Using
the ANSI/ISA-S5.1 Format)

4.2.5 Feedwater Flow Control

Feedwater flow shall be controlled by varying the speed of the boiler feedpump(s) and/or by varying the position of the feedwater control valve(s).

4.2.5.1 Variable-Speed Pump Control

When feedwater flow is controlled by the use of a hydraulic coupling, or a by a variable-speed motor, or by varying the speed of a turbine-driven pump, the following features shall be provided:

1. In the automatic mode the output of the flow controller shall have an adjustable limit. This limit is a direct function of the operating pressure. Since the output of the flow controller is speed demand (position demand of the hydraulic coupling), the limit will be a pump demand low limit. The low limit should be set to maintain minimum pump flow.

2. When using a flow controller with a speed controller, the flow controller sets pump flow demand. The speed controller shall serve in a cascade configuration to linearize the flow response of the pump.

4.2.5.2 Feedwater Control Valve Control Requirements

When feedwater flow is controlled by valve(s), a single or multiple feedwater control valve(s) may be used. Choice of valve configuration shall be based on a consideration of the pressure drop across the valve, rangeability, cavitation, and excessive valve wear with the valve nearly closed during low-load operation. If multiple valves are used, the control

S77.42
Fossil Fuel Power Plant Feedwater
Control System — Drum-Type

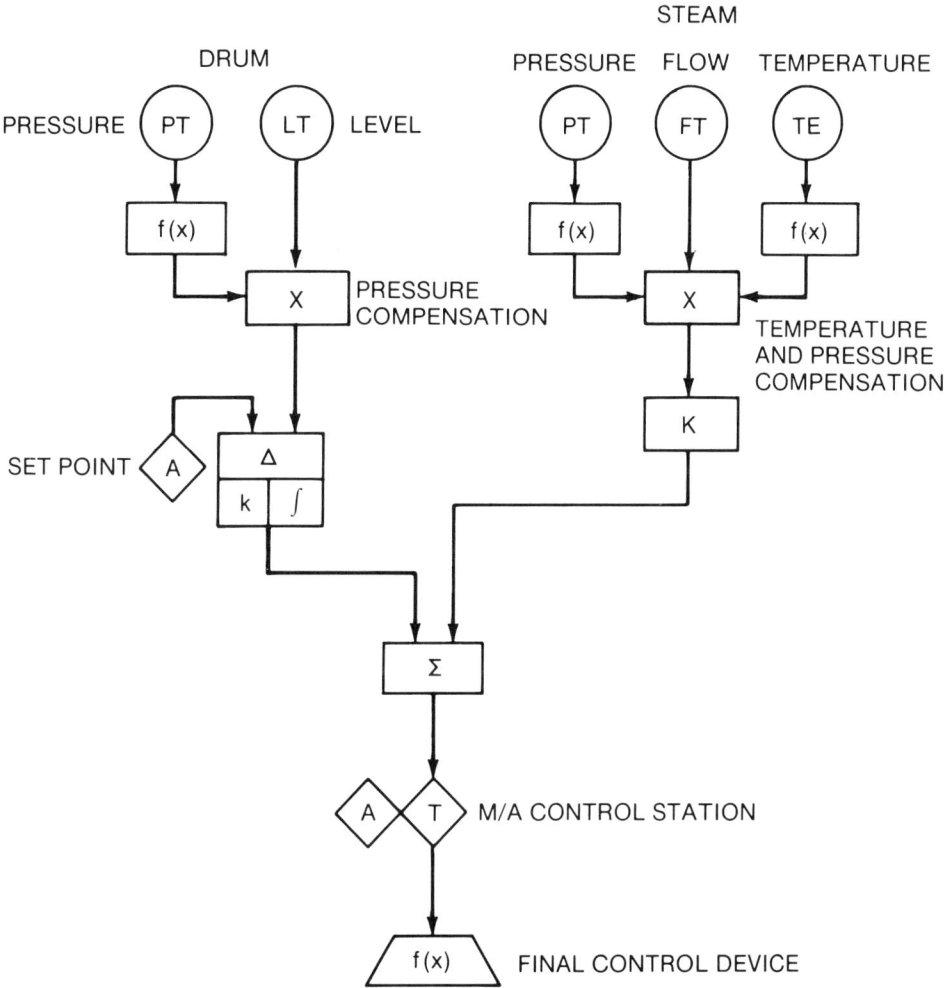

Typical single-drive control system. For simplicity, redundant transmitters have not been shown on this typical control drawing. See Figure 3A for ANSI/ISA-S5.1-1984 format.

**Figure 3. Two-Element Feedwater Control
(Functional Control Diagram)**

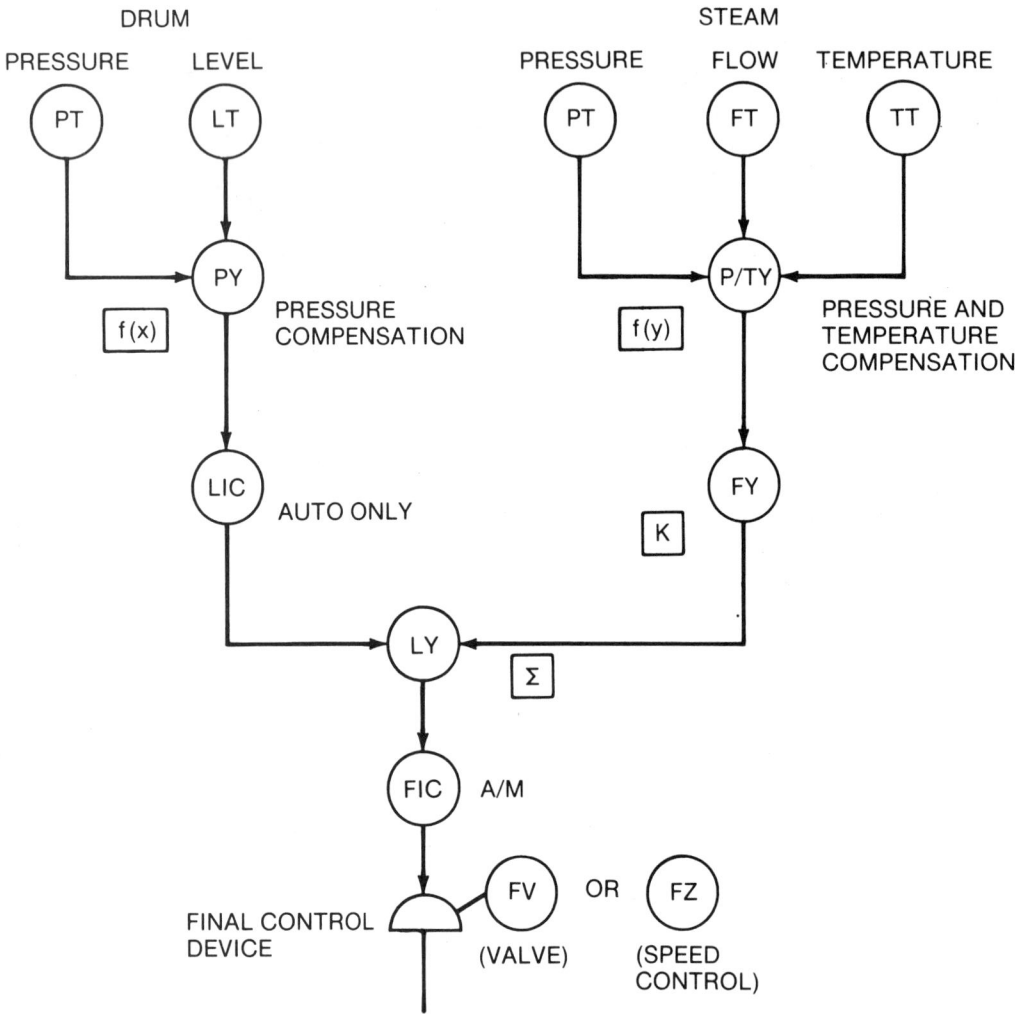

**Figure 3A. Two-Element Feedwater Control
(Functional Control Diagram Using
the ANSI/ISA-S5.1-1984 Format)**

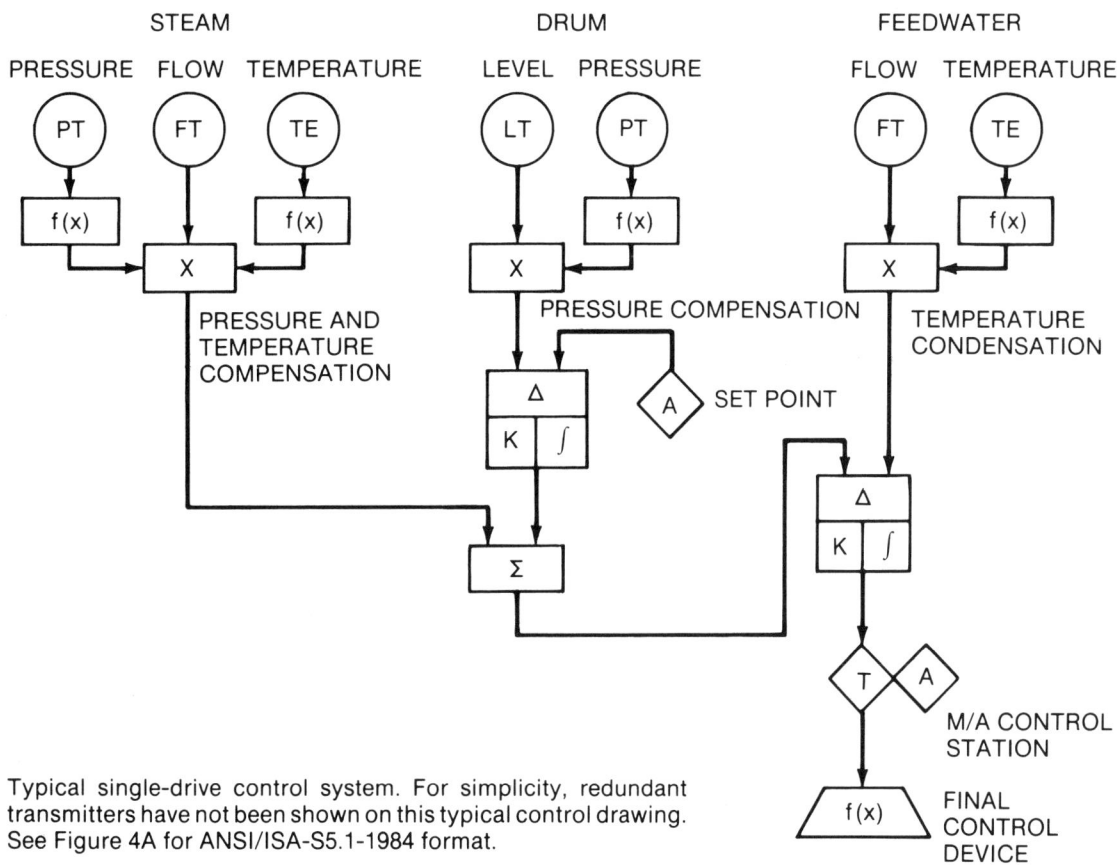

**Figure 4. Three-Element Feedwater Control
(Functional Control Diagram)**

system shall provide for a smooth transition during the crossover to each additional valve, both on flow increases and decreases.

To minimize wear across the main control valves and to improve controllability, the control system should be configured to close off the start-up valve(s) as the main control valve(s) assumes the load.

4.2.6 Boiler Feedpump Minimum Flow

Minimum flow recirculation is required for pump protection — based on the manufacturer's requirements. Control can be modulating or open/close and provided by remotely controlled or self-contained automatic regulating valves. As a minimum design, separate dedicated flow-monitoring systems consisting of flow elements, transmitters, and/or switches shall be provided for each individual boiler feedpump.

When flow switches are used, an adjustable dead band shall be provided. When an operator's manual override station is used, setting the control to "manual" shall always open the valve.

When minimum flow recirculation for a high-pressure booster stage is required by the pump manufacturer, it shall be provided using the same criteria as the main recirculation flow, but with its own flow measurement device.

4.3 Final Control Device Requirements

All final control devices shall be designed to fail safe on loss of demand signal or motive power — i.e., open, close, or lock in place. The fail-safe position shall be determined by the user based upon the specific application. Minimum flow recirculation valve(s) shall open on any failure in the minimum flow control system.

4.4 System Reliability and Availability

4.4.1 In order to establish minimum criteria, the feedwater control system specification shall include the following as part of the system design base:

1. Maximum unit load/steaming capacity

2. Normal operating load range

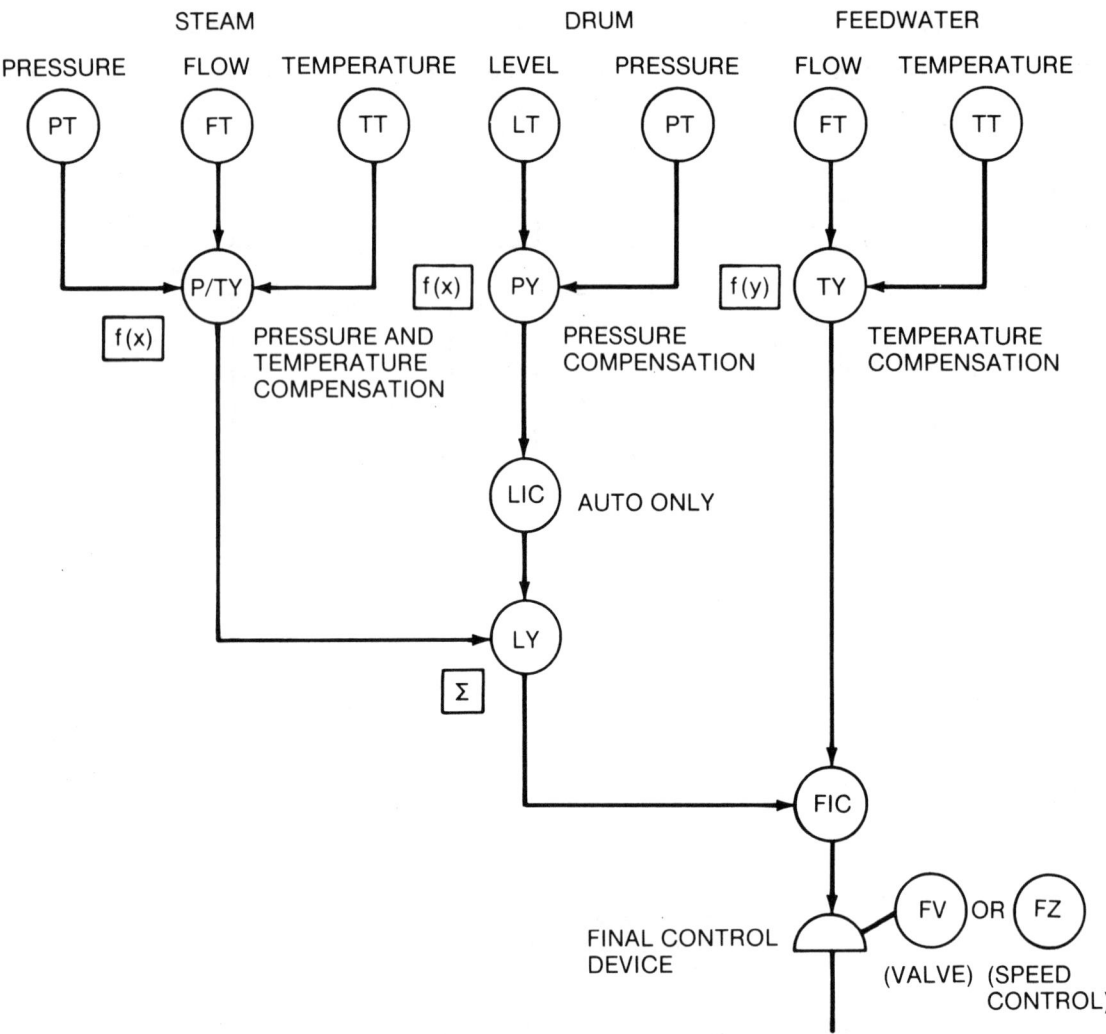

**Figure 4A. Three-Element Feedwater Control
(Functional Control Diagram Using
the ANSI/ISA-S5.1-1984 Format)**

3. Anticipated load changes (transients)

4. Start-up and shut-down frequency

5. Degree of automation

6. Boiler feedpump maximum and minimum capacity

4.4.2 All control transmitters shall be redundant. The following conditions apply:

1. When two transmitters are employed, excessive deviation between the transmitters shall be alarmed and the associated control loop shall be transferred to manual.

2. When three transmitters are employed, excessive deviation between the transmitters shall be alarmed. A transmitter select scheme shall be used for control purposes.

4.5 Minimum Alarm Requirements

Minimum alarm requirements shall include the following:

1. High and low drum level

2. Loss of control power

3. Loss of final drive power

4. Control loop trip-to-manual

5. Feedwater flow/system flow deviation (three-element control)

6. Loss of control transmitter (arming of the trip circuit)

4.6 Operator Interface

4.6.1 The following information used in the feedwater control system shall be made available to the operator. These shall include the following:

1. Drum level

2. Drum pressure

3. Feedwater flow

4. Feedwater temperature

5. Steam flow

6. Steam temperature

7. All alarms

8. Automatic/manual control loop status

9. Main steam pressure (where applicable)

4.6.2 In addition to the above, the following information should be made available to the operator:

1. Final drive position(s)

2. Valve position(s)

3. Pump speed(s)

4. Single- or three-element control status

5. Individual boiler feedpump flow

6. Drum level set point

4.6.3 The system shall include capabilities for the automatic/manual control of each individual final device (except for the boiler feedpump minimum flow valve[s] as discussed within Section 4.2.7).

5 REFERENCES

1. *ACVS Drive 1983 Catalogue*, Reliance Electric Company, Cleveland, Ohio, 1983.

2. ANSI/IEEE Std. 268-1982, "Metric Practice," American National Standards Institute, 1430 Broadway, New York, N.Y. 10018.

3. ANSI/ISA-S5.1-1984, "Instrumentation Symbols and Identification," Instrument Society of America, 67 Alexander Drive, P.O. Box 12277, Research Triangle Park, N.C. 27709.

4. ANSI/ISA-S5.4-1976 (R 1981), "Instrument Loop Diagrams," Instrument Society of America.

5. ANSI/ISA-S51.1-1979, "Process Instrumentation Terminology," Instrument Society of America.

6. API RP550 Pt. IV-1984, "Manual on Installation of Refinery Instruments and Control Systems: Steam Generators," American Petroleum Institute, 1220 L Street, NW, Wash., D.C. 20005.

7. *ASME Boiler and Pressure Vessel Code*, Section 1, American Society of Mechanical Engineers, 345 East 437th St., New York, N.Y. 10017, 1983.

8. ASME MFC-2M-1983, "Measurement Uncertainty for Fluid Flow in Closed Conduits," American Society of Mechanical Engineers.

9. Babcock & Wilcox, *Steam — Its Generation and Use*, 39th ed., New York, N.Y., 1978.

10. ISA-RPA-1a-1970, "Style Manual," Instrument Society of America.

11. Kempers, Gene, "Experience Report — Drum Level Indication Problems: George Neal Units 3 & 4," Edison Electric Institute's Prime Movers' Committee Report, Feb. 1-3, 1982, Wash., D.C.

12. Lockwood, Bernard H., "NPSH and How it is Calculated," Morris Pumps, Inc. (now Goulds Pumps Inc.), Baldwinsville, N.Y., May 1974 (internally published document).

13. PMC 20.1-1973, "Process Measurement and Control Terminology," Scientific Apparatus Makers Association, 1101 16th St., NW, Suite 300, Wash., D.C. 20036.

14. PMC 22.1-1981, "Functional Diagramming of Instrument and Control Systems," Scientific Apparatus Makers Association.

15. Power Technologies, Inc., *Drum Boiler Feedwater Controls*, Chap. IV, Electric Power Research Institute, Palo Alto, Calif.

16. "Recommendations for Location of Instrument and Control Connections for the Operation and Control of Watertube Boilers," SAMA/ABMA/IGCI's Recommended Standard: *Instrument Connections Manual*, jointly published by the Scientific Apparatus Makers Association, the American Boiler Manufacturers Association,

and Industrial Gas Cleaning Institute, Inc., 1981. ABMA's address is 950 North Glebe Road, Suite 180, Arlington, Va. 22203; IGCI's is 700 North Fairfax St., Suite 304, Alexandria, Va. 22314.

17. Singer, Joseph G., *Combustion: Fossil Power Systems*, 3d ed., Combustion Engineering, Inc., 1981.

18. UltraSystems Inc. (for Naval Civil Engineering Laboratory), "Theory of Boiler Control Systems — Operation Manual," N62474-81-C-9388, Irvine, Calif., 142 pp.

19. Waltz, R. J., ASME Paper 81-JPGC-Pwr.-21, "Drum Level Instrumentation for Reliable Boiler Operation," American Society of Mechanical Engineers.

APPENDIX A
FEEDWATER CONTROL

This appendix is included for informational purposes only and is not a part of ISA Standard S77.42.

A.1 PURPOSE

The purpose of this appendix is to provide tutorial information on the philosophy underlying this standard and to assist the user of this standard in specifying and applying feedwater control schemes.

A.2 INTRODUCTION

A.2.1 Design Specification Requirements

To adequately specify a feedwater control strategy, the following three fundamental questions must be addressed:

1. What are the anticipated process operational requirements — e.g., steady-state or cyclic operations, rates of change, etc.?

2. What equipment and operating parameters are required to properly interface the control system?

3. What characteristics must the control system possess to maintain the desired performance?

The extent to which these questions are answered will directly determine how well the control system is fitted to the design and operating requirements. A misapplication, at the least, could result in poor operating performance, and, at worst, could result in extensive boiler damage. The following subsections are intended to supplement good engineering judgment with a consistent means of communicating design requirements to suppliers, designers, or constructors.

A.2.2 Summary of Process Performance Requirements

A significant factor to consider in control system selection is the intended boiler usage. Since the operating requirements of the boiler define the required control system capabilities, design specifications must address the following unit characteristics:

1. Unit load/steaming capacity

2. Normal operating load range

3. Anticipated load changes (transients)

4. Start-up and shut-down frequency

5. Degree of automation

A complete description of the anticipated load characteristics will allow the engineer/supplier to properly evaluate the system and propose a control strategy. When the control strategy is preselected, these characteristics should still be defined as part of the design basis.

Table A.1 provides a general comparison of typical control systems for the engineer's use in specification development and evaluation. This table is not intended to be all-inclusive; rather, it is a summary of commonly used control strategies. The important conclusion to be drawn from the table is that all control systems are not the same, and therefore selection of a specific system requires careful consideration of design parameters.

TABLE A.1

SUMMARY OF TYPICAL CONTROL SYSTEMS

	Single-element	Two-element	Three-element
Prerequisite	Slow rate of change of steam flow	Constant feedwater pressure	Feedwater flow/steam signal available
Steady-state operability	Good	Good	Good
Transient operability	Poor	Good	Good
Response to load change	Slow	Fast	Fast
Control response type	Feedback	Feedforward	Feedforward
Compensation for drum shrink and swell	None	Partial	Effective
Potential for flow imbalance during load change	Probable	Dependent on final drive linearity and repeatability through the load range	Minimal

A.2.3 Single-Element Control (See Figure 2)

Single-element control requires a pressure-compensated drum level signal (if the instruments used to measure drum level are sensitive to density variation) and requires a desired set point signal. Proportional-plus-integral-action controllers maintain level by sending an output to the final control device. Single-element control in fossil-fired power plants should be used for start-up control before steam flow is delivered to the process. Single-element control is the minimum feedwater control system and is applied where steam flow is constant or at low loads when steam flow measurements are not available. When single-element control is combined with two- or three-element control, the mode selection may be automatically selected or operator-selected.

A.2.4 Two-Element Control (See Figure 3)

Two-element feedwater control requires (1) a pressure-compensated drum level signal if the instruments used to measure drum level are sensitive to density variations, and (2) a desired set point signal for level, along with a feedforward signal from a temperature-compensated steam flow transmitter. The error from the proportional-plus-integral-action level controller is summed with the steam flow signal to determine the demand to the final drive element.

A.2.5 Two-Element Control (See Figure 4)

Three-element feedwater control requires a pressure-compensated drum level signal (if the instruments used to measure drum level are sensitive to density variations), a desired set point signal for level, a feedforward signal from a temperature-compensated steam flow transmitter, and a signal from a feedwater flow transmitter. The feedwater flow transmitter should be temperature-compensated if the measurement is affected by feedwater temperature. The three-element control shall make feedwater flow follow steam flow and use the deviation in level as a resetting action to bring the required water inventory back to balance. Feedwater demand shall be derived from the error between the feedwater/steam flow error and the drum level. Consideration should be given to the inclusion of superheater spray flow in the total measurement of feedwater flow.

A.2.6 System Description and Interface Requirements

To achieve the performance objectives, the control system interface with the process must be considered carefully. At a minimum, a detailed process description should be provided which includes the following:

1. An instrumentation drawing defining all process design parameters such as temperatures, pressures, and normal flows.

2. Final drive descriptions of sufficient detail that a control strategy could be selected to provide an appropriate control action. Specifically, system head curves should be provided regardless of the method of feedwater regulation. Rated flows and minimum flows should be noted, in addition to pump recirculation flow requirements.

3. Boiler interface requirements defining drum level trip points, unmetered flow requirements (to include blowdown and superheat sprays), as well as any other boiler/feedwater interface requirements.

4. Existing process measurement interfaces, dimensioned sketches or diagrams. New flow element installations should be designed in accordance with the ASME MFC Series Standards. Drum level measurements should be in accordance with Section 1 of ASME's *Boiler and Pressure Vessel Code*. All measurements should be taken where vibrations, pulsations, and other flow disturbances are at a minimum.

5. A description of available electrical power and pneumatic supplies.

6. Control interlocks, set points, and alarm points.

7. Instrument loop diagrams, as defined by ISA S5.4-1976 (R1981).

A.3 BOILER DRUM

The drum of a subcritical boiler serves the following functions: (1) to maintain an adequate water level in the boiler tubes to prevent thermal damage; (2) to separate steam from the mixture of water and steam discharged into it; and (3) to house equipment to dry the drum steam after it is separated from the water.

The quantity of water contained in the boiler below the water level is relatively small compared to the total steam output. Primarily, the space required to accommodate steam-separating and -drying equipment determines the size of the drum.

The weight of the water in the mixture delivered to the drum for separation depends on the temperature and/or pressure and may range from less than two to over 25 times the weight of steam. To reduce this water to the small fraction found in the steam requires a high efficiency of water separation.

The factors that affect the separation of water from steam are:

1. The density of water with respect to steam

2. The available pressure drop

3. The amount of water in the mixture delivered to the steam drum

4. The quantity or total through-put of the water and steam to be separated

5. Viscosity, surface tension, and other such factors affected by pressure

6. Water level in the drum

7. The concentration of boiler water solids

A.4 DENSITY

The ratio of the density of steam to the density of water increases as pressure increases. This relationship is shown in Figure 5, which is a plot of the ratio of the density of water to the density of steam as a function of pressure. The density of water at 1200 psig (8.3 Pa) is approximately 16 times that of steam. At 2800 psig (19.1 Pa), the density of water is approximately three times that of steam.

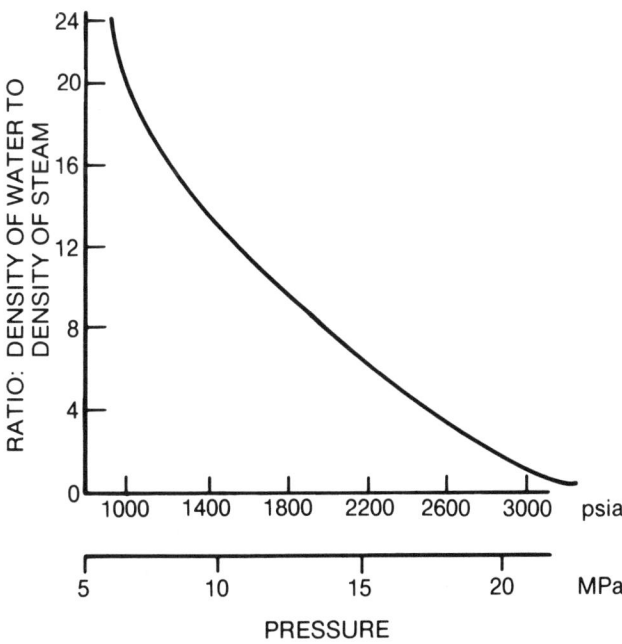

Figure 5. Water/Steam Density Relationship

The difficulty of measuring water level in a drum because of the variable density is shown in Figure 6. This drawing shows the relationship of the drum water with steam bubbles, the water level in a hot-gauge glass, the water level in a cold-gauge glass, and a method used to measure uncompensated drum level using a differential pressure transmitter. The level in the hot-gauge glass represents an undisturbed column of water, with saturated steam above that level.

A.5 CARRYOVER

As operating pressures increase, the steam phase exhibits greater solvent capabilities for the salts that may be present in the water phase. These salts will be partitioned in an equilibrium between the steam and water, known as "vaporous carryover." The phenomenon will contribute additional boiler water solids directly to the steam, independent of the efficiency of steam-water separation components.

A more serious problem with boilers below 2600 psig (17.9 Pa) is carryover occasioned by priming. Priming occurs when the water level is carried excessively high in the steam drum. The high water level can impair the discharge of the steam-water separators and result in water being carried through the driers.

Most materials that form boiler deposits originate in the preboiler system. Adherence to recommended operating procedures during start-up, normal operation, shutdown, and outages of a power plant is vital to minimize corrosion. The rate of deposition of preboiler corrosion products increases with increasing heat flux. Deposition is substantially greater on the hot side of the tube where boiling occurs. To minimize carryover, accurate drum level indication and controls should be primary considerations in the design of a feedwater control system.

A.6 SYSTEM HYDRAULICS

A system head curve represents the relationship between flow and hydraulic losses in a system. The representation (see Figures 7, 8, and 9) is in a graphical form — i.e., friction loss is shown to vary as the square of the flow rate; thus, the system curve is of parabolic shape.

Hydraulic losses in piping systems are comprised of pipe friction losses, valves, elbows, and other fitting losses — including losses from changes in pipe size. The parabolic shape of the system curve is determined by the friction losses through the system — including all bends, fittings, and valves. The static head does not affect the shape of the system curve or its steepness, but it does dictate the head of the system curve at zero flow rate. The operating point is at the intersection of the system curve and the pump curve. The flow rate can be reduced by throttling a discharge valve (see Figure 7), or by reducing pump speed (see Figure 10).

Pumps are usually operated in parallel when the heads are relatively low, but flow rates vary considerably. In this case, the pumps take their suction from a common supply and discharge into the same header. They may be required to operate independently or in combination. When shown superimposing the system curves, the pump performance curves clearly indicate the flow rates that can be expected and the heads at which each of the pumps would be operating (see Figure 8).

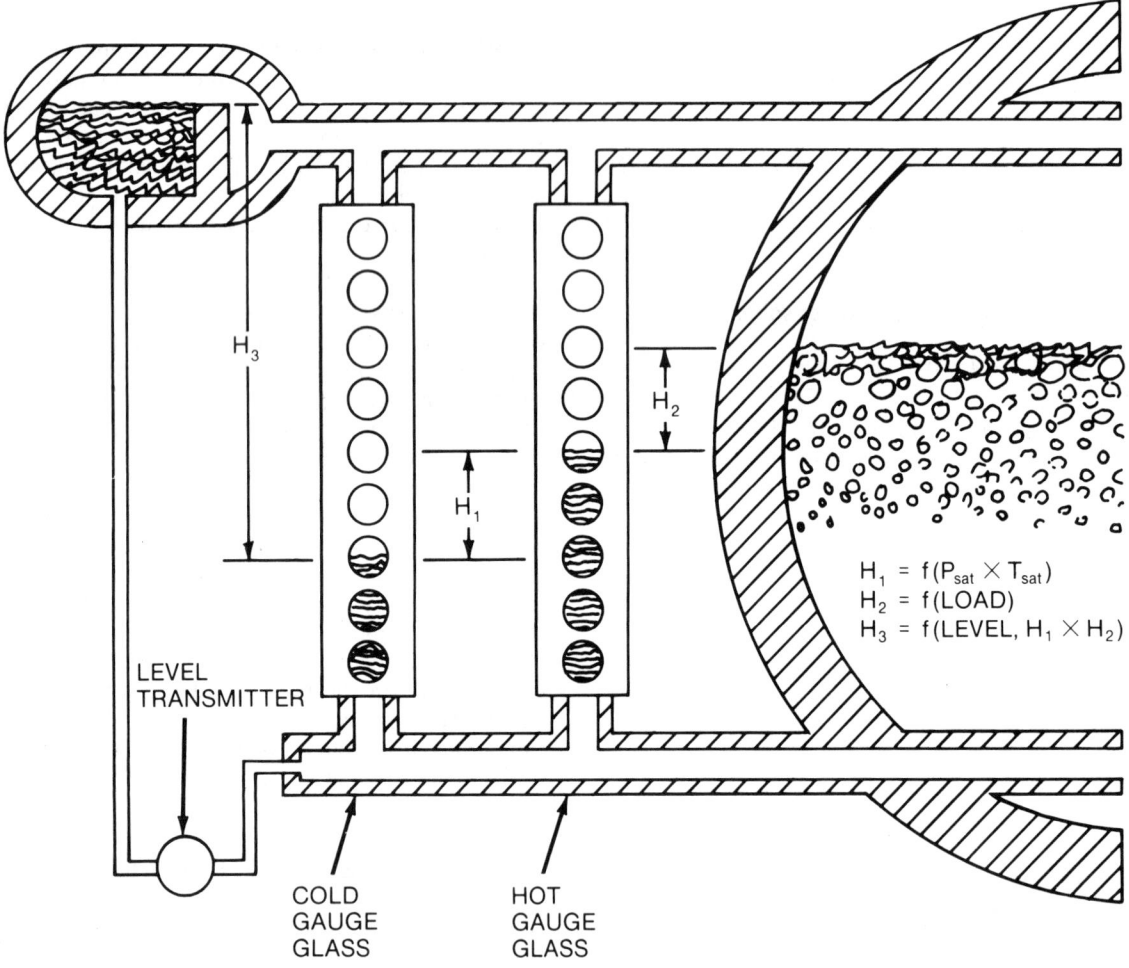

Figure 6. Level Measurements and Corrections

In some installations, the static system losses produce a head too high for one pump to obtain the required flow rate. In this case, two or more pumps are positioned in series, where the second and subsequent units take their suction directly from the discharge of the pump preceding it. The pumps are usually identical in size, speed, and impeller diameter. The combined pump performance curve is produced from the addition of heads for each pump at given flows (see Figure 9).

The slope of the system head curve can be adjusted by varying the friction loss in the pipeline. This can be accomplished either by throttling with a valve or (the more common way) by adjusting the capacity flow rate of the pump by reducing the pump speed. Variable speed is the most effective way of varying the flow rate (see Figure 10).

A.7 PUMP DRIVE SPEED CONTROL

A.7.1 Turbine Drives

For automatic control of turbine speed in response to boiler feedwater demand, a control system signal is sent to the pump turbine speed controller.

Below the minimum governing speed, the turbine motor control unit's manual selector switch allows the turbine to be controlled manually, similar to the way a valve-positioning device can be controlled. To operate the feedwater control on the pump turbine over the entire speed range from minimum to maximum speed, the manual motor speed changer must be in the high-speed stop position.

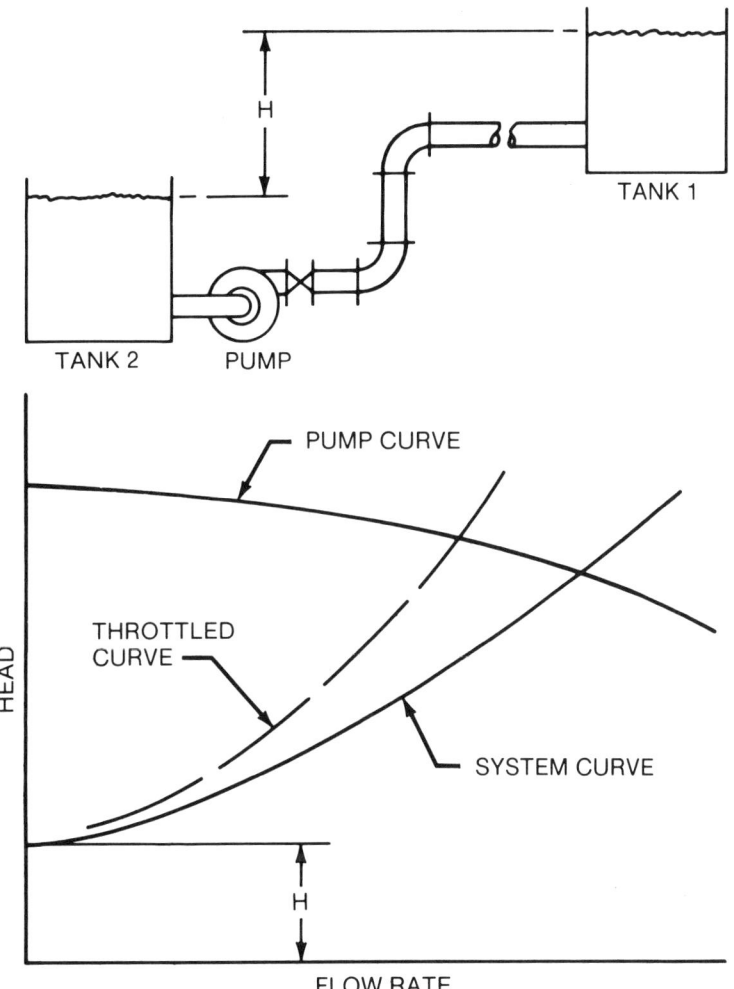

Figure 7. Positive Static Head Curves

The pump turbine speed control should be designed to be compatible with the feedwater control system signal. The minimum signal represents the minimum operating speed, and the maximum signal represents maximum operating speed, with the turbine developing its maximum specified capability.

A.7.2 Variable-Torque Drives

Variable-torque drives vary pump speed from minimum to full speed. This is usually accomplished by a mechanical positioning of the drive regulator.

A process controller interface package must be specified to provide the necessary interface circuitry between the process control system and the drive regulator. The interface package matches the drive operation's speed range with the process control system's signal range.

A.7.3 Variable-Speed Motor Drives

Variable-speed ac drives vary pump speed from minimum flow to full flow.

A.8 BOILER FEEDPUMP MINIMUM FLOW

The boiler feedpump is required to pass a minimum flow for internal cooling any time the pump is running. The minimum flow rate GPM (m^3/s) is available from the pump manufacturer.

Normally, recirculation flow is taken from the pump discharge and returned to the deaerator via the minimum flow line. A valve in the minimum flow line is actuated either by modulating control or on/off control.

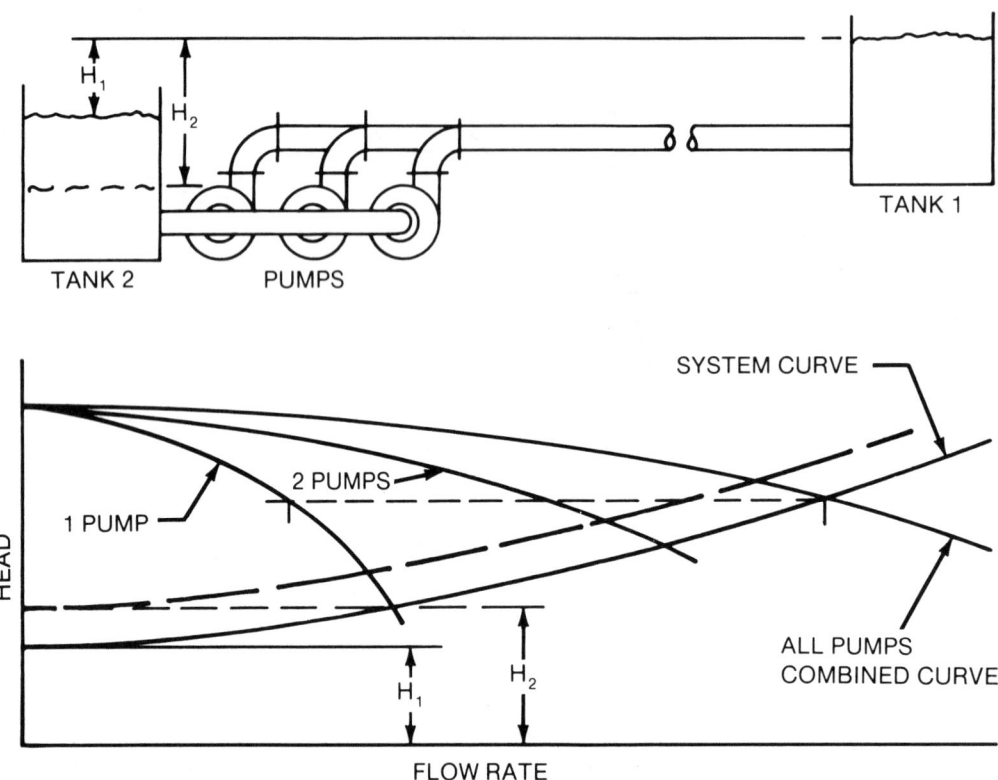

**Figure 8. Positive Static Head Curves —
Pumps in Parallel**

A flow transmitter is used in the boiler feedpump suction or discharge line to measure flow through the pump. Any time this flow drops below the minimum rate specified by the pump manufacturer for adequate cooling flow, a controller actuates the valve in an opening direction to assure adequate flow.

A.9 REDUNDANCY

Redundancy is employed when system reliability will be seriously affected by a component failure. Redundancy also permits on-line maintenance of components. For maximum availability, redundancy should always be considered. Deviation alarms and automatic failure detection/transfer should be considered in order to maximize the usefulness of the application of redundancy.

A.10 DRUM LEVEL OSCILLATIONS

Under certain operating conditions, drum level oscillations in the form of wave action or standing waves in the drum can occur. This should not be confused with the shrink-and-swell phenomenon. On forced circulation units, combinations of pumps in service may also cause wave action. This oscillation can cause variations in drum level readings when they are taken at both ends of the drum. Boiler manufacturers should be consulted for recommendations on the expected magnitude of such oscillations.

S77.42

Fossil Fuel Power Plant Feedwater
Control System — Drum-Type

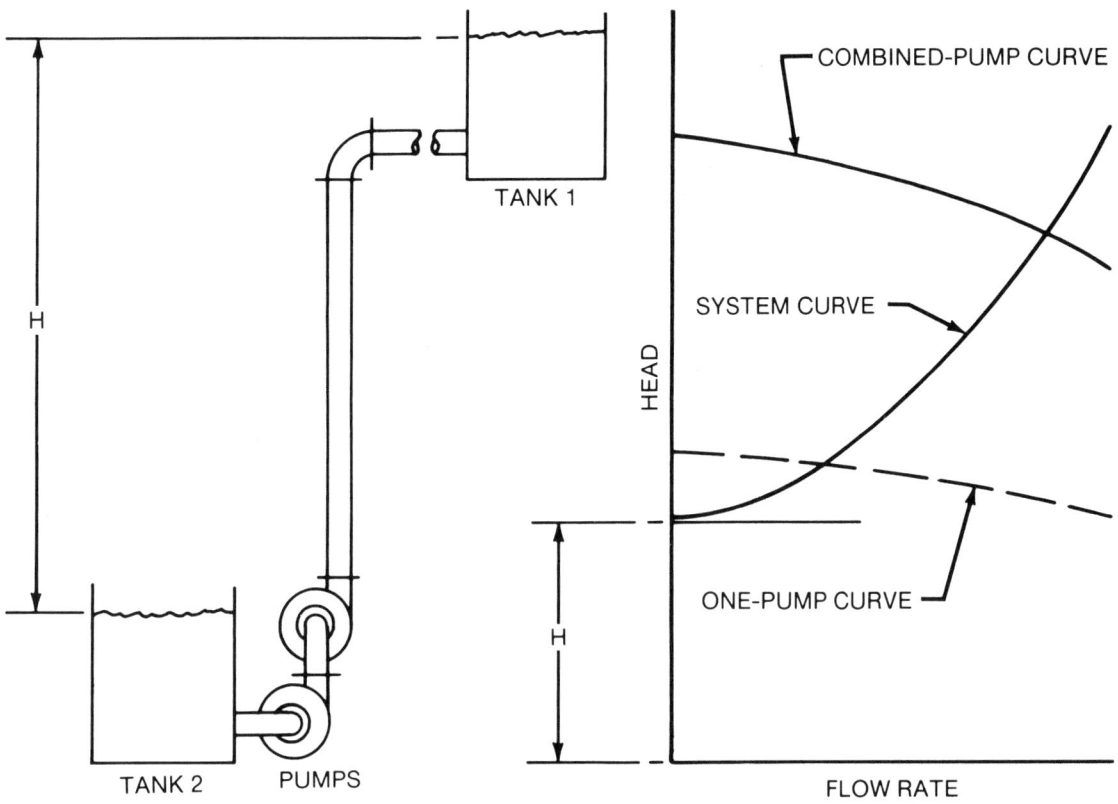

**Figure 9. Positive Static Head Curves —
Pumps in Series**

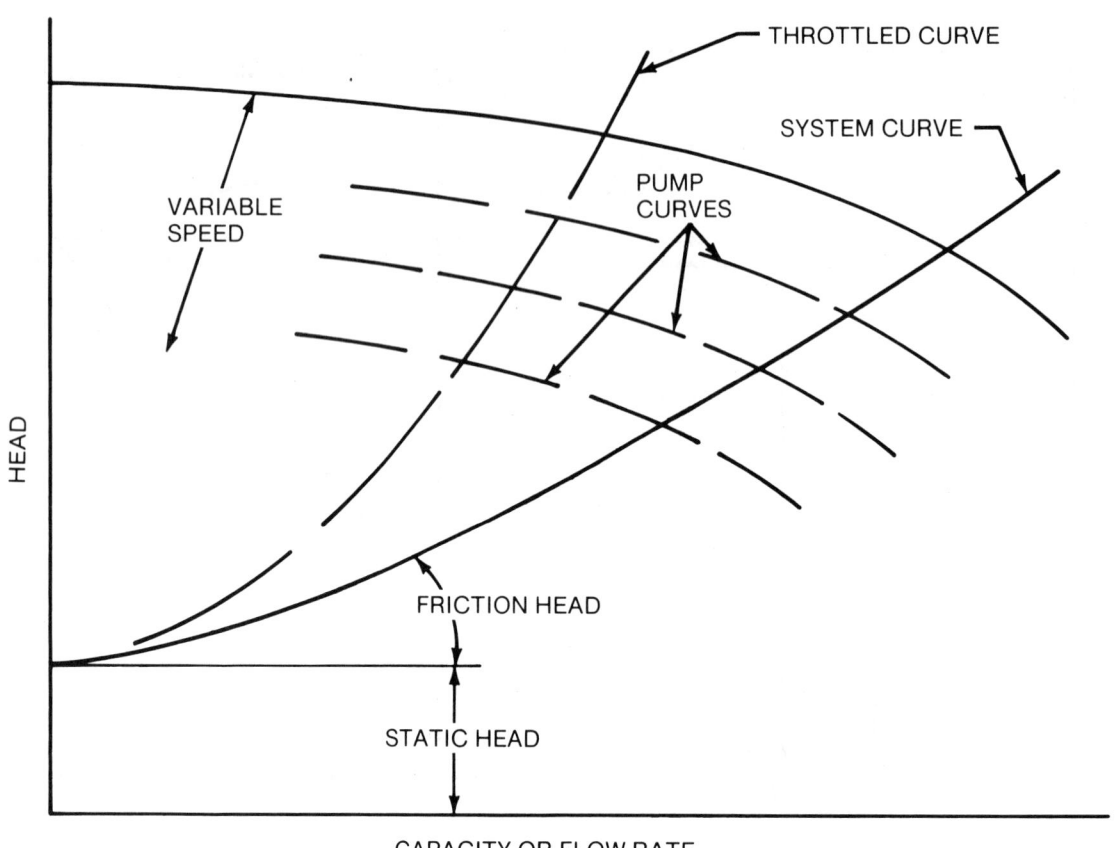

Figure 10. Diagram Depicting Effects of Throttling and Speed Reduction

ANSI/ISA-S82.01-1988
Approved February 18, 1988
(Partial Revision and Redesignation of ANSI C39.5-1974)

American National Standard

Safety Standard for Electrical and Electronic Test, Measuring, Controlling and Related Equipment

GENERAL REQUIREMENTS

Instrument Society of America

Instrument Society of America

ISBN 1-55617-100-5

ISA-S82.02 Safety Standard for Electrical and Electronic Test, Measuring, Controlling, and Related Equipment

Copyright ©1988 by the Instrument Society of America. All rights reserved. Printed in the United States of America. No part of this publication may be reproduced, stored in a retrieval system, or transmitted, in any form or by any means (electronic, mechanical, photocopying, recording, or otherwise), without the prior written permission of the publisher.

INSTRUMENT SOCIETY OF AMERICA
67 Alexander Drive
P.O. Box 12277
Research Triangle Park, North Carolina 27709

PREFACE

This preface is included for information purposes and is not part of ISA-S82.01.

This standard has been prepared as part of the service of the Instrument Society of America (ISA) toward a goal of uniformity in the field of instrumentation. To be of real value, this document should not be static, but should be subject to periodic review. Toward this end, the Society welcomes all comments and criticisms, and asks that they be addressed to the Secretary, Standards and Practices Board, Instrument Society of America, 67 Alexander Drive, P.O. Box 12277, Research Triangle Park, NC 27709, telephone (919) 549-8411.

The ISA Standards and Practices Department is aware of the growing need for attention to the metric system of units in general, and the International System of Units (SI) in particular, in the preparation of instrumentation standards. The Department is further aware of the benefits to U.S.A. users of ISA standards of incorporating suitable references to the SI (and the metric system) in their business and professional dealings with other countries. Toward this end, this Department will endeavor to introduce SI-acceptable metric units in all new and revised standards to the greatest extent possible. The Metric Practice Guide, which has been published by the Institute of Electrical and Electronics Engineers as ANSI/IEEE Std. 268-1982, and future revisions will be the reference guide for definitions, symbols, abbreviations, and conversion factors.

It is the policy of the Instrument Society of America to encourage and welcome the participation of all concerned individuals and interests in the development of ISA standards. Participation in the ISA standards-making process by an individual in no way constitutes endorsement by the employer of that individual, of the Instrument Society of America, or of any of the standards that ISA develops.

The information contained in the preface, footnotes, and appendices is included for information only and is not a part of the standard.

The development of this standard dates back to 1958. The need for the standard arose from requests that recording instruments meet the requirement of the American National Standard, the National Electrical Code. At that time there was also the beginning of the need for examination of measuring instruments by testing laboratories. The initial version of this standard was approved in 1964 after several years of careful development, review, and trial use.

Work began on safety requirements for electrical measuring instruments at the international level at about the same time. While consideration of specific safety requirements for measuring apparatus was begun in 1965, resulting first editions of IEC Publications 348 and 414 became available in 1971 and 1973, respectively.

American National Standard C39.5-1964 was reaffirmed in 1969 and revised in 1974. In 1982, responsibility for C39.5-1964 was transferred to the Instrument Society of America and the document was renumbered ISA-S82.01.

This 1988 edition reflects four major considerations:

- The scope has been expanded to include more equipment types. The scope and organization now readily allow future inclusion of more types of related equipment.

- A major goal was harmonization--to the extent possible--with worldwide standards, especially North American and IEC Standards. Not only were the IEC clause and multi-part formats adopted, but also many of the IEC requirements; many requirements were further defined and developed from the work of IEC Subcommittee 66E.

- Digital and microelectronic innovations have narrowed the one-time separation between digital and analog technologies and prompted commonality of safety requirements.

- Significant technical contributions were made by industry organizations and independent certifying laboratories.

Instrument Society of America

The following are significant features incorporated in this ongoing edition:

- Constructional requirements.

- Measurement of conformance to requirements is through the use of specific compliance statements. This allows the user of the standard to determine product compliance and to better understand the nature and intent of the requirements.

- Expansion of requirements for protection against fire and personal injury.

The following people served as members of ISA Subcommittee SP82.02, which prepared this standard:

NAME	COMPANY
Bruce J. Feikle, Chairman	OTIS
William Calder, Manager Director	The Foxboro Company
Abraham Abramowitz	Retired
David Braudaway	Sandia Laboratories
Joseph E. Devine	IBM Instruments, Inc.
Charles G. Gross	Gen Rad Inc.
Robert G. Harris	Underwriters Laboratories
Walter F. Hart	John Fluke Mfg. Co., Inc.
W.R. Howard	LFE Corporation
William H. Hulse, Jr.	Material Safety Eng. Div.
Donald S. Ironside	Biddle Instruments
A. Walter Jacobson	Retired
Al Kanode	Hewlett-Packard Company
Roger P. Lutfy	Factory Mutual Research
Donald A. Mader	Underwriters Laboratories, Inc.
Ernest C. Magison	Honeywell, Inc.
Richard C. Masek	Bailey Controls Co.
Frank J. McGowan	The Foxboro Company
Richard Nute	Tektronix, Inc.
Phillip A. Painchaud	Retired
William A. Phillips	Leeds & Northrup Company
William E. Rich	Westinghouse Electric Corp.
John N. Stanley	Leeds & Northrup Company
Robert Wallace	Tektronix, Inc.

The following people served as members of ISA Committee SP82 during the review of this standard:

NAME	COMPANY
David Braudaway, Co-Chairman	Sandia Laboratories
Frank McGowan, Co-Chairman	The Foxboro Company
William Calder, Manager Director	The Foxboro Company
Abraham Abramowitz	Retired
Norm Belecki	National Bureau of Standards
Bruce Feikle	OTIS

---

*Employer at time of participation

Name	Company
Robert G. Harris	Underwriters Laboratories, Inc.
W.R. Howard	LFE Corporation
A. Walter Jacobson	Retired
Donald A. Mader	Underwriters Laboratories, Inc.
Ivano Mazza	Casella Postale No. 11
Hubert O'Neil	Havti
Phillip A. Painchaud	Retired
Peter E. Perkins	Tektronix, Inc. DS 78-531
William E. Rich	Westinghouse Electric Corp.
DeWayne B. Sharp, P.E.	IBM Corporation
Jack V. Stegenga	Yokogawa Corp. of America

This standard was approved for publication by the ISA Standards and Practices Board in January 1988.

NAME	COMPANY
D. E. Rapley, Chairman	Rapley Engineering Services
D. N. Bishop	Chevron U.S.A. Inc.
W. Calder III	The Foxboro Company
N. Conger	Fisher Controls International, Inc.
R. S. Crowder	Ship Star Associates
C. R. Gross	Eagle Technology
H. S. Hopkins	Utility Products of Arizona
R. B. Jones	Dow Chemical Company
R. T. Jones	Philadelphia Electric Company
R. Keller	Consultant
A. P. McCauley	Chagrin Valley Controls, Inc.
E. M. Nesvig	ERDCO Engineering Corporation
C. W. Reimann	Moore Products Company
R. H. Reimer	Allen-Bradley Company
J. Rennie	Factory Mutual Research Corporation
W. C. Weidman	Gilbert/Commonwealth Inc.
K. A. Whitman	Fairleigh Dickinson University
* P. Bliss	Consultant
* B. A. Christensen	Continental Oil Company
* L. N. Combs	Consultant
* R. L. Gailey	Consultant
* T. J. Harrison	Florida State University
* O. P. Lovett, Jr.	Consultant
* E. C. Magison	Honeywell, Inc.
* W. B. Miller	Moore Products Company
* J. W. Mock	Bechtel Western Power Corporation
* G. Platt	Consultant
* J. R. Williams	Stearns Catalytic Corporation

* Director Emeritus

- SPECIFIC CLAUSE TREATMENTS -

The **SCOPE** includes environmental conditions envisioned for the equipment. **DEFINITIONS** have extensive additions to reduce misinterpretations. **MARKINGS** are expanded and clarified by inclusion of internationally recognized symbols. X-radiation considerations are added while several other **EMANATIONS** are identified as "under consideration" due to lack of recognized limits and standardized measurements. The clause for **PROTECTION FROM ELECTRIC SHOCK** deletes requirements for resistance tests, clarifies leakage current measurement procedures, details dieletric tests and includes requirements for measuring terminals and routine tests. **TESTING UNDER FAULT CONDITIONS** is new. **PROTECTION FROM FIRE** is an approach to measuring the likelihood of spread of fire. **EXTRA LOW VOLTAGE AND POWER LIMITED CIRCUITS** defines the circuits which could cause electric shock or the likelihood of fire and for which protection requirements of the standard apply.

Understanding the Appendix material of this standard under Clause 9 is also necessary for this edition.

TABLE OF CONTENTS

Section Title   Page

1. General 9
    1.1 Scope 9
2. Definitions 10
3. General Information 12
    3.1 Object 12
    3.2 Compliance 12
    3.3 Equipment construction 12
    3.4 Components 13
    3.5 Units of measurement 13
    3.6 Combination of protection means 13
4. General Test Requirements 13
    4.1 Conduct of tests 13
    4.2 Reference test conditions 13
    4.3 Fault conditions 14
5. Marking 14
    5.1 General requirements 14
    5.2 Equipment-identification markings 15
    5.3 Equipment-rating markings 15
    5.4 Warning and caution markings 16
    5.5 Operator-control markings 17
    5.6 Terminal markings 17
    5.7 Protective insulation 18
6. Equipment Emanations 18
    6.1 Ionizing radiation 19
    6.2 Non-ionizing radiation 19
    6.3 Ozone 19
    6.4 Ultrasonic pressure 19
7. Heating 19
    7.1 General 19
    7.2 Temperature limits 19
    7.3 Polymeric enclosure and insulating material thermal conditioning 22
8. Protection from Implosion and Explosion 22
    8.1 Implosion protection 22
    8.2 Explosion protection 23
9. Protection from Electric Shock 24
    9.1 Accessible parts 24
    9.2 Live part 25
    9.3 Exterior of equipment 25
    9.4 Construction requirements 27
    9.5 Protective grounding 27
    9.6 Protective insulation 29
    9.7 Protective impedance 29
    9.8 Spacing 29
    9.9 Polarization of primary circuits 31
    9.10 Humidity conditioning 32
    9.11 Leakage current 32
    9.12 Dielectric voltage-withstand tests 36
10. Testing under Fault Conditions 43
    10.1 Equipment preparation 43
    10.2 Compliance tests 44
    10.3 Fault conditions 44

11.	Mechanical Requirements	45
	11.1 Mechanical strength	45
	11.2 Handle loading test	49
	11.3 Tip-stability test	49
	11.4 Sharp edges	49
12.	Protection from Fire	49
	12.1 Enclosure materials	49
	12.2 Polymeric insulating materials	49
	12.3 Printed wiring boards	49
13.	Component Requirements and Applications	49
	13.1 Motors	49
	13.2 Supply-circuit and safety switches	49
	13.3 Protective devices	49
	13.4 Resistors	51
	13.5 Supply-circuit connectors	51
	13.6 Printed wiring	51
	13.7 Wire insulation and insulating tubing	51
	13.8 Connectors	51
	13.9 Receptacles	51
	13.10 Component applications	51
14.	Extra-Low Voltage and Power-Limited Circuits	52
	14.1 Extra-low voltage circuit	52
	14.2 Power-limited circuit	53
15.	Terminal Devices	54
	15.1 Supply-circuit connectors	54
16.	External Cords	54
	16.1 Attachment	54
	16.2 Flexible power-supply cords	55
	16.3 Interconnection cables	55
17.	Equipment Instructions	55
	17.1 Warnings and cautions	55

APPENDIX 57
Part 1   General Information: Manufacturing and Production Tests (Routine Tests)
Part 2   Recommended International Symbols

IN THIS STANDARD:

- Requirements proper are printed in roman type.
- Explanatory matter regarding the requirements is printed in smaller roman type.
- Test specifications are printed in italic type.
- Explanatory matter regarding the test specifications is printed in smaller italic type.
- Further explanatory matter regarding the requirements or specific subject marked by an asterisk can be found in the Appendix.

CAUTION

MANY TESTS REQUIRED BY THIS STANDARD ARE INHERENTLY HAZARDOUS. ADEQUATE SAFEGUARDS FOR PERSONNEL AND PROPERTY SHOULD BE EMPLOYED IN CONDUCTING SUCH TESTS.

*1. General

1.1 Scope

1.1.1 This standard applies to electrical, electronic, or electromechanical equipment designed to:

(1) measure, or
    observe, and
    indicate quantities of
    electrical, or
    electronic
    phenomena, or

(2) supply
    electrical, or
    electronic
    quantities for measuring or indicating purposes, or

(3) measure or control
    directly, or
    indirectly
    an industrial process, or

(4) measure or indicate electrical analogs of non-electrical phenomena.

*1.1.2 Constraints. This standard applies to equipment that:

(1) is rated for connection to supply circuits which exceed extra-low voltage and are not power-limited in accordance with Clause 14 and which do not exceed:

   (a) 600 volts between phases for three-phase supply circuits or

   (b) 250 volts rms, single-phase, or dc;

(2) is rated for connection to an extra-low voltage or power-limited circuit and internally derives from that circuit a voltage or power in excess of those limitations according to Clause 14;

(3) is rated for measuring, testing, or otherwise connecting to circuits operating in excess of extra-low voltage, or not power limited in accordance with Clause 14;

(4) is rated for connection to, or utilizing, extra-low voltage or power-limited circuits.

Some requirements of this standard do not apply to extra-low voltage and/or power-limited circuits.

1.1.3 Auxiliary or accessory equipment. This standard also applies to auxiliary equipment and accessories rated for use with electrical, electronic, or electro-mechanical equipment where auxiliary and accessory equipment is separate equipment which does not itself perform the desired function but is used in addition to or as a supplement to equipment according to Sub-clause 1.1.1.

1.1.4 Equipment covered. This standard forms one of a series of specific parts.

The General Requirements set forth in ISA-S82.01 apply to all electrical and electronic equipment or accessories in relation to Sub-clauses 1.1.1 and 1.1.3.

Requirements which apply specifically to certain types or classes of electrical and electronic equipment are set forth in the following standards:

ISA-S82.02 Electrical and Electronic Test and Measuring Equipment

ISA-S82.03 Electrical and Electronic Process Measurement and Control Equipment

In the event of conflict between the general requirements of this standard, ISA-S82.01, and another standard in this series, the requirements of the specific standard (i.e., ISA-S82.02 or ISA-S82.03) relative to a certain type or class of equipment, shall apply.

1.1.5 Equipment not covered. These requirements do not apply to the following types of equipment:

(1) Medical and laboratory equipment

(2) Watt-hour meters and associated equipment installed by electric utility companies for measuring electrical energy and related quantities

(3) General use battery chargers, auxiliary supply sources, substitute power supplies, or laboratory-type power supplies not specifically rated for use with measuring or testing equipment

Specifically rated is defined as being:

(a) packaged with the equipment, or

(b) referenced in an equipment marking, or

(c) referenced in the operating instructions packed with the equipment.

1.1.6 Environmental conditions. Equipment according to this standard is intended for use in enclosed locations where air temperature is controlled and normally only nonconductive pollution occurs.

The basic climatic conditions for heated or cooled enclosed locations are (IEC Publication 654-1, Class B location):

(1) Temperature------------From 5°C (41°F) to 40°C (14°F)

Equipment may occasionally be subjected to temperatures between -10°C (14°F) and +5°C (41°F) for short time periods.

(2) Altitude--------------Up to 2200 meters (7,218 feet)

(3) Relative Humidity----Up to 85%, with

Wet Bulb Temperature--Not to exceed 27°C (81°F)

2. Definitions

For the purposes of this standard, the following definitions apply:

2.1 ACCESSIBLE PART: A part that can be touched during normal use or operator servicing. See Sub-clause 9.1.

2.2 APPLIANCE COUPLER: See Sub-clause 13.5.1.

2.3 ATTACHMENT PLUG: A connecting device for a flexible cord that, by insertion into a receptacle, establishes supply-circuit connections between the flexible cord and the receptacle.

2.4 AWG: Wherever it appears in this standard, the abbreviation AWG means American Wire Gage.

2.5 BENCH TOP EQUIPMENT: Equipment designed to be used on and supported by a bench, table, stand, etc., but is neither fixed nor portable as determined by the following:

(1) It has at least one handle and the weight exceeds 20 kilograms (44 pounds), or

(2) It has no handle and the weight exceeds 5 kilograms (11 pounds), or

(3) It is not mobile (does not have casters, wheels, rollers, etc., nor is it provided with a cart).

2.6 BONDING: An electrically conductive connection between metallic parts of the equipment that are required to be grounded, and some other part of the equipment to which a grounding conductor is connected.

2.7 BRANCH CIRCUIT: That portion of permanently installed wiring between the final overcurrent protective device and the attachment-plug receptacle or outlet, or point of connection to the fixed equipment.

2.8 BY HAND: Denotes that an operation does not require the use of a tool, coin, or any other object that may serve as a tool.

2.9 CIRCUIT-TO-GROUND VOLTAGE: The rated value of voltage with respect to earth ground.

2.10 CLEARANCE DISTANCE: The shortest distance measured in air between conductive parts.

2.11 CREEPAGE DISTANCE: The shortest distance measured over the surface of insulation between conductive parts. Air gaps shorter than 1.0 millimeter (0.04 inch) are not considered to interrupt the surface path.

2.12 CONDUCTIVELY CONNECTED: A part is conductively connected to another part if the current between the parts, with the equipment at reference test conditions, exceeds the limit for leakage current.

2.13 CORD-CONNECTED EQUIPMENT: Equipment that connects to a supply circuit receptacle by means of a permanently attached flexible power-supply cord and attachment plug or by means of a detachable power-supply cord.

2.14 ELECTRONIC DEVICE: A part or an assembly of parts that employs electron or hole conduction in a semiconductor or electron conduction in a vacuum or gas.

2.15 EXTRA-LOW VOLTAGE CIRCUIT: See Sub-clause 14.1.

2.16 EQUIPMENT: An assembly of electrical or electronic components or circuits intended to perform a complete function apart from being a substructure of a system.

2.17 FIXED EQUIPMENT: Equipment which is designed to be fastened or otherwise physically secured to a supporting device.

2.18 GROUND: A conducting connection, whether intentional or accidental, between an electrical circuit or electrical equipment and either the earth or some other conducting body that serves in place of the earth.

2.19 GROUNDING: The act of establishing a conductive connection, whether intentional or accidental, between an electrical circuit or electrical equipment and either the earth or some other conducting body that serves in place of the earth.

2.20 HAND-HELD EQUIPMENT: Any piece of equipment either designed to be or indicated by the manufacturer that it can be (that is, in advertising literature or in the operating instructions) held in one hand during any phase of normal operation, regardless of its weight.

2.21 INDIRECT PROTECTIVE GROUNDING: See Sub-clause 9.5.9.

2.22 INSULATION:

(1) Basic: Insulation applied to live parts to provide basic protection against electric shock.

Where the term basic insulation is used in this standard, it means basic insulation rated for the application and applies to the insulation manufacturing ratings.

(2) Supplementary: Independent insulation applied in addition to basic insulation in order to provide protection against electric shock in the event of a failure of basic insulation.

(3) Protective (Double): Insulation comprising both basic insulation and supplementary insulation.

(4) Reinforced: A single insulation system applied to live parts, which provides protection against electric shock equivalent to double insulation.

The term "insulation system" does not imply that the insulation must be one homogeneous piece. It may comprise several layers which cannot be tested singly as supplementary or basic insulation.

2.23 ISOLATED CIRCUIT: A circuit in which the current, with the equipment at reference-test conditions, to any other circuit or conductive part does not exceed the limit for leakage current.

2.24 LIVE PART: A part which is considered capable of rendering an electric shock. See Sub-clause 9.2.

2.25 MEASURING-CIRCUIT VOLTAGE: The voltage between two terminals of a measuring circuit or between one of these terminals and ground.

2.26 OPERATIONAL MAINTENANCE: Any maintenance activity intended to be performed by the Operator, and which is required in order for the equipment to serve its intended purpose. Such activities include: interchanging plug-in modules, adjusting the response of a high-frequency probe, correcting "zero" on a panel instrument, changing charts, marking records, adding ink and the like. Such activities are expected to be performed by a person(s) not familiar with the risks of electrical shock, likelihood of fire, or personal injury. See Sub-clause 9.1.

2.27 OPERATOR: A person using or operating the equipment.

2.28 OPERATOR CONTROL: An operator-accessible control, usually a knob, push button, lever, or the like, provided to enable the operator to cause the equipment to perform its intended function and to serve its intended purpose.

2.29 PERMANENTLY CONNECTED EQUIPMENT: Equipment connected to a supply circuit by field wiring terminals or by means of separate installed leads.

2.30 PORTABLE EQUIPMENT: Equipment specifically designed to be carried by hand from one location to another as determined by the following:

(1) It is provided with at least one handle and does not exceed 20 kilograms (44 pounds); or

(2) it has no handle and does not exceed 5 kilograms (11 pounds).

2.31 POLYMERIC MATERIAL: A compound formed by molecular bonding (polymerizing) of two or more simple molecules (monomers). This material is commonly referred to as "plastic."

2.32 POWER-LIMITED CIRCUIT: See Sub-clause 14.2.

2.33 POWER-SUPPLY CORD: A flexible cord with attachment plug provided to connect equipment to a supply circuit receptacle.

2.34 PRIMARY CIRCUIT: Wiring and components of the equipment supply circuit which are at the voltage of or carry the current of the branch circuit.

2.35 RATED SUPPLY VOLTAGE: The supply voltage, or range of voltages, for which the manufacturer has designed the equipment.

2.36 RATED VALUE: The value or one of the values assigned by the equipment manufacturer or component manufacturer.

2.37 RECEPTACLE: A connector device permanently connected to the supply circuit (usually a branch circuit) into which an attachment plug is inserted.

2.38 SAFETY SWITCH: A switch employed for the purpose of providing protection according to this standard. Such a switch may be known as an interlock.

2.39 SPACING: (See Table 9-1):

(1) Basic: Physical separation between conductive parts which provides basic protection against electric shock.

(2) Supplementary: Physical separation used in addition to basic insulation in order to provide protection against electric shock in the event of a failure of the basic insulation.

(3) Protective: Increased physical separation between conductive parts which provides protection against electric shock equivalent to double insulation.

2.40 SUBSTITUTE POWER SUPPLY: Supply equipment that may be used instead of a battery supply for the equipment.

2.41 SUPPLY CIRCUIT: The circuit supplying electrical energy to the equipment from a branch circuit, a battery, or a power supply.

2.42 SUPPLY EQUIPMENT: Equipment that takes energy from an electrical supply, generally a branch circuit, and supplies it in a modified form to energize other equipment.

2.43 TERMINAL DEVICE: A part used to facilitate the making of external connections.

The part may contain several terminal contacts (a connector).

2.44 TERMINALS:

(1) Field wiring terminal: Any terminal to which a supply circuit wire is intended to be connected by an installer in the field.

(2) Measuring or testing terminal: An external terminal or connector of the equipment to which connection is made to serve the equipment's function.

(3) Measuring or testing grounded terminal: An external terminal of a grounded measuring or testing circuit which is internally connected to the equipment's protective grounding system and which is intended to be connected to the grounded side of the external circuit to which it is attached.

(4) Protective-grounding terminal: A terminal connected to the equipment protective-grounding system and intended to be externally connected to earth ground.

2.45 TESTS:

(1) Routine test: A test that is performed on each piece of equipment during the production process.

(2) Type test: A test that is performed on one or more pieces of equipment, representative of a type, to determine whether the design, construction, and manufacturing methods comply with the requirements according to this standard.

3. General Information

3.1 Object. The object of this standard is to specify:

(1) protections for operator and equipment;

(2) tests demonstrating compliance with these protections;

(3) terminology used in this standard.

3.2 Compliance. Compliance with the requirements is determined by performing inspections and tests according to the Clauses applicable to the particular equipment.

The equipment is expected to remain in compliance with these requirements after storage and transportation under the conditions specified by the manufacturer.

3.3 Equipment construction. The equipment construction shall be such that:

(1) fire, electric shock, or personal injury is not likely to occur during normal use or operational maintenance; and

(2) the components and materials are used within their rated values, such as electrical, mechanical, and temperature limits, and other limitations of use; see Sub-clause 3.4; and

(3) the components, wiring, and other internal parts are protected from being displaced or damaged.

Unless indicated otherwise, the components, parts and materials to which reference is made above and elsewhere in this standard are those involving or associated with

- the risk of fire, or
- the risk of electric shock, or
- the risk of personal injury.

3.4 Components. Where component or material rated values are required in this standard, such rated values shall be according to applicable component or material requirements.

3.5 Units of measurement. When a value for measurement as given in this standard is followed by an equivalent value in parentheses in other units, the first-stated value shall be regarded as the requirement. A given equivalent value may only be approximate.

For example: 25 millimeters (1 inch)
 Requirement (Approximate Equivalent Value)

3.6 Combination of protection means. Protection means may be combined in one equipment. Equipment as a whole shall meet all of the requirements for at least one of the protection means specified in this standard.

4. General Test Requirements

4.1 Conduct of tests

4.1.1 *Tests specified in this standard are type tests.*

Exception: Where indicated otherwise.

If in this standard it is required that tests on components or parts of equipment are to be made in accordance with their individual relevant specifications, such tests need not necessarily be performed during the type test specified in this standard.

4.1.2 *Type tests shall be carried out on one and the same piece of equipment or on identical multiple samples.*

Type tests specified in this standard shall be performed on equipment in the condition as shipped from the factory or as repaired to the same condition.

Where it is necessary to perform destructive tests, additional samples of identical components or equipment may be used.

When dimensions or mass make it impossible to perform out particular tests on a whole piece of equipment, separate tests on subassemblies are allowed.

4.1.3 *Unless indicated otherwise, for voltage measurements, the voltmeter sensitivity should not be less than 20,000 ohms per volt dc or 5,000 ohms per volt ac for potentials of 500 volts or less, and the resistance should not be less than 10 megohms for ac or dc potentials of more than 500 volts. Voltmeters having a higher input resistance should be used wherever it is warranted by the resistance of the circuit under test.*

4.1.4 *Cheesecloth specified in this standard shall be untreated cotton cloth running 26-28 square meters per kilogram (14-15 square yards per pound) and otherwise.*

4.1.5 *Airflow, such as drafts, shall not affect the test results.*

4.1.6 *The terms "voltage" and "current" indicate rms or dc values.*

Exception: Where specified otherwise.

4.2 Reference test conditions (reference conditions for test purposes). *The following conditions shall prevail in the location in which the equipment is being tested:*

- *Temperature within the range of 15-35°C (59-95°F)*

- *Relative humidity less than 80%*

- *Altitude up to 2200 meters (7,218 feet)*

- *Environment free of hoarfrost, dew, percolating water, rain, solar radiation, and the like*

4.2.1 *Each test shall be carried out under the most unfavorable combination of:*

(1) position
The equipment shall be in any position of normal use and mounted according to the manufacturer's instruction.

Exception: Where specified otherwise.

(2) supply circuit
For equipment rated for a nominal supply voltage, the applied supply voltage shall be 0.9 to 1.1 times the rated nominal supply voltage for which the equipment may be set.

For equipment rated for a range of supply voltage, that range of voltage may be used.

The maximum supply voltage applied shall not be less than the highest rated voltage of the supply to which the equipment may be connected.

For example: The maximum applied supply voltage for equipment rated for 105-115 volts ac, but intended for connection to a 120 volt ac nominal supply, shall not be less than 120 volts ac.

The applied supply-circuit frequency shall be any rated supply frequency of the equipment.

The applied supply voltage for ac and dc rated equipment shall be ac or dc.

Equipment that may be operated either from an ac or dc supply may be connected to the worst-case dc input specified if such operation imposes more severe conditions than if operated from the ac supply.

The applied primary voltage for single-phase ac rated equipment may be correctly polarized or reversed.

(3) measuring circuits
Measuring circuits shall be energized by any value between zero and the rated input.

Isolated measuring circuits shall be connected to any voltage within the rated value of circuit isolation.

Where the measuring circuits of equipment are to be connected to specified probes or transducers, they shall be connected to the probes and transducers recommended by the manufacturer; alternatively, equivalent networks for testing purposes may be used.

(4) ground terminals
Protective-grounding terminals shall be connected to ground.

Exception: Where specified otherwise in the test Clauses.

Measuring-circuit ground terminals shall be connected to ground when specified by the manufacuter.

(5) operating controls
Accessible controls shall be set to any position.

Exception: Supply-circuit voltage-setting devices shall be set according to the manufacturer's instructions.

Remote-control (operating) devices shall either be connected or not.

This condition should not cause the operating device to be set to cause abnormal operation of the equipment.

(6) normal use
The equipment shall be properly connected for its intended purpose, or not connected.

(7) motors and motor-driven parts
Motors and motor-driven parts of the equipment shall be loaded according to the intended purpose.

When testing motor-driven parts, other parts of the equipment that are intended to operate at the same time should remain connected.

(8) testing (output) circuits
Equipment supplying electrical quantities (voltage, current, and/or wave shapes) shall:

(a) Be operated in such a way as to provide the maximum rated output; and,

(b) If isolated, be connected to any voltage between zero and the rated circuit-to-ground voltage.

(9) substitute power supplies
Substitute power supplies shall be tested in or with the equipment for which they are intended, and according to the manufacturer's instructions.

(10) accessories
Noninterchangeable accessories shall be connected according to the manufacturer's instructions, or not connected.

Interchangeable accessories and accessories of limited interchangeability shall be subjected to separate tests relating to their own characteristics.

(11) ventilation
Normal ventilation shall not be impeded by artificial means.

4.3 Fault conditions

Testing under fault conditions denotes that, in addition to the reference test conditions mentioned in Sub-clause 4.2, the failure of particular components is simulated one after the other. Testing under fault conditions is according to Clause 10.

When performing tests under simulated fault conditions, multiple concurrent faults are not initially induced but are considered if they result from the initial induced fault.

5. Marking

The equipment shall be marked according to this Clause.

5.1 General requirements. The markings required by this Clause shall:

(1) resist the deleterious effects of handling, cleaning agents as specified by the manufacturer, etc., expected in normal use; and,

(2) be clear and legible under conditions of normal use.

Compliance for durability of markings on the outside of the equipment is checked by inspection and by performing the following test:

The markings are to be rubbed by hand, without undue pressure, for 15 seconds with a cloth soaked with the specified cleaning agent, or if not specified, for 15 seconds with water and then for 15 seconds with a cloth soaked with alcohol.

The markings shall be clearly legible after the above test. Adhesive labels shall not have worked loose or become curled at the edges.

5.2 Equipment identification markings. Equipment shall bear identification markings visible from the exterior according to this Sub-clause.

The markings should preferably be on the exterior of the equipment excluding the bottom. However, when space is limited, such identification markings may be on the bottom.

For equipment intended for rack or panel mounting, such identification markings may be on any surface where visible upon removing the equipment from the rack or panel.

The markings may be located where they are visible upon removing a cover or opening a door by hand that is intended to be removed or opened by an operator.

For permanently connected equipment the markings may be located where they are visible upon removing a cover or opening a door by means of a tool, if an external permanent notice identifying the location of the markings is provided on the cover or door.

Compliance is checked by inspection.

5.2.1 Manufacturer's identification. The equipment shall bear the name, trade name, or trademark of the manufacturer, or other distinctive marking of the manufacturer, or of the organization responsible for the equipment.

Compliance is checked by inspection.

5.2.2 Model identification. The equipment shall bear a distinctive designation, such as a model number, that distinguishes it from other equipment made by the same manufacturer.

Compliance is checked by inspection.

5.2.3 Factory identification. If equipment bearing the same distinctive designation (model number) is manufactured at more than one location, equipment from each manufacturing location shall bear an identification of such location.

Such identification may be in code and need not be on the equipment exterior.

Compliance is checked by inspection.

5.2.4 Accessory, plug-in module, and auxiliary-equipment identification. Equipment intended to work with specific accessory, plug-in module, or auxiliary equipment shall bear markings or be provided with a means for identifying each such accessory, plug-in module, or auxiliary equipment.

Accessory, plug-in module, and auxiliary equipment intended to work with specific equipment shall bear markings or be provided with a means for identifying such equipment.

The identification markings may be located on the equipment, on the accessory, plug-in module, or auxiliary equipment, or in the literature provided with either.

Compliance is checked by inspection.

5.3 Equipment rating markings. Equipment shall bear rating markings visible from the exterior according to this Sub-clause.

A rating is a value assigned to the equipment by the manufacturer. The markings should preferably be on the exterior of the equipment excluding the bottom. However, when space is limited, such rating markings may be on the bottom.

For equipment intended for rack or panel mounting, such rating markings may be on any surface where visible upon removing the equipment from the rack or panel.

The markings may be located where they are visible upon removing a cover or opening a door that is intended to be removed or opened by an operator without the use of tools.

For permanently connected equipment the markings may be located where they are visible upon removing a cover or opening a door by means of a tool, if an external permanent notice identifying the location of the markings is provided on the cover or door.

Compliance is checked by inspection.

5.3.1 Primary circuit type and frequency rating

(1) Equipment intended to operate from a direct current supply shall bear markings indicating that the supply shall be direct current.

The symbol ═══ (IEC Publication 417, see Appendix) may be used for this marking.

(2) Equipment intended to operate from an alternating current supply shall bear markings indicating that the supply be alternating current.

The symbol ∿ (IEC Publication 417, see Appendix) may be used for this marking.

The markings shall include the equipment primary-circuit frequency or primary-circuit frequency-range rating (hertz or Hz).

(3) Equipment intended to operate from either direct or alternating current supplies shall bear markings indicating that the supply may be either direct current or alternating current.

The symbol ⎓∿ (IEC Publication 417; see Appendix) may be used for this marking. The markings shall include the equipment primary-circuit frequency or primary-circuit frequency-range rating (hertz or Hz).

Compliance is checked by inspection.

5.3.2 Polyphase supply-circuit rating. Equipment intended for operation on a polyphase supply-circuit shall bear markings designating the number of phases and number of wires, including the grounded conductor and excepting the protective grounding conductor, if any.

The symbol 3 ∿ (IEC Publication 335.1, see Appendix) may be used to denote three-phase alternating current, and symbol 3N ∿ (IEC Publication 335.1, see Appendix) may be used to denote three-phase alternating current with neutral conductor.

Compliance is checked by inspection.

5.3.3 Supply-circuit voltage rating. Equipment shall bear markings designating the supply-circuit voltage(s) or voltage-range(s) rating(s) of the equipment that can be applied without the adjustment of a voltage-setting device.

Cord connected multiple-voltage equipment shall bear markings indicating the supply-circuit voltage or voltage range for which the equipment is set.

Alternatively, if the equipment is constructed so that the supply-circuit voltage-selector setting can be changed by the operator, the action of changing the voltage-selector setting shall also change the supply-circuit voltage indication.

Compliance is checked by inspection.

5.3.4 Supply-circuit current or power rating. Equipment shall bear markings designating the equipment supply-circuit current or power rating in amperes, volt-amperes, or watts. Such rating shall include the maximum current or power of any accessory or auxiliary equipment that is supplied through and operated simultaneously with the equipment.

The measured supply-circuit current or power shall not exceed 110 percent of the marked supply-circuit current or power rating.

Compliance is checked by inspection and by measuring the equipment supply-circuit current or power consumption under reference test conditions (Sub-clause 4.2).

5.3.5 Supply-circuit receptacle rating. Equipment incorporating operator-accessible supply-circuit receptacles for supplying operating energy to other equipment shall bear markings adjacent to such receptacles indicating the available voltage, frequency, and either current or power.

Compliance is checked by inspection.

5.3.6 Fuse ratings. Equipment incorporating fuses providing protection against the likelihood of fire shall bear markings adjacent to the fuse or fuseholder designating the fuse current and voltage ratings.

Exception: For fuses not accessible to the operator, the voltage rating markings may be omitted.

Compliance is checked by inspection.

5.3.7 Equipment duty-cycle rating. Equipment intended for short-term or intermittent use and, if operated continuously, would exceed the temperature rises according to Clause 7 shall bear a duty-cycle rating.

Such duty-cycle rating shall be visible from the operating position, and, if applicable, shall be adjacent to or otherwise related to the duty-cycle controls.

Compliance is checked by inspection.

5.4 Warning and caution markings. Equipment shall bear warning and caution markings according to this Sub-clause.

Alternately, the symbol ⚠ (IEC Publication 417, see Appendix) may be used to refer the operator to an explanation in the equipment instructions.

A warning or caution marking required in this Sub-clause shall

(1) be permanently attached; and

(2) not be attached to parts likely to be replaced during maintenance or servicing;

Exception: If the marking is integral with the replacement part.

(3) have lettering in which

- the precautionary signal word shall be at least 2.75 millimeters (0.11 inch) high;
- the text shall be at least 1.5 millimeters (0.06 inch) high and contrasting in color to the background; or
- if molded or stamped in a material, the text shall be at least 2.0 millimeters (0.08 inch) high and, if not contrasting in color, a depth or raised height of at least 0.5 millimeter (or 0.02 inch).

Warning and caution markings may be combined under the signal word "WARNING."

Compliance is checked by inspection.

5.4.1 Operational maintenance warning marking. Equipment in which live parts may become accessible as a result of operational maintenance shall bear a warning marking on the equipment surface through which the operator can gain access to the interior or such parts.

The text shall be prefaced by the signal word "WARNING," and

(1) identify the live parts or their location and state that such parts could render an electric shock, and

(2) instruct that the equipment is to be disconnected from the supply circuit or that the measuring terminals are to be disconnected prior to removal of the part.

The marking shall be clearly visible prior to operator removal of the equipment parts.

Compliance is checked by inspection.

5.4.2 High leakage current warning marking. Cord-connected equipment incorporating electromagnetic-suppression filtering and having a leakage current (see Sub-clause 9.5.6) exceeding 0.5 milliampere shall bear a warning marking. The symbol ⚠ shall not be used in lieu of this marking.

Such marking shall be prefaced by the signal word "WARNING" and shall

(1) state that the protective earth-grounding conductor provides protection from electric shock, and

(2) instruct that the equipment must be earth-grounded.

Such marking shall be located on the equipment adjacent to the equipment appliance coupler or from where the power supply cord emerges.

Compliance is checked by inspection.

5.5 Operator control markings. Equipment operator controls that are accessible shall bear markings according to this Sub-clause.

5.5.1 General marking requirements

(1) Identification of operator controls
Operator controls shall be identified as to their intended purpose.

Markings for operator controls may also indicate the sequence of operation and/or the direction of movement to be followed for operation of the equipment.

Exception: Where the intended purpose of the operator control does not affect compliance with this Standard.

(2) Location of markings
Markings shall be adjacent to or related to the controls.

Compliance is checked by inspection.

5.5.2 "ON/OFF" switch. The "ON" position, or the "OFF" position, or both, of the principal supply circuit switch controlling equipment operating energy shall be identified (see Sub-clause 13.10.1 and 13.10.2).

The symbol ○ (IEC Publication 417; see Appendix) for power "OFF" disconnection from the principal supply circuit and the symbol | (IEC Publication 417; see Appendix) for power "ON" connection to the principal supply circuit may be used for this marking.

Identification by illumination alone is not acceptable.

Switches that appear to be in the same position when "ON" or "OFF" shall not be used for the principal supply-circuit switch.

Compliance is checked by inspection.

5.6 Terminal marking. Equipment terminals shall bear markings according to this Sub-clause.

5.6.1 General marking requirements

(1) Identification of terminals
Equipment terminals shall be identified such that the equipment may be connected and operated as intended by the manufacturer, so that electric shock, fire, or personal injury are not likely.

If the equipment terminals are not expressly identified on the equipment, the terminals shall be identified in the operator instructions or literature provided with the equipment.

(2) Terminal function, use, or rating
If the equipment terminals are not expressly marked with function, use, or ratings, the applicable marking as to the

intent of the terminal shall be in the operator instructions or literature provided with the equipment.

A terminal or terminal set used for more than one function shall be marked to identify the function, use, or rating appropriate for each function. Abbreviations and symbols may be incorporated in such markings.

Such markings may include other functions, uses, or ratings pertinent to equipment operation.

Terminal function, use, or ratings may be incorporated into operator-control markings

(3) Terminal markings and ratings requiring explanation

If, to preserve the protections afforded by the equipment, further explanations beyond the terminal identification and ratings marked on the equipment are necessary, the marking shall include a reference to the operating instructions.

The symbol ⚠ may be used for this purpose.

The reference shall be adjacent to or related to the marking or rating requiring explanation.

Compliance is checked by inspection.

5.6.2 Protective Grounding Terminal. A protective grounding terminal which must be connected to earth ground prior to making any other connections to the equipment shall be identified with the word "Ground," or the symbol ⏚ (IEC Publication 417; see Appendix.).

When used, the symbol shall be defined in the operating instructions provided with the equipment.

Exception: Where protective grounding is an integral part of the power supply connection.

Alternatively, a green-colored terminal screw with a hexagonal head, a green-colored hexagonal terminal nut, or a green-colored pressure wire connector shall be used for this identification.

Compliance is checked by inspection.

5.7 Protective Insulation. Protective Insulated Equipment shall be marked with the symbol ▢ (IEC Publication 417; see Appendix.)

For equipment employing multiple protection means, care should be applied when using this marking. (See Figure 5-1).

Compliance is checked by inspection.

*6. Equipment Emanations

The equipment shall be constructed to limit the levels of ionizing and microwave radiation, the liberation of ozone, and of ultrasonic pressure.

Exception: For equipment which incorporates radioactive substances, additional requirements in other safety standards apply.

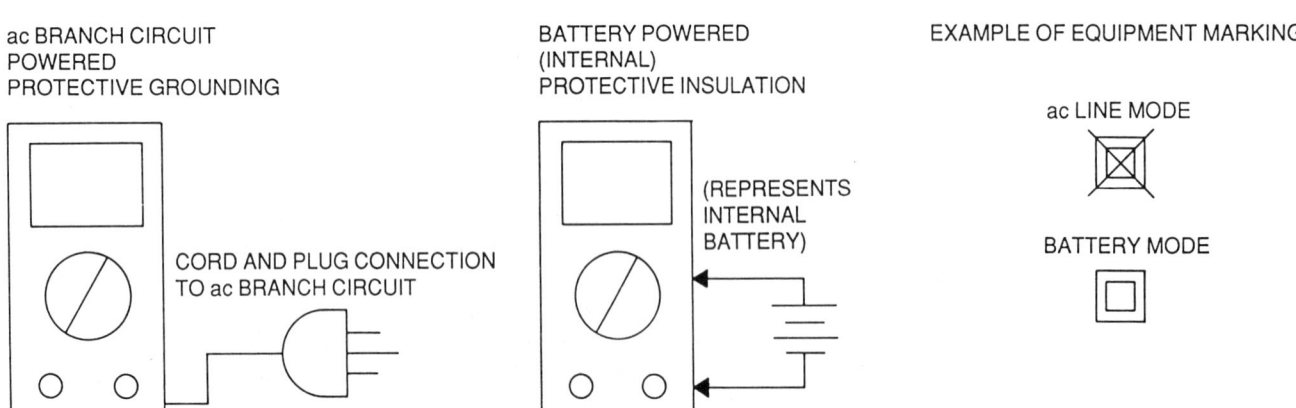

Figure 5-1. Marking for Equipment Employing Multiple Protection Means

(Example Only)

Compliance is checked according to Sub-clauses 6.1 to 6.4.

6.1 Ionizing radiation

6.1.1 X-radiation measurement. The exposure-rate limit averaged over an area of 10 square centimeters (1.55 square inches) shall not exceed 0.5 milliroentgen per hour (36 pico amperes per kilogram mass or 129 nanocoulombs per kilogram mass per hour) 5 centimeters (2 inches) from the surface of the equipment or surface of the part of the equipment being measured. The method of determining the amount of radiation shall be effective over the range of possible radiation energies.

The equipment shall be prepared and adjusted such that:

(1) the equipment under test shall be complete. Covers or doors that are intended to be removed or opened by the operator shall be removed or opened.

(2) the supply-circuit voltage to the equipment under test shall be:

(a) adjusted to be within the rated range(s) according to Sub-clause 4.2.1 (2), or

(b) 130 volts if the supply-circuit voltage rating is within the range of 105-130 volts, and 260 volts if the supply-circuit voltage rating is within the range of 210-260 volts,

whichever produces the maximum X-radiation.

(3) measurements on multifunction equipment shall be made with the equipment in all modes of normal operation, and with applicable signals applied to all input terminals in whatever manner produces the maximum X-radiation.

(4) cathode-ray-tube equipment shall display a pattern from each beam not exceeding 30 by 30 millimeters (1.19 by 1.19 inches) or the smallest possible display, whichever is larger. The display(s) shall be positioned to maximize X-radiation.

(5) to simulate conditions that may occur during repair of the equipment, the equipment under test shall be adjusted to produce maximum X-radiation, and all parts whose opening or removal is normal during service or repair shall be opened or removed during the test.

For example: Covers, CRT bezels, panels, windows, shields, barriers, or chassis.

Compliance is checked by measuring the X-radiation under the conditions of this Sub-clause in a direction normal to the surface of the equipment or surface of the part of the equipment being measured.

*6.2 Non-ionizing radiation. (Reserved for future use - see Appendix)

*6.3 Ozone. (Reserved for future use - see Appendix)

*6.4 Ultrasonic pressure. (Reserved for future use - see Appendix)

7. Heating

7.1 General. When equipment is operated under reference test conditions (Sub-clause 4.2), the temperature rises in Table 7-1 shall not be exceeded.

For equipment having special operating conditions, such as short-term or intermittent use, this Clause applies as far as compatible with such special operating conditions.

Compliance is checked by performing the tests according to this Sub-clause.

7.2 Temperature limits

Compliance with the temperature-rise limits is checked by measuring the temperature rise under reference test conditions after thermal equilibrium has been attained.

7.2.1 Thermal equilibrium

Thermal equilibrium shall be considered attained when three successive readings taken at equal intervals of

(1) 5 minutes, or

(2) ten percent of the total test time elapsed previous to the start of the first interval,
whichever is longer, indicate that there is no temperature change of the part.

7.2.2 Thermocouples. *Thermocouples, if used, shall be:*

(1) No. 28-32 AWG iron and constantan (Type J), or

(2) in accordance with the American National Standard for Temperature-Measurement Thermocouples, ANSI MC96.1, or IEC 584-1.

A thermocouple junction and lead wire shall be held in

thermal contact with the material being measured.

Taping or cementing the thermocouple in place is acceptable.

7.2.3 Test conditions. Each heating test shall be performed under the most unfavorable combination of the conditions described in this Sub-clause.

(1) Position general
The equipment shall be in any position of normal use or positioned and mounted according to the manufacturer's instructions. Normal ventilation shall not be impeded by artificial means.

All doors and covers that may be closed during operation of the equipment shall be closed.

Exception: Consideration may be given to the actual conditions of normal operation wherein doors and covers must by opened after each cycle of operation.

(2) Portable and bench top equipment
Rubber and other materials subject to deterioration shall be removed from feet and other supports of the equipment if deterioration of the material might result in the equipment or the supporting surface attaining higher temperatures.

Exception: If the arrangement of ventilation or other cooling means is such that the temperature rise is not affected by such placement or if the equipment design is such that a greater spacing has to be provided.

If the equipment lends itself to such placement, the heating test shall be conducted with the equipment against the wall in a right-angle corner, or in an alcove, even though such placement results in restricted ventilation. If the manufacturer's instructions specify minimum ventilation spacings, those spacings shall be provided between the equipment and the test surfaces. Walls shall be formed by black-painted vertical sheets of plywood not less than 9.5 mm (3/8 inch) thick and having such width and height that they extend not less than 0.6 m (2 ft) beyond the physical limits of the equipment.

(3) Rack- or panel-mounted and incomplete equipment
Equipment intended for rack or panel mounting or that is provided without a complete enclosure and is intended for a protected installation shall be tested for temperature rise in:

(a) an enclosure according to the manufacutrer's instructions, or

(b) a built-up enclosure constructed of wood not less than 9.5 millimeters (0.375 inch) thick and of dimensions

Table 7-1
Maximum Temperature Rise (1) Under Reference Test Conditions

Parts of the Equipment	Temperature Rise, °Celsius (°Fahrenheit) See note 5	
Accessible parts:		
A. surfaces of enclosures	35 (63)	
B. small areas and easily discernible heat sinks (not likely to be touched in normal use)	65 (117)	
Operating devices and handles:		
A. metallic	20 (36)	
B. nonmetallic	30 (54)	
Enclosure interior surfaces:		
A. wood	65 (117)	
B. insulating material	(See note 2)	
Insulating materials:		
A. polymeric	(See note 2)	
B. varnished cloth	60 (108)	
C. fiber	65 (117)	
D. wood and similar material	65 (117)	
	Method	
Insulating systems:	Resistance	Thermocouple
Class 105 windings of:		
A. transformers	75 (135)	65 (117)
B. relays, electromagnets, solenoids, etc.	85 (153)	65 (117)

Table 7-1 (cont.)
Maximum Temperature Rise (1) Under Reference Test Conditions

Parts of the Equipment	Temperature Rise, °Celsius	(°Fahrenheit) See note 5
C. motors: dc, universal, and ac with frame diameter larger than 178 millimeters (7 inches) (see note 3)		
i. open motors	75 (135)	65 (117)
ii. enclosed motors	80 (144)	70 (126)
D. ac motors with frame diameter of 178 millimeters (7 inches) or less (see note 3)		
i. open motors	75 (135)	75 (135)
ii. enclosed motors	80 (144)	80 (144)
E. vibrator coils	75 (135)	75 (135)
Class 130 windings of:		
A. transformers	95 (171)	85 (153)
B. relays, electromagnets, solenoids, etc.	105 (189)	85 (153)
C. motors: dc, universal and ac motors with frame diameter larger than 178 millimeters (7 inches) (see note 3)		
i. open motors	95 (171)	85 (153)
ii. enclosed motors	100 (180)	100 (162)
D. ac motors with frame diameter of 178 millimeters (7 inches) or less (see note 3)		
i. open motors	95 (171)	95 (171)
ii. enclosed motors	100 (180)	90 (180)
E. vibrator coils	95 (171)	95 (171)
Capacitors: (see note 4)		
A. electrolytic		40 (72)
B. other types		65 (117)
Fuses (see note 4)		65 (117)
Sealing compound		40 (72) (less than melting point)
Selenium rectifiers (see note 4)		50 (90)
Terminal box (on fixed equipment)		65 (117)
Surface on which fixed equipment might be mounted in service, and surfaces that might be adjacent to the unit when it is so mounted		65 (117)
Wires and cords (see note 4)		35 (63)

Notes

(1) The heating test can be conducted at any room temperature between 15°C and 35°C (59°F and 95°F) and the observed temperatures corrected to a room temperature of 25°C (77°F)

(2) Polymeric material shall be rated with respect to temperature.

(3) The diameter, measured in the plane of the laminations, of the circle circumscribing the stator frame, excluding lugs, boxes, etc., used solely for motor-mounting assembly, or connection.

(4) Does not apply if rated for a higher temperature.

(5) Equivalent degrees Fahrenheit values are an absolute temperature: therefore, 32°F should not be added to the value.

that provide 50 millimeters (2 inches) clearance from the top, sides, bottom, and back of the equipment with openings in front, back, etc., to expose operator controls, cables, jacks, connections, leads, and the like.

(4) Fixed equipment
Fixed equipment shall be installed according to the manufacturer's instructions, or as close to the wall or corner as the construction makes possible.

7.3 Polymeric enclosure and insulating material thermal conditioning. The equipment shall withstand thermal conditioning.

7.3.1 Thermal conditioning shall be conducted in either of two ways:

(1) One sample of the complete equipment, or that part of the equipment being evaluated, is to be placed in a full-draft circulating-air oven for at least 7 hours and maintained at a uniform temperature not less than 10°C (18°F) higher than the maximum operating temperature of the material, measured during the heating test (see Sub-clause 7.1), but not less than 70°C (158°F). The equipment shall not be operated during this thermal conditioning; or

(2) One sample of the complete equipment is to be placed in a test chamber for at least 7 hours. The circulation of air within the chamber is to simulate normal room conditions. The air temperature within the chamber, as measured at the supporting surface of the equipment, is to be maintained at 60.0 + 2.0°C (140.0 + 3.6°F). The equipment shall be operated in the same way as for the heating test (see Sub-clause 7.1).

7.3.2 *Compliance is checked after the thermal conditioning of Sub-clause 7.3.1 to determine that:*

(1) There is no interference with the operational maintenance of the equipment.

(2) The equipment complies with the requirements in Sub-clause 9.1.

(3) The equipment spacings according to Sub-clause 9.8 have been retained.

(4) The equipment dielectric voltage-withstand capabilities according to Sub-clause 9.12 have been retained.

(5) The equipment complies with the power-supply-cord strain-relief requirements.

(6) The integrity of the enclosure provides the same mechanical protection to internal parts of the equipment as prior to the conditioning.

8. Protection from Implosion and Explosion

8.1 Implosion protection. The equipment enclosure, or any direct-view high vacuum device such as a cathode-ray tube, shall provide protection with respect to the effects of implosion.

Implosion protection shall be provided for direct-view high-vacuum devices with a face plate dimension exceeding 160 millimeters (6.30 inches) in diameter, or with a face-plate area exceeding 20,000 square millimeters (31 square inches), whichever is greater.

Implosion protection shall be provided for both collapse of the vacuum-device structure and mechanical impact.

The integrity of the enclosure and any protective means shall not be impaired by an implosion.

Compliance is checked after implosion according to

(1) Sub-clause 8.1.1(1) (Thermal-shock method) or

(2) Sub-clause 8.1.1(2) (High-energy-impact method), and

(3) Sub-clause 8.1.2 (Mechanical impact) by

   (a) measuring the mass of expelled particles and their distance from the equipment, and

Exception: Shale-like particles, slivers, and glass dust are not considered glass particles.

   (b) inspecting the equipment enclosure for mechanical integrity.

The equipment shall be complete, with all covers and other normally openable or removable parts in place. A window that is not intended for implosion protection shall be removed. The equipment shall be in a normal operating position on a support with a height of 75 ± 5 centimeters (29.5 ± 2 inches) and horizontal dimensions not larger than those of the equipment. Equipment that normally stands on the floor shall be tested standing on the floor.

The floor shall be covered with a nonskid or nonbounce material. Two barriers each 13 millimeters (0.50 inch) thick, 250 millimeters (9.50 inches) high, and at least 2 meters (approximately 6 feet) long shall be placed on the floor located 0.5 meter (approximately 1.50 feet) and 2.0 meters *(approximately 6 feet) from the projection of the front of the equipment enclosure.*

After implosion is induced:

(1) no glass particle with a mass greater than 2 grams (0.07 ounce) shall pass the 0.5 meter (1.50 feet) barrier;

(2) the total mass of all of the glass particles between the two barriers shall not be more than 42.5 grams (1.50 ounces);

(3) no glass particle of any mass shall pass the 2.0 meter (6 feet) barrier;

(4) no glass particle of any mass shall be expelled from the sides, top, or rear of the enclosure.

Two implosion methods are described in Sub-clause 8.1.1. The thermal shock method is preferred; however, if an implosion cannot be induced by the thermal shock method, the high-energy impact method shall be used.

Two implosion tests shall be conducted on each protection system.

8.1.1 Implosion by collapse of the structure. An implosion that stimulates collapse of the structure shall be induced in the vacuum device.

Either method may be employed for inducing such an implosion provided that the method does not affect the measurement.

(1) Thermal shock method

An area shall be scratched with a glass cutter or diamond stylus in one of the patterns, according to Figure 8-1, on the funnel as near as possible to the juncture of the funnel and faceplate rim. Implosion shall be induced by repeated applications of the end of a 9.5 millimeter (0.375 inch) diameter glass rod, heated until nearly fluid, to the scratch pattern. Alternatively, implosion shall be induced by repeated applications of liquid nitrogen, or the like, to the scratch pattern.

(2) High-energy impact method

A 25 millimeter (1 inch) diameter pin shall be inserted through a hole in the equipment enclosure and rested on the funnel, as near as possible to the juncture of the funnel and faceplate rim. A means shall be provided to limit the maximum displacement of the pin to 15 millimeters (0.594 inch). The pin shall receive the impact energy of a 5 kilogram (11 pound) mass falling from a height of 1500 millimeters (5 feet) (Figure 8-2). The mass or height shall be increased, as necessary, to induce implosion.

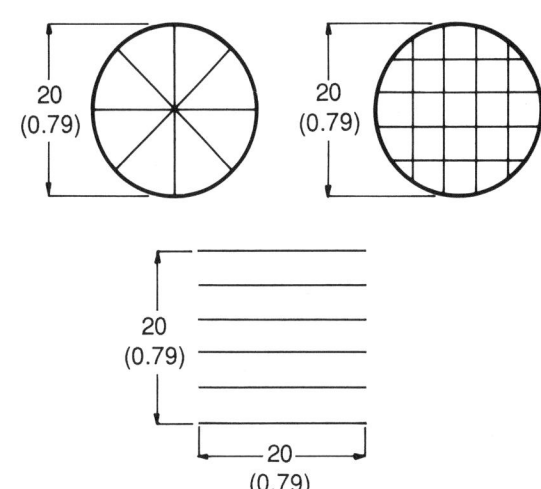

Figure 8-1. Scratch Patterns for Implosion by the Thermal Shock Method
Dimensions in millimeters (inches)

8.1.2 Implosion by mechanical impact. The display surface shall be subjected to the impact according to Table 8-1. The impact shall be produced by the specified spherical steel mass falling through the required distance as a pendulum.

Table 8-1
Impact Levels for Implosion by Mechanical Impact

Display surface dimension in millimeters (inches)	Steel ball diameter in millimeters (inches) (See Note 1)	Impact level in joules (foot-pounds force)
Exceeding 160 (6.30) and less than 400 (15.75)	40 (1.50) or 50.8 (2)	4.0 (3)
Equal to and exceeding 400 (15.75)	40 (1.50) or 50.8 (2)	5.5 (4)

Note 1: Either diameter of steel ball may be used as it is not expected that the difference in diameter can affect the results.

8.2 Explosion protection. The equipment enclosure or configuration shall provide protection with respect to explosion of a component within the enclosure. The equipment enclosure or configuration shall also provide protection with respect to a discharge.

For example: A discharge may occur from an electrolytic capacitor pressure-relief device or from a battery electrolyte leak.

Compliance is checked by inspection following the tests according to Clause 10.

*9. Protection from Electric Shock

9.1 Accessible parts. Accessible parts:

(1) shall not render an electric shock (shall not be live); and

(2) shall not be susceptible to becoming live in the event of a fault.

Parts of equipment which become accessible upon removing a cover, opening a door, adjusting a control, setting a supply-circuit voltage mechanism, replacing a fuse, attaching and detaching an interconnecting cable assembly, etc., and are intended for access by the operator during normal use, shall not render an electric shock.

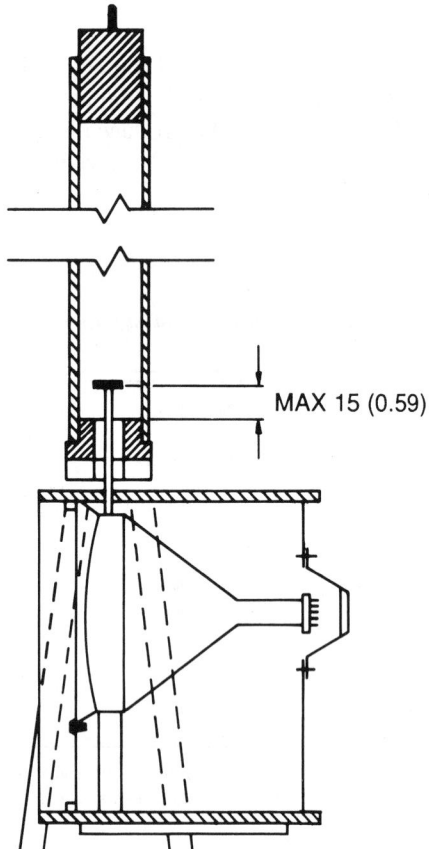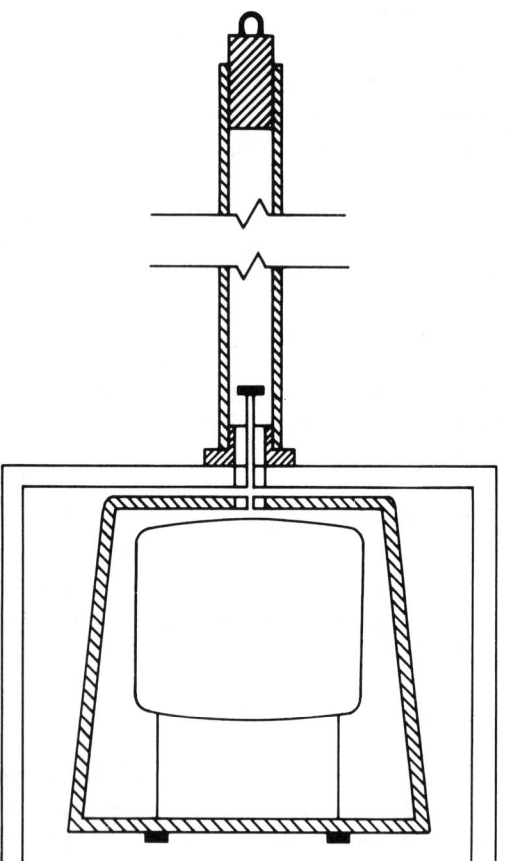

Figure 8-2. International Electrotechnical Commission (IEC) Test Apparatus for Implosion by the High-Energy Impact Method
Dimensions are given in millimeters (inches)

Exceptions:

1. Replacement of lamps that result in live parts of the lamp or socket being accessible after the lamp is removed. (As allowed for replacement of lamps in the house.)

2. Exchange of plug-in modules which do not require or intend for the operator to enter or contact parts accessible in the opening left by removal of the plug-in module when accessible live parts are recessed in the opening a distance greater than the major stop plate on the test finger of Figure 9-1 (180 millimeters).

3. When operator access to live parts is required by the basic nature of the equipment and a suitable warning is provided according to Sub-clause 5.4.1. Care should be used to provide guards where possible to prevent unintentional contact with such parts when the normal sequence of operational maintenance is being performed.

For measuring or testing terminals (see definition Sub-clause 2.44) refer to Sub-clause 9.3.5.

Because the insulation may be readily damaged, insulation consisting only of enamel, lacquer, oxides, anodic films, nonimpregnated paper, fiber and fibrous materials, wood, and the like shall not be regarded as providing protection from electric shock. Such insulation that is accessible shall be short-circuited during the tests according to Sub-clauses 9.1.1 and 9.2.

Exception: Wood may be used as enclosure material provided that such an enclosure is not in contact with live circuits.

An electronic device or an electrolytic capacitor shall not be regarded as providing protection from electric shock. Such devices that limit the voltage or current to an accessible part shall be short-circuited or open-circuited, one at a time, whichever creates the more unfavorable conditions, during the tests according to Sub-clauses 9.1.1 and 9.2.

Compliance is checked by performing the tests according to Sub-clauses 9.1.1 and 9.2.

9.1.1 Accessibility

A part is accessible when:

(1) the IEC articulate accessibility probe (Figure 9-1) applied in every possible position to the exterior or exposed surfaces, including the bottom; or

(2) the IEC rigid accessibility probe (Figure 9-2) applied with a maximum force of 30 newtons (6.75 pounds-force) in every possible position to the exterior or exposed surfaces, including the bottom,

touches the part.

Exception: For rack- or panel-mounted and incomplete equipment, the IEC articulate accessibility probe and IEC rigid accessibility probe shall be applied only to the exterior or exposed surfaces remaining after installation according to the manufacturer's instructions.

9.2 Live part

A live part is a part where under reference test conditions:

(1) the potential between the part and ground or any other simultaneously accessible part, exceeds

(a) 30 volts rms (42.4 volts peak), or

(b) 60 volts dc, or

(c) 24.8 volts dc interrupted at a rate of 10-200 hertz, and from which the leakage current exceeds the limits according to Sub-clause 9.11, or, at which,

(2) the potential between the part and ground or any other simultaneously accessible part exceeds

(a) 42.4 volts peak and not more than 450 volts peak, and the capacitance between the parts exceeds 0.1 microfarad; or

(b) 450 volts peak and not more than 15 kilovolts peak, and the product of capacitance in microfarads times the potential in volts exceeds 45 microcoulombs; or

(c) 15 kilovolts peak and one-half the product of the capacitance in microfarads times the square of the potential in volts exceeds 350 millijoules.

9.3 Exterior of equipment

9.3.1 Operating shafts. Operating shafts that are accessible after removing knobs, handles, and the like shall not be live.

Exception: Insulating knobs, handles, and the like that are captive to and enclose the shaft. The means for removing such insulating knobs, handles, and the like shall not be exterior to the equipment.

Compliance is checked according to Sub-clauses 9.1 and 9.2 after removing knobs, handles, and the like.

9.3.2 Openings over parts. Parts of equipment that can be touched by a freely suspended foreign body introduced into the equipment, while the equipment is in any of its normal operating positions, shall not be live.

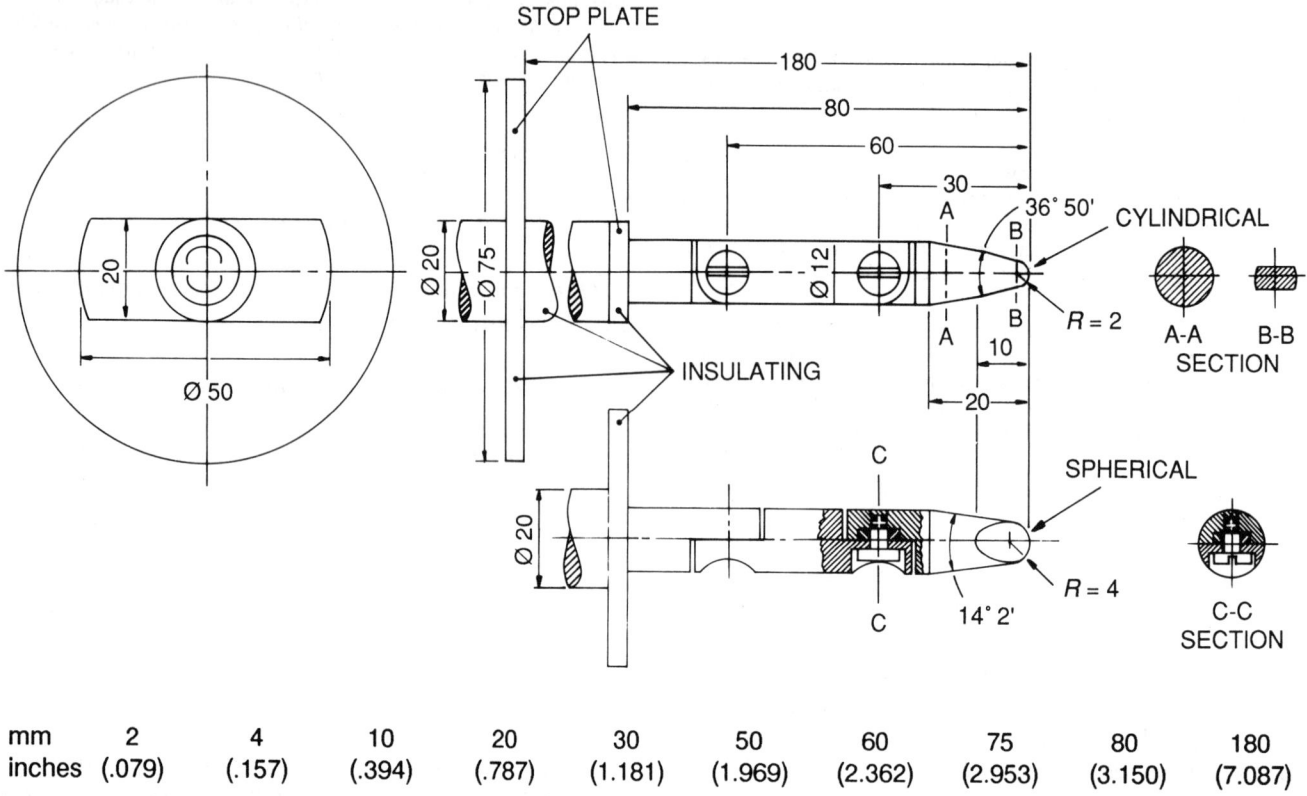

mm	2	4	10	20	30	50	60	75	80	180
inches	(.079)	(.157)	(.394)	(.787)	(1.181)	(1.969)	(2.362)	(2.953)	(3.150)	(7.087)

Figure 9-1. International Electrotechnical Commission (IEC) Articulate Accessibility Probe with Stop Plate
All dimensions in millimeters (inches)

Compliance is checked according to Sub-clause 9.2 after a freely suspended test pin, 4 millimeters (0.156 inch) in diameter penetrates an opening to a depth of 100 millimeters (4 inches).

9.3.3 Adjustment openings. Parts of equipment that can be touched by a tool introduced into the equipment through an adjustment opening shall not be live.

Compliance is checked according to Sub-clause 9.2 after inserting the tool specified or supplied by the manufacturer, or a test pin 3 millimeters (0.109 inch) in diameter, whichever creates the more unfavorable conditions, through the adjustment opening to a depth of three times the adjustment-device depth and at all possible angles made possible by the construction.

9.3.4 Screws. A screw that fixes enclosure covers and, when replaced by a longer screw, would reduce a creepage distance or clearance according to Sub-clause 9.8 shall be captive.

Compliance is checked by inspection.

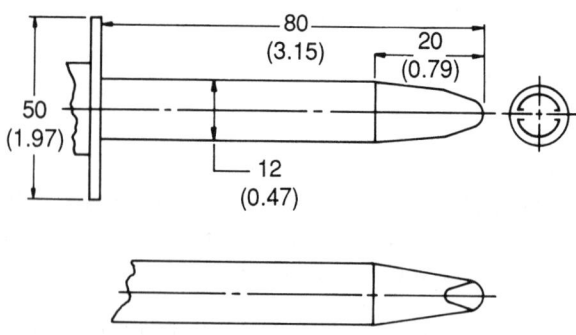

Figure 9-2. International Electrotechnical Commission (IEC) Rigid Accessibility Probe
All dimensions in millimeters (inches)

9.3.5 Terminals and connectors for circuits other than supply circuits

(1) Terminals and connectors, when fully mated as intended, and under reference test conditions, shall not have any accessible parts that could be live.

Exception: Live parts of measuring or testing terminals and connectors that for operational reasons must be accessible. Live parts that need not be exposed shall be covered, recessed, or otherwise made inaccessible.

Compliance is checked according to Sub-clauses 9.1 and 9.2 with the terminals and connectors fully mated as intended.

(2) An accessible terminal charged by an internal capacitor shall not be live 10 seconds after interruption of the supply.

Compliance is checked according to Sub-clause 9.2, 10 seconds after interruption of the supply.

9.3.6 Supply-circuit connector terminals.
The supply-circuit connector terminals of cord-connected equipment shall not be live 10 seconds after disconnection of the supply with the equipment supply circuit switch(es) in the "OFF" position.

Compliance is checked according to Sub-clause 9.2, 10 seconds after disconnection of the supply. Ideally, disconnection should occur at the supply-voltage peak. If disconnection at the supply-voltage peak cannot be assured, then the test shall be performed ten (10) times.

9.4 Construction requirements.
The construction requirements according to this Sub-clause are applicable to circuits which are considered live (according to Sub-clause 9.2) or where power is not limited according to Sub-clause 14.2.

(1) Conductors and basic insulation shall be arranged so that inadvertent loosening of wire connections, screws, etc., shall not reduce the spacings according to Sub-clause 9.8, or cause accessible parts to become live.

Compliance is checked according to Clause 11.

(2) An insulated conductor shall be restrained from abrasion of its basic insulation or from subjecting its basic insulation to a voltage or temperature greater than the voltage or temperature rating of the insulated conductor.

Wires should be routed away from sharp edges, screw threads, burrs, moving parts, etc. Holes through which wires are routed should have smooth, well-rounded surfaces, or shall have a bushing.

Clamps and guides used for routing or wiring should have smooth, well-rounded edges. Pressures exerted by such clamps should not cause cold-flow or otherwise deform the basic insulation. Supplementary insulation should be provided under a conductive clamp that restrains thermoplastic-insulated wires having 0.8 millimeter (0.031 inch) or less insulation.

(3) The basic insulation on each wire shall be rated for at least the maximum voltage to which the wire is connected, and for at least the temperature it attains according to Sub-clause 7.2.

(4) Insulating tubing, sleeving, and tape shall be rated for at least the maximum voltage against which it insulates, and for at least the temperature it attains according to Sub-clause 7.2.

(5) Where breaking of the bond between the conductor and base material of a printed-wiring board would render an electric shock or could result in a fire, the printed wiring shall have a rated bonding strength according to printed wiring board standards.

See, for example, ANSI/UL796, "Printed Circuits," for appropriate requirements.

(6) Where breaking or loosening of a circuit connection would render an electric shock or could result in a fire, such connection shall be made mechanically secure.

Mechanical security of connections may be provided by crimped, closed ring or flanged lugs, or a wrapping that forms at least an open U or by cable clamps, or by cable lacing, insulating tubing, or similar means. For printed-wiring boards, mechanical security is provided when wire leads, including those of a component, pass through a hole in the board and are soldered.

Compliance is checked by inspection.

9.5 Protective grounding.
Protective grounding is a system for bonding to ground those accessible conductive parts susceptible to becoming live in the event of a fault.

Equipment provided with a means for grounding (other than the measuring ground) shall comply with the applicable requirements in this Sub-clause, even though such equipment does not itself require protective grounding.

9.5.1 Accessible parts.
Accessible conductive parts

(1) separated from parts conductively connected to a live circuit by basic insulation or basic spacing shall be grounded according to this Sub-clause; or,

(2) shall be separated from parts conductively connected to a live circuit by

    (a) a conductive part that is grounded according to this Sub-clause; or,

(b) a protective insulating system according to Sub-clause 9.6.

Compliance is checked by inspection and by the tests according to this Sub-clause.

9.5.2 Grounded parts. Parts intended to be protectively grounded shall be bonded to the grounding terminal of the equipment. The resistance shall not be more than 0.1 ohm.

Exception: Accessible conductive parts not intended to be grounded, such as isolated trim, handles, nameplates, fastening screws, terminals, etc., must comply with Sub-clause 9.1.

Compliance is checked by inspection or, in case of doubt, by measuring the resistance between a grounded accessible conductive part and the point of connection of the protective-grounding means, excluding the protective-grounding conductor of the power-supply cord.

The type-test resistance shall be determined by measuring the voltage when a current of 25 amperes at supply-circuit frequency is conducted for one minute.

Exception: Control shafts

The method of measurement is to be determined by the user.

Compliance for the routine test is checked by verifying continuity of the grounding connection by any suitable means between a representative part required to be grounded and the attachment-plug grounding terminal.

9.5.3 Interconnected equipment. Where protective grounding between electrically interconnected pieces of equipment is required, such protective grounding shall be by means of at least one discrete conductor included in the interconnecting cable. A metal braid of such interconnecting cable shall not be used as the sole protective-grounding conductor.

Exception: Electrically interconnected equipment, each of which is separately grounded as, for example, through its supply-circuit connection.

Compliance is checked by inspection and according to Sub-clause 9.5.2.

9.5.4 Protective-grounding conductor

(1) The protective-grounding conductor of a power-supply cord shall be green, or green with one or more yellow stripes.

(2) The protective-grounding conductor of a power-supply cord or interconnecting cable shall be at least the size of the largest circuit conductor in the power-supply cord or interconnecting cable.

(3) The protective-grounding conductor shall be separately connected to the protective grounding terminal or the equipment frame or enclosure.

An example of an acceptable connection is by means of a screw and lockwasher used solely for that purpose.

(4) Connection of the protective-grounding conductor shall not depend on solder alone.

(5) The protective-grounding conductor of an attached power-supply cord shall, for the condition of breakage of the strain-relief mechanism, be the last conductor to break.

Compliance is checked by inspection.

9.5.5 Protective-grounding terminals

(1) The protective-grounding terminal of a detachable power-supply cord shall make first and break last with respect to the other connector terminals.

(2) A terminal provided on the equipment for the purpose of protective earth grounding shall be used solely for such purpose.

Exception: Panel-mounted parts may be bonded to the equipment grounding system by means of a conductive mounting stud or a metallic fixing device.

(3) Parts of a protective-grounding terminal (except parts to which the terminal is attached) shall be of corrosion-resistant material, or protected by a non-corrosive protective finish, and the contact surface shall be of metal.

(4) A terminal solely for connection of a protective-grounding conductor shall be capable of securing a conductor not smaller than the largest supply circuit conductor.

(5) A protective-grounding terminal shall be fastened in place to prevent turning of the terminal.

Compliance is checked by inspection.

9.5.6 High leakage-current equipment. Cord-connected equipment requiring an electromagnetic-suppression filter for functional performance and with leakage current exceeding 0.5 milliampere shall employ protective grounding.

Compliance is checked by measurement according to Sub-clause 9.11 and by inspection.

9.5.7 Location of supply-circuit-interrupting devices. A switch, overcurrent device, etc., shall not interrupt the equipment-protective-grounding system.

Compliance is checked by inspection.

9.5.8 Supply-circuit receptacles. Supply-circuit receptacles employed on grounded equipment shall be grounding types.

Compliance is checked by inspection.

9.5.9 Indirect protective grounding

Indirect protective grounding is a system where an intermediate device completes the protective-grounding circuit in the event of a fault.

9.5.10 Protective grounding system capacity

Reserved for future use.

9.6 Protective Insulation. Protective insulation is a system for additionally insulating noncurrent-carrying conductive parts susceptible to becoming live in the event of a fault.

The term protective insulation as used in this standard is synonymous with the commonly used term double insulation.

9.6.1 Accessible parts. Accessible conductive parts shall be separated from parts conductively connected to a live circuit by a protective insulation system comprised of at least:

(1) basic insulation plus supplementary insulation; or

(2) reinforced insulation; or

(3) basic insulation plus supplementary spacings; or

(4) basic spacings plus supplementary insulation; or

(5) protective spacings (Table 9.1).

Accessible conductive parts shall not be conductively connected to any secondary circuit, measuring circuit or testing circuit.

Exception: Secondary circuits, measuring circuits and testing circuits that are not live according to Sub-clause 9.2 under a single fault condition.

Spacings shall be according to Table 9.1.

Compliance is checked by inspection.

9.6.2 Supplementary insulation. Supplementary insulation shall be independent from and shall be rated at least equal to the relevant basic insulation.

Compliance is checked by inspection and according to Sub-clause 9.12.

9.6.3 Reinforced insulation. Reinforced insulation shall be a single insulating material rated at least twice the relevant ratings for basic insulations.

Compliance is checked by inspection and tested to twice the required voltage according to Sub-clause 9.12.

9.7 Protective impedance. Protective impedance is a system which limits an accessible voltage or current or charge or energy such that the values stated in Sub-clause 9.2 are not exceeded under normal operating conditions or in the event of a fault.

A protective impedance can be a combination of components connected between live parts and accessible conductive parts not connected to a protective grounding terminal.

Compliance is checked according to Subclause 9.2 and by application of fault conditions according to Clause 10.

*9.8 Spacing. Spacing is a system for separating

(1) uninsulated conductive parts, or

(2) insufficiently insulated conductive parts, or

(3) conductive parts attached to an insulating structure

from other such parts or from ground or from both.

9.8.1 Primary circuits. The spacings between an uninsulated part conductively connected to the primary circuit and

(1) an uninsulated part conductively connected to another pole of the primary circuit;

(2) another uninsulated part of any other circuit;

(3) a grounded part;

(4) accessible conductive parts; and

(5) the point of closest approach of either IEC accessi-

bility probe when applied according to Sub-clause 9.1.1 to any opening of the enclosure (see Figure 9-3)

shall be at least those according to Table 9-1.

Spacing requirements are not applied to the internal spacings of electronic devices or inherent spacings of other components. The acceptability of a component should be according to Clause 13 and applicable component requirements (see Sub-clause 3.4 and 4.1.1).

Compliance is checked by first applying and removing a 2 newton (0.5 pound) force against any wire or uninsulated part followed by measuring the creepage distance or clearance.

Table 9.1
Spacings[e]

		Basic Spacings or Supplementary Spacings				Protective Spacings			
		Distance-Millimeter (Inch)				Distance-Millimeter (Inch)			
Max. Circuit Voltage[b] (Volts)		Through Air (Clearance)[c]		Over Surface (Creepage)[d]		Through Air (Clearance)[c]		Over Surface (Creepage)[d]	
Sinusoidal ac (rms)	dc, ac peak, or Mixed Voltage	In Equipment	Printed Ckts.[a]	In Equipment	Printed Ckts.[a]	In Equipment	Printed Ckts.[a]	In Equipment	Printed Ckts.[a]
Up to 30	Up to 42	1.0 (0.040)	0.5 (0.020)	1.0 (0.040)	0.5 (0.020)	2.0 (0.080)	1.0 (0.040)	2.0 (0.080)	1.0 (0.040)
Over 30 up to 60	Over 42 up to 85	2.0 (0.080)	1.0 (0.040)	2.0 (0.080)	1.0 (0.040)	3.0 (0.120)	2.0 (0.080)	3.0 (0.120)	2.0 (0.080)
Over 60 up to 150	Over 85 up to 212	2.5 (0.100)	1.5 (0.060)	2.5 (0.100)	1.5 (0.060)	3.5 (0.140)	2.5 (0.100)	3.5 (0.140)	2.5 (0.100)
Over 150 up to 300	Over 212 up to 424	3.0 (0.120)	2.0 (0.080)	3.0 (0.120)	2.0 (0.080)	4.0 (0.160)	3.0 (0.120)	4.0 (0.160)	3.0 (0.120)
Over 300 up to 450	Over 424 up to 630	3.5 (0.140)		4.5 (0.180)		5.0 (0.200)		7.0 (0.280)	
Over 450 up to 650	Over 630 up to 920	4.0 (0.160)		6.0 (0.240)		6.0 (0.240)		9.0 (0.360)	
Over 650 up to 1000	Over 920 up to 1400	5.5 (0.220)		9.0 (0.360)		8.0 (0.320)		13.0 (0.520)	
Over 1000, not over 1500	Over 1400, not over 2100	10.0 (0.400)		12.0 (0.480)					
Over 1500, not over 2000	Over 2100, not over 2800	12.0 (0.480)		14.0 (0.560)					
Over 2000, not over 2500	Over 2800, not over 3600	14.0 (0.560)		15.5 (0.620)					

[a] Values apply to miniature-type components (for example, micromodules, etc.) and are accepted only where the spacings are rigidly maintained by constructional means and cannot be reduced during assembly of the component or part into the equipment.
[b] Voltage measured according to sub-clause 14.1.2.
[c] See definitions sub-clause 2.10
[d] See definitions sub-clause 2.11
[e] The type of spacing to be used (i.e., basic, supplementary, or protective) is determined by the Protection Class. (See Sub-Clauses 9.5, 9.6, and 9.8).

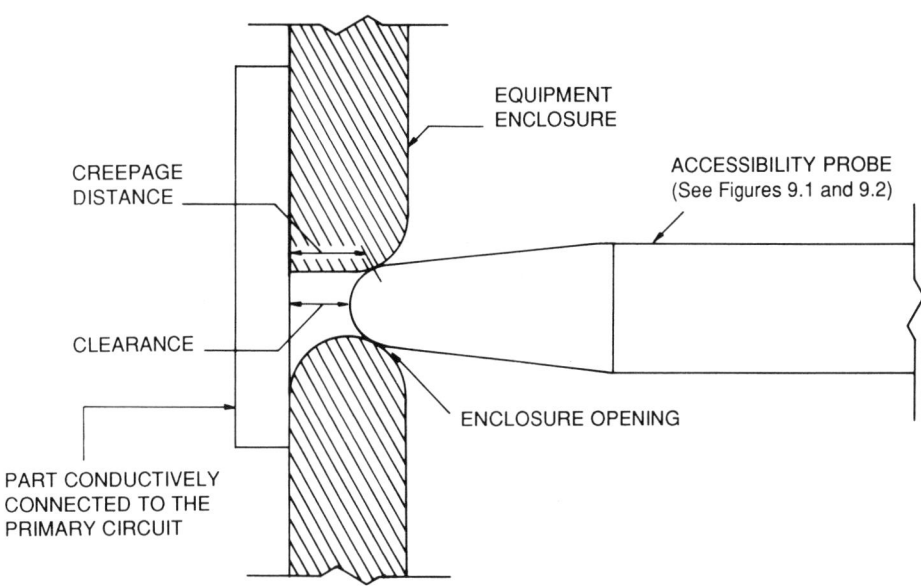

Figure 9-3. Measuring the Distance of Closest Approach, through an Enclosure Opening, to an Uninsulated Part Conductively Connected to the Primary Circuit Using Either of the IEC Accessibility Probes

9.8.2 Other circuits. The spacings in other circuits are evaluated according to Sub-clauses 9.12.3 to 9.12.5 (dielectric voltage-withstand tests).

Exception: Circuits employing spacing according to Table 9-1.

9.8.3 Field wiring terminal parts. The creepage and clearance distances between supply-circuit field wiring terminal parts of opposite polarity and between supply- circuit field wiring terminal parts and the enclosure in fixed equipment shall be at least those specified in Table 9-2.

Exception: Supply circuits rated for extra-low voltage and power limited according to Sub-clause 14.2

Compliance is checked by first applying and removing a 2-newton (0.5 pound) force against any wire or uninsulated part followed by measuring the applicable creepage distance or clearance.

9.9 Polarization of primary circuits. Equipment intended for use on a polarized branch circuit shall comply with the applicable requirements of this Sub-clause, even though such equipment does not itself rely upon polarization.

9.9.1 Primary circuit attachment plug and cord connector body

(1) Cord-connected polarized equipment shall employ a polarized attachment plug.

(2) Cord-connected polarized equipment employing a detachable power-supply cord shall also employ a polarized cord connector body.

Compliance is checked by inspection.

9.9.2 Primary-circuit receptacles. Primary-circuit receptacles employed on polarized equipment shall be polarized.

Compliance is checked by inspection.

9.9.3 Location of primary-circuit interrupting device. A single pole primary switch, overcurrent device; etc., employed in polarized equipment shall not interrupt the grounded conductor.

Exception: Where all primary-circuit conductors are interrupted simultaneously.

Compliance is checked by inspection.

9.9.4 Reverse polarization. Protection from electric shock in polarized equipment shall be provided for the condition of reversed polarization of the primary circuit.

Compliance is checked according to Sub-clause 9.11 (leakage current).

9.9.5 Accessible parts. Accessible conductive parts, including accessible conductive parts of terminals of polarized equipment, shall not be connected to the grounded primary-circuit conductor.

Compliance is checked by inspection.

9.9.6 Polarization check (routine). Equipment subject to any requirement in this Sub-clause shall be routinely inspected for polarization.

Compliance is checked by tracing the grounded conductor from the attachment plug or cord connector to the point of connection to each device in the equipment primary circuit.

9.10 Humidity conditioning. The equipment insulation shall not have leakage currents exceeding the limits according to Sub-clause 9.11 following humidity conditioning.

Humidity conditioning is not required for routine tests.

Compliance is checked by first conditioning the equipment according to this Sub-clause and then testing it according to Sub-clause 9.11.

9.10.1 Preconditioning. Equipment shall not be subjected to moisture condensation when subjected to humidity conditioning.

To avoid condensation, the equipment shall be temperature preconditioned by storing in a chamber whose temperature is 40.0-44.0°C (104.0-111.2°F) and whose relative humidity does not exceed 50 percent. The equipment shall be maintained in the chamber for a minimum of four hours and until the equipment temperature has stabilized.

Temperature stabilization may be determined by placing a thermocouple on the most massive component in the system. Temperature stabilization may be indicated by successive temperature measurements of that component that show that no additional temperature rise is apparent.

9.10.2 Conditioning

Covers or doors that are intended to be removed or opened by the operator for the purposes of operational maintenance shall be removed or opened during conditioning. Immediately following temperature preconditioning, according to Sub-clause 9.10.1, the equipment shall be conditioned in a chamber whose temperature is 38.0-40.0°C (100.4-104.0°F) and whose relative humidity is 90-95 percent. The equipment shall be maintained in the chamber for 48 hours.

The equipment shall not be operated during the humidity conditioning.

Before testing according to Sub-clause 9.11, the equipment shall have been removed from the humidity chamber.

Leakage current testing shall commence immediately after completion of the conditioning test.

*9.11 Leakage current. If the open-circuit potential between an accessible part and either ground or any other accessible part exceeds the limits according to Sub-clause 9.2, leakage currents between those parts shall not be more than the limits specified in this Sub-clause.

Leakage current refers to all currents, including capacitively coupled currents, that may be conveyed between exposed surfaces of the equipment and ground or other exposed surfaces of the equipment.

Leakage current shall be measured between equipment parts indicated in Table 9-3.

The leakage current measuring circuits shall be according to Figures 9-6 to 9-12.

Leakage current limits shall be according to Table 9-4 and Figure 9-5.

The specified leakage current measuring instrument shall be according to Figure 9-4.

Compliance is checked by measuring the current with the specified measuring instrument, or the equivalent.

The specified leakage current measuring instrument is one:

Table 9-2
Spacings
Field Wiring Terminal Parts

Terminal Parts	Vrms*	Creepage Distance mm (inches)	Clearance Distance mm (inches)
Supply Circuits	0-50	3.2 (0.13)	3.2 (0.13)
	51-250	6.4 (0.25)	6.4 (0.25)
	251-480	12.7 (0.50)	12.7 (0.50)

* For dc voltages multiply the Vrms value by $\sqrt{2}$.

(1) having an impedance of 1500 ohms resistance shunted by a capacitance of 0.15 microfarad; and

(2) for direct currents, indicating the average direct current and which can follow low-frequency alternating currents and indicate the peak value for frequencies up to 1 Hertz, and

(3) for alternating currents, indicating 1.11 times the average of the full-wave rectified composite waveform of current in milliamperes through the resistor; and, having the frequency response (ratio of indicated to actual value of current) over a frequency range of 1 hertz-100 kilohertz that is equal to the ratio of the impedance of a 1500-ohm resistor shunted by a 0.15- microfarad capacitor to 1500 ohms, and;

(4) having an error of not more than 5 percent for dc measurements of 2 milliamperes dc and 2 milliamperes peak ac at frequencies up to 1 hertz. The ac measure-

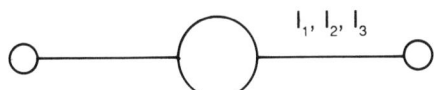

SPECIFIED LEAKAGE CURRENT MEASURING INSTRUMENT AS DEPICTED IN FIGURES 9.6 - 9.12.

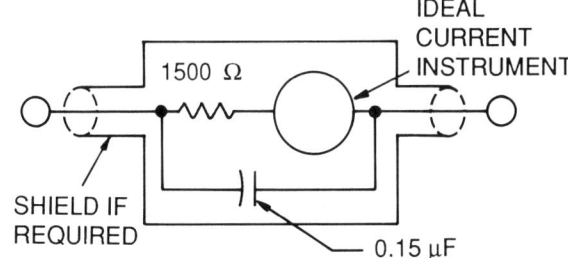

EQUIVALENT CIRCUIT OF THE SPECIFIED LEAKAGE CURRENT MEASURING INSTRUMENT

Figure 9-4. Specified Leakage Current Measuring Instrument

Table 9-3
Equipment Parts between Which the Open-Circuit Potential and Leakage Currents Shall Be Measured

Equipment Parts	Cord-Connected Equipment	Self-Powered Equipment	Permanently Connected Equipment
The open grounding conductor (including parts connected to the grounding terminal)	Yes	Yes [3]	No
The enclosure (or the foil wrapped about the enclosure) [1]	Yes	Yes	Yes
Other accessible parts	Yes	Yes	Yes
	AND		
The grounded pole of the primary circuit	Yes	No	No
Any other simultaneously accessible part [2]	Yes	Yes	Yes

Notes

[1] See Leakage Current Measuring Circuit, Note 8, Sub-Clause 9.11.1

[2] Parts are considered simultaneously accessible if they can be touched by one or both hands. Parts within a 100 millimeter (4 inch) by 200 millimeter (8 inch) rectangle are considered accessible to one hand; parts within a distance of 1.8 meters (6 feet) are considered accessible to both hands.

[3] If provided with a grounding terminal.

ments above 1 hertz shall not be in error more than 5 percent at the limits indicated in Figure 9-5.

The actual leakage current measuring instrument used need not have all the attributes of the specified leakage current measuring instrument if equivalent readings are obtained.

Above commercial power frequencies, tactile perception requires higher current. The frequency response of the specified leakage current measuring instrument approximates the experimentally determined effect on a person's hand holding a wire over a frequency range of dc to 100 kilohertz; therefore, the indicated current limit is applicable over this frequency range.

9.11.1 Leakage current measurement notes

Sub-clause 9.11 states that leakage current measuring circuits shall be according to Figures 9-6 to 9-12. Each of the

Table 9-4
Leakage Current Limits [f], [g]

Equipment Conditions		Equipment Employing Protective Grounding			Equipment Not Employing Protective Insulation	
		I_1 [a]	I_2 [b]	I_3 [c]	I_1 [d]	I_3 [e]
Leakage Current Due to Primary Circuit;	No EMI Filtering or EMI Filtering Removed	2 mA dc 0.5 mA ac	2 mA dc 0.5 mA ac		2 mA dc 0.5 mA ac	
Figure 9.6 or Figure 9.10	EMI Filtering Installed if Applicable	2 mA dc 3.5 mA ac	2 mA dc 0.5 mA ac		2 mA dc 0.5 mA ac	
Leakage Current due to Primary Circuit plus Isolation Voltage;	No EMI Filtering	2 mA dc 0.5 - 3.5 mA ac Figure 9.5	2 mA dc 0.5 mA ac		2 mA dc 0.5 mA ac	
Figures 9.7 or 9.11	EMI Filtering Installed if Applicable	2 mA dc 3.5mA ac	2 mA dc	0.5mA ac	2 mA dc	0.5mA ac
Leakage Current between Conductive Parts	EMI Filtering Installed if Applicable; Figure 9.8 or 9.12			2 mA dc 0.5 mA ac		2 mA dc 0.5 mA ac
	No EMI Filtering or EMI Filtering Removed; Figure 9.9			2 mA dc 0.5 - 3.5 mA ac; Figure 9.5		
	EMI Filtering Installed if Applicable; Figure 9.9			2 mA dc 3.5 mA ac		

Notes:
[a] Parts connected to the protective grounding conductor plus isolated parts
[b] Parts isolated from the protective grounding conductor
[c] Simultaneously accessible parts
[d] Isolated parts
[e] Simultaneously accessible parts
[f] Leakage current measurements shall be made for each condition that applies to the equipment under investigation
[g] The leakage current limits specified are the maximum acceptable indications of the specified leakage current measuring instrument over a frequency range of dc to 100 kilohertz.

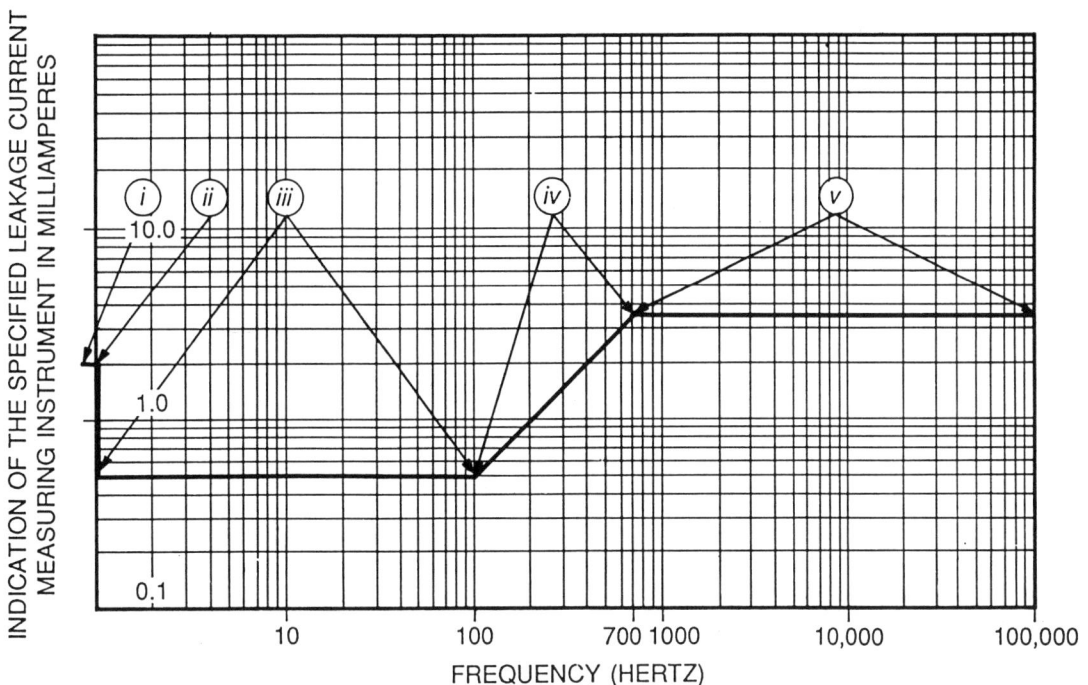

Leakage current limits applicable to Figure 9-5:

(i) 2.0 milliampers dc
(ii) 2.0 milliampers ac peak for frequencies up to 1 hertz
(iii) 0.5 milliamperes rms for frequencies of 1 hertz - 100 hertz
(iv) 0.5 milliamperes rms at 100 hertz linearly increasing to 3.5 milliamperes rms at 700 hertz
(v) 3.5 milliamperes rms for frequencies of 700 hertz to 100 kilohertz

Figure 9-5. Indication of the Specified Leakage Current Measuring Instrument Resulting from Primary Circuit and Isolation Voltage Plotted as a Function of Frequency

circuit diagrams makes reference to certain explanatory notes itemized in this Sub-clause. The following are the notes applicable to Figures 9-6 to 9-12.

1. The primary circuit can be connected directly to switch S-1 without transformer T-1 if one pole of the primary circuit is grounded as normally is the case in the U.S.A.

2. For single phase cord-connected equipment, leakage current shall be measured with switch S-1 "normal" and with switch S-1 "reversed." For dc equipment, and for permanently connected equipment, the supply conductors are not reversed for leakage current measurement.

3. High or signal terminals normally need not be connected. When there is doubt about their effect on leakage current, high or signal terminals may also be shorted or loaded as appropriate within the intent of equipment design and specifications.

4. The common, or reference, or low terminals and the guard or shield terminals of isolated terminal sets shall be connected to circuit point "A" as indicated.

5. The common, or reference, or low terminals of each isolated terminal set, input or output, and each guard or shield terminal shall be connected to its maximum rated circuit-to-ground voltage. Terminals rated to other terminals but not rated to ground are not connected to a voltage.

6. All accessories that contribute to leakage current are connected according to the manufacturer's ratings, instructions and specifications.

7. Normally, I1 and I2 can be used simultaneously without affecting the results; however, for referee conditions, one meter at a time shall be used with the other meter replaced by a short circuit.

8. *If the equipment is wholly or partially encased in insulation, accessible parts shall include*

(a) a 200 by 200 millimeter (8 by 8 inch) square of conductive foil wrapped anywhere about the equipment; or

(b) two 100 by 200 millimeter (4 by 8 inch) conductive foil pieces wrapped about the equipment at any two locations not more than 1.8 meters (6 feet) apart; or

(c) a conductive plane surface upon which any part of the equipment may rest.

9.11.2 Leakage current including an electromagnetic-suppression filtering

(1) If equipment requires an electromagnetic-suppression filtering for functional performance, the current measured according to Sub-clause 9.11 shall not cause an indication of the specified leakage current measuring instrument greater than 3.5 milliamperes rms.

(2) With the electromagnetic-suppression filtering removed from the equipment, the current measured according to Sub-clause 9.11 shall not cause an indication of the specified leakage current measuring instrument greater than those according to Sub-clause 9.12.

*9.12 Dielectric voltage-withstand tests. Insulation and spacings between conductors shall sustain a voltage stress, according to this Sub-clause, without electrical breakdown that would result in electric shock or fire.

Dielectric voltage-withstand tests need be performed only in those circuits where it can be demonstrated that a breakdown of insulation or spacings between conductors would result in electric shock or fire.

A breakdown which might result in electric shock or the likelihood of fire is one that precludes compliance to Sub-clause 10.2 (Compliance Tests - Fault Conditions). Clause 10 includes fault conditions normally applied to equipment; however, it may be necessary to perform

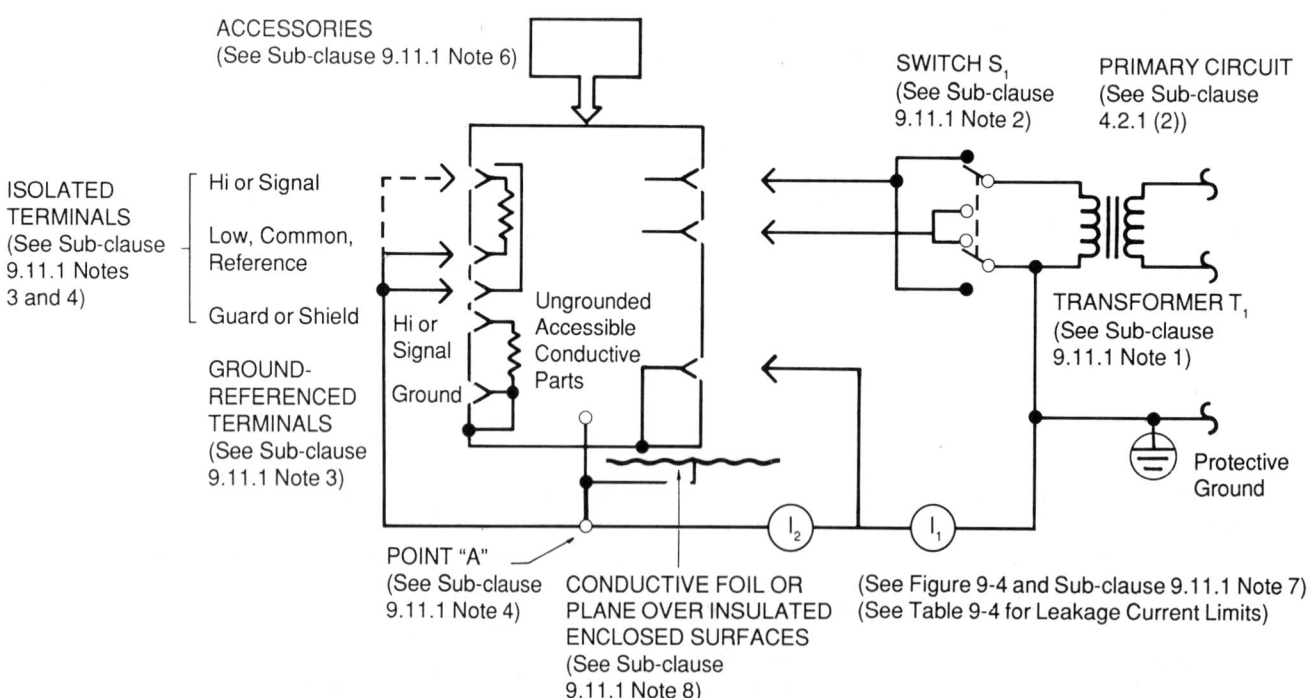

Figure 9-6. Leakage Current Measurement Circuit for Equipment Employing Protective Grounding -- Leakage Current Due to Primary Circuit

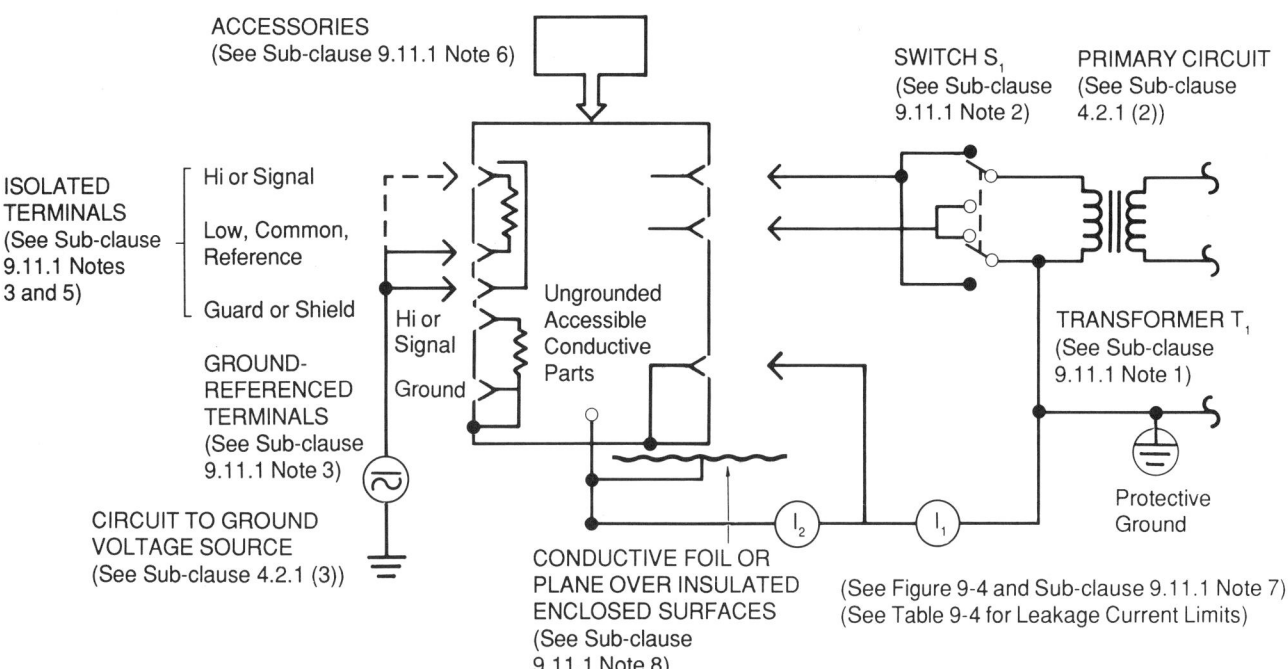

Figure 9-7. Leakage Current Measurement Circuit for Equipment Employing Protective Grounding -- Leakage Current due to Primary Circuit plus Isolation Voltage

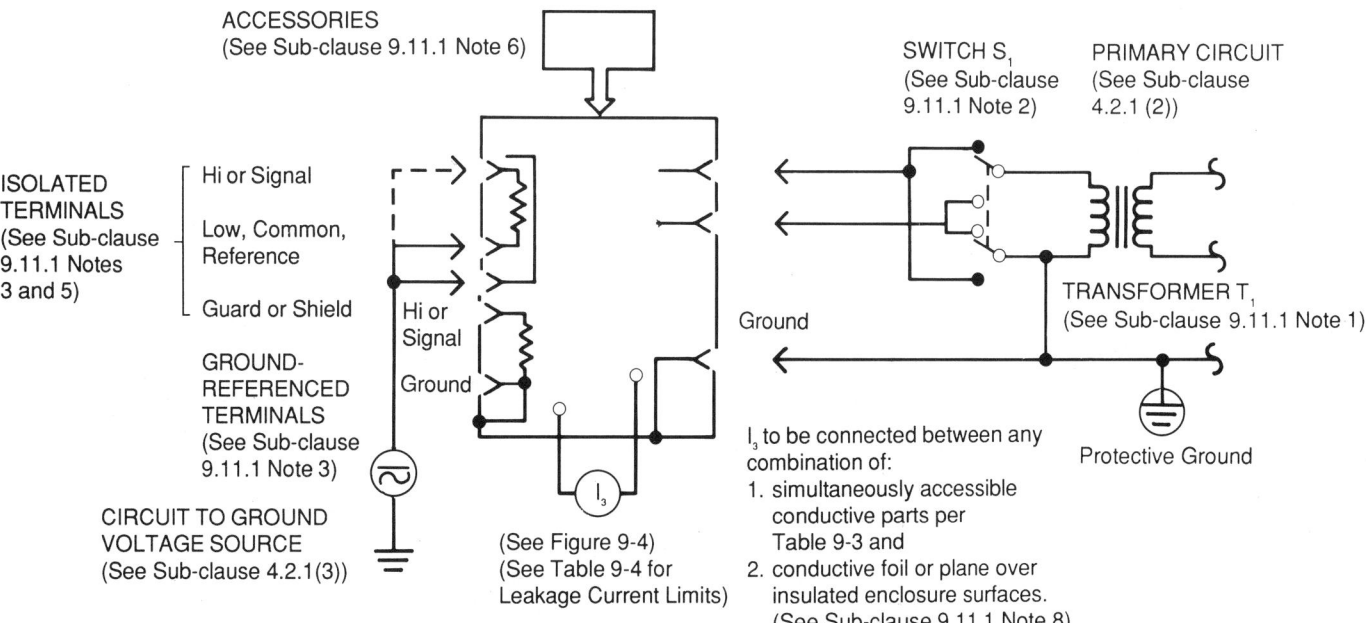

Figure 9-8. Leakage Current Measurement Circuit for Equipment Employing Protective Grounding -- Leakage Current between Conductive Parts with Ground Connected

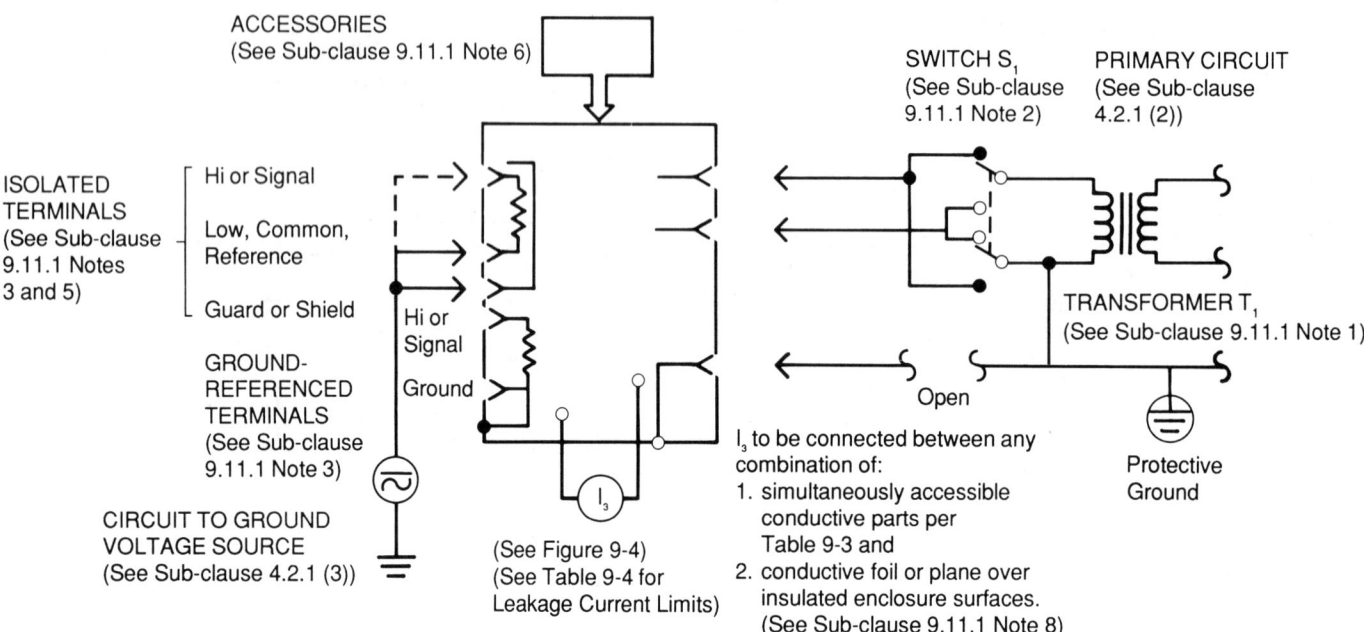

Figure 9-9. Leakage Current Measurement Circuit for Equipment Employing Protective Grounding -- Leakage Current due to Primary Circuit plus Isolation Voltage between Simultaneously Accessible Conductive Parts with Ground Open

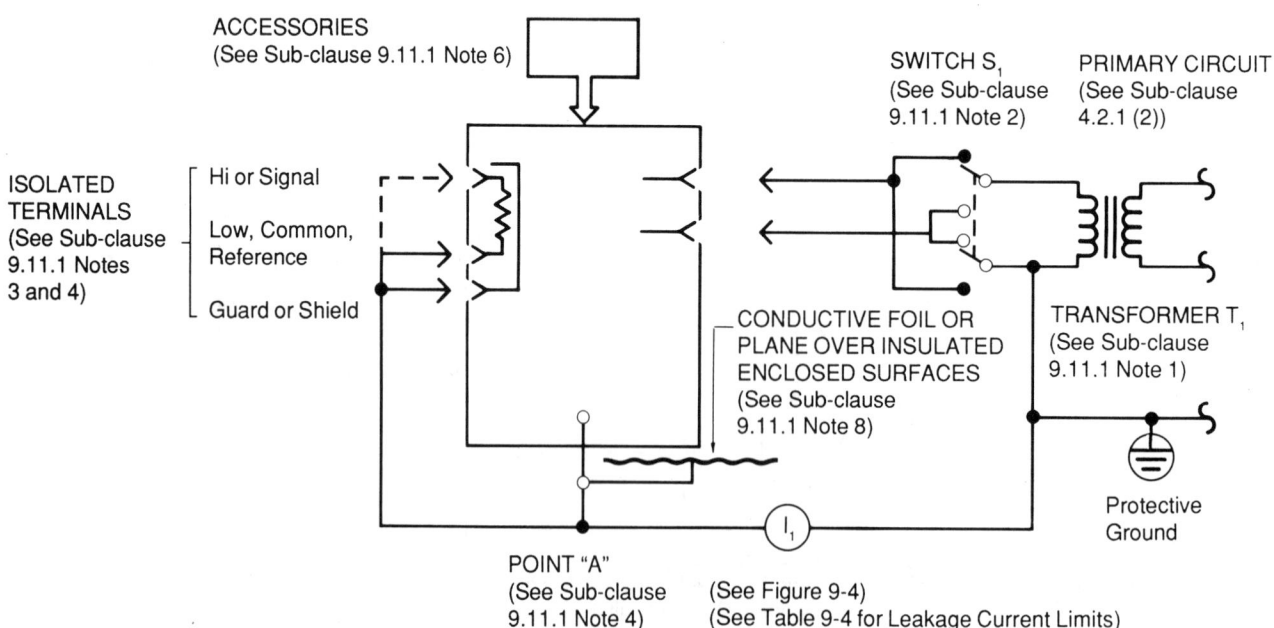

Figure 9-10. Leakage Current Measurement Circuit for Equipment Employing Protective Insulation -- Leakage Current due to Primary Circuit

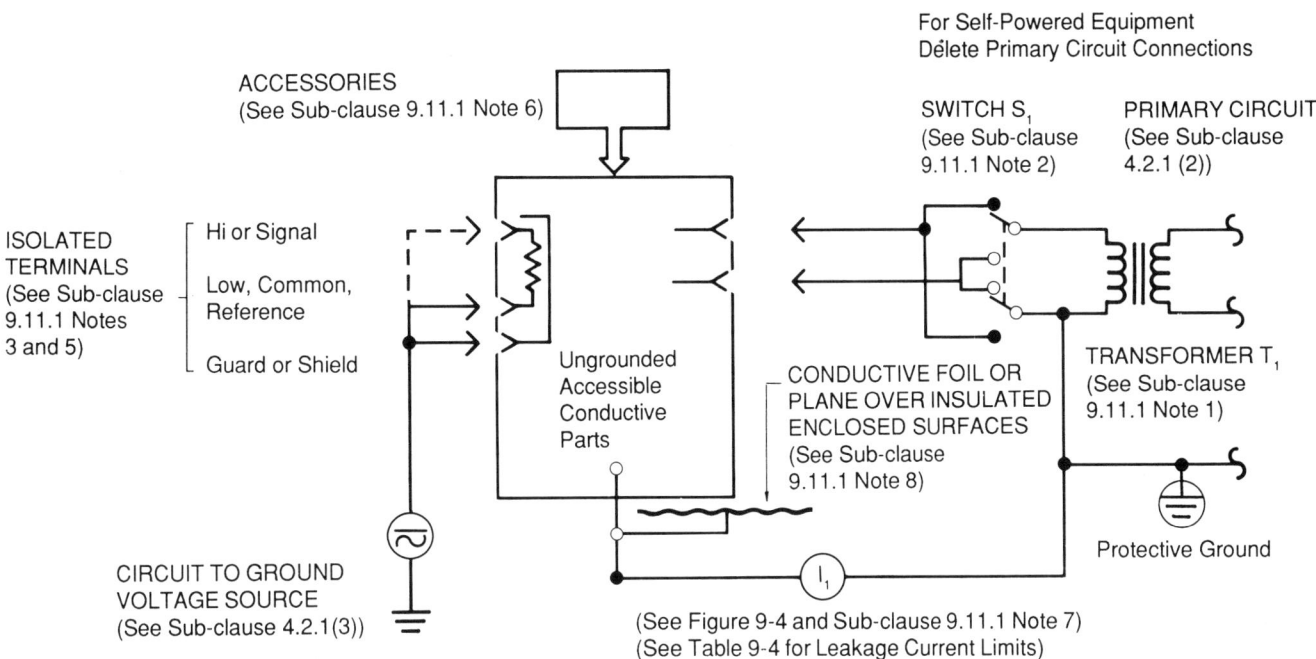

**Figure 9-11. Leakage Current Measurement Circuit for Self-Powered Equipment and Equipment Employing Protective Insulation --
Primary Circuit plus Isolation Voltage**

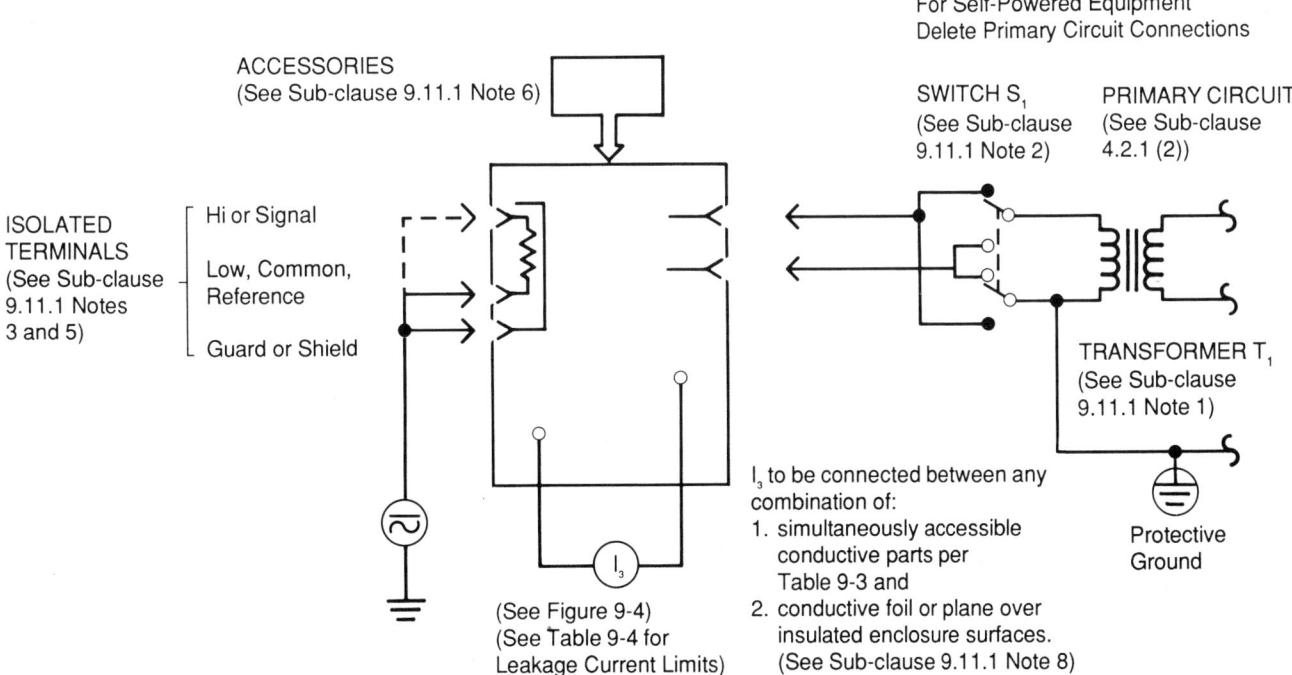

**Figure 9-12. Leakage Current Measurement Circuit for Self-Powered Equipment and Equipment Employing Protective Insulation --
Leakage Current due to Primary Circuit plus Isolation Voltage between Simultaneously Accessible Conductive Parts**

additional tests to examine the effects of the breakdown. It would be reasonable to consider such fault conditions as:

(1) a direct short circuit between the points of breakdown, and

(2) a continued application of the test potential to result in repeated breakdown until terminated according to Sub-clause 10.1(1) or Sub-clause 10.2.3.

Other abnormal phenomena which may result from the breakdown may need to be evaluated through application of other fault conditions to determine that protection is maintained according to the intent of this Standard (see Sub-clauses 3.1, 3.2, and 3.3).

Compliance is checked according to this Sub-clause by applying a test voltage as indicated in Table 9-5 between conductors. The test voltage shall be raised gradually and smoothly to its specified value and maintained for 1 minute.

Breakdown is often indicated by an abrupt decrease or nonlinear advance of voltage as the voltage is increased. Similarly, a breakdown is often indicated by an abrupt increase in current. Partial discharge (corona) and similar phenomena are disregarded during application of the test voltage.

9.12.1 Preparation of circuits for dielectric voltage-withstand type tests

For type tests: power-dissipating component parts, electronic devices, and capacitors located between the circuits under test shall be removed or disconnected such that the spacings and insulation rather than such component parts are subjected to the full dielectric voltage-withstand test potential. Switches and other controls, whether accessible or not, shall be set or adjusted such that all conductors intended to be tested are connected to the circuit under test.

9.12.2 Primary circuits

(1) Type test

The insulation and spacings between parts conductively connected to all conductors of the primary circuit, connected together, and

(a) the protective grounding terminal,

(b) the equipment enclosure, or conductive foil wrapped about the enclosure;

(c) terminals, whether operator-accessible or not, for connection of external circuits, not conductively connected to the primary circuit;

(d) other accessible conductive parts; and

(e) other circuits not conductively connected to the primary circuit

shall be type tested.

The parts may be tested separately or connected together.

The test voltage for basic or supplementary insulation and spacings shall be calculated according to Table 9-5, using the maximum rated voltage of the equipment. (See Sub-clause 4.2.1 (2)).

NOTE: *The test voltage for equipment employing a protective insulation system shall be twice the value calculated for basic insulation.*

(2) Routine test

Each piece of equipment shall withstand, without electrical breakdown, as a routine production-line test, the application of a test voltage between the primary circuit and accessible conductive parts.

The test potential shall be 1200 volts rms, at primary circuit frequency or 1700 volts dc, applied for one second. Alternatively, the test voltage shall be 1000 volts rms, at primary circuit frequency, or 1420 volts dc, applied for one minute.

The test potential for equipment employing a protective system insulation shall be twice the values specified in Table 9-5.

The equipment may be in a heated or unheated condition for the test.

The test shall be conducted when the equipment is complete (fully assembled) and with the primary switch in the "ON" position. It is not intended that the equipment be unwired, modified, or disassembled for the test.

Exception: Parts such as snap covers or friction-fit knobs that would interfere with performing the test need not be in place.

Exception: The test may be performed before final assembly if the test represents that for the completed equipment.

If the output of the routine dielectric voltage-withstand test equipment is less than 500 volt-amperes, it shall include a voltmeter or other equivalent indicator in the output circuit to directly indicate the test voltage, and an audible or visual indication of breakdown. In the event of breakdown, manual reset of an external switch is required or an automatic reject of the unit under test is to result.

If the output of the routine dielectric voltage-withstand test equipment is 500 volt-amperes or larger, the test voltage may be indicated by acceptable voltmeter or other equivalent indicator in another circuit, or by a selector switch marked to indicate the test voltage where there is only a single test-voltage output. When marking is used without an indicating voltmeter or other equivalent indicator, the routine dielectric voltage-withstand test equipment shall include a positive means, such as a power-on lamp, to indicate that the manual-reset switch has been reset following a tripout.

Routine dielectric voltage-withstand test equipment other than that according to this Sub-clause may be used if found to accomplish the intended factory control equivalent.

During the test, both sides of the primary circuit of the equipment are to be connected together and to one terminal of the dielectric test equipment; the other dielectric test equipment terminal is to be connected to the accessible conductive parts.

Electromagnetic-suppression capacitors connected to the primary circuit shall not be disconnected during such tests.

9.12.3 Transformers. The insulation and spacings between windings and parts of a transformer conductively connected to the primary circuit, or connected to a power circuit not limited according to Sub-clause 14.2, shall be type tested.

The windings and parts to be tested include:

(1) primary to core, shield, and each secondary.

These windings and parts may be tested separately or connected together.

The test voltage shall be calculated according to Table 9-5 using the maximum rated primary-circuit voltage.

The test voltage for equipment employing protective insulation shall be twice the value calculated according to Table 9-5 using the maximum rated primary-circuit voltage

(2) Each secondary to core, shield, each other secondary and primary.

These windings and parts may be tested separately or connected together.

The test voltage for each secondary shall be calculated according to Table 9-5 using its maximum rated voltage or its maximum rated isolation voltage, whichever is greater.

The test voltage for each secondary for equipment employing protective insulation shall be twice the value calculated according to Table 9-5 using its maximum rated voltage or its maximum rated isolation voltage, whichever is greater.

Table 9-5. Test Voltage for Dielectric Voltage-Withstand Type Tests

\	Voltage between conductors (U) [a]	\	Test voltage [b, e]	\
dc	ac$_{rms}$	ac$_{peak}$ or mixed ac$_{peak}$ + dc	ac$_{peak}$ [c] or dc [d]	ac$_{rms}$ [c]
0-60.0	0 to 30	0 to 42.4	707	500
	above 30		2.828 U$_{rms}$ + 1414	2U$_{rms}$ + 1000
		above 42.4	2U$_{peak}$ + 1414	1.414U$_{peak}$ + 1000
above 60.0			2U$_{dc}$ + 1414	1.414U$_{dc}$ + 1000

Notes:

(a) Voltage measured according to Sub-clause 14.1.2

(b) The test voltage specified as dc, ac$_{rms}$ and ac$_{peak}$ are considered equivalent for dielectric voltage-withstand test purposes. Either ac or dc voltages may be used for dielectric voltage-withstand tests.

(c) The ac test voltage waveform shall be an essentially sinusoidal voltage with a frequencey of 45-65 hertz.

(d) The test dc voltage shall have no more than 3 percent peak-to-peak ripple.

(e) The test voltage for equipment employing a protective insulation system shall be twice the values specified.

9.12.4 Other circuits (except primary and measuring circuits)

Dielectric voltage-withstand tests need be performed only in those other circuits where it can be demonstrated that a breakdown of insulation or spacings between conductors would result in electric shock or fire.

(1) Where spacings do not equal or exceed those according to Table 9-1, the insulation and spacings between each circuit determined to be live according to Sub-clause 9.2 or not power-limited according to Sub-clause 14.2 and

- *(a) the protective grounding terminal,*
- *(b) the equipment enclosure, or conductive foil wrapped about the equipment,*
- *(c) other accessible conductive parts, and*
- *(d) each other circuit*

shall be type tested.

The parts may be tested separately or connected together.

The test voltage for circuits not power limited shall be calculated according to Table 9-5 using the voltage according to Sub-clause 14.1.2.

The test voltage for live circuits of equipment employing protective insulation shall be twice the value calculated according to Table 9-5 using the voltage according to Sub-clause 14.1.2.

(2) Where spacings do not equal or exceed those according to Table 9-1, the insulation and spacings between parts of opposite polarity within each circuit not power-limited according to Sub-clause 14.2 shall be type tested.

The test voltage shall be calculated according to Table 9-5 using the voltage according to Sub-clause 14.1.2.

9.12.5 Measuring circuits

Dielectric voltage-withstand tests need be performed only in those measuring circuits where it can be demonstrated that a breakdown of insulation or spacings between conductors would result in electric shock or fire.

(1) Where spacings are less than those according to Table 9-1, the insulation and spacings between the circuitry and parts conductively connected to each ungrounded measuring terminal rated for connection to circuits exceeding extra-low voltage and not power-limited according to Clause 14 and

- *(a) the protective grounding terminal,*
- *(b) the equipment enclosure, or conductive foil wrapped about the equipment,*
- *(c) other accessible conductive parts,*
- *(d) the primary circuit, and*
- *(e) each other circuit*

shall be type tested.

The parts may be tested separately or connected together.

The test voltage shall be calculated according to Table 9-5 using the maximum rated circuit-to-ground voltage for the terminal.

(2) For equipment employing protective insulation where protective spacings are less than those according to Table 9-1, the insulation and spacings between an isolated measuring terminal or terminal set rated for connection to a live circuit and

- *(a) accessible conductive parts,*
- *(b) the enclosure (or foil wrapped about the enclosure),*
- *(c) each other isolated circuit, and*
- *(d) the primary circuit*

shall be type tested. The test voltage shall be twice that calculated according to Table 9-5 using the maximum rated circuit-to-ground voltage for the terminal or terminal set.

The parts may be tested separately or connected together.

(3) Where spacings are less than those according to Table 9-1, the insulation and spacings between voltage measuring terminals rated for connection to circuits exceeding extra-low voltage or not power-limited according to Clause 14 shall be type tested.

The test voltage shall be two times the maximum between-the-terminals voltage rating, but not less than 500 volts rms for circuits other than those designed for measurements on branch circuits. For circuits designed for measurement on branch circuits, the test voltage shall be calculated according to Table 9-5 using the maximum rated voltage. Whether the test voltage is dc, ac rms, or ac peak is determined according to the nature of the between-the-terminals rating(s).

Alternatively, when the test voltage is ac, and where it is desirable to conduct the test with all circuit components in place (to limit the power delivered to the circuits of the voltage-measuring terminals under test by the dielectric voltage-withstand test equipment), the test voltage shall consist of single half-cycle pulses separated in time. The pulses shall be sinusoidal, starting alternatively at 0 and, at a later time, at 180°. See Figure 9-13. The test potential shall be applied at least 3600 full cycles.

The peak value of the pulse shall be the peak value of the test voltage calculated above.

The test voltage may be limited to the breakdown voltage of a rated transient suppressive device, or a combination of such devices, if

(1) the transient suppressive device(s) is (are) rated for the purpose; and,

(2) the transient suppressive device(s) shunt the measuring circuit being protected; and

(3) the current through the measuring circuit and the transient suppressive device(s) is limited (such as by an overcurrent protector, fixed impedance, etc.) to the current rating of the transient suppressive device(s).

9.12.6 High Voltage Circuits

In addition to the dielectric voltage-withstand tests according to Sub-clause 9.12.4, the insulation and spacings between the circuitry and parts of a high-voltage source over 2500 volts peak shall be type tested by operating at the potential necessary so that the source output is 1.25 U + 1750 where U is the high-voltage source output voltage measured according to Sub-clause 14.1.2 without electrical breakdown that results in risk of fire.

The high-voltage circuit dielectric voltage-withstand test does not apply to measuring circuits.

10. Testing under Fault Conditions

When the equipment is operated under fault conditions according to this Clause:

(1) accessible parts according to Sub-clause 9.1 shall not render an electric shock according to Sub-clause 9.2;

(2) there shall be no spread of fire according to Sub-clause 10.2.1; and

(3) no other condition shall result in infringement of protection according to this standard.

Equipment having features not contemplated in the test procedures according to this Clause may be tested as necessary to meet the intent of this Clause.

Examination of the equipment and its circuit diagram will generally show the fault conditions that may result in infringement according to this standard. Circuits that are live according to Sub-clause 9.2 or not power-limited according to Sub-clause 14.2 should be examined to identify those parts to which faults are applied.

10.1 Equipment preparation. During fault conditions:

(1) the equipment shall be operated until further change as a result of the fault is not likely;

In most cases, continuous operation for four hours demonstrates that further change because of the fault is unlikely.

(2) multiple simultaneous faults shall not be applied (see Sub-clause 4.3);

(3) the equipment supply circuit shall be connected to a rated supply and in series with a nontime-delay fuse of the

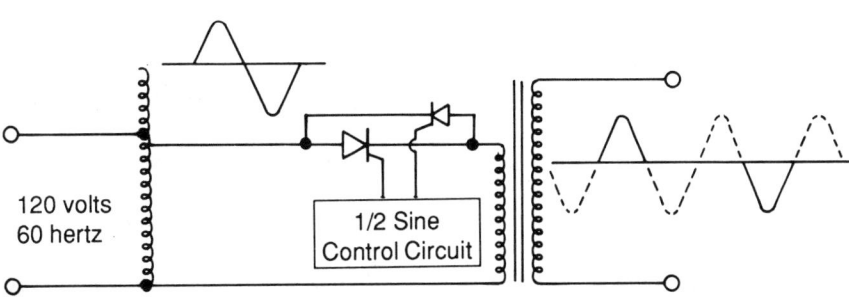

Figure 9-13. Half-Sine Pulse Dielectric Voltage-Withstand Test Circuit

maximum current rating that can be accommodated by the branch-circuit for which the equipment may be connected;

(4) ventilated equipment shall be:

(a) wrapped in a single layer of cheesecloth (or equivalent flame detection means shall be employed); and

(b) the equipment placed on a white tissue paper-covered softwood surface;

(5) each test shall be conducted under the least favorable combination of reference test conditions (see Sub-clause 4.2.1);

(6) the normal protective ground means shall be employed;

(7) parts of equipment that are intended to be removed or opened by the operator shall be removed or opened.

10.2 Compliance tests

Compliance is checked according to this Sub-clause after introduction of faults according to Sub-clause 10.3.

10.2.1 Fire

There shall be no emission of flame or molten metal that causes the cheesecloth or tissue paper to flame, char, or glow.

10.2.2 Flammability of liberated gases.
Reserved for future use.

10.2.3 Electric shock

The equipment is considered to render an electric shock if:

(1) the protective grounding means fails according to Sub-clause 9.5, or

(2) insulation breaks down when tested according to

(a) Sub-clause 9.12.2 (2), (a) - (e); and

(b) Sub-clause 9.12.3 (1); and

(c) when the test involves measuring circuits, Sub-clause 9.12.5 (1); (a) - (c); or

(3) accessible parts comply with Sub-clause 9.2; or

(4) spacings are reduced below those required in Sub-clause 9.8.

10.2.4 Termination of fault condition tests

The opening of:

(1) the supply circuit fuse; or

(2) the equipment protective device; or

(3) any other circuit component,

before any condition according to this Sub-clause occurs, is an acceptable termination of a fault condition.

10.3 Fault conditions

10.3.1 Component Fault Test

An electrolytic capacitor or electronic device with any terminal connected to a circuit not power-limited according to Sub-clause 14.2 shall be:

(1) short-circuited terminal-to-terminal one pair at a time; or

(2) open-circuited, one terminal at a time.

10.3.2 Temperature control device disabling test

The following parts shall be disabled:

(1) an automatic temperature-regulating or -limiting control; or

(2) a thermal protective device; or

(3) a temperature-controlled or other fan; and

(4) any other similar ventilating control.

Exception: Components complying with Clause 13.

10.3.3 Motor test

Motors shall be stopped or prevented from starting.

Exception: Motors rated for blocked conditions.

10.3.4 Continuous operation test

An electromagnetic device such as a motor, relay, solenoid, etc., used for short-term or intermittent operation shall be operated continuously if continuous operation may occur inadvertently.

Exception: An electromagnetic device rated for continuous operation.

10.3.5 Motor capacitor short-circuit test

A capacitor connected to the auxiliary winding circuit of a primary-circuit-connected motor shall be short-circuited.

Exception: A capacitor rated as a "self-healing" type.

10.3.6 Transformer short-circuit tests

A transformer conductively connected to the supply circuit or to a circuit not power-limited, according to Sub-clause 14.2, shall be subjected to a short-circuit test according to this Sub-clause.

The other secondary windings may be connected, or not connected, as may occur in the equipment, unless one condition produces more unfavorable results.

Transformer overload tests are under consideration.

Each secondary winding shall be separately short-circuited.

10.3.7 Substitute power supplies

The operator accessible output(s) shall be short-circuited to both ground and common (if not ground). Other output arrangements shall be short-circuited according to the intent of this Sub-clause.

10.3.8 Battery-circuit reverse-polarization test

Battery-operated circuits and battery-charging circuits shall be tested with the battery installed for reverse polarization if such installation is possible without mechanically damaging, modifying, or altering the equipment or the battery.

Exception: The battery-circuit reverse-polarization test is not required when the battery is not accessible to the operator and is not intended for replacement by the operator.

10.3.9 Battery protection tests

Equipment incorporating a battery(ies) shall be subjected to separate tests that maximize the battery discharge rate and, if a charging circuit is provided, that maximize the battery charge rate according to this Sub-clause.

(1) Discharge test.
The single electrolytic capacitor or electronic device fault which maximizes the battery discharge rate shall be introduced.

(2) Charge test.
The single electrolytic capacitor or electronic device fault which maximizes the battery charge rate shall be introduced.

Compliance shall be according to Sub-clause 10.2 and, in addition if the enclosure does not provide protection according to Sub-clause 8.2, the battery shall not explode.

11. Mechanical Requirements

The equipment shall have mechanical strength and stability and accessible parts shall not have sharp edges.

Compliance is checked by inspection and by performing the tests according to Table 11-1.

11.1 Mechanical strength. The equipment enclosure, or parts of the enclosure, required to be in place to comply with the requirements in this standard for:

(1) protection from electric shock, spread of fire, and personal injury; and,

(2) protection of internal parts and wiring; and,

(3) external-cord and cable-assembly strain relief

shall have mechanical strength.

Equipment that meets more than one of the equipment-type definitions according to Clause 2 shall be subjected to the applicable tests specified for each of the relevant equipment types.

For example: equipment that is classed as being both hand-held and bench top.

Exception: Impact and pressure tests need not be performed on enclosure and enclosure parts that do not provide such protection.

Impact and pressure tests need not be performed on cast metal enclosures according to Table 11-2 and sheet-metal enclosures according to Table 11-3.

Compliance is checked by inspection and by the requirements of Clause 9 after performing the tests according to Sub-clauses 11.1.1 to 11.1.4.

11.1.1 Drop test

(1) Hand-held equipment.

Each of three samples of equipment intended to be hand-held or hand-supported during operation is to be subjected three separate times to the impact that results from the equipment being dropped through a distance of 1.0 meter (or 3 feet) to strike a hardwood surface in the positions most likely to produce adverse results. Each of the impacts for a given sample shall be arranged so that the equipment strikes the hardwood surface in a different position for each of the three drops.

(2) Bench-top or portable equipment.

The equipment, standing in its normal position of use on a smooth, hardwood surface, is tilted about one bottom edge so that the distance between the opposite edge and the test surface is 100 millimeters (3.94 inches) or so that the angle made by the bottom and the test surface is 45°, whichever condition is the less severe.

The equipment is then allowed to fall freely onto the test surface.

The equipment shall be subjected to one drop about each of the four bottom edges.

The equipment shall not be allowed to topple onto an adjacent face instead of falling back as intended.

11.1.2 Impact test

The equipment shall be held firmly against a rigid support and shall be subjected to sets of three blows from a spring-operated impact hammer as shown in Figure 11-1. The hammer shall be applied to any external part that when broken is likely to expose live parts, including handles, levers, knobs, and the like, by pressing the release nose perpendicular to the surface of that part.

A window of an indicating device shall withstand an impact of 0.085 newton-meter (0.753 pound force-inch) from a hollow steel impact sphere 50.8 mm (2 inches) in diameter and an approximate mass of 113.4 grams (4 ounces).

Exception: Parts accessible during operational maintenance; such parts are subjected to the test according to Sub-clause 9.1.1.

Window impact tests need not be conducted where mechanical protection (guarding) is an integral part of the enclosure and the mechanical protection prevents impacting of the window.

11.1.3 Pressure test

A force of 90 newtons (20 pounds) shall be applied from a metal rod 12.7 millimeters (0.50 inch) in diameter, the end of which is rounded to a 12.7 millimeter (0.50 inch) diameter hemisphere. The force shall be applied for one minute to any point on the overall enclosure except the bottom. A force of 65 newtons (15 pounds) shall be applied for one minute to any point on the enclosure bottom.

For portable equipment having less than 4.5 kilograms mass (10 pounds mass), the 90 newton (20 pound) force shall be applied to all surfaces of the enclosure including the bottom.

Table 11-1. Equipment Tests

Sub-Clause Number	Requirement	Hand-Held	Portable and Bench Top	Fixed
11.1.1(1)	Drop	X		
11.1.1(2)	Drop		X	
11.1.2	Impact	X	X	X
11.1.3	Pressure	X	X	X
11.2	Handle Loading	X	X	
11.3	Tip Stability			X
11.4	Sharp Edge	X	X	X

Exception: A window covering the face of a meter indicator need withstand a force of only 35 newtons (8 pounds).

The apparatus consists of three main parts: the body, the striking element and the spring-loaded release nose.

The body comprises the housing, the striking element guide, the release mechanism and all parts rigidly fixed thereto. The mass of this assembly is 1.25 kilograms (2.756 pounds mass).

The striking element comprises the hammer head, the hammer shaft and the cocking knob. The mass of this assembly is 250 grams (8.81 ounces mass).

The hammer head has a hemispherical face of radius 10 millimeters (0.391 inch) and is of polyamide having a Rockwell hardness of R100; it is fixed to the hammer shaft in such a way that the distance from its tip to the plane of the front of the nose, when the striking element is on the point of release, is 20 millimeters (0.781 inch).

The nose has a mass of 60 grams (2.1 ounces mass) and the nose spring is such that it exerts a force when the release jaws are on the point of releasing the striking element of 20 newtons (4.5 pounds force).

The hammer spring is adjusted so that the product of the compression, in millimeters, and the force exerted, in newtons, equals 1000, (in pound-inches, 8.85) the compression being approximately 20 millimeters (0.781 inch). With this adjustment, the impact energy is 0.5 + 0.05 newton meter (4.43 + 0.443 pound force-inch).

The release mechanism springs are adjusted so that they exert just sufficient pressure to keep the release jaws in the closed position.

Ref.: Sub-clause 11.1.2

The apparatus is cocked by pulling the cocking knob back until the release jaws engage with the groove in the hammer function.

The blows are applied by pushing the release nose against the sample in a direction perpendicular to the surface at the point to be tested. The pressure is slowly increased so that the nose moves back until it is in contact with the release

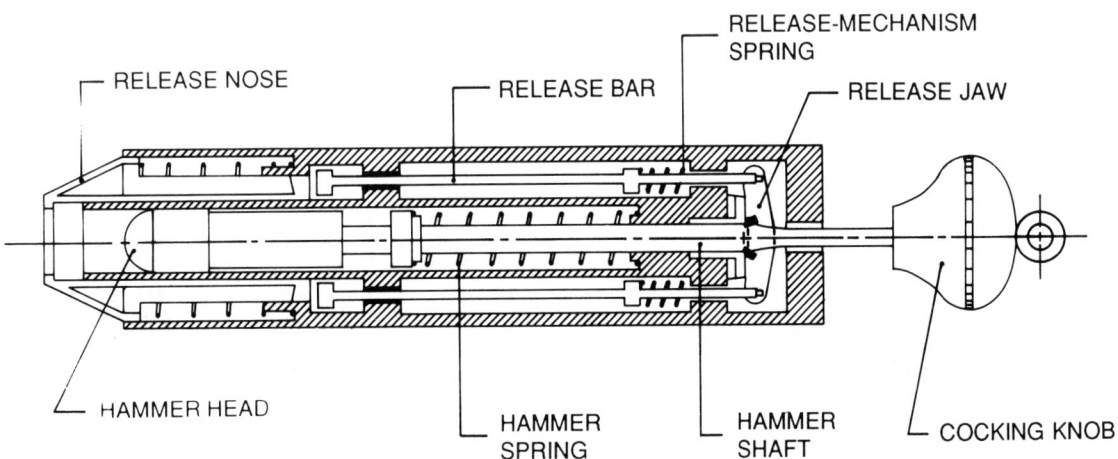

Figure 11-1. Spring-Operated Impact Hammer

Table 11-2. Thickness of Cast Metal Enclosures

	Minimum Average Thickness			
	Unreinforced Flat Surfaces		Curved, Ribbed, or Otherwise Reinforced	
	Millimeters	Inches	Millimeters	Inches
Cast metal enclosures	3.2	1/8	2.4	3/32
Malleable-iron enclosures	2.4	3/32	1.6	1/16
Die-cast metal enclosures	2.0	5/64	1.2	3/64

Table 11-3. Thickness of Sheet Metal Enclosures

Maximum Dimensions of Enclosure		Minimum Average Thickness of Sheet Metal[a]					
		Steel				Copper, Brass or Aluminum	
		Without Supporting Frame		With Supporting Frame or Equivalent Reinforcing			
Length or Width	Area	Zinc-Coated	Uncoated	Zinc-Coated	Uncoated	Without Supporting Frame	With frame or Equivalent Reinforcing
76.2 (3)	39 (6)[b]	0.58 (0.023)	0.51 (0.020)	0.58 (0.023)	0.51 (0.020)	0.58 (0.023)	0.58 (0.023)
203 (8)	232 (36)	0.74 (0.029)	0.66 (0.026)	0.58 (0.023)	0.51 (0.020)	0.91 (0.036)	0.74 (0.029)
305 (12)	581 (90)	0.86 (0.034)	0.81 (0.032)	0.58 (0.023)	0.51 (0.020)	1.14 (0.045)	0.74 (0.029)
457 (18)	871 (135)	1.14 (0.045)	1.07 (0.042)	0.86 (0.034)	0.81 (0.032)	1.47 (0.058)	1.14 (0.045)
610 (24)	2323 (360)	1.42 (0.056)	1.35 (0.053)	1.14 (0.045)	1.07 (0.042)	1.91 (0.075)	1.47 (0.058)
1219 (48)	7742 (1200)	1.78 (0.070)	1.70 (0.067)	1.42 (0.056)	1.35 (0.053)	2.41 (0.095)	1.91 (0.075)
1524 (60)	9678 (1500)	2.46 (0.097)	2.36 (0.093)	1.42 (0.056)	1.35 (0.053)	3.10 (0.122)	1.91 (0.075)
More than 1524 (60)	More than 9678 (1500)	3.20 (0.126)	3.12 (0.123)	1.42 (0.056)	1.35 (0.053)	3.89 (0.153)	1.91 (0.075)

[a] Dimensions and areas are noted in mm and cm^2 with corresponding inches and $inches^2$ in parentheses ( ).

[b] Volume of enclosure not more than 196 cm^2 (12 $inches^3$).

bars, which then move to operate the release mechanism and allow the hammer to strike.

11.2 Handle loading test. Carrying handles or grips supplied with equipment shall be capable of withstanding a force of four times the weight of the equipment.

Compliance shall be checked by subjecting the equipment handle or grip to the weight of the equipment plus a test weight of three times the weight of the equipment applied uniformly over a 75 millimeter (3 inch) width at the center of the handle without clamping. The test weight shall be applied from zero so that the full load is attained in 5-10 seconds and maintained for one minute. The handle or grip shall not break loose from the equipment nor shall there be any permanent distortion, cracking or other evidence of the handle or grip not being able to perform its intended function.

If more than one handle is furnished on the equipment, the test weight shall be distributed to each handle proportionately to the measured load on each handle in the normal carrying position of the equipment.

If the equipment is furnished with more than one handle but can be carried using only one handle, each handle shall be tested separately for its capability of sustaining the total test weight.

11.3 Tip stability test. Equipment having a weight of 11 kilograms (24 pounds) or more shall not tip over when placed at the center of an inclined plane that makes an angle of 10 degrees with the horizontal and then turned to the position most likely to cause tip-over.

Compliance shall be checked by placing the equipment in its intended position with all doors, drawers, covers, and other openable and sliding parts in the position that causes the least stability and facing the position on the 10 degree incline most likely to cause it to overturn.

Feet and other means of support may be blocked to prevent the equipment from sliding during the test.

Drawers and other openable and sliding parts are not loaded for this test (only extended), unless manufacturer-supplied equipment is intended to be in or on them or a load rating is assigned.

11.4 Sharp edges. An accessible edge, projection, or corner of an enclosure, opening, frame, guard, handle, or the like shall be smooth and well-rounded, and shall not cause a cut-type injury during normal use of the equipment.

Compliance is checked by inspection.

12. Protection from Fire

The equipment shall be constructed to provide protection against fire.

Compliance is checked according to this Clause and Clause 10.

12.1 Enclosure materials. Polymeric enclosure materials shall be rated for flammability and ignition according to Table 12-1.

Compliance is checked by inspection.

12.2 Polymeric insulating materials. Insulating materials shall be rated for flammability and ignition according to Table 12.1.

Compliance is checked by inspection.

12.3 Printed wiring boards. Printed-wiring boards shall be rated for flammability according to Table 12-1.

Compliance is checked by inspection.

13. Component Requirements and Applications

Components shall be employed and rated according to applicable component requirements and according to this Clause.

13.1 Motors. A supply-circuit connected motor shall be rated for the voltage and load to which it is connected.

Compliance is checked by inspection and measurement of the motor supply current while the motor is under load.

13.2 Supply-circuit and safety switches. A supply-circuit or safety switch shall have a rating for the voltage to which it is connected and not less than the load it controls.

Compliance is checked by inspection and by measurement of the current controlled by the switch under reference test conditions.

13.3 Protective devices

13.3.1 Overcurrent-protection and thermal-limiting devices shall be rated according to applicable device requirements.

Table 12-1. Polymeric Material Requirement[a]

Application	Flammability Classification[b,e]	Properties, Resistance to Ignition From Hot Wire[c,e]	High Ampere Arc[d,e]
In contact with parts conductively connected to the primary circuit			
(A) Enclosures and Insulating Materials	HB	30 sec	60 arcs
	V-2	30 sec	30 arcs
	V-1	15 sec	30 arcs
	V-0	10 sec	15 arcs
	5-V	10 sec	15 arcs
(B) Printed Wiring Boards	V-1, V-0	----	----
(C) Connectors	V-2, V-1, V-0	----	----
In contact with parts conductively connected to power circuits not limited[g] (other than primary circuits and measuring or testing circuits)			
(D) Enclosures and Insulating Materials[f]	HB	15 sec	30 arcs
	V-2, V-1		
	V-0, 5-V		
(E) Printed Wiring Boards[f]	V-1, V-0	----	----
(F) Connectors[f]	V-2, V-1, V-0	----	----
(G) Enclosures materials (required parts) used in applications other than those covered in lines (A) and (D)	HB	15 sec	30 arcs
	V-2, V-1		
	V-0, 5-V		
(H) Insulating materials, printed wiring boards and connectors used in measuring or testing circuits or in applications other than those covered in lines (B), (C), (E) and (F)	----	----	----

NOTES TO TABLE 12-1
General Requirements

a   These requirements do not apply to the internal insulating systems of components or where component requirements exist.

b   The flammability classifications V-0, V-1, V-2, 5-V and HB are to be determined by the test. Refer to ANSI/UL-94 "Tests for Flammability" for testing.

For enclosures, a material classified using 3.2 millimeter (0.125 inch) thick bar specimens can be accepted in lesser thicknesses in the end product. For other parts, a material classified using 1.6 millimeter (0.063 inch) thick bar specimens can be accepted in lesser thicknesses in the end product.

c   Hot-wire resistance to ignition-hot-wire ignition performance is expressed as the number of seconds needed to ignite standard specimens that are wrapped with resistance wire that dissipates a specified level of electrical energy. Bar samples are to be used for this test. For enclosures, a material classified using 3.2 millimeter (0.125 inch) thick bar specimens can be accepted in less thicknesses in the end product. For other parts, a material classified using 1.6 millimeter (0.063 inch) thick bar specimens can be accepted in lesser thicknesses in the end product.

d   High-ampere arc resistance to ignition: High-ampere arc ignition performance is expressed as the number of arc rupture exposures (standardized as to electrode type and shape and electrical circuits) that are necessary to ignite a material when they are applied at a standard rate on the surface of the material. Bar samples are to be used for this test. For enclosures, a material classified using 3.2 millimeter (0.125) thick bar specimens can be accepted in lesser thicknesses in the end product. For other parts, a material classified using 1.6 millimeter (0.063 inch) thick bar specimens can be accepted in lesser thicknesses in the end product.

e   For an assembly, samples for the test parameters according to Table 12-1 can consist of the assembly and can be tested as finished parts, or test samples can be cut from finished parts. In the case of small parts that might be consumed before the test is completed, large samples of the same material can be tested provided they represent the same or lesser thickness than the part in question. None of the larger samples is to be entirely consumed. Samples that consist of an assembly or a section thereof that are not flat stock samples are to be positioned in what is considered to be the worst position in the application.

f   The requirements that materials, used in applications covered by lines (D) through (F) of Table 12-1, possess certain flammability and ignition ratings do not apply to small parts. For the purpose of these requirements, a small part is one:

(1) in which the maximum dimension does not exceed 30 millimeters (1.18 inches), and

(2) in which the volume does not exceed 8000 cubic millimeters (0.488 cubic inches), and

(3) that is located where it could not act as a bridge between a source of arcing or ignition and other ignitable parts.

g   Power not limited according to Sub-clause 14.2.

A thermal-limiting device shall be rated for the load it controls.

Compliance is checked by inspection and performing the heating test according to Clause 7.

13.3.2 A protective, limiting, regulating, or similar component, or circuit, that operates as a result of testing according to this standard and is relied upon for compliance to this standard shall be rated for such overcurrent or thermal-limiting protection.

For example: Fuses, thermostats, regulators, circuit breakers, fusible resistors, limiting controls, regulating controls and like components and devices.

Compliance is checked by inspection.

13.4 Resistors. A resistor used for fixed impedance limiting according to Sub-clause 14.2 shall be rated for the power dissipated under short circuit or maximum transferable power conditions, whichever is greater.

Compliance is checked by inspection.

13.5 Supply circuit connectors. Attachment plugs for connection of the equipment to the supply circuit and receptacles for providing supply-circuit energy to other equipment shall be rated according to applicable requirements.

Compliance is checked by inspection.

13.5.1 Rating compatibility. The configuration and rating of the attachment plug, power cord, receptacle, if used, and appliance coupler, if used, shall be compatible with the equipment supply-circuit ratings.

Compliance is checked by inspection.

13.5.2 Attachment plug rating. The attachment plug shall be rated at a current not less than 100 percent of the equipment rated current, and rated at a voltage appropriate for the equipment rated voltage(s).

Compliance is checked by inspection.

13.6 Printed wiring. Printed wiring shall have a rated bonding strength according to Sub-clause 9.4(5).

Compliance is checked by inspection.

13.7 Wire insulation and insulating tubing. The basic insulation on each wire shall be rated for the maximum voltage to which the wire is connected.

The basic insulation on all wires shall be rated for the maximum temperature which the insulation attains under reference test conditions.

Insulating tubing, sleeving, and tape shall be rated for the maximum voltage against which it insulates and for the temperature it attains under reference test conditions.

Compliance is checked by inspection.

13.8 Connectors. Connectors that are:

(1) in live circuits according to Sub-clause 9.2 and are accessible to the operator; or

(2) in live circuits according to Sub-clause 9.2 which in the event of a fault could make accessible parts live; or

(3) in a circuit not power-limited according to Sub-clause 14.2,

shall be rated according to applicable requirements.

Compliance is checked by inspection.

13.9 Receptacles. Supply-circuit receptacles incorporated on nonpolarized equipment shall be nonpolarized.

Supply-circuit receptacles incorporated on ungrounded equipment shall be a nongrounding type.

Compliance is checked by inspection.

13.10 Component applications

13.10.1 Supply circuit switches. Equipment shall incorporate a switch in the supply circuit or a means for disconnecting the equipment from the supply circuit shall be provided.

Exception: A supply-circuit switch is not required:

(1) for fixed equipment, where the branch circuit provides means of disconnection; or
(2) for auxiliary and accessory equipment where continuous operation is required; or
    For example: battery recharging circuits, timing clocks, crystal temperature ovens, etc.
(3) where exterior processes are dependent on continuous operation of the equipment; or
(4) where equipment is connected to the source of supply by flexible cords having either an attachment plug or appliance coupler, the attachment plug or appliance coupler receptacle may serve as the disconnect.

Compliance is checked by inspection.

Supply-circuit switches shall disconnect all parts of the equipment from all poles of the supply circuit.

Exception: Fuses and electromagnetic interference suppression components and their interconnection means within the equipment need not be disconnected.

A supply-circuit switch in polarized equipment need only interrupt the ungrounded supply-circuit conductor.

Supply-circuit switches in equipment employing protective grounding shall be located according to Sub-clause 9.5.7.

Compliance is checked by inspection.

13.10.2 Safety switches. Safety switches employed in a supply circuit shall comply with the requirements according to Sub-clause 13.10.1.

A safety switch employed in any other circuit shall interrupt the supply to all accessible parts which could render an electric shock.

Compliance is checked by inspection.

13.10.3 Interlock mechanism. An interlock, when used, shall render parts that become accessible to the operator free from electric shock, emanation, excessive temperature, or driven movement (that could result in physical injury).

Exception: An interlock mechanism need not be provided if there are not parts of the equipment accessible to the operator which could render an electric shock during normal use or operational maintenance according to Sub-clause 9.1.

Compliance is checked by inspection.

13.10.4 Batteries. Equipment operating from a battery circuit not power-limited (see Sub-clause 14.2) shall employ overcurrent protection.

Such protection shall be in the conductor opposite the circuit common conductor, or in the common conductor if equivalent protection is provided, and located in or adjacent to the battery connecting means.

Equipment construction shall protect the battery terminals so they are not subject to inadvertent short circuiting while installed or during removal or installation of the battery.

Compliance is checked by inspection.

13.10.5 Transformers. A transformer conductively connected to the supply circuit or to a circuit exceeding extra-low voltage or not power-limited according to Clause 14 shall be:

(1) housed within its own enclosure; or

(2) housed within the equipment enclosure.

Compliance is checked by inspection.

13.10.6 Overcurrent protection and thermal-limiting devices. Overcurrent protection and thermal-limiting devices employed in equipment shall not operate under reference test conditions.

Compliance is checked by inspection with the equipment operating under reference test conditions.

Exception: Rated supply voltage is to be used.

13.10.7 Voltage setting switches. The equipment shall be so constructed that changing of the setting from one voltage to another or from one nature of supply to another cannot occur unintentionally.

Compliance is checked by inspection.

14. Extra-Low Voltage and Power-Limited Circuits

To determine those circuit locations where specific tests are applied according to Clauses 9 and 10, and to determine the requirements for materials in contact with current-carrying parts according to Clause 12, equipment circuits are divided into groups according to magnitudes of voltage and power.

14.1 Extra-low voltage circuit. An extra-low voltage circuit is one in which the measured voltage is not more than:

(1) 30 volts rms (42.4 volts peak), or

(2) 60 volts dc, or

(3) 24.8 volts dc interrupted at a rate of 10-200 hertz.

Compliance is checked by measuring the voltage according to Sub-clause 14.1.2.

14.1.2 The voltage measurement shall be made between the conductor or part in question and its associated circuit common with:

(1) the equipment connected to a supply circuit of maximum rated voltage, and

(2) the equipment operating under reference test conditions, and

(3) any combination of electron tubes and fuses removed, and

(4) any operator-accessible connectors, or similar parts, either connected or disconnected.

14.2 Power-limited circuit

14.2.1 A power-limited circuit is one in which the power capable of being delivered to an external resistor connected in parallel to the circuit load is not more than 150 watts, measured after one minute.

Compliance is checked by circuit analysis or by measuring the power according to Sub-clause 14.2.2.

14.2.2 The power measurement shall be made between the conductor or part in question and its associated circuit common with:

(1) the equipment connected to a supply circuit of maximum rated voltage; and

(2) the equipment operating under reference test conditions; and

(3) an external resistive load and wattmeter connected in parallel with the circuit load; and

(4) the circuit load shall be adjusted for minimum circuit power consumption under normal use.

The combination of the resistive load and wattmeter is connected according to Figure 14-1. The first power measurement is made at the power source; subsequent measurements, if necessary, are made at additional points between the power source and the load(s) until the power-limited point is identified.

Before energizing the equipment, the external resistor is adjusted for maximum resistance. The equipment is then energized and the resistor adjusted for maximum power or 150 watts dissipation as indicated by the wattmeter.

The supply side of the circuit under evaluation, at the point of measurement, is considered to be connected to a source of limited power when:

(1) the supply to the circuit under evaluation cannot deliver 150 watts to the external resistor; or

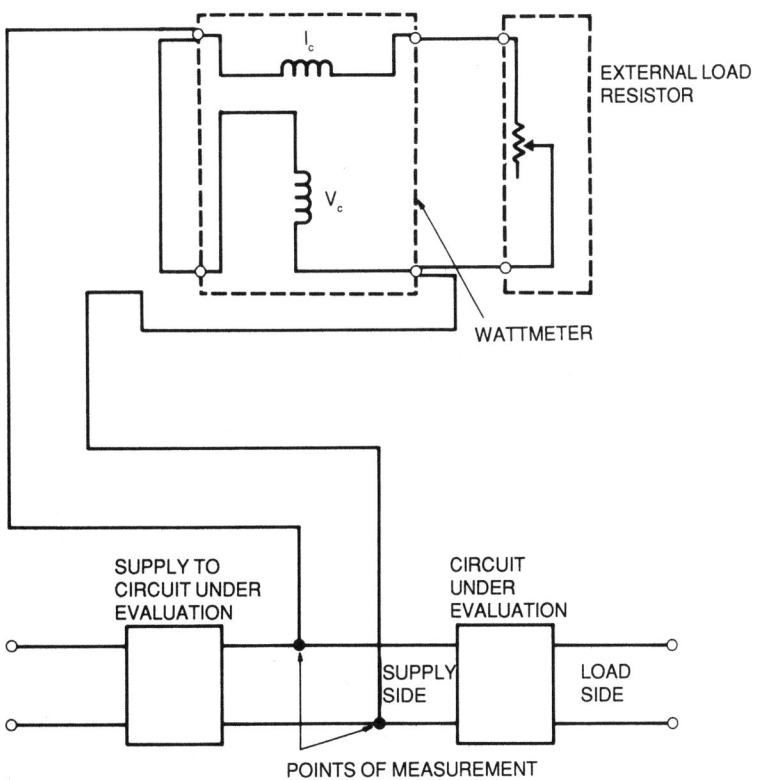

Figure 14-1. Connection of Wattmeter for Determining a Power-Limited Circuit

(2) a fixed impedance or a regulating network limits the delivery of power to the external resistor from the supply to the circuit under evaluation prior to reaching 150 watts; or

(3) an overload protector or circuit component opens to interrupt the delivery of power to the external resistor from the supply to the circuit under evaluation prior to reaching 150 watts.

15. Terminal Devices

15.1 Supply-circuit connectors. Attachment plugs and appliance couplers for connection of the equipment to the supply circuit and receptacles for providing supply-circuit energy to other equipment shall be rated according to applicable requirements.

Compliance is checked by inspection.

15.1.1 Rating compatibility. The configuration and rating of the attachment plug, power-supply cord, receptacle, (if used), and appliance coupler (if used) shall be compatible with the equipment supply-circuit ratings as set by the manufacturer.

Compliance is checked by inspection.

15.1.2 Attachment-plug rating. The attachment plug shall be rated at a current not less than 100 percent of the equipment rated current, and rated at a voltage applicable to the equipment rated voltage or voltage range.

Compliance is checked by inspection.

15.1.3 Receptacles. Supply-circuit receptacles incorporated on nonpolarized equipment shall be of a nonpolarized type.
Supply-circuit receptacles incorporated on polarized equipment shall be of a polarized type.
Supply-circuit receptacles incorporated on ungrounded equipment shall be of a nongrounding type.
Supply-circuit receptacles incorporated on earth-grounded equipment shall be of an earth-grounding type.

Compliance is checked by inspection.

16. External Cords

External cords and cable assemblies, including interconnecting cables, optional cables, and assemblies used in circuits which are live according to Sub-clause 9.2, or not power-limited according to Sub-clause 14.2, shall be rated and tested according to this Clause.

Exception: Measuring and testing leads.

External cords and cable assemblies may be permanently attached to the equipment or may be separable.

16.1 Attachment. External cords and cables permanently attached to the equipment shall withstand a force applied to the cord or cable according to this Sub-clause without transmitting any strain to the internal connections.

Compliance is checked by applying a steady pull of 156 newtons (35 pounds).

The force may be obtained from hanging a 16 kilogram (35 pound) weight on the external cord.

With the chassis in the cabinet in the normal manner, the force shall be applied from any angle possible. Three samples shall be tested. The minimum average time of holding shall be 15 seconds; and the minimum holding time shall be 5 seconds.

Alternatively, one sample can be tested if the 156 newton (35 pound) force is applied for a period of 1 minute.

Following the test:

(1) The insulation or covering on the cord or cable assembly shall not be cut or torn; and

(2) the bushing shall not slide through the hole in the chassis or enclosure; and

(3) the cord shall not slide in the bushing.

16.1.1 The attachment of an external cord shall be such that:

(1) the conductors are prevented from twisting; and

(2) the insulation is protected from abrasion, piercing and sharp bends and from temperatures and voltages exceeding its ratings, if pushed back inside the equipment enclosure.

Compliance is checked by inspection.

16.1.2 Where an external cord or cable assembly emerges from the enclosure and may be subjected to strain or motion, the equipment shall employ an insulating bushing attached to the enclosure.

Exception: When a sheathed polyvinyl chloride flexible cord rated not less than the designation SJT (227IEC53) is employed.

Compliance is checked by inspection.

16.2 Flexible power-supply cords

16.2.1 Type and rating. A flexible power-supply cord shall be of a type rated for the particular application.

(1) A flexible cord subjected to continuous abuse by virtue of its length, enabling it to be located on a floor, shall not be rated less than sheathed polyvinyl chloride flexible cord, designation SJT (227IEC53).

(2) A flexible cord partially protected from continuous abuse, by virtue of its short length not enabling it to be located on a floor, shall not be rated less than non-sheathed polyvinyl chloride flexible cord, designation SPT-1 (227IEC42) or sheathed polyvinyl chloride flexible cord, designation SVT (227IEC52).

Compliance is checked by inspection.

A flexible cord less than 2.5 meters (approximately 8 feet, 3 inches) in length is considered partially protected and not likely to be subjected to continuous abuse by being located on a floor.

A flexible cord equal to or exceeding 2.5 meters (approximately 8 feet, 3 inches) in length is considered subjected to continuous abuse by being located on a floor.

The length of a flexible cord is measured from the face of the attachment plug to the point at which the cord emerges from the equipment.

An equivalent or heavier type of flexible cord may be used. Refer to IEC Publication 541 for equivalency data with particular attention to other types identified therein for those countries that are still in the process of harmonizing with IEC Publications 227 and 245.

16.2.2 Conductor rating. The power-supply cord conductors shall have a cross section such that, if a short circuit occurs at the equipment end of the cord, the supply-circuit overcurrent protective device(s) operates before the cord-insulation temperature rating is exceeded.

The flexible cord shall have an ampacity (assigned current-carrying capacity) not less than the current rating of the equipment.

Compliance is checked by inspection.

16.3 Interconnection cables. Conductors of an external cable assembly shall have a cross section such that, if a fault occurs in the equipment, the temperature rise of the cable assembly shall not exceed the cable assembly temperature rating.

17. Equipment Instructions

Instructions for operating and servicing the equipment shall be according to this Clause and to ISA-S82.02 and ISA-S82.03.

17.1 Warnings and cautions. The equipment instructions shall include an explanation of those warnings and cautions which appear on the equipment and associated with operator and equipment protection within the intent of this standard.

Compliance is checked by inspection.

The following appendix is not a part of ISA 82.01, but is included to facilitate its use.

APPENDIX

General Information Part 1

Clause 1.

This standard, ISA-S82.01, is patterned from IEC Publications 348, 414, and 536 and the work of IEC Subcommittee 66E.

1.1.2 Generally, equipment which contains only circuits which are power-limited according to Sub-clause 14.2 and not live according to Sub-clause 9.2 are subject to Clauses 5, 6, 8, 12, and 17 only.

For equipment which contains such circuits together with other circuits not limited, only the circuits not limited are subject to all requirements. In such equipment, the insulations and spacing between limited and not limited circuits are subject to the applicable requirements.

Clause 6.

Relative to Sub-clauses 6.2, 6.3, and 6.4 below, the following currently available information is provided. The levels or measuring methods are under consideration for acceptability. When the questions and concerns for acceptability of the levels and measuring methods are fully resolved, the Sub-clauses will be completed. Until that time, these Sub-clauses are reserved.

6.2 Non-ionizing radiation

The power density of microwave radiation emitted shall not exceed 10 watts per square meter (1 milliwatt per square centimeter) (8.36 watts per square yard) (6.45 milliwatts per square inch) at any point 5 centimeters (2 inches) or more from the surface of the equipment or surface of the part of the equipment.

This requirement applies to spurious radiation at frequencies between 10 megahertz and 100 gigahertz. It does not apply to parts of the equipment where microwave radiation propagates intentionally, such as at microwave output ports.

Compliance is checked by measuring power density.

Power density is best determined by measuring separately: the Mean Squared Electric Field Strength (E) [4,000 volts squared per square meter (63 volts per meter) maximum], and the Mean Squared Magnetic Field Strength (H) [0.025 ampere squared per square meter (0.16 ampere per meter) maximum].

6.3 Ozone

During normal operation, the equipment shall not produce ozone exceeding:

(1) an average concentration of 0.1 part per million, and

(2) a maximum concentration of 0.3 part per million, measured during an 8 hour period.

Compliance is checked by measuring the ozone in a closed room of approximately 28.3 cubic meters (1000 cubic feet), at reference test conditions. Eight measurements shall be made at 1 hour intervals.

6.4 Ultrasonic pressure

During normal operation, the equipment shall not produce a sound-pressure level exceeding 110 decibels above a reference level of 20 micronewtons per square meter (2×10^{-4} microbar or 2 pascals).

Compliance is checked by measuring the sound pressure level over a frequency range of 20-100 kilohertz.

Test equipment may be designed in accordance with the American National Standard Specification for Sound Level Meters, ANSI S1.4-1983, using a microphone that covers the frequency range of 20-100 kilohertz. The measuring equipment shall have an accuracy of not less than + 1 decibel.

Clause 9.

Electric Shock: The effects of electric current and capacitative discharge on the body.

Electric shock results from contact with a live electrical circuit. Studies have been made to establish the level of voltage, current, and stored charge to determine the meaning of live, and to establish that level at which effects on the body occur. The following discussion and illustrations are taken from recent reports.

It is generally accepted that current through the body determines the reaction. The value of current depends on the body impedance and the voltage across it. Voltage below 60 Vdc, 30 Vac, and 42 V peak is not considered dangerous under ordinary conditions because body impedance limits the current. Body impedance varies over a wide range and consists of the series combination of skin impedance at the points of contact and the internal resistance of the body.

A resistance of 500 ohms is commonly used as the internal body resistance between major extremities and 1500 or 2000 ohms is accepted as the resistance between the normal perspiring hands of a worker for reaction current measurements. Skin impedance depends on the condition of the skin and the areas of contact. It may be as high as 500,000 ohms resistive but under moist conditions can be as low as 1,000 to 2,000 ohms. The skin also acts as a capacitor dielectric between the internal body resistance and the external contact. Capacitance through the skin, in parallel with the skin resistance, depends on contact area and skin thickness. Values from 0.01 μF to 0.2 μF are typical for finger contact and hand gripping, respectively. Breaks in the skin at both points of contact or high frequency can reduce skin impedance to near zero. Under this condition the only limitation to current flow is the internal resistance of the body. Leakage current measurements are normally made with a resistor-capacitor network which compensates for the effects of frequency on body reaction.

A series of illustrations is provided to further explain the effects of electric current and capacitative discharge on the human body. See Figures A-1 through A-4.

Effects of 60 Hz electric current on the body

1. Effect of electric current applied directly to the heart (micro shock).

Currents as little as 50 μA caused by voltages as small as 5 mV can cause fibrillation of the human heart.

2. Effects of 60-Hz electric current on an average human through the body trunk (macroshock).

Current Intensity, mA (1-second contact)	Physiological Effect
1	Threshold of perception
5	Accepted as maximum harmless current intensity
10-20	"Let-go" current before sustained muscular contraction
50	Pain, possible fainting, exhaustion, mechanical injury; heart and respiratory functions continue
100-300	Ventricular fibrillation will start, but respiratory center remains intact
6,000	Sustained myocardial contraction, followed by normal heart rhythm. Temporary respiratory paralysis. Burns if current density is high.

Bibliography for Electrical Shock

1. Safe current Limits for Electromedical Apparatus. ANSI/AAMI SCL 12/78.

2. Hewlett Packard. Application Note AN718, "Patient Safety."

3. Friedlander, Gordon D. "Electricity in hospitals: elimination of lethal hazards." IEEE *Spectrum*. 8(9): 40-51; 1971 September.

4. Dalziel, Charles F. "Electric shock hazard." IEEE *Spectrum*. 9 (2): 40-50; 1972 February.

5. "Effects of current passing through the human body." IEC Report Part I, 1984; Part II, 1986.

6. General Requirements for Electronic Equipment. Mil Std. 454K. 1986 February 15.

7. IEC Standard 348, 1971, Clause 9 (not available from ANSI)

8. Graphical symbols for use in equipment. Index, survey and compilation of the single sheets. IEC Std. 417, 1973.

9. General requirements: Safety of household and similar electrical appliances. IEC Std. 335.1, 1970.

10. Underwriters Laboratories. "What constitutes a shock hazard from capacitor discharge." Subject 855, 1985.

12. Safety requirements for electronic measuring apparatus. IEC Std. 348, 1978.

13. Enclosures for Electrical Equipment (1000 volts max). ANSI/NEMA-250, 1985.

14. Classification of degrees of protection provided by enclosure. IEC Publication 529, 1976.

All IEC Standards are available in U.S. from ANSI:

ANSI
IEC Publications
1430 Broadway
New York, N.Y. 10018

9.8   Spacings

The upper limits specified for circuit voltages given in Table 9.1 take into consideration the application of tolerances that may be applied to normal (rated) voltages (i.e., Sub-clause 4.2.12)

IEC Subcommittee 28A "Insulation Coordination for Low-Voltage Equipment" has been charged by the IEC Committee of Action to develop guides for insulation coordination in low-voltage systems and equipment. These guides are intended to be used by IEC technical committees in an effort to harmonize spacing requirements throughout the IEC publications for low-voltage equipment.

The IEC has approved for publication IEC Report 664-"Insulation co-ordination within low-voltage systems including clearance and creepage distances for equipment." Additionally, IEC Report 664A-First supplement to Publication 664, has been approved for publication.

Individual IEC Technical Committees have begun to consider the recommendations of SC28A for the dimensioning of clearance and creepage distances of low-voltage equipment. It is expected that future publications of these committees will reflect these recommendations and establish a harmonization of spacing requirements for all low-voltage equipment.

In the future, ISA-S82.01, ISA-S82.02, and ISA-S82.03 will need to consider the IEC work to establish international harmonization.

9.11   Leakage current

Rationale for Not Connecting High or Signal Terminals during Leakage Current Measurements

This document includes a major departure from former requirements, namely, that high or signal terminals need not be connected to maximize the leakage current measurement. Refer to ISA-S82.01 Sub-clause 9.11.1, Measurement Note 3.

For the purposes of explanation, leakage current is assumed to be the sum of those currents due to the impedances (usually distributed capacitances and insulation resistances) between the ungrounded supply circuit conductor and other conductive parts where a potential difference exists between those parts and the ungrounded supply circuit conductor. These impedances are identified in the diagram of leakage current paths as Z_1 and Z_2 (see Figure A-5). The leakage current paths, I_1 and I_2, resulting from the impedances Z_1 and Z_2, respectively, are shown by the two broken lines.

In earlier drafts of ISA-S82.01-1986, R_1 was assumed large compared to Z_2 and was therefore considered to be the element controlling the magnitude of current I_2. Because R_1 is between the terminals, any external resistance or impedance, R_2 connected between the terminals and in parallel with R_1 would reduce the impedance between Z_2 and the chassis. The worst-case situation would occur when R_1 is zero, hence the requirement that high or signal terminals be shorted at low or ground when measuring leakage current.

The ISA-S82.01-1988 revision contains a new method for leakage current measurements whereby the high or signal terminals may remain unconnected. During the in-depth analysis of this measurement technique, the sources of leakage currents were carefully defined and considered. The assumption that R_1 is large compared to Z_2 was challenged.

It is hypothesized that, in most equipment construction, Z_2 would be very large compared to R_1. This is a reasonable hypothesis because Z_2 would be a result of the proximity of supply-circuit wiring to the "high" or signal terminal. Most manufacturers would keep the supply highly segregated from any high-impedance secondary circuits, especially signal input circuits.

Some calculations were made to confirm the hypothesis: Assuming a 120 volt supply, the minimum impedance to ground for a leakage current of 0.5 milliampere is 240 kilohms.

If 90% of the total leakage current is via Z_1 and 10% is via Z_2, then the value of $Z_2 + R_1$ must be at least 2.4 megohms. If R_1 is one megohm, then Z_2 is 1.4 megohms. Furthermore, the potential across R_1 is 50 volts! Ob-

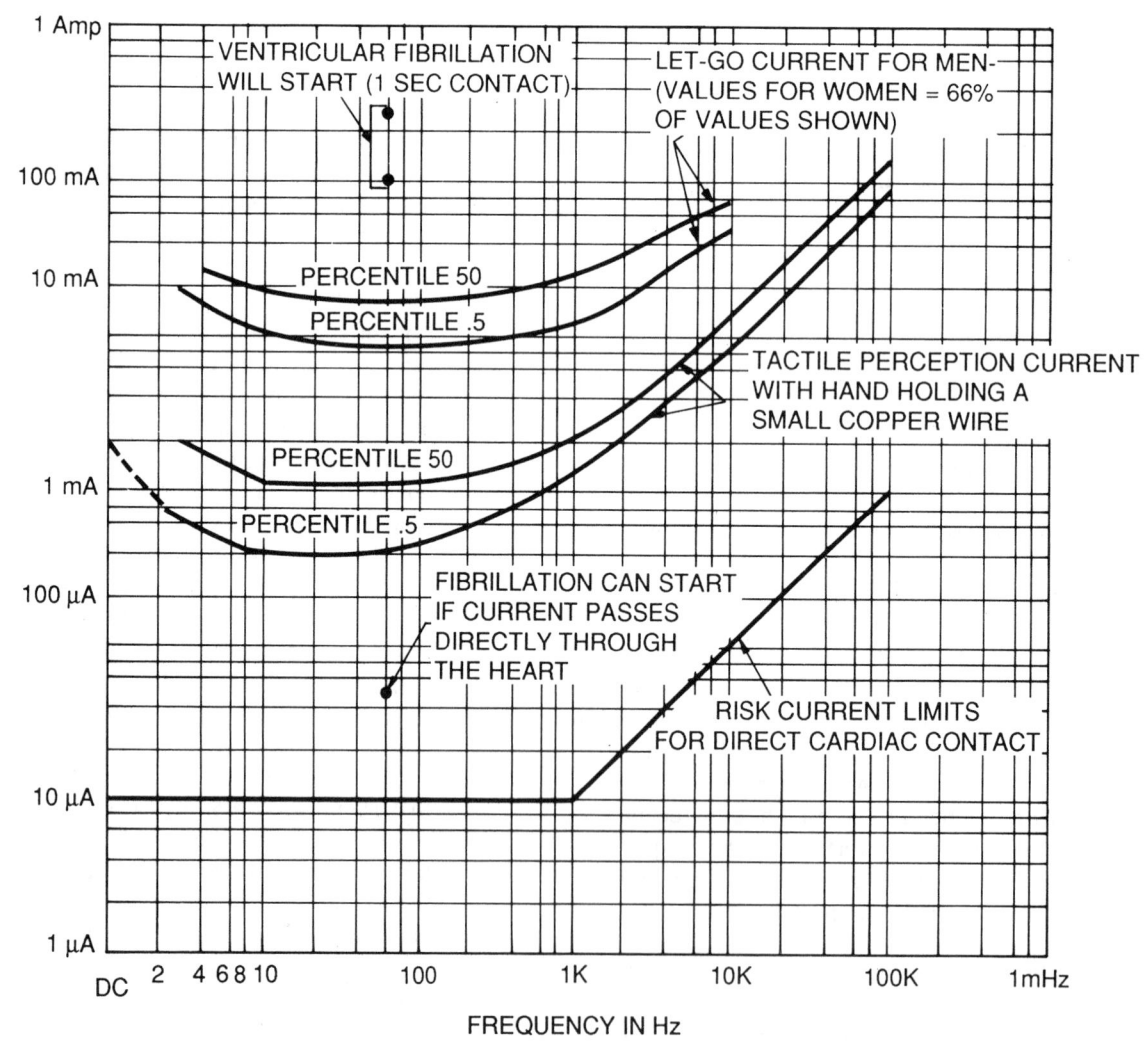

Figure A-1. Effects on the Body of Electric Current vs. Frequency

viously, the measuring equipment would then measure its own leakage and be useless for any other measurement.

Furthermore, it is hypothesized that the leakage current I_2 must not interfere with the equipment's function. For a 3-1/2-digit digital multimeter rated 0.2 volts full scale, the smallest measurable voltage is about 0.1 millivolt. With an input impedance of 10 megohms, the leakage current I_2 must be less than 12 picoamperes, and the leakage impedance greater than 12,000 gigohms. Similar calculations can be made for the other input impedances with the same conclusion: The value of Z_2 must be very much greater than R_1.

It can also be shown that, as resistance between the terminals goes lower, it is a smaller and smaller part of the total impedance ($Z_2 + R_1$) and has negligible effect upon the leakage current.

Therefore, the high or signal terminal need not be connected to maximize leakage current.

Since this standard does not require an entire enclosure of insulating material to be wrapped in foil for leakage current measurements as is required by IEC Publication 348 and previous editions of C39.5, an explanation is hereby presented. It was found that the increased leakage currents, due to the new requirement of this standard which includes rated circuit-to-ground terminal voltage in the measurement, could be unreasonably high, especially at frequencies between 1 kHz and 100 kHz, when an enclosure of insulating material is entirely wrapped in foil.

An example to illustrate the condition is an ac voltmeter with isolated input terminals in which the operator has made the signal error of reversing the input leads (see Figure A-6).

The internal shields, when driven by the 1000 V ac, 10 kHz source, couple a leakage current through the shield to foil capacitance. In order to keep the leakage current below 0.5 mA, the capacitance must be less then 112 pF. A value of 224 pF would be required for a 500-volt signal under the same conditions. Because of this new requirement, some equipment must be designed with lower capacitance from circuits to foil through the enclosure. It was also realized that it is unrealistic to pre-

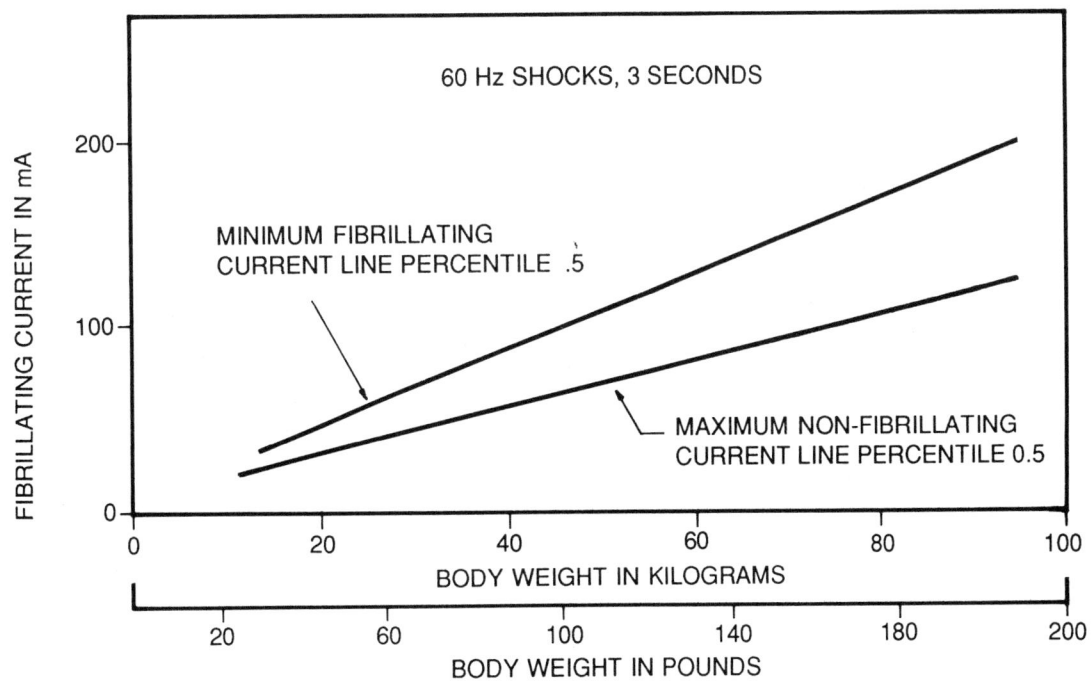

Figure A-2. Fibrillation Currents vs. Body Weight

sume that an enclosure would be completely enclosed by a conductive foil in normal use; therefore, a more realistic approach was taken.

The 200 by 200 millimeter (8 by 8 inch) foil represents the maximum area that can be covered by an operator with both hands.

The two 100 by 200 (4 by 8 inch) millimeter pieces of foil represent the fact that an operator can grasp the equipment with both hands at different locations.

The plane surface includes the conditions whereby the equipment may be sitting on another ungrounded or floating conductive equipment enclosure or another equipment may be resting on top of the equipment producing the leakage current.

The standard instructs the tester to position the foil or plane surface by trial and error in the location that produces the most leakage current. It should also be noted that other safety investigations such as dielectric voltage-withstand tests may still require the enclosure to be completely wrapped in foil.

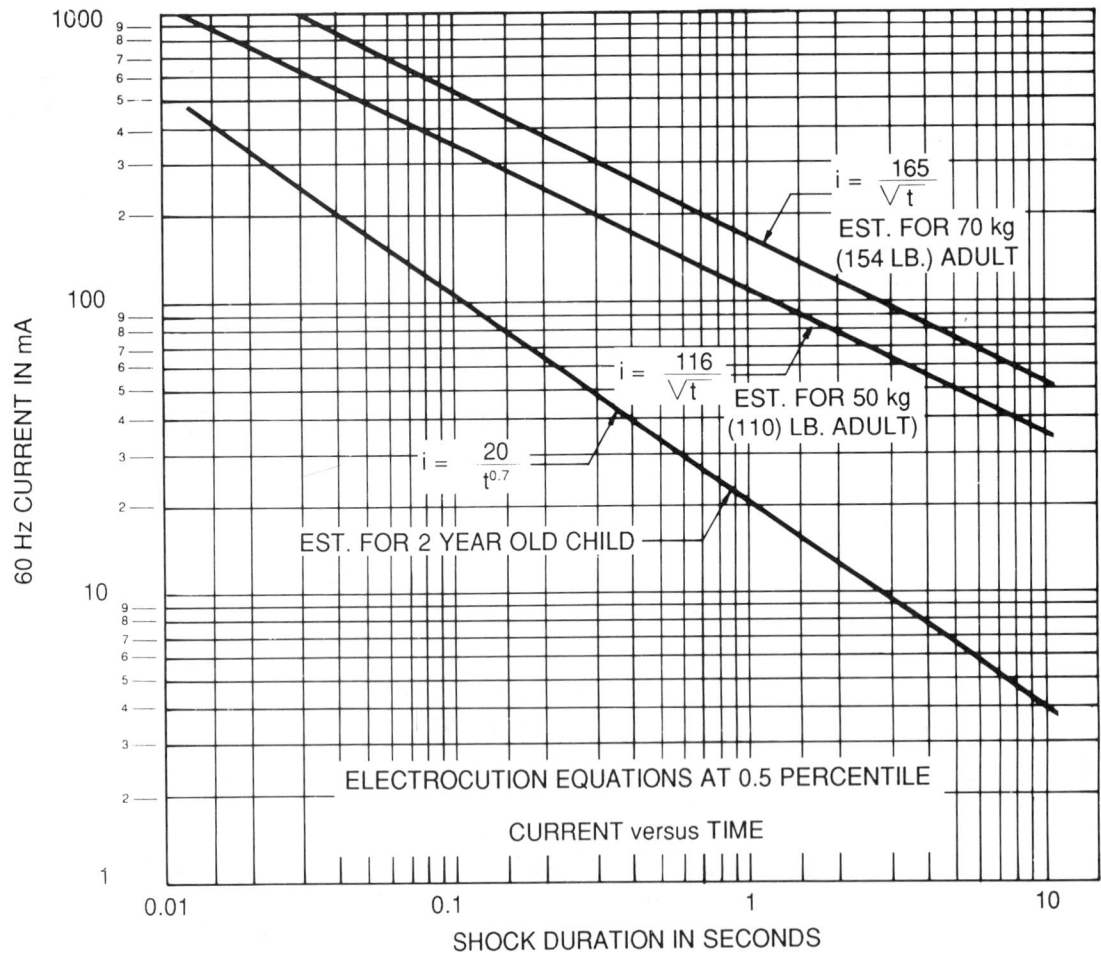

Figure A-3. Electrocution Equations, Current vs. Time

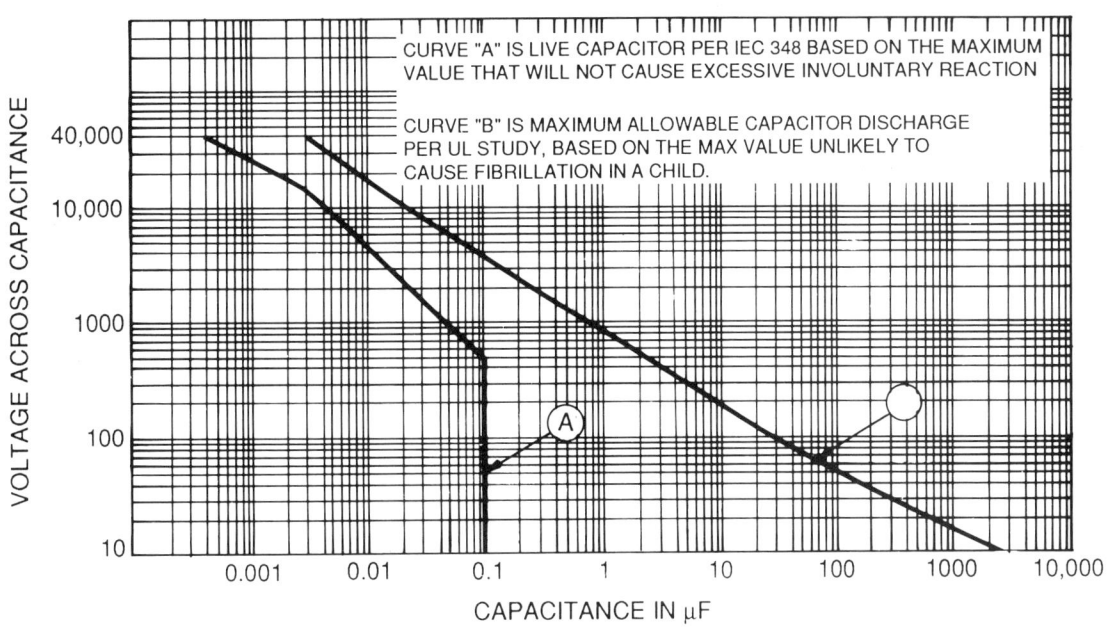

Figure A-4. Shock Hazard from Capacitor Discharge

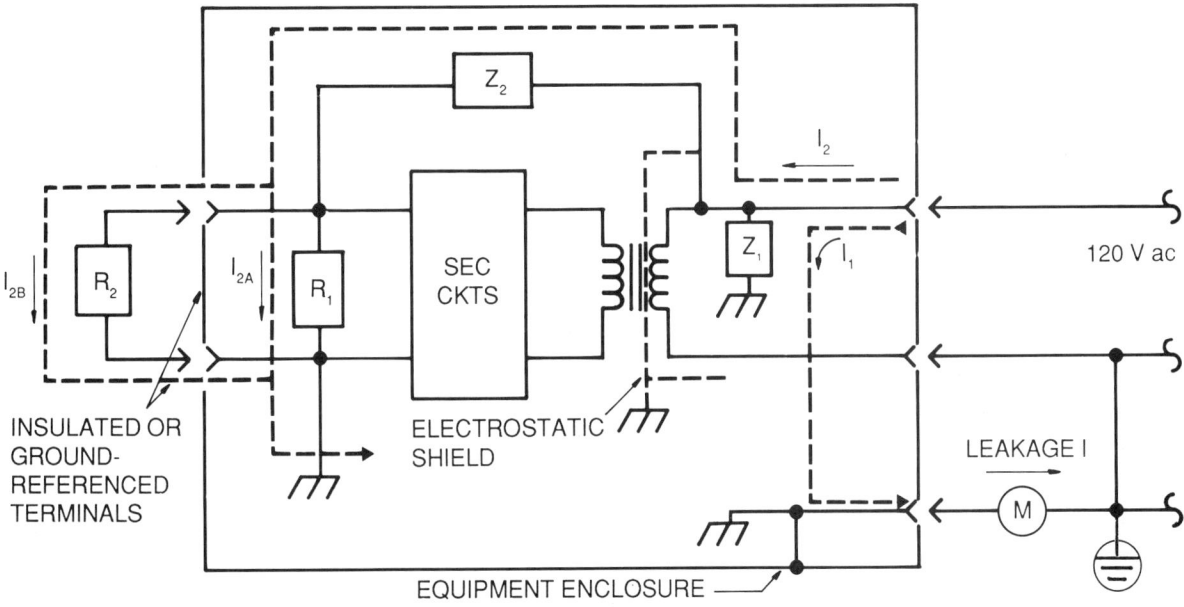

Figure A-5. Use of Conductive Foil During Leakage Current Measurement

Rationale for the Use of Conductive Foil or Plane during Leakage Current Measurements.

For leakage current measurements, ISA-S82.01 sub-clause 9.11.1, Note 8 states the following:

If the equipment is wholly or partially encased in insulation, accessible parts shall include:

(a) a 200 by 200 millimeter (8 by 8 inch) square of conductive foil wrapped anywhere about the equipment;

(b) two 100 by 200 millimeter (4 by 8 inch) conductive foil pieces wrapped about the equipment at any two locations not more than 1.8 meters (6 feet) apart; or

(c) a conductive plane surface upon which any part of the equipment may rest.

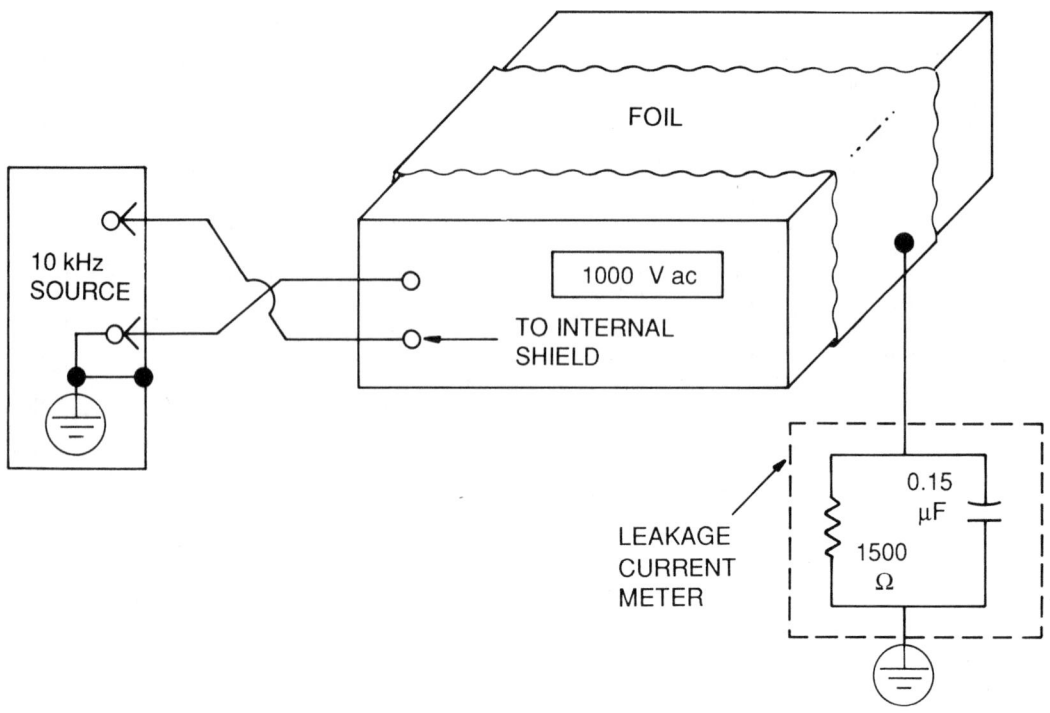

Figure A-6. Use of Conductive Foil During Leakage Current Measurements

9.12 Dielectric-Voltage-Withstand-Test

Circuits	Circuit Power					
	Power-Limited, Sub-Clause 14.2			Not Power-Limited, Sub-Clause 14.2		
	To Opposite Polarity	To Accessible Parts	Other Circuits	To Opposite Polarity	To Accessible Parts	Other Circuits
Circuits Not Live, Sub-Clause 9.2	No	No	No	Yes	No	Yes
Live Circuits, Sub-Clause 9.2	No	Yes	Yes	Yes	Yes	Yes

The table above is offered to clarify the applicability of dielectric voltage-withstand tests.

S82.01 Safety Standard for Electrical and Electronic Test, Measuring, Controlling, and Related Equipment
General Requirements

RECOMMENDED INTERNATIONAL SYMBOLS

SYMBOL		SYMBOL DEFINITION
IEC417	⏚	PROTECTIVE GROUNDING TERMINAL: A terminal which must be connected to earth ground prior to making any other connections to the equipment.
GREEN COLOR American National Standard National Electrical Code ANSI C1-1975	⬡	SUPPLY CIRCUIT PROTECTIVE GROUNDING TERMINAL: (1) A green-colored terminal screw with a hexagonal head; (2) a green-colored pressure wire connector.
IEC417	⏚	GROUNDED TERMINAL: A grounded terminal which, as far as the operator is concerned, is already grounded by means of an earth grounding system.
IEC417	∼	A terminal to which or from which an alternating (sine wave) current or voltage may be applied or supplied.
IEC417	⎓	A terminal to which or from which a direct current or voltage may be applied or supplied.
IEC417	∼	A terminal to which or from which an alternating and direct current or voltage may be applied or supplied.
IEC417	⚠	EXPLANATION: This marking indicates that the operator must refer to an explanation in the operating instructions.
RED COLOR IEC417	⚡	HIGH VOLTAGE TERMINAL: A terminal at which a voltage with respect to another terminal or part exists or may be adjusted to 1000 volts or more.
IEC417	I	ON: Power: connection to the principal supply circuit.
IEC417	O	OFF: Power: disconnection from the principal supply circuit.
IEC335.1	3 ∼	A terminal to which or from which a three-phase alternating (sine-wave) current or voltage may be applied or supplied.
IEC335.1	3N ∼	A terminal to which or from which a three-phase alternating (sine-wave) current or voltage and neutral conductor may be applied or supplied.

ANSI/ISA-S82.02-1988
Approved February 18, 1988
(Partial Revision and Redesignation of ANSI C39.5-1974)

American National Standard

Safety Standard for Electrical and Electronic Test, Measuring, Controlling and Related Equipment

ELECTRICAL AND ELECTRONIC TEST AND MEASURING EQUIPMENT

Instrument Society of America

Instrument Society of America

ISBN 1-55617-101-3

ISA-S82.02 Safety Standard for Electrical and Electronic Test, Measuring, Controlling, and Related Equipment

Copyright ©1988 by the Instrument Society of America. All rights reserved. Printed in the United States of America. No part of this publication may be reproduced, stored in a retrieval system, or transmitted, in any form or by any means (electronic, mechanical, photocopying, recording, or otherwise), without the prior written permission of the publisher.

INSTRUMENT SOCIETY OF AMERICA
67 Alexander Drive
P.O. Box 12277
Research Triangle Park, North Carolina 27709

PREFACE

This preface is included for information purposes and is not part of ISA-S82.02.

This standard has been prepared as part of the service of the Instrument Society of America (ISA) toward a goal of uniformity in the field of instrumentation. To be of real value, this document should not be static, but should be subject to periodic review. Toward this end, the Society welcomes all comments and criticisms, and asks that they be addressed to the Secretary, Standards and Practices Board, Instrument Society of America, 67 Alexander Drive, P.O. Box 12277, Research Triangle Park, NC 27709, telephone (919) 549-8411.

The ISA Standards and Practices Department is aware of the growing need for attention to the metric system of units in general, and the International System of Units (SI) in particular, in the preparation of instrumentation standards. The Department is further aware of the benefits to U.S.A. users of ISA standards of incorporating suitable references to the SI (and the metric system) in their business and professional dealings with other countries. Toward this end, this Department will endeavor to introduce SI-acceptable metric units in all new and revised standards to the greatest extent possible. The Metric Practice Guide, which has been published by the Institute of Electrical and Electronics Engineers as ANSI/IEEE Std. 268-1982, and future revisions will be the reference guide for definitions, symbols, abbreviations, and conversion factors.

It is the policy of the Instrument Society of America to encourage and welcome the participation of all concerned individuals and interests in the development of ISA standards. Participation in the ISA standards-making process by an individual in no way constitutes endorsement by the employer of that individual, of the Instrument Society of America, or of any of the standards that ISA develops.

The information contained in the preface, footnotes, and appendices is included for information only and is not a part of the standard.

The development of this standard dates back to 1958. The need for the standard arose from requests that recording instruments meet the requirement of the American National Standard, the National Electrical Code. At that time there was also the beginning of the need for examination of measuring instruments by testing laboratories. The initial version of this standard was approved in 1964 after several years of careful development, review, and trial use.

Work began on safety requirements for electrical measuring instruments at the international level at about the same time. While consideration of specific safety requirements for measuring apparatus was begun in 1965, resulting first editions of IEC Publications 348 and 414 became available in 1971 and 1973, respectively.

American National Standard C39.5-1964 was reaffirmed in 1969 and revised in 1974. In 1982, responsibility for C39.5-1964 was transferred to the Instrument Society of America and the document was renumbered ISA-S82.01.

This 1988 edition reflects four major considerations:

- The scope has been expanded to include more equipment types. The scope and organization now readily allow future inclusion of more types of related equipment.

- A major goal was harmonization--to the extent possible--with worldwide standards, especially North American and IEC Standards. Not only were the IEC clause and multi-part formats adopted, but also many of the IEC requirements; many requirements were further defined and developed from the work of IEC Subcommittee 66E.

- Digital and microelectronic innovations have narrowed the one-time separation between digital and analog technologies and prompted commonality of safety requirements.

- Significant technical contributions were made by industry organizations and independent certifying laboratories.

Instrument Society of America

The following are significant features incorporated in this ongoing edition:

- Constructional requirements.

- Measurement of conformance to requirements is through the use of specific compliance statements. This allows the user of the standard to determine product compliance and to better understand the nature and intent of the requirements.

- Expansion of requirements for protection against fire and personal injury.

The following people served as members of ISA Subcommittee SP82.02, which prepared this standard:

NAME	COMPANY
Bruce J. Feikle, Chairman	OTIS
William Calder, Manager Director	The Foxboro Company
Abraham Abramowitz	Retired
David Braudaway	Sandia Laboratories
Joseph E. Devine	IBM Instruments, Inc.
Charles G. Gross	Gen Rad Inc.
Robert G. Harris	Underwriters Laboratories
Walter F. Hart	John Fluke Mfg. Co., Inc.
W.R. Howard	LFE Corporation
William H. Hulse, Jr.	Material Safety Eng. Div.
Donald S. Ironside	Biddle Instruments
A. Walter Jacobson	Retired
Al Kanode	Hewlett-Packard Company
Roger P. Lutfy	Factory Mutual Research
Donald A. Mader	Underwriters Laboratories, Inc.
Ernest C. Magison	Honeywell, Inc.
Richard C. Masek	Bailey Controls Co.
Frank J. McGowan	The Foxboro Company
Richard Nute	Tektronix, Inc.
Phillip A. Painchaud	Retired
William A. Phillips	Leeds & Northrup Company
William E. Rich	Westinghouse Electric Corp.
John N. Stanley	Leeds & Northrup Company
Robert Wallace	Tektronix, Inc.

The following people served as members of ISA Committee SP82 during the review of this standard:

NAME	COMPANY
David Braudaway, Co-Chairman	Sandia Laboratories
Frank McGowan, Co-Chairman	The Foxboro Company
William Calder, Manager Director	The Foxboro Company
Abraham Abramowitz	Retired
Norm Belecki	National Bureau of Standards
Bruce Feikle	OTIS

*Employer at time of participation

Name	Company
Robert G. Harris	Underwriters Laboratories, Inc.
W.R. Howard	LFE Corporation
A. Walter Jacobson	Retired
Donald A. Mader	Underwriters Laboratories, Inc.
Ivano Mazza	Casella Postale No. 11
Hubert O'Neil	Havti
Phillip A. Painchaud	Retired
Peter E. Perkins	Tektronix, Inc. DS 78-531
William E. Rich	Westinghouse Electric Corp.
DeWayne B. Sharp, P.E.	IBM Corporation
Jack V. Stegenga	Yokogawa Corp. of America

This standard was approved for publication by the ISA Standards and Practices Board in January 1988.

NAME	COMPANY
D. E. Rapley, Chairman	Rapley Engineering Services
D. N. Bishop	Chevron U.S.A. Inc.
W. Calder III	The Foxboro Company
N. Conger	Fisher Controls International, Inc.
R. S. Crowder	Ship Star Associates
C. R. Gross	Eagle Technology
H. S. Hopkins	Utility Products of Arizona
R. B. Jones	Dow Chemical Company
R. T. Jones	Philadelphia Electric Company
R. Keller	Consultant
A. P. McCauley	Chagrin Valley Controls, Inc.
E. M. Nesvig	ERDCO Engineering Corporation
C. W. Reimann	Moore Products Company
R. H. Reimer	Allen-Bradley Company
J. Rennie	Factory Mutual Research Corporation
W. C. Weidman	Gilbert/Commonwealth Inc.
K. A. Whitman	Fairleigh Dickinson University
* P. Bliss	Consultant
* B. A. Christensen	Continental Oil Company
* L. N. Combs	Consultant
* R. L. Gailey	Consultant
* T. J. Harrison	Florida State University
* O. P. Lovett, Jr.	Consultant
* E. C. Magison	Honeywell, Inc.
* W. B. Miller	Moore Products Company
* J. W. Mock	Bechtel Western Power Corporation
* G. Platt	Consultant
* J. R. Williams	Stearns Catalytic Corporation

* Director Emeritus

- SPECIFIC CLAUSE TREATMENTS -

The **SCOPE** includes environmental conditions envisioned for the equipment. **DEFINITIONS** have extensive additions to reduce misinterpretations. **MARKINGS** are expanded and clarified by inclusion of internationally recognized symbols. X-radiation considerations are added while several other **EMANATIONS** are identified as "under consideration" due to lack of recognized limits and standardized measurements. The clause for **PROTECTION FROM ELECTRIC SHOCK** deletes requirements for resistance tests, clarifies leakage current measurement procedures, details dieletric tests and includes requirements for measuring terminals and routine tests. **TESTING UNDER FAULT CONDITIONS** is new. **PROTECTION FROM FIRE** is an approach to measuring the likelihood of spread of fire. **EXTRA LOW VOLTAGE AND POWER LIMITED CIRCUITS** defines the circuits which could cause electric shock or the likelihood of fire and for which protection requirements of the standard apply.

Understanding the Appendix material of ISA-S82.01 under Clause 9 is also necessary for this edition.

TABLE OF CONTENTS

Section Title   Page

1. General 9
   1.1 Scope 9
   1.2 Measuring or testing probe assemblies, connectors and terminals 9
2. Definitions 9
   2.1 Probe assembly, measuring or testing 9
3. General Information 10
   3.1 Object 10
   3.2 Compliance 10
4. General Test Requirements 11
   4.1 General 11
5. Marking 11
   5.1 General 11
   5.2 Accessible terminal ratings 11
   5.3 Measuring and testing probes 11
6. Equipment Emanations 12
   6.1 General 12
7. Heating 12
   7.1 General 12
8. Protection from Implosion and Explosion 12
   8.1 General 12
9. Protection from Electrical Shock 12
   9.1 General 12
   9.2 Measuring or testing probe assemblies and connectors 12
10. Testing Under Fault Conditions 14
    10.1 General 14
    10.2 Multi-function and overrange test 14
    10.3 Testing equipment 14
11. Mechanical Requirements 14
    11.1 General 14
    11.2 Measuring and testing probe assemblies 14
12. Protection from Fire 14
    12.1 General 14
13. Component Requirements and Applications 14
    13.1 General 14
    13.2 Range and function switches 14
14. Extra-Low Voltage and Power-Limited Circuits 17
    14.1 General 17
15. Terminal Devices 17
    15.1 General 17
    15.2 Terminal mounting 17
16. External Cords 17
    16.1 General 17
    16.2 Measuring and testing probe assemblies 17
17. Equipment Instructions 18
    17.1 Separation of information 18
    17.2 Operating instructions 18

IN THIS STANDARD

Requirements proper are printed in roman type.

Explanatory matter regarding the requirements is printed in smaller roman type.

Test specifications are printed in italic type.

Explanatory matter regarding the test specifications is printed in smaller italic type.

Further explanatory matter regarding the requirements or specific subject marked by an asterisk can be found in the Appendix of ISA-S82.01.

CAUTION

MANY TESTS REQUIRED BY THIS STANDARD ARE INHERENTLY HAZARDOUS. ADEQUATE SAFEGUARDS FOR PERSONNEL AND PROPERTY SHOULD BE EMPLOYED IN CONDUCTING SUCH TESTS.

1. General

1.1 Scope. This standard applies to electrical, electronic and electromechanical measuring and testing equipment, and to the terminals, connectors, wiring and probes rated for use in the interface between such equipment and

(1) the electrical or physical phenomena being measured, or

(2) that to which electrical or electronic quantities are being supplied for measuring purposes.

See Figure 1-1.

1.2 Measuring or testing probe assemblies, connectors and terminals

1.2.1 Measuring or testing probe assembly. A measuring or testing probe assembly packaged with the equipment and rated for connection to a live circuit according to Sub-clause 9.2 of ISA-S82.01 shall be marked, constructed, and tested according to the applicable requirements of this standard.

1.2.2 Accessories and adapters. This standard also applies to accessories and adapters rated for use with measuring or testing probe assemblies, connectors or terminals where accessories and adapters are separate items which do not themselves perform the desired function but are used in addition to or as a supplement to items according to Sub-clause 1.1.

1.2.3 Items covered. The requirements in this standard apply to measuring or testing probe assemblies, connectors, terminals, wiring and parts of such items that an end-use equipment operator might use and handle to cause

· electrical or electronic measuring equipment to measure or

· electrical or electronic testing equipment to supply electrical or electronic quantities (signals)

as intended.

Measuring or testing probe assemblies, connectors, terminals, wiring, accessories and adapters include, but are not necessarily restricted to:

(1) adapters (such as: binding post-to-banana, binding post-to-BNC, binding post-to-pin tip, BNC-to-banana, BNC-to-UHF, etc.),

(2) binding posts,

(3) connectors (such as: BNC, N, SMA, TNC; UHF, etc.),

(4) jacks and plugs (such as: banana, pin tip; etc.),

(5) probes (such as: clamp-on current, oscilloscope, voltmeter, etc.),

and assemblies of such items.

Portions of probe assemblies such as amplifiers and supply circuits are subject to the applicable requirements of ISA-S82.01, depending upon the design, construction, and features provided.

1.2.4 Constraints. This standard applies to measuring or testing probe assemblies, connectors and terminals, that are rated for measuring or testing:

(1) branch circuits up to 1000 volts or

(2) ac or dc voltages up to 40 kilovolts incorporated within electrical end-product equipment circuits.

1.2.5 Items not covered. This standard does not apply to measuring or testing probe assemblies, connectors or terminals that:

(1) are intended for use with equipment other than that according to the scope of ISA-S82.01,

(2) are rated for use exclusively in extra-low voltage and power-limited applications (see Clause 14 of ISA-S82.01),

(3) are intended primarily for equipment-to-equipment interconnection (for example, IEEE interface bus).

2. Definitions

For the purposes of this standard the definitions of Clause 2 of ISA-S82.01 and the following definitions shall apply.

2.1 Probe Assembly, Measuring or Testing: An assembly consisting of a probe(s), a remote terminal(s), and a connecting lead(s) or shielded cable for conveniently extend-

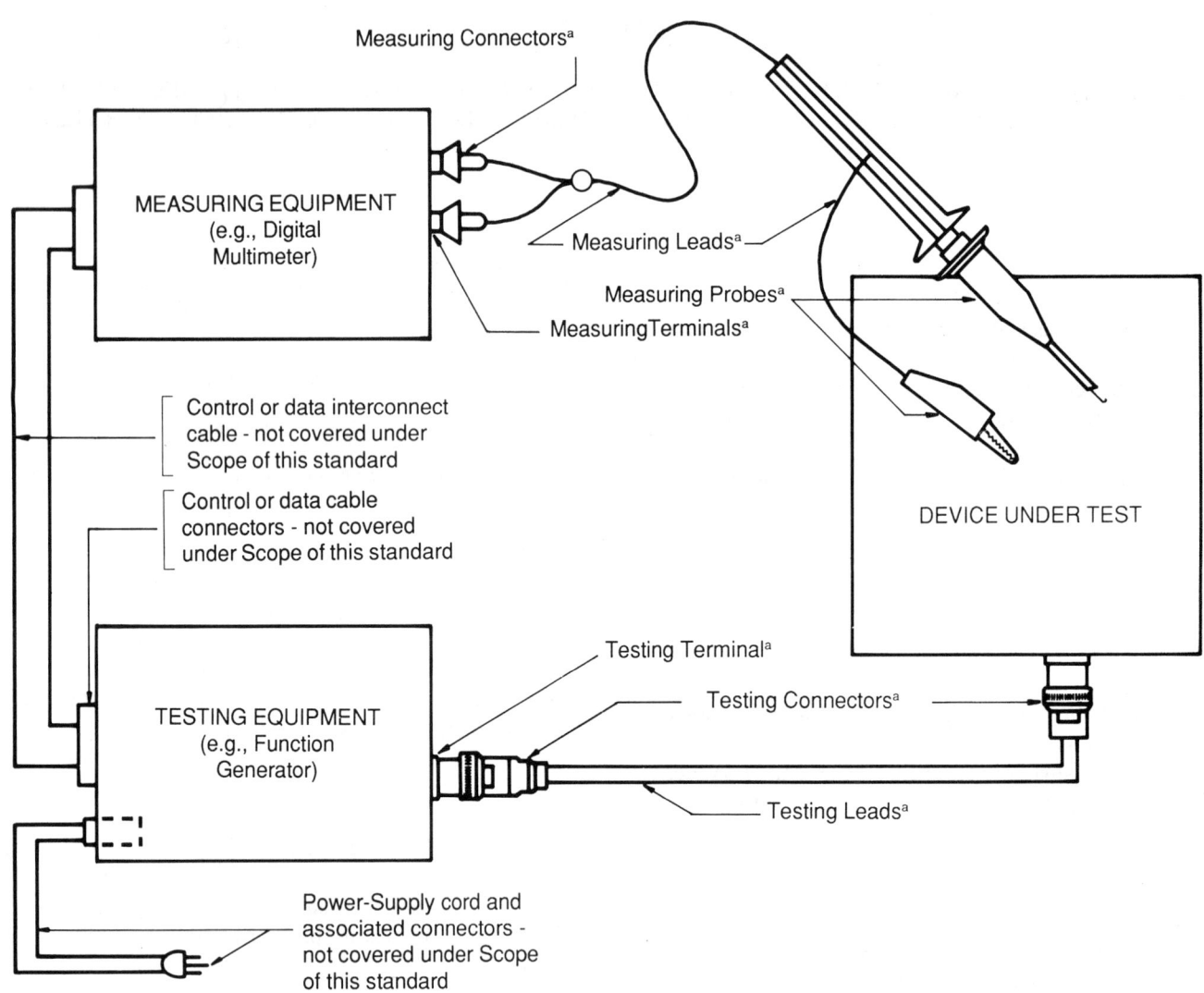

[a] This standard applies to the terminals, connectors, wiring (leads), and probes used in the interface between the measuring and testing equipment and that which is under test. See Sub-clause 1.1.

Figure 1-1: Illustration of Devices Related to the Scope

ing the equipment input or output circuits to the external circuit in which measurement or testing is desired. Refer to Figure 2-1 for a typical example.

The leads may be permanently connected to the equipment or may be terminated with a connector(s) to mate with the equipment measuring or testing terminals.

The probe assembly may contain components or circuits that modify the characteristics of the voltage or current passing through the probe assembly or the input or output characteristics of the equipment.

3. General Information

3.1 Object. This standard supplements or amends the requirements of ISA-S82.01.

3.2 Compliance. Where protections are adequately covered by ISA-S82.01 and identified by cross reference, the equipment shall comply with the requirements of the referenced Clauses of ISA-S82.01.

Compliance with the requirements of ISA-S82.02 is

determined by performing inspections and tests according to the Clauses applicable to the particular equipment.

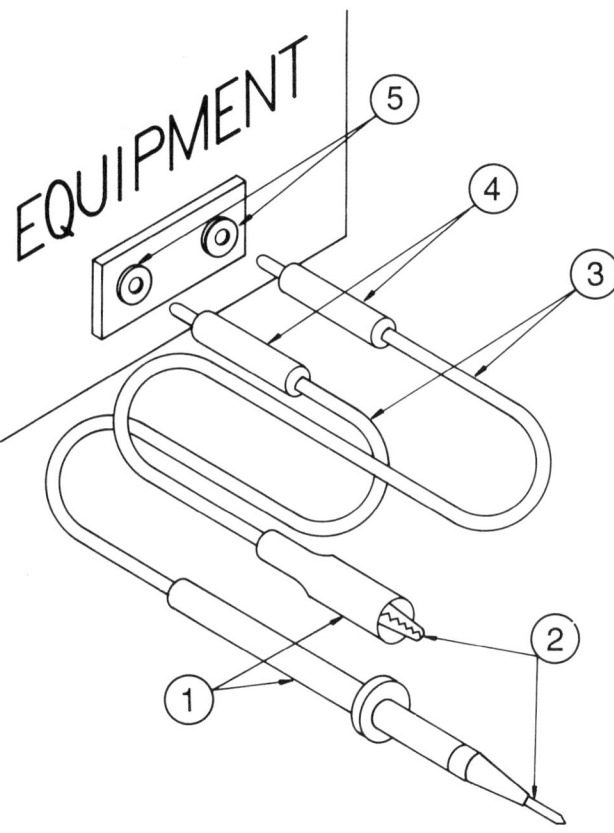

1 PROBE: The handle, body, boot, etc., used to support and insulate the remote terminal and to facilitate its application to the circuit in which measurement or testing is desired.

2 REMOTE TERMINAL: The clip, pin, grip or other connecting or contacting device that makes the electromechanical connection to the external circuit in which measurement or testing is desired.

3 LEADS: The insulated leads or shielded cable interconnecting the remote terminal(s) with the equipment input or output circuits.

4 CONNECTOR: The connector device, if any, that terminates the leads or cable with a means to make the electromechanical connection to the equipment input or output terminals or connector.

5 MEASURING OR TESTING TERMINALS: See Sub-clause 2.44(2) of ISA-S82.01.

Figure 2-1: Typical Probe Assembly
(Illustration only for defining generic parts)

4. General Test Requirements

4.1 General. The requirements of Clause 4 of ISA-S82.01 shall apply.

5. Marking

5.1 General. The requirements of Clause 5 of ISA-S82.01 shall apply and the requirements according to Sub-clauses 5.2 and 5.3 below.

5.2 Accessible terminal ratings

5.2.1 Isolated terminal rating marking. An isolated terminal or terminal set that is intended to be connected to an energized external source shall be marked with an isolated rating.

Compliance is checked by inspection.

5.2.2 Voltage terminal identification and rating marking. A terminal or terminal set intended for connection to an external voltage source or that supplies a voltage(s) not limited according to Sub-clause 14.2 of ISA-S82.01 shall be identified and bear a rating marking adjacent or otherwise related to the terminal(s).

Alternately, the related control markings, indicators, equipment configuration, equipment name, or other identifiers shall specify the terminal(s) rating or use, whichever is appropriate to the intent of the terminal(s).

Compliance is checked by inspection.

5.2.3 Current terminal marking. A terminal or terminal set intended for connection to an external current source or that supplies current shall bear a rating marking adjacent to or otherwise related to the terminal or terminal set.

Alternatively, the related controls, indicators, or other identifiers shall specify the terminal rating or use, whichever is appropriate to the intent of the terminal.

Compliance is checked by inspection.

5.2.4 High voltage terminal marking. A measuring or testing terminal to which or from which a voltage equal to or exceeding 1000 volts may be applied or supplied shall be marked with the flash symbol ⚡ in red (IEC Publication 417; see Appendix, S82.01).

Compliance is checked by inspection.

5.3 Measuring and testing probes. A measuring or testing probe assembly provided with a remote terminal shall be marked according to this Sub-clause.

5.3.1 Identification. A detachable measuring or testing probe assembly shall be provided with a means for identifying:

(1) name or trade name, or other descriptive marking, of the manufacturer, or the organization responsible for the measuring or testing probe assembly; and,

(2) a distinctive configuration or designation, such as a model number.

The identification means may be in the literature packed with the assembly.

Compliance is checked by inspection.

5.3.2 Probe assembly-to-equipment identification. A detachable (and interchangeable) measuring or testing probe assembly intended for use with a specific model of equipment shall be marked or be provided with a means for identifying such equipment.

The marking may be located on the probe assembly or on the literature packed with it.

Compliance is checked by inspection.

5.3.3 Rating marking. A detachable and interchangeable measuring or testing probe assembly shall bear a marking or be provided with a means for identifying the maximum rated value and nature of circuit-to-ground voltage.

The marking may be located on the measuring or testing probe assembly or in the literature packed with it.

Compliance is checked by inspection.

6. Equipment Emanations

6.1 General. The requirements of Clause 6 of ISA-S82.01 shall apply.

7. Heating

7.1 General. The requirements of Clause 7 of ISA-S82.01 shall apply.

8. Protection from Implosion and Explosion

8.1 General. The requirements of Clause 8 of ISA-S82.01 shall apply.

9. Protection from Electrical Shock

9.1 General. The requirements of Clause 9 of ISA-S82.01 shall apply and the requirements according to Sub-clause 9.2 below.

9.2 Measuring or testing probe assemblies and connectors. A measuring- or testing-probe assembly shall be constructed to provide protection from electrical shock.

9.2.1 Lead insulation. The basic insulation on each measuring- or testing-lead wire (Figure 2-1, item 3) shall be rated for at least the maximum voltage to which the wire is to be connected.

Compliance is checked by inspection.

9.2.2 Probe insulation. The basic insulation on each measuring or testing probe (Figure 2-1, item 1) shall be constructed so that uninsulated parts that could render an electrical shock are not accessible.

Exception: The remote terminal (Figure 2-1, item 2) need not be insulated but guarding according to the intent of Figure 9-1 shall be provided for voltages greater than 1200 volts peak. Other constructions shall be evaluated according to the intent of these requirements.

Compliance is checked according to Sub-clauses 9.1 and 9.2 of ISA-S82.01 and by inspection.

Table 9-1
Leakage Current Limits

Figure	S1	S2	I1	I2
9-2 A	--	--	0.5 mA ac 2 mA dc	--
9-2 B	--	--	0.5 mA ac 2 mA dc	--
9-2 C	--	Normal & Rev.	0.5 mA ac 2 mA dc	--
9-2 D	Open	Normal & Rev.	0.5 mA ac 2 mA dc	Limits of Fig. 9-5 of ISA-S82.01
9-2 D	Closed	Normal & Rev.	0.5 mA ac 2 mA dc	0.5 mA ac 2 mA dc

S82.02 Electrical and Electronic Test and Measuring Equipment

A. ALLIGATOR CLIP WITH INSULATING BOOT

B. PENCIL TYPE PROBE WITH INSULATED HANDLE

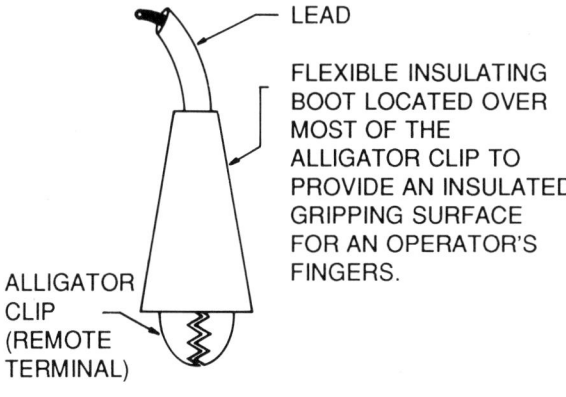
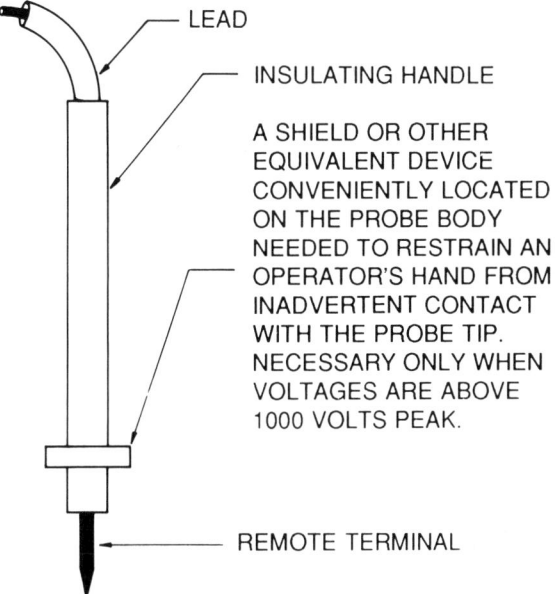

C. PROBE WITH INDIRECT ACTION THAT KEEPS OPERATOR'S HAND AWAY FROM PROBE TIP.

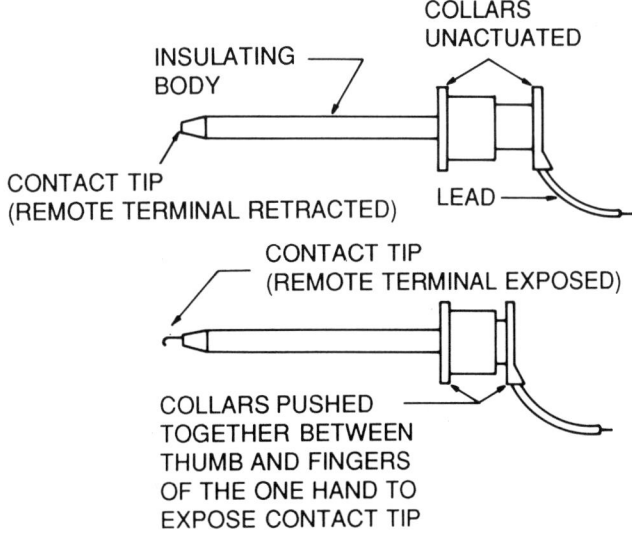

Figure 9-1. Insulated Measuring and Testing Devices

(Illustration only for depicting the intent)

9.2.3 General leakage current for probes and probe assemblies. If the open-circuit potential between an accessible part and either ground or any other accessible part exceeds the limits according to Sub-clause 9.2 of ISA-S82.01, the leakage current between those parts shall not cause an indication of the specified measuring instrument that exceeds the limits specified in Table 9-1.

Compliance is checked by measuring the leakage current with the specified measuring instrument, or the equivalent, for the connections shown in Figure 9-2.

10. Testing under Fault Conditions

10.1 General. The requirements of Clause 10 of ISA-S82.01 shall apply and the requirements according to Sub-clauses 10.2 and 10.3 below.

10.2 Multi-function and overrange test

Range and function switches shall be:

(1) operated, and

(2) set

for any combination of conditions. Terminals shall be connected to any maximum rated voltage or current or, for multiple-terminal equipment, any combination of rated voltages and currents. For equipment provided with more than one probe set where the interchange of probes, or probe sets, accomplishes a change of function or range, any combination of the foregoing conditions shall be applicable to any combination of probes, or probe sets. The test source connected to the equipment-measuring terminals during the multi-function and overrange test shall be limited to 3600 volt-amperes.

Unlike the other fault condition test, no specific minimum or maximum time duration criteria are specified as to how long the multi-function and overrange test is to be continued. The test duration is generally based upon the intended use of the equipment, its design and its construction with a recognition that the intent of the multi-function and overrange fault-condition test is to detect the likelihood of fire, glowing, arcing, explosion, electrical shock and similar abnormal phenomena which might occur shortly after an incorrect range or function switch has been actuated or a terminal misconnection made. When it has become obvious that the equipment is no longer capable of performing its intended function, the test may be terminated.

10.3 Testing equipment

The operator accessible output(s) shall be short-circuited to both ground and common (if not ground). Other output arrangements shall be short-circuited according to the intent of this Sub-clause.

11. Mechanical Requirements

11.1 General. The requirements of Clause 11 of ISA-S82.01 shall apply as well as the requirements of Sub-clause 11.2 below.

11.2 Measuring and testing probe assemblies. A measuring or testing probe assembly shall have mechanical strength according to this Sub-clause.

Compliance is checked by inspection and by performing the tests according to this Sub-clause. A measuring or testing probe assembly shall comply with the requirements of Sub-clause 9.1 and 9.12 of ISA-S82.01 after testing according to this Sub-clause.

11.2.1 Drop test

A measuring or testing probe assembly shall be subjected to a drop test according to Sub-clause 11.1.1(1) of ISA-S82.01.

11.2.2 Impact

(1) Each of three samples shall be subjected three separate times to the impact that results from the assembly swinging as a pendulum by its own wire or cable against a hardwood surface. The height of the drop shall be the length of the wire or cable with a maximum height of 1.83 meters (6 feet). The hardwood surface shall be vertical and positioned on a vertical line through the pendulum hinge point. See Figure 11-1.

(2) Alternatively, parts of the probe assembly shall be subjected three separate times at different locations to a preload force of 20 newtons (4.5 pounds force) and an impact of 0.5 newton-meter (70.81 pound force-inch). Such impacts shall be at the locations most likely to produce adverse effects. (See Figure 11-1 of ISA-S82.01.)

12. Protection from Fire

12.1 General. The requirements of Clause 12 of ISA-S82.01 shall apply.

Exception: Hand-held measuring and testing probes.

13. Component Requirements and Applications

13.1 General. The requirements of Clause 13 of ISA-S82.01 shall apply and the requirements according to Sub-clause 13.2 below.

13.2 Range and function switches. A range-changing switch, function-selector switch or similar control device

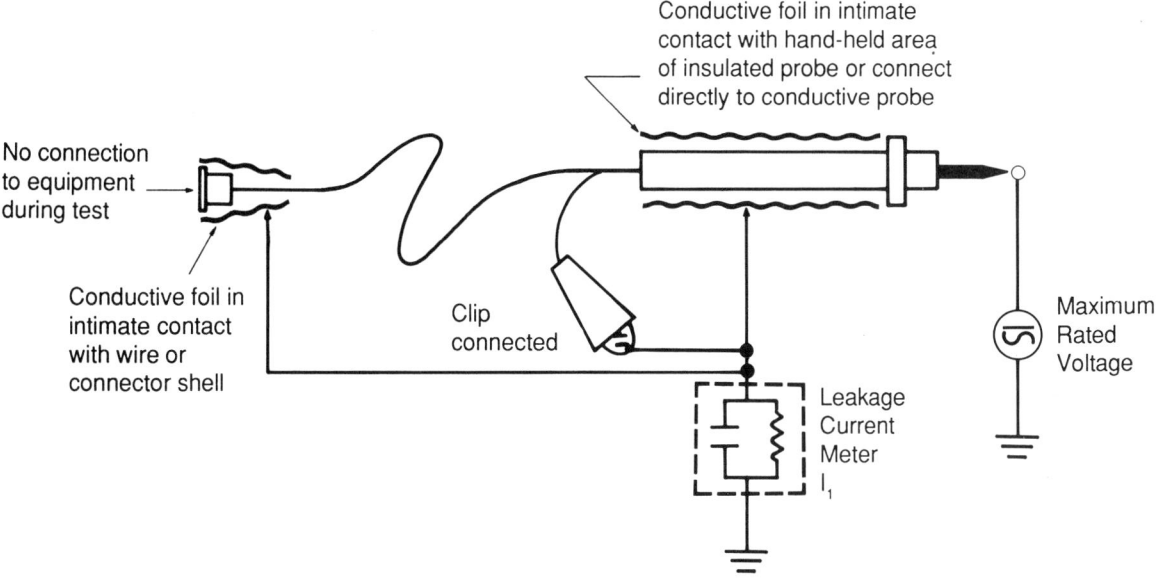

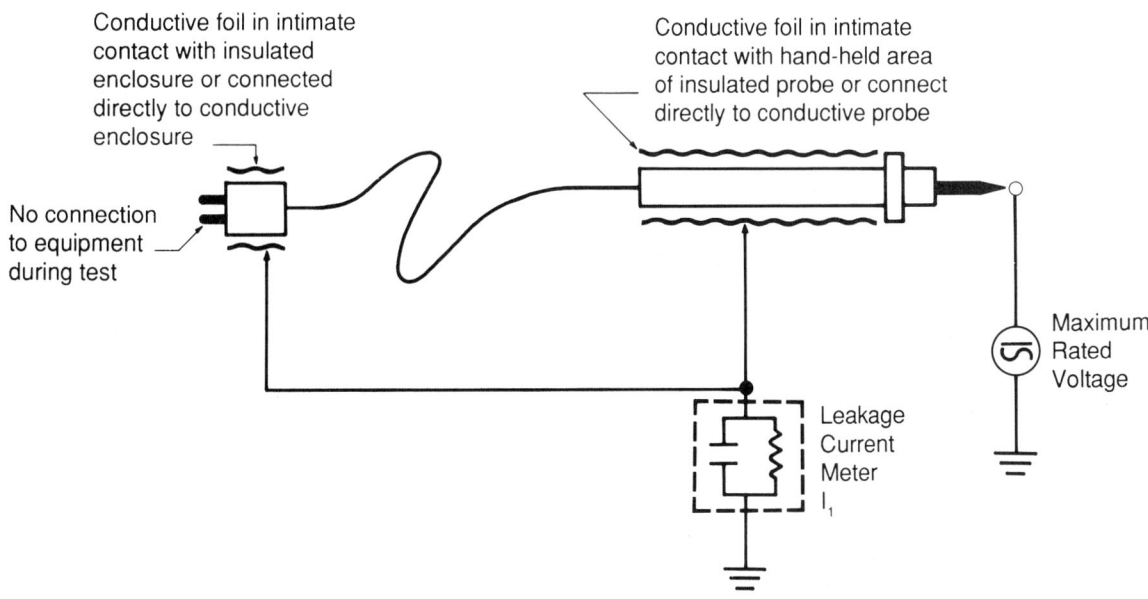

Figure 9-2. Leakage Current Measurement Circuits for Probes and Probe Assemblies

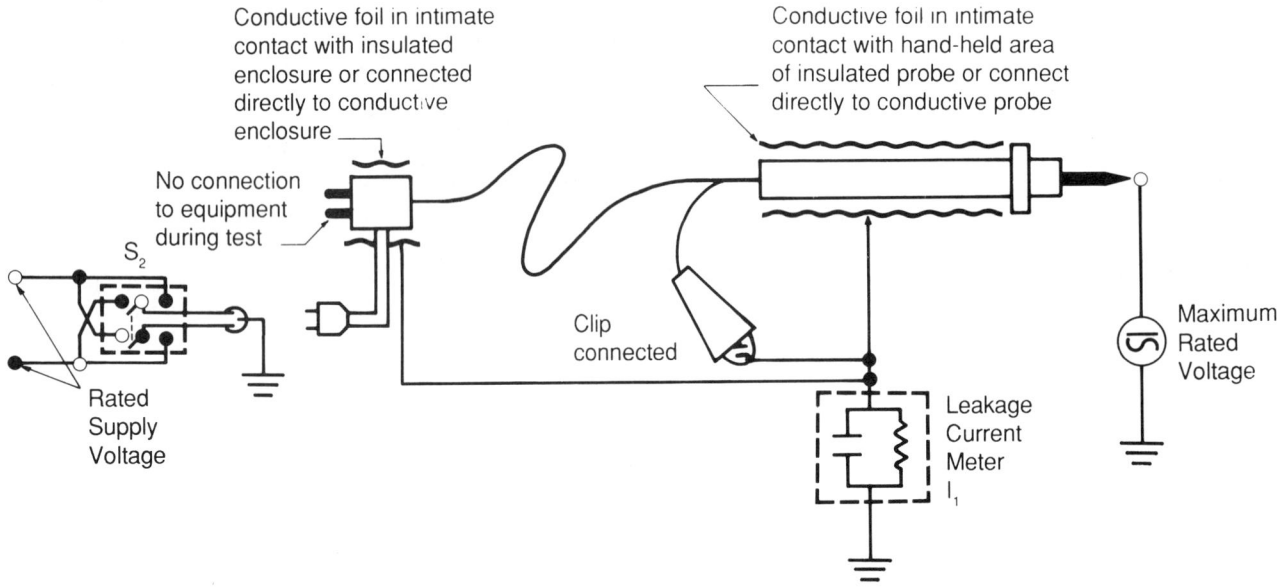

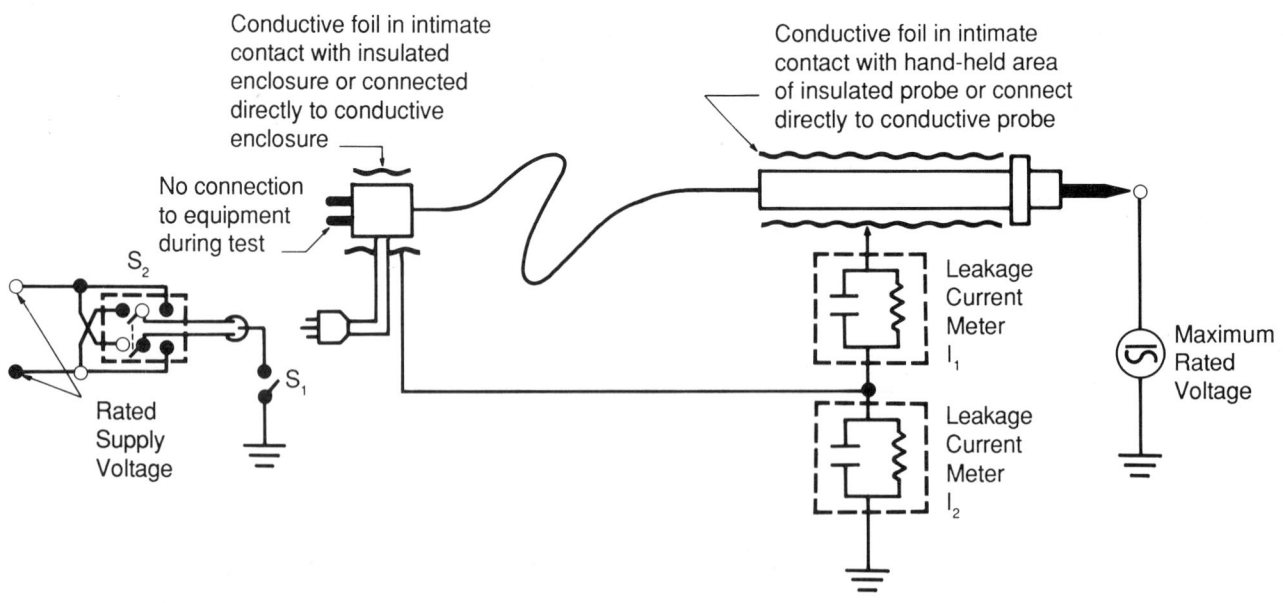

Figure 9-2, continued. Leakage Current Measurement Circuits for Probes and Probe Assemblies

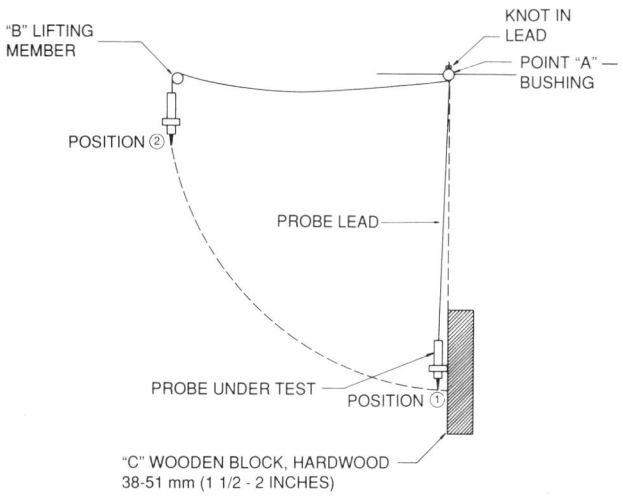

Figure 11-1: Pendulum Test

located in the equipment current-measuring circuit(s) shall be so designed that the current is not interrupted when changing current ranges.

Exception: When the equipment construction provides equivalent protection or the current measuring circuit in which the switch is connected is rated for a maximum value of no more than 2 amperes.

Compliance is checked by inspection.

14. Extra-Low Voltage and Power-Limited Circuits

14.1 General. The requirements of Clause 14 of ISA-S82.01 shall apply.

15. Terminal Devices

15.1 General. The requirements of Clause 15 of ISA-S82.01 shall apply and the requirements according to Sub-clause 15.2 below.

15.2 Terminal mounting. An operator accessible measuring or testing terminal, when used for its intended purpose and in its intended manner, shall be so anchored that the terminal mounting, its parts, and the internal and external connections are not loosened.

Exception: When the construction complies with the applicable requirements of Clause 9 following the applicable tests of Clause 11 with the terminals loosened to the maximum extent made possible by the construction.

Compliance is checked by inspection.

16. External Cords

16.1 General. The requirements of Clause 16 of ISA-S82.01 shall apply as well as the requirements of Sub-clause 16.2 below.

16.2 Measuring and testing probe assemblies. Measuring and testing cables and assemblies used in circuits that are live according to Sub-clause 9.2 of ISA-S82.01 or not power-limited according to Sub-clause 14.2 of ISA-S82.01 shall be rated and tested according to this Sub-clause.

16.2.1 Attachment. The attachment of a lead of a probe assembly to a connector or to a probe shall withstand a force according to this Sub-clause without transmitting any strain to the internal connections.

Compliance is checked by applying a force to the lead attachment means for one minute.

The insulation on the lead shall not be cut or torn, and the lead shall not slide in a bushing.

The force shall be:

- *for a friction-fit connector, four times the connector withdrawal force; and*

- *for a probe, four times the weight of the probe assembly; or*

- *36 newtons (8 pounds force)*

whichever is greater

16.2.2 Where a probe cord or cable emerges from an enclosure and may be subjected to strain or motion, the equipment shall employ an insulating bushing attached to the enclosure, and the strain or motion shall not be transmitted to the internal connections.

Compliance is checked by inspection.

17. Equipment Instructions

Operating and servicing instructions provided with the equipment shall include applicable information according to this Clause.

Additional safeguard information considered applicable by the manufacturer may be included.

Instructions according to this Clause may be supplemented, but not replaced, by illustrations.

17.1 Separation of information.
Operating and operational-maintenance instructions shall be separated from servicing instructions.

Where servicing requires access to parts that could render electrical shock, servicing instructions shall be preceded by a warning.

The warning shall be prefaced by the signal word "WARNING" and

(1) state that servicing requires access to parts that could render an electrical shock, and

(2) refer servicing to qualified personnel only.

For example:

> WARNING - THESE SERVICING INSTRUCTIONS ARE FOR USE BY QUALIFIED PERSONNEL ONLY. TO AVOID ELECTRICAL SHOCK, DO NOT PERFORM ANY SERVICING OTHER THAN THAT CONTAINED IN THE OPERATING INSTRUCTIONS UNLESS YOU ARE QUALIFIED TO DO SO.

Compliance is checked by inspection

17.2 Operating instructions.
With regard to equipment protection, the operating instructions shall:

(1) describe the equipment installation, including specifically:

    (a) assembly, if required, and mounting;

    (b) protective grounding means, if employed;

    (c) polarization of supply circuit, if employed; and

    (d) ventilation considerations.

(2) explain equipment markings, including specifically:

    (a) symbols;

    (b) controls; and

    (c) terminal ratings.

(3) identify and describe interconnections with:

    (a) auxiliary and accessory equipment;

    (b) measuring and testing circuits; and

    (c) other equipment.

(4) explain the operation and application of the equipment, specifically in regard to:

    (a) maintaining the protective grounding, if employed, during measuring and testing procedures; and

    (b) precautions when using probes that have accessible remote terminals.

Compliance is checked by inspection.

ANSI/ISA-S82.03-1988
Approved February 18, 1988
(Partial Revision and Redesignation of ANSI C39.5-1974)

American National Standard

Safety Standard for Electrical and Electronic Test, Measuring, Controlling and Related Equipment

ELECTRICAL AND ELECTRONIC PROCESS
MEASUREMENT AND CONTROL EQUIPMENT

Instrument Society of America

Instrument Society of America

ISBN 1-55617-102-1

ISA-S82.03 Safety Standard for Electrical and Electronic Test, Measuring, Controlling, and Related Equipment

Copyright ©1988 by the Instrument Society of America. All rights reserved. Printed in the United States of America. No part of this publication may be reproduced, stored in a retrieval system, or transmitted, in any form or by any means (electronic, mechanical, photocopying, recording, or otherwise), without the prior written permission of the publisher.

INSTRUMENT SOCIETY OF AMERICA
67 Alexander Drive
P.O. Box 12277
Research Triangle Park, North Carolina 27709

S82.03   Safety Standard for Electrical and Electronic Test, Measuring, Controlling and Related Equipment: Electrical and Electronic Process Measurement and Control Equipment

PREFACE

This preface is included for information purposes and is not part of ISA-S82.03.

This standard has been prepared as part of the service of the Instrument Society of America (ISA) toward a goal of uniformity in the field of instrumentation. To be of real value, this document should not be static, but should be subject to periodic review. Toward this end, the Society welcomes all comments and criticisms, and asks that they be addressed to the Secretary, Standards and Practices Board, Instrument Society of America, 67 Alexander Drive, P.O. Box 12277, Research Triangle Park, NC 27709, telephone (919) 549-8411.

The ISA Standards and Practices Department is aware of the growing need for attention to the metric system of units in general, and the International System of Units (SI) in particular, in the preparation of instrumentation standards. The Department is further aware of the benefits to U.S.A. users of ISA standards of incorporating suitable references to the SI (and the metric system) in their business and professional dealings with other countries. Toward this end, this Department will endeavor to introduce SI-acceptable metric units in all new and revised standards to the greatest extent possible. The Metric Practice Guide, which has been published by the Institute of Electrical and Electronics Engineers as ANSI/IEEE Std. 268-1982, and future revisions will be the reference guide for definitions, symbols, abbreviations, and conversion factors.

It is the policy of the Instrument Society of America to encourage and welcome the participation of all concerned individuals and interests in the development of ISA standards. Participation in the ISA standards-making process by an individual in no way constitutes endorsement by the employer of that individual, of the Instrument Society of America, or of any of the standards that ISA develops.

The information contained in the preface, footnotes, and appendices is included for information only and is not a part of the standard.

The development of this standard dates back to 1958. The need for the standard arose from requests that recording instruments meet the requirement of the American National Standard, the National Electrical Code. At that time there was also the beginning of the need for examination of measuring instruments by testing laboratories. The initial version of this standard was approved in 1964 after several years of careful development, review, and trial use.

Work began on safety requirements for electrical measuring instruments at the international level at about the same time. While consideration of specific safety requirements for measuring apparatus was begun in 1965, resulting first editions of IEC Publications 348 and 414 became available in 1971 and 1973, respectively.

American National Standard C39.5-1964 was reaffirmed in 1969 and revised in 1974. In 1982, responsibility for C39.5-1964 was transferred to the Instrument Society of America and the document was renumbered ISA-S82.01.

This 1988 edition reflects four major considerations:

- The scope has been expanded to include more equipment types. The scope and organization now readily allow future inclusion of more types of related equipment.

- A major goal was harmonization--to the extent possible--with worldwide standards, especially North American and IEC Standards. Not only were the IEC clause and multi-part formats adopted, but also many of the IEC requirements; many requirements were further defined and developed from the work of IEC Subcommittee 66E.

- Digital and microelectronic innovations have narrowed the one-time separation between digital and analog technologies and prompted commonality of safety requirements.

- Significant technical contributions were made by industry organizations and independent certifying laboratories.

Instrument Society of America

The following are significant features incorporated in this ongoing edition:

- Constructional requirements.

- Measurement of conformance to requirements is through the use of specific compliance statements. This allows the user of the standard to determine product compliance and to better understand the nature and intent of the requirements.

- Expansion of requirements for protection against fire and personal injury.

The following people served as members of ISA Subcommittee SP82.02, which prepared this standard:

NAME	COMPANY
Bruce J. Feikle, Chairman	OTIS
William Calder, Manager Director	The Foxboro Company
Abraham Abramowitz	Retired
David Braudaway	Sandia Laboratories
Joseph E. Devine	IBM Instruments, Inc.
Charles G. Gross	Gen Rad Inc.
Robert G. Harris	Underwriters Laboratories
Walter F. Hart	John Fluke Mfg. Co., Inc.
W.R. Howard	LFE Corporation
William H. Hulse, Jr.	Material Safety Eng. Div.
Donald S. Ironside	Biddle Instruments
A. Walter Jacobson	Retired
Al Kanode	Hewlett-Packard Company
Roger P. Lutfy	Factory Mutual Research
Donald A. Mader	Underwriters Laboratories, Inc.
Ernest C. Magison	Honeywell, Inc.
Richard C. Masek	Bailey Controls Co.
Frank J. McGowan	The Foxboro Company
Richard Nute	Tektronix, Inc.
Phillip A. Painchaud	Retired
William A. Phillips	Leeds & Northrup Company
William E. Rich	Westinghouse Electric Corp.
John N. Stanley	Leeds & Northrup Company
Robert Wallace	Tektronix, Inc.

The following people served as members of ISA Committee SP82 during the review of this standard:

NAME	COMPANY
David Braudaway, Co-Chairman	Sandia Laboratories
Frank McGowan, Co-Chairman	The Foxboro Company
William Calder, Manager Director	The Foxboro Company
Abraham Abramowitz	Retired
Norm Belecki	National Bureau of Standards
Bruce Feikle	OTIS

*Employer at time of participation

Robert G. Harris	Underwriters Laboratories, Inc.
W.R. Howard	LFE Corporation
A. Walter Jacobson	Retired
Donald A. Mader	Underwriters Laboratories, Inc.
Ivano Mazza	Casella Postale No. 11
Hubert O'Neil	Havti
Phillip A. Painchaud	Retired
Peter E. Perkins	Tektronix, Inc. DS 78-531
William E. Rich	Westinghouse Electric Corp.
DeWayne B. Sharp, P.E.	IBM Corporation
Jack V. Stegenga	Yokogawa Corp. of America

This standard was approved for publication by the ISA Standards and Practices Board in January 1988.

NAME	COMPANY
D. E. Rapley, Chairman	Rapley Engineering Services
D. N. Bishop	Chevron U.S.A. Inc.
W. Calder III	The Foxboro Company
N. Conger	Fisher Controls International, Inc.
R. S. Crowder	Ship Star Associates
C. R. Gross	Eagle Technology
H. S. Hopkins	Utility Products of Arizona
R. B. Jones	Dow Chemical Company
R. T. Jones	Philadelphia Electric Company
R. Keller	Consultant
A. P. McCauley	Chagrin Valley Controls, Inc.
E. M. Nesvig	ERDCO Engineering Corporation
C. W. Reimann	Moore Products Company
R. H. Reimer	Allen-Bradley Company
J. Rennie	Factory Mutual Research Corporation
W. C. Weidman	Gilbert/Commonwealth Inc.
K. A. Whitman	Fairleigh Dickinson University
* P. Bliss	Consultant
* B. A. Christensen	Continental Oil Company
* L. N. Combs	Consultant
* R. L. Gailey	Consultant
* T. J. Harrison	Florida State University
* O. P. Lovett, Jr.	Consultant
* E. C. Magison	Honeywell, Inc.
* W. B. Miller	Moore Products Company
* J. W. Mock	Bechtel Western Power Corporation
* G. Platt	Consultant
* J. R. Williams	Stearns Catalytic Corporation

* Director Emeritus

- SPECIFIC CLAUSE TREATMENTS -

The **SCOPE** includes environmental conditions envisioned for the equipment. **DEFINITIONS** have extensive additions to reduce misinterpretations. **MARKINGS** are expanded and clarified by inclusion of internationally recognized symbols. X-radiation considerations are added while several other **EMANATIONS** are identified as "under consideration" due to lack of recognized limits and standardized measurements. The clause for **PROTECTION FROM ELECTRIC SHOCK** deletes requirements for resistance tests, clarifies leakage current measurement procedures, details dieletric tests and includes requirements for measuring terminals and routine tests. **TESTING UNDER FAULT CONDITIONS** is new. **PROTECTION FROM FIRE** is an approach to measuring the likelihood of spread of fire. **EXTRA LOW VOLTAGE AND POWER LIMITED CIRCUITS** defines the circuits which could cause electric shock or the likelihood of fire and for which protection requirements of the standard apply.

Understanding the Appendix material of ISA-S82.01 under Clause 9 is also necessary for this edition.

TABLE OF CONTENTS

Section Title	Page
1. General	9
1.1 Scope	9
1.2 Environment	9
2. Definitions	10
2.1 Control equipment	10
2.2 Electronic measuring equipment	10
2.3 Indicating measuring equipment	10
2.4 Outdoor locations	10
2.5 Sheltered locations	10
3. General Information	10
3.1 Object	10
3.2 Compliance	10
4. General Test Requirements	10
4.1 General	10
4.2 Reference test conditions	10
5. Marking	11
5.1 General	11
5.2 Terminals	11
5.3 Enclosures for extended environments	11
6. Equipment Emanations	11
6.1 General	11
7. Heating	11
7.1 General	11
7.2 Maximum temperature rise	11
8. Protection from Implosion and Explosion	11
8.1 General	11
9. Protection from Electrical Shock	11
9.1 General	11
9.2 Live parts	12
9.3 Construction requirements - extended environments	12
9.4 Spacings - extended environments	12
9.5 Dielectric voltage-withstand tests	12
10. Testing under Fault Conditions	13
10.1 Application of fault conditions	13
11. Mechanical Requirements	13
11.1 General	13
12. Protection from Fire	13
12.1 General	13
13. Component Requirements and Applications	13
13.1 General	13
13.2 Protective coatings	13
14. Extra-Low Voltage and Power-Limited Circuits	14
14.1 General	14
15. Terminal Devices	14
15.1 General	14
15.2 Terminal screws	14
16. External Cords	14
16.1 General	14

Section Title	Page
17. Equipment Instructions	14
17.1 General	14
17.2 Form of equipment instructions	14
17.3 Equipment information and instructions	14
18. Pressure-Actuated Equipment	15
18.1 General	15
18.2 Hydrostatic test	15
19. Equipment Enclosures	17
19.1 General	17

APPENDIX

Comparison between NEMA Enclosure Type Numbers and IEC Enclosure Classification Designations 19

IN THIS STANDARD
Requirements proper are printed in roman type.
Explanatory matter regarding the requirements is printed in smaller roman type.
Test specifications are printed in italic type.
Explanatory matter regarding the test specifications is printed in smaller italic type.
Further explanatory matter regarding the requirements or specific subject marked by an asterisk can be found in the Appendix.

CAUTION

MANY TEST REQUIRED BY THIS STANDARD ARE INHERENTLY HAZARDOUS. ADEQUATE SAFEGUARDS FOR PERSONNEL AND PROPERTY SHOULD BE EMPLOYED IN CONDUCTING SUCH TESTS.

1. General

1.1 Scope. This standard applies to electrical, electronic (analog/digital) and electromechanical process measurement and control equipment which:

(1) measures and controls directly or indirectly an industrial process through a final control element (or elements);

(2) is intended to be connected to supply circuits which do not exceed 250 volts rms, single phase, or dc; and

(3) is rated for use in either indoor, outdoor or sheltered locations.

This equipment includes but is not necessarily restricted to:

· indicating, integrating or recording equipment with or without a control function,

· transmitters,

· transducers,

· analyzers,

· supervisory or telemetry equipment,

· accessories used with any of the above equipment.

1.2 Environment

1.2.1 Indoor locations. Equipment according to this standard is intended for use where air temperature is controlled within specified limits.

Exception: where specified otherwise.

The basic climatic conditions for heated and/or cooled locations at an atmospheric pressure of 86 to 106 kPa (12.5 to 15.4 psi) are according to Table 1-1.

Normally, only non-conductive pollution occurs. Occasionally, however, a temporary conductivity caused by condensation must be expected. In air-conditioned locations no pollution or only dry, non-conductive pollution occurs and has no influence.

Table 1-1
Heated and/or Cooled Indoor Locations[a]

Location Class	Temperature (°C)	Relative Humidity	Maximum Moisture Content kg/kg Dry Air
B1	+15 to +30	10 to 75%	0.020
B2	+5 to +40	10 to 75%	0.020
B3	+5 to +40	5 to 95%	0.028
BX	Equipment designed and rated for higher limits, however, intended only for use in indoor locations		

[a] The environment inside an equipment enclosure is considered to be the same as that external to the equipment enclosure when the enclosure does not provide adequate protection against the environmental stresses.

1.2.2 Sheltered locations. Equipment according to this standard intended for use in sheltered locations shall comply with the applicable requirements according to Clause 19 of ISA-S82.03.

Exception: where specified otherwise.

The basic climatic conditions for sheltered locations at an atmospheric pressure of 86 to 106 kPa (12.5 to 15.4 psi) are according to Table 1-2.

Normally, conductive pollution occurs, or dry, non-conductive pollution occurs which becomes conductive due to condensation which is expected.

Table 1-2
Sheltered Locations[a]

Location Class	Temperature (°C)	Relative Humidity	Maximum Moisture Content kg/kg Dry Air
C1	-25 to +55	5 to 100%[b]	0.028
C2	-40 to +70	5 to 100%[b]	0.028

[a] The environment inside an equipment enclosure is considered to be the same as that external to the equipment enclosure when the enclosure does not provide adequate protection against the environmental stresses.

[b] Including condensation

1.2.3 Outdoor locations. Equipment according to this standard intended for use in outdoor locations shall comply with the applicable requirements according to Clause 19 of ISA-S82.03.

Exception: where specified otherwise.

The basic climatic conditions for outdoor locations at an atmospheric pressure of 86 to 106 kPa (12.4 to 15.4 psi) are according to Table 1-3.

Normally, the pollution generates high and persistent conductivity caused, for instance, by conductive dust or by rain or snow.

Table 1-3
Outdoor Locations[a]

Location Class	Temperature (°C)	Relative Humidity	Maximum Moisture Content kg/kg Dry Air
D1	-25 to +70	5 to 100%[b]	0.050
D2	-40 to +85	5 to 100%[b]	0.050

[a] The environment inside an equipment enclosure is considered to be the same as that external to the equipment enclosure when the enclosure does not provide adequate protection against the environmental stresses.

[b] Including condensation and direct wetness.

2. Definitions

2.1 CONTROL EQUIPMENT: Equipment which controls one or more output quantities to specific values, each value being determined by manual setting, local or remote programming, or by one or more input variables.

2.2 ELECTRONIC MEASURING EQUIPMENT: Equipment which, by means of incorporating electronic devices, serves to measure or to observe quantities or to supply electrical quantities for measuring purposes.

Electronic devices are parts or assemblies of parts which use electron or hole conduction in semiconductors, gases, or a vacuum.

2.3 INDICATING MEASURING EQUIPMENT: Equipment which indicates the value of the measured quantity.

Process measurement and control equipment may combine one or more of the functions of Sub-clauses 2.1, 2.2, or 2.3 above.

2.4 OUTDOOR LOCATION: Locations where neither air temperature nor humidity are controlled and the equipment is exposed to outdoor atmospheric conditions such as direct sunshine, wind, rain, hail, sleet, snow and icing.

Sensors, transmitters, final control elements and actuators and some indicators separate from controllers are often located in outdoor locations.

2.5 SHELTERED LOCATIONS: Locations where neither air temperature nor humidity is controlled and equipment is protected against direct exposure to such climatic elements as direct sunlight, fall of rain and other precipitation, and full wind pressure.

Indoor locations which are neither heated nor cooled are sheltered locations.

Transmitters, final control elements, and some indicators separate from controllers are often located in sheltered locations when frequent operator attention is not important.

3. General Information

3.1 Object. ISA-S82.03 supplements or amends the requirements of ISA-S82.01.

3.2 Compliance

Where protections are adequately covered by ISA-S82.01 and identified by cross reference, the equipment shall comply with the requirements of the referenced Clauses of ISA-S82.01.

Compliance with the requirements of ISA-S82.03 is determined by performing inspections and tests according to the Clauses applicable to the particular equipment.

4. General Test Requirements

4.1 General

The requirements of Clause 4 of ISA-S82.01 shall apply and the requirements according to Sub-clause 4.2 below.

4.2 Reference test conditions

Reference test conditions shall be the same as those in Sub-

clause 4.2 of ISA-S82.01 for equipment ratings not exceeding those according to Sub-clause 1.2.1 of ISA-S82.03.

Equipment

(1) with ambient temperature ratings exceeding, or

(2) rated for sheltered or outdoor use, where environmental conditions exceed

those specified in Sub-clause 1.2.1 of ISA-S82.03, shall be tested at the extended environmental ratings if hazards may result at the more severe environmental conditions.

Exception: Unless otherwise specified.

Testing at higher ambient temberatures will ensure, among other factors, that absolute temperatures will not exceed the safe limits for insulating materials used.

See Clause 19 of ISA-S82.03 for extended environmental enclosure requirements and tests.

5. Marking

5.1 General. The requirements of Clause 5 of ISA-S82.01 shall apply and the requirements according to Sub-clauses 5.2 and 5.3 below.

5.2 Terminals. Terminals for external connection to switch and relay contact inputs and outputs shall be marked with the voltage and current rating of the contacts adjacent to the terminals or on the equipment name or data-plate.

Compliance is checked by inspection.

5.3 Enclosures for extended environments. In addition to such markings as may be required for electrical equipment within an enclosure rated for use in sheltered or outdoor locations, there shall be a marking to identify the conditions for which the enclosure is rated.

The enclosure type designation marking (see Clause 19) shall be a permanent marking (engraved, etched, die stamped, or equivalent) on the enclosure, or on a suitable nameplate on the enclosure.

Exception: Where this requirement conflicts with other equipment safety standards, the enclosed rating information may be provided in the equipment instructions.

Compliance is checked by inspection.

6. Equipment Emanations

6.1 General. The requirements of Clause 6 of ISA-S82.01 shall apply.

7. Heating

7.1 General. The requirements of Clause 7 of ISA-S82.01 shall apply and the requirements according to Sub-clause 7.2 below.

Exception: Surface temperature rise of an enclosure not operator accessible may exceed the maximum temperature rise specified according to Table 7-1 of ISA-S82.01.

7.2 Maximum temperature rise. See Table 7-1 (ISA-S82.01)

NOTES:

(1) The temperature rises specified in Table 7-1 are based on environmental conditions according to Sub-clause 1.1.6 of ISA-S82.01 and which may be as high as 40°C occasionally and for brief periods. If equipment is intended specifically for use in a prevailing ambient temperature constantly more than 25°C, the test of the equipment shall be made with that higher ambient temperature, and the allowable temperature rises specified in Table 7-1 shall be reduced by the amount by which the higher ambient exceeds 25°C.

7.2.1 Operator accessible parts. Operator accessible parts exceeding the allowable temperature rise according to Table 7-1 of ISA-S82.01 shall be marked with a warning marking according to Sub-clause 5.4 of ISA-S82.01.

Compliance is checked by inspection.

8. Protection from Implosion and Explosion

8.1 General. The requirements of Clause 8 of ISA-S82.01 shall apply.

9. Protection from Electrical Shock

9.1 General. The requirements of Clause 9 of ISA-S82.01 shall apply and the requirements according to Sub-clauses 9.2 through 9.5 below.

9.2 Live part

9.2.1 Accessible parts shall not be live.

A live part is a part where under reference test conditions the potential difference between the part and any other accessible part including ground either:

(1) exceeds those values specified in Sub-clause 9.2 of ISA-S82.01, or

(2) under damp conditions (as in certain outdoor environments, tropical regions, etc.) exceeds

(a) 15 volts rms (12.2 volts peak), or

(b) 30 volts dc, or

(c) 12.4 volts dc interrupted at a rate of 10-200 hertz

and from which the leakage current according to Sub-clause 9.11 of ISA-S82.01 is exceeded.

9.3 Construction requirements - extended environments.
Equipment intended to be used outdoors, or in sheltered locations under extended environmental conditions, according to Sub-clauses 1.2.2 and 1.2.3 shall have an enclosure rating according to Clause 19 of ISA-S82.03.

Equipment intended to be used indoors, where falling dirt, liquids, splashing water, or dust is expected, may require additional protections according to Clause 19 of ISA-S82.03.

Compliance is checked by inspection and by the tests specified in Clause 19 of this standard.

9.4 Spacing - extended environments

9.4.1 Supply circuits. The spacings between an uninsulated part conductively connected to the supply circuit and

(1) an uninsulated part conductively connected to another pole of the supply circuit;

(2) another uninsulated part of any other circuit;

(3) a grounded part;

(4) accessible conductive parts; and

(5) the point of closest approach of either accessibility probe, when applied according to Sub-clause 9.1.1 of ISA-S82.01, to any opening of the enclosure (see Figure 9-3 of ISA-S82.01)

shall be at least those

(1) according to Table 9-1 of ISA-S82.01, and

(2) additionally protected by at least an equipment enclosure type according to Clause 19 of this standard for use in sheltered locations, or

(3) additionally protected

(a) by a protective coating in compliance with Sub-clause 13.2 of this standard, and

(b) by at least a Type 4 equipment enclosure type according to Clause 19 of this standard for use in sheltered or outdoor locations; or

Exception: Supply circuits that are power-limited according to Sub-clause 14.2 of ISA-S82.01 and not live according to Sub-clause 9.2.1 above need not comply with the spacing requirements.

Spacing requirements are not applied to the internal spacings of electronic devices.

Compliance is checked by first applying and removing a 2-newton (0.5-pound) force against any wire or uninsulated part followed by measuring the creepage distance or clearance.

9.4.2 Other circuits. The spacings in other circuits shall be additionally protected by a protective coating in compliance with Sub-clause 13.2 of this standard.

Exception: Where spacings are:

(1) at least those according to Table 9-1 of ISA-S82.01, or

(2) additionally protected by an equipment enclosure according to Clause 19 of this standard suitable for the intended environment.

The spacings in other circuits are evaluated according to Sub-clause 9.12 (dielectric voltage-withstand tests) of ISA-S82.01.

Exception: Circuits that are power-limited according to Sub-clause 14.2 of ISA-S82.01 and not live according to Sub-clause 9.2 of this standard.

9.4.3 Field wiring terminal parts. The requirements of Sub-clause 9.8.3 of ISA-S82.01 shall apply.

Exception: Circuits that are powered-limited according to Sub-clause 14.2 of ISA-S82.01 and not live according to Sub-clause 9.2. of this standard.

9.5 Dielectric voltage-withstand tests
9.5.1 Supply circuits

(1) Type test.

The requirements of Sub-clause 9.12.2 (1) of ISA-S82.01 shall apply.

Exception: Circuits that are power-limited according to Sub-clause 14.2 of ISA-S82.01, and not live according to Sub-clause 9.2.1 of this standard.

(2) Routine test.

The requirements of Sub-clause 9.12.2(2) of ISA-S82.01 shall apply except for the test potentials.

The test potential, for supply circuits rated greater than extra-low voltage according to Sub-clause 14.1 of ISA-S82.01, shall be 1200 V rms at supply circuit frequency, or 1700 V dc, applied for one second. Alternately, the test potential shall be 1000 V rms at supply circuit frequency, or 1420 V dc, applied for one minute.

The test potential for extra-low voltage supply circuits according to Sub-clause 14.1 of ISA-S82.01 shall be 600 V rms at supply circuit frequency, or 860 V dc, applied for one second. Alternately, the test potential shall be 500 V rms at supply circuit frequency, or 707 V dc applied for one minute.

Exception: Circuits that are power-limited according to Sub-clause 14.2 of ISA-S82.01 and not live according to Sub-clause 9.2. of this standard.

10. Testing under Fault Conditions

10.1 Application of fault conditions. Fault conditions shall be applied to the equipment in the manner described in Clause 10 of ISA-S82.01.

11. Mechanical Requirements

11.1 General. Equipment enclosures, or parts of the enclosure, required to be in place to comply with the requirements in this standard shall have mechanical strength and stability necessary to resist the abuse to which they may be subjected in their intended uses. The requirements of Clause 11 of ISA-S82.01 shall apply.

Additional factors to be considered when evaluating process measurement and control equipment enclosures with regard to their intended uses include but are not limited to:

(1) resistance to corrosion,

(2) resistance to harmful solvents and gases.

12. Protection from Fire

12.1 General. The requirements of Clause 12 of ISA-S82.01 shall apply.

13. Component Requirements and Applications

13.1 General. The requirements of Clause 13 of ISA-S82.01 shall apply and the requirements according to Sub-clause 13.2 below.

13.2 Protective coatings. Where protective coatings are used according to Sub-clause 9.4 of this standard the suitability of the coating shall be determined by subjecting the equipment or the coated part of the equipment being evaluated to the tests specified in Sub-clauses 13.2.1 through 13.2.4. The equipment shall not be operated during aging and humidity conditioning.

13.2.1 Aging

The equipment (or the coated part of the equipment) being evaluated shall be aged by maintaining it at 90° C ± 1° C (194°F ± 2° F) for 96 hours. The test sample shall be removed from the aging chamber and subjected to a dielectric voltage-withstand test according to Sub-clause 13.2.3 of this standard.

13.2.2 Humidity conditioning

Following aging according to Sub-clause 13.2.1 of this standard, the equipment or the coated part of the equipment being evaluated shall be conditioned by maintaining it at 23° C ± 1° C (73° F ± 2° F) and 96 ± 2 percent relative humidity for 96 hours. The test sample shall be removed from the conditioning chamber and subjected to a dielectric voltage-withstand test according to Sub-clause 13.2.3 of this standard.

13.2.3 Dielectric voltage-withstand test

A test voltage according to Table 9-5 of ISA-S82.01 shall be applied for one minute without breakdown between adjacent circuits where coated spacings exist.

13.2.4 Adhesion

Following the tests of Sub-clause 13.2.1 and 13.2.2 above, the equipment or the coated part of the equipment being evaluated shall be investigated for adhesion of the coating, where coated spacings exist, by scraping or cutting. The coating shall not flake.

14. Extra-Low Voltage and Power-Limited Circuits

14.1 General. The requirements of Clause 14 of ISA-S82.01 shall apply.

15. Terminal Devices

15.1 General. The requirements of Clause 15 of ISA-S82.01 shall apply and the requirements according to Sub-clause 15.2 below.

15.2 Terminal screws. Wire-binding screws for field wiring terminals shall be selected according to Table 15-1.

**Table 15-1
Wire-Binding Screws**

Wire Size (AWG)	10	12	14
Minimum Binding Screw Size	10	8	6

Compliance is checked by inspection.

16. External Cords

16.1 General. The requirements of Clause 16 of ISA-S82.01 shall apply.

17. Equipment Instructions

17.1 General. The requirements of Clause 17 of ISA-S82.01 shall apply and the requirements according to Sub-clauses 17.2 and 17.3 below.

Additional applicable safeguard information considered appropriate by the manufacturer may be included.

Instructions according to this Clause may be supplemented, but not replaced, by illustrations.

Compliance is checked by inspection.

17.2 Form of equipment instruction. Equipment instructions shall be in the form of either:

(1) a pamphlet, a folder, or data sheet(s) containing operator instructions and operational maintenance information; or

(2) a combined operators' and servicing manual; or

(3) an operators' manual and a separate servicing manual.

17.3 Equipment information and instructions. Operating and operational maintenance instructions shall be separated from servicing instructions within the equipment instructions.

17.3.1 Operating instructions. With regard to equipment protections, the operating instructions shall:

(1) describe the equipment installation, including specifically:

   (a) assembly, if required, and mounting,

   (b) protective grounding means, if employed,

   (c) intended supply circuit and connection of the equipment to the supply circuit,

   (d) ventilation consideration,

   (e) the environmental conditions under which the equipment can be operated, and

   (f) additional protective means, if any, for equipment operated in sheltered or outdoor locations;

(2) explain equipment markings, including specifically

   (a) symbols

   (b) controls

   (c) terminal ratings;

(3) identify and describe interconnection with

   (a) auxiliary and accessory equipment

(b) other equipment;

(4) where operating instructions and servicing instructions are combined into a single manual and where servicing requires access to parts which could render electrical shock or replacement of a like component(s) is depended upon to provide extra-low voltage and power-limited circuits according to Clause 14 of ISA-S82.01, servicing instructions shall be preceded by a warning.

The warning shall be prefaced by the signal word "WARNING", and

(a) state that servicing requires access to parts which could render an electrical shock, and refer servicing only to qualified personnel;

(b) state that extra-low voltage and power-limited circuits are dependent upon replacement of live components;

(5) include operational maintenance information.

17.3.2 Servicing instructions. With regard to equipment protections, the servicing instructions shall include specifically:

(1) equipment calibration other than factory calibration;

(2) equipment maintenance;

(3) equipment repair, including any limitations thereto.

18. Pressure-Actuated Equipment

18.1 General. The requirements of this Clause are intended to:

(1) provide a means by which the pressure protections of fluid-pressure-actuated equipment can be verified (type tested);

(2) apply to fluid-pressure-actuated equipment employing flexible-metal bellows, diaphragms, Bourdon tubes, or the like that are rated for pressures between 2000 kPa and 175,000 kPa (300 and 25,000 psig); and

(3) apply to use of fluid-pressure-actuated equipment with ordinary benign fluids.

Equipment rated above 175,000 kPa (25,000 psig) or actuated by hazardous fluids is special-purpose equipment outside the scope of this standard.

Compliance is checked by inspection and by performing the tests according to Sub-clause 18.2 below.

The maximum rated operating pressure used in conjunction with pressure test multipliers is that which is marked on the equipment. In the case of differential pressure equipment, this marked rating would be the maximum static (working) pressure.

The maximum rated overrange pressure (maximum pressure which may be applied without permanent change in performance) is used in conjunction with pressure test multipliers only when marked on the equipment.

18.2 Hydrostatic test

The part of the equipment that is normally subjected to the actuating fluid pressure shall be:

(1) filled with a suitable liquid, such as water, to exclude air, and

(2) connected to a hydraulic pump, and the pressure shall be raised gradually to the hydrostatic pressure indicated in the Sub-clauses below.

Those portions of the equipment which normally receive indirect pressure loading, as in hydraulically coupled systems, shall be simultaneously subjected to the hydrostatic test pressure either through the original hydraulic filling fluid or, in its absence, by filling with the test liquid.

18.2.1 Initial hydrostatic test

The following test sequence shall be followed:

(1) The equipment shall withstand a hydrostatic pressure for one minute applied in accordance with Column II of Table 18-1.

No visible leakage shall occur.

(2) The equipment shall withstand a hydrostatic pressure for one minute applied in accordance with Column III of Table 18-1.

No rupture or failure which results in flying fragments outside of the equipment shall occur.

Leakage may occur because of splits in Bourdon tubes, diaphragms, or bellows or because of joint or gasket failure. These are not considered test failures if the hydrostatic pressure can be maintained for one minute.

18.2.2 Hydrostatic test-modifications to minimize leakage. If excessive leakage occurs in tests according to

Sub-clause 18.2.1(2) above such that the hydrostatic pressure cannot be maintained for one minute, certain equipment modifications shall be permitted:

(1) External fittings may be modified to eliminate leakage.

(2) A leaking gasket or flexible seal member (not part of the measuring element) which serves as a structural partition (barrier) between that part of the equipment normally subjected to the actuating fluid pressure and the external components (enclosure) may be replaced by a stronger non-functional member.

The modified equipment shall then comply with the requirements of Sub-clause 18.2.1(2) above.

If modifications are made in the structural partition, the modified equipment shall also comply with the requirements of Sub-clause 18.2.4 below.

18.2.3 Hydrostatic tests--under conditions of excessive leakage. Where excessive leakage cannot be successfully reduced according to the modifications of Sub-clause 18.2.2 above so that the leakage, in fact, serves as a pressure relief mechanism, the equipment complies with this Clause when tested according to the following requirements:

(1) Where no enclosure is provided, the equipment shall withstand, for one minute, a hydrostatic pressure applied in accordance with Column IV of Table 18-1.

No rupture or failure which results in flying fragments outside of the equipment shall occur.

(2) Where an enclosure is provided, the equipment shall withstand, for one minute, a hydrostatic pressure applied in accordance with Column IV of Table 18-1.

No rupture or failure which results in flying fragments outside of the equipment shall occur.

The equipment shall also comply with the requirements of Sub-clause 18.2.4 below.

An enclosure is an unpressured case, cover or housing, which may enclose all or part of the pressure-actuated equipment. The enclosure is not subjected to the actuating fluid pressure under normal operation.

18.2.4 Supplementary tests--under conditions of excess leakage. One of the following requirements shall be met for equipment tested according to Sub-clauses 18.2.2 or 18.2.3(2) above:

(1) The enclosure shall leak at a rate sufficient to prevent a pressure buildup without rupture or failure which results in flying fragments outside of the equipment; or

"Sufficient" means a leakage rate at least equal to the leakage rate of the primary structure.

(2) The enclosure shall withstand, without rupture or failure which results in flying fragments outside of the equipment, a pressure equal to the maximum operating pressure of the equipment; or

(3) The unaltered structural partition shall withstand a hydrostatic pressure in accordance with Column IV of Table 18-1.

No visible leakage shall occur.

Table 18.1
Test Pressure

Col. I	Col. II	Col. III	Col. IV
Marked Maximum Operating Pressure Rating	Test Pressure for Sub-clause 18.2.1 (1)	Test Pressure for Sub-clauses 18.2.1 (2) & 18.2.2	Test Pressure for Sub-clauses 18.2.1 (1) & (2) & 18.2.4 (3)
2100-14,000 kPa (300 - 2,000 psig)	(a) 2.0 Times Rated Pressure	(a) 3 Times Rated Pressure	(a) 2.5 Times Rated Pressure
Over 14,000 - 70,000 kPa (over 2,000 - 10,000 psig)	1.75 Times Rated Pressure plus 3500 kPa (500 psig)	2.5 Times Rated Pressure plus 7000 kPa (1,000 psig)	2.0 Times Rated Pressure plus 7,000 kPa (1000 psig)
Over 70,000 - 175,000 kPa (over 10,000 - 25,000 psig)	1.3 Times Rated Pressure plus 35,000 kPa (5,000 psig)	2 Times Rated Pressure plus 42,000 kPa (6,000 psig)	1.5 Times Rated Pressure plus 42,000 kPa (6,000 psig)

(a) Marked Maximum Operating Pressure or Maximum Overrange Rating

19. Equipment Enclosures for Use in Extended Environmental Conditions

19.1 General. This Clause specifies additional protections for equipment in an enclosure intended to be used outdoors or in sheltered locations under the environmental conditions according to Sub-clauses 1.2.2 and 1.2.3 of ISA-S82.03.

Enclosures intended to be used under the environmental conditions according to Sub-clauses 1.2.2 and 1.2.3 of ISA-S82.03 shall be a NEMA enclosure type according to Table 19-1 and shall comply with the applicable requirements of ANSI/NEMA-250, "Enclosures for Electrical Equipment" for the specific enclosure type selected.

Enclosures intended to be used under extended environmental conditions are designated by a type number which is intended to indicate the environmental conditions for which they are suitable. The higher the type number, the higher the degree of protection provided by the enclosure.

Other ANSI/NEMA-250 type enclosures may be suitable for application in sheltered and outdoor locations provided the requirements of Clause 19 of ISA-S82.03 have been satisified.

Equipment which, due to transport, may occasionally be subjected to condensation is not considered to require protections for extended environmental conditions.

Equipment intended to be used indoors where falling dirt, liquids, splashing water, or dust is expected, may require additional protections according to Table 19-1.

Compliance is checked by performing the tests according to ANSI/NEMA-250 for the specified enclosure type.

Equipment is tested in a non-energized state unless otherwise specified.

Table 19-1
Enclosure Types Suitable for Use in Sheltered and Outdoor Locations

Environment	3	3R	3S	4	4X	5
Sheltered locations (see sub-clause 1.2.2 of ISA-S82.03)				X	X	X
Outdoor (see sub-clause 1.2.3 of ISA-S82.03)	X	X	X	X	X	

*See ANSI/NEMA-250-1986.

APPENDIX

COMPARISON BETWEEN NEMA ENCLOSURE TYPE NUMBERS AND IEC ENCLOSURE CLASSIFICATION DESIGNATIONS

IEC Publication 529 *Classification of Degrees of Protection by Enclosures* provides a system for specifying the enclosures of electrical equipment on the basis of the degree of protection provided by the enclosure. IEC 529 does not specify degrees of protection against mechanical damage of equipment, risk of explosions, or conditions such as moisture (produced, for example, by condensation), corrosive vapors, fungus, or vermin. NEMA Standards Publication 250 does test for environmental conditions such as corrosion, rust, icing, oil, and coolants. For this reason, and because the tests and evaluations for other characteristics are not identical, the IEC Enclosure Classification Designations cannot be exactly equated with NEMA Enclosure Type Numbers.

The IEC designation consists of the letters IP followed by two numerals. The first characteristic numeral indicates the degree of protection provided by the enclosure with respect to persons and solid foreign objects entering the enclosure. The second characteristic numeral indicates the degree of protection provided by the enclosure with respect to the harmful ingress of water.

Table A-1 provides an equivalent conversion from NEMA Enclosure Type Numbers to IEC Enclosure Classification Designations. The NEMA Types meet or exceed the test requirements for the associated IEC Classifications; for this reason Table A-1 cannot be used to convert from IEC Classifications to NEMA Types.

Table A-1
Conversion of NEMA Type Numbers to IEC Classification Designations
(Cannot be used to convert IEC Classification Designations to NEMA Type Numbers)

NEMA Enclosure Type Number	IEC Enclosure Classification Designation
1	IP10
2	IP11
3	IP54
3R	IP14
3S	IP54
4 and 4X	IP56
5	IP52
6 and 6P	IP67
12 and 12K	IP52
13	IP54

NOTE: This comparison is based on tests specified in IEC Publication 529.

*Reprinted from ANSI/NEMA-250-1986, "Enclosures for Electrical Equipment," with permission of the copyright holder, National Electrical Manufacturer's Association.

ANSI/MC96.1-1982
(Revision of ANSI-MC96.1-1975)

American National Standard

Temperature Measurement Thermocouples

Instrument Society of America

AMERICAN NATIONAL STANDARD

An American National Standard implies a consensus of those substantially concerned with its scope and provisions. An American National Standard is intended as a guide to aid the manufacturer, the consumer, and the general public. The existence of an American National Standard does not in any respect preclude anyone, whether he has approved the standard or not, from manufacturing, marketing, purchasing, or using products, processes, or procedures not conforming to the standard. American National Standards are subject to periodic review and users are cautioned to obtain the latest editions.

CAUTION NOTICE: This American National Standard may be revised or withdrawn at any time. The procedures of the American National Standards Institute require that action be taken to reaffirm, revise, or withdraw this standard no later than five years from the date of publication. Purchasers of American National Standards may receive current information on all standards by calling or writing the American National Standards Institute.

This standard is the revision of American National Standard, C96.1 "Temperature Measurement Thermocouples." C96.1, sponsored by the Instrument Society of America, originally approved by the United States of America Standards Institute on June 9, 1964, and reaffirmed without change by the American National Standards Institute in 1969. Subsequently, a revised version was approved by ANSI in 1975 with the designation ANSI-MC96.1-1975. This current revision was approved by ANSI on August 12, 1982 with the designation ANSI-MC96.1 - 1982 (Revision of ANSI-MC96.1-1975).

ISBN 0-87664-708-5

ANSI-MC96.1-1982 Temperature
Measurement Thermocouples

Copyright © by the instrument Society of America 1982. All rights reserved. Printed in the United States of America. No part of this publication may be reproduced, sorted in a retrieval system, or transmitted, in any form or any means electronic, mechanical, photocopying, recording or otherwise without the prior written permission of the publisher.

INSTRUMENT SOCIETY OF AMERICA
67 Alexander Drive
P.O. Box 12277
Research Triangle Park, North Carolina 27709

Copyright © by the Instrument Society of America 1982

ACKNOWLEDGMENTS

ANSI - C96 COMMITTEE

EDWARD D. ZYSK - Chairman
Engelhard Minerals & Chemicals Corporation

LOIS M. FERSON - Secretary
Instrument Society of America

ORGANIZATIONAL LIAISONS

AMERICAN SOCIETY OF MECHANICAL ENGINEERS

ROBERT P. BENEDICT - Liaison
Westinghouse Electric Corporation

K. WOODFIELD - Alternate
General Motors Institute

AMERICAN SOCIETY FOR TESTING AND MATERIALS

DONALD I. FINCH - Consultant (Deceased)

IEC/TC65-WG5

GEORGE J. CHAMPAGNE - Liaison
The Foxboro Company

EDWARD D. ZYSK - Alternate
Engelhard Minerals & Chemicals Corporation

NATIONAL BUREAU OF STANDARDS

GEORGE W. BURNS - Liaison
National Bureau of Standards

OIML PS12/RS5

EDWARD D. ZYSK - Technical Advisor
Engelhard Corp.

SOCIETY OF AUTOMOTIVE ENGINEERS
R. B. CLARK - Liaison
General Electric Company

UNITED STATES AIR FORCE

JAMES E. ORWIG - Liaison
USAF Aerospace Guidance &
Metrology Center

INSTRUMENT SOCIETY OF AMERICA

PHILIP BLISS - Liaison
Consultant

EDWARD Z. ZYSK - Alternate
Engelhard Minerals & Chemicals Corporation

INDIVIDUAL MEMBERS

ROY F. ABRAHMSEN
Combustion Engineering, Inc.

J. A. BARD
Matthey Bishop, Inc.

ALEX H. CLARK
Leeds & Northrup Co.

CLINTON R. DODD
Driver-Harris Co.

A. E. GEALT
Honeywell

WILEY W. JOHNSTON, JR
Consultant

HENRY L. KURTZ
Driver-Harris Co.

EDWIN L. LEWIS
Consultant

JOHN D. MITILINEOS
Sigmund Cohn Corporation

LLOYD J. PICKERING
Claud S. Gordon Company

R. A. PUSTELL
General Electric Co.

T. P. WANG
Wilber B. Driver Company

J. D. WILLIAMS
Claud S. Gordon Company

TABLE OF CONTENTS

Section	Title	Page
	Foreword	7
1.	Coding of Thermocouple Wire and Extension Wires	9
2.	Terminology, Wire Size, Upper Temperature Limit, and Initial Calibration Tolerance for Thermocouples and Extension Wire	12
	2.1 Scope and Purpose	12
	2.2 Terminology and Symbols	12
	2.3 Wire Sizes	14
	2.4 Upper Temperature Limits	14
	2.5 Tolerance of Initial Calibration	15
3.	Non-Ceramic Insulation of Thermocouple and Extension Wires	16
4.	Temperature-EMF Tables for Thermocouples	17
	4.1 Scope and Purpose	17
	4.2 Introduction	17
	4.3 Use of Temperature-EMF Tables	17

Appendices

A	Bare Thermocouple Element Fabrication	37
	A.1 General	37
	A.2 Thermocouple Wires	37
	A.3 Joining Thermocouple Wires	37
B	Sheathed Thermocouple Element Fabrication	
	B.1 General	40
	B.2 Special Equipment	40
	B.3 General Precautions	40
	B.4 Measuring Junction Fabrication	40
C	Thermocouples and Thermocouple Extension Wires - Selection, Assembly, and Installation	
	C.1 Scope and Purpose	41
	C.2 Types and Uses	41
	C.3 Assembly	42
	C.4 Installation Considerations for Thermocouples	44
	C.5 Installation of Extension Wires	45
D	Thermocouples - Checking Procedures	46
	D.1 General	46
	D.2 Scope and Purpose	46
	D.3 Procedure	46
	Bibliography	47

LIST OF ILLUSTRATIONS

Figure	Title	Page
1	Thermocouple Elements	12
2	Thermocouple Element with Terminal Block	12
3	Thermocouple Element with Connection Head	13
4	Connection Head	13
5	Protecting Tube	13
6	Protecting Tube with Mounting Bushing	13
7	Protecting Tube with Mounting Flange	13
8	Thermocouple Element with Protecting Tube and Connection Head	13
9	Open End Protecting Tube	13
10	Well	13
11	Thermocouple Assembly with Well	13
12	Immersion and Insertion Lengths for Thermocouple Assembly with Thermowell	14

LIST OF ILLUSTRATIONS (continued)

Figure	Title	Page
A-1.	Method of Twisting Wires for Gas and Electric Arc Welding	37
A-2.	Method of Forming Metal Wires for Resistance Welding	37
A-3.	Formed Butt Welded Thermocouple	38
A-4.	Method of Forming Metal Wires for Electric Arc Welding	38
A-5.	Neutral Flame for Gas Welding	38
B-1.	Typical Exposed Junction	40
B-2.	Fixturing For Weldment	40
B-3.	Cutaway View of Grounded Junction	41
B-4.	Cutaway View of Ungrounded Junction	41

LIST OF TABLES

Table	Title	Page
1	Thermocouple Type Letter Designations	9
2	Symbols for Types of Thermocouple Wire	10
3	Symbols for Types of Extension Wire	10
4	Color Code - Duplex Insulated Thermocouple Wire	11
5	Color Code - Single Conductor Insulated Thermocouple Extension Wire	11
6	Color Code - Duplex Insulated Thermocouple Extension Wire	11
7	Recommended Upper Temperature Limits for Protected Thermocouples Upper Temperature Limit for Various Wire Sizes	15
8	Initial Calibration Tolerances for Thermocouples	15
9	Initial Calibration Tolerances for Thermocouple Extension Wires	16
10	Initial Calibration Tolerances for Thermocouple Compensating Extension Wire	16
11	Temperature-EMF for Type B Thermocouples, Reference Junction at 0°C	18
12	Temperature-EMF for Type E Thermocouples, Reference Junction at 0°C	21
13	Temperature-EMF for Type J Thermocouples, Reference Junction at 0°C	23
14	Temperature-EMF for Type K Thermocouples, Reference Junction at 0°C	25
15	Temperature-EMF for Type R Thermocouples, Reference Junction at 0°C	28
16	Temperature-EMF for Type S Thermocouples, Reference Junction at 0°C	31
17	Temperature-EMF for Type T Thermocouples, Reference Junction at 0°C	34

FOREWORD

(This Foreword is included for information purposes and is not part of ANSI MC96.1)

The development of this American National Standard has resulted from the work of the American National Standards Committee on Temperature Measurement, MC96. The Committee was organized in 1946 under the sponsorship of the Instrument Society of America, the scope of the Committee being designated as follows:

Requirements for temperature measurement thermocouples, including terminology, fabrication, wire sizes, installation, color codes of thermocouple and thermocouple extension wire, Temperature-EMF tables and tolerances have been coordinated with the International Electrotechnical Commission (IEC).

Credit must be given to the National Bureau of Standards and to Committee E20 on Temperature Measurement of the American Society for Testing and Materials for the development of the temperature-EMF tables and for recommendations as to the maximum recommended temperature of the various materials. Special credit must also be given to G. W. Burns, NBS-Washington, D.C., and Dr. Robert Powell, formerly with NBS-Boulder, for providing the thermocouple reference tables.

This Standard has been prepared as a part of the service of the Instrument Society of America toward a goal of uniformity in the field of instrumentation. To be of real value this document should not be static, but should be subjected to periodic review. Toward this end the Society welcomes all comments and criticisms, and asks that they be addressed to the Standards and Practices Board Secretary, Instrument Society of America, P.O. Box 12277, Research Triangle Park, N.C. 27709.

In 1821, Seebeck discovered that, in a closed circuit made up of wires of two dissimilar metals, electric current will flow if the temperature of one junction is elevated above that of the other. In 1886, Le Chatelier introduced a thermocouple consisting of one wire of platinum and the other of 90 percent platinum-10 percent rhodium. This combination, Type S, is still the international standard for purposes of calibration and comparison, and defines the International Practical Temperature Scale of 1968 from the antimony to the gold point. This type of thermocouple was made and sold by W. C. Heraeus, GmbH of Hanau, Germany, and is sometimes called the Heraeus Couple. Somewhat later, it was learned that a thermoelement composed of 87 percent platinum and 13 percent rhodium, Type R, would give a somewhat higher EMF output. This type is frequently used in industry. In 1954 a thermocouple was introduced in Germany whose positive leg is an alloy of platinum and 30 percent rhodium. Its negative leg is also an alloy of platinum and 6 percent rhodium. This combination, Type B, gives somewhat greater physical strength and greater stability, and can withstand somewhat higher temperature than types R and S.

In an effort to find less costly metals for use in thermocouples, a number of combinations were tried. Iron and nickel were useful and inexpensive. Pure nickel, however, becomes very brittle upon oxidation; and it was learned that an alloy of about 55 percent copper, 45 percent nickel originally known as constantan would eliminate this problem. This alloy combination, iron-constantan, has since been widely used and is designated Type J. The present calibration for Type J was established by the National Bureau of Standards (see NBS Monograph 125).

In an effort to find a couple useful to higher temperatures than the iron versus copper-nickel combination, a 90 percent nickel-10 percent chromium alloy as a positive wire, and a 95 percent nickel-5 percent aluminum, manganese, silicon alloy as a negative wire was developed. This combination (originally called Chromel-Alumel) is known as Type K. Similar alloys for specific applications have since become available, to the same curve.

Another combination, copper versus copper-nickel, Type T, is used particularly at below-zero temperatures. The temperature-EMF Reference Table was prepared by the National Bureau of Standards in 1938 and revised in NBS Monograph 125.

The Type E Thermocouple, 90 percent nickel-10 percent chromium versus copper-nickel, is receiving increasing attention and use where corrosion of small diameter iron wire is a problem and a higher EMF output is desirable.

Further information on the letter designated type thermocouples is given in Appendix C.

Several combinations using tungsten, rhenium and their binary alloys are widely used at high temperatures in inert or reducing atmospheres, and are nearing acceptance as standard.

For additional information on temperature measurement thermocouples, reference may be made to NBS Special Publication 300, Volume II, "Precision Measurement and Calibration-Temperature," 1968 and to NBS Monographs 124 and 125, published by United States Department of Commerce, National Bureau of Standards. Specific attention is called to the reference categories on Thermoelectric Theory and Calibration, and Thermoelectric Devices. Additional information is in STP-470B, "Manual on the Use of Thermocouples," 1981, published by the American Society for Testing and Materials.

For many years, letter designations have been assigned by ANSI Committee MC96 and endorsed by international standards as a device to identify certain common types without using proprietary trade names, and to associate them with temperature-emf relationships established by the National Bureau of Standards. Color codes for the insulation of letter-designated wires are also assigned by MC96 to facilitate identification in the field. The assignment of a letter designation and/or color code by MC96 constitutes an acknowledgment of an existing recognition by NBS of a defining temperature-emf relationship and an existing general usage, and does not constitute an endorsement of the thermocouple type by ISA, ANSI, and NBS. The letter designation applies only to the temperature-emf relationship and not to the material. Other material, having different temperature-emf relationships, may well be equivalent or superior in some applications.

The use of the letter X to indicate thermocouple extension wire appeared obvious. The use of the term lead wire, or compensating lead wire, is to be discouraged because it frequently is confused with the term lead (element).

Much discussion was involved in the use of the color red to designated polarity, since red is used popularly in electrical circuits to indicate positive. No nationally-accepted code known to the committee covered this point. Research into manufacturers' records showed that, in thermocouple circuits, the red negative had been in use for more than forty years.

The colors used to designate the various compositions and combination of thermocouple and extension wire were originally selected upon an almost arbitrary basis. Colors which had been used by large manufacturers were given very careful consideration and comparison so that as few changes as possible would be required to establish uniformity. Millions of miles of wire with these color codes are presently in use.

In ANSI-MC96.1 thermocouple and thermocouple extension wires are designated by letters. This has been done primarily to eliminate the use of proprietary names. The designations are given in Table 1 of the text.

The ISA Standards and Practices Department is aware of the growing need for attention to the metric system of units in general, and the International System of Units (SI) in particular, in the preparation of instrumentation standards. The Department is further aware of the benefits to USA users of ISA Standards of incorporating suitable references to the SI (and the metric system) in their business and professional dealings with other countries. Toward this end this Department will endeavor to introduce SI and SI-acceptable metric units as optional alternatives to English units in all new and revised standards to the greatest extent possible. The ASTM "Metric Practice Guide," endorsed and published as National Bureau of Standards Handbook 102 and as ANSI Z210.1, is the reference guide for definitions, symbols, abbreviations and conversion factors.

1. CODING OF THERMOCOUPLE WIRE AND EXTENSION WIRES

This standard applies to thermocouples and extension wires.

Its purpose is to establish uniformity in the designation of thermocouples and extension wires and to provide, by means of the color of its insulation, an identification of its type or composition as well as its polarity when used as part of a thermocouple system.

TABLE 1

THERMOCOUPLE TYPE LETTER DESIGNATIONS

Type	Nominal Temperature Range	Temperature-EMF Relationship Data	Material Identification* (Positive Material in Caps)**
B	0 to 1820°C	Refer to Table 11	PLATINUM-30 PERCENT RHODIUM versus platinum-6 percent rhodium
E	−270 to 1000°C	Refer to Table 12	NICKEL-10 PERCENT CHROMIUM† versus copper-nickel
J	−210 to 760°C	Refer to Table 13	IRON versus copper-nickel
K	−270 to 1372°C	Refer to Table 14	NICKEL-10 PERCENT CHROMIUM † versus nickel-5 percent (aluminum, silicon) ††
R	−50 to 1768°C	Refer to Table 15	PLATINUM-13 PERCENT RHODIUM versus platinum
S	−50 to 1768°C	Refer to Table 16	PLATINUM-10 PERCENT RHODIUM versus platinum
T	−270 to 400°C	Refer to Table 17	COPPER versus copper-nickel

* Any combination of thermocouple materials having EMF-temperature relationships within the tolerances for any of the above-mentioned tables shall bear that table's appropriate type letter designation.

** The indicated polarity of the thermocouple materials applies for conditions when the measuring junction is at higher temperatures than the reference junction.

† It should not be assumed that thermoelements used with more than one thermocouple type are interchangeable or have the same millivolt limits of error.

†† Silicon, or aluminum and silicon may be present in combination with other elements.

TABLE 2

SYMBOLS FOR TYPES OF THERMOCOUPLE WIRE

Type*	Thermoelements	
	Positive	Negative
B	BP	BN
E	EP	EN
J	JP	JN
K	KP	KN
R	RP	RN
S	SP	SN
T	TP	TN

* Any thermocouple material having temperature-EMF relationships within the tolerances for any of the above-mentioned tables shall bear that table's appropriate ''type-letter'' designation. Identification of some typical materials is contained in Appendix C (Table C-1).

TABLE 3

SYMBOLS FOR TYPES OF EXTENSION WIRE

Type	Combination	Positive	Negative
B	BX**	BPX	BNX
E	EX	EPX	ENX
J	JX	JPX	JNX
K	KX	KPX	KNX
R or S	SX*	SPX	SNX
T	TX	TPX	TNX

* Both Type R or S Thermocouples use the same SX compensating extension wire.

** Special compensating extension wires are not required for reference junction temperatures up to 100°C. Generally copper conductors are used. However, proprietary alloys may be obtained for use at higher reference junction temperatures.

NOTE: Identification of some typical materials is contained in Appendix C (Table C-3).

TABLE 4
COLOR CODE - DUPLEX INSULATED THERMOCOUPLE WIRE

Thermocouple			Color of Insulation		
Type	Positive	Negative	Overall*	Positive*	Negative
E	EP	EN	Brown	Purple	Red
J	JP	JN	Brown	White	Red
K	KP	KN	Brown	Yellow	Red
T	TP	TN	Brown	Blue	Red

* A tracer color of the positive wire code color may be used in the overall braid.

TABLE 5
COLOR CODE - SINGLE CONDUCTOR INSULATED THERMOCOUPLE EXTENSION WIRE

Extension Wire Type			Color of Insulation	
Type	Positive	Negative	Positive	Negative*
B	BPX	BNX	Gray	Red-Gray Trace
E	EPX	ENX	Purple	Red-Purple Trace
J	JPX	JNX	White	Red-White Trace
K	KPX	KNX	Yellow	Red-Yellow Trace
R or S	SPX	SNX	Black	Red-Black Trace
T	TPX	TNX	Blue	Red-Blue Trace

* The color identified as a trace may be applied as a tracer, braid, or by any other readily identifiable means.

NOTE OF CAUTION: In the procurement of random lengths of single conductor insulated extension wire, it must be recognized that such wire is commercially combined in matching pairs to conform to established temperature-EMF curves. Therefore, it is imperative that all single conductor insulated extension wire be procured in pairs, at the same time, and from the same source.

TABLE 6
COLOR CODE - DUPLEX INSULATED THERMOCOUPLE EXTENSION WIRE

Extension Wire Type			Color of Insulation		
Type	Positive	Negative	Overall	Positive	Negative*
B	BPX	BNX	Gray	Gray	Red
E	EPX	ENX	Purple	Purple	Red
J	JPX	JNX	Black	White	Red
K	KPX	KNX	Yellow	Yellow	Red
R or S	SPX	SNX	Green	Black	Red
T	TPX	TNX	Blue	Blue	Red

* A tracer having the color corresponding to the positive wire code color may be used on the negative wire color code.

2. TERMINOLOGY, WIRE SIZE, UPPER TEMPERATURE LIMIT, AND INITIAL CALIBRATION TOLERANCE FOR THERMOCOUPLES AND EXTENSION WIRE

2.1 Scope and Purpose

This section applies to thermocouples and extension wire.

This section establishes terminology, symbols, normal wire size, recommended upper temperature limit, and tolerance for thermocouples and extension wire.

2.2 Terminology and Symbols

2.2.1 Thermoelement

A thermoelement is one of the two dissimilar electrical conductors comprising a thermocouple.

2.2.2 Thermocouple

A thermocouple is two dissimilar thermoelements so joined as to produce a thermal emf when the measuring and reference junctions are at different temperatures.

1. *Measuring Junction.* The measuring junction is that junction of a thermocouple which is subjected to the temperature to be measured.

2. *Reference Junction.* The reference junction is that junction of a thermocouple which is at a known temperature or which is automatically compensated for its temperature.

NOTE: In normal industry practice the thermocouple element is terminated at the connection head. However, the Reference Junction is not ordinarily located in the connection head but is transferred to the instrument by the use of thermocouple extension wire.

2.2.3 Extension Wire

Extension wire is a pair of wires having such temperature-emf characteristics relative to the thermocouple with which the wires are intended to be used that, when properly connected to the thermocouple, the reference junction is transferred to the other end of the wires.

NOTE: Extension wires which are basically different in chemical composition from the thermocouple wires with which they are to be used are sometimes referred to as compensating extension wire. In this context, type SX and BX wires would be compensating extension wire and types TX, JX, EX, and KX wires would be extension wire.

2.2.4 Tolerances

The tolerance of a thermocouple or extension wire is the maximum allowable deviation in degrees from the standard emf-temperature values for the type of thermocouple in question when the reference junction temperature is at the ice point and the measuring junction is at the temperature to be measured.

2.2.5 Thermocouple Element

A thermocouple element is a pair of bare or insulated thermoelements joined at one end to form a measuring junction and intended for use as a thermocouple or as a part of a thermocouple assembly. (See Figure 1.)

The thermocouple element length is the overall length of the thermocouple element and is assigned the symbol A.

The thermocouple element diameter is the maximum transverse dimension of the insulated portion of the thermocouple element and is assigned the symbol Y.

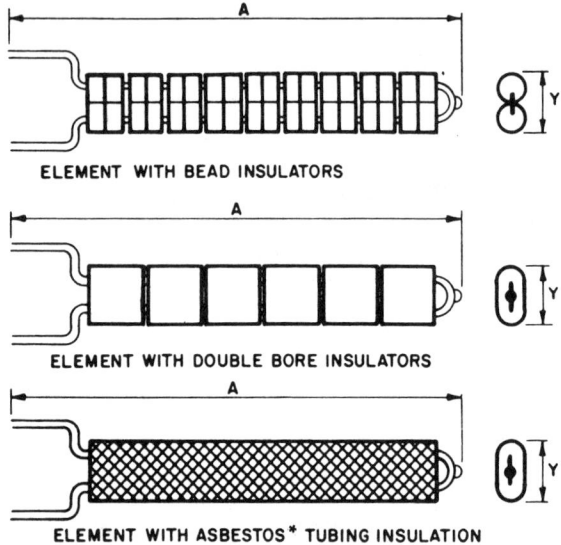

Figure 1. Thermocouple Elements

2.2.6 Thermocouple Assembly

A thermocouple assembly is an assembly consisting of a thermocouple element and one or more associated parts such as terminal block, connection head, and protecting tube.

1. *Terminal Block.* A terminal block is a block of insulating material that is used to support and join the terminations of conductors. (See Figure 2.)

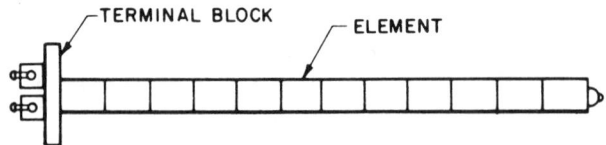

Figure 2. Thermocouple Element with Terminal Block

2. *Connection Head.* A connection head is a housing enclosing a terminal block for an electrical temperature sensing device and usually provided with threaded openings for attachment to a protecting tube and for attachment of conduit. (See Figures 3 and 4.)

---

*Asbestos is being replaced with safer high-temperature materials.

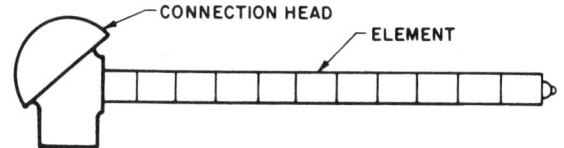

Figure 3. Thermocouple Element with Connection Head

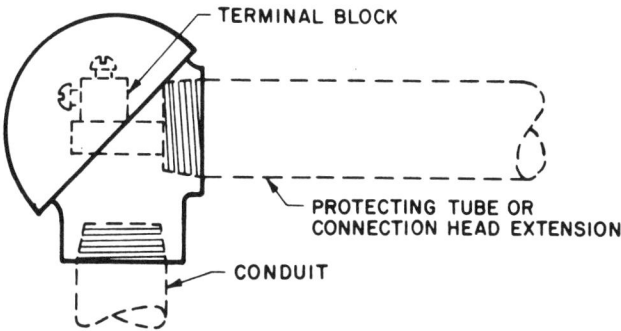

Figure 4. Connection Head

3. *Connection Head Extension.* A connection head extension is a threaded fitting or an assembly of fittings extending between the thermowell or angle fitting and the connection head.

The connection head extension length is the overall length of the connection head extension and is assigned the symbol N. (See Figure 11.)

4. *Protecting Tube.* A protecting tube is a tube designed to enclose a temperature sensing device and protect it from the deleterious effects of the environment. It may provide for attachment to a connection head but is not primarily designed for pressure-tight attachment to a vessel. A bushing or flange may be provided for the attachment of a protecting tube to a vessel. (See Figures 5, 6, 7, and 8.)

The protecting tube length is the overall length of a protecting tube and is assigned the symbol P. (See Figure 5.)

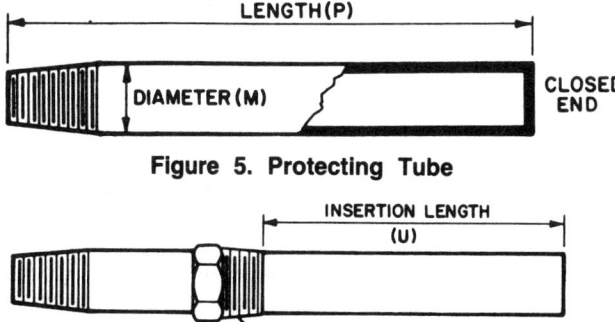

Figure 5. Protecting Tube

Figure 6. Protecting Tube with Mounting Bushing

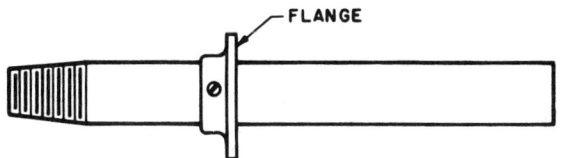

Figure 7. Protecting Tube with Mounting Flange

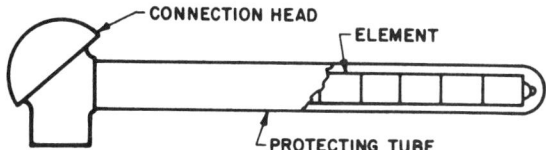

Figure 8. Thermocouple Element with Protecting Tube and Connection Head

The protecting tube diameter is the outside diameter of a protecting tube and is assigned the symbol M.

A protecting tube has one end closed unless it is specified as open end. (See Figure 9.)

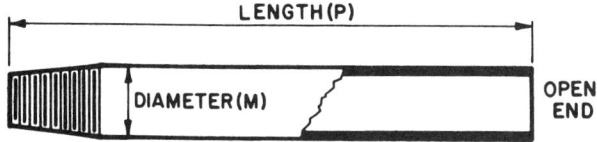

Figure 9. Open End Protecting Tube

5. *Thermowell.* A thermowell is a pressure-tight receptacle adapted to receive a temperature sensing element and provided with external threads or other means for pressure-tight attachment to a vessel.

A lagging extension is that portion of a thermowell above the threads, intended to extend through the lagging of a vessel. The lagging extension length is the length from the lower end of the external threads of the well to the outer end of the portion intended to extend through the lagging of a vessel, less one inch allowance for threads, and is assigned the symbol T. (See Figure 10.)

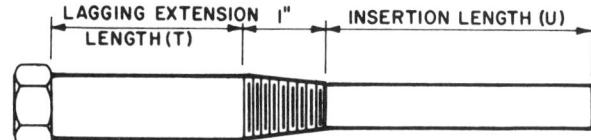

Figure 10. Well

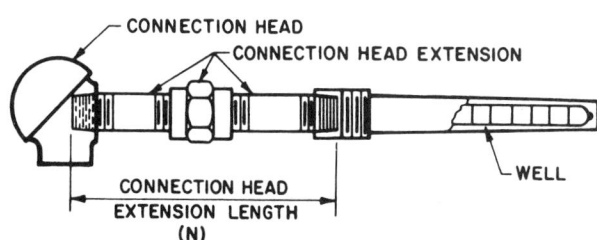

Figure 11. Thermocouple Assembly with Thermowell

The immersion length of a thermowell, protecting tube, or thermocouple element is the length from the free end to the point of immersion in the medium which is being measured and is assigned the symbol R. (See Figure 12.)

The insertion length of a thermowell, protecting tube or thermocouple element is the length from the free end to, but not including, the external threads or other means of attachment to a vessel and is assigned the symbol U. (See Figures 10 and 12.)

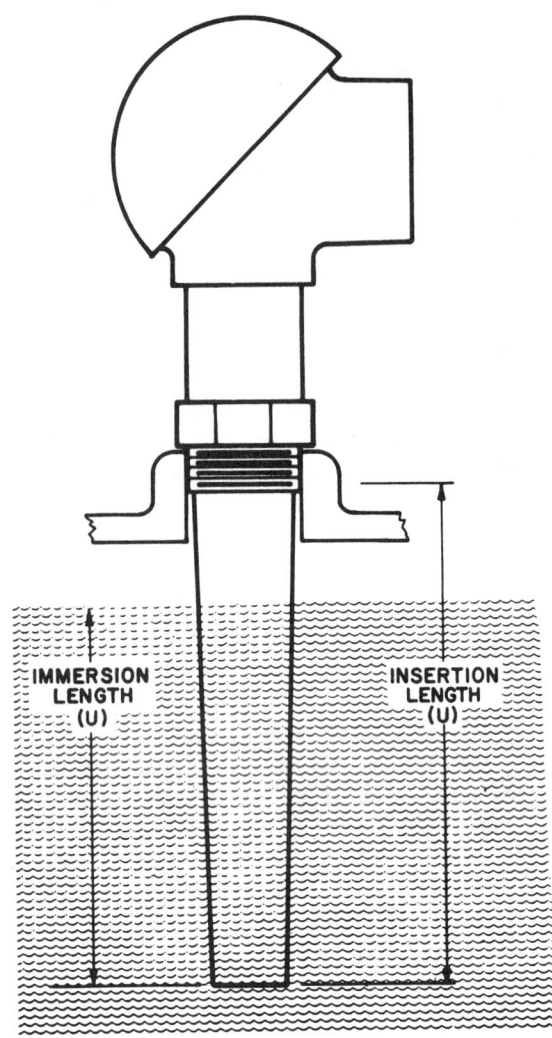

Figure 12. Immersion and Insertion Lengths for Thermocouple Assembly with Thermowell

2.2.7 Angle Type Thermocouple Assembly

An angle type thermocouple assembly is an assembly consisting of a thermocouple element, a protecting tube, an angle fitting, a connection head extension, and a connection head.

2.2.8 Other Forms of Thermocouples and Thermocouple Elements

1. *Coaxial Thermocouple Element.* A coaxial thermocouple element consists of a thermoelement in wire form within a thermoelement in tube form and electrically insulated from the tube except at the measuring junction.

2. *Sheathed Thermocouple.* A sheathed thermocouple is a thermocouple having its thermoelements, and sometimes its measuring junction, embedded in mineral oxide insulation compacted within a metal protecting tube.

SYMBOLS

	Symbols
Thermocouple element length	A
Thermocouple element diameter	Y
Connection head extension length	N
Protecting tube length	P
Protecting tube diameter	M
Lagging extension length	T
Immersion length	R
Insertion length	U

2.3 WIRE SIZES

2.3.1 Thermocouples

The wire sizes normally used for non-sheathed thermocouples are as follows:

For J, K, and E: 8, 14, 20, 24 and 28 AWG*

For T: 14, 20, 24, and 28 AWG*

For R, S and B: 24 AWG* only

2.3.2 Extension wires

The wire sizes normally used for extension wire, either singly or in pairs, are 14, 16, and 20 AWG*. Sixteen (16) gage is most commonly used. Twenty (20) gage and smaller may be used when bundled and reinforced to provide strength for pulling. These sizes apply to all types of extension wires.

2.4 UPPER TEMPERATURE LIMITS

Table 7 gives the recommended upper temperature limits for the various thermocouples and wire sizes. These limits apply to protected thermocouples in conventional closed-end protecting tubes. They do not apply to sheathed thermocouples having compacted mineral oxide insulation.

In any general recommendation of thermocouple temperature limits, it is not practical to take into account special cases. In actual operation, there may be instances where the temperature limits recommended can be exceeded. Likewise, there may be applications where satisfactory life will not be obtained at the recommended temperature limits. However, in general, the temperature limits listed are such as to provide satisfactory thermocouple life when the wires are operated continuously at these temperatures. Various factors affecting thermocouple life are discussed in Appendix C.

---

* American Wire Gage, also known as B&S (Brown & Sharpe)

2.5 TOLERANCE OF INITIAL CALIBRATION

Table 8, 9 and 10 give the standard and special tolerance of initial calibration for thermocouples and thermocouple extension wires. The tolerance of initial calibration is defined as the allowable deviation of the thermocouple and extension wire in its initial condition as supplied by the manufacturer from the standard emf-temperature tables. Once the thermocouple is in use its calibration will change. The magnitude and direction of the change are dependent on temperature, time and environmental conditions affecting the thermocouple and may not be accurately predicted. The tolerances for each type of thermocouple apply only over the temperature range for which the wire size in question is recommended (see Table 7). These tolerances should be applied only to standard wire sizes. The same tolerances may not be obtainable in special sizes. These tolerances do not include installation or system errors. See Appendix C, paragraph C4.1 for the thermocouple installations and errors.

Where tolerances are given in percent, in Table 8, the percentage applies to the temperature being measured. For example, the standard tolerance of Type J over the temperature range 277° to 760°C is ±3/4 percent. If the temperature being measured is 538°C, the tolerance is ±3/4 percent of 538, or ±4.0°C. To determine the tolerance in degrees Fahrenheit, multiply the tolerance in degrees Celsius times 1.8.

TABLE 7
RECOMMENDED UPPER TEMPERATURE LIMITS FOR PROTECTED THERMOCOUPLES. UPPER TEMPERATURE LIMIT FOR VARIOUS WIRE SIZES (AWG), deg C

Thermocouple Type	No. 8 Gage	No. 14 Gage	No. 20 Gage	No. 24 Gage	No. 28 Gage
B				1700	
E	870	650	540	430	430
J	760	590	480	370	370
K	1260	1090	980	870	870
R & S				1480	
T		370	260	200	200

TABLE 8
INITIAL CALIBRATION TOLERANCES FOR THERMOCOUPLES
Reference Junction 0°C

Thermocouple Type	Temperature Range, °C	TOLERANCES Standard (whichever is greater)	TOLERANCES Special (whichever is greater)
B	870 to 1700	±0.5%	
E	0 to 900	±1.7°C or ±0.5%	±1°C or ±0.4%
J	0 to 750	±2.2°C or ±0.75%	±1.1°C or ±0.4%
K	0 to 1250	±2.2°C or ±0.75%	±1.1°C or ±0.4%
R or S	0 to 1450	±1.5°C or ±0.25%	±0.6°C or ±0.1%
T	0 to 350	±1°C or ±0.75%	±0.5°C or ±0.4%
	Cryogenic Ranges		
E*	–200 to 0	±1.7°C or ±1%	**
K*	–200 to 0	±2.2°C or ±2%	**
T*	–200 to 0	±1°C or ±1.5%	**

* Thermocouples and thermocouple material are normally supplied to meet the tolerances specified in the table for the normal specified range. The same materials, however, may not fall within the cryogenic tolerances in the second section of the table. If materials are required to meet the cryogenic tolerances, the purchase order must so state. Selection of materials usually will be required. Tolerances indicated in this table are not necessarily an indication of the accuracy of temperature measurements in use after initial heating of the materials.

** Little information is available to justify establishing special tolerances for cryogenic temperatures. Limited experience suggests the following tolerances for types E and T thermocouples:
  Type E –200 to 0°C ±1°C or ±0.5% (whichever is greater)
  Type T –200 to 0°C ±0.5°C or ±0.8% (whichever is greater)

These tolerances are given only as guide for discussion between purchaser and supplier. Due to the characteristics of the materials, cryogenic tolerances for Type J thermocouples and special cryogenic tolerances for Type K thermocouples are not listed.

TABLE 9
INITIAL CALIBRATION TOLERANCES FOR THERMOCOUPLE EXTENSION WIRES
Reference Junction 0°C

Extension Wire Type	Temperature Range, °C	Tolerances Standard	Special
EX	0 to 200°C	±1.7°C	-
JX	0 to 200°C	±2.2°C	±1.1°C
KX	0 to 200°C	±2.2°C	-
TX	0 to 100°C	±1.0°C	±0.5°C

3. NON-CERAMIC INSULATION OF THERMOCOUPLE AND EXTENSION WIRES

The normal function of thermocouple and extension wire insulation is to provide electrical insulation. If this function is not provided or is compromised in any way, the indicated temperature may be in error. Insulation of this type (non-ceramic) may be affected adversely by moisture, abrasion, flexing, temperature extremes, chemical attack, and nuclear radiation. Each type of insulation has its own limitations. a knowledge of these limitations is essential if accurate and reliable measurements are to be made.

A number of coatings are presently available commercially. The strong points as well as limitations are discussed in ASTM Special Technical Publication STP-470B, "Manual on the Use of Thermocouples in Temperature Measurement."

In summary, this type of insulation should be selected only after considering possible exposure temperatures and heating rates, the number of temperature cycles, mechanical movement, moisture, routing of the insulated wire, and chemical deterioration.

TABLE 10
INITIAL CALIBRATION TOLERANCES FOR THERMOCOUPLE COMPENSATING EXTENSION WIRE
Reference Junction 0°C

Thermocouple Type	Compensating Wire Type	Temperature Range, °C	Tolerances
B	BX***	0 to 100	$\begin{Bmatrix} +0.000 & (+0°C*) \\ -0.033 & (-3.7°C*) \end{Bmatrix}$ mV
R, S	SX**	0 to 200	±.057 mV (±5°C*)

* Due to the non-linearity of the Types, R, S, and B temperature-emf curves, the error introduced into a thermocouple system by the compensating wire will be variable when expressed in degrees. The degree C tolerances given in parentheses are based on the following measuring junction temperatures:

Type Wire	Measuring Junction Temperature
BX	Greater than 1000°C
SX	Greater than 870°C

** Copper (+) versus copper nickel alloy (−).
***Copper versus copper compensating extension wire, usable to 100°C with maximum errors as indicated, but with no significant error over 0 to 50°C range. Matched proprietary alloy compensating wire is available for use over the range 0 to 200°C with claimed tolerances of ±0.033 mV. (±3.7°C*)

4. TEMPERATURE-EMF TABLES FOR THERMOCOUPLES

4.1 Scope and Purpose

This section applies to the temperature-emf relationships of materials used for temperature measurement thermocouples.

Its purpose is to provide reference tables of temperature-amf values for Type B,E,J,K,R,S, and T thermocouples, in a form convenient for industrial and laboratory use.

4.2 Introduction

The values in these tables are based upon the International Practical Temperature Scale of 1968 (IPTS-68) and the U.S. legal electrical units. All the data in Tables 11 to 17 have been extracted from "Thermocouple Reference Tables Based on the IPTS-68," National Bureau of Standards Monograph 125. These tables differ slightly from previous tables for the following reasons: improved measurements and data analysis techniques, slight changes in commercial thermocouple materials, and also changes in the temperature scale and electrical units. The significance of these factors, as well as the origin of each of the tables, is discussed in the NBS reference noted above, and it should be consulted for details.

These tables give values of EMF to three decimal places (0.001 mV) for one degree Celsius (°C) temperature intervals. If greater precision is required. the NBS reference noted above should be consulted. It includes tables giving values of EMF to four decimal places (0.0001 mV) and analytical functions for each thermocouple type that allow a direct and precise calculation of the EMF-temperature relationship.

Tables for each type of thermocouple giving values of EMF as a function of temperature in degrees Fahrenheit (°F) can be found in ANSI/ASTM Standard E230, "Temperature-Electromotive Force (EMF) Tables for Thermocouples." Tables giving EMF-temperature values (in both °C and °F) for single-leg thermoelements referenced to platinun (NBS Pt-67) are also given in the above ANSI/ASTM standard.

4.3 Use of Temperature-EMF Tables

These Temperature-EMF reference tables serve two very useful purposes in that they provide a means for converting the generated EMF of certain thermocouple material combinations into equivalent temperatures, and they enable the calibration and checking of thermocouples and thermocouple extension wire.

If the reference junction is maintained at 0°C, the appropriate temperature or EMF data may be read directly from the tables. When it is not practical to maintain the reference junction temperature at 0°C, these tables may still be used by applying an appropriate correction. The value of the correction may be obtained from these tables. An example to illustrate how to obtain and apply this correction follows.

Let us suppose a Type J thermocouple was used in an installation to determine the temperature of a fluid medium and an EMF output of 18.070 mV was observed. Also, a mercury thermometer in close proximity to the thermocouple reference junction produced a reading of 20°C.

To use the Type J Table to obtain a value for the temperature of the fluid medium, the observed EMF output of the thermocouple must first be corrected to compensate for the difference between the reference junction temperature actually used and 0°C. The correction is the EMF value given by the Type J Table at the reference junction temperature actually used (20°C). As shown below, this EMF value (1.019 mV) is algebraically added to the observed EMF output to obtain the value of EMF that the thermocouple would produce if the reference junction were at 0°C.

Observed Type J Thermocouple Output:	18.070mV
Correction Factor (Table value at reference temperature actually used) for reference junction at 20°C:	1.019mV
Corrected Output:	19.089mV

The corrected output of 19.089 mV is then used to determine from the Type J Table the equivalent temperature value of 350°C.

TABLE 11

TEMPERATURE-EMF FOR TYPE B THERMOCOUPLES

TEMPERATURES IN DEGREES CELSIUS (IPTS-68)   **REFERENCE JUNCTIONS AT 0°C**

DEG C	0	1	2	3	4	5	6	7	8	9	10	DEG C
					THERMOELECTRIC VOLTAGE IN MILLIVOLTS							
0	0.000	-0.000	-0.000	-0.001	-0.001	-0.001	-0.001	-0.001	-0.002	-0.002	-0.002	0
10	-0.002	-0.002	-0.002	-0.002	-0.002	-0.002	-0.002	-0.002	-0.003	-0.003	-0.003	10
20	-0.003	-0.003	-0.003	-0.003	-0.003	-0.002	-0.002	-0.002	-0.002	-0.002	-0.002	20
30	-0.002	-0.002	-0.002	-0.002	-0.002	-0.001	-0.001	-0.001	-0.001	-0.001	-0.000	30
40	-0.000	-0.000	-0.000	0.000	0.000	0.001	0.001	0.001	0.002	0.002	0.002	40
50	0.002	0.003	0.003	0.003	0.004	0.004	0.004	0.005	0.005	0.006	0.006	50
60	0.006	0.007	0.007	0.008	0.008	0.009	0.009	0.010	0.010	0.011	0.011	60
70	0.011	0.012	0.012	0.013	0.014	0.014	0.015	0.015	0.016	0.017	0.017	70
80	0.017	0.018	0.019	0.020	0.020	0.021	0.022	0.022	0.023	0.024	0.025	80
90	0.025	0.026	0.026	0.027	0.028	0.029	0.030	0.031	0.031	0.032	0.033	90
100	0.033	0.034	0.035	0.036	0.037	0.038	0.039	0.040	0.041	0.042	0.043	100
110	0.043	0.044	0.045	0.046	0.047	0.048	0.049	0.050	0.051	0.052	0.053	110
120	0.053	0.055	0.056	0.057	0.058	0.059	0.060	0.062	0.063	0.064	0.065	120
130	0.065	0.066	0.068	0.069	0.070	0.071	0.073	0.074	0.075	0.077	0.078	130
140	0.078	0.079	0.081	0.082	0.083	0.085	0.086	0.088	0.089	0.091	0.092	140
150	0.092	0.093	0.095	0.096	0.098	0.099	0.101	0.102	0.104	0.106	0.107	150
160	0.107	0.109	0.110	0.112	0.113	0.115	0.117	0.118	0.120	0.122	0.123	160
170	0.123	0.125	0.127	0.128	0.130	0.132	0.133	0.135	0.137	0.139	0.140	170
180	0.140	0.142	0.144	0.146	0.148	0.149	0.151	0.153	0.155	0.157	0.159	180
190	0.159	0.161	0.163	0.164	0.166	0.168	0.170	0.172	0.174	0.176	0.178	190
200	0.178	0.180	0.182	0.184	0.186	0.188	0.190	0.192	0.194	0.197	0.199	200
210	0.199	0.201	0.203	0.205	0.207	0.209	0.211	0.214	0.216	0.218	0.220	210
220	0.220	0.222	0.225	0.227	0.229	0.231	0.234	0.236	0.238	0.240	0.243	220
230	0.243	0.245	0.247	0.250	0.252	0.254	0.257	0.259	0.262	0.264	0.266	230
240	0.266	0.269	0.271	0.274	0.276	0.279	0.281	0.284	0.286	0.289	0.291	240
250	0.291	0.294	0.296	0.299	0.301	0.304	0.307	0.309	0.312	0.314	0.317	250
260	0.317	0.320	0.322	0.325	0.328	0.330	0.333	0.336	0.338	0.341	0.344	260
270	0.344	0.347	0.349	0.352	0.355	0.358	0.360	0.363	0.366	0.369	0.372	270
280	0.372	0.375	0.377	0.380	0.383	0.386	0.389	0.392	0.395	0.398	0.401	280
290	0.401	0.404	0.406	0.409	0.412	0.415	0.418	0.421	0.424	0.427	0.431	290
300	0.431	0.434	0.437	0.440	0.443	0.446	0.449	0.452	0.455	0.458	0.462	300
310	0.462	0.465	0.468	0.471	0.474	0.477	0.481	0.484	0.487	0.490	0.494	310
320	0.494	0.497	0.500	0.503	0.507	0.510	0.513	0.517	0.520	0.523	0.527	320
330	0.527	0.530	0.533	0.537	0.540	0.544	0.547	0.550	0.554	0.557	0.561	330
340	0.561	0.564	0.568	0.571	0.575	0.578	0.582	0.585	0.589	0.592	0.596	340
350	0.596	0.599	0.603	0.606	0.610	0.614	0.617	0.621	0.625	0.628	0.632	350
360	0.632	0.636	0.639	0.643	0.647	0.650	0.654	0.658	0.661	0.665	0.669	360
370	0.669	0.673	0.677	0.680	0.684	0.688	0.692	0.696	0.699	0.703	0.707	370
380	0.707	0.711	0.715	0.719	0.723	0.727	0.730	0.734	0.738	0.742	0.746	380
390	0.746	0.750	0.754	0.758	0.762	0.766	0.770	0.774	0.778	0.782	0.786	390
400	0.786	0.790	0.794	0.799	0.803	0.807	0.811	0.815	0.819	0.823	0.827	400
410	0.827	0.832	0.836	0.840	0.844	0.848	0.853	0.857	0.861	0.865	0.870	410
420	0.870	0.874	0.878	0.882	0.887	0.891	0.895	0.900	0.904	0.908	0.913	420
430	0.913	0.917	0.921	0.926	0.930	0.935	0.939	0.943	0.948	0.952	0.957	430
440	0.957	0.961	0.966	0.970	0.975	0.979	0.984	0.988	0.993	0.997	1.002	440
450	1.002	1.006	1.011	1.015	1.020	1.025	1.029	1.034	1.039	1.043	1.048	450
460	1.048	1.052	1.057	1.062	1.066	1.071	1.076	1.081	1.085	1.090	1.095	460
470	1.095	1.100	1.104	1.109	1.114	1.119	1.123	1.128	1.133	1.138	1.143	470
480	1.143	1.148	1.152	1.157	1.162	1.167	1.172	1.177	1.182	1.187	1.192	480
490	1.192	1.197	1.202	1.206	1.211	1.216	1.221	1.226	1.231	1.236	1.241	490
500	1.241	1.246	1.252	1.257	1.262	1.267	1.272	1.277	1.282	1.287	1.292	500
510	1.292	1.297	1.303	1.308	1.313	1.318	1.323	1.328	1.334	1.339	1.344	510
520	1.344	1.349	1.354	1.360	1.365	1.370	1.375	1.381	1.386	1.391	1.397	520
530	1.397	1.402	1.407	1.413	1.418	1.423	1.429	1.434	1.439	1.445	1.450	530
540	1.450	1.456	1.461	1.467	1.472	1.477	1.483	1.488	1.494	1.499	1.505	540
550	1.505	1.510	1.516	1.521	1.527	1.532	1.538	1.544	1.549	1.555	1.560	550
560	1.560	1.566	1.571	1.577	1.583	1.588	1.594	1.600	1.605	1.611	1.617	560
570	1.617	1.622	1.628	1.634	1.639	1.645	1.651	1.657	1.662	1.668	1.674	570
580	1.674	1.680	1.685	1.691	1.697	1.703	1.709	1.715	1.720	1.726	1.732	580
590	1.732	1.738	1.744	1.750	1.756	1.762	1.767	1.773	1.779	1.785	1.791	590
600	1.791	1.797	1.803	1.809	1.815	1.821	1.827	1.833	1.839	1.845	1.851	600
610	1.851	1.857	1.863	1.869	1.875	1.882	1.888	1.894	1.900	1.906	1.912	610
620	1.912	1.918	1.924	1.931	1.937	1.943	1.949	1.955	1.961	1.968	1.974	620
630	1.974	1.980	1.986	1.993	1.999	2.005	2.011	2.018	2.024	2.030	2.036	630
640	2.036	2.043	2.049	2.055	2.062	2.068	2.074	2.081	2.087	2.094	2.100	640
DEG C	0	1	2	3	4	5	6	7	8	9	10	DEG C

TABLE 11 (Continued)

TEMPERATURE-EMF FOR TYPE B THERMOCOUPLES

TEMPERATURES IN DEGREES CELSIUS (IPTS-68) **REFERENCE JUNCTIONS AT 0°C**

DEG C	0	1	2	3	4	5	6	7	8	9	10	DEG C
				THERMOELECTRIC VOLTAGE IN MILLIVOLTS								
650	2.100	2.106	2.113	2.119	2.126	2.132	2.139	2.145	2.151	2.158	2.164	650
660	2.164	2.171	2.177	2.184	2.190	2.197	2.203	2.210	2.216	2.223	2.230	660
670	2.230	2.236	2.243	2.249	2.256	2.263	2.269	2.276	2.282	2.289	2.296	670
680	2.296	2.302	2.309	2.316	2.322	2.329	2.336	2.343	2.349	2.356	2.363	680
690	2.363	2.369	2.376	2.383	2.390	2.396	2.403	2.410	2.417	2.424	2.430	690
700	2.430	2.437	2.444	2.451	2.458	2.465	2.472	2.478	2.485	2.492	2.499	700
710	2.499	2.506	2.513	2.520	2.527	2.534	2.541	2.548	2.555	2.562	2.569	710
720	2.569	2.576	2.583	2.590	2.597	2.604	2.611	2.618	2.625	2.632	2.639	720
730	2.639	2.646	2.653	2.660	2.667	2.674	2.682	2.689	2.696	2.703	2.710	730
740	2.710	2.717	2.724	2.732	2.739	2.746	2.753	2.760	2.768	2.775	2.782	740
750	2.782	2.789	2.797	2.804	2.811	2.818	2.826	2.833	2.840	2.848	2.855	750
760	2.855	2.862	2.869	2.877	2.884	2.892	2.899	2.906	2.914	2.921	2.928	760
770	2.928	2.936	2.943	2.951	2.958	2.966	2.973	2.980	2.988	2.995	3.003	770
780	3.003	3.010	3.018	3.025	3.033	3.040	3.048	3.055	3.063	3.070	3.078	780
790	3.078	3.086	3.093	3.101	3.108	3.116	3.124	3.131	3.139	3.146	3.154	790
800	3.154	3.162	3.169	3.177	3.185	3.192	3.200	3.208	3.215	3.223	3.231	800
810	3.231	3.239	3.246	3.254	3.262	3.269	3.277	3.285	3.293	3.301	3.308	810
820	3.308	3.316	3.324	3.332	3.340	3.347	3.355	3.363	3.371	3.379	3.387	820
830	3.387	3.395	3.402	3.410	3.418	3.426	3.434	3.442	3.450	3.458	3.466	830
840	3.466	3.474	3.482	3.490	3.498	3.506	3.514	3.522	3.530	3.538	3.546	840
850	3.546	3.554	3.562	3.570	3.578	3.586	3.594	3.602	3.610	3.618	3.626	850
860	3.626	3.634	3.643	3.651	3.659	3.667	3.675	3.683	3.691	3.700	3.708	860
870	3.708	3.716	3.724	3.732	3.741	3.749	3.757	3.765	3.773	3.782	3.790	870
880	3.790	3.798	3.806	3.815	3.823	3.831	3.840	3.848	3.856	3.865	3.873	880
890	3.873	3.881	3.890	3.898	3.906	3.915	3.923	3.931	3.940	3.948	3.957	890
900	3.957	3.965	3.973	3.982	3.990	3.999	4.007	4.016	4.024	4.032	4.041	900
910	4.041	4.049	4.058	4.066	4.075	4.083	4.092	4.100	4.109	4.117	4.126	910
920	4.126	4.135	4.143	4.152	4.160	4.169	4.177	4.186	4.195	4.203	4.212	920
930	4.212	4.220	4.229	4.238	4.246	4.255	4.264	4.272	4.281	4.290	4.298	930
940	4.298	4.307	4.316	4.325	4.333	4.342	4.351	4.359	4.368	4.377	4.386	940
950	4.386	4.394	4.403	4.412	4.421	4.430	4.438	4.447	4.456	4.465	4.474	950
960	4.474	4.483	4.491	4.500	4.509	4.518	4.527	4.536	4.545	4.553	4.562	960
970	4.562	4.571	4.580	4.589	4.598	4.607	4.616	4.625	4.634	4.643	4.652	970
980	4.652	4.661	4.670	4.679	4.688	4.697	4.706	4.715	4.724	4.733	4.742	980
990	4.742	4.751	4.760	4.769	4.778	4.787	4.796	4.805	4.814	4.824	4.833	990
1000	4.833	4.842	4.851	4.860	4.869	4.878	4.887	4.897	4.906	4.915	4.924	1000
1010	4.924	4.933	4.942	4.952	4.961	4.970	4.979	4.989	4.998	5.007	5.016	1010
1020	5.016	5.025	5.035	5.044	5.053	5.063	5.072	5.081	5.090	5.100	5.109	1020
1030	5.109	5.118	5.128	5.137	5.146	5.156	5.165	5.174	5.184	5.193	5.202	1030
1040	5.202	5.212	5.221	5.231	5.240	5.249	5.259	5.268	5.278	5.287	5.297	1040
1050	5.297	5.306	5.316	5.325	5.334	5.344	5.353	5.363	5.372	5.382	5.391	1050
1060	5.391	5.401	5.410	5.420	5.429	5.439	5.449	5.458	5.468	5.477	5.487	1060
1070	5.487	5.496	5.506	5.516	5.525	5.535	5.544	5.554	5.564	5.573	5.583	1070
1080	5.583	5.593	5.602	5.612	5.621	5.631	5.641	5.651	5.660	5.670	5.680	1080
1090	5.680	5.689	5.699	5.709	5.718	5.728	5.738	5.748	5.757	5.767	5.777	1090
1100	5.777	5.787	5.796	5.806	5.816	5.826	5.836	5.845	5.855	5.865	5.875	1100
1110	5.875	5.885	5.895	5.904	5.914	5.924	5.934	5.944	5.954	5.964	5.973	1110
1120	5.973	5.983	5.993	6.003	6.013	6.023	6.033	6.043	6.053	6.063	6.073	1120
1130	6.073	6.083	6.093	6.102	6.112	6.122	6.132	6.142	6.152	6.162	6.172	1130
1140	6.172	6.182	6.192	6.202	6.212	6.223	6.233	6.243	6.253	6.263	6.273	1140
1150	6.273	6.283	6.293	6.303	6.313	6.323	6.333	6.343	6.353	6.364	6.374	1150
1160	6.374	6.384	6.394	6.404	6.414	6.424	6.435	6.445	6.455	6.465	6.475	1160
1170	6.475	6.485	6.496	6.506	6.516	6.526	6.536	6.547	6.557	6.567	6.577	1170
1180	6.577	6.588	6.598	6.608	6.618	6.629	6.639	6.649	6.659	6.670	6.680	1180
1190	6.680	6.690	6.701	6.711	6.721	6.732	6.742	6.752	6.763	6.773	6.783	1190
1200	6.783	6.794	6.804	6.814	6.825	6.835	6.846	6.856	6.866	6.877	6.887	1200
1210	6.887	6.898	6.908	6.918	6.929	6.939	6.950	6.960	6.971	6.981	6.991	1210
1220	6.991	7.002	7.012	7.023	7.033	7.044	7.054	7.065	7.075	7.086	7.096	1220
1230	7.096	7.107	7.117	7.128	7.138	7.149	7.159	7.170	7.181	7.191	7.202	1230
1240	7.202	7.212	7.223	7.233	7.244	7.255	7.265	7.276	7.286	7.297	7.308	1240
1250	7.308	7.318	7.329	7.339	7.350	7.361	7.371	7.382	7.393	7.403	7.414	1250
1260	7.414	7.425	7.435	7.446	7.457	7.467	7.478	7.489	7.500	7.510	7.521	1260
1270	7.521	7.532	7.542	7.553	7.564	7.575	7.585	7.596	7.607	7.618	7.628	1270
1280	7.628	7.639	7.650	7.661	7.671	7.682	7.693	7.704	7.715	7.725	7.736	1280
1290	7.736	7.747	7.758	7.769	7.780	7.790	7.801	7.812	7.823	7.834	7.845	1290
DEG C	0	1	2	3	4	5	6	7	8	9	10	DEG C

TABLE 11 (Continued)

TEMPERATURE-EMF FOR TYPE B THERMOCOUPLES

TEMPERATURES IN DEGREES CELSIUS (IPTS-68)   **REFERENCE JUNCTIONS AT 0°C**

DEG C	0	1	2	3	4	5	6	7	8	9	10	DEG C

THERMOELECTRIC VOLTAGE IN MILLIVOLTS

DEG C	0	1	2	3	4	5	6	7	8	9	10	DEG C
1300	7.845	7.855	7.866	7.877	7.888	7.899	7.910	7.921	7.932	7.943	7.953	1300
1310	7.953	7.964	7.975	7.986	7.997	8.008	8.019	8.030	8.041	8.052	8.063	1310
1320	8.063	8.074	8.085	8.096	8.107	8.118	8.128	8.139	8.150	8.161	8.172	1320
1330	8.172	8.183	8.194	8.205	8.216	8.227	8.238	8.249	8.261	8.272	8.283	1330
1340	8.283	8.294	8.305	8.316	8.327	8.338	8.349	8.360	8.371	8.382	8.393	1340
1350	8.393	8.404	8.415	8.426	8.437	8.449	8.460	8.471	8.482	8.493	8.504	1350
1360	8.504	8.515	8.526	8.538	8.549	8.560	8.571	8.582	8.593	8.604	8.616	1360
1370	8.616	8.627	8.638	8.649	8.660	8.671	8.683	8.694	8.705	8.716	8.727	1370
1380	8.727	8.738	8.750	8.761	8.772	8.783	8.795	8.806	8.817	8.828	8.839	1380
1390	8.839	8.851	8.862	8.873	8.884	8.896	8.907	8.918	8.929	8.941	8.952	1390
1400	8.952	8.963	8.974	8.986	8.997	9.008	9.020	9.031	9.042	9.053	9.065	1400
1410	9.065	9.076	9.087	9.099	9.110	9.121	9.133	9.144	9.155	9.167	9.178	1410
1420	9.178	9.189	9.201	9.212	9.223	9.235	9.246	9.257	9.269	9.280	9.291	1420
1430	9.291	9.303	9.314	9.326	9.337	9.348	9.360	9.371	9.382	9.394	9.405	1430
1440	9.405	9.417	9.428	9.439	9.451	9.462	9.474	9.485	9.497	9.508	9.519	1440
1450	9.519	9.531	9.542	9.554	9.565	9.577	9.588	9.599	9.611	9.622	9.634	1450
1460	9.634	9.645	9.657	9.668	9.680	9.691	9.703	9.714	9.726	9.737	9.748	1460
1470	9.748	9.760	9.771	9.783	9.794	9.806	9.817	9.829	9.840	9.852	9.863	1470
1480	9.863	9.875	9.886	9.898	9.909	9.921	9.933	9.944	9.956	9.967	9.979	1480
1490	9.979	9.990	10.002	10.013	10.025	10.036	10.048	10.059	10.071	10.082	10.094	1490
1500	10.094	10.106	10.117	10.129	10.140	10.152	10.163	10.175	10.187	10.198	10.210	1500
1510	10.210	10.221	10.233	10.244	10.256	10.268	10.279	10.291	10.302	10.314	10.325	1510
1520	10.325	10.337	10.349	10.360	10.372	10.383	10.395	10.407	10.418	10.430	10.441	1520
1530	10.441	10.453	10.465	10.476	10.488	10.500	10.511	10.523	10.534	10.546	10.558	1530
1540	10.558	10.569	10.581	10.593	10.604	10.616	10.627	10.639	10.651	10.662	10.674	1540
1550	10.674	10.686	10.697	10.709	10.721	10.732	10.744	10.756	10.767	10.779	10.790	1550
1560	10.790	10.802	10.814	10.825	10.837	10.849	10.860	10.872	10.884	10.895	10.907	1560
1570	10.907	10.919	10.930	10.942	10.954	10.965	10.977	10.989	11.000	11.012	11.024	1570
1580	11.024	11.035	11.047	11.059	11.070	11.082	11.094	11.105	11.117	11.129	11.141	1580
1590	11.141	11.152	11.164	11.176	11.187	11.199	11.211	11.222	11.234	11.246	11.257	1590
1600	11.257	11.269	11.281	11.292	11.304	11.316	11.328	11.339	11.351	11.363	11.374	1600
1610	11.374	11.386	11.398	11.409	11.421	11.433	11.444	11.456	11.468	11.480	11.491	1610
1620	11.491	11.503	11.515	11.526	11.538	11.550	11.561	11.573	11.585	11.597	11.608	1620
1630	11.608	11.620	11.632	11.643	11.655	11.667	11.678	11.690	11.702	11.714	11.725	1630
1640	11.725	11.737	11.749	11.760	11.772	11.784	11.795	11.807	11.819	11.830	11.842	1640
1650	11.842	11.854	11.866	11.877	11.889	11.901	11.912	11.924	11.936	11.947	11.959	1650
1660	11.959	11.971	11.983	11.994	12.006	12.018	12.029	12.041	12.053	12.064	12.076	1660
1670	12.076	12.088	12.099	12.111	12.123	12.134	12.146	12.158	12.170	12.181	12.193	1670
1680	12.193	12.205	12.216	12.228	12.240	12.251	12.263	12.275	12.286	12.298	12.310	1680
1690	12.310	12.321	12.333	12.345	12.356	12.368	12.380	12.391	12.403	12.415	12.426	1690
1700	12.426	12.438	12.450	12.461	12.473	12.485	12.496	12.508	12.520	12.531	12.543	1700
1710	12.543	12.555	12.566	12.578	12.590	12.601	12.613	12.624	12.636	12.648	12.659	1710
1720	12.659	12.671	12.683	12.694	12.706	12.718	12.729	12.741	12.752	12.764	12.776	1720
1730	12.776	12.787	12.799	12.811	12.822	12.834	12.845	12.857	12.869	12.880	12.892	1730
1740	12.892	12.903	12.915	12.927	12.938	12.950	12.961	12.973	12.985	12.996	13.008	1740
1750	13.008	13.019	13.031	13.043	13.054	13.066	13.077	13.089	13.100	13.112	13.124	1750
1760	13.124	13.135	13.147	13.158	13.170	13.181	13.193	13.204	13.216	13.228	13.239	1760
1770	13.239	13.251	13.262	13.274	13.285	13.297	13.308	13.320	13.331	13.343	13.354	1770
1780	13.354	13.366	13.378	13.389	13.401	13.412	13.424	13.435	13.447	13.458	13.470	1780
1790	13.470	13.481	13.493	13.504	13.516	13.527	13.539	13.550	13.562	13.573	13.585	1790
1800	13.585	13.596	13.607	13.619	13.630	13.642	13.653	13.665	13.676	13.688	13.699	1800
1810	13.699	13.711	13.722	13.733	13.745	13.756	13.768	13.779	13.791	13.802	13.814	1810
1820	13.814											1820

DEG C	0	1	2	3	4	5	6	7	8	9	10	DEG C

TABLE 12

TEMPERATURE-EMF FOR TYPE E THERMOCOUPLES

TEMPERATURES IN DEGREES CELSIUS (IPTS-68)   **REFERENCE JUNCTIONS AT 0°C**

DEG C	0	1	2	3	4	5	6	7	8	9	10	DEG C
					THERMOELECTRIC VOLTAGE IN MILLIVOLTS							
-270	-9.835											-270
-260	-9.797	-9.802	-9.808	-9.813	-9.817	-9.821	-9.825	-9.828	-9.831	-9.833	-9.835	-260
-250	-9.719	-9.728	-9.737	-9.746	-9.754	-9.762	-9.770	-9.777	-9.784	-9.791	-9.797	-250
-240	-9.604	-9.617	-9.630	-9.642	-9.654	-9.666	-9.677	-9.688	-9.699	-9.709	-9.719	-240
-230	-9.455	-9.472	-9.488	-9.503	-9.519	-9.534	-9.549	-9.563	-9.577	-9.591	-9.604	-230
-220	-9.274	-9.293	-9.313	-9.332	-9.350	-9.368	-9.386	-9.404	-9.421	-9.438	-9.455	-220
-210	-9.063	-9.085	-9.107	-9.129	-9.151	-9.172	-9.193	-9.214	-9.234	-9.254	-9.274	-210
-200	-8.824	-8.850	-8.874	-8.899	-8.923	-8.947	-8.971	-8.994	-9.017	-9.040	-9.063	-200
-190	-8.561	-8.588	-8.615	-8.642	-8.669	-8.696	-8.722	-8.748	-8.774	-8.799	-8.824	-190
-180	-8.273	-8.303	-8.333	-8.362	-8.391	-8.420	-8.449	-8.477	-8.505	-8.533	-8.561	-180
-170	-7.963	-7.995	-8.027	-8.058	-8.090	-8.121	-8.152	-8.183	-8.213	-8.243	-8.273	-170
-160	-7.631	-7.665	-7.699	-7.733	-7.767	-7.800	-7.833	-7.866	-7.898	-7.931	-7.963	-160
-150	-7.279	-7.315	-7.351	-7.387	-7.422	-7.458	-7.493	-7.528	-7.562	-7.597	-7.631	-150
-140	-6.907	-6.945	-6.983	-7.020	-7.058	-7.095	-7.132	-7.169	-7.206	-7.243	-7.279	-140
-130	-6.516	-6.556	-6.596	-6.635	-6.675	-6.714	-6.753	-6.792	-6.830	-6.869	-6.907	-130
-120	-6.107	-6.149	-6.190	-6.231	-6.273	-6.314	-6.354	-6.395	-6.436	-6.476	-6.516	-120
-110	-5.680	-5.724	-5.767	-5.810	-5.853	-5.896	-5.938	-5.981	-6.023	-6.065	-6.107	-110
-100	-5.237	-5.282	-5.327	-5.371	-5.416	-5.460	-5.505	-5.549	-5.593	-5.637	-5.680	-100
-90	-4.777	-4.824	-4.870	-4.916	-4.963	-5.009	-5.055	-5.100	-5.146	-5.191	-5.237	-90
-80	-4.301	-4.350	-4.398	-4.446	-4.493	-4.541	-4.588	-4.636	-4.683	-4.730	-4.777	-80
-70	-3.811	-3.860	-3.910	-3.959	-4.009	-4.058	-4.107	-4.156	-4.204	-4.253	-4.301	-70
-60	-3.306	-3.357	-3.408	-3.459	-3.509	-3.560	-3.610	-3.661	-3.711	-3.761	-3.811	-60
-50	-2.787	-2.839	-2.892	-2.944	-2.996	-3.048	-3.100	-3.152	-3.203	-3.254	-3.306	-50
-40	-2.254	-2.308	-2.362	-2.416	-2.469	-2.522	-2.575	-2.628	-2.681	-2.734	-2.787	-40
-30	-1.709	-1.764	-1.819	-1.874	-1.929	-1.983	-2.038	-2.092	-2.146	-2.200	-2.254	-30
-20	-1.151	-1.208	-1.264	-1.320	-1.376	-1.432	-1.487	-1.543	-1.599	-1.654	-1.709	-20
-10	-0.581	-0.639	-0.696	-0.754	-0.811	-0.868	-0.925	-0.982	-1.038	-1.095	-1.151	-10
-0	0.000	-0.059	-0.117	-0.176	-0.234	-0.292	-0.350	-0.408	-0.466	-0.524	-0.581	-0
0	0.000	0.059	0.118	0.176	0.235	0.295	0.354	0.413	0.472	0.532	0.591	0
10	0.591	0.651	0.711	0.770	0.830	0.890	0.950	1.011	1.071	1.131	1.192	10
20	1.192	1.252	1.313	1.373	1.434	1.495	1.556	1.617	1.678	1.739	1.801	20
30	1.801	1.862	1.924	1.985	2.047	2.109	2.171	2.233	2.295	2.357	2.419	30
40	2.419	2.482	2.544	2.607	2.669	2.732	2.795	2.858	2.921	2.984	3.047	40
50	3.047	3.110	3.173	3.237	3.300	3.364	3.428	3.491	3.555	3.619	3.683	50
60	3.683	3.748	3.812	3.876	3.941	4.005	4.070	4.134	4.199	4.264	4.329	60
70	4.329	4.394	4.459	4.524	4.590	4.655	4.720	4.786	4.852	4.917	4.983	70
80	4.983	5.049	5.115	5.181	5.247	5.314	5.380	5.446	5.513	5.579	5.646	80
90	5.646	5.713	5.780	5.846	5.913	5.981	6.048	6.115	6.182	6.250	6.317	90
100	6.317	6.385	6.452	6.520	6.588	6.656	6.724	6.792	6.860	6.928	6.996	100
110	6.996	7.064	7.133	7.201	7.270	7.339	7.407	7.476	7.545	7.614	7.683	110
120	7.683	7.752	7.821	7.890	7.960	8.029	8.099	8.168	8.238	8.307	8.377	120
130	8.377	8.447	8.517	8.587	8.657	8.727	8.797	8.867	8.938	9.008	9.078	130
140	9.078	9.149	9.220	9.290	9.361	9.432	9.503	9.573	9.644	9.715	9.787	140
150	9.787	9.858	9.929	10.000	10.072	10.143	10.215	10.286	10.358	10.429	10.501	150
160	10.501	10.573	10.645	10.717	10.789	10.861	10.933	11.005	11.077	11.150	11.222	160
170	11.222	11.294	11.367	11.439	11.512	11.585	11.657	11.730	11.803	11.876	11.949	170
180	11.949	12.022	12.095	12.168	12.241	12.314	12.387	12.461	12.534	12.608	12.681	180
190	12.681	12.755	12.828	12.902	12.975	13.049	13.123	13.197	13.271	13.345	13.419	190
200	13.419	13.493	13.567	13.641	13.715	13.789	13.864	13.938	14.012	14.087	14.161	200
210	14.161	14.236	14.310	14.385	14.460	14.534	14.609	14.684	14.759	14.834	14.909	210
220	14.909	14.984	15.059	15.134	15.209	15.284	15.359	15.435	15.510	15.585	15.661	220
230	15.661	15.736	15.812	15.887	15.963	16.038	16.114	16.190	16.266	16.341	16.417	230
240	16.417	16.493	16.569	16.645	16.721	16.797	16.873	16.949	17.025	17.101	17.178	240
250	17.178	17.254	17.330	17.406	17.483	17.559	17.636	17.712	17.789	17.865	17.942	250
260	17.942	18.018	18.095	18.172	18.248	18.325	18.402	18.479	18.556	18.633	18.710	260
270	18.710	18.787	18.864	18.941	19.018	19.095	19.172	19.249	19.326	19.404	19.481	270
280	19.481	19.558	19.636	19.713	19.790	19.868	19.945	20.023	20.100	20.178	20.256	280
290	20.256	20.333	20.411	20.488	20.566	20.644	20.722	20.800	20.877	20.955	21.033	290
300	21.033	21.111	21.189	21.267	21.345	21.423	21.501	21.579	21.657	21.735	21.814	300
310	21.814	21.892	21.970	22.048	22.127	22.205	22.283	22.362	22.440	22.518	22.597	310
320	22.597	22.675	22.754	22.832	22.911	22.989	23.068	23.147	23.225	23.304	23.383	320
330	23.383	23.461	23.540	23.619	23.698	23.777	23.855	23.934	24.013	24.092	24.171	330
340	24.171	24.250	24.329	24.408	24.487	24.566	24.645	24.724	24.803	24.882	24.961	340
DEG C	0	1	2	3	4	5	6	7	8	9	10	DEG C

TABLE 12 (Continued)

TEMPERATURE-EMF FOR TYPE E THERMOCOUPLES

TEMPERATURES IN DEGREES CELSIUS (IPTS-68)   **REFERENCE JUNCTIONS AT 0°C**

DEG C	0	1	2	3	4	5	6	7	8	9	10	DEG C
					THERMOELECTRIC VOLTAGE IN MILLIVOLTS							
350	24.961	25.041	25.120	25.199	25.278	25.357	25.437	25.516	25.595	25.675	25.754	350
360	25.754	25.833	25.913	25.992	26.072	26.151	26.230	26.310	26.389	26.469	26.549	360
370	26.549	26.628	26.708	26.787	26.867	26.947	27.026	27.106	27.186	27.265	27.345	370
380	27.345	27.425	27.504	27.584	27.664	27.744	27.824	27.903	27.983	28.063	28.143	380
390	28.143	28.223	28.303	28.383	28.463	28.543	28.623	28.703	28.783	28.863	28.943	390
400	28.943	29.023	29.103	29.183	29.263	29.343	29.423	29.503	29.584	29.664	29.744	400
410	29.744	29.824	29.904	29.984	30.065	30.145	30.225	30.305	30.386	30.466	30.546	410
420	30.546	30.627	30.707	30.787	30.868	30.948	31.028	31.109	31.189	31.270	31.350	420
430	31.350	31.430	31.511	31.591	31.672	31.752	31.833	31.913	31.994	32.074	32.155	430
440	32.155	32.235	32.316	32.396	32.477	32.557	32.638	32.719	32.799	32.880	32.960	440
450	32.960	33.041	33.122	33.202	33.283	33.364	33.444	33.525	33.605	33.686	33.767	450
460	33.767	33.848	33.928	34.009	34.090	34.170	34.251	34.332	34.413	34.493	34.574	460
470	34.574	34.655	34.736	34.816	34.897	34.978	35.059	35.140	35.220	35.301	35.382	470
480	35.382	35.463	35.544	35.624	35.705	35.786	35.867	35.948	36.029	36.109	36.190	480
490	36.190	36.271	36.352	36.433	36.514	36.595	36.675	36.756	36.837	36.918	36.999	490
500	36.999	37.080	37.161	37.242	37.323	37.403	37.484	37.565	37.646	37.727	37.808	500
510	37.808	37.889	37.970	38.051	38.132	38.213	38.293	38.374	38.455	38.536	38.617	510
520	38.617	38.698	38.779	38.860	38.941	39.022	39.103	39.184	39.264	39.345	39.426	520
530	39.426	39.507	39.588	39.669	39.750	39.831	39.912	39.993	40.074	40.155	40.236	530
540	40.236	40.316	40.397	40.478	40.559	40.640	40.721	40.802	40.883	40.964	41.045	540
550	41.045	41.125	41.206	41.287	41.368	41.449	41.530	41.611	41.692	41.773	41.853	550
560	41.853	41.934	42.015	42.096	42.177	42.258	42.339	42.419	42.500	42.581	42.662	560
570	42.662	42.743	42.824	42.904	42.985	43.066	43.147	43.228	43.308	43.389	43.470	570
580	43.470	43.551	43.632	43.712	43.793	43.874	43.955	44.035	44.116	44.197	44.278	580
590	44.278	44.358	44.439	44.520	44.601	44.681	44.762	44.843	44.923	45.004	45.085	590
600	45.085	45.165	45.246	45.327	45.407	45.488	45.569	45.649	45.730	45.811	45.891	600
610	45.891	45.972	46.052	46.133	46.213	46.294	46.375	46.455	46.536	46.616	46.697	610
620	46.697	46.777	46.858	46.938	47.019	47.099	47.180	47.260	47.341	47.421	47.502	620
630	47.502	47.582	47.663	47.743	47.824	47.904	47.984	48.065	48.145	48.226	48.306	630
640	48.306	48.386	48.467	48.547	48.627	48.708	48.788	48.868	48.949	49.029	49.109	640
650	49.109	49.189	49.270	49.350	49.430	49.510	49.591	49.671	49.751	49.831	49.911	650
660	49.911	49.992	50.072	50.152	50.232	50.312	50.392	50.472	50.553	50.633	50.713	660
670	50.713	50.793	50.873	50.953	51.033	51.113	51.193	51.273	51.353	51.433	51.513	670
680	51.513	51.593	51.673	51.753	51.833	51.913	51.993	52.073	52.152	52.232	52.312	680
690	52.312	52.392	52.472	52.552	52.632	52.711	52.791	52.871	52.951	53.031	53.110	690
700	53.110	53.190	53.270	53.350	53.429	53.509	53.589	53.668	53.748	53.828	53.907	700
710	53.907	53.987	54.066	54.146	54.226	54.305	54.385	54.464	54.544	54.623	54.703	710
720	54.703	54.782	54.862	54.941	55.021	55.100	55.180	55.259	55.339	55.418	55.498	720
730	55.498	55.577	55.656	55.736	55.815	55.894	55.974	56.053	56.132	56.212	56.291	730
740	56.291	56.370	56.449	56.529	56.608	56.687	56.766	56.845	56.924	57.004	57.083	740
750	57.083	57.162	57.241	57.320	57.399	57.478	57.557	57.636	57.715	57.794	57.873	750
760	57.873	57.952	58.031	58.110	58.189	58.268	58.347	58.426	58.505	58.584	58.663	760
770	58.663	58.742	58.820	58.899	58.978	59.057	59.136	59.214	59.293	59.372	59.451	770
780	59.451	59.529	59.608	59.687	59.765	59.844	59.923	60.001	60.080	60.159	60.237	780
790	60.237	60.316	60.394	60.473	60.551	60.630	60.708	60.787	60.865	60.944	61.022	790
800	61.022	61.101	61.179	61.258	61.336	61.414	61.493	61.571	61.649	61.728	61.806	800
810	61.806	61.884	61.962	62.041	62.119	62.197	62.275	62.353	62.432	62.510	62.588	810
820	62.588	62.666	62.744	62.822	62.900	62.978	63.056	63.134	63.212	63.290	63.368	820
830	63.368	63.446	63.524	63.602	63.680	63.758	63.836	63.914	63.992	64.069	64.147	830
840	64.147	64.225	64.303	64.380	64.458	64.536	64.614	64.691	64.769	64.847	64.924	840
850	64.924	65.002	65.080	65.157	65.235	65.312	65.390	65.467	65.545	65.622	65.700	850
860	65.700	65.777	65.855	65.932	66.009	66.087	66.164	66.241	66.319	66.396	66.473	860
870	66.473	66.551	66.628	66.705	66.782	66.859	66.937	67.014	67.091	67.168	67.245	870
880	67.245	67.322	67.399	67.476	67.553	67.630	67.707	67.784	67.861	67.938	68.015	880
890	68.015	68.092	68.169	68.246	68.323	68.399	68.476	68.553	68.630	68.706	68.783	890
900	68.783	68.860	68.936	69.013	69.090	69.166	69.243	69.320	69.396	69.473	69.549	900
910	69.549	69.626	69.702	69.779	69.855	69.931	70.008	70.084	70.161	70.237	70.313	910
920	70.313	70.390	70.466	70.542	70.618	70.694	70.771	70.847	70.923	70.999	71.075	920
930	71.075	71.151	71.227	71.303	71.380	71.456	71.532	71.608	71.683	71.759	71.835	930
940	71.835	71.911	71.987	72.063	72.139	72.215	72.290	72.366	72.442	72.518	72.593	940
950	72.593	72.669	72.745	72.820	72.896	72.972	73.047	73.123	73.199	73.274	73.350	950
960	73.350	73.425	73.501	73.576	73.652	73.727	73.802	73.878	73.953	74.029	74.104	960
970	74.104	74.179	74.255	74.330	74.405	74.480	74.556	74.631	74.706	74.781	74.857	970
980	74.857	74.932	75.007	75.082	75.157	75.232	75.307	75.382	75.458	75.533	75.608	980
990	75.608	75.683	75.758	75.833	75.908	75.983	76.058	76.133	76.208	76.283	76.358	990
1000	76.358											1000
DEG C	0	1	2	3	4	5	6	7	8	9	10	DEG C

TABLE 13

TEMPERATURE-EMF FOR TYPE J THERMOCOUPLES

TEMPERATURES IN DEGREES CELSIUS (IPTS-68) **REFERENCE JUNCTIONS AT 0°C**

DEG C	0	1	2	3	4	5	6	7	8	9	10	DEG C
					THERMOELECTRIC VOLTAGE IN MILLIVOLTS							
-210	-8.096											-210
-200	-7.890	-7.912	-7.934	-7.955	-7.976	-7.996	-8.017	-8.037	-8.057	-8.076	-8.096	-200
-190	-7.659	-7.683	-7.707	-7.731	-7.755	-7.778	-7.801	-7.824	-7.846	-7.868	-7.890	-190
-180	-7.402	-7.429	-7.455	-7.482	-7.508	-7.533	-7.559	-7.584	-7.609	-7.634	-7.659	-180
-170	-7.122	-7.151	-7.180	-7.209	-7.237	-7.265	-7.293	-7.321	-7.348	-7.375	-7.402	-170
-160	-6.821	-6.852	-6.883	-6.914	-6.944	-6.974	-7.004	-7.034	-7.064	-7.093	-7.122	-160
-150	-6.499	-6.532	-6.565	-6.598	-6.630	-6.663	-6.695	-6.727	-6.758	-6.790	-6.821	-150
-140	-6.159	-6.194	-6.228	-6.263	-6.297	-6.331	-6.365	-6.399	-6.433	-6.466	-6.499	-140
-130	-5.801	-5.837	-5.874	-5.910	-5.946	-5.982	-6.018	-6.053	-6.089	-6.124	-6.159	-130
-120	-5.426	-5.464	-5.502	-5.540	-5.578	-5.615	-5.653	-5.690	-5.727	-5.764	-5.801	-120
-110	-5.036	-5.076	-5.115	-5.155	-5.194	-5.233	-5.272	-5.311	-5.349	-5.388	-5.426	-110
-100	-4.632	-4.673	-4.714	-4.755	-4.795	-4.836	-4.876	-4.916	-4.956	-4.996	-5.036	-100
-90	-4.215	-4.257	-4.299	-4.341	-4.383	-4.425	-4.467	-4.508	-4.550	-4.591	-4.632	-90
-80	-3.785	-3.829	-3.872	-3.915	-3.958	-4.001	-4.044	-4.087	-4.130	-4.172	-4.215	-80
-70	-3.344	-3.389	-3.433	-3.478	-3.522	-3.566	-3.610	-3.654	-3.698	-3.742	-3.785	-70
-60	-2.892	-2.938	-2.984	-3.029	-3.074	-3.120	-3.165	-3.210	-3.255	-3.299	-3.344	-60
-50	-2.431	-2.478	-2.524	-2.570	-2.617	-2.663	-2.709	-2.755	-2.801	-2.847	-2.892	-50
-40	-1.960	-2.008	-2.055	-2.102	-2.150	-2.197	-2.244	-2.291	-2.338	-2.384	-2.431	-40
-30	-1.481	-1.530	-1.578	-1.626	-1.674	-1.722	-1.770	-1.818	-1.865	-1.913	-1.960	-30
-20	-0.995	-1.044	-1.093	-1.141	-1.190	-1.239	-1.288	-1.336	-1.385	-1.433	-1.481	-20
-10	-0.501	-0.550	-0.600	-0.650	-0.699	-0.748	-0.798	-0.847	-0.896	-0.945	-0.995	-10
- 0	0.000	-0.050	-0.101	-0.151	-0.201	-0.251	-0.301	-0.351	-0.401	-0.451	-0.501	- 0
0	0.000	0.050	0.101	0.151	0.202	0.253	0.303	0.354	0.405	0.456	0.507	0
10	0.507	0.558	0.609	0.660	0.711	0.762	0.813	0.865	0.916	0.967	1.019	10
20	1.019	1.070	1.122	1.174	1.225	1.277	1.329	1.381	1.432	1.484	1.536	20
30	1.536	1.588	1.640	1.693	1.745	1.797	1.849	1.901	1.954	2.006	2.058	30
40	2.058	2.111	2.163	2.216	2.268	2.321	2.374	2.426	2.479	2.532	2.585	40
50	2.585	2.638	2.691	2.743	2.796	2.849	2.902	2.956	3.009	3.062	3.115	50
60	3.115	3.168	3.221	3.275	3.328	3.381	3.435	3.488	3.542	3.595	3.649	60
70	3.649	3.702	3.756	3.809	3.863	3.917	3.971	4.024	4.078	4.132	4.186	70
80	4.186	4.239	4.293	4.347	4.401	4.455	4.509	4.563	4.617	4.671	4.725	80
90	4.725	4.780	4.834	4.888	4.942	4.996	5.050	5.105	5.159	5.213	5.268	90
100	5.268	5.322	5.376	5.431	5.485	5.540	5.594	5.649	5.703	5.758	5.812	100
110	5.812	5.867	5.921	5.976	6.031	6.085	6.140	6.195	6.249	6.304	6.359	110
120	6.359	6.414	6.468	6.523	6.578	6.633	6.688	6.742	6.797	6.852	6.907	120
130	6.907	6.962	7.017	7.072	7.127	7.182	7.237	7.292	7.347	7.402	7.457	130
140	7.457	7.512	7.567	7.622	7.677	7.732	7.787	7.843	7.898	7.953	8.008	140
150	8.008	8.063	8.118	8.174	8.229	8.284	8.339	8.394	8.450	8.505	8.560	150
160	8.560	8.616	8.671	8.726	8.781	8.837	8.892	8.947	9.003	9.058	9.113	160
170	9.113	9.169	9.224	9.279	9.335	9.390	9.446	9.501	9.556	9.612	9.667	170
180	9.667	9.723	9.778	9.834	9.889	9.944	10.000	10.055	10.111	10.166	10.222	180
190	10.222	10.277	10.333	10.388	10.444	10.499	10.555	10.610	10.666	10.721	10.777	190
200	10.777	10.832	10.888	10.943	10.999	11.054	11.110	11.165	11.221	11.276	11.332	200
210	11.332	11.387	11.443	11.498	11.554	11.609	11.665	11.720	11.776	11.831	11.887	210
220	11.887	11.943	11.998	12.054	12.109	12.165	12.220	12.276	12.331	12.387	12.442	220
230	12.442	12.498	12.553	12.609	12.664	12.720	12.776	12.831	12.887	12.942	12.998	230
240	12.998	13.053	13.109	13.164	13.220	13.275	13.331	13.386	13.442	13.497	13.553	240
250	13.553	13.608	13.664	13.719	13.775	13.830	13.886	13.941	13.997	14.052	14.108	250
260	14.108	14.163	14.219	14.274	14.330	14.385	14.441	14.496	14.552	14.607	14.663	260
270	14.663	14.718	14.774	14.829	14.885	14.940	14.995	15.051	15.106	15.162	15.217	270
280	15.217	15.273	15.328	15.383	15.439	15.494	15.550	15.605	15.661	15.716	15.771	280
290	15.771	15.827	15.882	15.938	15.993	16.048	16.104	16.159	16.214	16.270	16.325	290
300	16.325	16.380	16.436	16.491	16.547	16.602	16.657	16.713	16.768	16.823	16.879	300
310	16.879	16.934	16.989	17.044	17.100	17.155	17.210	17.266	17.321	17.376	17.432	310
320	17.432	17.487	17.542	17.597	17.653	17.708	17.763	17.818	17.874	17.929	17.984	320
330	17.984	18.039	18.095	18.150	18.205	18.260	18.316	18.371	18.426	18.481	18.537	330
340	18.537	18.592	18.647	18.702	18.757	18.813	18.868	18.923	18.978	19.033	19.089	340
DEG C	0	1	2	3	4	5	6	7	8	9	10	DEG C

TABLE 13 (Continued)

TEMPERATURE-EMF FOR TYPE J THERMOCOUPLES

TEMPERATURES IN DEGREES CELSIUS (IPTS-68)   **REFERENCE JUNCTIONS AT 0°C**

DEG C	0	1	2	3	4	5	6	7	8	9	10	DEG C
					THERMOELECTRIC VOLTAGE IN MILLIVOLTS							
350	19.089	19.144	19.199	19.254	19.309	19.364	19.420	19.475	19.530	19.585	19.640	350
360	19.640	19.695	19.751	19.806	19.861	19.916	19.971	20.026	20.081	20.137	20.192	360
370	20.192	20.247	20.302	20.357	20.412	20.467	20.523	20.578	20.633	20.688	20.743	370
380	20.743	20.798	20.853	20.909	20.964	21.019	21.074	21.129	21.184	21.239	21.295	380
390	21.295	21.350	21.405	21.460	21.515	21.570	21.625	21.680	21.736	21.791	21.846	390
400	21.846	21.901	21.956	22.011	22.066	22.122	22.177	22.232	22.287	22.342	22.397	400
410	22.397	22.453	22.508	22.563	22.618	22.673	22.728	22.784	22.839	22.894	22.949	410
420	22.949	23.004	23.060	23.115	23.170	23.225	23.280	23.336	23.391	23.446	23.501	420
430	23.501	23.556	23.612	23.667	23.722	23.777	23.833	23.888	23.943	23.999	24.054	430
440	24.054	24.109	24.164	24.220	24.275	24.330	24.386	24.441	24.496	24.552	24.607	440
450	24.607	24.662	24.718	24.773	24.829	24.884	24.939	24.995	25.050	25.106	25.161	450
460	25.161	25.217	25.272	25.327	25.383	25.438	25.494	25.549	25.605	25.661	25.716	460
470	25.716	25.772	25.827	25.883	25.938	25.994	26.050	26.105	26.161	26.216	26.272	470
480	26.272	26.328	26.383	26.439	26.495	26.551	26.606	26.662	26.718	26.774	26.829	480
490	26.829	26.885	26.941	26.997	27.053	27.109	27.165	27.220	27.276	27.332	27.388	490
500	27.388	27.444	27.500	27.556	27.612	27.668	27.724	27.780	27.836	27.893	27.949	500
510	27.949	28.005	28.061	28.117	28.173	28.230	28.286	28.342	28.398	28.455	28.511	510
520	28.511	28.567	28.624	28.680	28.736	28.793	28.849	28.906	28.962	29.019	29.075	520
530	29.075	29.132	29.188	29.245	29.301	29.358	29.415	29.471	29.528	29.585	29.642	530
540	29.642	29.698	29.755	29.812	29.869	29.926	29.983	30.039	30.096	30.153	30.210	540
550	30.210	30.267	30.324	30.381	30.439	30.496	30.553	30.610	30.667	30.724	30.782	550
560	30.782	30.839	30.896	30.954	31.011	31.068	31.126	31.183	31.241	31.298	31.356	560
570	31.356	31.413	31.471	31.528	31.586	31.644	31.702	31.759	31.817	31.875	31.933	570
580	31.933	31.991	32.048	32.106	32.164	32.222	32.280	32.338	32.396	32.455	32.513	580
590	32.513	32.571	32.629	32.687	32.746	32.804	32.862	32.921	32.979	33.038	33.096	590
600	33.096	33.155	33.213	33.272	33.330	33.389	33.448	33.506	33.565	33.624	33.683	600
610	33.683	33.742	33.800	33.859	33.918	33.977	34.036	34.095	34.155	34.214	34.273	610
620	34.273	34.332	34.391	34.451	34.510	34.569	34.629	34.688	34.748	34.807	34.867	620
630	34.867	34.926	34.986	35.046	35.105	35.165	35.225	35.285	35.344	35.404	35.464	630
640	35.464	35.524	35.584	35.644	35.704	35.764	35.825	35.885	35.945	36.005	36.066	640
650	36.066	36.126	36.186	36.247	36.307	36.368	36.428	36.489	36.549	36.610	36.671	650
660	36.671	36.732	36.792	36.853	36.914	36.975	37.036	37.097	37.158	37.219	37.280	660
670	37.280	37.341	37.402	37.463	37.525	37.586	37.647	37.709	37.770	37.831	37.893	670
680	37.893	37.954	38.016	38.078	38.139	38.201	38.262	38.324	38.386	38.448	38.510	680
690	38.510	38.572	38.633	38.695	38.757	38.819	38.882	38.944	39.006	39.068	39.130	690
700	39.130	39.192	39.255	39.317	39.379	39.442	39.504	39.567	39.629	39.692	39.754	700
710	39.754	39.817	39.880	39.942	40.005	40.068	40.131	40.193	40.256	40.319	40.382	710
720	40.382	40.445	40.508	40.571	40.634	40.697	40.760	40.823	40.886	40.950	41.013	720
730	41.013	41.076	41.139	41.203	41.266	41.329	41.393	41.456	41.520	41.583	41.647	730
740	41.647	41.710	41.774	41.837	41.901	41.965	42.028	42.092	42.156	42.219	42.283	740
750	42.283	42.347	42.411	42.475	42.538	42.602	42.666	42.730	42.794	42.858	42.922	750
760	42.922											760
DEG C	0	1	2	3	4	5	6	7	8	9	10	DEG C

TABLE 14

TEMPERATURE-EMF FOR TYPE K THERMOCOUPLES

TEMPERATURES IN DEGREES CELSIUS (IPTS-68) **REFERENCE JUNCTIONS AT 0°C**

DEG C	0	1	2	3	4	5	6	7	8	9	10	DEG C
					THERMOELECTRIC VOLTAGE IN MILLIVOLTS							
-270	-6.458											-270
-260	-6.441	-6.444	-6.446	-6.448	-6.450	-6.452	-6.453	-6.455	-6.456	-6.457	-6.458	-260
-250	-6.404	-6.408	-6.413	-6.417	-6.421	-6.425	-6.429	-6.432	-6.435	-6.438	-6.441	-250
-240	-6.344	-6.351	-6.358	-6.364	-6.371	-6.377	-6.382	-6.388	-6.394	-6.399	-6.404	-240
-230	-6.262	-6.271	-6.280	-6.289	-6.297	-6.306	-6.314	-6.322	-6.329	-6.337	-6.344	-230
-220	-6.158	-6.170	-6.181	-6.192	-6.202	-6.213	-6.223	-6.233	-6.243	-6.253	-6.262	-220
-210	-6.035	-6.048	-6.061	-6.074	-6.087	-6.099	-6.111	-6.123	-6.135	-6.147	-6.158	-210
-200	-5.891	-5.907	-5.922	-5.936	-5.951	-5.965	-5.980	-5.994	-6.007	-6.021	-6.035	-200
-190	-5.730	-5.747	-5.763	-5.780	-5.796	-5.813	-5.829	-5.845	-5.860	-5.876	-5.891	-190
-180	-5.550	-5.569	-5.587	-5.606	-5.624	-5.642	-5.660	-5.678	-5.695	-5.712	-5.730	-180
-170	-5.354	-5.374	-5.394	-5.414	-5.434	-5.454	-5.474	-5.493	-5.512	-5.531	-5.550	-170
-160	-5.141	-5.163	-5.185	-5.207	-5.228	-5.249	-5.271	-5.292	-5.313	-5.333	-5.354	-160
-150	-4.912	-4.936	-4.959	-4.983	-5.006	-5.029	-5.051	-5.074	-5.097	-5.119	-5.141	-150
-140	-4.669	-4.694	-4.719	-4.743	-4.768	-4.792	-4.817	-4.841	-4.865	-4.889	-4.912	-140
-130	-4.410	-4.437	-4.463	-4.489	-4.515	-4.541	-4.567	-4.593	-4.618	-4.644	-4.669	-130
-120	-4.138	-4.166	-4.193	-4.221	-4.248	-4.276	-4.303	-4.330	-4.357	-4.384	-4.410	-120
-110	-3.852	-3.881	-3.910	-3.939	-3.968	-3.997	-4.025	-4.053	-4.082	-4.110	-4.138	-110
-100	-3.553	-3.584	-3.614	-3.644	-3.674	-3.704	-3.734	-3.764	-3.793	-3.823	-3.852	-100
-90	-3.242	-3.274	-3.305	-3.337	-3.368	-3.399	-3.430	-3.461	-3.492	-3.523	-3.553	-90
-80	-2.920	-2.953	-2.985	-3.018	-3.050	-3.082	-3.115	-3.147	-3.179	-3.211	-3.242	-80
-70	-2.586	-2.620	-2.654	-2.687	-2.721	-2.754	-2.788	-2.821	-2.854	-2.887	-2.920	-70
-60	-2.243	-2.277	-2.312	-2.347	-2.381	-2.416	-2.450	-2.484	-2.518	-2.552	-2.586	-60
-50	-1.889	-1.925	-1.961	-1.996	-2.032	-2.067	-2.102	-2.137	-2.173	-2.208	-2.243	-50
-40	-1.527	-1.563	-1.600	-1.636	-1.673	-1.709	-1.745	-1.781	-1.817	-1.853	-1.889	-40
-30	-1.156	-1.193	-1.231	-1.268	-1.305	-1.342	-1.379	-1.416	-1.453	-1.490	-1.527	-30
-20	-0.777	-0.816	-0.854	-0.892	-0.930	-0.968	-1.005	-1.043	-1.081	-1.118	-1.156	-20
-10	-0.392	-0.431	-0.469	-0.508	-0.547	-0.585	-0.624	-0.662	-0.701	-0.739	-0.777	-10
-0	0.000	-0.039	-0.079	-0.118	-0.157	-0.197	-0.236	-0.275	-0.314	-0.353	-0.392	-0
0	0.000	0.039	0.079	0.119	0.158	0.198	0.238	0.277	0.317	0.357	0.397	0
10	0.397	0.437	0.477	0.517	0.557	0.597	0.637	0.677	0.718	0.758	0.798	10
20	0.798	0.838	0.879	0.919	0.960	1.000	1.041	1.081	1.122	1.162	1.203	20
30	1.203	1.244	1.285	1.325	1.366	1.407	1.448	1.489	1.529	1.570	1.611	30
40	1.611	1.652	1.693	1.734	1.776	1.817	1.858	1.899	1.940	1.981	2.022	40
50	2.022	2.064	2.105	2.146	2.188	2.229	2.270	2.312	2.353	2.394	2.436	50
60	2.436	2.477	2.519	2.560	2.601	2.643	2.684	2.726	2.767	2.809	2.850	60
70	2.850	2.892	2.933	2.975	3.016	3.058	3.100	3.141	3.183	3.224	3.266	70
80	3.266	3.307	3.349	3.390	3.432	3.473	3.515	3.556	3.598	3.639	3.681	80
90	3.681	3.722	3.764	3.805	3.847	3.888	3.930	3.971	4.012	4.054	4.095	90
100	4.095	4.137	4.178	4.219	4.261	4.302	4.343	4.384	4.426	4.467	4.508	100
110	4.508	4.549	4.590	4.632	4.673	4.714	4.755	4.796	4.837	4.878	4.919	110
120	4.919	4.960	5.001	5.042	5.083	5.124	5.164	5.205	5.246	5.287	5.327	120
130	5.327	5.368	5.409	5.450	5.490	5.531	5.571	5.612	5.652	5.693	5.733	130
140	5.733	5.774	5.814	5.855	5.895	5.936	5.976	6.016	6.057	6.097	6.137	140
150	6.137	6.177	6.218	6.258	6.298	6.338	6.378	6.419	6.459	6.499	6.539	150
160	6.539	6.579	6.619	6.659	6.699	6.739	6.779	6.819	6.859	6.899	6.939	160
170	6.939	6.979	7.019	7.059	7.099	7.139	7.179	7.219	7.259	7.299	7.338	170
180	7.338	7.378	7.418	7.458	7.498	7.538	7.578	7.618	7.658	7.697	7.737	180
190	7.737	7.777	7.817	7.857	7.897	7.937	7.977	8.017	8.057	8.097	8.137	190
200	8.137	8.177	8.216	8.256	8.296	8.336	8.376	8.416	8.456	8.497	8.537	200
210	8.537	8.577	8.617	8.657	8.697	8.737	8.777	8.817	8.857	8.898	8.938	210
220	8.938	8.978	9.018	9.058	9.099	9.139	9.179	9.220	9.260	9.300	9.341	220
230	9.341	9.381	9.421	9.462	9.502	9.543	9.583	9.624	9.664	9.705	9.745	230
240	9.745	9.786	9.826	9.867	9.907	9.948	9.989	10.029	10.070	10.111	10.151	240
250	10.151	10.192	10.233	10.274	10.315	10.355	10.396	10.437	10.478	10.519	10.560	250
260	10.560	10.600	10.641	10.682	10.723	10.764	10.805	10.846	10.887	10.928	10.969	260
270	10.969	11.010	11.051	11.093	11.134	11.175	11.216	11.257	11.298	11.339	11.381	270
280	11.381	11.422	11.463	11.504	11.546	11.587	11.628	11.669	11.711	11.752	11.793	280
290	11.793	11.835	11.876	11.918	11.959	12.000	12.042	12.083	12.125	12.166	12.207	290
300	12.207	12.249	12.290	12.332	12.373	12.415	12.456	12.498	12.539	12.581	12.623	300
310	12.623	12.664	12.706	12.747	12.789	12.831	12.872	12.914	12.955	12.997	13.039	310
320	13.039	13.080	13.122	13.164	13.205	13.247	13.289	13.331	13.372	13.414	13.456	320
330	13.456	13.497	13.539	13.581	13.623	13.665	13.706	13.748	13.790	13.832	13.874	330
340	13.874	13.915	13.957	13.999	14.041	14.083	14.125	14.167	14.208	14.250	14.292	340
DEG C	0	1	2	3	4	5	6	7	8	9	10	DEG C

TABLE 14 (Continued)

TEMPERATURE-EMF FOR TYPE K THERMOCOUPLES

TEMPERATURES IN DEGREES CELSIUS (IPTS-68) **REFERENCE JUNCTIONS AT 0°C**

DEG C	0	1	2	3	4	5	6	7	8	9	10	DEG C
					THERMOELECTRIC VOLTAGE IN MILLIVOLTS							
350	14.292	14.334	14.376	14.418	14.460	14.502	14.544	14.586	14.628	14.670	14.712	350
360	14.712	14.754	14.796	14.838	14.880	14.922	14.964	15.006	15.048	15.090	15.132	360
370	15.132	15.174	15.216	15.258	15.300	15.342	15.384	15.426	15.468	15.510	15.552	370
380	15.552	15.594	15.636	15.679	15.721	15.763	15.805	15.847	15.889	15.931	15.974	380
390	15.974	16.016	16.058	16.100	16.142	16.184	16.227	16.269	16.311	16.353	16.395	390
400	16.395	16.438	16.480	16.522	16.564	16.607	16.649	16.691	16.733	16.776	16.818	400
410	16.818	16.860	16.902	16.945	16.987	17.029	17.072	17.114	17.156	17.199	17.241	410
420	17.241	17.283	17.326	17.368	17.410	17.453	17.495	17.537	17.580	17.622	17.664	420
430	17.664	17.707	17.749	17.792	17.834	17.876	17.919	17.961	18.004	18.046	18.088	430
440	18.088	18.131	18.173	18.216	18.258	18.301	18.343	18.385	18.428	18.470	18.513	440
450	18.513	18.555	18.598	18.640	18.683	18.725	18.768	18.810	18.853	18.895	18.938	450
460	18.938	18.980	19.023	19.065	19.108	19.150	19.193	19.235	19.278	19.320	19.363	460
470	19.363	19.405	19.448	19.490	19.533	19.576	19.618	19.661	19.703	19.746	19.788	470
480	19.788	19.831	19.873	19.916	19.959	20.001	20.044	20.086	20.129	20.172	20.214	480
490	20.214	20.257	20.299	20.342	20.385	20.427	20.470	20.512	20.555	20.598	20.640	490
500	20.640	20.683	20.725	20.768	20.811	20.853	20.896	20.938	20.981	21.024	21.066	500
510	21.066	21.109	21.152	21.194	21.237	21.280	21.322	21.365	21.407	21.450	21.493	510
520	21.493	21.535	21.578	21.621	21.663	21.706	21.749	21.791	21.834	21.876	21.919	520
530	21.919	21.962	22.004	22.047	22.090	22.132	22.175	22.218	22.260	22.303	22.346	530
540	22.346	22.388	22.431	22.473	22.516	22.559	22.601	22.644	22.687	22.729	22.772	540
550	22.772	22.815	22.857	22.900	22.942	22.985	23.028	23.070	23.113	23.156	23.198	550
560	23.198	23.241	23.284	23.326	23.369	23.411	23.454	23.497	23.539	23.582	23.624	560
570	23.624	23.667	23.710	23.752	23.795	23.837	23.880	23.923	23.965	24.008	24.050	570
580	24.050	24.093	24.136	24.178	24.221	24.263	24.306	24.348	24.391	24.434	24.476	580
590	24.476	24.519	24.561	24.604	24.646	24.689	24.731	24.774	24.817	24.859	24.902	590
600	24.902	24.944	24.987	25.029	25.072	25.114	25.157	25.199	25.242	25.284	25.327	600
610	25.327	25.369	25.412	25.454	25.497	25.539	25.582	25.624	25.666	25.709	25.751	610
620	25.751	25.794	25.836	25.879	25.921	25.964	26.006	26.048	26.091	26.133	26.176	620
630	26.176	26.218	26.260	26.303	26.345	26.387	26.430	26.472	26.515	26.557	26.599	630
640	26.599	26.642	26.684	26.726	26.769	26.811	26.853	26.896	26.938	26.980	27.022	640
650	27.022	27.065	27.107	27.149	27.192	27.234	27.276	27.318	27.361	27.403	27.445	650
660	27.445	27.487	27.529	27.572	27.614	27.656	27.698	27.740	27.783	27.825	27.867	660
670	27.867	27.909	27.951	27.993	28.035	28.078	28.120	28.162	28.204	28.246	28.288	670
680	28.288	28.330	28.372	28.414	28.456	28.498	28.540	28.583	28.625	28.667	28.709	680
690	28.709	28.751	28.793	28.835	28.877	28.919	28.961	29.002	29.044	29.086	29.128	690
700	29.128	29.170	29.212	29.254	29.296	29.338	29.380	29.422	29.464	29.505	29.547	700
710	29.547	29.589	29.631	29.673	29.715	29.756	29.798	29.840	29.882	29.924	29.965	710
720	29.965	30.007	30.049	30.091	30.132	30.174	30.216	30.257	30.299	30.341	30.383	720
730	30.383	30.424	30.466	30.508	30.549	30.591	30.632	30.674	30.716	30.757	30.799	730
740	30.799	30.840	30.882	30.924	30.965	31.007	31.048	31.090	31.131	31.173	31.214	740
750	31.214	31.256	31.297	31.339	31.380	31.422	31.463	31.504	31.546	31.587	31.629	750
760	31.629	31.670	31.712	31.753	31.794	31.836	31.877	31.918	31.960	32.001	32.042	760
770	32.042	32.084	32.125	32.166	32.207	32.249	32.290	32.331	32.372	32.414	32.455	770
780	32.455	32.496	32.537	32.578	32.619	32.661	32.702	32.743	32.784	32.825	32.866	780
790	32.866	32.907	32.948	32.990	33.031	33.072	33.113	33.154	33.195	33.236	33.277	790
800	33.277	33.318	33.359	33.400	33.441	33.482	33.523	33.564	33.604	33.645	33.686	800
810	33.686	33.727	33.768	33.809	33.850	33.891	33.931	33.972	34.013	34.054	34.095	810
820	34.095	34.136	34.176	34.217	34.258	34.299	34.339	34.380	34.421	34.461	34.502	820
830	34.502	34.543	34.583	34.624	34.665	34.705	34.746	34.787	34.827	34.868	34.909	830
840	34.909	34.949	34.990	35.030	35.071	35.111	35.152	35.192	35.233	35.273	35.314	840
850	35.314	35.354	35.395	35.435	35.476	35.516	35.557	35.597	35.637	35.678	35.718	850
860	35.718	35.758	35.799	35.839	35.880	35.920	35.960	36.000	36.041	36.081	36.121	860
870	36.121	36.162	36.202	36.242	36.282	36.323	36.363	36.403	36.443	36.483	36.524	870
880	36.524	36.564	36.604	36.644	36.684	36.724	36.764	36.804	36.844	36.885	36.925	880
890	36.925	36.965	37.005	37.045	37.085	37.125	37.165	37.205	37.245	37.285	37.325	890
900	37.325	37.365	37.405	37.445	37.484	37.524	37.564	37.604	37.644	37.684	37.724	900
910	37.724	37.764	37.803	37.843	37.883	37.923	37.963	38.002	38.042	38.082	38.122	910
920	38.122	38.162	38.201	38.241	38.281	38.320	38.360	38.400	38.439	38.479	38.519	920
930	38.519	38.558	38.598	38.638	38.677	38.717	38.756	38.796	38.836	38.875	38.915	930
940	38.915	38.954	38.994	39.033	39.073	39.112	39.152	39.191	39.231	39.270	39.310	940
950	39.310	39.349	39.388	39.428	39.467	39.507	39.546	39.585	39.625	39.664	39.703	950
960	39.703	39.743	39.782	39.821	39.861	39.900	39.939	39.979	40.018	40.057	40.096	960
970	40.096	40.136	40.175	40.214	40.253	40.292	40.332	40.371	40.410	40.449	40.488	970
980	40.488	40.527	40.566	40.605	40.645	40.684	40.723	40.762	40.801	40.840	40.879	980
990	40.879	40.918	40.957	40.996	41.035	41.074	41.113	41.152	41.191	41.230	41.269	990
DEG C	0	1	2	3	4	5	6	7	8	9	10	DEG C

TABLE 14 (Continued)

TEMPERATURE-EMF FOR TYPE K THERMOCOUPLES

TEMPERATURES IN DEGREES CELSIUS (IPTS-68) **REFERENCE JUNCTIONS AT 0°C**

DEG C	0	1	2	3	4	5	6	7	8	9	10	DEG C
					THERMOELECTRIC	VOLTAGE IN	MILLIVOLTS					
1000	41.269	41.308	41.347	41.385	41.424	41.463	41.502	41.541	41.580	41.619	41.657	1000
1010	41.657	41.696	41.735	41.774	41.813	41.851	41.890	41.929	41.968	42.006	42.045	1010
1020	42.045	42.084	42.123	42.161	42.200	42.239	42.277	42.316	42.355	42.393	42.432	1020
1030	42.432	42.470	42.509	42.548	42.586	42.625	42.663	42.702	42.740	42.779	42.817	1030
1040	42.817	42.856	42.894	42.933	42.971	43.010	43.048	43.087	43.125	43.164	43.202	1040
1050	43.202	43.240	43.279	43.317	43.356	43.394	43.432	43.471	43.509	43.547	43.585	1050
1060	43.585	43.624	43.662	43.700	43.739	43.777	43.815	43.853	43.891	43.930	43.968	1060
1070	43.968	44.006	44.044	44.082	44.121	44.159	44.197	44.235	44.273	44.311	44.349	1070
1080	44.349	44.387	44.425	44.463	44.501	44.539	44.577	44.615	44.653	44.691	44.729	1080
1090	44.729	44.767	44.805	44.843	44.881	44.919	44.957	44.995	45.033	45.070	45.108	1090
1100	45.108	45.146	45.184	45.222	45.260	45.297	45.335	45.373	45.411	45.448	45.486	1100
1110	45.486	45.524	45.561	45.599	45.637	45.675	45.712	45.750	45.787	45.825	45.863	1110
1120	45.863	45.900	45.938	45.975	46.013	46.050	46.088	46.126	46.163	46.201	46.238	1120
1130	46.238	46.275	46.313	46.350	46.388	46.425	46.463	46.500	46.537	46.575	46.612	1130
1140	46.612	46.649	46.687	46.724	46.761	46.799	46.836	46.873	46.910	46.948	46.985	1140
1150	46.985	47.022	47.059	47.096	47.134	47.171	47.208	47.245	47.282	47.319	47.356	1150
1160	47.356	47.393	47.430	47.468	47.505	47.542	47.579	47.616	47.653	47.689	47.726	1160
1170	47.726	47.763	47.800	47.837	47.874	47.911	47.948	47.985	48.021	48.058	48.095	1170
1180	48.095	48.132	48.169	48.205	48.242	48.279	48.316	48.352	48.389	48.426	48.462	1180
1190	48.462	48.499	48.536	48.572	48.609	48.645	48.682	48.718	48.755	48.792	48.828	1190
1200	48.828	48.865	48.901	48.937	48.974	49.010	49.047	49.083	49.120	49.156	49.192	1200
1210	49.192	49.229	49.265	49.301	49.338	49.374	49.410	49.446	49.483	49.519	49.555	1210
1220	49.555	49.591	49.627	49.663	49.700	49.736	49.772	49.808	49.844	49.880	49.916	1220
1230	49.916	49.952	49.988	50.024	50.060	50.096	50.132	50.168	50.204	50.240	50.276	1230
1240	50.276	50.311	50.347	50.383	50.419	50.455	50.491	50.526	50.562	50.598	50.633	1240
1250	50.633	50.669	50.705	50.741	50.776	50.812	50.847	50.883	50.919	50.954	50.990	1250
1260	50.990	51.025	51.061	51.096	51.132	51.167	51.203	51.238	51.274	51.309	51.344	1260
1270	51.344	51.380	51.415	51.450	51.486	51.521	51.556	51.592	51.627	51.662	51.697	1270
1280	51.697	51.733	51.768	51.803	51.838	51.873	51.908	51.943	51.979	52.014	52.049	1280
1290	52.049	52.084	52.119	52.154	52.189	52.224	52.259	52.294	52.329	52.364	52.398	1290
1300	52.398	52.433	52.468	52.503	52.538	52.573	52.608	52.642	52.677	52.712	52.747	1300
1310	52.747	52.781	52.816	52.851	52.886	52.920	52.955	52.989	53.024	53.059	53.093	1310
1320	53.093	53.128	53.162	53.197	53.232	53.266	53.301	53.335	53.370	53.404	53.439	1320
1330	53.439	53.473	53.507	53.542	53.576	53.611	53.645	53.679	53.714	53.748	53.782	1330
1340	53.782	53.817	53.851	53.885	53.920	53.954	53.988	54.022	54.057	54.091	54.125	1340
1350	54.125	54.159	54.193	54.228	54.262	54.296	54.330	54.364	54.398	54.432	54.466	1350
1360	54.466	54.501	54.535	54.569	54.603	54.637	54.671	54.705	54.739	54.773	54.807	1360
1370	54.807	54.841	54.875									1370
DEG C	0	1	2	3	4	5	6	7	8	9	10	DEG C

TABLE 15

TEMPERATURE-EMF FOR TYPE R THERMOCOUPLES

TEMPERATURES IN DEGREES CELSIUS (IPTS-68) **REFERENCE JUNCTIONS AT 0°C**

DEG C	0	1	2	3	4	5	6	7	8	9	10	DEG C

THERMOELECTRIC VOLTAGE IN MILLIVOLTS

DEG C	0	1	2	3	4	5	6	7	8	9	10	DEG C
-50	-0.226											-50
-40	-0.188	-0.192	-0.196	-0.200	-0.204	-0.207	-0.211	-0.215	-0.219	-0.223	-0.226	-40
-30	-0.145	-0.150	-0.154	-0.158	-0.163	-0.167	-0.171	-0.175	-0.180	-0.184	-0.188	-30
-20	-0.100	-0.105	-0.109	-0.114	-0.119	-0.123	-0.128	-0.132	-0.137	-0.141	-0.145	-20
-10	-0.051	-0.056	-0.061	-0.066	-0.071	-0.076	-0.081	-0.086	-0.091	-0.095	-0.100	-10
-0	0.000	-0.005	-0.011	-0.016	-0.021	-0.026	-0.031	-0.036	-0.041	-0.046	-0.051	-0
0	0.000	0.005	0.011	0.016	0.021	0.027	0.032	0.038	0.043	0.049	0.054	0
10	0.054	0.060	0.065	0.071	0.077	0.082	0.088	0.094	0.100	0.105	0.111	10
20	0.111	0.117	0.123	0.129	0.135	0.141	0.147	0.152	0.158	0.165	0.171	20
30	0.171	0.177	0.183	0.189	0.195	0.201	0.207	0.214	0.220	0.226	0.232	30
40	0.232	0.239	0.245	0.251	0.258	0.264	0.271	0.277	0.283	0.290	0.296	40
50	0.296	0.303	0.310	0.316	0.323	0.329	0.336	0.343	0.349	0.356	0.363	50
60	0.363	0.369	0.376	0.383	0.390	0.397	0.403	0.410	0.417	0.424	0.431	60
70	0.431	0.438	0.445	0.452	0.459	0.466	0.473	0.480	0.487	0.494	0.501	70
80	0.501	0.508	0.515	0.523	0.530	0.537	0.544	0.552	0.559	0.566	0.573	80
90	0.573	0.581	0.588	0.595	0.603	0.610	0.617	0.625	0.632	0.640	0.647	90
100	0.647	0.655	0.662	0.670	0.677	0.685	0.692	0.700	0.708	0.715	0.723	100
110	0.723	0.730	0.738	0.746	0.754	0.761	0.769	0.777	0.784	0.792	0.800	110
120	0.800	0.808	0.816	0.824	0.831	0.839	0.847	0.855	0.863	0.871	0.879	120
130	0.879	0.887	0.895	0.903	0.911	0.919	0.927	0.935	0.943	0.951	0.959	130
140	0.959	0.967	0.975	0.983	0.992	1.000	1.008	1.016	1.024	1.032	1.041	140
150	1.041	1.049	1.057	1.065	1.074	1.082	1.090	1.099	1.107	1.115	1.124	150
160	1.124	1.132	1.140	1.149	1.157	1.166	1.174	1.183	1.191	1.200	1.208	160
170	1.208	1.217	1.225	1.234	1.242	1.251	1.259	1.268	1.276	1.285	1.294	170
180	1.294	1.302	1.311	1.319	1.328	1.337	1.345	1.354	1.363	1.372	1.380	180
190	1.380	1.389	1.398	1.407	1.415	1.424	1.433	1.442	1.450	1.459	1.468	190
200	1.468	1.477	1.486	1.495	1.504	1.512	1.521	1.530	1.539	1.548	1.557	200
210	1.557	1.566	1.575	1.584	1.593	1.602	1.611	1.620	1.629	1.638	1.647	210
220	1.647	1.656	1.665	1.674	1.683	1.692	1.702	1.711	1.720	1.729	1.738	220
230	1.738	1.747	1.756	1.766	1.775	1.784	1.793	1.802	1.812	1.821	1.830	230
240	1.830	1.839	1.849	1.858	1.867	1.876	1.886	1.895	1.904	1.914	1.923	240
250	1.923	1.932	1.942	1.951	1.960	1.970	1.979	1.988	1.998	2.007	2.017	250
260	2.017	2.026	2.036	2.045	2.054	2.064	2.073	2.083	2.092	2.102	2.111	260
270	2.111	2.121	2.130	2.140	2.149	2.159	2.169	2.178	2.188	2.197	2.207	270
280	2.207	2.216	2.226	2.236	2.245	2.255	2.264	2.274	2.284	2.293	2.303	280
290	2.303	2.313	2.322	2.332	2.342	2.351	2.361	2.371	2.381	2.390	2.400	290
300	2.400	2.410	2.420	2.429	2.439	2.449	2.459	2.468	2.478	2.488	2.498	300
310	2.498	2.508	2.517	2.527	2.537	2.547	2.557	2.567	2.577	2.586	2.596	310
320	2.596	2.606	2.616	2.626	2.636	2.646	2.656	2.666	2.676	2.685	2.695	320
330	2.695	2.705	2.715	2.725	2.735	2.745	2.755	2.765	2.775	2.785	2.795	330
340	2.795	2.805	2.815	2.825	2.835	2.845	2.855	2.866	2.876	2.886	2.896	340
350	2.896	2.906	2.916	2.926	2.936	2.946	2.956	2.966	2.977	2.987	2.997	350
360	2.997	3.007	3.017	3.027	3.037	3.048	3.058	3.068	3.078	3.088	3.099	360
370	3.099	3.109	3.119	3.129	3.139	3.150	3.160	3.170	3.180	3.191	3.201	370
380	3.201	3.211	3.221	3.232	3.242	3.252	3.263	3.273	3.283	3.293	3.304	380
390	3.304	3.314	3.324	3.335	3.345	3.355	3.366	3.376	3.386	3.397	3.407	390
400	3.407	3.418	3.428	3.438	3.449	3.459	3.470	3.480	3.490	3.501	3.511	400
410	3.511	3.522	3.532	3.542	3.553	3.563	3.574	3.584	3.595	3.605	3.616	410
420	3.616	3.626	3.637	3.647	3.658	3.668	3.679	3.689	3.700	3.710	3.721	420
430	3.721	3.731	3.742	3.752	3.763	3.774	3.784	3.795	3.805	3.816	3.826	430
440	3.826	3.837	3.848	3.858	3.869	3.879	3.890	3.901	3.911	3.922	3.933	440
450	3.933	3.943	3.954	3.964	3.975	3.986	3.996	4.007	4.018	4.028	4.039	450
460	4.039	4.050	4.061	4.071	4.082	4.093	4.103	4.114	4.125	4.136	4.146	460
470	4.146	4.157	4.168	4.178	4.189	4.200	4.211	4.222	4.232	4.243	4.254	470
480	4.254	4.265	4.275	4.286	4.297	4.308	4.319	4.329	4.340	4.351	4.362	480
490	4.362	4.373	4.384	4.394	4.405	4.416	4.427	4.438	4.449	4.460	4.471	490
500	4.471	4.481	4.492	4.503	4.514	4.525	4.536	4.547	4.558	4.569	4.580	500
510	4.580	4.591	4.601	4.612	4.623	4.634	4.645	4.656	4.667	4.678	4.689	510
520	4.689	4.700	4.711	4.722	4.733	4.744	4.755	4.766	4.777	4.788	4.799	520
530	4.799	4.810	4.821	4.832	4.843	4.854	4.865	4.876	4.888	4.899	4.910	530
540	4.910	4.921	4.932	4.943	4.954	4.965	4.976	4.987	4.998	5.009	5.021	540
550	5.021	5.032	5.043	5.054	5.065	5.076	5.087	5.099	5.110	5.121	5.132	550
560	5.132	5.143	5.154	5.166	5.177	5.188	5.199	5.210	5.221	5.233	5.244	560
570	5.244	5.255	5.266	5.278	5.289	5.300	5.311	5.322	5.334	5.345	5.356	570
580	5.356	5.367	5.379	5.390	5.401	5.413	5.424	5.435	5.446	5.458	5.469	580
590	5.469	5.480	5.492	5.503	5.514	5.526	5.537	5.548	5.560	5.571	5.582	590

DEG C	0	1	2	3	4	5	6	7	8	9	10	DEG C

TABLE 15 (Continued)

TEMPERATURE-EMF FOR TYPE R THERMOCOUPLES

TEMPERATURES IN DEGREES CELSIUS (IPTS-68) **REFERENCE JUNCTIONS AT 0°C**

DEG C	0	1	2	3	4	5	6	7	8	9	10	DEG C

THERMOELECTRIC VOLTAGE IN MILLIVOLTS

DEG C	0	1	2	3	4	5	6	7	8	9	10	DEG C
600	5.582	5.594	5.605	5.616	5.628	5.639	5.650	5.662	5.673	5.685	5.696	600
610	5.696	5.707	5.719	5.730	5.742	5.753	5.764	5.776	5.787	5.799	5.810	610
620	5.810	5.821	5.833	5.844	5.856	5.867	5.879	5.890	5.902	5.913	5.925	620
630	5.925	5.936	5.948	5.959	5.971	5.982	5.994	6.005	6.017	6.028	6.040	630
640	6.040	6.051	6.063	6.074	6.086	6.098	6.109	6.121	6.132	6.144	6.155	640
650	6.155	6.167	6.179	6.190	6.202	6.213	6.225	6.237	6.248	6.260	6.272	650
660	6.272	6.283	6.295	6.307	6.318	6.330	6.342	6.353	6.365	6.377	6.388	660
670	6.388	6.400	6.412	6.423	6.435	6.447	6.458	6.470	6.482	6.494	6.505	670
680	6.505	6.517	6.529	6.541	6.552	6.564	6.576	6.588	6.599	6.611	6.623	680
690	6.623	6.635	6.647	6.658	6.670	6.682	6.694	6.706	6.718	6.729	6.741	690
700	6.741	6.753	6.765	6.777	6.789	6.800	6.812	6.824	6.836	6.848	6.860	700
710	6.860	6.872	6.884	6.895	6.907	6.919	6.931	6.943	6.955	6.967	6.979	710
720	6.979	6.991	7.003	7.015	7.027	7.039	7.051	7.063	7.074	7.086	7.098	720
730	7.098	7.110	7.122	7.134	7.146	7.158	7.170	7.182	7.194	7.206	7.218	730
740	7.218	7.231	7.243	7.255	7.267	7.279	7.291	7.303	7.315	7.327	7.339	740
750	7.339	7.351	7.363	7.375	7.387	7.399	7.412	7.424	7.436	7.448	7.460	750
760	7.460	7.472	7.484	7.496	7.509	7.521	7.533	7.545	7.557	7.569	7.582	760
770	7.582	7.594	7.606	7.618	7.630	7.642	7.655	7.667	7.679	7.691	7.703	770
780	7.703	7.716	7.728	7.740	7.752	7.765	7.777	7.789	7.801	7.814	7.826	780
790	7.826	7.838	7.850	7.863	7.875	7.887	7.900	7.912	7.924	7.937	7.949	790
800	7.949	7.961	7.973	7.986	7.998	8.010	8.023	8.035	8.047	8.060	8.072	800
810	8.072	8.085	8.097	8.109	8.122	8.134	8.146	8.159	8.171	8.184	8.196	810
820	8.196	8.208	8.221	8.233	8.246	8.258	8.271	8.283	8.295	8.308	8.320	820
830	8.320	8.333	8.345	8.358	8.370	8.383	8.395	8.408	8.420	8.433	8.445	830
840	8.445	8.458	8.470	8.483	8.495	8.508	8.520	8.533	8.545	8.558	8.570	840
850	8.570	8.583	8.595	8.608	8.621	8.633	8.646	8.658	8.671	8.683	8.696	850
860	8.696	8.709	8.721	8.734	8.746	8.759	8.772	8.784	8.797	8.810	8.822	860
870	8.822	8.835	8.847	8.860	8.873	8.885	8.898	8.911	8.923	8.936	8.949	870
880	8.949	8.961	8.974	8.987	9.000	9.012	9.025	9.038	9.050	9.063	9.076	880
890	9.076	9.089	9.101	9.114	9.127	9.140	9.152	9.165	9.178	9.191	9.203	890
900	9.203	9.216	9.229	9.242	9.254	9.267	9.280	9.293	9.306	9.319	9.331	900
910	9.331	9.344	9.357	9.370	9.383	9.395	9.408	9.421	9.434	9.447	9.460	910
920	9.460	9.473	9.485	9.498	9.511	9.524	9.537	9.550	9.563	9.576	9.589	920
930	9.589	9.602	9.614	9.627	9.640	9.653	9.666	9.679	9.692	9.705	9.718	930
940	9.718	9.731	9.744	9.757	9.770	9.783	9.796	9.809	9.822	9.835	9.848	940
950	9.848	9.861	9.874	9.887	9.900	9.913	9.926	9.939	9.952	9.965	9.978	950
960	9.978	9.991	10.004	10.017	10.030	10.043	10.056	10.069	10.082	10.095	10.109	960
970	10.109	10.122	10.135	10.148	10.161	10.174	10.187	10.200	10.213	10.227	10.240	970
980	10.240	10.253	10.266	10.279	10.292	10.305	10.319	10.332	10.345	10.358	10.371	980
990	10.371	10.384	10.398	10.411	10.424	10.437	10.450	10.464	10.477	10.490	10.503	990
1000	10.503	10.516	10.530	10.543	10.556	10.569	10.583	10.596	10.609	10.622	10.636	1000
1010	10.636	10.649	10.662	10.675	10.689	10.702	10.715	10.729	10.742	10.755	10.768	1010
1020	10.768	10.782	10.795	10.808	10.822	10.835	10.848	10.862	10.875	10.888	10.902	1020
1030	10.902	10.915	10.928	10.942	10.955	10.968	10.982	10.995	11.009	11.022	11.035	1030
1040	11.035	11.049	11.062	11.076	11.089	11.102	11.116	11.129	11.143	11.156	11.170	1040
1050	11.170	11.183	11.196	11.210	11.223	11.237	11.250	11.264	11.277	11.291	11.304	1050
1060	11.304	11.318	11.331	11.345	11.358	11.372	11.385	11.399	11.412	11.426	11.439	1060
1070	11.439	11.453	11.466	11.480	11.493	11.507	11.520	11.534	11.547	11.561	11.574	1070
1080	11.574	11.588	11.602	11.615	11.629	11.642	11.656	11.669	11.683	11.697	11.710	1080
1090	11.710	11.724	11.737	11.751	11.765	11.778	11.792	11.805	11.819	11.833	11.846	1090
1100	11.846	11.860	11.874	11.887	11.901	11.914	11.928	11.942	11.955	11.969	11.983	1100
1110	11.983	11.996	12.010	12.024	12.037	12.051	12.065	12.078	12.092	12.106	12.119	1110
1120	12.119	12.133	12.147	12.161	12.174	12.188	12.202	12.215	12.229	12.243	12.257	1120
1130	12.257	12.270	12.284	12.298	12.311	12.325	12.339	12.353	12.366	12.380	12.394	1130
1140	12.394	12.408	12.421	12.435	12.449	12.463	12.476	12.490	12.504	12.518	12.532	1140
1150	12.532	12.545	12.559	12.573	12.587	12.600	12.614	12.628	12.642	12.656	12.669	1150
1160	12.669	12.683	12.697	12.711	12.725	12.739	12.752	12.766	12.780	12.794	12.808	1160
1170	12.808	12.822	12.835	12.849	12.863	12.877	12.891	12.905	12.918	12.932	12.946	1170
1180	12.946	12.960	12.974	12.988	13.002	13.016	13.029	13.043	13.057	13.071	13.085	1180
1190	13.085	13.099	13.113	13.127	13.140	13.154	13.168	13.182	13.196	13.210	13.224	1190
1200	13.224	13.238	13.252	13.266	13.280	13.293	13.307	13.321	13.335	13.349	13.363	1200
1210	13.363	13.377	13.391	13.405	13.419	13.433	13.447	13.461	13.475	13.489	13.502	1210
1220	13.502	13.516	13.530	13.544	13.558	13.572	13.586	13.600	13.614	13.628	13.642	1220
1230	13.642	13.656	13.670	13.684	13.698	13.712	13.726	13.740	13.754	13.768	13.782	1230
1240	13.782	13.796	13.810	13.824	13.838	13.852	13.866	13.880	13.894	13.908	13.922	1240

DEG C	0	1	2	3	4	5	6	7	8	9	10	DEG C

TABLE 15 (Continued)

TEMPERATURE-EMF FOR TYPE R THERMOCOUPLES

TEMPERATURES IN DEGREES CELSIUS (IPTS-68) **REFERENCE JUNCTIONS AT 0°C**

DEG C	0	1	2	3	4	5	6	7	8	9	10	DEG C
					THERMOELECTRIC VOLTAGE IN MILLIVOLTS							
1250	13.922	13.936	13.950	13.964	13.978	13.992	14.006	14.020	14.034	14.048	14.062	1250
1260	14.062	14.076	14.090	14.104	14.118	14.132	14.146	14.160	14.174	14.188	14.202	1260
1270	14.202	14.216	14.230	14.244	14.258	14.272	14.286	14.301	14.315	14.329	14.343	1270
1280	14.343	14.357	14.371	14.385	14.399	14.413	14.427	14.441	14.455	14.469	14.483	1280
1290	14.483	14.497	14.511	14.525	14.539	14.554	14.568	14.582	14.596	14.610	14.624	1290
1300	14.624	14.638	14.652	14.666	14.680	14.694	14.708	14.722	14.737	14.751	14.765	1300
1310	14.765	14.779	14.793	14.807	14.821	14.835	14.849	14.863	14.877	14.891	14.906	1310
1320	14.906	14.920	14.934	14.948	14.962	14.976	14.990	15.004	15.018	15.032	15.047	1320
1330	15.047	15.061	15.075	15.089	15.103	15.117	15.131	15.145	15.159	15.173	15.188	1330
1340	15.188	15.202	15.216	15.230	15.244	15.258	15.272	15.286	15.300	15.315	15.329	1340
1350	15.329	15.343	15.357	15.371	15.385	15.399	15.413	15.427	15.442	15.456	15.470	1350
1360	15.470	15.484	15.498	15.512	15.526	15.540	15.555	15.569	15.583	15.597	15.611	1360
1370	15.611	15.625	15.639	15.653	15.667	15.682	15.696	15.710	15.724	15.738	15.752	1370
1380	15.752	15.766	15.780	15.795	15.809	15.823	15.837	15.851	15.865	15.879	15.893	1380
1390	15.893	15.908	15.922	15.936	15.950	15.964	15.978	15.992	16.006	16.021	16.035	1390
1400	16.035	16.049	16.063	16.077	16.091	16.105	16.119	16.134	16.148	16.162	16.176	1400
1410	16.176	16.190	16.204	16.218	16.232	16.247	16.261	16.275	16.289	16.303	16.317	1410
1420	16.317	16.331	16.345	16.360	16.374	16.388	16.402	16.416	16.430	16.444	16.458	1420
1430	16.458	16.472	16.487	16.501	16.515	16.529	16.543	16.557	16.571	16.585	16.599	1430
1440	16.599	16.614	16.628	16.642	16.656	16.670	16.684	16.698	16.712	16.726	16.741	1440
1450	16.741	16.755	16.769	16.783	16.797	16.811	16.825	16.839	16.853	16.867	16.882	1450
1460	16.882	16.896	16.910	16.924	16.938	16.952	16.966	16.980	16.994	17.008	17.022	1460
1470	17.022	17.037	17.051	17.065	17.079	17.093	17.107	17.121	17.135	17.149	17.163	1470
1480	17.163	17.177	17.192	17.206	17.220	17.234	17.248	17.262	17.276	17.290	17.304	1480
1490	17.304	17.318	17.332	17.346	17.360	17.374	17.388	17.403	17.417	17.431	17.445	1490
1500	17.445	17.459	17.473	17.487	17.501	17.515	17.529	17.543	17.557	17.571	17.585	1500
1510	17.585	17.599	17.613	17.627	17.641	17.655	17.669	17.684	17.698	17.712	17.726	1510
1520	17.726	17.740	17.754	17.768	17.782	17.796	17.810	17.824	17.838	17.852	17.866	1520
1530	17.866	17.880	17.894	17.908	17.922	17.936	17.950	17.964	17.978	17.992	18.006	1530
1540	18.006	18.020	18.034	18.048	18.062	18.076	18.090	18.104	18.118	18.132	18.146	1540
1550	18.146	18.160	18.174	18.188	18.202	18.216	18.230	18.244	18.258	18.272	18.286	1550
1560	18.286	18.299	18.313	18.327	18.341	18.355	18.369	18.383	18.397	18.411	18.425	1560
1570	18.425	18.439	18.453	18.467	18.481	18.495	18.509	18.523	18.537	18.550	18.564	1570
1580	18.564	18.578	18.592	18.606	18.620	18.634	18.648	18.662	18.676	18.690	18.703	1580
1590	18.703	18.717	18.731	18.745	18.759	18.773	18.787	18.801	18.815	18.828	18.842	1590
1600	18.842	18.856	18.870	18.884	18.898	18.912	18.926	18.939	18.953	18.967	18.981	1600
1610	18.981	18.995	19.009	19.023	19.036	19.050	19.064	19.078	19.092	19.106	19.119	1610
1620	19.119	19.133	19.147	19.161	19.175	19.188	19.202	19.216	19.230	19.244	19.257	1620
1630	19.257	19.271	19.285	19.299	19.313	19.326	19.340	19.354	19.368	19.382	19.395	1630
1640	19.395	19.409	19.423	19.437	19.450	19.464	19.478	19.492	19.505	19.519	19.533	1640
1650	19.533	19.547	19.560	19.574	19.588	19.602	19.615	19.629	19.643	19.656	19.670	1650
1660	19.670	19.684	19.698	19.711	19.725	19.739	19.752	19.766	19.780	19.793	19.807	1660
1670	19.807	19.821	19.834	19.848	19.862	19.875	19.889	19.903	19.916	19.930	19.944	1670
1680	19.944	19.957	19.971	19.985	19.998	20.012	20.025	20.039	20.053	20.066	20.080	1680
1690	20.080	20.093	20.107	20.120	20.134	20.148	20.161	20.175	20.188	20.202	20.215	1690
1700	20.215	20.229	20.242	20.256	20.269	20.283	20.296	20.309	20.323	20.336	20.350	1700
1710	20.350	20.363	20.377	20.390	20.403	20.417	20.430	20.443	20.457	20.470	20.483	1710
1720	20.483	20.497	20.510	20.523	20.537	20.550	20.563	20.576	20.590	20.603	20.616	1720
1730	20.616	20.629	20.642	20.656	20.669	20.682	20.695	20.708	20.721	20.734	20.748	1730
1740	20.748	20.761	20.774	20.787	20.800	20.813	20.826	20.839	20.852	20.865	20.878	1740
1750	20.878	20.891	20.904	20.916	20.929	20.942	20.955	20.968	20.981	20.994	21.006	1750
1760	21.006	21.019	21.032	21.045	21.057	21.070	21.083	21.096	21.108			1760
DEG C	0	1	2	3	4	5	6	7	8	9	10	DEG C

TABLE 16

TEMPERATURE-EMF FOR TYPE S THERMOCOUPLES

TEMPERATURES IN DEGREES CELSIUS (IPTS-68) **REFERENCE JUNCTIONS AT 0°C**

DEG C	0	1	2	3	4	5	6	7	8	9	10	DEG C
					THERMOELECTRIC VOLTAGE IN MILLIVOLTS							
-50	-0.236											-50
-40	-0.194	-0.199	-0.203	-0.207	-0.211	-0.215	-0.220	-0.224	-0.228	-0.232	-0.236	-40
-30	-0.150	-0.155	-0.159	-0.164	-0.168	-0.173	-0.177	-0.181	-0.186	-0.190	-0.194	-30
-20	-0.103	-0.108	-0.112	-0.117	-0.122	-0.127	-0.132	-0.136	-0.141	-0.145	-0.150	-20
-10	-0.053	-0.058	-0.063	-0.068	-0.073	-0.078	-0.083	-0.088	-0.093	-0.098	-0.103	-10
-0	0.000	-0.005	-0.011	-0.016	-0.021	-0.027	-0.032	-0.037	-0.042	-0.048	-0.053	-0
0	0.000	0.005	0.011	0.016	0.022	0.027	0.033	0.038	0.044	0.050	0.055	0
10	0.055	0.061	0.067	0.072	0.078	0.084	0.090	0.095	0.101	0.107	0.113	10
20	0.113	0.119	0.125	0.131	0.137	0.142	0.148	0.154	0.161	0.167	0.173	20
30	0.173	0.179	0.185	0.191	0.197	0.203	0.210	0.216	0.222	0.228	0.235	30
40	0.235	0.241	0.247	0.254	0.260	0.266	0.273	0.279	0.286	0.292	0.299	40
50	0.299	0.305	0.312	0.318	0.325	0.331	0.338	0.345	0.351	0.358	0.365	50
60	0.365	0.371	0.378	0.385	0.391	0.398	0.405	0.412	0.419	0.425	0.432	60
70	0.432	0.439	0.446	0.453	0.460	0.467	0.474	0.481	0.488	0.495	0.502	70
80	0.502	0.509	0.516	0.523	0.530	0.537	0.544	0.551	0.558	0.566	0.573	80
90	0.573	0.580	0.587	0.594	0.602	0.609	0.616	0.623	0.631	0.638	0.645	90
100	0.645	0.653	0.660	0.667	0.675	0.682	0.690	0.697	0.704	0.712	0.719	100
110	0.719	0.727	0.734	0.742	0.749	0.757	0.764	0.772	0.780	0.787	0.795	110
120	0.795	0.802	0.810	0.818	0.825	0.833	0.841	0.848	0.856	0.864	0.872	120
130	0.872	0.879	0.887	0.895	0.903	0.910	0.918	0.926	0.934	0.942	0.950	130
140	0.950	0.957	0.965	0.973	0.981	0.989	0.997	1.005	1.013	1.021	1.029	140
150	1.029	1.037	1.045	1.053	1.061	1.069	1.077	1.085	1.093	1.101	1.109	150
160	1.109	1.117	1.125	1.133	1.141	1.149	1.158	1.166	1.174	1.182	1.190	160
170	1.190	1.198	1.207	1.215	1.223	1.231	1.240	1.248	1.256	1.264	1.273	170
180	1.273	1.281	1.289	1.297	1.306	1.314	1.322	1.331	1.339	1.347	1.356	180
190	1.356	1.364	1.373	1.381	1.389	1.398	1.406	1.415	1.423	1.432	1.440	190
200	1.440	1.448	1.457	1.465	1.474	1.482	1.491	1.499	1.508	1.516	1.525	200
210	1.525	1.534	1.542	1.551	1.559	1.568	1.576	1.585	1.594	1.602	1.611	210
220	1.611	1.620	1.628	1.637	1.645	1.654	1.663	1.671	1.680	1.689	1.698	220
230	1.698	1.706	1.715	1.724	1.732	1.741	1.750	1.759	1.767	1.776	1.785	230
240	1.785	1.794	1.802	1.811	1.820	1.829	1.838	1.846	1.855	1.864	1.873	240
250	1.873	1.882	1.891	1.899	1.908	1.917	1.926	1.935	1.944	1.953	1.962	250
260	1.962	1.971	1.979	1.988	1.997	2.006	2.015	2.024	2.033	2.042	2.051	260
270	2.051	2.060	2.069	2.078	2.087	2.096	2.105	2.114	2.123	2.132	2.141	270
280	2.141	2.150	2.159	2.168	2.177	2.186	2.195	2.204	2.213	2.222	2.232	280
290	2.232	2.241	2.250	2.259	2.268	2.277	2.286	2.295	2.304	2.314	2.323	290
300	2.323	2.332	2.341	2.350	2.359	2.368	2.378	2.387	2.396	2.405	2.414	300
310	2.414	2.424	2.433	2.442	2.451	2.460	2.470	2.479	2.488	2.497	2.506	310
320	2.506	2.516	2.525	2.534	2.543	2.553	2.562	2.571	2.581	2.590	2.599	320
330	2.599	2.608	2.618	2.627	2.636	2.646	2.655	2.664	2.674	2.683	2.692	330
340	2.692	2.702	2.711	2.720	2.730	2.739	2.748	2.758	2.767	2.776	2.786	340
350	2.786	2.795	2.805	2.814	2.823	2.833	2.842	2.852	2.861	2.870	2.880	350
360	2.880	2.889	2.899	2.908	2.917	2.927	2.936	2.946	2.955	2.965	2.974	360
370	2.974	2.984	2.993	3.003	3.012	3.022	3.031	3.041	3.050	3.059	3.069	370
380	3.069	3.078	3.088	3.097	3.107	3.117	3.126	3.136	3.145	3.155	3.164	380
390	3.164	3.174	3.183	3.193	3.202	3.212	3.221	3.231	3.241	3.250	3.260	390
400	3.260	3.269	3.279	3.288	3.298	3.308	3.317	3.327	3.336	3.346	3.356	400
410	3.356	3.365	3.375	3.384	3.394	3.404	3.413	3.423	3.433	3.442	3.452	410
420	3.452	3.462	3.471	3.481	3.491	3.500	3.510	3.520	3.529	3.539	3.549	420
430	3.549	3.558	3.568	3.578	3.587	3.597	3.607	3.616	3.626	3.636	3.645	430
440	3.645	3.655	3.665	3.675	3.684	3.694	3.704	3.714	3.723	3.733	3.743	440
450	3.743	3.752	3.762	3.772	3.782	3.791	3.801	3.811	3.821	3.831	3.840	450
460	3.840	3.850	3.860	3.870	3.879	3.889	3.899	3.909	3.919	3.928	3.938	460
470	3.938	3.948	3.958	3.968	3.977	3.987	3.997	4.007	4.017	4.027	4.036	470
480	4.036	4.046	4.056	4.066	4.076	4.086	4.095	4.105	4.115	4.125	4.135	480
490	4.135	4.145	4.155	4.164	4.174	4.184	4.194	4.204	4.214	4.224	4.234	490
500	4.234	4.243	4.253	4.263	4.273	4.283	4.293	4.303	4.313	4.323	4.333	500
510	4.333	4.343	4.352	4.362	4.372	4.382	4.392	4.402	4.412	4.422	4.432	510
520	4.432	4.442	4.452	4.462	4.472	4.482	4.492	4.502	4.512	4.522	4.532	520
530	4.532	4.542	4.552	4.562	4.572	4.582	4.592	4.602	4.612	4.622	4.632	530
540	4.632	4.642	4.652	4.662	4.672	4.682	4.692	4.702	4.712	4.722	4.732	540
550	4.732	4.742	4.752	4.762	4.772	4.782	4.792	4.802	4.812	4.822	4.832	550
560	4.832	4.842	4.852	4.862	4.873	4.883	4.893	4.903	4.913	4.923	4.933	560
570	4.933	4.943	4.953	4.963	4.973	4.984	4.994	5.004	5.014	5.024	5.034	570
580	5.034	5.044	5.054	5.065	5.075	5.085	5.095	5.105	5.115	5.125	5.136	580
590	5.136	5.146	5.156	5.166	5.176	5.186	5.197	5.207	5.217	5.227	5.237	590
DEG C	0	1	2	3	4	5	6	7	8	9	10	DEG C

TABLE 16 (Continued)

TEMPERATURE-EMF FOR TYPE S THERMOCOUPLES

TEMPERATURES IN DEGREES CELSIUS (IPTS-68) **REFERENCE JUNCTIONS AT 0°C**

THERMOELECTRIC VOLTAGE IN MILLIVOLTS

DEG C	0	1	2	3	4	5	6	7	8	9	10	DEG C
600	5.237	5.247	5.258	5.268	5.278	5.288	5.298	5.309	5.319	5.329	5.339	600
610	5.339	5.350	5.360	5.370	5.380	5.391	5.401	5.411	5.421	5.431	5.442	610
620	5.442	5.452	5.462	5.473	5.483	5.493	5.503	5.514	5.524	5.534	5.544	620
630	5.544	5.555	5.565	5.575	5.586	5.596	5.606	5.617	5.627	5.637	5.648	630
640	5.648	5.658	5.668	5.679	5.689	5.700	5.710	5.720	5.731	5.741	5.751	640
650	5.751	5.762	5.772	5.782	5.793	5.803	5.814	5.824	5.834	5.845	5.855	650
660	5.855	5.866	5.876	5.887	5.897	5.907	5.918	5.928	5.939	5.949	5.960	660
670	5.960	5.970	5.980	5.991	6.001	6.012	6.022	6.033	6.043	6.054	6.064	670
680	6.064	6.075	6.085	6.096	6.106	6.117	6.127	6.138	6.148	6.159	6.169	680
690	6.169	6.180	6.190	6.201	6.211	6.222	6.232	6.243	6.253	6.264	6.274	690
700	6.274	6.285	6.295	6.306	6.316	6.327	6.338	6.348	6.359	6.369	6.380	700
710	6.380	6.390	6.401	6.412	6.422	6.433	6.443	6.454	6.465	6.475	6.486	710
720	6.486	6.496	6.507	6.518	6.528	6.539	6.549	6.560	6.571	6.581	6.592	720
730	6.592	6.603	6.613	6.624	6.635	6.645	6.656	6.667	6.677	6.688	6.699	730
740	6.699	6.709	6.720	6.731	6.741	6.752	6.763	6.773	6.784	6.795	6.805	740
750	6.805	6.816	6.827	6.838	6.848	6.859	6.870	6.880	6.891	6.902	6.913	750
760	6.913	6.923	6.934	6.945	6.956	6.966	6.977	6.988	6.999	7.009	7.020	760
770	7.020	7.031	7.042	7.053	7.063	7.074	7.085	7.096	7.107	7.117	7.128	770
780	7.128	7.139	7.150	7.161	7.171	7.182	7.193	7.204	7.215	7.225	7.236	780
790	7.236	7.247	7.258	7.269	7.280	7.291	7.301	7.312	7.323	7.334	7.345	790
800	7.345	7.356	7.367	7.377	7.388	7.399	7.410	7.421	7.432	7.443	7.454	800
810	7.454	7.465	7.476	7.486	7.497	7.508	7.519	7.530	7.541	7.552	7.563	810
820	7.563	7.574	7.585	7.596	7.607	7.618	7.629	7.640	7.651	7.661	7.672	820
830	7.672	7.683	7.694	7.705	7.716	7.727	7.738	7.749	7.760	7.771	7.782	830
840	7.782	7.793	7.804	7.815	7.826	7.837	7.848	7.859	7.870	7.881	7.892	840
850	7.892	7.904	7.915	7.926	7.937	7.948	7.959	7.970	7.981	7.992	8.003	850
860	8.003	8.014	8.025	8.036	8.047	8.058	8.069	8.081	8.092	8.103	8.114	860
870	8.114	8.125	8.136	8.147	8.158	8.169	8.180	8.192	8.203	8.214	8.225	870
880	8.225	8.236	8.247	8.258	8.270	8.281	8.292	8.303	8.314	8.325	8.336	880
890	8.336	8.348	8.359	8.370	8.381	8.392	8.404	8.415	8.426	8.437	8.448	890
900	8.448	8.460	8.471	8.482	8.493	8.504	8.516	8.527	8.538	8.549	8.560	900
910	8.560	8.572	8.583	8.594	8.605	8.617	8.628	8.639	8.650	8.662	8.673	910
920	8.673	8.684	8.695	8.707	8.718	8.729	8.741	8.752	8.763	8.774	8.786	920
930	8.786	8.797	8.808	8.820	8.831	8.842	8.854	8.865	8.876	8.888	8.899	930
940	8.899	8.910	8.922	8.933	8.944	8.956	8.967	8.978	8.990	9.001	9.012	940
950	9.012	9.024	9.035	9.047	9.058	9.069	9.081	9.092	9.103	9.115	9.126	950
960	9.126	9.138	9.149	9.160	9.172	9.183	9.195	9.206	9.217	9.229	9.240	960
970	9.240	9.252	9.263	9.275	9.286	9.298	9.309	9.320	9.332	9.343	9.355	970
980	9.355	9.366	9.378	9.389	9.401	9.412	9.424	9.435	9.447	9.458	9.470	980
990	9.470	9.481	9.493	9.504	9.516	9.527	9.539	9.550	9.562	9.573	9.585	990
1000	9.585	9.596	9.608	9.619	9.631	9.642	9.654	9.665	9.677	9.689	9.700	1000
1010	9.700	9.712	9.723	9.735	9.746	9.758	9.770	9.781	9.793	9.804	9.816	1010
1020	9.816	9.828	9.839	9.851	9.862	9.874	9.886	9.897	9.909	9.920	9.932	1020
1030	9.932	9.944	9.955	9.967	9.979	9.990	10.002	10.013	10.025	10.037	10.048	1030
1040	10.048	10.060	10.072	10.083	10.095	10.107	10.118	10.130	10.142	10.154	10.165	1040
1050	10.165	10.177	10.189	10.200	10.212	10.224	10.235	10.247	10.259	10.271	10.282	1050
1060	10.282	10.294	10.306	10.318	10.329	10.341	10.353	10.364	10.376	10.388	10.400	1060
1070	10.400	10.411	10.423	10.435	10.447	10.459	10.470	10.482	10.494	10.506	10.517	1070
1080	10.517	10.529	10.541	10.553	10.565	10.576	10.588	10.600	10.612	10.624	10.635	1080
1090	10.635	10.647	10.659	10.671	10.683	10.694	10.706	10.718	10.730	10.742	10.754	1090
1100	10.754	10.765	10.777	10.789	10.801	10.813	10.825	10.836	10.848	10.860	10.872	1100
1110	10.872	10.884	10.896	10.908	10.919	10.931	10.943	10.955	10.967	10.979	10.991	1110
1120	10.991	11.003	11.014	11.026	11.038	11.050	11.062	11.074	11.086	11.098	11.110	1120
1130	11.110	11.121	11.133	11.145	11.157	11.169	11.181	11.193	11.205	11.217	11.229	1130
1140	11.229	11.241	11.252	11.264	11.276	11.288	11.300	11.312	11.324	11.336	11.348	1140
1150	11.348	11.360	11.372	11.384	11.396	11.408	11.420	11.432	11.443	11.455	11.467	1150
1160	11.467	11.479	11.491	11.503	11.515	11.527	11.539	11.551	11.563	11.575	11.587	1160
1170	11.587	11.599	11.611	11.623	11.635	11.647	11.659	11.671	11.683	11.695	11.707	1170
1180	11.707	11.719	11.731	11.743	11.755	11.767	11.779	11.791	11.803	11.815	11.827	1180
1190	11.827	11.839	11.851	11.863	11.875	11.887	11.899	11.911	11.923	11.935	11.947	1190
1200	11.947	11.959	11.971	11.983	11.995	12.007	12.019	12.031	12.043	12.055	12.067	1200
1210	12.067	12.079	12.091	12.103	12.116	12.128	12.140	12.152	12.164	12.176	12.188	1210
1220	12.188	12.200	12.212	12.224	12.236	12.248	12.260	12.272	12.284	12.296	12.308	1220
1230	12.308	12.320	12.332	12.345	12.357	12.369	12.381	12.393	12.405	12.417	12.429	1230
1240	12.429	12.441	12.453	12.465	12.477	12.489	12.501	12.514	12.526	12.538	12.550	1240
DEG C	0	1	2	3	4	5	6	7	8	9	10	DEG C

TABLE 16 (Continued)

TEMPERATURE-EMF FOR TYPE S THERMOCOUPLES

TEMPERATURES IN DEGREES CELSIUS (IPTS-68) **REFERENCE JUNCTIONS AT 0°C**

DEG C	0	1	2	3	4	5	6	7	8	9	10	DEG C

THERMOELECTRIC VOLTAGE IN MILLIVOLTS

DEG C	0	1	2	3	4	5	6	7	8	9	10	DEG C
1250	12.550	12.562	12.574	12.586	12.598	12.610	12.622	12.634	12.647	12.659	12.671	1250
1260	12.671	12.683	12.695	12.707	12.719	12.731	12.743	12.755	12.767	12.780	12.792	1260
1270	12.792	12.804	12.816	12.828	12.840	12.852	12.864	12.876	12.888	12.901	12.913	1270
1280	12.913	12.925	12.937	12.949	12.961	12.973	12.985	12.997	13.010	13.022	13.034	1280
1290	13.034	13.046	13.058	13.070	13.082	13.094	13.107	13.119	13.131	13.143	13.155	1290
1300	13.155	13.167	13.179	13.191	13.203	13.216	13.228	13.240	13.252	13.264	13.276	1300
1310	13.276	13.288	13.300	13.313	13.325	13.337	13.349	13.361	13.373	13.385	13.397	1310
1320	13.397	13.410	13.422	13.434	13.446	13.458	13.470	13.482	13.495	13.507	13.519	1320
1330	13.519	13.531	13.543	13.555	13.567	13.579	13.592	13.604	13.616	13.628	13.640	1330
1340	13.640	13.652	13.664	13.677	13.689	13.701	13.713	13.725	13.737	13.749	13.761	1340
1350	13.761	13.774	13.786	13.798	13.810	13.822	13.834	13.846	13.859	13.871	13.883	1350
1360	13.883	13.895	13.907	13.919	13.931	13.943	13.956	13.968	13.980	13.992	14.004	1360
1370	14.004	14.016	14.028	14.040	14.053	14.065	14.077	14.089	14.101	14.113	14.125	1370
1380	14.125	14.138	14.150	14.162	14.174	14.186	14.198	14.210	14.222	14.235	14.247	1380
1390	14.247	14.259	14.271	14.283	14.295	14.307	14.319	14.332	14.344	14.356	14.368	1390
1400	14.368	14.380	14.392	14.404	14.416	14.429	14.441	14.453	14.465	14.477	14.489	1400
1410	14.489	14.501	14.513	14.526	14.538	14.550	14.562	14.574	14.586	14.598	14.610	1410
1420	14.610	14.622	14.635	14.647	14.659	14.671	14.683	14.695	14.707	14.719	14.731	1420
1430	14.731	14.744	14.756	14.768	14.780	14.792	14.804	14.816	14.828	14.840	14.852	1430
1440	14.852	14.865	14.877	14.889	14.901	14.913	14.925	14.937	14.949	14.961	14.973	1440
1450	14.973	14.985	14.998	15.010	15.022	15.034	15.046	15.058	15.070	15.082	15.094	1450
1460	15.094	15.106	15.118	15.130	15.143	15.155	15.167	15.179	15.191	15.203	15.215	1460
1470	15.215	15.227	15.239	15.251	15.263	15.275	15.287	15.299	15.311	15.324	15.336	1470
1480	15.336	15.348	15.360	15.372	15.384	15.396	15.408	15.420	15.432	15.444	15.456	1480
1490	15.456	15.468	15.480	15.492	15.504	15.516	15.528	15.540	15.552	15.564	15.576	1490
1500	15.576	15.589	15.601	15.613	15.625	15.637	15.649	15.661	15.673	15.685	15.697	1500
1510	15.697	15.709	15.721	15.733	15.745	15.757	15.769	15.781	15.793	15.805	15.817	1510
1520	15.817	15.829	15.841	15.853	15.865	15.877	15.889	15.901	15.913	15.925	15.937	1520
1530	15.937	15.949	15.961	15.973	15.985	15.997	16.009	16.021	16.033	16.045	16.057	1530
1540	16.057	16.069	16.080	16.092	16.104	16.116	16.128	16.140	16.152	16.164	16.176	1540
1550	16.176	16.188	16.200	16.212	16.224	16.236	16.248	16.260	16.272	16.284	16.296	1550
1560	16.296	16.308	16.319	16.331	16.343	16.355	16.367	16.379	16.391	16.403	16.415	1560
1570	16.415	16.427	16.439	16.451	16.462	16.474	16.486	16.498	16.510	16.522	16.534	1570
1580	16.534	16.546	16.558	16.569	16.581	16.593	16.605	16.617	16.629	16.641	16.653	1580
1590	16.653	16.664	16.676	16.688	16.700	16.712	16.724	16.736	16.747	16.759	16.771	1590
1600	16.771	16.783	16.795	16.807	16.819	16.830	16.842	16.854	16.866	16.878	16.890	1600
1610	16.890	16.901	16.913	16.925	16.937	16.949	16.960	16.972	16.984	16.996	17.008	1610
1620	17.008	17.019	17.031	17.043	17.055	17.067	17.078	17.090	17.102	17.114	17.125	1620
1630	17.125	17.137	17.149	17.161	17.173	17.184	17.196	17.208	17.220	17.231	17.243	1630
1640	17.243	17.255	17.267	17.278	17.290	17.302	17.313	17.325	17.337	17.349	17.360	1640
1650	17.360	17.372	17.384	17.396	17.407	17.419	17.431	17.442	17.454	17.466	17.477	1650
1660	17.477	17.489	17.501	17.512	17.524	17.536	17.548	17.559	17.571	17.583	17.594	1660
1670	17.594	17.606	17.617	17.629	17.641	17.652	17.664	17.676	17.687	17.699	17.711	1670
1680	17.711	17.722	17.734	17.745	17.757	17.769	17.780	17.792	17.803	17.815	17.826	1680
1690	17.826	17.838	17.850	17.861	17.873	17.884	17.896	17.907	17.919	17.930	17.942	1690
1700	17.942	17.953	17.965	17.976	17.988	17.999	18.010	18.022	18.033	18.045	18.056	1700
1710	18.056	18.068	18.079	18.090	18.102	18.113	18.124	18.136	18.147	18.158	18.170	1710
1720	18.170	18.181	18.192	18.204	18.215	18.226	18.237	18.249	18.260	18.271	18.282	1720
1730	18.282	18.293	18.305	18.316	18.327	18.338	18.349	18.360	18.372	18.383	18.394	1730
1740	18.394	18.405	18.416	18.427	18.438	18.449	18.460	18.471	18.482	18.493	18.504	1740
1750	18.504	18.515	18.526	18.536	18.547	18.558	18.569	18.580	18.591	18.602	18.612	1750
1760	18.612	18.623	18.634	18.645	18.655	18.666	18.677	18.687	18.698			1760

DEG C	0	1	2	3	4	5	6	7	8	9	10	DEG C

TABLE 17

TEMPERATURE-EMF FOR TYPE T THERMOCOUPLES

TEMPERATURES IN DEGREES CELSIUS (IPTS-68) **REFERENCE JUNCTIONS AT 0°C**

DEG C	0	1	2	3	4	5	6	7	8	9	10	DEG C

THERMOELECTRIC VOLTAGE IN MILLIVOLTS

DEG C	0	1	2	3	4	5	6	7	8	9	10	DEG C
-270	-6.258											-270
-260	-6.232	-6.236	-6.239	-6.242	-6.245	-6.248	-6.251	-6.253	-6.255	-6.256	-6.258	-260
-250	-6.181	-6.187	-6.193	-6.198	-6.204	-6.209	-6.214	-6.219	-6.224	-6.228	-6.232	-250
-240	-6.105	-6.114	-6.122	-6.130	-6.138	-6.146	-6.153	-6.160	-6.167	-6.174	-6.181	-240
-230	-6.007	-6.018	-6.028	-6.039	-6.049	-6.059	-6.068	-6.078	-6.087	-6.096	-6.105	-230
-220	-5.889	-5.901	-5.914	-5.926	-5.938	-5.950	-5.962	-5.973	-5.985	-5.996	-6.007	-220
-210	-5.753	-5.767	-5.782	-5.795	-5.809	-5.823	-5.836	-5.850	-5.863	-5.876	-5.889	-210
-200	-5.603	-5.619	-5.634	-5.650	-5.665	-5.680	-5.695	-5.710	-5.724	-5.739	-5.753	-200
-190	-5.439	-5.456	-5.473	-5.489	-5.506	-5.522	-5.539	-5.555	-5.571	-5.587	-5.603	-190
-180	-5.261	-5.279	-5.297	-5.315	-5.333	-5.351	-5.369	-5.387	-5.404	-5.421	-5.439	-180
-170	-5.069	-5.089	-5.109	-5.128	-5.147	-5.167	-5.186	-5.205	-5.223	-5.242	-5.261	-170
-160	-4.865	-4.886	-4.907	-4.928	-4.948	-4.969	-4.989	-5.010	-5.030	-5.050	-5.069	-160
-150	-4.648	-4.670	-4.693	-4.715	-4.737	-4.758	-4.780	-4.801	-4.823	-4.844	-4.865	-150
-140	-4.419	-4.442	-4.466	-4.489	-4.512	-4.535	-4.558	-4.581	-4.603	-4.626	-4.648	-140
-130	-4.177	-4.202	-4.226	-4.251	-4.275	-4.299	-4.323	-4.347	-4.371	-4.395	-4.419	-130
-120	-3.923	-3.949	-3.974	-4.000	-4.026	-4.051	-4.077	-4.102	-4.127	-4.152	-4.177	-120
-110	-3.656	-3.684	-3.711	-3.737	-3.764	-3.791	-3.818	-3.844	-3.870	-3.897	-3.923	-110
-100	-3.378	-3.407	-3.435	-3.463	-3.491	-3.519	-3.547	-3.574	-3.602	-3.629	-3.656	-100
-90	-3.089	-3.118	-3.147	-3.177	-3.206	-3.235	-3.264	-3.293	-3.321	-3.350	-3.378	-90
-80	-2.788	-2.818	-2.849	-2.879	-2.909	-2.939	-2.970	-2.999	-3.029	-3.059	-3.089	-80
-70	-2.475	-2.507	-2.539	-2.570	-2.602	-2.633	-2.664	-2.695	-2.726	-2.757	-2.788	-70
-60	-2.152	-2.185	-2.218	-2.250	-2.283	-2.315	-2.348	-2.380	-2.412	-2.444	-2.475	-60
-50	-1.819	-1.853	-1.886	-1.920	-1.953	-1.987	-2.020	-2.053	-2.087	-2.120	-2.152	-50
-40	-1.475	-1.510	-1.544	-1.579	-1.614	-1.648	-1.682	-1.717	-1.751	-1.785	-1.819	-40
-30	-1.121	-1.157	-1.192	-1.228	-1.263	-1.299	-1.334	-1.370	-1.405	-1.440	-1.475	-30
-20	-0.757	-0.794	-0.830	-0.867	-0.903	-0.940	-0.976	-1.013	-1.049	-1.085	-1.121	-20
-10	-0.383	-0.421	-0.458	-0.496	-0.534	-0.571	-0.608	-0.646	-0.683	-0.720	-0.757	-10
-0	0.000	-0.039	-0.077	-0.116	-0.154	-0.193	-0.231	-0.269	-0.307	-0.345	-0.383	-0
0	0.000	0.039	0.078	0.117	0.156	0.195	0.234	0.273	0.312	0.351	0.391	0
10	0.391	0.430	0.470	0.510	0.549	0.589	0.629	0.669	0.709	0.749	0.789	10
20	0.789	0.830	0.870	0.911	0.951	0.992	1.032	1.073	1.114	1.155	1.196	20
30	1.196	1.237	1.279	1.320	1.361	1.403	1.444	1.486	1.528	1.569	1.611	30
40	1.611	1.653	1.695	1.738	1.780	1.822	1.865	1.907	1.950	1.992	2.035	40
50	2.035	2.078	2.121	2.164	2.207	2.250	2.294	2.337	2.380	2.424	2.467	50
60	2.467	2.511	2.555	2.599	2.643	2.687	2.731	2.775	2.819	2.864	2.908	60
70	2.908	2.953	2.997	3.042	3.087	3.131	3.176	3.221	3.266	3.312	3.357	70
80	3.357	3.402	3.447	3.493	3.538	3.584	3.630	3.676	3.721	3.767	3.813	80
90	3.813	3.859	3.906	3.952	3.998	4.044	4.091	4.137	4.184	4.231	4.277	90
100	4.277	4.324	4.371	4.418	4.465	4.512	4.559	4.607	4.654	4.701	4.749	100
110	4.749	4.796	4.844	4.891	4.939	4.987	5.035	5.083	5.131	5.179	5.227	110
120	5.227	5.275	5.324	5.372	5.420	5.469	5.517	5.566	5.615	5.663	5.712	120
130	5.712	5.761	5.810	5.859	5.908	5.957	6.007	6.056	6.105	6.155	6.204	130
140	6.204	6.254	6.303	6.353	6.403	6.452	6.502	6.552	6.602	6.652	6.702	140
150	6.702	6.753	6.803	6.853	6.903	6.954	7.004	7.055	7.106	7.156	7.207	150
160	7.207	7.258	7.309	7.360	7.411	7.462	7.513	7.564	7.615	7.666	7.718	160
170	7.718	7.769	7.821	7.872	7.924	7.975	8.027	8.079	8.131	8.183	8.235	170
180	8.235	8.287	8.339	8.391	8.443	8.495	8.548	8.600	8.652	8.705	8.757	180
190	8.757	8.810	8.863	8.915	8.968	9.021	9.074	9.127	9.180	9.233	9.286	190
200	9.286	9.339	9.392	9.446	9.499	9.553	9.606	9.659	9.713	9.767	9.820	200
210	9.820	9.874	9.928	9.982	10.036	10.090	10.144	10.198	10.252	10.306	10.360	210
220	10.360	10.414	10.469	10.523	10.578	10.632	10.687	10.741	10.796	10.851	10.905	220
230	10.905	10.960	11.015	11.070	11.125	11.180	11.235	11.290	11.345	11.401	11.456	230
240	11.456	11.511	11.566	11.622	11.677	11.733	11.788	11.844	11.900	11.956	12.011	240
250	12.011	12.067	12.123	12.179	12.235	12.291	12.347	12.403	12.459	12.515	12.572	250
260	12.572	12.628	12.684	12.741	12.797	12.854	12.910	12.967	13.024	13.080	13.137	260
270	13.137	13.194	13.251	13.307	13.364	13.421	13.478	13.535	13.592	13.650	13.707	270
280	13.707	13.764	13.821	13.879	13.936	13.993	14.051	14.108	14.166	14.223	14.281	280
290	14.281	14.339	14.396	14.454	14.512	14.570	14.628	14.686	14.744	14.802	14.860	290
300	14.860	14.918	14.976	15.034	15.092	15.151	15.209	15.267	15.326	15.384	15.443	300
310	15.443	15.501	15.560	15.619	15.677	15.736	15.795	15.853	15.912	15.971	16.030	310
320	16.030	16.089	16.148	16.207	16.266	16.325	16.384	16.444	16.503	16.562	16.621	320
330	16.621	16.681	16.740	16.800	16.859	16.919	16.978	17.038	17.097	17.157	17.217	330
340	17.217	17.277	17.336	17.396	17.456	17.516	17.576	17.636	17.696	17.756	17.816	340

DEG C	0	1	2	3	4	5	6	7	8	9	10	DEG C

TABLE 17 (Continued)

TEMPERATURE-EMF FOR TYPE T THERMOCOUPLES

TEMPERATURES IN DEGREES CELSIUS (IPTS-68)   **REFERENCE JUNCTIONS AT 0°C**

DEG C	0	1	2	3	4	5	6	7	8	9	10	DEG C
					THERMOELECTRIC VOLTAGE IN MILLIVOLTS							
350	17.816	17.877	17.937	17.997	18.057	18.118	18.178	18.238	18.299	18.359	18.420	350
360	18.420	18.480	18.541	18.602	18.662	18.723	18.784	18.845	18.905	18.966	19.027	360
370	19.027	19.088	19.149	19.210	19.271	19.332	19.393	19.455	19.516	19.577	19.638	370
380	19.638	19.699	19.761	19.822	19.883	19.945	20.006	20.068	20.129	20.191	20.252	380
390	20.252	20.314	20.376	20.437	20.499	20.560	20.622	20.684	20.746	20.807	20.869	390
400	20.869											400
DEG C	0	1	2	3	4	5	6	7	8	9	10	DEG C

APPENDICES

APPENDIX A

BARE THERMOCOUPLE ELEMENT FABRICATION

A1. General

While completely fabricated thermocouples are available commercially, this Appendix is intended to assist those who desire to fabricate their own thermocouples.

A2. Thermocouple Wires

Carefully selected and tested pairs of thermocouple wires are available commercially in standard AWG diameters. When purchased as a pair simultaneously from a single supplier, the pair will conform to the specified calibration limits and be referred to as a matched pair.

1. Interchange of a common wire between two types of thermocouples (e.g. copper-nickel from Type J to T) or even between different matched pairs of the same type may yield a thermocouple that will not conform to the specified calibration limits.

2. See Appendix D for checking procedure and a reference on calibration.

A3. Joining Thermocouple Wires

A3.1 General

The dissimilar wires of a thermocouple must be joined at the temperature measuring junction by a joint of good electrical and thermal conductivity, without destroying the mechanical and metallurgical properties of the thermocouple wires at this joint.

1. For use below 500°C (1000°F) most base metal thermocouple wires may be silver soldered using borax as a flux.

2. Above 500°C (1000°F) experience has shown that properly welded thermocouple junctions provide long life and excellent thermal and electrical properties. Welded thermocouple junctions are used in practically all industrial applications today. Noble metal thermocouples should always be joined by welding. Common methods of welding thermocouples are gas, electric arc, resistance, tungsten-inert-gas and plasma-arc welding.

A3.2 Preparation of Wires

1. Often the matched wires must be straightened prior to joining to facilitate stringing of insulators in the final thermocouple assembly, but where possible, excessive bending of thermocouple wires should be avoided because cold working may alter the EMF output of thermocouple wire. Hammering, stretching and excessive twisting should be avoided for the same reason.

2. The thermocouple wires are cut to the length desired allowing for one or two attempts at welding and for any forming that must be done at the junction.

3. All thermocouple wire should be cleaned carefully with a suitable solvent such as Freon TF*, Methyl-Ethyl-Ketone, or Alcohol (such as Isopropyl) prior to welding.

4. Simple jigs and fixtures are usually used to shape the wires prior to welding, except for butt welded thermocouples which are often bent around a mandrel after welding. Care must be taken to avoid nicking or damaging the wire during the forming operation as damage to the wire or wire surface may shorten thermocouple life. The wires should be spaced to permit free insertion into insulators.

A3.3 Gas or Arc Welding Types, E, J, K and T Thermocouples

In preparation for welding, the wires may be twisted as shown in Figure A-1 or positioned in a "V" as shown in Figure A-4. The twisted construction adds strength and facilitates welding.

For twisted AWG sizes 8 and 14, one inch of each wire should be prepared by removing any oxide or other surface finish with abrasive paper or by very careful filing or grinding. For twisted AWG 20, 24, and 28, the prepared length need be only one-half inch. The prepared ends are either twisted together to yield one and one-half turns as shown in Figure A-1 or positioned in a "V" as shown in Figure A-4 and then welded.

Figure A-1. Method of Twisting Wires for Gas and Electric Arc Welding

A3.4 Resistance Welding Types J and K Thermocouples

This method is recommended only for the 8 and 14 AWG wires. Approximately one-half inch of each wire should be sanded, in preparation for welding, with abrasive paper or by very careful filing or grinding.

Figure A-2. Method of Forming Metal Wires for Resistance Welding

---
*Trade Name

The sanded ends should be formed to produce longitudinal contact as shown in Figure A-2.

A3.5 Butt Resistance Welding Types E, J, and K Thermocouples

This method is recommended for 8 through 20 AWG wires and requires a good, commercially available, wire butt welder of suitable current capacity for the gage wire being welded. Approximately 0.5 inch of each wire should be sanded with abrasive paper in preparation for welding.

The sanded ends of the straight wires are butted together in the spring loaded butt welder jaws and spring pressure applied to the jaws. The weld is performed and the flash is removed by grinding. The wires are then bent as shown in Figure A-3.

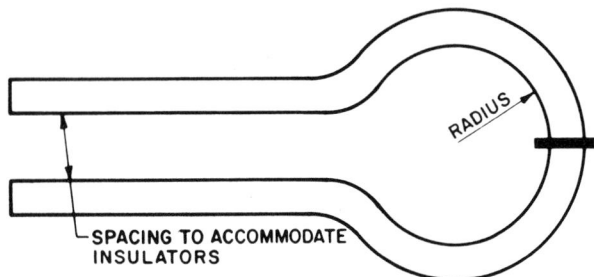

Figure A-3. Formed Butt Welded Thermocouple

A3.6 Resistance Welding Types B, R, and S Thermocouples

Extreme care should be taken to avoid cold working and contamination by oils, perspiration, dirt, etc. Sanding is not required.

The ends should be formed to produce a longitudinal contact of about one-eighth inch as shown in Figure A-2.

A3.7 Arc Welding Types B, R, and S Thermocouples

Extreme care should be taken to avoid cold working and contamination by oils, perspiration, dirt, etc. Sanding is not required.

The ends of the wires are positioned as shown in Figure A-4.

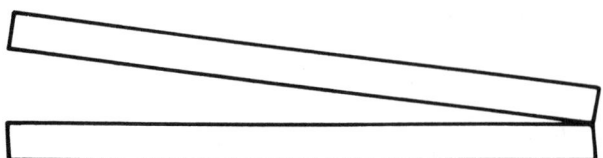

Figure A-4. Method of Forming Metal Wires for Electric Arc Welding

A3.8 Gas Welding

The character of the gas flame is the primary consideration of gas welding. A neutral flame as shown in Figure A-5 is essential. The neutral flame is obtained by increasing the oxygen until the excess gas flame - shown dashed in Figure A-5 - just vanishes. Overshooting the vanishing point gives an oxidizing flame. AN OXIDIZING FLAME IS INJURIOUS AND SHOULD NEVER BE USED.

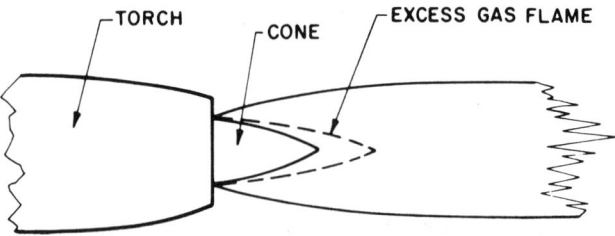

Figure A-5. Neutral Flame for Gas Welding

The smallest tip, that will readily heat the wires to fusion temperature, should be used. Continued heating at welding temperatures yields a poor weld.

Heat the ends of the twisted or "V" positioned wires to redness with the tip of the cone and plunge them into the flux. Reheat the wires to fusion temperature simultaneously, and rotate the weld to form a ball at the tip. Quench the welded junction in water to remove excess flux.

Attainment of the simultaneous fusion requires that the heating of the lower melting wire (see Table A-1) be delayed.

The junction should be examined with a low power magnifier for smoothness. A weld with a pitted surface must be rejected because it is burned (i.e. overheated to the point of incipient melting or intergranular oxidation). Repair of the weld is not feasible. An unsatisfactory weld must be cut off and the procedure repeated.

A3.9 Electric Arc Welding

Welded junctions may also be produced by an arc between two soft carbon electrodes or between one carbon electrode and the thermocouple wires as the second electrode. Only 8 and 14 AWG thermocouples should be used as the second electrode. Finer wires require two carbon electrodes. Direct current (dc) or alternating current (ac) may be used, but direct current is preferred. With direct current and a single carbon electrode, the thermocouple is connected to the positive lead. The ends of the twisted or "V" positioned wires are moistened, dipped into flux, and clamped upright in a vise with copper jaws. (Avoid surface damage in clamping.)

The two leads from the variable electric current source are connected either to the two carbon electrodes or one to the carbon electrode and the other to the vise clamping the thermocouple.

The size of the pure, soft carbon electrodes should be proportional to the wire size.

An arc is struck between the carbon electrodes or carbon electrode and thermocouple by momentary shorting. For an 8 AWG thermocouple 30 to 45 volts is typical, and this value should be proportionately reduced for smaller sizes. With the single carbon electrode, a 1/16 inch arc gap minimizes oxidation and nitrogen absorption.

A brief welding cycle is best, since excessive current will result in burning. The bead should be small and solid. Bridges or gaps between the wires are weak and unsatisfactory. If an unsatisfactory weld results, it must be cut off and the procedure repeated.

A3.10 Electric Resistance Welding

Heating and fusion of the wires are accomplished by resistance heating of the wires and by contact resistance at their junction. This method is recommended only by Types B, R, and S, and the 8 and 14 AWG sizes of Types J and K. The junction of Figure A-2 is placed between the electrodes of a resistance welder. A suitable pressure-current-time cycle must be established by trial-and-error on identical scrap wires or by experience. Visual and destructive examination are required to establish proper welding conditions. Excessive pressure will produce a good looking weld but only peripheral fusion. An unsatisfactory weld must be cut off the the procedure repeated.

A3.11 TIG and Plasma Arc Welding

The tungsten-inert gas (TIG) welding process and plasma-arc welding process are rapidly gaining in importance for welding thermocouple junctions. These welding processes use an inert gas envelope to protect the weld from oxidation in lieu of a flux. Welding using the TIG or plasma arc processes is done following the same routine as welding with one carbon electrode, except a flux is not used. The plasma arc has distinct advantages such as: no tungsten inclusions in the weld, extremely high temperatures in the plasma arc, a more controllable arc, and a constant pilot arc which can actually be used for fine welding, as well as a guide light to position the torch prior to starting the main arc, to name but a few. These processes are especially recommended for welding junctions in sheathed thermocouple wire. Procedures outlined above may also be used on types B, R, and S.

TABLE A-1
SUMMARY OF METHODS FOR JOINING OF BARE WIRE THERMOCOUPLES

Type of T/C*	Lower Melting	Silver Brazing	Welding Flux**	Welding Gas	Welding Arc	Resistance Welding	Plasma Arc or TIG	Butt Welding
B	BN	N.R.	None	A3.1.2	A.3.7 & A.3.9	A3.6 & A3.9	A3.11	-
E	EN	A3.1.1	Fluorspar	A3.3 & A3.8	A3.3 & A3.9	N.R.	A3.11	A3.5
J	JN	A3.1.1	Borox	A3.3 & A3.8	A3.3 & A3.9	A3.4 & A3.10	A3.11	A3.5
K	KN	A3.1.1	Fluorspar	A3.3 & A3.8	A3.3 & A3.9	A3.4 & A3.10	A3.11	A3.5
R	RN	N.R.	None	A3.1.2	A3.7 & A3.9	A3.6 & A3.9	A3.11	-
S	SN	N.R.	None	A3.1.2	A3.7 & A3.9	A3.6 & A3.9	A3.11	-
T	TP	A3.1.1	Borax	A3.3 & A3.8	A3.3 & A3.9	N.R.	A3.11	N.R.

N.R. —Not recommended
* See Table 1 of Standard for typical alloys
** Boric Acid also recommended for Types J, E, and K
NOTE: Numbers in body of table refer to paragraphs in Appendix A where the procedure for that type of joining is covered.

APPENDIX B
SHEATHED THERMOCOUPLE ELEMENT FABRICATION

B1. General

Sheathed thermocouple elements may be fabricated from commercially available sheathed thermocouple wire described in Chapter 5 of ASTM Special Technical Publication 470B. Fabricating such thermocouples successfully requires a higher degree of skill, special equipment and techniques, and virtually clean room conditions compared to fabricating customary bare wire thermocouples. Although this appendix is intended to assist those who desire to fabricate their own sheathed thermocouple elements, only general fabrication procedures are outlined below. A study of the literature on sheathed thermocouples and Chapter 5 of ASTM Special Technical Publication 470B, plus refinement of the procedures outlined below by practical experience are preerequisites to successful fabrication of sheathed thermocouple elements.

B2. Special Equipment

Although it is possible to remove the sheath by grinding and filing, special sheath stripping tools are commercially available and are highly recommended.

Welding is generally done using the Tungsten-Inert-Gas (TIG) process, and therefore a TIG welder is recommended. A plasma-arc welder is also excellent for this purpose.

It is often desirable to remove the insulant from around the thermocouple wires. Although this can be accomplished by tedious picking with a needle or other sharp instrument a miniature sandblaster is far superior.

A clean, dry and well lighted work area is essential to creating a finished element of high integrity.

Ovens capable of continuous operation at a minimum temperature of 200°F are suggested for storage of unsealed sheathed thermocouple wire and thermocouple elements during even short periods of delay where the compacted insulation might be exposed to air-borne moisture or other contaminants.

Special holding vises or fixtures are recommended for junctioning and capping sheathed thermocouple wire. The jaws may be made of commercial copper, and should be grooved to accept the various diameters of sheathed wire with which the fabricator plans to work.

B3. General Precautions

The crushed mineral oxide insulation in all sheathed thermocouple wire will rapidly absorb moisture. Thus the cable should be purchased with the ends closed by welding or suitably sealed in some other manner. Unsealed cable should be stored in an oven at 200°F or higher to reduce moisture pick-up.

All fixtures, vises and other tools brought into contact with the ceramic in the sheathed wire should be surgically clean to prevent contamination of the ceramic.

B4. Measuring Junction Fabrication

Exposed Junction (Wires not encased within sheath end closure.) The sheathed wire is cut to the desired length allowing for reference junction and measuring junction sheath removal. The sheath is removed from both ends as required, and the wires are cleaned by sandblasting or other suitable means. Wires are positioned and lightly clamped in a fixture such as a special copper jaws vise so that approximately 1/32 in. of wires are exposed above the vise jaws. The wires must touch one another at the surface of the jaws. The junction is fused by an electrical arc using the TIG welder with the ground lead connected to the vise jaws. The finished junction should look like Figure B-1. Both ends of the thermocouple element must be sealed with a suitable sealer to prevent contamination of the insulation, or the thermocouple element can be stored in an oven at 200°F until it is used.

Figure B-1. Typical Exposed Junction

Grounded Junction (Wires encased within sheath end closure.) The sheathed wire is cut to the desired length allowing for the reference end sheath removal.

B4.1 On sheath diameters of 0.125 in. and smaller, the sheath and wires of the squared end can be simple welded over thus captivating both the sheath and the wires into the cap weld, creating a sound mechanical joint as well as electrical contact between the wires without diluting the cap weld metal with metal from the thermoelements. This weld is performed by positioning the squared end in a fixture such as copper jaws in a vise with closely fitting grooves for each size of sheathed wire. See Figure B-2.

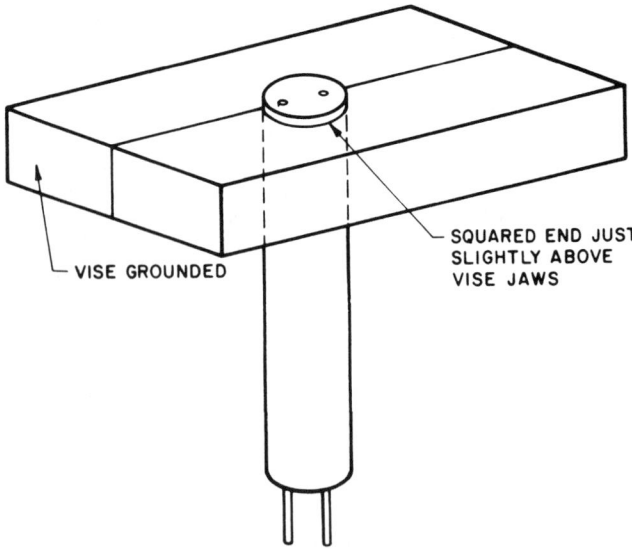

Figure B-2. Fixturing for Weldment

A TIG welder or plasma arc welder are best suited for welding, adding filler metal compatible with the sheath material as required.

B4.a Sheath diameters larger than 0.125 in. require slightly different methods. After squaring the end, a portion of the insulation is removed to a depth equal to approximately the inside diameter of the sheath. The reference junction end must have the sheath removed and the wires exposed to which the ground clamp from the welder is attached. The unit is positioned in the vise jaws in the same fashion as for forming the cap weld. An arc is struck on the wires thereby fusing them into a neat ball. The end is then capped as on smaller sheathed elements bringing the weld metal down into intimate contact with the welded wires, but not re-melting them. The general configuration of a grounded junction is shown in Figure B-3.

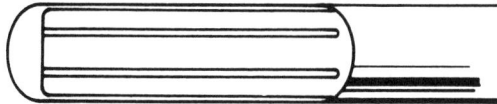

Figure B-3. Cutaway View of Grounded Junction

B4.3 Ungrounded Junction (Wires encased within but not touching sheath end closure.) The procedure for producing this form of junction is not too unlike that of the grounded junction in sheathed wire larger than 0.125 in. O.D. After preparing the ends by squaring and removing the sheath, the insulation is removed on the measuring junction end to a level which will allow the wires to be welded together to the proper depth from the squared end. Care must be taken not to create a weld ball as large as the sheath I.D. which will ground the wires to the sheath and defeat the purpose of this type of junction. After the wires are joined, an insulation powder equivalent to that within the sheath is packed tightly around the welded wires and out flush to the squared end of the sheath. Capping is then done as before. Figure B-4 is typical of an Ungrounded Junction

Figure B-4. Cutaway View of Ungrounded Junction

B4.4 The hygroscopic nature of most compacted insulations demands extreme care to avoid moisture pickup. It is very important to seal all ends where insulation is exposed as quickly and completely as possible.

APPENDIX C
THERMOCOUPLES AND THERMOCOUPLE EXTENSION WIRES - SELECTION, ASSEMBLY, AND INSTALLATION

C.1. Scope and Purpose

This section gives general information on the selection, assembly, and installation of commonly used thermocouples and associated thermocouple extension wires for specific applications.

C2. Types and Uses

There are seven thermocouple types now in general use. The material designations are shown in C-1. The positive element of each couple is listed first.

Each of these types has individual characteristics that are desirable for some applications and unsuitable for others. In addition to the usual insulated forms all couples are available as compacted ceramic insulated thermocouples (See Section 2.2.8, page 14.)

Type T may be used in a vacuum, inert, oxidizing or reducing atmosphere. It commonly is used for sub-zero temperatures and has an upper temperature limit of 370°C (700°F) for wires at least 0.064 in. (1.63 mm) in diameter, in conventional closed-end protecting tubes. Its copper positive element is preferred to the iron positive element of the Type J type for sub-zero use because of its superior corrosion resistance in moist atmospheres.

Type J may be used in a vacuum, inert, oxidizing, or reducing atmosphere. With elements of at least 0.128 in. (3.25 mm) in diameter, in conventional closed-end protecting tubes, it has an upper temperature limit of 760°C (1400°F). At high temperatures it should not be used in certain hydrogen containing atmospheres due to possible embrittling effects on the iron element. While this couple is occasionally used for sub-zero temperature measurements, the possible rusting or embrittlement of the iron under these conditions is at times undesirable.

Type E may be used in a vacuum, inert, oxidizing atmosphere and for temperatures up to 870°C (1600°F) for element wires at least 0.128 in. (3.25 mm) in diameter in closed-end protecting tubes. At sub-zero temperatures, the couple is not subject to corrosion. This thermocouple has the highest EMF output of any standard metal thermocouple.

Type K nominally is used to an upper temperature limit of 1260°C (2300°F) for wires at least 0.128 in. (3.25 mm) in diameter in conventional closed-end protecting tubes. It should be used either in inert or oxidizing atmospheres. It has a short life in atmospheres that are marginally oxidizing, alternately oxidizing and reducing, or reducing atmospheres, particularly in the temperature range of 820°C to 1010°C (1500°F to 1850°F). An oxidizing atmosphere inside the protecting tube may be obtained by providing adequate ventilating. The use of a sufficiently large protecting tube and an open head will be of material assistance. Type K thermocouples should not be used for accurate temperature measurements below 480°C (900°F) after prolonged exposure above 760°C (1400°F).

TABLE C-1
THERMOCOUPLE TYPES

Type	Thermocouple Element	Thermocouple Wire Materials*
B	BP BN	Platinum - 30% rhodium Platinum - 6% rhodium
E	EP EN	Nickel Chromium Copper Nickel
J	JP JN	Iron Copper Nickel
K	KP KN	Nickel Chromium Nickel Aluminum Silicon
R	RP RN	Platinum - 13% rhodium Platinum
S	SP SN	Platinum - 10% rhodium Platinum
T	TP TN	Copper Copper Nickel

* These thermocouple materials are defined by their EMF characteristics. Alloy compositions may vary from lot to lot.

Types R and S may be used for temperatures up to 1480°C (2700°F) and Type B to 1700°C (3100°F) for wires at least 0.020 in. (0.51 mm) in diameter in conventional closed-end protecting tubes. These thermocouples are easily contaminated and should always be used in a protecting tube. The protecting tube also should be non-metallic and silica free, since the thermocouple can be contaminated by metallic vapors, reduced oxides, or other impurities at high temperatures. These elements may be used in inert or oxidizing atmospheres and for short periods of time in a vacuum - longer periods of time in vacuum if, for example, gas tight ceramic tubes or metal sheaths are used. They should not be used in a reducing atmosphere. Type S thermocouples are frequently used for calibration or checking since this type is one of the primary standards.

C3. Assembly

The fabrication of thermocouples requires special techniques as described in Appendix A. If the equipment and skill required to fabricate thermocouples properly are not available, the user should purchase fabricated thermocouples, since improper techniques can result in significant errors in temperature measurements.

Wire for making thermocouples preferably should be purchased in matched pairs in order to insure accuracy within standard limits of error of Section 3. However, positive and negative wires for types E, K, and T thermocouples, purchased at different times or from different suppliers, can be combined interchangeably. Wires to special limits of error given in Section 3 are obtained by selection and always should be purchased in matched pairs.

It is essential that the thermocouple have the same calibration as the instrument with which it is to be used.

Hard-fired ceramic insulators are used on most bare thermocouple elements. Insulators are available in single, double or multi-bore and a variety of shapes, sizes and lengths.

For types B, R, and S thermocouples it is recommended that insulators be of aluminum oxide and be of one piece, full length construction, to provide maximum protection from contamination. The insulator should also be light in weight or the assembly designed to minimize mechanical stresses on the noble metal wire.

For base metal thermocouples, insulation of braided glass or other fabric is sometimes used. Such materials should not be used with noble metal elements, as they will contaminate the thermocouple. Thermocouples may be made from insulated thermocouple wire, provided the insulation is suitable for the exposure temperature and intended service and will not contaminate the thermocouple or environment.

Protecting tubes are used for most thermocouple installations to prevent contamination of the thermocouple and provide mechanical protection and support. The minimum diameter of the protecting tube must be such as to accommodate the thermocouple element. However, larger diameter tubes are often required for (a) strength, (b) to permit insertion of a checking thermocouple alongside the service thermocouple and (c) to provide an adequate diameter to length ratio to assist in maintaining an oxidizing atmosphere for type K or E thermocouples. One half, three fourths, and one inch pipe size tubes are commonly used.

The length of the protecting tube (and thermocouple element) should be such as to place the measuring junction of the thermocouple well into the medium, the temperature of which is to be measured. A minimum immersion length of 8 to 10 tube diameters is recommended in order to minimize conduction errors.

Protection tubes must be internally clean and free of sulphur bearing compounds, oil, oxides, and sulphur-bearing compounds.

A wide range of metal and ceramic protecting tubes is available. Depending upon the application, the protecting tube should have some or all of the following properties:

1. Mechanical strength to withstand pressure and resist sagging at high temperatures.

2. Temperature resistance to withstand the temperature being measured; thermal shock resistance so that sudden temperature changes will not damage the tube.

3. Corrosion resistance to avoid chemical action with the medium in which the tube is immersed.

4. Erosion resistance.

5. Low porosity at operating temperature. This is especially true of protecting tubes installed in furnaces since furnace gases are generally damaging to thermocouples.

Some of the common protection tube materials and the maximum operating temperatures are given in Table C2.

TABLE C-2

PROTECTION TUBES

Protecting Tube Materials	Maximum Operation Temperature	
	Deg. C	Deg. F
Carbon Steel	540	1000
Wrought Iron	700	1300
Cast Iron	700	1300
304 Stainless Steel	870	1600
316 Stainless Steel	870	1600
446 Stainless Steel	980	1800
Nickel	980	1800
75 Nickel - 15 Chromium-Iron	1150	2100
Porcelain	1650	3000*
Silicon Carbide	1650	3000
Alumina-Silica	1650	3000*
Aluminum Oxide	1750	3200*

* Horizontal tubes should receive additional support above 1480°C (2700°F)

In addition to the conventional thermocouple construction described in this Appendix, sheathed, compacted, ceramic insulated thermocouple material is in common use. This material consists of one or more thermoelements encased in ceramic insulating material (usually magnesium oxide) which is firmly compacted within a metallic sheath. The nature of this construction is such as to require special fabricating techniques, and it is therefore recommended that the user purchase such thermocouples completely fabricated.

1. Sheathed, compacted thermocouple material can be obtained in (a) all calibration types, (b) a variety of sheath diameters ranging from 0.010 in. to 0.500 in. diameter and (c) a choice of sheath materials to withstand specific environments. Insulating materials, other than magnesium oxide, are also available.

2. The choice of sheath diameter will depend upon such factors as life expectancy, speed of response requirements and space limitations. Large diameter material will provide longer life but will have a slower response. For a more complete discussion of sheathed, compacted, ceramic insulated thermocouples, the user is referred to ASTM "Manual on the Use of Thermocouples in Temperature Measurement," STP 470B.

C4. Installation Considerations for Thermocouples

C4.1 In installing thermocouples it must be always borne in mind that the EMF produced depends upon the difference in temperature between the measuring and reference junctions. With a fixed or known reference junction, the thermocouple thermometer is capable only of indicating the temperature attained by its measuring junction. It is thus necessary in a particular process to insure that the measuring junction is at the same temperature within the accuracy desired as the medium to be measured. The errors discussed elsewhere in this standard are negligible compared with those that may result by not making the installation in such a manner that the measuring junction attains the temperature to be measured.

The measuring junction temperature actually obtained in an installation is a result of the net heat supplied to that junction by the conventional modes of heat transfer; i.e., conduction, convection and radiation. Where protection tubes or wells are necessary in an installation, the problem is only aggravated. Among the many factors which influence the measuring junction temperature of a particular installation are:

1. Temperature of the surroundings
2. Velocity and properties of the fluid
3. Emissivity of the exposed surface
4. Thermal conductivity of thermocouple and well materials
5. Ratio of heat-transfer areas

Under installation conditions where the surrounding (duct wall) temperatures are appreciably different from the fluid temperatures in the case of gases, heat exchange will take place by the mechanism of radiation by the thermocouple and its surroundings. In addition, heat will flow from or to the thermocouple by the mechanism of conduction, and heat will be transferred by convection. Depending upon whether the surrounding temperatures are higher or lower than the gas temperature, the thermocouple will indicate higher or lower temperature. Where great differences in temperature exist between the gas and the surroundings, publications on heat transfer should be consulted treating the thermocouple thermometer as a "thin-rod" type of problem.

The thermocouple thermometer should be located in a position where the mass velocity is as high as practicable to assure good heat transfer by convection; however, if the velocity is in excess of 300 ft/sec. then a specially-designed stagnation-type probe should be used. When a thermocouple must be installed in a location where the velocity is very low, then it may be necessary to induce a flow of gases past the junction. Several aspirating types of pyrometers are available for this purpose.

C4.2 A thermocouple connection head is recommended to provide positive connections between the thermocouple and the extension wire. The head also permits easy replacement of the thermocouple.

C4.3 The protecting tube should extend beyond the outer surface of the vessel furnace or processing equipment so that the temperature of the connection head approximates the ambient atmospheric temperature. This is especially true for Types B, R, and S thermocouples using compensating extension wires. The connection head temperature should never exceed the temperature limits given in Section 3 for thermocouple extension wires.

C4.4 After all the steps outlined above have been carried out, the actual installation of a thermocouple still requires some care. Both the thermocouple and the extension wire should be cleaned before fastening in the terminal block to assure good electrical contacts. Color-coded insulation identifies the positive and negative elements of the extension wire. It is necessary to have the thermocouple wire tagged or otherwise identified as to polarity. The following information can be used to determine polarity in the field.

C4.4.1 For extension wire having insulation color-coded in accordance with Section 1, *the negative wire insulation is always colored red.*

C4.4.2 For Type E, the negative wire is silver in appearance. It has a lower resistance in ohms/foot than the positive element for the same size wire.

C4.4.3 For Type J, the positive element is frequently rusty and is magnetic. It has a lower resistance in ohms/foot for the same size wire.

C4.4.4 For Type K, the negative element is slightly magnetic. It has a lower resistance in ohms/foot for the same size wire.

C4.4.5 For Type R and S, the negative wire is softer than the positive wire. The negative wire also has a lower resistance in ohms/foot for the same size wire.

C4.4.6 For Type T, the positive wire is copper-colored and the negative wire is silver in appearance. The positive wire has lower resistance in ohms/foot for the same size wire.

C4.5 Bottoming of the thermocouple in the protecting tube is often practiced to improve the response to temperature change. Bottoming consists of having the thermocouple junction pressed tightly against the end or "bottom" of the protecting tube. However, bottoming may ground the thermocouple which, with some types of installations causes difficulties.

C4.6 It must be borne in mind that zero error is unattainable. In addition to the instrument error, the thermocouple and the extension wire will introduce errors. In Section 3 are tabulated the initial tolerances that can be expected in new materials. The installed components may deteriorate with use, and methods of checking the installation are given in Appendix D.

C5 Installation of Extension Wires

C5.1 Types of extension wires for use with various types of thermocouples are listed in Table C3.

TABLE C-3
TYPES OF EXTENSION WIRE

Thermocouple Type	Extension Wire	Extension Wire Elements*
B	BX	BPX Copper BNX Copper
E	EX	EPX Nickel Chromium ENX Copper Nickel
J	JX	JPX Iron JNX Copper Nickel
K	KX	KPX Nickel Chromium KNX Nickel Aluminum
R or S	SX	SPX Copper SNX Copper Nickel Alloy
T	TX	TPX Copper TNX Copper Nickel

*These thermocouple materials are defined by their EMF characteristics. Alloy compositions may vary from lot to lot.

C5.2 Potentiometer type instruments are not critical as to extension wire resistance. However, single wires and pairs smaller than 16-gage are not recommended for use in conduit, as they do not have sufficient strength for pulling. Twenty-gage and smaller wire may be used when assembled in suitable reinforced bundles to provide pulling strength. Where extension wire smaller than 16-gage is required, insulated thermocouple wire, instead of extension wire, may be used. The total resistance of the extension wire is important when used with galvanometer millivoltmeter-type instruments. Some millivoltmeters require a definite resistance in the extension wire.

Many of the more recent millivoltmeters are made with a high internal resistance. The extension wire used with these instruments does not have to be calibrated but the total resistance should be kept to approximately the value given on the instrument scale. Therefore, it is usually necessary to use a large size wire. Sizes 14- and 16-gage are recommended.

The resistance of electronic instruments is often high enough to place no restraint on the resistance of the extension wire.

Due to the fast response of such instruments, fluctuations due to noise may require noise suppression techniques.

C5.3 The insulation used on extension wires may be divided into four general classifications: waterproof, moisture resistant, heat resistant, and radiation resistant. Materials used for insulating extension wire are selected to perform a variety of functions. These include: physical protection, bonding, mechanical separation, and electrical insulation.

In a permanently dry location these functions can be performed by non-conducting substances such as cotton, glass, asbestos fibers, paper tapes, and ceramic beads. Where moisture may be present, more or less impervious barriers are required. These may be enamel coatings, asphalt or wax impregnations, plastics, rubber or lead sheaths. Where heat resistance is necessary, glass, asbestos fibers, and ceramics may be used. Where the extension wire may be exposed to varying degrees of heat and moisture, a combination of two or more of these materials may give a satisfactory insulation.

C5.4 Thermocouple extension wire should always be installed in the best manner to protect it from excessive heat, moisture, and mechanical damage. Wherever practicable it should be installed in conduit so that it is not subjected to excessive flexing or bending, which might change the thermoelectric characteristics.

The layout and arrangement of the conduits for a thermocouple system should be given considerable thought. Long radius bends should be used instead of elbows where possible; cold working of thermocouple elements can introduce inhomogeneity, and pulling of the extension wire through a number of elbows could work the wires unnecessarily.

While it is generally desirable to keep the length of extension wire as short as possible, it is often possible to have one conduit serve a number of thermocouples. The extent of conduit fill is not as significant with thermocouples as with power wiring because the lower current causes lower heating. (Conduit fill is the ratio between the cross-sectional area of all of the wires and the area of the conduit.)

For minimum error the extension wire should be run from the thermocouple connection head to the instrument terminal or reference junction in one continuous length; junction boxes may introduce errors and can be avoided by design of the conduit system with the proper pull points. When splices are unavoidable, they should be made by compressing the two wires to be joined with a mechanical device to obtain intimate contact. No other electrical wires should ever be run in the same conduit with extension wires.

When installing extension wire underground, always use waterproof insulation. Running extension wires parallel to or in close proximity to power lines should be avoided. When any connections are made, polarity must be strictly observed.

APPENDIX D
THERMOCOUPLES-CHECKING PROCEDURES

D1. General

New thermocouples, thermocouple materials, and thermocouple extension wire are controlled by the supplier to fit a published temperature-EMF table or curve within states tolerance limits. Thermocouple extension wire normally retains its original characteristics when used within recommended temperature limits, but thermocouples, which are exposed to high temperatures in various atmospheres, may change characteristics. To avoid the continued use of thermocouples with excessive deviations from the original characteristic due to such exposure or contamination, it is good practice to check the thermocouples at regular intervals.

New material, not previously exposed to temperature gradients, can be checked by techniques described in ASTM E-220-72.

D2. Scope and Purpose

Recommendations and suggestions are given below for simple and ordinarily adequate procedures for checking installed thermocouples. These are not intended to be completely self-sufficient, however, and usually it will be advantageous to consult more detailed treatments as well.

"Temperature Measurement," Part 3 of the Performance Test Codes Supplement on Instruments and Apparatus, PTC 19.3, published by the American Society of Mechanical Engineers, gives thorough coverage to the use and calibration of thermocouples. Calibrating procedures for new thermocouple materials are given in NBS Special Publication 300, Volume II, "Precision Measurement and Calibration-Temperature". This describes various testing methods and the precautions which must be observed in order to attain various degrees of accuracy. In particular, it describes in detail the methods developed and used at the National Bureau of Standards. Further information is available in E220-72, "Calibration of Thermocouples by Comparison Techniques," published by the American Society for Testing and Materials.

D3. Procedure

D3.1 The checking of installed thermocouples is complicated by the thermo-electric non-uniformity resulting from contamination or deterioration of the elements. The unheated terminals of the used thermocouple will normally be like new - the actual junction, contaminated or deteriorated, and the intermediate material affected to various degrees.

The output of a contaminated or deteriorated thermocouple will not be determined solely by the temperature of the heated junction, as with a new homogeneous thermocouple, but also by the temperature gradient between the measuring and reference ends and the pattern of contamination and deterioration in the temperature gradient zone. FOR THIS REASON A USED THERMOCOUPLE SHOULD NOT BE REMOVED FROM ITS INSTALLED LOCATION AND PLACED IN A CALIBRAZING FURNACE FOR CHECKING. IT IS HIGH-

LY IMPROBABLE THAT THE TEMPERATURE GRADIENTS IN THE TWO INSTALLATIONS WILL BE THE SAME.

A used thermocouple must be checked in its normal installed location. The purpose of checking an installed thermocouple is not to determine its temperature-EMF characteristics, but to determine the temperature error in actual service. This can most readily be done by temporarily installing a new or checking thermocouple along side the service thermocouple, or in its place, and comparing the readings. If the installed thermocouple is used to measure a wide range of temperatures, it should be checked at more than one temperature within the range of its use. Testing of a thermocouple at a single temperature yields some information, but it is not safe to assume that the changes in the EMF of the couple are proportional to the temperature or to the EMF.

D3.2 Where the protecting tube is large enough, a checking thermocouple may be inserted beside the service thermocouple. It is recommended that a separate checking instrument be used with the checking thermocouple to permit checking of the service instrument, as well as of the service thermocouple.

Where the protecting tube is not large enough to permit the insertion of an additional thermocouple, it is necessary to remove the service thermocouple and to replace it with a checking thermocouple. When this method is used, it is essential that stable temperature conditions be maintained. In general, the higher temperature or more contaminating the atmosphere, the more frequently checks should be made.

D3.3 Large temperature gradients can exist in commonly used furnaces and other devices, and points physically close together may be at surprisingly different temperatures. The procedure of checking a thermocouple installation by means of a checking thermocouple inserted through a furnace door or otherwise installed in a different part of the apparatus from the service thermocouple is not recommended, since the thermocouple reading may fail to agree and yet both may be correct.

D3.4 Checking thermocouples or secondary standards should be homogeneous and uncontaminated. Any new thermocouple may be used, but it should be checked against a primary standard and tagged with its deviation from the standard curve. If a user does not have the equipment and technique for doing this, calibrated and tagged thermocouples are available. The National Bureau of Standards or other standardizing laboratories will furnish a report on the temperature-EMF characteristics of a submitted thermocouple.

D3.5 The accuracy of a checking thermocouple or secondary standard will become questionable after use. Noble metal thermocouples may normally be relied upon for a considerable period of use, provided that the checking temperatures have not been avoided. Base metal thermocouples used for checking purposes should be checked frequently.

NOTE: Base metal couples should not be used for checking purposes below 480°C (900°F), if they are exposed between checks at temperatures above 760°C (1400°F).

BIBLIOGRAPHY

ASTM Standard E220, "Standard Method for Calibration of Thermocouples by Comparison Techniques", American Society for Testing and Materials

ASTM Special Technical Publication STP-470B, "Manual on the Use of Thermocouples in Temperature Measurement", American Society for Testing and Materials

ASTM Standard E-563, "Standard Recommended Practice for Preparation and Use of Freezing Point Reference Baths", American Society for Testing and Materials

NBS Monograph 124, "Reference Tables for Low Temperature Thermocouples", National Bureau of Standards

NBS Monograph 125, "Thermocouple Reference Tables Based on the IPTS-68 including Supplement 1", National Bureau of Standards

NBS Special Publication 300, Volume II, "Precision Measurement and Calibration Temperature", 1968.

NBS SP 373, "Bibliograph of Temperature Measurement", National Bureau of Standards, Jan. 1953-Dec. 1969

Roeser, W.F. and S.T. Lonberger, "Methods of Testing Thermocouples and Thermocouple Materials", Circular 590, National Bureau of Standards, 1956

"Temperature Measurement", Performance Test Codes Supplement on Instruments and Apparatus, Part 3. PTC 19.3-1974, American Society of Mechanical Engineers, 1974.

"Thermocouple Thermometers", PMC Standard No. 8-10-1963, SAMA-PMC Section Inc.

See also standards of ASTM Committee E-20 on Temperature Measurement.

INSTRUMENT SOCIETY of AMERICA
Research Triangle Park, North Carolina

ANSI C100.6-3-1984

American National Standard

Voltage or Current Reference Devices: Solid State Devices

Instrument Society of America

ANSI C100.6-3-1984

American National Standard Voltage or Current Reference Devices: Solid-State Devices

Secretariat

Scientific Apparatus Makers Association

Approved May 15, 1984

American National Standards Institute, Inc.

Abstract The comprehensive standard describes voltage or current reference devices used to maintain a unit of dc voltage or current having uncertainties of 100 ppm of output or less. This part of the standard covers solid-state devices. Sections are included on performance specifications, markup and symbols, and transport or interlaboratory standards. Appendixes describe recommended test circuits and test procedures.

ANSI
C100.6-3-1984

AMERICAN NATIONAL STANDARD

American National Standard

An American National Standard implies a consensus of those substantially concerned with its scope and provisions. An American National Standard is intended as a guide to aid the manufacturer, the consumer, and the general public. The existence of an American National Standard does not in any respect preclude anyone, whether he has approved the standard or not, from manufacturing, marketing, purchasing, or using products, processes, or procedures not conforming to the standard. American National Standards are subject to periodic review and users are cautioned to obtain the latest editions.

CAUTION NOTICE: This American National Standard may be revised or withdrawn at any time. The procedures of the American National Standards Institute require that action be taken to reaffirm, revise, or withdraw this standard no later than five years from the date of publication. Purchasers of American National Standards may receive current information on all standards by calling or writing the American National Standards Institute.

Published by

**Instrument Society of America
67 Alexander Drive
P.O. Box 12277
Research Triangle Park, North Carolina 27709**

Copyright © 1984 by the Instrument Society of America
All rights reserved.

No part of this publication may be reproduced in any form, in an electronic retrieval system or otherwise, without the prior written permission of the publisher.

Printed in the United States of America

VOLTAGE OR CURRENT REFERENCE
DEVICES: SOLID-STATE DEVICES

ANSI C100.6-3-1984

Foreword

(This Foreword is not a part of American National Standard for Voltage or Current Reference Devices: Solid-State Devices, C100.6-3-1984.)

This American National Standard was intentionally subdivided into three parts, each part describing a different reference device. This was done for the following reasons:

(1) The differences in characteristics are sufficiently great that one comprehensive standard would not be as useful as a standard subdivided and focused on one of three specific devices.

(2) With the standard subdivided in this manner, it is easier to append standards for new devices as they become available.

The first part of this standard, C100.6.-1, covers chemical-type standard cells outside of temperature-control enclosures. It is a direct adoption of IEC Publication 428.

The second part, C100.6-2, covers chemical standard cells in active or passive temperature-controlled enclosures.

The third part, C100.6-3, covers physical devices used to maintain the unit of dc voltage or current having uncertainties of 100 ppm of output or less, treating these devices from the standpoint of performance characteristics, not details of design.

American National Standards Committee on Electrical Reference Instruments and Devices, C100, under whose jurisdiction this standard was developed, has the following scope:

Definitions, classification, ratings, method of testing, performance requirements, and constructional details where necessary, for various types of electrical reference measuring apparatus and devices covering direct-current and the frequency range up to 1 MHz as used in electrical standardizing laboratories.

Suggestions for improvement of this standard will be welcome. They should be sent to the American National Standards Institute, 1430 Broadway, New York, N.Y. 10018.

American National Standards Committee C100 had the following members at the time it processed and approved this standard:

Kenneth J. Koep, Chairman

Organization Represented	Name of Representative
James G. Biddle Company	Donald S. Ironside
Epply Laboratories	A. R. Karoli
Guideline Instruments	Jack Sutcliffe
I.B.M. Corporation	A.E. Warwick
Julie Research Labs	Loebe Julie
National Bureau of Standards	Norman Belecki
	Woodward G. Eicke
Newark Air Force Station	Frank A. Palvolygi
Standard Reference Labs, Inc.	Kenneth J. Koep
U.S. Army Missile Command	Charles Brooks
Individual Members	A. E. Anderson
	D. Braudaway
	W. H. Shirk, Jr.

Contents

SECTION	PAGE
1. Scope	7
2. General Requirements	7
2.1 Manufacturers' Test Data	7
2.2 Performance Specifications	7
2.3 Markings and Symbols	8
2.4 Transport or Interlaboratory Standards	8
2.5 Optimization Characteristics	9
2.6 Test Circuits	9
2.7 Terms and Definitions	9

Appendixes

Appendix A Recommended Test Circuits and/or Test Procedures for Solid-State Voltage and Current Devices 11

A1	DC Output Voltage Measurement - Opposition Method	11
A2	Time Stability Test - Long Term	13
A3	DC Internal Resistance	13
A4	Ripple Voltage	13
A5	Dielectric Strength	13
A6	Warm-Up Time	13
A7	Ambient Temperature Effects - Normal Operating	13
A8	Ambient Relative Humidity	14
A9	Retrace - On/Off	14

List of Illustrations

Figure A1-a DC Output Voltage Measurement: Opposition Method (Output Nominally Equal) Where $\Delta E_1 = E_1 - E_2 + P$ 11

Figure A1-b DC Output Voltage Measurement: Opposition Method (Output Nominally Equal) Where $\Delta E_2 = E_2 - E_1 + P$ 11

Figure A2 DC Output Current Measurement: Opposition Method (Generated Output Voltage Nominally Equal to Reference Standard) 12

Figure A3 Test Device Output Voltage Higher Than Reference Device Output Voltage Level 12

American National Standard Voltage or Current Reference Devices: Solid-State Devices

1. Scope

This standard applies to physical devices used to maintain the unit of dc voltage of current having uncertainties of 100 ppm of output or less. It treats these devices from the standpoint of performance characteristics, but does not specify design or construction details or techniques. This part of the standard, C100.6-3, applies to solid state devices used to maintain the unit of dc voltage or current having uncertainties of 100 ppm of output or less.

2. General Requirements

In order that solid-state voltage or current reference devices can be useful as standards, the manufacturer shall supply an instruction manual or its equivalent, which shall contain as a minimum the following information, if applicable:

(1) Operating instructions
(2) Report of calibration
(3) Schematic/wiring diagram
(4) Shipping, storage, and handling instructions
(5) Troubleshooting procedures
(6) Spare parts list

2.1 Manufacturers' Test Data. The manufacturer shall also supply the following data relating to a specific serial number.

2.1.1 Output Voltage or Current. The output voltage shall be given for each solid-state voltage device as measured at its terminals under no load conditions and, after a suitable period of stabilization, under reference conditions. The output current shall be given for each solid-state current device as measured at its terminals under a specified load condition and, after a suitable period of stabilization, under reference conditions. See Section 2.2.6.

For multiple output devices, the output value of each output level shall be given as above.

2.1.2 Operating Temperature

2.1.2.1 Solid-State Reference Devices in Active Temperature-Controlled Enclosures. The measured ambient temperature of the reference element or elements during the measurements made to obtain the output voltage or current values in 2.1.1 shall be given. The stated temperature may be expressed in terms of the nominal enclosure working temperature.

2.1.2.2 Solid-State Reference Devices in Passive Temperature Controlled Enclosures. The nominal ambient temperature of the test enclosure containing the solid-state devices during the test measurement period shall be given.

2.1.3 Accuracy of Device Output (Voltage or Current). The uncertainties of the value or values in 2.1.1 shall be given. The uncertainties given are obtained by directly adding the random error of the mean, at the 99 percent confidence level, with a systematic error of x ppm associated with the working standard.

2.1.4 Manufacturers' Test Conditions. The following minimum manufacturers' test conditions shall be given:

(1) Ambient temperature and range
(2) Ambient humidity and range
(3) Line voltage and regulation (or power supply voltage and regulation)
(4) Line frequency
(5) Ground and/or shielding conditions
(6) Recommended measurement circuit, if applicable
(7) Other physical and/or electrical conditions having a significant effect on the principal characteristics of the device.

2.2 Performance Specifications (Relating to a Type of Device or a Specific Model). The expected change in the output voltage(s) for voltage reference devices or the output current reference devices shall be specified in all of the following ways and shall be given for reference conditions (as specified in 2.2.7) and for normal operating conditions (as specified in 2.2.6) or as specified by the manufacturer. It shall be stated clearly whether the stability figures given require the operator to correct for drift or for any other correctable operating temperature fluctuations in temperature-controlled enclosures.

2.2.1 Output Voltage or Current Stability as a Function of Time

2.2.1.1 Long Term. The bounds within which the output voltage or current is expected to remain for a specified time interval — for which the duration is a year or longer—shall be stated.

2.2.1.2 Short Term. The bounds within which the output voltage or current is expected to remain for a specified time interval — for which the duration is in the range of 60-90 days—shall be stated.

2.2.1.3 Shortest Term. The bounds within

which the output voltage or current is expected to remain for a period of one hour shall be stated.

2.2.1.4 Other Terms. Other definitions of stability may be used in addition to those given, but they must be clearly defined by the manufacturer.

2.2.2 Stabilization Characteristics. The length of time required for active temperature control enclosures from the time after their temperature controller is energized until the internal devices perform within their initial accuracy rating shall be stated. An indication of the degree of output stabilization achieved 24 hours, or any other specified shorter time interval after the temperature controller is energized, shall also be given.

The magnitude of the on/off retrace effect shall also be given for all units that must be de-energized during transit. (Note: all solid-state references do not necessarily reach a condition of a stable output level when the device environment has reached a stable condition.)

2.2.3 Temperature Control and Monitoring Circuitry. For active control enclosures either with or without an internal temperature monitor.

2.2.3.1 Temperature Stability (Control). The bounds within which the temperature control is expected to remain for a period of one year and also for a period of one hour shall be stated.

2.2.3.2 Temperature Monitoring Stability (If Applicable). The meaningful resolution and repeatability of the temperature monitoring device shall be stated.

The bounds within which the temperature monitor is expected to remain stable shall be stated, as well as the confidence level of this expectation.

The accuracy at a stated level of confidence with which the monitor can make meaningful measurements of temperature changes shall also be stated. (Note: If a part of the adjustment or control means is readily accessible, the control means must be calibrated in order for the device to meet this standard.)

2.2.4 Power Requirements. The characteristics of the source of power necessary to maintain the proper operation of the unit within the specifications previously set forth shall be clearly stated. For devices that can be energized from more than one type of source (i.e., ac or dc), each set of characteristics shall be given. As a minimum, the upper and lower limits of excitation voltage, frequency, and maximum power demand shall be given.

2.2.5 Normal Operating Conditions. Unless otherwise stated by the manufacturer, the normal operating conditions shall be those given below:

Ambient temperature	15 to 40°C
Ambient relative humidity	15 to 80%
Ambient barometric pressure	76 to 107 kPa

2.2.6 Reference Operating Conditions. Unless otherwise stated by the manufacturer, the reference operating conditions shall be those given below:

Ambient temperature	23 ± 1°C
Ambient relative humidity	15 to 55%
Ambient barometric pressure	76 to 107 kPa
Input power	± 1%

The following reference conditions shall be specified by the manufacturer in addition to those in the preceding paragraph

(1) Position with respect to normal (degrees)

(2) Operating power (line or battery)

(3) Line voltage or battery voltage regulation (ppm output per percent output)

(4) Line voltage ripple (magnitude and frequency)

(5) Line voltage frequency range (of line-operated power supplies)

(6) Grounding and/or guarding requirements

Failure to specify any of the above will be taken to mean that the effect on the device output will cause less than 20 percent change in the uncertainty level.

2.2.7 Insulation Resistance. The leakage path resistance of the device output, including its associated leads and connections to any conducting element, shall be given. Insulation resistance shall be sufficiently high that its effect will not cause greater than a 20 percent change in the uncertainty rating of the device due to leakages.

2.2.8 Current Reference Devices. The load resistance range that the current reference device can drive while remaining within its uncertainty rating as specified in 2.1.1 shall be given. The characteristics of the load required shall also be clearly stated, e.g., its inductive and capacitive impedances, its stability, and its type and number of terminals.

2.3 Marking and Symbols. Each enclosure shall be marked. Where size does not permit direct marking, an attached tag may be used. The following minimum information is to be clearly and indelibly marked:

(1) Manufacturer's name, symbol, or initials

(2) Polarity (Note: This applies to both the output and input, as applicable)

(3) Power requirement (both ac and dc, if applicable)

(4) Output levels for multiple output devices

(5) Ground and/or guard terminals, if applicable

(6) Serial number of enclosure

(7) Model number of enclosure

(8) All switches, terminals, and controls, with their respective functions

2.4 Transport or Interlaboratory Standards. Transport or interlaboratory standards are subjected to conditions in transit that are more severe than those encountered in most reference or normal operating conditions. Therefore, ratings of operation on standby battery power, temperature control under transit conditions and other characteristics may be required to ensure their proper use as reference standards. In addition to the requirements outlined in Sections 2.1 and 2.2, transport or interlaboratory standards shall be supplied with the following data by the manufacturer.

2.4.1 Transport Standards without Self-Contained Batteries

(1) Input power requirements under reference conditions

(2) Input requirements under normal operating conditions

(3) Input power requirements under shipping conditions

2.4.2 Transport Standards with Self-Contained Batteries

(1) Duration of self-sustaining operation under reference conditions

(2) Duration of self-sustaining operation under normal operating conditions

(3) Duration of self-sustaining operation under shipping conditions with device in shipping container

2.4.3 Temperature Control under Extreme Cold Conditions. The lowest external temperature and time interval which the standards can sustain.

2.4.4 Temperature Control under Extreme Hot Conditions. The maximum external temperature and the time interval which the standard can sustain.

2.5 Optimization Characteristics. Under ideal conditions it is possible to improve on the operating characteristics of certain devices to optimize their operation and achieve greater precision and accuracy.

Data in addition to that required above, when supplied to the user, may permit improvement in the performance of a specific device.

The following data shall be supplied, if applicable, when this ability is required:

(1) Temperature coefficient of temperature controlled enclosure

(2) Output resistance of voltage reference devices (Note: Specify test resistance)

(3) Output voltage drive range of current reference devices

(4) Effects of operating on each power mode for multiple type devices, e.g., ac or dc

(5) Effects of short-circuiting voltage device outputs or open-circuiting current device outputs

2.6 Test Circuits

Recommended test circuits and/or methods are given in Appendix A. Other test circuits and/or methods can be used that demonstrate the proper capability to make the required measurement.

2.6.1 DC Output Voltage or Current Measurement. Three output voltage test methods are shown in Appendix A1. The preferred method is Method Number 1, where two nominally equal levels are compared and the difference level measured. This method is capable of achieving a higher degree of accuracy as long as a reference standard approximately equal to the test device is available. See Figures A1-a and A1-b.

The other two voltage test methods, Method 3 and Method 4, can be used to compare a test device against a reference device where both do not have a similar output voltage level. Method 2 shows a test method of comparing a current device output level by converting it to a voltage test using a four-terminal resistor. Once a test voltage level is available, the other three test methods can be applied.

2.6.2 Stability Test - Long Term. A recommended test method is given in Appendix A2.

2.6.3 DC Internal Resistance Test. A recommended test method is given in Appendix A3.

2.6.4 Ripple Voltage. A recommended method is given in Appendix A4.

2.6.5 Dielectric Strength (Breakdown Voltage — Insulation). A recommended test method is given in Appendix A5.

2.6.6 Warm-Up Time. A recommended test method is given in Appendix A6.

2.6.7 Ambient Temperature Effects - Normal Operating. A recommended test method is given in Appendix A7.

2.6.8 Ambient Relative Humidity - Normal Operating. A recommended test method is given in Appendix A8.

2.6.9 On/Off Retrace. A recommended test method is given in Appendix A9.

2.6.10 Other Characteristics. Other characteristics may be specified by the manufacturer of the voltage or current device for special design and application purposes. In this case, the manufacturer shall define the test procedure.

2.7 Terms and Definitions. The definitions of terms used in C100.6-2, Clause 2.7, are supplemented below.

2.7.1 Zener Diode (Semiconductor). This is a class of silicon diodes that exhibit in the avalanche-breakdown region a change in reverse current over a very narrow range of reverse voltage. (Note: This characteristic permits a highly stable reference voltage to be maintained across the diode despite a relatively wide range of current through the diode.)

2.7.2 Operating Time. This refers to any time interval of continuous energization of the device starting at the instant of application of operating power.

2.7.3 Retrace (On/Off). This refers to the change in reference output level expressed in parts-per-million, ppm, or percent that occurs when the device is energized and de-energized as a function of time. (Note: The time required prior to successive measurements must be stated.)

2.7.4 Output Ripple Voltage. This refers to the portion of the output voltage harmonically related in frequency to the input voltage and arising solely from the input voltage. (Note: Unless otherwise specified, percent ripple is the ratio of root-mean-square (RMS) value of ripple voltage to the average value of the total voltage expressed in percent.)

2.7.5 Excitation Voltage Influence. Voltage regulation is a measure of the change in the output of the device when the energizing potential is changed. The magnitude of the input energization level change is stated in percent and must be maintained long enough for a steady state value to be accomplished at the new set level before the output level is measured. The effect is stated in percent or parts-per-million (ppm) change in output relative to the percent or ppm change in input, e.g., 2ppm/10 percent.

VOLTAGE OR CURRENT REFERENCE
DEVICES: SOLID-STATE DEVICES

ANSI C100.6-3-1984

APPENDIX A

This appendix describes recommended test circuits and/or test procedures for solid-state voltage and current devices. (Note: All tests are conducted under reference conditions except for the parameter under test as spelled out in the test procedure.)

A1. Method 1: DC Output Voltage Measurement: Opposition Method (Output Nominally Equal)

In Method 1 the small difference ΔE between a test device and a reference device connected in series opposition will be measured. The potentiometer system used to measure the difference shall have a resolution of 0.1 ppm and be of the low thermal type.

The design proposed provides information on (1) the emf of the test device relative to the reference devices, (2) the left-right component, (3) the deviation of each observation from the predicted value, and (4) the standard deviation of a single observation.

All of these are estimated from the data of a single run and provide information on the measurement process. It has been shown that there are often small residual emf's in a measuring circuit which do not affect the quality of the measurements. Certain of these tend to remain constant and can be estimated.

Figure A1-a shows a measuring circuit with a small emf, P, in the circuit. If the difference ΔE is measured, it equals $\Delta E_1 = E_1 - E_2 + P$. Now if we reverse the position of the two devices in the circuit, that is, connect the test device to the terminals where the reference device had been connected, and connect the reference device to the terminals where the test device had been connected, we have the circuit shown in Figure A1-b. The difference ΔE is now $\Delta E_2 = E_2 - E_1 + P$. We can estimate P and eliminate the effect of P on the final results. If we add ΔE_1 to ΔE_2 we can estimate P where $P = \frac{1}{2}(\Delta E_1 + \Delta E_2)$. Subtracting, we obtain the difference $E_1 - E_2$, which is free of P where $E_1 - E_2 = \frac{1}{2}(\Delta E_1 - \Delta E_2)$. P is called the left-right component and is estimated by interchanging the position of the test and reference devices in the measuring circuit.

To standardize the relative position of the devices with

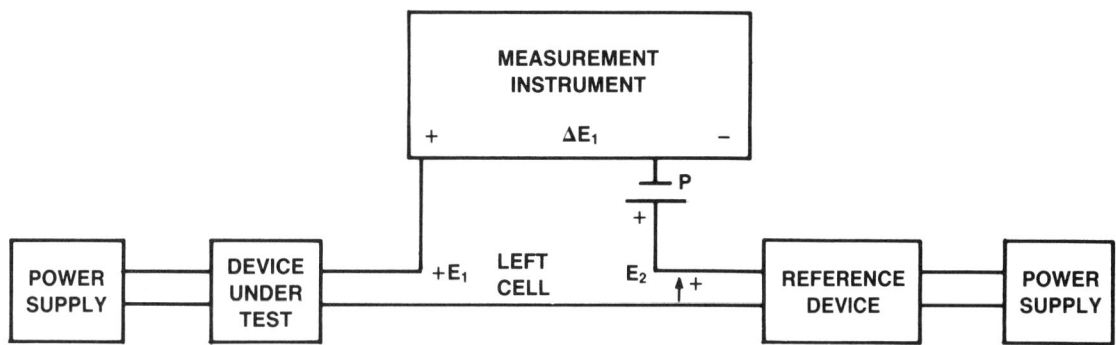

Fig. A1-a
DC Output Voltage Measurement: Opposition Method
(Output Nominally Equal) where $\Delta E_1 = E_1 - E_2 + P$

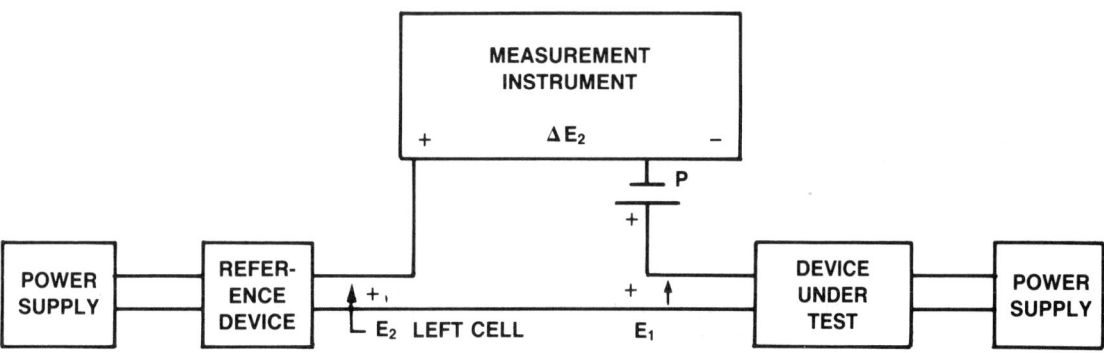

Fig. A1-b
DC Output Voltage Measurement: Opposition Method
(Output Nominally Equal) where $\Delta E_2 = E_2 - E_1 + P$

respect to the input terminals of the measuring instrument, the left cell is always connected to the potentiometer so that its polarity agrees with the polarity marked on the terminals of the measuring instrument.

The measurement described will result in positive and negative differences in emf being measured sequentially by the instrument. If there is a polarity switch on the instrument, use it. If there is no polarity switch, the current source to the instrument should be reversed. To accomplish the current reversal, use a low-contact reversing switch and do *not* standardize the instrument while the current source is reversed.

A1. Method 2: DC Output Current Measurement: Opposition Method (Generated Output Voltage Nominally Equal to Reference Standard)

This method is for current reference devices.

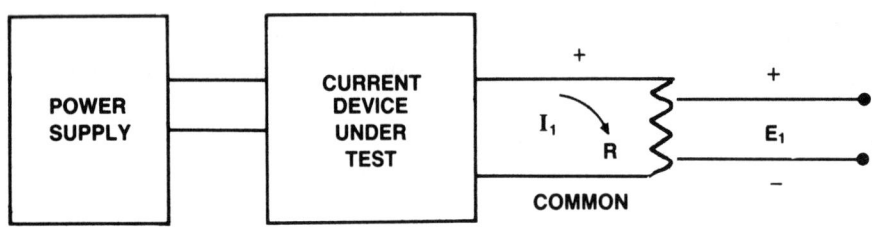

Fig. A2
DC Output Current Measurement: Opposition Method (Generated Output Voltage Nominally Equal to Reference Standard)

A1. Method 2.

The test method shown in Method 1 can be applied to current test devices after the output current level I_1 is converted to a voltage level E_1, using a Kelvin (4 terminal) resistor, R, as the transducer. The resistor must be of a quality that its uncertainty contribution to establishing the test voltage level, E_1, is less than 25 percent of the current device's uncertainty rating. This includes its initial accuracy rating plus self-heating effects, etc.

Once the E_1 test voltage level is determined, it is evaluated by Method 1. Note that $I_1 = \frac{E_1}{R}$. (Note: Methods 3 and 4 can also be applied when the magnitude of E_1 generated is larger or smaller than an available reference voltage value.)

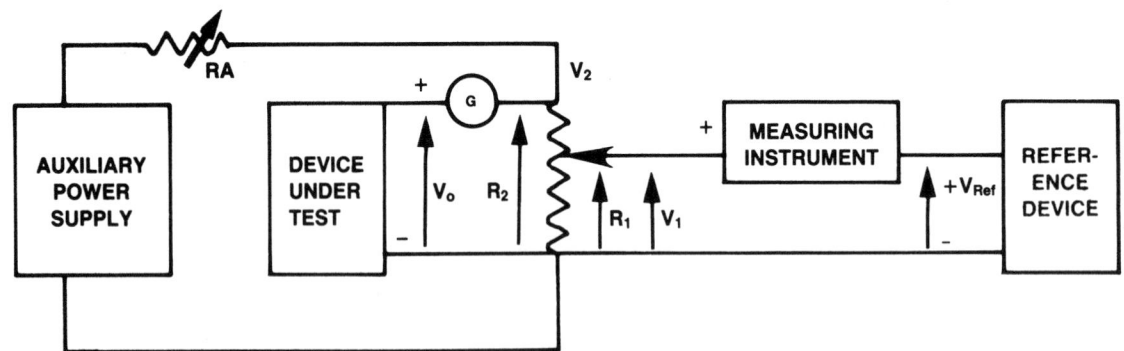

Fig. A3
Test Device Output Voltage Higher Than Reference Device Output Voltage Level

(1) Adjust RA for balance of G. At this condition, $V_O = V_2$

(2) Adjust R_1/R_2 ratio for null on measuring instrument.

$$V \text{ reference} = V_1 = \frac{R_1}{R_2}(V_2) = \frac{R_1}{R_2}(V_O)$$

Method 3: Test Device Output Voltage Higher Than Reference Device Output Voltage Level

In this method, a stable, auxiliary power supply is used to energize a resistive divider. The voltage developed across the divider is first adjusted by comparison to the test device. The divider ratio is then adjusted until the stepped-down voltage is equal to the reference voltage. If a fixed divider is used, the ratio must develop a voltage V_1 approximately equal to the reference level and the difference voltage measured on the measuring instrument.

A second reading should be made with the input to the measuring instrument reversed, if possible. The average of the two readings should be used to indicate the value of the test device output relative to the reference device output.

Method 4: Test Device Output Voltage Lower Than Reference Device Output Voltage Level

This method is the same as Method 3 except that the positions of the test device and the reference device are interchanged.

Note: In all four methods the following items should be evaluated prior to making output level measurements:
  (1) The effect of leakage between supplies and other parts of the circuit
  (2) The auxiliary supplies should be highly isolated, low ripple, well regulated dc sources or should be stabilized battery supplies
  (3) The resistive divider must be stable and of either a known fixed or variable ratio depending on type used
  (4) The circuit connections should be of the low thermal type

A2. Time Stability Test - Long Term

The test device shall be adjusted in accordance with the manufacturer's instructions at reference conditions, and no further adjustments shall be made for the duration of the test.

Each test device shall be energized for at least 24 hours at reference conditions prior to output level measurements. Each output level is then measured daily over a 10-day interval. Each day a minimum of 3 measurements is to be made using a test procedure according to one of the methods given above.

The average of the 30 (minimum) measurements made over the 10-day interval is to be used as the device's initial output value.

On 13-week (or shorter) intervals, measurements are to be made for 5 consecutive days, with a minimum of 3 measurements being made each day. The average value of the 15 measurements is compared to the initial output value previously calculated to indicate the value of drift with time. The test shall be continued for one year, with measurements being made at specified time intervals to determine long term stability.

A3. DC Internal Resistance (Voltage Source)

With a test load resistance, R_L, specified by the manufacturer connected to the output terminals, the output voltage, E_L, shall be measured. The load shall be removed and the output voltage, E_O, again measured. The source resistance, R_S is defined as $R_L (E_O - E_L) / E_L$.

A4. Ripple Voltage

With the test device adjusted in accordance with the conditions stated in 2.2.6 for reference conditions and the manufacturer's instructions regarding load, the harmonically related ac voltage at the output terminals shall be measured using an appropriate instrument responding to the rms value of the ripple voltage.

A5. Dielectric Strength (Breakdown Voltage-Insulation)

The test device shall be conditioned for this test for a minimum of 16 hours at a temperature of approximately 30°C and a relative humidity of 75 percent. The recommended test voltage for the dielectric withstand test is 500V. The test voltage should be applied to the test device as follows:
  (1) Between power terminals and the case, with all terminals of the output circuit tied to the case
  (2) Between the output circuit and the case, with all terminals of the output shorted.

To avoid voltage surges, the applied voltage shall be raised to its full value gradually and held at the rated value for 60 seconds. It shall then be reduced gradually.

Note: If the case is not conductive, the case is to be placed on a conductive plate and the above measurements made to the conductive plate instead of the case.

A6. Warm-Up Time (For Units Equipped Only with an AC Output)

The power shall be disconnected from the test device for a minimum period of 12 hours. The unit shall then be energized under reference conditions and the output measured by the A1 procedure (or equivalent) until the output is stabilized to the value given in 2.1.1, within the limits of uncertainty stated. The time required for the output to come within specification is the device warm-up time.

A7. Ambient Temperature Effects - Normal Operating

The test device shall be placed in a test chamber

at a temperature of 40 ±0.5°C for a period of 8 hours. At the end of that period the output of the unit shall be measured. The temperature shall then be reduced to 15 ±0.5°C for 8 hours and the output re-measured. The measured values at 40°C and 15°C compared to the value given in 2.1.1 at reference conditions indicate the temperature effect.

A8. Ambient Relative Humidity
The test device shall be placed in a test chamber at a relative humidity of 15 ±10% and in a temperature of 23 ± 3°C for a period of 16 hours. The output of the unit shall then be measured. The humidity shall then be increased to 80 ±10 percent and the test repeated. The output at 80% relative humidity and at 15% relative humidity shall be compared to the value measured at reference conditions.

A9. Retrace - On/Off
This test is only applicable to devices that must be de-energized for shipment, or where a standby power supply is not available to accomplish constant energization.

This test is to be performed at reference conditions as stated in 2.2.6. The test device shall be energized for a period of time specified by the manufacturer to allow the output to stabilize. The output value shall be recorded. The device shall then be de-energized for a minimum period of 12 hours. After this period, the device is again re-energized and allowed to stabilize according to the manufacturer's specifications. The change in the output level stated in ppm or percent is the "On/Off Retrace Error." An On/Off Retrace Error greater than 25 percent of the rating is not acceptable to meet this standard.

INSTRUMENT SOCIETY OF AMERICA
Research Triangle Park, North Carolina 27709

Titles, Abstracts, and Contact Information for Instrumentation and Control Standards

In this section representative standards that also cover instrumentation and control are presented as supplied from responding organizations. In most cases these listings provide only a sampling of the available material but help identify the general area of interest of each organization. They also provide contact information for complete listings, further information and ordering.

In addition to the abstracts of standards published by U.S. Organizations, titles of selected international standards published by the International Organization for Standardization (ISO) and the International Electrotechnical Commission (IEC) are included. The United States is represented in these organizations through its member body, the American National Standards Institute, Inc. (ANSI), formerly, USASI.

ACOUSTICAL SOCIETY OF AMERICA (ASA)

The following publications are examples of the many acoustical standards available from the Acoustical Society of America, 335 East 45th Street, New York, NY 10017. (212) 661-9404

ANSI: S1.1-1960 (R 1976), Acoustical Terminology (Including Mechanical Shock and Vibration), 62 pp.
Describes acoustical terminology in areas such as oscillation, vibration and shock, transmission, linear systems, transducers, recording, underwater sound, sonics, architectural acoustics, hearing and speech, and music.

ANSI: S1.4-1983, Specification for Sound Level Meters, 18 pp.
Covers definitions, characteristics, and tolerances, indication, internal noise and extraneous influences, and calibration of general-purpose sound level meters. Includes measurement of transient sound signals and permits use of digital techniques and displays.

ANSI: S1.6-1984, Preferred Frequencies, Frequency Levels, and Band Numbers for Acoustical Measurements, 3 pp.
Provides guide to the design of new acoustical equipment and the selection of frequencies to make comparison of the results of acoustical measurements most convenient.

ANSI: S1.8-1969 (R 1974), Preferred Reference Quantities for Acoustical Levels, 10 pp.
Concerned with the reference quantities and the definitions of some levels for acoustics, electroacoustics and mechanical vibrations. Purpose is to provide a preferred reference quantity of convenient magnitude for a given kind of acoustical level. (Sponsored by Acoustical Society of America and ASME).

ANSI: S1.10-1966 (R 1976), Method for the Calibration of Microphones, 35 pp.
Describes methods for performing absolute and comparison calibrations of laboratory standard microphones.

ANSI: S1.11-1986 (R 1976), Specification for Octave-Band and Fractional-Octave-Band Analog and Digital Filters
Provides performance requirements for fractional-octave-band bandpass filters, including, in particular, octave-band and one-third-octave-band filters. Basic requirements are given by equations with selected empirical constants to establish limits on the required performance. The requirements are applicable to passive or active analog filters that operate on continuous-time signals, to analog and digital filters that operate on discrete-time signals, and to fractional-octave-band analyses synthesized from narrow-band spectral components. An Appendix is included for reference to terminology used in digital signal processing.

ANSI: S1.12-1967 (R 1977), Specifications for Laboratory Standard Microphones, 11 pp.
Describes types of laboratory microphones suitable for calibration by an absolute method such as

Titles & Abstracts

the reciprocity technique described in ANSI S1.10-1966 (R 1976), Method for the Calibration of Microphones. These microphones are intended for use as acoustical measurement standards either in a free field or in conjunction with a variety of devices such as artificial voices and couplers for calibrating earphones or microphones.

ANSI: S1.20-1972 (R 1988), Procedures for Calibration of Underwater Electroacoustic Transducers, 40 pp.

This standard establishes measurement procedures for calibrating electroacoustic transducers and describes forms for presenting the resultant data. It is a revision of American National Standard S1.20-1972 (R1977). Both primary and secondary calibration procedures are specified for the frequency range from a few hertz to a few megahertz. Procedures are specified for determining the measurable characteristics of free-field sensitivity, transmitting response, directional response, impedance, dynamic range, equivalent noise pressure level, and overload pressure level. Equations are given for obtaining the derived characteristics directivity factor, directivity index, efficiency, theoretical noise pressure level, and quality factor (Q). A coordinate system and forms of data presentation are specified so that results may be readily compared and easily understood.

ANSI: S1.23-1976, Method for the Designation of Sound Power Emitted by Machinery and Equipment, 3 pp.

Describes a method for expressing the noise emissions of machinery and equipment in a convenient manner. Applies to all machinery and equipment which is essentially stationary in nature and for which a sound power spectrum may be determined. The designation described in this standard is based on the A-weighted sound power emitted by the source.

ANSI: S1.40-1984, Specification for Acoustical Calibrators

Specifies performance requirements for coupler-type acoustical calibrators. For each microphone type that may be used with the calibrator, requirements include the sound pressure level in the coupler, the frequency of the sound, and the determination of the influence of atmospheric pressure, temperature, humidity, and magnetic fields on the pressure level and frequency of the sound produced by the calibrator. Specifications are to be met within stated tolerances at each frequency and sound pressure level of operation.

ANSI: S1.42-1986, Design Response of Weighting Networks for Acoustical Measurements

Provides the design criteria for both the frequency domain response (amplitude and phase) and time domain response of the A-, B-, and C-weighting networks used in acoustically related measurements. The poles and zeros for each weighting network are given, along with equations for computing the amplitude and phase responses as functions of frequency and the impulse and step responses as functions of time. In the Appendix, similar information is provided for the D- and E-weighting networks.

ANSI: S2.4-1976 (R 1982), Method for Specifying the Characteristics of Auxiliary Analog Equipment for Shock and Vibration Measurements, 10 pp.

Provides a uniform technology and format for the presentation of the performance and other characteristics of auxiliary analog equipment for shock and vibration measurements.

ANSI: S2.9-1976 (R 1982), Nomenclature for Specifying Damping Properties of Materials, 8 pp.

Presents the preferred nomenclature (parameters, symbols, and definitions) for specifying the damping properties of uniform materials and uniform specimens where "uniform" implies homogeneity on a macroscopic scale.

ANSI: S2.11-1969 (R 1978), Selection of Calibration and Tests for Electrical Transducers Used for Measuring Shock and Vibration, 19 pp.

Includes considerations relevant to commonly employed electromechanical shock and vibration measurement transducers, but not to those transducers primarily designed for measurement of acoustic or pressure phenomena.

ANSI: S2.31-1979, Method for the Experimental Determination of Mechanical Mobility. Part I: Basic Definitions and Transducers, 17 pp.

Covers the experimental determination of mechanical mobility of structures by a variety of methods appropriate for different test situations. Part I covers basic concepts and definitions and serves as a guide for the selection, calibration and evaluation of the transducers and instruments used in mobility measurements. The material in Part I is common to most mobility measurement tasks. This document supersedes ANSI S2.6-1963 (R1976).

ANSI: S12.11-1987, Methods for the Measurement of Noise Emitted by Small Air-Moving Devices

This standard defines uniform methods for measuring and reporting the noise emitted by small air-moving devices. The sound power produced by the air mover is determined and reported in terms of the noise power emission level in bels. The standard covers specific methods for installing and mounting air movers, operation of the air mover during the tests, and environmental conditions during the measurements.

AIR-CONDITIONING AND REFRIGERATION INSTITUTE (ARI)

The following are examples of many HVAC standards that are available from the Air-Conditioning and Refrigeration Institute, 1501 Wilson Blvd., 6th Floor, Arlington, VA 22209. (703) 524-8800

ARI: STD. 130-88, Graphic Electrical/Electronic Symbols for Air-Conditioning and Refrigerating Equipment, 1988.
Contains over 100 electrical symbols commonly used for air-conditioning and refrigeration equipment.

ARI: STD. 575-87, Method of Measuring Machinery Sound Within Equipment Spaces, 1987, 12 pp.
Establishes a uniform method of measuring, recording and determining the sound pressure level of machinery installed in mechanical equipment spaces.

ARI: STD. 720-88, Standard for Refrigerant Access Valves and Hose Connectors, 1988.
Applies to 1/4 inch SAE flare refrigerant access valves and hose connectors. It has definitions of components and establishes location requirements of the valve core pin in the access valve body and the location of the hose connector gasket and core pin depressor. It includes drawings, dimensions and gaging procedures.

ARI: STD. 750-87, Thermostatic Refrigerant Expansion Valves, 1987, 4 pp.
Definitions; testing and rating requirements; a specification for minimum published data; recommended standard maximum operating pressures for pressure-limiting type valves; recommended refrigerant designation and color coding; and recommended standard connection sizes.

ARI: STD. 760-87 Solenoid Valves for Use with Volatile Refrigerants, 1987, 19 pp.
Definitions and classifications; testing and rating requirements; specifications for minimum published data; performance requirements; electrical specifications; recommended line connection sizes; design, construction and assembly recommendations for safety; nameplate data; and conformance conditions.

ARI: STD. 770-84, Refrigerant Pressure Regulating Valves, 1984, 11 pp.
Establish, for refrigerant pressure regulating valves: definitions; requirements for testing and rating; specifications, literature and advertising requirements; requirements for marking; and conformance conditions.

ARI: STD. 780-86, Definite Purpose Magnetic Contactors, 1986, 5 pp.
Establish, for definite purpose magnetic contactors within its scope: definitions; requirements for testing; performance requirements; and conformance conditions.

ARI: STD. 790-86, Definite Purpose Magnetic Contactors for Limited Duty, 1986, 4 pp.
Limited to magnetic contactors not intended for repeatedly establishing and interrupting an electrical power motor protection circuit or primary safety circuit (refrigeration or air conditioning). These contactros may be applicable to control fan motors, pump motors, resistive loads, etc.

AIR MOVEMENT AND CONTROL ASSOCIATION (AMCA)

The following standards are available from the Air Movement and Control Association, 30 W. University Drive, Arlington Heights, IL 60004. (312) 394-0150

AMCA: Standard 300-85; Reverberant Room Method for Sound Testing of Fans, 1985, 25 pp.
Detailed specifications of test setups, instrumentation, procedures, and calculations for determining sound power output of fans.

AMCA: Standard 210-85; Laboratory Methods of Testing Fans for Rating, 1985, 60 pp.
Detailed specifications of test setups, instrumentation, and procedures for measuring air performance of fans.

AMCA: Standard 500-83; Test Method for Louvers, Dampers, and Shutters, R1986, 28 pp.
Detailed specifications of instrumentation, calibration, test procedures and apparatus for determining pressure drop, leakage, and water penetration across ventilating louvers and dampers.

AMCA: Interim Standard 220-82; Test Methods for Air Curtain Units, 1982.
Establishes uniform methods for laboratory testing of Air Curtain units to determine performance in terms of flow rate, outlet velocity uniformity, power consumption, and velocity profile.

Titles & Abstracts

ALUMINUM ASSOCIATION (AA)

The following publications are available from the Aluminum Association, 900 19th St. NW, Washington, DC 20006. (202) 862-5100

ASD-1, Aluminum Standards and Data, 1988, 216 pp.
Issued biennially, this reference book contains data on chemical compositions, mechanical, physical and other properties, tolerances and aluminum products in general use. It also includes separate sections on sheet and plate, rolled rod and bar, extruded rod, bar and shapes, drawn and extruded tube, forgings, electrical conductors and other aluminum forms and shapes.

ASD-1-M, Aluminum Standards and Data Metric SI, 1986, 212 pp.
Metric version of Aluminum Standards and Data which is divided into three sections: general information, including typical physical and mechanical data; nomenclature unique to the aluminum industry; and chemical compositions, mechanical properties and tolerances.

AMERICAN ASSOCIATION OF TEXTILE CHEMISTS AND COLORISTS (AATCC)

The following annual publication is available from the American Association of Textile Chemists and Colorists, P.O. Box 12215, Research Triangle Park, NC 22709. (919) 549-8141

Technical Manual of the AATCC, 1989.
An annual publication which contains AATCC Test Methods, committee rosters and reports.

AMERICAN BOILER MANUFACTURERS ASSOCIATION (ABMA)

The following are available from the American Boiler Manufacturers Association, Suite 160, 950 No. Glebe Road, Arlington, VA 22203. (703) 522-7350

Boiler Water Limits and Steam Purity Recommendations for Watertube Boilers (1982, 3rd Edition)
These recommendations are to acquaint engineers and purchasers of steam generating equipment with ABMA's judgement as to the relationships of boiler water conditions to steam purity and are published for general guidance.

Boiler Water Requirements and Associated Steam Purity for Commercial Boilers (1984, 1st Edition)
This publication is to acquaint engineers, purchasers ad operators of commercial boilers with ABMA's judgement as to the relationship between boiler water and boiler performance. This document discusses the effect of various feedwater and condensate systems on the boiler operation. It also provides information on boiler water and steam testing as well as system care and maintenance.

Fluidized Bed Combustion Guidelines (1987, 1st Edition)
Provides architects, engineers, fuel producers and users with general information about fluid bed combustion systems for steam generation. Intended as a supplement to other sources of information furnished by the equipment manufacturers.

Guide to Clean and Efficient Operation of Coal Stoker Fired Boilers
This report was prepared as a set of guidelines for those in charge of operating coal stoker fired boilers. It explains and illustrates the different types of coal stokers in operation today. It explains the combustion process in simple terms, specifically as it relates to stoker coal combustion. It explains the various heat losses in stoker boilers and how they may be minimized. Most importantly it discusses ways in which coal stoker fired boilers may be operated at peak efficiency and with minimum pollutant emissions.

Guidelines for Industrial Boiler Performance Improvement
Provides general guidelines for use by industrial boiler operators to reduce stack emissions on nitrogen oxides and improve boiler operating efficiency. Deals primarily with boiler adjustments that are typically within the control of boiler operators and plant engineering personnel.

Lexicon — Boiler & Auxiliary Equipment (1987, 5th Edition)
This publication contains 1,200 terms covering such areas as fuels, fuel-burning equipment, combustion, boilers, superheaters and dust collecting equipment. A selection also covers boiler and fuel systems controls.

Matrix of Recommended Quality Control Requirements
This recommended matrix of the ABMA concerning items to be considered for quality control/assurance in connection with boiler manufacturer's procured items was prepared by the ABMA Committee on Quality Assurance/Control and represents those items a number of experts

Titles & Abstracts

believe should be considered when qualifying procured materials.

Operation and Maintenance Safety Manual
Prepared by the ABMA Product Safety Committee for the purpose of alerting boiler operators and maintenance personnel to some of the hazards of operating and maintaining boiler systems. Applies to utility, industrial and commercial boilers.

Procedure for the Measurement of Sound from Field-Erected Stationary Steam Generators (1973, 3rd Edition)
Sets forth a method for the measurement and recording on data sheets, the sound pressure levels of field-erected stationary steam generators.

Procedure for the Measurement of Sound from Boiler Units, Bottom Supported, Shop or Field-Erected (1973, 3rd Edition)
Provides a standard test method for the measurement of airborne sound from bottom supported steam or hot water generators (boilers), using water or other fluids and from liquid phase heaters.

Recommended Design Guidelines for Stoker Firing of Bituminous Coals (1st Edition)
Provides for the application of stokers firing bituminous coal. Contains coal sizing and carbon loss curves, which have been developed as an aid in boiler and stoker design. They provide empirical data covering projected stoker operation and carbon loss expectations.

Thermal Shock Damage to Hot Water Boilers as a Result of Energy Conservation Measures
Warns users about problems that can arise from incorrect application of energy management systems to boilers and related systems.

AMERICAN CHEMICAL SOCIETY (ACS)

The following publication is available from the American Chemical Society, Books and Journals' Division, 1155 Sixteenth Street NW, Washington, DC 20036. (202) 872-4600

Reagent Chemicals, ACS Specifications, 7th Edition, 1986, 675 pp.
ACS Specifications for 352 reagent chemicals; includes 50 pages of definitions, tests, and reagent solutions. Features flame and flameless atomic absorption methods; new polarographic and chromatographic procedures; and new colorimetric test for arsenic; and the chemical abstracts number for each standard.

AMERICAN CONFERENCE OF GOVERNMENTAL INDUSTRIAL HYGIENISTS (ACGIH)

The following publications are available from the American Conference of Governmental Industrial Hygienists, Committee on Industrial Ventilation, 6500 Glenway Avenue, Bldg. D-7, Cincinnati, OH 45211-4438. (513) 661-7881

Air Sampling Instruments Manual 7th Edition, 1989
Five sections:
Basics of Air Sampling
Sampling for Specific Purposes
Sampling Systems and Components
Sample Collectors
Direct Reading Instruments
Total of 22 Chapters and 7 instrument sections

Industrial Ventilation — A Manual of Recommended Practice, 20th Edition, 1988.
Ten chapters from Chapter I, "General Principles of Ventilation," to Chapter X "Specific Operations" discuss basic ventilation principles and provide useful information.

Ventilation System Testing
From Industrial Ventilation — A Manual of Recommended Practice.

AMERICAN GAS ASSOCIATION (AGA)

The following publications are available from Order and Billing Department, American Gas Association, 1515 Wilson Boulevard, Arlington, VA 22209. (703) 841-8400

ANSI Z21.2-1983, Z21.2a-1985 and Z21.2b-1987, Gas Hose Connectors for Portable Indoor Gas-Fired Equipment
Details test and examination criteria for gas hose connectors for use indoors with laboratory, shop or ironing equipment that requires mobility during operation. Such connectors may have end fittings of the slip end type or fittings provided with taper pipe threads; are limited to a maximum nominal inside diameter of 3/8 inch and a maximum nominal length of 6 feet; are intended for use only in unconcealed indoor locations and where they will not be likely to be subject to excessive temperatures (above 125 F); are for use only with natural, manufactured and mixed gases having a specific gravity less than 1.0; and are for use on gas piping systems having fuel gas pressures not in excess of 1/2 psig. This standard does not apply to gas appliance connectors covered under ANSI Z21.24,

Titles & Abstracts

Z21.45, Z21.54 or Z21.69.

ANSI Z21.13-1987 Gas-Fired Low-Pressure Steam and Hot Water Boilers

Details test and examination criteria for low-pressure steam and hot water boilers for use with natural, manufactured and mixed gases, liquefied petroleum gases, and LP gas-air mixtures. A low-pressure boiler is defined in the standard as a boiler operating at or below the following pressures or temperatures: steam heating boiler — 15 psig steam pressure; hot water heating or supply boiler — 160 psig water pressure, 250 F water temperature.

ANSI Z21.15-1979, Z21.15a-1981 and Z21.15b-1984 Manually Operated Gas Valves

Details test and examination criteria for manually operated gas valves which are substantially of the plug and body, or rotating disc type, and to valves of other types which will provide equivalent performance. The standard presents minimum levels for the substantial and durable construction, safe operation and acceptable performance of such valves.

ANSI Z21.18-1987 Gas Appliances Pressure Regulators

Details test and examination criteria for gas appliance pressure regulators for use with natural, manufactured and mixed gases, liquified petroleum gases and LP gas-air mixtures. Such devices, either individual or in combination with other controls, are intended to control selected outlet gas pressures to individual gas appliances.

ANSI Z21.21-1987 Automatic Valves for Gas Appliances

Details test and examination criteria for individual automatic valves, valves utilized as parts of automatic gas ignition systems, or the automatic valve functions of combination controls, which have maximum operating pressure ratings of 1/2 psi, 2 psi, 5 psi, or higher than 5 psi in 5 psi increments up to and including a maximum operating pressure of 60 psi.

ANSI Z21.22-1986 Relief Valves and Automatic Gas Shutoff Devices for Hot Water Supply Systems

Details test and examination criteria for: (1) temperature relief valves and combination temperature and pressure relief valves for use on storage tanks of hot water supply systems without heater input limitation; (2) valves having only pressure relief features for use on storage tanks of hot water supply systems with inputs up to and including 200,000 Btu per hour; (3) automatic gas shutoff valves and devices; and (4) vacuum relief valves.

ANSI Z21.23-1980 and Z21.23a-1985 Gas Appliance Thermostats

Details test and examination criteria for integral gas valve type and electric type thermostats which are used as integral parts of gas-burning appliances. It presents minimum levels or the substantial and durable construction, safe operation and acceptable performance for such thermostats. The standard does not apply to wall-mounted thermostats for comfort heating control.

ANSI Z21.24-1987 Metal Connectors for Gas Appliances

Details test and examination criteria for gas appliance connectors comprised of semi-rigid metal tubing and having a fitting at each end provided with taper pipe threads for connection to a gas appliance and to house piping, or consisting of corrugated tubing depending on all-metal construction for gastightness. Such connectors are suitable for connecting gas-fired appliances to fixed gas supply lines at pressure not in excess of 1/2 psig. These connectors are limited to a maximum nominal length of 6 feet and are not intended for continuous movement.

ANSI Z21.41-1978, Z21.41a-1981 and Z21.41b-1983 Quick-Disconnect Devices for Use With Gas Fuel

Details test and examination criteria for hand-operated devices which provide means for connecting and disconnecting gas-fired appliances or gas appliance connectors to gas supplies and which are for use under indoor or outdoor applications. These devices are equipped with automatic means to shut off gas flow when disconnected.

ANSI: Z21.45-1985 and Z21.45a-1987 Flexible Connectors of Other Than All-Metal Construction for Gas Appliances

Details test and examination criteria for gas appliance connectors consisting of flexible tubing dependent on other than all-metal construction for gastightness. Such connectors are suitable for connecting gas-fired appliances to fixed gas supply lines containing natural, manufactured or mixed gases, liquefied petroleum gases or LP gas-air mixtures at pressures not in excess of 1/2 psig. These connectors are limited to a maximum nominal inside diameter of 1 inch and a nominal length of 6 feet and are not intended for continuous movement. This standard does not apply to gas appliance connectors covered under ANSI Z21.2, Z21.24, Z21.54 or Z21.69.

ANSI Z21.70-1981 Earthquake Actuated Automatic Gas Shutoff Systems (new standard)

Details test and examination criteria for automatic gas shutoff systems consisting of, (1) a seis-

Titles & Abstracts

mic sensing means and, (2) an actuating means designed to automatically actuate a companion gas shutoff means installed in gas piping. Such systems are designed to automatically shut off the gas supply downstream of the gas shutoff means in the event of a seismic disturbance. The system may consist of separable components or may incorporate all functions in a single body.

ANSI Z21.71-1981 and Z21.71a-1985 Automatic Intermittent Pilot Ignition Systems for Field Installation

Details the construction and installation procedures for automatic intermittent pilot ignition systems designed to be adapted to continuous pilot burners on listed forced air heating appliances and boilers equipped with atmospheric burners. These systems, which consist of a pilot ignition device, pilot flame sensing means, and a necessary related controls and wiring, ignite the pilot burner gas on a call for heat, prove the presence of the pilot before allowing main burner gas flow, and shut off both main burner and pilot gas when the call for heat is satisfied.

ANSI Z83.3-1971, Z83.3a-1972, Z83.3b-1986 Gas Utilization Equipment in Large Boilers (reaffirmation)

Details criteria for the installation of gas utilization equipment in boilers having inputs over 400,000 Btu per hour per combustion chamber, except water-tube boilers having outputs of 10,000 pounds of steam per hour or more.

ANSI: Z223.1-1988 National Fuel Gas Code

A safety code for gas piping systems on consumers' premises, the installation of gas utilization equipment and accessories for use with fuel gases. Piping systems covered by this code are limited to a maximum operating pressure of 60 psig.

AMERICAN LEATHER CHEMISTS ASSOCIATION (ALCA)

ALCA is working with ASTM in developing uniform test procedures for leather processing. ALCA has accepted Standard Methods for the Examination of Water and Wastewater, 14th Edition, published by the American Wastewater Association, WPCF Association, and the American Public Health Association, 1015 18th Street, NW., Washington, DC 20036. This publication replaces ACLA's former M-series of methods. Other information on leather processing is available from the American Leather Chemists Association, c/o Campus Station, Location 14, Cincinnati, OH 45221. (513) 475-2707

AMERICAN NATIONAL STANDARDS INSTITUTE (ANSI)

The American National Standards Institute (ANSI) does not develop standards but rather approves standards developed by other organizations after ANSI has verified that the requirements for due process and consensus have been met. In addition ANSI coordinates the private sector standards activity within the United States and manages US participation in nongovernmental, international standards developing organizations, primarily the International Organization for Standardization (ISO) and the International Electrotechnical Commission (IEC). Copies of ANSI-approved standards may be purchased either from the standards developer or ANSI. ANSI sells IEC and ISO standards and serves as a clearinghouse for information on other international standards organizations. Contact: American National Standards Institute, 1430 Broadway, New York, New York 10018. (212) 354-3300

AMERICAN NUCLEAR SOCIETY (ANS)

The following ANS standards are available from the American Nuclear Society, 555 North Kensington Avenue, La Grange Park, IL 60521. (312) 352-6611

ANS: 10.2-1988, Recommended Programming Practices to Facilitate the Portability of Scientific Computer Programs, 1988.

This standard recommends programming practices to facilitate the portability of computer programs prepared for scientific and engineering computations.

ANS: 2.2-1988, Earthquake Instrumentation Criteria for Nuclear Power Plants, 1988.

This ANSI standard specified earthquake instrumentation for the site, structures, equipment, and piping. It is intended for use at water-cooled nuclear power plants, and may be used for guidance at other types of nuclear power plants.

ANS: 2.10-1979, Guidelines for Retrieval, Review, Processing and Evaluation of Records Obtained from Seismic Instrumentation, 1979.

Provides instructions for treatment of data from a variety of seismic instruments used in light water-cooled, land-based nuclear power plants. Defines the type and timing of plant owner activities, required in the event of an earthquake and includes specific procedures for the evaluation of records obtained from seismic instrumentation specified in ANS-2.2-1988.

Titles & Abstracts

ANS: 3.1-1987 Selection, Qualification and Training of Personnel for Nuclear Power Plants (Revision of ANSI/ANS: 3.1-1981), 1987.

This standard provides criteria for the selection, qualification, and training of personnel for stationery nuclear power plants. Qualifications, responsibilities, and training of personnel in operating and support organizations appropriate for the safe and efficient operation of nuclear power plants are addressed.

ANS: 4.5-1986, Criteria for Accident Monitoring Functions in Light-Water-Cooled Reactors, 1986.

Criteria are provided for determining the variable to be monitored by the control room operator of a light water reactor, as required for safety, during the course of an accident, including long-term stable shutdown. Also included are criteria for determining the requirements for the equipment used to monitor those variables.

ANS: 6.6.1-1987, Calculation and Measurement of Direct and Scattered Gamma Radiation from LWR Nuclear Power Plants, 1987.

This standard defines calculational requirements and discusses measurement techniques for estimates of dose rates near nuclear power plants due to direct and scattered gamma-rays from contained sources on site. It describes the considerations necessary to computer dose rates, including component self-shielding, shielding afforded by walls and structures, and scattered radiation.

ANS: 8.3-1986, Criticality Accident Alarm System, (Revision of ANSI/ANS: 8.3-1979), 1986.

This standard applies to all operations with plutonium, 233U, 235U, and other fissionable materials in which inadvertent criticality may occur and cause the exposure of personnel to unacceptable amounts of radiation. It is directed principally toward gamma-radiation rate-sensing systems. This revision incorporates relevant features of ANSI/ANS-8.3-1979 and ANS-N2.3-1979.

ANS: 10.3-1986, Guidelines for the Documentation of Digital Computer Programs (Revision of ANSI/N413-1974), 1986.

This standard presents guidelines for the documentation of digital computer programs prepared for scientific and engineering computations. The guidelines are designed to facilitate effective usage, transfer, conversion, and modification of computer programs.

ANS: 10.4-1987, Guidelines for the Verification and Validation of Scientific and Engineering Computer Programs for the Nuclear Industry

This provides guidelines for the verification and validation (V&V) of scientific and engineering computer programs developed for use by the nuclear industry.

ANS: 10.5-1986, Guidelines for Considering User Needs in Computer Program Development (Revision of ANSI/ANS-10.5-1979)

This standard provides guidelines for accommodating user needs in preparing computer programs for scientific and engineering applications. These guidelines will help ensure proper application and simplify the use of the computer programs as well as encourage development of a product that will be easily and correctly applied.

ANS: 59.3-1984, Safety Criteria for Control Air Systems, 1984.

This standard provides nuclear safety criteria for the control air system that furnishes compressed air to safety-related components in nuclear power plants.

It applies only to the air supply system and does not apply to air-operated devices.

AMERICAN PETROLEUM INSTITUTE (API)

The following API Publications, Recommended Practices, Standards, and Bulletins are available from the American Petroleum Institute, 1220 L Street, NW, Washington, DC 20005. (202) 682-8000

API: Bull 5T1, Bulletin on Nondestructive Testing Terminology, Seventh Edition, 1985.

Provides definitions in English, French, German, Italian, Japanese, and Spanish for a number of defects which commonly occur in steel pipe.

API: Spec. 6D, Specifications for Pipeline Valves, End Closures, Connectors and Swivels, Eighteenth Edition, 1982.

Covers materials, dimensions, and pressure ratings for flanged and welding-end pipeline valves.

Guide for Inspection of Refinery Equipment:
Chap. IV, Inspection Tools, Third Edition, 1983.

This chapter describes and illustrates both homemade and purchasable refinery inspection tools, such as hammer, calipers, gages, electronic and mechanical devices, and special tools. It also covers the use of tools to solve specific inspection problems.

Chap. XV, Instruments and Control Equipment, Third Edition, 1981.

This chapter is a guide for instrument inspection.

Titles & Abstracts

The inspection procedures given cover the majority of types of standard commercial instruments and associated equipment used in modern refineries.

API: RP 500A, Classification of Areas for Electrical Installations in Petroleum Refineries, Fourth Edition, 1982.
A guide that applies to refinery areas when flammable vapors and liquids are processed, stored, loaded, unloaded, or otherwise handled. It is intended to serve as a supplement to the **National Electrical Code.** 19 pages.

API: RP 500B, Recommended Practice for Classification of Areas for Electrical Installations at Drilling Rigs and Production Facilities on Land and on Marine Fixed and Mobile Platforms, Second Edition, 1973.
Classifies areas surrounding drilling rigs and production facilities on land and on marine fixed and mobile platforms for the safe installation of electrical equipment. 14 pages.

API: RP 500C, Classification of Areas for Electrical Installation, Petroleum, Second Edition, 1984.
Classified areas include pump stations, compressor stations, storage facilities, loading racks, and manifold and pipeline right-of-way areas where flammable liquids and gases are handled.

API: RP 520, Design and Installation of Pressure-Relieving Systems in Refineries, Parts I and II.
Part I — Design, Fourth Edition, 1976.
Applies to relieving devices and their discharge systems on refinery pressure vessels and equipment designed for maximum allowable working pressure of more than 15 psig.
Part II — Installation, Second Edition, 1963, (Reaffirmed 1973).
Applies to installation of pressure relief valves in gas, vapor, and liquid service. It includes information on inlet and discharge piping, valve location and position, valve setting, and handling and testing.

API: RP 521, Guide for Pressure Relief and Depressuring Systems, Second Edition, 1982.
Supplements the material set forth in API RP 520, Parts I and II. Guidelines are provided for examining the principal causes of over-pressure; for determining individual relieving rates; and for selecting and designing disposal systems.

API: Std. 526, (ANSI/API Std. 526-1984), Flanged Steel Safety Valves, Third Edition, 1984.
Specifies dimensions of carbon and alloy steel safety relief valves for the purpose of promoting interchangeability. Basic requirements are given for orifice designation and area, valve size and rating, materials, pressure-temperature limits, and center-to-face dimensions.

API: Std. 527, (ANSI/API Std. 527-1978), Commercial Seat Tightness of Safety Relief Valves with Metal-to-Metal Seats, Second Edition, 1978.
Describes a method of determining seat tightness of safety relief valves as covered in API Std. 526.

API: RP 550, Manual on Installation of Refinery Instruments and Control Systems, Parts I, II, III, and IV.

Part I — Process Instrumentation and Control.

Section 1 — Flow, Third Edition, 1977
Section 2 — Level, Fourth Edition, 1980 (Reaffirmed 1983)
Section 3 — Temperature, Fourth Edition, 1985
Section 4 — Pressure, Fourth Edition, 1980 (Reaffirmed 1983)
Section 5 — Controllers and Control Systems, Fourth Edition, 1985
Section 6 — Control Valves and Accessories, Fourth Edition, 1985
Section 7 — Transmission Systems, Third Edition, 1974
Section 8 — Seals, Purges, and Winterizing, Fourth Edition, 1980
Section 9 — Air Supply Systems, Fourth Edition, 1980
Section 10 — Hydraulic Systems, Fourth Edition, 1981
Section 11 — Electrical Power Supply, Third Edition, 1981
Section 12 — Control Centers, Third Edition, 1977
Section 13 — Alarms and Protective Devices, Fourth Edition, 1985
Section 14 — Process Computer Systems, First Edition, 1982

Part II — Process Stream Analyzers

Section 1 — Analyzers, Fourth Edition, 1985
Section 2 — Process Chromatographs, Fourth Edition, 1981
Section 4 — Moisture Analyzers, Fourth Edition, 1983
Section 5 — Oxygen Analyzers, Fourth Edition, 1983
Section 6 — Analyzers for the Measurement of Sulfur and Its Compounds, Fourth Edition, 1983
Section 7 — Electrochemical Liquid Analyzers, Fourth Edition, 1984
Section 9 — Water Quality Analyzers, Fourth Edition, 1984

Titles & Abstracts

Section 10 — Area Safety Monitors, Fourth Edition, 1983

Part III — Fired Heaters and Inert Gas Generators, Third Edition, 1985

Part IV — Steam Generators, Second Edition, 1984

API: Std. 598, Valve Inspection and Test, Fifth Edition, 1982.
Covers inspection and pressure test requirements for both resilient seated and metal-to-metal seated valves of the gate, globe, plug, check, ball, and butterfly types.

API: Std. 612, Special-Purpose Steam Turbines for Refinery Services, Third Edition, November 1987
Covers minimum requirements for special-purpose steam turbines for refinery services. Special-purpose steam turbines are those horizontal turbines used to drive equipment that is usually not spared, is relatively large in size (power), or is in critical service. This category is not limited by steam conditioners or turbine speed. These requirements include basic design, materials, and related lube-oil systems, controls, and auxiliary equipment.

API: Std. 617, Centrifugal Compressors for General Refinery Services, Fifth Edition, November 1988
This standard covers the minimum requirements for centrifugal compressors used in refinery services handling air or gas. It does not cover machines that may be classed as fans or blowers that develop less than 5 psig (0.34 bar effective) from atmospheric pressure, or packaged integrally geared centrifugal air compressors. 52 pages. Also see, 1984 Interpretations — Technical Publications.

API: Std. 618, Reciprocating Compressors for General Refinery Services, Third Edition, February 1986
This standard covers the minimum requirements for moderate- to low-speed reciprocating compressors used to handle process air or gas in refinery services. Both lubricated- and nonlubricated-cylinder designs are covered. Also covered are related lubricating systems, controls, instrumentation, intercoolers, aftercoolers, pulsation suppression devices and other auxiliary equipment. 111 pages.

API: Std. 619, Rotary-Type Positive Displacement Compressors for General Refinery Services, Second Edition, May 1985
This standard is a purchase specification for helical, spiral, and straight lobe compressors used for vacuum or pressure or both in refinery services. It is primarily intended for compressors that are in continuous duty on process units and generally unspared. It does not cover portable air compressors, liquid ring compressors, vane-type compressors, or compressors in oxygen-bearing gas service using flammable liquid for injection or flooding. 64 pages. Also see, 1985 Interpretations — Technical Publications.

API: Std. 670, Vibration, Axial-Position, and Bearing-Temperature Monitoring Systems, Second Edition, June 1986
This standard covers the minimum requirements for monitoring machinery shaft vibration, shaft axial position, and bearing temperature. It outlines a standardized system for monitoring these items and covers requirements, hardware, installation, calibration, and arrangement. This standard does not apply to accelerometer-based vibration monitoring systems, which are covered by Standard 678. 39 pages.

API: Std. 672, Packages, Integrally Geared, Centrifugal Plant and Instrument Air Compressors for General Refinery Services, Second Edition, 1988.
This standard establishes the minimum requirements for constant-speed, packaged, integrally geared centrifugal plant and instrument air compressors including their drivers and auxiliaries. 33 pages.

API: Std. 675, Positive Displacement Pumps — Controlled Volume, First Edition, March 1980, Reaffirmed December 1987
This standard covers the minimum requirements for controlled-volume positive displacement pumps for use in refinery service. Both packed-plunger and diaphragm types are included. Diaphragm pumps that use direct mechanical actuation are excluded. 21 pages.

API: Std. 676, Positive Displacement Pumps — Rotary, First Edition, September 1980, Reaffirmed December 1987
This standard covers the minimum requirements for rotary positive displacement pumps used in refinery services. Included are pumps consisting of a casting containing gears, screws, lobes, cams, vanes, plungers, or similar elements actuated by relative motion between the drive shaft and the casing. 23 pages.

API: Std. 678, Accelerometer-Based Vibration Monitoring System, First Edition, May 1981, Reaffirmed December 1987
This specification covers the minimum requirements for machinery monitoring systems using

piezoelectric vibratory acceleration transducers. It outlines a standardized system covering requirements for and the installation and arrangement of hardware (sensors and instruments). This system may be used separately or in conjunction with noncontacting, eddy current proximity devices as described in API Standard 670. 25 pages. Also see, 1984 Interpretations — Technical Publications.

API: Std. 680, Packaged Reciprocating Plant and Instrument Air Compressors for General Refinery Services, First Edition, October 1987

This standard covers the minimum requirements for double-acting reciprocating compressers of 600 brake horsepower (448 kilowatts) and below used in refinery services with lubricated or non-lubricated cylinders in utility air and/or instrument air service up to 125 psig. Driver, related lubricating systems, controls, instrumentation, intercooler, aftercooler, receiver, and other auxiliary equipment are covered. 80 pages.

Bull. 2509B, Shop Testing of Automatic Liquid-Level Gages, 1961.

This publication is not included in the manual. 49 pages.

API: Manual of Petroleum Measurement Standards (Complete Set)

This manual, which includes all subject matter found in API measurement publications, is an ongoing project.

Chapter 4, Proving Systems, First Edition, 1978 (ANSI/API MPMS 4-1978)

Serves as a guide for the design, installation, calibration, and operation of meter proving systems.

Chapter 5, Metering

Covers the dynamic measurement of liquid hydrocarbons, or metering. It is divided into subchapters.

Chapter 5.2, Measurement of Liquid Hydrocarbons by Displacement Meter Systems, First Edition, 1977. (ANSI/API MPMS 5.2-1977)

Describes and illustrates methods and practices that may be used to obtain optimum measurement of liquid hydrocarbons and maximum service life when using displacement meters.

Chapter 5.3, Turbine Meters, First Edition, 1976.

Specifies the characteristics of turbine meters and gives rules for applying appropriate considerations to the nature of the liquids to be measured. It also covers the installation of metering systems that use a turbine meter, and their performance, operation, and maintenance in liquid hydrocarbon service.

Chapter 5.4, Instrumentation or Accessory Equipment for Liquid Hydrocarbon Metering Systems, First Edition, 1976. (ANSI/API MPMS 5.4-1976)

Specifies the characteristics of available and necessary equipment that can be used to attain desired purposes when used in conjunction with volumetric hydrocarbon meters.

Chapter 5.5, Fidelity and Security of Flow Measurement Pulsed-Data Transmission Systems, First Edition, June 1982.

Provides a guide to the selection, operation, and maintenance of pulsed-data, cabled transmission systems for fluid metering systems to provide the desired level of fidelity and security of transmitted data.

Chapter 6, Metering Assemblies

Discusses the design, installation and operation of metering systems for coping with special situations in hydrocarbon measurement. Portions of Chapter 6 are in preparation.

Chapter 7, Temperature Determination

Covered the sampling, reading, averaging, and rounding of the temperature of liquid hydrocarbons in both the static and dynamic modes of measurement for volumetric purposes. Portions of Chapter 7 are in preparation.

Chapter 8, Sampling

Covers standardized procedures for sampling crude oil or its products.

Chapter 9, Density Determination

Describes the standard methods and apparatus used to determine the specific gravity of crude petroleum products normally handled as liquids. It is divided into subchapters as follows.

Chapter 10, Sediment and Water

Describes methods for determining the amount of sediment and water, either together or separately. Laboratory and field methods are covered as follows.

Chapter 12, Calculation of Petroleum Quantities

Describes the standard procedures for calculating net standard volumes, including the application of correction factors and the importance of significant figures. The purpose of standardizing the calculation procedure is to achieve the same result regardless of what person or computer does the calculating.

Chapter 13, Statistical Aspects of Measuring and Sampling

Covers the application of statistical methods to

Titles & Abstracts

petroleum measurement and sampling. Chapter 13 is in preparation.

Chapter 14, Natural Gas Fluids Measurement
Standardizes practices for measuring, sampling, and testing natural gas fluids. Chapter 14 is in preparation.

Chapter 16, Measurement of Petroleum by Weight
Provides references to model regulations promulgated by NCWM regarding commercial weighing, tolerances, and other technical requirements and to the recognized practices of the petroleum industry when products are handled on a weight basis. Chapter 16 is in preparation.

AMERICAN SOCIETY FOR QUALITY CONTROL (ASQC)

The following ASQC and ANSI standards are available from the American Society of Quality Control, 310 West Wisconsin Avenue, Milwaukee, WI 53203. (414) 272-8575

ANSI/ASQC: A1-1978, Definitions, Symbols, Formulas and Tables for Control Charts.
A standardization of the symbols, concepts, terms, and procedures relation to Shewhart control charts, control charts with warning limits, moving averages and ranges, exponentially smoothed averages and ranges, exponentially smoothed averages, Cusum charts, multi-variate control, trend control, and the acceptance control chart.

ANSI/ASQC: A2-1978, Terms, Symbols and Definitions for Acceptance Sampling.
Covers the major forms of acceptance sampling schemes for both attributes and variables measures, including extensive comments, explanations, and comparison of the various acceptance sampling approaches.

ANSI/ASQC: A3-1978, Quality Systems Terminology.
Presents basic definitions dealing with quality assurance, quality control, quality programs, and quality systems for general use within U.S. commerce and industry.

ANSI/ASQC: C1-1985 (ANSI Z1.8-1971), Specifications of General Requirements for a Quality Program.
This standard concerns the establishment and maintenance of a quality program by a contractor to assure compliance requirements in the areas of quality management, design information, procurement, manufacture, acceptance, and documentation.

ANSI/ASQC: Z1.15-1979, Generic Guidelines for Quality Systems.
Describes the significant elements that should be considered in the quality system of a manufactured product, including quality policy, design assurance, purchased materials control, quality control at various production stages, field performance, and product liability.

ANSI/ASQC: Z1.4-1980, Sampling Procedures and Tables for Inspection by Attributes.
This standard, which corresponds to MIL-STD-105D, establishes sampling plans and procedures for inspection by attributes. Its tables and procedures are completely compatible with MIL-STD-105D. It is also compatible and interchangeable with ANSI/ASQC Z1.9-1980 for variables inspection.

ANSI/ASQC: Z1.9-1980, Sampling Procedures and Tables for Inspection by Variables for Percent Nonconforming.
This standard, establishing sampling plans and procedures for inspection by variables, corresponds to the military standard MIL-STD-414 and is interchangeable with ISO/DIS3951. It contains tables and procedures of MIL-STD-414, suitably modified to achieve correspondence with ISO/DIS 3951 and matching with MIL-STD-105D and ANSI/ASQC Z1.4-1980.

ANSI/IEEE: 730-1981, Software Quality Assurance Plans.
This standard assists in the preparation of quality assurance plans for the development and maintenance of critical software. It provides developers, users, and the public with criteria against which such plans can be prepared and assessed.

ANSI/ASQC: E2-1984, Guide to Inspection Planning.
This standard describes the significant elements that should be considered in the development of inspection activities. It provides generic guidelines for planning and applying a product/process inspection system for construction, manufacturing, operating, or service functions.

ANSI/ASQC B1, B2 and B3-1985 (ANSI Z1.1-1958, Z1.2-1958 and Z1.3-1958 Revised 1975) Guide for Quality Control, Control Chart Method of Analyzing Data, and Control Chart Method of Controlling Quality During Production.
Contains three standards; a guide for handling problems concerning the economic control of quality of materials and manufactured products with

particular reference to methods of collecting, arranging, and analyzing inspection and test records to detect lack of uniformity of quality; a guide to the control chart method of analyzing a collection of data, with particular reference to quality data resulting from inspections and tests of materials and manufactured products; and a guide outlining the control chart method of identifying and eliminating causes of trouble in repetitive production processes, in order to reduce variations in the quality of manufactured product and materials.

ANSI/ASQC Q1-1986, Generic Guidelines for Auditing of Quality Systems.

Describes the significant elements that should be considered in the planning and execution of audits. The standard is intended to provide generic guidelines for internal and external Audits of Quality Systems. Not intended to be applied in all specific situations, but is to be used in developing and describing criteria for effective and efficient auditing.

ANSI/ASQC S1-1987, An Attribute Skip-Lot Sampling Program.

Provides a procedure for reducing the inspection effort on products submitted by those suppliers who have demonstrated their ability to control, in an effective manner, all facets of product quality and consistently produce superior quality material. Shall not be applied to the inspection of product characteristics which involve the safety of personnel.

AMERICAN SOCIETY FOR TESTING AND MATERIALS (ASTM)

The following ASTM Standards are available from the American Society for Testing and Materials, 1916 Race Street, Philadelphia, PA 19103. (215) 299-5400. Standards are available from either organization.

ASTM: Index to Standards; The 1985 Annual Book of ASTM Standards, Volume 00.01, 1988.

This book is a key reference volume. It lists by number designation, and cross-indexed title every one of over 8500 ASTM Standards. The index refers the reader to the exact volume of the 67-volume Annual Book of ASTM Standards where a particular standard may be found.

ASTM: A 105/A 105M-87a, Standard Specification for Forgings, Carbon Steel, for Piping Components, 7 pp.

Included are flanges, fittings, valves, and similar parts to specified dimensions or to dimensional standards such as ANSI and API specifications.

ASTM: A 181/A 181M-87, Standard Specification for Forgings, Carbon Steel for General Purpose Piping, 5 pp.

Two Grades of material are covered, designated as grades I and II, respectively, and are classified in accordance with their chemical and physical properties.

ASTM: A 182/A 182M-88, Standard Specification for Forged or Rolled Alloy-Steel Pipe Flanges, Forged Fittings, and Valves and Parts for High-Temperature Service, 14 pp.

Twenty-five grades are covered including eleven ferritic steels and fourteen austenitic steels. Selection will depend upon design and service conditions, mechanical properties, and the high-temperature characteristics.

ASTM: A 522/A 522M-87, Standard Specification for Forged or Rolled 8 and 9 Percent Nickel Alloy Steel Flanges, Fittings, Valves, and Parts for Low-Temperature Service, 5 pp.

The specification is applicable to forgings with maximum section thickness of 33 in. (76.2 mm) in the double normalized and tempered condition and 5 in. (127.0 mm) in the quenched and tempered condition. Forgings to this specification are intended for service at operating temperatures not lower than -320 F (-196 C) or higher than 250 F (121 C).

ASTM: B 106-84, Standard Test Methods for Flexivity of Thermostat Metals, 8 pp.

Intended for determining the flexure-temperature characteristics of thermostat metals tested in the form of flat strips 0.012 in. or over in thickness and in the form of spiral coils less than 0.012 in. in thickness.

ASTM: B 223-85, Standard Test Method for Modulus of Elasticity of Thermostat Metals (Cantilever Beam Method), 7 pp.

Covers the procedures for determining the modulus of elasticity of thermostat metals at any temperature between -300 and -1000°F by mounting the specimen as a cantilever beam and measuring the deflection when subjected to a mechanical load.

ASTM: B 244-79, Standard Method for Measurement of Thickness of Anodic Coatings on Aluminum and of Other Nonconductive Coatings on Nonmagnetic Basis Metals with Eddy-Current Instruments, 1 pp.

Describes procedures for measuring by nondestructive means the thickness of anodic coatings on aluminum using eddy-current instruments.

ASTM-B305-56 (1978), Standard Test Method for Maximum Loading Stress at Temperature of

Titles & Abstracts

Thermostat Metals (Cantilever Beam Method), 5 pp.

Intended for the evaluation of the maximum stress at temperatures that can be applied to thermostat metals by a static or dead load before the combination of thermal and mechanical stresses causes displacement of the metal in excess of the elastic limit of the material. In this test the specimen is mounted as a cantilever beam.

ASTM: B 362-86, Standard Test Method for Mechanical Torque Rate of Spiral Coils of Thermostat Metal, 3 pp.

Covers the principles of determining the mechanical torque rate of spiral coils of thermostat metal.

ASTM: B 388-86, Standard Specification for Thermostat Metal Sheet and Strip, 1974, 9 pp.

Covers thermostat metals in the form of sheet or strip which are used for the temperature-sensitive elements of devices for controlling, compensating, or indicating temperature and is intended to supply acceptance requirements to purchasers ordering this material by type designation.

ASTM: B 389-81 (1986), Standard Test Method for Thermal Deflection Rate of Spiral and Helical Coils of Thermostat Metal, 6 pp.

Covers a procedure for determining thermal deflection rate of spiral and helical coils for thermostat metal.

ASTM: B 430-84, Standard Test Method for Particle Size Distribution of Refractory Metal-Type Powders by Turbidimetry. 6 pp.

Covers the procedure for the determination of particle size distribution of refractory metal powder with a turbidimeter.

ASTM: B 462-87, Standard Specification for Forged or Rolled Chromium-Nickel-Iron-Molybdenum-Copper-Columbium Stabilized Alloy (UNS N08020) Pipe Flanges, Forged Fittings, and Valves and Parts for Corrosive High-Temperature Service, 1970, 4 pp.

ASTM: C 115-86, Standard Test Method for Fineness of Portland Cement by the Turbidimeter, 10 pp.

Describes the Wagner turbidimeter apparatus and procedure for determining the fineness of portland cement. Includes details for construction, calibration, and application of the turbidimeter.

ASTM: C 518-85, Standard Test Method for Steady-State Heat Flux Measurements and Thermal Transmission Properties by Means of the Heat Flow Meter Apparatus, 32 pp.

Covers the determination of, by means of a heat flow meter, the thermal conductivity of homogeneous insulating, building, and other materials whose thermal conductivities do not exceed 2.0 Btu/h; $\times$ ft $2 \times °F$ (1.13 mW/cm2x°C).

ASTM: C 604-86, Standard Test Method for True Specific Gravity of Refractory Materials by Gas-Comparison Pycnometer, 3 pp.

Covers the determination of the true specific gravity of solid materials, and is particularly useful for easily hydrateable materials which are not suitable for test with ASTM Method C 135, Test for True Specific Gravity of Refractory Materials.

ASTM: D 240-87, Standard Test Method for Heat of Combustion of Liquid Hydrocarbon Fuels by Bomb Calorimeter, 10 pp.

Describes procedures for determining heat of combustion. It is applicable to a variety of substances but particularly to liquid hydrocarbon fuels of both low and high volatility.

ASTM: D 287-87 - (API: STD 2544), Standard Test Method for API Gravity of Crude Petroleum and Petroleum Products (Hydrometer Method), 4 pp.

Covers the determination by means of a glass hydrometer of the API gravity of crude petroleum and petroleum products normally handled as liquids and having a Reid vapor pressure of 26 lb. or less.

ASTM: D 941-83, Standard Test Method for Density and Relative Density (Specific Gravity) of Liquids by Lipkin Bicapillary Pycnometer, 6 pp.

Intended for the measurement of the density of any hydrocarbon material that can be handled in a normal fashion as a liquid at the specified test temperatures of 20 and 25°C. Its application is restricted to liquids having vapor pressures less than 600 mm Hg (approximately 0.8 atm) and having viscosities less than 15 cSt at 20°C.

ASTM: D 1071-83, Standard Method for Measurement of Gaseous Fuel Samples, 19 pp.

Are applicable to the measuring of gaseous fuel samples, including, liquefied petroleum gases, in the gaseous state at normal temperature and pressures.

ASTM: D 1085-65 (1984) - (API: STD 2545), Standard Method of Gaging Petroleum and Petroleum Products, 39 pp. Published as reprint only.

Describes the procedure for gaging crude petroleum and its liquid products in various types of tanks, containers, and carriers.

ASTM: D 1142-86, Standard Test Method for Water Vapor Content of Gaseous Fuels by Measurement of Dew-Point Temperature. 13 pp.

Covers the determination of the water vapor content of gaseous fuels by measurement of the dewpoint temperature and the calculation therefrom of the water vapor content.

ASTM: D 1145-80, Standard Method of Sampling Natural Gas, 14 pp.

Covers the procedures for the sampling of (1) natural gases containing primarily hydrocarbons and nitrogen, (2) natural gases containing hydrogen, (3) natural gases containing hydrogen sulfide, or organic sulfur compounds, or other sulfur contaminants, (4) natural gas containing carbon dioxide, (5) natural gas containing gasoline and condensables.

ASTM: D 1186-87, Standard Method for Nondestructive Measurement of Dry Film Thickness of Nonmagnetic Coatings Applied to a Ferrous Base, 5 pp.

Covers the measurement of film thickness of nonmagnetic dried films of paint, varnish, lacquer, and related products applied over a magnetic base material.

ASTM: D 1217-86, Standard Test Method for Density and Relative Density (Specific Gravity) of Liquids by Bingham Pycnometer, 8 pp.

Intended for the measurement of the density of pure hydrocarbons or petroleum distillates boiling between 194 and 230°F (90 and 110°C) that can be handled in a normal fashion as a liquid at the specified test temperatures of 68 and 77°F (20 and 25°C). The method was developed especially for the reference fuels n-heptane and iso-octane and is designed to provide values having an accuracy of 0.00003 g/mL.

ASTM: D 1238-88, Standard Test Method for Measuring Flow Rates of Thermoplastics by Extrusion Plastometer, 13 pp.

Covers measurement of the rate of extrusion of molten resins through an orifice of a specified length and diameter under prescribed conditions of temperature and pressure.

ASTM: D 1247-80, Standard Method of Sampling Manufactured Gas, 11 pp.

Covers the procedures for securing representative samples of manufactured gas, and correlates the size or type of sample with the analysis to be done subsequently on that sample.

ASTM: D 1298-85, Standard Test Method for Density, Relative Density, (Specific Gravity), or API Gravity of Crude Petroleum and Liquid Petroleum Products by Hydrometer Method, 7 pp.

Covers the laboratory determination, using a glass hydrometer, of the density, specific gravity, or API gravity of crude petroleum, petroleum products, or mixtures of petroleum and non-petroleum products normally handled as liquids, and having a Reid vapor pressure [ASTM Method D 323, Test for Vapor Pressure of Petroleum Products (Reid Method) or IP 69] of 26 lb. or less.

ASTM: D 1356-73(1979), Standard Definitions of Terms Relating to Atmospheric Sampling and Analysis, 6 pp.

Includes definitions relating to atmospheric conditions, sampling devices, and methods of analysis.

ASTM: D 1408-65 (1984), Standard Methods for Measurement and Calibration of Spherical and Spheroidal Tanks, 1 pg. Available as separate reprint only.

Describes the procedures for calibrating spherical and spheroidal tanks which are used as liquid containers.

ASTM: D 1410-65 (1984), Standard Method for Measurement and Calibration of Stationary Horizontal Tanks, 1 pg. Available as separate reprint only.

Describes external measurement procedures for calibrating horizontal aboveground stationary tanks larger than a barrel or drum.

ASTM: D 1480-86, Standard Test Method for Density and Relative Density (Specific Gravity) of Viscous Materials by Bingham Pycnometer, 9 pp.

Describes two procedures for the measurement of the density of materials which are fluid at the desired test temperature. Its application is restricted to liquids of vapor pressures below 600 mm Hg and viscosities below about 400 cSt at the test temperature.

ASTM: D 1481-86, Standard Test Method for Density and Relative Density (Specific Gravity) of Viscous Materials by Lipkin Bicapillary Pycnometer, 9 pp.

Intended for determining the density of oils more viscous than 15 cSt at 20°C (68°F), and of viscous oils and melted waxes at elevated temperatures, but not at temperatures at which the sample would have a vapor pressure of 100 mm Hg or above.

ASTM: D 1605-60 (1979), Standard Recommended Practices for Sampling Atmospheres for Analysis of Gases and Vapors, 22 pp.

Covers two types of sampling methods for the sampling of atmospheres for analysis of gases and vapors. Includes a description of procedures, apparatus, and methods for the determination of performance.

ASTM: D 1657-88, Standard Test Method for

Titles & Abstracts

Density or Relative Density of Light Hydrocarbons by Pressure Hydrometer, 5 pp.

Covers a procedure for determining the specific gravity 60/60°F of light hydrocarbons including liquefied petroleum hydrocarbons, LPG, and butadiene.

ASTM: D 1826-88, Standard Test Method for Calorific Value of Gases in Natural Gas Range by Continuous Recording Calorimeter, 15 pp.

Covers a procedure for determining with the continuous recording calorimeter the total calorific value of fuel gas produced or sold in the natural gas range of 900 to 1200 Btu per standard cubic foot. Includes definitions, description of apparatus, methods of installation, and operation of the calorimeter.

ASTM: D 1890-81, Standard Test Method for Beta Particle Radioactivity of Water, 7 pp.

Covers the measurement of beta particle activity of water by means of several types of instruments composed of a detecting device and combined amplifier, power supply, and scaler.

ASTM: D 1943-81, Standard Test Method for Alpha Particle Radioactivity of Water, 5 pp.

Covers the measurement of alpha particle activity of water. It is applicable to alpha emitters having energies above 3.9 MeV and at activity levels above 9.5 pCi/mL of radioactivity homogeneous water. The method is not applicable to samples containing alpha-emitting radio-elements that are volatile under conditions of the analysis.

ASTM: D 1945-81, Standard Method for Analysis of Natural Gas by Gas Chromatography, 15 pp.

Describes the method determination of the complete chemical composition of reformed gases and similar gaseous mixtures containing the following components: hydrogen, oxygen, nitrogen, carbon monoxide, carbon dioxide, methane, ethane, and ethylene. Includes definitions, measurement variables, description of apparatus, sampling, standardization and calibration, and calculations.

ASTM: D 1946-82, Standard Method for Analysis of Reformed Gas by Gas Chromatography, 6 pp.

Describes the method determination of the complete chemical composition of reformed gases and similar gaseous mixtures containing the following components: hydrogen, oxygen, nitrogen, carbon monoxide, carbon dioxide, methane, ethane, and ethylene. Includes definitions, measurement variables, description of apparatus, sampling, standardization and calibration, and calculations.

ASTM: D 2009-65 (1979), Recommended Practice for Collection by Filtration and Determination of Mass, Number, and Optical Sizing of Atmospheric Particulates, 8 pp.

Covers the collection and measurement of mass particle size and particle size distribution of atmospheric material. Applies to both solid and liquid particles.

ASTM: D 2124-70 (1984), Standard Method for Analysis of Components in Poly (Vinyl Chloride) Compounds Using an Infrared Spectrophotometric Technique, 7 pp.

Provides for the infrared identification of resins, plasticizers, stabilizers, and fillers in poly (vinyl chloride) (PVC) compounds.

ASTM: D 2162-86, Standard Method for Basic Calibration of Master Viscometers and Viscosity Oil Standards, 9 pp.

Covers procedures for calibrating master viscometer and viscosity oil standards both of which may be used to calibrate routine viscometers as described in ASTM Method D 445, Test for Kinematic Viscosity of Transparent and Opaque Liquids (and the Calculation of Dynamic Viscosity).

ASTM: D 2163-87, Standard Method for Analysis of Liquefied Petroleum (LP) Gases and Propane Concentrates by Gas Chromatography, 6 pp.

Covers the determination of the composition of liquefied petroleum (LP) gases. It is applicable to analysis of propane, propylene, and butane in all concentration ranges 0.1% and above.

ASTM: D 2168-80, Standard Methods for Calibration of Laboratory Mechanical-Rammer Soil Compactors, 7 pp.

This method of calibration is intended for use only in the calibration of mechanical compactors equipped with rammers striking directly the surface of the soil.

ASTM: D 2186-84, Standard Test Methods for Deposit-Forming Impurities in Steam, 9 pp.

Covers the determination of the amount of deposit-forming impurities in steam by the evaporative electrical conductivity and sodium tracer methods. Special techniques for silica and certain metal oxides are also presented.

ASTM: D2389-83, Standard Test Method for Minimum Pressure for Vapor Phase Ignition of Monopropellants, 15 pp.

Covers the determination of the minimum pressure for the ignition of monopropellants in the vapor phase. This measure can lead to the determination of minimum ignition energy.

ASTM: D2597-88, Standard Method for Analysis of Natural Gas-Liquid Mixtures by Gas Chromatography, 7 pp.

Titles & Abstracts

Covers the analysis of wide-range natural gas-liquid (NGL) mixtures, such as commercial de-ethanized and depropanized natural gasoline mixtures, that cannot readily be entered into the chromatograph as a liquid by syringe or as a vapor at atmospheric pressure because of both highly volatile and heavy-end components.

ASTM: D 2600-87, Standard Test Method for Aromatic Traces in Light Saturated Hydrocarbons by Gas Chromatography, 11 pp.

Covers the determination of benzene, toluene, and C8 aromatics in light saturate hydrocarbon samples. The method is limited by aromatic selectivity of the stationary liquid to samples containing n-decane as the highest boiling compound.

ASTM: D2650-83, Standard Test Method for Chemical Composition of Gases by Mass Spectrometry, 10 pp.

Covers the quantitative analysis of gases containing specific combinations of the following components; hydrogen, hydrocarbons with up to six carbon atoms per molecule; carbon monoxide; carbon dioxide, mercaptans with one or two carbon atoms per molecule; hydrogen sulfide; and air (nitrogen, oxygen, and argon).

ASTM: E1-88, Standard Specification for ASTM Thermometers, 61 pp.

Covers specifications for 184 etched-stem liquid-in-glass thermometers graduated in Celsius or Fahrenheit degrees.

ASTM: E70-77 (1986), Standard Test Method for pH of Aqueous Solutions with the Glass Electrode, 7 pp.

Covers the definition of pH and the apparatus and procedures for the electrometric measurement of pH values of aqueous solutions or extracts with the glass electrode.

ASTM: E74-83, Standard Practice of Calibration of Force Measuring Instruments for Verifying the Load Indication of Testing Machines, 9 pp.

Covers procedures for the verification of calibration devices suitable for calibrating testing machines in accordance with the requirements of the Methods of Load Verification of Testing Machines (ASTM Designation: E4).

ASTM: E77-84, Standard Method for Verification and Calibration of Liquid-In-Glass Thermometers, 18 pp.

Describes the principles, apparatus, and procedures, for visual and dimensional insection, test for permanency of pigment, test for bulb stability, and test for scale accuracy to be used in the verification and calibration of etched-stem liquid-in-glass thermometers.

ASTM: E83-85, Standard Practice for Verification and Classification of Extensometers, 8 pp.

Covers procedures for the verification and classification of extensometers. The method applies only to instruments that indicate or record values which are proportional to changes in length. Extensometers are classifed on the basis of the magnitude of their errors.

ASTM: E94-88a, Standard Guide for Radiographic Testing, 13 pp.

Provides a guide for satisfactory radiographic testing. Statements about preferred practices are given without discussion of the technical reasons leading to the preference.

ASTM: E116-81 (1986), Standard Practice for Photographic Photometry in Spectrochemical Analysis, 31 pp.

Provides a practical guide to the preparation and use of emulsion calibration curves for determining spectral line intensity ratios in spectrochemical analysis.

ASTM: E131-84, Standard Definitions of Terms and Symbols Relating to Molecular Spectroscopy, 9 pp.

ASTM: E135-87, Standard Definitions of Terms and Symbols Relating to Emission Spectroscopy, 7 pp.

ASTM: E137-(82) 87, Standard Practice for Evaluation of Mass Spectrometers for Chemical Analysis, 4 pp.

Provides means for evaluation of the suitability of mass spectrometers for use in ASTM mass spectrometric methods of chemical analysis. Also includes discussion of tests that are generally helpful in evaluating the performance of a particular mass spectrometer as used in a particular ASTM method of analysis.

ASTM: E168-67 (1977), Standard Recommended Practices for General Techniques of Infrared Quantitative Analysis, 9 pp.

Provides general information on the various techniques most often used in infrared quantitative analysis. Includes definitions and symbols, theory, apparatus, calculation methods, and special techniques.

ASTM: E170-84b, Terminology Relating to Radiation Measurements and Dosimetry, 9 pp.

Wherever possible, these definitions are the same as, or similar to, those recommended by the International Commission on Radiological Units and Measurements (ICRU) as presented in the National Bureau of Standards Handbook 62.

Titles & Abstracts

ASTM: E172-85, Standard Practice for Describing and Specifying the Excitation Source in Emission Spectrochemical Analysis, 5 pp.
Provides general recommendations for the description of the various types of sources used in spectrographic analysis and for the specification of source parameters.

ASTM: E177-86, Standard Practice for Use of the Terms Precision and Bias in ASTM Test Methods, 16 pp.
The purpose of this recommended practice is to outline some general concepts regarding the terms "precision" and "accuracy", to provide some standard usages for ASTM committees in reference to precision and accuracy, and to illustrate some important features of the experimental determination of precision.

ASTM: E179-81, Standard Practice for Selection of Geometric Conditions for Measurement of Reflectance and Transmittance, 8 pp.
Intended for use in selecting terminology, measurement scales, and instrumentation for describing or evaluating such appearance characteristics as glossiness, opacity, lightness, transparency, and haziness, in terms of reflected or transmitted light.

ASTM: E189-63 (1975), Standard Recommended Practice for Determining Temperature-Electrical Resistance Characteristics (EMF) of Metallic Materials, 5 pp.
Covers procedures determining the temperature versus electrical resistance of emf characteristics of metallic materials.

ASTM: E230-87, Standard Temperature Electromotive Force (EMF) Table for Standardized Thermocouples, 100 pp.
Consists of temperature-emf tables for thermocouple types B, E, J, K, R, S, and T; standard and special limits of error and upper temperature limits All intervals are 1 degree.

ASTM: E235-82, Standard Specification for Thermocouples, Sheathed, Type K, for Nuclear or Other High-Reliability Applications, 7 pp.
Presents the material, operating, and environmental requirements two-wire thermocouples intended for nuclear service. Provisions for temperatures up to 900 degrees C (1650 degrees F) are covered.

ASTM: E261-77, Standard Method for Determining Neutron Flux, Fluence, and Spectra by Radioactivation Techniques, 12 pp.
Covers the determination of neutron flux in a radiation field from the radioactivity that is induced in a detector specimen. The description is directed toward the need for characterization of the magnitude and energy distribution of neutron flux in connection with radiation effects on materials.

ASTM: E275-83, Standard Practice for Describing and Measuring Performance of Ultraviolet, Visible, and Near Infrared Spectrophotometers, 16 pp.
Covers the description of requirements of spectrophotometric performance especially for ASTM methods, and the testing of the adequacy of available equipment for a specific method.

ASTM: E304-81 (87), Standard Practice for Use and Evaluation of Spark Source Mass Spectrometers for the Analysis of Solids, 6 pp.
Provides guidelines for evaluation of suitability of solids mass spectrometers for use in ASTM methods for analysis of solids which specify the use of such apparatus. This practice is restricted to those instruments in which ions are produced in an electrical discharge directly from the solid.

ASTM: E306-71 (1976), Standard Method for Absolute Calibration of Reflectance Standards, 5 pp.
Describes the requirements for an auxiliary-integrating sphere to be used with a spectrophotometer already equipped with a sphere to measure diffusely-reflected flux, and describes the calibration of an instrument standard on a scale on which the perfectly-reflecting, perfectly-diffusing specimen is assigned a value of unity.

ASTM: E317-85, Standard Practice for Evaluating Performance Characteristics of Ultrasonic Pulse-Echo Testing Systems Without the Use of Electronic Measurement Instruments, 17 pp.
Describes procedures for determining some important performance characteristics that establish the capabilities of pulse-echo ultrasonic testing systems in which test results are displayed on an A-scan cathode-ray tube screen.

ASTM: E334-81, Standard Practices for General Techniques in Infrared Microanalysis, 8 pp.
These recommended practices cover general information on techniques that are of general use in securing and subsequently analyzing (by infrared spectrophotometric techniques) microgram quantities of solid or liquid samples.

ASTM: E337-84 — (ANSI: L14/19-1963), Standard Test Method for Measuring Humidity With a Psychrometer (The Measurement of Wet-Bulb and Dry-Bulb Temperatures), 24 pp.
Covers the procedure for determining relative humidity of atmospheric air by means of wet- and dry-bulb temperature readings.

Titles & Abstracts

ASTM: E355-77 (1983), Standard Recommended Practice for Gas Chromatography Terms and Relationships, 9 pp.
List of operating parameters and relationships that applies in most cases only to steady-gas elution chromatography.

ASTM: E380-86, Standard Metric Practice (Complete in Vol. 14.02 Only; Excerpts in Related Material Section of All Other Volumes), 42 pp.
Gives guidance for application of the International System of Units (SI). Includes information on SI, a limited list of non-SI units recognized for use with SI, a list of conversion factors from non-SI to SI units, and general guidance on proper style and usage.

ASTM: E386-78 (1984), Standard Definitions of Terms, Symbols, Conventions, and References Relating to High-Resolution Nuclear Magnetic Resonance (NMR) Spectroscopy, 12 pp.

ASTM: E425-85, Standard Definitions of Terms Relating to Leak Testing, 8 pp.

ASTM: E432-71 (1984), Standard Guide for the Selection of a Leak Testing Method, 4 pp.
Intended as a guide for the selection of a leak testing method. The type of item to be tested or the test system and the method considered for either leak measurement or location are related in order of increasing sensitivity.

ASTM: F305-70 (1980), Standard Practice for Sampling Particulates from Reservoir-Type Pressure-Sensing Instruments by Fluid Flushing, 2 pp.
Covers the sampling of reservoir-type pressure-sensing instruments which enclose a volume that has dubious drainage capabilities.

AMERICAN SOCIETY OF AGRICULTURAL ENGINEERS (ASAE)

The following publications are available from the American Society of Agricultural Engineers, 2950 Niles Rd., St. Joseph, MI 49085. (616) 429-0300

ASAE: S313.2, Soil Cone Penetrometer, revised 1985, 1 p.
Described is the soil cone penetrometer, recommended as a measuring device to provide a standard uniform method characterizing the penetration resistance of soils.

ASAE: D271.2, Psychrometric Data, reconfirmed 1983, 5 charts.
Convenient reference psychrometric charts that yield data for a dry bulb temperature range of -35 degrees to 600 degrees F.

ASAE: D272.2, Resistance to Airflow of Grains, Seeds, Other Agricultural Products, and Perforated Metal Sheets, 5 pp.
Provides an estimate of airflow resistance that can be used as the basis for the design of systems to aerate grain and seed.

ASAE: D293.1 Dielectric Properties of Grain and Seed, revised 1984, 8 pp.
These data are intended to provide a basis for design of equipment and application of radio-frequency energy for treatment of grain and seed, or possible electrical measurement of moisture content, and development of capacitive type of RF-energy absorption sensing devices.

ASAE: S368.1 Compression Test of Food Materials of Convex Shape, reconfirmed 1987.
This standard is intended for use in determining mechanical attributes of food texture, resistance of mechanical injury and force deformation behavior of food materials of convex shape, such as fruits and vegetables, seeds and grains, and manufactured food materials.

ASAE: D309.1, Wet-Bulb Temperatures and Wet-Bulb Depressions, revised June 1987, 7 pp.
Explanation and accompanying set of maps show mean wet-bulb temperatures, mean wet-bulb depressions, and their standard deviations in both Celsius and Fahrenheit.

AMERICAN SOCIETY OF HEATING, REFRIGERATING, AND AIR-CONDITIONING ENGINEERS, INC. (ASHRAE)

The following ASHRAE Standards may be obtained from the American Society of Heating, Refrigerating, and Air-Conditioning Engineers, Inc., 1791 Tullie Circle N.E., Atlanta, GA 30329. (404) 636-8400

ASHRAE: Std. 12-75, (ANSI: B53.1-1974), Refrigeration Terms and Definitions, 1975, 33 pp.
Intended to provide authoritative definitions of words and terms employed in all phases of activity connected with refrigeration and air-conditioning.

ASHRAE: Std. 28-78, Methods of Testing Flow Capacity of Refrigerant Capillary Tubes, 1978, 4 pp.
Covers air flow capacity tests for tubes used in refrigerant metering.

Titles & Abstracts

ASHRAE: Std. 41.1-74, Part I, Standard Measurements Guide: Section on Temperature Measurements, 1974, 18 pp.
Provides methods for accurate temperature measurement for the particular needs of heating, refrigeration, and air conditioning. The rates of heat flow, both to and from moving volatile and non-volatile fluids, in the range of -40 to 400 degrees F are covered. The use of thermometers, thermocouples, and thermistors and the effect of changes in enthalpy are discussed.

ASHRAE: Std. 41.2-1987, Standard Methods for Laboratory Air Flow Measurement

ANSI/ASHRAE: 41.4-1984, Standard Method of Measurement of Proportion of Oil in Liquid Refrigerant, 4 pp.
Intended to apply only where it is known that the sample is from a single-phase solution of oil in liquid refrigerant. Does not apply to measurement of oil concentrations so low that the criterion of 6.1 is not met.

ASHRAE: 41.5-1975, Standard Measurement Guide: Engineering Analysis of Experimental Data, 15 pp.
To provide recommended practices for reporting of uncertainty in results for data obtained from an experiment. Defines terms and sets forth procedures for applying statistical methods to experimental data.

ANSI/ASHRAE: 41.6-1982, Standard Method for Measurement of Moist Air Properties, 24 pp.
Sets forth recommended practices and procedures for the measurement and calculation of moist air properties in order to promote accurate measurement methods for specific use in the preparation of other ASHRAE standards.

ASHRAE: 41.7-1984, (reaffirmed, supersedes 41.7-78), Standard Method for Measurement of Flow of Gas, 7 pp.
Provide recommended practices for the measurement of the flow of dry gases for use in the preparation of ASHRAE Standards.

ASHRAE: 41.8-78, Standard Methods of Measurement of Flow of Fluids — Liquids, 14 pp.
Establish recommended practices for the measurement of flow of fluids as liquids. It shall also establish the standard technique to be used for the calibration of other instruments more convenient to use. This standard is not intended to be used as a replacement for the calibration of flow meters by facilities traceable to NBS nor restrict the use of such facilities that do not incorporate the methods outlined below.

ASHRAE: Std. 41.9-1988, A Standard Calorimeter Test Method for Flow Measurement of a Volatile Refrigerant

ASHRAE: Std. 74-73, Method of Measuring Solar-Optical Properties of Materials, 1973, 7 pp.
The purpose of this standard is to develop a standard method of measuring and reporting the solar-optical properties of materials.

ANSI/ASHRAE: 86-1983, Methods of Testing Floc Point of Refrigeration Grade Oils, 3 pp.
The test for floc point is intended to determine the waxing tendency of refrigeration grade oils at low temperatures and is based on evaluation of wax precipitation tendency of a mixture of 90% Refrigerant 12 and 10% of oil being tested, the results of which can be used to compare several different oils.

ANSI/ASHRAE: 97-1983, Sealed Glass Tube Method to Test the Chemical Stability of Material for Use within Refrigerant Systems, 10 pp.
Establishes a procedure using sealed glass tubes for the evaluation of materials to be used in refrigerant systems. Detailed safety precautions are included.

ANSI/ASHRAE: 101-1981, Application of Infrared Sensing Devices to the Assessment of Building Heat Loss Characteristics, 27 pp.
Describe acceptable procedures and specifications for the applied use of infrared radiation sensing devices for assessment of building heat loss characteristics and interpretation of data resulting therefrom.

AMERICAN SOCIETY OF MECHANICAL ENGINEERS (ASME)

The following ASME and ANSI Standards may be obtained from The American Society of Mechanical Engineers, 345 East 47th Street, New York, NY 10017. (212) 705-7722

ANSI/ASME: MFC-1M-1979 (R-1986), Glossary of Terms Used in the Measurement of Fluid Flow in Pipes, 1979, 177 pp., Bk. No. J00065
Presents a collection of definitions of those terms which pertain to the measurement of fluid flow in pipes.

ANSI/ASME: MFC-2M-1983 (R-1988), Measurement Uncertainty for Fluid Flow in Closed Conduits, 1983, pp. 71, Bk. No. K00112
Presents a working outline detailing and illustrating the techniques for estimating measurement uncertainty for fluid flow in closed conduits.

Titles & Abstracts

ASME: MFC-3M-1985, Measurement of Fluid Flow in Pipes Using Orifice, Nozzle, and Venturi, 1985, pp. 63, Bk. No. K000113

Specifies the geometry and method of use (installation and flowing conditions) for orifice plates, nozzles, and venturi tubes when they are inserted in a conduit running full, to determine the rate of the fluid flowing. It also gives necessary information for calculating flow rate and its associated uncertainty.

ANSI/ASME: MFC-4M-1986, Measurement of Gas Flow by Turbine Meters, 1986, pp. 18, Bk. No. K0018

Applies to the measurement of gas by a turbine meter; the meter's construction, installation, operation, performance characteristics, data computation and presentation, calibration, field checking, and other related considerations of the meter.

ANSI/ASME: MFC-5M-1985, Measurement of Liquid Flow in Closed Conduit Using Transit-Time Ultrasonic Flowmeters, 1985, pp. 14, Bk. No. K0015

Provides a description of the operating principles employed by the ultrasonic flow meters covered, a description of error sources and performance verification procedures, and a common set of terminology, symbols, definitions, and specifications.

ASME/ANSI: MFC-6M-1987, Measurement of Fluid Flow in Pipes Using Vortex Flow Meters, 1987, pp. 11, Bl. No. K00117

Describes vortex shedding fluid flow meters in which a von Karman vortex sheet is produced by one or more struts installed in a closed conduit and features needed for the user to select a flowmeter satisfactory for the application. It also provides general information generic to a vortex shedding flow meter, a glossary, a set of engineering equations useful in specifying performance.

ANSI/ASME: MFC-7M-1987, Measurement of Gas Flow by Means of Critical Flow Venturi Nozzles, 1987 pp. 32, Bk. No. K00119

Specifies the geometry and method of use (installation and operating conditions) of critical flow venturi nozzles inserted in a system to determine the mass flow rate of the gas flowing through the system. It also gives necessary information for calculating the flow rate and its associated uncertainty.

ASME/ANSI: MFC-8M-1988, Fluid Flow Measurement in Closed Conduits — Connections for Pressure Signal Transmissions Between Primary and Secondary Devices, 1988, pp. 39, Bk. No. K12188

Describes means whereby a pressure signal from a primary device can be transmitted by known techniques to a secondary device in such a way that the value of the signal is not distorted or modified. It relates to the types of pressure difference primary devices for flow measurement, described in ASME MFC-3M.

ASME-10M-1988, Method for Establishing Installation Effects on Flowmeters, 1988, pp. 7

Establishes methods for determining the influence of installation conditions or flow patterns on the performance of flowmeters in closed conduits running full.

ASME: PTC 6A, Appendix A to Test Code for Steam Turbines, 1982, 56 pp. Bk. No. C00029.

This appendix provides numerical examples of various turbine test calculations.

ASME: PTC 6R, Guidance for Evaluation of Measurement Uncertainty in Performance Tests of Steam Turbines, 1969, 32 pp. Bk. No. D00041.

This report provides guidance to establish the degree of uncertainty with test results when there are deviations from the requirements of Performance Test Code No. 6 on Steam Turbines.

ASME: PTC 6S, Simplified Procedures for Routine Performance Tests of Steam Turbines, 1970, 4 pp. Bk. No. D00042.

Provides for the testing of steam turbines operating predominantly within the moisture region for the purpose of determining the level of performance with minimum uncertainty.

ASME: PTC 9, Displacement Compressors, Vacuum Pumps, and Blowers, 1970, 41 pp. Bk. No. C00009.

Applies to tests for determining the performance of positive displacement, compressors, blowers, and vacuum pumps whether reciprocating or rotating.

ASME: PTC 19.2, Instruments and Apparatus: Pressure Measurement, 1964, 58 pp. Bk. No. D00029.

Discusses the technology of pressure measurement: general considerations and definitions, pressure connections, liquid-level gages, deadweight gages and testers, elastic gages, and low-pressure measurement.

ASME: PTC 19.3, Instruments and Apparatus: Temperature Measurement, (R 1979), 118 pp., Bk. No. C00035.

Presents a revision, expansion, and consolidation of all earlier pamphlets on temperature measurement instruments, with particular emphasis on

Titles & Abstracts

basic sources of errors and means of coping with them.

ASME: PTC 19.5, Interim Supplement on Instruments and Apparatus: Application, Part II of Fluid Meters. Sixth Edition, 1972, 140 pp., Bk. No. G00018.

Presents the recommended conditions, procedures and data for measuring the flow of fluids, particularly with the three principal differential pressure meters: the orifice, the flow nozzle, and the venturi tube.

ASME: PTC 19.5.1, Instruments and Apparatus: Weighing Scales, 1964, 17 pp., Bk. No. D00028.

Discusses weighing scales suitable for quantity measurement of materials in connection with tests of power equipment.

ASME: PTC 19.6, Electrical Measurements in Power Circuits, 1965, 40 pp., Bk. No. D00007.

The methods given include measurements made with either indicating or integrating instruments of power, voltage and current in direct-current and alternating-current single-phase and poly-phase rotating machinery, transformers induction apparatus, arc and resistance heating equipment, and mercury are rectifiers.

ASME: PTC 19.7, Measurement of Shaft Horsepower, 1980, 29 pp., Bk. No. D00009.

Shows how measurement of shaft horsepower of rotating machines can be accomplished either by the direct method of utilizing dynamometers or the indirect method of using calibrated motors or generators, heat balance, or heat exchangers.

ASME: PTC 19.8, Measurement of Indicated Horsepower, 1970, 29 pp., Bk. No. D00008.

This supplement of the Performance Test Codes treats the direct measurement of indicated power of piston engines and compressors by use of the engine indicator.

ASME: PTC 19.10, Flue and Exhaust Gas Analyses, Instruments and Apparatus — V Part 10, 1981, 32 pp., Bk. No. C00031.

This supplement provides descriptions of methods, apparatus, and calculations which are used in conjunction with Performance Test Codes to determine quantitatively the constituents of the gases resulting from combustion of carbonaceous or hydrocarbon fuel in the solid, liquid, or gaseous form.

ASME: PTC 19.11, Water and Steam in the Power Cycle (Purity and Quality, Leak Detection, and Measurement), 1970, 98 pp., Bk. No. D00011.

Specifies and discusses the methods of instrumentation for testing boiler feedwater, steam, and condensate in relation to performance testing in the power cycle and for leak detection and leakage measurement for surface condensers and other cycle components.

ASME: PTC 19.12, Measurement of Time, 1958, 12 pp., Bk. No. D00012.

General purpose clocks, chronometers, clocks or regulators for indicating time to the nearest second, astronomical clocks, watches, stop watches, timers, chronographs, and oscillographs are the types of timekeepers described.

ASME: PTC 19.13, Measurement of Rotary Speed, 1961, 17 pp., Bk. No. D00013.

Covers commonly used instruments and methods, and discusses characteristics and limitations of commercially available instruments used for testing rotating machinery, turbines, blowers, or electric motors.

ASME: PTC 19.14, Linear Measurements, 1958, 14 pp., Bk. No. D00014.

Coverage includes tapes, rules and scales, calipers, and dividers, slide calipers, depth gages, vernier calipers, vernier depth gages, micrometer calipers, micrometer depth gages, internal micrometers, telescoping and small hole gages, dial indicators, dial bore gages, dial caliper gages, thickness gages, and gage blocks.

ASME: PTC 19.16, Density Determinations of Solids and Liquids, 1965, 12 pp., Bk. No. D00016.

Considers available methods for determining specific gravity.

ASME: PTC 19.17, Determination of the Viscosity of Liquids, 1965, 16 pp., Bk. No. D00017.

Gives procedures for and information on various types of viscometers and their applications.

ASME: PTC 32.1, Nuclear Steam Supply Systems, 1969, 34 pp., Bk. No. C00012.

Contains instructions for the performance testing of nuclear steam supply systems. It establishes procedures for conducting tests to determine the thermal performance of a nuclear steam supply system as a unit.

The following Standards, identified by their ANSI number, are ANSI approved, but published by ASME. They are available from either ASME or ANSI.

ANSI: B40.1-1980, Gauges — Pressure and Vacuum, Indicating Dial Type — Elastic Element, 1980, 15 pp., Bk. No. K00015.

Titles & Abstracts

The scope is confined to dial type indicating gauges which indicate pressure or vacuum by means of a pointer and a graduated scale, utilizing an elastic element for measuring the pressure or vacuum. It does not include dead weight types, mercury-floated piston types, or other special constructions which do not utilize an elastic element.

ANSI: MC88.1-1972 (R 1978), Guide for Dynamic Calibration of Pressure Transducers, 29 pp., Bk. No. L00042.
Provides calibration techniques and methods for use with pressure transducers.

ANSI: Y1.1-1972, Abbreviations for Use on Drawings and in Text, Bk. No. J00003.
Provides list of basic abbreviations.

ANSI: Y14.1-1980, Drawing Sheet Size and Format, Bk. No. N00001.
Established standard drawing sheet sizes.

ANSI: Y14.26M-1981, Engineering Drawing and Related Practices — Digital Representation for Communication of Product Definition Data, Bk. No. N00099.
Establishes data required to describe and communicate the essential engineering characteristics of physical objects as manufactured products. Establishes information structures to be used for the digital representation and communication of products definition data.

ANSI: Y32.10-1967 (R 1979), Graphic Symbols for Fluid Power Diagrams, 1967, 22 pp., Bk. No. N00022.
Presents a system of graphic symbols for fluid power diagrams. Elementary forms of symbols are: Circles, Squares, Rectangles, Triangles, Arcs, Arrows, Lines, Dots, and Crosses.

ANSI: Y32.11-1961, Graphical Symbols for Process Flow Diagrams in Petroleum and Chemical Industries, Bk. No. K00040.
A preliminary set of standard symbols, developed for use on the basic process flow diagrams in order to represent the major items of equipment used by the petroleum and chemical industries.

ANSI: Z32.2.3-1949 (R 1953), Graphical Symbols for Pipe Fittings, Valves and Piping, 12 pp., Bk. No. K00006.

ANSI: Z32.2.4-1949 (R 1953), Graphical Symbols for Heating, Ventilating, and Air Conditioning, 16 pp., Bk. No. K00005.

ANSI: Z32.2.6-1950 (R 1956), Graphical Symbols for Heat-Power Apparatus, 8 pp., Bk. No. K00004.

ANSI: B88.2-1974 (R 1981), Procedure for Bench Calibration of Tank Level Gauging Tapes and Sounding Rules, (R 1981), 4 pp., Bk. No. L00043.
Procedure applies to any gauging tape or sounding rule using a graduated scale to determine level of liquid in tanks. Procedures for both linear and non-linear scales are provided.

ANSI: B89.1.9-1973 (R 1980), Precision Inch Gauge Blanks for Length Measurement (Thru 20 Inches), (R 1980), 16 pp., Bk. No. L00044.
Covers specifications for gauge blocks up to and including 20 inches in length, including physical properties, general dimensions, tolerance grades, flatness, parallelism and surface texture requirements.

ANSI: B89.3.1-1972, (R 1979), Measurement of Out-Of-Roundness, 1972, 27 pp., Bk. No. L00020.
Covers the specification and measurement of out-of-roundness of a surface of revolution. Deals primarily with precision spindle instruments for out-of-roundness measurement and polar chart presentation.

ANSI: B89.6.2-1973 (R 1979), Temperature and Humidity Environment for Dimensional Measurement, 1973, 31 pp., Bk. No. L00047.
Covers methods of describing and testing temperature-controlled environments for dimensional measurement and ensuring adequate temperature control for the calibration of measuring equipment as well as the manufacture and acceptance of work-pieces.

ANSI: C85.1-1981, Terminology for Automatic Control, 1981, Bk. No. N00036.
Terminology pertaining to systems such as: automatic process control, feedback control, regulating, and other related systems not requiring human intervention as a part of the regulating procedure.

AMERICAN VACUUM SOCIETY (AVS)

The following tentative Standards are available from the American Vacuum Society, 335 East 45th Street, New York, NY 10017. (212) 661-9404

AVS: 2.1, Calibration of Leak Detectors of the Mass Spectrometer Type, 1973, 10 pp.
Prescribes procedures to be used for calibrating leak detectors of the mass spectrometer type, that is, for determining a sensitivity figure for such leak detectors.

AVS: 2.2, Method for Vacuum Leak Calibration, 1968, 4 pp.

Titles & Abstracts

This standard describes an apparatus for measuring the leak rate of vacuum leaks, in the range of 10^{-5} to 10^{-3} atm cm^3/s, and a procedure for using the apparatus to determine such leak rates.

AVS: 2.3-1972, Procedure for Calibrating Gas Analyzers of the Mass Spectrometer Type, 1972, 15 pp.
Concerned with calibration procedures for determining the minimum detectable partial pressure of gas analyzers of the mass spectrometer type. Procedures are also given for evaluating the resolution of the analyzer.

AVS: 5.3, Method for Measuring Pumping Speed of Mechanical Vacuum Pumps for Permanent Gases, 1967, 3 pp.
Describes procedures for determining the pumping speed for permanent gases of all positive displacement mechanical vacuum pumps.

AVS: 6.2-1969, Procedure for Calibrating Vacuum Gages of the Thermal Conductivity Type, 1969, 4 pp.
Procedures are given and apparatus described for calibrating vacuum gages of the thermal conductivity type of direct comparison with measurements made with an absolute reference instrument such as the McLeod gage. The pressure range considered is of the order of 10^{-4} to several Torr.

AVS: 6.4-1969, Procedure for Calibrating Hot Filament Ionization Gauges Against a Reference Manometer in the Range of 10^{-2}–10^{-5} Torr, 1969, 5 pp.
Procedures are given for the calibration of hot cathode ionization gauges and gauge tube by direct comparison against a McLeod gauge or other absolute manometer in the pressure range of 10^{-2}–10^{-5} Torr.

AVS: 6.5-1971, Procedures for the Calibration of Hot Filament Ionization Gauge Controls, 1971, 7 pp.
Guidelines and methods are provided for the electrical calibration of hot cathode ionization gauge controls.

AVS: 7.1, Graphic Symbols in Vacuum Technology, 1966, 4 pp.
A uniform system of graphic symbols to be used in vacuum technology.

AMERICAN WATER WORKS ASSOCIATION (AWWA)

The following AWWA Publications are available from the American Water Works Association, 6666 West Quincy Avenue, Denver, CO 80235. (303) 794-7711

AWWA: C700-77, Cold-Water Meters — Displacement Type, 1977, 16 pp.
Covers the various types and classes of cold-water displacement meters in sizes 5/8 in. through 6 in., and the materials and workmanship employed in their fabrication.

AWWA: C701-88, Cold-Water Meters — Turbine Type for Customer Service, 1988, 20 pp.
Covers the various types and classes of cold-water turbine meters in sizes 1-1/2 in.–12 in. for water works customer service, and the material and workmanship employed in their fabrication.

AWWA: C702-86, Cold-Water Meters — Compound Type, 1986.
Covers the various types and classes of cold-water compound meters in sizes 2 in.–/10 in. and covers the materials and workmanship employed in their fabrication.

AWWA: C703-86, Cold-Water Meters — Fire Service Type, 1986.
Covers the various types and classes of cold-water fire service type meters in sizes 3 in. through 10 in., and the materials and workmanship employed in their fabrication.

AWWA: C704-70 (R 84), Cold-Water Meters — Propeller Type for Main Line Application, 1970, Reaffirmed without revision, 1984, 12 pp.
Covers the various types and classes of cold-water propeller meters in sizes 2 in. to 36 in., for main line applications, and the materials and workmanship employed in their fabrication.

AWWA: C706-86, Direct-Reading Remote Registration Systems for Cold-Water Meters, 1986, 12 pp.
Covers direct-reading remote registration systems for use on cold-water meters for water utility customer service, and the materials and workmanship employed in their fabrication and assembly.

AWWA: C707-82, Encoder-Type Remote Registration Systems for Cold-Water Meters, 1982, 16 pp.
Covers encoder-type remote-registration systems for use on cold-water meters for water-utility customer service and the materials and workmanship employed in their fabrication and assembly.

AWWA: C708-82, Cold-Water Meters Multi-Jet Type for Customer Service, 1982, 16 pp.
Covers the various types and classes of cold-water multi-jet meters in sizes 5/8 in. through 2 in., and

Titles & Abstracts

the materials and workmanship employed in their fabrication.

AWWA: C710-88, Cold-Water Meters — Displacement Type, Plastic Main Case, 1988, 15 pp (estimated).
Covers the various types and classes of plastic main case cold-water displacement meters in sizes ⅝" through 2", and the materials and workmanship employed in their fabrication.

AWWA: C500-86, Gate Valves — 3 in. through 48 in. — for Water and Other Liquids, 1986.
Covers iron-body, bronze-mounted non-rising stem gate valves, 3 in. through 48 in. in diameter with either double disc gates having parallel or inclined seats, or solid-wedge gates.

AWWA: C501-80, Sluice Gates, 1980, 28 pp.
Covers wall thimble, vertically-mounted sluice gates designed for either seating head or unseating head or both, in ordinary water supply or wastewater service.

AWWA: C504-80, Rubber-Seated Butterfly Valves, 1980, 24 pp.
Covers cast-iron and ductile-iron body, rubber-seated tight-closure butterfly valves, 3–72 in. in size with four body types for fresh water having a pH greater than 6 and temperatures generally less than 125 degrees F.

AWWA: C506-78 (R 83), Backflow Prevention Devices — Reduced Pressure Principle and Double Check Valve Types, 1978, 20 pp.
Covers two types of backflow prevention devices designed for operation on cold-water lines (maximum 110 degrees F) at 150 psi operating pressure.

AWWA: C507-85, Ball Valves, Shaft- or Trunion-Mounted — 6 in. through 48 in. — for Water Pressures up to 300 psi, 1985, 16 pp.
Covers cast-iron, ductile-iron, and cast-steel flanged end, tight shut-off, shaft- or trunion-mounted, full port, double- and single-seated ball valves for use in fresh water with a pH greater than 6 and temperature generally less than 125 degrees F.

AWWA: C508-82, Swing-Check Valves for Waterworks Service, 2 in. through 24 in. NPS, 1982, 16 pp.
Covers iron-body, bronze-mounted swing-check valves, 2–24 in. NPS, for normal horizontal installation in water systems.

AWWA: C509-87, Resilient-Seated Gate Valves, 3 through 12 NPS, for Water and Sewage Systems, 1987, 20 pp.
Covers iron-body resilient-seated gate valves with non-rising stems (NRS) and outside screw-and-yoke (OS&Y) rising stems for installation in water and sewage systems.

AWWA: C540-87, Power Actuating Devices for Valves and Sluice Gates, 1987, 26 pp.
Covers power actuating devices for valves 3-in. in diameter and larger and sluice gates in ordinary water service.

APHA, AWWA, WPCF: Standard Methods for the Examination of Water and Wastewater, 1985, 1268 pp, 16th ed.
"Standard Methods" presents the best current practice of American water analysis in connection with the ordinary water purification, sewage disposal and sanitary investigations.

APHA, AWWA, WPCF: Standard Methods for the Examination of Water and Wastewater, Supplement to the 16th ed., 1988, 161 pp.
Supplement to the 16th edition of Standard Methods.

AMERICAN WELDING SOCIETY (AWS)

The following publications are available from the American Welding Society, P.O. Box 351040, Miami, FL 33135. (305) 443-9353

AWS: A4.2-86, Standard Procedures for Calibrating Magnetic Instruments to Measure the Delta Ferrite Content of Austenitic Stainless Steel Weld Metal, 1974, 16 pp.
Prescribes procedures for the calibration and maintenance of calibration of magnetic instruments for measuring the delta ferrite content of austenitic stainless steel weld metals in terms of their Ferrite Number.

AWS: F6.1-78, Method for Sound Level Measurement of Manual Arc Welding and Cutting Processes, 1978, 8 pp.
Describes the equipment and procedure to be used in measuring sound levels of manual arc welding and cutting processes. The procedure described allows the user to measure the sound level associated with specific processes in a reproducible manner that permits comparison with other selected processes. This method is not applicable to the determination of operator exposure to process sound.

Titles & Abstracts

ANTI-FRICTION BEARING MANUFACTURERS ASSOCIATION, INC. (AFBMA)

The following documents are available from the Anti-Friction Bearing Manufacturers Association, Inc., 1101 Connecticut Ave., NW, Suite 700, Washington, DC 20036-4303. (202) 429-5155

ANSI/AFBMA Std. 12.1-1985, Instrument Ball Bearings, Metric Design, 1985, 43 pp.

ANSI/AFBMA Std. 12.2-1985, Instrument Ball Bearings, Inch Design, 1985, 43 pp.

ANSI/AFBMA Std. 13-1987, Roller Bearing Vibration and Noise (Methods of Measuring), 1987, 10 pp.

These Standards for Instrument Ball Bearings have been established by The Anti-Friction Bearing Manufacturers Association, Inc., for the purpose of defining the characteristics of these bearings, such as boundary dimensions, tolerances, classification used for selective assembly, radial internal clearance values, recommended gaging practices, mounting practices and starting torque values.

ASSOCIATION FOR THE ADVANCEMENT OF MEDICAL INSTRUMENTATION (AAMI)

The following publication and other related standards are available from Association for the Advancement of Medical Instrumentation, Suite 602, 1901 N. Fort Myer Drive, Arlington, VA 22209. (703) 525-4890

AAMI: ANSI/AAMI ES1-1985, Safe Current Limits for Electromedical Apparatus, 1978, 20 pp.
Provides designers and consumers with limits and measuring techniques for risk currents of electromedical apparatus as a function of frequency, the characteristics of the apparatus, and the nature of the intentional contact with the patient.

AAMI: ANSI/AAMI DF2 5/81, Standard for Cardiac Defibrillator Devices, 1982, 20 pp.
Provides minimum labeling, performance and safety requirements for cardiac defibrillator devices. Also included are referee test methods by which compliance can be verified.

AAMI: ANSI/AAMI RD 5 6/81, Standard for Hemodialysis Systems, 1982, 35 pp.
Establishes requirements for materials, components, monitors, accessories, maintenance and labeling for hemodialysis systems. The quality of water and hemodialysis bath concentrate used in the system is also defined. Additionally, a guideline for the user of the device, with particular emphasis on water purity assurance and monitoring, is appended to this standard.

AAMI: ANSI/AAMI AT6 6/81, Standard for Autotransfusion Devices, 1982, 12 pp.
Describes labeling requirements sufficient to assure that adequate information is available to the clinician to choose the appropriate device for his or her particular application, requirements for safety and performance of the device and test methods to verify that the labeling and performance requirements of the standard have been met.

AAMI: ANSI/AAMI EC11—1982, American National Standard for Diagnostic Electrocardiographic Devices, 1983, 36 pp.
Establishes minimum safety and performance requirements for ECG systems with direct writing devices which are intended for use in EGC contour analysis for diagnostic purposes.

AAMI: ANSI/AAMI EC12—1983, American National Standard for Pregelled EGC Disposable Electrodes, 1984, 16 pp.
Contains minimum labeling and electrical performance requirements, test methods, and terminology for any pregelled electrocardiographic disposable electrode in which the electrolyte has been placed in contact with the sensing element by the manufacturer.

AAMI: ANSI/AAMI EC13—1983, American National Standard for Cardiac Monitors, Heart Rate Meters and Alarms, 1984, 40 pp.
Establishes minimum safety and performance requirements for cardiac monitors, heart rate meters, and alarms which are used to acquire and/or display ECG signals with the primary purpose of continuous detection of cardiac rhythm.

AAMI: ANSI/AAMI NS14—1984, American National Standard for Implantable Spinal Cord Stimulators, 1984, 16 pp.
Establishes minimum labeling, safety, and performance requirements for implantable spinal cord stimulators intended for use in the relief of chronic, severe pain.

AAMI: ANSI/AAMI NS15—1984, American National Standard for Implantable Peripheral Nerve Stimulators, 1984, 16 pp.
Establishes minimum labeling, safety, and per-

formance requirements for implantable peripheral nerve stimulators intended for use in the relief of chronic, severe pain.

AAMI: ECGC—5/83, Standard for ECG Connectors, 1983, 16 pp.
Establishes a preferred connector set for cable and panel and certain cable requirements to achieve universal defibrillation protection and in order to allow cable and apparatus interchangeability of ECG monitoring systems with isolated patient connections.

AAMI: ANSI/AAMI BP22—1986, Blood Pressure Transducers, General, 1986, 20 pp.
Establishes safety and performance requirements for isolated transducers, including cables, designed for blood pressure measurements through an indwelling catheter or direct puncture. Also included are disclosure requirements to permit the user to determine compatibility between the transducer and the blood pressure monitor.

AAMI: ANSI/AAMI BP23—1986, Blood Pressure Transducers, Interchangeability and Performance of Resistive Bridge Type, 1986, 20 pp.
Establishes interchangeability and performance requirements for resistive bridge-type blood pressure transducers, including the connector interface to the blood pressure monitor. The standard addresses compatibility between transducers and monitors regardless of the manufacturer.

AAMI: ANSI/AAMI HF18—1986, Electrosurgical Devices, 1986, 42 pp.
Provides minimum labeling, performance and safety requirements to help ensure safe and effective use of electrosurgical devices. Devices included are electrosurgical high frequency generators, directly related accessories, including active electrodes and cables, dispersive electrodes and cables, and footswitches and other operator-controlled mechanisms for activating the generator output.

AAMI: ANSI/AAMI NS4—1985, Transcutaneous Electrical Nerve Stimulators, 1986, 18 pp.
Establishes certain requirements for portable, battery-powered, transcutaneous electrical nerve stimulators that are used in the treatment of pain syndromes, that are intended for use on intact skin and mucous membranes, and that do not require surgical intervention or violation of the skin surface.

AAMI: ANSI/AAMI SP9—1985, Sphygmomanometers, Non-Automated, 1986, 20 pp.
Provides safety and performance requirements for pneumatic or other non-automated sphygmomanometers that are used with an occluding cuff for the indirect determination of blood pressure. Aneroid and mercury gravity sphygmomanometers are included, used in conjunction with a stethoscope or other manual methods for detecting Korotkoff sounds.

ASSOCIATION OF OFFICIAL ANALYTICAL CHEMISTS (AOAC)

The following publication is available from the Association of Official Analytical Chemists, 1111 N. 19th Street, Arlington, VA 22209. (703) 522-3032

Official Methods of Analysis of the AOAC, 1984, 1141 pp.
Methods of analysis, adopted after collaborative studies to demonstrate that they are reliable, have convenient practical application, and give reproducible results in the hands of professional analytical chemists. These standardized methods are used by regulatory agencies of federal, state, and municipal governments and by the regulated industries; in specifications to government, private, and research workers in agriculture and public health.

THE CHLORINE INSTITUTE (CI)

The following publications are available from The Chlorine Institute, Inc., 2001 L Street, NW, Washington, DC 20036. (202) 775-2790

CI: No. 6, Piping Systems for Dry Chlorine (11th Ed.), 1985, 14 pp.
Contains information on selected pipe, valves and fittings suitable for use with dry chlorine (gas or liquid) within the range — 20 F and 300 F.

CI: No. 39, Maintenance Instructions for Chlorine Institute Standard Safety Valves, Type 1-1/2 JQ (8th Ed.), 1982, 32 pp.
Illustrated instructions for maintenance of Crosby style 1-1/2JQ chlorine safety valves.

CI: No. 40, Maintenance Instructions for Chlorine Institute Standard Angle Valve (4th Ed.), 1982, 19 pp.
Illustrated instructions on maintenance of standard chlorine angle valve.

CI: No. 41, Maintenance Instructions for Chlorine Institute Standard Safety Valves, Type 4 JQ (3rd Ed.), 1975, 20 pp.
Illustrated instructions on maintenance of Crosby style 4 JQ chlorine safety valves.

Titles & Abstracts

CI: No. 42, Maintenance Instructions for Chlorine Institute Standard Excess Flow Valves (3rd Ed.), 1981, 16 pp.
Illustrated instructions on maintenance of standard chlorine excess flow valves.

COOLING TOWER INSTITUTE (CTI)

The following documents are available from the Cooling Tower Institute, P.O. Box 73383, Houston, TX 77273. (713) 583-4087

CTI: ATC-105, Acceptance Test Code for Industrial Water-Cooling Towers, 1982, 43 pp (Addendum 1986).
This Code covers the determination of the capability of water-cooling towers. The purpose is to describe instrumentation and test procedures which will yield accuracy and uniformity in calculated tower capacity.

CTI: Gear Speed Reducers, 1979.
Rating practice and operating considerations for use with propellor-type fans: includes AGMA-approved service factors; recommendations for lubrication, alignment and protection during shutdown.

CTI: ACT-128, Code for Measurement of Sound from Water Cooling Towers, 1981, 3 pp.
Applies to mechanical and natural draft towers. Test and measurement procedures, operating conditions, and instrumentation are specified.

CTI: STD-201, Certification Standard for Commercial Water-Cooling Towers, 1981.
This standard sets forth procedures whereby the Cooling Tower Institute may certify that a line of water cooling towers being offered for sale by a specific manufacturer, will perform in accordance with the manufacturers' published ratings. The certification thus provided will assure customers of specified water cooling tower capacity without costly field testing on an individual tower basis.

DAIRY AND FOOD, 3-A SANITARY STANDARDS COMMITTEES (DFSSC)

The following 3-A Sanitary Standards were formulated by the cooperative effort of industry and regulatory groups as represented by the Dairy Industry Committee, International Association of Milk, Food, and Environmental Sanitarians, U.S. Public Health Service, U.S. Department of Agriculture, Dairy and Food Industries Supply Association, and Poultry and Egg Institute of America. 3-A and E-3-A Standards and Practices are published in Dairy and Food Sanitation, P.O. Box 701, Ames, IA 50010. (301) 984-1444

3-A: 09-07, Sanitary Standards for Instrument Fittings and Connections Used on Milk and Milk Products Equipment. Part One: Text; Part Two: Drawings.
These standards cover the sanitary aspects of instrument fittings and connections for milk and milk product equipment and on lines which hold or convey milk or milk products.

3-A: 28-01, Sanitary Standards for Flow Meters for Milk and Liquid Milk Products, 1972, 3 pp.
Covers the sanitary aspects of flow meters for milk and liquid milk products, and includes that portion of any device integral with the meter such as strainers, temperature sensors and density sensors, which is in contact with the flowing product.

3-A: 37-00, Sanitary Standards for Pressure and Level Sensing Devices, 1978, 3 pp.
Covers the sanitary aspects of elements used on milk and milk products equipment for sensing pressure and/or product level.

ELECTRONIC INDUSTRIES ASSOCIATION (EIA)

The following EIA Standards are available from the Electronics Industries Association, Standards Sales Office, 2001 Eye Street, NW, Washington, DC 20006. (202) 457-4900. In addition, EIA, NEMA and EEI have jointly prepared several standards. The joint standards may also be listed under EEI or NEMA and may be ordered from any of the three.

EIA: 156-B, (R 1978) (R 1982), Battery Socket Patterns.
This standard is intended to provide clearance dimensions on the mating plates used in battery sockets.

EIA: 186-E, Standard Test Methods for Passive Electronic Component Parts, 1978.
Establishes uniform methods for testing electronic component parts. The methods provide a number of test conditions of varying degrees of severity so that appropriate test conditions may be selected for any component. Included in this base document are definitions and general instructions, as well as an index of the test methods which are available individually.

Titles & Abstracts

EIA: RS:186-1E, Method 1: Humidity (Steady State).
This test is intended to evaluate the effect of absorption and diffusion of moisture and moisture vapor on component parts.

EIA: 186-2E, Method 2: Moisture Resistance (Cycling).
This test is intended to evaluate in an accelerated manner, the resistance of component parts to deterioration resulting from high humidity and heat conditions typical of tropical environments.

EIA: 186-3E, Method 3: Humidity (Steady State-Sealed Container).
This test method provides a means for performing humidity testing without the need for specialized humidity test equipment. It is more severe than Method I.

EIA: 186-4E, Method 4: Dielectric Test (Withstanding Voltage).
This dielectric test is performed for the purpose of determining the ability of component parts to withstand a potential at sea level or at a specified altitude.

EIA: 186-5E, Method 5: Salt Spray (Corrosion).
The salt spray test is performed for the purpose of determining the adequacy of protective coatings or finishes, and has been widely used to evaluate the resistance of metals to corrosion in marine service or in exposed shore locations.

EIA: RS186-6E, Method 6: Mechanical Robustness of Terminals.
These tests are provided to cover various significant characteristics and are intended to determine the ability of terminals to withstand the usual stresses which may be applied during assembly or disassembly operations.

EIA: 186-7E, Method 7: Vibration Fatigue Test (Low Frequency, 10 to 55 Hertz).
This vibration fatigue test is performed for the purpose of determining the ability of component parts and their mountings to withstand vibration in the low frequency range of 10 to 55 Hertz.

EIA: 186-8E: Method 8: Vibration, High Frequency.
This sinusoidal high frequency vibration test is performed for the purpose of determining the effect on component parts of vibration in the frequency range of 10 to 500 Hz or 10 to 2000 Hz as may be encountered in aircraft, missiles, space or automotive vehicles.

EIA: 186-9E, Method 9: Solderability.
The purpose of this test standard is to determine the solderability of solid lead wires, terminals, and other terminations which are normally joined by means of soft solder.

EIA: RS:186-10E, Method 10: Effect of Soldering.
This test is performed to determine the effect of normal soldering operations on component parts.

EIA: 186-11E, Method 11: Thermal Shock in Air.
This test is conducted for the purpose of determining the resistance of a component part to exposures at extremes of high and low temperatures in air, and to the shock of alternate exposures to these extremes.

EIA: 186-12E, Method 12: Heat-Life Test.
This test is performed to determine the effect of storing or operating component parts at elevated temperatures for various time periods.

EIA: 186-13E, Method 13: Insulation Resistance Test.
This test is to measure the resistance offered by the insulating members of a component part to an impressed direct voltage tending to produce a leakage of current through or on the surface of these members.

EIA: 186-14E, Method 14: Panel Seal Test.
This test is intended to determine the effectiveness of panel seals on electronic components which are intended for mounting through holes in panels or enclosures. The panel seals are exposed to water under pressure and observed for leakage.

EIA: 232-D, Interface Between Data Terminal Equipment and Data Communication Equipment Employing Serial Binary Data Interchange, 1987.
Applicable to the interconnection of data terminal equipment (DTE) and data communication equipment (DCE) employing serial binary data interchange. It defines Electrical Signal Characteristics, Interface Mechanical Characteristics, Function Description of Interchange Circuits and Solid Interfaces for Selected Communication System Configurations. Included are thirteen specific interface configurations intended to meet the needs of fifteen defined system applications. (A companion document to EIA-232-D is Industrial Electronics Bulletin No. 9. It provides the application notes to EIA-232-C.)

EIA: 236-C (Also NEMA SK 502-1968), Color Coding of Semiconductor Devices, 1986.
Describes a color code system for marking semiconductor devices with their JEDEC assigned type numbers by means of color bands.

**EIA: 267-B (ANSI/EIA-RS-267-B-83), Axis and

Titles & Abstracts

Motion Nomenclature for Numerically Controlled Machines, 1983.
Intended to simplify programming, to simplify the training of programmers, and to facilitate the interchangeability of control tapes; it applies to all numerically controlled machines.

EIA: 272. Definition and Measurement of Voltage Jump — for Voltage Regulator and Reference Tubes, 1963 (R 1979).
Provides a definition of voltage jump and establishes a general test method including the necessary circuit constants.

EIA: 275-A (ANSI/EIA-275-A-72), Thermistor Definitions and Test Methods, 1971 (R 1976) (R 1981).
Cover definitions of terms and test methods for measurement of the performance characteristics of thermistors. The following are defined: Zero Power Temperature Coefficient, Maximum Operating Temperature, Dissipation Constant, Zero Power Resistance Temperature Characteristic, Temperature Wattage Characteristic, Current-time Characteristic, Resistance Ratio, Beta, Stability, and Maximum Power.

EIA: 281-B, Electrical and Construction Standards for Numerical Machine Control, 1979.
The provisions of this standard apply when equipment of one of the following categories is furnished as part of a numerically controlled machine system: (1) electronic control; (2) electromechanical low power logic and switching equipment (small relays, stepping switches, etc.); (3) electro-mechanical transducers (synchros, tachometers, position sensors); (4) other servo components.

EIA: 309 (ANSI/EIA-309-68), (R 1977), General Specifications for Thermistors, Insulated and Non-Insulated, 1965 (R 1970) (R 1976) (R 1981).
Covers insulated and non-insulated thermistor disks and rods with leads.

EIA: 310-C, (ANSI/EIA) (RS-310-C-1977), Racks, Panels and Associated Equipment, 1977.
Establishes those dimensions which are critical in ensuring compatibility between racks (open and enclosed), panels, and the equipment or apparatus installed thereon. It is intended as a guide to equipment manufacturers and designers. Three cabinet and rack widths to accommodate each of the three standard panel widths 19", 24" and 30" are covered.

EIA: 313-B, Thermal Resistance Measurements of Conduction Cooled Power Transistors.
Described are conditions and methods for thermal resistance measurements on conduction cooled power transistors.

EIA: 337, (ANSI: C83.28-1968), (R 1977), General Specification for Glass Coated Thermistor Beads and Thermistor Beads in Glass Probes and Glass Rods (Negative Temperature Coefficient).
Covers definitions and test methods, characteristics and requirements of glass coated thermistor beads and thermistor beads in glass probes and glass rods.

EIA: 359 (ANSI/EIA RS-359-A-84), EIA Standard Colors for Color Identification and Coding, 1984.
Supersedes EIA GEN-101-A, "Color Coding for Numerical Values," and associated color chips. All of the nominal and limit colors of GEN-101-A fall within this new standard with the exception of dark limits of yellow and orange which could easily have been confused with orange or brown respectively. The colors defined in this standard are intended to be applied to the marking of electronic components, wires, terminals, and circuit functions, based on the visual color attributes of hue, value, and chroma of the Munsell System of color notation.

EIS: RS-364 (ANSI: C83.63-1971), Standard Test Procedures for Low Frequency (Below 3 MHz) Electrical Connectors.
Provides uniform procedures for testing a wide range of electrical and mechanical parameters of electrical connectors at frequencies below 3 MHz.

EIA: 364-1, Addendum No. 1 to RS-364 (ANSI C83.63a-1972), 1970 (R 1976).

EIA: 364-2, Addendum No. 2 to RS-364 (ANSI C83.63b-1972, 1971 (R 1976).

EIA: 364-3, Addendum No. 3 to RS-364 (ANSI C83.63c-1973), 1974.

EIA: 364-4, Addendum No. 4 to RS-364 (ANSI C83.63d-1975), 1975.

EIA: 364-7, Addendum No. 7 (ANSI/EIARS-364-7-1978), 1978.

EIA: 364-04 (ANSI/EIA-364-04-87), Normal Force Test Procedure for Electrical Connectors, 1988.
This procedure determines the magnitude of normal force being generated by a contact system at any given deflection within its normal operating levels. This data and its relationship to contact pressure allows the electrical integrity and stability of the contact interface to be evaluated in

Titles & Abstracts

proper perspective when integrated with other monitored attributes. The procedure is not considered a destructive test and is not intended for acceptance testing.

EIA: 364-37A (ANSI/EIA-364-37A-87), Contact Engagement and Separation Force Test Procedure for Electrical Connectors, 1988.

The object of this test is to detail standard test methods to measure the force required to fully engage or separate standard test pins, blades, or mating components with individual contacts.

EIA: 394 (ANSI: C83.71-1972), Recorded Tape Formats for 7, 14 and 21 Tracks on 1/2 Inch Magnetic Tape and 14, 28 and 42 Tracks on 1 Inch Magnetic Tape for Instrumentation Recording.

EIA: 405, (ANSI C83.99-1973), Flutter Measurement of Instrumentation Magnetic Tape Recorder/Reproducers — Recommended Test Method.

Covers acceptable instrumentation and procedures for the measurement of flutter in instrumentation magnetic recording equipment. The purpose is to promote interchangeability and to eliminate misunderstandings between manufacturers and users by specifying standardized and reproducible flutter measurement techniques. In addition, it is intended to help the user ascertain the suitability of magnetic recording equipment for his requirements.

EIA: 408, Interface Numerical Control Equipment and Data Terminal Equipment Employing Parallel Binary Data Interchange, 1973.

This standard applies to the interconnection of data terminal equipment and numerical control equipment at the tape reader interface. The data terminal would typically be connected to a remote data source/sink such as a computer.

EIA: 413 (ANSI: C83.94-1973), Recommended Test Method — Timing Error Measurements of Instrumentation Magnetic-Tape Recorder/Reproducers.

This Standard covers acceptable instrumentation and procedures for the measurement of Time Base Error, Intertrack Time Displacement Error, Composite Time Base Error, and Pulse-to-Pulse Jitter in instrumentation magnetic-recording equipment.

EIA: 414-A, Simulated Shipping Test for Consumer Electronic Products.

This standard covers all consumer electronic products, as packaged for shipping. It recommends test procedures and minimum performance requirements.

EIA: 422-A, Electrical Characteristics of Balanced Voltage Digital Interface Circuits.

Specifies the electrical characteristics of the balanced voltage digital interface circuit normally implemented in integrated circuit technology.

EIA: 423-A, Electrical Characteristics of Unbalanced Voltage Digital Interface Circuits.

Specifies the electrical characteristics of the unbalanced digital interface circuit normally implemented in integrated circuit technology.

EIA: 455-33A (ANSI/EIA-455-33A-87), Fiber Optic Cable Tensile Loading and Bending Test, 1988.

This test is intended to verify the ability of the fiber optic cable to satisfactorily perform as required by Detail Specifications (a) while undergoing tensile and bending forces and (b) after undergoing tensile and bending forces.

EIA: 455-62 (ANSI/EIA-455-62-88), Optical Fiber Macrobend Attenuation, 1988.

This new Test Method describes procedures for determining the bending loss of uncabled multimode or single-mode fiber. Various bend radio and numbers of turns can be accommodated.

EIA: 455-65 (ANSI/EIA-455-65-87), Optical Fiber Flexure Test, 1988.

This test procedure is intended to determine the ability of an optical fiber to withstand repeated flexing at a specified (usually low) temperature.

EIA: 455-78 (ANSI/EIA-455-78-87), Spectral-Attenuation Cutback Measurement for Single-Mode Optical Fibers, 1988.

This Test Method describes a procedure for measuring the spectral attenuation of single-mode optical fibers. The procedure is restricted to non-polarization sensitive fibers at wavelengths greater than or equal to the cutoff wavelength of the fiber.

EIA: 455-82A (ANSI/EIA-455-82A), Fluid Penetration Test for Fluid-Blocked Fiber Optic Cable, 1988.

This test is intended to assure that a fluid-blocked fiber optic cable is filled and flooded sufficiently to restrict or prohibit the penetration and flow of water or other fluid within the core or along the cable sheath interfaces.

EIA: 455-83A, Cable to Interconnecting Device Axial Compressive Loading, 1987.

The intent of this test is to evaluate the mechanical and optical compatibility of fiber optic cables and interconnecting devices when they are subject to axial compressive loading.

Titles & Abstracts

EIA: 455-101A (previously published as EIA-455-101), Accelerated Oxygen Aging of Fiber Optic Cables and Devices, 1988.

The intent of this test is to accelerate the aging process of fiber optic cables and fiber optic devices (such as connectors) which are made of certain materials that are known to be, susceptible to oxygen (e.g., materials such as synthetic rubber, polychlorphene, and chlorosulfonated polyethylene).

EIA: 455-102 (ANSI/EIA-455-102-87), Water Pressure Cycling, 1988.

The intent of this procedure is to determine the response of a fiber optic cable or interconnecting device to repeated exposure to high water pressure.

EIA: 455-104 (ANSI/EIA-455-104-87), Fiber Optic Cable Cyclic Flexing Test, 1988.

This method describes a procedure for determining the effects of repeated flexings on a fiber optic cable. The procedure is designed to measure optical transmittance changes and requires an assessment of any damage occurring to other cable components.

EIA: 455-179 (ANSI/EIA-455-179-88), Inspection of Cleaved Fiber End Faces by Interferometry, 1988.

The intent of this method is to delineate one means of comparing cleaved optical fiber end faces with an ideal surface that is smooth, flat and perpendicular to the fiber axis.

EIA: 492B000 (ANSI/EIA-492B000-88), Sectional Specification for Class IV Single-Mode Optical Waveguide Fibers, 1988.

This new Specification was formulated to provide a document setting forth engineering and use requirements as necessary for optimum use of Class IV single-mode optical fibers. Use of this document is intended to eliminate misunderstanding or confusion between the supplier and user with respect to product performance requirements and test procedures.

EIA: 492BA00 (ANSI/EIA-492BA00-88), Blank Detail Specification for Class IVa Dispersion-Unshifted Single-Mode Optical Waveguide Fibers, 1988.

This Blank Detail Specification is a supplementary document to Sectional Specification EIA-492B000 and contains requirements for style, layout, and minimum content of Detail Specification.

EIA: 548 (ANSI/EIA-548-88), (formerly published as EIA-IS-44), Electronic Design Interchange Format (EDIF) Version 2 0 0, 1988.

The development of the Electronic Design Interchange Format came about as a result of industry's need for a more standard approach to applying Design Automation techniques to automation in manufacturing. EDIF was created to facilitate and standardize the transport of IC design date. EDIF Version 2 0 0 expands the communication of all levels of digital and analog design information and support a wide variety of data types. A major advantage gained through the use of the EDIF is that design date from incompatible CAE/CAD systems can be transferred using EDIF.

JEDEC Standard 77, Recommendations for Letter Symbols, Abbreviations, Terms and Definitions for Semi-conductor Device Data Sheets and Specifications.

The purpose of this publication is to provide supplemental information which will facilitate the use of Joint Electron Device Engineering Council (JEDEC) registration formats. Registration formats contain a number of technical terms and symbols for which definitions should be of assistance to the writers and users of specifications.

JEDEC Standard 77-1, Addendum No. 1 to JEDEC Standard No. 77, March 1988.

FACTORY MUTUAL SYSTEM (FM)

The following Approval Standards and Data Sheets, although prepared for the information of manufacturers and Factory Mutual engineers and policyholders, are available for help on specific problems from Factory Mutual Research Corporation, 1151 Boston-Providence Turnpike, Norwood, Massachusetts 02062. (617) 762-4300

FM: Data Sheet 5-0/14-1, Graphical Symbols for Electrical Diagrams — Device Numbers — Functions, 1977, 12 pp.

The symbols shown in the bulletin, based on American Standard Y32.2, are those which might be encountered when studying electrical power or control diagrams.

FM: Data Sheet 5-1, Electrical Equipment in Hazardous Locations, 1976, 10 pp.

Discusses the classification of hazardous (classified) locations for electrical installations, and the types of electrical equipment that should be provided in Class I (flammable gases or vapors) and Class II (combustible dusts) hazardous locations.

FM: Data Sheet 5-10/14-10, Protective Grounding of Electrical Power Systems and Equipment, 1984, 19 pp.

Describes the various methods used for grounding

Titles & Abstracts

electrical systems and the non-current carrying metal parts of electrical wiring systems and equipment. The advantages and disadvantages of the different grounding methods, and the means employed to safeguard property from arc damage and fire are also discussed.

FM: Data Sheet 5-32, Electronic Computer Systems, 1978, 11 pp.
Covers protection of computer equipment, records and tapes for protection of electrical equipment, occupancy and location for electronic computer systems. Also covers protection under raised floors for wiring.

FM: Data Sheet 14-6, Insulation-Resistance Tests, April 1980, 10 pp.
Presented are insulation-resistance tests which are of considerable value in detecting grounds, damp windings, carbonized or damaged insulation, foreign deposits, current leakage to ground and other conditions that cause or contribute to electrical breakdowns.

FM: Data Sheet 17-1, Nondestructive Flaw-Detection Methods, 1978, 11 pp.
Gives nondestructive flaw detection methods, readily adaptable to Code procedure and loss prevention including: magnetic particle flaw detection; liquid penetrant inspection; eddy-current, ultrasonic and radiographic testing.

FM: Approval Standard, Intrinsically Safe Apparatus and Associated Apparatus for use in Class I, II and III Division I Hazardous Locations, 1979, 59 pp.
Covers requirements for approval of intrinsically safe apparatus and associated apparatus. These requirements are based to a large extent on NFPA 493 and previous work done by the Instrument Society of America and the International Electrotechnical Commission.

FM: Approval Standard, Explosionproof Electrical Equipment, 1979, 21 pp.
Covers safety requirements for approval of explosionproof (flameproof) electrical equipment.

FM: Approval Standard, Electrical Utilization Equipment, 1979, 35 pp.
Covers safety requirements for approval of ordinary location electrical equipment. States construction criteria and test requirements for protection against shock and fire and for electrical equipment. Requirements are based to a large extent on ANSI C39.5 and IEC 348.

FM: Approval Standard, Fire Safe Valves, 1981, 5 pp.
Covers requirements for approval of valves designed for flammable liquid service.

FM: Approval Standard, Less Flammable Transformer Fluids, 1979, 3 pp.
Covers fire point, fluid properties and convection and radiative heat release rates for approval of less flammable fluid. Tests to a large degree are based on ASTM procedures.

FM: Approval Standard, Flame Radiation Detectors for Automatic Fire Alarm Signaling, 1977, 4 pp.
Covers requirements for approval of flame radiation detectors used to sense accidental fire and to actuate alarm or extinguishing systems.

FM: Approval Standard, Smoke Actuated Detectors for Automatic Fire Alarm Signaling, 1976, 5 pp.
Covers requirements for approval of detectors sensitive to airborne products of combustion responding to particulates or gaseous products.

FM: Approval Standard, Electrostatic Finishing Equipment, 1974, 6 pp.
Covers safety requirements for approval of electrostatic paint or powder finishing equipment.

FM: Approval Standard, Electric Interlocking Fuel Gas and Fuel Oil Cocks, 1973, 4 pp.
Covers requirements for approval of electric interlocking cocks which are manually operated to shut off the fuel supply to gas or oil fired burner equipment.

FM: Approval Standard, Airflow Interlocking Switches and Pressure Supervisory Switches for Fuel Oil, Fuel Gas and Ventilation or Combustion Air, 1974, 4 pp.
Covers requirements for approval of switches that electrically interlock air or fuel pressure with other combustion control equipment.

FM: Approval Standard, Nonprogramming and Programming Single or Multi-Burner Combustion Safeguards of the Industrial Gas and/or Fuel Oil Flame Supervising Types, 1970, 6 pp.
Covers requirements for approval of combustion safeguards for oil or gas fired industrial heating equipment. Units sense the presence of flame and cause the fuel to be shut off in the event of accidental flame failure.

FM: Approval Standard, Fuel Gas and Oil Safety Shutoff Valves, 1976, 3 pp.
Covers requirements for approval of shutoff valves installed in the fuel supply of industrial heating equipment.

FM: Approval Standard, Combustible Gas Detectors, 1982, 20 pp.
Covers proposed requirements for approval of

Titles & Abstracts

fixed and portable combustible gas detectors. Requirements are based to a large extent on work done by the Instrument Society of America and the Canadian Standards Association.

FM: Approval Standard, Electrical Equipment for Use in Class I, Division 2, Class II, Division 2 and Class III, Division 1 and 2 Hazardous Locations, 1986, 18 pp.

Covers requirements for approval of electrical equipment having Nonincendive circuits, field wiring and/or components. These requirements are based to a large extent on ISA-S12.12.

FLUID CONTROLS INSTITUTE, INC. (FCI)

The following FCI publications are available from the Fluid Controls Institute, Inc., P.O. Box 9036, 31 South Street, Morristown, NJ 07960. (201) 829-0990

FCI: 65-3, Standards for Determining Industrial Steam Trap Capacity Rating, 1965, 16 pp.

A brief description of the operating principles of various types of traps and the temperature depression below saturation on which capacities are based. This information provides an authoritative source of reference and makes possible a comparison of the relative capacities of traps of different types and manufacture.

FCI: 68-1, Procedure in Rating Flow and Pressure Characteristics of Solenoid Valves for Gas Service, 1968, 12 pp.

A recognized procedure under which solenoid valves intended for gas service may be rated for flow and pressure characteristics.

FCI: 68-2, Procedure in Rating Flow and Pressure Characteristics of Solenoid Valves for Liquid Service, 1968, 12 pp.

A recognized procedure for rating flow and pressure characteristics of solenoid valves for liquid service.

ANSI/FCI 69-1-1977, Pressure Rating Standard for Steam Traps, 1977, 7 pp.

Provides the minimum requirements for the design, fabrication, pressure rating, marking and testing of pressure-containing housing for steam traps.

FCI: 70-1, Standard Terminology and Definition for Filled Thermal Systems of Remote Sensing Temperature Regulators, 1970, 10 pp.

The purpose is to establish uniform terminology and definition for filled thermal system of remote sensing temperature regulators.

ANSI B16.104-1976 (FCI-70-2), Quality Control Standard for Control Valve Seat Leakage, 1976, 7 pp.

Establishes a series of seat leakage classes for control valves and defines the test procedures.

FCI: 71-1, Standard Terminology for Regulators, 1971, 6 pp.

The terminology in this standard is intended to identify by preferred current usage the different parts in a regulator.

FCI: 73-1, Pressure Rating Standard for "Y" Type Strainers, 1973, 6 pp.

Provides the minimum requirements for the design, fabrication, pressure rating, marking and testing of pressure containing housings for FCI approved "Y" Type Strainers for use with pipe conforming to dimensions specified in ANSI B36.10 and ANSI B36.19.

ANSI/FCI 74-1-1979, Silent Check Valve Standard, 1979.

A standard establishing uniform methods for design, rating and testing silent check valves.

ANSI/FCI 75-1-1979, Test Conditions and Procedures for Measuring Electrical Characteristics of Solenoid Valves, 1979.

A guide for manufacturers, users or others interested in test conditions and procedures for measuring electrical characteristics of solenoid valves.

FCI: 78-1, Pressure Rating Standard for Pipeline Strainers Other than "Y" Type, 1978.

A standard defining uniform methods for design, rating and testing of pipeline strainers other than "Y" Type.

FCI: 79-1, Standard for Proof of Pressure Ratings for Pressure Reducing Regulators, 1979.

Establishes guidelines for proof-of-design testing of pressure reducing regulators.

FCI: 81-1, Standard for Proof of Pressure Rating for Temperature Regulators, 1981.

Establishes guidelines for proof-of-design testing of temperature regulators.

FCI: 82-1, Recommended Methods for Testing and Classifying the Water Hammer Characteristics of Electrically Operated Valves.

A design guide to establish test procedures and to set performance criteria for valves which cause a shock wave in water media on shut-off.

FCI: 84-1, Metric Definition of the Valve Flow Coefficient Cv.

A standard defining the valve flow coefficient Cv commonly used for the purpose of calculating

Titles & Abstracts

valve flow capacity or leakage in metric terminology.

FCI: 85-1, Standard for Production and Performance Tests for Steam Traps.
A standard to assist manufacturers, users and specifiers to the product to comply with production and performance characteristics of automatic steam traps.

FCI: 86-2, Regulator Terminology
The terminology in this standard is intended to establish a common and preferred usage of terms as they apply to regulators.

FCI: 87-1 (Supercedes FCI 65-3).

FCI-87-2 Power Signal Standard for Spring-Diaphragm Actuated Control Valves.

GAS PROCESSORS ASSOCIATION (GPA)

The following are available from the Gas Processors Association, 6526 East 60th Street, Tulsa, OK 74145. (918) 493-3872.

GPA: Standard 2140-86, Liquefied Petroleum Gas Specifications and Test Methods, 1986, 52 pp.
Covers definitions and specifications for liquefied petroleum gases, vapor pressure, test, specific gravity test (pycnometer method), specific gravity test (hydrometer method), corrosion test (copper strip method), volatile sulphur test (lamp method), dryness test — commercial propane (dew point method) residue test — commercial propane (mercury freeze method), weathering test, residue test (end point index method), and sampling methods.

GPA: Standard 2142-57, Standard Factors for Volume Correction and Specific Gravity Conversion of Liquified Petroleum Gases.

GPA: Standard 2145-88, Physical Constants for the Paraffin Hydrocarbons and Other Components of Natural Gas Data are given in both English and SI units.
Compilation of authoritative numerical values for the paraffin hydrocarbons and other compounds occurring in natural gas and natural gas liquids.

GPA: Standard 2261-86, Method for Natural Gas Analysis by Gas Chromatography, 1986, 13 pp.
Covers the chromatographic analysis of natural gas to oxygen, nitrogen, helium, carbon dioxide, and hydrocarbons, methane through hexanes, and heavier. This method is applicable to natural gases and similar mixtures.

GPA: Publication 2165-75, Method for Analysis of Natural Gas Liquid Mixtures by Gas Chromatography, 1975, 6 pp.
This method is intended for the analysis of wide-range NGL mixtures, such as commercial de-ethanized and de-propanized natural gasoline mixtures, that cannot readily be entered into the chromatograph as a liquid by syringe or as a vapor at atmospheric pressure because of the presence of both highly volatile and heavy-end components. This method is intended to determine ethane and heavier hydrocarbons. Hydrogen sulfide, carbon dioxide, air, and methane are not determined.

GPA: Publication 2166-86, Methods for Obtaining Natural Gas Samples for Analysis by Gas Chromatography, 1986, 9 pp.
Describes procedures for obtaining representative "spot" samples of natural gas from vacuum or pressure sources in containers which are suitable for transporting the gas to a laboratory. Comparison of the degree of accuracy of eight methods.

GPA: Standard 2172, Method for Calculation of Gross Heating Value, Specific Gravity and Compressibility of Natural Gas Mixtures from Compositional Analysis.
This method outlines prcedures for calculating, at base conditions from compositional analyses, the following properties of natural gas mixtures: gross heating value (gross calorific value, combustion), relative density and compressibility factor.

GPA: Publication 2174-83, Method for Obtaining Hydrocarbon Fluid Samples Using A Floating Piston Cylinder, 1983, 5 pp.
Describes an alternate procedure for obtaining a representative sample of a hydrocarbon fluid and the subsequent preparation of that sample for laboratory analysis.

GPA: Standard 2186-86, Tentative Method for the Extended Analysis of Hydrocarbon Liquid Mixtures Containing Nitrogen and Carbon Dioxide by Temperature Programmed Gas Chromatography, 1986, 20 pp.
Intended for the compositional analysis of natural gas liquid streams where precise physical property data of the hexanes and heavier fraction are required.

GPA: Standard 2286-86, Tentative Method of Extended Analysis for Natural Gas and Similar Gaseous Mixtures by Temperature Programmed Gas Chromatography, 1986, 17 pp.
Covers determination of the chemical composition of natural gas and similar gaseous mixtures. In-

Titles & Abstracts

tended for use with rich gas systems and in situations where the heptanes plus compositional breakdown is desired.

GPA/API Standard 8182-84, Tentative Standard for the Mass Measurement of Natural Gas Liquids
This publication serves as a reference for the selection, design, installation, operation, and maintenance of single phase liquid mass measurement systems which operate in the 0.3 to 0.7 gm/cc density range.

GPA: Standard 8185-85, Orifice Metering of Natural Gas and Other Related Hydrocarbon Fluids. This is the revised AGA #3 now issued as API/ANSI 2530 and GPA 8185. These are identical publications.
This standard provides a procedure for the measurement of natural gas, hydrocarbons, and other related fluid flows using flange tap and pipe tap orifice meters.

INDUSTRIAL FASTENERS INSTITUTE (IFI)

The following standards are available from the Industrial Fasteners Institute, 1505 East Ohio Building, Cleveland, OH 44114. (216) 241-1482

IFI: Metric Standards, 1983, 500 pp.
Covers screw threads, materials, bolts, screws, studs, slotted and recessed screws, nuts, non-threaded fasteners. Provides data in metric only.

IFI: Fastener Standards, Fifth Edition, 1970, 500 pp.
Provides standards on all inch fasteners. Information is arranged by product group and contains over 100 pages of supporting technical data.

IFI: Capability Guide, 1986, 100 pp.
Covers fastener types, specifications, materials, bolt nut compatibility, platings, bolted joint design and tightening.

INDUSTRIAL RISK INSURERS (IRI)

The following are available from the Communications Department, Industrial Risk Insurers, 85 Woodland St., Hartford, CT 06102. (203) 520-7300

IRI: Loss Prevention and Protection for Chemical and Petrochemical Plants, 2nd edition, 28 pp.
Highlights the problems of insuring chemical and petrochemical plants and outlines prevention and protection guidelines aimed at preventing or minimizing losses in these properties.

IRI: Recommended Good Practice for the Protection of Electronic Data Processing and Computer-Controlled Industrial Processes, 2nd edition, Revised 1983, 62 pp.
Data processing and automatic control equipment, even though not directly involved in a fire, can be severely damaged or rendered inoperative by even a moderate exposure to heat, smoke, or corrosive fumes. Discussion or preferred fire prevention and protection measures to reduce potential loss or associated effects.

Pulp and Paper Mills: Loss Prevention and Protection, 2nd edition, 1983, 36 pp.
Discusses loss experience within the pulp and paper industry with recommendations for reducing and controlling losses.

INSTITUTE OF ELECTRICAL AND ELECTRONICS ENGINEERS (IEEE)

Publications of the former American Institute of Electrical Engineers (AIEE) and Institute of Radio Engineers (IRE) are identified by the IEEE number. All of these publications are available from the Institute of Electrical and Electronics Engineers, 345 East 47th Street, New York, 10017. (212) 705-7900

IEEE: STD 74-1958 (REAFF 1974), Test Code for Industrial Control (600 Volts or Less), 20 pp.
Establishes procedures for tests on a representative sample of an industrial control device or apparatus in order to substantiate conformance of that type of device or apparatus with a recognized standard of performance.

ANSI/IEEE: STD 81-1983, Guide for Measuring Earth Resistivity, Ground Impedence, and Earth Surface Potentials of a Ground System, 1983, 42 pp.
Describes and discusses the present state of the technique of measuring ground impedance, earth resistivity, potential gradients from currents in the earth, and the prediction of the magnitudes of ground resistance and potential gradients from scale model tests.

IEEE: STD 85-1973, (REAFF 1986), Test Procedure for Airborne Noise Measurements on Rotating Electric Machinery, 34 pp.
Outlines practical techniques and procedures which can be followed for conducting and reporting tests on rotating electrical machines of all

Titles & Abstracts

sizes to determine the airborne noise characteristics under steady state conditions.

ANSI/IEEE: STD 91-1984, Graphic Symbols for Logic Diagrams (Two-State Devices), 1984, 146 pp.

Sets forth principles governing the formation of graphical symbols for logic diagrams in which connections between symbols are generally shown with lines. Definitions of logic functions, the graphical representations of these functions, and examples of their application are given.

IEEE: STD 94-1970, Definition of Terms for Automatic Generation Control on Electric Power Systems, 8 pp.

Covers terms applicable to power system operation, governors, automatic control elements and action, telemetering, and load economics, includes nonstandard and alphabetical cross-indexes.

ANSI/IEEE: STD 100-1988, Standard Dictionary of Electrical and Electronic Terms — Fourth Edition.

IEEE: STD 108-1955, Proposed Recommended Guide for Specification of Electronic Voltmeters, 6 pp.

Provides a common basis of comparison among electronic voltmeters offered for general purpose applications.

IEEE: STD 118-1978, Standard Master Test Code for Resistance Measurements, 31 pp.

Provides instructions for those measurements of electric resistance which are commonly needed in determining the performance characteristics of electric machinery and equipment.

IEEE: STD 120-1955 (REAFF 1980), Master Test Code for Electrical Measurements in Power Circuits, 40 pp.

Gives instructions for those measurements of electrical quantities which are commonly needed in determining the performance characteristics of electric machinery and equipment.

ANSI/IEEE: STD 125-1988, Recommended Practice for Preparation of Equipment Specifications for Speed-Governing of Hydraulic Turbines.

IEEE: STD 139-1953, Recommended Practice for Measurement of Field Intensity Above 300 MHz from Radio-Frequency Industrial, Scientific and Medical Equipments, 14 pp.

Includes information on methods of measurement, associated measuring equipment and measurement precautions and factors affecting accuracy of measurement.

ANSI/IEEE: STD 141-1986, Recommended Practice for Electric Power Distribution for Industrial Plants, 608 pp.

Includes chapters on: System Planning; Voltage Considerations; Surge Voltage Protection; Application and Coordination of System Protective Devices; Fault Calculations; Grounding; Power Factors and Related Considerations; Power Switching; Transformation, and Motor-Control Apparatus; Instruments and Meters; Cable Systems; Busways; Electrical Energy Conservation; and Cost Estimating of Industrial Power Systems.

ANSI/IEEE: STD 142-1982, Recommended Practice for Grounding of Industrial and Commercial Power Systems, 136 pp.

Includes chapters on: Systems Grounding; Equipment Grounding; Static and Lightning Protection Grounding; and Connection to Earth.

ANSI/IEEE: STD 146-1980, Definitions of Fundamental Waveguide Terms, 1980, 10 pp.

Gives standard definitions of fundamental waveguide terms placing emphasis on waveguides (hollow uniconductor transmission lines).

ANSI/IEEE: STD 147-1979, Definitions for Waveguide Components, 1979, 7 pp.

Contains definitions for 41 of the more general, basic, and established terms for waveguide components in which the waveguide is treated as a generic term covering both transmission line and uniconductor waveguide.

IEEE: STD 148-1959, Waveguide and Waveguide Component Measurements, 1959 (REAFF 1971), 16 pp.

Provides general techniques for measurement of quantities which characterize a waveguide. References are included to cover specific procedures and equipment. Contents are limited to linear, reciprocal systems.

ANSI/IEEE: STD 162-1963, Definitions of Terms for Electronic Digital Computers, 2nd edition, 1963, (REAFF 1984), 12 pp.

Contains nearly all important terms related to electronic digital computers. Analog and programming terms are not included because they are covered in other standards.

ANSI/IEEE: STD 165-1977, (REAFF 1984), Definitions of Terms for Analog Computers, 1977, 12 pp.

Contains definitions for 172 terms common to analog computing techniques and associated hardware.

**ANSI/IEEE: STD 166-1977, (REAFF 1984), Defi-

Titles & Abstracts

nitions of Terms for Hybrid Computer Linkage Components, 1977.

ANSI/IEEE: STD 181-1977, Pulse Measurement and Analysis by Objective Techniques, 16 pp.
Provides definitions and descriptions of the techniques and procedures for time domain pulse measurements.

IEEE: STD 182-1961, Definitions of Terms for Radio Transmitters, 6 pp.
Defines 36 terms basic to the design of radio transmitters.

ANSI/IEEE: STD 186-1948, (REAFF 1976) Standard Methods of Testing Amplitude — Modulation Broadcast Receivers and ANSI/IEEE STD 189-1955 (REAFF 1976) Standard Method of Testing Receivers Employing Ferrite Core Loop Antennas, 28 pp.
Describes test assemblies and adjustments of input and output, operating condition, and radio receiver adjustments as applied to any type of receiver. Recommends procedures for measuring sensitivity, selectivity, and fidelity and other characteristics.

ANSI/IEEE: STD 206-1960, (REAFF 1978), Television: Measurement of Differential Gain and Differential Phase, 12 pp.
Describes a method of measuring differential gain and differential phase in video transmission systems.

IEEE: STD 216-1960, Definitions of Semiconductor Terms, 1960, (REAFF 1980), 4 pp.
Provides definitions of important terms relating to the physical aspects of semiconductor of materials and basic components developed from them.

ANSI/IEEE: STD 242-1986, Recommended Practice for Protection and Coordination of Industrial and Commercial Power Systems, 588 pp.
Includes chapters on: Calculation of Short-Circuit Currents; Instrument Transformers; Selection and Application of Protective Relays; Fuses; Low-Voltage Circuit Breakers; Ground Fault Protection; Conductor Protection; Motor Protection; Transformer Protection; Generator Protection; Bus and Switchgear Protection; Service Supply Line Protection; Overcurrent Coordination; and Maintenance Testing and Calibration.

ANSI/IEEE: STD 251-194, Proposed Heat Procedure for Direct-Current Tachometer Generators, 15 pp.
Covers instructions for conducting and reporting the more generally applicable and acceptable tests to determine the performance characteristics of direct-current tachometer generators.

IEEE: STD 261-1965, Letter Symbols for Thermoelectric Devices, 1965, 3 pp.
Presents standard letter symbols for quantities used in the field of thermoelectric devices.

ANSI/IEEE: STD 268-1982, Standard for Metric Practice, 48 pp.
Provides guidance for application of the modernized metric system and includes sections on: SI units and symbols; Application of the metric system; Recommendations concerning units; style and usage; Rules for conversion and rounding; and terminology.

IEEE: STD 284-1968, Standards Report on State-of-the-Art of Measuring Field Strength, Continous Wave, Sinusoidal, 1968, 7 pp.
Report on the state-of-the-art of measuring field strength of radio-frequency electromagnetic waves, with respect to available and desirable accuracies.

ANSI/IEEE: STD 290-180, Standard for Electric Couplings, 19 pp.
Covers the more generally applicable performance characteristics and conducting and reporting of the test for determining the performance characteristics of electric couplings.

ANSI/IEEE: STD 300-1982, Standard Test Procedure for Semiconductor Charge Particle Detectors, 28 pp.
Establishes standard test procedures for semiconductor charged-particle detectors. These detectors are in widespread use for the detection and high-resolution spectroscopy of charged particles.

ANSI/IEEE: STD 301-1976, Standard Test Procedures for Amplifiers and Preamplifiers for Semiconductor Radiation Detectors for Ionizing Radiation, 28 pp.
Includes sections on: Noise Linewidth Measurements, Preamplifier Noise; Pulse-Height Linearity; Count Rate Effects; Overload Effects; Pulse-Height Dependence on Rise Time; Pulse-Height Stability; and Crossover Walk.

IEEE: STD 302-1969, (REAFF 1981), Methods for Measuring (Below 1000 MHz) Electromagnetic Field Strength for Frequencies Below 1000 MHz in Radiowave Propagation, 1969, 15 pp.
Most measurements with which radio wave propagation are concerned involve the measurement of field strength. Standard methods for the measurement of this fundamental quantity for frequencies below 1000 megahertz are described.

IEEE: STD 306-1969, (REAFF 1981), Test Procedure for Charging Inductors, 1969, 57 pp.
This document pertains to the methods of mea-

surement of the electrical characteristics of charging inductors used in radar transmitters, linear particle accelerators; and similar equipment.

ANSI/IEEE: STD 308-1980, Standard Criteria for Class IE Power Systems for Nuclear Power Generating Stations, 1980, 24 pp.
Applies to the Class 1E portions of the following systems and equipment in single unit and multi-unit nuclear power generating stations: vital instrumentation and control power systems; alternating current power systems; and direct current power systems.

ANSI/IEEE: STD 309-1970, (REAFF 1984), Test Procedure for Geiger-Muller Counters, 8 pp.
Presents standard test procedures for Geiger-Muller counter radiation detectors.

IEEE: STD 311-1970, Standard Specification of General-Purpose Laboratory Cathode-Ray Oscilloscopes, 16 pp.
This standard documents the minimum information that users of general-purpose laboratory cathode-ray oscilloscopes typically need; provides potential purchasers and others with a common means for making comparisons between instruments; and provides uniformity of information from manufacturers.

IEEE: STD 314-1971, Standard Report on State of the Art of Measuring Unbalanced Transmission Line Impedance, 8 pp.
Applies to reporting the state-of-the-art of measuring impedance in distributed parameter coaxial waveguide systems, propagating a TEM wave.

IEEE: STD 316-1971, Requirements for Direct Current Instrument Shunts, 7 pp.
Applies to shunts for use in direct current circuits to extend the range of instruments or other measuring devices.

ANSI/IEEE: STD 325-1971, (REAFF 1982), Test Procedures for Germanium Gamma-Ray Detectors, 19 pp.
Presents test procedure for germanium gamma-ray detectors used for the detection and analysis of gamma-radiation.

ANSI/IEEE: STD 336-1985, Installation, Inspection, and Testing Requirements for Class 1E Instrumentation and Electric Equipment at Nuclear Power Generating Stations, 1980, 12 pp.
Sets forth the requirements for installation, inspection, and testing of Class 1E electric power, instrumentation, and control equipment and systems during the construction phase of a nuclear power generating station.

IEEE: STD 337-1972, Specifications Format Guide and Test Procedure for Linear, Single-Axis, Pendulous, Analog Torque Balance Accelerometer, 1972, (REAFF 1978), 47 pp.
Defines the requirements for a linear, single-axis, pendulous, analog torque balance accelerometer. The instrument is equipped with a permanent magnet torquer and is used as a sensing element to provide an electrical signal proportional to acceleration.

ANSI/IEEE: STD 338-1987, Standard Criteria for the Periodic Surveillance Testing of Nuclear Power Generating Station Safety Systems, 1987, 15 pp.
Establishes specific criteria for the periodic testing required to ensure operational availability of protection systems utilizing the capability called for in sections of the Criteria for Nuclear Power Generating Station Protection Systems.

ANSI/IEEE: STD 344-1975, (REAFF 1980), Guide for Seismic Qualification of Class 1E Equipment for Nuclear Power Generating Stations, 1975, 24 pp.
Provides direction for establishing procedures that will yield data which verify that the Class I electric equipment can meet its performance requirements during and following a design basis earthquake.

ANSI/IEEE: STD 376-1975, (REAFF 1986), Measurement of Impulse Strength and Impulse Bandwidth, 16 pp.
Provides basic information relating to the use of the impulse generator and interpretation of measurements made using instruments based on it.

ANSI/IEEE: STD 379-1988, Standard Application of the Single-Failure Criterion to Nuclear Power Generating Station Safety Systems.

ANSI/IEEE: STD 381-1977, (REAFF 1984), Criteria for Type Tests of Class 1E Modules, Used in Nuclear Power Generating Stations, 27 pp.
Describes the basic requirements of a type test program with the objective of verifying that a module used as Class IE Equipment in a nuclear power generating station meets or exceeds its design specifications.

IEEE: Std 389-1979, Recommended Practice for Testing Electronic Transformers and Inductors. 60 pp.
Presents a number of tests for use in determining the significant parameters and performance characteristics of electronic transformers and inductors. Included are tests for: electric strength, DC resistance, power losses, balance, equivalent impedances, terminated impedances, and other transformer properties.

Titles & Abstracts

ANSI/IEEE: Std 398-1972, Standard Test Procedures for Photomultipliers for Scintillation Counting and Glossary for Scintillation Counting Field. 28 pp.

Includes sections on: Photomultiplier characteristics, testing of photomultiplier characteristics, test conditions for photomultipliers, test instrumentation, and glossary for scintillation counting field.

ANSI/IEEE: Std 399-1980, Recommended Practice for Industrial and Commercial Power System Analysis. 224 pp.

Chapters include: Applications of Power System Analysis; Analytical Procedures; System Modeling; Load Flow Studies; Motor Starting Studies; Harmonic Studies; Switching Transients Studies; Reliability Studies; Grounding Mat Studies; and Computer Services.

ANSI/IEEE: Std 415-1986 Guide for Planning of Preoperational Testing Programs for Class 1E Power Systems for Nuclear Power Generating Stations. 15 pp.

Provides direction for establishing an acceptable preoperational testing program for Class 1E Power Systems.

ANSI/IEEE: Std 416-1984, ATLAS Test Language, 1984, 448 pp.

Defines the Abbreviated Test Language for All Systems (ATLAS). The term "all" was substituted for "avionics" in recognition of the wider application of the language. Used in preparation and documentation of test procedures which can be implemented either manually or with automatic or semi-automatic test equipment.

ANSI/IEEE: STD 446-1987, Emergency and Standby Power Systems for Industrial and Commercial Applications, 1987, 272 pp.

Provides commercial facility designers, operators and owners with guidelines for assuring uninterrupted power.

ANSI/IEEE: Std 449-1984, Standard for Ferroresonant Voltage Regulators. 26 pp.

Pertains to ferroresonant voltage regulators which operate at relatively constant frequencies and provide substantially constant ouput voltages in spite of relatively large changes in input voltage, and to controlled ferroresonant regulators which maintain substantially constant output voltages regardless of variations, within limits, of input voltage, temperature, frequency, and output load.

IEEE: STD 457-1982, Definitions of Terms for Nonlinear, Active, and Nonreciprocal Waveguide Components, 1982.

IEEE: STD 467-1980, Quality Assurance Program Requirements for the Design and Manufacture of Class 1E Instrumentation and Electric Equipment for Nuclear Power Generating Stations, 1980, 12 pp.

Sets forth the quality assurance program requirements for the design and manufacture of Class 1E instrumentation and electric equipment in a nuclear power generating station.

IEEE: STD 470-1972, Application Guide for Bolometric Power Meters, 26 pp.

Applies to bolometric power meters as complete instruments and to their constituent parts: bolometric detectors, bolometer units, and bolometer elements.

ANSI/IEEE: STD 474-1973, (REAFF 1982), Specifications and Test Methods for Fixed and Variable Attenuators, DC to 40 GHz, 1973, 27 pp.

Covers absorptive and reflective attenuators, both fixed as well as continuously variable or variable in fixed steps, both manual and programmable types. It does not cover electronic or solid-state-type attenuators.

ANSI/IEEE: Std 475-1983, Standard Measurement Procedure for Field-Disturbance Sensors (rf Intrusion Alarms). 16 pp.

Defines a test procedure for field-disturbance sensors to measure radio-frequency (rf), radiated field strength of the fundamental frequency, including second and third harmonics, and of any nonharmonic spurious emission within the frequency range from 0.3 GHz to 40.0 GHz.

ANSI/IEEE: STD 488-1978, Standard Digital Interface for Programmable Instrumentation, 83 pp.

Applies to interface systems used to interconnect both programmable and nonprogrammable electronic measuring apparatus with other apparatus and accessories necessary to assemble instrumentation systems.

ANSI/IEEE: STD 488.1-1987, Standard Digital Interface for Programmable Instrumentation.

ANSI/IEEE: STD 488.2-1987, Standard Codes, Formats, Protocols, and Common Commands for Use with ANSI/IEEE Std 488.1-1987, IEEE Standard Digital Interface for Programmable Instrumentation.

ANSI/IEEE: STD 497-1981, Criteria for Accident Monitoring Instrumentation for Nuclear Power Generating Stations, 1981, 14 pp.

Establishes the minimum design criteria for accident monitoring instrumentation.

Titles & Abstracts

ANSI/IEEE: STD 498-1985, Supplementary Requirements for the Calibration and Control of Measuring and Test Equipment Used in the Construction and Maintenance of Nuclear Power Generating Stations, 8 pp.

Establishes the requirements for a calibration program to control and verify the accuracy of M&TE (measuring and test equipment) which is used to assure that important parts of nuclear power generating stations are in conformance with prescribed technical requirements. This standard is intended to be used in conjunction with ANSI N45.2-1971.

ANSI/IEEE: STD 500-1984, Reliability Data Manual, 1984, 1424 pp.

Contains comprehensive and up-to-date collection of reliability data including maintenance rates as well as failure rates, failure rate ranges, failure modes and environmental factor information on generic components actually or potentially in use in nuclear power generating stations.

IEEE: STD 510-1983, Recommended Practices for Safety in High-Voltage and High-Power Testing, 1983, 19 pp.

Provide guidance for those who are establishing or revising their safety rules. Cites some of the hazards present in making high-voltage and high-power measurements, and some of the procedures and equipment which can be used to reduce personnel hazards.

ANSI/IEEE: STD 518-1982, Guide for the Installation of Electrical Equipment to Minimize Electrical Noise Inputs to Controllers from External Sources, 160 pp.

Develops a guide for the installation and operation of industrial controllers to minimize the disturbing effects of electrical noise on these controllers.

ANSI/IEEE: STD 525-1987, Guide for the Design and Installation of Cable Systems in Substations.

IEEE: STD 544-1975, Electrothermic Power Meters, 1975, 26 pp.

Provides a common means for making comparisons between instruments and to provide a uniformity of information among manufacturers. Also documents the minimum information that users of electrothermic power meters may need.

ANSI/IEEE: STD 583-1982, Modular Instrumentation and Digital Interface Systems (CAMAC), 1982, 81 pp.

Serves as a basis for a range of modular instrumentation capable of interfacing transducers and other devices to digital controllers for data and control. Specifies a data bus (dataway) by means of which instruments can communicate with each other with peripherals, computers and other external controllers.

ANSI/IEEE: STD 595-1982, Serial Highway Interface Systems (CAMAC), 1982, 85 pp.

Defines a serial highway system using byte-organized messages, and configured as a uni-directional loop to which are connected a system controller and up to sixty-two CAMAC crate assemblies or other control devices.

ANSI/IEEE: STD 596-1982, Parallel Highway Interface System (CAMAC), 1982, 42 pp.

Defines the CAMAC parallel highway interface system for interconnecting up to seven CAMAC crates (or other devices) and a system controller.

ANSI/IEEE: Std 622A-1984, Recommended Practice for the Design and Installation of Electric Pipe Heating Control and Alarm Systems for Power Generating Stations. 24 pp.

Provides recommended practices for designing and installing electric pipe heating control and alarm systems as applied to mechanical piping systems that require heat. Includes selection of control and alarm systems, accuracy considerations, local control usage, centralized control usage, qualification criteria of controls and alarms, and calibration and testing of controls and alarms.

IEEE: Std 625-1979, Recommended Practices to Improve Electrical Maintenance and Safety in the Cement Industry. 24 pp.

Recommendations apply to all electrical equipment such as, substations, power transformers, motor controls, generators, distribution systems, instruments and storage batteries commonly used in cement plants.

ANSI/IEEE: STD 645-1977, (REAFF 1982), Test Procedures for High-Purity Germanium Detectors for Ionizing Radiation, 11 pp.

Standard is a supplement to ANSI/IEEE STD 325-1971 (REAFF 1977) and provides additional test procedures required for high-purity germanium detectors.

ANSI/IEEE: STD 649-1980, Qualifying Class 1 E Motor Control Centers for Nuclear Power Generating Stations, 1980, 22 pp.

Describes the basic principles, requirements, and methods for qualifying Class 1E motor control centers for outside containment applications in nuclear power generating stations.

ANSI/IEEE: STD 675-1982, Multiple Controllers

Titles & Abstracts

in a CAMAC Crate, 1982, 32 pp.
Provides for the use of auxiliary controllers in a CAMAC crate to extend the capabilities and fields of application of the CAMAC modular instrumentation and interface system.

ANSI/IEEE: STD 683-1976 (REAFF 1987), Recommended Practice for Block Transfers in CAMAC Systems, 1976, 19 pp.
Presents recommended algorithms to encourage uniformity in design of CAMAC modules and controllers with resulting increased compatibility.

ANSI/IEEE: STD 690-1984, Standard for the Design and Installation of Cable Systems for Class 1E Circuits in Nuclear Power Generating Stations, 30 pp.
Provides direction for the design and installation of safety related electrical cable systems, including associated circuits. Also provided is guidance for the design and installation of those non-safety related cable systems that may affect the function of safety related systems.

IEEE: 696-1983, 696 Interface Devices, 1983, 40 pp.
Defines a general-purpose interface system for designers of new computer system components that will ensure their compatibility with present and future IEEE Std. 696 computer systems.

ANSI/IEEE: STD 716-1985, C/ATLAS Test Language, 1985, 438 pp.
Describes the language to be used for the writing of test programs for Units Under Test (UUTs), so that these programs can operate on various makes and models of Automatic Test Equipment (ATE).

ANSI/IEEE: STD 726-1982, Real-Time BASIC for CAMAC, 1982, 20 pp.
Defines the declarations and real-time statements for use with CAMAC hardware.

ANSI/IEEE: STD 728-1982, Recommended Practice for Code and Format Conventions, 1982, 52 pp.
Elaborates a family of codes and formats to be generated, processed, and interpreted by the device functions of apparatus operating in concert with interface functions.

ANSI/IEEE: STD 729-1983, Glossary of Software Engineering Terminology, 1983, 38 pp.
Defines more than 450 terms in general use in the software engineering field.

ANSI/IEEE: STD 730-1984, Software Quality Assurance Plans, 1984, 12 pp.
Provides uniform, minimum acceptable requirements for preparation and content of Software Quality Assurance Plans and provides a standard against which such plans can be assessed.

ANSI/IEEE: STD 739-1984, Bronze Book, 1984, 160 pp.
Energy conservation and cost-effective planning are addressed in the areas of engineering, design, applications, utilization, and to some extent the operation and maintenance of electric power systems to provide for the optimal use of electrical energy.

ANSI/IEEE: STD 746-1984, Standard for Performance Measurements of A/D and D/A Converters for PCM Television Video Circuits, 28 pp.
Describes methods for measuring the performance of uniformly coded analog-to-digital (A/D) converters and digital-to-analog (D/A) converters for pulse code modulation (PCM) television video signals.

IEEE: STD 748-1979, Spectrum Analyzers, 1979, 20 pp.
Provides a reference for specification comparison allowing a user to evaluate different instruments from a common base.

ANSI/IEEE: STD 758-1979 (REAFF 1987), Subroutines for CAMAC, 1979, 24 pp.
Presents a recommended set of software subroutines for use with CAMAC modular instrumentation and interface system.

ANSI/IEEE: STD 759-1984, Standard Test Procedures for Semiconductor X-Ray Energy Spectrometers, 39 pp.
Presents standard test procedures for semiconductor X-Ray energy spectrometers. Such systems consist of a semiconductor radiation detector assembly and signal processing electronics interfaced to a pulse height analyzer/computer.

ANSI/IEEE: STD 771-1984, Guide to the Use of ATLAS Test Language, 1984, 318 pp.
Defines ATLAS, Abbreviated Test Language for All Systems, which is standardized test language for expressing test specifications and test procedures.

ANSI/IEEE: STD 796-1983, Microcomputer System Bus, 1983, 46 pp.
Prepared for those users who evaluate or design products that will be compatible with the IEEE Std 796 system bus structure. Deals only with the interface characteristics of microcomputer devices, not with design specification, performance requirements, and safety requirements of modules.

Titles & Abstracts

ANSI/IEEE: STD 802.2-1985, Logical Link Control, 1985, 111 pp.
Provides a description of the peer-to-peer protocol procedures that are defined for the transfer of information and control between any pair of Data Link Layer service access points on a local area network.

ANSI/IEEE: STD 802.3-1985, Carrier Sense Multiple Access with Collision Detection, 1985, 143 pp.
Encompasses several media types and techniques for signal rates from 1 Mb/s to 20 Mb/s. Provides the necessary specifications and related parameter values for a 10 Mb/s baseband implementation.

ANSI/IEEE: STD 802.4-1985, Token-Passing Bus Access Method and Physical Layer Specifications, 1985, 238 pp.
Deals with all elements of the token-passing bus access method and its associated physical signalling and media technologies. Deals exclusively with the broadcast type.

ANSI/IEEE: STD 802.5-1985, Local Area Networks: Token Ring Access Method and Physical Layer Specifications, 89 pp.
Defines: the frame format and introduces medium access control frames, timers and priority stacks; medium access control protocol; physical layer functions of symbol encoding and decoding, symbol timing, and latency buffering; and the 1 and 4 Mb/s, shielded twisted pair attachment of the station to the medium including the definition of the medium interface connector.

ANSI/IEEE: STD 803-1983, Principles and Definitions, 1983.
Presents principles, definitions, and a procedure whereby systems/structures and component functions of power plant projects and related facilities can be uniquely identified.

ANSI/IEEE: STD 803A-1983, Component Function Identifiers, 1983, 20 pp.
Provides component function identifiers which have been selected considering the functional levels within a power plant which are significant to design, procurement, construction, operation, and maintenance applications.

ANSI/IEEE: STD 804-1983, Implementation, 1983, 15 pp.
Provides instruction for the implementation of the EIIS and includes information for establishing codes, code structure, and identification codes on equipment and design documents, and for reporting.

ANSI/IEEE: STD 805-1984, System Identification in Nuclear Power Plants, 1984, 177 pp.
Source of nuclear power plant system description concentrating on system function and includes internal detail where necessary to clearly support the system function description. System descriptions and diagrams represent typical systems.

ANSI/IEEE: STD 810-1987, Standard for Hydraulic Turbine and Generator Integrally Forged Shaft Couplings and Shaft Runout Tolerances.

ANSI/IEEE: STD 828-1983, Software Configuration Management Plans, 1983, 11 pp.
Provides the minimum requirements for preparation and content of Software Configuration Management (SCM) Plans and is applicable to the entire life cycle of critical software. Identifies essential items that appear in SCM plans, and provides examples to enhance clarity and promote understanding.

ANSI/IEEE: STD 829-1983, Software Test Documentation, 1983, 48 pp.
Describes a set of basic test documents associated with the dynamic aspects of software testing and defines the purpose, outline and content of each basic document.

ANSI/IEEE: STD 830-1984, Software Requirements Specifications, 1984, 24 pp.
Describes approaches to good practice in the specification of software requirements.

ANSI/IEEE: STD 896.1-1987, Standard Backplane Bus Specification for Multiprocessor Architectures: Futurebus.

ANSI/IEEE: STD 960-1986, Standard FASTBUS Modular High-Speed Data Acquisition and Control System, 232 pp.
Standardized modular data-bus system for data acquisition, data processing and control. The system consists of multiple bus segments that can operate independently, but link together for passing data and other information.

IEEE: STD 961-1987, Standard for an 8-Bit Microcomputer Bus System: STD Bus.

IEEE: 1003.1-1988, Standard Portable Operating System Interface for Computer Environments.

IEEE: 1051-1988, Recommended Practice for Parameters to Characterize Digital Loop Performance.

43

Titles & Abstracts

ANSI/IEEE/ANS: STD 7432-1982, Application Criteria for Programmable Digital Computer Systems in Safety Systems of Nuclear Power Generating Stations, 1982, 5 pp.
Establishes application criteria for programmable digital computer systems of nuclear power generating stations.

ANSI/IEEE: C57.13-1978, Standard Requirements for Instrument Transformers, 62 pp.
Covers certain electrical, dimensional and mechanical characteristics, and takes into consideration certain safety features of current and inductively coupled voltage transformers of types generally used in the measurement of electricity and the control equipment associated with the generation, transmission and distribution of alternating current.

IEEE: C57.13.2-1988, Conformance Test Procedures for Instrument Transformers, American National Standard for SH12005.

IEEE: C62-1988, Complete 1988 Edition: Guides and Standards for Surge Protection.

ANSI: C63.2-1980, Specifications for Electromagnetic Noise and Field Strength Instrumentation, 10 kHz to 1 GHz, 24 pp.
Describes requirements for instruments measuring quasi-peak, peak, rms, and average values.

ANSI: C95.3-1973 (Reaff 1979) Standard Techniques and Instrumentation for the Measurement of Potentially Hazardous Electromagnetic Radiation at Microwave Frequencies, 36 pp.
Sections included on: electromagnetic environment, theoretical calculations, methods of measurement and instrumentation requirements and calibration.

ANSI: C95.5-1981, Recommended Practice for the Measurement of Hazardous Electromagnetic Fields — RF and Microwave, 35 pp.
Specifies techniques and instrumentation for measurement of potentially hazardous electromagnetic fields both in the near field and far field of the electromagnetic source. Management techniques and instruments also apply to fields in the neighborhood of flammable materials and explosive devices.

ANSI: N13.4-1971 (Reaff 1983) Standard for the Specification of Portable X- or Gamma-Radiation Survey Instruments, 10 pp.
Provides the means of stating or describing performance as applicable to portable X- or gamma-radiation survey instruments, which includes a detector, and visual, analog and/or digital type readout.

ANSI: N42.4-1971 (Reaff 1985) Standard for High Voltage Connectors for Nuclear Instruments, 8 pp.
Applicable to coaxial high voltage connectors on nuclear instruments for dc applications up to 3500 volts rms at 60 Hz. The connectors are of the "safe" type in that the pin and socket contacts are well and securely recessed in the connector housing.

ANSI: N42.14-1978, (Reaff 1985), Calibration and Usage of Germanium Detectors for Measurement of Gamma-Ray Emission of Radionuclides, 1978, 13 pp.
Covers the energy and full energy peak efficiency calibration as well as the determination of gamma-ray energies in the 0.06 to 2 MeV energy region and is designed to yield gamma-ray emission rates with an uncertainty of plus/minus 3 percent.

ANSI: N42.18-1980 (Reaff 1985), Specification and Performance of On-Site Instrumentation for Continuously Monitoring Radioactivity in Effluents, 1980, 16 pp.
Applies to continuous monitors that measure normal releases, detect inadvertent releases, show general trends, and annunciate radiation levels that have exceeded predetermined values.

ANSI: N317-1980 (Reaff 1985), Standard Performance Criteria for Instrumentation Used for Inplant Plutonium Monitoring, 16 pp.
Presents performance criteria for radiation protection instrumentation essential to inplant plutonium monitoring. Plutonium radiations are also characterized. Criteria is limited to instruments capable of measuring: photon radiations within the energy range of 0.010 to 1.25 MeV; neutron radiations within the energy range from thermal to 10 MeV; and alpha radiations within the emitted energy range of 4.5 to 7.5 MeV.

ANSI: N320-1979 (Reaff 1985), Standard Performance Specifications for Reactor Emergency Radiological Monitoring Instrumentation, 15 pp.
Defines the essential performance parameters, and general placement of emergency instrumentation, radiological instrumentation systems, systems for monitoring conditions within the reactor facility, instrumentation systems for detecting and quantifying the release to the environs, installed systems for monitoring conditions in the environs, and portable instrumentation for monitoring the release of radionuclides associated with a postulated serious accident at a reactor facility.

ANSI: N323-1978 (Reaff 1983), Standard Radiation Protection Instrumentation Test and Calibration, 23 pp.
Establishes calibration methods for portable radi-

Titles & Abstracts

ation protection instruments used for detection and measurement of levels of ionizing radiation fields or levels of radioactive surface contamination, and includes conditions, equipment, and techniques for calibration as well as the degree of precision and accuracy required.

INSTITUTE OF ENVIRONMENTAL SCIENCES (IES)

The following IES Standards are available from Institute of Environmental Sciences, 940 East N.W. Hwy., Mt. Prospect, IL 60056. (312) 255-1561

IES: RP-CC-001-86, Recommended Practice for HEPA Filters, 1986, 10 pp.
Covers basic requirements for HEPA (high efficiency particulate air) filter units as a basis for agreement between buyer and seller.

IES: RP-CC-002-86, Recommended Practice for Laminar Flow Clean Air Devices, 1986, 10 pp.
Covers definitions, procedures for evaluating performance, and major requirements of laminar flow clean air devices.

IES: RP-CC-006-84-T, Recommended Practice for Testing Clean Rooms, 1984, 16 pp.
Covers testing methods for characterizing the performance of clean rooms.

IES: RP-CC-008-84, Recommended Practice for Gas-Phase Adsorber Cells, 1984, 8 pp.
Covers the design and testing of modular gas-phase adsorber cells for use where high efficiency removal of gaseous contaminants is required.

IES: CC-009-84, Compendium of Standards, Practices, Methods, Relating to Contamination Control, 1984, 53 pp.
Lists standards, practices, methods, technical orders, specifications, and similar documents developed by government, industry, and technical societies in the United States and other countries, which are related to the field of contamination control.

IES: CC-011-85-T, A Glossary of Terms and Definitions Related to Contamination Control, 16 pp.
This document defines more than 180 terms related to contamination control, contains charts on Characteristics of Particles and Particle Dispersoids, chart on the Sizes of Air-Borne Contaminants, Useful Equivalent and Conversion Table, and list of frequently used Acronyms and Initials. Prepared by RP-11 Working Group.

IES: RP-CC-013-86-T, Recommended Practice for Equipment Calibration or Validation Procedures.
Covers definitions and procedures for calibrating instruments used for testing clean rooms and clean air devices, and for determining intervals of calibration.

INSTITUTE FOR INTERCONNECTING AND PACKAGING ELECTRONIC CIRCUITS (IPC)

The following documents are available from the Institute for Interconnecting and Packaging Electronic Circuits, 7380 N. Lincoln Ave., Lincolnwood, IL 60646. (312) 677-2850

Each of the manuals listed is constantly being expanded and updated. Your initial purchase entitles you to receive all new information developed for these publications for two full years, plus the balance of the months during the year the publication is purchased.

IPC: Assembly-Joining Handbook.
The new Assembly and Joining Handbook contains practical and useful information regarding various approaches and techniques for the interconnection of electronic components. This material, developed by experts in the field of assembly and joining techniques and approved by the IPC committee on assembly and joining, is divided into eight sections. These sections contain information on: **Introductory Material; Printed Wiring Boards; Component/Lead types; Joining Materials; Component Mounting; Joining Techniques; Packaging; Quality Assurance and Testing.**
It is recognized that ideas and techniques discussed in the initial release will be subject to improvement and change. As changes occur in the technology the material will be revised and the changes or new information sent to subscribers of this handbook.

IPC: D-390A, Automated Design Guidelines.
This document is a general overview of computer aided design and its processes, techniques, considerations and problem areas with respect to printed circuit design. The guidelines describe the CAD design process from the initial input package requirements through to engineering change.

IPC: TM-650, Printed Circuit Test Methods Manual.
Provides a comprehensive compendium of pertinent information on test methods that will be useful to manufacturers and users of printed circuit boards. Designed to provide specific informa-

Titles & Abstracts

tion on test methods it does not attempt to establish acceptability levels for performance. Most of the test methods have been taken from previously approved standards and specifications of the IPC. Other test methods have been taken from either government standards or from other standards that have extensive use in the printed circuit industry. Contains over 100 test methods, and is divided into five major sections. A separate Table of Contents is shown at the beginning of each section.

IPC: Printed Wiring Design Guide.
Contains a compilation of valuable and useful data pertinent to the problems encountered in the application of printed wiring design principles. Sections include: General Printed Wiring Information, Single-Sided Printed Wiring (Rigid), Double-Sided Printed Wiring (Rigid), Multilayer Printed Wiring (Rigid), Flexible Printed Wiring, Printed Wiring Assemblies, Documentation, Quality Assurance and Testing, Computer Aided Design, plus Appendix. Update supplements available.

IPC: Technical Manual.
Divided into three volumes, it includes copies of all IPC (more than 50) Specifications and Standards.

INSTRUMENT SOCIETY OF AMERICA (ISA)

The following ISA Standards and Recommended Practices are available from the Instrument Society of America, 67 Alexander Drive, P.O. Box 12277, Research Triangle Park, NC 27709. A complete copy of each Standard is in Section III of this book. (919) 549-8411

ISA-RP2.1, Manometer Tables, Reaffirmed 1978, 31 pp.
Presents abbreviations and fundamental conversion factors commonly used in manometry, recommended definitions of pressure in terms of a column of mercury and water, and for a large number of liquids, tables of pressures indicated by, or equivalent to, heights of columns at various temperatures.

ISA-S5.1, Instrumentation Symbols and Identification (Formerly ANSI Y32.20), (ANSI/ISA-1984), 1984, 52 pp.
Establishes a uniform means of designating instruments and instrumentation systems used for measurement and control. The differing established procedural needs of various organizations are recognized (where not inconsistent with the objectives of the standard) by providing alternative symbolism methods. A number of options are provided for adding information or simplifying the symbolism, if desired. Includes additional information on symbolism for function blocks, function designations, computer functions, and programmable logic control.

ISA-S5.2, (ANSI/ISA-1976, R1981), Binary Logic Diagrams for Process Operations, R1981, 19 pp.
Provides symbols, both basic and non-basic, for binary operating functions. Intended to symbolize the binary operating functions of a system in a manner that can be applied to any class of hardware, whether electronic, electrical, fluidic, pneumatic, hydraulic, mechanical, manual, optical, or other.

ISA-S5.3, Graphic Symbols for Distributed Control/Shared Display Instrumentation, Logic and Computer Systems, 1982, 14 pp.
Establishes documentation for that class of instrumentation consisting of computers, programmable controllers, minicomputers and microprocessor based systems that have shared control, shared display or other interface features. Symbols are provided for interfacing field instrumentation, control room instrumentation and other hardware to the above.

ISA-S5.4, (ANSI/ISA-1976, R1981), Instrument Loop Diagrams, 1981, 11 pp.
Provides a method and practice for the preparation and use of instrument loop diagrams in the design, construction, checkout, startup, operation, maintenance, and reconstruction of instrument systems in industrial plants.

ISA-S5.5, Graphic Symbols for Process Displays, (ANSI/ISA-1985), 1986, 40 pp.
Provides a system of graphic symbols for conveying information on visual display units (VDUs) used for process monitoring and control. Intended to ensure compatibility of symbols on process VDUs with related symbols used in other disciplines. The standard applies to computers, distributed control systems, etc., and covers both color and monochromatic displays. It supplements ISA-S5.1 and ISA-S5.3.

ISA-RP7.1 Pneumatic Control Circuit Pressure Test, 1956, 6 pp.
Intended to provide a satisfactory procedure for the testing of pneumatic control circuits for leaks together with reasonable criteria for acceptance of work done and suitable aids for performance.

ISA-S7.3 (ANSI/ISA-1975, R1981), Quality Standard for Instrument Air, R1981, 6 pp.
Establishes a maximum allowable moisture content at which the instruments will function satisfactorily; a maximum entrained particle size

which will avoid plugging and wear/erosion of air passages and orifices; a maximum allowable oil content which will avoid malfunction due to clogging and wear of the components; an awareness of a possible source of corrosive or toxic contamination entering the air system; through the compressor suction, plant air system cross connections, or instrument air connections directly connected to processes.

ISA-S7.4, (ANSI/ISA-1981), Air Pressures for Pneumatic Controllers, Transmitters, and Transmission Systems, 1981, 4 pp.

Purpose is to establish standard operating pressure ranges for pneumatic intelligence transmission systems; and standard air supply pressures (with limit values) for operation of pneumatic controllers and pneumatic intelligence transmission systems.

ISA-RP7.7 Recommended Practice for Producing Quality Instrument Air, 1984, 16 pp.

Establishes general equipment guidelines for producing instrument air of the quality defined in ANSI/ISA-S7.3-1975 (R1981). This document enumerates equipment characteristics to include types, range, and performance of the components necessary to meet air quality requirements of ANSI/ISA-S7.3. This recommended practice lists tests of system, and components where applicable, to check performance of instrument air supply system to ANSI/ISA-S7.3 requirements.

ISA-RP12.1, Electrical Instruments in Hazardous Atmospheres, 1960, 7 pp.

Provides general guidance for the safe installation of electrical instruments using appropriate means to prevent ignition of flammable gases and vapors.

ISA-S12.4, Instrument Purging for Reduction of Hazardous Area Classification, 1970, 14 pp.

Covers a technique for reducing the hazard classification by the continuous addition of an air or inert gas within a general purpose enclosure. Refers only to hazards created by gases or vapors, and is concerned only with those system design criteria related to electrical ignition of a hazardous gas or vapor.

ISA-RP12.6 (ANSI/ISA-1977), Installation of Intrinsically Safe Instrument Systems in Class I Hazardous Locations, 10 pp.

Provides guidance for the design and installation of field installed wiring in non-hazardous locations and for the layout and wiring of panels which contain intrinsically safe wiring. Intended for use in conjuction with nationally recognized codes covering wiring practices, such as NEC Code NFPA 70 (ANSI C1) and Canadian Electrical Code, Part 1.

ISA-S12.10, Area Classification in Hazardous Dust Locations, 1973, 24 pp.

Evaluates the degree of dust hazard in locations made hazardous by the presence of a cloud or blanket of dust. Is in conformance with (and attempts to expand and clarify) the National Electrical Code and the Canadian Electrical Code.

ISA-S12.11, Electrical Instruments in Hazardous Dust Locations, 1973, 12 pp.

Summarizes the requirements for safe and economical installation of electrical instruments in locations made hazardous by a cloud or blanket of dust. (See ISA S12.10 for classification of such areas.) Is in conformance with (and attempts to expand and clarify) the National Electrical Code and the Canadian Electrical Code.

ISA-S12.12 Electrical Equipment for Use in Class I, Division 2 Hazardous (Classified) Locations, (ANSI/ISA-1984), 1984, 24 pp.

Provides requirements for the design, construction, and marking of electrical equipment, or parts of such equipment used in Class I, Division 2 locations. This document establishes uniformity in test methods for determining the suitability of the equipment and associated circuits and components as they are related to their ability to ignite a specified flammable gas or vapor-in-air mixture. This standard applies only to equipment, circuits, or components designed and assessed specifically for use in Class I, Division 2, hazardous locations as defined by the National Electrical Code NFPA No. 70, Articles 500 and 501, or the Canadian Electrical Code (Part I), C22.1, Section 18.

ISA-S12.13, Part I, Performance Requirements, Combustible Gas Detectors, (ANSI/ISA-S12.13, Part I-1986), 1986, 24 pp.

Improves the level of electrical safety and safety-oriented performance of combustible gas detection instruments used in hazardous (classified) locations. Covers the details of construction, performance, and test for portable, mobile, and stationary electrical instruments for sensing the presence of combustible gas or vapor concentrations in ambient air; parts of these instruments may be installed or used in Class I hazardous locations and gaseous mines in accordance with codes specified by authorities having jurisdiction. This standard does **not** cover gas detection instruments of the laboratory- or scientific-type used for analysis or measurement, instruments used for process control and process monitoring purposes, or instruments used for residential purposes.

ISA-RP12.13, Part II, Installation, Operation, and Maintenance of Combustible Gas Detection Instruments, 1987, 152 pp.

Establishes user criteria for the installation, oper-

Titles & Abstracts

ation, and maintenance of combustible gas detection instruments. Covers storage, user recordkeeping, maintenance, checkout procedures, calibration, external power supply systems, etc. Provides a substantial list of references.

ISA-RP16.1, 2, 3, Terminology, Dimensions, and Safety Practices for Indicating Variable Area Meters (Rotameters, Glass Tube, Metal Tube, Extension Type Glass Tube), 1959, 6 pp.

Combined RP16.1, 16.2 and 16.3 — intended to (a) establish uniformity of connection dimensions to permit interchangeability of one manufacturer's meters with another manufacturer's meters of the same size; (b) provide a common ground of understanding of the terminology, use, and component parts and accuracies of these meters; and (c) to provide a reference for the safe working pressures of these meters.

ISA-RP16.4, Nomenclature and Terminology for Extension Type Variable Area Meters (Rotameters), 1960, 3 pp.

Defines the nomenclature and terminology of various types of extensions applicable to 5 in. (125 mm) glass and metal tube variable area meters (rotameters) covered in ISA-RP16.1, 2, 3.

ISA-RP16.5, Installation, Operation, Maintenance Instructions for Glass Tube Variable Area Meters (Rotameters), 1961, 6 pp.

Covers the general considerations, important to the installation, operation and maintenance of meters to obtain the most reliable results.

ISA-RP16.6, Methods and Equipment for Calibration of Variable Area Meters (Rotameters), 1961, 7 pp.

Describes the methods and equipment used for calibrating the glass and metal metering tube area meters (rotameters) covered in RP16.1, 2, 3.

ISA-S18.1 (ANSI/ISA-1979, R 1985), Annunciator Sequences and Specifications, 36 pp.

Covers electrical annunciators that call attention to abnormal process conditions by the use of individual illuminated visual displays and audible devices. Sequence designations provided can be used to describe basic annunciator sequences and also many sequence variations.

ISA-S20, Specification Forms for Process Measurement and Control Instruments, Primary Elements and Control Valves, R1981, 72 pp.

These forms are intended to assist the specification writer to present the basic information. In this sense they are "short-form" specifications or "check sheets" and may not include all necessary engineering data or definitions of application requirements. While the types of instruments described by these forms are more common to the process industries the forms should also prove useful in the other areas if special requirements are defined elsewhere.

ISA-S26 (ANSI MC4.1-1975), Dynamic Response Testing of Process Control Instrumentation, 25 pp.

Incorporating four revised ISA recommended practices, the standard establishes the basis for dynamic response testing of measurement and control equipment with pneumatic output and electric output, and for closed loop actuators for externally actuated control valves and other final control elements. Pulse testing techniques as well as methods for sine wave, step, and pulse-type signals are included.

ISA-RP31.1 (ANSI/ISA RP31.1-1977), Specification, Installation, and Calibration of Turbine Flowmeters, 21 pp.

Establishes minimum ordering information, recommended acceptance and qualification test methods including calibration techniques, uniform terminology and drawing symbols, and recommended installation techniques for volumetric turbine flow transducers having an electrical output.

ISA-S37.1 (ANSI/ISA-1975, R1982), Electrical Transducer Nomenclature and Terminology, R1982, 15 pp.

Establishes uniform nomenclature for transducers and uniform simplified terminology for transducer characteristics.

ISA-RP37.2, Guide for Specifications and Tests for Piezoelectric Acceleration Transducers for Aerospace Testing, R1982, 19 pp.

Covers piezoelectric acceleration transducers, primarily those used in aerospace test instrumentation. Terminology used in this document follows ISAS37.1, "Electrical Transducer Nomenclature and Terminology," except that additional terms considered applicable to piezoelectric vibration transducers are defined.

ISA-S37.3 (ANSI/ISA-1975, R, 1982), Specifications and Tests for Strain Gage Pressure Transducers, R 1982, 22 pp.

Establishes for strain gage pressure transducers; uniform minimum specifications for design and performance characteristics; uniform acceptance and qualification test methods, including calibration techniques; uniform presentation of minimum test data; and a drawing symbol for use in electrical schematics.

ISA-S37.5 (ANSI/ISA-1975, R 1982), Specifications and Tests for Strain Gage Linear Acceler-

ation Transducers, R1982, 18 pp.

Establishes uniform minimum specifications for design and performance characteristics, uniform acceptance and qualification test methods including calibration techniques, uniform presentation of minimum test data, and a drawing symbol for use in electrical schematics for strain gage linear acceleration transducers.

ISA-S37.6 (ANSI/ISA-1976, R1982), Specifications and Tests of Potentiometric Pressure Transducers, R1982, 27 pp.

Establishes for potentiometric pressure transducers; uniform minimum specifications for design and performance characteristics; uniform acceptance and qualification test methods, including calibration techniques; uniform presentation of minimum test data; and a drawing symbol for use in electrical schematics.

ISA-S37.8 (ANSI/ISA-1977, R1982), Specifications and Tests for Strain Gage Force Transducers, R1982, 16 pp.

Outlines uniform general specifications, acceptance and qualification methods, methods for data presentation, and includes a drawing symbol used in electrical schematics for tension, compression and combination tension/compression transducers.

ISA-S37.10 (ANSI/ISA-1975, R1982), Specifications and Tests for Piezoelectric Pressure and Sound-Pressure Transducers, R1982, 22 pp.

Establishes uniform specifications for describing design and performance characteristics, acceptance and qualification test methods and calibration techniques, and procedures for presenting test data for piezoelectric (including ferro-electric) pressure and sound-pressure transducers.

ISA-S37.12 (ANSI/ISA-1977, R1982), Specification and Test for Potentiometric Displacement Transducers, R1982, 21 pp.

Covers potentiometric displacement transducers, primarily those used in measuring systems. The specifications are not intended to cover transducers used in hazardous locations as specified in the National Electrical Code nor are all requirements covered for transducers used in nuclear power plants.

ISA-RP42.1 Nomenclature for Instrument Tube Fittings, 1982, 8 pp.

Defines nomenclature for tubing fittings most commonly used in instrumentation. It is not intended as a substitute for manufacturer's catalog numbers, nor does it apply to special fittings. It is intended to apply to a mechanical fitting rather than a sweat fitting.

ISA-S50.1 (ANSI/ISA-1975, R1982), Compatibility of Analog Signals for Electronic Industrial Process Instruments, R1982, 11 pp.

This standard applies to analog dc signals used in process control and monitoring systems to transmit information between subsystems or separated elements of systems. Its purpose is to provide for compatibility between the several subsystems or separated elements of given systems.

ISA-S51.1 (ANSI/ISA 1979), Process Instrumentation Terminology, 1979, 41 pp.

Intended to include all specialized terms used to describe the use and performance of the instrumentation and instrument systems used for measurement, control or both in the process industries.

ISA-RP52.1, Recommended Environments for Standard Laboratories, 1975, 18 pp.

Recommendations for three levels of standardization are presented — from the more general National Bureau of Standards, through commercial, industrial and government laboratories. Requirements for nine environmental factors are discussed.

ISA-RP55.1 (ANSI: MC8.1-1975), Hardware Testing of Digital Process Computers, R1983, 54 pp.

Establishes a basis for evaluating functional hardware performance of digital process computers. Covers general recommendations applicable to all hardware performance testing, specific tests for pertinent subsystems and system parameters. Includes a brief glossary of terms used.

ISA-RP60.3, Human Engineering for Control Centers, 16 pp.

Assists the design engineer in establishing concepts which accommodate physical and mental capabilities of the operator while recognizing the operator's limitations. This recommended practice is limited to those aspects of human engineering that will affect the layout of and equipment selection for the control center. It is recognized that some of the human factors discussed in this document are also used in the design and manufacture of instruments.

ISA-RP60.6, Nameplates, Labels and Tags for Control Centers, 1984, 24 pp.

Assists the designer or engineer in choosing and specifying the method of identifying items mounted on a control center or associated with a control center facility. This recommended practice summarizes identification methods and suggests the use of nameplates, labels, and tags. Examples are included for guidance in preparing drawings and specifications. This recommended practice also covers functional definitions associated with nameplates, labels, and tags.

Titles & Abstracts

ISA-RP60.8, Electrical Guide for Control Centers, 1978, 6 pp.
Assists the design engineer in establishing the electrical requirements of a control center; it is also intended to comply with the provisions of the NEC. Special considerations which may apply to particular devices or circuits are not taken into account in this recommended practice.

ISA-RP60.9, Piping Guide for Control Centers, 1981, 11 pp.
Assists the design engineer in defining the piping requirements for pneumatic signals and supplies in control centers.

ISA-S61.1 (ANSI/ISA-S61.1-1977), Industrial Computer System FORTRAN Procedures for Executive Functions, Process Input-Output and Bit Manipulation, 1977, 11 pp.
Presents external procedure references for use in industrial computer control systems. These external procedure references permit interface with executive programs, process input and output functions and allow manipulation of bit strings. The FORTRAN statements described in this standard conform to the ANSI:X3.9-1966 Standard FORTRAN. No changes to standard FORTRAN syntax are intended.

ISA-S61.2 (ANSI/ISA-S61.2-1978), Industrial Computer System FORTRAN Procedures for File Access and the Control of File Contention, 1978, 7 pp.
Presents external procedure references for use in industrial computer control systems. These external procedure references provide means for accessing files, and also provide means for resolving problems of file access contention in a multi-programming multiprocessing environment.

ISA-S67.01, (ANSI/ISA-S67.01-1981, R1986), Transducer and Transmitter Installation for Nuclear Safety Applications, R1986, 14 pp.
Covers the installation of transducers for nuclear-safety-related applications, excepting those for measurands of liquid metals. It establishes requirements and recommendations for the installation of transducers and auxiliary equipment for nuclear power plant applications outside of the main reactor vessel.

ISA-S67.02, Nuclear-Safety-Related Instrument Sensing Line Piping and Tubing Standards for Use in Nuclear Power Plants, (ANSI/ISA-1980), 1981, 14 pp.
Covers design, protection and installation of nuclear-safety-related instrument sensing lines for light water cooled nuclear power plants. The standard covers the pressure boundary requirements for piping, capillary tubing, and tubing lines up to and including one inch (25.4 mm) outside diameter or three-quarter inch nominal pipe.

ISA-S67.03, Standard for Light Water Reactor Coolant Pressure Boundary Leak Detection, (ANSI/ISA-1982), 1982.
Defines design criteria that are intended to insure that adequate Reactor Coolant Pressure Boundary leak detection capabilities are provided to the nuclear plant operator and to meet the Code of Federal Regulations.

ISA-S67.04, Setpoints for Nuclear Safety-Related Instrumentation Used in Nuclear Power Plants, 1982, 15 pp.
Develops a basis for establishing setpoints for actions determined by the design basis for protection systems and to account for instrument errors and drift in the channel from the sensor through and including the bistable trip device.

ISA-S67.06, Response Time Testing of Nuclear-Safety-Related Instrument Channels in Nuclear Power Plants, (ANSI/ISA-1984), 1984, 20 pp.
Delineates requirements and methods for determining the response time characteristics of nuclear-safety-related instrument channels. The standard applies only to those instrument channels whose primary sensors measure pressure, temperature, or neutron flux. This document provides the nuclear power industry with requirements and acceptable methods for response time testing nuclear-safety-related instrument channels.

ISA-S67.10, Sample-Line Piping and Tubing Standard for Use in Nuclear Power Plants, (ANSI/ISA-1986), 1986, 24 pp.
Covers design, protection, and installation of sample lines connecting nuclear-safety related power plant processes with sampling instrumentation. The standard applies to light-water-cooled nuclear power plants, covering the pressure boundary requirements for piping and tubing. It applies to the areas from the process tap to the upstream side of the sample panel, bulkhead fitting, or analyzer shut-off valve, and it includes in-line sample probes.

ISA-S67.14, Qualifications and Certification of Instrumentation and Control Technicians in Nuclear Power Plants, (ANSI/ISA-1983), 1983, 16 pp.
Identifies the criteria for certification of instrumentation and control technicians in nuclear power plants. These criteria address qualifications based on education, experience, training, and job performance.

**ISA-S71.01, Environmental Conditions for Pro-

Titles & Abstracts

cess Measurement and Control Systems: Temperature and Humidity, (ANSI/ISA-1985), 1985, 18 pp.

Establishes uniform classifications of the environmental conditions of temperature and humidity as they relate to industrial process measurement and control equipment. The standard is compatible with IEC Publication 654-1, 1979, **Operating Conditions for Industrial-Process Measurement and Control Equipment, Part I: Temperature, Humidity and Barometric Pressure.**

ISA-S71.04, Environmental Conditions for Process Measurement and Control Systems: Airborne Contaminants, (ANSI/ISA-1985), 1985, 16 pp.

Classifies airborne contaminants that may affect process measurement and control instruments. This classification system provides a means of specifying the type and concentration of airborne contaminants to which a specified instrument may be exposed. This standard is limited to airborne contaminants and biological influences only, covering contamination influences that affect industrial process measurement and control systems.

ISA-S72.01, PROWAY-LAN Industrial Data Highway, (ANSI/ISA-1985), 1985, 200 pp.

Specifies those elements which are required for compatible interconnection of stations by way of a Local Area Network (LAN) using the Token Bus access method in an industrial environment. The standard is compatible with (but more restrictive than) the IEEE 802.2 and 802.4 standards for general LANS.

ISA-RP74.01, Application and Installation of Continuous-Belt Weighbridge Scales, 1984, 28 pp.

Furnishes design criteria inducive to simplified specifications and provides recommendations for installation, calibration, and maintenance of continuous-belt, weigh-bridge type scales. This recommended practice provides an effective base of comparison of scale suppliers, establishes minimum values, and ensures that a scale specification and purchase incorporates the essentials to satisfy a particular weighing job. It permits early belt conveyor design, with the full knowledge of the weight scale configuration, regardless of the manufacturer.

ISA-S75.01, Flow Equations for Sizing Control Valves, (ANSI/ISA-1985), 1985, 34 pp.

Establishes equations for predicting the flow of compressible and incompressible fluids through control valves. The equations are not intended for use when mixed-phase fluids, dense slurries, dry solids, or non-Newtonian liquids are encountered. The prediction of cavitation, noise, or other effects is not a part of this standard.
The equations are not, however, intended for use with mixed phases.

ISA-S75.02 (ANSI/ISA-S75.02-1988), Control Valve Capacity Test Procedure, 1988, 16 pp.

This standard provides a test procedure for obtaining the following factors for sizing control valves: valve flow coefficient (C); liquid pressure recovery factors (F) and (F); Reynolds Number factor (F); liquid critical pressure ratio factor (F); piping geometry factor (F); and pressure drop ratio factors (X and X). The standard is intended for control valves used in flow control of process fluids and is not intended to apply to fluid power components as defined in the National Fluid Power Association Standard NFPA T.3.5.28-1977.

ISA-S75.03, Face-to-Face Dimensions for Flanged Globe-Style Control Valve Bodies (Formerly ISA-S4.01.1), (ANSI/ISA-1985), 1984, 12 pp.

Applies to flanged globe-style control valves, sizes ½ inch through 16 inches, having top, top and bottom, port, or cage guiding. This standard aids users in the piping design by providing ANSI Class 125, flat face, and ANSI Classes 150, 250, 300 and 600, raised face, flanged control valve dimensions, without giving special consideration to the equipment manufacturer to be used.

ISA-S75.04, Face-to-Face Dimensions for Flangeless Control Valves (Formerly ISA-S4.01.2), (ANSI/ISA-1985), 1984, 8 pp.

Applies to flangeless control valves, sizes ¾ inch through 16 inches for ANSI Classes 150 through 600. This standard aids users in their piping designs for flangeless control valves without giving special consideration to the equipment manufacturer to be used. This standard applies to flangeless ball control valves utilizing a full ball or a segment of a ball and other rotary-stem or sliding-stem flangeless control valves. It does not apply to weld-end valves, butterfly valves, or other rotary-stem valves that may be covered by other standards.

ISA-S75.05, Control Valve Terminology, (ANSI/ISA-1983), 1983, 33 pp.

Provides terminology for control valves of seven different types and also for common types of actuators used with these valves. This standard names individual valve parts, defines assemblies of parts, and provides terminology for part and assembly functions.

ISA-RP75.06, (Formerly ISA-RP4.2), Control Valve Manifold Designs, 1981, 20 pp.

Presents six control valve manifold types with space estimates for various sizes. Each of these six

Titles & Abstracts

types consists of a straight through globe control valve, isolation upstream and downstream block valves and bypass piping with a manually activated valve.

ISA-S75.07, Laboratory Measurement of Aerodynamic Noise Generated by Control Valves, 1987, 16 pp.

Defines equipment, methods, and procedures for the laboratory testing and measurement of airborne sound radiated by a compressible fluid flowing through a control valve and its associated piping, including fixed-flow restrictions. The test may be conducted under any conditions agreed upon by the user and the manufacturer. Although this standard is designed for measurement of noise radiated from the piping downstream of the valve, other test variations are optional, including the use of insulation and nonstandard piping. Applications of this standard to control valves discharging directly to atmosphere are excluded.

ISA-S75.08, Installed Face-to-Face Dimensions for Flanged Clamp or Pinch Valves, (ANSI/ISA-S75.08-1985), 1985, 12 pp.

Applies to clamp or pinch valves sizes 1 inch through 8 inches. The purpose of this standard is to aid users in their piping design by providing installed face-to-face dimensions for control valves, incorporating clamp or pinch elements, which have flanges that mate with ANSI B16.1 Class 125 (PN20) and/or ANSI B16.5 Class 125 (PN20) flanges, without giving special consideration to the manufacturer of the equipment to be used.

ISA-S75.11, Inherent Flow Characteristic and Rangeability of Control Valves, (ANSI/ISA-1985), 1984, 16 pp.

Defines the statement of typical control valve inherent flow characteristics and inherent rangeabilities and establishes criteria for adherence to manufacturer-specified flow characteristics. This standard uses the basic definitions from ISA-S75.05 and also defines specific terms related to flow characteristic and rangeability. A table listing inherent flow characteristic deviations and sample plots of relative flow coefficient versus relative travel are also given.

ISA-S75.12, Face-to-Face Dimensions for Socket Weld-End and Screwed-End Globe-Style Control Valves, (ANSI classes 150, 300, 600, 900, 1500, and 2500), (ANSI/ISA-S75.12-1987), 1986, 12 pp.

Applies to socket weld-end globe-style control valves, sizes ½ inch through 4 inches, and screwed-end globe-style control valves, size ½ inch through 2½ inches, having top, top and bottom, port, or cage guiding. This standard aids users in their piping designs by providing ANSI Classes 150 through 2500 socket weld-end control valve dimensions and ANSI Classes 150 through 600 screwed-end control valve dimensions, without giving special considerations to the equipment manufacturer to be used.

ISA-S75.14, Face-to-Face Dimensions for Buttweld-End Globe-Style Control Valves, (ANSI/ISA-1984), 1984, 8 pp.

Applies to buttweld-end globe-style control valves, sizes ½ inch through 8 inches, having top and cage guiding. This standard aids users in their piping designs by providing ANSI Class 4500 buttweld-end control valve dimensions, without giving special consideration to the equipment manufacturer to be used.

ISA-S75.15, Face-to-Face Dimensions for Buttweld-End Globe-Style Control Valves (ANSI/ISA Classes 150, 300, 600, 900, 1500, and 2500), (ANSI/ISA-S75.15-1987), 1986, 12 pp.

Applies to buttweld-end globe-style control valves, sizes ½ inch through 18 inches, for ANSI Classes 150 through 2500, having top, top and bottom, port, or cage guiding. This standard aids users in their piping designs by providing buttweld-end control valve dimensions, without giving special consideration to the equipment manufacturer to be used.

ISA-S75.16, Face-to-Face Dimensions for Flanged Globe-Style Control Valve Bodies (ANSI Classes 900, 1500, and 2500), (ANSI/ISA-S75.16-1987), 1986, 12 pp.

Applies to flanged globe-style control valves, sizes ½ inch through 18 inches, having top, top and bottom, port, or cage guiding. This standard aids users in their piping designs by providing ANSI Classes 900, 1500, and 2500 raised-face, flanged control valve dimensions, without giving special consideration to the equipment manufacturer to be used.

ISA-S77.42, Fossil-Fuel Plant Feedwater Control System—Drum-Type, 1987, 32 pp.

Establishes minimum criteria for the control of levels, pressures, and flow for the safe and reliable operation of drum-type feedwater systems in fossil power plants. Aids in the development of design specifications covering the measurement and control of feedwater systems. The following requirements are defined for minimum system design: (1) process measurement requirements; (2) control and logic requirements; (3) final control device requirements; (4) system reliability and availability; (5) alarm requirements; and (6) operator interface. The safe physical containment of the feedwater shall be in accordance with applicable piping codes and standards and is beyond the scope of this standard.

ISA-S82.01, Safety Standard for Electrical and Electronic Test, Measurement, Controlling and

Titles & Abstracts

Related Equipment — General Requirements, 1988, 65 pp.

This standard applies to electrical and electronic test, measuring, controlling and related equipment. This standard applies to equipment that is rated for connection to supply circuits which exceed extra-low voltage and which do not exceed 480 volts rms, between phases for three-phase supply circuits or 250 volts rms, single-phase or dc. This standard does not apply to: (1) medical and laboratory equipment; (2) watt-hour meters and associated equipment installed by electrical utility companies for measuring electrical energy and related quantities; and (3) general use battery charges, auxiliary supply sources, substitute power supplies, or laboratory-type power supplies not specifically rated for use with measuring or testing equipment.

ISA-S82.02, Safety Standard for Electrical and Electronic Test, Measuring, Controlling, and Related Equipment — Electrical and Electronic Test and Measuring Equipment, 1988, 18 pp.

This standard applies to electrical and electronic and electromechanical measuring and testing equipment, and to the terminals, connectors, wiring and probes used in the interface between. This standard also applies to accessories and adaptors rated for use with measuring or testing probe assemblies, connectors or terminals. This standard applies to measuring or testing probe assemblies, connectors and terminals that are rated for measuring or testing branch circuits up to 1000 volts or ac or dc voltages up to 40 kilovolts incorporated within electrical end-product equipment circuits. This standard does not apply to equipment that is rated for use exclusively in extra-low voltage and power limited applications or equipment intended primarily for equipment-to-equipment interconnection (for example: IEEE interface bus, etc.)

ISA-S82.03, Safety Standard for Electrical and Electronic Test, Measuring, Controlling, and Related Equipment — Electrical and Electronic Process Measurement and Control Equipment, 1988, 20 pp.

This standard applies to electrical, electronic (analog/digital) and electromechanical process measurement and control equipment which: (1) measures and controls directly or indirectly an industrial process through a final control device; (2) is intended to be connected to supply circuits which do not exceed 250 volts rms, single phase, or dc; and (3) is rated for use in either indoor, outdoor, or sheltered locations. This equipment includes but is not necessarily restricted to: (1) integrating, indicating, or recording equipment with or without a control function; (2) transmitters; (3) transducers; (4) analyzers; (5) supervisory or telemetry equipment; and (6) accessories used with any of the above equipment.

The following standard, identified by its ANSI number, was sponsored and published by ISA, approved by ANSI, and is available from either ISA or ANSI.

MC96.1, American National Standard for Temperature Measurement Thermocouples, 1982, 48 pp.

Covers coding of thermocouple and extension wire; coding of insulated duplex thermocouple extension wires; terminology, limits of error and wire sizes for thermocouples and thermocouple extension wires; temperature EMF tables for thermocouples; plus appendices that cover fabrication, checking procedures, selection, and installation.

ANSI C100.6-3, American National Standard for Voltage or Current Reference Devices; Solid State Devices, 1984, 12 pp.

Applies to physical devices used to maintain the unit of dc voltage or current having uncertainties of 100 ppm of output or less. This standard treats these devices from the standpoint of performance characteristics, but does not specify design or construction details or techniques. This part of the standard, C100.6-3, applies to solid state devices used to maintain the unit of dc voltage or current having uncertainties of 100 ppm of output or less.

INSULATED CABLE ENGINEERS ASSOCIATION, INC. (ICEA)

The following publications are available from the Insulated Cable Engineers Association, Inc. (Prior to March 7, 1979, the Insulated Power Cable Engineers Association), P.O. Box 9, South Yarmouth, MA 02664. (617) 394-4424. Standards developed by NEMA-ICEA are available from NEMA, and standards developed by IEEE-ICEA are available from IEEE.

IPCEA: T-22-294, (R 1983), Test Procedures for Extended Time-Testing of Wire and Cable Insulation for Service in Wet Locations, 1983, 3 pp.

Describes procedures for extended time testing of extruded wire and cable insulations for service in wet locations.

IPCEA: T-24-380, Guide for Partial-Discharge Test Procedures, 1980, 6 pp.

Applies to the detection and measurement of partial discharges occurring in the insulation of single-conductor shielded cables and assemblies thereof and multiple-conductor cables with individually shielded conductors.

IPCEA: T-25-425, Guide for Establishing Stabil-

Titles & Abstracts

ity of Volume Resistivity for Conducting Polymeric Components of Power Cables, 1981, 4 pp.
Applies to testing of extruded conducting polymeric components of power cables with extruded insulation. It describes a method of demonstrating the stability over a period of time of the volume resistivity (calculated from longitudinal resistance) of these components at temperatures up to the emergency operating temperature of the cable.

IPCEA: T-28-562, (R 1983), Test Method for Measurement of Hot Creep of Polymeric Insulations, 1983, 3 pp.
Provides procedure for determining the relative degree of crosslinking of polymeric electrical cable insulations.

IPCEA: T-29-520, Vertical Cable Tray Flame Tests @ 210,000 Btu, 1986.

IPCEA: T-30-520, Vertical Cable Tray Flame Tests @ 70,000 Btu, 1986.

INTERNATIONAL ASSOCIATION OF PLUMBING AND MECHANICAL OFFICIALS (IAPMO)

The following documents are available from the International Association of Plumbing and Mechanical Officials, 5032 Alhambra Avenue, Los Angeles, CA 90032. (714) 595-8449

IAPMO: PS 8-77, 1977, 2 pp.
Implements Section 209 of the Uniform Plumbing Code published by the International Association of Plumbing and Mechanical Officials (IAPMO) for backwater valves.

IAPMO: PS 10-84, Globe-Type Loglighter Valves Angle or Straight Pattern, 1984, 4 pp.
Establishes a generally acceptable standard for globe-type loglighter valves, angle or straight pattern.

IAPMO: PS 15-77, Pressure Reducing and Regulating Valves for Installation on Domestic Water Supply Lines, 1977, 3 pp.
Serves to supplement the provisions of the Uniform Plumbing Code, Section 1007 for pressure reducing and regulating valves as required on domestic water supply lines; to prescribe minimum standards for materials in the construction of such valves and providing for test standards.

IAPMO: PS 31-77, Backflow Prevention Devices, 1977, 20 pp.
Covers material requirements, dimensions, and design and other specific properties, in addition to general description of materials.

Uniform Solar Energy Code, 1985, 74 pp.
Provisions apply to the erection, installation, alteration, addition, repair, relocation, replacement, maintenance or use of any solar system except as otherwise provided for in this Code.

INTERNATIONAL CONFERENCE OF BUILDING OFFICIALS (ICBO)

The following handbook is available from the International Conference of Building Officials, 5360 South Workman Mill Road, Whittier, CA 90601. (213) 699-0541

Uniform Building Code Standards, 1988, 1340 pp.
A collection of building code standards consisting of specifications developed by ANSI, ASTM, UL and other organizations. Some materials testing standards are included.

INTERNATIONAL ELECTROTECHNICAL COMMISSION (IEC)

The American National Standards Institute has administrative and technical affiliation with the U.S. National Committee of the IEC. This committee, in turn, represents the U.S. in the IEC. The IEC is composed of 43 National Committees that collectively represent about 80 percent of the world's population that produces and consumes 95 percent of all electrical energy. The IEC holds the international responsibility for the coordination and unification of all national electrotechnical standards and it is affiliated with the ISO. It also acts as the coordinating body for the activities of other international organizations whose responsibilities relate to or overlap the electrotechnical field.

The following representative IEC standards are available from ANSI, 1430 Broadway, New York, NY 10018. (212) 354-3300

IEC: 27, Letter Symbols to be Used in Electrical Technology:

IEC: 27-1 (1971), Part 1: General, incorporating Amendments No. 1 (1974) and No. 2 (1977). Supplement 27-1A (1976): Time-Dependent Quantities.

IEC: 27-2 (1972), Part 2: Telecommunications and Electronics. Supplement 27-2A (1975).

Titles & Abstracts

IEC: 27-3 (1974), Part 3: Logarithmic Quantities and Units.

IEC: 38 (1983), IEC Standard Voltages, including Amendment No. 1 (1977).

IEC: 50, International Electrotechnical Vocabulary.
A glossary of the terms, with their definitions in English and French, used in electrical engineering. The equivalent terms only are given in Dutch, German, Italian, Polish, Swedish and Spanish. A separate index is given for each of the eight languages. The vocabulary is issued in the form of separate booklets, each dealing with a specific field.

IEC: 50(00) (1979), International Electrotechnical Vocabulary, General Index.

IEC: 50(08) (1960), Electro-Acoustics.

IEC: 50(12) (1955), Transductors.

IEC: 68, Basic Environmental Testing Procedures:
Describes a standard general procedure for climatic and mechanical robustness tests, designed to assess the durability, under various conditions of use, transport and storage, of components used in equipment for radio-communication and in electronic equipment employing similar techniques.

IEC: 68-1 (1988), Part 1: General.

IEC: 68-2, Part 2: Tests.
This part describes the different tests in detail. Each test is identified by a letter of the alphabet and is issued in the form of a separate booklet.

IEC: 79, Electrical Apparatus for Explosive Gas Atmospheres:

IEC: 79-0 (1983), Part 0: General Introduction.

IEC: 79-1 (1971), Part 1: Construction and Test of Flameproof Enclosures of Electrical Apparatus, including Supplement 79-1a (1975): Appendix D: Method of Test for Ascertainment of Maximum Experimental Safe Gap.

IEC: 79-2 (1983), Part 2: Pressurized Enclosures.

IEC: 79-3 (1972), Part 3: Spark Test Apparatus for intrinsically-Safe Circuits.

IEC: 79-4 (1975), Part 4: Method of Test for Ignition Temperature, including Supplement 79-4A (1970).

IEC: 79-5 (1967), Part 5: Sand-Filled Apparatus.

IEC: 79-6 (1968), Part 6: Oil-Immersed Apparatus.

IEC: 79-7 (1969), Part 7: Construction and Test of Electrical Apparatus, Type of Protection "e".

IEC: 79-8 (1969), Part 8: Classification of Maximum Surface Temperatures. WD — See 79-0

IEC: 79-9 (1970), Part 9: Marking. WD — See 79-0

IEC: 79-10 (1986), Part 10: Classification of Hazardous Area.

IEC: 113, Diagrams, Charts, Tables.

IEC: 113-1 (1971), Part 1: Definitions and Classification.

IEC: 113-2 (1971), Part 2: Item Designation.

IEC: 117, Recommended Graphical Symbols; Graphical Symbols.

IEC: 117-0 (1973), Part O: General Index.

IEC: 117-1 (1960), Part 1: Kind of Current, Distribution Systems, Methods of Connection and Circuit Elements, incorporating Amendments No. 1 (1966), No. 2 (1967) and No. 3 (1973) and including Supplement 117-1A (1976).

IEC: 117-2 (1960), Part 2: Machines, Transformers, Primary Cells and Accumulators, Transductors and Magnetic Amplifiers, Inductors, incorporating Amendments No. 1 (1966), No. 2 (1971), No. 3 (1973) and first supplement (1974).

IEC: 117-3 (1977), Part 3: Switching and Protective Devices, superseding first edition (1963), Amendments No. 1 (1966), No. 2 (1972), No. 3 (1973), No. 4 (1974), and Supplement 117-3A (1970) and 117-3B (1972).

IEC: 117-4 (1963), Part 4: Measuring Instruments and Electric Clocks, incorporating Amendments No. 1 (1971) and including Amendments No. 2 (1973) and No. 3 (1974) and Supplement 117-4A (1974).

IEC: 150 (1963), Testing and Calibration of Ultrasonic Therapeutic Equipment.

IEC: 185 (1987), Current Transformers.
Applies to newly manufactured current transformers for use with electrical measuring instruments and electrical protective devices at frequencies from 15 Hz to 100 Hz. Although the require-

Titles & Abstracts

ments relate basically to transformers with separate windings they are also applicable, where appropriate, to auto-transformers.

IEC: 186 (1987), Voltage Transformers.
Applies to new voltage transformers for use with electrical measuring instruments and electrical protective devices at frequencies from 15 Hz to 100 Hz. The general requirements apply to all voltage transformers, but, for certain types, e.g. capacitor voltage transformers, the requirements are subject to the modifications stated in the appropriate chapter. Although the requirements relate basically to transformers with separate windings, they are also applicable, where appropriate, to auto-transformers. The standard does not apply to transformers for use in laboratories.

IEC: 271 (1974), List of Basic Terms, Definitions and Related Mathematics for Reliability, including Supplement 271A (1978).

IEC: 272 (1968), Preliminary Reliability Considerations.

IEC: 278 (1968), Documentation to be Supplied with Electronic Measuring Apparatus, including Supplement 278A (1974).

IEC: 284 (1968), Rules of Behavior with Respect to Possible Hazards when Dealing with Electronic Equipment and Equipment Employing Similar Techniques, including Amendment No. 1 (1972).

IEC: 319 (1978), Presentation of Reliability Data on Electronic Components (or parts).

IEC: 351, Expression of the Properties of Cathode-Ray Oscilloscopes.

IEC: 351-1 (1976), Part 1: General, superseding 351 (1971).

IEC: 351-2 (1976), Part 2: Storage Oscilloscopes.

IEC: 359 (1987), Expression of the Performance of Electrical and Electronic Measuring Equipment.
Applies to the specification of the performance of electrical and electronic equipment and instruments and the accessories used with them which measure electrical quantities (indicating and recording instruments), or supply measured electrical quantities (supply instruments), such as signal generators and some power supplies, or measure non-electrical quantities using electrical means, excluding any parts, such as transducers. The expression of performance covers only the electrical or electronic part. This standard provides methods for ensuring uniformity in the specification and measurement of errors of the above equipment.

IEC: 393, Potentiometers.

IEC: 393-1 (1973), Part 1: Terms and Methods of Test, including Supplements 393-1A (1977) and 393-1B (1978).

IEC: 393-2 (1976), Part 2: Sectional Specification: Lead-Screw Actuated Preset Potentiometers. Selection of Methods of Test and General Requirements.

IEC: 393-3 (1977), Part 3: Sectional Specification: Single-Turn Rotary Wirewound and Non-Wirewound Potentiometers. Selection of Methods of Test and General Requirements.

IEC: 393-4 (1978), Part 4: Sectional Specifications: Single-Turn Rotary Power Potentiometers. Selection of Methods of Test and General Requirements.

IEC: 393-5 (1978), Part 5: Sectional Specification: Single-Turn Rotary Low-Power Wirewound and Non-Wirewound Potentiometers. Selection of Methods of Test and General Requirements.

IEC: 405 (1972), Nuclear Instruments: Constructional Requirements to Afford Personal Protection Against Ionizing Radiation.

IEC: 414 (1973), Safety Requirements for Indicating and Recording Electrical Measuring Instruments and their Accessories.

IEC: 416 (1988), General Principles for the Formulation of Graphical Symbols, including Amendment No. 1 (1978).

IEC: 473 (1974), Dimensions for Panel-Mounted Indicating and Recording Electrical Measuring Instruments.

IEC: 477 (1974), Laboratory D.C. Resistors.

IEC: 482 (1975), Dimensions of Electronic Instrument Modules (for Nuclear Electronic Instruments).

IEC: 484 (1974), Indirect Acting Electrical Measuring Instruments.

IEC: 485 (1974), Digital Electronic D.C. Voltmeters and D.C. Electronic Analogue-to-Digital Converters.

**IEC: 529 (1976), Classification of Degrees of Pro-

Titles & Abstracts

tection Provided by Enclosures, including Amendment No. 1 (1978).

IEC: 534 Industrial-Process Control Valves.

IEC: 534-1 (1987), Part 1: Control Valve Terminology and General Considerations.
Applies to all types of industrial-process control valves. Establishes a basic component and functional terminology list and gives guidance on the use of the other parts of this publication. Gives overall design requirements, test requirements and prediction methods.

IEC: 534-2 (1978), Part 2: Flow Capacity. Section One — Sizing Equations for Incompressible Fluid Flow under Installed Conditions.
Deals with a general sizing equation for incompressible fluid flow under installed conditions. Factors are introduced which permit accurate sizing when choked conditions exist due to either cavitation or flashing. Procedures and factors are incorporated to handle installations where the valve is installed between fittings such as reducers, expanders, etc. A Reynolds number factor allows for applications where non-turbulent flow conditions exist caused by low differential pressure and/or high viscosity fluids. Numerical constants are presented permitting the use of three recognized flow coefficients: A_v, K_v, and C_v.

IEC: 534-2-2 (1980), Part 2: Flow Capacity, Section Two — Sizing Equations for Compressible Fluid Flow under Installed Conditions.
Covers equations suitable for use in sizing industrial process control valves when the flowing media are compressible fluids. The equations are for use with gas or vapor and are not intended for use with multiphase streams.

IEC: 534-2-3 (1983), Section Three — Test Procedures.
Applies to industrial-process control valves and provides the flow capacity test procedures for determining some of the variables used in the equations given in Publication 534-2.

IEC: 534-3 (1976), Part 3: Dimensions. Section One: Face-to-Face Dimensions for Flanged, Two-Way, Globe-Type Control Valves.
Covers nominal sizes between 20 mm and 400 mm.

IEC: 534-3-2 (1984), Part 3: Dimensions. Section Two — Face-to-Face Dimensions for Flangeless Control Valves Except Wafer Butterfly Valves.
Gives the overall lengths of following types: segmental ball, eccentric rotary plug, and barstock globe; wafer butterfly valves are excluded.

IEC: 534-4 (1982), Part 4: Inspection and Routine Testing.
Defines the methods of inspection, the pressure and leak tests and performance tasks for valves with pressure ratings up to PN 100 (Class 600).

IEC: 536 (1976), Classification of Electrical and Electronic Equipment with Regard to Protection Against Electric Shock.

IEC: 539 (1976), Directly Heated Negative Temperature Coefficient Thermistors.

IEC: 540 (1982), Test Methods for Insulation and Sheaths of Electric Cables and Cords (elastomerica and thermoplastic compounds), superseding 330 (1970).

IEC: 544-1 (1977), Guide for Determining the Effects of Ionizing Radiation on Insulating Materials Part 1: Radiation Interaction.

IEC: 546 Controllers with Analogue Signals for Use in Industrial-Process Control Systems.

IEC: 546-1 (1987), Part 1: Methods of Evaluating the Performances.
Applies to pneumatic and electric industrial-process controllers using analogue continuous input and output signals. Specifies uniform methods of test for evaluating the performances of such controllers.

IEC: 546-2 (1987), Part 2: Guidance for Inspection and Routine Testing.
Provides technial guidance for inspection and routine testing of controllers, for instance, as acceptance tests or after repair.

IEC: 552 (1977), CAMAC — Organization of Multi-Crate Systems, Specification of the Branch-Highway and CAMAC Crate Controller Type A1.

IEC: 561 (1976), Electro-Acoustical Measuring Equipment for Aircraft Noise Certification.

IEC: 601-1 (1977), Safety of Medical Electrical Equipment Part 1: General Requirements.

IEC: 748-4 (1987), Part 4: Interface Integrated Circuits.
Gives standards for the following categories of sub-categories of interface integrated circuits: Category I — Sub-category A: Line circuits (transmitters and receivers); Sub-category B: Sense amplifiers; Sub-category C: Peripheral drivers (including memory drivers) and level shifters; Sub-category D: Voltage comparators.

Titles & Abstracts

Category II: Linear and non-linear analogue-to-digital and digital-to-analogue converters.

IEC: 793 Optical Fibers.

IEC: 793-1 (1987), Part 1: Generic Specification.
Applies to primary coated or primary buffered optical fibers for use in telecommunication equipment and in devices employing similar techniques. Establishes uniform requirements for the geometrical, optical, transmission, mechanical and environmental properties of optical fibers. This entirely revised edition contains new measuring methods which were previously under consideration, such as refracted near field, optical fiber proof test and temperature cycling.

IEC: 794 Optical Fiber Cables.

IEC: 794-1 (1987), Part 1: Generic Specification.
Applies to optical fiber cables for use with telecommunication equipment and devices employing similar techniques and to cables having a combination of both optical fibers and electrical conductors. Presents requirements for the geometrical, transmission, mechanical and climatic characteristics of optical fiber cables, and electrical requirements. This entirely revised edition contains new measuring methods which were previously under consideration, such as torsion, cnatch, kink, cable bend, temperature cycling, sheath integrity and water penetration.

IEC: 794-2 (1987), Part 2: Product Specifications.
Provides product specifications for single fiber optical cables for indoor use with applications such as transmission equipment, telephone equipment, data processing equipment and communication and transmission networks.

IEC: 821 (1987), Bus — Microprocessor System Bus for 1 to 4 Byte Data.
Describes a high performance backplane bus for use in microprocessor bases systems. This parallel bus supports single and block transfer cycles on a 32-bit non-multiplexed address and data highway. Transmission is governed by an asynchronous handshaken protocol. The bus allocation provides for multi-processor architectures. This bus also supports inter-module interrupts for facilitating quick response to internal and external events. The mechanics of the boards and chassis are based on IEC Publication 297: Dimensions of Panels and Racks. (Note: This bus is similar to the VME bus.)

IEC: 874 (1987), Part 1: Generic Specification.
This publication also bears a QC number. This number, QC 210000 is the specification number in the IEC Quality Assessment System for Electronic Components (IECQ). Contains standard optical, mechanical and environmental tests and measuring methods for connectors for optical fibers and cables.

IEC: 902 (1987), Industrial-Process Measurement and Control Terms and Definitions.
Standardizes terms and definitions in the field of industrial process measurement and control. Includes conventional and specialized terms to describe the physical nature, the use, and the performance of instrumentation and instrument systems for measurement and control in the process industry. This publication has the status of a report.

IEC: 912 (1987), ECL (Emitter Coupled Logic) Front Panel Interconnections in Counter Logic.
Defines ECL front panel interconnections (signals, cables, connectors, terminators, etc.) in counter logic for modular instruments used in nuclear instrumentation and other applications.

IEC: 210000 (1987), Part 1: Connectors for Optical Fibers and Cables.
This publication also bears the number 874-1. Contains standard optical, mechanical and environmental tests and measuring methods for connectors for optical fibers and cables.

Titles & Abstracts

INTERNATIONAL ORGANIZATION FOR STANDARDIZATION (ISO)

The American National Standards Institute is the Member Body representing the United States in the International Organization for Standardization (ISO). Sixty-nine national standards bodies comprise the world membership and cooperate in formulating the technical program in which each member maintains a status as a participant or observer in accordance with the interest of the member in the specific standard under consideration.

The following representative ISO standards are available from ANSI, 1430 Broadway, New York, NY 10018. (212) 354-3300

ISO: 1-1975, Standard Reference Temperature for Industrial Length Measurements.

ISO: 31/I-1978, Quantities and Units of Space and Time.

ISO: 31/II-1978, Quantities of Units of Periodic and Related Phenomena.

ISO: 31/IV-1978, Quantities and Units of Heat.

ISO: R31/V-1965, Quantities and Units of Electricity and Magnetism.

ISO: 31/VI-1973, Quantities and Units of Light and Related Electromagnetic Radiations.

ISO: 31/VII-1978, Quantities and Units of Acoustics.

ISO: 31/IX-1973, Quantities and Units of Atomic and Nuclear Physics.

ISO: 31/X-1973, Quantities and Units of Nuclear Reactions and Ionizing Radiations.

ISO: 91-1982, Petroleum Measurement Tables.

ISO: 128-1982, Engineering Drawing, Principles of Presentation.

ISO: 140/I-1978, Part I: Requirements for Laboratories.

ISO: 140/2-1978, Part II: Statement of Precision Requirements.

ISO: 140/3-1978, Part III: Laboratory Measurements of Airborne Sound Insulation of Building Elements.

ISO: 228/I-1982, Pipe Threads where Pressure-Tight Joints are not made on Threads Part I: Designation Dimension and Tolerances.

ISO: 228/2-1987, Pipe Threads where Pressure-Tight Joints are not made on Threads — Part 2: Verification by Means of Limit Gauges.

ISO: 261-1973, ISO General Purpose Metric Screw Threads, General Plan.

ISO: 386-1977, Liquid-in-Glass Laboratory Thermometers — Principles of Design, Construction, and Use.

ISO: 226-1987, Acoustics Relation Between Sound Pressure Levels of Narrow Bands of Noise in a Diffuse Field and in a Frontally-Incident Free Field of Equal Loudness.

ISO: R495-1966, General Requirements for the Preparation of Test Codes for Measuring the Noise Emitted by Machines.

ISO: R508-1966, Identification Colors for Pipes Conveying Fluids in Liquid or Gaseous Condition in Land Installations and on Board Ships.

ISO: 532-1975, Acoustics — Method for Calculating Loudness Level.

ISO: R541-1967, Measurement of Fluid Flow by Means of Orifice Plates and Nozzles.

ISO: 554-1967, Standard Atmospheres for Conditioning and/or Testing Specifications.

Titles & Abstracts

ISO: 555, Liquid Flow Measurements in Open Channels-Dilution Method for Measurements of Steady Flow.

ISO: 555/2-1987, Liquid Flow Measurement in Open Channels — Dilution Methods for the Measurement of Steady Flow — Part II: Integration Method.

ISO: 605-1977, Pulses: Methods of Test.

ISO: 651-1975, Solid-Stem Calorimeter Thermometers.

ISO: 652-1975, Enclosed-Scale Calorimeter Thermometers.

ISO: 748-1979, Liquid Flow Measurement in Open Channels by Velocity Area Methods.

ISO: 772-1988, Liquid Flow Measurement in Open Channels — Vocabulary and Symbols, Bilingual Edition.

ISO: 921-1972, (E/F/R) Nuclear Energy Glossary.

ISO: 1028-1973, Information Processing Flowchart Symbols.

ISO: 1070-1973, Liquid Flow Measurement in Open Channels by Slope Area Method.

ISO: R1087-1969, Vocabulary of Terminology.

ISO: 1100-1 (1981), Liquid Flow Measurement in Open Channels: Establishment and Operation of a Gauging Station and Determination of the Stage-Discharge Relation.

ISO: 1607-1970, Methods of Measurements of the Performance Characteristics of Positive-Displacement Vacuum Pumps, Part 1: Measurement of the Volume Rate of Flow (Pumping Speed).

ISO: R1608-1970, Methods of Measurement of the Performance Characteristics of Vapor Vacuum Pumps, Part I: Measurement of the Volume Rate of Flow (Pumping Speed).

ISO: 1660-1987, Technical Drawings — Tolerances of form and of Position — Part III: Dimensioning and Tolerancing of Profiles.

ISO: R1661-1971, Technical Drawings — Tolerances of Form and of Position — Part IV: Practical Examples of Indications on Drawings.

ISO: 1709-1975, Nuclear-Energy Fissile Materials — Principles of Criticality Safety in Handling and Processing.

ISO: 1999-1975, Acoustics — Assessments of Occupational Noise Exposure for Hearing Conservation Purposes.

ISO: 2186-1973, Fluid Flow in Closed Conduits — Connections for Pressure Signal Transmissions between Primary and Secondary Elements.

ISO: 2373-1987, Mechanical Vibration of Certain Rotating Electrical Machinery with Shaft Heights between 80 and 400 mm — Measurement and Evaluation of the Vibration Severity.

ISO: 2382/3-1987, Information Processing Systems — Vocabulary — Part III: Equipment Technology.

ISO: 2382/4-1987, Information Processing Systems — Vocabulary — Part IV: Organization of Data.

ISO: 2382/6-1987, Information Processing Systems — Vocabulary — Part VI: Preparation and Handling of Data.

ISO: 2382/11-1987, Information Processing Systems — Vocabulary — Part XI: Processing Units.

ISO: 2382/18-1987, Information Processing Systems — Vocabulary — Part XVIII: Distributed Data Processing.

ISO: 2636-1973, Information Processing-Conventions for Incorporating Flowchart Symbols in Flowcharts.

ISO: 2975, Measurement of Water Flow in Closed Circuits — Tracer Method.

ISO: 2975/1-1974, Part I: General.

ISO: 2975/2-1975, Part II: Constant Rate Injection Method Using Non-Radioactive Tracers.

ISO: 2975/3-1976, Part III: Constant Injection Method Using Radioactive Tracers.

ISO: 2975/6-1977, Part VI: Transit Time Method Using Non-Radioactive Tracers.

ISO: 2975/7-1977, Part VII: Transit Time Method Using Radioactive Tracers.

ISO: 3313-1974, Measurement of Pulsating Fluid Flow in a Pipe by Means of Orifice Plates, Nozzles, or Venturi Tubes, in Particular in the Case of Sinusoidal or Square Wave Intermittent Periodic-Type Fluctuations.

ISO: 3354-1988, Measurement of Clean Water

Titles & Abstracts

Flow in Closed Conduits — Velocity Area Method Using Current-Meters.

ISO: 3966-1977, Measurement of Fluid Flow in Closed Conduits — Velocity Area Method Using Pitot Static Tubes.

ISO: 4006-1977, Measurement of Fluid Flow in Closed Conduits — Vocabulary and Symbols, Bilingual Edition.

ISO: 4335-1987, Information Processing Systems — Data Communication — High-Level Data Link Control Element of Procedures.

ISO: 5198-1987, Centrifugal, Mixed Flow and Axial Pumps Code for Hydraulic Performance Tests — Precision Grade.

ISO: 7478-1987, Information Processing Systems — Data Communication — Multilink Procedures.

ISO: 8316-1987, Measurement of Liquid Flow in Closed Conduits — Method by Collection of the Liquid in a Volumetric Tank.

ISO: 8348-1987, Information Processing Systems — Data Communications — Network Service Definition.

ISO: 8471-1987, Information Processing Systems — Data Communication — High-Level Data Link Control Balanced Classes of Procedures — Data-Link Layer Address Resolution/Negotiation in Switched Environments.

MANUFACTURERS STANDARDIZATION SOCIETY OF THE VALVE AND FITTINGS INDUSTRY (MSS)

The following MSS Standards publications can be obtained from the Manufacturers Standardization Society of the Valve and Fitting Industry, 127 Park St., NE, Vienna, VA 22180. (703) 281-6613

MMS: SP-6-1985, Standard Finishes for Contact Faces of Pipe Flanges and Connecting-End Flanges of Valves and Fittings.

MMS: SP-9-1984, Spot Facing for Bronze, Iron and Steel Flanges.

MMS: SP-25-1978 (R 1988), Standard Marking System for Valves, Fittings, Flanges and Unions.

MMS: SP-42-1985, Class 150 Corrosion Resistant Gate, Globe, Angle and Check Valves with Flanged and Butt-Weld Ends.

MMS: SP-43-1982 (R 1986), Wrought Stainless Steel Butt-Welding Fittings.

MMS: SP-44-1985, Steel Pipeline Flanges.

MMS: SP-45-1982, By-pass and Drain Connection Standard.

MMS: SP-51-1986, Class 150LW Corrosion Resistant Cast Flanges and Flanged Fittings.

MMS: SP-53-1985, Quality Standard for Steel Castings and Forgings for Valves, Flanges, and Fittings and Other Piping Components — Magnetic Particle Examination Method.

MMS: SP-54-1985, Quality Standard for Steel Castings for Valves, Flanges and Fittings and Other Piping Components — Radiographic Examination Method.

MMS: SP-55-1985, Quality Standard for Steel Castings for Valves, Flanges and Fittings and Other Piping Components — Visual Method.

ANSI/MMS: SP-58-1983, Pipe Hangers and Supports — Materials, Design and Manufacture.

ANSI/MMS: SP-60-1982 (R 1986), Connecting Flange Joint Between Tapping Sleeves and Tapping Valves.

ANSI/MMS: SP-61-1985, Pressure Testing of Steel Valves.

ANSI/MMS: SP-65-1988, High Pressure Chemical Industry Flanges and Threaded Stubs for Use with Lens Gaskets.

ANSI/MMS: SP-67-1983, Butterfly Valves.

ANSI/MMS: SP-68-1984, High Pressure-Offset Seat Butterfly Valves.

ANSI/MMS: SP-69-1983, Pipe Hangers and Supports — Selection and Application.

ANSI/MMS: SP-70-1984, Cast Iron Gate Valves, Flanged and Threaded Ends.

ANSI/MMS: SP-71-1984, Cast Iron Swing Check Valves, Flanged and Threaded Ends.

ANSI/MMS: SP-72-1970, Ball Valves with Flanged or Butt-Welding Ends for General Service.

Titles & Abstracts

ANSI/MMS: SP-73-1986, Brazing Joints for Wrought and Cast Copper Alloy Solder Joint Pressure Fittings.

ANSI/MMS: SP-75-1988, Specification for High Test Wrought Butt-Welding Fittings.

ANSI/MMS: SP-77-1984, Guidelines for Pipe Support Contractual Relationships.

ANSI/MMS: SP-78-1977, Cast Iron Plug Valves, Flanged and Threaded Ends.

ANSI/MMS: SP-79-1980, Socket-Welding Reducer Inserts.

ANSI/MMS: SP-80-1979, Bronze Gate, Globe, Angle and Check Valves.

ANSI/MMS: SP-81-1981 (R 1986), Stainless Steel, Bonnetless, Flanged, Knife Gate Valves.

ANSI/MMS: SP-82-1976 (R 1981, 1986), Valve Pressure Testing Methods.

ANSI/MMS: SP-83-1976, Carbon Steel Pipe Unions, Socket-Welding and Threaded.

ANSI/MMS: SP-84-1985, Steel Valves — Socket Welding and Threaded Ends.

ANSI/MMS: SP-85-1985, Cast Iron Globe & Angle Valves, Flanged and Threaded Ends.

ANSI/MMS: SP-86-1981, Guidelines for Metric Data in Standards for Valves, Flanges, and Fittings.

ANSI/MMS: SP-87-1982 (R 1986), Factory-Made Butt-Welding Fittings for Class 1 Nuclear Piping Applications.

ANSI/MMS: SP-88-1983 (R 1988), Diaphragm Type Valves.

ANSI/MMS: SP-89-1985, Pipe Hangers and Supports — Fabrication and Installation Practices.

ANSI/MMS: SP-90-1986, Guidelines on Terminology for Pipe Hangers and Supports.

ANSI/MMS: SP-91-1984, Guidelines for Manual Operation of Valves.

ANSI/MMS: SP-92-1980, MSS Valve User Guide.

ANSI/MMS: SP-93-1982, Quality Standards for Steel Castings and Forgings for Valves, Flanges, and Fittings and Other Piping Components — Liquid Penetrant Examination Method.

ANSI/MMS: SP-94-1982, Quality Standards for Ferritic and Martensitic Steel Castings and Forgings for Valves, Flanges, and Fittings and Other Piping Components — Ultrasonic Examination Method.

ANSI/MMS: SP-95-1986, Swage(d) Nipples and Bull Plugs.

ANSI/MMS: SP-96-1986, Guidelines on Terminology for Valves and Fittings.

ANSI/MMS: SP-97-1987, Forged Carbon Steel Branch Outlet Fittings — Socket Welding, Threaded and Buttwelding Ends.

ANSI/MMS: SP-98-1987, Protective Epoxy Coatings for the Interior of Valves and Hydrants.

ANSI/MMS: SP-99-1987, Instrument Valves.

ANSI/MMS: SP-100-1988, Qualification Requirements for Elastomer Diaphragm for Nuclear Service Diaphragm Type Valves.

METAL POWDER INDUSTRIES FEDERATION (MPIF)

The following publication is available from the Metal Powder Industries Federation, 105 College Rd. East, Princeton, NJ 08540. (609) 452-7700

Standard Test Methods for Metal Powders and Powder Metallurgy Products, 1985-86 edition, 95 pp.
Bound edition of MPIF standards covering five categories: P/M Nomenclature, Powder Testing Standards, P/M Material Standards, Material Testing Standards, and Safety Standards.

Materials Standards for P/M Structural Parts, 1987-1988 Edition, 20 pp.

Materials Standards for P/M Self-Lubricating Bearings, 1986-1987 Edition, 14 pp.

NATIONAL ASSOCIATION OF PIPE COATING APPLICATORS (NAPCA)

The following publications are available from NAPCA, 717 Commercial National Bank Building, Shreveport, LA 71101. (318) 227-2769

1-65-87, Recommended Specification Designations for Enamel Coatings.

2-66-87, Standard Applied Pipe Coating Weights for NAPCA Coating Specifications.

3-67-87, External Application Procedures for Hot Applied Coal Tar and Asphalt Enamel Coatings to Steel Pipe.

5-69-87, NAPCA Specifications Pipeline Felts.

6-69-83-1, Suggested Procedures to Hand Wrap Field Joints Using Hot Enamel.

6-69-83-2, Suggested Procedures for Coating of Girth Welds With Fusion Bonded Epoxy.

6-69-83-3, Suggested Procedures for Coating Field Joints, Fittings, Connections and Pre-Fabricated Sections Using Tape Coatings.

6-69-83-4, Suggested Procedures for Field Joint Application Using Mastic Mix and Field Mold.

6-69-83-5, Suggested Application Procedures for Coating Field Joints Using Heat Shrinkable Materials.

12-78-83, Application Specifications Mill Applied Fusion Bonded Epoxy Coatings (Rev. April 1983).

12-78-87, External Application Procedures for Plant Applied Fusion Bonded Epoxy (F.B.E.) Coatings to Steel Pipe.

13-79-87, Application Specifications for Coal Tar Epoxy Protective Coatings.

14-83-87, Application Specifications for Polyolefin Pipe Coating Applied by the Cross Head Extrusion Method or the Side Extrusion Method.

15-83-87, Plant Applied Tape Coating Application Specification for the Exterior of Steel Pipe.

Pocket Edition of National Association of Pipe Coating Applicators Specifications and Plant Coating Guide (1-65-83 through 15-83, Rev. April 1983).

NATIONAL ASSOCIATION OF RELAY MANUFACTURERS (NARM)

The following publications are available from the National Association of Relay Manufacturers, P.O. Box 1505, Elkhart, IN 46515. (219) 264-9421

Engineer's Relay Handbook, 3rd Edition, 1980.
Purpose is to bring together information that simplifies and clarifies the specifying and obtaining of correct relays. Information is directed at individuals responsible for specifying the correct type of relays for a given application; it isn't intended for designers and manufacturers or relays.

Definitions of Relay Terms, 1980.
A glossary of words and terms in common use by relay users and manufacturers, includes symbols.

NATIONAL BOARD OF BOILER AND PRESSURE VESSEL INSPECTORS (NBBI)

The following publications are available from NBBI, 1055 Crupper Avenue, Columbus, OH 43229. (614) 888-8320

NB-23, National Board Inspection Code, Revision 6, 1987, 237 pp.
This manual for inspectors, owners and users of boilers and pressure vessels presents rules and guidelines for inspection after installation, repairs, alterations and rerating; thereby, helping to ensure that these objects may continue to be safely used.

NB-18, Pressure Relief Device Certifications, Revision 3, 1987, 328 pp., plus addenda.
Lists safety and safety relief valves by manufacturer and model number and provides the relieving capacities of these valves as determined by test and certified by the National Board. Also lists names of companies which have been authorized by the National Board to repair safety and safety relief valves.

NB-169, Making It With Metric, 1st Edition, 1985, 89 pp.
Provides a basis for metric communication in a coordinated and orderly fashion.

NATIONAL INSTITUTE OF STANDARDS AND TECHNOLOGY (NIST)

The Office of Standards Code and Information (OSCI) Program of the NBS Office of the Associate Director for Industry and Standards maintains the National Center for Standards and Certification Information (NCSCI), the central repository of standards-related information in the United States. NCSCI provides access to

Titles & Abstracts

more than 240,000 titles of standards and related documents published by U.S. professional and technical organizations, U.S. Federal and State Government groups, and foreign national and international organizations. For further information concerning the Office of Standards Code and Information Program contact SCI Program, A629 Administration, National Institute of Standards and Technology, Gaithersburg, MD 20899. (301) 975-4038

NBS: SP-681, Standards Activities of Organizations in the United States.
Lists over 750 U.S. organizations and describes their standards-related activities. The standards activities covered include those of the Federal government, the private sector, and state procurement offices. The directory also lists sources of standards information and documents.

NIST: SP-649, Directory of International and Regional Organizations Conducting Standards-Related Activities.
This directory contains information on 272 international and regional organizations which conduct standardization, certification, laboratory accreditation, or other standards-related activities. It describes their work in these areas, as well as the scope of each organization, national and international.

KWIC Index (Computer Output Microform (COM) produced).
The KWIC Index contains the titles of more than 25,000 U.S. voluntary product and engineering standards. A standard can be located by means of any significant or key word in the title. Key words are arranged alphabetically.

NATIONAL CABLE TELEVISION ASSOCIATION (NCTA)

The following documents are available from NCTA, 1724 Massachusetts Avenue, NW, Washington, DC 20036. (202) 775-3550

NCTA: Standard Graphic Symbols for Cable Television Systems, 16 pp.
This standard provides a list of graphic symbols for the designation of electrical, electronic, and pole line devices for layout drawings of cable television systems.

NCTA: Recommended Practices for Measurements on Cable Television Systems, 1983, 120 pp. (supplements issued 1985).
Provides informative, readily updated descriptions of good engineering practices required for making test measurements to the head end and distribution system; current satellite transmission practices, NTC no. 7.

NATIONAL COUNCIL OF RADIATION PROTECTION AND MEASUREMENTS (NCRP)

The following, and other publications on radiation monitoring and protection, are available from the National Council of Radiation Protection and Measurements, 7910 Woodmont Avenue, Suite 1016, Bethesda, MD 20814. (301) 657-2652

NCRP: Report No. 23, Measurement of Neutron Flux and Spectra for Physical and Biological Applications, 1960, 92 pp.
NCRP Report No. 23 presents a discussion of neutron flux and spectra, compares various methods of neutron source calibration and presents the results of the intercomparisons. There is a discussion of methods of measurement of the emission rate of neutron sources, thermal neutron flux, intermediate neutron flux, fast neutron flux, and neutron spectra. Neutron radiation instruments for area survey and personnel monitoring involving flux and spectrum measurements are discussed also. Typical spectra of various neutron sources are shown. (Published in 1960 as National Bureau of Standards Handbook 72).

NCRP: Report No. 25, Measurement of Absorbed Dose of Neutrons and of Mixtures of Neutrons and Gamma Rays, 1961, 86 pp.
NCRP Report No. 25 represents a summary of methods for determining energy absorption in matter as a result of its interaction with neutrons. Since neutrons are almost invariably accompanied by gamma radiation, mixtures of gamma radiation and neutrons are discussed. Discussions are general wherever possible; however, most of the detailed examples have been drawn from the fields of health physics and radiobiology.

NCRP: Report No. 33, Medical X-Ray and Gamma-Ray Protection for Energies up to 10 MeV — Equipment Design and Use, 1968, 66 pp.
NCRP Report No. 33 is concerned with radiation protection in connection with the medical use of x and gamma rays of energies up to 10 MeV. This report presents recommendations pertaining to equipment design, use and operating conditions, and to radiation protection surveys and personnel monitoring. It includes sections for the guidance of the physician and his associates; the equipment designer and manufacturer; and the radiological

Titles & Abstracts

physicist concerned with calibration procedures, equipment inspection and survey measurements.

NCRP: Report No. 47, Tritium Measurement Techniques, 1976, 97 pp.
Describes and discusses methods for the measurement of tritium in a variety of media. Included are most of the important methods for the measurement of tritium and information on their advantages and disadvantages. Step-by-step procedures and detailed descriptions of equipment are not included.

NCRP: Report No. 50, Environmental Radiation Measurements, 1976, 246 pp.
Presents information on the properties of widely-distributed radionuclides and of typical radiation fields in the environment and on methods for their measurement. Emphasis is placed on the role of measurements in the realistic assessment of dose to man. Techniques applicable to routine monitoring programs during normal operation of nuclear facilities are described.

NCRP: Report No. 51, Radiation Protection Design Guidelines for 0.1-100 MeV Particle Accelerator Facilities, 1977, 159 pp.
Provides design guidelines for radiation protection in particle-accelerator facilities and describes one or more methods by which this protection may be achieved.

NCRP: Report No. 57, Instrumentation and Monitoring Methods for Radiation Protection, 1978, 177 pp.
Describes techniques, instruments, and practices applicable to all types of institutions concerned with radiation or radioactive materials. The first section presents information of a general character related to radiation surveys and instrumentation. Subsequent sections contain discussions of specific installations and types of measurement.

NCRP: Rpt. No. 58, A Handbook of Radioactivity Measurements Procedures, 2nd edition, 1985.
Updates progress made in the field of radionuclide metrology. Includes material dealing with liquid scintillation counting and the latest data from the Oak Ridge data banks.

NCRP: Rpt. No. 68, Radiation Protection in Pediatric Radiology, 1981.
Offers practical information on how to conduct the radiological examinations of children to reduce the radiation dose to the children and those responsible for their care.

NCRP: Rpt. No. 69, Dosimetry of X-Ray and Gamma-Ray Beams for Radiation Therapy in the Energy Range 10 keV to 50 MeV, 1982.
Describes a dosimetric process that will allow the delivery of a prescribed absorbed dose from x-ray and gamma-ray sources to a uniform phantom within the accuracy needed for radiation therapy.

NCRP: Rpt. No. 72, Radiation Protection and Measurements for Low Voltage Neutron Generators, 1983.
Provides information on the radiation protection problems in the use of generators that operate at voltages below a few hundred kilovolts and produce neutrons chiefly by the T(d,n) reaction.

NCRP: Rpt. No. 73, Protection in Nuclear Medicine and Ultrasound Diagnostic Procedures in Children, 1983.
Provides information on the manner of conducting nuclear medicine studies in children to reduce the dose to these patients and those responsible for their care. Also addresses the application of ultrasound to children and discusses the factors that need to be considered to insure continued safe use in clinical practice.

NCRP: Rpt. No. 79, Neutron Contamination From Medical Electron Accelerators, 1985.
Reviews the source of neutrons generated from medical electron accelerators and provides an examination of the transport of the neutrons in the protective housing of the accelerator, as well as in structural shielding barriers when equipment is operated at energies above 10 MeV.

NATIONAL ELECTRICAL MANUFACTURERS ASSOCIATION (NEMA)

The following NEMA Standards publications are available from the National Electrical Manufacturers Association, 2101 L St. NW, Washington, DC 20037. (202) 457-8400

NEMA: DC 2-1982, Quick-Connect Terminals.
Specifies dimensional construction of flat quick-connect terminals and performance characteristics of female connectors.

NEMA: DC 4-1986, Warm Air Limit and Fan Controls.
Describes the characteristics of temperature actuated electric devices intended to control the temperature of air through air-heating equipment intended primarily for residential use.

NEMA: DC 10-1983, Temperature Limit Controls for Electric Baseboard Heaters.
Defines basic construction standards and perfor-

Titles & Abstracts

mance characteristics of temperature limit controls and control systems for use with electric baseboard heaters.

NEMA: DC 12-1985, Hot-Water Immersion Controls.
Defines basic construction standards and performance characteristics of electric-switch-type hot-water immersion controls intended primarily for use with hot-water boilers and heaters used in residential heating.

NEMA: DC 13-1979 (R 1985), Line-Voltage Integrally-Mounted Thermostats for Electric Heaters.
Defines basic construction standards and performance characteristics of integrally-mounted thermostats.

NEMA: DC 22-1977 (R 1982), Load Control for Use on Central Electric Heating Systems.
Defines classifications, ratings, and other characteristics of load controls for use on central electric heating systems intended primarily for residential use.

NEMA: CC 1-1984, Electrical Power Connectors for Substations.
Defines manufacturing, rating, and testing standards.

ANSI/NEMA: CC 3-1973 (R 1978, 1983), Connectors for Use Between Aluminum or Aluminum-Copper Overhead Conductors.
Defines performance requirements and testing procedures.

NEMA: CC 4-1986, 8.3 kV and 8.3/14.4 kV Probe for Separable Insulated Loadbreak Connectors.
Establishes dimensions, design tests, test conditions, and interchangeable construction features for probes of loadbreak separable insulated connectors rated 8.3 kV and 8.3/14.4 kV AC, 200 Amperes.

ANSI/NEMA: ICS 1-1983, General Standards for Industrial Control and Systems.
Provides practical general information concerning ratings, construction, testing, performance, and manufacture of industrial control and systems equipment, terminal blocks, and resistance welding controls. This publication is recommended for use in conjunction with other NEMA ICS publications.

NEMA: ICS 1.1-1984, Safety Guidelines for the Application Installation, Maintenance of Solid State Control.

NEMA: ICS 1.3-1986, Preventive Maintenance of Industrial and Systems Equipment.
Covers fundamental principles, safety precautions and common guidelines for preventive maintenance of equipment within the scope of the NEMA Industrial Control and Systems Section. (Also published as Part 1-115 of ICS 1.)

NEMA: ICS 2-1983, Industrial Control Devices, Controllers and Assemblies.
Provides practical information concerning ratings, construction, testing, performance, and manufacture of industrial control devices and equipment.

NEMA: ICS 2.2-1983, Maintenance of Motor Controllers After a Fault Condition.
Covers the procedures to be followed in order to return to service a motor controller which has been subjected to a short circuit or ground fault.

NEMA: ICS 2.3-1983, Instructions for the Handling, Installation, Operation, and Maintenance of Motor Control Centers.
Guide to practical information containing instructions for the handling, installation, operation, and maintenance of motor control centers rated 600 volts or less.

ANSI/NEMA: ICS 3-1983, Industrial Systems.
Provides practical information concerning ratings, construction, testing, performance, and manufacture of industrial systems equipment.

NEMA: ICS 3.1-1983, Safety Standards for Construction and Guide for Selection, Installation and Operation of Adjustable-Speed Drive Systems.
These standards apply to all industrial equipment electrical components and wiring which are part of the electrical drive system, commencing at the point of input power.

ANSI/NEMA: ICS 4-1983, Terminal Blocks for Industrial Use.
Covers terminal blocks with screw type or screwless clamping units intended for industrial use, to provide electrical and mechanical connection for conductors having a cross section of 24 AWG to 2000 MCM.

NEMA: ICS 5-1983, Resistance Welding Control.
Defines construction standards, performance requirements, and safety standards for resistance welding control equipment.

ANSI/NEMA: ICS 6-1983, Enclosures for Industrial Controls and Systems.
This publication is recommended for use in conjunction with other NEMA ICS publications. It should also be used in conjunction with NEMA

Titles & Abstracts

Standards Publication No. 250. (Unless otherwise specified, ICS 6 is shipped with a copy of NEMA Standards Publication No. 250.)

ACR/NEMA 300-1985, Digital Imaging and Communications Standard.
Specifies a standard method for communicating between digital diagnostic imaging devices and associated equipment.

NEMA: PB 1-1984, Panelboards.
Covers single panelboards or groups of panel units suitable for assembly in the form of single panelboards, including buses, and with or without switches or automatic overload protective devices (fuses or circuit breakers), or both. These units are used in the distribution of electricity for light, heat, and power at: 600 volts and less, 1600-ampere mains and less, and 1200-ampere branch circuits and less.

NEMA: SM 23-1985, Steam Turbines for Mechanical Drive Service.
Defines construction standards and testing and performance requirements for single and multistage turbines intended to drive equipment such as pumps, fans, compressors, and generators.

NEMA: WC 55-1986, Instrumentation Cables and Thermocouple Wire (ICEA S-82-552).
Applies to materials, constructions and testing of multiconductor instrumentation cables including thermocouple extension cables.

NATIONAL ENVIRONMENTAL BALANCING BUREAU (NEBB)

The following manuals are available from NEBB, 8224 Old Courthouse Road, Vienna, VA 22180. (703) 734-3840

NEBB: Procedural Standards for Testing, Adjusting, and Balancing of Environmental Systems, 4th edition, 1983, 136 pp.
Contains a comprehensive reference on instruments and measurement accuracies, including a description of each TAB instrument required by NEBB for certification, its recommended uses, limitations, precision of readings and calibration required, U.S. unit and metric unit equations, and HUAC system air and hydronic design tables.

NEBB: Procedural Standards for Measuring Sound and Vibration, 1st edition, 1977, 96 pp.
Covers NEBB specifications, instruments and accuracy, conditions required for sound tests, sound measurement procedures, vibration isolation devices and systems, field inspection and measurement of vibration, sound and vibration table, charts and equations, reduced copies of NEBB certified report forms, and other topics.

NEBB: Environmental Systems Technology, 1st Edition, 1984.
Incorporating HVAC system history and fundamentals, engineering principles, system design, equipment, components and installation testing and balancing, controls, acoustics, and an extensive glossary and set of engineering tables.

NEBB: Testing, Adjusting, Balancing Manual for Technicians, 1st Edition, 1986.
A basic educational text on testing, adjusting and balancing work, as well as a comprehensive reference manual.

NATIONAL FIRE PROTECTION ASSOCIATION (NFPA)

The following NFPA standards and codes are not instrumentation standards as such; however, their content is important to anyone planning measurement and automatic control systems. They are special applications of the long-accepted guides for protective systems. Attention is drawn to them, but abstracts are not included. Copies of them may be obtained from the National Fire Protection Association, Batterymarch Park, Quincy, MA 02269. (617) 770-3000

NFPA: Electrical Installations in Hazardous Locations.
In one coherent, easy-to-use volume, you get the latest information and practical advice on solving the many complex problems encountered with electrical installations in hazardous locations. Electrical Installations in Hazardous Locations explains the background and reasoning behind many code and standard requirements for equipment and installation in hazardous locations to help you understand and apply these provisions correctly in your work. Plus, it includes diagrams, references, and information on the differences between U.S. and foreign requirements.

NFPA: 54, National Fuel Gas Code
Provides requirements for the safe design, installation, operation, and maintenance of gas piping and for installation and venting gas appliances in residential, commercial, and industrial applications.

NFPA: 70-84, National Electrical Code, 1984.
Contains most widely adopted set of electrical safety requirements for electricians, inspectors,

67

Titles & Abstracts

contractors, electrical manufacturers, architects, builders, and consulting engineers.

NFPA: 70A, Electrical Code for One- and Two-Family Dwellings, 1984.
All you need to know to meet Code regulations when installing electrical services in dwellings. Excerpted and edited from the 1984 National Electrical Code.

NFPA: 70B, Electrical Equipment Maintenance, 1983.
A recommended practice confined to preventive maintenance for industrial-type electrical systems and equipment to reduce hazards that result from their failure.

NFPA: 70E, Electrical Safety Requirements for Employee Workplaces.
Includes installation requirements, work practices, and maintenance requirements necessary to provide practical and safe working areas for employees.

NFPA: 71, Central Station Signaling Systems, 1977.
Provides a standard for a system, or group of systems, maintained and supervised from an approved central station controlled and operated by a person, firm, or corporation whose principal business is furnishing and maintaining a supervised signaling service.

NFPA: 72A, Local Protective Signaling Systems, 1979.
Describes fire alarm or supervisory signals within the protected premises primarily for the protection of life and secondarily for the protection of property.

NFPA: 72B, Auxiliary Protective Signaling Systems, 1979.
A standard on protection of an individual occupancy or building or group of buildings of a single occupancy where the municipal fire alarm facilities are utilized to transmit an alarm to the fire department.

NFPA: 72C, Remote Station Protective Signaling Systems, 1982.
Provides a standard for employing a direct circuit connection between signaling devices at protected premises and signal receiving equipment in a remote station.

NFPA: 72D, Proprietary Protective Signaling Systems, 1981.
Provides a standard for systems having their operation under the control or domination of the owner or others interested in the property to be protected.

NFPA: 72E, Automatic Fire Detectors, 1984.
Covers minimum performance, location, mounting, testing, and maintenance requirements of automatic fire detectors.

NFPA: 72H, Testing Procedures for Local, Auxiliary, Remote Station, and Proprietary Protective Signaling Systems
Provides procedures for acceptance and periodic testing of installed protective signaling systems.

NFPA: 75, Protection of Electronic Computer/Data Processing Equipment, 1981.
State requirements for installations needing fire protection or special building construction, rooms, areas or operating environment.

NFPA: 77, Recommended Practice on Static Electricity, 1977.
Assists in reducing the fire hazard of static by discussing its nature and origin, methods of mitigation, and dissipation.

NFPA: 214, Water-Cooling Towers
Provides requirements and useful data on fire protection systems for field-erected water-cooling towers.

NFPA: 493, Intrinsically Safe Apparatus in Division 1 Hazardous Locations, 1978.
Provides standard for the construction and evaluation of equipment of limited energy for use in hazardous locations.

NFPA: 496, Purged and Pressurized Enclosures for Electrical Equipment, 1982.
Provides standard for the design of equipment to eliminate or reduce a hazardous atmosphere.

NFPA: 497, Classification of Class 1 Hazardous Locations for Electrical Installations in Chemical Plants, 1975.
Recommended practice for classifying the zones of potential hazards from flammable atmospheres in chemical plants.

NFPA: 654, Prevention of Fire and Dust Explosions in the Chemical, Dye, Pharmaceutical, and Plastics Industries.
Specifies methods for reducing the risk of fire and explosion in manufacturing, handling and processing operations involving dusts other than agricultural commodities.

NFPA: 802, Nuclear Research Reactors
Provides recommendations for the safe design,

construction, operation, and protection of nuclear research facilities.

NFPA: 803, Standard for Fire Protection for Nuclear Power Plants, 1978.
Covers the protection of nuclear electric generating facilities from the consequences of fire, including safety to life, protection of property, and continuity of production.

NATIONAL FLUID POWER ASSOCIATION (NFLDP)

A comprehensive publication catalog listing all available NFPA standards or copies of the following are available from the National Fluid Power Association, 3333 N. Mayfair Rd., Milwaukee, WI 53222. (414) 778-3344

National Fluid Power Standards, 10 Volume Edition, 1987, over 2500 pp.
This bound, 10-volume set is the most complete compilation of NFPA and ANSI Standards relating to fluid power. Contains over 130 fluid power standards. The 10-volume set includes the following: A - Communications Standards; B - Pressure Rating Standards; C - Pump, Motor, Power Unit and Reservoir Standards; D - Filtration and Contamination Standards; E - Conductor and Associated Component Standards; F - Control Products Standards; G - Cylinder and Accumulator Standards; H - Fluid, Lubricant and Sealing Device Standards; I - Testing Standards; and J - Bibliographies Standards. Volumes may be purchased individually.

Volume A — Communications

ANSI/B93.2-1986, American National Standard Fluid Power Systems and Products — Glossary.

NFPA/T2.10.1M-1978, Metric Units for Fluid Power Applications.

NFPA/T2.10.2M-1977, Survey of Metric Language Usage by the U.S. Fluid Power Industry.

ISO 1000-1981, SI Units and Recommendations for the Use of Their Multiples and of Certain Other Units.

ISO 1219-1976, Fluid Power Systems and Components — Graphic Symbols.

ISO 2944-1974, Fluid Power Systems and Components — Nominal Pressures.

Volume B — Pressure Rating

ANSI/B93.2-1986, Excerpts Extracted From American National Standard Fluid Power Systems and Products — Glossary.

NPFA/T2.6.12-1974 (R1982), Method for Verifying the Fatigue and Static Pressure Ratings of the Pressure Containing Envelope of a Metal Fluid Power Component.

NFPA/T3.4.7M-1975 (R1980), Method for Establishing and Verifying the Fatigue and Static Pressure Ratings and Conducting Production Testing of the Pressure Containing Envelope of a Metal Fluid Power Accumulator.

NFPA/T3.6.29M-1976 (R1981), Method for Establishing and Verifying the Fatigue and Static Pressure Ratings of the Metal Pressure Containing Envelope of a Tie Rod or Bolted Fluid Power Cylinder.

NFPA/T2.6.1M S7 (T3.6.31M)-1976 (R1981), Telescopic Cylinders and Cylinder of Non-Bolted End Construction Pressure Rating Supplement No. 7 to NFPA Recommended Standard for Verifying the Fatigue and Static Pressure Ratings of the Pressure Containing Envelope of Metal Fluid Power Component.

NFPA/T3.9.22-1982, Pump/Motor Pressure Rating Supplement No. 6 to NFPA Recommended Standard Method for Verifying the Fatigue and Static Pressure Ratings of the Pressure Containing Envelope of a Metal Fluid Power Component.

NFPA/T2.6.1M S4 (T3.12.10M)-1976 (R1981), Air Line Filter, Regulator and/or Lubricator Pressure Rating Supplement No. 4 to NFPA Recommended Standard for Verifying the Fatigue and Static Pressure Ratings of the Pressure Containing Envelope of a Metal Fluid Power Components.

NFPA/T2.6.1M S8 (T3.16.8M)-1975 (R1982), Hydraulic Reservoir Pressure Rating Supplement No. 8 to NFPA Recommended Standard for Verifying the Fatigue and Static Pressure Ratings of the Pressure Containing Envelope of a Metal Fluid Power Component — Part 1 — Static Ratings.

NFPA/T2.6.1M S5 (T3.20.8M-1975) (R1981), Quick Action Couplings Pressure Rating Supplement No. 5 to NFPA Recommended Standard for Verifying the Fatigue and Static Pressure Ratings of the Pressure Containing Envelope of a Metal Fluid Power Component.

Titles & Abstracts

NFPA/T2.6.1M S3 (T3.29.2M)-1976 (R1982), Pressure Switch Pressure Rating Supplement No. 3 to NFPA Recommended Standard for Verifying the Fatigue and Static Pressure Ratings of the Pressure Containing Envelope of a Metal Fluid Power Component.

Volume C — Pumps, Motors, Power Units and Reservoirs

ANSI/B93.71M-1986, American National Standard Hydraulic Fluid Power — Pumps — Test Code for the Determination of Airborne Noise Levels.

ANSI/B93.73M-1986, American National Standard Hydraulic Fluid Power — Motors — Test Code for the Determination of Airborne Noise Levels.

ANSI/B93.6M-1972 (R1981), American National Standard Dimensions and Identification Code for Mounting Flanges and Shafts for Positive Displacement Hydraulic Fluid Power Pumps and Motors.

NFPA/T3.9.13-1982, Hydraulic Fluid Power — Pumps and Motors — Glossary.

ANSI/B93.27M-1973 (R1979), American National Standard Method of Testing and Presenting Basic Performance Data for Positive Displacement Hydraulic Fluid Power, Pumps and Motors.

NFPA/T3.9.18M R1-1978, Method of Establishing the Flow Degradation of Fixed Displacement Hydraulic Fluid Power Pumps When Exposed to Particulate Contaminant.

NFPA/T3.9.25M-1977 (R1982), Method of Establishing the Speed Degradation of Hydraulic Fluid Power Motors When Exposed to Particulate Contaminant.

ANSI/B93.57M-1982, American National Standard Hydraulic Fluid Power — Pumps and Motors — Geometric Displacement.

ANSI/B93.18M-1973 (R1980), American National Standard Non-Integral Industrial Fluid Power Hydraulic Reservoirs.

ANSI/B93.41M-1976 (R1982), American National Standard Requirements of Non-Integral Industrial Fluid Power Hydraulic Power Units.

ANSI/B93.12M-1971 (R1977), American National Standard Method of Rating for Mechanical Vacuum Pumps.

NFPA/T3.9.21-1978, Bibliography of Fluid Power Pump/Motor Standards.

NFPA/T3.16.9-1977 (R1982), Bibliography of Hydraulic Fluid Power Reservoirs and Power Units Standards.

Volume D — Filtration and Contamination

NFPA/T1.21.1-1978 (R1983), Procedure for Self-Certification by Fluid Power Manufacturers.

NFPA/T2.6.1M S1 (T3.10.5.1M)-1976 (R1981), Hydraulic Filter/Separator Housing Pressure Rating Supplement No. 1 to NFPA Recommended Standard for Verifying the Fatigue and Static Pressure Ratings of the Pressure Containing Envelope of a Metal Fluid Power Component.

ANSI/B93.19M-1972 (R1980), American National Standard Method for Extracting Fluid Samples from the Lines of an Operating Hydraulic Fluid Power System (for Particulate Contamination Analysis).

ANSI/B93.20M-1972 (R1980), American National Standard Procedure for Qualifying and Controlling Cleaning Methods for Hydraulic Fluid Power Fluid Sample Containers.

ANSI/B93.30M-1980, American National Standard — Hydraulic Fluid Power — Contamination Analysis Data — Reporting Method.

NFPA/T2.9.5M-1976 (R1980), Hydraulic Fluid Power — Calibration Method — To Count and Measure Computer Assisted Image Analysis Systems — Particles in the 1-10 μm Range.

ANSI/B93.28M-1973 (R1980), American National Standard Hydraulic Fluid Power — Calibration of Liquid Automatic Particle-Count Instruments — Method Using Air Cleaner Fine Test Dust Contaminant.

ANSI/B93.54M-1981, American National Standard Hydraulic Fluid Power — Assembled Systems — Method for Achieving Roll-off Cleanliness.

ANSI/B93.44M-1978 (R1986), American National Standard Method for Extracting Fluid Samples

Titles & Abstracts

from a Reservoir of an Operating Hyrdaulic Fluid Power System.

ANSI/B93.73M-1986, American National Standard Hydraulic Fluid Power — In-line Liquid Automatic Particle Counting Systems — Method of Validation.

NFPA/T3.10.4M-1968 (R1980), Graphic Symbols for Hydraulic Fluid Power Filters and Separators.

ANSI/B93.21M-1972 (R1980), American National Standard End Load Test Method for a Hydraulic Fluid Power Filter Element.

ANSI/B93.22M-1972 (R1979), American National Standard Hydraulic Fluid Power — Filter Elements — Determination of Fabrication Integrity.

ANSI/B93.25M-1972 (R1980), American National Standard Hydraulic Fluid Power — Filter Elements — Verification of Collapse/Burst Resistance.

ANSI/B93.23M-1972 (R1980), American National Standard Hydraulic Fluid Power — Filter Elements — Verification of Material Compatibility Fluids.

ANSI/B93.46M-1978 (R1986), American National Standard Method for Determining the Pore Size of a Cleanable Surface Type Hydraulic Fluid Power Filter Element.

NFPA/T3.10.8.18M-1977 (R1982), Multi-Pass Method for Evaluating the Filtration Performance of a Coarse Hydraulic Fluid Power Filter Element.

NFPA/T3.10.12 R1-1983, NFPA Information Report — Bibliography of Existing Standards Relating to Filtration and Contamination.

Volume E — Conductors and Associated Products

NFPA/T1.21.1-1978 (R1983), Procedure for Self-Certification by Fluid Power Manufacturers.

NFPA/T3.8.11-1977, A Bibliography of Fluid Power Tube Fittings and Conductors Standards.

ANSI/B93.59M-1982, American National Standard Fluid Power Systems and Products — Connectors and Associated Components — Outside Diameters of Tubes and Inside Diameters of Hoses.

ANSI/B93.60M-1982, American National Standard Fluid Power Systems and Products — Connectors and Associated Components — Nominal Pressures.

ANSI/B93.4M-1981, American National Standard Hydraulic Fluid Power — Line Tubing — Electric Resistance Welded, Mandrel Drawn.

ANSI/B93.11M-1981, American National Standard Hydraulic Fluid Power — Line Tubing — Seamless Low Carbon Steel.

ANSI/B93.42M-1977 (R1983), American National Standard Method for Testing Hydraulic Fluid Power Quick Action Couplings.

ANSI/B93.51M-1980, American National Standard Pneumatic Fluid Power — Quick Action Couplings — Test Conditions and Procedures.

NFPA/T3.20.1-1973 (R1981), Glossary of Terms For Fluid Power Quick Action Couplings.

NFPA/T3.20.7 R1-1983, A Bibliography of Fluid Power Quick Action Coupling Standards.

NFPA/T2.6.1M S5 (T3.20.8M)-1975 (R1981), Quick Action Couplings Pressure Rating Supplement No. 5 to NFPA Recommended Standard for Verifying the Fatigue and Static Pressure Ratings of the Pressure Containing Envelope of a Metal Fluid Power Component.

ANSI/B93.68M-1983, American National Standard Hydraulic Fluid Power — Quick Action Couplings — Surge Flow Test (Short Duration Flow).

ANSI/B93.69M-1983, American National Standard Hydraulic Fluid Power — Quick Action Couplings — Surge Flow Test (Long Duration Flow).

NFPA/T3.26.1 R1-1977, A Bibliography of Fluid Power Hose, Hose Fittings and Hose Assemblies.

Volume F — Control Products

NFPA/T3.27.1-1972 (R1981), Glossary of Terms for Compressed Air Dryers.

NFPA/T1.21.1-1978 (R1983), Procedure for Self-Certification by Fluid Power Manufacturers.

NFPA/T2.6.1M S2 (T3.21.4M)-1977 (R1982), Pneumatic Valve Pressure Rating Supplement No. 2 to NFPA Recommended Standard for Verifying the Fatigue and Static Pressure Ratings of the

Titles & Abstracts

Pressure Containing Envelope of a Metal Fluid Power Component.

NFPA/T2.6.1M S9 (T3.5.26)-1977 (R1982), Hydraulic Valve Pressure Rating Supplement No. 9 to NFPA Recommended Standard for Verifying the Fatigue and Static Pressure Ratings of the Pressure Containing Envelope of a Metal Fluid Power Component.

ANSI/B93.7M-1986, American National Standard Hydraulic Fluid Power — Valves — Mounting Interfaces.

ANSI/B93.9M-1969 (R1981), American National Standard Symbols for Marking Electrical Leads and Ports on Fluid Power Valves.

ANSI/B93.40M-1976 (R1982), American National Standard Series of Mounting Interfaces for 4567 Maximum psi (315 bar) Four-Port Hydraulic Fluid Power Directional Valves.

ANSI/B93.66M-1983, American National Standard Hydraulic Fluid Power — Directional Control Valve — Method for Determining the Metering Characteristics.

ANSI/B93.55M-1981, American National Standard Hydraulic Fluid Power — Solenoid-Piloted Industrial Valves — Interface Dimensions for Electrical Connectors.

NFPA/T3.5.33M-1985, Hydraulic fluid power — Cylinder Actuator Mounted Valves — Standard Dimensions for Mounting Surfaces.

ANSI/B93.65M-1983, American National Standard Hydraulic Fluid Power — Code for Identification of Valve Mounting Surfaces.

NFPA/T3.7.2M-1968 (R1980), Graphic Symbols for Fluidic Devices and Circuits.

ANSI/B93.14M-1971 (R1979), American National Standard Methods for Presenting Basic Performance Data for Fluidic Devices.

ANSI/B93.39M-1978 (R1986), American National Standard Requirements for Presentation of Catalog Data, Fluid Compatibility, Cleaning Media, Markings and Dimensional Identification Codes and Pressure Drop Characteristics for Fluid Power Air Line Filters.

ANSI/B93.13M-1981, American National Standard Pneumatic Fluid Power — Pressure Regulators — Industrial Type.

ANSI/B93.33M-1974 (R1981), American National Standard Interfaces for 4-Way General Purpose Industrial Pneumatic Directional Control Valves.

NFPA/T3.21.7M-1976 (R1981), Defining Interface Surfaces for each Pneumatic Valve Interface in NFPA Recommended Standard T3.21.1-1973.

NFPA/T3.21.9M-1976 (R1981), Definition of Port Communication for the Fluid Power Pneumatic Valve Interface to NFPA Recommended Standard T3.21.1 with the valve in Position in Response to a Remote Pilot Signal or Electrical Energization.

ANSI/B93.67M-1983, American National Standard Pneumatic Fluid Power — Five-Port Directional Control Valves — Mounting Surfaces — Optional Electrical Connector — Dimensions and Requirements.

ANSI/B93.45M-1982, American National Standard Pneumatic Fluid Power — Compressed Air Dryers — Methods for Rating and Testing.

ANSI/B93.38-1976 (R1981), American National Standard Method of Diagramming for Moving Parts Fluid Controls.

NFPA/T3.5.27-1976 (R1982), A Bibliography of Hydraulic Valve Standards and Test Procedures.

NFPA/T3.12.9-1977, A Bibliography of Fluid Power Pneumatic FRL Standards.

NFPA/T3.21.5-1978, A Bibliography of Fluid Power Pneumatic Valve Standards.

NFPA/T3.27.4-1979, A Bibliography of Compressed Air Dryers Standards.

NFPA/T3.28.11-1982, A Bibliography of Fluid Logic Devices.

Volume G — Cylinders and Accumulators

ANSI/B93.2-1986, Excerpts extracted from American National Standard Fluid Power Systems and Products — Glossary.

NFPA/T3.4.7M-1975 (R1980), Method for Establishing and Verifying the Fatigue and Static Pressure Ratings and Conducting Production Testing of the Pressure Containing Envelope of a Metal Fluid Power Accumulator.

ANSI/B93.3-1984, American National Standard Fluid Power Systems and Products — Cylinder

Bores and Piston Rod Diameters — Inch Series.

ANSI/B93.1M-1964 (R1982), American National Standard Dimension Identification Code for Fluid Power Cylinders.

ANSI/B93.8M-1968 (R1986), American National Standard Bore and Rod Size Combinations and Rod End Configurations for Cataloged Square Head Industrial Fluid Power Cylinders.

ANSI/B93.15-1981, American National Standard Fluid Power Systems and Products — Square Head Industrial Cylinders — Mounting Dimensions.

ANSI/B93.29M-1986, American National Standard Fluid Power Systems — Cylinders — Dimensions for Accessories for Catalogued Square Head.

ANSI/B93.34M-1973 (R1979), American National Standard Bore and Rod Size Combinations, Rod End Configurations, Dimensional Identification Code, and Mounting Dimensions for 3/4, 1 and 1 1/8 inch Bore Cataloged Square Head Tie Rod Type Industrial Fluid Power Cylinders.

NFPA/T3.6.17M-1971 (R1980), Port Nominal Pipe Sizes for Merged Inch and Metric Series Cataloged Square Head Industrial Pneumatic Fluid Power Cylinders.

ANSI/B93.52M-1981, American National Standard Fluid Power Systems and Products — Cylinder Bores and Piston Rod Diameters — Metric Series.

NFPA/T3.6.34-1979, Fluid Power Systems and Components — Cylinder Bores and Piston Rod Diameters — Inch Series.

ANSI/B93.53M-1981, American National Standard Fluid Power Systems and Products — Cylinder — Nominal Pressures.

ANSI/B93.56M-1982, American National Standard Fluid Power Systems and Products — Cylinders — Basic Series of Piston Strokes.

ANSI/B93.61M-1982, American National Standard Fluid Power Systems and Products — Cylinders — Piston Rod and Thread Dimensions and Types.

NFPA/T3.6.54M-1986, Hydraulic Fluid Power — Cylinder Ports — SAE Straight Thread O-ring and 4-bolt Flange Ports — Heavy Duty and Light Duty Cylinders.

NFPA/T3.6.36-1978 (R1984), A Bibliography of Fluid Power Cylinder Standards.

Volume H — Fluid, Lubricant and Sealing Devices

ANSI/B93.50M-1979, American National Standard Pneumatic Fluid Power — Use of Synthetic Lubricants — Guidelines.

ANSI/B93.5M-1979, American National Standard Practice for the Use of Fire Resistant Fluids in Industrial Hydraulic Fluid Power Systems.

NFPA/T2.13.2 R2-1980, Hydraulic Fluid Power — Fire Resistant Fluids — Information Report on Company Trade Names.

NFPA/T2.13.3-1979, Index of Non-proprietary Hydraulic Fluid Specifications and Selected Recommended Practices.

ANSI/B93.63M-1984, American National Standard Hydraulic Fluid Power — Petroleum Fluids — Prediction of Bulk Moduli.

NFPA/T3.19.4M R1-1985, Hydraulic Fluid Power — Seal Housings — Dimensions and Tolerances — Cylinder Rod and Piston Seals for Reciprocating Applications — Nominal Series.

ANSI/B93.17M-1979 (R1986), American National Standard Fluid Power Systems And Components — Multiple Lip Packing Sets — Methods for Measuring Stack Heights.

ANSI/B93.35M-1978 (R1986), American National Standard Cavity Dimensions For Fluid Power Exclusion Devices (Inch Series).

ANSI/B93.36M-1973 (R1986), American National Standard Groove Dimensions For Floating Type Metallic and Non-Metallic Fluid Power Piston Rings.

ANSI/B93.62M-1982, American National Standard Hydraulic Fluid Power — Reciprocating Dynamic Sealing Devices in Linear Actuators — Method of Testing, Measuring, and Reporting Leakage.

ANSI/B93.32M-1973 (R1986), American National Standard Groove Dimensions for Fluid Power Radial Compression Type Piston Rings.

Titles & Abstracts

ANSI/B93.58M-1982, American National Standard Fluid Systems — O-rings — Inside Diameters, Cross Sections, Tolerances and Size Identification Code.

NFPA/T3.19.22-1982, A Bibliography of Fluid Power Sealing Devices Standards.

Volume I — Testing

ANSI/B93.2-1986, Excerpts extracted from American National Standard Fluid Power Systems and Products — Glossary.

NFPA/T1.21.1-1978 (R1983), Procedure for Self-Certification by Fluid Power Manufacturers.

NFPA/T2.12.5-1983, National Fluid Power Association Information Report Fluid Power Laboratory Guidelines.

ANSI/B93.49M-1980, American National Standard Hydraulic Fluid Power — Valves — Pressure Differential-Flow Characteristic — Method of Measuring and Reporting.

ANSI/B93.10M-1969 (R1982), American National Standard Static Pressure Rating Methods of Square Head Fluid Power Cylinders.

ANSI/B93.48M-1979, American National Standard Pneumatic Fluid Power, Applications — Metal Separable Tube Fittings — Qualifications Test.

ANSI/B93.71M-1986, American National Standard Hydraulic Fluid Power — Pumps — Test Code for the Determination of Airborne Noise Levels.

ANSI/B93.72M-1986, American National Standard Hydraulic Fluid Power — Motors — Test Code for the Determination of Airborne Noise Levels.

ANSI/B93.27M-1973 (R1979), American National Standard Method of Testing and Presenting Basic Performance Data for Positive Displacement Hydraulic Fluid Power Pumps and Motors.

ANSI/B93.22M-1972 (R1979), American National Standard Method for Determining the Fabrication Integrity of a Hydraulic Fluid Power Filter Element.

ANSI/B93.24M-1972 (R1980), American National Standard Method for Verifying the Flow Fatigue Characteristics of a Hydraulic Fluid Power Filter Element.

ANSI/B93.31M-1973 (R1981), American National Standard Multi-Pass Method for Evaluating the Filtration Performance of a Fine Hydraulic Fluid Power Filter Element.

NFPA/T3.10.8.18M-1977 (R1982), Multi-Pass Method for Evaluating the Filtration Performance of a Coarse Hydraulic Fluid Power Filter Element.

ANSI/B93.42M-1977 (R1983), American National Standard Method for Testing Hydraulic Fluid Power Quick Disconnect Couplings.

ANSI/B93.51M-1980, American National Standard Pneumatic Fluid Power — Quick Action Couplings — Test Conditions and Procedures.

ANSI/B93.12-1971 (R1977), American National Standard Method of Rating for Mechanical Vacuum Pumps.

PIPE FABRICATION INSTITUTE (PFI)

The following standards are available from PFI, P.O. Box 173, Springdale, PA 15144. (412) 274-4722

PFI: ES-1, Internal Machining and Solid Machined Backing Rings for Circumferential Butt Welds, 1983.

PFI: ES-2, Method of Dimensioning Piping Assemblies, Revised 1984.

PFI: ES-3, Fabricating Tolerances, Reaffirmed 1984.

PFI: ES-4, Hydrostatic Testing of Fabricated Piping, Revised 1985.

PFI: ES-5, Cleaning of Fabricated Piping, Revised 1984.

PFI: ES-7, Minimum Length and Spacing for Welded Nozzles, 1984.

PFI: ES-11, Permanent Marking of Piping Materials, 1974 (R 1984), 2 pp.
  Covers recommended identification of piping materials welder's symbols or other data. Methods of marking only are involved.

PFI: ES-16, Access Holes and Plugs for Radiographic Inspection of Pipe Welds, Revised 1985.

PFI: ES-20, Wall Thickness Measurement by Ultrasonic Examination, Revised 1985.

Titles & Abstracts

PFI: ES-21, Internal Machining and Fit-up of GTAW Root Pass Circumferential Butt Welds, 1983.

PFI: ES-22, Recommended Practice for Color Coding of Piping Materials, 1974 (R 1984), 2 pp.
Provides identification of piping materials by a general material classification. It is not intended that the color coding will distinguish between the various grades of a particular material (A106B vs. A106A) or between specifications of the same materials (as ASTM A53, A106 or A155 pipe), or representing different manufacturing methods (such as seamless and seam welded).

PFI: ES-24, Pipe Bending Tolerances — Minimum Bending Radii — Minimum Tangents, Revised 1984.

PFI: ES-25, Random Radiography of Pressure Retaining Girth Butt Welds, Revised 1985.

PFI: ES-26, Welded Load Bearing Attachments to Pressure Retaining Piping Materials, 1984.

PFI: ES-27, Visual Examination — The Purpose, Meaning and Limitation of the Term, Reaffirmed 1984.

PFI: ES-29, Abrasive Blast Cleaning of Ferritic Piping Materials, Reaffirmed 1984.

PFI: ES-30, Random Ultrasonic Examination of Butt Welds, 1979.

PFI: ES-31, Standard for Protection of Ends of Fabricated Piping Assemblies, Revised 1985.

PFI: ES-32, Tool Calibration, Reaffirmed 1985.

PFI: ES-33, Circumferential Butt Welds in The Arc of Pipe Bends, Reaffirmed 1985.

PFI: ES-34, Painting of Fabricated Piping, 1983.

PFI: ES-35, Nonsymmetrical Bevels and Joint Configurations for Butt Welds, 1984.

PFI: TB1-1974, Pressure Temperature Ratings of Seamless Pipe Used in Power Plant Piping Systems, Reaffirmed 1984, 40 pp.
The information is derived from formulae and stress values contained in Section 1 Power Boiler and Pressure Vessel Code, and the Power Piping Section of the USA Standard Code for Pressure Piping ANSI: B31.1.0 Usage of these as maximum values is mandatory for piping systems within the jurisdiction of either of these Codes.

PFI: TB2-1983, Reinforcement Tables for 45 Degrees and 90 Degrees Branch Connections of Seamless Steel Pipe.

PFI: TB3-Revised 1985, Guidelines Clarifying Relationships and Design Engineering Responsibilities Between Purchasers' Engineers & Pipe Fabricator or Pipe Fabricator Erector.

PLUMBING AND DRAINAGE INSTITUTE (PDI)

The following publication is available from the Plumbing and Drainage Institute, 5342 Boulevard Place, Indianapolis, IN 46208. (317) 251-5298

PDI: WH201, Water Hammer Arresters, 1977, 24 pp.
Covers certification, sizing, placement and reference data on water hammer arresters used for reduction of noise, vibration and destruction in piping systems, valves, meters, etc.

PDI: G101, Testing and Rating Procedure for Grease Interceptors, with Appendix of Sizing and Installation Data, 1981, 12 pp.
This project includes the design and construction of the testing equipment, preliminary research and testing, the development of a certification test procedure and the development of a standard method of rating the flow capacities and grease capacities of grease interceptors.

PDI: Code Guide 302, Glossary of Industry Terms, 1979, 17 pp.
Describes plumbing devices which are normally used in plumbing and drainage systems and should be included when plumbing codes are written and adopted.

RADIO TECHNICAL COMMISSION FOR AERONAUTICS (RTCA)

The following publications are available from the RTCA Secretary, 1425 K Street, NW, Suite 500, Washington, DC 20005. (202) 682-0266

RTCA: DO-52, Calibration Procedures for Signal Generators Used in the Testing of VOR and ILS Receivers, 1953.
Recommends procedures for testing and calibrating signal generators used in the servicing of airborne VOR and ILS receivers. The accuracy of the components of simulated VOR and ILS signals is stated for signal generators calibrated as described.

Titles & Abstracts

RTCA: DO-56, VOR Test Signals, 1954.
Describes methods for determining, in an aircraft, the accuracy of VOR bearing indications. The causes of VOR bearing error due to VOR receiver malfunctioning are analyzed.

RTCA: DO-62, Calibration Procedures — Test Standard Omni-Bearing Selector Test Sets, 1954.
Recommends procedures to aid operators of aircraft radio service stations in the Calibration of Test Standard Omni-Bearing Selectors and Omni-Bearing Selector Test Sets used in testing and adjusting VOR receivers and their associated Omni-Bearing Selectors.

RTCA: DO-88, Altimetry, 1959.
Reports on studies of the problems associated with the measurement of aircraft altitude. States requirement that would permit all aircraft to maintain assigned heights within specific limits as related to terrain clearance and the safe vertical separation of aircraft in flight. Appendix I reports on Meteorological Aspects of Pressure Altimetry.

RTCA: DO-117, Standard Adjustment Criteria for Airborne Localizer and Glide Slope Receivers, 1963.
Recommends procedures for adjustment of Airborne Glide Slope and Localizer Receivers.

RTCA: DO-119, Interference to Aircraft Electronic Equipment from Devices Carried Aboard, 1963.
Recommends limits of permissible radiation of RF energy from portable electronic equipment used aboard aircraft in flight, including test procedures for the measurement thereof. Also recommends regulatory actions relating to the operation and identification of passenger-operated devices.

RTCA: DO-127, Standard Procedure for the Measurement of the Radio-Frequency Radiation from Aviation Radio Receivers Operating Within the Radio-Frequency Range of 30-890 Megacycles, 1965.
Recommends standards and test procedures for use by manufacturers of aviation receivers in making necessary radiation measurements using the Far-Field method. In addition, the report discusses the alternative methods of performing such measurement using the Near-Field method.

RTCA: DO-136, Universal Air-Ground Digital Communication System Standard, 1968.
Recommends universal digital standards for linking aircraft into the ground communications and data processing environment of the air traffic control system and airlines and military management information systems.

RTCA: DO-143, Minimum Performance Standards — Airborne Radio Marker Receiving Equipment Operating on 75 Megahertz, 1970.
Recommends standards and test procedures for Airborne Radio Marker Receiving Equipment. Coordinated with EUROCAE.

RTCA: DO-144, Minimum Operational Characteristics — Airborne ATC Transponder Systems, 1970.
Part I defines the concepts, philosophy and development of MOCs for airborne systems, and Part II covers the MOCs for Airborne ATC Transponder Systems, including system characteristics; provides information for demonstration of compliance, and guidance material.

RTCA: DO-148, A New Guidance System for Approach and Landing, 1970.
Defines a system concept and technical description (signal format) for a new precision instrument approach and landing guidance system (LGS) intended to satisfy the varied operational needs of different classes of aviation users, civil and military, in the United States and abroad. Volume 1 is an 80-page summary of findings and recommendations. Volume II is a 400-page compilation of the milestone Special Committee Reports, including the Tentative Operational Requirements, the Report of the Techniques Assessment Team and the Report of the Signal Format Development Team.

RTCA: DO-152, Minimum Operational Characteristics — Vertical Guidance Equipment Used in Airborne Volumetric Navigational Systems, 1922/Appendix 1974.
Part I defines the concepts, philosophy and development of MOCs for airborne systems, and Part II covers the MOCs for vertical guidance equipment used in airborne volumetric navigation systems, including system characteristics; provide information for demonstration of compliance; and guidance accuracy analysis. Appendix D provides a VOR/DME/Altimeter vertical guidance analysis.

RTCA: DO-154, Recommended Basic Characteristics for Airborne Radio Homing and Alerting Equipment for Use with Emergency Locator Transmitters (ELT), 1973.
Recommends basic system characteristics and provides test and guidance material for Airborne Radio Homing and Alerting Equipment.

RTCA: DO-155, Minimum Performance Standards — Airborne Low-Range Radar Altimeters, 1974.
Recommends standards and test procedures for those characteristics of an Airborne Low-Range Altimeter which are essential for its operation in

Titles & Abstracts

application to provide measured height above terrain for obstruction clearance and landing. Coordinated with EUROCAE.

RTCA: DO-158, Minimum Performance Standards — Airborne Doppler Radar Navigation Equipment, 1975.

Recommends standards and test procedures for Airborne Doppler Radar Navigation Equipment. Appendices include conditions of testing and detailed test procedures. Coordinated with EUROCAE.

RTCA: DO-160B, Environmental Conditions and Test Procedures for Airborne Equipment, 1984.

Standard procedures and environmental test criteria for testing airborne equipment for the whole spectrum of aircraft from light general aviation aircraft and helicopters through the "Jumbo Jets" and SST categories of aircraft. Coordinated with EUROCAE, RTCA DO-160B and EUROCAE/ED-14B are identically worded. Endorsed by the International Organization for Standardization (ISO) as de facto international standard ISO 7137.

RTCA: DO-161A, Minimum Performance Standards — Airborne Ground Proximity Warning Equipment, 1976.

Recommends standards and test procedures for Ground Proximity Warning Equipment. This is a revision of DO-161 and includes changes (1 & 2) to that document and other improvements suggested by operating experience. Appendices include envelopes of conditions for warning, conditions for testing, and detailed test procedures.

RTCA: DO-162, Report on Air-Ground Communications — Operational Considerations for 1980 and Beyond, 1975.

Provides an analysis of air-ground communications requirements anticipated for the post-1980 time frame. Includes definitions of U.S. aviation system and future trends, requirements, systems concepts, and recommendations. This is a companion report to RTCA Paper No. 128-72/EC-671 issued 8-18-72, entitled Proposed U.S. National Aviation Standard for the VHF A/G Communications System.

RTCA: DO-163, Minimum Performance Standards — Airborne HF Radio Communications Transmitting and Receiving Equipment Operating Within the Radio-Frequency Range of 1.5 to 3.0 Megahertz, 1976/Errata.

Recommends standards and test procedures for HF/SSB receivers and transmitters designed to operate in a 3 kHz channel environment. Also includes standards for the provision of AM equivalent mode of operation. Appendices include conditions for testing and detailed test procedures.

RTCA: DO-164A, Minimum Performance Standards — Airborne Omega Receiving Equipment, 1979.

Recommends standards and test procedures for Airborne Omega Navigation Receivers, Systems Sensors, and Navigation Systems. Also included are operational characteristics. Appendices include conditions for testing, detailed test procedures, and a description of Omega Error Mechanisms.

RTCA: DO-165, Initial Report of Civil Aviation Frequency Spectrum Requirements — 1980–2000, 1976.

Provides a comprehensive report on civil aviation's frequency requirements. Appendices recommend revisions to the ITU Table of Allocations and the footnotes thereto; provides justification for stated operational requirements; provides an aviation forecast.

RTCA: DO-166, Microwave Landing System (MLS) Implementation, 1977.

Reports on a study to develop user recommendations for a national implementation policy for MLS as the primary landing system in service by the year 2000. Volume I provides recommendations on how best to transition from ILS to MLS; recommends implementation strategy and a national implementation policy, which are summarized in a findings and recommendation chapter. Volume II includes six appendices which are the reports of working groups in special categories such as Benefits, Airborne Systems Operational Capabilities, and Civil System Costs.

RTCA: DO-167, Airborne Electronics and Electrical Equipment Reliability, 1977.

Provides a tutorial discussion of reliability related to aircraft accidents. Discusses airborne electronic equipment failures and means of reducing failures.

RTCA: DO-169, VHF Air-Ground Communication Technology and Spectrum Utilization, 1979.

Reports on VHF (118–136 MHz) spectrum utilization including the investigation of modulation techniques and reduced channel separation. Identifies problem areas and recommends, among other things, use of reduced channel spacing on a selective basis.

RTCA: DO-170, Audio Systems Characteristics and Minimum Performance Standards — Aircraft Microphones (Except Carbon), Aircraft Headsets and Speakers, Aircraft Audio Selector Panels and Amplifiers, 1980.

Titles & Abstracts

Part I discusses audio systems response characteristics affecting the intelligibility of air-ground voice communications and recommends means for improvement by users and designers of communication equipment. Part II recommend standards and test procedures for microphones (except carbon), headsets and speakers, and audio and interphone amplifiers for use in aircraft. Coordinated with EUROCAE.

RTCA: DO-171, Recommendations on Policies and Procedures for Off-the-Shelf Electronic Test Equipment Acquisition and Support, 1980.

Provides rationale and recommendations for various conditions and procedures that could provide major benefits to those responsible for drafting legislation, policies, procedures and guidelines for the acquisition and support of electronic test equipment.

RTCA: DO-172, Minimum Operational Performance Standards for Airborne Radar Approach and Beacon Systems for Helicopters, 1980.

Postulates operational goals and applications, and recommends standards and test procedures for Airborne Radar Approach (ARA) systems for helicopters, particularly when operating under IFR, IMC conditions or at night, including standards for the ground-based radar beacon. Includes test conditions and procedures for installed equipment performance, and operational characteristics with test procedures.

RTCA: DO-173, Minimum Operational Performance Standards for Airborne Weather and Ground Mapping Pulsed Radars, 1980.

Postulates operational goals and applications, and recommends standards and test procedures for airborne weather and ground mapping pulsed radars. It takes into account new radar technology and is applicable to both large aircraft and general aviation aircraft systems. Includes test conditions and procedures for installed equipment performance, and operational characteristics with test procedures.

RTCA: DO-174, Minimum Operational Performance Standards for Optional Equipment Which Displays Non-Radar Derived Data on Weather and Ground Mapping Radar Indicators, 1981.

Postulates operational goals and applications, and recommends standards and test procedures for use of weather and ground mapping radar indicators for display of non-radar graphic and/or alphanumeric data. Includes test conditions and procedures for installed equipment performance and operational characteristics with test procedures.

RTCA: DO-175, Minimum Operational Performance Standards for Ground-Based Automated Weather Observation Equipment, 1981.

Postulates operational goals and applications, and recommends standards and test procedures for ground-based automated weather observation equipment. Provides system characteristics for users, designers, manufacturers and installers of such equipment — of interest to various users, including airfield operators, meteorological services, aviation administrations, airplane and helicopter operators, etc. Includes test conditions and procedures for installed system performance, and operational characteristics with test procedures.

RTCA: DO-176, FM Broadcast Interference Related to Airborne ILS, VOR and VHF Communications, 1981.

Reviews the various aspects of the problem of commercial FM broadcast stations contributing to the interference of airborne systems. Recommends improved intra-governmental coordination procedures; and recommends steps to limit growth of the problem, to reduce the problem with installed receivers, and to minimize the problem with new receivers or installations.

RTCA: DO-177, Minimum Operational Performance Standards for Microwave Landing System (MLS) Airborne Receiving Equipment, 1981.

Postulates operational goals and applications, and recommends standards and test procedures for use of Microwave Landing Systems (MLS) airborne receiving equipment. Includes test conditions and procedures for installed equipment performance and operational characteristics with test procedures. Coordinated with EUROCAE.

RTCA: DO-178A, Software Considerations in Airborne Systems and Equipment Certification, 1981.

Describes techniques and methods that may be used for the orderly development and management of software for airborne digital computer-based equipment and systems. Provides guidance to both industry and regulators for use in the certification process. Coordinated with EUROCAE. EUROCAE/ED-12A is identically worded.

RTCA: DO-179, Minimum Operational Performance Standards for Automatic Direction Finding (ADF) Equipment, 1982.

Postulates operational goals and applications, and recommends standards and test procedures for airborne automatic direction finding equipment. Includes test conditions and procedures for installed equipment performance and operational characteristics with test procedures. Coordinated with EUROCAE.

RTCA: DO-180, Minimum Operational Performance Standards for Airborne Area Navigation Equipment Using VOR/DME Reference Facility Sensor Inputs, 1982.

Postulates operational goals and applications, and recommends standards and test procedures for airborne area navigation equipment (2S and 3D) using VOR/DME reference facility sensor inputs. Includes test conditions and procedures for installed equipment performance and operational characteristics with test procedures. Coordinated with EUROCAE.

RTCA: DO-181, Minimum Operational Performance Standards for Air Traffic Control Radar Beacon System/Mode Select (ATCRBS/Mode S) Airborne Equipment, 1983.

Postulates operational goals and applications, and recommends standards and test procedures for ATCRBS and Mode S airborne equipment. Includes test conditions and procedures for installed equipment performance and operational characteristics with test procedures. Coordinated with EUROCAE.

RTCA: DO-182, Emergency Locator Transmitter (ELT) Equipment Installation and Performance, 1982.

Provides analyses of ELT performance in regard to false alarms and activations in crash environments; provides criteria and guidelines for placement and installation of ELTs in aircraft; reports on ELT system performance in a variety of typical installations; and provides specific recommendations on all of the above standards.

RTCA: DO-183, Minimum Operational Performance Standards for Emergency Locator Transmitters — Automatic Fixed — ELT (AF), Automatic Portable — ELT (AP), Automatic Deployable — ELT (AD), Survival — ELT(s) Operating on 121.5 and 243.0 Megahertz, 1983.

Postulates operational goals and applications, and recommends standards and test procedures for emergency locator transmitters. Includes test conditions and procedures for installed equipment performance and operational characteristics with test procedures.

RTCA: DO-184, Traffic Alert and Collision Avoidance Systems (TCAS) I Functional Guidelines, 1983.

Sets forth minimum requirements and describes the various elements of TCAS I. Discusses both passive and active TCAS I applications. Provides the minimum performance requirements for electromagnetic compatibility for an active TCAS I, and test procedures for both active and passive systems. Appendix A addresses cross-link advisories.

RTCA: DO-185, Minimum Operational Performance Standards for Traffic Alert and Collision Avoidance Systems (TCAS) Airborne Equipment, 1984.

Volume I sets forth operational goals and applications and recommends standards and test procedures for airborne traffic alert and collision avoidance equipment intended primarily for use on transport category aircraft. Includes test conditions and procedures for ensuring installed equipment performance. This baseline TCAS is described as having omnidirectional interrogation/reception capability. Additional features are described which will provide sectorized interrogations for operations in higher density airspace and, separately, will provide bearing estimation measurements. When all features are included, the system will meet FAA Minimum TCAS II requirements for collision avoidance. Volume II contains the required Collision Avoidance Algorithms. The algorithms are presented as high-level pseudocode to convey functional design, and as low-level pseudocode to serve as detailed specification. Coordinated with EUROCAE.

RTCA: DO-186, Minimum Operational Performance Standards for Airborne Radio Communications Equipment Operating Within the Radio Frequency Range 117.975-137.000 MHZ, 1984.

Postulates operational goals and applications, and recommends standards and test procedures for airborne VHF communications receivers and transmitters. Includes test conditions and procedures for installed equipment performance, and operational characteristics with test procedures. Coordinated with EUROCAE.

RTCA: DO-187, Minimum Operational Performance Standards for Airborne Area Navigation Equipment Using Multi-Sensor Inputs.

Postulates operational goals and applications, and recommends standards and test procedures for airborne area navigation equipment (2D and 3D) using multisensor inputs. Includes test conditions and procedures for installed equipment performance and operational characteristics with test procedures. Coordinated with EUROCAE.

RTCA: DO-189, Minimum Operational Performance Standards for Airborne Distance Measuring Equipment (DME) Operating Within the Radio Frequency Range of 960-1215 Megahertz.

Postulates operational goals and applications, and recommends standards and test procedures for airborne distance measuring equipment (DME). It updates the former DME operational characteristics and performance standards for airborne

Titles & Abstracts

equipment that operate with conventional DME (DME/N) ground facilities and establishes standards for airborne equipment which will operate with both DME/N and precision DME (DME/P) ground facilities. Coordinated with EUROCAE.

RANGE COMMANDERS COUNCIL (RCC)

The following Interrange Instrumentation Group (IRIG) and Range Commanders Council (RCC) Standards have been prepared and published by the RCC for use by government agencies and industries under contract to them who have an interest in missiles, rockets and associated equipment. Limited copies of these publications are available to authorized government agencies and contractors with active government contracts from the Secretariat, Range Commanders Council, STEWS-SA-R, U.S. Army White Sands Missile Range, NM 88002. (505) 678-2121. Others may obtain copies from the Defense Logistics Agency, Defense Technical Information Center, ATTN: FDRA, Cameron Station, Alexandria, VA 22304-7633. (703) 892-5620. Use the AD/AO numbers that follow each listing below when ordering from DTIC.

IRIG: 106-86, Telemetry Standards, 1986, 132 pp.
Provides development and coordination agencies with the necessary criteria on which to base equipment designs and modification. The standards are intended to ensure efficient spectrum and interference-free operation of the radio link for telemetry systems at the RCC member ranges.

RCC: 118-79, Test Methods for Telemetry Systems and Subsystems, Volume I, End-to-End Methods for Telemetry Systems, 1979, 97 pp.
Addresses the methodology employed to accomplish solar calibrations and transducer-based system calibrations.

Volume II, Test Methods for Telemetry RF Subsystems, 1979, 183 pp.
Contains test procedures for telemetry antenna systems, RF preamps, multicouplers, receivers, and diversity combiners.

Volume III, Test Methods for Recorder/Reproducer Systems and Magnetic Tape, 1979, 136 pp.
Provides test methodologies for tape heads, tape speed, speed variation and timing error, direct record systems, FM systems, and serial high density digital systems.

Volume IV, Test Methods for Data Multiplex Equipment, 1979, 164 pp.
Contains test procedures for frequency division multiplexing, time division multiplex systems, subcarrier oscillators, and bit synchronizers.

RCC: 118-82, Test Methods for Telemetry Systems and Subsystems, Volume V, Test Methods for Vehicle Telemetry Systems, 120 pp.
Addresses test methodologies which apply to transducers, charge amplifiers, differential dc amplifiers, power supplies, and telemetry transmitters.

IRIG: 152-83, IRIG Standard for Distributing Vector Acquisition Data, 1983, 14 pp.
Defines a standard procedure for distributing interrange vector acquisition data to remote tracking instruments.

IRIG: 154-71, IRIG Standards for Distributing Raw Radar Antenna Data, 1971, 22 pp.
Defines a standard procedure for distributing raw antenna data from remote tracking sites via teletype.

IRIG: 161-65, IRIG Standard Data Format for Interrange Transmission of Tracking Data from Computer to Computer (2400 BPS Synchronous EFG Format O), 1985, 12 pp.
Provides a standard format for interrange transmission of tracking data from a computer on one range to a computer at another range.

IRIG: 200-70, IRIG Standard Time Formats, 1970, 26 pp.
Describes standard instrumentation timing formats suitable for recording on magnetic tape, recording oscillographs, film, and real-time transmission, which meet both manual and automatic data reduction requirements. Definitions of terms relating to the standard formats are included.

RCC: 203-64, Standard Format for Interrange Distribution of Visual Count Status Information, 1964, 18 pp.
Specifies a format to be used for the transfer of visual count status information between elements of a global range.

IRIG: 205-77, IRIG Standard Parallel Binary Time Code Formats, 1977, 12 pp.
Presents the "ground rules" for attaining maximum compatibility between present and future time generation equipment and the user interface. The selected combination codes included contain most, if not all, of the various parallel binary time codes.

IRIG: 206-77, IRIG Standard for VHF Time Transmission, 1977, 5 pp.

Provides guidance and criteria for procurement and evaluation of projected systems which transmit signals via VHF at various Department of Defense installations.

IRIG: 208-85, IRIG Standard for UHF Command Systems, 1985, 16 pp.

Serves as a guide for the orderly implementation and application of UHF command systems for test ranges. Provides development and coordination agencies with the necessary criteria on which to base equipment designs and modifications.

IRIG: 251-80, IRIG Standard for Pulse Repetition Frequencies and Reference Oscillator Frequency for C-Band Radars, 1980, 3 pp.

Contains standards established to accommodate existing C-band instrumentation radars.

IRIG: 252-74, IRIG Tracking Radar Compatibility and Design Standards for G-Band (4 to 6 GHz) Radars, 1974, 18 pp.

Outlines the minimum noncoherent G-band radar compatibility standards for interrange use with emphasis on transmitter and receiver characteristics. Also provides guidelines for the design of new G-band noncoherent instrumentation radars.

IRIG: 253-65, IRIG Standard Coordinate System and Data Formats for Antenna patterns, 1965, 100 pp.

Defines a vehicle antenna coordinate system and antenna pattern formats recommended for use at the National and Service Ranges. Also addresses antenna polarization considerations and includes a number of definitions associated with vehicle antenna pattern representations.

RCC: 257-86, Coherent C-Band Transponder Standards, 1986, 25 pp.

Defines minimum transponder parameters in such a manner that any C-band instrumentation radar on any test range may use the transponder.

IRIG: 303-64, Standardized Test Procedures for UHF Flight Termination Receivers, revised 1968, 45 pp.

Provides standardized interrange test procedures for the UHF Flight Termination Receivers for use in determining their adequacy for range safety applications and their compatibility with existing ground instrumentation systems.

IRIG: 352-72, IRIG Standards for Range Meteorological Data Reduction Part I — Rawinsonde, 1972, 129 pp.

Establishes a standard method for reducing rawinsonde data and defines all formulae and computation routines needed to carry out this data reduction process.

IRIG: 352-85, IRIG Standards for Range Meteorological Data Reduction Part I — Rocketsonde, 1985, 437 pp.

Contains computer program documentation for two different computer systems used for the reduction of range meteorological data. Section A addresses a program used by a large central computer to process data from either non-transponder or transponder rocketsondes. Section B deals with the program for the minicomputer (NOVA-3/12) used in the Meteorological Sounding System to process data from transponder rocketsondes.

RCC: 452-86, Video Standards and Formats, 1986, 50 pp.

Provides users of standard and high resolution monochrome and standard color closed circuit television equipment with the criteria essential for the interchange and compatibility of equipment, tape recordings and live signals. Applies to locally generated signals as well as metric video applications involving video tape recorders, video disc recorders and data insertion equipment.

RCC: 503-82, Standard for Data Labels and Data Annotation Procedures, 1982, 31 pp.

Establishes a standard data labeling process to simplify and expedite the annotation of selected data (items) generated from test operations conducted by the various RCC member and associate member ranges and facilities.

RESISTANCE WELDER MANUFACTURERS ASSOCIATION (RWMA)

The following publication and others dealing with special welding applications (piping, exotic materials, etc.) are available from the Resistance Welder Manufacturers Association, 1900 Arch Street, Philadelphia, PA 19103. (215) 564-3484

RWMA: Bulletin No. 16, Resistance Welding Equipment Standards, 1984, 105 pp.

Contains 10 standards including such topics as nomenclature and definitions, butt welding, electrical standards and fluid power standards.

SCIENTIFIC APPARATUS MAKERS ASSOCIATION (SAMA)

The following publications are available from SAMA, 1101 16th Street, NW, Suite 300, Washington, DC 20036. (202) 223-1360

Titles & Abstracts

SAMA: AI 1.1, Recommended Test Procedures for Glass pH and Reference Electrodes, 1974.
Provides a uniform means for users of glass pH electrodes to evaluate and calibrate these sensors. Procedures for percent theoretical slope, sodium error, zero potential pH, isopotential point, and time response are provided.

SAMA: AI 2.5, Instrumental Specifications for Atomic Absorption Spectrophotometers, 1978, 6 pp.
Provides a set of performance specifications which users of atomic absorption spectrophotometers can expect to obtain from manufacturers. These specifications are classified by function and relative importance.

SAMA: AI 2.1a, Guidelines for Purity and Handling of Gases Used in Atomic Absorption Spectroscopy, 1978, 6 pp.
Describes the purity requirements for gases used in atomic absorption spectroscopy. Contains information on handling and gas volume requirements for acetylene, hydrogen, nitrous oxide, argon, and nitrogen.

SAMA: AI 2.4B, Terminology for Atomic Absorption Spectrophotometers, 1978, 6 pp.
Provides readily usable definitions for 34 terms significant in atomic absorption spectroscopy. Usage is consistent with ASTM where applicable.

SAMA: PMC 4-1-1962, Bimetallic Thermometers.

SAMA: PMC 5-10-1963, Resistance Thermometers.

SAMA: PMC 6-10-1963, Filled System Thermometers.

SAMA: PMC 8-10-1963, Thermocouple Thermometers (Pyrometers).

SAMA: PMC 17-10-1963, Bushings and Wells for Temperature Sensing Elements.

SAMA: PMC 18-10, Markings for Adjustable Means in Automatic Controllers, 2nd edition, 1965, 2 pp.
Establishes standard terms and units for the marking of adjustment means of controllers capable of producing one or more of the following control actions — proportional action, reset or integral action, and rate of derivative action.

SAMA: PMC 19-10, Tubing Connection Markings for Pneumatic Instruments, 1963, 1 pp.
Establishes a uniform system of marking tubing connection to simplify the inter-connection of pneumatic instruments in their application to industrial processes.

SAMA: PMC 20.1-1973, Process Measurement and Control Terminology, 1973, 44 pp.
Applies to terminology associated with industrial process instrumentation used in industries such as chemical, petroleum, metallurgical, power, food, textile and paper. It includes terms relating to measurement and control, and the static and dynamic performance of indicators, recorders, controllers, indicating controllers, recording controllers, transmitters and transducers.

SAMA: PMC 21.4-1966, Temperature-Resistance Valves for Resistance Thermometer Elements for Platinum, Nickel and Copper.

SAMA: PMC 22.1-1981, Functional Diagramming of Instrument and Control Systems, 1981, 20 pp.
Presents both symbols and diagramming format for use in representing measuring, controlling and computing systems as used in industrial practice. The purpose is to establish uniformity of symbols and practices in diagramming such systems in their basic functional form, exclusive of their operating media or specific equipment detail.

SAMA: PMC 23-2-1971, Hydrostatic Testing of Control Valves, 1971, 6 pp.
This standard applies to ferrous, including stainless steel, control valves. The purpose of the hydrostatic test is to prove the structural integrity and liquid tightness of the valve.

SAMA: PMC 27.1-1980, Pressure Safety for Pressure and Differential Pressure Process Control Devices, 1980.

SAMA: PMC 28.1-1973, Dimensions of Wide Chart Recorders, 1973, 3 pp.
Covers mounting of wide chart recorders in panels, racks, and rack panels. The purpose is to establish mounting dimensions so that panel mounting configurations and rack configurations shall be compatible.

SAMA: PMC 28.2-1976, Dimensions for Panel and Rack Mounted Industrial Process Measurement and Control Instruments.
Covers dimensions for panel and rack mounted industrial process measurement and control instruments.

SAMA: PMC 31.1-1980, Generic Test Methods for the Testing and Evaluation of Process Control Instrumentation, 1980.
Describes the conditions and procedures for the testing under static and dynamic conditions of

process control instrumentation with analog input and output signals, using the manufacturer's specifications and instructions for installation and operation.

SAMA: 31.2-1983, Guidelines for Presenting Specifications of Analog Process Measurement and Control Instruments.

SAMA: PMC 32.0-1981, Process Instrumentation Reliability Terminology Handbook.
Defines process instrumentation and process instrumentation reliability terms.

SAMA: PMC 32.1-1976, Process Instrumentation Reliability Terminology.
Establishes terminology for use in process instrumentation reliability.

SAMA: PMC 33.1-1978, Electromagnetic Susceptibility of Process Control Instrumentation.
Discusses electromagnetic susceptibility as it relates to process control instrumentation.

SAMA-ABMA: Recommended Standard Instrument Connections Manual, 1981 Edition, 148 pp.
The recommendations represent the continuation of joint efforts between the Process Measurement and Control Section (PMC) SAMA and members of the American Boiler Manufacturers Association (ABMA) for the purpose of standardizing identification and location of instrument and control connections for water-tube boilers. No attempt has been made to incorporate connections used specifically for environmental monitoring control.

SOCIETY OF AUTOMOTIVE ENGINEERS, INC. (SAE)

The following and numerous other SAE publications are available from the Society of Automotive Engineers, 400 Commonwealth Drive, Warrendale, PA 15096. (412) 776-4841

SAE/J 254, Instrumentation and Techniques for Exhaust Gas Emissions Measurement, 1984.
Establishes uniform laboratory techniques for the continuous and bag-sample measurement of various constituents in the exhaust gas of the gasoline engines installed in passenger cars and light-duty trucks. Concentrates on the measurement of the following components in exhaust gas: hydrocarbons (HC), carbon monoxide (CO), carbon dioxide (CO_2), oxygen (O_2), and nitrogen oxides (NO_x).

SAE/J 247, Instrumentation for Measuring Acoustic Impulses Within Vehicles, 1980.
Provides guidelines for selection and application of instrumentation for proper measurement of acoustic impulses within vehicles, as typified by those generated during the deployment of a passive restraint system. The objective is to achieve uniformity in instrumentation practice and reporting of test measurements. Use should provide a basis for meaningful comparisons of test results from different sources.

SAE/J 211, Instrumentation for Impact Tests, 1980.
Provides guidelines for instrumentation used in automotive safety impact tests. The aim is to achieve uniformity in instrumentation practice and in reporting test results, without imposing undue restrictions on the performance characteristics of the individual elements in an instrumentation or data analysis system. Provides a basis for meaningful comparisons of test results from different sources.

SAE/AS 942, Pressure Altimeter System — Minimum Safe Performance Standard.
Specifies the requirements for minimum safe performance of an altimeter system in its normal mode of operation on subsonic aircraft. The instrument system specified shall accept an input of the static pressure and in some equipment other inputs that contribute altitude information to provide a visual indication of pressure altitude. If equipped with an automatic correction mechanism, it shall indicate by a positive means when the automatic correction mechanism is not in use. If the static source pressure error compensating mechanism is operational it shall be functional throughout the required operating envelope of the particular aircraft. Each aircraft type has its own static source error data which shall be obtained from the airframe manufacturer's certified data. When a Central Air Data Computer is used in the altimeter system, the CADC shall be certified to its own governing document and the altimeter system (CADC and display) shall comply with the requirements of this document. NOTE: The instrument system specified herein does not include the aircraft pressure lines and pressure sources.

SAE/AS 793, Total Temperature Measuring Instruments (Turbine Powered Subsonic Aircraft).
Establishes essential minimum safe performance requirements for total temperature measuring instruments, primarily for use with turbine-powered subsonic transport aircraft, the operation of which may subject the instruments to the environmental conditions specified in this report. Covers three basic types of total temperature measuring instruments used as a means of determining the total temperature developed by adia-

Titles & Abstracts

batic heating of the air due to motion of the aircraft through the air.

SAE/J 1045, Instrumentation and Techniques for Vehicle Refueling Emissions Measurement.

Describes a procedure for measuring the hydrocarbon emissions occurring during the refueling of passenger cars and light trucks. It can be used as a method for investigating the effects of temperatures, fuel characteristics, etc., on refueling emissions in the laboratory. It also can be used for determining the reduction in emissions achieved with emission control hardware. For this latter use, standard temperatures, fuel volatility, and fuel quantities are specified.

SAE/AS 407B, Fuel Flowmeters.

To specify minimum requirements for Fuel Flowmeters for use primarily in reciprocating engine powered civil transport aircraft, the operation of which may subject the instruments to the environmental conditions specified in Section 3.3. This Aeronautical Standard covers two basic types of instruments, or combinations thereof, intended for use in indicating fuel consumption of aircraft engines as follows: TYPE I — Measure rate of flow of fuel used. TYPE II — Totalize amount of fuel consumed or remaining.

SAE/AS 406, Flight Directors (Turbine-Powered Subsonic Aircraft).

This standard establishes the essential minimum safe performance requirements for flight director instruments, primarily for use with turbine-powered transport aircraft, the operation of which may subject the instruments to the environmental conditions specified in paragraph 3.3. This standard covers flight directors for use on aircraft to indicate to the pilot, by visual means, the correct control application for the operation of an aircraft in accordance with a preselected flight plan.

SAE/AS 404B, Electric Tachometer: Magnetic Drag (Indicator and Generator).

To specify minimum requirements for Electric Tachometers primarily for use in reciprocating engine powered civil transport aircraft, the operation of which may subject the instruments to the environmental conditions specified in Section 3.3. This Aeronautical Standard covers magnetic drag tachometers with or without built-in synchroscopes.

SAE/AS 394A, Rate of Climb Indicator, Pressure Actuated (Vertical Speed Indicator).

To specify minimum requirements for pressure, actuated Climb Indicators for use in aircraft, the operation of which may subject the instruments to the environmental conditions specified in paragraph 3.3. This Aeronautical Standard covers four (4) basic types of direct indicating instruments as follows: TYPE I — Range 0-2000 feet per minute climb and descent, TYPE II — Range 0-3000 feet per minute climb and descent, TYPE III — Range 0-4000 feet per minute climb and descent, TYPE IV — Range 0-4000 feet per minute climb and descent.

SAE/AS 392C, Altimeter, Pressure Actuated Sensitive Type.

To specify minimum requirements for Pressure Actuated Sensitive Altimeters for use in aircraft, the operation of which may subject the instrument to the environmental conditions specified in paragraph 3.3.

SAE/AS 391C, Airspeed Indicator (Pitot Static) (Reciprocating Engine Powered Aircraft).

To establish the essential minimum safe performance standards for pitot static pressure type of airspeed indicators, primarily for use with reciprocating engine power transport aircraft, the operation of which may subject the instruments to the conditions specified in Section 3.3.

SAE/AS 1104, Specification for Single-Degree-of-Freedom Spring-Restrained Rate Gyros.

This specification covers that gyroscopic instrument normally defined as a "subminiature rate gyro." The rate gyro, when subjected to an angular rate abut its input axis, provides an AC output voltage proportional to the angular rate. The subminiature size category generally includes gyro instruments of one (1) inch diameter or less and three and one-half (3½) inches length or less. This specification defines the requirements for a subminiature spring-restrained, single-degree-of-freedom rate gyro for aircraft, missile, and spacecraft applications.

SAE/ARP 416, Directional Indicating System (Turbine Powered).

To recommend the essential minimum safe performance requirements for gyroscopically stabilized Directional Indicating System, primarily for use with turbine powered subsonic transport aircraft, the operation of which may subject the instruments to the environmental conditions specified in paragraph 3.3. This recommended practice covers the requirements for gyroscopically stabilized Directional Indicating Systems, which will operate as a 1 degree/hour latitude corrected, free directional gyro or as a slaved gyro, magnetic compass with ½ degree accuracy.

SAE/ARP 1278, Oscilloscopic Method of Measuring Spark Energy.

This report provides specific information on instrumentation and procedure for the measurement of capacitance discharge spark energy using

Titles & Abstracts

an oscilloscope. This report describes basic method for measurement of spark energy on all types of capacitance discharge exciters. Reference is made to other methods which may be used if limitations are observed.

SAE/ARP 1267, Electromagnetic Interference Measurement Impulse Generators; Standard Calibration Requirements and Techniques.

This Aerospace Recommended Practice (ARP) describes a standard method and means for measuring or calibrating the "Spectrum Amplitude" output of an impulse generator. This ARP also outlines the method for the measurement of EMI instruments impulse bandwidth.

SAE/ARP 1217, Instrumentation Requirements for Turboshaft Engine Performance Measurements.

This Aerospace Recommended Practice (ARP) defines the measurement parameters that may be used by a pilot or operator to monitor the thermodynamic health of a turboshaft engine in a helicopter and the measurement system accuracies desired.

SAE/ARP 24B, Determination of Hydraulic Pressure Drop.

This ARP is intended to serve as an instrument to determine hydraulic pressure drop, utilizing the best known practices for accessories in the hydraulic, fuel, oil, and coolant systems to aerospace vehicles.

SAE/AIR 1255, Spectrum Analyzers for Electromagnetic Interference Measurements.

This AIR was prepared to inform the aerospace industry about the electromagnetic interference measurement capability of spectrum analyzers. The spectrum analyzers considered are of the wide dispersion type which are electronically tuned over an octave or wider frequency range. The reason for limiting the AIR to this type of spectrum analyzer is that several manufacturers produce them as general-purpose instruments, and their use for EMI measurement will give significant time and cost savings. The objective of the AIR is to give a description of the spectrum analyzers, consider the analyzer parameters, and describe how the analyzers are usable for collection of EMI data. The operator of a spectrum analyzer should be thoroughly familiar with the analyzer and the technical concepts reviewed in the AIR before performing EMI measurements.

SAE/AIR 1092, High Tension Exciter Output Voltage Measurement Using Cathode-Ray Oscilloscope.

The purpose of this report is to provide specific information on instrumentation and procedure for the measurement of high tension exciter output voltage. This report describes a method of voltage measurement using a cathode-ray oscilloscope and high-voltage probe, with emphasis on calibration.

ASTM/B, Test Method for Indentation Hardness of Aluminum Alloys by Means of a Newage Portable Non-Caliper-Type Instrument.

Aluminum alloys — indentation hardness — portable non-caliper-type hardness instrument, test; Hardness (indentation) — aluminum alloys — portable non-caliper-type hardness instrument, test; Portable non-caliper-type hardness method — indentation hardness (of aluminum alloys), by portable non-caliper-type hardness instrument, test.

SAE/J 209, Instrument Face Design and Location for Construction and Industrial Equipment, 1980.

The instrument design criteria and grouping described are recommended to manufacturers of construction and industrial equipment for all new designs. Adherence to these recommendations will promote improved performance and ease of machine and operation, protect machine and operator, and simplify instrument design and production.

SAE/AE 8021, Minimum Performance Standards for Direction Instrument, Non-Magnetic (Gyroscopically Stabilized).

This Aerospace Standard (AS) defines minimum performance requirements under standard and environmental conditions for Gyroscopically Stabilized Non-Magnetic Direction Instruments for use in aircraft. This document establishes the minimum requirements for design and qualification of equipment identified as Gyroscopically Stabilized Non-Magnetic Direction Instruments.

SAE/AS 8016, Vertical Velocity Instrument (Rate of Climb).

This Aerospace Standard (AS) establishes the minimum performance standards for vertical velocity instruments for aircraft use.

SAE/AS 8013, Minimum Performance Standard for Direction Instrument, Magnetic (Gyroscopically Stabilized).

This Aerospace Standard (AS) defines minimum performance requirements under standard and environmental conditions for Gyroscopically Stabilized Magnetic Direction Instruments for use in aircraft. This document establishes the minimum requirements for design and qualification of equipment identified as Gyroscopically Stabilized Magnetic Direction Instruments.

SAE/AS 8004, Minimum Performance Standard

Titles & Abstracts

for Turn and Slip Instruments.

This standard establishes the minimum performance standards for turn and slip instruments for aircraft use.

SAE/AS 439, Stall Warning Instrument (Turbine Powered Subsonic Aircraft).

This Aerospace Standard establishes the essential minimum safe performance standards for stall warning instruments primarily for use with turbine powered subsonic transport aircraft, the operation of which may subject the instruments to the environmental conditions specified in paragraph 3.4. This standard covers stall warning instruments to provide positive warning to the pilot of an impending stall. Stall, as defined for the purpose of this standard, is the minimum steady flight speed at which the airplane is controllable.

SAE/AS 403A, Stall Warning Instrument.

To specify minimum requirements for stall warning instruments for use in aircraft, the operation of which may subject the instrument to environmental conditions specified in Section 3.3.

SAE/AS 399A, Direction Instrument, Magnetic, (Stabilized Type).

To specify minimum requirements for gyroscopically stabilized i Magnetic Direction Instruments for use in aircraft, the operation of which may subject the instruments to the environmental conditions specified in Paragraph 3.3 This Aeronautical Standard covers minimum requirements for gyroscopically stabilized Magnetic Direction Instruments for use in aircraft.

SAE/AS 398A, Direction Instrument, Magnetic, Non-Stabilized Type (Magnetic Compass).

To specify minimum requirements for non-stabilized magnetic direction instruments for use in aircraft, the operation of which may subject the instrument to the environmental conditions specified in Paragraph 3.3 This Aeronautical Standard covers two basic types of instruments: Type I — Direct Reading, Type II — Remote Indicating.

SAE/AS 397A, Direction Instrument, Non-Magnetic, Stabilized Type (Directional Gyro).

To specify minimum requirements for non-magnetic gyroscopically stabilized direction indicators for use in aircraft, the operation of which may subject the instruments to the environmental conditions specified in Paragraph 3.3. This Aeronautical Standard covers two basic types: Type I — Air Operated, Type II — Electrically Operated.

SAE/AIR 818C, Aircraft Instrument Standards: Wording, Terminology, Phraseology, and Environmental and Design Standards.

Provides the sponsors of Aerospace Standards, (AS), with standard wording, formatting, and minimum environment and design requirements for use in the preparation of their document.

SAE/AS 802B, Powerplant Fire Detection Instruments, Thermal & Flame Contact Types (Reciprocating and Turbine Engine Powered Aircraft).

This Standard establishes minimum requirements for powerplant fire detection instruments primarily for use in reciprocating and turbine engine powered aircraft.

SAE/AS 8019, Airspeed Instruments.

This standard establishes minimum performance standards for total and static pressure actuated airspeed instruments.

SAE/AS 8005, Minimum Performance Standard — Temperature Instruments.

This Aerospace Standard (AS) establishes the essential minimum performance requirements for electrical type temperature instruments primarily for use on aircraft which may subject the instruments to environmental conditions specified herein.

SAE/AS 431A, True Mass Fuel Flow Instruments.

Establishes the essential minimum safe performance standards for True Mass Fuel Flow Instruments primarily for use with turbine powered, subsonic transport aircraft, the operation of which may subject the instruments to the environmental conditions specified in Section 3.3. This Aerospace Standard covers three basic types of true mass flow indicating instruments. Each may consist of an indicator, transmitter and other auxiliary means such as a power supply or amplifier as required.

SAE/AS 428, Exhaust Gas Temperature Instruments.

This standard establishes the essential minimum safe performance standards for exhaust gas temperature instruments primarily for use with turbine powered, subsonic aircraft, the operation of which may subject the instruments to the environmental conditions specified in paragraph 3.3 et seq. The exhaust gas temperature instruments covered by this standard are of the electrical servonull balance type, actuated by varying emf output or one or more parallel connected Chromel-Alumel thermocouples.

SAE/AS 414A, Temperature Instruments (Turbine Powered Subsonic Aircraft).

Establishes the essential minimum safe performance standards for electrical type temperature instruments primarily for use with turbine powered subsonic transport aircraft, the operation of which may subject the instruments to the environmental conditions specified in Section 3.4. Covers basic types of temperature instruments: TYPE I: Ratiometer type, actuated by changes in electrical resistance of a temperature sensing electrical resistance element; TYPE II: Millivoltmeter type, operated and actuated by varying EMF input to the instrument being obtained by temperature changes of the temperature sensing thermocouple.

SAE/AS 413B, Temperature Indicator.

Establishes the minimum safe performance standards for electrical type temperature instruments primarily for use with reciprocating engine powered transport aircraft, the operation of which may subject the instruments to the environmental conditions specified in Section 3.4. Covers two basic types of temperature instruments: TYPE I: Ratiometer type, actuated by changes in electrical resistance of a temperature sensing electrical resistance element; TYPE II: Millivoltmeter type, operated and actuated by varying E.M.F. output of a thermocouple.

SAE/AS 412A, Carbon Monoxide Detection Instruments.

To specify minimum requirements for carbon monoxide detector instruments for use in aircraft, the operation of which may subject the instrument to the environmental conditions specified in Paragraph 3.3. Not intended to cover fire detectors. Covers the basic type of carbon monoxide detector instrument used to determine toxic concentrations of carbon monoxide by the measurement of heat changes through catalytic oxidation.

SAE/AS 411A, Manifold Pressure Indicating Instruments.

Establishes the essential minimum safe performance standards for manifold pressure instruments primarily for use with reciprocating engine powered transport aircraft, the operation of which may subject the instruments to the environmental conditions specified in Section 3.3. Covers two basic types of manifold pressure instruments: TYPE I — Direct Indicating; TYPE II — Remote Indicating.

SAE/AS 408B, Pressure Instruments — Fuel, Oil, and Hydraulic (Reciprocating Engine Powered Aircraft).

Establishes the essential minimum safe performance standards for fuel, oil and hydraulic pressure instruments primarily for use with reciprocating engine powered transport aircraft, the operation of which may subject the instruments to the environmental conditions specified in Section 3.3. Covers two basic types of fuel, oil and hydraulic pressure instruments: TYPE I — Direct Indicating; TYPE II — Remote Indicating.

SAE/AS 405B, Fuel and Oil Quantity Instruments.

To specify minimum requirements for Fuel and Oil Quantity Instruments for use in aircraft, the operation of which may subject the instruments to the environmental conditions specified in Paragraph 3.3. Covers two basic types of instruments: Type I — Float Instruments; Type II — Capacitance Instruments.

SAE/ARP 427, Pressure Ratio Instruments.

To recommend requirements for electrical Pressure Ratio Indicating Instruments for use in aircraft, the operation of which may subject the instruments to the environmental conditions specified in Para. 3.3. Covers two types of two unit Pressure Ratio Instruments each of which consists of a Transducer and an Indicator.

SAE/ARP 1254, Fluidics Test Methods and Instrumentation.

Establishes acceptable methods, procedures and instrumentation required for testing fluidic devices. The tests described include only those necessary to predict the performance of a device when used in a system of circuit. The term "instrumentation" is understood to include all laboratory instrumentation (electrical, fluidic or mechanical, etc.).

THE SOCIETY OF NAVAL ARCHITECTS AND MARINE ENGINEERS (SNAME)

The following publications are available from SNAME, 601 Pavonia Avenue, Jersey City, NJ 07306. (201) 798-4800

SNAME: Code C-1, Code for Shipboard Vibration Measurements, 1975, 35 pp.

Establishes standard procedures for gathering and interpreting data on hull vibrations in single-screw commercial ships. These data are needed to compare the vibration characteristics of different ships of a given class, to establish vibration reference levels, and to provide a basis for the improvement of individual ships.

SNAME: Code C-2, Code for Sea Trials, 1973.

Includes a section which describes the instruments and apparatus commonly used for making measurements of the performance of various items of machinery in ship's trials.

Titles & Abstracts

SNAME: Code C-4, Local Shipboard Structures and Machinery Vibration Measurements, 1976, 28 pp.
Establishes standard procedures for gathering and presenting data on vibrations measured on structural elements of ships.

SNAME: Code C-5, Acceptable Vibration of Marine Steam and Gas Turbine Noise and Auxiliary Machinery Plants, 1976, 16 pp.
Provides criteria for mechanical vibration and serves as a refence standard in establishing ships' specifications and procurement documents for new marine equipment.

SNAME: T&R Bulletin 3-29, Guide for Centralized Control and Automation of Ship's Gas Turbine Propulsion Plant, 1978, 55 pp.
Provides technical guidance in the development of centralized control and automation of geared or electric drive gas turbine ship propulsion plants.

SNAME: T&R Bulletin 3-37, Design Guide for Shipboard Airborne Noise Control, 1983, 357 pp.
Contains technical guidance in the control of shipboard noise and acoustical design practice.

SNAME: T&R Bulletin 3-41, Guide for Centralized Control and Automation of Ship's Steam Propulsion Plant, 1986, 68 pp.
Gives technical guidance to establishing the desired degree and methods for employing centralized control and for automating a ship's steam propulsion plant.

SNAME: T&R Bulletin 3-42, Guidelines for the use of Vibration Monitoring for Preventive Maintenance, 1987, 98 pp.
Contains technical suggestions for implementing a vibration monitoring program for use in the maintenance of ship machinery.

SNAME: T&R Bulletin 4-18, Propulsion Monitoring Instrumentation for Shipboard Energy Conservation, 1984, 24 pp.
Reviews state of the art machinery instrumentation and provides guidance to ship operators on the selection of appropriate hardware to optimize propulsion plant performance.

SPRING MANUFACTURERS INSTITUTE, INC. (SMI)

The following handbook is available from the Spring Manufacturer's Institute, Inc., 1211 West 22nd Street, Oak Brook, IL 60521. (312) 520-3290

SMI: Handbook of Spring Design, 1977, 34 pp.
Gives specifications of material and various test procedures and standards for commercial springs including compression, extension, torsion, flat and hot wound. Contains glossary of spring and related testing terminology.

TECHNICAL ASSOCIATION OF THE PULP AND PAPER INDUSTRY (TAPPI)

The following and other TAPPI publications are available from TAPPI, Technology Park/Atlanta, P.O. Box 105113, Atlanta, GA 30348. (404) 446-1400

TAPPI: T 210 hm-86, Weighing, Sampling and Testing Pulp for Moisture, 1986, 4 pp.
The methods selected for moisture testing are specified for each given kind of pulp.

TAPPI: T 656 hm-83, Measuring, Sampling, and Analyzing White Waters, 1983, (Historical Method), 4 pp.
Presents methods of white water evaluation so that different mills may use substantially the same procedures and thus establish a common basis of comparison.

Titles & Abstracts

TAPPI: T 1206 rp-86, Precision Statement for Test Methods, 1986, 5 pp.

This recommended practice defines terms and describes how to estimate precision from available data.

TAPPI: T 1209 rp-87, Identification of Instrumental Methods of Color or Color Difference Measurement, 1987, 4 pp.

Provides brief, yet specific, recommendations for identification of instrumentation methods for the measurement of color or color difference.

TAPPI: TIS 0414-01, Instrument Symbols and Nomenclature, 1981, 7 pp.

Provides for the pulp and paper industry modifications to ISA Standard 5.1 and recommendations for symbolizing recent innovations in instrument hardware.

TAPPI: TIS 0414-02, Instrument Air Tubing and Piping Materials Recommends, 1982, 3 pp.

Assists in the proper specification and application of instrument air tubing in the pulp and paper mill.

TAPPI: TIS 0804-06, Photometric Linearity of Optical Properties Instruments, 1981, 2 pp.

Describes a test for linearity of optical properties instruments used in several TAPPI optical test methods.

ULTRASONIC INDUSTRY ASSOCIATION, INC. (UIA) (Formerly, Ultrasonic Manufacturers Association, Inc., UMA)

The following publication is available from the Ultrasonic Industry Association, c/o Penn Michael Management, P.O. Drawer F, Jamesburg, NJ 08831. (201) 521-4441

UIA: Recommended Standard Rating for Electric Generators, 1965, 1 pg.

Provides ratings for ultrasonic electrical generators on the basis of the average power output developed into a pure resistance or simulated load. The effects of peak pulse power and application of intermittent loads are also discussed.

UNDERWRITERS' LABORATORIES, INC. (UL)

The following "Standards for Safety" are available from Underwriters' Laboratories, Inc., Publications Stock, 333 Pfingsten Road, Northbrook, IL 60062. Offices and testing stations also located in Melville, LI, NY; Santa Clara, CA; and Research Triangle Park, NC. (312) 272-8800. Standards approved by ANSI have ANSI number in parentheses and are available from either UL or ANSI.

UL 25 (ANSI/UL25-1979 R 1985), Standard for Meters for Flammable and Combustible Liquids and LP-Gas 1979, 11 pp.

These requirements cover meters for measuring flammable and combustible liquids such as gasoline, kerosene, fuel oil, and similar petroleum products, and liquefied-petroleum gas in the liquid state. Liquid-measuring meters of the designs covered by this standard are commonly used in the assembly of dispensing equipment, tank trucks, and other low- or medium-pressure applications where there is a need for measurement of the product transferred.

UL 132, Standard for Relief Valves for Anhydrous Ammonia and LP-Gas, 1984, 12 pp.

These requirements cover safety valves and hydrostatic relief valves for anhydrous ammonia and liquefied petroleum gas (LP-Gas) for use in nonrefrigerated systems in facilities covered by the following American National and other Standards: ANSI K61.1, ANSI Z106.1, NFPA No. 58, and NFPA No. 59.

Titles & Abstracts

UL 144 (ANSI/UL 144-1977), Standard for Pressure Regulating Valves for LP-Gas, 1985, 14 pp.
These requirements cover pressure regulators for use with LP-Gas equipment other than in automotive and marine applications or gas-welding and cutting operations. They are also not intended to cover regulators for use in chemical, petrochemical, petroleum, or utility power plants; nor pipeline or marine terminals; nor related storage facilities at such plants or terminals.

UL 268, (ANSI/UL 268-1981), Standard for Smoke Detectors for Fire-Protective Signaling Systems, 1981, 83 pp.
These requirements cover smoke detectors to be employed in ordinary indoor locations in accordance with the following Standards of the National Fire Protection Association NPFA No. 72E-1978 and 74-1980.

UL 180 (ANSI/UL1.80-1980 R 1985), Standard for Liquid-Level Indicating Gauges and Tank-Filling Signals for Petroleum Products, 1980, 11 pp.
These requirements cover liquid-level indicating gauges and tank filling signals for use with vented tanks for the storage of petroleum products, such as gasoline, kerosene, fuel, oil, and other similar petroleum products.

UL 252 (ANSI/UL 252-1984), Standard for Compressed Gas Regulators, 1984, 11 pp.
These requirements cover pressure regulators which reduce the storage cylinder or line gas pressure to the use pressure. Regulators covered by these requirements are intended for use with air, inert gases, and fuel gases.

UL 404, Standard for Indicating Pressure Gauges for Compressed Gas Service, 1979, 7 pp.
These requirements cover indicating pressure gauges of the elastic element type usually employed in the high-pressure side of regulators or reducing valves used on compressed gas containers or cylinders of oxygen, hydrogen, nitrogen, and other gases. Such gauges usually have pressure ranges of 0-1500, 0-2000, 0-3000, or 0-4000 psi.

UL 187 (ANSI/UL 187-1985), Standard for X-Ray Equipment, 1983, 41 pp.
These requirements cover X-ray equipment for medical commercial, and industrial use, to be employed in accordance with the National Electrical Code.

UL 429, Standard for Electrically Operated Valves, 1982, 45 pp.
These requirements cover electrically general purpose and safety operated valves for the control of fluids such as air, gases, oils, refrigerants, steam, water, etc. Electrically operated valves covered by these requirements are intended to be employed in ordinary locations in accordance with the National Electrical Code.

UL 466 (ANSI: C 33.16-1976), Standard for Electrically Illuminated Scales, 1976, 8 pp.
These requirements cover portable, electrically illuminated counter scales rated at 250 volts or less and ordinarily of the computing type, intended for the measurement of weight and to be employed in accordance with the National Electrical Code.

UL 508, Standard for Industrial Control Equipment, 1984, 125 pp.
These requirements cover industrial control equipment for use in ordinary locations in accordance with the National Electrical Code. Includes apparatus and the devices immediately accessory thereto for starting, stopping, regulating, controlling, or protecting electric motors.

UL 521, (ANSI/UL 521-1987), Standard for Heat Detectors for Fire Protective Signaling Systems, 1988, 23 pp.
These requirements cover heat detectors for fire protective signaling systems intended to be installed in ordinary indoor and outdoor locations in accordance with the Standard for Automatic Fire Detectors, NFPA No. 72E.

Titles & Abstracts

UL 565, Standard for Liquid-Level Gages and Indicators for Anhydrous Ammonia and LP-Gas, 1986, 7 pp.

These requirements cover liquid-level gages and indicators for anhydrous ammonia and liquefied petroleum gas (LP-Gas) for use with pressure vessels in nonrefrigerated systems in installations covered by the following American National and other Standards: ANSI: K61.1, ANSI: Z106.1, NFPA No. 58, and NFPA No. 59.

UL 632 (ANSI/UL 632-1980), Standard for Electrically Actuated Transmitters, 1980, 31 pp.

These requirements cover electrically actuated transmitters intended for permanent installation and use in ordinary indoor locations. They do not cover manually actuated signaling boxes.

UL 698 (ANSI/UL 698-1985), Standard for Industrial Control Equipment for Use in Hazardous Classified Locations, 1984, 28 pp.

These requirements cover industrial control equipment for installation and use in hazardous locations, Class I, Groups, A, B, C, and D, and Class II, Groups, E, F, and G, in accordance with the National Electrical Code. They do not cover intrinsically safe electrical circuits of industrial control equipment for use in hazardous locations.

UL 873 (ANSI/UL 873-1981), Standard for Electrical Temperature-Indicating and Regulating Equipment, 1979, 90 pp.

These requirements cover general-use field-installation equipment and controls intended to be factory installed on or in certain appliances as safety, limiting, or operating controls. These controls respond directly or indirectly to changes in temperature, humidity, or pressure to effect control for equipment or appliance operation, etc.

UL 894-1972 (ANSI/UL 894-1977), Standard for Switches for Use in Hazardous Classified Locations, 1986, 18 pp.

These requirements cover snap and similar type switches rated at 60 amperes or less at 250 volts or less; switches 30 amperes or less at 600 volts or less; switches rated at 2 horsepower or less at 600 volts or less; Class I, Groups A, B, C, and D, and Class II, Groups, E, F, and G as defined in Article 500 of the National Electrical Code.

UL 913, (ANSI/UL/NFPA 4913-1979), Standard for Intrinsically Safe Apparatus and Associated Apparatus for Use in Class I, II, and III, Division I, Hazardous Locations, 1979, 42 pp.

These requirements shall apply to: Apparatus or parts of apparatus in Class I, II, or III; Division 1 locations. Those parts of apparatus located outside of the Class I, II, or III; Division 1 location whose design and construction may influence the intrinsic safety of an electrical circuit within the Class I, II, or III; Division 1 location.

UL 1002 (ANSI/UL 1002-1977), Standard for Electrically Operated Valves for Use in Hazardous Locations, Class I, Groups A, B, C, and D, and Class II, Groups E, F, and G, 1984, 18 pp.

These requirements cover electrically operated valves for installation and use in hazardous locations, Class I, Division I, Groups A, B, C, and D, and Class II, Division 1, Groups E, F, and G, in accordance with the National Electrical Code. They are not intended to cover valves for a fluid power system, which is a system that transmits and controls power through use of a pressurized fluid within an enclosed circuit.

UL 1244, Electrical and Electronic Measuring and Testing Equipment, 1980, 61 pp.

Applies to electrical, electronic, or electro-mechanical measuring or testing equipment designed to measure or observe and indicate quantities of electrical, or electronic phenomena (measuring equipment).

UL 1262, Laboratory Equipment, 1981.

These requirements cover cord-connected equipment rated 250 volts or less and permanently connected equipment rated 600 volts or less intended for use on interior wiring systems in accordance with the National Electrical Code.

Titles & Abstracts

UL 1437, Electrical Analog Instruments — Panel Board Types, 1979, 32 pp.

These requirements cover electrical and electrically operated indicating and recording instruments of the analog type that are powered only from the measured parameter and are intended for ordinary use in panel boards and the like.

UL 1481 Power Supplies for Fire Protective Signalling Systems, 1987.

Fixed and stationary commercial power supply units, rectifier units, transformers, and battery charger units.

UL 1604 Electrical Equipment for Use in Hazardous (Classified) Locations, 1988.

For equipment used in hazardous locations at ambient temperatures of 5 to 40°C, oxygen concentrations not greater than 21 percent, and at a nominal pressure of one atmosphere.

Agents for Standards Throughout the World

(Outside of the United States)

National standardizing bodies, especially the Member Bodies of the International Organization for Standardization (ISO), serve as national representatives for one another. Each acts within its own country as sole sales agent and information center for the standards of the other national standardizing bodies. Hence standards issued by the American National Standards Institute, the United States Member Body of ISO, may be purchased abroad from the respective national standardizing body in each of the following countries.

Algeria
 INAPI, Institut Algerien de Normalisation et de Propriete Industrielle, 5 Rue Abou Hamou Moussa, B.P. 1021, Centre de Tri, Alger

Argentina
 IRAM, Instituto Argentino de Racionalizacion de Materiales, Chile 1192, Buenos Aires

Australia
 SAA, Standards Association of Australia, Standards House, 80–86 Arthur Street, North Sidney, N.S.W. 2060

Austria
 ON, Oesterreichisches Normungsinstitut, Leopoldsgasse 4, Postfach 130, A 1020 Wien 2

Bangladesh
 BDSI, Bangladesh Standards Institution, 3-DIT (Extension) Avenue, Motijheel Commercial Area, Dhaka-2

Barbados
 BNSI, Barbados National Standards Institution, "Flodden" Culloden Road, St. Michael

Belgium
 IBN, Institut Belge de Normalisation, Av. de la Brabanconne, 29, B-1040 Bruxelles

Brazil
 ABNT, Associacao Brasileira de Normas Tecnicas, Av. 13 de Maio, Andar 13-28°, Caixa Postal 1680, CEP 20.003, Rio de Janeiro

Bulgaria
 BDS, State Committee for Science and Technical Progress, 21, 6th September Street, Sofia

Canada
 SCC, Standards Council of Canada, International Standardization Branch, 2000 Argentia Road, Suite 2-401, Mississauga, Ontario L5N 1V8

Chile
 INN, Instituto Nacional de Normalizacion, Matias Cousino 64-6°, Casilla 995-Correo 1, Santiago

China (Peking, Peoples Republic of China)
 CAS, China Association for Standardization, P.O. Box 820 Beijing

China (Republic of China)
 National Bureau of Standards, Ministry of Economic Affairs, 102 Kwang-Fu S. Rd, Taipei, Taiwan, 105

Columbia
 ICONTEC, Instituto Colombiano de Normas Tecnicas, Carrera 37 No. 52–95, P.O. Box 14237, Bogota

Costa Rica
 Instituto Centroamericano de Investigaciones y Tecnologia Industrial, 4a Calle y Avenida la Reforma, Zona 10, Guatemala City, Guatemala

Cyprus
 CYS, Cyprus Organization for Standards and Control of Quality, Ministry of Commerce and Industry, Nicosia

Czechoslovakia
 CSN, Urad pro Normalizaci a Mereni, Vaclavske Namesti 19, 113 47 Praha 1

Agents for Standards

Denmark
DS, Dansk Standardiseringsraad, Aurehojvej 12 and 15, Postbox 77, DK-2900 Hellerup

Dominican Republic
DIGENOR, Direccion General de Normas y Sistemas de Calidad, Secretaria de Industria y Comercio, Av. Mexico No. 30, Santo Domingo

Egypt
EOS, Egyptian Organization for Standardization, 2 Latin America Street, Garden City, Cairo

El Salvador
Instituto Centroamericano de Investigaciones y Tecnologia Industrial, 4a Calle y Avenida la Reforma, Zona 10, Guatemala City, Guatemala

Ethiopia
ESI, Ethiopian Standards Institution, P.O. Box 2310, Addis Ababa

Finland
SFS, Suomen Standardisoimisliitto r.y., P.O. Box 205, SF-00121 Helsinki 12

France
AFNOR, Association Francaise de Normalisation, Tour Europe, Cedex 7, 92080 Paris La Defense

Germany F.R.
DIN, Deutsches Institut fur Normung, Burggrafenstrasse 4–10, Postfach 1107, D-1000 Berlin 30

Ghana
GSB, Ghana Standards Board, P.O. Box M.245, Accra

Greece
ELOT, Hellenic Organization for Standardization, Didotou 15, 106 80 Athens

Guatemala
ICAITI, Instituto Centroamericano de Investigaciones y Tecnologia Industrial, 4a Calle y Avenida la Reforma, Zona 10, Apartado Postal 1552, Guatemala City

Honduras
Instituto Centroamericano de Investigaciones y Tecnologia Industrial, 4a Calle y Avenida la Reforma, Zona 10, Guatemala City, Guatemala

Hong Kong
Hong Kong Standards and Testing Centre, Dai Wang Street, Taipo Industrial Estate, Taipo, N.T., Hong Kong

Hungary
MSZH, Magyar Szabvanyugyi Hivatel, Postafiok 24, 1450 Budapest

Iceland
Technological Institute of Iceland, Division of Standards, Skipholt 37, IS-110 Reykjavik

India
ISI, Indian Standards Institution, Manak Bhavan, 9 Bahadur Shah Zafar Marg, New Delhi 110002

Indonesia
YDNI, Badan Kerjasama Standardisasi LIPI-YDNI (LIPI-YDNI Joint Standardization Committee), Jln, Teuku Chik ditiro 43, P.O. Box 250, Jakarta

Iran
ISIRI, Institute of Standards and Industrial Research of Iran, Ministry of Industries and Mines, P.O. Box 2937, Teheran

Iraq
COSOC, Central Organization for Standardization and Quality Control Planning Board, P.O. Box 13032, Aljadiria, Baghdad

Ireland
IIRS, IS, Institute for Industrial Research and Standards, Ballymun Road, Dublin-9

Israel
SII, Standards Institution of Israel, 42 University Street, Tel Aviv 69977.

Italy
UNI, Ente Nazionale Italiano de Unificazione, Piazza Armando Diaz 2, 1 20123 Milano

Jamaica
JBS, Bureau of Standards, 6 Winchester Road, P.O. Box 113, Kingston 10

Japan
JSA, Japanese Standards Association, 1-24 Akaska 4, Minato-ku 107, Tokyo

Jordan
Directorate of Standards, Ministry of Industry and Trade, P.O. Box 2019, Amman

Kenya
KEBS, Kenya Bureau of Standards, Off Mombasa Road Behind Belle Vue Cinema, P.O. Box 54974, Nairobi

Agents for Standards

Korea
KBS, Bureau of Standards, Industrial Advancement Administration, Yongdeungpo-Dong, Seoul

Korea
KSA, Korea Standards Association, 105-153, Kongduck-dong, Mapo-ku, Seoul

Kuwait
Standards and Metrology Department, Ministry of Commerce and Industry, Post Box No. 2944, Kuwait

Lebanon
LIBNOR, Institut Libanais de Normalisation, B.P. 195144, Beyrouth

Liberia
Ministry of Commerce, Industry and Transportation, Division of Standards, Monrovia

Libya
Standards and Specifications Section, Department of Industrial Organization, Secretariat of Industry, Tripoli

Madagascar
Ministere du Developpement Rural et de la Reforme Agraire, Direction l'agriculture, Service du Controle des Qualites et du Conditionnement, B.P. 1.316, Antananarivo

Malawi
Malawi Bureau of Standards, P.O. Box 946, Blantyre

Malaysia
SIRIM, Standards and Industrial Research Institute of Malaysia, Lot 10810, Phase 3, Federal Highway, P.O. Box 35, Shah Alam, Selangor

Mauritius
Mauritius Standards Bureau, Ministry of Commerce and Industry, Reduit

Mexico
DGN, DGN, Direccion General de Normas, Tuxpan No. 2, Mexico 7, D.F.

Morocco
SNIMA, Service de Normalisation Industrielle Marocaine, Direction de l'Industrie, Ministere du Commerce, de l'Industrie, des Mines et de la Marine Marchande, Rabat

Netherlands
NNI, NEN, Nederlands normalisatie-Instituut, Polakweg 5, P.O. Box 5810, 2280 HV Rijswijk ZH

New Zealand
SANZ, Standards Association of New Zealand, Private Bag, Wellington

Nicaragua
Instituto Centroamericano de Investigaciones y Tecnologia Industrial, 4a Calle y Avenida la Reforma, Zona 10, Guatemala City, Guatemala

Nigeria
NSO, Nigerian Standards Organization, Federal Ministry of Industries, 11 Kofo Abayomi Road, Victoria Island, Lagos

Norway
NSF, NS, Norges Standardiseringsforbund, Haakon VII's gt. 2, N-Oslo 1

Oman
Directorate General for Specifications and Measurements, Ministry of Commerce and Industry, P.O. Box 550, Muscat

Pakistan
PSI, PS, Pakistan Standards Institution, 39 Garden Road, Saddar, Karachi-3

Panama
Instituto Centroamericano de Investigaciones y Tecnologia Industrial, 4a Calle y Avenida la Reforma, Zona 10, Guatemala City, Guatemala

Peru
ITINTEC, Instituto de Investigacion Tecnologia, Industrial y de Normas Tecnicas, Jr. Morelli—2da Cuadra, Urbanizacion San Borja—Surquillo, Lima 34

Philippines
PS, Philippines Bureau of Standards, TML Commercial Bldg, 100 Quezon Avenue, Quezon City, P.O. Box 3719, Manila

Poland
PKNiM, Polski Komitet Normalizacji i Miar, UI, Electoraina 2, 00–139 Warszawa

Portugal
IGPAI, NP, Reparticao de Normalizacao, Avenida de Berna 1, Lisboa-1

Republic of South Africa
SABS, SABS, South African Bureau of Standards, Private Bag X191, Pretoria 0001

Rhodesia
Standards Association of Central Africa, Coventry Road, Workington P.O. Box 2259, Salisbury 4

Agents for Standards

Romania
IRS, STAS, Institutul Roman de Standardizare, Casuta Postala 6214, Bucarest 1

Saudi Arabia
SASO, Saudi Arabian Standards Organization, Airport Street, P.O. Box 3437, Riyadh

Singapore
SISIR, Singapore Institute of Standards and Industrial Research, 179 River Valley Road, P.O. Box 2611, Singapore 6

Spain
IRANOR, Instituto Nacional de Racionalizacion y Normalizacion, Serrano 150, Madrid 6

Sri Lanka
BCS, Bureau of Ceylon Standards, 53 Dharmapala Mawatha, Colombo 3

Sweden
SIS, SIS—Standardiseringskommissionen i Sverige, Tegnergatan 11, Box 3 295, S-103 66 Stockholm

Switzerland
SNV, SNV, *Association Suisse de Normalisation, Kirchenweg 4, Postfach, 8032 Zurich

Syria
Industrial Testing and Research Centre (Syrian Standards Organization), P.O. Box 845, Damascus

Thailand
TISI, Thai Industrial Standards Institute, Department of Science, Ministry of Industry, Rama VI Street, Bangkok 4

Trinidad and Tobago
Trinidad and Tobago Bureau of Standards, Room 318, Salvatori Bldg, Frederick Street, P.O. Box 288, Port of Spain

Tunisia
Ministere de l'Economie Nationale, Tunis

Turkey
TSE, TS, Turk Standardlari Enstitusu, Necatibey Caddesi 112, Bakanliklar, Ankara

United Kingdom
BSI, BS, British Standards Institution, 2 Park Street, London W1A 2BS, England

USSR
GOST, GOST, Gosudarstvennyj Komitet Standartov, Soveta Ministrov S.S.S.R, Leninsky Prospekt 9, Moskva 117049

Venezuela
COVENIN, Comision Venezolana de Normas industriales, Av. Boyaca (Cota Mil), Edf. Fundacion La Salle Piso 5, Caracas 105

Yugoslavia
JZS, Jugoslovenski zavod za Standardizaciju, Slobodana Penezica-Krcuna br. 35, Post Pregr. 933, 11000 Beograd

Zambia
Zambia Standards Institute, P.O. Box RW 259, Lusaka

Zimbabwe
Standards Association of Zimbabwe, 17, Coventry Road, P.O. Box 2259, Workington, Harare, Zimbabwe

SUBJECT INDEX

This index was designed to help the reader quickly identify, by subject area, the standards listed in the abstract section. All the standard titles were searched for key words and key subject terms. The index is based upon those words and terms and includes: subject terms listed in alphabetical order; the titles of standards containing a specific term or a closely related term under the subject term along with the acronym of the sponsoring organization; and the page within the abstract section that contains additional information pertaining to each standard.

The reader, when seeking standards dealing with a subject area, should search the index under all possible terms that may be related to the subject. Standards tend to have narrow and concise titles and the index terms are not grouped by general subject areas; therefore, related standards may be listed under a number of entries.

This index covers:

ABSORPTION
ACCELERATION
ACOUSTICS
AIR
AIRCRAFT INSTRUMENTATION
ALARMS
ALUMINUM
ANALOG
ANALYZERS AND ANALYSIS
ATMOSPHERE
ATTENUATORS

BEARINGS
BOILERS

CABLES
CALIBRATION
CALORIMETRY
CERTIFICATION
CHROMATOGRAPHY
CIRCUITS
COATINGS
CODES AND CODING
COILS
COMMUNICATIONS
CONNECTORS
CONTAMINATION
CONTROL CENTERS
CONTROL AND CONTROLLERS
CONVERTERS
COOLING TOWERS
COUPLINGS
CYLINDERS

DATA
DENSITY
DIAGRAMS
DIGITAL
DOSIMETRY
DRAWINGS
DUST

ELECTRICAL CONNECTORS
ELECTRICAL EQUIPMENT
ELECTRICAL GENERATION
ELECTRICAL MEASUREMENTS
ELECTROMAGNETIC
  RADIATION
ELECTRONIC COMPONENTS
ELECTRONIC HAZARDS
EMISSION MEASUREMENTS
ENCLOSURES
EXPLOSIVE SERVICE

FASTENERS
FIBER OPTICS
FILTERS
FITTINGS
FLANGES
FLOW MEASUREMENT
FLUID POWER
FLUIDICS
FREQUENCY
FUELS

GAGES
GASES
GEARS

GENERAL
GENERATORS
GRAPHICS
GRAVITY, SPECIFIC

HAZARDOUS LOCATIONS
HEAT MEASUREMENTS
HUMIDITY MEASUREMENTS
HUMIDITY, RELATIVE
HYDROCARBONS
HYDROMETERS

IMPULSE
INDEX
INDUCTANCE
INFRARED TECHNIQUES
INSPECTION
INSULATION
INTRINSIC SAFETY

LABORATORY
LEAK DETECTION
LENGTH MEASUREMENT
LEVEL MEASUREMENT
LINEAR MEASUREMENT
LOSS CONTROL

MAGNETIC TECHNIQUES
MAINTENANCE
MANOMETERS
MEASUREMENT
MEDICAL APPARATUS
METEOROLOGICAL
  MEASUREMENT

METERS AND METERING
METRIFICATION
MOISTURE MEASUREMENTS

NOISE MEASUREMENTS
NOMENCLATURE

OSCILLOSCOPES

PANELS
PETROLEUM
PETROLEUM APPLICATIONS
pH MEASUREMENTS
PHOTOMETRY
PHOTOMULTIPLIERS
PIEZOELECTRIC DEVICES
PIPES AND PIPING
PILOT TUBES
POTENTIOMETERS
POWDERS MEASUREMENTS
POWER
POWER GENERATION
PRECISION
PRESSURE CONTROL &
  MEASUREMENT
PROGRAMMABLE
  CONTROLLERS/
  INSTRUMENTATION
PSYCHROMETRY
PULSES
PUMPS AND PUMPING
PYCNOMETER

Continued next page

Subject Index

This index covers *continued from previous page*

QUALITY CONTROL	SENSORS	TELEMETRY	TURBIDITY
	SHIPBOARD APPLICATIONS	TEMPERATURE MEASURE-	TURBINES
RADIATION	SHOCK & VIBRATORY	MENT & CONTROL	
RADIO FREQUENCY	MEASUREMENTS	TERMINALS	ULTRASONIC TECHNIQUES
RECEIVERS	SHUNTS	TEXTILES	
RECORDERS AND RECORDING	SIGNALING SYSTEMS	THERMAL MEASUREMENTS	VACCUUM APPLICATIONS
REFINERIES	SOFTWARE	THERMISTORS	VALVES
RELECTIONS	SOLAR	THERMOCOUPLES	VENTILATING SYSTEMS
REFRIGERATION	SOLID STATE DEVICES	THERMOELECTRIC	VIBRATION MEASUREMENTS
APPLICATIONS	SOUND MEASUREMENTS	THERMOMETERS	VISCOMETRY
REGULATORS	SPECIFICATIONS	THERMOPLASTICS	VISCIOUS MATERIALS
RELAY	SPECTROMETRY &	THERMOSTATS	VOLTAGE MEASUREMENTS
RELIABILITY	SPECTROPHOTOMETRY	TIMING	
RESISTANCE	SPRINGS	TOLERANCES	WATER MEASUREMENTS
ROTAMETERS	STATIC ELECTRICITY	TRANSDUCERS	WATER AND WASTEWATER
ROUNDNESS	STEAM APPLICATIONS	TRANSFORMERS	WAVE CHARACTERISTICS
	STRESS MEASUREMENT	TRANSISTORS	WEIGHING
SAFETY CONSIDERATIONS	SYMBOLS	TRANSMISSION DEVICES &	WELDS AND WELDING
SAMPLING		SYSTEMS	WIRES & WIRING
SCALES	TACHOMETERS	TRANSMITTERS	
SEISMIC MEASUREMENTS	TANKS	TRANSPONDER STANDARDS	
SEMICONDUCTOR DEVICES	TAPE	TUBING	

A

Absorption

		PAGE
Instrumental Specifications for Atomic Absorption Spectrophotometers	SAMA	82
Guidelines for Purity and Handling of Gases Used in Atomic Absorption Spectroscopy	SAMA	82
Terminology for Atomic Absorption Spectrophotometers	SAMA	82

Acceleration

Specifications Format Guide and Test Procedure for Linear, Single-Axis, Pendulous, Analog Torque Balance Accelerometer	IEEE	39
Specifications and Tests for Strain Gage Linear Acceleration Transducers	ISA	48

Acoustics

Acoustical Terminology	ASA	1
Preferred Frequencies, Frequency Levels, and Band Numbers for Acoustical Measurements	ASA	1
Preferred Reference Quantities for Acoustical Levels	ASA	1
Method for the Calibration of Microphones	ASA	1
Specification for Octave-Band and Fractional-Octave-Band Analog and Digital Filters	ASA	1
Specification for Laboratory Standard Microphones	ASA	1
Procedures for Calibration of Underwater Electroacoustic Transducers	ASA	2
Specification for Acoustical Calibrators	ASA	2
Design Response of Weighing Networks for Acoustical Measurements	ASA	2

Subject Index

		PAGE
Nomenclature for Specifying Damping Properties of Materials	ASA	2
Electro-Acoustical Measuring Equipment for Aircraft Noise Certification	IEC	57
Quantities and Units of Acoustics	ISO	59
Acoustics Relation Between Sound Pressure Levels of Narrow Bands of Noise in a Diffuse Field and in a Frontally-Incident Free Field of Equal Loudness	ISO	59
Acoustics — Method for Calculating Loudness Level	ISO	59
Acoustics — Assessments of Occupational Noise Exposure for Hearing Conservation Purposes	ISO	60
Instrumentation for Measuring Acoustic Impulses Within Vehicles	SAE	83

Air

Laboratory Methods of Testing Fans for Rating	AMCA	3
Test Method for Louvers, Dampers, and Shutters	AMCA	3
Test Methods for Air Curtain Units	AMCA	3
Safety Criteria for Control Air Systems	ANS	8
Resistance to Airflow through Grains, Seeds, and Perforated Metal Sheets	ASAE	19
Standard Method for Measurement of Moist Air Properties	ASHRAE	20
Standard Methods for Laboratory Air Flow Measurement	ASHRAE	20
Graphical Symbols for Heating, Ventilating, and Air Conditioning	ASME	23
Method 11: Thermal Shock in Air	EIA	29
Approval Standard, Airflow Interlocking Switches and Pressure Supervisory Switches for Fuel Oil, Fuel Gas and Ventilation or Combustion Air	FM	33
Recommended Practice for Laminar Flow Clean Air Devices	IES	45
Recommended Practice for Testing Clean Rooms	IES	45
Quality Standard for Instrument Air	ISA	46
Air Pressures for Pneumatic Controllers, Transmitters, and Transmission Systems	ISA	47
Recommended Practice for Producing Quality Instrument Air	ISA	47
Environmental Conditions for Process Measurement and Control Systems: Airborne Contaminants	ISA	51
Warm Air Limit and Fan Controls	NEMA	65
Air Line Filter, Regulator and/or Lubricator Pressure Rating Supplement	NFLDP	69
Fluid Power Industrial Type Air Line Pressure Regulators	NFLDP	72
Glossary of Terms for Compressed Air Dryers	NFLDP	71
Pneumatic Fluid Power — Compressed Air Dryers — Methods for Rating and Testing	NFLDP	72

Subject Index

		PAGE
Aircraft Instrumentation		
Electro-Acoustical Measuring Equipment for Aircraft Noise Certification	IEC	57
Calibration Procedures for Signal Generators Used in the Testing of VOR and ILS Receivers	RTCA	75
VOR Test Signals	RTCA	76
Calibration Procedures — Test Standard Omni-Bearing Selector Test Sets	RTCA	76
Altimetry	RTCA	76
Standard Adjustment Criteria for Airborne Localizer and Glide Slope Receivers	RTCA	76
Interference to Aircraft Electronic Equipment from Devices Carried Aboard	RTCA	76
Standard Procedure for the Measurement of the Radio-Frequency Radiation from Aviation Radio Receivers Operating Within the Radio-Frequency Range of 30-890 Megacycles	RTCA	76
Universal/Air-Ground Digital Communication System Standard	RTCA	76
Minimum Performance Standards — Airborne Radio Marker Receiving Equipment Operating on 75 Megahertz	RTCA	76
Minimum Operational Characteristics — Airborne ATC Transponder Systems	RTCA	76
A New Guidance System for Approach and Landing	RTCA	76
Minimum Operational Characteristics — Vertical Guidance Equipment Used in Airborne Volumetric Navigational Systems	RTCA	76
Recommended Basic Characteristics for Airborne Radio Homing and Alerting Equipment for Use with Emergency Locator Transmitters (ELT)	RTCA	76
Minimum Performance Standards — Airborne Doppler Radar Navigation Equipment	RTCA	77
Minimum Performance Standards — Airborne Low-Range Radar Altimeters	RTCA	76
Environmental Conditions and Test Procedures for Airborne Equipment	RTCA	77
Minimum Performance Standards — Airborne Ground Proximity Warning Equipment	RTCA	77
Report on Air-Ground Communications — Operational Considerations for 1980 and Beyond	RTCA	77
Minimum Performance Standards — Airborne HF Radio Communications Transmitting and Receiving Equipment Operating Within the Radio-Frequency Range of 1.5 to 3.0 Megahertz	RTCA	77
Minimum Performance Standards — Airborne Omega Receiving Equipment	RTCA	77
Initial Report of Civil Aviation Frequency Spectrum Requirements — 1980-2000	RTCA	77
Microwave Landing System (MLS) Implementation	RTCA	77
Airborne Electronics and Electrical Equipment Reliability	RTCA	77

Subject Index

		PAGE
VHF Air-Ground Communication Technology and Spectrum Utilization	RTCA	77
Audio Systems Characteristics and Minimum Performance Standards — Aircraft Microphones, Aircraft Headsets and Speakers, Aircraft Audio Selector Panels and Amplifiers	RTCA	77
Recommendations on Policies and Procedures for Off-the-Shelf Electronic Test Equipment Acquisition and Support	RTCA	78
Minimum Operational Performance Standards for Airborne Radar Approach and Beacon Systems for Helicopters	RTCA	78
Minimum Operational Performance Standards for Airborne Weather and Ground Mapping Pulsed Radars	RTCA	78
Minimum Operational Performance Standards for Optional Equipment Which Displays Non-Radar Derived Data on Weather and Ground Mapping Radar Indicators	RTCA	78
Minimum Operational Performance Standards for Ground-Based Automated Weather Observation Equipment	RTCA	78
FM Broadcast Interference Related to Airborne ILS, VOR and VHF Communications	RTCA	78
Minimum Operational Performance Standards for Microwave Landing System (MLS) Airborne Receiving Equipment	RTCA	78
Software Considerations in Airborne Systems and Equipment Certification	RTCA	78
Minimum Operational Performance Standards for Automatic Direction Finding (ADF) Equipment	RTCA	78
Minimum Operational Performance Standards for Airborne Area Navigation Equipment Using VOR/DME Reference Facility Sensor Inputs	RTCA	79
Minimum Operational Performance Standards for Air Traffic Control Radar Beacon System/Mode Select (ATCRBS/Mode S) Airborne Equipment	RTCA	79
Emergency Locator Transmitter (ELT) Equipment Installation and Performance	RTCA	79
Minimum Operational Performance Standards for Emergency Locator Transmitter — Automatic Fixed — ELT (AF), Automatic Portable — ELT (AP), Automatic Deployable — ELT (AD), Survival — ELT(s) Operating on 121.5 and 243.0 Megahertz	RTCA	79
Traffic Alert and Collision Avoidance Systems (TCAS) I Functional Guidelines	RTCA	79
Minimum Operational Performance Standards for Traffic Alert and Collision Avoidance System (TCAS) Airborne Equipment	RTCA	79
Minimum Operational Performance Standards for Airborne Radio Communications Equipment Operating Within the Radio Frequency Range 117.975-137.000 MHZ	RTCA	79
Minimum Operational Performance Standards for Airborne Area Navigation Equipment Using Multi-Sensor Inputs	RTCA	79
Minimum Operational Performance Standards for Airborne Distance Measuring Equipment (DME) Operating Within the Radio Frequency Range of 960-1215 Megahertz	RTCA	79

Subject Index

		PAGE
Pressure Altimeter System	SAE	83
Electromagnetic Interference Measurement Impulse Generators	SAE	85
Aircraft Instrument Standards: Wording, Terminology, Phraseology, Environment and Design Standards	SAE	86
Flight Directors	SAE	84
Rate of Climb Indicator, Pressure Actuated	SAE	84
Spectrum Analyzers for Electromagnetic Interference Measurements	SAE	85
Altimeter, Pressure Actuated Sensitive Type	SAE	84
Airspeed Indicator (Pitot Static)	SAE	84
Specification for Single-Degree-of-Freedom Spring-Restrained Rate Gyros	SAE	84
Directional Indicating System	SAE	84
Pressure Instruments — Fuel, Oil and Hydraulic	SAE	87
Determination of Hydraulic Drop	SAE	85
Minimum Performance Standards for Direction Instrument, Non-Magnetic	SAE	85
Powerplant Fire Detection Instruments — Thermal and Flame Contact Types	SAE	86
Vertical Velocity Instrument	SAE	85
Fuel Flowmeters	SAE	84
Carbon Monoxide Detection Instruments	SAE	87
Exhaust Gas Temperature Instruments	SAE	86
True Mass Fuel Flow Instruments	SAE	86
Electric Tachometer: Magnetic Drag	SAE	84
Total Temperature Measuring Instruments	SAE	83
Minimum Performance Standard for Direction Instrument, Magnetic	SAE	85
Stall Warning Instrument	SAE	86
Minimum Performance Standard for Turn and Slip Instruments	SAE	86
Direction Instrument Magnetic, Non-Magnetic	SAE	86
Temperature Instruments	SAE	86
Airspeed Instruments	SAE	86
Pressure Ratio Instruments	SAE	87
Manifold Pressure Indicating Instruments	SAE	87
Temperature Indicator	SAE	87
Fuel and Oil Quantity Instruments	SAE	87
Fluidic Test Methods and Instrumentation	SAE	87
Instrumentation Requirements for Turboshaft Engine Performance Measurements	SAE	85

Subject Index

		PAGE
Alarms		
American National Standard for Cardiac Monitors, Heart Rate Meters and Alarms	AAMI	26
Criticality Accident Alarm System	ANS	8
Approval Standard, Flame Radiation Detectors for Automatic Fire Alarm Signaling	FM	33
Approval Standard, Smoke Actuated Detectors for Automatic Fire Alarm Signaling	FM	33
Annunciator Sequences and Specifications	ISA	48
Automatic Fire Detectors	NFPA	68
Standard for Smoke Detectors for Fire-Protective Signaling Systems	UL	90
Aluminum		
Aluminum Standards and Data	AA	4
Aluminum Standards and Data Metric SI	AA	4
Standard Method for Measurement of Thickness of Anodic Coatings on Aluminum and of Other Nonconductive Coatings on Nonmagnetic Basis Metals with Eddy-Current Instruments	ASTM	13
Test Method for Indentation Hardness of Aluminum Alloys by Means of a Newage Portable Non-Caliper-Type Instrument	SAE	85
Analog		
Specification for Octave-Band and Fractional-Octave-Band Analog and Digital Filters	ASA	1
Method for Specifying the Characteristics of Auxiliary Analog Equipment for Shock and Vibration Measurements	ASA	2
Digital Electronic D.C. Voltmeters and D.C. Electronic Analogue-to-Digital Converters	IEC	56
Methods of Evaluating the Performances	IEC	57
Definitions of Terms for Analog Computers	IEEE	37
Specifications Format Guide and Test Procedure for Linear, Single-Axis, Pendulous, Analog Torque Balance Accelerometer	IEEE	39
Compatibility of Analog Signals for Electronic Industrial Process Instruments	ISA	49
Guidelines for Presenting Specifications of Analog Process Measurement and Control Instruments	SAMA	83
Electrical Analog Instruments — Panel Board Types	UL	92
Analyzers and Analysis		
Official Methods of Analysis of the AOAC	AOAC	27
Flue and Exhaust Gas Analyses, Instruments and Apparatus	ASME	22
Standard Practice for Photographic Photometry in Spectrochemical Analysis	ASTM	17
Standard Practice for Evaluation of Mass Spectrometers for Chemical Analysis	ASTM	17

Subject Index

		PAGE
Standard Recommended Practices for General Techniques of Infrared Quantitative Analysis	ASTM	17
Standard Practice for Use and Evaluation of Spark Source Mass Spectrometers for the Analysis of Solids	ASTM	18
Procedure for Calibrating Gas Analyzers of the Mass Spectrometer Type	AVS	24
Spectrum Analyzers	IEEE	42
Spectrum Analyzers for Electromagnetic Interference Measurements	SAE	85

Atmosphere

Standard Definitions of Terms Relating to Atmosphere Sampling and Analysis	ASTM	15
Standard Recommended Practices for Sampling Atmospheres for Analysis of Gases and Vapors	ASTM	15
Standard Method for Collection by Filtration and Determination of Mass, Number, and Optical Sizing of Atmospheric Particulates	ASTM	16
Electrical Apparatus for Explosive Gas Atmospheres	IEC	55
Standard Atmospheres for Conditioning and/or Testing Specifications	ISO	59

Attenuators

Specifications and Test Methods for Fixed and Variable Attenuators	IEEE	40

B

Bearings

Instrument Ball Bearings	AFBMA	26
Materials Standards for P/M Self-Lubricating Bearings	MPIF	62

Boilers

Boiler Water Requirements and Associated Steam Purity for Commercial Boilers	ABMA	4
Recommended Design Guidelines for Stoker Firing of Bituminous Coals	ABMA	5
Boiler Water Limits and Steam Purity Recommendations for Watertube Boilers	ABMA	4
Fluidized Bed Combustion Guidelines	ABMA	4
A Guide to Clean and Efficient Operation of Coal Stoker Fired Boilers	ABMA	4
Guidelines for Industrial Boiler Performance Improvement	ABMA	4
Operation and Maintenance Safety Manual	ABMA	5
Procedure for Measurement of Sound from Boiler Units	ABMA	5
Matrix of Recommended Quality Control Requirements	ABMA	4
Thermal Shock Damage to Hot Water Boilers	ABMA	5

Subject Index

		PAGE
Automatic Intermittent Pilot Ignition Systems for Field Installation	AGA	7
Gas Fired Low-Pressure Steam and Hot Water Boilers	AGA	6
Gas Utilization Equipment in Large Boilers	AGA	7
National Board Inspection Code	NBBI	63
Hot-Water Immersion Controls	NEMA	66

C

Cables

Vertical Cable Tray-Flame Tests @ 210,000 Btu	ICEA	54
Guide for Partial-Discharge Test Procedures	ICEA	53
Vertical Cable Tray Flame Tests @ 70,000 Btu	ICEA	54
Guide for Establishing Stability of Volume Resistivity for Conducting Polymeric Components of Power Cables	ICEA	53
Test Methods for Insulation and Sheaths of Electric Cables and Cords	IEC	57
Guide for the Design and Installation of Cable Systems in Substations	IEEE	41
Instrumentation Cables	NEMA	67

Calibration

Procedures for Calibration of Underwater Electroacoustic Transducers	ASA	2
Method for the Calibration of Microphones	ASA	1
Specifications for Laboratory Standard Microphones	ASA	1
Specification for Acoustical Calibrators	ASA	2
Selection of Calibration and Tests for Electrical Transducers Used for Measuring Shock and Vibration	ASA	2
Method for the Experimental Determination of Mechanical Mobility. Part I: Basic Definitions and Transducers	ASA	2
Guide for Dynamic Calibration of Pressure Transducers	ASME	23
Procedure for Bench Calibration of Tank Level Gauging Tapes and Sounding Rules	ASME	23
Standard Method for Measurement and Calibration of Spherical and Spheroidal Tanks	ASTM	15
Standard Method for Measurement and Calibration of Stationary Horizontal Tanks	ASTM	15
Standard Method for Basic Calibration of Master Viscometers and Viscosity Oil Standards	ASTM	16
Standard Method for Calibration of Laboratory Mechanical-Rammer Soil Compactors	ASTM	16
Standard Practice for Calibration of Force Measuring Instruments for Verifying the Load Indication of Testing Machines	ASTM	17
Standard Method for Verification and Calibration of Liquid-In-Glass Thermometers	ASTM	17

Subject Index

		PAGE
Standard Method for Absolute Calibration of Reflectance Standards	ASTM	18
Calibration of Leak Detectors of the Mass Spectrometer Type	AVS	23
Method for Vacuum Leak Calibration	AVS	23
Procedure for Calibrating Gas Analyzers of the Mass Spectrometer Type	AVS	24
Procedure for Calibrating Hot Filament Ionization Gauges Against a Reference Manometer in the Range of $10^{-2} - 10^{-5}$ Torr	AVS	24
Procedures for the Calibration of Hot Filament Ionization Gauge Controls	AVS	24
Procedure for Calibrating Vacuum Gages of the Thermal Conductivity Type	AVS	24
Standard Procedures for Calibrating Magnetic Instruments to Measure the Delta Ferrite Content of Austenitic Stainless Steel Weld Metal	AWS	25
Testing and Calibration of Ultrasonic Therapeutic Equipment	IEC	55
Supplementary Requirements for the Calibration and Control of Measuring and Test Equipment Used in the Construction and Maintenance of Nuclear Power Generating Stations	IEEE	41
Calibration and Usage of Germanium Detectors for Measurement of Gamma-Ray Emission of Radionuclides	IEEE	44
Recommended Practice for Equipment Calibration or Validation Procedures	IES	45
Methods and Equipment for Calibration of Variable Area Meters (Rotameters)	ISA	48
Specification, Installation, and Calibration of Turbine Flowmeters	ISA	48
Calibration Method for Computer Assisted Image Analysis Systems Used to Count and Measure Particles in the 1-10 μm Range in Fluid Power Systems	NFLDP	70
Method for Calibration of Liquid Automatic Particle Counters Using "AC" Fine Test Dust	NFLDP	70
Tool Calibration	PFI	75
Calibration Procedures for Signal Generators Used in the Testing of VOR and ILS Receivers	RTCA	75
Calibration Procedures — Test Standard Omni-Bearing Selector Test Sets	RTCA	76

Calorimetry

Standard Test Method for Heat of Combustion of Liquid Hydrocarbon Fuels by Bomb Calorimeter	ASTM	14
Standard Test Method for Calorific Value of Gases in Natural Gas Range by Continuous Recording Calorimeter	ASTM	16
Solid-Stem Calorimeter Thermometers	ISO	60
Enclosed-Scale Calorimeter Thermometers	ISO	60

Subject Index

		PAGE
Certification		
Qualification and Certification of Instrumentation and Control Technicians in Nuclear Power Plants	ISA	50
Directory of International and Regional Organizations Conducting Standards-Related Activities	NBS	64
Standards Activities of Organizations in the United States	NBS	64
Chromatography		
Standard Method for Analysis of Natural Gas by Gas Chromatography	ASTM	16
Standard Method for Analysis of Reformed Gas by Gas Chromatography	ASTM	16
Standard Method for Analysis of Liquefied Petroleum (LP) Gases and Propylene Concentrates by Gas Chromatography	ASTM	16
Standard Method for Analysis of Natural Gas-Liquid Mixtures by Gas Chromatography	ASTM	16
Standard Test Method for Aromatic Traces in Light Saturated Hydrocarbons by Gas Chromatography	ASTM	17
Standard Recommended Practice for Gas Chromatography Terms and Relationships	ASTM	19
Method for Natural Gas Analysis by Gas Chromatography	GPA	35
Method for Analysis of Natural Gas Liquid Mixtures by Gas Chromatography	GPA	35
Method for Obtaining Natural Gas Samples for Analysis by Gas Chromatography	GPA	35
Circuits		
Definite Purpose Magnetic Contactors for Limited Duty	ARI	3
Part 4: Interface	IEC	57
Printed Circuit Test Methods Manual	IPC	45
Coatings		
Method 5: Salt Spray (Corrosion)	EIA	29
Recommended Specification Designations for Enamel Coatings	NAPCA	62
Standard Applied Pipe Coating Weights for NAPCA Coating Specifications	NAPCA	63
External Application Procedures for Hot Applied Coal Tar and Asphalt Enamel Coatings to Steel Pipe	NAPCA	63
Suggested Procedures to Hand Wrap Field Joints Using Hot Enamel	NAPCA	63
Suggested Procedures for Coating of Girth Welds With Fusion Bonded Epoxy	NAPCA	63
Suggested Procedures for Coating Field Joints, Fittings, Connections and Pre-Fabricated Sections Using Tape Coatings	NAPCA	63
Suggested Procedures for Field Joint Application Using Mastic Mix and Field Mold	NAPCA	63

Subject Index

		PAGE
Suggested Application Procedures for Coating Field Joints Using Heat Shrinkable Materials	NAPCA	63
Application Specifications Mill Applied Fusion Bonded Epoxy Coatings (Rev. April 1983)	NAPCA	63
External Application Procedures for Plant	NAPCA	63
Application Specifications for Coal Tar Epoxy Protective Coatings	NAPCA	63
Application Specifications for Polyolefin Pipe Coating Applied by the Cross Head Extrusion Method or the Side Extrusion Method	NAPCA	63
Plant Applied Tape Coating Application Specification for the Exterior of Steel Pipe	NAPCA	63

Codes and Coding

National Fuel Gas Code	AGA	7
EIA Standard Colors for Color Identification and Coding	EIA	30
PS 8-77, Uniform Plumbing Code	IAPMO	54
Uniform Solar Energy Code	IAPMO	54
Uniform Building Code Standards	ICBO	54
Test Code for Industrial Control	IEEE	36
Recommended Practice for Code and Format Conventions	IEEE	42
National Board Inspection Code	NBBI	63
Dimensions and Identification Code for Mounting Flanges and Shafts for Positive Displacement Hydraulic Fluid Power Pumps and Motors	NFLDP	70
Hydraulic Fluid Power — Code for the Identification of Valve Mounting Surfaces	NFLDP	72
Requirements for Presentation of Catalog Data, Fluid Compatibility, Cleaning Media, Markings and Dimensional Identification Codes, and Pressure Drop Characteristics for Fluid Power Air Line Filters	NFLDP	72
Dimension Identification Code for Fluid Power Cylinders	NFLDP	73
National Electrical Code	NFPA	67
Electrical Code for One- and Two-Family Dwellings	NFPA	68
Code Guide 302, Glossary of Industry Terms	PDI	75
Recommended Practice for Color Coding of Piping Materials	PFI	75
IRIG Standard Parallel Binary Time Code Formats	RCC	80
Code for Shipboard Vibration Measurements	SNAME	87
Code for Sea Trials	SNAME	87
Local Shipboard Structures and Machinery Vibration Measurements	SNAME	88
Acceptable Vibration of Marine Steam and Gas Turbine Noise and Auxiliary Machinery Plants	SNAME	88

Subject Index

		PAGE
Coils		
Test Method for Mechanical Torque Rate of Spiral Coils of Thermostat Metal	ASTM	14
Standard Test Method for Thermal Deflection Rate of Spiral and Helical Coils of Thermostat Metal	ASTM	14
Communications		
Interface Between Data Terminal Equipment and Data Communication Equipment Employing Serial Binary Data Interchange	EIA	29
Information Processing Systems — Data Communication — Multilink Procedures	ISO	61
Information Processing Systems — Data Communication — High-Level Data Line Control Elements of Procedures	ISO	61
Information Processing Systems — Data Communications — Network Service Definition	ISO	61
Information Processing Systems — Data Communication — High-Level Data Link Control Balanced Classes of Procedures — Data-Link Layer Address Resolution/Negotiation in Switched Environments	ISO	61
Standard Format for Interrange Distribution of Visual Count Status Information	RCC	80
IRIG Standard for UHF Command Systems	RCC	81
IRIG Tracking Radar Compatibility and Design Standards for G-Band Radars	RCC	81
IRIG Standard Coordinate System and Data Formats for Antenna Patterns	RCC	81
Compressors		
Packaged Reciprocating Plant and Instrument Air Compressors for General Refinery Services	API	11
Reciprocating Compressors for General Refinery Services	API	10
Rotary-Type Positive Displacement Compressors for General Refinery Services	API	10
Centrifugal Compressors for General Refinery Services	API	10
Packages, Integrally Geared, Centrifugal Plant and Instrument Air Compressors for General Refinery Services	API	10
Displacement Compressors, Vacuum Pumps, and Blowers	ASME	21
Computers		
Guidelines for the Verification and Validation of Scientific and Engineering Computer Programs for the Nuclear Industry	ANS	8
Recommended Programming Practices to Facilitate the Portability of Scientific Computer Programs	ANS	7
Guidelines for Considering User Needs in Computer Program Development	ANS	8
Guidelines for the Documentation of Digital Computer Programs	ANS	8

Subject Index

		PAGE
Electronic Computer Systems	FM	33
Definitions of Terms for Electronic Digital Computers	IEEE	37
Definitions of Terms for Analog Computers	IEEE	37
Standard for an 8-Bit Microcomputer Bus System: STD Bus	IEEE	43
Definitions of Terms for Hybrid Computer Linkage Components	IEEE	38
Standard Portable Operating System Interface for Computer Environments	IEEE	43
Serial Highway Interface System (CAMAC)	IEEE	41
Parallel Highway Interface System (CAMAC)	IEEE	41
Multiple Controllers in a CAMAC Crate	IEEE	42
Recommended Practice for Block Transfers in CAMAC Systems	IEEE	42
Interface Devices	IEEE	42
Recommended Practice for Code and Format Conventions	IEEE	42
Glossary of Software Engineering Terminology	IEEE	42
Software Quality Assurance Plans	IEEE	42
Subroutines for CAMAC	IEEE	42
Microcomputer System Bus	IEEE	42
Application Criteria for Programmable Digital Computer Systems in Safety Systems of Nuclear Power Generating Stations	IEEE	44
Local Area Networks: Token Ring Access Method and Physical Layer Specifications	IEEE	43
Standard FASTBUS Modular High-Speed Data Acquisition and Control System	IEEE	43
Electronic Data Processing and Computer-Controlled Industrial Processes	IRI	36
Graphic Symbols for Distributed Control/Shared Display Instrumentation, Logic and Computer Systems	ISA	46
Hardware Testing of Digital Process Computers	ISA	49
Industrial Computer System FORTRAN Procedures for Executive Functions, Process Input-Output and Bit Manipulation	ISA	50
Industrial Computer System FORTRAN Procedures for File Access and the Control of File Contention	ISA	50
Calibration Method for Computer Assisted Image Analysis Systems Used to Count and Measure Particles in the 1-10 %m Range in Fluid Power Systems	NFLDP	70
Protection of Electronic Computer/Data Processing Equipment	NFPA	68
Standards for Range Meteorological Data Reduction	RCC	81
IRIG Standard Parallel Binary Time Code Formats	RCC	80
Interrange Transmission of Tracking Data from Computer to Computer	RCC	80
IRIG Standard Parallel Binary Time Code Formats	RCC	80

Subject Index

		PAGE
Connectors		
Gas Hose Connectors for Portable Indoor Gas-Fired Equipment	AGA	5
Flexible Connectors of Other Than All-Metal Construction for Gas Appliances	AGA	6
Metal Connectors for Gas Appliances	AGA	6
Quick-Disconnect Devices for Use with Gas Fuel	AGA	6
Specifications for Pipeline Valves, End Closures, Connectors and Swivels	API	8
Sanitary Standards for Instrument Fittings and Connections Used on Milk and Milk Products Equipment	DFSSC	28
Contact Engagement and Separation Force Test Procedure for Electrical Connectors	EIA	30
Normal Force Test Procedure for Electrical Connectors	EIA	30
Connectors for Optical Fibres and Cables	IEC	58
ECL (Emitter Coupled Logic) Front Panel Interconnections in Counter Logic	IEC	58
High Voltage Connectors for Nuclear Instruments	IEEE	44
Assembly-Joining Handbook	IPC	45
Bypass and Drain Connection Standard	MSS	61
Electric Power Connectors	NEMA	66
Probes for Separable Insulated Loadbreak Connectors	NEMA	66
Connectors for Use Between Aluminum or Aluminum Copper Overhead Conductors	NEMA	66
Fluid Power Systems and Components — Connectors and Associated Components — Nominal Pressures	NFLDP	73
Glossary of Terms for Fluid Power Quick Action Couplings	NFLDP	71
Hydraulic Fluid Power Solenoid-Piloted Industrial Valves — Interface Dimensions for Electrical Connectors	NFLDP	72
Pneumatic Fluid Power — Five-Port Directional Control Valves — Mounting Surfaces — Optional Electrical Connector — Dimensions and Requirements	NFLDP	72
Tubing Connection Markings for Pneumatic Instruments	SAMA	82
Recommended Standard Instrument Connections Manual	SAMA	83
Contamination		
Compendium of Standards, Practices, Methods, Relating to Contamination Control	IES	45
Neutron Contamination from Medical Electron Accelerators	NCRP	65
Control Centers		
Human Engineering for Control Centers	ISA	49
Nameplates, Labels and Tags for Control Centers	ISA	49
Electrical Guide for Control Centers	ISA	50
Piping Guide for Control Centers	ISA	50

Subject Index

		PAGE
Instructions for the Handling, Installation, Operation, and Maintenance of Motor Control Centers	NEMA	66

Control and Controllers

Definitions, Symbols, Formulas and Tables for Control Charts	ASQC	12
Guide for Quality Control, Control Chart Method of Analyzing Data, and Control Method of Controlling Quality During Production	ASQC	12
Control Chart Method of Controlling Quality During Production	ASQC	12
Procedures for the Calibration of Hot Filament Ionization Gauge Controls	AVS	24
Electrical and Construction Standards for Numerical Machine Control	EIA	30
Controllers with Analogue Signals for Use in Industrial Process Control Systems — Part 1: Methods of Evaluating the Performance; Part 2: Guidance for Inspection and Routine Testing.	IEC	57
Methods of Evaluating the Performance of Controllers with Analogue Signals for Use in Industrial Process Control	IEC	57
CAMAC — Organization of Multi-Crate Systems, Specification of the Branch-Highway and CAMAC Crate Controller Type A1	IEC	57
Guide for the Installation of Electrical Equipment to Minimize Electrical Noise Inputs to Controllers from External Sources	IEEE	41
Parallel Highway Interface System (CAMAC)	IEEE	41
Serial Highway Interface System (CAMAC)	IEEE	41
Recommended Practice for Parameters to Characterize Digital Loop Performance	IEEE	44
Multiple Controllers in a CAMAC Crate	IEEE	42
Pneumatic Control Circuit Pressure Test	ISA	46
Air Pressures for Pneumatic Controllers, Transmitters, and Transmission Systems	ISA	47
Specification Forms for Process Measurement and Control Instruments, Primary Elements and Control Valves	ISA	48
Dynamic Response Testing of Process Control Instrumentation	ISA	48
General Standards for Industrial Control Systems	NEMA	66
Industrial Control Devices, Controllers and Assemblies	NEMA	66
Maintenance of Motor Controllers After a Fault Condition	NEMA	66
Preventive Maintenance of Industrial Systems and Equipment	NEMA	66
Markings for Adjustable Means in Automatic Controllers	SAMA	82
Pressure Safety for Pressure and Differential Pressure Process Control Devices	SAMA	82
Standard for Industrial Control Equipment	UL	90
Standard for Industrial Control Equipment for Use in Hazardous Classified Locations	UL	91

Subject Index

		PAGE
Converters		
Digital Electronic D.C. Voltmeters and D.C. Electronic Analogue-to-Digital Converters	IEC	56
Performance Measurements of A/D and D/A Converters for PCM Television Video Circuits	IEEE	42
Cooling Towers		
Acceptance Test Code for Industrial Water-Cooling Towers	CTI	28
Code for Measurement of Sound from Water Cooling Towers	CTI	28
Certification Standard for Commercial Water-Cooling Towers	CTI	28
Water-Cooling Towers	NFPA	68
Couplings		
Standard for Electric Couplings	IEEE	38
Glossary of Terms for Fluid Power Quick Action Couplings	NFLDP	71
Bibliography of Fluid Power Quick Action Coupling Standards	NFLDP	71
Testing Hydraulic Fluid Power Quick Action Couplings	NFLDP	71
Pneumatic Fluid Power — Quick Action Couplings — Test Conditions and Procedures	NFLDP	71
Hydraulic Fluid Power — Quick Action Couplings — Surge Flow Test (Short Duration Flow)	NFLDP	71
Hydraulic Fluid Power — Quick Action Couplings — Surge Flow Test (Long Duration Flow)	NFLDP	71
Cylinders		
Fluid Power Systems and Products — Cylinder Bores and Piston Rod Diameters — Inch Series	NFLDP	73
Dimension Identification Code for Fluid Power Cylinders	NFLDP	73
Bore and Rod Size Combinations and Rod End Configurations for Cataloged Square Head Industrial Fluid Power Cylinders	NFLDP	73
Mounting Dimensions for Square Head Industrial Fluid Power Cylinders	NFLDP	73
Dimensions for Accessories for Cataloged Square Head Tie Rod Type Industrial Fluid Power Cylinders	NFLDP	73
Port Nominal Pipe Sizes for Merged Inch and Metric Series Cataloged Square Head Industrial Pneumatic Fluid Power Cylinders	NFLDP	73
Fluid Power Systems and Products — Cylinder Bores and Piston Rod Diameters — Metric Series	NFLDP	73
Fluid Power Systems and Components — Cylinder Bores and Piston Rod Diameters — Inch Series	NFLDP	73
Fluid Power Systems and Products — Cylinder — Nominal Pressures	NFLDP	73
Fluid Power Systems and Products — Cylinders — Basic Series of Piston Strokes	NFLDP	73

Subject Index

		PAGE
D		
Data		
Interface Between Data Terminal Equipment and Data Communication Equipment Employing Serial Binary Data Interchange	EIA	29
Interface Between Numerical Control Equipment and Data Terminal Equipment Employing Parallel Binary Data Interchange	EIA	31
Data Sheet 5-1, Electrical Equipment in Hazardous Locations	FM	32
Microprocessor System Bus for 1 to 4 Byte Data	IEC	58
Reliability Data Manual	IEEE	41
Logical Link Control	IEEE	43
Protection of Electronic Data Processing and Computer-Controlled Industrial Processes	IRI	36
PROWAY-LAN, Industrial Data Highway	ISA	51
Guidelines for Metric Data in Standards for Valves, Flanges and Fittings	MSS	62
Method of Testing and Presenting Basic Performance Data for Positive Displacement Hydraulic Fluid Power Pumps and Motors	NFLDP	74
Protection of Electronic Computer/Data Processing Equipment	NFPA	68
Test Methods For Data Multiplex Equipment	RCC	80
Distributing Vector Acquisition Data	RCC	80
Raw Radar Antenna Data	RCC	80
Data Labels and Data Annotation Procedures	RCC	81
Density		
Standard Test Method for Density and Relative Density (Specific Gravity) of Liquids by Lipkin Bicapillary Pycnometer	ASTM	14
Standard Test Method for Density and Relative Density (Specific Gravity) of Liquids by Bingham Pycnometer	ASTM	15
Standard Test Method for Density, Relative Density, (Specific Gravity) or API Gravity of Crude Petroleum and Liquid Petroleum Products by Hydrometer Method	ASTM	15
Standard Test Method for Density and Relative Density (Specific Gravity) of Viscous Materials by Bingham Pycnometer	ASTM	15
Standard Test Method for Density and Relative Density (Specific Gravity) of Viscous Materials by Lipkin Bicapillary Pycnometer	ASTM	15
Standard Test Method for Density or Relative Density of Light Hydrocarbons by Pressure Hydrometer	ASTM	16
Density Determinations of Solids and Liquids	ASME	22
Diagrams		
Graphic Symbols for Fluid Power Diagrams	ASME	23

Subject Index

		PAGE
Graphical Symbols for Process Flow Diagrams in Petroleum and Chemical Industries	ASME	23
Graphical Symbols for Electrical Diagrams — Device Numbers — Functions	FM	32
Diagrams, Charts, Tables	IEC	55
Graphic Symbols for Logic Diagrams	IEEE	37
Binary Logic Diagrams for Process Operations	ISA	46
Instrument Loop Diagrams	ISA	46
Methods of Diagramming for Moving Parts Fluid Controls	NFLDP	72
Functional Diagramming of Instrument and Control Systems	SAMA	82

Digital

Specification for Octave-Band and Fractional-Octave-Band Analog and Digital Filters	ASA	1
Engineering Drawing and Related Practices — Digital Representation for Communication of Product Definition Data	ASME	23
Electrical Characteristics of Balanced Voltage Digital Interface Circuits	EIA	31
Electrical Characteristics of Unbalanced Voltage Digital Interface Circuits	EIA	31
Digital Electronic D.C. Voltmeters and D.C. Electronic Analogue-to-Digital Converters	IEC	56
Definitions of Terms for Electronic Digital Computers	IEEE	37
Standard Backplane Bus Specification for Multiprocessor Architectures: Futurebus	IEEE	43
Standard Digital Interface for Programmable Instrumentation	IEEE	40
Modular Instrumentation and Digital Interface Systems (CAMAC)	IEEE	41
Serial Highway Interface System (CAMAC)	IEEE	41
Parallel Highway Interface System (CAMAC)	IEEE	41
Multiple Controllers in a CAMAC Crate	IEEE	42
Recommended Practice for Block Transfers in CAMAC Systems	IEEE	42
Interface Devices	IEEE	42
Recommended Practice for Code and Format Conventions	IEEE	42
Glossary of Software Engineering Terminology	IEEE	42
Software Quality Assurance	IEEE	42
Subroutines for CAMAC	IEEE	42
Microcomputer System Bus	IEEE	42
Carrier Sense Multiple Access with Collision Detection	IEEE	43
Token-Passing Bus Access Method and Physical Layer Specifications	IEEE	43
Application Criteria for Programmable Digital Computer Systems in Safety Systems of Nuclear Power Generating Stations	IEEE	44

Subject Index

		PAGE
Automated Design Guidelines	IPC	45
Hardware Testing of Digital Process Computers	ISA	49
Information Processing Systems — Vocabulary — Part IV: Organization of Data	ISO	60
Information Processing Systems — Vocabulary — Part VI: Preparation and Handling of Data	ISO	60
Information Processing Systems — Vocabulary — Part XI: Processing Units	ISO	60
Information Processing Systems — Vocabulary — Part XVIII: Distributed Data Processing	ISO	60
Information Processing Systems — Data Communication — High-Level Data Link Control Elements of Procedures	ISO	61
Information Processing Systems — Data Communication — Multilink Procedures	ISO	61
Information Processing Systems — Data Communication — Network Service Definition	ISO	61
Information Processing Systems — Data Communication — High-Level Data Link Control Balanced Classes of Procedures — Data-Link Layer Address Resolution/Negotiation in Switched Environments	ISO	61
Digital Imaging and Communications	NEMA	67
Universal Air-Ground Digital Communication System Standard	RTCA	76

Dosimetry

Terminology Relating to Radiation Measurements and Dosimetry	ASTM	17

Drawings

Abbreviations for Use on Drawings and in Text	ASME	23
Drawing Sheet Size and Format	ASME	23
Engineering Drawing and Related Practices — Digital Representation for Communication of Product Definition Data	ASME	23
Engineering Drawing, Principles of Presentation	ISO	59
Technical Drawings — Tolerances of Form and of Position: Dimensioning and Tolerancing of Profiles	ISO	60
Technical Drawings — Tolerances of Form and of Position: Practical Examples of Indications on Drawings	ISO	60

Dust

Area Classification in Hazardous Dust Locations	ISA	47
Electrical Instruments in Hazardous Dust Locations	ISA	47
Method for Calibration of Liquid Automatic Particle-Count Instruments Using Air Cleaner Fine Test Dust Contaminant	NFLDP	70

Subject Index

PAGE

E

Electrical Connectors

Standard Test Procedures for Low Frequency (Below 3 MHz) Electrical Connectors	EIA	30
Data Sheet 5-10/14-10, Protective Grounding of Electrical Circuits and Equipment	FM	32
Electric Power Connectors	NEMA	66

Electrical Equipment

Electrosurgical Devices	AAMI	27
Transcutaneous Electrical Nerve Stimulators	AAMI	27
Classification of Areas for Electrical Installation in Petroleum Refineries	API	9
Areas for Electrical Installation at Drilling Rigs and Production Facilities	API	9
Classification of Areas for Electrical Installation, Petroleum	API	9
Electrical Equipment for Use in Class I, Div. 2, Class II, Div. 2 and Class III, Div. 1 and 2 Hazardous Locations	FM	34
Classification of Electrical and Electronic Equipment with Regard to Protection Against Electronic Shock	IEC	57
Safety of Medical Electrical Equipment	IEC	57
Electrical Equipment for Use in Hazardous (Classified) Locations	UL	92

Electrical Generation

Bronze Book	IEEE	42
Recommended Standard Rating for Electric Generators	UIA	89

Electrical Measurements

Dielectric Properties of Grain and Seeds	ASAE	19
Electrical Measurements in Power Circuits	ASME	22
Approval Standard, Electrical Utilization Equipment	FM	33
Standard Voltages	IEC	55
Expression of the Performance of Electrical and Electronic Measuring Equipment	IEC	56
Documentation to be Supplied with Electronic Measuring Apparatus	IEC	56
Safety Requirements for Indicating and Recording Electrical Measuring Instruments and their Accessories	IEC	56
Dimensions for Panel-Mounted Indicating and Recording Electrical Measuring Instruments	IEC	56
Indirect Acting Electrical Measuring	IEC	56
Master Test Code for Electrical Measurements in Power Circuits	IEEE	37
Electrical Equipment for Use in Class I, Division 2 Hazardous (Classified) Locations	ISA	47

Subject Index

		PAGE
Electrical Instruments in Hazardous Dust Locations	ISA	47
Electrical Guide for Control Centers	ISA	50
Quantities and Units of Electricity and Magnetism	ISO	59
Load Control for Use on Central Electric Heating Systems	NEMA	66
Electrical and Electronic Measuring and Testing Equipment	UL	91

Electromagnetic Radiation

Techniques and Instrumentation for the Measurement of Potentially Hazardous Electromagnetic Radiation at Microwave Frequencies	IEEE	44
Measurement of Hazardous Electromagnetic Fields	IEEE	44
Quantities and Units of Light and Related Electromagnetic Radiations	ISO	59
Spectrum Analyzers for Electromagnetic Interference Measurements	SAE	85
Electromagnetic Susceptibility of Process Control Instrumentation	SAMA	83

Electronic Components

Method 10: Effect of Soldering	EIA	29
Electronic Design Interchange Format Version 200	EIA	32
Battery Socket Patterns	EIA	28
Standard Test Methods for Passive Electronic Component Parts	EIA	28
Method 14: Panel Seal Test	EIA	29
Electrical and Construction Standards for Numerical Machine Control	EIA	30
Simulated Shipping Test for Consumer Electronic Products	EIA	31
Electronic Computer Systems	FM	33
Interface Integrated Circuits	IEC	57
Generic Specification	IEC	58
Proposed Recommended Guide for Specification of Electronic Voltmeters	IEEE	37
Compatibility of Analog Signals for Electronic Industrial Process Instruments	ISA	49
Recommendations on Policies and Procedures for Off-the-Shelf Electronic Test Equipment Acquisition and Support	RTCA	78

Electronic Hazards

Rules of Behavior with Respect to Possible Hazards when Dealing with Electronic Equipment and Equipment Employing Similar Techniques	IEC	56

Emission Measurements

Instrumentation and Techniques for Vehicle Refueling Emissions Measurement	SAE	84

Subject Index

		PAGE
Instrumentation and Techniques for Exhaust Gas Emissions Measurement	SAE	83

Enclosures

Classification of Degrees of Protection Provided by Enclosures	IEC	56
Enclosures for Industrial Controls and Systems	NEMA	66
Purged and Pressurized Enclosures for Electrical Equipment	NFPA	68

Explosive Service

Approval Standard, Explosion-proof Electrical Equipment	FM	33
Prevention of Fire and Dust Explosions in the Chemical, Dye, Pharmaceutical, and Plastics Industries	NFPA	68

F

Fasteners

Metric Standards	IFI	36
Fastener Standards	IFI	36
Capability Guide	IFI	36

Fiber Optics

Fiber Optic Cable Tensile Loading and Bending Test	EIA	31
Optical Fiber Macrobend Attenuation	EIA	31
Optical Fiber Flexure Test	EIA	31
Spectral-Attenuation Cutback Measurement for Single-Mode Optical Fibers	EIA	31
Fluid Penetration Test for Fluid-Blocked Fiber Optic Cable	EIA	31
Cable to Interconnecting Device Axial Compressive Loading	EIA	31
Accelerated Oxygen Aging of Fiber Optic Cables and Devices	EIA	32
Water Pressure Cycling	EIA	32
Fiber Optic Cable Cycling Flexing Test	EIA	32
Inspection of Cleaved Fiber End Faces by Interferometry	EIA	32
Sectional Specification for Class IV Single-Mode Optical Waveguide Fibers	EIA	32
Blank Detail Specification for Class IVa Dispersion — Unshifted Single-Mode Optical Waveguide Fibers	EIA	32
Optical Fibres. Part 1: Generic Specification	IEC	58
Optical Fibre Cables — Part 1: Generic Specification; Part 2: Production Specifications	IEC	58
Connectors for Optical Fibres and Cables — Part 1: Generic Specification	IEC	58

Filters

Specification for Octave-Band and Fractional-Octave-Band Analog and Digital Filters	ASA	1
Recommended Practice for HEPA Filters	IES	45

Subject Index

		PAGE
Hydraulic Filter Separator Housing Pressure Rating Supplement	NFLDP	70
Filtration and Contamination	NFLDP	70
Graphic Symbols for Hydraulic Fluid Power Filters and Separators	NFLDP	71
End Load Test Method for a Hydraulic Fluid Power Filter Element	NFLDP	71
Method for Determining the Fabrication Integrity of a Hydraulic Fluid Power Filter Element	NFLDP	71
Method for Verifying the Collapse/Burst Resistance of a Hydraulic Fluid Power Filter Element	NFLDP	71
Method for Verifying the Material Compatibility of a Hydraulic Fluid Power Filter Element	NFLDP	71
Method for Verifying the Flow Fatigue Characteristics of a Hydraulic Fluid Power Filter Element	NFLDP	74
Multi-Pass Method for Evaluating the Filtration Performance of a Fine Hydraulic Fluid Power Filter Element	NFLDP	71
Multi-Pass Method for Evaluating the Filtration Performance of a Coarse Hydraulic Fluid Filter	NFLDP	71
Requirements for Presentation of Catalog Data, Fluid Compatibility, Cleaning Media, Markings and Dimensional Identification Codes, and Pressure Drop Characteristics for Fluid Power Air Line Filters	NFLDP	72

Fittings

Specification for Forged or Rolled Alloy-Steel Pipe Flanges, Forged Fittings and Valves and Parts for High-Temperature Service	ASTM	13
Specification for Forged or Rolled 8 and 9 Percent Nickel Alloy Steel Flanges, Fittings, Valves, and Parts for Low-Temperature	ASTM	13
Standard Specification for Forged or Rolled Chromium-Nickle-Iron-Molybdenum-Copper-Columbium Stabilized Alloy Pipe Flanges, Forged Fittings, and Valves and Parts for Corrosive High Temperature	ASTM	14
Nomenclature for Instrument Tube Fittings	ISA	49
Brazing Joints for Wrought and Cast Copper Alloy Solder Joint Pressure Fittings	MSS	62
Standard Marking System for Valves, Fittings, Flanges, and Unions	MSS	61
Wrought Stainless Steel Butt-Welding Fittings	MSS	61
Guidelines for Metric Data in Standards for Valves, Flanges and Fittings	MSS	62
Factory-Made Butt-Welding Fittings for Class 1 Nuclear Piping Applications	MSS	62
Swage(d) Nipples and Bull Plugs	MSS	62
Forged Carbon Steel Branch Outlet Fittings — Socket Welding, Threaded and Buttwelding Ends	MSS	62
Specification for High Test Wrought Butt-Welding Fittings	MSS	62

… Subject Index

		PAGE
Pneumatic Fluid Power Applications — Metal Separable Tube Fittings — Qualifications Test	NFLDP	72

Flanges

Specifications for Forged or Rolled Alloy-Steel Pipe Flanges, Forged Fittings and Valves and Parts for High-Temperature Service	ASTM	13
Specification for Forged or Rolled 8 and 9 Percent Nickel Alloy Steel Flanges, Fittings, Valves, and Parts for Low-Temperature	ASTM	13
Standard Specification for Forged or Rolled Chromium-Nickel-Iron-Molybdenum-Copper-Columbium Stabilized Alloy Pipe Flanges, Forged Fittings, and Valves and Parts for Corrosive High Temperature	ASTM	14
Connecting Flange Joint Between Tapping Sleeves and Tapping Valves	MSS	61
Standard Marking System for Valves, Fittings, Flanges, and Unions	MSS	61
High Pressure Chemical Industry Flanges	MSS	61
Corrosion Resistant Cast Flanges and Flanged Fittings	MSS	61
Cast Iron Gate Valves, Flanges and Threaded Ends	MSS	61
Spot Facing for Bronze, Iron and Steel Flanges	MSS	61
Guidelines for Metric Data in Standards for Valves, Flanges and Fittings	MSS	62
Standard Finishes for Contact Faces of Pipe Flanges	MSS	61
Dimensions and Identification Code for Mounting Flanges and Shafts for Positive Displacement Hydraulic Fluid Power Pumps and Motors	NFLDP	70

Flow Measurement

Standard Methods for Laboratory Air Flow Measurement	ASHRAE	20
A Standard Colorimeter Test Method for Flow Measurement of a Volatile Refrigerant	ASHRAE	20
Method of Testing Flow Capacity of Refrigerant Capillary Tubes	ASHRAE	19
Standard Method for Measurement of Flow of Fluids	ASHRAE	20
Standard Method for Measurement of Flow of Gas	ASHRAE	20
Glossary of Terms Used in Measurement of Fluid Flow in Pipes	ASME	20
Graphical Symbols for Process Flow Diagrams in Petroleum and Chemical Industries	ASME	23
Measurement Uncertainty for Fluid Flow in Closed Conduits	ASME	20
Maintenance Instructions for Chlorine Institute Standard Excess Flow Valves	CI	28
Sanitary Standards for Flow Meters for Milk and Liquid Milk Products	DFSSC	28
Procedure in Rating Flow and Pressure Characteristics of Solenoid Valves for Gas Service	FCI	34

Subject Index

		PAGE
Procedure in Rating Flow and Pressure Characteristics of Solenoid Valves for Liquid Service	FCI	34
Backflow Prevention Devices	IAPMO	54
Industrial Process Control Valves — Flow Capacity	IEC	57
Industrial-Process Control Valves — Test Procedures	IEC	57
Specification, Installation, and Calibration of Turbine Flowmeters	ISA	48
Flow Equations for Sizing Control Valves	ISA	51
Inherent Flow Characteristic and Rangeability of Control Valves	ISA	52
Measurement of Fluid Flow by Means of Orifice Plates and Nozzles	ISO	59
Liquid Flow Measurements in Open Channels—Dilution Method for Measurements of Steady Flow	ISO	60
Liquid Flow Measurement in Open Channels by Velocity Area Methods	ISO	60
Liquid Flow Measurement in Open Channels — Vocabulary and Symbols, Bilingual Edition	ISO	60
Information Processing Flowchart Symbols	ISO	60
Liquid Flow Measurement in Open Channels by Slope Area Method	ISO	60
Liquid Flow Measurement in Open Channels: Establishment and Operation of a Gauging System	ISO	60
Methods of Measurements of the Performance Characteristics of Positive-Displacement Vacuum Pumps: Measurement of the Volume Rate of Flow	ISO	60
Methods of Measurements of the Performance Characteristics of Vapor Vacuum Pumps: Measurement of the Volume Rate of Flow	ISO	60
Fluid Flow in Closed Conduits — Connections for Pressure Signal Transmissions between Primary and Secondary Elements	ISO	60
Measurement of Liquid Flow in Closed Conduits — Method by Collection of the Liquid in a Volumetric Tank	ISO	61
Information Processing—Conventions for Incorporating Flowchart Symbols in Flowcharts	ISO	60
Measurement of Pulsating Fluid Flow in a Pipe by Means of Orifice Plates, Nozzles, or Venturi Tubes, in Particular in the Case of Sinusoidal or Square Wave Intermittent Periodic-Type Fluctuations	ISO	60
Measurement of Clean Water Flow in Closed Conduits — Velocity Area Method Using Current-Meters	ISO	60

Subject Index

		PAGE
Measurement of Fluid Flow in Closed Conduits — Velocity Area Method Using Pitot Static Tubes	ISO	61
Measurement of Fluid Flow in Closed Conduits — Vocabulary and Symbols	ISO	61
Liquid Flow Measurement in Open Channels — Dilution Methods for the Measurement of Steady Flow — Part 2: Integration Method	ISO	60
Method for Verifying the Flow Fatigue Characteristics of a Hydraulic Fluid Power Filter Element	NFLDP	74
Fuel Flowmeters	SAE	84
True Mass Fuel Flow Instruments	SAE	86

Fluid Power

Graphic Symbols for Fluid Power Diagrams	ASME	23
Method of Establishing the Flow Degradation of Fixed Displacement Hydraulic Fluid Power Pumps When Exposed to Particulate Contaminant	NFLDP	70
Methods of Establishing Speed Degradation of Hydraulic Fluid Power Motors When Exposed to Particulate Contaminant	NFLDP	70
Non-Integral Industrial Fluid Power Hydraulic Reservoirs	NFLDP	70
Survey of Metric Language Usage by U.S. Fluid Power Industry	NFLDP	69
Requirements for Non-Integral Industrial Fluid Power Hydraulic Power Units	NFLDP	70
Hydraulic Fluid Power-Pumps and Motors — Geometric Displacement	NFLDP	70
SI Units and Recommendations for Use	NFLDP	69
Method for Extracting Fluid Samples from the Lines of an Operating Hydraulic Fluid Power System (for Particulate Contamination Analysis)	NFLDP	70
Procedure for Qualifying and Controlling Cleaning Methods for Hydraulic Fluid Power Sample Containers	NFLDP	70
Hydraulic Fluid Power — Contamination Analysis Data — Reporting Method	NFLDP	70
Calibration Method for Computer Assisted Image Analysis Systems Used to Count and Measure Particles in the 1–10 μm Range in Fluid Power Systems	NFLDP	70
Hydraulic Fluid Power — Assembled Systems — Method for Achieving Roll-off Cleanliness	NFLDP	70
Graphic Symbols for Hydraulic Fluid Power Filters and Separators	NFLDP	70
End Load Test Method for a Hydraulic Fluid Power Filter Element	NFLDP	71
Method for Determining the Fabrication Integrity of a Hydraulic Fluid Power Filter Element	NFLDP	71
Method for Verifying the Collapse/Burst Resistance of a Hydraulic Fluid Power Filter Element	NFLDP	71

Subject Index

		PAGE
Method for Verifying the Material Compatibility of a Hydraulic Fluid Power Filter Element	NFLDP	71
Method for Verifying the Flow Fatigue Characteristics of a Hydraulic Fluid Power Filter Element	NFLDP	74
Multi-Pass Method for Evaluating the Filtration Performance of a Fine Hydraulic Fluid Power Filter Element	NFLDP	74
Method of Determining the Pore Size of a Cleanable Surface Type Hydraulic Fluid Power Element	NFLDP	71
Multi-Pass Method for Evaluating the Filtration Performance of a Coarse Hydraulic Fluid Filter	NFLDP	71
Pneumatic Fluid Power Applications — Metal Separable Tube Fittings — Qualifications Test	NFLDP	74
Fluid Power Systems and Products — Outside Diameters of Tubes and Inside Diameters of Hoses	NFLDP	71
Fluid Power Systems and Components — Nominal Pressures	NFLDP	69
Glossary of Terms for Fluid Power Quick Action Couplings	NFLDP	71
Pneumatic Fluid Power — Quick Action Couplings — Test Conditions and Procedures	NFLDP	71
Hydraulic Fluid Power — Quick Action Couplings — Surge Flow Test (Short Duration Flow)	NFLDP	71
Hydraulic Fluid Power — Quick Action Couplings — Surge Flow Test (Long Duration Flow)	NFLDP	71
Series of Mounting Interfaces for 4567 Maximum psi (315 bar) Four Port Hydraulic Fluid Power Directional Valves	NFLDP	72
Hydraulic Fluid Power — Directional Control Valve — Method for Determining the Metering Characteristic	NFLDP	72
Hydraulic Fluid Power Solenoid-Piloted Industrial Valves — Interface Dimensions for Electrical Connectors	NFLDP	72
Hydraulic Fluid Power — Code for the Identification of Valve Mounting Surfaces	NFLDP	72
Graphic Symbols for Fluidic Devices and Circuits	NFLDP	72
Methods of Presenting Basic Performance Data for Fluidic Devices	NFLDP	72
Requirements for Presentation of Catalog Data, Fluid Compatibility, Cleaning Media, Markings and Dimensional Identification Codes, and Pressure Drop Characteristics for Fluid Power Air Line Filters	NFLDP	72
Fluid Power Industrial Type Air Line Pressure Regulators	NFLDP	72
Pneumatic Fluid Power — Five-Port Directional Control Valves — Mounting Surfaces — Optional Electrical Connector — Dimensions and Requirements	NFLDP	72
Definition of Port Communication for the Fluid Power Valve Interface to NFPA Recommended Standard T3.21.1 with the Valve in Position in Response to a Remote Pilot Signal or Electrical Energization	NFLDP	72
Methods of Diagramming for Moving Parts Fluid Controls	NFLDP	72
Cylinder Bore and Piston Rod Sizes for Fluid Power Cylinders	NFLDP	73

Subject Index

		PAGE
Fluid Power Systems and Products — Cylinder Bores and Piston Rod Diameters — Inch Series	NFLDP	72
Dimension Identification Code for Fluid Power Cylinders	NFLDP	73
Bore and Rod Size Combinations and Rod End Configurations for Cataloged Square Head Industrial Fluid Power Cylinders	NFLDP	73
Mounting Dimensions for Square Head Industrial Fluid Power Cylinders	NFLDP	73
Dimensions for Accessories for Cataloged Square Head Industrial Fluid Power Cylinders	NFLDP	73
Fluid Power Systems and Products — Cylinder Bores and Piston Rod Diameters — Metric Series	NFLDP	73
Fluid Power Systems and Products — Cylinder Bores and Piston Rod Diameters — Inch Series	NFLDP	73
Fluid Power Systems and Products — Cylinder — Nominal Pressures	NFLDP	73
Fluid Power Systems and Products — Cylinder — Basic Series of Piston Strokes	NFLDP	73
Recommended Guidelines for the Use of Synthetic Lubricants in Pneumatic Fluid Power Systems	NFLDP	73
Fluid and Lubricants, Sealing Devices	NFLDP	73
Practice for the Use of Fire Resistant Fluids for Industrial Hydraulic Fluid Power Systems	NFLDP	73
Company Trade Names for Hydraulic Fluid Power Fire Resistant Fluids	NFLDP	73
Hydraulic Fluid Power — Petroleum Fluids — Prediction of Bulk Moduli	NFLDP	73
Cavity Dimensions for Fluid Power Exclusion Devices (Inch Series)	NFLDP	73
Groove Dimensions for Floating Type Metallic and Non-Metallic Fluid Power Piston Rings	NFLDP	73
Groove Dimensions for Fluid Power Radial Compression Type Piston Rings	NFLDP	73
Fluid Systems — O-Rings — Inside Diameters, Cross-Sections, Tolerances and Size Identification	NFLDP	74
Information Report — Fluid Power Laboratory Guidelines	NFLDP	74

Fluidics

Standard Method for Measurement of Flow of Fluids	ASHRAE	20
Approval Standard, Less Flammable Transformer Fluids	FM	33
National Fluid Power Standards	NFLDP	69
Graphic Symbols for Fluid Power Systems and Components	NFLDP	69
Metric Units for Fluid Power Applications	NFLDP	69
Method for Verifying the Fatigue of Static Pressure Ratings of the Pressure Containing Envelope of a Metal Fluid Power Component	NFLDP	69

Subject Index

		PAGE
Method for Establishing and Verifying the Fatigue and Static Pressure Ratings and Conducting Production Testing of the Pressure Containing Envelope of a Metal Fluid Power Accumulator	NFLDP	69
Static Pressure Rating Methods of Square Head Fluid Power Cylinders	NFLDP	74
Method for Establishing and Verifying the Fatigue and Static Pressure Ratings of the Metal Pressure Containing Envelope of a Tie Rod or Bolted Fluid Power Cylinder	NFLDP	69
Dimensions and Identification Code for Mounting Flanges and Shafts for Positive Displacement Hydraulic Fluid Power Pumps and Motors	NFLDP	70
Hydraulic Fluid Power-Glossary-Pumps and Motors	NFLDP	70
Method of Testing and Presenting Basic Performance Data for Positive Displacement Hydraulic Fluid Power Pumps and Motors	NFLDP	70

Frequency

Standard for Pulse Repetition Frequencies and Reference Oscillator Frequency for C-Band Radars	RCC	81
Preferred Frequencies, Frequency Levels, and Band Numbers for Acoustical Measurements	ASA	1
Measurement of Field Intensity above 300 MHz from Radio Frequency Equipment	IEEE	37

Fuels

Quick-Disconnect, Devices for Use with Gas Fuel	AGA	6
National Fuel Gas Code	AGA	7
Standard Test Method for Heat of Combustion of Liquid Hydrocarbon Fuels by Bomb Calorimeter	ASTM	14
Standard Method for Measurement of Gaseous Fuel Samples	ASTM	14
Standard Test Method for Water Vapor Content of Gaseous Fuels by Measurement of Dew-Point Temperature	ASTM	14
Standard Method for Sampling Manufactured Gas	ASTM	15
Approval Standard, Electric Interlocking Fuel Gas and Fuel Oil Cocks	FM	33
Approval Standard, Airflow Interlocking Switches and Pressure Supervisory Switches for Fuel Oil, Fuel Gas and Ventilation or Combustion Air	FM	33
Approval Standard, Nonprogramming and Programming Single or Multi-Burner Combustion Safeguards of the Industrial Gas and/or Fuel Oil Flame Supervising Types	FM	33
Approval Standard, Fuel Gas and Oil Safety Shutoff Valves	FM	33
Petroleum Measurement Tables	ISO	59
Fuel Flowmeters	SAE	84
True Mass Fuel Flow Instruments	SAE	86
Standard for Liquid-Level Indicating Gauges and Tank-Filling Signals for Petroleum Products	UL	90

Subject Index

G

Gages

		PAGE
Shop Testing of Automatic Liquid-Level Gages	API	11
Linear Measurements	ASME	22
Gauges — Pressure and Vacuum, Indicating Dial Type — Elastic Element	ASME	22
Procedure for Bench Calibration of Tank Level Gauging Tapes and Sounding Rules	ASME	23
Precision Inch Gauge Blanks for Length Measurement	ASME	23
Standard Method of Gaging Petroleum and Petroleum Products	ASTM	14
Procedure for Calibrating Vacuum Gages of the Thermal Conductivity Type	AVS	24
Procedure for Calibrating Hot Filament Ionization Gages Against a Reference Manometer in the Range of 10^{-2} — 10^{-5} Torr	AVS	24
Procedures for the Calibration of Hot Filament Ionization Gauge Controls	AVS	24
Specifications and Tests for Strain Gage Pressure Transducers	ISA	48
Specifications and Tests for Strain Gage Linear Acceleration Transducers	ISA	49
Specifications and Tests for Strain Gage Force Transducers	ISA	49
Pipe Threads Where Pressure-Tight Joints are not made on Threads — Part 2: Verification by Means of Limit Gauges	ISO	59
Standard for Liquid-Level Indicating Gauges and Tank-Filling Signals for Petroleum Products	UL	90
Standard for Indicating Pressure Gauges for Compressed Gas Service	UL	90
Standard for Liquid-Level Gages and Indicators for Anhydrous Ammonia and LP-Gas	UL	91

Gases

National Fuel Gas Code	AGA	7
Gas-Fired Low-Pressure Steam and Hot Water Boilers	AGA	6
Manually Operated Gas Valves	AGA	6
Automatic Valves for Gas Appliances	AGA	6
Relief Valves, Automatic Gas Shutoff Devices	AGA	6
Quick-Disconnect Devices	AGA	6
Earthquake Actuated Automatic Gas Shutoff Systems	AGA	6
Metal Connectors for Gas Appliances	AGA	6
Quick-Disconnect, Devices for Use with Gas Fuel	AGA	6
Flexible Connectors of Other Than All-Metal Construction for Gas Appliances	AGA	6
Standard Method for Measurement of Flow of Gas	ASHRAE	20
Flue and Exhaust Gas Analyses, Instruments and Apparatus	ASME	22

Subject Index

		PAGE
Standard Recommended Practices for Sampling Atmospheres for Analysis of Gases and Vapors	ASTM	15
Standard Test Method for Calorific Value of Gases in Natural Gas Range by Continuous Recording Calorimeter	ASTM	16
Standard Method for Analysis of Natural Gas by Gas Chromatography	ASTM	16
Standard Method for Analysis of Reformed Gas by Gas Chromatography	ASTM	16
Standard Method for Analysis of Liquefied Petroleum (LP) Gases and Propylene Concentrates by Gas Chromatography	ASTM	16
Standard Method for Analysis of Natural Gas-Liquid Mixtures by Gas Chromatography	ASTM	16
Standard Test Method for Aromatic Traces in Light Saturated Hydrocarbons by Gas Chromatography	ASTM	17
Standard Test Method for Chemical Composition of Gases by Mass Spectrometry	ASTM	17
Method for Measuring Pumping Speed of Mechanical Vacuum Pumps for Permanent Gases	AVS	24
Procedure for Calibrating Gas Analyzers of the Mass Spectrometer Type	AVS	24
Procedure for Calibrating Vacuum Gages of the Thermal Conductivity Type	AVS	24
Piping Systems for Dry Chlorine	CI	27
Maintenance Instructions for Chlorine Institute Standard Safety Valves, Type 1-1/2JQ	CI	27
Maintenance Instructions for Chlorine Institute Standard Safety Valves, Type 4 JQ	CI	27
Maintenance Instructions for Chlorine Institute Standard Excess Flow Valves	CI	28
Procedure in Rating Flow and Pressure Characteristics of Solenoid Valves for Gas Service	FCI	34
Approval Standard, Electric Interlocking Fuel Gas and Fuel Oil Cocks	FM	33
Approval Standard, Airflow Interlocking Switches and Pressure Supervisory Switches for Fuel Oil, Fuel Gas and Ventilation or Combustion Air	FM	33
Approval Standard, Nonprogramming and Programming Single or Multi-Burner Combustion Safeguards of the Industrial Gas and/or Fuel Oil Flame Supervising Types	FM	33
Approval Standard, Fuel Gas and Oil Safety Shutoff Valves	FM	33
Approval Standard, Combustible Gas Detectors	FM	33
Method for Calculation of Gross Heating Value, Specific Gravity and Compressibility of Natural Gas Mixtures from Compositional Analysis	GPA	35
Liquefied Petroleum Gas Specifications and Test Methods	GPA	35
Method for Natural Gas Analysis by Gas Chromatography	GPA	35

Subject Index

		PAGE
Tentative Method of Extended Analysis for Natural Gas and Similar Gaseous Mixtures	GPA	35
Tentative Method for the Extended Analysis of Hydrocarbon Liquids Containing Nitrogen and Carbon Dioxide	GPA	35
Standard Factors for Volume Correction and Specific Gravity Conversion of Liquified Petroleum Gases	GPA	35
Method for Analysis of Natural Gas Liquid Mixtures by Gas Chromatography	GPA	35
Tentative Standard for the Mass Measurement of Natural Gas Liquids	GPA	36
Method for Obtaining Natural Gas Samples for Analysis by Gas Chromatography	GPA	35
Orifice Metering of Natural Gas and Other Related Hydrocarbon Fluids	GPA	36
Physical Constants for the Paraffin Hydrocarbons and Other Components of Natural Gas	GPA	35
Electrical Apparatus for Explosive Gas Atmospheres	IEC	55
Recommended Practice for Gas-Phase Adsorber Cells	IES	45
Performance Requirements, Combustible Gas Detectors	ISA	47
Installation, Operation, and Maintenance of Combustible Gas Detection Instruments	ISA	47
National Fuel Gas Code	NFPA	67
Exhaust Gas Temperature Instruments	SAE	86
Standard for Meters for Flammable and Combustible Liquids and LP-Gas	UL	89
Standard for Relief Valves for Anhydrous Ammonia and LP-Gas	UL	89
Standard for Pressure Regulating Valves for LP-Gas	UL	90
Standard for Compressed Gas Regulators	UL	90
Standard for Indicating Pressure Gauges for Compressed Gas Service	UL	90
Standard for Liquid-Level Gages and Indicators for Anhydrous Ammonia and LP-Gas	UL	91
Gears		
Gear Speed Reducers	CTI	28
General		
Standards Activities of Organizations in the United States	NBS	64
Industrial Systems	NEMA	66
Directory of International and Regional Organizations Conducting Standards Related Activities	NIST	64
KWIC Index	NIST	64
Materials Standards for P/M Structural Parts	MPIF	62
Instrument Face Design and Location for Construction and Industrial Equipment	SAE	85

Subject Index

		PAGE
Generators		
Procedure for Measurement of Sound from Field-Erected Stationary Steam Generators	ABMA	5
Proposed Heat Procedure for Direct Current Tachometer Generators	IEEE	38
Rating for Electric Generators	UIA	90
Graphics		
Recommended Graphical Symbols	IEC	55
General Principles for the Formulation of Graphical Symbols	IEC	56
Gravity Specific		
Density Determinations of Solids and Liquids	ASME	22
Standard Test Method for True Specific Gravity of Refractory Materials by Gas-Comparison Pycnometer	ASTM	14
Standard Test Method for API Gravity of Crude Petroleum and Petroleum Products (Hydrometer Method)	ASTM	14
Standard Test Method for Density and Relative Density (Specific Gravity) of Liquids by Lipkin Bicapillary Pycnometer	ASTM	14
Standard Test Method for Density and Relative Density (Specific Gravity) of Liquids by Bingham Pycnometer	ASTM	15
Test Method for Density, Relative Density, (Specific Gravity) or API Gravity of Crude Petroleum and Liquid Petroleum Products by Hydrometer Method	ASTM	15
Standard Test Method for Density and Relative Density (Specific Gravity) of Viscous Materials by Bingham Pycnometer	ASTM	15
Standard Test Method for Density and Relative Density (Specific Gravity) of Viscous Materials by Lipkin Bicapillary Pycnometer	ASTM	15
Standard Test Method for Density or Relative Density of Light Hydrocarbons by Pressure Hydrometer	ASTM	16
Liquefied Petroleum Gas Specifications and Test Methods	GPA	35

H

Hazardous Locations		
Electrical Equipment for Use in Class I, Division 2, Class II, Division 2 and Class III, Division 1 and 2 Hazardous Locations	FM	34
Intrinsically Safe Apparatus and Associated Apparatus for use in Class I, II and III Division I Hazardous Locations	FM	33
Data Sheet 5-1, Electrical Equipment in Hazardous Locations	FM	32
Electrical Instruments in Hazardous Atmospheres	ISA	47
Instrument Purging for Reduction of Hazardous Area Classification	ISA	47
Installation of Intrinsically Safe Instrument Systems in Class I Hazardous Locations	ISA	47
Area Classification in Hazardous Dust Locations	ISA	47

Subject Index

		PAGE
Electrical Equipment for Use in Class I, Division 2 Hazardous (Classified) Locations	ISA	47
Classification of Class 1 Hazardous Locations for Electrical Installations in Chemical Plants	NFPA	68
Electrical Installations in Hazardous Locations	NFPA	67
Standard for Industrial Control Equipment for Use in Hazardous Locations	UL	91
Electrical Equipment for Use in Hazardous (Classified) Locations	UL	92
Standard for Switches for Use in Hazardous (Classified) Locations	UL	91
Standard for Intrinsically Safe Apparatus and Associated Apparatus for Use in Class I, II, III, Division I, Hazardous Locations	UL	91

Heat Measurements

Application of Infrared Sensing Devices to the Assessment of Building Heat Loss Characteristics	ASHRAE	20
Standard Test Measurements for Steady State Heat Flux Measurements and Thermal Transmission Properties	ASTM	14
Quantities and Units of Heat	ISO	59
Load Control for Use on Central Electric Heating Systems	NEMA	66

Humidity Measurements

Temperature and Humidity Environment for Dimensional Measurement	ASME	23
Method 1: Humidity (Steady State)	EIA	29
Method 3: Humidity (Steady State-Sealed Container)	EIA	29
Environmental Conditions for Process Measurement and Control Systems: Temperature and Humidity	ISA	51

Humidity, Relative

Standard Test Method for Measuring Humidity With a Psychrometer	ASTM	18

Hydrocarbons

Manual on Installation of Refinery Instruments and Control Systems	API	9
API Manual of Petroleum Measurement Standards	API	11
Standard Test Method for Heat of Combustion of Liquid Hydrocarbon Fuels by Bomb Calorimeter	ASTM	14
Standard Method for Measurement of Gaseous Fuel Samples	ASTM	14
Standard Method for Gaging Petroleum and Petroleum Products	ASTM	14
Standard Test Method for Water Vapor Content of Gaseous Fuels by Measurement of Dew-Point Temperature	ASTM	14
Standard Method of Sampling Natural Gas	ASTM	15

Subject Index

		PAGE
Test Method for Density, Relative Density, (Specific Gravity) or API Gravity of Crude Petroleum and Liquid Petroleum Products by Hydrometer Method	ASTM	15
Standard Test Method for Density or Relative Density of Light Hydrocarbons by Pressure Hydrometer	ASTM	16
Standard Method for Analysis of Liquefied Petroleum (LP) Gases and Propylene Concentrates by Gas Chromatography	ASTM	16
Standard Test Method for Aromatic Traces in Light Saturated Hydrocarbons by Gas Chromatography	ASTM	17
Method for Obtaining Hydrocarbon Fluid Samples Using a Floating Piston Cylinder	GPA	35
Hydraulic Fluid Power — Petroleum Fluids — Prediction of Bulk Moduli	NFLDP	73

Hydrometers

Standard Test Method for API Gravity of Crude Petroleum and Petroleum Products (Hydrometer Method)	ASTM	14
Test Method for Density, Relative Density, (Specific Gravity) or API Gravity of Crude Petroleum and Liquid Petroleum Products by Hydrometer Method	ASTM	15
Standard Test Method for Density or Relative Density of Light Hydrocarbons by Pressure Hydrometer	ASTM	16

I

Impulse

Measurement of Impulse Strength and Impulse Bandwidth	IEEE	39

Index

ASTM Index, The 1985 Annual Book of ASTM Standards	ASTM	13
KWIC Index	NBS	64

Inductance

Test Procedure for Charging Inductors	IEEE	38

Infrared Techniques

Standard Method for Analysis of Components in Poly (Vinyl Chloride) Compounds Using an Infrared Spectrophotometric Technique	ASTM	16

Inspection

Sampling Procedures and Tables for Inspection by Attributes	ASQC	12
Sampling Procedures and Tables for Inspection by Variables for Percent Nonconforming	ASQC	12
Guide to Inspection Planning	ASQC	12

Insulation

Method 13: Insulation Resistance Test	EIA	29
Test Method for Measurement of Hot Creep of Polymeric Insulations	ICEA	54

Subject Index

		PAGE
Intrinsic Safety		
Intrinsically Safe Apparatus and Associated Apparatus for use in Class I, II and III Division I Hazardous Locations	FM	33
Installation of Intrinsically Safe Instrument Systems in Class I Hazardous Locations	ISA	47
Intrinsically Safe Apparatus in Division 1 Hazardous Locations	NFPA	68
Standard for Intrinsically Safe Apparatus and Associated Apparatus for Use in Class I, II, III, Division I, Hazardous Locations	UL	91

L

Laboratory		
Recommended Environments for Standard Laboratories	ISA	49
Requirements for Laboratories	ISO	59
Laboratory Measurements of Airborne Sound Insulation of Building Elements	ISO	59
Liquid-in-Glass Laboratory Thermometers	ISO	59
Information Report — Fluid Power Laboratory Guidelines	NFLDP	
Laboratory Equipment	UL	91
Leak Detection		
Water and Steam in the Power Cycle (Purity and Quality, Leak Detection, and Measurement)	ASME	22
Standard Definitions of Terms Relating to Leak Testing	ASTM	19
Standard Guide for the Selection of a Leak Testing Method	ASTM	19
Calibration of Leak Detectors of the Mass Spectrometer Type	AVS	23
Method for Vacuum Leak Calibration	AVS	23
Method 14: Panel Seal Test	EIA	29
Pneumatic Control Circuit Pressure Test	ISA	46
Standard for Light Water Reactor Coolant Pressure Boundary Leak Detection	ISA	50
Length Measurement		
Precision Inch Gauge Blanks for Length Measurement	ASME	23
Standard Method for Verification and Classification of Extensometers	ASTM	17
Standard Reference Temperature for Industrial Length Measurements	ISO	59
Level Measurement		
Sanitary Standards for Pressure and Level Sensing Devices	DFSSC	28
Linear Measurement		
Linear Measurements	ASME	22
Specifications Format Guide and Test Procedure for Linear, Single-Axis, Pendulous, Analog Torque Balance Accelerometer	IEEE	39

Subject Index

		PAGE
Loss Control		
Pulp and Paper Mills: Loss Prevention and Protection	IRI	36

M

Magnetic Techniques		
Definite Purpose Magnetic Contactors	ARI	3
Definite Purpose Magnetic Contactors for Limited Duty	ARI	3
Standard Method for Nondestructive Measurement of Dry Film Thickness of Nonmagnetic Coatings Applied to a Ferrous Base	ASTM	15
Standard Procedures for Calibrating Magnetic Instruments to Measure the Delta Ferrite Content of Austenitic Stainless Steel Weld Metal	AWS	25
Quantities and Units of Electricity and Magnetism	ISO	59
Maintenance		
Recommended Practices for Measurements on Cable Television Systems	NCTA	64
Maintenance of Motor Controllers After a Fault Condition	NEMA	66
Preventive Maintenance of Industrial and Systems Equipment	NEMA	66
Manometers		
Procedure for Calibrating Hot Filament Ionization Gauges Against a Reference Manometer in the Range of 10^{-2} — 10^{-5} Torr	AVS	24
Manometer Tables	ISA	46
Measurement		
Procedures for Calibration of Underwater Electroacoustic Transducers	ASA	2
Design Response of Weighing Networks for Accoustical Measurements	ASA	2
Methods for the Measurement of Noise Emitted by Small Air-Moving Devices	ASA	2
Standard Measurement Guide — Section on Temperature Measurement	ASHRAE	20
Standard Measurement Guide — Engineering Analysis of Experimental Data	ASHRAE	20
Orifice Metering of Natural Gas and Other Related Hydrocarbon Fluids	GPA	36
Tentative Standard for the Mass Measurement of Natural Gas Liquids	GPA	36
Industrial-Process Measurement and Control Terms and Definitions	IEC	58
Quantities and Units of Periodic and Related Phenomena	ISO	59
Making It With Metric	NBBI	63
Identification of Instrumental Methods of Color or Color Difference Measurement	TAPPI	89

Subject Index

		PAGE
Medical Apparatus		
Safe Current Limits for Electromedical Apparatus	AAMI	26
Standard for Cardiac Defibrillator Devices	AAMI	26
Standard for Hemodialysis Systems	AAMI	26
Standard for Autotransfusion Devices	AAMI	26
American National Standard for Diagnostic Electrocardiographic Devices	AAMI	26
American National Standard for Pregelled ECG Disposable Electrodes	AAMI	26
American National Standard for Cardiac Monitors, Heart Rate Meters and Alarms	AAMI	26
American National Standard for Implantable Spinal Cord Stimulators	AAMI	26
American National Standard for Implantable Peripheral Nerve Stimulators	AAMI	26
Standard for ECG Connectors	AAMI	27
Safety of Medical Electrical Equipment	IEC	57
Meteorological Measurement		
Standards for Range Meteorological Data Reduction	RCC	81
Meters and Metering		
American National Standard for Cardiac Monitors, Heart Rate Meters and Alarms	AAMI	26
API Manual of Petroleum Measurement Standards	API	11
Specification for Sound Level Meters	ASA	1
Interim Supplement on Instruments and Apparatus: Application, Part II of Fluid Meters	ASME	22
Cold-Water Meters — Displacement Type, Plastic Main Case	AWWA	25
Cold-Water Meters — Displacement Type	AWWA	24
Cold-Water Meters — Displacement Type, Plastic Main Case	AWWA	25
Cold-Water Meters — Turbine Type for Customer Service	AWWA	24
Cold-Water Meters — Compound Type	AWWA	24
Cold-Water Meters — Fire Service Type	AWWA	24
Cold-Water Meters — Propeller Type for Main Line Application	AWWA	24
Direct-Reading Remote Registration Systems for Cold-Water Meters	AWWA	24
Encoder-Type Remote Registration Systems for Cold-Water Meters	AWWA	24
Cold-Water Meters Multi-Jet Type for Customer Service	AWWA	24
Application Guide for Bolometric Power Meters	IEEE	40
Electrothermic Power Meters	IEEE	41

Subject Index

		PAGE
Terminology, Dimensions, and Safety Practices for Indicating Variable Area Meters (Rotameters, Glass Tube, Metal Tube, Extension Type Glass Tube)	ISA	48
Nomenclature and Terminology for Extension Type Variable Area Meters (Rotameters)	ISA	48
Installation, Operation, Maintenance Instructions for Glass Tube Variable Area Meters (Rotameters)	ISA	48
Methods and Equipment for Calibration of Variable Area Meters (Rotameters)	ISA	48
Measurement of Clean Water Flow in Closed Conduits — Velocity Area Method Using Current-Meters	ISO	61
Hydraulic Fluid Power — Directional Control Valve — Method for Determining the Metering Characteristic	NFLDP	72
Standard for Meters for Flammable and Combustible Liquids and LP-Gas	UL	89

Metrification

Aluminum Standards and Data Metric SI	AA	4
Standard Metric Practice	ASTM	19
Standard for Metric Practice	IEEE	38
Metric Standards	IFI	36
General Purpose Metric Screw Threads	ISO	59
Guidelines for Metric Data in Standards for Valves, Flanges and Fittings	MSS	62
Making It With Metric	NBBI	63
Metric Units for Fluid Power Applications	NFLDP	69
Port Nominal Pipe Sizes for Merged Inch and Metric Series Cataloged Square Head Industrial Pneumatic Fluid Power Cylinders	NFLDP	73
Fluid Power Systems and Components — Cylinder Bores and Piston Rod Diameters — Metric Series	NFLDP	73

Moisture Measurements

Dielectric Properties of Grain and Seeds	ASAE	19
Standard Method for Measurement of Moist Air Properties	ASHRAE	20
Standard Test Method for Measuring Humidity With a Psychrometer	ASTM	18
Method 2: Moisture Resistance (Cycling)	EIA	29
Weighing, Sampling and Testing Pulp for Moisture	TAPPI	88

N

Noise Measurements

Methods for the Measurement of Noise Emitted by Small Air-Moving Devices	ASA	2
Electro-Acoustical Measuring Equipment for Aircraft Noise Certification	IEC	57

Subject Index

		PAGE
Test Procedure for Airborne Noise Measurements on Rotating Electrical Machinery	IEEE	36
Guide for the Installation of Electrical Equipment to Minimize Electrical Noise Inputs to Controllers from External Sources	IEEE	41
Specifications for Electromagnetic Noise and Field Strength Instrumentation	IEEE	44
Acoustics Relation Between Sound Pressure Levels of Narrow Bands of Noise in a Diffuse Field and in a Frontally-Incident Free Field of Equal Loudness	ISO	59
General Requirements for the Preparation of Test Codes for Measuring the Noise Emitted by Machines	ISO	59
Acoustics — Assessment of Occupational Noise Exposure for Hearing Conservation Purposes	ISO	60
Acceptable Vibration of Marine Steam and Gas Turbine Noise and Auxiliary Machinery Plants	SNAME	88
Design Guide for Shipboard Airborne Noise Control	SNAME	88

Nomenclature

Lexicon — Boiler & Auxiliary Equipment	ABMA	4
Bulletin on Nondestructive Testing Terminology	API	8
Acoustical Terminology	ASA	1
Nomenclature for Specifying Damping Properties of Materials	ASA	1
Method for the Experimental Determination of Mechanical Mobility. Part I: Basic Definitions and Tranducers	ASA	2
Refrigeration Terms and Definitions	ASHRAE	19
Terminology for Automatic Control	ASME	23
Glossary of Terms Used in the Measurement of Fluid Flow in Pipes	ASME	20
Definitions, Symbols, Formulas and Tables for Control Charts	ASQC	12
Terms, Symbols and Definitions for Acceptance Sampling	ASQC	12
Quality Systems Terminology	ASQC	12
Standard Definitions of Terms Relating to Atmosphere Sampling and Analysis	ASTM	15
Standard Definitions of Terms and Symbols Relating to Molecular Spectroscopy	ASTM	17
Standard Definitions of Terms and Symbols Relating to Emission Spectroscopy	ASTM	17
Standard Practice for Use of the Terms Precision and Bias	ASTM	18
Standard Recommended Practice for Gas Chromatography Terms and Relationships	ASTM	19
Standard Definitions of Terms Relating to Definition of Terms, Symbols, Conventions, and References Relating to High-Resolution Nuclear Magnetic Resonance (NMR) Spectroscopy	ASTM	19
Standard Definitions of Terms Relating to Leak Testing	ASTM	19
Axis and Motion Nomenclature for Numerically Controlled Machines	EIA	30

Subject Index

		PAGE
Thermistor Definitions and Test Methods	EIA	30
Recommendations for Letter Symbols, Abbreviations, Terms and Definitions for Semiconductor Device Data Sheets and Specifications	EIA	32
Standard Terminology and Definitions for Filled Thermal Systems of Remote Sensing Temperature Regulators	FCI	34
Standard Terminology for Regulators	FCI	34
Regulator Terminology	FCI	35
Industrial-Process Measurement and Control Terms and Definitions	IEC	58
Programs, Charts, Tables	IEC	55
International Electrotechnical Vocabulary	IEC	55
Letter Symbols to be Used in Electrical Technology	IEC	54
List of Basic Terms, Definitions and Related Mathematics for Reliability	IEC	56
Industrial-Process Control Valves	IEC	57
Control Valve Terminology and General Considerations	IEC	57
Industrial-Process Measurement and Control Terms and Definitions	IEC	58
Guidelines on Terminology for Valves and Fittings	IEC	62
Definition of Terms for Automatic Generation Control on Electric Power Systems	IEEE	37
Dictionary of Electrical and Electronics Terms	IEEE	37
Definitions of Fundamental Waveguide Terms	IEEE	37
Definitions for Waveguide Components	IEEE	37
Definitions of Terms for Electronic Digital Computers	IEEE	37
Definitions of Terms for Analog Computers	IEEE	37
Definitions of Terms for Hybrid Computer Linkage Components	IEEE	38
Definitions of Terms for Radio Transmitters	IEEE	38
Definitions of Semiconductor Terms	IEEE	38
ATLAS Test Language	IEEE	40
Definitions of Terms for Nonlinear, Active, and Nonreciprocal Waveguide Components	IEEE	40
C/ATLAS Test Language	IEEE	42
Glossary of Software Engineering Terminology	IEEE	42
Guide to the Use of ATLAS Test Language	IEEE	42
Principles and Definitions	IEEE	43
A Glossary of Terms and Definitions Related to Contamination Control	IES	45
Terminology, Dimensions, and Safety Practices for Indicating Variable Area Meters (Rotameters, Glass Tube, Metal Tube, Extension Type Glass Tube)	ISA	48

Subject Index

		PAGE
Nomenclature and Terminology for Extension Type Variable Area Meters (Rotameters)	ISA	48
Electrical Transducer Nomenclature and Terminology	ISA	48
Nomenclature for Instrument Tube Fittings	ISA	49
Compatibility of Analog Signals for Electronic Industrial Process Instruments	ISA	49
Process Instrumentation Terminology	ISA	49
Control Valve Terminology	ISA	51
Graphic Symbols for Distributed Control/Shared Display Instrumentation, Logic, and Computer Systems	ISA	46
Graphic Symbols for Process Displays	ISA	46
Instrument Loop Diagrams	ISA	46
Instrumentation Symbols and Identification	ISA	46
Nameplates, Labels and Tags For Control Centers	ISA	49
Nomenclature for Instrument Tubing Fittings	ISA	49
Process Instrumentation Terminology	ISA	49
Terminology, Dimensions and Safety Practices for Indicating Variable Area Meters (Rotameters) (Glass, Metal, and Extension Type Glass Tubes)	ISA	48
Information Processing Systems — Vocabulary	ISO	60
Liquid Flow Measurement in Open Channels — Vocabulary and Symbols, Bilingual Edition	ISO	60
Nuclear Energy Glossary	ISO	60
Vocabulary of Terminology	ISO	60
Measurement of Fluid Flow in Closed Conduits — Vocabulary and Symbols	ISO	61
Guidelines on Terminology for Pipe Hangers and Supports	MSS	62
Definitions of Relay Terms	NARM	63
Glossary of Terms for Fluid Power Systems and Products	NFLDP	69
Hydraulic Fluid Power-Glossary-Pumps and Motors	NFLDP	70
Glossary of Terms for Fluid Power Quick Action Couplings	NFLDP	71
Glossary of Terms for Compressed Air Dryers	NFLDP	71
Visual Examination — The Purpose, Meaning and Limitation of the Term	PFI	76
Code Guide 302, Glossary of Industry Terms	PDI	75
Terminology for Atomic Absorption Spectrophotometers	SAMA	82
Functional Diagramming of Instrument and Control Systems	SAMA	82
Process Measurement and Control Terminology	SAMA	82
Process Instrumentation Reliability Terminology Handbook	SAMA	83
Process Instrumentation Reliability Terminology	SAMA	83
Instrument Symbols and Nomenclature	TAPPI	89

Subject Index

		PAGE
Nuclear		
Earthquake Instrumentation Criteria for Nuclear Power Plants	ANS	7
Guidelines for Retrieval, Review, Processing and Evaluation of Records Obtained from Seismic Instrumentation	ANS	7
Selection, Qualification and Training of Personnel for Nuclear Power Plants	ANS	8
Criteria for Accident Monitoring Functions in Light-Water-Cooled Reactors	ANS	8
Calculation and Measurement of Direct and Scattered Gama Radiation from LWR Nuclear Power Plants	ANS	8
Criticality Accident Alarm System	ANS	8
Safety Criteria for Control Air Systems	ANS	8
Nuclear Steam Supply Systems	ASME	22
Standard Specification for Thermocouples, Sheathed, Type K, for Nuclear or Other High-Reliability Applications	ASTM	18
Standard Method for Measuring Neutron Flux, Fluence, and Spectra by Radioactivation Techniques	ASTM	18
Standard Definitions of Terms Relating to Definition of Terms, Symbols, Conventions, and References Relating to High-Resolution Nuclear Magnetic Resonance (NMR) Spectroscopy	ASTM	19
Nuclear Instruments: Constructional Requirements to Afford Personal Protection Against Ionizing Radiation	IEC	56
Dimensions of Electronic Instrument Modules (for Nuclear Electronic Instruments)	IEC	56
ECL (Emitter Coupled Logic) Front Panel Interconnections in Counter Logic	IEC	58
Installation, Inspection, and Testing Requirements for Class 1E Instrumentation and Electric Equipment at Nuclear Power Generating Stations	IEEE	39
Criteria for Class 1E Power Systems for Nuclear Power Generating Stations	IEEE	39
Standard Application of the Single-Failure Criterion to Nuclear Power Generating Station Safety Systems	IEEE	39
Standard Criteria for the Periodic Surveillance Testing of Nuclear Power Generating Station Safety Systems	IEEE	39
Guide for Seismic Qualification of Class 1E Equipment for Nuclear Power Generating Stations	IEEE	39
Criteria for Type Tests of Class 1E Modules, Used in Nuclear Power Generating Stations	IEEE	39
Quality Assurance Program Requirements for the Design and Manufacture of Class 1E Instrumentation and Electric Equipment for Nuclear Power Generating Stations	IEEE	40
Criteria for Accident Monitoring Instrumentation for Nuclear Power Generating Stations	IEEE	40
Supplementary Requirements for the Calibration and Control of Measuring and Test Equipment Used in the Construction and Maintenance of Nuclear Power Generating Stations	IEEE	41

Subject Index

		PAGE
Qualifying Class 1E Motor Control Centers for Nuclear Power Generating Stations	IEEE	41
Design, Installation of Cable Systems for Class 1 Circuits in Nuclear Power Generating Stations	IEEE	42
System Identification in Nuclear Power Plants	IEEE	43
Application Criteria for Programmable Digital Computer Systems in Safety Systems of Nuclear Power Generating Stations	IEEE	44
Performance Specifications for Reactor Emergency Radiological Monitoring Instrumentation	IEEE	44
High Voltage Connectors for Nuclear Instruments	IEEE	44
Transducer and Transmitter Installation for Nuclear Safety Applications	ISA	50
Nuclear-Safety-Related Instrument Sensing Line Piping and Tubing Standards for Use in Nuclear Power Plants	ISA	50
Standard for Light Water Reactor Coolant Pressure Boundary Leak Detection	ISA	50
Setpoints for Nuclear Safety-Related Instrumentation Used in Nuclear Power Plants	ISA	50
Response Time Testing of Nuclear-Safety-Related Instrument Channels in Nuclear Power Plants	ISA	50
Qualifications and Certification of Instrumentation and Control Technicians in Nuclear Power Plants	ISA	50
Sample-Line Piping and Tubing for Nuclear Power Plants	ISA	50
Quantities and Units of Atomic and Nuclear Physics	ISO	59
Quantities and Units of Nuclear Reactions and Ionizing Radiations	ISO	59
Nuclear Energy Glossary	ISO	60
Nuclear-Energy Fissile Materials — Principles of Criticality Safety in Handling and Processing	ISO	60
Protection in Nuclear Medicine and Ultrasound Diagnostic Procedures in Children	NCRP	65
Standard for Fire Protection for Nuclear Power Plants	NFPA	69
Nuclear Research Reactors	NFPA	68

O

Oscilloscopes

Expression of the Properties of Cathode-Ray Oscilloscopes	IEC	56
Specification for General-Purpose Cathode-Ray Oscilloscopes	IEEE	39
High Tension Exciter Output Voltage Measurement Using Cathode-Ray Oscilloscope	SAE	85
Oscilloscope Method of Measuring Spark Energy	SAE	84

Subject Index

		PAGE

P

Panels

Racks, Panels, and Associated Equipment	EIA	30
Dimensions for Panel-Mounted Indicating and Recording Electrical Measuring Instruments	IEC	56
Standards for Panelboards	NEMA	67
Audio Systems Characteristics and Minimum Performance Standards — Aircraft Microphones, Aircraft Headsets and Speakers, Aircraft Audio Selector Panels and Amplifiers	RTCA	77
Dimensions for Panel and Rack Mounted Industrial Process Measurement and Control Instruments	SAMA	82
Electrical Analog Instruments — Panel Board Types	UL	92

Petroleum Applications

API Manual of Petroleum Measurement Standards	API	11

pH Measurements

Standard Test Method for pH of Aqueous Solutions with the Glass Electrode	ASTM	17
Recommended Test Procedures for Glass pH and Reference Electrodes	SAMA	82

Photometry

Standard Recommended Practices for Photographic Photometry in Spectrochemical Analysis	ASTM	17
Photometric Linearity of Optical Properties Instruments	TAPPI	89

Photomultipliers

Standard Test Procedures for Photomultipliers for Scintillation Counting	IEEE	40

Piezoelectric Devices

Guide for Specifications and Tests for Piezoelectric Acceleration Transducers for Aerospace Testing	ISA	48
Specifications and Tests for Piezoelectric Pressure and Sound-Pressure Transducers	ISA	49

Pipes and Piping

Specifications for Pipeline Valves, End Closures, Connectors and Swivels	API	8
Classification of Areas for Electrical Installation in Petroleum Refineries	API	9
Nondestructive Testing Terminology	API	8
Measurement of Fluid Flow in Pipes Using Orifice, Nozzle, and Venturi	ASME	21
Measurement of Fluid Flow in Pipes Using Vortex Flow Meters	ASME	21
Graphical Symbols for Pipe Fittings, Valves and Piping	ASME	23

Subject Index

		PAGE
Specification for Forgings, Carbon Steel, for Piping Components	ASTM	13
Specification for Forgings, Carbon Steel for General Purpose Piping	ASTM	13
Piping Systems for Dry Chlorine	CI	27
Pressure Rating Standard for Pipeline Strainers Other than "Y" Type	FCI	34
Piping Guide for Control Centers	ISA	50
Nuclear-Safety-Related Instrument Sensing Line Piping and Tubing Standards for Use in Nuclear Power Plants	ISA	50
Sample-Line Piping and Tubing Standard for use in Nuclear Power Plants	ISA	50
Pipe Threads where Pressure Tight Joints are made on Threads — Part 2: Verification by Means of Limit Gauges	ISO	59
Pipe Threads where Pressure-Tight Joints are not made on Threads — Part I: Designation Dimension and Tolerance	ISO	59
Identification Colors for Pipes Conveying Fluids in Liquid or Gaseous Condition in Land Installation and on Board Ships	ISO	59
Measurement of Pulsating Fluid Flow in a Pipe by Means of Orifice Plates, Nozzles, or Venturi Tubes, in Particular in the Case of Sinusoidal or Square Wave Intermittent Periodic-Type Fluctuations	ISO	60
Pipe Hangers and Supports — Materials, Design and Manufacture	MSS	61
Pipe Hangers and Supports — Selection and Application	MSS	61
Guide Lines for Pipe Support Contractual Relationship	MSS	62
Carbon Steel Pipe Unions, Socket Welding and Threaded	MSS	62
Pipe Hangers and Supports — Fabrication and Installation Practices	MSS	62
Guidelines on Terminology for Pipe Hangers and Supports	MSS	62
NAPCA Specifications Pipeline Felts	NAPCA	63
External Application Procedures for Plant Applied Fusion Bonded Epoxy (F.B.E.) Coatings to Steel Pipe	NAPCA	63
Water Hammer Arresters	PDI	75
Testing and Rating Procedure for Grease Interceptors	PDI	75
Code Guide 302, Glossary of Industry Terms	PDI	75
Visual Examination — The Purpose, Meaning and Limitation of the Term	PFI	75
Method of Dimensioning Piping Assemblies	PFI	74
Hydrostatic Testing of Fabricated Piping	PFI	74
Cleaning of Fabricated Piping	PFI	74
Permanent Marking on Piping Materials	PFI	74
Access Holes and Plugs for Radiographic Inspection of Pipe Welds	PFI	74
Recommended Practice for Color Coding of Piping Materials	PFI	75

Subject Index

		PAGE
Pipe Bending Tolerances — Minimum Bending Radii — Minimum Tangents	PFI	75
Welded Load Bearing Attachments to Pressure Retaining Piping Materials	PFI	75
Abrasive Blast Cleaning of Ferritic Piping Materials	PFI	75
Standard for Protection of Ends of Fabricated Piping Assemblies	PFI	75
Circumferential Butt Welds in the Arc of Pipe Bends	PFI	75
Painting of Fabricated Piping	PFI	75
Pressure Temperature Ratings of Seamless Pipe Used in Power Plant Piping Systems	PFI	75
Reinforcement Tables for 45 Degrees and 90 Degrees Branch Connections of Seamless Steel Pipe	PFI	75
Guidelines Clarifying Relationships and Design Engineering Responsibilities Between Purchasers' Engineers & Pipe Fabricator or Pipe Fabricator Erector	PFI	75
Instrument Air Tubing and Piping Materials Recommends	TAPPI	89

Pitot Tubes

Measurement of Fluid Flow in Closed Conduits — Velocity Area Method Using Pitot Static Tubes	ISO	61

Potentiometers

Potentiometers	IEC	56
Specifications and Tests of Potentiometric Pressure Transducers	ISA	49
Specifications and Tests for Potentiometric Displacement Transducers	ISA	49

Powders Measurements

Standard Test Methods for Metal Powders and Powder Metallurgy Products	MPIF	62

Power

Method for Designation of Sound Power Emitted by Machinery and Equipment	ASA	2
Electrical Measurements in Power Circuits	ASME	22
Measurement of Shaft Horsepower	ASME	22
Measurement of Indicated Horsepower	ASME	22
Water and Steam in the Power Cycle (Purity and Quality, Leak Detection, and Measurement)	ASME	22
Graphical Symbols for Heat-Power Apparatus	ASME	23
Thermal Resistance Measurements of Conduction Cooled Power Transistors	EIA	30
Power Signal Standard for Spring-Diaphragm Actuated Controls Valves	FCI	35
Definition of Terms for Automatic Generation Control on Electric Power Systems	IEEE	37

Subject Index

		PAGE
Complete 1988 Edition: Guides and Standards for Surge Protection	IEEE	44
Master Test Code for Electrical Measurements in Power Circuits	IEEE	37
Electric Power Distribution for Industrial Plants	IEEE	37
Protection and Coordination of Industrial and Commercial Power Systems	IEEE	38
Grounding of Industrial and Commercial Power Systems	IEEE	37
Emergency and Standby Power Systems for Industrial and Commercial Applications	IEEE	40
Industrial and Commercial Power System Analysis	IEEE	40
Component Function Identifiers	IEEE	43
Pumps, Motors, Power Units and Reservoirs	NFLDP	70

Power Generation

Installation, Inspection, and Testing Requirements for Class 1E Instrumentation and Electric Equipment at Nuclear Power Generating Stations	IEEE	39
Criteria for the Periodic Surveillance Testing of Nuclear Power Generating Station Safety Systems	IEEE	39
Guide for Seismic Qualification of Class 1E Equipment for Nuclear Power Generating Stations	IEEE	39
Criteria for Type Tests of Class 1E Modules, Used in Nuclear Power Generating Stations	IEEE	39
Quality Assurance Program Requirements for the Design and Manufacture of Class 1E Instrumentation and Electric Equipment for Nuclear Power Generating Stations	IEEE	40
Criteria for Accident Monitoring Instrumentation for Nuclear Power Generating Stations	IEEE	40
Supplementary Requirements for the Calibration and Control of Measuring and Test Equipment Used in the Construction and Maintenance of Nuclear Power Generating Stations	IEEE	41
Planning of Preoperational Test Programs for Class 1E Power Systems for Nuclear Power Generating Stations	IEEE	40
Qualifying Class 1E Motor Control Centers for Nuclear Power Generating Stations	IEEE	41
Application Criteria for Programmable Digital Computer Systems in Safety Systems of Nuclear Power Generating Stations	IEEE	44
Design and Installation of Electric Pipe Heating Control and Alarm Systems for Power Generating Stations	IEEE	41
Design and Installation of Cable Systems for Class 1E Circuits in Nuclear Power Generating Stations	IEEE	42
Nuclear-Safety-Related Instrument Sensing Line Piping and Tubing Standards for Use in Nuclear Power Plants	ISA	50
Setpoints for Nuclear Safety-Related Instrumentation Used in Nuclear Power Plants	ISA	50

Subject Index

		PAGE
Response Time Testing of Nuclear-Safety-Related Instrument Channels in Nuclear Power Plants	ISA	50
Qualifications and Certification of Instrumentation and Control Technicians in Nuclear Power Plants	ISA	50
Standard for Fire Protection for Nuclear Power Plants	NFPA	69
Pressure Temperature Ratings for Seamless Pipe Used in Power Plant Piping Systems	PFI	75

Precision

Statement of Precision Requirements	ISO	59
Precision Statement for Test Methods	TAPPI	89

Pressure Control & Measurement

Gas-Fired Low-Pressure Steam and Hot Water Boilers	AGA	6
Design and Installation of Pressure-Relieving Systems in Refineries	API	9
Guide for Pressure Relief and Depressuring Systems	API	9
Instruments and Apparatus: Pressure Measurement	ASME	21
Gauges — Pressure and Vacuum, Indicating Dial Type — Elastic Element	ASME	22
Guide for Dynamic Calibration of Pressure Transducers	ASME	23
Fluid Flow in Closed Conduits—Connections for Pressure Signal Transmissions Between Primary and Secondary Devices	ASME	21
Standard Test Methods for Minimum Pressure for Vapor Phase Ignition of Monopropellants	ASTM	16
Standard Test Method for Density or Relative Density of Light Hydrocarbons by Pressure Hydrometer	ASTM	16
Standard Practice for Sampling Particulates from Reservoir-Type Pressure-Sensing Instruments by Fluid Flushing	ASTM	19
Backflow Prevention Devices — Reduced Pressure Principle and Double Check Valve Types	AWWA	25
Ball Valves, Shaft- or Trunion-Mounted — 6 in. through 48 in. — for Water Pressures up to 300 psi	AWWA	25
Sanitary Standards for Pressure and Level Sensing Devices	DFSSC	28
Standards for Determining Industrial Steam Trap Capacity Rating	FCI	34
Procedure in Rating Flow and Pressure Characteristics of Solenoid Valves for Gas Service	FCI	34
Procedure in Rating Flow and Pressure Characteristics of Solenoid Valves for Liquid Service	FCI	34
Pressure Rating Standard for Steam Traps	FCI	34
Pressure Rating Standard for "Y" Type Strainers	FCI	34
Standard for Proof of Pressure Ratings for Pressure Reducing Regulators	FCI	34
Specifications and Tests for Strain Gage Pressure Transducers	ISA	48

Subject Index

		PAGE
Specifications and Tests of Potentiometric Pressure Transducers	ISA	49
Specifications and Tests for Strain Gage Force Transducers	ISA	49
Specifications and Tests for Piezoelectric Pressure and Sound-Pressure Transducers	ISA	49
Fluid Flow in Closed Conduits — Connections for Pressure Signal Transmissions between Primary and Secondary Elements	ISO	60
Pressure Relief Device Certifications	NBBI	63
Method for Verifying the Fatigue of Static Pressure Ratings of the Pressure Containing Envelope of a Metal Fluid Power Component	NFLDP	70
Method for Establishing and Verifying the Fatigue and Static Pressure Ratings and Conducting Production Testing of the Pressure Containing Envelope of a Metal Fluid Power Accumulator	NFLDP	72
Static Pressure Rating Methods of Square Head Fluid Power Cylinders	NFLDP	74
Hydraulic Filter Separator Housing Pressure Rating Supplement	NFLDP	70
Air Line Filter, Regulator and/or Lubricator Pressure Rating Supplement	NFLDP	69
Hydraulic Reservoir Pressure Rating Supplement, Part 1—Static Ratings	NFLDP	69
Quick Action Coupling Pressure Rating Supplement	NFLDP	69
Pressure Temperature Ratings of Seamless Pipe Used in Power Plant Piping Systems	PFI	75
Pressure Instruments — Fuel, Oil and Hydraulic	SAE	87
Pressure Ratio Instruments	SAE	87
Pressure Safety for Pressure and Differential Pressure Process Control Devices	SAMA	82

Programmable Controllers/Instrumentation

Standard Digital Interface for Programmable Instrumentation	IEEE	40
Standard Codes, Formats, Protocols, and Common Commands for Use with ANSI/IEEE Std. 488.1-1987, IEEE Standard Digital Interface for Programmable Instrumentation	IEEE	40
Industrial Systems	NEMA	66

Programming Languages

CAMAC — Organization of Multi-Crate Systems, Specification of the Branch-Highway and CAMAC Crate Controller Type A1	IEC	57
Modular Instrumentation and Digital Interface Systems (CAMAC)	IEEE	41
Serial Highway Interface System (CAMAC)	IEEE	41
Parallel Highway Interface System (CAMAC)	IEEE	41
Multiple Controllers in a CAMAC Crate	IEEE	42

Subject Index

		PAGE
Recommended Practice for Block Transfers in CAMAC Systems	IEEE	42
C/ATLAS Test Language	IEEE	42
Real-Time BASIC for CAMAC	IEEE	42
Subroutines for CAMAC	IEEE	42
ATLAS Test Language	IEEE	40
Microcomputer System Bus	IEEE	42
Token-Passing Bus Access Method and Physical Layer Specifications	IEEE	43

Psychrometry

Psychrometric Data	ASAE	19

Pulses

Pulse Measurement and Analysis by Objective Techniques	IEEE	38
Pulses: Methods of Text	ISO	60
Measurement of Pulsating Fluid Flow in a Pipe by Means of Orifice Plates, Nozzles, or Venturi Tubes, in Particular in the Case of Sinusoidal or Square Wave Intermittent Periodic-Type Fluctuations	ISO	60
Pulse Repetition Frequencies and Reference Oscillator Frequency for C-Band Radars	RCC	81

Pumps and Pumping

Positive Displacement Pumps—Controlled Volume	API	10
Positive Displacement Pumps—Rotary	API	10
Displacement Compressors, Vacuum Pumps, and Blowers	ASME	21
Method for Measuring Pumping Speed of Mechanical Vacuum Pumps for Permanent Gases	AVS	24
Methods of Measurements of the Performance Characteristics of Positive-Displacement Vacuum Pumps: Measurement of the Volume Rate of Flow	ISO	60
Centrifugal, Mixed Flow and Axial Pumps — Code for Hydraulic Performance Tests — Precision Grade	ISO	61
Methods of Measurements of the Performance Characteristics of Vapor Vacuum Pumps: Measurement of the Volume Rate of Flow	ISO	60
Pumps, Motors, Power Units and Reservoirs	NFLDP	70
Hydraulic Fluid Power-Glossary-Pumps and Motors	NFLDP	70
Method of Testing and Presenting Basic Performance Data for Positive Displacement Hydraulic Fluid Power Pumps and Motors	NFLDP	70
Method of Establishing the Flow Degradation of Fixed Displacement Hydraulic Fluid Power Pumps When Exposed to Particulate Contaminant	NFLDP	70
Method of Rating for Mechanical Vacuum Pumps	NFLDP	70
Hydraulic Fluid Power-Pumps and Motors — Geometric Displacement	NFLDP	70

Subject Index

		PAGE
Pycnometer		
Standard Test Method for True Specific Gravity of Refractory Materials by Gas-Comparison Pycnometer	ASTM	14
Standard Test Method for Density and Relative Density (Specific Gravity) of Liquids by Lipkin Bicapillary Pycnometer	ASTM	14
Standard Test Method for Density and Relative Density (Specific Gravity) of Liquids by Bingham Pycnometer	ASTM	15
Standard Test Method for Density and Relative Density (Specific Gravity) of Viscous Materials by Bingham Pycnometer	ASTM	15
Standard Test Method for Density and Relative Density (Specific Gravity) of Viscous Materials by Lipkin Bicapillary Pycnometer	ASTM	15
Liquefied Petroleum Gas Specifications and Test Methods	GPA	35

Q

Quality Control		
Matrix of Recommended Quality Control Requirements	ABMA	4
Quality Systems Terminology	ASQC	12
Specifications of General Requirements for a Quality Program	ASQC	12
Generic Guidelines for Quality Systems	ASQC	12
Software Quality Assurance Plans	ASQC	12
Guide for Quality Control	ASQC	12
Generic Guidelines for Auditing of Quality Systems	ASQC	13
An Attribute Skip-Lot Sampling Program	ASQC	13
Quality Control Standard for Control Valve Seat Leakage	FCI	34
Quality Assurance Program Requirements for the Design and Manufacture of Class 1E Instrumentation and Electric Equipment for Nuclear Power Generating Stations	IEEE	40
Software Quality Assurance Plans	IEEE	42

R

Radiation		
Application of Infrared Sensing Devices to the Assessment of Building Heat Loss Characteristics	ASHRAE	20
Standard Test Method for Beta Particle Radioactivity of Water	ASTM	16
Standard Test Method for Alpha Particle Radioactivity of Water	ASTM	16
Standard Guide for Radiographic Testing	ASTM	17
Standard Recommended Practices for General Techniques of Infrared Quantitative Analysis	ASTM	17
Terminology Relating to Radiation Measurements and Dosimetry	ASTM	17
Standard Method for Measuring Neutron Flux, Fluence, and Spectra by Radioactivation Techniques	ASTM	18

Subject Index

		PAGE
Standard Practice for Describing and Measuring Performance of Ultraviolet, Visible, and Near Infrared Spectrophotometers	ASTM	18
Standard Practices for General Techniques in Infrared Microanalysis	ASTM	18
Approval Standard, Flame Radiation Detectors for Automatic Fire Alarm Signalling	FM	33
Nuclear Instruments: Constructional Requirements to Afford Personal Protection Against Ionizing Radiation	IEC	56
Guide for Determining the Effects of Ionizing Radiation on Insulating Materials	IEC	57
Test Procedure for Geiger-Muller Counters	IEEE	39
Test Procedures for Germanium Gamma-Ray Detectors	IEEE	39
Test Procedures for High-Purity Germanium Detectors for Ionizing Radiation	IEEE	41
Calibration and Usage of Germanium Detectors for Measurement of Gamma-Ray Emission of Radionuclides	IEEE	44
Specification and Performance of On-Site Instrumentation for Continuously Monitoring Radioactivity in Effluents	IEEE	44
Specification of Portable X- or Gamma-Radiation Survey Instruments	IEEE	44
Performance Criteria for Instrumentation for Plutonium Monitoring	IEEE	44
Standard Radiation Protection Instrumentation Test and Calibration	IEEE	44
Quantities and Units of Light and Related Electromagnetic Radiations	ISO	59
Quantities and Units of Nuclear Reactions and Ionizing Radiations	ISO	59
Measurement of Neutron Flux and Spectra for Physical and Biological Applications	NCRP	64
Measurement of Absorbed Dose of Neutrons and of Mixtures of Neutrons and Gamma Rays	NCRP	64
Medical X-Ray and Gamma-Ray Protection for Energies up to 10 MeV — Equipment Design and Use	NCRP	64
Tritium Measurement Techniques	NCRP	65
Environmental Radiation Measurements	NCRP	65
Radiation Protection Design Guidelines for 0.1-100 MeV Particle Accelerator Facilities	NCRP	65
Instrumentation and Monitoring Methods for Radiation Protection	NCRP	65
A Handbook of Radioactivity Measurements Procedures	NCRP	65
Radiation Protection in Pediatric Radiology	NCRP	65
Dosimetry of X-Ray and Gamma-Ray Beams for Radiation Therapy in the Energy Range 10 keV to 50 MeV	NCRP	65
Radiation Protection and Measurements for Low Voltage Neutron Generators	NCRP	65

Subject Index

		PAGE
Neutron Contamination From Medical Electron Accelerators	NCRP	65
Access Holes and Plugs for Radiographic Inspection of Pipe Welds	PFI	74
Random Radiography of Pressure Retaining Girth Butt Welds	PFI	75
Standard Procedure for the Measurement of the Radio-Frequency Radiation from Aviation Radio Receivers Operating Within the Radio-Frequency Range of 30-890 Megacycles	RTCA	77
Guidelines for Purity and Handling of Gases Used in Atomic Absorption Spectroscopy	SAMA	82
Instrumental Specifications for Atomic Absorption Spectrophotometers	SAMA	82
Terminology for Atomic Absorption Spectrophotometers	SAMA	82
Standard for X-Ray Equipment	UL	90

Radio Frequency

Dielectric Properties of Grain and Seeds	ASAE	19
Standards Report on State-of-the-Art of Measuring Field Strength, Continuous Wave, Sinusoidal	IEEE	38
Methods for Measuring (Below 1000 MHz) Electromagnetic Field Strength for Frequencies Below 1000 MHz in Radiowave Propagation	IEEE	38
Recommended Practice for Measurement of Field Intensity Above 300 MHz from Radio-Frequency Industrial, Scientific and Medical Equipment	IEEE	37
Standard Procedure for the Measurement of the Radio-Frequency Radiation from Aviation Radio Receivers Operating Within the Radio-Frequency Range of 30-890 Megacycles	RTCA	76

Receivers

Testing Amplitude-Modulation Broadcast Receivers	IEEE	38
Standardized Test Procedures for UHF Flight Termination Receivers	RCC	81
Calibration Procedures for Signal Generators Used in the Testing of VOR and ILS Receivers	RTCA	75
Standard Adjustment Criteria for Airborne Localizer and Glide Slope Receivers	RTCA	76
Minimum Performance Standards — Airborne Radio Marker Receiving Equipment Operating on 75 Megahertz	RTCA	76
Airborne HF Radio Communications Transmitting and Receiving Equipment Operating Within the Radio-Frequency Range of 1.5 to 3.0 Megahertz	RTCA	77
Airborne Omega Receiving Equipment	RTCA	77

Recorders and Recording

Recorded Tape Formats for 7, 14 and 21 Tracks on 1/2 Inch Magnetic Tape and 14, 18 and 42 Tracks on 1 Inch Magnetic Tape for Instrumentation Recording	EIA	31

Subject Index

		PAGE
Flutter Measurement of Instrumentation Magnetic Tape Recorder/Reproducers — Recommended Test Method	EIA	31
Recommended Test Method — Timing Error Measurements of Instrumentation Magnetic-Tape Recorder/Reproducers	EIA	31
Test Methods for Recorder/Reproducer Systems and Magnetic Tape	RCC	80
Dimensions of Wide Chart Recorders	SAMA	82

Refineries

Guide for Inspection of Refinery Equipment	API	8
Classification of Areas for Electrical Installations in Petroleum Refineries	API	9
Design and Installation of Pressure-Relieving Systems in Refineries	API	9
Manual on Installation of Refinery Instruments and Control Systems	API	9

Reflections

Standard Recommended Practice for Selection of Geometric Conditions for Measurement of Reflectance and Transmittance	ASTM	18
Standard Method for Absolute Calibration of Reflectance Standards	ASTM	18

Refrigeration Applications

Graphic Electrical/Electronic Symbols for Air-Conditioning and Refrigerating Equipment	ARI	3
Standard for Refrigerant Access Valves and Hose Connectors	ARI	3
Thermostatic Refrigerant Expansion Valves	ARI	3
Solenoid Valves for Use with Volatile Refrigerants and Water	ARI	3
Refrigerant Pressure Regulating Valves	ARI	3
Refrigeration Terms and Definitions	ASHRAE	19
Method of Testing Flow Capacity of Refrigerant Capillary Tubes	ASHRAE	19
Method of Measurement of Proportion of Oil in Liquid Refrigerant	ASHRAE	20
Methods of Testing Floc Point of Refrigeration Grade Oils	ASHRAE	20
Sealed Glass Tube Method to Test the Chemical Stability of Material for Use Within Refrigerant Systems	ASHRAE	20

Regulators

Gas Appliance Pressure Regulators	AGA	6
Standard Terminology for Regulators	FCI	34
Standard for Proof of Pressure Ratings for Pressure Reducing Regulators	FCI	34
Standard for Proof of Pressure Ratings for Temperature Regulators	FCI	34
Ferroresonant Voltage Regulators	IEEE	40

Subject Index

		PAGE
Standard for Pressure Regulating Valves for LP-Gas	UL	90
Standard for Compressed Gas Regulators	UL	90
Standard for Temperature-Indicating and Regulating Equipment	UL	91

Relay

Engineers' Relay Handbook	NARM	63
Definitions of Relay Terms	NARM	63

Reliability

Standard Specification for Thermocouples, Sheathed, Type K, for Nuclear or Other High-Reliability Applications	ASTM	18
List of Basic Terms, Definitions and Related Mathematics for Reliability	IEC	56
Preliminary Reliability Considerations	IEC	56
Presentation of Reliability Data on Electronic Components	IEC	56
Reliability Data Manual	IEEE	41
Airborne Electronics and Electrical Equipment Reliability	RTCA	79
Process Instrumentation Reliability Terminology Handbook	SAMA	83
Process Instrumentation Reliability Terminology	SAMA	83

Resistance

Standard Recommended Practice for Determining Temperature-Electrical Resistance Characteristics (EMF) of Metallic Materials	ASTM	18
Soil Cone Penetrometer	ASAE	19
Resistance to Airflow of Grains, Seeds, and Perforated Metal Sheets	ASAE	19
Compression Test of Food Materials of Convex Shape	ASAE	19
Method 2: Moisture Resistance (Cycling)	EIA	29
Method 5: Salt Spray (Corrosion)	EIA	29
Method 13: Insulation Resistance Test	EIA	29
Thermal Resistance Measurements of Conduction Cooled Power Transistors	EIA	30
Insulation-Resistance Tests	FM	33
Laboratory D.C. Resistors	IEC	56
Guide for Measuring Earth Resistivity, Ground Impedance, and Earth Surface Potentials of a Ground System	IEEE	36
Standard Master Test Code for Resistance Measurement	IEEE	37
Electric Resistance Welded Mandrel Drawn Hydraulic Line Tubing	NFLDP	71

Rotameters

Nomenclature and Terminology for Extension Type Variable Area Meters (Rotameters)	ISA	48

Subject Index

		PAGE
Installation, Operation, Maintenance Instructions for Glass Tube Variable Area Meters (Rotameters)	ISA	48
Methods and Equipment for Calibration of Variable Area Meters (Rotameters)	ISA	48

Roundness

Measurement of Out-Of-Roundness	ASME	23

S

Safety Considerations

Operation and Maintenance Safety Manual	ABMA	5
Definite Purpose Magnetic Contactors for Limited Duty	ARI	3
National Fuel Gas Code	AGA	7
Relief Valves and Automatic Gas Shutoff Devices	AGA	6
Earthquake Actuated Gas Shutoff Devices	AGA	6
Criticality Accident Alarm System	ANS	8
Safety Criteria for Control Air Systems	ANS	8
Classification of Areas for Electrical Installation, Petroleum	API	9
Flanged Steel Safety Valves	API	9
Maintenance Instructions for Chlorine Institute Standard Safety Valves, Type 1-1/2JQ	CI	27
Maintenance Instructions for Chlorine Institute Standard Safety Valves, Type 4 JQ	CI	27
EIA Standard Colors for Color Identification and Coding	EIA	30
Approval Standard, Explosion-proof Electrical Equipment	FM	33
Approval Standard, Fire Safe Valves	FM	33
Approval Standard, Flame Radiation Detectors for Automatic Fire Alarm Signalling	FM	33
Approval Standard, Smoke Actuated Detectors for Automatic Fire Alarm Signalling	FM	33
Approval Standard, Nonprogramming and Programming Single or Multi-Burner Combustion Safeguards of the Industrial Gas and/or Fuel Oil Flame Supervising	FM	33
Approval Standard, Fuel Gas and Oil Safety Shutoff Valves	FM	33
Approval Standard, Combustible Gas Detectors	FM	33
Data Sheet 5-1, Electrical Equipment in Hazardous Locations	FM	32
Data Sheet 5-10/14-10, Protective Grounding of Electrical Circuits and Equipment	FM	32
Basic Environmental Testing Procedures	IEC	55
Nuclear Instruments: Constructional Requirements to Afford Personal Protection Against Ionizing Radiation	IEC	56
Safety Requirements for Indicating and Recording Electrical Measuring Instruments and their Accessories	IEC	56
Classification of Degrees of Protection Provided by Enclosures	IEC	57

Subject Index

		PAGE
Classification of Electrical and Electronic Equipment with Regard to Protection Against Electronic Shock	IEC	57
Guide for Determining the Effects of Ionizing Radiation on Insulating Materials	IEC	57
Safety of Medical Electrical Equipment	IEC	57
Guide for Measuring Earth Resistivity, Ground Impedance, and Earth Surface Potentials of a Ground System	IEEE	36
Standard Criteria for Periodic Suveillance Testing of Nuclear Power Generating Station Safety Systems	IEEE	39
Guide for Seismic Qualifications of Class 1E Equipment for Nuclear Power Generating Stations	IEEE	39
Criteria for Accident Monitoring Instrumentation for Nuclear Power Generating Stations	IEEE	40
Recommended Practices for Safety in High-Voltage and High-Power Testing	IEEE	41
Implementation	IEEE	43
Application Criteria for Programmable Digital Computer Systems in Safety Systems of Nuclear Power Generating Stations	IEEE	44
Improve Electrical Maintenance and Safety in Cement Industry	IEEE	41
Loss Prevention and Protection for Chemical and Petrochemical Plants	IRI	36
Electrical Instruments in Hazardous Atmospheres	ISA	47
Instrument Purging for Reduction of Hazardous Area Classification	ISA	47
Area Classification in Hazardous Dust Locations	ISA	47
Electrical Equipment for Use in Class I, Division 2 Hazardous (Classified) Locations	ISA	47
Electrical Instruments in Hazardous Dust Locations	ISA	47
Terminology, Dimensions, and Safety Practices for Indicating Variable Area Meters (Rotameters, Glass Tube, Metal Tube, Extension Type Glass Tube)	ISA	48
Annunciator Sequences and Specifications	ISA	48
Recommended Environments for Standard Laboratories	ISA	49
Nameplates, Labels and Tags for Control Centers	ISA	49
Transducer and Transmitter Installation for Nuclear Safety Applications	ISA	50
Nuclear-Safety-Related Instrument Sensing Line Piping and Tubing Standards for Use in Nuclear Power Plants	ISA	50
Setpoints for Nuclear Safety-Related Instrumentation Used in Nuclear Power Plants	ISA	50
Response Time Testing of Nuclear-Safety-Related Instrument Channels in Nuclear Power Plants	ISA	50
Environmental Conditions for Process Measurement and Control Systems: Temperature and Humidity	ISA	51

Subject Index

		PAGE
Environmental Conditions for Process Measurement and Control Systems: Airborne Contaminants	ISA	51
Nuclear-Energy Fissile Materials — Principles of Criticality Safety in Handling and Processing	ISO	60
Instrumentation and Monitoring Methods for Radiation Protection	NCRP	65
Radiation Protection in Pediatric Radiology	NCRP	65
Radiation Protection and Measurements for Low Voltage Neutron Generators	NCRP	65
Protection in Nuclear Medicine and Ultrasound Diagnostic Procedures in Children	NCRP	65
Environmental Systems Technology	NEBB	67
Testing, Adjusting, Balancing Manual for Technicians	NEBB	67
Procedural Standards for Testing-Adjusting-Balancing of Environmental Systems	NEBB	67
Safety Standard for Construction and Guide for Selection, Installation and Operation of Adjustable-Speed Drive Systems	NEMA	66
Safety Guidelines for the Application, Installation, Maintenance of Solid State Control	NEMA	66
Resistance Welding Control	NEMA	66
Practice for the Use of Fire Resistant Fluids for Industrial Hydraulic Fluid Power Systems	NFLDP	73
Company Trade Names for Hydraulic Fluid Power Fire Resistant Fluids	NFLDP	73
National Electrical Code	NFPA	67
Electrical Code for One- and Two-Family Dwellings	NFPA	68
Electrical Equipment Maintenance	NFPA	68
Electrical Safety Requirements for Employee Work Places	NFPA	68
Local Protective Signaling Systems	NFPA	68
Auxiliary Protective Signaling Systems	NFPA	68
Proprietary Protective Signaling Systems	NFPA	68
Automatic Fire Detectors	NFPA	68
Protection of Electronic Computer/Data Processing Equipment	NFPA	68
Recommended Practice on Static Electricity	NFPA	68
Purged and Pressurized Enclosures for Electrical Equipment	NFPA	68
Prevention of Fire and Dust Explosions in the Chemical, Dye, Pharmaceutical, and Plastics Industries	NFPA	68
Classification of Class 1 Hazardous Locations for Electrical Installations in Chemical Plants	NFPA	68
Standard for Fire Protection for Nuclear Power Plants	NFPA	69
Recommended Practice for Color Coding of Piping Materials	PFI	75
Emergency Locator Transmitter — Automatic Fixes — ELT (AF), Automatic Portable — ELT (AP), Automatic Deployable — ELT (AD), Survival — ELT(s) Operating on 121.5 and 243.0 Megahertz	RTCA	79

Subject Index

		PAGE
Traffic Alert and Collision Avoidance System (TCAS) I Functional Guidelines	RTCA	79
Traffic Alert and Collision Avoidance System (TCAS) Airborne Equipment	RTCA	79
Powerplant Fire Detection Instruments — Thermal and Flame Contact Types	SAE	86
Carbon Monoxide Detection Instruments	SAE	87
Instrumentation for Impact Tests	SAE	83
Pressure Safety for Pressure and Differential Pressure Process Control Devices	SAMA	82
Identification of Instrumental Methods of Color or Color Difference Measurements	TAPPI	89
Standard for Electrically Operated Valves for Use in Hazardous Locations	UL	91

Sampling

Terms, Symbols and Definitions of Acceptance Sampling	ASQC	12
Sampling Procedures and Tables for Inspection by Attributes	ASQC	12
Sampling Procedures and Tables for Inspection by Variables for Percent Nonconforming	ASQC	12
Attribute Skip-Lot Sampling Program	ASQC	13
Standard Method of Sampling Natural Gas	ASTM	15
Standard Method for Sampling Manufactured Gas	ASTM	15
Standard Definitions of Terms Relating to Atmosphere Sampling and Analysis	ASTM	15
Standard Recommended Practices for Sampling Atmospheres for Analysis of Gases and Vapors	ASTM	15
Standard Practice for Sampling Particulates from Reservoir-Type Pressure-Sensing Instruments by Fluid Flushing	ASTM	19
Procedure for Qualifying and Controlling Cleaning Methods for Hydraulic Fluid Power Sample Containers	NFLDP	70
Method for Extracting Fluid Samples from a Reservoir of an Operating Hydraulic Fluid Power System	NFLDP	71
Weighing, Sampling and Testing Pulp for Moisture	TAPPI	88
Measuring, Sampling and Analyzing White Waters	TAPPI	88

Scales

Standard for Electrically Illuminated Scales	UL	90

Seismic Measurements

Earthquake Instrumentation Criteria for Nuclear Power Plants	ANS	7
Guidelines for Retrieval, Review, Processing and Evaluation of Records Obtained from Seismic Instrumentation	ANS	7

Semiconductor Devices

Color Coding of Semiconductor Devices	EIA	29

Subject Index

		PAGE
Recommendations for Letter Symbols, Abbreviations, Terms and Definitions for Semiconductor Device Data Sheets and Specifications	EIA	32
Standard Test Procedure for Semiconductor Charged Particle Detectors	IEEE	38
Test for Amplifiers and Preamplifiers for Semiconductor Radiation Detectors	IEEE	38
Definitions of Semiconductor Terms	IEEE	38

Sensors

Measurement Procedure for Field-Disturbance Sensors	IEEE	40
Airborne Area Navigation Equipment Using VOR/DME Reference Facility Sensor Inputs	RTCA	80

Shipboard Applications

Identification Colors for Pipes Conveying Fluids in Liquid or Gaseous Condition in Land Installations and on Board Ships	ISO	59
Code for Shipboard Vibration Measurements	SNAME	87
Code for Sea Trials	SNAME	87
Local Shipboard Structures and Machinery Vibration Measurements	SNAME	88
Acceptable Vibration of Marine Steam and Gas Turbine Noise and Auxiliary Machinery Plants	SNAME	88
Guide for Centralized Control and Automation of Ship's Steam Propulsion Plant	SNAME	88
Guide for Centralized Control and Automation of Ship's Gas Turbine Propulsion Plant	SNAME	88
Propulsion Monitoring Instrumentation for Shipboard Energy Conservation	SNAME	88

Shock & Vibration Measurements

Earthquake Instrumentation Criteria for Nuclear Power Plants	ANS	7
Guidelines for Retrieval, Review, Processing and Evaluation of Records Obtained from Seismic Instrumentation	ANS	7
Acoustical Terminology	ASA	1
Method for Specifying the Characteristics of Auxiliary Analog Equipment for Shock and Vibration Measurements	ASA	2
Selection of Calibration and Tests for Electrical Transducers Used for Measuring Shock and Vibration	ASA	2
Method 11: Thermal Shock in Air	EIA	29

Shunts

Requirements for Direct Current Instrument Shunts	IEEE	39

Signaling Systems

Carrier Sense Multiple Access with Collision Detection	IEEE	43
Compatibility of Analog Signals for Electronic Industrial Process Instruments	ISA	49

Subject Index

		PAGE
Fluid Flow in Closed Conduits — Connections for Pressure Signal Transmissions between Primary and Secondary Elements	ISO	60
Testing Procedures for Local, Auxiliary, Remote Station, and Proprietary Protective Signaling Systems	NFPA	68
Central Station Signaling Systems	NFPA	68
Local Protective Signaling Systems	NFPA	68
Auxiliary Protective Signaling Systems	NFPA	68
Remote Station Protective Signaling Systems	NFPA	68
Proprietary Protective Signaling Systems	NFPA	68
Calibration Procedures for Signal Generators Used in the Testing of VOR and ILS Receivers	RTCA	75
Standard for Heat Detectors for Fire Protection Signalng Systems	UL	90
Standard for Smoke Detectors for Fire-Protective Signaling Systems	UL	90
Power Supplies for Fire Protection Signaling Systems	UL	92

Software

Glossary of Software Engineering Terminology	IEEE	42
Software Quality Assurance	IEEE	42
Software Configuration Management Plans	IEEE	43
Software Test Documentation	IEEE	43
Software Requirements Specifications	IEEE	43
Software Considerations in Airborne Systems and Equipment Certification	RTCA	78

Solar

Method of Measuring Solar-Optical Properties of Materials	ASHRAE	20
Uniform Solar Energy Code	IAPMO	54

Solid State Devices

American National Standard for Voltage or Current Reference Devices; Solid State Devices	ISA	53

Sound Measurements

Sphygmomanometers	AAMI	27
Procedure for Measurement of Sound from Field-Erected Steam Generators	ABMA	5
Procedure for Measurement of Sound from Boiler Units, Bottom Supported, Shop or Field-Erected	ABMA	5
Reverberant Room Method for Sound Testing of Fans	AMCA	3
Method of Measuring Machinery Sound Within Equipment Spaces	ARI	3
Specification for Sound Level Meters	ASA	1

Subject Index

		PAGE
Method for Sound Level Measurement of Manual Arc Welding and Cutting Processes	AWS	25
Code for Measurement of Sound from Water Cooling Towers	CTI	28
Procedural Standards for Measuring Sound and Vibration	NEBB	67
Laboratory Measurements of Airborne Sound Insulation of Building Elements	ISO	59
Acoustics Relation Between Sound Pressure Levels of Narrow Bands of Noise in a Diffuse Field and in a Frontally-Incident Free Field of Equal Loudness	ISO	59

Specifications

Specification for Sound Level Meters	ASA	1
Specification for Octave-Band and Fractional-Octave Band Analog and Digital Filters	ASA	1
Specification for Acoustical Calibrators	ASA	2
Specification for Laboratory Standard Microphones	ASA	1
Thermostatic Refrigerant Expansion Valves	ARI	3
Laboratory Methods of Testing Fans for Rating	AMCA	3
Reagent Chemicals, ACS Specifications, 6th Edition	ACS	5
Specifications for Pipeline Valves, End Closures, Connectors and Swivels	API	8
Specifications of General Requirements for a Quality Program	ASQC	12
Specification for Forgings, Carbon Steel, for Piping Components	ASTM	13
Specification for Forgings, Carbon Steel, for General Purpose Piping	ASTM	13
Specification for Forged or Rolled Alloy-Steel Pipe Flanges, Forged Fittings and Valves and Parts for High-Temperature Service	ASTM	13
Specification for Forged or Rolled 8 and 9 Percent Nickel Alloy Steel Flanges, Fittings, Valves, and Parts for Low-Temperature	ASTM	13
Standard Specification for Thermostat Metal Sheet and Strip	ASTM	14
Standard Specification for Forged or Rolled Chromium-Nickel-Iron-Molybdenum-Copper-Columbium Stabilized Alloy Pipe Flanges, Forged Fittings, and Valves and Parts for Corrosive High Temperature Service	ASTM	14
Standard Specification for ASTM Thermometers	ASTM	17
Uniform Building Code Standards	ICBO	54
Token-Passing Bus Access Method and Physical Layer Specifications	IEEE	43
Specification and Performance of On-Site Instrumentation for Continuously Monitoring Radioactivity in Effluents	IEEE	44
Technical Manual	IPC	46
Annunciator Sequences and Specifications	ISA	48
Specification Forms for Process Measurement and Control Instruments, Primary Elements and Control Valves	ISA	48

Subject Index

		PAGE
Specification, Installation, and Calibration of Turbine Flowmeters	ISA	48
Guide for Specifications and Tests for Piezoelectric Acceleration Transducers for Aerospace Testing	ISA	48
Specifications and Tests for Strain Gage Pressure Transducers	ISA	48
Specifications and Tests for Strain Gage Linear Acceleration Transducers	ISA	49
Specifications and Tests of Potentiometric Pressure Transducers	ISA	49
Specifications and Tests for Strain Gage Force Transducers	ISA	49
Specifications and Tests for Piezoelectric Pressure and Sound-Pressure Transducers	ISA	49
Specification and Test for Potentiometric Displacement Transducers	ISA	49
Index on Non-Proprietary Hydraulic Fluid Specifications and Selected Recommended Practices	NFLDP	74
Video Standards and Formats	RCC	81
Generic Test Methods for the Testing and Evaluation of Process Control Instrumentation	SAMA	82

Spectrometry & Spectrophotometry

Standard Method for Analysis of Components in Poly (Vinyl Chloride) Compounds Using an Infrared Spectrophotemetric Technique	ASTM	16
Standard Test Method for Chemical Composition of Gases by Mass Spectrometry	ASTM	17
Standard Practice for Photographic Photometry in Spectrochemical Analysis	ASTM	17
Standard Definitions of Terms and Symbols Relating to Molecular Spectroscopy	ASTM	17
Standard Definitions of Terms and Symbols Relating to Emission Spectroscopy	ASTM	17
Standard Practice for Evaluation of Mass Spectrometers for Chemical Analysis	ASTM	17
Standard Recommended Practice for Describing and Specifying the Excitation Source in Emission Spectrochemical Analysis	ASTM	18
Standard Practice for Describing and Measuring Performance of Ultraviolet, Visible, and Near Infrared Spectrophotometers	ASTM	18
Standard Practice for Use and Evaluation of Spark Source Mass Spectrometers for the Analysis of Solids	ASTM	18
Standard Recommended Practices for General Techniques Infrared Microanalysis	ASTM	17
Standard Definitions of Terms Relating to Definition of Terms, Symbols, Conventions, and References Relating to High-Resolution Nuclear Magnetic Resonance (NMR) Spectroscopy	ASTM	19
Calibration of Leak Detectors of the Mass Spectrometer	AVS	23
Procedure for Calibrating Gas Analyzers of the Mass Spectrometer Type	AVS	24

Subject Index

		PAGE
Test for Semiconductor X-Ray Energy Spectrometers	IEEE	42
Instrumental Specifications for Atomic Absorption Spectrophotometers	SAMA	82
Guidelines for Purity and Handling of Gases Used in Atomic Absorption Spectroscopy	SAMA	82
Terminology for Atomic Absorption Spectrophotometers	SAMA	82

Springs

Handbook of Spring Design	SMI	88

Static Electricity

Approval Standard, Electrostatic Finishing Equipment	FM	33

Steam Applications

Gas-Fired Low-Pressure Steam and Hot Water Boilers	AGA	6
Appendix A to Test Code for Steam Turbines	ASME	21
Guidance for Evaluation of Measurement Uncertainty in Performance Tests of Steam Turbines	ASME	21
Simplified Procedures for Routine Performance Tests of Steam Turbines	ASME	21
Water and Steam in the Power Cycle (Purity and Quality, Leak Detection, and Measurement)	ASME	22
Nuclear Steam Supply Systems	ASME	22
Standard Test Methods for Deposit-Forming Impurities in Steam	ASTM	16
Pressure Rating Standard for Steam Traps	FCI	34
Production and Performance Tests for Steam Traps	FCI	35
Guide for Centralized Control and Automation of Ship's Steam Propulsion Plant	SNAME	88

Stress Measurement

Standard Test Method for Maximum Loading Stress at Temperature of Thermostat Metals	ASTM	13
Method 6: Mechanical Robustness of Terminals	EIA	29

Symbols

Graphic Electrical/Electronic Symbols for Air-Conditioning and Refrigerating Equipment	ARI	3
Graphic Symbols for Fluid Power Diagrams	ASME	23
Graphical Symbols for Process Flow Diagrams in Petroleum and Chemical Industries	ASME	23
Graphical Symbols for Pipe Fittings, Valves and Piping	ASME	23
Graphical Symbols for Heating, Ventilating, and Air Conditioning	ASME	23
Graphical Symbols for Heat-Power Apparatus	ASME	23
Definitions, Symbols, Formulas and Tables for Control Charts	ASQC	12
Terms, Symbols and Definitions for Acceptance Sampling	ASQC	12

Subject Index

		PAGE
Graphic Symbols in Vacuum Technology	AVS	24
Recommendations for Letter Symbols, Abbreviations, Terms and Definitions for Semiconductor Device Data Sheets and Specifications	EIA	32
Graphical Symbols for Electrical Diagrams — Device Numbers — Functions	FM	32
Letter Symbols to Be Used in Electrical Technology	IEC	54
Graphic Symbols for Logic Diagrams	IEEE	37
Letter Symbols for Thermoelectric Devices	IEEE	38
Instrumentation Symbols and Identification	ISA	46
Graphic Symbols for Distributed Control/Shared Display Instrumentation, Logic and Computer Systems	ISA	46
Graphic Symbols for Process Displays	ISA	46
Liquid Flow Measurement in Open Channels — Vocabulary and Symbols, Bilingual Edition	ISO	60
Information Processing Flowchart Symbols	ISO	60
Information Processing-Conventions for Incorporating Flowchart Symbols in Flowcharts	ISO	60
Measurement of Fluid Flow in Closed Conduits — Vocabulary and Symbols	ISO	61
Standard Graphic Symbols for Cable Television	NCTA	64
Graphic Symbols for Hydraulic Fluid Power Filters and Separators	NFLDP	71
Graphic Symbols for Fluidic Devices and Circuits	NFLDP	73
Instrument Symbols and Nomenclature	TAPPI	89

T

Tachometers

Proposed Heat Procedure for Direct Current Tachometer Generators	IEEE	38

Tanks

Procedure for Bench Calibration of Tank Level Gauging Tapes and Sounding Rules	ASME	23
Standard Method for Measurement and Calibration of Spherical and Spheroidal Tanks	ASTM	15
Standard Method for Measurement and Calibration of Stationary Horizontal Tanks	ASTM	15
Standard for Liquid-Level Indicating Gauges and Tank-Filling Signals for Petroleum Products	UL	90

Tape

Recorded Tape Formats for 7, 14 and 21 Tracks on 1/2 Inch Magnetic Tape and 14, 18 and 42 Tracks on 1 Inch Magnetic Tape for Instrumentation Recording	EIA	31
Flutter Measurement of Instrumentation Magnetic Tape Recorder/Reproducers — Recommended Test Method	EIA	31

Subject Index

		PAGE
Recommended Test Method — Timing Error Measurements of Instrumentation Magnetic-Tape Recorder/Reproducers	EIA	31

Telemetry

Wet-Bulb Temperatures and Wet-Bulb Depressions	ASAE	19
Test Methods for Recorder/Reproducer Systems	RCC	80
Test Methods for Telemetry Systems and Subsystems	RCC	80
Telemetry Standards	RCC	80

Temperature Measurement & Control

Wet-Bulb Temperatures and Wet-Bulb Depressions	ASAE	19
Standard Measurements Guide: Section on Temperature Measurements	ASHRAE	20
Instruments and Apparatus: Temperature Measurement	ASME	21
Graphical Symbols for Heating, Ventilating, and Air Conditioning	ASME	23
Graphical Symbols for Heat-Power Apparatus	ASME	23
Temperature and Humidity Environment for Dimensional Measurement	ASME	23
Standard Test Method for Flexivity of Thermostat Metals	ASTM	13
Standard Test Method for Modulus of Elasticity of Thermostat Metals	ASTM	13
Standard Test Method for Maximum Loading Stress at Temperature of Thermostat Metals	ASTM	13
Test Method for Mechanical Torque Rate of Spiral Coils of Thermostat Metal	ASTM	14
Standard Specification for Thermostat Metal Sheet and Strip	ASTM	14
Standard Test Method for Thermal Deflection Rate of Spiral and Helical Coils of Thermostat Metal	ASTM	14
Standard Specification for Forged or Rolled Chromium-Nickel-Iron-Molybdenum-Copper-Columbium Stabilized Alloy Pipe Flanges, Forged Fittings, and Valves and Parts for Corrosive High Temperature	ASTM	14
Standard Test Method for Heat of Combustion of Liquid Hydrocarbon Fuels by Bomb Calorimeter	ASTM	14
Standard Test Method for Water Vapor Content of Gaseous Fuels by Measurement of Dew-Point Temperature	ASTM	14
Standard Recommended Practice for Determining Temperature-Electrical Resistance Characteristics (EMF) of Metallic Materials	ASTM	18
Standard Temperature Electromotive Force (EMF) Tables for Standardized Thermocouples	ASTM	18
Standard Test Method for Measuring Humidity With a Psychrometer	ASTM	18
Method 12: Heat-Life Test	EIA	29
Standard Terminology and Definition for Filled Thermal Systems of Remote Sensing Temperature Regulators	FCI	34

Subject Index

		PAGE
Environmental Conditions for Process Measurement and Control Systems: Temperature and Humidity	ISA	51
American National Standard for Temperature Measurement Thermocouples	ISA	53
Standard Reference Temperature for Industrial Length Measurements	ISO	59
Warm Air Limit and Fan Controls	NEMA	65
Temperature Limit Controls for Electric Baseboard Heaters	NEMA	65
Pressure Temperature Ratings of Seamless Pipe Used in Power Plant Piping Systems	PFI	76
Exhaust Gas Temperature Instruments	SAE	86
Total Temperature Measuring Instrumentation	SAE	83
Temperature Instruments	SAE	86
Temperature Indicator	SAE	87
Bushings and Wells for Temperature Sensing Elements	SAMA	82
Standard for Temperature-Indicating and Regulating Equipment	UL	91

Terminals

Method 9: Solderability	EIA	29
Quick Connect Terminals	NEMA	65
Terminal Blocks for Industrial Use	NEMA	66

Textiles

Technical Manual of the AATCC	AATCC	4

Thermal Measurements

Thermal Transmission Properties	ASTM	14
Thermal Resistance Measurements of Conduction Cooled Power Transistors	EIA	30
Standard Terminology and Definition for Filled Thermal Systems of Remote Sensing Temperature Regulators	FCI	34
Standard Test Measurement for Steady-State Heat Flow Measurements and Thermal Transmission Properties	ASTM	14

Thermistors

Thermistor Definitions and Test Methods	EIA	30
General Specifications for Thermistors, Insulated and Non-Insulated	EIA	30
Specification for Glass Coated Thermistor Beads and Thermistor Beads in Glass Probes and Glass Rods (Negative Temperature Coefficient)	EIA	30
Directly Heated Negative Temperature Coefficient Thermistors	IEC	57

Thermocouples

Standard Temperature Electromotive Force (EMF) Tables for Standardized Thermocouples	ASTM	18

Subject Index

		PAGE
Standard Specification for Thermocouples, Sheathed, Type K, for Nuclear or Other High-Reliability Applications	ASTM	18
American National Standard for Temperature Measurement Thermocouples	ISA	53
Thermocouple Thermometers (Pyrometers)	SAMA	84

Thermoelectric

Letter Symbols for Thermoelectric Devices	IEEE	38

Thermometers

Standard Specification for ASTM Thermometers	ASTM	17
Standard Method for Verification and Calibration of Liquid-In-Glass Thermometers	ASTM	17
Liquid-in-Glass Laboratory Thermometers	ISO	59
Solid-Stem Calorimeter Thermometers	ISO	60
Enclosed-Scale Calorimeter Thermometers	ISO	60
Bimetallic Thermometers	SAMA	82
Resistance Thermometers	SAMA	82
Filled System Thermometers	SAMA	82
Thermocouple Thermometers (Pyrometers)	SAMA	82
Temperature-Resistance Valves for Resistance Thermometer Elements for Platinum, Nickel and Copper	SAMA	82

Thermoplastics

Standard Test Method for Measuring Flow Rates of Thermoplastics by Extrusion Plastometer	ASTM	15

Thermostats

Gas Appliance Thermostats	AGA	6
Thermostatic Refrigerant Expansion Valves	ARI	3
Standard Test Method for Flexivity of Thermostat Metals	ASTM	13
Standard Test Method for Modulus of Elasticity of Thermostat Metals	ASTM	13
Standard Test Method for Maximum Loading Stress at Temperature of Thermostat Metals	ASTM	13
Standard Test Method for Mechanical Torque Rate of Spiral Coils of Thermostat Metal	ASTM	14
Standard Specification for Thermostat Metal Sheet and Strip	ASTM	14
Standard Test Method for Thermal Deflection Rate of Spiral and Helical Coils of Thermostat Metal	ASTM	14
Line Voltage Integrally-Mounted Thermostats for Electric Heaters	NEMA	66

Timing

Measurement of Time	ASME	22
Quantities and Units of Space and Time	ISO	59

Subject Index

		PAGE
IRIG Standard Time Formats	RCC	80
VHF Time Transmission	RCC	80

Tolerances

Fabricating Tolerances	PFI	76

Transducers

Blood Pressure Transducers, General	AAMI	27
Blood Pressure Transducers, Inter-changeability and Performance	AAMI	27
Guide for Dynamic Calibration of Pressure Transducers	ASME	23
Procedures for Calibration of Underwater Electroacoustic Transducers	ASA	2
Selection of Calibration and Tests for Electrical Transducers Used for Measuring Shock and Vibration	ASA	2
Method for the Experimental Determination of Mechanical Mobility. Part I: Basic Definitions and Transducers	ASA	2
Electrical Transducer Nomenclature and Terminology	ISA	48
Guide for Specifications and Tests for Piezoelectric Acceleration Transducers for Aerospace Testing	ISA	48
Specifications and Tests for Strain Gage Pressure Transducers	ISA	48
Specifications and Tests for Strain Gage Linear Acceleration Transducers	ISA	49
Specifications and Tests of Potentiometric Pressure Transducers	ISA	49
Specifications and Tests for Strain Gage Force Transducers	ISA	49
Specifications and Tests for Piezoelectric Pressure and Sound-Pressure Transducers	ISA	49
Specification and Test for Potentiometric Displacement Transducers	ISA	49
Transducer and Transmitter Installation for Nuclear Safety Applications	ISA	50

Transformers

Approval Standard, Less Flammable Transformer Fluids	FM	33
Current Transformers	IEC	55
Voltage Transformers	IEC	56
Testing Electronic Transformers and Inductors	IEEE	39
Standard Requirements for Instrument Transformers	IEEE	44
Conformance Test Procedures for Instrument Transformers	IEEE	44
Power Supplies for Fire Protective Signaling Systems	UL	92

Transistors

Thermal Resistance Measurements of Conduction Cooled Power Transistors	EIA	30

Subject Index

		PAGE
Transmission Devices & Systems		
Standard Recommended Practice for Selection of Geometric Conditions for Measurement of Reflectance and Transmittance	ASTM	18
Television: Measurement of Differential Gain and Differential Phase	IEEE	38
Measuring Unbalanced Transmission Line Impedance	IEEE	39
Air Pressures for Pneumatic Controllers, Transmitters, and Transmission Systems	ISA	47
Fluid Flow in Closed Conduits — Connections for Pressure Signal Transmissions between Primary and Secondary Elements	ISO	60
IRIG Standard for VHF Time Transmission	RCC	81
Transmitters		
Definitions of Terms for Radio Transmitters	IEEE	38
Air Pressures for Pneumatic Controllers, Transmitters, and Transmission Systems	ISA	47
Transducer and Transmitter Installation for Nuclear Safety Applications	ISA	50
Emergency Locator Transmitter (ELT) Equipment Installation and Performance	RTCA	79
Standard for Electrically Actuated Transmitters	UL	91
Transponder Standards		
Coherent C-Band Transponder Standards	RCC	81
Minimum Operational Characteristics — Airborne ATC Transponder Systems	RTCA	78
Tubing		
Nomenclature for Instrument Tube Fittings	ISA	49
Nuclear-Safety-Related Instrument Sensing Line Piping and Tubing Standards for Use in Nuclear Power Plants	ISA	50
Pneumatic Fluid Power Applications — Metal Separable Tube Fittings — Qualifications Test	NFLDP	74
Electric Resistance Welded Mandrel Drawn Hydraulic Line Tubing	NFLDP	71
Seamless Low Carbon Steel Hydraulic Line Tubing	NFLDP	71
Tubing Connection Markings for Pneumatic Instruments	SAMA	82
Turbidity		
Standard Test Method for Particle Size Distribution of Refractory Metal-Type Powders by Turbidemetry	ASTM	14
Standard Test Method for Fineness of Portland Cement by the Turbidimeter	ASTM	14
Turbines		
Special-Purpose Steam Turbines for Refinery Services	API	10
Appendix A to Test Code for Steam Turbines	ASME	21

Subject Index

		PAGE
Measurement of Gas Flow by Turbine Meters	ASME	21
Simplified Procedures for Routine Performance Tests of Steam Turbines	ASME	21
Guidance for Evaluation of Measurement Uncertainty in Performance Tests of Steam Turbines	ASME	21
Measurement of Rotary Speed	ASME	22
Cold-Water Meters — Turbine Type for Customer Service	AWWA	24
Preparation of Equipment Specifications for Speed-Governing of Hydraulic Turbines Intended to Drive Electric Generators	IEEE	37
Standard for Hydraulic Turbine and Generator Integrally Forged Shaft Couplings and Shaft Runout Tolerances	IEEE	43
Specification, Installation, and Calibration of Turbine Flowmeters	ISA	48
Steam Turbines for Mechanical Drive Service	NEMA	67
Acceptable Vibration of Marine Steam and Gas Turbine Noise and Auxiliary Machinery Plants	SNAME	88
Guide for Centralized Control and Automation of Ship's Gas Turbine Propulsion Plant	SNAME	88

U

Ultrasonic Techniques

Measurement of Liquid Flow in Closed Conduits Using Transit-Time Ultrasonic Flowmeters	ASME	21
Standard Practice for Evaluating Performance Characteristics of Ultrasonic Pulse-Echo Testing Systems Without the Use of Electronic Measurement Instruments	ASTM	18
Nondestructive Flaw-Detection Methods	FM	33
Testing and Calibration of Ultrasonic Therapeutic Equipment	IEC	55
Wall Thickness Measurement by Ultrasonic Examination	PFI	74
Random Ultrasonic Examination of Butt Welds	PFI	75

V

Vacuum Applications

Displacement Compressors, Vacuum Pumps, and Blowers	ASME	21
Gauges — Pressure and Vacuum, Indicating Dial Type — Elastic Element	ASME	22
Method for Vacuum Leak Calibration	AVS	23
Method for Measuring Pumping Speed of Mechanical Vacuum Pumps for Permanent Gases	AVS	24
Procedure for Calibrating Vacuum Gages of the Thermal Conductivity Type	AVS	24
Graphic Symbols in Vacuum Technology	AVS	24
Methods of Measurements of the Performance Characteristics of Positive-Displacement Vacuum Pumps: Measurement of the Volume Rate of Flow	ISO	60

Subject Index

		PAGE
Methods of Measurements of the Performance Characteristics of Vapor Vacuum Pumps: Measurement of the Volume Rate of Flow	ISO	60
Method of Rating for Mechanical Vacuum Pumps	NFLDP	74
Valves		
Manually Operated Gas Valves	AGA	6
Automatic Valves for Gas Appliances	AGA	6
Relief Valves and Automatic Gas Shutoff Devices	AGA	6
Specifications for Pipeline Valves, End Closures, Connectors and Swivels	API	8
Flanged Steel Safety Valves	API	9
Valve Inspection and Test	API	10
Commercial Seat Tightness of Safety Relief Valves with Metal-to-Metal Seats	API	9
Design and Installation of Pressure-Relieving Systems in Refineries	API	9
Guide for Pressure Relief and Depressuring Systems	API	9
Standard for Refrigerant Access Valves and Hose Connectors	ARI	3
Thermostatic Refrigerant Expansion Valves	ARI	3
Solenoid Valves for Use with Volatile Refrigerants and Water	ARI	3
Refrigerant Pressure Regulating Valves	ARI	3
Graphical Symbols for Pipe Fittings, Valves and Piping	ASME	23
Specification for Forged or Rolled Alloy-Steel Pipe Flanges, Forged Fittings and Valves and Parts for High-Temperature Service	ASTM	13
Specification for Forged or Rolled 8 and 9 Percent Nickel Alloy Steel Flanges, Fittings, Valves, and Parts for Low-Temperature	ASTM	13
Standard Specification for Forged or Rolled Chromium-Nickel-Iron-Molybdenum-Copper-Columbium Stablized Alloy Pipe Flanges, Forged Fittings, and Valves and Parts for Corrosive High Temperature	ASTM	14
Gate Valves — 3 in. through 48 in. — for Water and Other Liquids	AWWA	25
Power Actuating Devices for Valves and Sluice Gates	AWWA	25
Rubber-Seated Butterfly Valves	AWWA	25
Backflow Prevention Devices — Reduced Pressure Principle and Double Check Valve Types	AWWA	25
Ball Valves, Shaft- or Trunion-Mounted — 6 in. through 48 in. — for Water Pressures up to 300 psi	AWWA	25
Swing-Check Valves for Waterworks Service, 2 in. through 24 in. NPS	AWWA	25
Resilient-Seated Gate Valves, 3 through 12 NPS, for Water and Sewage Systems	AWWA	25
Maintenance Instructions for Chlorine Institute Standard Safety Valves, Type 1-1/2JQ	CI	27

Subject Index

		PAGE
Maintenance Instructions for Chlorine Institute Standard Angle Valve	CI	27
Maintenance Instructions for Chlorine Institute Standard Safety Valves, Type 4 JQ	CI	27
Maintenance Instructions for Chlorine Institute Standard Excess Flow Valves	CI	28
Procedure in Rating Flow and Pressure Characteristics of Solenoid Valves for Gas Service	FCI	34
Procedure in Rating Flow and Pressure Characteristics of Solenoid Valves for Liquid Service	FCI	34
Quality Control Standard for Control Valve Seat Leakage	FCI	34
Silent Check Valve Standard	FCI	34
Test Conditions and Procedures for Measuring Electrical Characteristics of Solenoid Valves	FCI	34
Testing and Classifying the Water Hammer Characteristics of Electrically Operated Valves	FCI	34
Metric Definition of the Valve Flow Coefficient Cv	FCI	34
Approval Standard, Fire Safe Valves	FM	33
Approval Standard, Fuel Gas and Oil Safety Shutoff Valves	FM	33
PS 8-77, Uniform Plumbing Code	IAPMO	54
Globe-Type Loglighter Valves Angle or Straight Pattern	IAPMO	54
Pressure Reducing and Regulating Valves for Installation on Domestic Water Supply Line	IAPMO	54
Control Valve Terminology and General Considerations	IEC	58
Industrial-Process Control Valves	IEC	57
Industrial-Process Control Valves — Part 3: Dimensions, Section One: Face-to-Face Dimensions for Flanged, Two-Way, Globe-Type Control Valves	IEC	57
Industrial-Process Control Valves Part 4: Inspection and Routine Testing	IEC	57
Industrial-Process Control Valves — Part 3: Dimensions, Section One: Face-to-Face Dimensions for Flangeless Control Valves Except Wafer Butterfly Valves	IEC	57
Specification Forms for Process Measurement and Control Instruments, Primary Elements and Control Valves	ISA	48
Flow Equations for Sizing Control Valves	ISA	51
Control Valve Capacity Test Procedure	ISA	51
Face-to-Face Dimensions for Flanged Globe-Style Control Valve Bodies	ISA	51
Face-to-Face Dimensions for Flangless Control Valves	ISA	51
Control Valve Terminology	ISA	51
Control Valve Manifold Designs	ISA	51
Inherent Flow Characteristic and Rangeability of Control Valves	ISA	52

Subject Index

		PAGE
Face-to-Face Dimensions for Socket Weld-End and Screwed-End Globe-Style Control Valves	ISA	52
Face-to-Face Dimensions for Buttweld-End Globe-Style Control Valves	ISA	52
Laboratory Measurement of Aerodynamic Noise Generated by Control Valves	ISA	52
Installed Face-to-Face Dimensions for Flanged Clamp or Pinch Valves	ISA	52
Face-to-Face Dimensions for Flanged Globe-Style Control Valve Bodies (ANSI Classes 900, 1500 and 2500)	ISA	52
Standard Marking System for Valves, Fittings, Flanges, and Unions	MSS	61
Corrosion-Resistant Cast Flanged Valves	MSS	61
Corrosion Resistant Gate, Globe, Angle and Check Valves	MSS	61
Butterfly Valves	MSS	61
Quality Standard for Steel Castings and Forgings for Valves	MSS	61
Cast Iron Gate Valves, Flanges and Threaded Ends	MSS	61
Cast Iron Swing Check Valves, Flanged and Threaded Ends	MSS	61
Ball Valves with Flanged or Butt-Welding Ends for General Service	MSS	61
Pressure Testing of Steel Valves	MSS	61
Cast Iron Plug Valves	MSS	62
Bronze Gate, Globe, Angle and Check Valves	MSS	62
High Pressure-Offset Seat Butterfly Valves	MSS	61
Valve Pressure Testing Methods	MSS	62
Stainless Steel, Bonnetless, Flanged, Knife Gate Valves	MSS	62
Instrument Valves	MSS	62
Diaphram Type Valves	MSS	62
Steel Valves, Socket Welding and Threaded Ends	MSS	62
Cast Iron Globe and Angle Valves, Flanged and Threaded Ends	MSS	62
Qualification Requirements for Elastomer Diaphragm for Nuclear Service Diaphragm Type Valves	MSS	62
Guidelines for Metric Data in Standards for Valves, Flanges and Fittings	MSS	62
Protective Epoxy Coatings for the Interior of Valves and Hydrants	MSS	62
Guidelines for Manual Operation of Valves	MSS	62
Valve User Guide	MSS	62
Pressure Relief Device Certifications	NBBI	63
Hydraulic Valve Pressure Rating Supplement	NFLDP	69
Pneumatic Valve Pressure Rating Supplement	NFLDP	69
Series of Mounting Interfaces for 4567 Maximum psi (315 bar) Four Port Hydraulic Fluid Power Directional Valves	NFLDP	72

Subject Index

		PAGE
Hydraulic Fluid Power — Directional Control Valve — Method for Determining the Metering Characteristics	NFLDP	72
Hydraulic Fluid Power Solenoid-Piloted Industrial Valves — Interface Dimensions for Electrical Connectors	NFLDP	72
Hydraulic Fluid Power — Code for the Identification of Valve Mounting Surfaces	NFLDP	72
Interfaces for 4-Way General Purpose Industrial Pneumatic Directional Control Valves	NFLDP	72
Pneumatic Fluid Power — Five-Port Directional Control Valves — Mounting Surfaces — Optional Electrical Connector — Dimensions and Requirements	NFLDP	72
Defining Interface Surfaces for Each Pneumatic Valve Interface in NFPA Recommended Standard T3.2.1	NFLDP	72
Definition of Port Communication for the Fluid Power Valve Interface to NFPA Recommended Standard T3.21.1 with the Valve in Position in Response to a Remote Pilot Signal or Electrical Energization	NFLDP	72
Temperature-Resistance Valves for Resistance Thermometer Elements for Platinum, Nickel and Copper	SAMA	82
Hydrostatic Testing of Control Valves	SAMA	82
Standard for Relief Valves for Anhydrous Ammonia and LP-Gas	UL	89
Standard for Electrically Operated Valves	UL	90

Ventilating Systems

Industrial Ventilation — A Manual of Recommended Practice, 20th Edition	ACGIH	5
Air Sampling Instruments Manual, 7th Edition	ACGIH	5
Ventilation System Testing	ACGIH	5
Test Method for Louvers, Dampers, and Shutters	AMCA	3
Graphical Symbols for Heating, Ventilating, and Air Conditioning	ASME	23
Approval Standard, Airflow Interlocking Switches and Pressure Supervisory Switches for Fuel Oil, Fuel Gas and Ventilation or Combustion Air	FM	33

Vibration Measurements

Roller Bearing Vibration and Noise Measurements	AFBMA	26
Vibration, Axial-Position, and Bearing Temperature Monitoring Systems	API	10
Accelerometer-Based Vibration Monitoring System	API	10
Acoustical Terminology	ASA	1
Selection of Calibration and Tests for Electrical Transducers Used for Measuring Shock and Vibration	ASA	2
Method for Specifying the Characteristics of Auxiliary Analog Equipment for Shock and Vibration Measurements	ASA	2
Method 7: Vibration Fatigue Test (Low Frequency, 10 to 55 Hertz)	EIA	29

Subject Index

		PAGE
Method 8: Vibration, High Frequency	EIA	29
Procedural Standards for Measuring Sound and Vibration	NEBB	67
Mechanical Vibration of Certain Rotating Electrical Machinery with Shaft Heights between 80 and 400mm — Measurement and evaluation of the Vibration Severity	ISO	60
Water Hammer Arresters	PDI	77
Code for Shipboard Vibration Measurements	SNAME	87
Local Shipboard Structures and Machinery Vibration Measurements	SNAME	88
Acceptable Vibration of Marine Steam and Gas Turbine Noise and Auxiliary Machinery Plants	SNAME	88
Guidelines for the Use of Vibration Monitoring for Preventive Maintenance	SNAME	88

Viscometry

Determination of the Viscosity of Liquids	ASME	22
Standard Method for Calibration of Master Viscometers and Viscosity Oil Standards	ASTM	16

Viscous Materials

Standard Test Method for Density and Relative Density (Specific Gravity) of Viscous Materials by Bingham Pycnometer	ASTM	15
Standard Test Method for Density and Relative Density (Specific Gravity) of Viscous Materials by Lipkin Bicapillary Pycnometer	ASTM	15

Voltage Measurements

Method 4: Dielectric Test (Withstanding Voltage)	EIA	20
Definition and Measurement of Voltage Jump — for Voltage Regulator and Reference Tubes	EIA	30
Voltage Transformers	IEC	56

W

Water Measurements

Sluice Gates	AWWA	25
Uniform Plumbing Code	IAPMO	54
Pressure Reducing and Regulating Valves for Installation on Domestic Water Supply Line	IAPMO	54
Measurement of Water Flow in Closed Circuits — Tracer Method	ISO	60
Measurement of Clean Water Flow in Closed Conduits — Velocity Area Method Using Current-Meters	ISO	60
Water Hammer Arresters	PDI	75
Measuring, Sampling and Analyzing White Waters	TAPPI	88

Water and Wastewater

Standard Methods for the Examination of Water & Wastewater	AWWA	25

Subject Index

		PAGE
Swing-Check Valves for Waterworks Service, 2 in. through 24 in. NPS	AWWA	25
Resilient-Seated Gate Valves, 3 through 12 NPS, for Water and Sewage Systems	AWWA	25

Wave Characteristics

Definitions of Fundamental Waveguide Terms	IEEE	37
Definitions for Waveguide Components	IEEE	37
Waveguide and Waveguide Component Measurements	IEEE	37
Standards Report on State-of-the-Art of Measuring Field Strength, Continuous Wave, Sinusoidal	IEEE	38
Methods of Measuring (Below 1000 MHz) Electromagnetic Field Strength for Frequencies Below 1000 MHz in Radiowave Propagation	IEEE	38
Definitions of Terms for Nonlinear, Active, and Nonreciprocal Waveguide Components	IEEE	40
Measurement of Pulsating Fluid Flow in a Pipe by Means of Orifice Plates, Nozzles, or Venturi Tubes, in Particular in the Case of Sinusoidal or Square Wave Intermittent Periodic-Type Fluctuations	ISO	60

Weighing

Instruments and Apparatus: Weighing Scales	ASME	22
Application and Installation of Continuous-Belt Weighbridge Scales	ISA	51
Standard for Electrically Illuminated Scales	UL	90

Welds and Welding

Specifications for Pipeline Valves, End Closures, Connectors and Swivels	API	8
Standard Procedures for Calibrating Magnetic Instruments to Measure the Delta Ferrite Content of Austenitic Stainless Steel Weld Metal	AWS	25
Method for Sound Level Measurement of Manual Arc Welding and Cutting Processes	AWS	25
Method 10: Effect of Soldering	EIA	29
Socket Welding Reducer Inserts	MSS	62
Carbon Steel Pipe Unions, Socket Welding and Threaded	MSS	62
Steel Valves, Socket Welding and Threaded Ends	MSS	62
General Standards for Industrial Control and Systems	NEMA	66
Resistance Welding Control	NEMA	66
Electric Resistance Welded Mandrel Drawn Hydraulic Line Tubing	NFLDP	71
Internal Machining and Solid Machined Backing Rings for Circumferential Butt Welds	PFI	74
Minimum Length and Spacing for Welded Nozzles	PFI	74

Subject Index

		PAGE
Access Holes and Plugs for Radiographic Inspection of Pipe Welds	PFI	74
Internal Machining and Fit-up of GTAW Root Pass Circumferential Butt Welds	PFI	75
Random Radiography of Pressure Retaining Girth Butt Welds	PFI	75
Welded Load Bearing Attachments to Pressure Retaining Piping Materials	PFI	75
Random Ultrasonic Examination of Butt Welds	PFI	75
Circumferential Butt Welds in the Arc of Pipe Bends	PFI	75
Nonsymmetrical Bevels and Joint Configurations for Butt Welds	PFI	75
Resistance Welding Equipment Standards	RWMA	81

Wires & Wiring

Method 9: Solderability	EIA	29
Test Procedures for Extended Time-Testing of Wire and Cable Insulation for Service in Wet Locations	ICEA	53
Printed Wiring Design Guide	IPC	46